Lecture Notes in Computer Science 5732

Ian P. Gent (Ed.)

Principles and Practice of Constraint Programming – CP 2009

15th International Conference, CP 2009
Lisbon, Portugal, September 20-24, 2009
Proceedings

Springer

Volume Editor

Ian P. Gent
University of St. Andrews
School of Computer Science
North Haugh, St Andrews
Fife KY16 9SX, Scotland, UK
E-mail: ipg@cs.st-andrews.ac.uk

Library of Congress Control Number: 2009933414

CR Subject Classification (1998): D.1.6, D.3, F.3, G.2, D.3.2, F.4.1, G.1.6

LNCS Sublibrary: SL 2 – Programming and Software Engineering

ISSN 0302-9743
ISBN-10 3-642-04243-0 Springer Berlin Heidelberg New York
ISBN-13 978-3-642-04243-0 Springer Berlin Heidelberg New York

springer.com

Printed in Germany

Typesetting: Camera-ready by author, data conversion by Scientific Publishing Services, Chennai, India
Printed on acid-free paper SPIN: 12753741 06/3180 5 4 3 2 1 0

Preface

This volume contains the papers presented at CP 2009: The 15th International Conference on Principles and Practice of Constraint Programming. It was held from September 20–24, 2009 at the Rectory of the New University of Lisbon, Portugal. Everyone involved with the conference thanks our sponsors for their support.

There were 128 submissions to the research track, of which 53 were accepted for a rate of 41.4%. Each submission was reviewed by three reviewers, with a small number of additional reviews obtained in exceptional cases. Each review was either by a Programme Committee member, or by a colleague invited to help by a committee member thanks to their particular expertise. Papers submitted as long papers were accepted at full length or not at all. It is important to note that papers submitted as short papers were held to the same high standards of quality as long papers. There is thus no distinction in these proceedings between long and short papers, except of course the number of pages they occupy. As it happens, the acceptance rates of short and long papers were very similar indeed.

There were 13 submissions to the application track, of which 8 were accepted, for a rate of 61.5%. Papers underwent the same review process as regular papers, and there was not a separate committee for reviewing application track papers. However, papers in the application track were not required to be original or novel research, but to be original and novel as an application of constraints.

The programme included three invited talks from distinguished scientists, Carla Gomes, Philippe Baptiste, and Barbara Smith. They have each provided a brief abstract of their talk. Additionally, three tutorials were given during the main programme. Affiliated workshops were held on September 20, 2009. Details of tutorials and workshops are given on a separate page.

To help with the review process and constructing the proceedings, I used the EasyChair conference management system. On another issue of process, while I did not submit any papers myself to the conference, a number of colleagues from St Andrews University did. I thank Fahiem Bacchus for overseeing these papers, assigning PC members and making final decisions on them.

I would like to thank the Association for Constraint Programming for inviting me to be Programme Chair of this conference. A conference proceedings such as this is the work of literally hundreds of people, including authors, Programme Committee, local organizers, tutorial and workshop chairs, the doctoral programme chairs, and additional reviewers. I would like to thank all of them for their hard work, especially remembering those authors of papers which were rejected.

Apart from all those who have played some formal role, many people have provided special help with a lot of the day-to-day, and often dull, tasks that arise while preparing the proceedings. For this, I would especially like to thank

Chris Jefferson, Lars Kotthoff, Angela Miguel, Ian Miguel, Neil Moore, Peter Nightingale, Karen Petrie, Andrea Rendl, and Judith Underwood.

July 2009 Ian Gent

Prize-Winning Papers

A small number of colleagues from the Programme Committee helped me decide on the papers deserving of being recognized as of the highest standard of those submitted. As well as prizes, a small number of other papers were designated as runners-up.

– *Ian Gent*

Best Paper (Application Track)

A Hybrid Constraint Model for the Routing and Wavelength Assignment Problem, by Helmut Simonis

Best Paper (Research Track)

Edge Finding Filtering Algorithm for Discrete Cumulative Resources in $O(kn \log n)$, by Petr Vilím

***Runners-Up** (in alphabetical order)*

- *Conflict Resolution*, by Konstantin Korovin, Nestan Tsiskaridze, Andrei Voronkov
- *Failed Value Consistencies for Constraint Satisfaction*, by Christophe Lecoutre, Olivier Roussel
- *On the Power of Clause-Learning SAT Solvers with Restarts*, by Knot Pipatsrisawat, Adnan Darwiche

Best Student Paper[1]

On the Power of Clause-Learning SAT Solvers with Restarts, by Knot Pipatsrisawat, Adnan Darwiche

Runner-Up

- *Using Relaxations in Maximum Density Still Life*, by Geoffrey Chu, Peter Stuckey, Maria Garcia de la Banda

[1] Student papers were those papers declared by the authors to be mainly the work (both in research and writing) of PhD or other students.

Workshops and Tutorials

Workshops

A range of workshops affiliated with the conference took place the day before the main conference, on September 20, 2009. The workshops accepted by the Workshop Chairs were as follows:

- Bin Packing and Placement Constraint (BPPC 2009)
- Local Search Techniques in Constraint Satisfaction (LSCS 2009)
- Mathematical Foundations of Constraint Programming
- Constraint Modelling and Reformulation (ModRef 2009)
- Symmetry and Constraint Satisfaction Problems (SymCon 2009)
- Interval Analysis and Constraint Propagation for Applications (IntCP 2009)
- Constraint-Based Methods for Bioinformatics (WCB 2009)
- Constraint Reasoning and Optimization for Computational Sustainability (CROCS 2009)

Tutorials

Three tutorial presentations were given during the main programme of the conference. These were as follows:

- *Amortized and Expected Case Analysis of Constraint Propagation Algorithms*, by Chris Jefferson, Meinolf Sellmann
- *Soft Global Constraints*, by Willem-Jan van Hoeve
- *Exploiting Fixed-Parameter Tractability in Satisfiability and Constraint Satisfaction*, by Barry O'Sullivan and Igor Razgon

Conference Organization

Programme Chair

Ian Gent	University of St Andrews, UK

Conference Chair

Pedro Barahona	New University of Lisbon, Portugal

Tutorial and Workshop Chairs

Ian Miguel	University of St Andrews, UK
Patrick Prosser	University of Glasgow, UK

Doctoral Programme Chairs

Karen Petrie	University of Oxford, UK
Olivia Smith	University of Melbourne, Australia

Publicity Chair

Francisco Azevedo	New University of Lisbon, Portugal

Sponsorship Chair

Jorge Cruz	New University of Lisbon, Portugal

Organizing Committee

Inês Lynce	Technical University of Lisbon, Portugal
Vasco Manquinho	Technical University of Lisbon, Portugal
Ludwig Krippahl	New University of Lisbon, Portugal

Sponsors

ACP - Association for Contraint Programming
NICTA - National Information and Communications Technology Australia
FCT - Foundation for Science and Technology
CENTRIA - Centre for Artificial Intelligence
APPIA - Portuguese Association for Artificial Intelligence

4C - Cork Constraint Computation Centre
TAP - Air Portugal
Widescope: Optimization Solutions

Programme Committee

Fahiem Bacchus	University of Toronto, Canada
Pedro Barahona	New University of Lisbon, Portugal
Peter van Beek	University of Waterloo, Canada
Frédéric Benhamou	University of Nantes, France
Christian Bessiere	Université Montpellier, CNRS, France
Lucas Bordeaux	Microsoft Research, Cambridge, UK
Ken Brown	University College Cork, Ireland
Andrei Bulatov	Simon Fraser University, Canada
Mats Carlsson	Swedish Institute of Computer Science, Sweden
Hubie Chen	Universitat Pompeu Fabra, Spain
Martin Cooper	University of Toulouse, France
Victor Dalmau	Universitat Pompeu Fabra, Spain
Jeremy Frank	NASA, USA
Enrico Giunchiglia	Università di Genova, Italy
Simon de Givry	INRA Biometrics and Artificial Intelligence, France
Alexandre Goldsztejn	Centre National de la Recherche Scientifique, France
Brahim Hnich	IEU, Turkey
Christopher Jefferson	University of St Andrews, UK
Ulrich Junker	ILOG, France
Jimmy Lee	The Chinese University of Hong Kong
Ines Lynce	Technical University of Lisbon, Portugal
Felip Manya	IIIA-CSIC, Spain
Joao Marques-Silva	University College Dublin, Ireland
Pedro Meseguer	IIIA-CSIC, Spain
Ian Miguel	University of St Andrews, UK
Michela Milano	Università di Bologna, Italy
David Mitchell	Simon Fraser University, Canada
Barry O'Sullivan	University College Cork, Ireland
Patrick Prosser	University of Glasgow, UK
Claude-Guy Quimper	Google, Canada
Ashish Sabharwal	Cornell University, USA
Meinolf Sellmann	Brown University, USA
Paul Shaw, ILOG	France
Kostas Stergiou	University of the Aegean, Greece

Peter Stuckey	University of Melbourne, Australia
Michael Trick	Carnegie Mellon University, USA
Kristen Brent Venable	University of Padova, Italy
Gérard Verfaillie	ONERA, France
Mark Wallace	Monash University, Australia
Toby Walsh	NICTA and UNSW, Australia
Roland Yap	National University of Singapore, Singapore
Weixiong Zhang	Washington University in St. Louis, USA

Additional Reviewers

Magnus Ågren
Josep Argelich
Francisco Azevedo
Yoram Bachrach
Thanasis Balafoutis
Mauro Bampo
Ralph Becket
Ramon Bejar
Nicolas Beldiceanu
Stefano Bistarelli
Eric Bourreau
Simone Bova
Sebastian Brand
Pascal Brisset
Hadrien Cambazard
Catarina Carvalho
Martine Ceberio
Kenil C.K. Cheng
Raphael Chenouard
Marc Christie
David Cohen
Remi Coletta
Marco Correia
Jorge Cruz
Jessica Davies
Renaud Dumeur
Redouane Ezzahir
Helene Fargier
Thibaut Feydy
Raphael Finkel
Pierre Flener
Jeremy Frank
Marco Gavanelli
Mirco Gelain
Laurent Granvilliers
Andrew Grayland
Diarmuid Grimes
Yunsong Guo
Tarik Hadzic
Fang He
Emmanuel Hebrard
Martin Henz
Willem Jan van Hoeve
Alan Holland
Eric Hsu
Ruoyun Huang
Frank Hutter
Dejan Jovanović
Valentine Kabanets
Serdar Kadioglu
George Katsirelos
Tom Kelsey
Emil Keyder
Zeynep Kiziltan
Lars Kotthoff
Lukas Kroc
Philippe Laborie
Arnaud Lallouet
Javier Larrosa
Christophe Lecoutre
Ho-fung Leung
Olivier Lhomme
Chu-Min Li
Wenkai Li
Wei Li
Chavalit Likitvivatanavong
Lengning Liu
Michele Lombardi

Yuri Malitsky
Marco Maratea
Paolo Marin
Carles Mateu
Eric Monfroy
Neil Moore
Massimo Narizzano
Nina Narodytska
Jorge Navas
Bertrand Neveu
Peter Nightingale
Gustav Nordh
Alexandre Papadopoulos
Karen Petrie
Maria Silvia Pini
Jordi Planes
Steven Prestwich
Anthony Przybylski
Riccardo Pucella
Luca Pulina
Luis Quesada
Igor Razgon
Pierre Regnier
Andrea Rendl
Juan Antonio Rodríguez-Aguilar
Andrea Roli
Emma Rollon
Emanuele Di Rosa
Francesca Rossi
Tyrel Russell
Andras Salamon
Horst Samulowitz
Frédéric Saubion
Thomas Schiex
Joachim Schimpf
Christian Schulte
Charles F.K. Siu
Evgeny Skvortsov
Barbara Smith
Paul Strooper
Guido Tack
Eugenia Ternovska
Kevin Tierney
Guillaume Verger
Petr Vilím
Richard Wallace
Philipp Weis
Tomas Werner
Mark Weyer
Christoph Wintersteiger
May H.C. Woo
Michal Wrona
William Yeoh
Makoto Yokoo
Evangeline Young
Alessandro Zanarini
Standa Zivny

Association for Constraint Programming

The Association for Constraint Programming aims at promoting constraint programming in every aspect of the scientific world, by encouraging its theoretical and practical developments, its teaching in academic institutions, its adoption in the industrial world, and its use in application fields.

The ACP is a non-profit association, which uses the profit of the organized events to support future events or activities. At any given time, members of the ACP are all attendees of a CP conference in the past five years, and all members of the Programme Committee of the current CP conference.

Executive Committee

President: Barry O'Sullivan (Elected 2008-2012, President until the end of 2009)
Secretary: Jimmy H.M. Lee (Elected 2006-2009, Secretary until end of 2009)
Treasurer: Thomas Schiex (Elected 2007-2010, Treasurer until end of 2009)
Conference Coordinator: Pedro Meseguer (Elected 2007-2010, Conference coordinator until end of 2009)
Other Members:

- Christian Bessiere (Non-Elected Member 2007-2009, Programme Chair of CP2007)
- John Hooker (Elected 2008-2012)
- Karen Petrie (Elected 2008-2012)
- Christian Schulte (Elected 2006-2009)
- Peter Stuckey (Elected 2006-2012)
- Michael Trick (Elected 2006-2009)
- Roland Yap (Elected 2008-2012)

Table of Contents

Invited Talks

Application Track Papers

Research Track Papers

Constraint-Based Schedulers, Do They Really Work?

Philippe Baptiste

CNRS LIX, Ecole Polytechnique,
91128 Palaiseau, France
Philippe.Baptiste@polytechnique.fr

Constraint programming has been widely applied in the area of scheduling, enabling the implementation of flexible scheduling systems. Over the last ten years, the flexibility offered by CP has been combined with the efficiency of specialized Operations Research algorithms. As a result, CP tools dedicated to scheduling are now claimed to be fast, efficient and easy to use.

In this talk, I will show that this claim is partially true and that, indeed, constraint-based schedulers behave very well on a rather large class of problems. However, there are several scheduling situations in which CP does not work. I will try to indentify the key ingredients that make scheduling problems hard for CP. I will also introduce new mixed integer formulations that could compete/cooperate with CP for such problems.

I.P. Gent (Ed.): CP 2009, LNCS 5732, p. 1, 2009.

Challenges for Constraint Reasoning and Optimization in Computational Sustainability

Carla P. Gomes

Cornell University
Ithaca, NY, USA
gomes@cs.cornell.edu

Abstract. Computational Sustainability is a new emerging research field with the overall goal of studying and providing solutions to computational problems for balancing environmental, economic, and societal needs for a sustainable future. I will provide examples of challenge problems in Computational Sustainability, ranging from wildlife preservation and biodiversity, to balancing socio-economic needs and the environment, to large-scale deployment and management of renewable energy sources, highlighting overarching computational themes in constraint reasoning and optimization and interactions with machine learning, and dynamical systems. I will also discuss the need for a new approach to study such challenging problems in which computational problems are viewed as "natural" phenomena, amenable to a scientific methodology in which principled experimentation, to explore problem parameter spaces and hidden problem structure, plays as prominent a role as formal analysis.

Extended Abstract

Humanity's use of Earth's resources is threatening our planet and the livelihood of future generations. The dramatic growth in our use of natural resources over the past century is reaching alarming levels. Our Common Future [3], the seminal report of the United Nations World Commission on Environment and Development, published in 1987, raised environmental concerns and introduced the notion of sustainable development: "development that meets the needs of the present without compromising the ability of future generations to meet their needs." Our Common Future also stated the urgency of policies for sustainable development. The United Nations Environment Program in its fourth Global Environmental Outlook report published in October of 2007 [4] and the United Nations Intergovernmental Panel on Climate Change (IPCC) [2] reiterated the concerns raised in Our Common Future. For example, the fourth Global Environmental Outlook report stated that "there are no major issues raised in Our Common Future for which the foreseeable trends are favorable" [4].

Key sustainability issues translate into decision making and policy making problems concerning the management of our natural resources involving significant computational challenges that fall into the realm of computing and information science and related discilines, even though in general they are not studied

I.P. Gent (Ed.): CP 2009, LNCS 5732, pp. 2–4, 2009.

by researchers in those disciplines. In fact, the impact of modern information technology has been highly uneven, mainly benefiting large firms in profitable sectors, with little or no benefit in terms of the environment. It is therefore imperative and urgent that we turn our attention to computational problems that arise in the context of the environment and the way we use our natural resources.

Our vision is that computer scientists can — and should — play a key role in increasing the efficiency and effectiveness of the way we manage and allocate our natural resources. Furthermore, we argue for the establishment of critical mass in the new emerging field of *Computational Sustainability*. Computational Sustainability is an interdisciplinary field that aims to apply techniques from computer science and related fields, namely information science, operations research, applied mathematics, and statistics, to balance environmental, economic, and societal needs for sustainable development. The range of problems that fall under Computational Sustainability is rather wide, encompassing computational challenges in disciplines as diverse as ecology, natural resources, atmospheric science, and biological and environmental engineering. Research in Computational Sustainability is therefore necessarily an interdisciplinary endeavor, where scientists with complementary skills must work together in a collaborative process. The focus of Computational Sustainability is on developing computational and mathematical models, methods, and tools for decision making concerning the management and allocation of resources for sustainable development.

In this talk I will provide examples of computational sustainability challenge domains ranging from wildlife preservation and biodiversity, to balancing socio-economic needs and the environment, to large-scale deployment and management of renewable energy sources. I will discuss how computational sustainability problems offer challenges but also opportunities for the advancement of the state of the art of computing and information science and related fields, highlighting some overarching computational themes in constraint reasoning and optimization, machine learning, and dynamical systems. I will also discuss the need for a new approach to study such challenging problems in which computational problems are viewed as "natural" phenomena, amenable to a scientific methodology in which principled experimentation, to explore problem parameter spaces and hidden problem structure, plays as prominent a role as formal analysis [1]. Such an approach differs from the traditional computer science approach, based on abstract mathematical models, mainly driven by worst-case analyses. While formulations of real-world computational tasks lead frequently to worst-case intractable problems, often such real world tasks contain hidden structure enabling scalable methods. It is therefore important to develop new approaches to identify and exploit real-world structure, combining principled experimentation with mathematical modeling, that will lead to scalable and practically effective solutions.

In summary, the new field of Computational Sustainability brings together computer scientists, operation researchers, applied mathematicians, biologists, environmental scientists, and economists, to join forces to study and provide solutions to computational problems concerning sustainable development, offering challenges but also opportunities for the advancement of the state of the art of computing and information science and related fields.

Acknowledgments

The author is the lead Principal Investigator of an NSF Expedition in Computing grant on Computational Sustainability (Award Number: 0832782). The author thanks the NSF Expeditions in Computing grant team members for their many contributions towards the development of a vision for Computational Sustainability, in particular, Chris Barrett, Antonio Bento, Jon Conrad, Tom Dietterich, John Gunckenheimer, John Hopcroft, Ashish Sabharwhal, Bart Selman, David Shmoys, Steve Strogatz, and Mary Lou Zeeman.

References

[1] Gomes, C., Selman, B.: The science of constraints. Constraint Programming Letters 1(1) (2007)
[2] IPCC. Fourth assessment report (AR4). Technical report, United Nations Intergovernmental Panel on Climate Change, IPCC (2007)
[3] UNEP. Our common future. Published as annex to the General Assembly document A/42/427, Development and International Cooperation: Environment. Technical report, United Nations Environment Programme, UNEP (1987)
[4] UNEP. Global environment outlook 4 (GEO4). Technical report, United Nations Environment Programme, UNEP (2007)

Observations on Symmetry Breaking

Barbara M. Smith

School of Computing, University of Leeds
Leeds LS2 9JT
United Kingdom
b.m.smith@leeds.ac.uk

A common way to exploit symmetry in constraint satisfaction problems is to transform the symmetric CSP instance by adding constraints in such a way the new CSP has at least one solution from each symmetry equivalence class of solutions in the original CSP, and ideally only one. Crawford, Ginsberg, Luks and Roy, in a 1996 paper, gave a standard procedure for deriving so-called lex-leader constraints in SAT problems that has subsequently been adapted for the general CSP, principally for variable symmetries. The lex-leader constraint for a given element of the symmetry group excludes any solution that is lexicographically larger than its symmetric equivalent, given a ordering of the variables of the CSP instance.

Ensuring that there is only one solution in the transformed CSP instance for every symmetry equivalence class requires in principle a lex-leader constraint for every element of the symmetry group. Where it is impracticable to generate so many constraints, we can resort to partial symmetry breaking, and generate constraints for only a subset of the symmetry group.

Partial symmetry breaking using lex-leader constraints requires us to choose a subset of the symmetries: I discuss which symmetries might lead to the best symmetry-breaking constraints. Moreover, a balance has to be struck. The more constraints we construct, the fewer solutions there will be to the reduced CSP instance and the less search we can expect to do to solve it, but at the same time, more constraints mean longer propagation time. For some example CSPs, I show where the best balance between less search and longer run-time seems to lie. In constructing lex-leader constraints, we have to choose a variable ordering and I discuss the effect of changing the ordering and why it might be efficient to use the same variable ordering for search. I also discuss including auxiliary variables in the construction of lex-leader constraints.

I.P. Gent (Ed.): CP 2009, LNCS 5732, p. 5, 2009.

Generating Optimal Stowage Plans for Container Vessel Bays

Alberto Delgado[1], Rune Møller Jensen[1], and Christian Schulte[2]

[1] IT University of Copenhagen, Denmark
{alde,rmj}@itu.dk
[2] KTH - Royal Institute of Technology, Sweden
cschulte@kth.se

Abstract. Millions of containers are stowed every week with goods worth billions of dollars, but container vessel stowage is an all but neglected combinatorial optimization problem. In this paper, we introduce a model for stowing containers in a vessel bay which is the result of probably the longest collaboration to date with a liner shipping company on automated stowage planning. We then show how to solve this model efficiently in - to our knowledge - the first application of CP to stowage planning using state-of-the-art techniques such as extensive use of global constraints, viewpoints, static and dynamic symmetry breaking, decomposed branching strategies, and early failure detection. Our CP approach outperforms an integer programming and column generation approach in a preliminary study. Since a complete model of this problem includes even more logical constraints, we believe that stowage planning is a new application area for CP with a high impact potential.

1 Introduction

More than 60% of all international cargo is carried by liner shipping container vessels. To satisfy growing demands, the size vessel has increased dramatically over the last two decades. This in turn has made the traditional manual stowage planning of the vessels very challenging. A container vessel stowage plan assigns containers to slots on the vessel. It is hard to generate good stowage plans since containers cannot be stacked freely due to global constraints like stability and bending forces and many interfering local stacking rules over and under deck.

Despite of the importance of stowage planning, the amount of previous work is surprisingly scarce. In the last two decades, less than 25 scientific publications have been made on the topic and there only exists two patents. The early approaches were "flat" in the sense that they introduced a decision variable or similar for each possible slot assignment of the containers (e.g.,[1],[2]). None of these scale beyond small feeder vessels of a few hundred 20-foot containers. Approaches with some scalability are heuristic (e.g., [3],[4],[5]) in particularly by decomposing the problem hierarchically (e.g., [6],[7],[8],[9]). None of these techniques, though, have been commercialized. They are either too slow or neglect important aspects of the problem due to little contact with industry experts.

I.P. Gent (Ed.): CP 2009, LNCS 5732, pp. 6–20, 2009.

We have since 2005 collaborated closely with a large liner shipping company that has developed an efficient hierarchical stowage planning algorithm using a more accurate domain model than any published work. An important sub-problem of this algorithm and other hierarchical algorithms is to assign a set of containers in a vessel bay. One of our objectives has been to compare different optimization techniques for this sub-problem. To this end, we have first defined the complete set of constraints and objectives used by our industrial partner and then constructed a simplified problem with a representative subset of these for an under deck bay. We have investigated incomplete methods based on local search (e.g., [10]) and evaluated these using complete methods.

In this paper, we introduce an optimal CP approach which to our knowledge is the first application of CP to container vessel stowage planning, to solve the sub-problem mentioned above. We present our stowage model and show how to solve it efficiently using Gecode [11]. State-of-the-art modeling techniques are considered including: different viewpoints to achieve better propagation, extensive use of global constraints to avoid modeling with boolean variables, and static and dynamic symmetry breaking. In addition, we use a branching strategy that takes advantage of the structure of the problem and a set of early failure detection algorithms that determines whether a partial assignment is inconsistent. The CP approach presented in this paper has been successfully tested on industrial data. Our experimental evaluation shows that the modeling decisions we made, in particularly the ones related to early failure detection, improve computation times substantially. The definition of the problem and model introduced in this paper have been slightly modified in order to make them easy to understand. We consider, though, that this simplification does not make less relevant the results here presented. Interestingly, preliminary results on the original version of the problem shows that a less elaborated CP model outperforms two optimal approaches based on Integer Programming (IP) [12] and Column Generation (CG). We believe this to be due to the logical nature of local stacking rules and objectives of stowage planning that mathematical programming is unable to handle efficiently. Thus, we consider CP to be the most efficient general technique to solve these problems optimally and, it gave us the main motivation to upgrade the initial CP model to the one presented here.

The remainder of the paper is organized as follows. Section 2 describes stowage planning problems. Section 3 defines our CP model. Section 4 describes why we believe CP outperforms mathematical programming on this problem. Finally, Section 5 presents experimental results, and Section 6 draws conclusions and discusses directions for future work.

2 The Container Stowage Problem for an under Deck Location

A container vessel is divided in sub-sections called *bays*, each bay of a container vessel consists of over and under deck *stacks* of containers. A *location* is a set of stacks that can be over or under deck. These stacks are not necessarily consecutive, but all stacks in the set are either over or under deck. Left figure in Fig.1

depicts a bay. The stacks 3, 4 and 5 under deck form a location, stacks 1, 2, 6 and 7 under deck form another location. The same stacks form two extra locations in the over deck section. This paper focuses on the under deck locations. The vertical alignments of cells in a location are called *tiers*. Left figure in Fig.1 shows how the tiers are enumerated for each section of the bay.

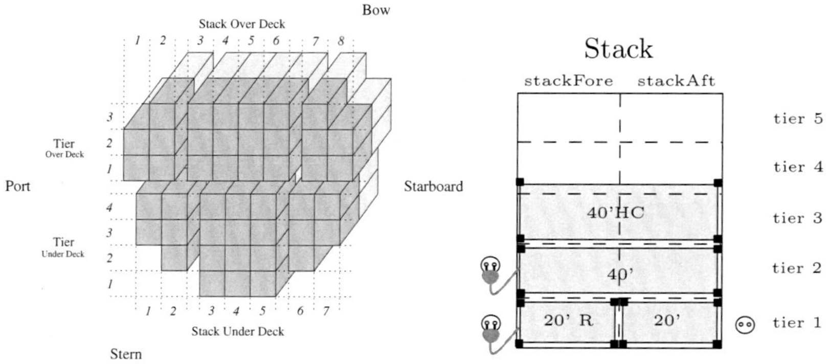

Fig. 1. To the left a back view of a bay, to the right a side view of a partially loaded stack. Each plug in the right figure represents a reefer slot, reefer containers are the ones with electric cords.

Each stack has a weight and height limit that must be satisfied by the containers allocated there. A stack can be seen as a set of piled *cells* one on top of the other. Each of these cells is divided in two *slots*, *Fore* and *Aft*. It is also possible to refer to the *Fore* and *Aft* part of a stack, i.e. *stackFore* refers to the *Fore* slots of all cells in a stack . Some slots have a plug to provide electricity to the containers, in case their cargo needs to be refrigerated. Such slots are called *reefer* slots. Right figure in Fig.1 shows the structure of a stack. As depicted in left figure in Fig.1, it is common for stacks not to have all slots physically available, they must fit into the layout of the vessel and some of the slots must be taken away to do so. These slots can be located either in the bottom or in the top of the stack and we refer to them as *Blocked slots.*

A *container* is a box where goods are stored. Each container has a weight, height, length, port where it has to be unloaded (discharge port) and indicates whether it11 needs to be provided with electric power (reefer). In an under deck location, containers have 20 or 40-foot length and 8'6" or 9'6" height. The weight is limited according to the length of the container and the discharge port depends on the route of the vessel. Containers that are 9'6" high are called *high-cube* containers, and according to the definition of the problem all high-cube containers are 40-foot long. Each cell in a stack can hold one 40-foot container or two 20-foot containers.

In order to generate stowage plans for complete vessels an efficient hierarchical algorithm has been developed. This algorithm decomposes the process of generating stowage plans into solving two derived problems: a master and a sub problem. The master problem focuses on constraints over the complete vessel i.e. stability constraints, bending forces, etc. and distributes containers in the different locations of the vessel but it does not assign them to a specific slot. Here, all containers to be loaded in the actual port are considered, together with forecasting information of further ports on the route of the vessel.

The sub-problem finds stowage plans for the locations of the vessel according to the distribution of containers made by the master problem. The constraints here are mostly stack wise and each container is assigned to an specific slot. There are two main types of locations in a vessel, over and under deck. Besides their position on the vessel, they differ in the constraints that the containers allocated there must fulfilled. As mentioned before, this paper focuses on finding stowage plans for the under deck locations.

The Container Stowage Problem for an Under Deck Location (*CSPUDL* is defined by the following constraints and objectives. A feasible stowage plan for an under deck location must satisfy the following constraints:

1. Assigned cells must form stacks (containers stand of top of each other in the stacks. They can not hang in the air).
2. 20-foot containers can not be stacked on top of 40-foot containers.
3. A 20-foot reefer container must be placed in a reefer slot. A 40-foot reefer container must be placed in a cell with at least one reefer slot, either Fore or Aft.
4. The sum of the heights and weights of the containers allocated in a stack are within the stack limits.

Every allocation plan that satisfies these constraints is valid, but since the problem we are solving here is to find the best allocation plan possible, a set of objectives must be defined to evaluate the quality of the solutions:

1. **Minimize overstows.** A container is stored above another container if it is stored in a cell with a higher tier number. A container A overstows a container B in a stack, if A is stored above B and the discharge port of A is after the one of B, such that A must be removed in order to unload B. A cost is paid for each container overstowing any other containers below.
2. **Keep stacks empty if possible.** A cost is paid for every new stack used.
3. **Avoid loading non reefer container into reefer cells.** A cost is paid for each non reefer container allocated in a reefer cell.

The first objective is directly related to the economical costs of a stowage plan. The second and third are rules of thumb of the shipping industry with respect to generating allocation plans for further ports in the route of a vessel. Using as few stacks as possible increases the available space in a location and reduce the possibility of overstowage in further ports. Minimizing the reefer objective allows reefer containers to be loaded also in further ports.

A feasible solution to the *CSPUDL* satisfies the constraints above. An optimal solution is a feasible solution that has minimum cost.

As a requirement from the industry, generating a stowage plan for a vessel should not take more than 10 minutes. Since a big vessel can have an average of one hundred locations, solving the *CSPUDL* as fast as possible is mandatory for this problem. An average of one second or less per location has been set as goal for solving the *CSPUDL*.

3 The Model

We present here a constraint programming model to find the optimal allocation plan for a set of containers *Containers* in a location l. In order to make the description of the model clearer, some constants are defined: *Slots* and *Stacks* are the set of slots and stacks of location l. $stackFore_i$ and $stackAft_i$ are the set of Fore and Aft slots in stack i. The weight and height limit of a stack i are represented by $stack_i^w$ and $stack_i^h$. Without loss of generality and in order to simplify some of the constraints, $stack_i$ refers to the set of cells from stack i. Every time a constraint is posted over a cell of $stack_i$, the constraint is actually posted over the Aft and Fore slots of the cell.

The set of decision variables of the problem is defined first. To improve propagation, we implement two different viewpoints[13] and channel them together such that both sets of variables contain the same information all the time.

The first viewpoint is the set of variables S, where each variable corresponds to a $slot_i \in Slots$ from location l. The second one is the set of variables C, where each variable represents a container from *Containers*.

$$S = \{s_i | i \in Slots\} \wedge s_i \in \{Containers\}, \forall i \in Slots$$
$$C = \{c_i | i \in Containers\} \wedge c_i \in \{Slots\}, \forall i \in Containers$$

In order to connect the two viewpoints, it is necessary to define a set of channeling constraints. Since the number of containers is not guaranteed to be equal to the number of slots in location l, we consider two possible alternatives to modeling the channeling. The first one is to declare a new set of boolean variables $C \times S = \{c \times s_{ij} | i \in Slots, j \in Containers\}$, where $c \times s_{ij} \leftrightarrow c_j = i \wedge s_i = j$, that channels set S and C, and add to the domain of each variable in S the value 0 to represent an empty slot. The second one is to extend the number of containers to match the number of slots of l, and define a single global channeling constraint in order to propagate information from one model to the other.

The main difference between these two approaches is how and when the information is propagated among the two viewpoints. Since the first approach uses boolean variables to channel the two models, the flow of information is limited to reflect assignment of variables from one viewpoint to the other.

In the second approach, since there is a channeling constraint connecting the two viewpoints, any update in the domains of the variables are propagated among viewpoints as they occur, increasing the levels of propagation. An extra advantage of using the second approach is that the alldifferent constraint is implied here, all containers must be allocated in a different slot, and all slots must hold different containers.

Our model implements the second approach. Artificial containers are added to the original set of containers to match the number of slots in l. Since a 40-foot container occupies two slots, these containers are split into two smaller containers of the size of a slot each: Aft40 and Fore40. All 40-foot containers are removed from the original set of containers and replaced by the new Aft40 and Fore40 containers. Empty containers are also added, they will be allocated in valid slots where no container is placed. Finally, it is necessary to add some extra containers that will be allocated in the blocked slots. Once the number of containers matches the slots of l, a global channeling constraint is used to connect the two view points, i.e. $channeling(S, N)$.

The first two constraints from the previous section describe how containers can be stacked according to physical limitations of the problem. They define the valid patterns the containers in stacks can form. We assign a code to each type of container, i.e. 0 to blocked containers, 1 to 20-foot containers, 2 to 40-foot containers and, 3 to empty containers, and define a regular expression that recognizes all the well-formed stacks according to the first two constraints: $R = \{r | r \in 0^*1^*2^*3^*0^*\}$. Then we define a constraint just allowing stacks accepted by R. In order to do this, a new set of auxiliary variables must be defined. These variables will represent the type of the container allocated in a slot, and their domain is the set of possible types for the containers: $T = \{t_i | i \in Slots\}$, $t_i \in \{0, 1, 2, 3\}$. To bind this new set of variables with one of the viewpoints, it is necessary to declare an array of integers $types$ representing the type associated to each container, and use element constraints such that: $types[s_i] = t_i, \forall i \in Slots$.

With the new set of auxiliary variables defined, we proceed to declare the constraints that will just allow well-formed stacks. A regular constraint [14] is declared for each Aft and Fore stack, together with the regular expression R that defines the well-formed stacks. In this constraint $stack_i$ refers to the subset of variables from T in stack i.

$$regular(stack_i, R), \forall\ i \in Stacks$$

For the reefer constraint two subsets of containers are defined: $\neg RC$ and $\neg 20RC$. $\neg RC$ is the subset of non-reefer containers and $\neg 20RC$ is a subset containing 40-foot, 40-foot reefer containers and 20-foot containers. The purpose of these two subsets is to restrict the domain of some of the slots of location l to allocate just the allowed containers. The first subset of slots is $\neg RS$, which are the slots that are non-reefer and that are not part of reefer cells. The second subset is RCS, which are slots that are non-reefer but that are part of a reefer cell. Then we remove the reefer containers from slots where it is not possible to allocate any reefer containers at all, and remove the 20-foot reefer containers from slots where it is possible to allocate part of a reefer container.

$$s_i \in \neg RC, \forall\ i \in \neg RS \wedge s_i \in \neg 20RC, \forall\ i \in RCS$$

Some extra sets of auxiliary variables are used in order to model the height and weight limit constraints for each stack in l. H is a set of variables where each h_i represents the height of the container allocated in s_i, W represents the same as

H but with respect to the weight of the container. Both sets of auxiliary variables are bound to S with element constraints, as it was previously explained for T. An extra set of variables is also declared here: $HS = \{hs_i | i \in Stacks\}$, $hs_i \in \{0, ..., stack_i^h\}, \forall i \in Stacks$, representing the height of each stack in location l. The constraints restricting the height and weight load of each stack in l are:

$$\sum_{j \in stackAft_i} h_j \leq hs_i, \forall i \in Stacks$$

$$\sum_{j \in stackFore_i} h_j \leq hs_i, \forall i \in Stacks$$

$$\sum_{j \in stack_i} w_j \leq stack_i^w, \forall i \in Stacks$$

3.1 Objectives

The first objective is overstowage. It is based on a feature of the containers that is not related to any previous constraint, the discharge port. A new set P of auxiliary variables is introduced here, where each p_i represents the discharge port of the container allocated in s_i. A new function is defined: $bottom : Stack \rightarrow Slots$, which associates each stack with its bottom slot. The finite domain variable Ov captures the number of overstows in location l.

$$Ov = \sum_{i \in Stacks} \sum_{j \in stack_i - \{bottom(i)\}} \Big(\sum_{k=bottom(i)}^{j-1} (P_j > P_k) > 0 \Big)$$

Since empty and blocked containers have discharge port 0, we use the previously declared set of auxiliary variables P to determine the number of stacks used in a solution. When a stack i is empty, the sum of the values assigned to the subset of P variables in i should be 0, otherwise the stack is been used. A finite domain variable Us captures the number of used stacks in location l.

$$Us = \sum_{i \in Stacks} \Big(\big(\sum_{j \in stack_i} P_j \big) > 0 \Big)$$

A check over the reefer slots is performed, the reefer objective increases its value if a non-reefer container is allocated in a reefer slot. A finite domain variable Ro captures the number of non-reefer containers allocated in reefer slots.

$$Ro = \sum_{i \in RS} (s_i \in \neg RC)$$

3.2 Branch and Bound

The relevance of the objectives defined for this problem is given by the relation $Ov > Us > Ro$. A lexicographic order constraint is used for the branch and bound search procedure to prune branches not leading to any better solution. Our model does not rely on an objective function to measure the quality of the solutions but on an order over the objective variables, which provides us with a stronger propagation. The branch and bound approach constraints the objective value of the next solution to be better than the one found so far. When this objective value is calculated by a mathematical function, the only constraint branch and bound posts is a relational constraint over this objective value, considerably reducing the amount of backwards propagation that can be achieved. In cases where an order determines the quality of the solution, lexicographic constraints can be used, which in most of the cases, propagate directly over each objective variable if necessary, increasing the level of backwards propagation achieved.

3.3 Branching Strategies

In our branching strategy we take advantage of the structure of the problem and the set of auxiliary variables defined in the model in order to find high-quality solutions as early as possible. We decompose the branching in three sub-branchings: the first one focuses on finding high-quality solutions, the second one in feasibility with respect to a problematic constraint and the third one finds a valid assignment for the decision variables.

Since two of the three objectives defined for this problem rely on the discharge port of the containers allocated in the slots of l, we start by branching over the set of variables P. Slots with discharge ports less or equal than the one assigned to the slot right below are selected, which decreases the probability of overstowage. The slots from stacks already used are considered first to reduce the used stack objective. When it is necessary to select a slot from an empty stack, the highest discharge port possible for the slot is selected.

After assigning all variables from P, we branch over a new set of auxiliary variables involved in one of the most problematic constraints: H. The height limits of the stacks are usually more strict than the weight limits and therefore, finding allocation plans that respect these limits become a difficult task in itself. No variable selection heuristic is involved for variables, we fill up stacks bottom-up and select the maximal height possible for a container to be allocated there.

At last, we branch over the set of variables S in order to generate an allocation plan. By this time, the discharge port and the height of the container to be allocated in each slot has been decided, and it is most likely that the objective value that any possible solution to be generated from this point is already known. Here we try to allocate slots from bottom-up in each stack, selecting the maximal possible container in the domain of the slot.

The decomposition of the branching plays along with the branch and bound strategy. The domain size of variables in P are considerable smaller than the ones from any of the viewpoints, making the process of finding valid assignments for P easier. Once the first valid allocation plan is found, most of the time the backtracking algorithm backtracks directly to the variables of the first branching in order to find a solution with a better objective value. Therefore, most of the search is concentrated in a considerable smaller sub-problem, branching over the two other sets of variables just when the possibilities of finding a better solution are almost certain.

3.4 Improving the Model

Some Extra Constraints. Here some redundant constraints are declared in order to improve propagation and reduce search time. The first constraint is an *alldifferent* constraint over the set S of slots, reinforcing the fact that it is not possible to allocate one container in more than one slot. This constraint is forced in the model by the *channeling* constraint between S and C.

A second constraint deals with a sub-problem presented in the model. Since all containers have a height, they must be allocated in the stacks from l and each stack has a height limit, the problem of finding a stack for each container to be allocated without violating the height capacity can be seen as a bin-packing problem. A global constraint for this problem is introduced in [15], where the load of each bin and the position of each item are given as variables. Here a new set of auxiliary variables CS representing the stack where a container is allocated is necessary. We use element constraints to bind this new set of variables with the set C. A modified implementation of [15] is considered to model this sub-problem, in order to tighten the height limits of each stack and not allow unfeasible assignments of containers based on their height.

Symmetries on Containers. The weight of the containers make each of them almost unique, limiting the possibility of applying symmetry breaking constrains. It is possible, though, to use these constraints on the artificial containers that were added to the model. First we focus on the set of empty containers, this set is split into two equal subsets that become the empty containers to be allocated in each part of the location. By doing this we avoid any set of equivalent solutions where empty containers are swapped between Aft and Fore slots. Then, a non-increasing order is applied over each of the subsets mentioned before in order to avoid any symmetrical solution.

Splitting up all 40-foot containers into two smaller containers, Aft40 and Fore40, also provoke symmetrical solutions. All Fore40 containers are removed from Aft slots and all Aft40 from Fore slots in location l.

Symmetries on Slots. The first subset of slots that we consider for symmetry breaking is the cells: swapping the containers allocated in Aft and Fore slots of a cell generates equivalent solutions in several cases. Therefore, when the Aft and Fore slot of a cell can allocate containers with the same features, a non-increasing ordering constraint is used indicating that the id of the container allocated in the Aft slot of the cell has to be greater than the one allocated in the Fore slot. It is not possible to apply an order to the slots in a stack since the tier of the slot where a container is allocated is related to the overstowage objective. There are some cases, though, where ordering can be applied. The first case is when all containers to load in location l have the same discharge port. In this case, it is possible to use a non-decreasing ordering constraint over all the slots in a stack that can allocate containers with the same features. In cases where there are different discharge ports it is not possible to sort the containers from the beginning. Here we take advantage of the different branching steps described in the previous section and select the subsets of slots in each stack where an ordering constraint can be used after the branching over the set of auxiliary variables P is finished. These subsets are defined by all the slots where containers with the same discharge port and the same features are allocated. Then a non-decreasing order constraint is used in each of this subsets.

At last, the possible symmetries between identical stacks are considered. In a pre-processing stage, stacks with the same features are grouped together: same slots capacity, reefer capacity, height limit and weight limit. When two stacks are in the same group, a lexicographic order is applied between them. The lexicographic order is also applied on the set of auxiliary variables P since this set of variables is assigned first than the set S.

Discussion on Symmetries. There is one relevant issue about the symmetry breaking constraints described in this section, more specifically on the lexicographic order constraints posted over stacks grouped together. Since these constraints are ordering the discharge ports, the height and the id of the containers in each stack, any assignment of values to these variables that does not follow such order will be considered invalid. It is necessary to sort the containers at a pre-processing stage to avoid any conflict among symmetry-breaking constraints over the different set of variables. The set of containers $Containers$ is sorted such that containers with the highest discharge port have associated the highest id. This sorting avoids the lexicographic constraints posted over the set S and set P of variables in identical stacks to conflict with each other.

Estimators. Three estimators have been defined to determine whether a complete valid solution can be generated from a partial solution, leading to early pruning of branches from the search tree with no future. Two of the estimators are simple algorithms that compute lower bounds of objectives from relaxed versions of the problem, while the third estimator is an early termination detector for the height constraint.

The first estimator finds the minimum number of stacks necessary to allocate all containers in a location from a partial solution. It greedily solves a simplified version of the allocation problem, where the only constraint considered is the height limit of the stacks and all containers not yet allocated are considered as normal height containers. The estimator starts by assigning containers to used stacks, no new penalization is paid to do so. Once all used stacks have been totally filled up, the remaining containers are allocated in the empty stacks, which are sorted by capacity before the estimator fills them up. By sorting the empty stacks we guarantee that the number of stacks used to allocate the remaining containers will be the smallest possible.

Formally, let $\prec^\rho$ be a total pre-order defined over the set of stacks:

$$\begin{aligned} k \prec^\rho m \Leftrightarrow \quad & (k, m \in Stacks_{used}) \vee \\ & (k \in Stacks_{used} \wedge m \in Stacks_{empty}) \vee \\ & (k, m \in Stacks_{empty} \wedge cap(k) \geq cap(m)), \end{aligned}$$

where the capacity $cap(k)$ is the remaining number of free slots in stack k. Let C^N denote the number of containers not assigned yet in a partial solution

$$C^N = |\{C_i \mid i \in Containers, |C_i| > 1, i \notin Empty\}|.$$

A recursive function calculating a lower bound of the number of used stacks is then given by:

$$\mu_\rho(c, \prec^\rho_j, \sigma) = \begin{cases} j & : \text{ if } c = 0 \\ \mu_\rho(c-1, \prec^\rho_j, \sigma - 1) & : \text{ if } c > 0 \wedge \sigma > 0 \\ \mu_\rho(c, \prec^\rho_{j+1}, cap(\prec^\rho_{j+1})) & : \text{ if } c > 0 \wedge \sigma = 0 \end{cases}$$

where c is the number of remaining containers to be placed, $\prec^\rho_j$ is the jth stack in the ordering, and σ is the free capacity of this stack. The estimated number of used stacks for any partial solution is then given by:

$$ES^U = \mu_\rho(C^N, \prec^\rho_1, cap(\prec^\rho_1))$$

For the reefer objective, let s_R, c_R, and c_E denote the number of unassigned reefer slots, unassigned reefer containers, and unassigned empty containers. A lower bound of the number of non-reefer containers placed in reefer slots Ro is:

$$ES^R = \begin{cases} 0 & : \text{ if } s_R \leq c_R + c_E \\ s_R - c_R - c_E & : \text{ otherwise} \end{cases}$$

To achieve improved propagation, we restrict the reefer and used stack objective to be greater or equal to the estimated lower bound: $Ro \geq ES^R \wedge Us \geq ES^U$.

The third estimator detects inconsistency of the height constraint. Since stacks are filled bottom-up, a stack j for some partial solution p has some free height $h(j)$ at the top. Let M^N_j and M^{HC}_j denote the maximum number of normal and high-cube containers that can be placed in stack j, respectively. We have:

$$\begin{aligned} M^N_j &= \lfloor h(j)/h(N) \rfloor, \\ M^{HC}_j &= \lfloor h(j)/h(HC) \rfloor \end{aligned}$$

where $h(N)$ and $h(HC)$ denote the height of normal and high-cube containers. Let C^N and C^{HC} denote the number of unassigned normal and high-cube containers of p. For the height constraint to be consistent for p, we then must have:

$$\sum_{j \in Stacks} M^N_j \geq C^N \quad \wedge \quad \sum_{j \in Stacks} M^{HC}_j \geq C^{HC}.$$

4 Why CP

Despite the fact that several of the capacity constraints of the *CSPUDL* are linear, it is non-trivial to represent logical constraints and objectives like no 20-foot over 40-foot and overstowage using mathematical programming. Moreover, the under deck stowage problem considered in industry includes more rule-based constraints and may even affect containers in adjacent stacks. Two of these constraints are due to pallet-wide containers and IMO containers with hazardous goods. The former takes up the limited space between stacks and therefore can not be placed in adjacent stacks, while the latter may require packing patterns where no IMO containers are placed next to each other in any direction.

We have made a preliminary investigation of two optimal mathematical programming approaches for solving a slightly different version of the problem, where one extra objective related to clustering containers with the same discharge port and pre-placed containers in locations are considered, and the objective function is a linear inequality with different weights for each objective.

The first of these approaches is described in [12] and uses an IP model with binary decision variables c_{jki} indicating whether container i is placed in cell k in stack j. The results are shown in table 1. Despite adding several specialized cuts and exploiting the general optimization techniques of the CPLEX solver, this approach only performs significantly better than CP in two instances, 4 and 5, and slightly better (a matter of few milliseconds) in instances 11 and 12.

The second approach uses column generation. The idea is to let each variable of the master LP problem represent a particular packing of a stack. The dual variables of the master problem are used by the slave problem to find a packing with negative reduced price wrt. the current set of candidate packings. In our preliminary experiments, IP was used to solve the pricing problem. The approach was implemented in GAMS using CPLEX. As depicted in table 1 the results are much worse than for IP and CP. Moreover, the LP variables of the master problem become fractional, which actually means that lower bounds rather than optimal feasible solutions are found.

The CP model from table 1 was our first attempt to use constraint programming to tackle the *CSPUDL*. It heavily relies on boolean variables for modeling, the use of global constraints is limited and not all estimation algorithms were implemented. The results obtained with this model were promising enough to continue our work with CP. Four out of seventeen instances were notoriously performing over the time limit established as goal (one second), all instances were solved to optimality and, in just four of the instances IP outperformed CP.

Table 1. Preliminary results comparing three optimal approaches to a slightly different version of the *CSPUDL*. All the results are in seconds.

Method (time)	1	2	3	4	5	6	7	8	9	10	11	12	13	14	15	16	17
CP(ms)	0.02	0.03	2	30.51	1.32	0.03	0.01	20	5.32	0.01	0.07	0.04	0.02	0.96	8.34	1.7	20.43
IP(ms)	0.06	0.15	-	1.75	0.160	0.12	0.19	0.15	-	0.01	0.01	0.01	0.1	1.17	21.89	4.94	40.05
CG(s)	14	22	1688	15588	-	17	22	22	352	7	12	28	14	27	18	11	37

Table 1 shows the results of solving to optimality seventeen instances of the *CSPUDL*. The first row shows the problem number. For these experiments, we used the same industrial problems as described in next section. The next three rows shows the computation time for the constraint programming (CP), integer programming (IP), and column generation (CG) approach. The dash in table 1 represents a timeout, the given time for this was thirty minutes.

5 Experimental Evaluation

A representative set of instances from different real-life vessels of different size, with different configurations of containers and discharge ports, with capacity location from 16 to 144 slots were gathered for the experiments here presented.

In table 2 the first eight columns of each row gives an overview of the features of each instance: The first column shows the instance number, the second one the total number of containers to be allocated, the third and fourth columns represent the number of 40 and 20-foot containers. The fifth and sixth columns are the number of reefer and high-cube containers, seventh column is the number of slots available and, the level of occupancy of the location is in the eight column. It is important to notice that a 40-foot container requires two slots to be allocated. All the experiments were run in a Ubuntu Linux machine with a Core 2 Duo 1.6 GHz and 1GB of RAM. The implementation of the constraint model took place in the C++ constraint library Gecode [11], version 3.0.2. The dash in table 2 represents the same as in table 1.

Table 2 shows the results of solving the set of locations using the CP model presented in this paper. Our main goal here is to present the impact of the estimator algorithms presented in section 3.4 and how they help to close the gap between the execution time of the CP approach and the time limit for this problem. The NS(s) and NS(nodes) columns show the response time in seconds and explored nodes of solving the instances without estimation algorithms. The E(s) and E(nodes) columns show the results of solving all instances to optimality including estimation algorithms in the model, time and explored nodes respectively. The branching strategy defined in section 3.3 was the one used.

From the results in table 2 it can be observed that estimation does have an important effect on the response time of the solver. Time is reduced in all instances but one, and in most of the cases the explored nodes also decrease. In instances where the explored nodes are equal but the time response has decreased, estimation algorithms are making nodes fail faster, avoiding unnecessary propagation. It is hard to determine the order of magnitude of the impact of the estimators, since it seems to vary from instance to instance, e.g. instance 4, 11, 13 and 16. The instance where it was not possible to prove optimality before has been solved in reasonable time, and instances 3 and 9 had a considerable reduction in their response time. It is also important to notice from the results in table 2, that our CP approach is getting closer to the goal described in the introduction, since just three instances out of seventeen remain with an execution time over one second.

With respect to the results presented in table 1, a small overhead can be seen in instances where finding the optimal solution usually takes less than 0.2 seconds, e.g. instances 1, 2, 6, etc. However, the substantial reductions in time response in instances 3, 4, 9 and 15 compensates the overhead.

Table 2. Problem instances and CP experimental results. Each row represents an instance. The first eight columns are general information of the instance, the extra four are the response time and explore nodes of the experiments described in this section.

Inst	Conts	C^{20}	C^{40}	C^{R}	C^{HC}	Slots	Full(%)	NS(s)	NS(nodes)	E(s)	E(nodes)
1	23	0	23	0	2	62	74	0.15	117	0.17	105
2	38	0	38	0	38	86	88	0.24	159	0.19	139
3	75	10	65	0	9	144	97	366.78	120461	0.59	213
4	40	0	40	34	34	90	89	14.14	4405	4.06	2391
5	28	0	28	24	28	90	62	0.79	271	0.31	187
6	28	0	28	0	10	60	93	0.28	119	0.07	61
7	35	0	35	0	8	72	97	0.39	153	0.19	153
8	34	0	34	0	7	70	97	0.34	147	0.17	147
9	53	0	53	0	5	108	98	37.19	9015	0.36	199
10	4	0	4	0	0	16	50	0.06	21	0.03	21
11	7	0	7	0	7	40	35	0.14	53	0.07	51
12	42	0	42	0	42	88	95	1.12	279	0.52	259
13	24	0	24	0	0	90	53	0.47	157	0.20	141
14	23	0	23	0	23	108	42	2.23	553	0.26	151
15	34	0	34	0	8	90	75	2.34	639	1.15	639
16	19	0	19	0	19	90	42	0.84	289	0.40	275
17	37	0	37	1	34	116	63	-	-	22.16	10153

6 Conclusion

In this paper we have introduced a model for stowing containers in an under deck storage area of a container vessel bay. We have shown how to solve this model efficiently using CP and compared our approach favorably with an integer programming and a column generation approach. CP is not widely used to solve problems to optimality. The estimation algorithms introduced in this paper, however, improves the performance of the branch and bound dramatically, good lower bounds are generated from partial solutions and unpromising branches are pruned in early stages without discarding any optimal solution.

We consider that the main reason of CP outperforming IP in most of the cases presented in this paper is the non-linear nature of some of the constraints and objectives of this problem, i.e. no 20-foot on top of 40-foot container, over-stowage. The logical nature of these constraints makes their linearization with 0-1 variables a non trivial task, and since further constraints to be included in this problem have the same logical nature as the ones mentioned before, i.e. IMO and pallet-wide containers, the CP approach will be most likely to keep outperforming an IP implementation.

An important objective of our future work is to make instances of stowage planning problems available to the CP community. We also plan to develop CP-based LNS stowage algorithms and investigate whether CP can be used to solve the pricing problem of column generation methods for this problem.

Acknowledgements. We would like to thank the anonymous reviewers and the following collaborators: Thomas Stidsen, Kent Hj Andersen, Trine Hyer Rose, Kira Janstrup, Nicolas Guilbert, Benoit Paquin and Mikael Lagerkvist. This research was partly funded by The Danish Council for Strategic Research, within the programme "Green IT".

References

1. Botter, R., Brinati, M.: Stowage container planning: A model for getting an optimal solution. In: Proceedings of the Seventh International Conference on Computer Applications in the Automation of Shipyard Operation and Ship Design (1992)
2. Giemesh, P., Jellinhaus, A.: Optimization models for the containership stowage problem. In: Proceedings of the International Conference of the German Operations Research Society (2003)
3. Ambrosino, D., Sciomachen, A., Tanfani, E.: Stowing a conteinership: the master bay plan problem. Transportation Research 38 (2004)
4. Avriel, M., Penn, M., Shpirer, N., Witteboon, S.: Stowage planning for container ships to reduce the number of shifts. Annals of Operations Research 76(55-71) (1998)
5. Dubrovsky, O., Penn, M.: A genetic algorithm with a compact solution encoding for the container ship stowage problem. Journal of Heuristics 8(585-599) (2002)
6. Ambrosino, D., Sciomachen, A., Tanfani, E.: A decomposition heuristics for the container ship stowage problem. Journal of Heuristics 12(3) (2006)
7. Kang, J., Kim, Y.: Stowage planning in maritime container transportation. Journal of the Operations Research society 53(4) (2002)
8. Wilson, I., Roach, P.: Principles of combinatorial optimisation applied to container-ship stowage planning. Journal Heuristics 1(5) (1999)
9. Gumus, M., Kaminsky, P., Tiemroth, E., Ayik, M.: A multi-stage decomposition heuristic for the container stowage problem. In: Proceedings of 2008 MSOM Conference (2008)
10. Pacino, D., Jensen, R.: A local search extended placement heuristic for stowing under deck bays of container vessels. In: Proceedings of ODYSSEUS 2009 (2009)
11. Gecode Team: Gecode: Generic constraint development environment (2006), `http://www.gecode.org`
12. Rose, H.T., Janstrup, K., Andersen, K.H.: The Container Stowage Problem. Technical report, IT University of Copenhagen (2008)
13. Smith, B.: Modelling. In: Rossi, F., van Beek, P., Walsh, T. (eds.) Handbook of Constraint Programming. Elsevier, Amsterdam (2006)
14. Pesant, G.: A regular language membership constraint for finite sequences of variables. In: Wallace, M. (ed.) CP 2004. LNCS, vol. 3258, pp. 482–495. Springer, Heidelberg (2004)
15. Paul, S.: A constraint for bin packing. In: Wallace, M. (ed.) CP 2004. LNCS, vol. 3258, pp. 648–662. Springer, Heidelberg (2004)

Real-Time Tabu Search for Video Tracking Association

Ivan Dotu[1], Pascal Van Hentenryck[1], Miguel A. Patricio[2], A. Berlanga[2], Jose García[2], and Jose M. Molina[2]

[1] Brown University, Box 1910, Providence, RI 02912
[2] Universidad Carlos III, 22, 28270 Colmenarejo (Madrid)

Abstract. Intelligent Visual Surveillance (IVS) systems are becoming a ubiquitous security component as they aim at monitoring, in real time, persistent and transcient activities in specific environments. This paper considers the data association problem arising in IVS systems, which consists in assigning blobs (connected sets of pixels) to tracks (objects being monitored) in order to minimize the distance of the resulting scene to its prediction (which may be obtained with a Kalman filter). It proposes a tabu-search algorithm for this multi-assignment problem that can process more than 11 frames per seconds on standard IVS benchmarks, thus significantly outperforming the state of the art.

1 Introduction

Intelligent Visual Surveillance (IVS) systems are becoming a key component in ensuring security at buildings, commercial areas, public transportation, parking, ports, etc. [13,26,28]. The primary aims of these systems are the real-time monitoring of persistent and transient objects and the understanding of their activity within a specific environment.

IVS Systems track all the objects (*tracks*) moving within their local field of view, giving rise to the so-called Multi-target Joint Estimation (MTJE) problem. MTJE is concerned with the estimation of the number of objects in a scene, together with their instantaneous locations, kinematic states, and other relevant characteristics. These entities, which are used for tracking, are called *track states.* A frequently used track state $\mathbf{x}$ consists of associating a vector $[x, y, \dot{x}, \dot{y}, w, h, s]$ with each track, representing the x and y coordinates of the centroid, the x and y velocities, the width, the height and scale of the object [15]. In general, the number of objects present in the scene is unknown and time-dependent, as is the state of each object. The tracking problem over time consists of estimating the sequence $\hat{\mathbf{X}}_k^n = \{\hat{N}_k; \hat{\mathbf{x}}_k^n\}_{n=1}^{N_k}$, where k is the time index, using the observations available up to that point. The observations or measurements concern pixels belonging to moving objects at time k. These pixels are organized in blobs and a blob is a connected set of pixels in an image, which is locally distinguishable from pixels not part of the blob. *One of the main tasks of an IVS system is to associate blobs with tracks, a process which is visualized in Figure 1.*

I.P. Gent (Ed.): CP 2009, LNCS 5732, pp. 21–34, 2009.

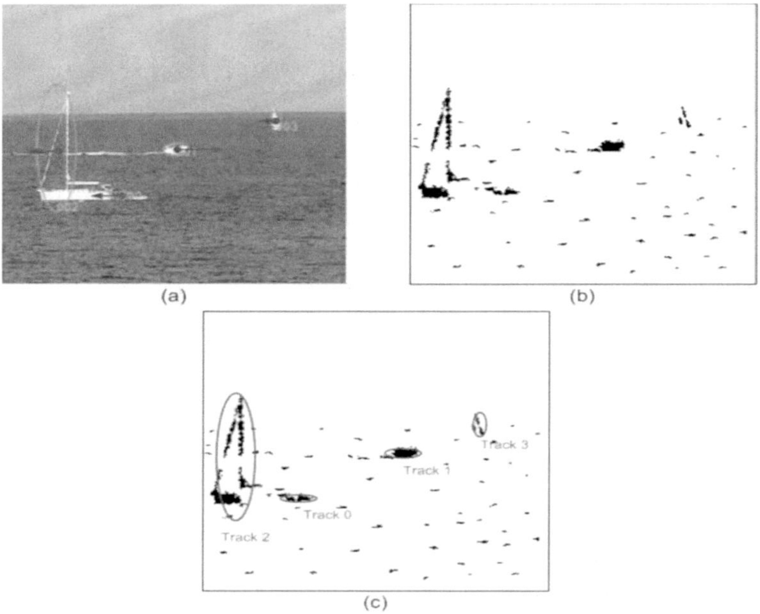

Fig. 1. (a) Four tracks (from 0 to 3) and their current state (position, size and velocity) at frame $k-1$; (b) Observation/detection of moving objects at frame k, we call *blob* to each connected group of detected pixels; (c) Association output at frame k

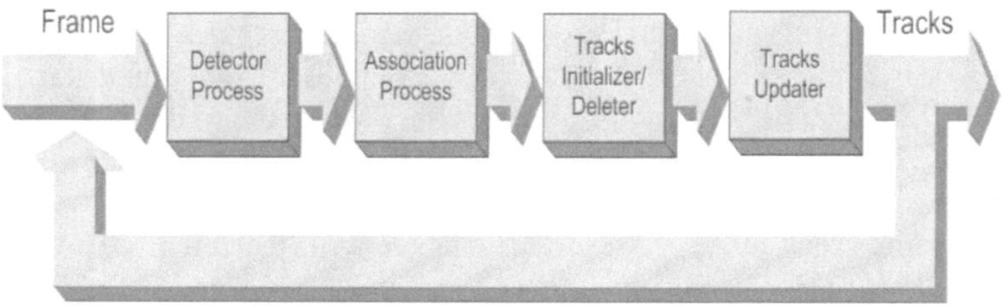

Fig. 2. The Architecture of an IVS System

Figure 2 depicts a standard architecture for an IVS system. First, a detector process produces the list of blobs found in a frame, with their positions and sizes. Then, the association process solves the blob-to-track multi-assignment problem, which assigns blobs to tracks, allowing a given blob to be assigned to several tracks. The goal in this step is to find a multi-assignment maximizing the similarity between the resulting tracks and those produced by the prediction module. Typically, a Kalman filter is used to predict the track state from the previous observations [31]. The blob initializer/deleter module eliminates those blobs not associated with any track (they are considered noise), and initializes new blobs. Finally, the tracks are updated and the process is repeated.

This paper focuses on the association problem. It proposes a tabu-search algorithm which solves the multi-assignment problem in real time (more than 11 frames per second), thus improving the performance of existing algorithms by an order of magnitude, while preserving solution quality.

The rest of the paper is organized as follows. Section 2 presents the related work on the association problem. Section 3 describes the problem formally. Section 4 presents the tabu-search algorithm and Section 5 presents the experimental results. Section 6 concludes the paper.

2 Related Work

The performance of a IVS depends on two strictly coupled subtasks: the data association method used for assigning observations to tracks and the model selected to estimate the movement of an object. Tracking algorithms can be formulated as the correspondence of detected objects represented by tracks across frames. Probabilistic methods are one of the most used algorithms. They consider the object measurement and the uncertainties to establish the correspondence. The target-tracking community has usually formulated the total process as an optimal state estimation and a data association problem, with a Bayesian approach to recursively filter the observations coming from the sensors. Only in the case that a single target appears, with no false alarms, there is no association problem and optimal Bayesian filters can be applied, for instance, Kalman filters under ideal conditions, or suboptimal Bayesian filters such as Multiple Models (IMM) [32] for realistic maneuvering situations and Particle Filters (PF) [2,12] in non-Gaussian conditions.

In the Kalman filter tracker, data association is a sequential frame-by-frame decision process based on current detections and the result of previous assignments. It can be formulated as maximizing the likelihood function of current observations (detections) conditioned on detections and associations from previous frames. With the Kalman filter one can estimate the track position based on previous assignments. The best assignment solution is the one that best matches the estimation of the Kalman filter.

Let us know review algorithms in the literature that belong to the first subtask mention: the data association. The most relevant approaches for solving the data association problem are different variants of evolutionary algorithms and the Mean-Shift and Particle Filtering ([8]).

Evolutionary Algorithms. Angus et al [1] applied Genetic Algorithms (GAs) to the data association problem in the radar contex and on a single scan scenario. GAs were also used by Hillis [17] for the multi-scan data association problem. Reference [25] proposed and analysed a family of efficient evolutionary computation algorithms designed specifically to achieve a fast video process, the performance of a family of which is analyzed. This novel approach is called Estimation of Distribution Algorithms (EDAs) [21] and these algorithms do not implement crossover or mutation operators. The new population is generated by sampling the probability distribution, which is estimated from the information of several selected individuals of the previous generations. The estimation of the joint probability distribution associated to the selected individuals is the most complex task and is approached in different ways by different authors. For instance, UMDA [22] assumes linear problems with independent variables, PBIL (Population-Based Incremental Learning) [6,3] uses a vector of probabilities instead of a population and also assumes variable independence, MIMIC [4] searches for permutations of variables in each generation in order to find the probability distribution using the Kullback-Leibler distance (two-order dependencies), and FDA [23] factorizes the joint probability distribution by combining an evolutionary algorithm and simulated annealing.

To evaluate our research, we implemented three different EDAs approaches: UMDA, PBIL, and the Compact Genetic Algorithm (cGA) [16]. In the UMDA algorithm [22], the joint probability distribution is estimated as the relative frequencies of the variable values stored in a data set. UMDA has been shown to work almost perfectly with linear problems and rather well when the dependencies are not very significant. PBIL mixes genetic algorithms with competitive learning. It applies a Hebbian rule to update the vector of probabilities. cGA simulates the performance of a canonical genetic algorithm with uniform crossover. It is the simplest EDA and only needs a parameter: the population size.

Mean-Shift and Particle Filtering. An implementation of the Mean-Shift algorithm together with a Particle Filtering algorithm is one of the most powerful tracking system in the vision community [7]. The Mean-Shift algorithm was proposed by Comaniciu and Meer [8] as an image segmentation technique. The algorithm finds clusters in the joint spatial & color space. It is initialized with a large number of hypothesized cluster centers, which are chosen, randomly, from a particular image. Then, each cluster center is moved to the mean of the data lying inside the multidimensional ellipsoid centered on the cluster center. Clusters can also get merged in some iterations. Mean-shift clustering has been used in various applications such as edge detection, image regularization [8], and tracking [9].

Tabu Search. As mentioned, this paper tackles the data association problem using tabu search [14]. The most related applications of tabu search include [20] acoustic control, [24] VLSI design, [10] image registration, and [18] 2D object tracking. This last reference focuses on different issues and does not describe the local search algorithm.

3 Problem Formalization and Modeling

Let us now formalize the specific problem we will be dealing with from the constraint-based local search standpoint: Given a set of blobs with information about their widths, heights, and centroids (measured in a 2-D grid of pixels) and a set of tracks with *predicted* information on their widths, heights, and *minimize*

$$\sum_{t \in T}(t_d + t_s)$$

subject to

$$t_d = \frac{|\widehat{t_x} - t_x|}{\widehat{t_w}} + \frac{|\widehat{t_y} - t_y|}{\widehat{t_h}} \quad (t \in T)$$

$$t_s = \frac{|\widehat{t_w} - t_w|}{\widehat{t_w}} + \frac{|\widehat{t_h} - t_h|}{\widehat{t_h}} \quad (t \in T)$$

$$t_{west} = \min_{b \in B: a[b,t]=1}(b_x - b_w/2) \quad (t \in T)$$
$$t_{east} = \max_{b \in B: a[b,t]=1}(b_x + b_w/2) \quad (t \in T)$$
$$t_{north} = \min_{b \in B: a[b,t]=1}(b_y - b_h/2) \quad (t \in T)$$
$$t_{south} = \max_{b \in B: a[b,t]=1}(b_y + b_h/2) \quad (t \in T)$$
$$t_w = t_{east} - t_{west} \quad (t \in T)$$
$$t_h = t_{south} - t_{north} \quad (t \in T)$$
$$t_x = t_{west} - t_w/2 \quad (t \in T)$$
$$t_y = t_{south} - t_h/2 \quad (t \in T)$$

$$0 \leq a[b,t] \leq 1 \quad (b \in B \,\&\, t \in T)$$

Fig. 3. The Model for the Data Association Problem

centroids, assign blobs to tracks to minimize the distance between the predicted tracks and the tracks specified by the assignment. Additionally, we may have at our disposal a hypothesis matrix that discards some blob-to-track assignments.

More precisely, we are given a set of blobs B and a set of tracks T. Each blob $b \in B$ is characterized by its centroid (b_x, b_y), its height b_h, and its width b_w. Each track $t \in T$ is given a prediction which is also characterized by its centroid $(\widehat{t_x}, \widehat{t_y})$, its height $\widehat{t_h}$, and its width $\widehat{t_w}$. The goal is to assign the blobs to the tracks or, more precisely, to determine whether a blob b is assigned to a track t. The decision variables are thus 0/1 variables of the form $a[b,t]$ which is equal to 1 when blob b is assigned to track t and 0 otherwise. This assignment determines the shape of each track t as the smallest box that encloses all its blobs. This shape is once again characterized by its centroid (t_x, t_y), its height t_h, and its width t_w. Note that t_x, t_y, t_h, and t_w are variables since they depend on the assignment variables in matrix a.

The objective function consists in maximizing the similarity with the proposed tracks and their predictions. This is expressed as the minimization of

$$F = \sum_{t \in T}(t_d + t_s) \tag{1}$$

where t_d is the normalized distance between the proposed and predicted track t:

$$t_d = \frac{|\widehat{t_x} - t_x|}{\widehat{t_w}} + \frac{|\widehat{t_y} - t_y|}{\widehat{t_h}} \tag{2}$$

and t_s is the normalized size similarity between the proposed and predicted track t:

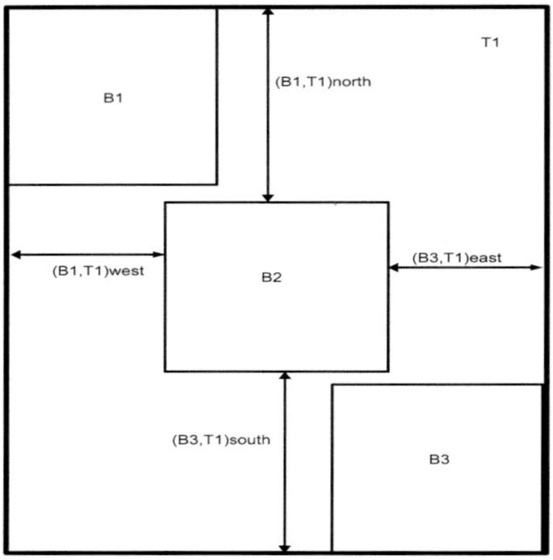

Fig. 4. Auxiliary Information Maintained For Each Blob

$$t_s = \frac{\left|\widehat{t_w} - t_w\right|}{\widehat{t_w}} + \frac{\left|\widehat{t_h} - t_h\right|}{\widehat{t_h}}. \tag{3}$$

The complete model is specified in Figure 3. The first set of constraints describe the components of the objective function. The second set of constraints compute the centroids, widths, and heights of the tracks, using the positions of the blobs. The first four equations in this group computes the west, east, north, and south of a track t from the blob assigned to t. The remaining equations specify the centroid, width, and height of the track from these values. Additional constraints can be easily added for excluding the possibility of assigning a blob b to a track t. Observe also that a blob may be assigned to multiple tracks. At this level of abstraction, when two tracks collide, it may be impossible to decide whether the blob belongs to one track or another. Forcing a blob to belong to one single track in these situations may result in the disappearance of one of the objects in the collision, which is obviously undesirable. Moreover, a blob does not necessarily need to be assigned to any track, since it is possible that it is only noise. In our modeling a track cannot be empty but in the pipeline implementation this can happen and other modules are responsible for losing the track in subsequent frames. Finally, it is worth mentioning that this model is independent of the prediction technique, although the Kalman filter is traditionally used but see [25] for further details.

4 The Tabu Search Algorithm

We now describe the tabu search algorithm for the data association problem. The neighborhood of the algorithm consists of three moves: removing a blob

Table 1. Removing Blob b from Track t. $(b,t)_\lambda$ to denote coordinate λ of blob b with respect to track t.

Attribute	Formula
x coordinate	$t_x + 0.5 * (b,t)_{west} - 0.5 * (b,t)_{east}$
y coordinate	$t_y + 0.5 * (b,t)_{north} - 0.5 * (b,t)_{south}$
width	$t_w - (b,t)_{west} - (b,t)_{east}$
height	$t_h - (b,t)_{north} - (b,t)_{south}$

from a track, adding a blob to a track, and swapping the tracks of two blobs. The algorithm maintains two tabu lists. The first list contains pairs of the form $\langle b,t\rangle$ to prevent a blob b from being added or removed from track t. The second list maintains triplets of the form $\langle b_1,b_2,t\rangle$ to avoid swapping blob b_1 in track t with blob b_2. The tabu tenure is dynamically and randomly generated in the interval $[4,100]$. The main issue in the implementation of the algorithm is how to perform and evaluate the moves efficiently, since the algorithm should run in real time. The problem comes from the fact that the cost function cannot be calculated directly. It is only when the blobs are assigned to tracks that the shapes of the tracks, and thus the distance to the prediction, can be computed.

4.1 Incremental Data Structures

To perform and evaluate moves efficiently, the tabu-search algorithm maintains additional information for each blob. For each blob b and each track t to which it is assigned, the algorithm maintains the following information:

- **North:** the number of pixels blob b adds to track t from the centroid toward a virtual north coordinate.
- **South:** the number of pixels blob b adds to track t from the centroid toward a virtual south coordinate.
- **East:** the number of many pixels blob b adds to track t from the centroid toward a virtual east coordinate.
- **West:** the number of many pixels blob b adds to track t from the centroid toward a virtual west coordinate.

Figure 4 illustrates these concepts and shows only the non-zero values. This figure indicates that removing block B1 would change the track size from the North and the West, while block B3 from the South and the East. Removing B2 would not change the track size.

This information allows the algorithm to compute how many pixels a track t would lose in each dimension if the blob is removed. It is initialized at the beginning of the algorithm and then updated after each move. It is also used to compute the costs of possible moves. The algorithm also maintains information about each track

Table 2. Adding blob b to track t. $t_{*'}$ denotes the value of characteristic $*$ of track t after the move and $(b,t)_\lambda$ to denote coordinate λ of blob b with respect to track t.

Attribute	Formula
x coord.	$min(b_{left}, t_x - 0.5 * t_w) + 0.5 * t_{w'}$
y coord.	$min(b_{top}, t_y - 0.5 * t_h) + 0.5 * t_{y'}$
width	$t_w + max(0, b_{right} - (t_x + 0.5 * t_w)) + max(0, (t_x - 0.5 * t_w) - b_{left})$
height	$t_h + max(0, b_{bottom} - (t_y + 0.5 * t_h)) + max(0, (t_y - 0.5 * t_h) - b_{top})$
b_{top}	$b_y - 0.5 * b_h$
b_{bottom}	$b_y + 0.5 * b_h$
b_{left}	$b_x - 0.5 * b_w$
b_{right}	$b_x + 0.5 * b_w$

Table 3. Swapping blobs b in track t with blob $b2$. $t_{*'}$ denotes the value of characteristic $*$ of track t after the move and $(b,t)_\lambda$ to denote coordinate λ of blob b with respect to track t.

	Formula
x	$min(b2_{left}, t_x - 0.5 * t_w) + 0.5 * t_{w'} + 0.5 * b_w - 0.5 * b_e$
y	$min(b2_{top}, t_y - 0.5 * t_y) + 0.5 * t_{y'} + 0.5 * b_n - 0.5 * b_s$
w	$t_w + max(0, b2_{right} - (t_x + 0.5 * t_w)) + max(0, (t_x - 0.5 * t_w) - b2_{left}) - b_w - b_e$
h	$t_h + max(0, b2_{bottom} - (t_y + 0.5 * t_h)) + max(0, (t_y - 0.5 * t_h) - b2_{top}) - b_n - b_s$
b_n	$min((b,t)_{north}, max(0, b2_{top} - b1_{top}))$
b_s	$min((b,t)_{south}, max(0, b1_{bottom} - b2_{bottom}))$
b_e	$min((b,t)_{east}, max(0, b1_{right} - b2_{right}))$
b_w	$min((b,t)_{west}, max(0, b2_{left} - b1_{left}))$

- The **number** of blobs currently assigned to it.
- The **centroid** of the track in a 2-D matrix of pixels.[1]
- The **height** and the **width** in pixels.

The last three items are also maintained for each blob.

4.2 The Moves

Table 1 describes the effect of removing a blob b from a track t. It describes how to recompute its centroid (first two lines), its width, and its height in terms of the auxiliary information described earlier. Table 2 describes the effect of adding a blob b to a track t. Once again, the first four lines describe how to recompute its centroid (first two lines), its width, and its height in terms of the auxiliary information, while the last four lines compute some auxiliary information about the blobs (the top/bottom/left/right corrdinates of the blob). Finally, Table 3 describes the computation to swap blob b in track t with blob $b2$. These formulas allow the algorithm to compute the new shape of a track in constant time. As a consequence, all moves can be performed and evaluated in constant time.

[1] The north-west corner is position $(0, 0)$.

4.3 The Starting Point

The starting point of a local search may have a significant impact on its efficiency and solution quality. Two types of starting points were considered in the tabu-search algorithm.

- **Randomly** assign 0 or 1 to the $a[b,t]$ variables. This is equivalent to randomly assign blobs to tracks. Whenever a blob is not allowed to belong to a certain track (given by the hypothesis matrix mentioned in Sect. 3), the corresponding $a[b,t]$ variable is set to 0.
- **In Proximity** order to the predicted centroid of a track t, assign blobs to t until it surpasses either its predicted width or its predicted height. Repeat for every track.

The experimental results will indicate that the second choice brings significant benefits.

5 Results

This section reports the experimental results of the tabu-search algorithm for the data-association problem. Starting with a random or proximity-based initial solution, the algorithm iteratively chooses the best move possible in the neighborhood until a given number of iterations is reached. The algorithm maintains the two tabu lists described earlier for forbidden moves and implement the aspiration criterion to overwrite the tabu status whenever a new best solution would be obtained. This algorithm has been implemented in the language Comet [30] and experiments were run on a MacBook under Mac OS X with a 2.4 GHz Intel Core 2 Duo processor and 4 GB of memory. The results are given for several instances of four different videos:

- **CLEAR (Classification of Events, Activities and Relationships) [29].** This dataset is deployed to evaluate systems that are designed to recognize events, activities, and their relationships in interaction scenarios. 2 tracks and up to 50 blobs.
- **CANDELA [5] (Content Analysis and Network DELivery Architectures).** From 5 to 10 tracks and up to 17 blobs.
- **HANDBALL.** This video is from the publicly available CVBASE dataset [11]. It has 16 tracks and 23 blobs.
- **Football INMOVE.** This video belongs to the Performance Evaluation of Tracking and Surveillance (PETS) dataset [27]. From 8 to 10 tracks and up to 42 blobs.

The scenes were chosen based on their complexity. We selected scenes where there were a large number of blobs to be assigned to a high number of tracks. These problems become a challenging problem of blob-to-track association.

To establish the solution quality of our approach compared to other methods, we implemented our algorithm within the pipeline of an IVS system. Table 4

Table 4. Measures of quality of the algorithms applied to HANDBALL. Tracks per Frame (TPF), Frames per Second (FPS) and Lost Track Probability (LTP).

	mean TPF (ideal=14)	sd TPF	FPS	LTP
TABU	12.175	1.077	11.33	0.03
CGA	10.769	1.095	0.81	0.20
PBIL	11.639	1.220	0.66	0.20
MSPF	12.920	0.523	0.56	0.22
UMDA	7.353	1.014	0.31	0.71
GA	12.403	1.821	0.04	0.43

Table 5. Best Results from 20 Static Scenarios: 1-4 from **CLEAR**, 5-13 from **CANDELA**, 14-15 from **HANDBALL** and 16-20 from **INMOVE**

Scenario	Optimum	Proximity	Tabu	Tabu/Init	Tabu/Swap	SAGreedy	SA
1	1.021	1.024	1.021	1.021	1.021	1.021	1.021
2	0.369	1.294	0.369	0.369	0.369	0.369	0.369
3	0.385	1.074	0.385	0.385	0.385	0.385	0.438
4	0.372	0.588	0.372	0.372	0.372	0.382	0.372
5	1.403	1.569	1.403	1.403	1.403	1.514	1.403
6	1.040	6.401	1.040	1.040	1.040	1.040	1.040
7	0.893	2.435	0.893	0.893	0.893	0.893	0.893
8	1.807	6.055	3.120	1.807	1.807	1.807	1.807
9	1.784	5.684	1.784	1.784	1.784	1.784	1.784
10	1.023	5.431	1.023	1.023	1.023	1.023	1.023
11	2.587	6.867	2.587	2.587	2.587	2.891	2.587
12	2.853	9.122	5.362	2.853	2.853	2.853	3.735
13	4.162	8.933	5.462	4.162	4.162	4.162	5.274
14	7.177	10.199	13.745	7.177	7.177	7.264	7.264
15	8.067	11.668	21.144	8.067	8.067	8.067	8.067
16	2.532	2.642	10.830	2.635	2.532	2.642	2.635
17	2.823	9.040	9.603	2.823	2.823	2.823	2.823
18	3.433	4.838	9.970	3.433	3.433	3.433	3.433
19	4.805	6.761	6.267	4.805	4.805	4.977	4.977
20	4.080	5.190	7.295	4.108	4.080	4.108	4.080

shows a comparison of the tabu-search algorithm against the methods described in Section 2. It can be seen that the tabu-search algorithms compares well in terms of solution quality and outperforms the other approaches in efficiency by orders of magnitude. *In particular, it is the only method that can process video tracking in real-time (11 Frames Per Second, FPS).*

It is hard to reproduce the actual behavior of the algorithms without watching at the video scene itself, however, the measures used in this paper can give us an approximate idea of their quality. TPF (Tracks Per Frame) shows how many tracks appear on every frame, which should be close the actual number of tracks (ideal number of tracks). However, this does not mean that these tracks

Table 6. Average Results from 20 Static Scenarios: 1-4 from **CLEAR**, 5-13 from **CANDELA**, 14-15 from **HANDBALL** and 16-20 from **INMOVE**

Scenario	Opt.	Proximity	Tabu	Tabu/Init	Tabu/Swap	SAGreedy	SA
1	1.021	1.024	1.937	1.021	1.021	1.021	1.024
2	0.369	1.294	1.083	0.369	0.369	0.369	0.940
3	0.385	1.074	0.657	0.385	0.385	0.385	0.818
4	0.372	0.588	0.784	0.372	0.372	0.382	0.568
5	1.403	1.569	5.492	1.403	1.403	1.514	1.552
6	1.040	6.401	4.249	1.040	1.040	1.040	2.814
7	0.893	2.435	2.324	0.893	0.893	0.893	1.035
8	1.807	6.055	5.739	1.807	1.807	1.807	4.465
9	1.784	5.684	2.814	1.784	1.784	1.784	3.419
10	1.023	5.431	3.705	1.023	1.023	1.023	3.500
11	2.587	6.867	3.055	2.587	2.587	3.277	4.201
12	2.853	9.122	10.036	2.853	2.853	2.889	7.335
13	4.162	8.933	9.334	4.162	4.162	4.162	7.820
14	7.177	10.199	18.693	7.177	7.177	9.002	9.816
15	8.067	11.668	25.503	8.067	8.067	8.814	11.283
16	2.532	2.642	15.292	2.635	2.532	2.642	2.642
17	2.823	9.040	14.644	2.823	2.823	3.646	6.563
18	3.433	4.838	13.976	3.433	3.433	3.445	4.677
19	4.805	6.761	9.074	4.805	4.805	5.004	6.322
20	4.080	5.190	8.228	4.108	4.080	4.108	5.067

are the real tracks we would like to target (noise instead of objects). Thus, this measure should be taken into account along with the LTP (Lost Track Probability) measure. This measure tells us how likely it is that we lose a track using a given algorithm. Note that this measures has also been used in other frameworks ([19]).

Moreover, we have captured 20 different static scenarios from the 4 different videos introduced above. Table 5 shows the optimal and best results in terms of the value of the fitness function F described in section 3 for a variety of algorithms, while Table 6 depicts the average solution quality over 100 runs. The considered algorithms are the proximity heuristic (Proximity), tabu search with a random initialization and no swapping (Tabu), tabu search with the proximity-based initialization (Tabu + Init), tabu search with the proximity-based initialization and swaps (Tabu + Swap). We also report results of simulated annealing with and without greedy acceptance (SAgreedy and SA respectively) and using the proximity based initialization. The algorithms all have a time limit to achieve real-time performance (at least 7 FPS). Note that the data association is only one part of an IVS system, and other modules can take up to 90% of the process time. Thus the time limits for the different scenarios range from 30 to 62 miliseconds, depending on the number of blobs and tracks.

The experimental results indicate that the initialization function is key to efficiency. It allows tabu search to find the optimal solutions to all the instances except for 2, in the very limited amount of iterations allowed (around 50 iterations per track). The two remaining scenarios can also be solved optimally by tabu search but requires a more complex neighborhood that includes swaps. Overall, tabu search with the proximity-based initialization and a neighborhood including swaps finds the optimal solution consistently (i.e, it finds the optimal solution in every run). The simulated annealing algorithm cannot achieve the required solution quality in real-time: It can solve very few instances optimally in the average and its returned solution may be far from the optimal value, highlighting the importance of tabu search for this problem. Note that we provide the optimal fitness using Constraint Programming (CP); the times needed to solve these instances to optimality range from several hours up to a day.

Finally, we have fixed the number of iterations for all the algorithms and compared the computation times with and without the incremental data structures. The incremental data structures reduces the execution times by a factor 9 for the tabu search algorithms (with swaps), showing how critical the incremental data structures are to achieve real-time performance in video tracking.

In summary, the experimental results indicate that a tabu-search algorithm with the proximity-based heuristic and both assignment and swap moves produces the required solution in real time and is the only method with these functionalities. This is a significant contribution to intelligent visual surveillance systems.

6 Conclusions

This paper considered the data association problem arising in Intelligent Visual Surveillance (IVS) systems whose aim is to monitor, in real time, persistent and transcient activities in specific environments. The data association problem consists in assigning blobs (connected sets of pixels) to tracks (objects being monitored) in order to minimize the distance of the resulting scene to its prediction (which may be obtained with a Kalman filer). The paper has proposed a tabu-search algorithm for this multi-assignment problem that finds optimal solutions consistently and processes more than 7 frames per seconds on standard IVS benchmarks. The resulting algorithms significantly outperforms the state of the art by providing the first real-time algorithm delivering the required solution quality.

References

1. Angus, J., Zhou, H., Bea, C., Becket-Lemus, L., Klose, J., Tubbs, S.: Genetic algorithms in passive tracking. Technical report, Claremont Graduate School, Math Clinic Report (1993)
2. Arulampalam, M.S., Maskell, S., Gordon, N., Clapp, T.: A tutorial on particle filters for online nonlinear/non-gaussian bayesian tracking. IEEE Transactions on Signal Processing [see also IEEE Transactions on Acoustics, Speech, and Signal Processing] 50(2), 174–188 (2002)

3. Baluja, S.: Population-based incremental learning: A method for integrating genetic search based function optimization and competitive learning, Technical Report CMU-CS-94-163, CMU-CS, Pittsburgh, PA (1994)
4. de Bonet, J.S., Isbell Jr., C.L., Viola, P.: MIMIC: Finding optima by estimating probability densities. In: Mozer, M.C., Jordan, M.I., Petsche, T. (eds.) Advances in Neural Information Processing Systems, vol. 9, p. 424. The MIT Press, Cambridge (1997); Artech House, Inc. (1999)
5. Content analysis and network delivery architectures, `http://www.hitech-projects.com/euprojects/candela/`
6. Cestnik, B.: Estimating probabilities: A crucial task in machine learning. In: ECAI, pp. 147–149 (1990)
7. Chen, T.P., Haussecker, H., Bovyrin, A., Belenov, R., Rodyushkin, K., Kuranov, A., Eruhimov, V.: Computer vision workload analysis: Case study of video surveillance systems. j-INTEL-TECH-J 9(2), 109–118 (2005)
8. Comaniciu, D., Meer, P.: Mean shift: a robust approach toward feature space analysis. IEEE Transactions on Pattern Analysis and Machine Intelligence 24(5), 603–619 (2002)
9. Comaniciu, D., Ramesh, V., Meer, P.: Kernel-based object tracking. IEEE Trans. Pattern Anal. Mach. Intell. 25(5), 564–575 (2003)
10. Cordon, O., Damas, S.: Image registration with iterated local search. Journal of Heuristics 12(1-2), 73–94 (2006)
11. University of Ljubljana Machine Vision Group. In: Cvbase 2006 workshop on computer vision based analysis in sport environments (2001), `http://vision.fe.uni-lj.si/cvbase06/`
12. Djuric, P.M., Kotecha, J.H., Zhang, J., Huang, Y., Ghirmai, T., Bugallo, M.F., Miguez, J.: Particle filtering. IEEE Signal Processing Magazine, 19–38 (2003)
13. Ferryman, J.M., Maybank, S.J., Worrall, A.D.: Visual surveillance for moving vehicles. Int. J. Comput. Vision 37(2), 187–197 (2000)
14. Glover, F., Laguna, M.: Modern Heuristic Techniques for Combinatorial Problems. Blackwell Scientific Publishing, Malden (1993)
15. Han, M., Xu, W., Tao, H., Gong, Y.: An algorithm for multiple object trajectory tracking. In: CVPR 2004: Proceedings of the IEEE Computer Society Conference on Computer Vision and Pattern Recognition, vol. 01, pp. 864–871 (2004)
16. Harik, G.R., Lobo, F.G., Goldberg, D.E.: The compact genetic algorithm. IEEE Transactions on Evolutionary Computation 3(4), 287 (1999)
17. Hillis, D.B.: Using a genetic algorithm for multi-hypothesis tracking. In: ICTAI 1997: Proceedings of the 9th International Conference on Tools with Artificial Intelligence, Washington, DC, USA, p. 112. IEEE Computer Society, Los Alamitos (1997)
18. Huwer, S., Niemann, H.: 2d-object tracking based on projection-histograms. In: Burkhardt, H.-J., Neumann, B. (eds.) ECCV 1998. LNCS, vol. 1406, pp. 861–876. Springer, Heidelberg (1998)
19. Kan, W.Y., Krogmeier, J.V.: A generalization of the pda target tracking algorithm using hypothesis clustering. Signals, Systems and Computers 2, 878–882 (1996)
20. Kincaid, R.K., Laba, K.E.: Reactive tabu search and sensor selection in active structural acoustic control problems. Journal of Heuristics 4(3), 199–220 (1998)
21. Larraaga, P., Lozano, J.A.: Estimation of Distribution Algorithms: A New Tool for Evolutionary Computation. Kluwer Academic Publishers, Norwell (2001)
22. Mühlenbein, H.: The equation for response to selection and its use for prediction. Evolutionary Computation 5(3), 303–346 (1997)

23. Mühlenbein, H., Mahnig, T.: The factorized distribution algorithm for additively decompressed functions. In: 1999 Congress on Evolutionary Computation, pp. 752–759 (1999)
24. Pisinger, D., Faroe, O., Zachariasen, M.: Guided local search for final placement vlsi design. Journal of Heuristics 9(3), 269–295 (2003)
25. Patricio, M.A., García, J., Berlanga, A., Molina, J.M.: Video tracking association problem using estimation of distribution algorithms in complex scenes. In: Mira, J., Álvarez, J.R. (eds.) IWINAC 2007. LNCS, vol. 4528, pp. 261–270. Springer, Heidelberg (2007)
26. Regazzoni, C.S., Vernazza, G., Fabri, G. (eds.): Highway traffic monitoring. Kluwer Academic Publishers, Dordrecht (1998)
27. 4th IEEE International Workshop on Performance Evaluation of Tracking and Surveillance (PETS 2003), `http://www.cvg.cs.rdg.ac.uk/VSPETS/vspets-db.html`
28. Regazzoni, C.S., Vernazza, G., Fabri, G. (eds.): Security in ports: the user requirements for surveillance system. Kluwer Academic Publishers, Norwell (1998)
29. Stiefelhagen, R., Bernardin, K., Bowers, R., Rose, R.T., Michel, M., Garofolo, J.: The CLEAR 2007 Evaluation. In: Stiefelhagen, R., Bowers, R., Fiscus, J.G. (eds.) RT 2007 and CLEAR 2007. LNCS, vol. 4625, pp. 3–34. Springer, Heidelberg (2008)
30. Van Hentenryck, P., Michel, L.: Constraint-Based Local Search. The MIT Press, Cambridge (2005)
31. Xiao-Rong, L., Bar-Shalom, Y.: Multitarget-Multisensor Tracking. In: Principles and Techniques (1995)
32. Yeddanapudi, M., Bar-Shalom, Y., Pattipati, K.: Imm estimation for multitarget-multisensor air traffic surveillance. Proceedings of the IEEE 85, 80–96 (1997)

Pin Assignment Using Stochastic Local Search Constraint Programming

Bella Dubrov[1], Haggai Eran[1], Ari Freund[1], Edward F. Mark[2], Shyam Ramji[2], and Timothy A. Schell[2]

[1] IBM Haifa Research Lab, Haifa, Israel
{bella,haggaie,arief}@il.ibm.com
[2] IBM East Fishkill, Hopewell Junction, NY
{efmark,ramji,tschell}@us.ibm.com
http://www.haifa.ibm.com/projects/verification/csp/

Abstract. VLSI chips design is becoming increasingly complex and calling for more and more automation. Many chip design problems can be formulated as constraint problems and are potentially amenable to CP techniques. To the best of our knowledge, though, there has been little CP work in this domain to date. We describe a successful application of a CP based tool to a particular *pin-assignment* problem in which tens of thousands of pins (i.e., connection points) belonging to internal units on the chip must be placed within their units so as to satisfy certain constraints and optimize the wirability of the design. Our tool has been tested on real IBM designs and is now being integrated into IBM's chip development environment.

Keywords: Constraint Programming, Stochastic Local Search, EDA, ASIC, Chip Design, Pin Assignment.

1 Introduction

This paper explores the application of Constraint Programming (CP) for developing Computer Aided Design (CAD) tools for Integrated Circuit (IC) design and demonstrates the successful automation of *pin-assignment* during IC design.

We start with a brief background on the various phases involved in a typical IC design process. For illustration, we consider an Application Specific Integrated Circuit (ASIC) design flow [1]. ASIC chips are designed for specific applications or functions such as encoding and decoding digital data, embedded functions within a factory automation system and so on. Generally, ASIC designs adopt a standard cell design methodology wherein the circuit layout for primitive logic operations (AND, OR, NAND, XOR etc.) is available as a standard cell library that is then used to implement the chip logic function. As Figure 1 illustrates, starting with the design specification in a high-level language such as VHDL or Verilog, the logic synthesis phase generates a cell-level implementation, i.e., netlist (interconnected cells) that is presented to physical design to generate a layout mask for chip fabrication. The netlist is then partitioned into

I.P. Gent (Ed.): CP 2009, LNCS 5732, pp. 35–49, 2009.

blocks based on logic function, physical connectivity or other extraneous design constraints. At this point, the physical design implementation of the partitioned logic is considered subject to fabrication technology, chip package, IO cells, metal layer stack for interconnect, power distribution etc. Floorplanning is the phase where the circuit blocks (partitions) are assigned an area, shape and location along with chip IO cell placement. Once the large blocks are floorplanned, the standard cells are placed and logic optimized (physical synthesis) based on estimated interconnect length. Then the connections are routed along the shortest length using metal layers on a regular grid (per layer) to complete the chip implementation while meeting the design frequency targets. The generated layout mask captures the geometrical shapes which correspond to the metal, oxide or semiconductor layers that make up the components of the integrated circuit. The enormous design complexity and short time-to-market for ASIC chips has lead to evolution of CAD tools to automate various phases in this design flow.

The IC design flow presents several feasibility and optimization problems that demand efficient and effective algorithms to be developed in the CAD tools [2,3]. The growing complexity of VLSI designs in ultra-deep sub-micron technologies has also driven the need for hierarchical design methodologies to reduce the overall turn-around time. Each phase in the design flow exhibits a flavor of constrained optimization problem, as seen in a typical hierarchical IC design flow. For example, logic synthesis attempts to minimize cell area subject to available library cells (logic functions), delay, power etc.; partitioning divides the circuit into sub-blocks called macros with a defined physical boundary and macro pins serving as the interface for connections from the top (chip) level to the cells within the macros (i.e., connections between cells in different macros and connections between cells and the chip's external pins), where the objective is to minimize the cuts or interconnect crossings subject to arbitrary cell area balance criteria between the partitions; floorplanning attempts to shape and place blocks to minimize estimated interconnect length and chip layout area subject to constraints such as relative placement of large blocks with respect to IO cells, spacing of cells, macro pin assignment along the periphery of the blocks and alignment to the metal layer pitch; placement and routing of cells within the macro blocks and at the chip level with an objective to minimize the routed interconnect length subject to meeting design timing (frequency) and power constraints.

Given the inherently hard sub-problems, several optimization techniques such as integer linear programming, network flows, quadratic programming, simulated annealing, and combinatorial optimization integrated with effective heuristics have been explored in CAD for VLSI [3,4]. However, to the best of our knowledge, there has been little research on the application of constraint programming (CP) techniques to IC physical design problems. We explore this aspect in the present paper.

In particular, we address the macro *pin-assignment* problem during the chip floorplanning phase. As described earlier, in a hierarchical design approach, the partitioned macros have pins which are the logical interface between the internal

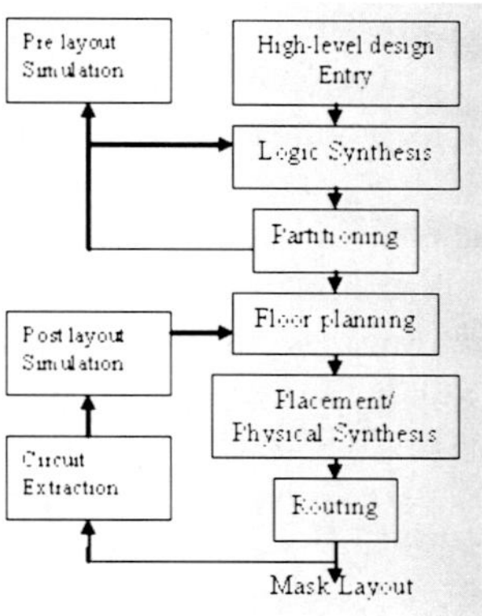

Fig. 1. ASIC design flow

(macro) level cells and the top (chip) level blocks or cells. Assigning pin locations, typically along the periphery of the macros as shown in Figure 2, in order to minimize the interconnect length subject to physical technology rules (spacing, alignment to grid etc.) is referred to as *pin-assignment*. In such a design flow, the macro pin assignment heavily influences the quality of cell placement and routing at both the macro and the chip level which in-turn impacts the overall chip physical design time. Therefore, a good (optimal) pin assignment is desirable for any hierarchical VLSI design flow.

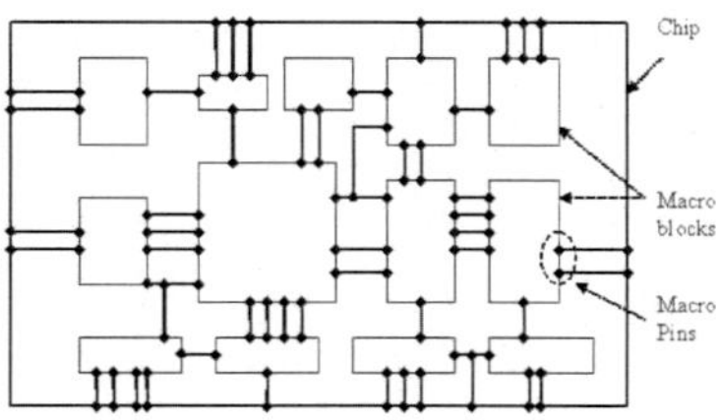

Fig. 2. Macro pin-assignment for a given VLSI chip floorplan

Pin-assignment in VLSI physical design has been researched over several decades with various techniques developed for different design styles ranging from printed circuit boards [5] to recent multi-million gate integrated circuits (IC) layouts [6,7]. Broadly, the existing macro pin-assignment algorithms can be classified as: a) pin placement on the macros to minimize the estimated top (chip) level wire length [8,9], and b) pin-assignment coupled with global routing on a net-by-net basis [10,6]. The term *net* refers to a set of pins (belonging to different cells) that must all be connected together. More recently, simultaneous pin-assignment and global routing for all 2-pin nets using network flow formulation [7], and pin-placement integrated with analytical cell placement using quadratic programming methods [11] have also been discussed. However, the known pin-assignment algorithms have limitations in that they either use greedy heuristics, or consider nets (pin-connections) sequentially leading to inferior solutions, or create an abstraction for continuous global optimization that ignores detailed pin-placement constraints such as pin-spacing, layer restriction and alignment. Additionally, current methods of macro pin-assignment require a good deal of manual intervention by a physical design engineer. This paper explores the use of CP techniques to model the general macro pin-assignment problem while considering all nets simultaneously. We describe an automated macro pin-assignment tool using CP techniques and demonstrate the results on real-world industrial chip designs. We believe that the area of electronic design automation has the potential to become a fertile application domain for CP methods since many of its problems have natural formulations in terms of known constraints. Also, the flexibility offered by CP, whereby constraints may be easily added or removed, is an important advantage in this domain where different problems (stemming from different design methodologies, different hardware technologies, different levels in the design hierarchy, etc.) are similar but not identical to each other.

1.1 Stochastic Local Search

Stochastic CSP solvers work by defining a non-negative cost function on the points of the search space (full assignments to all variables in the problem). The cost of a point is zero if the point is *feasible*, i.e., it satisfies all constraints, and is positive otherwise. Intuitively, the cost of a point corresponds to a measure of how far the point is from feasibility. Typically, the cost of a point is defined as the sum of the individual costs attached by the constraints to it, that is, there is a cost function (on the full assignments) associated with each constraint—intuitively measuring how violated the constraint is by the assignment, with 0 indicating satisfaction—and the cost of the point is the sum of these individual costs. It is the job of the problem modeler to provide these cost functions.

The solver then starts with a randomly (or otherwise) chosen full assignment for the CSP variables, and in each iteration examines a random sample of the "neighborhood" of the current assignment, and tries to find an assignment with lower cost. This is repeated until a feasible point is found or a timeout occurs. Of course, the solver may be augmented with various heuristics to accelerate the search, escape from local minima, etc.

This approach can also be extended to solve constrained optimization problems by similarly defining an optimization objective function which is used once feasibility has been attained.

Our tool is based on *Stocs*—a stochastic CSP solver developed in IBM [12] which has the ability to solve constraint optimization problems as well as pure feasibility ones. Stocs employs different strategies in its decision which neighboring points to examine at each iteration, ranging from random to user-defined. The choice of which strategy to use at each step is itself (non-uniformly) random. One powerful strategy involves *learning*. During the search Stocs tries to learn the inherent topography of the search space by remembering the directions and step sizes that proved most useful in past iterations. It uses this information to predict the most promising points to examine next. However, Stocs never abandons completely random attempts, which helps it escape from local minima. We remark that (in contrast with some other stochastic local search algorithms) Stocs never moves to a point with a higher cost than the current point. It continues to improve its position in this manner until it reaches a point with cost 0 (i.e., a feasible point). Stocs will also halt if it cannot find a better point within a user-defined number of attempts or allotted time.

Stocs provides for optimization problem solving by defining the cost function as an ordered pair (c_1, c_2), where c_1 is the total cost reported by the constraints and c_2 is the value of the objective function. Comparison between two cost pairs gives priority to c_1, so that a “more feasible” point is always better than a “less feasible” one, even if the value of the objective function is better in the latter. When solving an optimization problem, Stocs does not immediately halt upon reaching a point with $c_1 = 0$. Rather, it continues to search for better points (i.e., points with lower c_2) until it cannot improve further.

1.2 Paper Organization

The remainder of the paper is organized as follows. We define the pin-assignment problem and its modeling as a CSP within our prototype tool in Section 2; we present experimental results obtained with the tool in Section 3; and we conclude in Section 4.

2 The Pin Assignment Problem and Its Modeling

The particular design problem we solve is the following. Logic blocks are already placed at fixed locations on the floorplan, and the problem is to place the pins on the block edges. The overall objective is to place the pins such that wiring them (in later stages in the design process) will be feasible.

The pin assignment problem arises at a point in the design flow at which there is not enough data available to develop exact constraints that will capture wiring feasibility. In fact, it is premature to attempt this, since the design is expected to evolve quite significantly before the wiring stage is reached. Thus rather than trying to satisfy any number of detailed wiring constraints we substitute an optimization objective for the feasibility goal. Since the main wiring constraints typically translate into a wire length constraint, we attempt to minimize the total wire length. Of course the actual routing is done downstream in the workflow, so we can only estimate this value roughly. We employ the commonly used *half perimeter wire length* (HPWL) measure, which can be computed relatively quickly and is sufficiently accurate as an approximation. In this method, the total wire length of a single net is estimated as half the perimeter length of the smallest bounding rectangle containing the net's pins. Since wiring is done rectilinearly, the most economic routing is through the shortest rectilinear Steiner tree [13] connecting the pins. HPWL is an exact estimate of this length for nets connecting up to three pins, and a lower bound for larger nets.

In addition to the wiring objective, there are also various constraints on the pins' locations. Specifically, pins may not overlap (on the same metal layer), pins must be located on macro edges, and pins may not be placed in certain *blockage* areas (e.g., on the power or ground grids).

An important consideration with respect to pin-assignment is that macros are typically reused multiple times in a given design. Thus a design may contain several copies, refered to as *instances*, of a given macro, and moreover, these may be rotated or mirrored as well. The pertinent aspect of this is that pin-assignment is done per macro, not per instance, so that all copies of a macro share the same (relative) placement of pins. This makes the problem harder (at least for humans) because different instances of the same macro may benefit from different pin-assignments, but only one assignment is allowed, so a compromise must be struck.

2.1 Formalization

Pin locations. The most natural way, perhaps, to model pin locations is through their two-dimensional x-y coordinates. However, the requirement that pins must be placed on the edges of their macros allows us to use a one dimensional modeling wherein the location of a pin is given by its distance from the macro's origin (lower left corner) going clockwise through the macro's perimeter. This modeling has the advantage of reducing the number of variables and simultaneously obviating the pins-on-macro-edges constraint. On the other hand, calculating the HPWL objective function requires the conversion of relative one-dimensional pin locations to absolute two-dimensional coordinates. Nevertheless, we found that the advantage offered by the one-dimensional modeling far offsets the penalty of increased objective function computation time, so we use the one-dimensional modeling.

Note that pin locations are non-negative integers in a finite range since pins must be located on a specific pin placement grid whose lines are called *tracks*. Furthermore, if—as is the case in the designs we have encountered—the pin size is smaller than the distance between two adjacent grid points, then the no-overlap constraint can be simplified into an *all-different* constraint on the pin locations (of pins belonging to the same macro).

Variables and Domains. We denote the set of pins by P, and the set of macros by M. Let $Macro : P \to M$ be the function mapping pins to their macros (i.e., pin p belongs to macro $Macro(p)$). Also, for each macro $m \in M$, denote the macro's perimeter length in track units by $Perimeter\,(m) \in \mathbb{N}$. We define a CSP variable v_p for each pin $p \in P$, with domain $D_p = \{0, \ldots, Perimeter\,(Macro(p)) - 1\}$. Blockages are modeled by simply deleting forbidden pin locations from the corresponding domains. (In actuality, although this issue has been raised, we have not encountered yet problems with blockages and our tool does not support them.)

Constraints. Having modeled away the pins-on-edges and blockage constraints, the only remaining ones are the no-overlap constraints. For each macro m we introduce a constraint *AllDiff* $\{v_p \mid Macro\,(p) = m\}$. If pin sizes are greater than the grid dimension, we instead use a one-dimensional no-overlap constraint (with wraparound at the macro's origin).

Optimization Objective Function. In order to describe the objective function we first need to formalize the connectivity between pin instances in the design. (Note that pins are defined with respect to macros, hence each pin may have multiple instances in the design—one per instance of the corresponding macro.) Let I denote the set of all macro instances, and for each instance $i \in I$ let (by a slight abuse of notation) $Macro(i) \in M$ denote the corresponding macro. An instance can be placed anywhere on the grid, rotated by right angles, or mirrored. A *pin instance* is a pair $(i, p) \in I \times P$ such that $Macro(p) = Macro(i)$. The connectivity between pin instances is defined by a set of *nets* N. Each net $n \in N$ is a subset of $I \times P$, namely, the set of all pin instances connected by the net. A pin instance (i, p) may only appear in a single net $n \in N$, which we denote $Net(i, p)$.

For every macro m we define a function $Rel_m : \{0, \dots, Perimeter\,(m) - 1\} \to \mathbb{R}^2$ that converts a one-dimensional pin location on m's perimeter to the corresponding two-dimensional coordinate pair relative to the macro's origin. This function can be computed easily by first finding out on which edge the pin location falls (based on the macro's dimensions), then the relative position on that edge, and finally the relative position with respect to the macro's origin.

For each macro instance $i \in I$ we now define a function $Abs_i : \mathbb{R}^2 \to \mathbb{R}^2$ that transforms a coordinate pair relative to the macro's origin to absolute coordinates. This function is implemented by considering the placement, rotation and mirroring of the instance i, and converting the coordinates accordingly.

Finally, given a pin instance (i, p) and a pin location $x \in D_p$, the corresponding absolute two-dimensional coordinates are given by $Abs_i(Rel_{Macro(i)}(x))$.

The objective function is therefore

$$\sum_{n \in N} Hpwl(\{Abs_i(Rel_{Macro(i)}(v_p)) \mid (i, p) \in n\},$$

where $Hpwl(\cdot)$ is the half perimeter wire length function defined by

$$Hpwl(\{(x_1, y_1), \dots, (x_k, y_k)\}) = \frac{1}{2}(\max\{x_1, \dots, x_k\} - \min\{x_1, \dots, x_k\} + \max\{y_1, \dots, y_k\} - \min\{y_1, \dots, y_k\}).$$

Constraint Violation Cost. Recall that for a given point in the search space, i.e., an assignment of values to all variables, each constraint must be assigned a non-negative cost which (intuitively) reflects the degree to which the constraint is violated by the assignment. We use the natural cost function of total amount of overlap, defined as the sum over all pin locations in which pins are placed of the number of pins overlapping in that location minus one (i.e., a location containing one pin contributes 0 to the cost; a location containing two pins contributes 1; a location containing three pins contributes 2, etc.).

2.2 Grouping into Buses

For sufficiently small designs the model above results in good solutions (See Figure 4(a)). However, for designs with large numbers of pins, the resulting model is too large (in terms of variable count) for the solver to be able to deal with in a reasonable amount of time. Our solution is to solve a more constrained problem: instead of having each pin placed by the solver individually, we group sets of pins together to form *buses*[1] and have the solver place each bus as a whole. In a post-processing phase we then place the constituent pins of each bus contiguously within the space alloted to the bus.

Thus we model the position of each bus as a single CSP variable, and therefore dramatically reduce the number of variables. The semantics of a bus position variable is the same as before: it is the distance of the bus's center from the macro's origin, going clockwise through the macro's perimeter.

Formally, the set of buses is B. Each bus $b \in B$ is a set of pins belonging to the same macro. We denote the macro by $Macro(b)$. We denote the bus to which pin p belongs by $Bus(p)$. For each bus $b \in B$ we have a variable v_b whose domain is $D_b = \{0, \ldots, Perimeter\,(Macro(b)) - 1\}$. Since the CSP is defined in terms of buses, we also need to introduce the notion of *bus net*. Specifically, a *bus instance* is a pair $(i, b) \in I \times B$ such that $Macro(i) = Macro(b)$. The *bus net* corresponding to net $n \in N$ is defined as the set of bus instances

$$BusNet(n) = \{(i, Bus(p)) \mid (i, p) \in n\}\,.$$

Naturally, in the case of nets with identical connectivity relative to the respective buses (i.e., nets whose corresponding bus nets are identical), we only allow a single bus net. Thus a given bus net may correspond to multiple nets. We denote the number of nets for which bus net x substitutes by $Width\,(x)$, i.e.

$$Width\,(x) = |\{n \in N \mid BusNet\,(n) = x\}|\,.$$

The no-overlap constraints are now defined in terms of buses rather than pins. They ensure that no two buses overlap, which requires taking the bus sizes into account. (The size of a bus is proportional to the number of pins in it.)

Finally, we also modify the objective function to use weighted distances between bus center points rather than between pins. Specifically, let N_B denote the set of bus nets. Then the objective function is

$$\sum_{x \in N_B} Width\,(x)\, Hpwl\left(\left\{Abs_i\left(Rel_{Macro(i)}(v_b)\right) \mid (i, b) \in x\right\}\right).$$

Bus Grouping Method. We now describe how the set of pins of a given macro is partitioned into buses. A simple approach could have been to use the names given by the designer to the pins to group similarly named pins into buses. (Typically, pins are grouped by the designer into buses and their names reflect this,

[1] Some authors use the term *bus* to refer to a set of wires running in parallel and connecting what we refer to as buses.

i.e., the kth pin in bus X is called something like X<k>.) However, this method could cause problems when different pins in the same bus have different connectivity (e.g., the most significant bit of a data bus may connect identically as all other bits in the bus, but additionally to macros concerned exclusively with the sign). The extra constraint that buses must be placed as a whole would impact the solution's cost. An opposite problem occurs when designers, who typically define and name buses based on their functionality rather than connectivity, group pins with identical connectivity into multiple buses. In this case grouping by name would be finer than needed and would lead to unnecessary bloating of the CSP.

We therefore group buses according to their pin destinations. Pins of the same macro are grouped only when they are connected to the same set of macro instances. We also split buses in cases where the bus is connected to pins that would otherwise be grouped into a single bus in one instance, but into multiple buses in another. More specifically, our grouping algorithm is as follows.

1. Calculate for each pin instance (i, p) its *instance set*:
$$\{i' \in I \mid (i', p') \in Net\,(i, p)\}\,.$$
2. Group pin instances that have the same instance set and belong to the same macro instance.
3. For each pin group, check that all its pins are only connected to pins within the same pin group. If not split the groups accordingly.
Repeat this step until no such pin groups exist.
4. The resulting partitioning of the macro's pins defines the grouping of the macro pins into buses.

Implementation of the Constraint Violation Cost Functions. As we have mentioned, the no-overlap constraints must take bus sizes into account. Specifically, given any two buses b_1 and b_2 (belonging to the same macro) they must satisfy
$$First\,(b_1) > Last\,(b_2) \vee Last\,(b_1) \leq First\,(b_2)\,,$$
where $First\,(b)$ and $Last\,(b)$ are the two endpoints of the bus, calculated based on the value assigned to v_b and b's size. (Recall that bus locations are modeled one-dimensionally.) However, this simple condition is not valid for buses that wrap through the macro's origin, in which case a more elaborate condition is required. To sidestep this issue, whenever we detect such a situation we simply split the bus in two at the origin.

Our algorithm for computing the violation cost of the no-overlap constraint for a given macro instance is as follows.

1. Split any buses wrapping through the origin.
2. Sort together the starting and ending points of all buses. Maintain for each point its identity as a starting point or ending point.

3. Iterate through the sorted list of points keeping track of the number of starting points encountered, n_s, and the number of ending points encountered, n_e. Let x' be the previous point and x'' be the current point. Then $n_s - n_e$ is the number of buses passing through the interval between x' and x''. Thus add $\max\{0, l(n_s - n_e - 1)\}$ to the violation cost, where l is the length of the interval between x' and x''.

The running time of this algorithm is linear in the number of buses, except for the sorting step, which takes $O(n \log n)$ time.

Post-Processing. After solving the CSP, we still need to provide a placement for the individual pins. This is done in a short post-processing phase that orders the pins of each bus in the region allotted to that bus. The ordering of the pins within buses is done so that two buses connected at the ends of a bus net have the same relative pin ordering, thus avoiding wire crossovers (in the future routing stage). However, it is not always possible to eliminate all crossovers. Keeping in mind that all instances of the same macro share the same pin ordering, and that macro instances may be rotated or mirrored, it is easy to envision situations in which a particular ordering of pins in a given bus eliminates crossovers between macro instances X and Y but necessitates crossovers between X and Z, while reversing this order eliminates crossovers between X and Z but introduces crossovers between X and Y. It is also easy to see that the problem is only sensitive to order reversals—not the particulars of the order. (I.e., given an ordering of the pins in one of the buses, it is easy to find orderings of the other buses connected to it such that all crossovers are eliminated. However the constraint that all macro instances share the same pin placement might force some of these orderings to be reversed, and no amount of tweaking the order will prevent this.) Thus in our post-processing phase we first compute the pin order within each bus up to its clockwise/counter clockwise orientation, and then heuristically test both orientations for each bus independently and choose the one that yields lower cost. More specifically, We start with random orientations, iterate through the buses in random order, and for each bus, invert its orientation if doing so lowers the overall cost.

2.3 Model Improvements

When experimenting with the above model, some of the buses were occasionally placed close to the corners of their macros, wrapping around the corner. This was due to the fact that the objective function only considered the center of the bus, and did not care whether or not the entire bus was on the same edge as the center point. However, such a placement usually entails a greater cost (once the individual pins are laid out) than a placement in which the entire bus is located on the same edge. To alleviate the problem we use a modified version of the cost function which calculates for each bus net the average of two HPWL values, one for each of the two ends of the connected buses. This way, the cost of wrapping around a corner tends to be higher than not doing so, and the solver gravitates to solutions in which buses do not wrap around corners.

3 Experimental Results

We developed a prototype tool for automated pin-assignment implementing the ideas described in this paper. We applied the tool to five real-life IBM designs, with satisfying results. While there are still unresolved issues, the results were sufficiently encouraging that designers have expressed interest in using our tool. We are currently in the process of integrating the tool into *ChipBench* [14], the chip design environment in use across IBM hardware development labs.

In this section we describe the results we have obtained on the five designs with the solver running on a Pentium 4 3GHz Linux machine with 2GB of RAM.

Table 1 summarizes the characteristics of the five designs. As can be seen, while the number of blocks is quite small, the number of pins and buses is large.

Table 1. The hardware designs used in the experiments

	Macros	Instances	Pins	Buses
Design 1	4	6	1,956	1,610
Design 2	12	46	40,899	133
Design 3	20	20	87,492	254
Design 4	12	43	22,391	366
Design 5	9	10	14,623	55

We carried out the following experiments. For each of the five designs we ran the solver twice, once using a random point (i.e., random pin placement) as the search's starting point, and once using the solution obtained by ChipBench's current semi-automatic pin-assignment tool *mcport* as the starting point (thus using our tool to improve a given solution). We refer to the former as *random init* and to the latter as *manual init*. In each case we halted the solver after five hours. (However, for Designs 2 and 5 the engine's internal heuristics halted it much earlier, after it reached a local minimum from which it could not escape.) In addition, for the smallest design (Design 1) we also ran the solver without first grouping pins into buses (i.e., using *AllDiff* on the individual pins).

In order to compare our results with *mcport*'s one must understand *mcport*'s nature. *Mcport*'s work is comprised of three stages. The first stage is *constraint generation*. Constraint generation consists of grouping of pins based on internal (intra-macro) and external (inter-macro) connectivity, as well as pin names, and then generating constraints which assign each group of pins to a certain portion of one of the macro edges. This is done based on connectivity and an abstract floorplan provided by the designer by means of a graphical user interface. Following constraint generation *mcport* performs *pin spreading, legalization,* and *refinement*. Pin spreading generates an evenly spaced initial pin distribution respecting the pin constraints, based on the length and depth (number of layers) of the pin constraint. Individual pins within a group are ordered based on their name and other criteria. In this stage pins are placed along a macro edge without regard to design constraints, e.g., they may be placed on top of power shapes, other blockage shapes, or other fixed pins. Legalization then moves pins

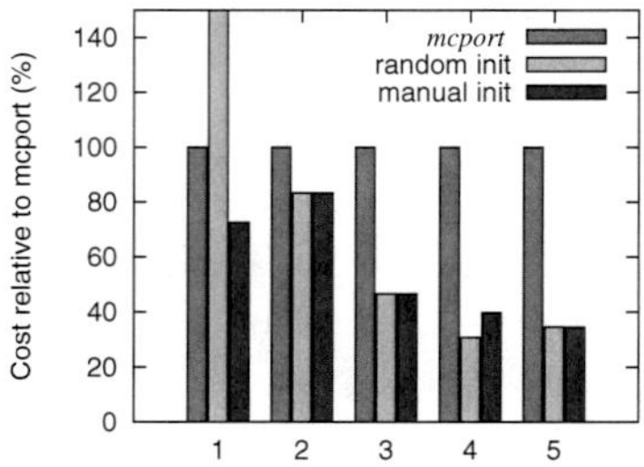

Fig. 3. Improvement over the semi-automatic *mcport* method. The *random init* experiment of Design 1 resulted in a higher ratio than the chart shows (280%) and was truncated in order to enhance the chart's clarity.

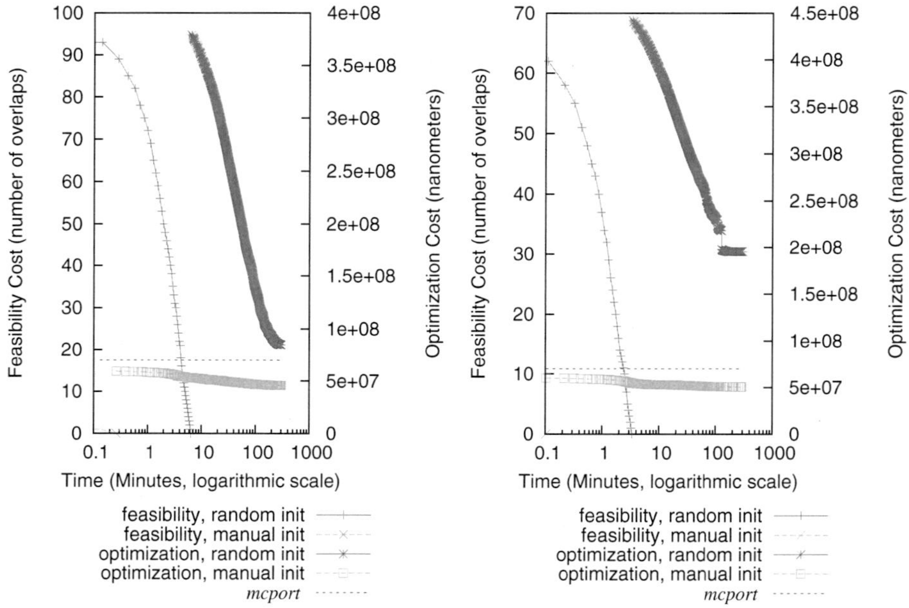

(a) Without grouping (b) With grouping

Fig. 4. Design 1

the minimum distance to legal locations that do not violate design rules. Finally, refinement may improve upon the existing pin placement, e.g., by swapping pin locations to improve the estimated net wire length between macros. *Mcport* allows a high degree of user intervention and control over its various stages, and is best described as a semi-automatic pin placement tool.

Figure 3 compares our results with *mcport*'s. (The data for Design 1 is with pin grouping.) In all cases except for Design 1 with random init, our tool achieved significant improvement. We also see that in the other cases, *manual init* did not offer an advantage over *random init* in terms of solution quality.

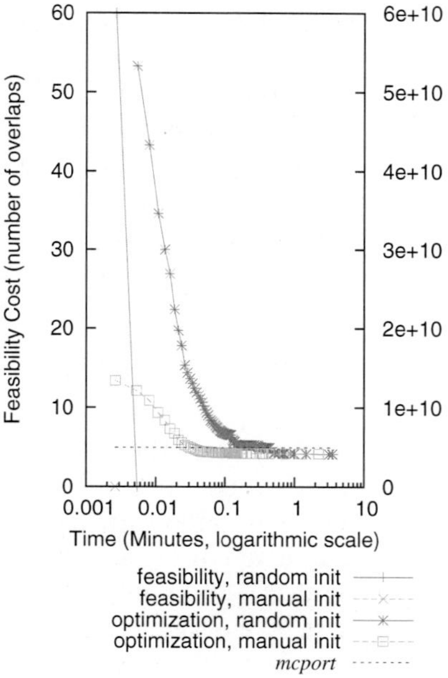

(a) Design 2

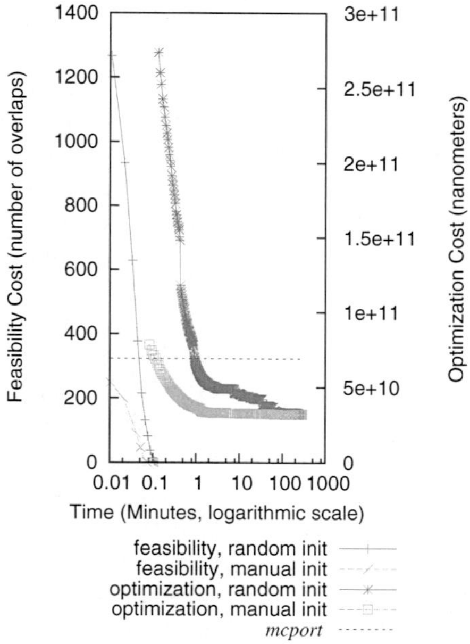

(b) Design 3

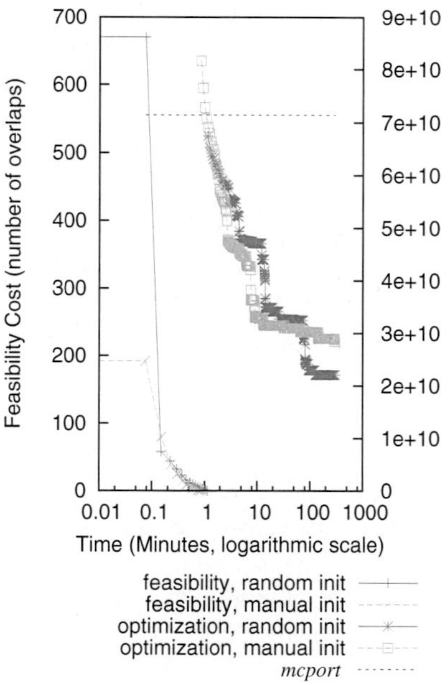

(c) Design 4

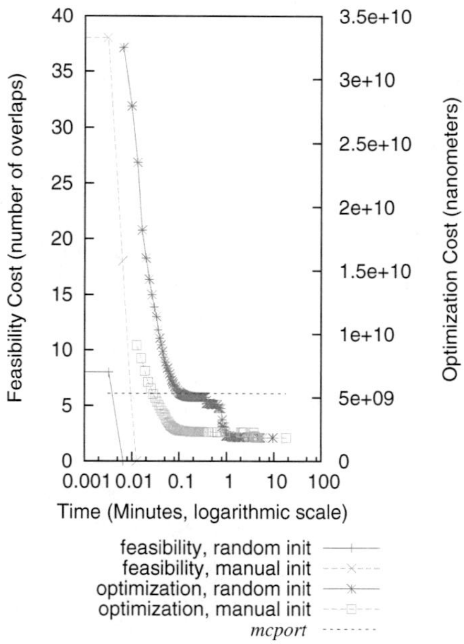

(d) Design 5

Fig. 5. Experimental Results

Figures 4(a)–5(d) show the solver's progress on each of the designs. To conserve space, each figure depicts the results of the two experiments (random init and manual init) for one design. For each experiment we plot two graphs (for a total of four graphs per figure): one depicting the decline with time of the total constraint violation cost until feasibility is reached (violation cost = 0), and the other depicting the decline with time of the optimization function (HPWL) once feasibility has been reached. The X axis corresponds to time. The left Y axis is labeled with constraint violation cost values, while the right Y axis is labeled with HPWL values. In addition, each figure contains a horizontal dotted line indicating the HPWL value of the solution found by *mcport*. Note the logarithmic scale used for time.

In the experiments in which our tool's starting point is *mcport*'s solution, one would expect that feasibility would be reached immediately (since *mcport*'s solution is feasible) and the HPWL graph would start at the dotted line (the cost of *mcport*'s solution). However, the graphs show that this is not the case. (The feasibility cost is shown in the graphs by the curve with × markers. Note that these curves are barely visible in the bottom left corners of Figures 4(a), 4(b) and 5(a).) The reason is that *mcport* places pins in different metal layers, whereas our tool does not. (Although we have written the code to accommodate multiple metal layers, it was not necessary in these particular designs, and so we have disabled it.) Thus the search's starting point was actually a projection of *mcport*'s solution onto one layer, which accounts for the initial infeasibility and slightly different cost. There were also some numerical inaccuracies introduced by rounding errors in the conversion of *mcport*'s solution (described in absolute floating point coordinates) into the coordinates used by our tool (relative integer coordinates).

The graphs show that *manual init* and *random init* both converge to approximately the same result, but, as expected, *manual init* does so much faster. We also see that for Design 1, not grouping pins into buses yielded better results (especially with *random init*), which is also to be expected since the solver then has more flexibility in assigning the pins. Of course, retaining individual pins is only possible for small designs due to the issue of scalability.

4 Conclusion

In this work we have demonstrated a fruitful application of constraint programming technology to automated chip design. Using CP allowed us to focus our work on the details of the problem, and not on the solvers and search algorithms.

An important lesson (re)learnt was to invest time working out the model and the choice of CSP variables. These had great impact on the running time relative to the constraint implementation.

Our future plans include enabling the support for multi-layer placement of pins, supporting blockages, improving the bus grouping algorithm, and improving the bus orientation post-processing algorithm. We believe that CP technology could be useful for other chip design problems as well. In fact we are currently developing a CP-based tool for floorplaning using a systematic CSP solver.

Acknowledgment. We thank Ariel Birnbaum for helpful ideas on bus grouping.

References

1. Smith, M.J.S.: Application Specific Integrated Circuits. VLSI Systems Series. Addison-Wesley, Reading (1997)
2. Sherwani, N.A.: Algorithms for VLSI Physcial Design Automation, 3rd edn. Kluwer Academic Publishers, Norwell (1998)
3. Lengauer, T.: Combinatorial algorithms for integrated circuit layout. John Wiley & Sons, Inc., New York (1990)
4. Drechsler, R.: Evolutionary algorithms for VLSI CAD, 2nd edn. Springer, Heidelberg (1998)
5. Koren, N.L.: Pin assignment in automated printed circuit board design. In: ACM/IEEE Design Automation Conference, pp. 72–79 (1972)
6. Liu, L., Sechen, C.: Multi-layer pin assignment for macro cell circuits. IEEE Trans. Computer-Aided Design 18, 1452–1461 (1999)
7. Xiang, H., Tang, X., Wong, D.F.: An algorithm for simultaneous pin assignment and routing. In: Proc. of International Conference on Computer Aided Design, pp. 232–238 (2001)
8. Brady, H.: An approach to topological pin assignment. IEEE Trans. Computer-Aided Design CAD-3, 250–255 (1984)
9. Yao, X., Yamada, M., Liu, C.L.: A new approach to pin assignment problem. In: Proc. of Design Automation Conference, pp. 566–572 (1988)
10. Wang, L., Lai, Y., Liu, B.: Simultaneous pin assignment and global wiring for custom vlsi design. In: Proc. IEEE International Symposium on Circuits and Systems, vol. 4, pp. 2128–2131 (1991)
11. Westra, J., Groeneveld, P.: Towards integration of quadratic placement and pin assignment. In: IEEE Proc. of ISVLSI, pp. 284–286 (2005)
12. Naveh, Y.: Guiding stochastic search by dynamic learning of the problem topography. In: Perron, L., Trick, M.A. (eds.) CPAIOR 2008. LNCS, vol. 5015, pp. 349–354. Springer, Heidelberg (2008)
13. Hwang, F.K., Richards, D.S.: Steiner tree problems. Networks 22(1), 55–89 (1992)
14. Sayah, J., Gupta, R., Sherlekar, D., Honsinger, P., Apte, J., Bollinger, S., Chen, H., DasGupta, S., Hsieh, E., Huber, A., Hughes, E., Kurzum, Z., Rao, V., Tabtieng, T., Valijan, V., Yang, D.: Design planning for high-performance asics. IBM Journal of Research and Development 40(4), 431–452 (1996)

Modelling Equidistant Frequency Permutation Arrays: An Application of Constraints to Mathematics

Sophie Huczynska[2], Paul McKay[1], Ian Miguel[1], and Peter Nightingale[1]

[1] School of Computer Science, University of St Andrews, UK
{ianm,pn}@cs.st-andrews.ac.uk, pgm9@st-andrews.ac.uk
[2] School of Mathematics and Statistics, University of St Andrews, UK
sophieh@mcs.st-andrews.ac.uk

Abstract. Equidistant Frequency Permutation Arrays are combinatorial objects of interest in coding theory. A frequency permutation array is a type of constant composition code in which each symbol occurs the same number of times in each codeword. The problem is to find a set of codewords such that any pair of codewords are a given uniform Hamming distance apart. The equidistant case is of special interest given the result that any optimal constant composition code is equidistant. This paper presents, compares and combines a number of different constraint formulations of this problem class, including a new method of representing permutations with constraints. Using these constraint models, we are able to establish several new results, which are contributing directly to mathematical research in this area.[1]

1 Introduction

In this paper we consider Equidistant Frequency Permutation Arrays (EFPAs), combinatorial objects of interest in coding theory. A frequency permutation array (introduced in [1]) is a special kind of constant composition code (CCC), in which each symbol occurs the same number of times in each codeword. CCCs have many applications, for example in powerline communications and balanced scheduling, and have recently been much studied (eg [2], [3]). The situation when CCCs are equidistant is of particular interest, since it is known that any CCC which is optimal must be equidistant. EFPAs are introduced in [4], where various bounds and constructions are obtained; other results on families of such codes can be found in [5].

Informally, the problem is to find a set (often of maximal size) of codewords, such that any pair of codewords are Hamming distance d apart. Each codeword (which may be considered as a sequence) is made up of symbols from the alphabet $\{1, \ldots, q\}$, with each symbol occurring a fixed number λ of times per codeword.

The problem has parameters v, q, λ, d and it is to find a set E of size v, of sequences of length $q\lambda$, such that each sequence contains λ of each symbol in the set $\{1, \ldots, q\}$. For each pair of sequences in E, the pair are Hamming distance d apart (i.e. there are d

[1] Sophie Huczynska is supported by a Royal Society Dorothy Hodgkin Research Fellowship, and Ian Miguel was supported by a UK Royal Academy of Engineering/EPSRC Research Fellowship. Peter Nightingale is supported by EPSRC grant EP/E030394/1.

I.P. Gent (Ed.): CP 2009, LNCS 5732, pp. 50–64, 2009.

Table 1. EFPA example with $v = 5$, $q = 3$, $\lambda = 2$, $d = 4$

c_1	0	0	1	1	2	2
c_2	0	1	0	2	1	2
c_3	0	1	2	0	2	1
c_4	0	2	1	2	0	1
c_5	0	2	2	1	1	0

places where the sequences disagree). For the parameters $v = 5$, $q = 3$, $\lambda = 2$, $d = 4$, Table 1 shows a set $E = \{c_1, c_2, c_3, c_4, c_5\}$.

Computers have long been used to assist in solving related mathematical problems. Slaney et al. found great success in using automated reasoning to attack quasigroup existence problems [6].

We model and solve EFPA in constraints, using the Minion solver [7,8] and the Tailor modelling assistant [9]. Constraint solving proceeds in two phases. First, the problem is *modelled* as a set of discrete decision variables, and a set of constraints (relations, e.g. $x_1 + x_2 \leq x_3$) on those variables that a solution must satisfy. Second, a constraint solver is used to search for solutions to the model: assignments of values to variables satisfying all constraints. Many constraint solvers (including Minion) interleave propagation and depth-first backtracking search. Propagation simplifies the problem by removing values from variable domains, and search assigns values to variables, searching for a solution. A successful model must propagate well, and will also include an ordering of the variables for search.

We investigate six different formulations, starting with two simple models based on two viewpoints (where a viewpoint is a choice of variables and domains sufficient to characterise the problem). A set of implied constraints is derived, which prove to be useful in experiments. Furthermore, we develop a novel method of modelling permutations with constraints.

The work described in this paper has direct application in mathematics. During the course of this work, we generated 24 EFPAs (21 of which were proven to be maximal) to assist with the mathematical research of one of the authors (Huczynska). Some of these EFPAs directly refuted a working conjecture, and others provided supporting evidence that a construction is maximal. This illustrates that, with careful modelling and the power of fast constraint solvers such as Minion and modelling assistants such as Tailor, constraint programming can contribute to research in other disciplines.

Experiments are performed to compare the different models, using 24 carefully chosen instances of EFPA.

2 Modelling the EFPA Problem

First we present two straightforward models based on two viewpoints. One represents the set of sequences explicitly in a two-dimensional table (like Table 1). The other is similar, but extends the two-dimensional table in a third dimension, expanding each original variable into a set of Boolean variables corresponding to each original value. The problem constraints and symmetry breaking constraints are quite different on these two models. The three-dimensional model is able to break symmetry in three planes rather than two, and the two-dimensional model is able to take advantage of the Global Cardinality Constraint (GCC) [10] to enforce the requirement that there are λ occurrences of each symbol. Both straightforward models perform reasonably well, which reflects well on constraint programming.

The two simple models are developed in various ways to give six variants in total. Firstly, the two models are channelled together, combining the symmetry breaking constraints on the three-dimensional Boolean model with problem constraints on the two-dimensional model. The channelled model is superior to the Boolean model in almost all cases in our experiments.

Secondly, a set of implied constraints are derived for the two-dimensional model. The first row of the table is fixed by the symmetry breaking constraints. All other rows are related to the first by the Hamming distance d, and we derive additional GCC constraints from this. The implied constraints are somewhat similar to those proposed for BIBD [11]. The implied constraints are beneficial in most cases in our experiments.

Thirdly, we explicitly model the permutation between each pair of sequences, using a representation of cycle notation. While permutation has been modelled frequently in CSP and SAT (e.g. [12,13]), this is to the best of our knowledge the first time cycle notation has been explicitly modelled using standard CSP integer variables. This approach greatly reduces the search space. Unfortunately this model has a large number of constraints and is not always superior in terms of solver time. However, it shows promise because of the reduction in search space.

To model the EFPA problem, we used the Tailor modelling assistant [9]. We formulated each model in the solver-independent modelling language ESSENCE′ and used Tailor v0.3 to flatten each instance for input to the constraint solver Minion 0.8.1. Tailor provides optional common subexpression elimination (CSE) [14]. Preliminary experiments revealed that CSE improved the speed of Minion by a small margin, without affecting the search tree explored. We use CSE throughout.

2.1 Boolean and Non-Boolean Models

In this section, we investigate two viewpoints and construct two models based on the viewpoints, and a third which channels them together. The three models are compared experimentally.

A Non-Boolean Model. Viewed abstractly, the problem is to find a fixed-size set of codewords, where a codeword is a permutation of a multiset of symbols. Therefore the decisions are how to represent the set, and how to represent the permutations. In the non-Boolean model, we use an explicit representation of the set [15] (i.e. each element is explicitly represented). For each codeword, we use the primal model of a permutation [12]. Each position in the codeword has one variable whose domain is the alphabet. We do not explore the dual model here, because it would complicate the Hamming distance constraints.

There is an additional set of Boolean variables representing where pairs of codewords differ. The variables are as follows.

- $\forall a \in \{1 \ldots v\}, \forall i \in \{1 \ldots q\lambda\} \ : \ c[a,i] \in \{1 \ldots q\}$ representing the set of sequences. a is the sequence number and i is the index into the sequence.
- $\forall a \in \{1 \ldots v\}, \forall b \in \{a+1 \ldots v\}, \forall i \in \{1 \ldots q\lambda\} \ : \ \mathit{diff}[a,b,i] \in \{0,1\}$ representing whether two sequences a and b differ at position i.

For each sequence, a GCC constraint is used to ensure that there are λ occurrences of each symbol in the alphabet. A reified disequality constraint is used to connect the *diff* variables to c, and a sum is used to enforce the Hamming distance d.

- $\forall a \in \{1 \ldots v\} \ : \ \mathrm{GCC}(c[a, 1 \ldots q\lambda], \langle 1 \ldots q \rangle, \langle \lambda \ldots \lambda \rangle)$
- $\forall a \in \{1 \ldots v\}, \forall b \in \{a+1 \ldots v\}, \forall i \in \{1 \ldots q\lambda\} \ : \ \mathit{diff}[a,b,i] \Leftrightarrow (c[a,i] \neq c[b,i])$
- $\forall a \in \{1 \ldots v\}, \forall b \in \{a+1 \ldots v\} \ : \ \sum_{i=1}^{q\lambda} \mathit{diff}[a,b,i] = d$ (any pair of sequences differ in d places)

The matrix c has a number of symmetries. In a solution, rows, columns and symbols of the alphabet may be freely permuted to create other solutions. To break some of the symmetry, we apply lexicographic ordering (lex-ordering) constraints to the rows and columns, following Flener et al. [16].

- $\forall a \in \{2 \ldots v\} \ : \ c[a-1, 1 \ldots q\lambda] \leq_{lex} c[a, 1 \ldots q\lambda]$ (rows are lex-ordered)
- $\forall b \in \{2 \ldots q\lambda\} \ : \ c[1 \ldots v, b-1] \leq_{lex} c[1 \ldots v, b]$ (columns are lex-ordered)

These two constraint sets do not explicitly order the symbols. It would be possible to order the symbols by using value symmetry breaking constraints [18]. However we leave this for future work.

These constraints are all found in Minion 0.8.1. The GCC constraint enforces GAC in this situation (with a fixed number of occurrences of each symbol). The reified not-equal constraint enforces Bounds($\mathbb{Z}$)-consistency [17]. The sum constraint above is decomposed into $\leq d$ and $\geq d$ constraints (sumleq and sumgeq in Minion) which also enforce Bounds($\mathbb{Z}$)-consistency. The lex ordering constraints enforce GAC.

The variable order is row-wise on c, in index order, as follows.

$$c[1,1], \ldots, c[1,q\lambda], c[2,1], \ldots, c[2,q\lambda], \ldots$$

The values are searched in ascending order.

A Boolean Model. In this section we consider another simple model for EFPA, based on a different viewpoint to the one above. The difference is in the representation of each symbol in each codeword using a vector of Boolean variables. The model uses a three-dimensional matrix m of Boolean variables to represent occurrences of the q symbols in the v codewords. The first dimension is the codeword $1 \ldots v$, the second is the symbol $1 \ldots q$ and the third is the codeword position $1 \ldots q\lambda$.

- $\forall i \in \{1 \ldots v\}, \forall j \in \{1 \ldots q\}, \forall k \in \{1 \ldots q\lambda\} \ : \ m[i,j,k] \in \{0,1\}$

Variable $m[i,j,k]$ is 1 iff the codeword i has symbol j at position k.

We must ensure that exactly one symbol appears at each position in each codeword. This is done with the following set of constraints.

- $\forall i \in \{1...v\}, \forall j \in \{1...q\lambda\} \ : \ \sum_{k=1}^{q} m[i,k,j] = 1$

To ensure that there are λ of each symbol in each codeword we post the following set of constraints.

– $\forall i \in \{1...v\}, \forall j \in \{1...q\} \,:\, \sum_{k=1}^{q\lambda} m[i,j,k] = \lambda$

The final problem constraint set states that the Hamming distance between any pair of codewords is exactly d. The two codewords are represented as planes in the matrix. For each position where the pair of codewords differ, the planes in m differ in two places corresponding to one symbol being removed and another inserted. Therefore the number of places where the two planes differ is $2d$.

– $\forall i \in \{1...v\}, \forall j \in \{i+1...v\} \,:\, [\sum_{k=1}^{q\lambda} \sum_{l=1}^{q} (m[i,l,k] \neq m[j,l,k])] = 2d$

Symbols, codewords and positions may all be freely permuted. In order to break some of this symmetry, we lexicographically order (lex order) planes of the matrix in all three dimensions, using the technique of Flener et al. [16]. (We rely on Tailor to vectorize the planes in a consistent manner.) This set of symmetry breaking constraints orders the symbols, in contrast to those of the non-Boolean model.

– $\forall i \in \{1...q\lambda - 1\} \,:\, m[1\ldots v, 1\ldots q, i] \leq_{lex} m[1\ldots v, 1\ldots q, i+1]$
– $\forall i \in \{1...q - 1\} \,:\, m[1\ldots v, i, 1\ldots q\lambda] \leq_{lex} m[1\ldots v, i+1, 1\ldots q\lambda]$
– $\forall i \in \{1\ldots v - 1\} \,:\, m[i, 1\ldots q, 1\ldots q\lambda] \leq_{lex} m[i+1, 1\ldots q, 1\ldots q\lambda]$

Preliminary experiments reveal that these three constraint sets drastically improve performance. In one instance the addition of symmetry breaking constraints improved the performance of the model by approximately 40 times. The variable ordering is as follows. For each i in ascending order: for each j in ascending order: for each k in ascending order: $m[i,j,k]$. To illustrate:

$$m[1,1,1], \ldots, m[1,1,q\lambda], m[1,2,1], \ldots, m[1,q,q\lambda], m[2,1,1], \ldots$$

For all variables, value 0 is branched on first.

Channelling Boolean and Non-Boolean Models. The Boolean model may have better symmetry breaking than the non-Boolean model, because the value symmetry of the non-Boolean model is transformed into variable symmetry [16] and broken using lex constraints. However, in the non-Boolean model, the first row is invariant because of the column lex constraints and therefore some of the value symmetry is broken there.

The non-Boolean model is able to exploit the GCC constraint on the rows, and also has a neater representation of the Hamming distance requirement. In this section we aim to gain the advantages of both models by connecting the two with channelling constraints, given below.

– $\forall i \in \{1\ldots v\}, \forall j \in \{1\ldots q\}, \forall k \in \{1\ldots q\lambda\} \,:\, m[i,j,k] \Leftrightarrow (c[i,k] = j)$

The symmetry breaking constraints in the non-Boolean model are removed, because they contradict those in the Boolean model.

The Boolean model has three sets of constraints other than the symmetry breaking constraints. Inspection of each set suggests that they will provide no useful propagation, because the non-Boolean representation of the same constraint set is equivalent or stronger. Preliminary experimentation on instance $d = \lambda = q = 4$, $v = 9$ showed that removing all three sets does not affect the node count (2,350,155) but does reduce the time taken from 86 s to 52 s. Therefore we do not include the three sets of constraints.

Preliminary experiments suggest that searching on the c (non-Boolean) variables is not effective when channelling, using either an ascending or descending value ordering (with the variable ordering described for c above). Therefore we search on m, using the same variable and value ordering as the standard Boolean model. Given that the constraints on the non-Boolean formulation appear to be stronger, we expect that the channelled model will improve on the Boolean model in terms of search nodes.

Empirical Evaluation. To compare the three models empirically, we picked twelve tuples $\langle d, \lambda, q \rangle$ with a range of different values of d, λ and q. For each parameter set, the usual task is to find the maximal set of codewords. This can be done by solving iteratively, increasing v until the instance is unsatisfiable. This provides a maximal set of codewords, and a proof that there is no larger set. Typically the unsatisfiable instance is much more difficult than the others, because of the need to exhaust the search space. Instances are identified by the tuple $\langle d, \lambda, q, v \rangle$.

For each parameter set, we use two consecutive values of v such that the smaller instance is satisfiable and the larger one is unsatisfiable or it takes longer than the time limit of 2 hours to solve. This provides 12 satisfiable instances, 11 unsatisfiable instances and one $\langle 6, 4, 4, 14 \rangle$ which is unknown[2]. We used Minion 0.8.1 on an Intel Xeon E5430 2.66 GHz 8-core machine, using all cores. The three models are named as follows.

Non-Boolean refers to the two-dimensional model.
Boolean refers to the three-dimensional model.
Channelled refers to the combined model described in the section above.

We enabled the SAC [19] preprocessing option of Minion to be consistent with our other experiments presented below. SAC preprocessing is cheap, taking less than 0.2 s on the largest instances.

Both the non-Boolean and Channelled models include some variables which are not necessary here, but are required for the implied constraints and permutation model described below[3]. For satisfiable instances, these spurious variables are set at the end of the search process. Assigning each variable takes 1 search node, and a very small amount

[2] The problem instances are available at
`http://minion.sourceforge.net/benchmarks.html`

[3] Several models were expressed in one Essence′ file for convenience. Essence′ allows constraints to be included or excluded as required, and we used this to generate the different models, however the set of variables remains the same.

of time. For unsatisfiable instances, these variables do not affect search or propagation in any way.

Figure 1 shows our results for the three models. Instance $\langle 6,4,4,14\rangle$ times out for all three models, and is the only time-out for the channelled model. The non-Boolean model times out on five instances. The channelled model improves upon the Boolean model in both nodes and time, except for the very easy satisfiable instance $\langle 4,4,3,7\rangle$. In this case, there are 327 spurious variables, and the channelled model explores 396 nodes. Therefore the spurious variables account for most of the nodes, and this instance should be disregarded for comparing Boolean and channelled models.

For the non-Boolean and channelled models, neither always dominates the other, either in nodes or time. The non-Boolean model is faster on 13 instances, and the channelled model on 10 instances. For the instance that timed out, we can observe that the channelled model explores fewer nodes than the other two. For this instance, the channelling has some overhead, as one would expect.

2.2 Extensions of the Non-Boolean Model

In this section we explore two extensions of the non-Boolean model, both of which exploit knowledge about the permutation of d elements between pairs of codewords.

Implied Constraints. It is possible to derive some implied constraints between pairs of sequences. Consider the sequence $\langle 1,1,1,2,2,2,3,3,3\rangle$, and assume that $d=4$. To construct another sequence with the appropriate Hamming distance, we can swap two 1's with two 2's: $\langle 1,2,2,1,1,2,3,3,3\rangle$. However, it is not possible to move all three 1's, since that would cause six disagreements. In general, for each symbol, the maximum number which can be moved is $\lfloor \frac{d}{2} \rfloor$.

If $\lfloor \frac{d}{2} \rfloor < \lambda$, then this observation allows us to add useful constraints to the model. Between any pair of sequences in the set, and for each symbol a, at least $\lambda - \lfloor \frac{d}{2} \rfloor$ instances of a must remain in the same place. We do not exploit this observation for every pair of rows, but only for the pairs containing the first row. This is because the first row is fixed and this makes the statement of the constraints considerably simpler.

The first row is fixed by the combination of column lex ordering constraints and the cardinality constraint. If $q=\lambda=3$ then the first row is $\langle 1,1,1,2,2,2,3,3,3\rangle$ in the non-Boolean model. It is arranged in q blocks of length λ, and each of the other sequences is divided into q blocks in the same way, as shown in the table below.

$c[1,1\ldots q\lambda]$	1 1 1	2 2 2	3 3 3
$c[2,1\ldots q\lambda]$	Block 1	Block 2	Block 3

To state the constraints, we have auxiliary variables $occ^a_{b,d} \in \{0\ldots\lambda\}$ representing the number of occurrences of value d in sequence a, block b (where the blocks are numbered $1\ldots q$ in index order). We post q GCC constraints to count the symbols in each block, as follows. Also, we constrain the occurrences of the relevant occ variables.

- $\forall a \in \{2\ldots v\}, \forall b \in \{1\ldots q\}\ :\ \mathrm{GCC}(c^a_{(b-1)\lambda+1\ldots b\lambda}, \langle 1\ldots q\rangle, occ^a_{b,1\ldots q})$
- $\forall a \in \{2\ldots v\}, \forall b \in \{1\ldots q\}\ :\ occ^a_{b,b} \geq \lambda - \lfloor \frac{d}{2} \rfloor$

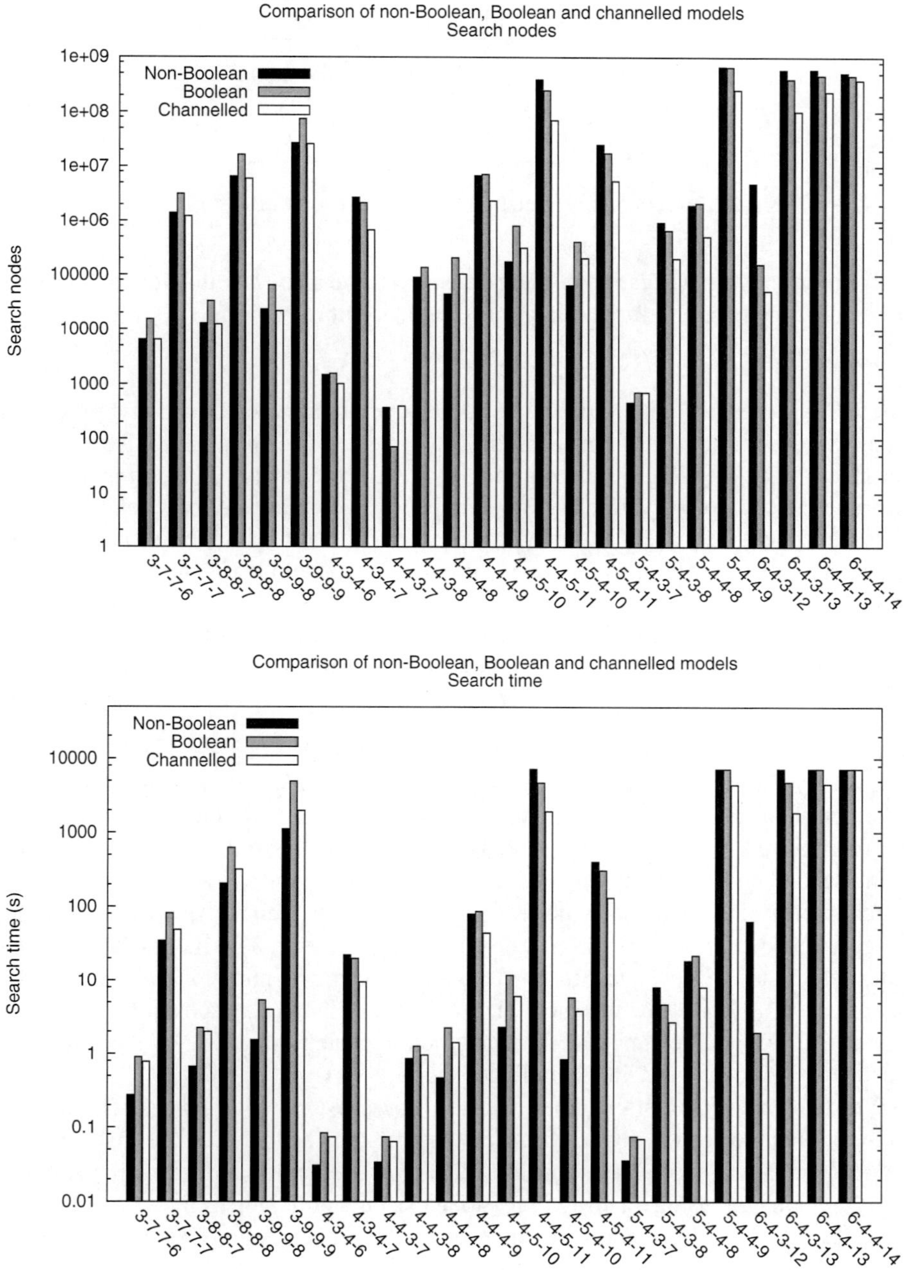

Fig. 1. Comparison of non-Boolean, Boolean and channelled models

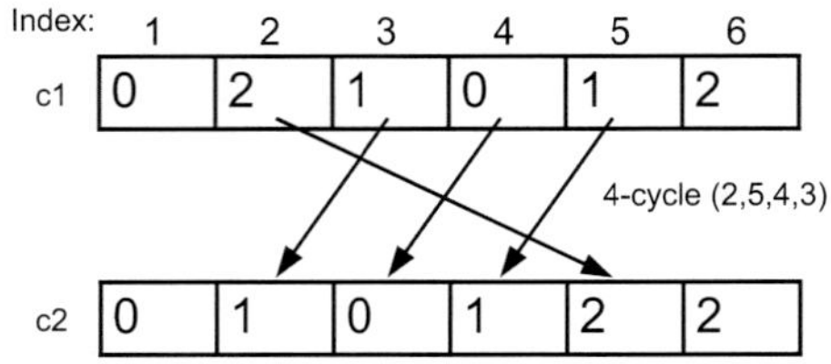

Fig. 2. The action of permutation (2,5,4,3) on codeword c_1 to form c_2

To improve the propagation of these constraints, we also state that for each symbol, the occurrence variables for all blocks in a sequence must sum to λ.

– $\forall a \in \{2 \dots v\}, \forall b \in \{1 \dots q\} \; : \; \sum occ^a_{1 \dots q,b} = \lambda$

The GCC constraint (with variables as its third argument) performs a hybrid consistency. It reads the bounds of the occurrence variables, and performs GAC over the target variables only, using Régin's algorithm [10]. Also, it uses a simple counting algorithm to prune the occurrence variables. For example, if two target variables are assigned to 0, and five variables have 0 in their domain, then the lower bound for occurrences of 0 is two, and the upper bound is five. This implementation of GCC is named gccweak in Minion.

Note that this constraint set cannot be applied to the channelled model, because it relies on the symmetry breaking constraints of the non-Boolean model. In the channelling model, the first row is in descending order.

Modelling Permutations. Between any pair of codewords c_1 and c_2, there are d indices where they differ. Since the two codewords have the same multiset of symbols, the difference between them can be represented as a permutation. In order to have d points of disagreement, d symbols in the first codeword must be moved to a different position in the second codeword. We consider permutations of the indices $1 \dots q\lambda$, specifying which indices of the first codeword are moved to form the second. Any permutation can be represented in *cycle notation* — for example, the cycle $(1, 3, 2)$ moves the symbol at index 1 to index 3, 3 to 2 and 2 to 1. We do not allow the permutation to permute index i to j if $c_1[i] = c_1[j]$, since this would leave c_1 and c_2 the same at position j.

Figure 2 shows an example of a permutation in cycle form acting on a codeword to create another codeword with Hamming distance $d = 4$.

From any sequence to any other, there is a permutation of d indices. The implied constraints in the section above make some use of this fact between the first sequence and all others. In this section we consider all pairs of sequences, and we explicitly model the cycle notation, using an array of d variables containing indices into the sequences. A further variable *cform* represents the form of the cycle notation. For example, when $d = 4$, there are two possible forms of the cycle notation: (p, q, r, s) and $(p, q)(r, s)$, therefore *cform* has two values. In fact we have only implemented the permutation model for $d = 4$. The subclass of EFPA where $d = 4$ is of interest to mathematicians as it is the smallest value of d where the precise upper bound for the size of an EFPA is not obvious (due to the fact that there is more than one possible cycle structure for a derangement of 4 points).

Table 2. Example of p

		1	2	3	4
1	$(1,3)(4,6)$	$s_1[3]=2$	$s_1[1]=1$	$s_1[6]=3$	$s_1[4]=2$
2	$(1,3,4,6)$	$s_1[6]=3$	$s_1[1]=1$	$s_1[3]=2$	$s_1[4]=2$

- $\forall e \in \{1 \dots d\} : perm[e] \in \{1 \dots q\lambda\}$
- $cform \in \{1 \dots cforms\}$ where *cforms* is the number of cycle forms.

To allow us to map from one sequence to another using the permutation, we introduce a table of variables p. There are *cforms* rows and d columns in p. The rows correspond to different forms of the cycle notation. Each row contains elements of the first sequence (those elements indexed by *perm*) permuted according to the form of the cycle notation.

- $\forall i \in \{1 \dots cforms\}, \forall j \in \{1 \dots d\} : p[i,j] \in \{1 \dots q\}$

For example, if the first sequence is $s_1 = \langle 1,1,2,2,3,3 \rangle$, $d = 4$ and *perm* is $\langle 1,3,4,6 \rangle$, then p is given in Table 2. In the first row, indices for each pair are swapped, and in the second row the indices are rotated according to the inverse of the 4-cycle.

For the second sequence s_2, positions 1,3,4 and 6 (i.e. the values of *perm*) must equal the appropriate value from p. *cform* is used to select the appropriate row in p. For position 1, $s_2[1] = p[cform, 1]$. Also, constraints are posted stating that s_1 and s_2 are equal at all positions not in *perm*. In this example, if $cform = 1$ then $s_2 = \langle 2,1,1,3,3,2 \rangle$, and if $cform = 2$ then $s_2 = \langle 3,1,1,2,3,2 \rangle$.

The basic set of constraints is given below.

- $\forall i \in \{1 \dots cforms\}, \forall j \in \{1 \dots d\} : p[i,j] = s_1[k]$ where k is the inverse mapping of j by the permutation (*perm* with cycle form i).
- $\forall i \in \{1 \dots d\} : s_2[perm[i]] = p[cform, i]$
- $\forall i \in \{1 \dots q\lambda\} : (\bigwedge_{j=1}^{d} perm[j] \neq i) \Leftrightarrow (s_1[i] = s_2[i])$ (If index i is not present in *perm*, then the value at position i remains the same, and vice versa)

We add symmetry breaking constraints to *perm*. In general, within each cycle, the smallest element is placed at the front, and cycles of equal length are ordered by their first element. The ordering constraints are shown below for $d = 4$. Only one of the constraints is conditional on *cform*; the other three are true for either cycle form.

- $(cform = 1) \Rightarrow (perm[3] < perm[4])$
- $(perm[1] < perm[2]) \wedge (perm[1] < perm[3]) \wedge (perm[1] < perm[4])$

As a special case for $d = 4$, we add an allDifferent to the values permuted by the 4-cycle form. The reason is that if the values are not all different, the 4-cycle can be reformulated as two 2-cycles.

- $(cform = 2) \Rightarrow \text{allDifferent}(p[2, 1 \dots d])$

The expression above is decomposed as a reified allDifferent and an implies constraint on two Boolean variables. GAC is enforced on both these constraints, which is equivalent to GAC on the original expression.

When *cform* $= 1$, the transposed values must be different, therefore we add the following constraints. These constraints are also true when *cform* $= 2$, so there is no need for them to be conditional.

- $p[1,1] \neq p[1,2] \wedge p[1,3] \neq p[1,4]$

Empirical Evaluation. In this section we compare the following four models, using the same set of instances and experimental details as in the previous experiment.

Non-Boolean refers to the two-dimensional model.

Implied is the non-Boolean model with additional constraints described in the section with heading Implied Constraints.

Permutation is the non-Boolean model with additional permutation constraints described in the Modelling Permutations section.

Implied+Perm is the non-Boolean model with both Implied and Permutation constraint sets.

All the above models have the same set of variables, including all variables used by the implied constraints and the permutation model. For solvable instances, the unused variables are enumerated at the end of the search process, adding a small constant to the node count. The permutation models are only used where $d = 4$, since they are not defined for other values.

We found it important to perform singleton consistency (SAC) [19] preprocessing. It is a very cheap preprocessing step which is nevertheless very important for the permutation model. Table 3 shows that SAC preprocessing is important on the instance where $d = \lambda = q = 4$ and $v = 9$, for two of the four models listed above. For both models that include the permutation constraint set, the preprocessing is vital. Therefore we use SAC preprocessing throughout.

Figure 3 shows our results for these models. It is clear that both the implied constraints and the permutation model are effective in reducing the number of search nodes, and in most cases the combined model gives a further reduction. The permutation model is particularly effective. For example, on the instance $\langle 4, 5, 4, 11 \rangle$ the non-Boolean model takes 25,271,680 nodes, the implied model takes 14,607,410 nodes and the permutation model takes 6,032,900 nodes, a reduction of 76%. Where $d = 4$, there is a clear ordering among the four models.

Table 3. Search nodes with and without SAC preprocessing, $d = \lambda = q = 4$, $v = 9$

Search nodes	Non-Boolean	Implied	Permutation	Implied+Perm
No preprocessing	6,788,586	4,032,510	6,021,363	3,471,775
SAC preprocessing	6,788,361	4,032,306	1,674,826	1,340,546

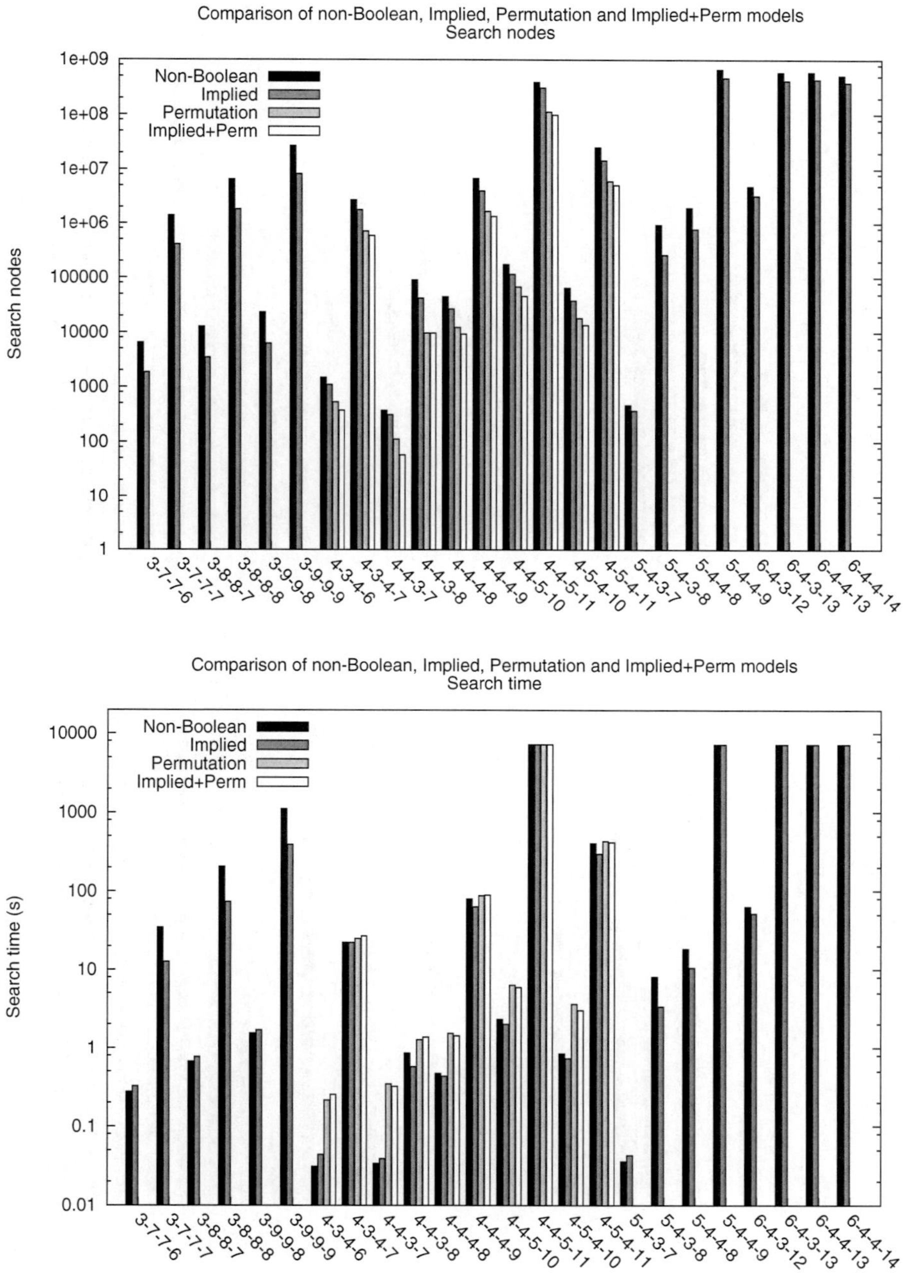

Fig. 3. Comparison of non-Boolean, Implied, Permutation and Implied+Perm models

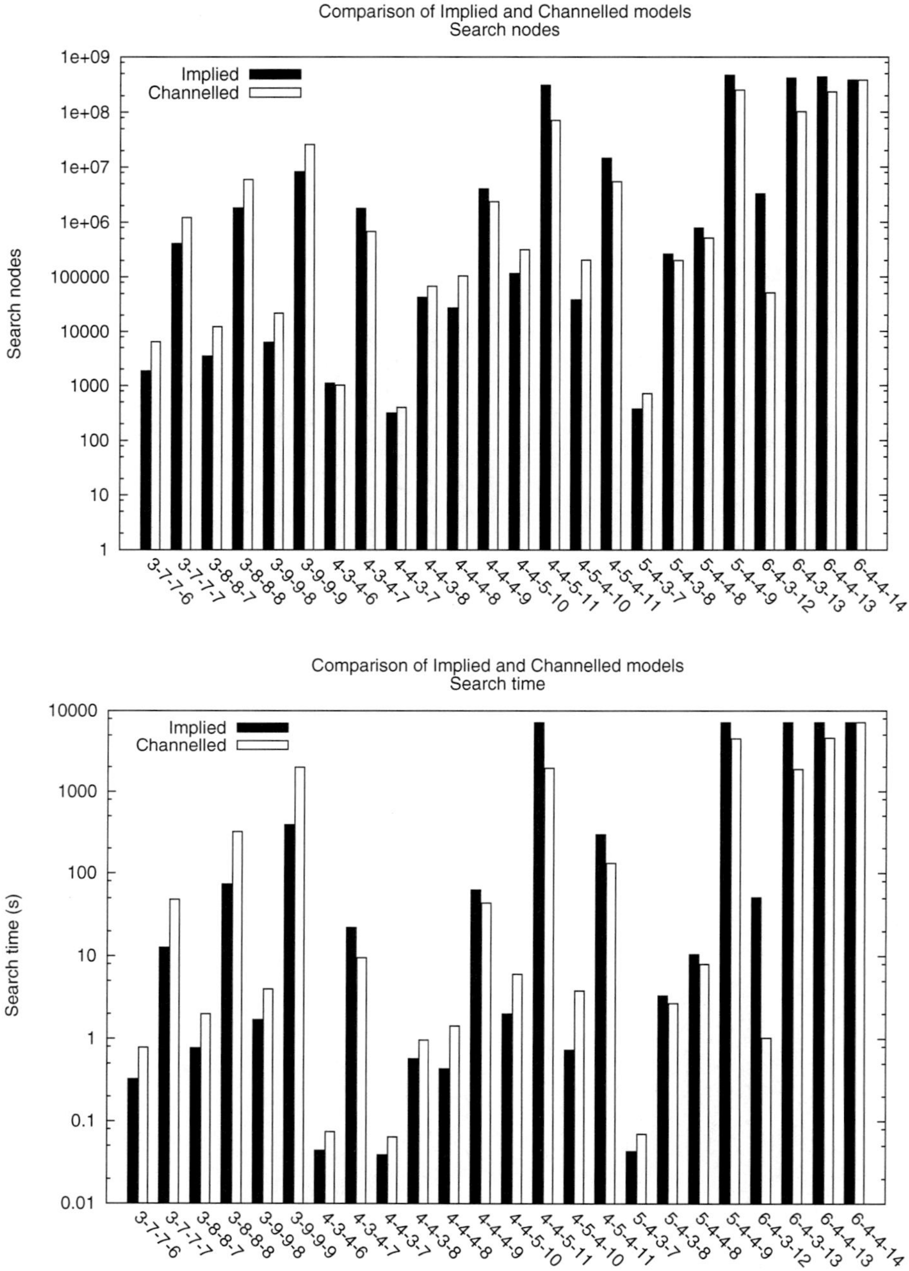

Fig. 4. Comparison of Implied and Channelled models

However, the search times are not so straightforward. In most cases, the implied constraints are worthwhile. However, the permutation model is never worthwhile, and likewise for the Implied+Perm model. It is clear that the permutation constraint set adds a considerable overhead to the search process, and therefore it takes longer to solve even though it is exploring many fewer nodes.

Finally, we compare the Implied model with the Channelled model from the previous experiment. These models use a different variable and value ordering as well as a different constraint set, so there is no reason to expect one to always dominate the other in terms of search nodes. The Implied model is the most efficient when searching on c variables, and likewise the channelled model is the most efficient when searching on m, therefore this makes an interesting final comparison. The data are plotted in Figure 4. Recall that the implied constraint set cannot be added to the channelling model, because it is incompatible with the symmetry breaking constraints.

The two models behave remarkably similarly, given their considerable differences. For the instance which times out, it can be seen that the two models explored a similar number of nodes, indicating a similar node rate. The implied model was faster for 13 instances, and the channelled model was faster for 10. However, the implied model timed out on five instances, and the channelled model timed out on one.

3 Conclusions

We have modelled the equidistant frequency permutation array problem using constraint programming, investigating a range of models. We devised a channelled model which exploits symmetry breaking constraints on one viewpoint and problem constraints on the other viewpoint. We invented a set of implied constraints and showed their benefit in most cases. This set of constraints may generalise to other problems involving fixed Hamming distance. As a potential item of future work for EFPA, the implied constraints could be reformulated to be compatible with the channelled model.

We gave a novel representation of cycle notation, modelling a permutation with a fixed number of move points. This was shown to be very effective for cutting down the search space, which indicates its potential. However, the overhead of the additional constraints negated the benefits. With a different formulation or a different constraint solver, the permutation model could prove to be beneficial. Also, it may apply to other problems involving fixed Hamming distance. It would be interesting to investigate this further.

Our work has direct application in mathematics. One of the authors (Huczynska) is a mathematician and is exploiting our novel results in her own theoretical investigations [4]. This illustrates that, with careful modelling and the power of fast constraint solvers such as Minion and modelling assistants such as Tailor, constraint programming can contribute to research in other disciplines.

References

1. Huczynska, S., Mullen, G.: Frequency permutation arrays. J. Combin. Des. 14, 463–478 (2006)
2. Chu, W., Colbourn, C., Dukes, P.: Constructions for permutation codes in powerline communications. Designs, Codes and Cryptography 32, 51–64 (2004)

3. Chu, W., Colbourn, C., Dukes, P.: On constant composition codes. Discrete Applied Math. 154, 912–929 (2006)
4. Huczynska, S.: Equidistant frequency permutation arrays and related constant composition codes. Designs, Codes and Cryptography (to appear, 2009)
5. Ding, C., Yin, J.: A construction of optimal constant composition codes. Designs, Codes and Cryptography 40, 157–165 (2006)
6. Slaney, J., Fujita, M., Stickel, M.: Automated reasoning and exhaustive search: Quasigroup existence problems. Computers and Mathematics with Applications 29, 115–132 (1995)
7. Gent, I.P., Jefferson, C., Miguel, I.: Minion: A fast, scalable, constraint solver. In: Proceedings 17th European Conference on Artificial Intelligence (ECAI 2006), pp. 98–102 (2006)
8. Gent, I.P., Jefferson, C., Miguel, I.: Watched literals for constraint propagation in minion. In: Benhamou, F. (ed.) CP 2006. LNCS, vol. 4204, pp. 182–197. Springer, Heidelberg (2006)
9. Gent, I.P., Miguel, I., Rendl, A.: Tailoring solver-independent constraint models: A case study with Essence′ and Minion. In: Miguel, I., Ruml, W. (eds.) SARA 2007. LNCS (LNAI), vol. 4612, pp. 184–199. Springer, Heidelberg (2007)
10. Régin, J.C.: Generalized arc consistency for global cardinality constraint. In: Proceedings 13th National Conference on Artificial Intelligence (AAAI 1996), pp. 209–215 (1996)
11. Frisch, A.M., Jefferson, C., Miguel, I.: Symmetry breaking as a prelude to implied constraints: A constraint modelling pattern. In: Proceedings ECAI 2004, pp. 171–175 (2004)
12. Hnich, B., Smith, B.M., Walsh, T.: Dual modelling of permutation and injection problems. JAIR 21, 357–391 (2004)
13. Velev, M., Gao, P.: Efficient SAT techniques for absolute encoding of permutation problems: Application to hamiltonian cycles. In: Proceedings SARA 2009 (to appear, 2009)
14. Rendl, A., Miguel, I., Gent, I.P., Jefferson, C.: Automatically enhancing constraint model instances during tailoring. In: Proceedings of the Eighth International Symposium on Abstraction, Reformulation and Approximation, SARA 2009 (to appear, 2009)
15. Jefferson, C.: Representations in Constraint Programming. PhD thesis, Computer Science Department, York University, UK (2007)
16. Flener, P., Frisch, A.M., Hnich, B., Kiziltan, Z., Miguel, I., Pearson, J., Walsh, T.: Breaking row and column symmetries in matrix models. In: Van Hentenryck, P. (ed.) CP 2002. LNCS, vol. 2470, pp. 462–476. Springer, Heidelberg (2002)
17. Choi, C.W., Harvey, W., Lee, J.H.M., Stuckey, P.J.: Finite domain bounds consistency revisited. In: Sattar, A., Kang, B.-h. (eds.) AI 2006. LNCS (LNAI), vol. 4304, pp. 49–58. Springer, Heidelberg (2006)
18. Walsh, T.: Breaking value symmetry. In: Proceedings of AAAI 2008 (2008)
19. Debruyne, R., Bessière, C.: Some practicable filtering techniques for the constraint satisfaction problem. In: Proceedings 15th International Joint Conference on Artificial Intelligence (IJCAI 1997), pp. 412–417 (1997)

Scheduling the CB1000 Nanoproteomic Analysis System with Python, Tailor, and Minion

Andrew Loewenstern

Andrew Loewenstern Consulting
andrew@gigagig.org

Abstract. An effective scheduler for parallel jobs on a robotic protein analysis system was created with Python, Tailor, and the Minion constraint solver. Tailor's implementation of the expressive Essence' constraint modeling language allowed the use of the powerful Minion solver by non experts. Constructing the model in Python allowed its use in several parts of the software, including generation of the Essence' model used by Minion, visualization of the schedule, and verification of the correct execution of the system.

1 Introduction

1.1 Cell Biosciences CB1000

The CB1000 is a commercial nanoproteomic analysis system that provides new analytic capabilites to life sciences researchers. It is produced by Cell Biosciences of Palo Alto California, USA.[1] The CB1000 possesses a robotic arm which moves glass capillaries containing cell samples from location to location within the instrument over the course of job. The embedded software controlling the system is primarily written in Python and runs on the Ubuntu platform.

The use of the robotic arm is the primary physical constraint in the system as it performs several vital functions but can be used by only one job at a time. In addition to manipulating capillaries and transferring them from location to location, the arm is also used for loading samples and reagents into the capillaries, and dispensing and cleaning up liquids.

The other constrained resource in the system is the separation chamber, which can also only be used by one job at a time but parallel to the arm. The separation chamber is used not only at the beginning of the job but also has the instruments for collecting data at the end.

The system contains a finite supply of capillaries, reagents, and other liquids used in the job. At this time only eight jobs can be executed before the system requires manual intervention by a human. The goal of the scheduler is to improve the throughput of the system, minimizing the amount of time it takes to process a run of up to eight jobs. This represents a major upgrade from the first generation sytems which are only capable of executing jobs sequentially.

I.P. Gent (Ed.): CP 2009, LNCS 5732, pp. 65–72, 2009.

1.2 Job

The job performed by the CB1000 produces results similar to a Western Blot but with new quantification capabilities and extremely small sample sizes.[5,6] It consists of a series of steps that are fixed but can vary in duration for each experiment. A simplified schematic of a typical job is shown in Fig.1. The black lines are robotic arm steps. The two steps labeled S and D, Separation and Detection respectively, use the separation chamber. There are two periods of incubation, labeled PI (Primary Incubation) and SI (Secondary Incubation), where the capillaries lay undisturbed in a drawer and provide opportunity for other jobs to run. This schematic does not show that most of the arm steps partially overlap the step preceeding or following it. Those arm steps are modeled as separate steps to optimize running time. Each step in the constraint model comprises a series of actual steps on the system that cannot be interrupted.

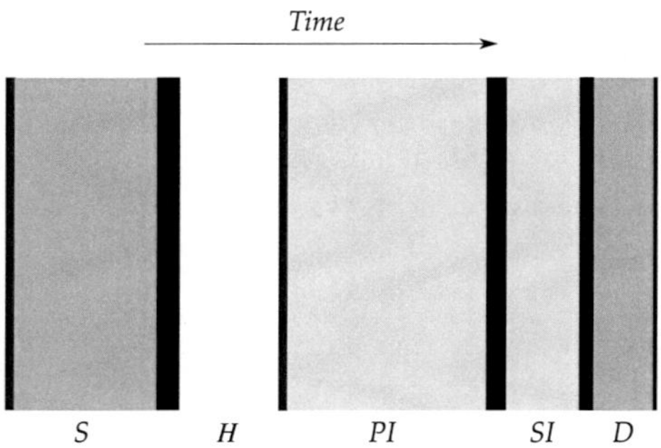

Fig. 1. Simplified schematic of a job. Arm steps are black. S = Separation, H = Hold, PI & SI = Incubation, D = Detection.

At the point labeled H the experiment is in a semi-stable state. The duration of this step may vary in order to optimize the utilization of constrained resources. Some users may not desire to have an additional variable in their experiment so this step may be skipped.

1.3 Interface

Users control the CB1000 from an application written in Java that runs on a regular PC positioned next to the system. Once the user has set the parameters for up to eight jobs the control application signals the embedded software in the CB1000 to start the run. The scheduler operates while the system prepares for the run of jobs.

2 Implementation

2.1 Tailor and Minion

Tailor and Minion are under active development, have a file based interface that integrates with any language through subprocesses, and possess permissive licenses. Key to the decision to use Tailor is the high level Essence modeling language.[2] Tailor implements a subset of Essence called Essence', provides an interactive compilation tool, and outputs models suitable for the Minion constraint solver.[3] Models constructed in Essence' are easily understandable to professional software developers who may not be experts in constraint programming. Minion, a powerful constraint solver with years of development, was treated as a "black box."[4]

2.2 Model

The goal of the scheduler is to find the start and end time of every step relative to the beginning of the run so the machine can execute them in the correct order. Since two jobs cannot start at the same time, each job is started after delay relative to the beginning of the run. This delay comprises one of the two objective variables to be found by the solver.

Step start and stop times are modeled as expressions representing the sums of the durations of all preceding steps in the job. Step durations are determined empirically and all are known prior to scheduling except for the delay from the beginning of the run prior to starting a job and the hold step at point H in Fig. 1 for each job. This hold step in the middle of the job comprises the second objective variable for each job that must be found by the solver.

Once the solver finds the start delay and hold times they must be plugged back into the schedule to determine the start times for each step. To avoid having duplicate models of the job in both Essence' and Python, the model is constructed once in Python. The same model is used to generate Essence' and subsequently to compute the complete schedule.

The job performed by the CB1000 has 12 separate steps using the arm. Since each arm step must end before or start after every other arm step for every other job in a run of up to eight jobs, there are 4032 disjunctions forming the constraints for the arm. The `forall` operator in Essence' makes writing out every constraint unnecessary but it would still be tedious. Generating Essence' from Python simplifies the creation of Essence' and allows is to be generated on the fly. The complete Python source for generating the model, including Essence' templates and lists of steps and names used by other parts of the software but not the solver, is only 250 lines.

Since the step marked S in Fig. 1 fixes the experiment and makes it more stable, the solver is instructed to minimize the delay prior to starting a job from the beginning of the run. This is accopmlished with a `minimizing` statement in Essence'.

Essence' Template. The Python code contains a string template template of the Essence' model leading up to the constraints. This portion of the Essence' has the definition of step durations that are referenced in the expressions representing the start and stop times for each step. The durations are defined with a series of `let` statements. At scheduling time, this template has all of the actual step durations inserted using Python string substitution. Then the constraints are generated and appended to the Essence' code.

Python. The model of the job itself consists of a set of Python variables that each represent either the start time or stop time of a step relative to the start of the run. Each of these variables is a string containg an expression that evaluates to either the start or stop time for a step. By a happy coincidence, these expressions are valid in either Essence' and Python. Each expression is simply the sum of the variables represting the start delay plus the durations of all steps preceeding the one being defined. The start of the first step is equal to start delay. The end of the first step is equal to the start of the first step plus the duration of the first step. The start of second step is equal to the end of the first, and the end of the second step is equal to its start plus its duration, and so forth.

Simplified Python code showing the definition of the first three steps in a job.

```
firstStepStart = "jobDelay"
firstStepEnd = "jobDelay + firstStepDuration"
secondStepStart = "jobDelay + firstStepDuration"
secondStepEnd = "jobDelay + firstStepDuration + secondStepDuration"
thirdStepStart = "jobDelay + firstStepDuration + secondStepDuration"
thirdStepEnd = "jobDelay + firstStepDuration + secondStepDuration
               + thirdStepDuration"
```

In this manner all of the steps in the model are constructed as strings in Python. These strings are be emitted in the constraints portion of the Essence' code to ensure that no two steps using the same resource overlap. Later, the strings are evaluated in a Python environment that contains all of the variables referenced by the expressions, including the ones found by the solver, to determine the actual start time of each step.

Constraints. Two sets of constraints ensure mutually exclusive access to the arm and separation chamber. This is done simply by ensuring all steps using the same resource do not overlap with any other step in any job using the same resource. For every job, each arm step must must be compared against the arm steps for every other job. If the step begins after the other step ends or the step ends before the other step begins, the mutually exclusive constraint for the pair of stepms is satisfied.

The Python listing below demonstrates the loop that generates mutually exclusive constraints for all steps for a constrained resource. `allsteps` below contains the string expressions representing the start and stop times for each step that uses the resource. The same code is used to output constraints for the usage of the separation chamber.

```
for a in range(jobCount):
  for b in range(jobCount):
    if b > a:
        for aStart, aFinish in allsteps:
          for bStart, bFinish in allsteps:
             emitEssence("(((%s) > (%s)) \/ ((%s) < (%s))),\n"
                        % (aStart, bFinish, aFinish, bStart)
```

The listing below shows the Essence' constraint for comparing the first step of the first job against the first step of the second job.

```
(((jobDelay1) > (jobDelay2 + firstStepDuration)) \/
  (jobDelay1 + firstStepDuration) < (jobDelay2)))
```

2.3 Scheduling

Once the durations for each step, which are dependent on user supplied parameters, are collected the Essence' model is emitted. Tailor is run on the Essence' model and a model suitable for Minion is generated. Tailor reports 26288 constraints and 15104 auxiliary variables in the output. Finally, Minion is run and the values for the two objective variable arrays are sent to standard output on two lines. Execution of Tailor and Minion is accomplished through the built in Python `subprocess` module.

Once the values for the objective variables are parsed, they are added to a Python dictionary containing the durations of each step. The keys in this dictionary are the same as the variable names used in the model expressions and in the Essence' code. This dictionary is used as an environment for evaluating the model strings specifying the start and stop times for each step. Each model string evaluates to an integer representing the elapsed time in seconds since the beginning of the run. In this way the start and stop times for steps relative to the start of the run can be computed using the same model as used in the constraint solver. Once the start time for each steps is known, scheduling is simply a matter of sorting by start time the Python object instances containing the code for the steps for all job and executing them in order.

2.4 Visualization

Once the schedule is computed the start and stop times are fed to Matplotlib and a horizontal bar chart is generated.[7] A set of plots are shown in Fig. 2, Fig. 3, and Fig. 4. As the system operates the start and stop time of each step is recorded. At the end of the run another plot is generated using the actual start and stop times. Code used to generate the Essence' model is also used to transform the data into a format acceptable to Matplotlib. The lists of steps in the Python model are also used to assign colors to the plots. These plots are used to quickly and informally verify the correct operation of the scheduler and the CB1000 itself.

3 Results

3.1 Modeling

Modeling with Essence' greatly eased the development of the scheduler and reduced the risk of creating code that would be very difficult to understand by someone unfamiliar with constraint programming. Being able to examine the intermediate Essence' code and execute it using Tailor's interactive facilities made debugging the model fairly simple. Conversely, increasing the complexity of the tool chain introduced an unknown risk of bugs resulting in explicable results or failure.

Development was done with an early version of Tailor, version 0.2.0. Generating Essence' fromPython code also made it easy avoid limitations with `foreach` expression in the early version of Tailor. Once a working model was constructed, however, the combination of Tailor and Minion never failed to find a correct solution.

Approximately one day was required to learn Essence' and construct a working model and scheduler. Much more time was subsequently spent refining the model, creating a visualization tool, and integrating the scheduler into a code base that had been written for serial execution of jobs.

Python and Essence' sharing syntax for array access is a happy coincidence that opened the door to powerful meta programming techniques. The same code specifying the start and stop times of steps in a job was used in several places. Reducing duplication reduced the chance of error and allowed the scheduler to get up and running quickly. Refinements and corrections to the model since the original development were easy to apply and did not produce any unexpected results.

3.2 Execution

On the CPU embedded in the CB1000, a 1.6 GHz Celeron, the entire process of computing the schedule takes approximately 12 seconds. This is done in parallel with the system initialization performed at the beginning of the run so it adds no additional time to the run. This performance is acceptable for the operation of the system. Adding substantially more arm steps, however, would likely dramatically increase the search time required to find a solution.

3.3 Throughput

The ratio of the duration of arm usage, separation chamber usage, and passive incubation of a typical job allows the scheduler to increase throughput dramatically. The first generation system performed the job sequentially, leading to long run times as shown in Fig.2.

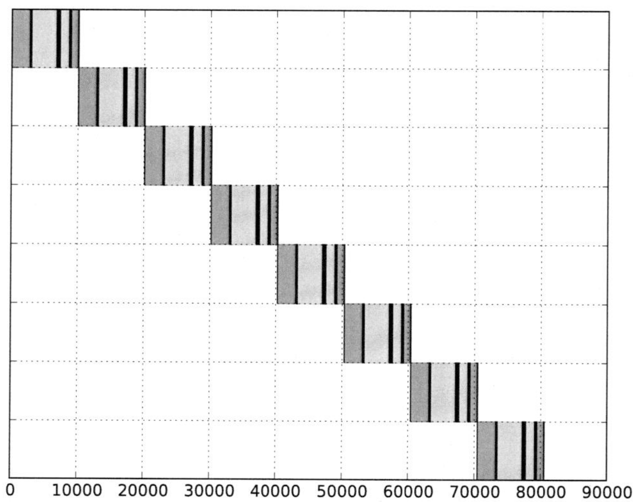

Fig. 2. Unscheduled run with sequential execution, time in seconds

Figure 3 shows the same run with the scheduler. In this case there is no hold step in the middle of the job. Even without that optimization, the scheduler decreases the time to complete the run by 30%

Figure 4 shows the same run with the scheduler using the hold step in the middle of the run. In this case the scheduler minimizes the delay before starting a job, which results in all of the separations being done first. With the hold step the scheduler is able to effectively double the throughput of the system.

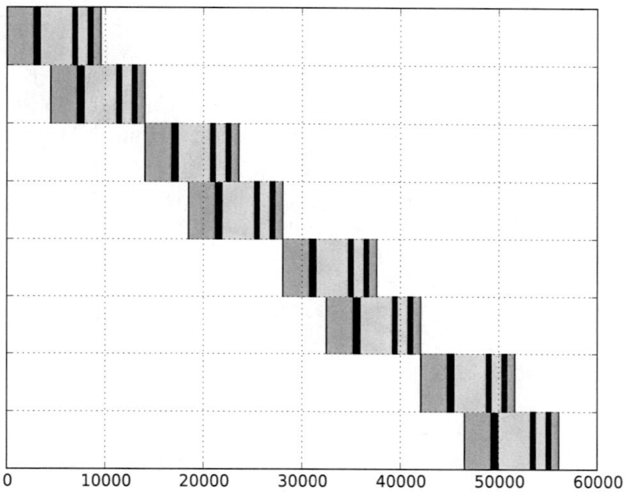

Fig. 3. Scheduled run with no delay before adding first reagent, time in seconds

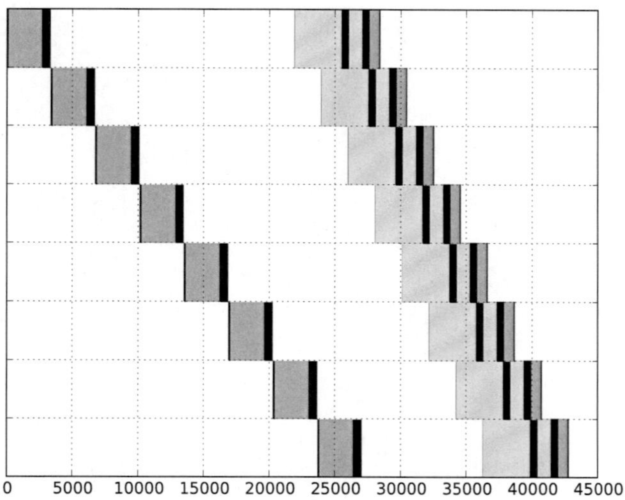

Fig. 4. Scheduled run with variable hold step in the middle of the job, time in seconds

References

1. Cell Biosciences web site, http://www.cellbiosciences.com/
2. Frisch, A.M., Harvey, W., Jefferson, C., Hernandez, B.M., Miguel, I.: ESSENCE: A Constraint Language for Specifying Combinatorial Problems. To appear in Constraints 13(3) (July 2008)
3. Gent, I.P., Miguel, I., Rendl, A.: Tailoring Solver-independent Constraint Models: A Case Study with Essence and Minion. In: Proceedings of SARA 2007 (2007)
4. Gent, I.P., Jefferson, C., Miguel, I.: MINION: A Fast, Scalable, Constraint Solver. In: The European Conference on Artificial Intelligence 2006, ECAI 2006 (2006)
5. O'Neill, R.A., Bhamidipati, A., Bi, X., Deb-Basu, D., Cahill, L., Ferrante, J., Gentalen, E., Glazer, M., Gossett, J., Hacker, K., Kirby, C., Knittle, J., Loder, R., Mastroieni, C., MacLaren, M., Mills, T., Nguyen, U., Parker, N., Rice, A., Roach, D., Suich, D., Voehringer, D., Voss, K., Yang, J., Yang, T., Vander Horn, P.B.: Isoelectric focusing technology quantifies protein signaling in 25 cells. PNAS 103(44), 16153 (2006)
6. Neal Burnette, W.: 'Western blotting': electrophoretic transfer of proteins from sodium dodecyl sulfate — polyacrylamide gels to unmodified nitrocellulose and radiographic detection with antibody and radioiodinated protein A. Analytical Biochemistry 112(2), 195–203 (1981)
7. Matplotlib web site, http://matplotlib.sourceforge.net/

Solving Nurse Rostering Problems Using Soft Global Constraints

Jean-Philippe Métivier, Patrice Boizumault, and Samir Loudni

GREYC (CNRS - UMR 6072) – Université de Caen
Campus II – Boulevard du Maréchal Juin
14000 Caen Cedex

Abstract. Nurse Rostering Problems (NRPs) consist of generating rosters where required shifts are assigned to nurses over a scheduling period satisfying a number of constraints. Most NRPs in real world are NP-hard and are particularly challenging as a large set of different constraints and specific nurse preferences need to be satisfied. The aim of this paper is to show how NRPs can be easily modelled and efficiently solved using soft global constraints. Experiments on real-life problems and comparison with ad'hoc OR approaches are detailed.

1 Introduction

Due to their complexity and importance in real world modern hospitals, Nurse Rostering Problems (NRPs) have been extensively studied in both Operational Research (OR) and Artificial Intelligence (AI) for more than 40 years [5,11]. Most NRPs in real world are NP-hard [16] and are particularly challenging as a large set of different rules and specific nurse preferences need to be satisfied to warrant high quality rosters for nurses in practice. Other wide range of heterogeneous and specific constraints makes the problem over-constrained and hard to solve efficiently [1,27].

NRPs consist of generating rosters where required shifts are assigned to nurses over a scheduling period satisfying a number of constraints [5,7]. These constraints are usually defined by regulations, working practices and preferences of nurses and are usually categorised into two groups: hard constraints and soft constraints (with their violation costs).

From a Constraint Programming (CP) point of view, global constraints are often key elements in successfully modelling and solving real-life problems due to their efficient filtering. Global constraints are particularly well suited [31] for modelling NRPs: sequence constraints on every nurse planning, daily capacity constraints, etc. But, for over-constrained problems as NRPs, such filtering can only be performed in very particular cases. Soft global constraints proposed by [30,26,33] take advantage from the semantics of a constraint and from the semantics of its violation to efficiently perform filtering. The aim of this paper is to show how NRPs can be modelled using soft global constraints and solved efficiently with solutions quality and computing times close to those obtained using ad'hoc OR methods.

I.P. Gent (Ed.): CP 2009, LNCS 5732, pp. 73–87, 2009.

Section 2 gives a synthetic overview of NRPs and describes the problem we selected as example for our presentation. Although this problem is fictional, hard and soft constraints it contains are representative of constraints encountered in most NRPs. Section 3 is devoted to soft global constraints and their filtering. In Section 4, we show how NRPs can be modelled in a concise and elegant way using soft global constraints.

The next two sections are devoted to the resolution of NRPs. First, we introduce (Section 5) the global constraint `regularCount` (and its soft version) which combines `regular` and `atleast/atmost` constraints in order to provide a more efficient filtering. Then, we motivate and present our resolution method VNS/LDS+CP [18] based on a Variable Neighborhood Search (VNS [22]) where the reconstruction step is performed using a Limited Discrepancy Search (LDS [13]) combined with filtering (CP) performed by soft global constraints.

The ASAP site (Automated Scheduling, optimization And Planning) of University of Nottingham (`http://www.cs.nott.ac.uk/~tec/NRP/`) records a large and various set of NRPs instances as well as the methods used to solve them. We performed experimentations over different instances we selected in order to be representative of the diversity and the size of NRPs. For each instance, we compare quality of solutions and computing times for our method with the best known method for solving it [14]. Experimentations show (Section 7) that, despite its genericity and flexibility, our method provides excellent results on small and middle size problems and very promizing results on large scale problems.

2 Nurse Rostering Problems

2.1 An Overview of NRPs

NRPs consist of generating rosters where required *shifts* are assigned to nurses over a scheduling period (*planning horizon*) satisfying a number of constraints [5,11]. These constraints are usually defined by regulations, working practices and nurses preference. Constraints are usually categorised into two groups: *hard* and *soft* ones. Hard constraints must be satisfied in order to obtain feasible solutions for use in practice. A common hard constraint is to assign all shifts required to the limited number of nurses. Soft constraints are not obligatory but are desired to be satisfied as much as possible. The violations of soft constraints in the roster are used to evaluate the quality of solutions. A common soft constraint in NRPs is to generate rosters with a balanced workload so that human resources are used efficiently.

Shift types are hospital duties which usually have a well-defined start and end time. Many nurse rostering problems are concerned with the three traditional shifts Morning, (7:00–15:00), Evening (15:00–23:00), and Night (23:00–7:00). *Shift constraints* express the number of personnel needed for every skill category and for every shift or time interval during the entire planning period. *Nurse constraints* refer to all the restrictions on personal schedules. All the personal requests, personal preferences, and constraints on balancing the workload among personnel belong to this category.

2.2 Example: A 3-Shifts NRP

The 3 shifts are Morning (M), Evening (E) and Night (N). Off (O) will represent repose. 8 nurses must be planned over a planning horizon of 28 days satisfying the below constraints.

1. **Hard Constraints:**
   (H1) From Monday to Friday, M, E, N shifts require respectively (2, 2,1) nurses. For weekend, the demand of all shifts is reduced to 1.
   (H2) A nurse must have at least 10 days off.
   (H3) A nurse must have 2 free Sundays.
   (H4) A nurse is not allowed to work more than 4 N shifts.
   (H5) The number of consecutive N shifts is at least 2 and at most 3.
   (H6) Shift changes must be performed respecting the order: M, E, N.
2. **Soft Constraints:**
   (S1) For a nurse, the number of M and N shifts should be within the range [5..10]. Any deviation δ is penalised by a cost $\delta \times 10$.
   (S2) The number of consecutive working days is at most 4. Any excess δ generates a penality of $\delta \times 1000$.
   (S3) Every isolated day off is penalised by a cost 100.
   (S4) Every isolated working day is penalised by a cost 100.
   (S5) Two working days must be separated by 16 hours of rest. Any violation generates a cost 100.
   (S6) An incomplete weekend has cost 200.
   (S7) Over a period of 2 weeks a nurse must have 2 days off during weekends. Any deviation δ is penalised by a cost $\delta \times 100$.
   (S8) A N shift on Friday before a free weekend is penalised by a cost 500.

Related CP Works. An operational system GYMNASTE using (hard) global constraints is described in [31]. Other practical systems are also mentionned. But, dealing with over-constrained problems is only discussed as perspectives. Three directions are indicated (Hierarchical CP [1], heuristics and interactions). The last two proposals are problem dependent. As quoted by [31], the main difficulty with the Hierarchical CP approach is that global constraints have to be extended to handle constraint violations. This is the aim of this paper.

3 Soft Global Constraints

3.1 Principles

Over-constrained problems are generally modelled as Constraint Optimization Problems (COP). A cost is associated to each constraint in order to quantify its violation. A global objective related to the whole set of costs is usually defined (for example to minimize the total sum of costs). Global constraints are often key elements in successfully modelling and solving real-life problems due to their efficient filtering. But, for over-constrained problems, such filtering can only be performed in very particular cases. Soft global constraints proposed by [30,26,33] take advantage from the semantics of a constraint and from the semantics of its violation to efficiently perform filtering.

Definition 1 (violation measure). *μ is a violation measure for the global constraint* $\mathtt{c}(X_1, ..., X_n)$ *iff μ is a function from $D_1 \times D_2 \times ... \times D_n$ to $\Re^+$ s.t. $\forall A \in D_1 \times D_2 \times ... \times D_n$, $\mu(A) = 0$ iff A satisfies $c(X_1, ..., X_n)$.*

To each soft global constraint c are associated a violation measure μ_c and a cost variable Z_c that measures the violation of c. So the COP is transformed into a CSP where all constraints are hard and the cost variable $Z = \sum_c Z_c$ will be minimized. If the domain of a cost variable is reduced during the search, propagation will be performed on domains of other cost variables.

Definition 2 (soft global constraint). *Let* $\mathtt{c}(X_1, ..., X_n)$ *be a global constraint, Z_c its cost variable, and μ a violation measure. The soft global constraint* Σ-$\mathtt{c}([X_1, ..., X_n], \mu, Z_c)$ *has a solution iff $\exists A \in D_1 \times D_2 \times ... \times D_n$ s.t. $min(D_{Z_c}) \leq \mu(A) \leq max(D_{Z_c})$.*

Example (decomposition based violation measure). Let $\mathtt{c}(X_1, ..., X_n)$ be a global constraint which can de decomposed as a conjunction of binary constraints $c_{i,j}$ over variables X_i and X_j. Let $\varphi_{i,j}$ be the violation cost of $c_{i,j}$; let $A \in D_1 \times D_2 \times ... \times D_n$ and $unsat(A)$ be the set of constraints $c_{i,j}$ unsatisfied by A. Then, $\mu_{dec}(A) = \sum_{c_{i,j} \in unsat(A)} \varphi_{i,j}$.

3.2 Relaxation of gcc

i) A Global Cardinality Constraint (gcc) on a sequence of variables specifies, for each value in the union of their domains, an upper and lower bound to the number of variables that are assigned to this value [28].

Definition 3. *Let $\mathcal{X}$=$\{X_1, \ldots, X_n\}$, $Doms$=$\cup_{X_i \in \mathcal{X}} D_i$. Let $v_j \in Doms$, l_j and u_j the lower and upper bounds for v_j.* $\mathtt{gcc}(\mathcal{X}, l, u)$ *has a solution iff $\exists A \in D_1 \times D_2 \times ... \times D_n$ s.t. $\forall v_j \in Doms$, $l_j \leq \mid \{X_i \in \mathcal{X} \mid X_i = v_j\} \mid \leq u_j$.*

Each constraint $\mathtt{gcc}(\mathcal{X}, l, u)$ can be decomposed as a conjunction of atleast and atmost constraints over values in *Doms*:

$$\mathtt{gcc}(\mathcal{X}, l, u) = \bigwedge_{v_j \in Doms} (\mathtt{atleast}(\mathcal{X}, v_j, l_j) \wedge \mathtt{atmost}(\mathcal{X}, v_j, u_j))$$

ii) Global constraint Σ-gcc is a soft version of gcc for the decomposition based violation measure μ_{dec}[20]. Σ-gcc allows the violation of the lower and/or upper bounds of values. To each value $v_j \in Doms$ are associated a *shortage* function $s(\mathcal{X}, v_j)$ measuring the number of missing assignments of v_j to satisfy $\mathtt{atleast}(\mathcal{X}, v_j, l_j)$, and an *excess* function $e(\mathcal{X}, v_j)$ measuring the number of assignments of v_j in excess to satisfy $\mathtt{atmost}(\mathcal{X}, v_j, u_j)$. Each constraint Σ-gcc is modelled by adding *violation arcs* to the network of gcc [28]. These violation arcs represent the shortage or the excess for each value of *Doms* [33,20].

Definition 4 (μ_{dec}). *For each value $v_j \in Doms$, let $\varphi_j^{atleast}$ be the violation cost of its lower bound l_j and φ_j^{atmost} the violation cost of its upper bound u_j,*

$$\mu_{dec}(\mathcal{X}) = \sum_{v_j \in Doms} \mu_{card}(\mathcal{X}, v_j)$$
$$where\ \mu_{card}(\mathcal{X}, v_j) = s(\mathcal{X}, v_j) \times \varphi_j^{atleast} + e(\mathcal{X}, v_j) \times \varphi_j^{atmost}$$

Property 1 (**filtering**). Let Z be a cost variable and Φ a violation structure, Σ-gcc$(\mathcal{X}, l, u, \mu_{dec}, \Phi, Z)$ is domain-consistent iff for every arc $a = (X_i, v_j)$, there exists an integer $s-t$ flow f of value $max(n, \sum_{v_j \in Doms} l_j)$ with $f(a) = 1$ and weight p s.t. $min(D_Z) \leq p \leq max(D_Z)$.

Worst case time complexity: $O(n^2 log(n) \times d)$ where $d = max(|D_i|)$.

3.3 Relaxation of `Regular`

i) Global Constraint `Regular` [25]

Definition 5. *Let M be a Deterministic Finite Automaton (DFA), $\mathcal{L}(M)$ the language defined by M, $\mathcal{X}$ a sequence of n variables.* `regular`*$(\mathcal{X}, M)$ has a solution iff $\exists A \in D_1 \times D_2 \times ... \times D_n$ s.t. $A \in \mathcal{L}(M)$.*

A DFA is defined by a 5-tuple $M = \{Q, \Sigma, \delta, q_0, F\}$ where Q is a finite set of states, Σ is an alphabet, $\delta : Q \times \Sigma \rightarrow Q$ is a transition function, q_0 is the initial state and $F \subseteq Q$ is the set of final (or accepting) states. A `regular` constraint over a sequence of n variables is modelled by a layered directed graph $\mathcal{G} = (V, U)$:

- vertex set $V = \{s\} \cup V_0 \cup \ldots \cup V_n \cup \{t\}$ where $\forall i \in [1..n], V_i = \{q_l^i \mid q_l \in Q\}$
- arc set $U = \{(s, q_0^0)\} \cup U_0 \cup \ldots U_n \cup \{(q_l^n, t) \mid q_l \in F\}$
where $\forall i \in [1..n], U_i = \{(q_l^i, q_m^{i+1}, v_j) \mid v_j \in D_i, \delta(q_l, v_j) = q_m\}$

Property 2. Solutions for `regular`$(\mathcal{X}, M)$ correspond to s-t paths in graph $\mathcal{G}$.

ii) Global Constraint `Cost-Regular` enables to model the fact that some transitions of an automaton may have a cost. To each `cost-regular` constraint is associated a directed weighted layered graph $\mathcal{G} = (V, U, \Phi)$, where each arc representing a transition is valued by the cost of this transition [10]. For an instantiation A, the measure $\mu_{reg}(A)$ is defined as the total sum of the transition costs for arcs belonging to the path associated to A.

Property 3 (**filtering**). Let Z be a cost variable and Φ a violation structure. `cost-regular`$(\mathcal{X}, M, \mu_{reg}, \Phi, Z)$ is domain-consistent iff for every arc $a = (X_i, v_j)$ there exists an s-t path of weight p s.t. $min(D_Z) \leq p \leq max(D_Z)$.

Worst case time complexity: $O(n \times |Q| \times |\Sigma|)$. For each layer i, each vertex q_l^i may have at most $|\Sigma|$ successors.

iii) `Cost-Regular` **used as a soft constraint for NRPs** every hard constraint of sequence c will be modelled using a DFA M_c (see Section 4.3). A soft constraint of sequence Σ-`c` will be modelled by adding new transitions to M_c as well as their costs. Let M_c' be this new DFA; then a `cost-regular` constraint over M_c' is stated. So, for modelling NRPs, `cost-regular` will be used as a soft version of `regular` $(\mathcal{L}(M_c) \subset \mathcal{L}(M_c'))$.

4 Modelling a 3-Shifts NRP

This section presents the modelling of the problem specified in Section 2.2 and shows how soft global constraints are well suited for modelling NRPs.

4.1 Variables and Domains

Let J=[1..28] be the scheduling period and I=[1..8] the set of nurses; variable X_j^i, with domain D_j^i=Doms=$\{M,E,N,O\}$, will represent the shift assigned to nurse i for day j. For gcc constraints, values will be ordered as in *Doms*.

4.2 Capacity Constraints

(H1) $\forall j \in [1,2,3,4,5,8,9,...,24,25,26], \texttt{gcc}([X_j^1,\ldots,X_j^8],[2,2,1,0],[2,2,1,8])$.
$\forall j \in [6,7,13,14,20,21,27,28], \texttt{gcc}([X_j^1,\ldots,X_j^8],[1,1,1,0],[1,1,1,8])$.
(H2) $\forall i \in I, \texttt{atleast}([X_1^i,\ldots,X_{28}^i],O,10)$.
(H3) $\forall i \in I, \texttt{atleast}([X_7^i,X_{14}^i,X_{21}^i,X_{28}^i],O,2)$.
(H4) $\forall i \in I, \texttt{atmost}([X_1^i,\ldots,X_{28}^i],N,4)$.
(S1) $\forall i \in I, \Sigma\texttt{-gcc}([X_1^i,\ldots,X_{28}^i],[5,5],[10,10],[10,10],[10,10],Z_i)$ for values M and E with $\varphi_M^{atleast}=\varphi_M^{atmost}=10$ (violation costs for E are the same).
(S7) is also modelled using Σ-gcc constraints.

(H2), (H4) and (S1) can be grouped together using Σ-gcc constraints:

$\forall i \in I$, $\Sigma\texttt{-gcc}([X_1^i,\ldots,X_{28}^i],[5,5,0,10],[10,10,4,28],[10,10,0,\infty],[10,10,\infty,0],Z_i)$.
As (H2) is a hard atleast constraint, then u_O=28, $\varphi_O^{atleast}=\infty$ and $\varphi_O^{atmost}=0$.
As (H4) is a hard atmost constraint, then l_N=0, $\varphi_N^{atleast}=0$ and $\varphi_N^{atmost}=\infty$.

4.3 Sequence Constraints

Let Σ be an alphabet, $\overline{x}$ will represent any symbol $y \in \Sigma$ s.t. $y \neq x$.

(H5) $\forall i \in I, \texttt{regular}([X_1^i,\ldots,X_{28}^i],A_1)$ (Figure 1)
(H6) states that shift changes must be performed respecting the order: M, E, N. (see automaton A_2 (Figure 2)). For modelling (S5), two arcs are added: one for transition (e_3,e_2,E) with cost 100 and one for transition (e_2,e_1,M) with cost 100. So, $\forall i \in I$, $\texttt{cost-regular}([X_1^i,\ldots,X_{28}^i],A_2,Z_i^{A2})$.
(S6) $\forall\ i{\in}I$, $\texttt{cost-regular}([X_6^i,X_7^i,X_{13}^i,X_{14}^i,...,X_{27}^i,X_{28}^i,],A_3,Z_i^{A3})$ (Figure 3).

Finally, (S2), (S3) and (S4) can be grouped together using cost-regular. (S8) can also be modelled by cost-regular.

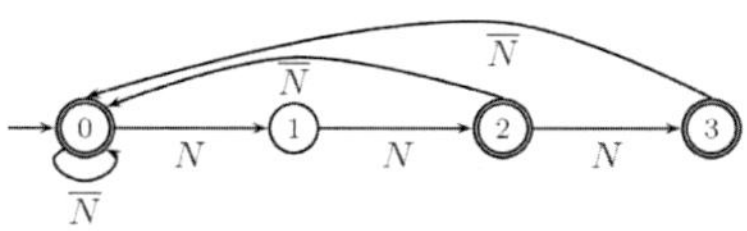

Fig. 1. Automaton A1 for (H5)

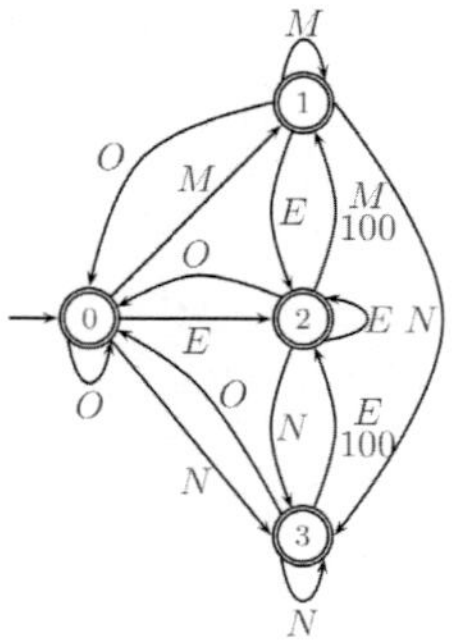

Fig. 2. Automaton A2 for (H6) and (S5)

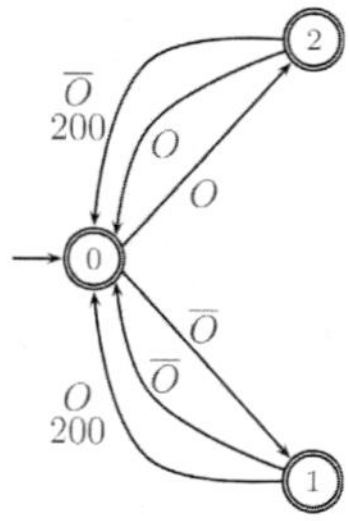

Fig. 3. Automaton A3 for (S6)

5 Interaction between Global Constraints

Despite the efficient filtering of each global constraint, the lack of communication between them reduces significantly the quality of the whole filtering. Indeed, each global constraint uses its internal representation (bipartite graph, network,...) and does not (or partially) exploit information deduced by other global constraints. In most NRPs, a great number of global constraints share a common set of variables (e.g. constraints over the entire planning of a nurse). Few works have been done on the interaction between global constraints:

- `cardinality matrix constraint` [29] combining several `gcc` as a matrix,
- `multi-cost-regular` [19] merging multiple `cost-regular`.

In this section, we propose the `regularCount` (resp. `cost-regularCount`) constraint which combines a `regular` (resp. `cost-regular`) constraint with several `atleast`/`atmost` constraints on a same value.

5.1 Motivating Example

Rule (H4) is modelled as: $\forall i \in I$, `atmost`$([X_1^i, \ldots, X_7^i], N, 4)$ and rule (H5) as: $\forall i \in I$, `regular`$([X_1^i, \ldots, X_7^i], A_1)$ (see Section 2.2 & Section 4.3). Let us consider the following reduced variable domains associated to the first week of nurse i: $D_1^i = D_2^i = D_3^i = \{N\}$, $D_4^i = \{O\}$ and $D_5^i = D_6^i = D_7^i = \{N, O\}$. Filtering separately `atmost` and `regular` will not detect that value N should be removed from D_5^i. Indeed, if $X_5^i = N$ then $X_6^i = N$ by (H5) but (H4) fails. For analogous reasons, value N should also be removed from D_6^i and D_7^i. This example illustrates the weakness of separate filterings.

5.2 `regularCount` Constraint

For an automaton and a particular value $v_j \in Doms$, a `regularCount` constraint will combine a `regular` constraint with several `atleast`/`atmost` constraints on value v_j.

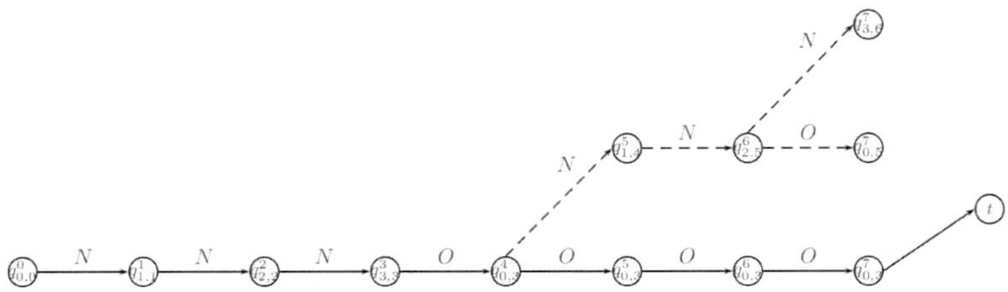

Fig. 4. Graph representation for regularCount($[X_1^i,\ldots, X_7^i]$,A_1,N,0,4)

Definition 6. *Let M=$\{Q, \Sigma, \delta, q_0, F\}$ be a DFA, $\mathcal{L}(M)$ its associated language, $\mathcal{X}$ a sequence of n variables, $v_j \in \Sigma$, l_j (resp. u_j) an upper (resp. lower) bound for v_j.* `regularCount`*($\mathcal{X}$,M,v_j,l_j,u_j) has a solution iff $\exists A \in D_1 \times \ldots \times D_n \mid A \in \mathcal{L}(M) \wedge l_j \leq \mid \{X_i \in \mathcal{X} \mid X_i = v_j\} \mid \leq u_j$.*

The `regularCount` constraint has been used for developping a track planner for a local Radio station. A new track has to be broadcast a bounded number of times into the daily planning which must respect musical transitions expressed as regular expressions. Other potential use for `regularCount` would be planning advertising for TV stations or planning maintenance periods for assembly-lines.

Graph Representation. As for `regular` (see Section 3.3), a constraint `regularCount` ($\mathcal{X}$,M,v_j,l_j,u_j) is modelled by a layered directed graph $\mathcal{G}'(V, U)$. For each layer i, states q_l are labelled both by the layer (q_l^i as for `regular`) and by k the number of occurrences of v_j found so far ($q_{l,k}^i$ for `atleast` and `atmost` constraints).

$$
\begin{aligned}
&V = \{s\} \cup V_0 \cup \ldots \cup V_n \cup \{t\} \\
&V_i = \{q_{l,k}^i \mid q_l \in Q, k \in [0 \ldots n]\},\ \forall i \in [1..n] \\
&U = \{(s, q_{0,0}^0)\} \cup U_0 \cup \ldots U_n \cup U_t \\
&U_t = \{(q_{l,k}^n, t) \mid q_l \in F \wedge (l_j \leq k \leq u_j)\} \\
&U_i = \{(q_{l,k}^i, q_{m,k'}^{i+1}, v) \mid v \in D_i, \delta(q_l, v) = q_m\},\ \forall i \in [1..n] \\
&\textit{where}\ \text{if}\,(v = v_j)\,\text{then}\,k' = k + 1\,\text{else}\,k' = k
\end{aligned}
$$

Example (Section 5.1) is modelled as: $\forall i \in I$, `regularCount`($[X_1^i,\ldots, X_7^i]$,A_1,N,0,4). Figure 4 describes its graph representation. `regularCount` filtering will remove value N from domains D_5^i, D_6^i and D_7^i.

Property 4 (**filtering**). Let M=$\{Q, \Sigma, \delta, q_0, F\}$ be a DFA, $v_j \in \Sigma$, l_j (resp. u_j) the upper (resp. lower) bound for v_j. `regularCount`($\mathcal{X}$,M,v_j,l_j,u_j) is domain-consistent iff for every arc $a = (X_i, v_j) \in U_i$ there exists an s-t path in $\mathcal{G}'(V, U)$.

Proof. There is an arc from $q_{l,k}^i$ to $q_{m,k'}^{i+1}$ iff there exists a value $v \in D_i$ such that $\delta(q_l, v)$= q_m. If $(v = v_j)$ then k'= $k + 1$ (i.e., the number of variables that are assigned to v_j is $k + 1$), otherwise k'=k. If an arc belongs to an s-t path, it belongs to a path from $q_{0,0}^0$ to $q_{l,k}^n$, with $q_l \in F$ and $l_j \leq k' \leq u_j$.

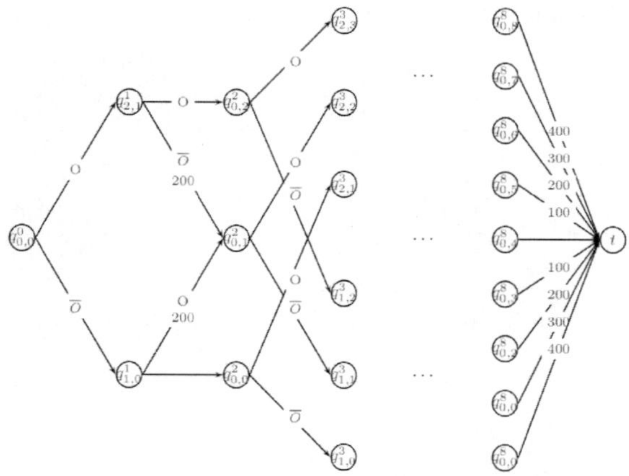

Fig. 5. Graph representation for `cost-regularCount`$([X_6^i,...,X_{28}^i],A_3,\Phi,O,2,2,Z_i^{A_3})$

Worst case time complexity: $O(n^2/2 \times |Q| \times |\Sigma|)$. For layer i, there are at most $i \times |Q|$ vertices as at most i occurrences of v_j may be assigned to variables $X_1, ..., X_i$. As for `regular`, each vertex may have at most $|\Sigma|$ successors. Summing for all layers leads to $\sum_{i=1}^{n} i \times |Q| \times |\Sigma|$.

5.3 `Cost-regularCount` Constraint

Let $M=\{Q, \Sigma, \delta, q_0, F\}$ be a DFA and $v_j \in \Sigma$. `cost-regularCount` is a soft version of `regularCount` for which some transitions may have a cost and which also allows the violation of lower/upper bounds for a value v_j.

Definition 7 (violation measure $\overline{\mu}_{reg}$). *Let $\varphi_j^{atleast}$ (resp. φ_j^{atmost}) be the violation cost associated to lower bound l_j (resp. upper bound u_j). $\forall v_j \in \Sigma$, $\overline{\mu}_{reg}(\mathcal{X}, v_j) = \mu_{reg}(\mathcal{X}) + \mu_{card}(\mathcal{X}, v_j)$.*

Definition 8. *Let M a DFA, $v_j \in \Sigma$, Z a cost variable and Φ a violation structure.* `cost-regularCount`*$(\mathcal{X}, M, \overline{\mu}_{reg}, \Phi, v_j, l_j, u_j, Z)$ has a solution iff $\exists A \in D_1 \times \ldots \times D_n$ s.t. $A \in \mathcal{L}(M) \wedge \min(D_Z) \leq \overline{\mu}_{reg}(A, v_j) \leq \max(D_Z)$.*

Graph Representation. There are two main differences between $\mathcal{G}''(V, U, \Phi)$ and $\mathcal{G}'(V, U)$: i) transition costs are associated to corresponding arcs (as for `cost-regular`), ii) arcs U_t are replaced by violation arcs $\tilde{U_t} = \{(q_{l,k}^n, t) \mid q_l \in F, k \in [0 \ldots n]\}$ which enable to model shortage or excess for a value v_j. To each violation arc $a=(q_{l,k}^n, t)$ is associated a cost $w(a)$:

$$w(a) = \begin{cases} (l_j - k) \times \varphi_j^{atleast} & \text{if } k < l_j \\ (k - u_j) \times \varphi_j^{atmost} & \text{if } k > u_j \\ 0 & \text{otherwise} \end{cases}$$

Figure 5 gives the graph representation of (S6) and (S7) modelled as:

$\forall i \in I$, `cost-regularCount`$([X_6^i, X_7^i, X_{13}^i, ..., X_{27}^i, X_{28}^i,], A_3, \overline{\mu}_{reg}, \Phi, O, 2, 2, Z_i^{A_3})$.

Table 1. Comparative results for Filtering. ($\star$) denotes optimal values.

Instance	$\|I\| \times \|J\|$	$\|D\|$	UB	Σ-Gcc & cost-regular		cost-regularCount	
				Time (s.)	♯backtracks	Time (s.)	♯backtracks
inst_01_07	28	2	3000*	1.4	1 559	**0.2**	**342**
inst_01_11	44	2	1500*	20.9	14 113	**6.7**	**6 002**
inst_01_14	56	2	2500*	380.1	193 156	**122.6**	**63 395**
inst_02_07	49	3	1100*	≥5 400	–	**3 303.1**	**2 891 874**
inst_02_14	98	3	100*	**73.6**	24 100	120.2	**16 587**
inst_02_21	147	3	100*	4 886.7	1 216 908	**940.5**	**107 612**

*Property 5 (***filtering***).* Let $M=\{Q,\Sigma,\delta,q_0,F\}$ be a DFA, $v_j \in \Sigma$, l_j (resp. u_j) the upper (resp. lower) bound for v_j, Φ a violation structure and Z a cost-variable. `cost-regularCount`$(\mathcal{X}, M, \overline{\mu}_{reg}, \Phi, v_j, l_j, u_j, Z)$ is domain consistent iff for every arc $a = (X_i, v) \in U_i$ there exists an s-t path in $\mathcal{G}''(V,,\Phi)$ of cost p s.t. $\min(D_z) \leq p \leq \max(D_z)$.

Proof. If a transition associated to arc $(q^i_{l,k}, q^{i+1}_{m,k'})$ uses v_j, then $k'=k+1$ else $k'=k$. An s-t path using a violation arc in $\tilde{U}_t$ corresponds to a solution with a shortage or an excess for value v_j and the cost to pay is reported on this violation arc. As transition costs are reported on their associated arcs in U_i, the cost of an s-t path A using a violation arc in $\tilde{U}_t$ corresponds exactly to $\overline{\mu}_{reg}(A, v_j)$.

Worst case time complexity: $O(n^2/2 \times |Q| \times |\Sigma|)$ as for `regularCount`.

Table 1 compares the efficiency of `cost-regularCount` filtering vs separate filterings in terms of computing times and number of backtracks. Experiments have been performed on small or medium NRPs instances using Depth First Branch and Bound and run on a 2.8 Ghz P4 processor. `cost-regularCount` filtering always performs better than separate filterings except for `inst_02_14` which is an instance easy to solve where filtering is not so crucial. The extra-cost comes from the fact that the complexity of `cost-regularCount` is slightly higher than those of `cost-regular` and Σ-`gcc`.

6 Variable Neighborhood Search

A great variety of approaches that have been proposed for solving NRPs are either ad'hoc OR methods (including preprocessing steps to reduce the problem size), or local search methods combining OR techniques to find an initial solution. NRPs seem to be well suited for defining large-scale neighborhoods (2-opt, swap and interchange of large portions of nurse plannings, ...).

Variable Neighborhood Search (VNS) [22] is a metaheuristic which systematically exploits the idea of large neighborhood change, both in descent to local minima and in escape from the valleys which contain them. Variable Neighborhood Decomposition Search (VNDS) [12] extends basic VNS within a successive

Algorithm 1. Pseudo-code for VNS/LDS+CP.

```
function VNS/LDS+CP(𝒳, 𝒞, k_init, k_max, δ_max)
begin
    s ← genInitialSol(𝒳)
    k ← k_init
    while (k < k_max) ∧ (not timeout) do
        𝒳_unaffected ← H_neighbor(𝒩_k, s)
        𝒜 ← s\{(x_i = a) s.t. x_i ∈ 𝒳_unaffected}
        s' ← NaryLDS(𝒜, 𝒳_unaffected, δ_max, 𝒱(s), s)
        if 𝒱(s') < 𝒱(s) then
            s ← s'
            k ← k_init
        else k ← k + 1
    return s
end
```

approximations method. For a solution of size n, all but k variables are fixed, and VNDS solves a sub-problem in the space of the k unfixed variables.

VNS/LDS+CP. [18] is a generic local search method based on VNDS [12]. Neighborhoods are obtained by unfixing a part of the current solution according to a neighborhood heuristic. Then the exploration of the search space related to the unfixed part of the current solution is performed by a partial tree search (LDS, [13]) with CP in order to benefit from the efficiency of global constraints filtering (See Algorithm 1). However, as the size of neighborhoods can quickly grow, the exploration of (very) larger neighborhoods may require a too expensive effort. That is why, in order to efficiently explore parts of the search space, LDS is used.

LDS+CP: Our *variable ordering* for LDS is `Dom/Deg` and our *value ordering* selects the value which leads to the lowest increase of the violation cost. CP is performed using soft global constraints filtering.

Neighborhood Heuristics. A lot of soft global constraints are stated over the whole planning of a nurse. So, all variables related to a nurse planning will be together unassigned. k will represent the number of nurse plannings to be unassigned (and not the number of variables as depicted in general Algorithm 1). In order to show the interest of soft global constraints filtering, only two "basic" heuristics have been considered: (i) `rand` which randomly selects a nurse planning, and (ii) `maxV` which selects the nurse planning having the highest violation cost.

7 Experimental Results

We performed experimentations over different instances we selected in order to be representative of the diversity and the size of NRPs. For each instance, **we always compare with the best method** for solving it [14]. As experiments have been run on various machines, we will report, for each instance, the original CPU time and the processor. For all instances, except the first three ones where the processor is unknown (they are noted in italic Table 1.), CPU times will be normalised[1] and denoted CPUN. **Some methods include a pre-treatment.** As CPU times for this step are not given in papers, reported CPU times concern in fact the second step. Finally, reported CPU times for our method always **include the computing time for obtaining the initial solution**.

Experimental Protocol. Each instance has been solved by VNS/LDS+CP, with a discrepancy varying from δ_{min}=3 to δ_{max}=8. k_{min} has been set to 2 and k_{max} to 66% of the total number of nurses. Timeout has been set according to the size of each instance. For the rand heuristic, a set of 10 runs per instance has been performed. VNS/LDS+CP has been implemented in C++. Experiments have been performed under Linux on a 2.8 Ghz P4 processor, with 1GB RAM.

Comparisons with ad'hoc Methods. Heuristic maxV, which is better informed that rand, provides the best performances except for instances LLR and Azaiez.

A) Ozkarahan *instance [24]: we find the optimum in less than* 1*s. using* maxV.

B) MILLAR *instance:* (2 methods)
B1) Network programming [21]: All feasible weekly shift patterns of length at most 4 days are generated. An ILP model is defined and solved using CPLEX.
B2) TS+B&B [15]: Nurse constraints are used to produce all feasible shift patterns for the whole scheduling period for each nurse (independently from shift constraints). Best combinations of these shift patterns are found using mathematical programming and Tabu Search.

With B1, a solution of cost 2.550 is found after 500 s. on an IBM RISC6000/340. With B2, a solution of cost 0 is obtained in 1 s. on a 1Ghz Intel P3 processor. *We find the optimum in less than* 1 *s. using* maxV.

C) Musa *instance [23]:* A solution of cost 199 is found in 28 s. on UNIVAC-1100. *We find the optimum (cost 175) in* 39 *s. using* maxV.

D) LLR *instance:* A hybrid AI approach (TS+CP), which combines CP techniques and Tabu Search is used in [17]. A solution of cost 366 is found after 96 s. on a PC/P-545MHz (CPUN 16 s.). *With* rand, *we obtain (on average) a solution of cost* 319 *after* 265 *s. The best solution (over the 10 runs) has a cost* 314 *(*79 *s.). The first solution (cost* 363*) is obtained in less than* 1 *s. Using* maxV, *a solution of cost* 326 *is found in* 186 *s.*

E) BCV-5.4.1 *instance:* (3 methods). All the results are obtained on a same machine (2.66GHz Intel P4 processor). Hybrid Tabu search [4] is the best of the 3 methods for this instance. VDS [6] finds the optimum after 135 s. (CPUN 128 s.). In [3], a solution of cost 200 is found after 16 s. (CPUN 15 s.). *With* maxV, *we obtain the optimum after* 180 *s.*

[1] For a machine κ times slower than ours, reported CPU times will be divided by κ.

Table 2. Comparative results. (⋆) denotes optimal values.

Instances	$\lvert I\rvert \times \lvert J\rvert$	$\lvert D\rvert$	Best Ub	Ad'hoc methods			VNS/LDS+CP	
				Algo.	**Cost**	**Time(s)**	**Cost**	**Time(s)**
Ozkarahan	14×7	3	0⋆	[24]	-	-	**0**	1
Millar	8×14	3	0⋆	Network	2550	*500*	**0**	1
				TS+B&B	**0**	*1*		
Musa	11×14	2	175⋆	[23]	199	*28*	**175**	39
LLR	26×7	4	301⋆	TS+CP	366	16	314	79
BCV-5.4.1	4×28	5	48⋆	Hybrid TS	**48**	5	**48**	180
				VDS	**48**	128		
				[3]	200	15		
Azaiez	13×28	3	0⋆	(0,1)-LGP	**0**	150	**0**	233
GPOST	8×28	3	3⋆	2-Phase	**3**	14	8	234
Valouxis	16×28	4	20⋆	VDS	120	4200	160	3780
Ikegami 3Shift-DATA1	25×30	4	6	TS+B&B	6	111060	63	671

F) `Azaiez` *instance:* An optimal solution is provided with the (0,1)-LGP method [2] after 600 s. on a PC/P-700MHz (`CPUN` 150 s.). `rand` *(resp.* `maxV`*) finds the optimum in* 233 *s. (resp.* 1.050 *s.).*

G) `GPOST` *instance* is solved using a 2 steps method [14]. First, all feasible schedules are enumerated for each nurse. Then, the final schedule is generated using CPLEX. An optimal solution of cost 3 is found in 8 s. on a 2.83GHz Intel Core2 Duo processor (`CPUN` 14 s.) without taking into account the time used in the first step [14]. *We find a solution of cost* 8 *in* 234 *s. using* `maxV`.

H) `Valouxis` *instance [32]:* In [6], Variable Depth Search (VDS) obtains a solution of cost 120 (6 workstretches of length 3) after 2.993 s. on a 2.66GHz Intel Core2 Duo processor (`CPUN` 4.200 s.). *We obtain a solution of cost* 160 *(*8 *workstretches of length 3) after* 3.780 *s. using* `maxV`.

I) `Ikegami-3shift-DATA1` *instance:* Experiments have been performed on a PENTIUM3-1Ghz. TS+B&B [15] finds a solution of cost 10 after 543 mns (`CPUN` 194 mns) with a timeout of 24h and a solution of cost 6 after 5.783 mns (`CPUN` 1.851 mns) with a timeout of 100h. `maxV` *provides a solution of cost* 63 *(where all unsatisfied constraints are of weight* 1*) after* 671 *s. with a timeout of* 1*h.*

Contrary to other instances, nurse constraints are hard and shift constraints are soft for `Ikegami`. So our neighborhood heuristics which unassign whole nurse plannings are irrelevant. If the timeout is increased, the solution quality is improved but it is not enough to bring the optimum. As it is more efficient to unassign variables related to soft constraints than hard ones, one may consider that basic heuristics unassigning shift constraints would be efficient. But it is not the case as it is very difficult to obtain a first solution: nurses constraints are larger than soft ones and more difficult to satisfy.

Conclusions. For each instance, we have compared our method with the best ad'hoc method for solving it [14]. Despite its genericity and flexibility, our method has obtained: (i) solutions of better quality and better computing times for `Ozkarahan`, `Millar`, `Musa` and `LLR`, (ii) solutions of equal quality with computing times close to those for `BCV541` and `Azaiez`, (iii) *very promising solution quality* on large scale instances as `GPOST` and `Valouxis`.

8 Conclusion

In this paper, we have shown how NRPs can be modelled in a concise and elegant way using soft global constraints. For each instance, we have compared quality of solutions and computing times for our method with the best known method for solving it. Experimentations show that, despite its genericity and flexibility, our method provides excellent results on small and middle size problems and very promizing results on large scale problems. For large instances or very specific ones like `Ikegami`, performances of our method could be greatly improved by using neighborhood heuristics especially designed for NRPs. In order to reduce the lack of communication between soft global constraints, it would be interesting to extend arc consistency for soft binary constraints [9,8].

References

1. Meyer auf'm Hofe, H.: Solving rostering tasks as constraint optimisation. In: Burke, E., Erben, W. (eds.) PATAT 2000. LNCS, vol. 2079, pp. 191–212. Springer, Heidelberg (2001)
2. Azaiez, M., Al Sharif, S.: A 0-1 goal programming model for nurse scheduling. Computers and Operations Research 32(3), 491–507 (2005)
3. Brucker, P., Burke, E., Curtois, T., Qu, R., Vanden Berghe, G.: A shift sequence based approach for nurse scheduling and a new benchmark dataset. J. of Heuristics (to appear, 2009)
4. Burke, E., De Causmaecker, P., Vanden Berghe, G.: A hybrid tabu search algorithm for the nurse rostering problem. In: McKay, B., Yao, X., Newton, C.S., Kim, J.-H., Furuhashi, T. (eds.) SEAL 1998. LNCS (LNAI), vol. 1585, pp. 187–194. Springer, Heidelberg (1999)
5. Burke, E., De Causmaecker, P., Vanden Berghe, G., Van Landeghem, H.: The state of the art of nurse rostering. Journal of Scheduling 7(6), 441–499 (2004)
6. Burke, E., Curtois, T., Qu, R., Vanden Berge, G.: A time predefined variable depth search for nurse rostering, TR 2007-6, University of Nottingham (2007)
7. Burke, E., Li, J., Qu, R.: A hybrid model of Integer Programming and VNS for highly-constrained nurse rostering problems. In: EJOR 2009 (to appear, 2009)
8. Cooper, M., de Givry, S., Sanchez, M., Schiex, T., Zytnicki, M.: Virtual arc consistency for Weighted CSP. In: AAAI 2008 (2008)
9. Cooper, M., Schiex, T.: Arc consistency for soft constraints. Artificial Intelligence 154(1-2), 199–227 (2004)

10. Demassey, S., Pesant, G., Rousseau, L.-M.: A `cost-regular` based hybrid column generation approach. Constraints 11(4), 315–333 (2006)
11. Ernst, A., Jiang, H., Krishnamoorthy, M., Sier, D.: Staff scheduling and rostering: A review of applications, methods and models. EJOR 153(1), 3–27 (2004)
12. Hansen, P., Mladenovic, N., Perez-Britos, D.: Variable neighborhood decomposition search. Journal of Heuristics 7(4), 335–350 (2001)
13. Harvey, W., Ginsberg, M.: Limited Discrepancy Search. In: IJCAI 1995, pp. 607–614 (1995)
14. `http://www.cs.nott.ac.uk/~tec/NRP/`
15. Ikegami, A., Niwa, A.: A subproblem-centric model and approach to the nurse scheduling problem. Mathematical Programming 97(3), 517–541 (2003)
16. Karp, R.: Reducibility among combinatorial problems. In: Complexity of Computer Computations, pp. 85–103. Plenum Press, New York (1972)
17. Li, H., Lim, A., Rodrigues, B.: A hybrid AI approach for nurse rostering problem. In: SAC, pp. 730–735 (2003)
18. Loudni, S., Boizumault, P.: Combining VNS with constraint programming for solving anytime optimization problems. EJOR 191(3), 705–735 (2008)
19. Menana, J., Demassey, S.: Sequencing and counting with the multicost-regular constraint. In: van Hoeve, W.J., Hooker, J.N. (eds.) CPAIOR 2009. LNCS, vol. 5547, pp. 178–192. Springer, Heidelberg (2009)
20. Métivier, J.-P., Boizumault, P., Loudni, S.: Softening `gcc` and `regular` with preferences. In: SAC 2009, pp. 1392–1396 (2009)
21. Millar, H., Kiragu, M.: Cyclic and non-cyclic sheduling of 12h shift nurses by network programming. EJOR 104(1), 582–592 (1996)
22. Mladenovic, N., Hansen, P.: Variable neighborhood search. Computers & OR 24(11), 1097–1100 (1997)
23. Musa, A., Saxena, U.: Scheduling nurses using goal-programming techniques. IIE transactions 16, 216–221 (1984)
24. Ozkarahan, I.: The zero-one goal programming model of a flexible nurse scheduling support system. In: Int. Industrial Engineering Conference, pp. 436–441 (1989)
25. Pesant, G.: A regular language membership constraint for finite sequences of variables. In: Wallace, M. (ed.) CP 2004. LNCS, vol. 3258, pp. 482–495. Springer, Heidelberg (2004)
26. Petit, T., Régin, J.-C., Bessière, C.: Specific filtering algorithms for over-constrained problems. In: Walsh, T. (ed.) CP 2001. LNCS, vol. 2239, pp. 451–463. Springer, Heidelberg (2001)
27. Qu, R., He, F.: A hybrid constraint programming approach for nurse rostering problems. In: 28th SGAI International Conference on AI, pp. 211–224 (2008)
28. Régin, J.-C.: Generalized arc consistency for global cardinality constraint. In: AAAI 1996, pp. 209–215 (1996)
29. Régin, J.-C., Gomes, C.: The cardinality matrix constraint. In: Wallace, M. (ed.) CP 2004. LNCS, vol. 3258, pp. 572–587. Springer, Heidelberg (2004)
30. Régin, J.-C., Petit, T., Bessière, C., Puget, J.-F.: An original constraint based approach for solving over-constrained problems. In: Dechter, R. (ed.) CP 2000. LNCS, vol. 1894, pp. 543–548. Springer, Heidelberg (2000)
31. Simonis, H.: Models for global constraint applications. Constraints 12(1), 63–92 (2007)
32. Valouxis, C., Housos, E.: Hybrid optimization techniques for the workshift and rest assignment of nursing personnel. A.I. in Medicine 20(2), 155–175 (2000)
33. van Hoeve, W., Pesant, G., Rousseau, L.-M.: On global warming: Flow-based soft global constraints. Journal of Heuristics 12(4-5), 347–373 (2006)

Online Selection of Quorum Systems for RAMBO Reconfiguration

Laurent Michel[2], Martijn Moraal[2], Alexander Shvartsman[2], Elaine Sonderegger[2], and Pascal Van Hentenryck[1]

[1] Brown University, Box 1910, Providence, RI 02912
[2] University of Connecticut, Storrs, CT 06269-2155

Abstract. RAMBO is the Reconfigurable Atomic Memory for Basic Objects, a formally specified algorithm that implements atomic read/write shared memory in dynamic, rapidly changing networking environments. RAMBO is particularly apt at dealing with volatile environments such as mobile networks. To maintain availability and consistency, even as hosts join, leave, and fail, RAMBO replicates objects and uses reconfigurable quorum systems. As the system dynamically changes, RAMBO installs new quorum configurations. This paper addresses the reconfiguration problem with three approaches based on a finite-domain model, an hybrid master-slave decomposition and a parallel composite to find optimal or near-optimal configurations. Current behaviors of RAMBO participants are observed, gossiped, and used as predictors for future behaviors, with the goal of finding quorum configurations that minimize read and write operation delays without affecting correctness and fault-tolerance properties of the system.

1 Introduction

Providing consistent shared objects in dynamic networked systems is one of the fundamental problems in distributed computing. Shared object systems must be resilient to failures and guarantee consistency despite the dynamically changing collections of hosts that maintain object replicas. RAMBO, which stands for Reconfigurable Atomic Memory for Basic Objects [7,4], is a formally specified distributed algorithm designed to support a long-lived atomic read/write memory service in such a rapidly changing network environment. To maintain availability and consistency of the memory service, RAMBO uses reconfigurable quorum systems, where each object is replicated at hosts that are quorum members, and where the intersection among quorum sets is used to guarantee atomicity. RAMBO specifications and algorithms provide a tailorable framework for implementing optimizable research testbeds [3] and robust enterprise storage systems [1].

The ability to rapidly reconfigure quorum systems in response to failures and delays is at the heart of the service. RAMBO allows for any configuration to be installed at any time, enabling it to adapt to changes in the set of participants

I.P. Gent (Ed.): CP 2009, LNCS 5732, pp. 88–103, 2009.

and tolerate failures while maintaining atomicity of the objects. However, deciding *when* reconfigurations should be done and *what* the new configurations should look like remain external to its specification.

This paper focuses on the task of determining what a new configuration should be and how it should be deployed. Note that a poorly crafted configuration may affect RAMBO's performance in negative ways: it may deteriorate the response time if the quorum members are over-burdened and slow to respond; it may also weaken fault-tolerance when failure-prone hosts are chosen to maintain object replicas. In response, RAMBO may need to perform additional reconfigurations, in the hope of installing a better quorum system, possibly thrashing between ill-chosen configurations. Applying optimization techniques to the design and deployment of sensible configurations addresses these issues. We pursue a methodic approach to producing strong configurations that will positively affect the performances of read and write operations, while balancing the workload and increasing the likelihood that the configuration will be long-lived.

The production of a new configuration is an activity that occurs *online* while RAMBO is running. Any participant may request a reconfiguration, after which consensus is used to agree on the new configuration. A participant ought to be able, based on historical observations, to propose a well-designed quorum configuration that is optimized with respect to relevant criteria, such as being composed of members who have been communicating with low latency, and consisting of quorums that will be well-balanced with respect to read and write operation loads. This paper investigates three optimization methods for producing high-quality configurations, namely: a CP method, a hybrid CBLS/CP master-slave similar in spirit to a Benders decomposition, and a parallel composite. The methods grow in their level of sophistication and offer non-trivial integrations of several optimization techniques. Results on instances with up to 16 hosts and half a dozen quorums indicate that excellent solutions can be found in a second or two, and proved optimal reasonably quickly (from a few seconds to a few minutes depending on the nature of the quorum system). This makes it possible to perform rapid online decisions for what quorum configurations to propose for deployment, substantially increasing the effectiveness of the reconfiguration in RAMBO. We believe our approach can be generalized to serve as a valuable tool for dynamic distributed systems in enabling swift online optimization decisions.

Section 2 presents RAMBO in more detail, including the reconfiguration problem and our approach. Section 3 introduces a high-level deployment model, and Section 4 presents the CP model. Section 5 presents the CBLS model, while Section 6 covers the hybridization. Section 7 reports the experimental results and analyzes the behavior of the models in detail. Section 8 concludes the paper.

2 The RAMBO System and Configuration Selection

The original RAMBO framework provides a formal specification for a reconfigurable atomic shared memory service, and its implementation as a distributed algorithm [7]. Basic read/write objects are replicated to achieve availability in the presence of failures. In order to maintain memory consistency in the presence of small and transient changes, the algorithm uses quorum *configurations*. In order to accommodate larger and more permanent changes, the algorithm supports dynamic *reconfiguration*, by which the quorum configurations are modified. RAMBO supports three activities, all concurrently: reading and writing objects, introducing new configurations, and removing obsolete configurations. Atomicity is guaranteed in all executions.

Each quorum configuration consists of a set of *members*, a set of *read-quorums*, and a set of *write-quorums*. Members are hosts for the replicated object. Quorums are subsets of members, with the requirement that every read-quorum has a non-empty intersection with every write-quorum. The algorithm performs read and write operations using a two-phase strategy. The first phase gathers information from at least one read-quorum of every active configurations, and the second phase propagates information to at least one write-quorum of every active configuration. The information propagates among the participants by means of background gossip. Because every read-quorum and write-quorum intersect, atomicity of the data object is maintained.

The distributed reconfiguration service uses consensus to agree on the successive configurations. Any member of the latest configuration may propose a new configuration at any time; different proposals are reconciled by an execution of consensus among the members of the latest configuration. The algorithm removes old configurations when their use is no longer necessary for maintaining consistency. This is done by "writing" the information about the new configuration and the latest object value to the new configuration.

RAMBO is designed as a flexible and optimizable service, in particular allowing for any configuration to be installed at any time. Exactly when to reconfigure and what new configuration to choose are left as decisions for an external service. Given the dynamic nature of the intended deployment environments, it is neither feasible nor desirable to pre-specify future configurations. The decision of what configuration to choose next ought to be made dynamically in response to external stimuli and observations about the performance of the service. Choosing sensible configurations is critical in maintaining good performance.

In this work we focus on the expedient determination of new configurations. We do not explicitly address the problem of when to reconfigure: we consider this to be an application-level decision made on the basis of the observations about the performance of the service and the suspected failures of object replicas.

We assume that the hosts have no knowledge of the underlying network, particularly as nodes join and leave. We further assume that the best available estimate of the network connections and host performance is the measurement of average round trip message delays. In particular, a measurement of round trip delay from host h_1 to host h_2 and back to h_1 can be used as an assessment of: (a) the communication "distances" between h_1 and h_2, (b) the communication loads on h_1 and h_2, (c) the processing loads at h_1 and h_2, and (d) the likelihood that host h_2 is failing, or even failed. The hosts can record delay measurements as running averages for the recent past, reasonably assuming that the measurements are quickly impacted by failures or slowdowns. Each host also measures the average frequency of read and write operations it initiates. The hosts share gathered operation frequencies and messaging delays by adding them to the gossip messages of RAMBO. Our overall guiding principle is that observations of current behaviors are the best available predictors of future behaviors.

Note that the RAMBO algorithms require minimal changes. The sole modification is the addition of local observations about system performance piggy-backed onto the gossip messages. Upon receipt of a gossip message, a host updates its local knowledge using the piggy-backed information and then delivers the message to RAMBO. Any participant can compute a new configuration using only the local knowledge about the behavior of the participants and submit it directly to the RAMBO reconfiguration service. This paper focuses exclusively on recommending a good or optimal candidate for reconfiguration.

The approach attempts to optimize the mapping of an abstract quorum system to a subset of hosts participating in the service. We assume that the participants have at their disposal an abstract specification of a quorum system, consisting of the members, the read-quorums, and write-quorums, and the problem to be solved is to assign each member to a participating host in such a way that delays to the members of the read and write quorums are minimized.

Once the mapping is computed, the resulting configurations may be augmented to include information recommending the best read and write quorums for each host to use, where use of best quorums will result in minimum delays for read and write operations and the most balanced load on replica hosts (until failures occur in those best quorums). Of course the use of this information is optional and a host can always use other quorums if it does not observe good responses from the recommended quorums.

Finally, it is worth amplifying that our approach uses observation about the recent and current behavior as the basis for optimization. Future behaviors may of course differ from the observed current behaviors. To avoid quorum configuration failure and performance degradation, the speed of reconfiguration is important. Consequently, there is substantial value in computing a good deployment mapping quickly and reconfiguring to it, even if it may end up not being the optimal mapping. Then, once an optimal mapping is determined, it is always possible to reconfigure one more time from the "good" to the "best" quorum configuration.

3 Modeling the RAMBO Configuration Selection

Model Parameters. The input data of the RAMBO deployment model consists of:

- The set of hosts H.
- For every host $h \in H$, the average frequency f_h of its read and write requests.
- For every pair of hosts $h_1, h_2 \in H$, the average round trip delay d_{h_1,h_2} of messages from h_1 to h_2.
- The abstract configuration c to be deployed on H, where c consists of:

  - The set of members M, each of which maintains a replica of the data.
  - The set of read quorums $R \subseteq \mathcal{P}(M)$.
  - The set of write quorums $W \subseteq \mathcal{P}(M)$.
- The load balancing factor α, restricting the spread of loads on the configuration members, assuming each host first tries to contact its fastest read and write quorums.

Decision Variables. A decision variable x_m with domain H is associated to each configuration member m. $x_m = h$ when replica m is deployed on host h. Each host h is also associated with two decision variables $readQ_h$ and $writeQ_h$ (with domains R and W) denoting, respectively, one read (write) quorum from the minimum average delay read(write)-quorums associated to h. Finally, an auxiliary variable $load_m$ represents the load of a configuration member m which is induced by the traffic between the hosts and their chosen read/write quorums.

The Objective. An optimal deployment minimizes

$$\sum_{h \in H} f_h \times \left(\min_{q \in R} \left(\max_{m \in q} d_{h,x_m} \right) + \min_{q \in W} \left(\max_{m \in q} d_{h,x_m} \right) \right)$$

where each term in the summation captures the time it takes in RAMBO for a host h to execute the read/write phase of the protocol. Indeed, a read (or a write) in RAMBO requires the client to contact all the members of at least one read quorum before it can proceed to a write round of communication with all the members of at least one write quorum to update the data item. The $\max_{m \in q} d_{h,x_m}$ reflects the time it takes to contact all the members of quorum q as one must wait for an answer from its slowest member. The outer $\min_{q \in R}$ reflects the fact that RAMBO must hear back from one quorum before it proceeds, and this happens when the "fastest" quorum replies.

The Constraints. A configuration is subject to the following constraints. First, all configuration members must be deployed on separate hosts.

$$\forall m, m' \in M : m \neq m' \Rightarrow x_m \neq x_{m'}$$

An implementation of RAMBO can use different strategies when contacting the read and write quorums. A conforming implementation might simply contact all the read quorums in parallel. Naturally, this does not affect the value of the objective, but it induces more traffic and work on the members of the quorum system. Another strategy for RAMBO is to contact what it currently perceives as the fastest read quorum first and fall back on the other read quorums if it does not receive a timely response. It could even select a read quorum uniformly at random. These strategies strike different trade-offs between the traffic they induce and the workload uniformity. The model presented below captures the greedy strategy, namely, RAMBO contacts its closest quorum first, and the model assumes that this quorum replies (unless a failure has occurred).

The model uses the $readQ_h$ and $writeQ_h$ variables of host h to capture which read (write) quorum RAMBO contacts. Recall that a read (write) quorum is a set and therefore the domain of $readQ_h$ is the set of read quorums (similarly for $writeQ_h$). More formally,

$$readQ_h = r \;\Rightarrow\; \max_{m\in r} d_{h,x_m} = \min_{q\in R}\left(\max_{m\in q} d_{h,x_m}\right)$$

$$writeQ_h = w \;\Rightarrow\; \max_{m\in w} d_{h,x_m} = \min_{q\in W}\left(\max_{m\in q} d_{h,x_m}\right)$$

In each equation, the first conjunct requires the chosen quorum to be a member of the possible quorum sets while the second requires the chosen quorum to induce a minimal delay. Note that several quorums might deliver the same minimal delay, so this constraint alone does not determine the ideal read (write) quorum.

The third constraint defines the load of a configuration member as

$$load_m = \sum_{h\in H}\left(\sum_{m\in readQ_h} f_h + \sum_{m\in writeQ_h} f_h\right)$$

Clearly, the load on m depends on which read (write) quorums are chosen among those that induce minimal delays. Finally, the load balancing constraint requires the maximum load on a member be within a factor α of the minimum load.

$$\max_{m\in M} load_m \;\leq\; \alpha \times \left(\min_{m\in M} load_m\right)$$

4 The CP Model

The COMET program for the CP model is shown in Figure 1. The data declarations in lines 2–8 correspond to the input data of the RAMBO model in Section 3. Line 9 declares an additional input used for breaking symmetries among the members of the quorum configuration. Lines 10–14 define derived data. Specifically, $nbrQ[m]$ is the number of quorums in which m appears, and $degree[h]$ is the number of neighbors of host h in the logical network graph. RQ and WQ are the index sets of the read and write quorums, respectively. The auxiliary matrices $readQC$ and $writeQC$ are encodings of the quorum membership, e.g., $readQC[i,j] = true \Leftrightarrow j \in R[i]$. Lines 15–20 declare the decision variables. Variable $x[m]$ specifies the host of configuration member m. Variables $readD[h,r]$ and $writeD[h,w]$ are the communication delays for host h to access read quorum r and write quorum w. The variables $readQ[h]$ and $writeQ[h]$ represent the read and write quorum selections for host h. Finally, the variable $load[m]$ represents the communication load on configuration member m, given the current deployment and quorum selections.

Line 22 specifies the objective function, which minimizes the total communication delay over all operations. Line 24 specifies the fault tolerance requirement,

```
Solver<CP> cp();
range M= ...; // The members of the quorum configuration
set{int}[] R = ...; // An array storing all the read quorums in the configuration
set{int}[] W = ...; // An array storing all the write quorums in the configuration
range H = ...; // The host nodes
int[] f = ...; // The frequency matrix
int[,] d = ...; // The delays matrix
int alpha = ...; // The load factor
set{tuple{int low; int high}} Order = ...; // The order of quorum members
int nbrQ[m in M] = .... ; // The number of quorums for each member
int degree[H] = ...; // The degree of a host (number of neighbors)
range RQ = R.getRange(); range WQ = W.getRange();
boolean readQC[RQ,M] = ...;
boolean writeQC[WQ,M] = ...;
var<CP>{int} x[M](cp,H);
var<CP>{int} readD[H,RQ](cp,0..10000);
var<CP>{int} writeD[H,WQ](cp,0..10000);
var<CP>{int} readQ[H](cp,RQ);
var<CP>{int} writeQ[H](cp,WQ);
var<CP>{int} load[M](cp,0..10000);
minimize <cp>
   sum(h in H) f[h] * (min(r in RQ) readD[h,r]+ min(w in WQ) writeD[h,w])
subject to {
   cp.post(alldifferent(x), onDomains);
   forall (o in Order) cp.post(x[o.low] < x[o.high]);
   forall (h in H, r in RQ)
      cp.post(readD[h,r] == max(m in R[r]) d[h,x[m]]);
   forall (h in H, w in WQ)
      cp.post(writeD[h,w] == max(m in W[w]) d[h,x[m]]);
   forall (h in H) {
      cp.post(readD[h,readQ[h]] == min(r in RQ) readD[h,r]);
      cp.post(writeD[h,writeQ[h]] == min(w in WQ) writeD[h,w]);
   }
   forall (m in M) cp.post(load[m] == sum(h in H) f[h] * (readQC[readQ[h],m]+
                                                           writeQC[writeQ[h],m])));
   cp.post(max(m in M) load[m] <= alpha * min(m in M) load[m]);
} using {
   while (sum(k in M) x[k].bound() <M.getSize())
      selectMax(m in M: !x[m].bound()) (nbrQ[m])
         tryall<cp>(h in H : x[m].memberOf (h)) by (- degree[h])
            cp.label(x[m], h);
         onFailure cp.diff(x[m], h);
   once<cp> phase2search(cp, readQ, writeQ);
}
function void phase2search (Solver s, int readQ[H], int writeQ[H]) {
   forall (h in H : !readQ[h].bound() || !writeQ[h].bound()) by (- f[h]) {
     label(readQ[h]);
     label(writeQ[h]);
   }
}
```

Fig. 1. The CP Model in COMET

namely, all members of the configuration must be deployed to distinct hosts. The `onDomains` annotation indicates that arc-consistency must be enforced. Line 25 breaks the variable symmetries among the configuration members [8].

Lines 26–36 constraint the auxiliary delay variables and quorum selection variables needed in the load-balancing constraint. The constraints on lines 27 and 29 capture the delays incurred by host h to use a read (write) quorum. Lines 30–33 require the quorums assigned to host h, namely $readQ[h]$ and $writeQ[h]$, to be among the quorums with minimum delay for that host. Lines 34-35 specify the communication load on m as the sum of the operation frequencies of each host for which m is a member of its assigned read and/or write quorum. Line 36 is the load-balancing constraint and requires the load on the most heavily loaded configuration member to be no more than $alpha$ times the load on the most lightly loaded configuration member.

The search procedure operates in two phases. The first phase (lines 38–42) assigns configuration members to hosts. The variable selection heuristic first focuses on variables that appear in many quorums. The value selection heuristic first considers hosts that have many neighbors 'close by' as these would be ideal locations for quorum members. The second phase, which finds an assignment of hosts to read and write quorums that satisfies the load-balancing constraint, is invoked on line 43. This second phase cannot impact the value of the objective function. Instead, its role is to decide which quorum among its best options each host should use to meet the load-balancing requirement. Clearly, only one such assignment is needed which explains the `once<cp>` annotation enclosing the second phase call. The phase two procedure, shown on lines 45–50, considers the most "talkative" hosts first (by decreasing frequencies) and attempts to assign one of the remaining legal (minimal) quorums from its domain.

5 An Hybrid CBLS/CP Master-Slave Algorithm

Determining a new configuration for RAMBO is, unarguably, an online problem. It must be solved quickly to submit a new proposal that RAMBO then puts up for consensus with the other participants. While an approach based on finite-domain is appealing based on its ability to prove optimality, it might not scale nicely or take a long time to establish optimality. This section investigates a hybrid CBLS/CP master-slave algorithm based on the assumption that local search can deliver high-quality solutions in short-order, a highly desirable property in an online setting. A first natural attempt recycles the finite-domain model and uses a search procedure with a neighborhood structure that re-assigns either the deployment or the quorum selection variables. Unfortunately, this *direct* approach is unsuccessful as load-balancing provides little to no guidance on the quorum selection until the deployment is fixed. The recognition of this difficulty suggests a second approach where a master local search is first used to find a deployment that minimizes the communication volume. For each candidate solution produced in the master, a slave model focuses on finding a quorum selection. Note that the slave does not affect the objective. Instead, it handles the

```
FunctionSum<LS> O(ls);
forall (h in H) O.post(f[h] * (min(r in RQ) (max(m in R[r]) d[h, x[m]]) +
                               min(w in WQ) (max(m in W[w]) d[h, x[m]])));
var{int} v(ls,0..1) := 1;
var{float} w(ls) := 0.5;
var{float} obj(ls) <- sqrt(O.value()^2 + (w*v)^2);
ls.close();
```

Fig. 2. The Constraint System for the CBLS Model in COMET

feasibility of the load-balancing constraint. Finding a feasible quorum selection is typically quite hard in its own right and this paper uses a finite-domain model to solve the slave. In summary, our hybrid uses a master local search to find the deployment and a slave CP model to establish feasibility and influence the master problem through a violation term in its objective function. The approach is reminiscent of Benders decompositions [2].

The hybrid master-slave model uses the same input parameters and decision variables as its finite-domain cousin. Namely, the core decision variable is an array $x[m]$ that associates with every member of the configuration the host on which it is deployed. Before delving into the modeling, it is useful to consider a few invariants that maintain key properties. Given a deployment x, the invariants

```
int readD[h in H,r in RQ] <- max(m in R[r]) d[h,x[m]]
set{int} bestReadQ[h in H] <- argMin(r in RQ) readD[h,r]
int readQMax[r in RQ] <- sum(h in H : member(r,bestReadQ[h])) f[h]
int readQMin[r in RQ] <- sum(h in H : card(bestReadQ[h]==1 &&
                                     member(r,bestReadQ[h])) f[h]
```

maintain the read delay, the set of read quorums that yield a minimal delay (*bestReadQ*), and upper and lower bounds on the communication volume for each quorum. Similar invariants are defined for the write quorums. The invariants

```
loadMin[m in M] <- sum(r in RQ : member(m,R[q])) readQMin[r] +
                   sum(r in WQ : member(m,W[q])) writeQMin[r]
loadMax[m in M] <- sum(r in RQ : member(m,R[q])) readQMax[r] +
                   sum(r in WQ : member(m,W[q])) writeQMax[r]
```

maintain lower and upper bounds on the load of any quorum member. The true load $load[m]$ for a configuration member m satisfies

$$loadMin[m] \leq load[m] \leq loadMax[m]$$

The `alldifferent` constraint on the deployment variables x that was needed in the finite-domain model can be avoided here by simply starting from a candidate assignment that satisfies it and relying on a neighborhood structure based on swaps that will not introduce violations. The load-balancing constraint is more delicate to handle as it cannot be verified until the model has selected which quorum (among its best quorums) each host will contact. This decision (picking

```
while (it < maxIt) {
   select (m in c.members, n in H : m !=n ) {
     float delta = (lookahead(ls, obj) v := makeMove(m, n);) − obj;
      if (distr.accept(−delta/t)) {
         v := makeMove(m, n);
         if (v == 0) bf = min(obj, bestFeasible);
         else bi = min(obj, bestInfeasible);
      }
   }
   it++; stableIt++; t = t ∗ 0.9995; w := w + 0.01;
   if (bf < bestFeasible || bi < bestInfeasible) updateBest();
   if (stableIt >= 1000) reheat();
   if (rounds >= 10) diversify();
}
function int makeMove(int m, int n) {
   x[m] :=: x[n];
   if ((O.value() < bestFeasible) && (max (m in M) loadMin[m] <= alpha
         ∗ min (m in M) loadMax[m])) {
      Solver<CP> cp();
      cp.limitFailures(1000);
      boolean feasible = false;
      var<CP>{int} readQ[h in H](cp, RQ);
      var<CP>{int} writeQ[h in H](cp, WQ);
      var<CP>{int} load[M](cp, 0..100000);
      solve<cp> {
         forall (h in H) {
            cp.post(readD[h, readQ[h]] == min(r in RQ) readD[h,r]);
            cp.post(writeD[h, writeQ[h]] == min(w in WQ) writeD[h,w]);
         }
         forall (m in M)
             cp.post(load[m] == sum(h in H) f[h] ∗ (readQC[readQ[h],m]+
                                                   writeQC[writeQ[h],m])));
         cp.post(max (m in M) load[m] <= alpha ∗ min (m in M) load[m]);
      } using { phase2search(cp, readQ, writeQ);feasible = true; }
      return (feasible) ? 0 : 1;
   } else return 1;
}
```

Fig. 3. The Hybrid Master-Slave Model in COMET

a quorum) used to be carried out during the phase 2 of the finite-domain model and cannot be easily modeled in the master. Note that it is possible to state an *approximation* of the load-balancing constraint. However this complicates the model and presentation and is not shown in the sake of brevity. Instead, the load-balancing constraint satisfaction is entrusted to the slave which now produces a simple Boolean output v indicating whether the load-balancing is satisfied.

The remainder of the model, shown in Figure 2, declares the objective function O which mimics the finite-domain model. Line 6 declares the objective obj as the combination of the objective value and the scaled measure of violations v established by the slave. This objective is reminiscent of the coloring objective in [5]. The scaling is carried out by a weight w which is used to shift the emphasis back and forth between the objective and the load-balancing constraint.

The Master. Figure 3 illustrates the implementation of a master-slave decomposition. The master (lines 1–14) uses simulated annealing [6] to find a deployment. It selects a random move on line 2, evaluates it on line 3, and accepts with probability $e^{-delta/t}$ on line 4. As is typical in simulated annealing, the temperature t is decreased during the search, so that many moves are accepted initially, but converges to a (local) minimum as the temperature drops. This effect is magnified by the weight w, which initially has a low value to place the emphasis on improving the objective and which is increased during the search to shift towards finding feasible solutions. If the selected move is accepted, it is performed on line 5 and both bi and bf are updated to reflect the best feasible and infeasible solutions thus far. After each iteration, line 10 updates the temperature and weight. Line 11 records the best feasible and infeasible solutions. Line 12 resets the temperature and weight when no improvements have been seen for 1000 iterations. After 10 reheats (resetting of temperature and weight) a diversification step is carried out on line 13 that replaces x by a random permutation of the hosts.

The Slave. The slave, shown in lines 15–37 of Figure 3, is invoked via a call to `makeMove`. It performs the swap and uses a CP model to determine the feasibility of the solution. It assigns the quorums and returns a Boolean to report feasibility w.r.t. the load balancing. The CP model can be costly and is only executed when there is some hope of finding a feasible solution. In particular, if the scaled minimum of the upper bounds on the load is smaller than the maximum of the lower bounds on the load, the load-balancing constraint is necessarily infeasible and line 36 returns 1 to report a violation. If the load-balancing constraint is potentially feasible, the body (lines 19–35) looks for a quorum assignment and returns 0 if it finds a solution. This CP model is essentially the same as the phase 2 model from Section 4. It contains the same decision variables ($readQ[h]$, $writeQ[h]$ and $load[m]$). Naturally, the read and write delays ($readD$ and $writeD$) are constants here. As before, lines 26–29 state the constraints on the quorum assignments $readQ[h]$ and $writeQ[h]$, lines 30–32 set up the load-defining constraints, and line 33 sets up the load-balancing constraint.

6 Parallel Composition

This model is a parallel composition of the two previous models. The two models run in separate threads and communicate through events [10].

```
thread t performCBLSmodel(center, benchmark, configuration);
performCPmodel(center, benchmark, configuration);
```

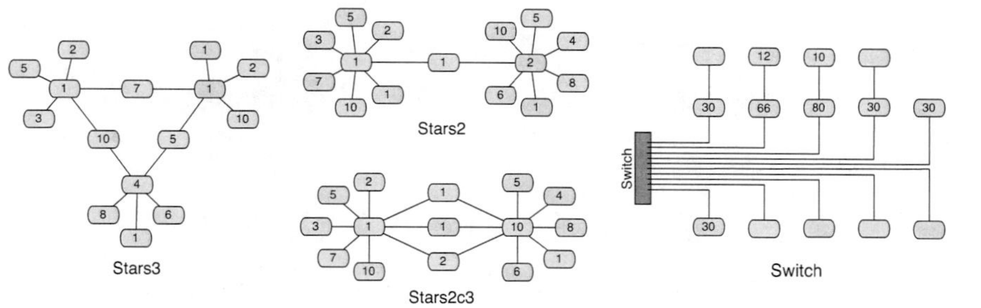

Fig. 4. Network Configuration Benchmarks `Stars3`, `Stars2`, `Stars2c3`, and `Switch`

The CBLS notifies the CP search each time it finds a new solution with

```
center.tellNewSolution(new Solution(ls, MinimizeIntValue(O.value())));
```

Finally, the snippet

```
whenever center@newSolution(Solution s)
  if (s.getObjectiveValue().compare(cp.getObjective().getPrimalBound()) < 0)
     cp.setPrimalBound(s.getObjectiveValue());
```

is added to the CP model to update the bound on the objective every time a new solution is received. Similarly (but not shown), whenever the CP model terminates, it notifies an event to ask the CBLS model to terminate.

7 Experimental Results

The Benchmarks. Six network configurations and four quorum systems are used as benchmarks, representing common networks and quorums. The first four networks, `Stars3`, `Stars2`, `Stars2c3`, and `Line`, use different interconnects for 15 hosts; `Stars3`, `Stars2`, and `Stars2c3` arrange the hosts in clusters, as illustrated in Figure 4, and `Line` arranges the hosts in a single line (a bus). `Hyper16` uses 16 hosts and an hypercube interconnect. `Switch`, which is shown in Figure 4, consists of 10 hosts on a switch and 4 other hosts hooked up via point-to-point links. The frequencies of the read/write operations for each host are given in Figure 4, and the delays are the number of "hops" between hosts. The benchmarks run equally well, but somewhat slower, when random noise is added to these frequencies and delays to more faithfully simulate network measurements.

Figure 5 shows the `3x2`, `3Step`, and `3x3` quorum systems, where horizontal groups are read quorums and vertical groups are write quorums. `Maj` uses majority quorums [9], with its six members grouped into four read quorums and four write quorums, each with four members.

Experimental Results for the CP Model. Table 1 reports the results for the CP model with COMET 1.1 (on a Core 2 @ 2.16 GHz) using $\alpha = 2$. The table provides

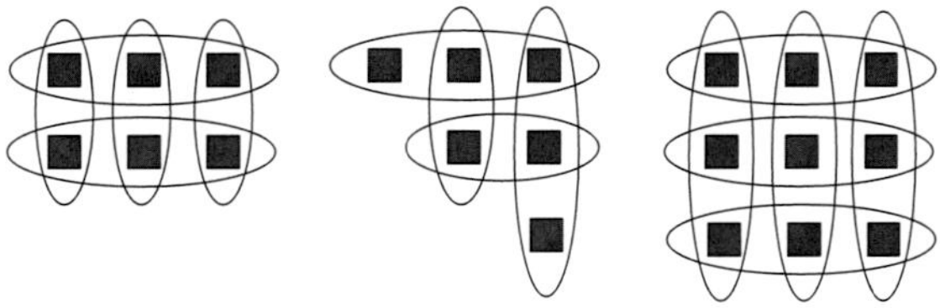

Fig. 5. Quorum System Benchmarks 3x2, 3Step, and 3x3

Table 1. Experimental Results for the CP Model with $\alpha = 2$

		3x2			3Step			3x3			Maj		
Benchmark		Opt	T_{end}	T_{opt}	Opt	T_{end}	T_{opt}	Opt	T_{end}	T_{opt}	Opt	T_{end}	T_{opt}
Stars3	μ	261	0.98	0.41	285	0.37	0.09	284	42.12	10.95	340	5.30	0.03
	σ		0.18	0.29		0.03	0.06		5.37	10.71		0.12	0.01
Stars2	μ	303	5.70	2.76	284	0.57	0.30	316	536.50	92.69	374	20.76	1.35
	σ		9.71	9.67		0.24	0.21		203.53	197.01		3.62	3.71
Stars2c3	μ	238	1.16	0.21	239	1.52	1.37	268	1414.09	914.20	270	8.07	0.03
	σ		0.15	0.14		1.79	1.79		402.47	609.09		0.17	0.01
Line	μ	485	4.02	2.76	479	2.92	2.59	517	445.28	319.96	607	26.04	6.91
	σ		0.57	0.80		1.88	2.02		109.01	155.78		0.87	0.48
Hyper16	μ	246	9.87	5.04	256	2.13	0.85	249	508.28	226.04	305	66.94	3.37
	σ		1.24	2.27		0.16	0.60		82.37	153.78		0.89	0.24
Switch	μ	610	1.59	0.71	620	0.46	0.30	620	59.03	45.06	620	0.23	0.08
	σ		0.80	0.70		0.58	0.57		69.38	69.67		0.05	0.05

two rows for each benchmark: the first reports averages and the second reports standard deviations. Columns are grouped by quorum system type. Within each group, column *Opt* gives the objective for the optimal solution found, T_{end} gives the time in seconds to find the optimum and prove optimality, and column T_{opt} reports the time in seconds to find the optimum. The results are the average and standard deviation over 50 runs. It is useful to review these results in more detail.

1. For all but two network configurations, the easiest quorum system to solve is `3Step`, followed by `3x2`, `Maj`, and then `3x3`.
2. There is a lot of variation in the time it takes to find optimal solutions. With `Maj`, the optimum is found quickly, sometimes as soon as one percent into the run. In contrast, `Stars2c3` and `Line` with `3Step` do not find the optimum until the very end.
3. The standard deviation for T_{Opt} and T_{end} tend to be highly dependent on the benchmarks. While the `Maj` quorum system induces small deviations, the `3x3` quorum system exhibits deviations that are often larger than the averages. A closer examination of the runs reveals that most runs are similar and a few outliers are significantly longer.

Table 2. Experimental Results for the Hybrid Master-Slave Model with $\alpha = 2$

	3x2			3Step			3x3			Maj		
Benchmark	T_{end}	T_{best}	$\#Opt$	T_{end}	T_{best}	$\#Opt$	T_{end}	T_{best}	$\#Opt$	T_{end}	T_{best}	$\#Opt$
Stars3	10.02	0.74	50	8.18	1.43	50	10.48	1.61	50	19.99	0.36	50
Stars2	9.37	2.24	50	7.37	0.38	50	11.03	1.79	50	17.85	0.29	50
Stars2c3	8.74	0.11	50	7.07	0.32	50	33.12	17.10	13	18.71	0.25	50
Line	10.83	1.24	50	8.52	2.23	48	11.25	5.57	16	22.53	0.70	50
Hyper16	12.46	2.97	48	9.69	2.80	46	13.35	2.98	3	23.61	8.37	48
Switch	16.26	0.10	50	7.26	0.03	50	9.95	0.09	50	18.66	0.12	50

Table 3. Experimental Results for the Composite Model with $\alpha = 2$

		3x3				Maj			
		CP		Composite		CP		Composite	
Benchmark		Tend	Topt	Tend	Topt	Tend	Topt	Tend	Topt
Stars3	μ	42.12	10.95	40.88	3.16	5.30	0.03	6.16	0.41
	σ	5.37	10.71	3.52	5.78	0.12	0.01	0.21	0.28
Stars2	μ	536.50	92.69	490.79	3.14	20.76	1.35	21.30	0.58
	σ	203.53	197.01	32.87	3.60	3.62	3.71	0.76	0.92
Stars2c3	μ	1414.09	914.20	779.81	175.85	8.07	0.03	9.07	0.27
	σ	402.47	609.09	59.78	221.15	0.17	0.01	0.34	0.32
Line	μ	445.28	319.96	257.86	75.15	26.04	6.91	26.95	0.78
	σ	109.01	155.78	27.09	78.05	0.87	0.48	0.80	0.48
Hyper16	μ	508.28	226.04	553.73	262.76	66.94	3.37	73.62	6.83
	σ	82.37	153.78	114.81	194.29	0.89	0.24	2.25	5.90
Switch	μ	59.03	45.06	15.14	0.10	0.23	0.08	0.28	0.13
	σ	69.38	69.67	3.14	0.10	0.05	0.05	0.05	0.08

The particular properties of network configurations and quorum systems that cause these variations in behavior need further study.

Experimental Results for the Hybrid Master-Slave. Table 2 reports the results for the hybrid master-slave with COMET 1.1 (on a Core 2 @ 2.16 GHz) using $\alpha = 2$. T_{end} reports the total runtime, T_{best} gives the time to the best solution, and $\#Opt$ indicates how frequent the best solution was found (out of 50 runs). Optimal solutions are found very reliably. The hardest benchmark is `Hyper16` on a `3x3` quorum system where the model was only able to find the optimum 3 times. Even here, though, the average best solution is only 1% away from the optimum. In most other instances, the model finds the optimum on all runs. Furthermore, most runs show a fast time to the best solution, pointing to a robust model. The standard deviations are consistently very small.

Although not shown, the results are consistent across α values with one notable exception. The `Stars2c3` with `3x3` and $\alpha = 2$ takes significantly more time than any other benchmark. This is due to a larger number of invocations of the slave search. In the other benchmarks most of the slave searches are eliminated through the conditions described in Section 5. Compared to a pure CP model, the hybrid finds the optimum faster on all but the easiest benchmarks.

Experimental Results for the Parallel Composition. Table 3 reports the results for the parallel composition on the hardest instances and compares them to the CP model. The composite was run with COMET 1.1 (on a Core 2 @ 2.16 GHz). T_{end} and T_{opt} carry the same meaning as in Table 1. The values from the composite are the average and standard deviation over 50 runs.

The composite clearly benefits from its local search component as it consistently delivers the optimum very early on and quite reliably. The time to complete the optimality proof offers a mixed set of results. For some instances (e.g., instances based on `Maj` quorums), there are no benefits to speak of. For others, the availability of the optimum early on translates into a shorter optimality proof and decreased standard deviation. The most dramatic instances in this respect are, perhaps, `Switch`, `Line`, and `Stars2c3` with `3x3` quorum systems.

8 Conclusion

This paper considered an online problem that arises during the execution of RAMBO, a distributed algorithm offering reconfigurable atomic memory for basic objects. The optimization problem consists of producing a new configuration for the quorum system used by RAMBO. The objective is to produce a configuration that minimizes communication traffic while retaining good fault-tolerance through load-balancing. Three approaches were considered, namely: a CP model, a hybrid CBLS/CP, and a parallel composition of the two. The approaches were evaluated on a suite of instances capturing different networks and quorum systems. The experimental results show that this problem is quite challenging and that a composite approach can deliver excellent solutions within the real-time requirements of an online setting.

References

1. Arif, S.F., Merchant, A., Saito, Y., Spence, S., Veitch, A.: Fab: enterprise storage systems on a shoestring. In: Operating Systems, Lihue, HI, May 18-21, pp. 133–138. USENIX Association (2003)
2. Benders, J.F.: Partitioning procedures for solving mixed variables programming problems. Numerische Mathematik 4, 238–252 (1962)
3. Georgiou, C., Musial, P.M., Shvartsman, A.A.: Long-lived rambo: Trading knowledge for communication. Theor. Comput. Sci. 383(1), 59–85 (2007)
4. Gilbert, S., Lynch, N.A., Shvartsman, A.A.: RAMBO II: Rapidly reconfigurable atomic memory for dynamic networks. In: DSN, pp. 259–268. IEEE Computer Society, Los Alamitos (2003)
5. Johnson, D., Aragon, C., McGeoch, L., Schevon, C.: Optimization by Simulated Annealing: An Experimental Evaluation; Part I, Graph Partitioning. Operations Research 37(6), 865–893 (1989)

6. Kirkpatrick, S., Gelatt, C., Vecchi, M.: Optimization by Simulated Annealing. Science 220, 671–680 (1983)
7. Lynch, N., Shvartsman, A.: RAMBO: A reconfigurable atomic memory service for dynamic networks. In: Proceedings of the 16th International Symposium on Distributed Computing, pp. 173–190 (2002)
8. Smith, B.M.: Sets of symmetry breaking constraints. In: Proc. of SymCon, vol. 5 (2005)
9. Thomas, R.H.: A majority consensus approach to concurrency control for multiple copy databases. ACM Trans. Database Syst. 4(2), 180–209 (1979)
10. Van Hentenryck, P., Michel, L.: Control Abstraction for Local Search. In: Rossi, F. (ed.) CP 2003. LNCS, vol. 2833, pp. 65–80. Springer, Heidelberg (2003)

A Hybrid Constraint Model for the Routing and Wavelength Assignment Problem

Helmut Simonis*

Cork Constraint Computation Centre
Department of Computer Science, University College Cork, Ireland
h.simonis@4c.ucc.ie

Abstract. In this paper we present a hybrid model for the demand acceptance variant of the routing and wavelength assignment problem in directed networks, an important benchmark problem in optical network design. Our solution uses a decomposition into a MIP model for the routing and optimization aspect, combined with a finite domain constraint model for the wavelength assignment. If a solution to the constraint problem is found, it provides an optimal solution to the overall problem. If the constraint problem is infeasible, we use an extended explanation technique to find a good relaxation of the problem which leads to a near optimal solution. Extensive experiments show that proven optimality is achieved for more than 99.8% of all cases tested, while run-times are orders of magnitude smaller than the best known MIP solution.

1 Introduction

The routing and wavelength assignment problem (RWA) [12,2,21] in optical networks considers a network where demands can be transported on different optical wavelengths through the network. Each accepted demand is allocated a path from its source to its sink, as well as a specific wavelength. Demands routed over the same link must be allocated different wavelengths, while demands whose paths are link disjoint may use the same wavelengths.

The RWA problem is a well studied, important problem in optical network design, for which many problem variants have been considered. Depending on the technology used, the network may be assumed to be *directed* or *undirected*. The *static design problem* considers the problem of allocating all given demands on a network topology, using the minimal number of frequencies. The *demand acceptance problem* considers a fixed, given number of frequencies on all links in the network. The objective is to accept the maximal number of demands in the network. In this paper we discuss the demand acceptance problem in a directed network, a companion paper [17] describes a solution for the easier static design problem variant.

More formally, we are considering a directed network $G = (N, E)$ of nodes N and edges E. A demand $d \in D$ is between source $s(d)$ and sink $t(d)$. We use the

* This work was supported by Science Foundation Ireland (Grant Number 05/IN/I886). Support from Cisco Systems and the Silicon Valley Community Foundation is gratefully acknowledged.

I.P. Gent (Ed.): CP 2009, LNCS 5732, pp. 104–118, 2009.

notation $\mathrm{In}(n)$ and $\mathrm{Out}(n)$ to denote all edges entering resp. leaving node n. The set Λ of available wavelengths is fixed and identical throughout the network. We can formulate a basic model of the problem with two sets of 0/1 integer variables. Variables y_d^λ denote whether demand d is accepted using wavelength λ, variables x_{de}^λ state whether edge e is used to transport demand d on wavelength λ.

$$\max \sum_{d\in D}\sum_{\lambda\in\Lambda} y_d^\lambda \tag{1}$$

s.t.

$$y_d^\lambda \in \{0,1\}, x_{de}^\lambda \in \{0,1\} \tag{2}$$

$$\forall d \in D: \quad \sum_{\lambda\in\Lambda} y_d^\lambda \le 1 \tag{3}$$

$$\forall e \in E, \forall \lambda \in \Lambda: \quad \sum_{d\in D} x_{de}^\lambda \le 1 \tag{4}$$

$$\forall d \in D, \forall \lambda \in \Lambda: \quad \sum_{e\in \mathrm{In}(s(d))} x_{de}^\lambda = 0, \quad \sum_{e\in \mathrm{Out}(s(d))} x_{de}^\lambda = y_d^\lambda \tag{5}$$

$$\forall d \in D, \forall \lambda \in \Lambda: \quad \sum_{e\in \mathrm{Out}(t(d))} x_{de}^\lambda = 0, \quad \sum_{e\in \mathrm{In}(t(d))} x_{de}^\lambda = y_d^\lambda \tag{6}$$

$$\forall d \in D, \forall \lambda \in \Lambda, \forall n \in N \setminus \{s(d), t(d)\}: \quad \sum_{e\in \mathrm{In}(n)} x_{de}^\lambda = \sum_{e\in \mathrm{Out}(n)} x_{de}^\lambda \tag{7}$$

Constraint (2) enforces integrality of the solution, constraint (3) states that a demand can use atmost one wavelength. The *clash* constraint (4) states that on each edge, only one demand may use any given wavelength. Constraints (5) and (6) link the x and y variables at the source (resp. sink) of each demand. Finally, constraint (7) enforces flow balance on all other nodes of the network.

The main contributions of this paper are

- a novel, two-step problem decomposition for the RWA problem into a MIP (Mixed Integer Programming) and finite domain constraint model,
- a new, very accurate upper bound to the RWA problem based on a resource-based relaxation of an existing, source aggregation MIP solution,
- the use of explanation techniques to understand infeasibility of the constraint model, suggesting good candidates for problem relaxation,
- extension of the model to handle parallel fibers without increasing problem size,
- experimental results showing that very high quality solutions are obtained by this method in seconds, outperforming the best MIP model by orders of magnitudes.

In the next section we will describe existing solutions to the problem, with special emphasis on complete MIP models. We then describe our decomposition strategy, presenting a resource-based MIP relaxation and the finite domain constraint model. In section 4, we describe how we can detect and explain infeasibility of the constraint model, and how we can relax the problem to obtain good, but possibly sub-optimal solutions.

We then extend our approach to allow parallel links in the network without increasing the size of the model. This is followed in section 6 by an extensive experimental evaluation of the proposed technique, before we consider possible further research in section 7.

2 Related Work

The RWA problem has been studied using many different solution approaches, see [4] for an overview. We can distinguish two main approaches. Greedy heuristics use local search techniques to accept demands incrementally, providing fast solutions for large problem cases, but without a formal guarantee of solution quality. Alternatively, complete methods, mainly based on ILP (Integer Linear Programming) techniques, can provide optimal solutions, but are restricted in the problem size handled.

The MIP formulation (1) does not scale well with increasing number of demands and network size. A major factor is the potential symmetry between all frequencies as well as additional symmetries due to multiple demands between the same source and sink. In [5], different ILP reformulations of the problem are considered, the best alternative uses a source aggregation, described below. In this model, one does not consider individual demands, but aggregates all demands starting in the same source node. We introduce integer variables y_{sd} to denote how many demands from a source s to a sink d are accepted. The upper limit for these variables is given by P_{sd}, the total number of requested demands between s and d. We then define variables x^{λ}_{se} to state whether wavelength λ on edge e is used to transport a demand originating in s, without identifying which demand is carried. We use the notation D_s for the set of destinations of requested demands originating in s.

$$\max \sum_{s \in N} \sum_{d \in D_s} y_{sd} \tag{8}$$

s.t.

$$y_{sd} \in \{0, 1 ... P_{sd}\}, x^{\lambda}_{se} \in \{0, 1\} \tag{9}$$

$$\forall e \in E, \forall \lambda \in \Lambda : \quad \sum_{s \in N} x^{\lambda}_{se} \leq 1 \tag{10}$$

$$\forall s \in N, \forall \lambda \in \Lambda : \quad \sum_{e \in \text{In}(s)} x^{\lambda}_{se} = 0 \tag{11}$$

$$\forall s \in N, \forall d \in D_s, \forall \lambda \in \Lambda : \quad \sum_{e \in \text{In}(d)} x^{\lambda}_{se} \geq \sum_{e \in \text{Out}(d)} x^{\lambda}_{se} \tag{12}$$

$$\forall s \in N, \forall d \in D_s : \quad \sum_{\lambda \in \Lambda} \sum_{e \in \text{In}(d)} x^{\lambda}_{se} = \sum_{\lambda \in \Lambda} \sum_{e \in \text{Out}(d)} x^{\lambda}_{se} + y_{sd} \tag{13}$$

$$\forall s \in N, \forall n \neq s, n \notin D_s, \forall \lambda \in \Lambda : \quad \sum_{e \in \text{In}(n)} x^{\lambda}_{se} = \sum_{e \in \text{Out}(n)} x^{\lambda}_{se} \tag{14}$$

Constraints (9) define the integrality conditions. Constraint (10) specifies the *clash* constraint between demands from different sources. Constraint (11) states that demands originating in s can not be routed through s, while constraints (12) and (13) consider the destinations of demands originating in s and state that the correct number of demands must be dropped in each node, linking the x and y variables. Finally, constraint (14) enforces flow balance at all other nodes of the network.

The solution of formulation (8) does not provide an immediate, unique solution for the demand acceptance problem. The path for each accepted demand must be extracted from the solution by a small procedure, which can also be used to eliminate loops in the paths at the same time.

A very nice property of the model is that the LP relaxation, replacing the integrality constraint (9) with continuous domain restrictions, provides a very tight upper bound for the ILP solution.

The aggregated model (8) performs much better than model (1), especially if the number of demands is increasing. But experiments in [5] show that obtaining optimal solutions even for small networks may still require several hours, if not days, of computation time. As network size increases, the number of sources to consider increases as well, leading to a dramatic performance loss.

The use of column generation has been considered [6] for a *path-based* [16] formulation of the problem. This increases the problem size that can be handled, but still requires solution times of several hours.

So far constraint programming has not been used to solve the RWA problem; a general overview of constraint applications in the network domain is given in [16]. Smith in [18] discusses a design problem for optical networks, but this is restricted to a ring topology, and minimizes the need for ADM multiplexers.

On the other hand, the RWA problem considered here is not too far removed from the demand acceptance problem in MPLS traffic engineering (MPLS-TE) in IP networks, which has been approached with multiple hybrid constraint solution techniques as described in [9,8,16]. The main difference is that demands in the MPLS-TE problem have integer sizes and overall link capacity limits are enforced instead of clash constraints. This motivated our solution approach to the RWA problem using a decomposition using a resource based relaxation, which we will describe in the next chapter.

3 Solution Approach

It seems unlikely that the complete optimization problem can be solved as a finite domain constraint program. In this section we describe our solution approach which is based on a simple decomposition strategy.

3.1 Problem Decomposition

A solution to the RWA problem must consider the following three activities:

1. Select demands to be accepted
2. Choose paths for each accepted demand
3. Assign wavelength

We choose a decomposition technique which handles the first two steps with a MIP program which is a relaxation of model (8), and a second step which consists of a finite

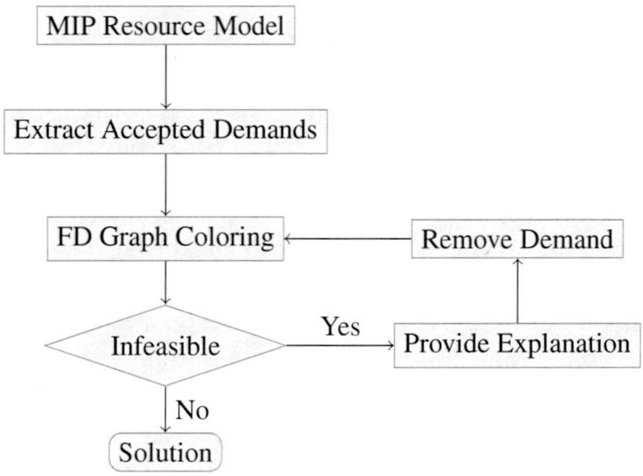

Fig. 1. Solution Approach

domain model for a resulting graph coloring problem, where each accepted demand is a node and there are *disequality* constraints between demands which are routed over the same link. The overall solution approach is shown in figure 1.

In the MIP model, we replace the clash constraints with simple capacity constraints which limit the number of demands that can be routed through each link. The optimal solution to this problem is an upper bound to the RWA problem. We then generate a graph coloring problem by imposing *disequality* constraints between all accepted demands which are routed over the same link. The number of available colors is limited by the available wavelengths. Using finite domain constraint programming, we search for a feasible solution to the problem. If we are successful, we have an optimal solution to the overall problem. If the constraint problem is infeasible, we drop some demands until a feasible solution is obtained. This solution may be sub-optimal, but usually is very close to the previously obtained upper bound. In order to choose which demand to drop, we have to understand why the constraint problem is infeasible. We use two techniques to find an explanation, one, structurally by detecting large cliques in the constraint graph, the other based on the QuickXplain [7] method. Once an explanation is obtained we heuristically choose one of its demands for relaxation in the overall problem. This creates a new graph coloring problem, which we recursively solve with our procedure.

Note that the suggested approach is not guaranteed to find an optimal solution, as a sub-optimal, infeasible solution to the second step only affects the second phase of the algorithm. To obtain a complete procedure, we would have to extend the feedback loop from the explanation generation back to the first phase MIP model. We will consider such an extension in future work.

3.2 Resource Allocation MIP Model

In order to simplify model (8), we relax the wavelength constraints and replace them with capacity constraints over all links. Each link has capacity $C = |\Lambda|$. We still use the integer variables y_{sd}, which state how many demands from s to d are accepted, but we replace the x_{se}^{λ} variables with integer z_{se} variables. These variables count how many demands originating in s are routed over link e. We use $T_s = \sum_{d \in D_s} P_{sd}$, the total number of requested demands starting in s, as the upper bound on the z_{se} variables.

$$\max \sum_{s \in N} \sum_{d \in D_s} y_{sd} \tag{15}$$

s.t.

$$y_{sd} \in \{0, 1 ... P_{sd}\}, z_{se} \in \{0, 1 ... T_s\} \tag{16}$$

$$\forall e \in E: \quad \sum_{s \in N} z_{se} \le C \tag{17}$$

$$\forall s \in N: \quad \sum_{e \in \text{In}(s)} z_{se} = 0 \tag{18}$$

$$\forall s \in N, \forall d \in D_s: \quad \sum_{e \in \text{In}(d)} z_{se} = \sum_{e \in \text{Out}(d)} z_{se} + y_{sd} \tag{19}$$

$$\forall s \in N, \forall n \neq s, n \notin D_s: \quad \sum_{e \in \text{In}(n)} z_{se} = \sum_{e \in \text{Out}(n)} z_{se} \tag{20}$$

Constraint (16) describes the integrality constraints, note that all variables have integer, not 0/1 domains. Constraint (17) is the link capacity constraint, which relaxes the clash constraints (10) for individual wavelengths. Constraint (18) limits the use of the source node, while constraint (19) describes the balance around the destination nodes, linking the y and z variables. Finally, constraint (20) imposes flow balance for all other nodes.

The optimal solution to model (15) provides an upper bound to the RWA problem. Perhaps surprisingly, this bound was not discussed in [5], although an equivalent relaxation based on a *path formulation* [16] was presented.

Similar to the situation for model (8), the solution does not directly tells us which demands are accepted and which paths they should take. A simple, but non-deterministic program is required to extract these elements, which are required in the second step of our procedure. At the same time, this procedure removes possible loops in the paths, which would cause problems in the graph coloring model. A simple example is given in figure 2. It shows the result obtained for one source (in green, marked **S**) and all possible sinks for demands starting in **S**. The thick, black links show the non-zero values of the z_{se} variables for this source, the nodes in blue show the non-zero values of the y_{sd} variables. The node marked **A** can be reached on two different paths. Which path is used for which demand is not defined in the MIP model, but will be selected by the solution extraction routine. The isolated, red nodes show demands which are not accepted.

3.3 Graph Coloring Model

After running the MIP model (15), and extracting the accepted demands and their paths, we can now try to allocate the available wavelengths with a simple finite domain constraint model. For each of the a accepted demands, a finite domain variable f_d ranging over all available wavelengths Λ is generated. For each link, we consider all demands

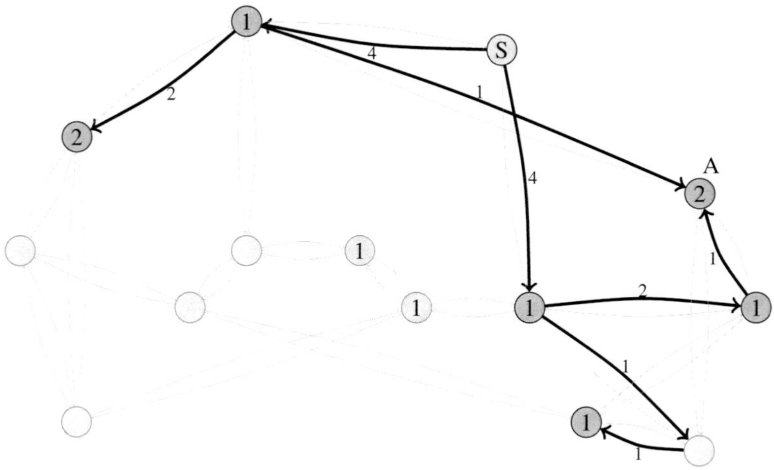

Fig. 2. Non-deterministic Solution Extraction

which are routed over it. The corresponding domain variables must be pairwise different, or, alternatively, an *alldifferent* constraint must hold between all of them. Considering all links together, we can express the constraints by either

- a set of binary *disequality* constraints between any two demands which are routed over a common link
- a set of *alldifferent* constraints [19], one for each link in the network
- a single *some-different* global constraint [14], expressing the complete disequality constraint network

More formally, using D^e to denote all accepted demands routed over link e, we generate the *alldifferent* constraints:

$$\forall e \in E : \text{alldifferent}(\{f_d | d \in D^e\}) \tag{21}$$

Alternatively, we can generate an equivalent set of *disequality* constraints:

$$\forall d_1, d_2 \in D, d_1 \neq d_2, \exists e \in E \text{ s.t. } d_1 \in D^e, d_2 \in D^e : \quad f_{d_1} \neq f_{d_2} \tag{22}$$

We initially use *bounds consistent* alldifferent constraints to provide a good compromise between computational effort required and resulting propagation.

In our search routine, we are interested only in finding a feasible solution quickly, not in proving infeasibility, as this is handled separately. We try three different search routines, each limited by a timeout (5 seconds) to restrict the effort expended:

- We first use an incomplete search routine, in our case *credit-based search* [3,1], which explores the top of the search tree completely, but allows only limited backtracking in lower parts of the tree. The variable selection method used is *first-fail*, and we use a randomized value selection method. We consecutively try credit amounts of a, a^2 and a^3 to improve chances of finding a solution quickly.

- In a second step, we try a complete search routine, still using bounds-consistent *alldifferent* constraints, keeping the same variable and value selection strategies. This might succeed when a larger amount of backtracking steps is required to find a solution.
- In a third step, we again use a complete search, but impose *domain consistent* versions of the *alldifferent* constraints, so that more propagation may detect failures which cause deep backtracking in earlier steps.

We could add specialized symmetry breaking techniques to handle the existing *value symmetries* between the available frequencies, as well as potential *variable symmetries* between demands sharing the same source, sink and path. These constraints will interfere with our method to detect infeasibility, and therefore were not implemented.

In the absence of additional symmetry breaking constraints, we can use a combination of shaving, impact analysis and symmetry breaking to preassign one of the large cliques to values. For each of the *alldifferent* constraints, we try to fix all their variables to values 1 to C, and measure the product of the domain sizes of the remaining variables in the problem. We fix the clique whose assignment will lead to the largest problem reduction before starting the overall search. We also observe that in a few cases this preassignment leads to a immediate failure, which we will exploit in the next section to explain infeasibility.

4 Handling Infeasibility

In the vast majority of examples tested, the finite domain solver quickly finds a feasible solution, which results in an optimal solution of the overall problem. In relatively few cases, the constraint problem is infeasible, which means we have to drop additional demands to obtain a complete RWA solution. We use two specialized methods to detect infeasibility and provide an explanation, which we can use to suggest likely demands to be rejected.

4.1 Clique Detection

A simple technique considers the structure of the constraint graph. For many links, the number of demands routed over them is equal to the overall capacity C, the number of available wavelengths. The variables corresponding to these demands form a clique in the disequality constraint graph. We can easily check each such clique if any additional variable (demand) must be different from each of its variables. This increases the size of the clique beyond the number of possible values and makes the problem infeasible. This can occur if demands routed over multiple links interact with each other. To obtain a feasible solution, we have to remove at least one of the demands in the infeasible clique from the problem.

As the graph coloring problem in this case is already given as a collection of cliques, many of which are already at limit size, we don't have to search for large cliques in the disequality constraint graph. This more generic method has been used in other constraint solvers for graph coloring problems.

4.2 Explanation

When we create the constraint variables and set up the *alldifferent* constraints, there will be no initial constraint propagation, if the size of the largest clique is equal to or smaller than the number of available frequencies. But we noticed that sometimes the preassignment of a single clique to initial values will lead to a failure, showing the problem to be inconsistent. At this point, we can use traditional explanation techniques like QuickXplain [7] to find a conflict set in the constraint graph after the preassignment. Together with the *alldifferent* constraint corresponding to the preassignment this explanation provides an (not necessarily minimal) explanation of the overall infeasibility. We run the explanation procedure at the level of individual *alldifferent* constraints, not their binary decomposition, in order to profit from the bounds consistent reasoning of the constraints. It might be possible to reduce the explanation further by removing some of the variables from the *alldifferent* constraints, we don't attempt this. Instead, we provide an explanation for each of the failed preassignments, and consider that set of possible explanations.

An explanation for the infeasibility in our system is a set of *alldifferent* constraints over variables which correspond to accepted demands. This means that this set of demands routed on the paths which were assigned in the first phase can not be allocated to the given wavelengths while satisfying the clash constraints. If (some of) the demands were allocated on different paths, such a solution might exist. We do not consider this possibility in our current approach, but resolve the infeasibility by rejecting one of the demands occurring in the explanation.

In order to suggest which demand should be removed to make the problem feasible, we count how often a demand occurs in an explanation and in how many explanations it is present. We order the demands by decreasing number of occurrences and try to remove the demands with the largest count first. This creates a new constraint problem, which we try to solve recursively with the same technique. The method will always terminate, as we remove one demand at each step, but fortunately will require only one or two steps in most cases before a feasible solution is found.

5 Extension to Multigraphs

An interesting extension of the problem considers the possibility of using more than one fiber between nodes in the network, e.g. replacing the directed graph with a multigraph. This is often done to increase capacity on connections which carry a lot of traffic, and can typically be achieved at little extra cost as cables between locations already carry many fibers in a single cable strand.

At first sight, we can use our existing model without change, modelling each parallel fiber as a separate edge in the multigraph, for which we generate variables in our MIP model and constraints in the finite domain graph coloring model. The disadvantage is that we increase the number of variables and introduce additional symmetries in our model, as the choice between the parallel fibers is unrestricted.

We can reduce the size of our models by changing some of the constraints slightly. In the MIP model (15), we only introduce one variable for every set $\hat{e} \in \hat{E}$ of $k_{\hat{e}}$ parallel fibers. We then adjust the capacity constraints (17) to

$$\forall \hat{e} \in \hat{E}: \quad \sum_{s \in N} z_{se} \leq k_{\hat{e}} * C \tag{23}$$

The rest of the model is not affected, and the number of constraints and variables is the same as in the case without multiple fibers.

The second phase of our decomposition is no longer a standard graph coloring problem, as on a set of parallel fibers we can use the same wavelength up to $k_{\hat{e}}$ times. Our constraint model is only slightly affected. We have to replace the *alldifferent* constraint with a global cardinality constraint (*gcc*)[13,11], which allows values to be used repeatedly. The variables and their domains don't need to be modified, and for every connection which uses multiple fibers we have to use *gcc* instead of *alldifferent* constraints. Note that we can no longer model the problem with *disequality* constraints alone, nor can we use a single *some-different* constraint. The explanation part of the program is not affected, the QuickXplain procedure works for any constraint network. In the experiments below we have not included scenarios with multiple fibers, as we did not find realistic network scenarios using them in the literature.

6 Experimental Results

Most of the published results on the RWA problem use randomly generated demands on a few given network structures. We also use this approach and generate given numbers of demands between randomly chosen source and sink nodes. Multiple demands between the same nodes are allowed, but source and sink must be different.

6.1 Fixed Network Structure

In the literature we found four actual optical network topologies used in experiments. Their size is quite small, ranging from 14 to 27 nodes.

nsf 14 nodes, 42 edges
eon 20 nodes, 78 edges
mci 19 nodes, 64 edges
brezil 27 nodes, 140 edges

We explored all combinations of number of demands (100-800 demands in increments of 50, 15 cases) and available wavelengths (5-50 in increments of 5, 10 cases) for the four networks, and created 100 random problems for each combination. This created 60000 problems ranging over a variety of typical scenarios. In many cases, all demands can be accepted, as there are enough frequencies available. At the other extreme, when there are many demands for few frequencies, most demands between far-distant nodes will be dropped, and only demands which can be satisfied with a short route will be used. This is an artifact of the objective function which does not reward the acceptance of long-distance demands. In between the two extreme cases, difficult optimization problems are found where most, but not all demands can be accepted and near-optimal solutions are important for customer satisfaction and revenue generation.

Table 1 shows the distribution of outcomes over all 60000 test cases considered. Only 88 (0.14%) are infeasible, the vast majority is solved to optimality. In more than 98%

of the cases the first search routine using only a units of credit finds a feasible solution. There are very few instances where complete search is required. As graph coloring problems, these instances do not seem to be very hard, given the structural information we can exploit.

Table 2 shows the result for selected, interesting parameter combinations. The entries summarize the results over 100 runs with the same parameters, but different random seeds. The column *Opt.* tells how many solutions were proven optimal. The columns *Avg LP*, *Avg MIP* and *Avg FD* show the average cost obtained by the LP relaxation of the MIP model (15), the MIP model itself and the final number of accepted demands obtained by the finite domain solver. The LP relaxation already is a very good approximation of the total cost, for the examples shown the LP and MIP cost coincide, and even on the full set of tests the LP cost is very tight, the MIP-LP gap never exceeds 0.94. The next column, *Max Gap*, shows the largest gap between MIP and FD solution,

Table 1. Overall Distribution of Solutions

Type	Technique	Count
Infeasible	clique	50
	preassign	38
Feasible	credit total	59962
	of that, credit a units	58861
	of that, credit a^2 units	940
	of that, credit a^3 units	161
	complete search, BC alldifferent	25
	complete search, GAC alldifferent	12

Table 2. Selected Examples (100 Runs Each)

Network	Dem.	λ	Opt.	Avg LP	Avg MIP	Avg FD	Max Gap	Avg Time	Max Time
brezil	500	15	98	483.86	483.86	483.84	1.00	0.92	1.34
brezil	600	20	100	590.96	590.96	590.96	0.00	1.00	1.34
brezil	700	20	100	672.53	672.53	672.53	0.00	1.19	1.78
brezil	800	25	99	781.39	781.39	781.37	2.00	1.44	11.47
eon	500	20	100	471.56	471.56	471.56	0.00	0.65	0.77
eon	600	25	100	574.80	574.80	574.80	0.00	0.82	1.13
eon	700	30	100	677.35	677.35	677.35	0.00	1.05	1.81
eon	800	35	100	779.17	779.17	779.17	0.00	1.28	1.94
mci	500	25	100	486.38	486.38	486.38	0.00	0.80	2.28
mci	600	30	100	585.18	585.18	585.18	0.00	1.27	29.81
mci	700	35	100	684.00	684.00	684.00	0.00	1.30	3.53
mci	800	40	100	782.86	782.86	782.86	0.00	1.68	5.21
nsf	500	35	100	495.20	495.20	495.20	0.00	0.50	0.60
nsf	600	40	100	588.63	588.63	588.63	0.00	0.66	0.98
nsf	700	45	100	678.44	678.44	678.44	0.00	0.86	1.35
nsf	800	45	100	727.15	727.15	727.15	0.00	0.95	1.56

i.e. the number of demands removed due to infeasibility of the graph coloring model. This value never exceeds 2 in the examples shown, it never exceeds 4 in any of the tests run. We then show average and maximal total run times on a Windows XP laptop with a 2.4GHz processor and 2GB of memory. Results were obtained using ECLiPSe 6.0 [20] with the eplex library [15] for the Coin-OR [10] CLP/CBC MIP solver.

In table 3 we compare our results (Hybrid Model) to the Full MIP model (8) presented in [5]. One can see that the difference in solution quality is minimal, but the times required for the full model (again, using ECLiPSe 6.0 with eplex and the Coin-OR CLP/CBC solver) are much higher, especially for larger network size and/or large number of demands. Note that for the brezil network with 700 demands and 25 frequencies (results are shown in italics), only 99 of the MIP models were solved. One problem instance did not terminate within 5 days of execution.

Table 3. Compared to MIP Model for Complete Problem

Network	Dem.	λ	Opt.	Hybrid Model Avg FD	Avg Time	Max Time	Full MIP Avg Opt	Avg Time	Max Time
brezil	500	15	98	483.84	0.92	1.34	483.86	1218.40	14103.84
brezil	600	20	100	590.96	1.00	1.34	590.96	6076.81	87767.95
brezil	700	25	98	695.48	1.01	1.80	*695.48*	*13623.15*	*78463.89*
brezil	800	25	99	781.37	1.44	11.47	781.39	7567.68	15456.50
eon	500	20	100	471.56	0.65	0.77	471.56	352.21	585.45
eon	600	25	100	574.80	0.82	1.13	574.80	1411.67	2877.88
eon	700	30	100	677.35	1.05	1.81	677.35	1727.52	3568.13
eon	800	35	100	779.17	1.28	1.94	779.17	2485.64	4116.11
mci	500	25	100	486.38	0.80	2.28	486.38	1023.16	1664.31
mci	600	30	100	585.18	1.27	29.81	585.18	1621.30	2895.88
mci	700	35	100	684.00	1.30	3.53	684.00	1987.23	3428.41
mci	800	40	100	782.86	1.68	5.21	782.86	2316.88	4402.44
nsf	500	35	100	495.20	0.50	0.60	495.20	82.85	173.19
nsf	600	40	100	588.63	0.66	0.98	588.63	155.90	373.63
nsf	700	45	100	678.44	0.86	1.35	678.44	205.82	586.61
nsf	800	45	100	727.15	0.95	1.56	727.15	173.53	410.97

6.2 Increasing Number of Demands

Table 4 shows results for another series of tests, where we increase the number of demands for the eon network, increasing at the same time the number of available frequencies so that over 90%, but below 100% of the demands can be accepted. We split the reported run-times into the average and maximum time needed for phase 1 (MIP), and phase 2 (FD). One can see that the time for phase 1 is not affected, as the MIP model (15) does depend neither on the number of demands, nor on the number of frequencies available. The average time for phase 2 slowly increase with problem size, while the number of optimal solutions stays very high (99-100%). The time for finding and exploiting the explanation in an infeasible scenario increases significantly, but not prohibitively.

Table 4. Increasing Number of Demands

Network	Dem.	λ	Opt.	Avg LP	Avg MIP	Avg FD	Max Gap	Avg MIP Time	Max MIP Time	Avg FD Time	Max FD Time
eon	800	30	100	741.78	741.78	741.78	0.00	0.15	0.17	0.83	1.61
eon	900	40	100	880.59	880.59	880.59	0.00	0.14	0.16	1.18	2.17
eon	1000	40	100	950.36	950.36	950.36	0.00	0.15	0.17	1.37	3.42
eon	1100	50	100	1082.61	1082.61	1082.61	0.00	0.14	0.16	1.71	2.83
eon	1200	50	100	1156.38	1156.38	1156.38	0.00	0.15	0.17	2.07	5.92
eon	1300	50	100	1219.82	1219.82	1219.82	0.00	0.16	0.17	2.22	5.24
eon	1400	60	100	1361.47	1361.47	1361.47	0.00	0.15	0.16	2.92	4.94
eon	1500	60	99	1428.78	1428.78	1428.77	1.00	0.15	0.17	4.22	106.97
eon	1600	70	100	1565.90	1565.90	1565.90	0.00	0.15	0.16	3.89	8.48
eon	1700	70	100	1637.47	1637.47	1637.47	0.00	0.16	0.17	4.58	13.59
eon	1800	80	100	1769.86	1769.86	1769.86	0.00	0.15	0.16	5.19	8.81
eon	1900	80	99	1844.46	1844.46	1844.45	1.00	0.15	0.17	7.23	163.41
eon	2000	90	100	1972.66	1972.66	1972.66	0.00	0.15	0.17	6.34	9.61

Table 5. Random Networks (Edge Density 0.25, 100 Runs Each)

Network	Dem.	λ	Opt.	Avg LP	Avg MIP	Avg FD	Avg MIP Time	Max MIP Time	Avg FD Time	Max FD Time
r30	500	30	100	391.82	391.82	391.82	0.45	0.55	0.12	0.16
r40	500	30	100	424.58	424.58	424.58	1.07	1.23	0.14	0.17
r50	500	30	100	437.69	437.69	437.69	2.13	2.38	0.09	0.13
r60	500	30	100	447.21	447.21	447.21	3.92	4.34	0.08	0.16
r70	500	30	100	453.41	453.41	453.41	6.78	7.50	0.10	0.17
r80	500	30	100	457.65	457.65	457.65	10.75	11.95	0.10	0.17
r90	500	30	100	464.69	464.69	464.69	16.08	17.45	0.08	0.22
r100	500	30	100	466.67	466.67	466.67	22.74	25.22	0.09	0.25

6.3 Random Networks

We also wanted to check the stability of the proposed method for larger network sizes. For this we generated random network structures with 30 to 100 nodes and an average edge density of 0.25. We tested these for randomly generated problems of 500 demands and 30 frequencies, with 100 instances for each problem case.

Table 5 shows the results, again splitting the execution times into the MIP and FD components. One can see that with increasing network size the MIP solution times start to dominate. Unfortunately, the LP solver times also increase proportionately, which means that just obtaining an upper bound with the model shown becomes expensive when we consider more than 100 nodes.

7 Future Work

Our current model works very well with large number of demands, but does not handle large network sizes well. In order to extend the problem size further, we will have to consider a further hybridization of the first phase of the algorithm, along the lines of [9], if we want to find proven optimal solutions.

We so far have only considered the directed network variant of the problem. It seems straightforward to extend the model to handle the undirected case discussed in [4] as well.

At the moment, we do not attempt to generate minimal explanations, or indeed a minimal set of explanations, as we are only interested in them to suggest a further relaxation of the demand acceptance problem. By creating a more compact representation, we might be able to feed them as no-good constraints into the first phase of the algorithm, generating a complete, hybrid algorithm for the RWA problem.

Another task will be a further study of the problem environment to determine which additional features are required to bring this method to real-life use.

8 Conclusion

In this paper we have described a hybrid combination of ILP and constraint programming to solve the demand acceptance variant of the routing and wavelength assignment (RWA) problem. We have shown that decomposing the problem into a resource-constrained, optimized routing problem and a graph coloring problem works very well, producing either proven optimal or near optimal solutions for all cases tested. This method outperforms a full MIP model by orders of magnitude, making the proposed method an efficient solution for realistic problem sizes.

Acknowledgment

We want to thank Paul Davern and Hadrien Cambazard for helpful comments on a draft of the paper.

References

1. Apt, K.R., Wallace, M.: Constraint Logic Programming using ECLiPSe. Cambridge University Press, New York (2007)
2. Banerjee, D., Mukherjee, B.: A practical approach for routing and wavelength assignment in large wavelength-routed optical networks. IEEE Journal on Selected Areas in Communications 14(5), 903–908 (1996)
3. Beldiceanu, N., Bourreau, E., Chan, P., Rivreau, D.: Partial search strategy in CHIP. In: 2nd International Conference on Metaheuristics MIC 1997, Sophia Antipolis, France (1997)
4. Jaumard, B., Meyer, C., Thiongane, B.: ILP formulations for the routing and wavelength assignment problem: Symmetric systems. In: Resende, M., Pardalos, P. (eds.) Handbook of Optimization in Telecommunications, pp. 637–677. Springer, Heidelberg (2006)
5. Jaumard, B., Meyer, C., Thiongane, B.: Comparison of ILP formulations for the RWA problem. Optical Switching and Networking 4(3-4), 157–172 (2007)
6. Jaumard, B., Meyer, C., Thiongane, B.: On column generation formulations for the RWA problem. Discrete Applied Mathematics 157, 1291–1308 (2009)

7. Junker, U.: Quickxplain: Conflict detection for arbitrary constraint propagation algorithms. In: IJCAI 2001 Workshop on Modelling and Solving problems with constraints (CONS-1), Seattle, WA, USA (August 2001)
8. Lever, J.: A local search/constraint propagation hybrid for a network routing problem. International Journal on Artificial Intelligence Tools 14(1-2), 43–60 (2005)
9. Liatsos, V., Novello, S., El Sakkout, H.: A probe backtrack search algorithm for network routing. In: Proceedings of the Third International Workshop on Cooperative Solvers in Constraint Programming, CoSolv 2003, Kinsale, Ireland (September 2003)
10. Lougee-Heimer, R.: The common optimization interface for operations research. IBM Journal of Research and Development 47, 57–66 (2003)
11. Quimper, C.-G.: Efficient Propagators for Global Constraints. PhD thesis, University of Waterloo (2006)
12. Ramaswami, R., Sivarajan, K.N.: Routing and wavelength assignment in all-optical networks. IEEE/ACM Trans. Netw. 3(5), 489–500 (1995)
13. Régin, J.-C.: Generalized arc consistency for global cardinality constraint. In: AAAI/IAAI, vol. 1, pp. 209–215 (1996)
14. Richter, Y., Freund, A., Naveh, Y.: Generalizing alldifferent: The somedifferent constraint. In: Benhamou, F. (ed.) CP 2006. LNCS, vol. 4204, pp. 468–483. Springer, Heidelberg (2006)
15. Shen, K., Schimpf, J.: Eplex: Harnessing mathematical programming solvers for constraint logic programming. In: van Beek, P. (ed.) CP 2005. LNCS, vol. 3709, pp. 622–636. Springer, Heidelberg (2005)
16. Simonis, H.: Constraint applications in networks. In: Rossi, F., van Beek, P., Walsh, T. (eds.) Handbook of Constraint Programming. Elsevier, Amsterdam (2006)
17. Simonis, H.: Solving the static design routing and wavelength assignment problem. In: CSCLP 2009, Barcelona, Spain (June 2009)
18. Smith, B.M.: Symmetry and search in a network design problem. In: Barták, R., Milano, M. (eds.) CPAIOR 2005. LNCS, vol. 3524, pp. 336–350. Springer, Heidelberg (2005)
19. van Hoeve, W.J.: The alldifferent constraint: A survey. CoRR, cs.PL/0105015 (2001)
20. Wallace, M., Novello, S., Schimpf, J.: ECLiPSe: A platform for constraint logic programming. ICL Systems Journal 12(1) (May 1997)
21. Zang, H., Jue, J.P., Mukherjee, B.: A review of routing and wavelength assignment approaches for wavelength-routed optical WDM networks. Optical Networks Magazine, 47–60 (January 2000)

Memoisation for Constraint-Based Local Search

Magnus Ågren

Swedish Institute of Computer Science
Box 1263, SE – 164 29 Kista, Sweden
magnus.agren@sics.se

Abstract. We present a memoisation technique for constraint-based local search based on the observation that penalties with respect to some interchangeable elements need only be calculated once. We apply the technique to constraint-based local search on set variables, and demonstrate the usefulness of the approach by significantly speeding up the penalty calculation of a commonly used set constraint.

1 Introduction and Background

Memoisation [1] is an optimisation technique often used to speed up function calls in programming languages. By caching the calculated results for inputs to a given function, subsequent calls for already seen inputs to the function do not need to be recalculated but can be looked up and returned directly. In this paper we apply memoisation to constraint-based local search.

In constraint-based local search [2], constraint measures are used to navigate in the search space and move towards (optimal) solutions. Given a constraint, such measures include penalties and variable conflicts, which are estimations on how far the constraint currently is from being satisfied and how much each variable contributes to that distance, respectively. Since a local search algorithm may perform many moves, and each move may mean evaluating the constraint measures with respect to a large number of configurations (complete assignments), the evaluation must be done efficiently. This is often achieved by using incremental algorithms (see for example [3]).

In [4] we presented constraint measures with such incremental algorithms for using monadic existential second-order logic (∃MSO) for modelling set constraints in local search. For example, the set constraint $S \subset T$ (strict subset) can be modelled in ∃MSO by:

$$\exists S \exists T((\forall x(x \notin S \vee x \in T)) \wedge (\exists x(x \notin S \wedge x \in T))) \quad (1)$$

We call such constraint models *∃MSO constraints*. Now, given a common universe $\mathbf{U}$ for all set variables and an element u of this universe, the penalty of a primitive constraint of the form $u \in S$ or $u \notin S$ is zero if it is satisfied, and one otherwise; the penalty of a conjunction (disjunction) is the sum (minimum) of the penalties of its conjuncts (disjuncts); and the penalty of a first-order universal (existential)

I.P. Gent (Ed.): CP 2009, LNCS 5732, pp. 119–126, 2009.

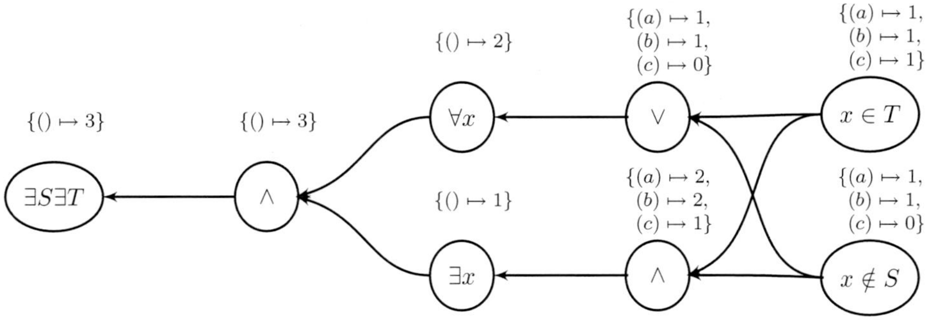

Fig. 1. Penalty dag of $\exists S \exists T((\forall x(x \notin S \vee x \in T)) \wedge (\exists x(x \notin S \wedge x \in T)))$

quantification is the sum (minimum) of the penalties of the quantified formula where the occurrences of the bound variable are replaced by each element of $\mathbf{U}$.

The practical relevance of using ∃MSO constraints in local search was demonstrated in [4], where a necessary built-in global constraint was assumed missing and replaced by a corresponding ∃MSO constraint, while still obtaining competitive results in terms of runtime and robustness.

In this paper we use penalty *dags* (directed acyclic graphs), namely attributed parse trees, for illustrative purposes only, and not as an implementation device for supporting incremental maintenance algorithms, as in [4]. For instance, the calculation of the penalty of (1) under the configuration $k = \{S \mapsto \{a, b\}, T \mapsto \emptyset\}$ is illustrated by the penalty dag in Fig. 1; the map $\{() \mapsto 3\}$ above the sink $\exists S \exists T$ indicates that the penalty of (1) is 3.

Interchangeable elements (or values) must often be identified in order to solve problems efficiently. The constraint programming community has traditionally done this in the context of symmetry breaking for complete search, where the aim is to *avoid* rediscovering (symmetrically) already encountered (non)solutions (see [5] for an early reference).

On the contrary, we here take advantage of interchangeable elements, and the dag in Fig. 1 can also be used to illustrate the key idea of this paper. Consider the penalty maps above the (rightmost) $\wedge$ and $\vee$ connectives in the dag which, for the corresponding subformulas rooted at those connectives, indicate the penalties with respect to each element of the universe $\mathbf{U} = \{a, b, c\}$. Note that, in both of these penalty maps, the penalties are the same for the elements a and b. This is not by chance but because a and b are interchangeable in the sense that they are both in S and not in T under the configuration k. In fact, *any* element (of the universe) in S and not in T would be interchangeable with a and b and would also share these same penalty maps in the dag. Hence, the penalty maps need only be calculated once for all interchangeable elements, and can then later simply be returned from a cache taking interchangeability into account. We show in the following that this can lead to a significant speedup of local search algorithms with ∃MSO constraints. Note that although we only consider penalties in this paper, all results can be generalised for variable-conflicts as well.

2 Memoising ∃MSO Penalties Using Signatures

Recall that the penalties of first-order quantifications are sums and minima of the penalties of the quantified subformula where the occurrences of the bound first-order variable are replaced by each element of the universe, and consider again (1) and $k = \{S \mapsto \{a,b\}, T \mapsto \emptyset\}$. *The key idea of this paper is based on the following observation:* since the elements a and b are both in S and not in T, the penalties of, for example, the quantified subformula $x \notin S \wedge x \in T$ are the same when x is replaced by a or b. Indeed, both a and b are bound to 2 in the map above the rightmost $\wedge$-node in Fig. 1. So a and b are *interchangeable* in the sense that it is only necessary to calculate the penalty of the subformula given one of the elements, and then reuse that value for the other element. We characterise such interchangeable elements by their *signatures.* The signature of an element $u \in \mathbf{U}$ with respect to a sequence of set variables $\langle S_1, \ldots, S_n \rangle$ under a configuration k is a bit string $b_1 \cdots b_n$ such that $b_i = 1$ if and only if $u \in k(S_i)$.

Example 1. The respective signatures of a, b, c with respect to $\langle S, T \rangle$ under $k = \{S \mapsto \{a,b\}, T \mapsto \emptyset\}$ are 10, 10, 00. So a and b share the same signature.

Using signatures to reason about interchangeable elements (or values) was done also in [6], but there in the context of symmetry breaking for complete search.

We will now present a penalty maintenance algorithm for ∃MSO constraints based on memoisation and element signatures. Given an ∃MSO constraint $\Phi = \exists S_1 \cdots \exists S_n \phi$, this algorithm operates on a data structure D with the following fields:

- *D.penalty*, the penalty of Φ under the current configuration;
- *D.signature*, an array indexed by the elements of the universe such that *D.signature*$[u]$ is the signature of u with respect to $\langle S_1, \ldots, S_n \rangle$ under the current configuration;
- *D.cache*, an array indexed by the (first-order) quantified subformulas ϕ of Φ such that *D.cache*$[\phi]$ is the penalty cache of ϕ (these penalty caches correspond to the penalty maps of the quantified subformulas in Fig. 1, but are based on element signatures and not on elements);
- *D.min*, an array indexed by the first-order existential quantifications $\exists x \phi$ of Φ such that *D.min*$[\exists x \phi]$ is a multiset of the penalties of ϕ under the current configuration, where the occurrences of x in ϕ is replaced by each element of $\mathbf{U}$. So the minimum value of *D.min*$[\exists x \phi]$ is the penalty of $\exists x \phi$.

Following the ideas in [7], the penalty maintenance algorithm consists of two parts: an *initialisation* part and an *update* part. Both of these parts call a generic function *proj_penalty* which is used to traverse the ∃MSO constraint. The intuition behind this is that a call *proj_penalty*(D, Φ, A) returns the penalty of Φ projected on some subset A of the universe. By initially setting A to $\mathbf{U}$, the penalty of Φ is obtained. By later setting A to $\{u\}$, for example, the penalty of Φ projected on $\{u\}$ is obtained. So given a move, for example, of the form *add*$(S,u)(k)$ (the result of adding u to S under k), the penalty change of Φ

Algorithm 1. Generic function for initialising and updating ∃MSO penalties

1: **function** $proj\_penalty(D, \Phi, A)$
2: **if** Φ is of the form $\forall x \phi$ **then**
3: $p \leftarrow 0$
4: **for all** $u \in A$ **do**
5: **if** $D.signature[u] \in D.cache[\phi]$ **then**
6: $p \leftarrow p + D.cache[\phi][D.signature[u]]$
7: **else**
8: $q \leftarrow proj\_penalty(D, \phi[u/x], \emptyset)$
9: $D.cache[\phi][D.signature[u]] \leftarrow q$
10: $p \leftarrow p + q$
11: **return** p
12: **if** Φ is of the form $\exists x \phi$ **then**
13: **for all** $u \in A$ **do**
14: **if** $D.signature[u] \in D.cache[\phi]$ **then**
15: $add(D.cache[\phi][D.signature[u]], D.min[\exists x \phi])$
16: **else**
17: $q \leftarrow proj\_penalty(D, \phi[u/x], \emptyset)$
18: $D.cache[\phi][D.signature[u]] \leftarrow q$
19: $add(q, D.min[\exists x \phi])$
20: **return** $\min(D.min[\exists x \phi])$
21: **if** Φ is of the form $\phi \wedge \psi$ **then**
22: **return** $proj\_penalty(D, \phi, A) + proj\_penalty(D, \psi, A)$
23: **if** Φ is of the form $\phi \vee \psi$ **then**
24: **return** $\min(proj\_penalty(D, \phi, A), proj\_penalty(D, \psi, A))$
25: **if** Φ is of the form $u \in S_i$ **then return** $1 - D.signature[u][i]$
26: **if** Φ is of the form $u \notin S_i$ **then return** $D.signature[u][i]$

can be obtained by two calls $proj\_penalty(D, \Phi, \{u\})$ before and after the move, the penalty change being the difference of the results of these two calls. *Note that this difference is the same as the difference of the results of two such calls where* $\{u\}$ *is replaced by* **U**. Also note that the current configuration would be a superfluous argument to *proj_penalty* since it is implicit from *D.signature*.

The function *proj_penalty* is shown in Algorithm 1. We here only discuss the quantifier cases on lines 2 to 20 as the other cases closely follow the penalty function described in the first section. For a call $proj\_penalty(D, \forall x \phi, A)$, the sum is calculated by looking up $D.cache[\phi]$ given the signature of each element of A. When a value is in the cache it can be directly used (line 6). Otherwise, the value is calculated by a recursive call where the occurrences of the bound variable x are replaced by the element $u \in A$, and stored in the cache for subsequent calls with the same signature (lines 8 to 10). A call $proj\_penalty(D, \exists x \phi, A)$ is similar, the only difference being that the minimum is calculated by first adding the penalty for the signature of each element of A to the multiset $D.min[\exists x \phi]$ (lines 15 and 19), of which the minimum value is then returned (line 20).

Note that *D.signature* is used to represent the current configuration. So before a call $proj\_penalty(D, \Phi, A)$, *D.signature* must be updated to reflect this.

Algorithm 2. Initialise and update procedures for $\exists$MSO penalties

1: **procedure** $\mathit{initialise}(D, \exists S_1 \cdots \exists S_n \phi, \mathbf{U})(k)$
2:     **for all** $u \in \mathbf{U}$ **do**
3:         $D.\mathit{signature}[u] \leftarrow$ the signature of u with respect to $\langle S_1, \ldots, S_n \rangle$ under k
4:     $D.\mathit{penalty} \leftarrow \mathit{proj\_penalty}(D, \phi, \mathbf{U})$
5: **procedure** $\mathit{update}(D, \Phi)(k, \ell)$
6:     **if** ℓ is of the form $\mathit{add}(S_i, u)(k)$ or $\mathit{drop}(S_i, u)(k)$ **then**
7:         **for all** subformulas $\exists x \phi$ of Φ **do**
8:             $\mathit{remove}(D.\mathit{cache}[\phi][D.\mathit{signature}[u]], D.\mathit{min}[\exists x \phi])$
9:         $p_0 \leftarrow \mathit{proj\_penalty}(D, \Phi, \{u\})$
10:         **for all** subformulas $\exists x \phi$ of Φ **do**
11:             $\mathit{remove}(D.\mathit{cache}[\phi][D.\mathit{signature}[u]], D.\mathit{min}[\exists x \phi])$
12:         flip the value $D.\mathit{signature}[u][i]$
13:         $p_1 \leftarrow \mathit{proj\_penalty}(D, \Phi, \{u\})$
14:         $D.\mathit{penalty} \leftarrow D.\mathit{penalty} + (p_1 - p_0)$
15:     **else**
16:         failure

Also note that, before a call $\mathit{proj\_penalty}(D, \Phi, A)$ on an already initialised D, the penalties in any multiset $D.\mathit{min}[\exists x \phi]$ corresponding to the elements of A, must be removed. This is necessary since projecting the penalty of an existential quantification on $A \subseteq \mathbf{U}$ still requires taking the penalties with respect to *all* elements of $\mathbf{U}$ into account (since it is a minimum value).

The procedure *initialise* is shown in Algorithm 2. A call $\mathit{initialise}(D, \Phi, \mathbf{U})(k)$ initialises the signatures of $D.\mathit{signature}$ to reflect the configuration k (lines 2 to 3) after which a call to *proj_penalty* is used to initialise $D.\mathit{penalty}$ (line 4).

The procedure *update* is also shown in Algorithm 2. Given a move ℓ of the form $\mathit{add}(S_i, u)(k)$ or $\mathit{drop}(S_i, u)(k)$ (the results of adding or dropping u from S_i under k), a call $\mathit{update}(D, \Phi)(k, \ell)$ must (twice) update each multiset in $D.\mathit{min}$ by removing one occurrence of the value corresponding to the signature of u (lines 7 to 8 and 10 to 11). (See also the note above concerning this.) The penalty change is then obtained as the difference of the results of two calls to *proj_penalty* (lines 9 and 13), before and after the move ℓ has been reflected on $D.\mathit{signature}$ (line 12). This penalty change is then used to update $D.\mathit{penalty}$ (line 14).

Example 2. Let Φ denote (1) and let $\forall x \phi$ and $\exists x \psi$ denote respectively the (first-order) universal and existential quantifications of Φ. Given $k = \{S \mapsto \{a, b\}, T \mapsto \emptyset\}$ and $\mathbf{U} = \{a, b, c\}$, the call $\mathit{initialise}(D, \Phi, \mathbf{U})(k)$ initialises the fields of the data structure D such that:

$$D.\mathit{penalty} = 3 \qquad D.\mathit{signature} = [a \mapsto 10, b \mapsto 10, c \mapsto 00]$$

$$D.\mathit{min} = [\exists x \psi \mapsto \{1, 2, 2\}] \qquad D.\mathit{cache} = \begin{bmatrix} \phi \mapsto [00 \mapsto 0, 10 \mapsto 1] \\ \psi \mapsto [00 \mapsto 1, 10 \mapsto 2] \end{bmatrix}$$

Considering now adding c to T, the subsequent call $\mathit{update}(D, \Phi)(k, \ell)$, where $\ell = \mathit{add}(T, c)(k)$, changes the fields of D such that:

$$D.penalty = 2 \qquad D.signature = [a \mapsto 10, b \mapsto 10, c \mapsto 01]$$

$$D.min = [\exists x\psi \mapsto \{0,2,2\}] \qquad D.cache = \begin{bmatrix} \phi \mapsto [00 \mapsto 0, 01 \mapsto 0, 10 \mapsto 1] \\ \psi \mapsto [00 \mapsto 1, 01 \mapsto 0, 10 \mapsto 2] \end{bmatrix}$$

The penalty is decreased to two since the constraint is now closer to being satisfied. This is calculated by:

1. removing 1 (the penalty cached for 00 in $D.cache[\psi]$) from $D.min[\exists x\psi]$, setting this multiset temporarily to $\{2,2\}$ (lines 7 to 8 of Algorithm 2);
2. obtaining $p_0 = 1$ by the call $proj\_penalty(D, \Phi, \{c\})$, which also adds 1 back to $D.min[\exists x\psi]$ (line 9);
3. removing 1 from $D.min[\exists x\psi]$ again (lines 10 to 11);
4. updating the signature of c to 01 (line 12);
5. obtaining $p_1 = 0$ by the call $proj\_penalty(D, \Phi, \{c\})$, which also adds 0 to $D.min[\exists x\psi]$, setting it to $\{0,2,2\}$ (line 13);
6. increasing $D.penalty$ by the difference $p_1 - p_0 = 0 - 1 = -1$ (line 14).

3 Evaluation

The algorithms of the previous section were implemented in *Objective Caml* (`http://caml.inria.fr`) and the experiments were performed on a 2.67 GHz Intel Core i7 Linux machine (using only one processor core).

In order to evaluate the memoisation-based penalty maintenance algorithm (called *memo* below) we have compared it to the incremental penalty maintenance algorithm of [4] (called *nomemo* below). We compared these two algorithms by measuring their speed in terms of average number of iterations per second when solving two given problems *subset* and *partition*. Both problems are stated on n set variables $\mathcal{S} = \{S_1, \ldots, S_n\}$ all with a common universe $\mathbf{U}$ of cardinality n such that:

- for *subset*, there is an $S_i \subset S_{i+1}$ constraint for each $1 \leq i < n$;
- for *partition*, there is a single $Partition(\mathcal{S})$ constraint.

While the $S_i \subset S_{i+1}$ constraints are modelled in ∃MSO as (1), the $Partition(\mathcal{S})$ constraint (requiring all set variables to be pairwise disjoint and their union to equal $\mathbf{U}$, where any set variable may be empty) is modelled in ∃MSO as:

$$\exists S_1 \cdots \exists S_n \left(\forall x \begin{pmatrix} (x \in S_1 \rightarrow (x \notin S_2 \wedge \cdots \wedge x \notin S_n)) \\ \wedge \\ (x \in S_2 \rightarrow (x \notin S_3 \wedge \cdots \wedge x \notin S_n)) \\ \wedge \cdots \wedge \\ (x \in S_{n-1} \rightarrow x \notin S_n) \\ \wedge \\ (x \in S_1 \vee \cdots \vee x \in S_n) \end{pmatrix} \right)$$

We chose both simple 2-ary constraints as well as a more complex n-ary constraint in order to compare the overhead versus the gain for *memo*. Intuitively,

Algorithm 3. A simple hill climber for evaluating *memo*

```
function HillClimber(V, C)
    k ← a random configuration for V
    while penalty(C)(k) > 0 do
        choose a possible move ℓ of the form add(S, u)(k) or drop(S, u)(k)
        minimising penalty(C)(ℓ) for
            k ← ℓ
    return k
```

the gain should be greater for more complex constraints (that is longer ∃MSO formulas), since each saved recalculation would have been more costly for such constraints.

The local search algorithm used for the experiments is a very simple hill climber, shown in Algorithm 3. After initialising the variables **V** to a random configuration, the hill climber greedily chooses an *add* or a *drop* move minimising the penalty of all constraints **C** as the next configuration. If a configuration satisfying all constraints is found (that is, if the penalty of all constraints is zero), this solution is returned. This hill climber serves our purposes simply since none of the problems *subset* or *partition* are particularly hard. This is however irrelevant as we are here *only* interested in measuring the *speed* of a memoisation-based penalty maintenance algorithm (that is *memo*) and comparing this speed with the speed of another incremental penalty maintenance algorithm (that is *nomemo*). *Solving open instances of hard problems or making comparisons with other solving approaches are thus not purposes of this paper.*

We ran the instances where $n = |\mathbf{U}| \in \{20, 25, 30, 35, 40, 45, 50, 55\}$ for both problems and the results are shown in Table 1. For each of the problems *subset* and *partition* and with respect to a given instance n, the columns labelled *memo* and *nomemo* indicate the number of iterations per second (in Algorithm 3, higher values are better) achieved by using the respective penalty maintenance

Table 1. Comparing *memo* with *nomemo* when solving the problems *subset* and *partition* for the instances in the column labelled n. For each problem, the columns labelled *memo* and *nomemo* indicate the number of iterations per second achieved by using the respective algorithms, and the column labelled *speedup* indicates the speedup of running *memo* compared with *nomemo*. All values are averages over ten runs.

	subset			*partition*		
n	*memo*	*nomemo*	*speedup*	*memo*	*nomemo*	*speedup*
20	665.7	893.4	0.7	1017.4	193.5	5.3
25	421.8	572.1	0.7	432.8	93.2	4.6
30	292.1	389.3	0.8	332.6	50.5	6.6
35	214.4	285.0	0.8	248.5	29.3	8.5
40	163.5	217.9	0.8	164.2	18.4	8.9
45	129.4	171.1	0.8	112.9	12.3	9.2
50	104.6	133.9	0.8	73.6	8.4	8.8
55	86.1	111.5	0.8	47.4	5.9	8.0

algorithms. The column labelled *speedup* indicates the speedup of running *memo* compared with *nomemo*. All values are averages over ten runs. The same random seeds were used when comparing the two different algorithms. Hence, the number of iterations (not reported here) as well as the solutions were the same for the two different algorithms.

On the one hand, using *memo* is slower for *subset* on all instances, although by a small (and seemingly constant) factor. On the other hand, using *memo* is significantly faster for *partition* on all instances. As suspected above, the overhead can be larger than the gain for simple 2-ary constraints since the cost for (re)calculating the penalty (or incrementally updating the same) is small for such constraints. This is not the case for more complex n-ary constraints, which is why the gain can be larger than the overhead for such constraints. This clearly shows the usefulness of memoisation-based penalty maintenance algorithms for local search with ∃MSO.

4 Conclusion

We have applied memoisation to the calculation of penalties for ∃MSO constraints. Our approach is based on identifying interchangeable elements in the first-order quantifications of the ∃MSO constraints, and characterising these elements by their signatures. Such interchangeable elements share penalties and need only be calculated and cached once, thereby lowering the number of necessary calculations as well as the number of cached penalties. Our results show that this can lead to a significant speedup when using ∃MSO constraints in local search.

Acknowledgements. I thank Pierre Flener and Justin Pearson for discussions and comments, as well as the anonymous referees for constructive reviews.

References

1. Michie, D.: Memo functions: a language feature with "rote-learning" properties. Research Memorandum MIP-R-29. Edinburgh: Department of Machine Intelligence & Perception (1967)
2. Van Hentenryck, P., Michel, L.: Constraint-Based Local Search. MIT Press, Cambridge (2005)
3. Van Hentenryck, P., Michel, L.: Differentiable invariants. In: Benhamou, F. (ed.) CP 2006. LNCS, vol. 4204, pp. 604–619. Springer, Heidelberg (2006)
4. Ågren, M., Flener, P., Pearson, J.: Generic incremental algorithms for local search. Constraints 12(3), 293–324 (2007); Collects the results of papers at CP-AI-OR 2005, CP 2005, and CP 2006
5. Freuder, E.C.: Eliminating interchangeable values in constraint satisfaction problems. In: Proceedings of AAAI 1991, pp. 227–233 (1991)
6. Sellmann, M., Van Hentenryck, P.: Structural symmetry breaking. In: Proceedings of IJCAI 2005, Professional Book Center, pp. 298–303 (2005)
7. Ågren, M.: Set Constraints for Local Search. PhD thesis, Uppsala University (2007)

On the Structure of Industrial SAT Instances*

Carlos Ansótegui[1], María Luisa Bonet[2], and Jordi Levy[3]

[1] Universitat de Lleida (DIEI, UdL)
[2] Universitat Politècnica de Catalunya (LSI, UPC)
[3] Artificial Intelligence Research Institute (IIIA, CSIC)

Abstract. During this decade, it has been observed that many real-world graphs, like the web and some social and metabolic networks, have a *scale-free* structure. These graphs are characterized by a big variability in the arity of nodes, that seems to follow a power-law distribution. This came as a big surprise to researchers steeped in the tradition of classical random networks.

SAT instances can also be seen as (bi-partite) graphs. In this paper we study many families of industrial SAT instances used in SAT competitions, and show that most of them also present this scale-free structure. On the contrary, random SAT instances, viewed as graphs, are closer to the classical random graph model, where arity of nodes follows a Poisson distribution with small variability. This would explain their distinct nature.

We also analyze what happens when we instantiate a fraction of the variables, at random or using some heuristics, and how the scale-free structure is modified by these instantiations. Finally, we study how the structure is modified during the execution of a SAT solver, concluding that the scale-free structure is preserved.

1 Introduction

The Satisfiability problem (SAT) is central in Computer Science. It was the first problem to be proven NP-Complete, and it is used extensively to encode many other problems into it. Therefore, finding good algorithms to solve SAT is of practical use in many areas of Computer Science. Even though the general SAT problem is NP-Complete, many very large industrial instances can be solved efficiently by modern solvers. The aim of this work is to study the body of industrial instances to detect general properties that are shared by the majority of instances. We focus on the structure of the instances viewed as bi-partite graphs, where nodes represent variables and clauses, and edges represent the presence of a variable in a clause. In particular, we try to detect the distribution on the frequencies of the variables and of the sizes of the clauses, in SAT instances used in the latest SAT Competitions and SAT Races. Our work was inspired by [BDIS05], where they suggest that industrial instances, as many other real-world graphs could have a scale-free structure.

* Research partially supported by the projects TIN2007-68005-C04-{01,03,04} and TIN2006-15662-C02-02 funded by the MEC.

I.P. Gent (Ed.): CP 2009, LNCS 5732, pp. 127–141, 2009.

The classical random graph model [ER59] was one of the best studied during the last century, and set the basis of graph theory. In [WS98], a new model of random graphs is proposed, called *small-world* to describe the structure of some social collectivities. In [AJB99], they show that the world wide web, viewed as a graph, has a structure than cannot be described by the classical random graph model. They propose a new model called *scale-free*. The name comes from the fact that, in this new model, the arity of nodes follows a power-law distribution $p(k) \sim k^{-\alpha}$, and these distributions are scale-free. However, the name also suggests that these graphs present some kind of self-similarity. In recent years it has been observed that many other real-world graphs, like some social and metabolic networks, also have a scale-free structure.

Power-law (zeta and Pareto) distributions are characterized by a big variability, consequence of a polynomially decreasing tail. A small fraction of the individuals is responsible for most of the average, in what is popularly known as the *80:20 rule* (i.e. 80% of the land is owned by the 20% of the population). Many other heterogeneous distribution are also called power-law or *heavy-tailed* when their tail decreases polynomially, in contrast with other classical distributions, like normal, Poisson, or binomial that have a exponentially decreasing tail. Experience tells us that power-law distributions are as frequent in nature, if not more frequent, as exponentially decreasing distributions. For instance, the CPU time of the different executions (with different random variable selection) of a solver on a formula follow a power-law distribution [GFSB04].

The topology of graphs have a major impact on the cost of solving search problems on these graphs. Gent et al. [GHPW99] analyze the impact of a small-world topology on the cost of coloring graphs, and Walsh [Wal01] does the same in the case of scale-free graphs. Therefore, we can expect that SAT solving, viewed as a search process of on a graph (the formula), will be affected by the topology of this graph.

It is well-known in the SAT community that classical random k-CNF formulas and industrial (or real-world) formulas have a distinct nature. This makes SAT solvers to specialize in one or the other kind of formulas. In the SAT competition there is a special track for each kind of formulas, whereas in the SAT Race competition, only industrial formulas are used to test the solvers. Random k-CNF formulas, as graphs, follow the Erdös-Rényi model. In the phase transition point for $k = 3$, for instance, most of the variables have a number of occurrences very close to 12.75.[1] In this paper, we show that most industrial instances are better modeled as scale-free graphs.

We think that the present study provides a step towards a theoretical explanation of why some SAT solvers perform better on industrial instances, and others on random SAT instances. Moreover, the better understanding of real-world instances could lead to the improvement of existing SAT solvers.

The paper can also serve as basis for new random SAT generation models that produce instances closer to real-world ones. This problem is distinguished as one

[1] The number 12.75 comes from multiplying the size of the clauses $k = 3$ by the clause/variable ratio $m/n = 4.25$ at the phase transition point.

of the 10 challenge problems in SAT [SKM97, Sel00, KS03, KS07]. Recently, in [ABL09], we have proposed some random SAT instance generators that produce formulas with variable frequencies following a power-law distribution. We show that solvers specialized on industrial instances perform better in these *random industrial-like instances* than solvers specialized on random formulas.

Another application of the study could be to evaluate which is the best family of solvers to use on a particular instance, by analyzing the distribution of variable frequencies or clause sizes. In particular, this could be use as one more selection criteria in a portfolio approach [XHHLB08].

The paper proceeds as follows. In Section 2, we present the study of the distributions that best represent the frequencies of variable occurrences and clause sizes. Also we describe the statistical techniques we use in our work. In Section 3, we study whether the scale-free nature is preserved under partial instantiations of variables. In Section 4, we analyze the structure of the formulas during the execution of complete SAT solver of different nature. We conclude in Section 5.

2 Analysis of Industrial SAT Instances

2.1 Methodological Background

Every SAT instance can be seen as a bi-partite graph, with a set of nodes $V \cup C$, where V represents the variables and C represents the clauses. The edges are the pairs $(v, c) \in V \times C$ such that variable v appears in clause c. In what follows, $n = |V|$ and $m = |C|$. In order to analyze if a bi-partite graph is scale-free, we have to study the arity of the nodes. Notice that the arity of a node $v \in V$ is the number of occurrences of the variable v, and the arity of $c \in C$ is the size of the clause c.

For every bi-partite graph we can compute $f_v^{real}(k)$ as the number of variables that have a number of occurrences equal to k, divided by n, and similarly, $f_c^{real}(k)$ is the number of clauses of size k divided by m. We add the label *real* to emphasize that these functions come from empirical data. We can also define the accumulative versions of these functions as $F_v^{real}(k) = \sum_{i \geq k} f_v^{real}(i)$ and $F_c^{real}(k) = \sum_{i \geq k} f_c^{real}(i)$. Notice that, assuming that there are no empty clauses and all variables occur somewhere, $F_v^{real}(1) = F_c^{real}(1) = 1$.

In the scale-free model, the arity of nodes is characterized by a random variable K that follows a power-law distribution $f^{pow}(k) = P(K{=}k) = c\,k^{-\alpha}$. The exponent α has typically values inside $[2, 3]$. This distribution diverges at zero, and there is a lower bound k_{min} for the values of k from where we get the power-law behavior or *heavy tail*. In the discrete case (the one that concerns us), the normalizing constant is $c = 1/\zeta(\alpha, k_{min}) = 1/\sum_{i=0}^{\infty}(i + k_{min})^{-\alpha}$, where ζ is the Hurwitz zeta function. For big values of k_{min} we can approximate this distribution using the continuous version. In this case the probability density function is $f^{pow}(k) = \frac{\alpha-1}{k_{min}} \left(\frac{k}{k_{min}}\right)^{-\alpha}$, and the cumulative function is $F^{pow}(k) = \left(\frac{k}{k_{min}}\right)^{-\alpha+1}$.

There is not a proper (formal) definition of what a scale-free graph is, but one of their basic properties –usually taken as a definition– is that the arity of nodes *seems* to follow a power-law distribution. Therefore, we must check if, for some values of α_v and α_c, we have $f_v^{real}(k) \approx c\,k^{-\alpha_v}$ and $f_c^{real}(k) \approx c\,k^{-\alpha_c}$. Notice that, applying logarithms to both sides, we get $\log f(k) = \log c - \alpha \log k$. Therefore, if $f_v^{real}(k)$ and $f_c^{real}(k)$ are power-law, representing them as a function of k with double-logarithmic axes, we should get closed to a straight line with slope $-\alpha$.

In some papers, the value α is calculated by linear regression of $\log f(k)$ as a function of $\log k$. In [LADW05, section 2.1.3] there is a discussion of why it is better to plot the cumulative logarithm $\log F(k)$, instead of $\log f(k)$, to compute the regression. But, in this case, the slope is $-\alpha + 1$. Following this argument, in Figure 1 we represent $F_v(k)$ and $F_c(k)$ versus k with double-logarithmic axes, for some families of industrial formulas.

We will follow the maximum likelihood method for computing an estimation of α, as described in [CSN07]. To estimate the value of α for a collection of empirical data $k_1, \dots, k_n$, we compute the value of α that maximizes the probability that the data were drawn from the model:

$$P(k_1, \dots, k_n \mid \alpha) = \prod_{i=1}^{n} \frac{\alpha - 1}{k_{min}} \left(\frac{k_i}{k_{min}} \right)^{-\alpha}$$

We take logarithms, since the maximum will be in the same place, then we take derivatives and make the function equal to zero:

$$\begin{aligned} &\frac{\partial}{\partial\alpha} \log P(x_1, \dots, x_n \mid \alpha) = \\ &\quad = \frac{\partial}{\partial\alpha} \left(n \log \frac{\alpha - 1}{k_{min}} - \alpha \sum_{i=1}^{n} \log \frac{k_i}{k_{min}} \right) = \\ &\quad = \frac{n}{\alpha - 1} - \sum_{i=1}^{n} \log \frac{k_i}{k_{min}} = 0 \end{aligned}$$

we get

$$\hat{\alpha} = 1 + \frac{n}{\sum_{i=1}^{n} \log(k_i / k_{min})}$$

For the discrete case, a good approximation for big values of k_{min} is

$$\hat{\alpha} = 1 + \frac{n}{\sum_{i=1}^{n} \log \frac{k_i}{k_{min} - 1/2}}$$

Notice that the estimated α depends on k_{min}. To compute the value of $\hat{k}_{min}$, we try to minimize the distance between the (experimental) cumulative distribution function $F^{real}(x)$ and the (theoretical) cumulative distribution function $F^{pow}(x; \alpha, k_{min})$. The distance between both distributions is calculated as the maximal difference between both functions. Then, we compute the value of k_{min} that minimizes this distance:

$$d = \min_{k_{min} \geq 1} \left\{ \max_{k \geq k_{min}} \left\{ \left| \frac{F^{real}(k)}{F^{real}(k_{min})} - F^{pow}(k; \hat{\alpha}, k_{min}) \right| \right\} \right\}$$

We get so the value of k_{min} and of d. The value of this distance d is an indicator of the fitness of the estimation.

When we say that arity of nodes *seems* to follow a power-law distribution, we emphasize the *seems* because it is obvious that SAT formulas, as well as the WWW and other scale-free graphs, are not randomly generated. Therefore, we do not expect the arity of nodes to follow exactly any distribution. However, we want to check if some distribution fits the data better than others. In particular, we have tried to fit, apart from a power-law distribution, an exponential distribution.

The probability density function for an exponential distribution has the form $c\,e^{-\beta\,x}$. Calculating the constant, for the discrete case, we get $f^{exp}(k;\beta,k_{min}) = (1-e^{-\beta})\,e^{-\beta\,(k-k_{min})}$ and its cumulative function $F^{exp}(k) = e^{-\beta(k-k_{min})}$. In this case the estimation of the β parameter by the method of maximum likelihood gives:

$$\begin{aligned}\frac{\partial}{\partial\beta}\log P(k_1,\dots,k_n\,|\,\beta) =& \\ &= \frac{\partial}{\partial\beta}\left(n\,\log(1-e^{-\beta}) - \beta\sum_{i=1}^{n}(k_i-k_{min})\right) = \\ &= \frac{n\,e^{-\beta}}{1-e^{-\beta}} - \sum_{i=1}^{n}(k_i-k_{min}) = 0\end{aligned}$$

Hence,

$$\hat{\beta} = \log\left(\frac{n}{\sum_{i=1}^{n}(k_i-k_{min})}+1\right)$$

The value of k_{min} is calculated as in the case of the power-law distribution.

For distinct families of industrial formulas, we have calculated $f_v^{real}(k)$and $f_c^{real}(k)$, as well as their cumulative functions. First, we have studied instances independently in each family, observing that they all have the same nature. Thus, we decide to group them by families, assuming that all formulas of the same family follow the same probability distribution. Therefore, for a family, $f_v^{real}(k)$ is the sum for every formula of the number of variables that have k occurrences, and similarly for $f_c^{real}(k)$. Notice that, under this assumption, the arity of a variable, independently of in which formula of the family it occurs, is an independent realization of the same random variable. Therefore, we can do this addition. Later, we have fitted a power-law distribution and an exponential distribution, and we have calculated the distance d^{pow} between $F_v^{real}(k)$ and the estimated $F_v^{pow}(k;\alpha,k_{min})$, and the distance d^{exp} between $F_v^{real}(k)$ and the estimated $F_v^{exp}(k;\beta,k_{min})$. When $d^{pow} < d^{exp}$, we say that the power-law distribution fits better than the exponential distribution. We use this criteria to state that a family of formulas has a scale-free structure. It is also important to compare the value of k_{min} obtained in each estimation, noted k_{min}^{pow} and k_{min}^{exp}. A big value of k_{min} means that we need to discard a lot of values of $F^{real}(k)$ to fit the distribution, and it must be taken as a point against the fitted distribution. Also a value of α far away from the interval $[2,3]$ must be read as a point against the scale-free structure.

Table 1. Most likelihood values of α and β estimated for a power-low and an exponential distribution. In bold we remark the smallest distance between the real and the fitted distributions. We also report the total number of variable occurrences n, mean $E[V]$ and variance $Var[V]$, and the respective values for clause sizes.

Variables (V)										
					Power-law			Exponential		
Family	#inst	n	$E[V]$	$Var[V]$	α	k_{min}^{pow}	d^{pow}	β	k_{min}^{exp}	d^{exp}
cmu	3	16678	7.95	12.11	3.49	5	**0.072**	0.224	4	0.176
een	12	739744	7.60	13.26	2.67	10	**0.043**	0.054	15	0.136
fuhs	2	73486	9.05	14.56	2.79	62	**0.075**	0.181	4	0.158
goldb	11	114038	21.02	88.71	2.05	21	**0.042**	0.003	100	0.204
grieu	9	6914	364.21	42.23	1.77	100	0.577	0.004	100	0.538
ibm	38	4985723	10.75	23.97	2.63	7	**0.027**	0.017	45	0.083
manol	59	7827736	6.93	16.24	2.95	57	0.059	0.017	76	**0.033**
mizh	13	725644	12.49	148.12	4.09	15	0.172	0.034	22	0.247
narai	6	9642548	9.72	16.38	3.85	5	0.152	0.109	1	0.347
palac	2	298266	10.82	60.55	1.84	20	0.087	0.003	100	0.074
post	10	12906872	7.90	44.15	2.57	12	0.132	0.135	1	0.334
schup	7	2196731	8.09	12.40	2.59	41	0.120	0.063	9	0.182
simon	12	798804	7.78	11.96	2.53	14	**0.028**	0.022	50	0.065
uts	10	1420464	13.01	74.70	1.76	69	0.111	0.003	75	**0.088**
velev	60	8442829	88.31	379.04	1.82	13	**0.030**	0.003	87	0.287
random	40	400000	12.75	3.57	18.65	24	0.019	0.777	25	**0.008**
SAT'08	100	27964721	13.30	113.48	2.29	12	**0.051**	0.003	73	0.254

Clauses (C)										
					Power-law			Exponential		
Family	#inst	m	$E[C]$	$Var[C]$	α	k_{min}^{pow}	d^{pow}	β	k_{min}^{exp}	d^{exp}
cmu	3	53769	2.46	1.21	5.35	3	0.126	1.778	3	**0.048**
een	12	2278059	2.47	0.69	3.80	4	**0.044**	2.420	3	0.046
fuhs	2	256742	2.59	0.82	4.89	5	0.041	2.182	3	**0.020**
goldb	11	710559	3.37	1.46	10.48	5	0.158	4.803	5	**0.008**
grieu	9	961030	2.62	0.76	8.54	26	0.108	3.878	3	**0.020**
ibm	38	21084555	2.54	1.57	3.77	6	**0.023**	0.375	4	0.032
manol	59	23244626	2.33	0.47						
mizh	13	3036234	2.98	0.91	1.58	1	0.328	0.408	1	0.334
narai	6	37639556	2.49	2.05	3.33	2	**0.088**	1.113	2	0.090
palac	2	1274356	2.53	9.33	1.71	4	0.116	1.055	2	0.116
post	10	42441234	2.40	1.39	3.33	2	0.143	2.884	33	**0.053**
schup	7	6947242	2.56	1.36	4.30	4	0.093	2.585	3	**0.046**
simon	12	2675233	2.32	0.90	3.76	4	0.033	0.498	5	**0.026**
uts	10	7101806	2.60	11.56	3.63	2	0.114	0.004	35	0.116
velev	60	253221473	2.94	9.01	3.35	72	0.042	0.021	28	**0.040**
random	40	1700000	3.00	0.00						
SAT'08	100	140942860	2.64	5.68	3.03	17	**0.054**	0.074	10	0.068

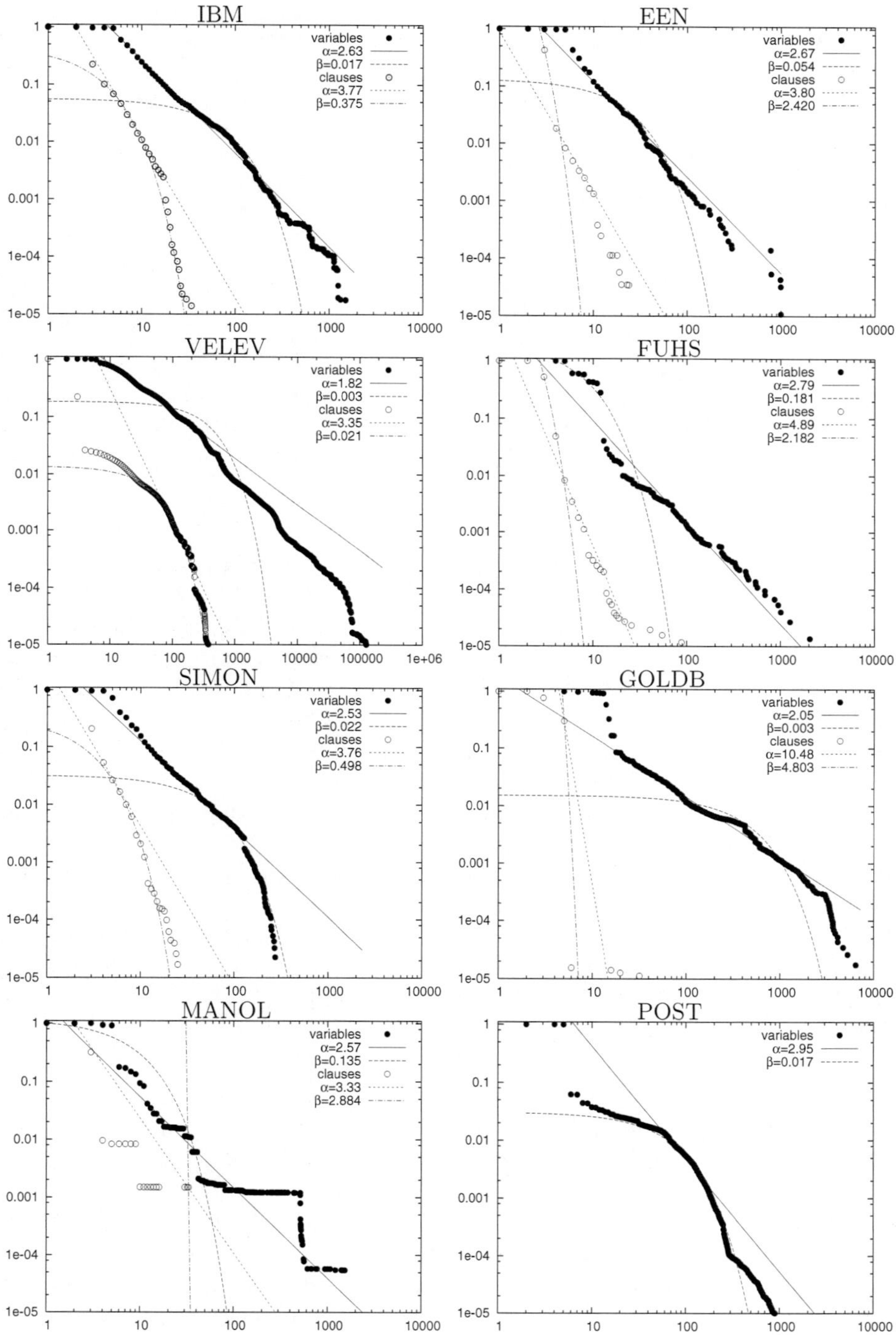

Fig. 1. Plotting of $F_v^{real}(k)$ and $F_c^{real}(k)$, and their respective power-law (characterized by α) and exponential (characterized by β) estimations, for some families of formulas. In families where all clauses are small, we have avoided the representation of $F_c^{real}(k)$.

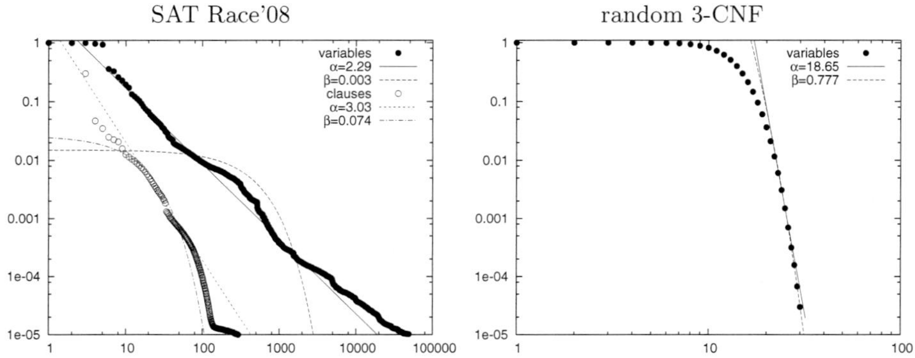

Fig. 2. Plotting of $F_v^{real}(k)$ and $F_c^{real}(k)$, and their respective power-law and exponential estimation, for the formulas of the SAT'08 Race and random 3CNF

2.2 Results of the Analysis

We have selected a set of families of formulas from the industrial category of the 2002–2005 and 2007 SAT Competitions, and the 2006 and 2008 SAT Races. For these families, Table 1 presents the estimations of the parameters of the distributions power-law and exponential for variables occurrences and clause sizes. We have also extended the study to a family of 40 random 3-CNF instances of 10^4 variables in the phase transition point; and to the *heterogeneous* family composed by the 100 instances used in the latest SAT Race 2008 competition. In Table 1 we also include information about the sum of the number of variables and clauses of all formulas of the family, and the average number of occurrences of variables and sizes of clauses, as well as their variance. For the computation of k_{min} (the value where the data starts to fit the distribution) we impose a limit value of 100. We consider that, if the distance d between the observed data and the distribution is smaller than 0.1, then it is plausible that the data follows that distribution. To conclude that the family follows a power-law distribution we also require that $d^{pow} < d^{exp}$ and the value of k_{min} to be small. For the families where all clauses have size at most 3, we obviate the study for the distribution of clause size.

In Figure 1 we plot the distributions of some families, as well as the estimated power-law and exponential distributions that best fit them. In Figure 2 we also plot the distributions for the heterogeneous family of the SAT Race 2008, and the random 3-CNF formulas.

We can conclude that for the families: CMU, EEN, FUHS, GOLDB, IBM, SIMON and VELEV, the number of variable occurrences follow a power-law distribution. In the case of clause size, only the families EEN, IBM and NARAI seem to follow a power-law distribution. Therefore, in general, the variable occurrences follows a power-law distribution in more families than the clause size. The value of α for variables is also smaller than the α for clauses, that tends to fall out of the interval [2, 3]. We think that the explanation for this phenomena is that, when the formulas are encoded, people try to avoid the use of very big clauses, since they weak the propagation power in SAT solvers. We also observe that some families, like MANOL, do not seem to follow a particular distribution.

In the random 3-CNF formulas, the exponential distribution fits better than the power-law, although the distance d^{pow} is surprisingly small. If we plot the distribution for each formula of the family, we see that it is very homogeneous, without the typical peeks that we find in industrial data. Moreover, the value of $\alpha = 18.65$ is big enough to discard a power-law distribution.

Looking at the plot of the SAT Race'08 heterogeneous family, we see that the data fits better the power-law distribution than other homogeneous families. In this case, we have to take into account that the addition of so many instances, by a kind of law of the big numbers, tends to make distributions smoother. The values of α that we get are $\alpha = 2.29$ for variable occurrence and $\alpha = 3.03$ for clause size. As in some homogeneous families, we observe that the value of α in the case of clause size is bigger than the value of α for variable occurrence, and falls in the limit of the interval [2, 3].

3 Instantiating Variables in Industrial Instances

Albert, Jeong & Barabási [AJB00] studied the effect of *failure* and *attack* actions in the diameter of an Erdös-Rényi graph and of a scale-free graph. The diameter is the average minimum distance between two nodes, failure consists in removing a certain percentage of randomly selected nodes, and attack consists in removing the nodes following a certain heuristic (e.g. those nodes with higher arity). They observed that failure and attack have the same effect on Erdös-Rényi graphs (after removing 5% of the nodes, the diameter increases in the same way independently of how nodes are chosen). However, while failure almost does not change the diameter of scale-free graphs, attack increases the diameter even more than in the case of an Erdös-Rényi graphs. Considering that Internet is a scale-free graph, they conclude that it is robust against random failures of the servers, but it is specially susceptible to *terrorism* attacks.

In the case of SAT solvers, the instantiation of variables removes nodes in the bi-partite graph representing the formula (e.g. the instantiation $v = true$ removes the variable-node v, and all those clause-nodes c, where c contains the literal v). Since classical random SAT instances are similar to Erdös-Rényi graphs, we can expect the same behavior on random formulas, when we instantiate variables randomly, as when we use some heuristics. However, in scale-free industrial instances, we expect a very different effect.

We have conducted a series of experiments where we instantiate up to 10% of the variables of some families of formulas, and we analyze the decrease in size of the formula. Notice that we only instantiate variables, i.e., we do not apply any local inference like unit propagation, and we do not discard the obtained subformula, even if it contains the empty clause. In Section 4 we perform a similar experiment using real SAT solvers.

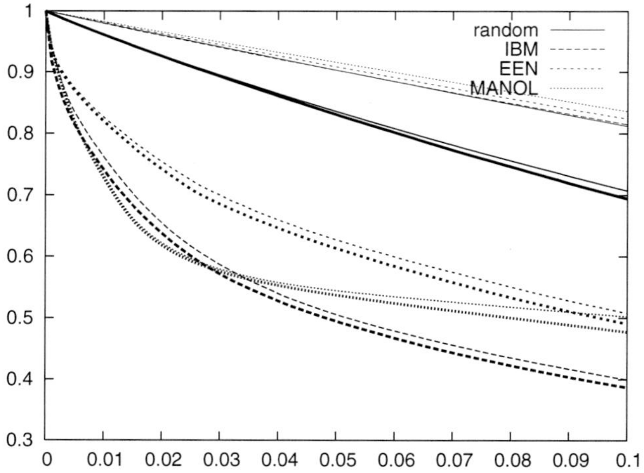

Fig. 3. Percentage of the formula-size decrease as a function of the percentage of instantiated variables. For the 3 lines of each family, the upper one corresponds to the random strategy, the middle to the Jeroslow-Wang, and the lower to the most-frequent strategy.

We experiment with the IBM and the EEN families –the ones with a more clear scale-free structure–, with the MANOL family –that does not seem to follow a neat distribution–, and with the random 3CNF set –that we know have an absolutely different structure–. Apart from the random selection of variables, we have analyzed the use of the most-frequent variable[2] and of the Jeroslow-Wang heuristics [JW90]. Results are shown in Figure 3. We observe that instantiating randomly selected variables has the same effect in all families: after instantiating 10% of the variables, the size of the formula decreases between 16% and 19%. The size-decrement seems to be proportional to the percentage of instantiated variables, i.e. the slope seems to be constant and the same in all families.

For the other two heuristics (most-frequent variable and Jeroslow-Wang), the size-decrease in random formulas is bigger, but not so much as in the industrial formulas: in random formulas, after instantiating 10% of the variables, the decrease is around 30%, whereas in industrial formulas the decrease is around 50%. Moreover, the size-decrease seems to be constant in the case of random formulas, whereas in industrial formulas, the use of these heuristics speeds up the size-decrease, but at a certain point, when we have instantiated around 1% or 2% of the variables, the slope decreases substantially. Both heuristics seem to have the same effect, although the most-frequent heuristic is always a little better (bigger decrease) than the Jeroslow-Wang heuristic.

The natural question is now: after instantiating a significant part of the variables, is the formula still scale-free? We have studied the formulas that we get after instantiating some variables of the IBM formulas following the three

[2] The most-frequent heuristic consists in selecting the variable with a higher number of occurrences, and the polarity with it appears more times.

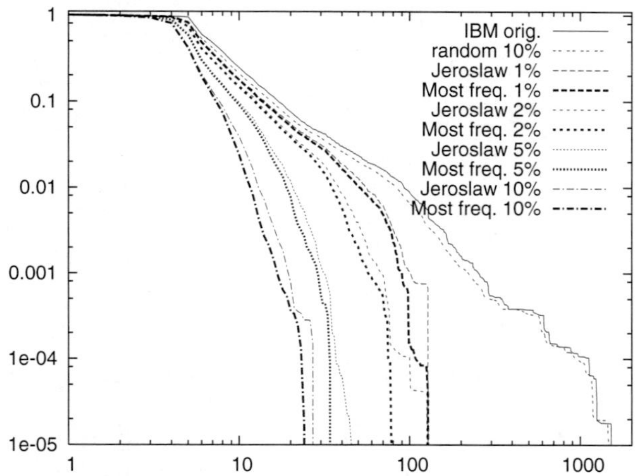

Fig. 4. Function $F_v(k)$ for IBM formulas where 1%, 2%, 5% and 10% of the variables have been instantiated using the random, Jeroslow-Wang and most-frequent strategies

Table 2. Analysis of the partially instantiated IBM formulas

	random				Jeroslow-Wang				Most freq.			
	Power-law		Exponential		Power-law		Exponential		Power-law		Exponential	
	α	d^{pow}	β	d^{exp}	α	d^{pow}	β	d^{exp}	α	d^{pow}	β	d^{exp}
0%	2.63	**0.027**	0.017	0.083	2.63	**0.027**	0.017	0.083	2.63	**0.027**	0.017	0.083
1%	2.56	**0.027**	0.017	0.078	2.72	**0.017**	0.046	0.036	2.79	**0.020**	0.052	0.034
2%	2.57	**0.025**	0.017	0.076	2.82	**0.015**	0.083	0.026	2.89	**0.012**	0.093	0.030
5%	2.59	**0.020**	0.018	0.075	3.27	0.029	0.218	**0.019**	3.39	0.029	0.250	**0.014**
10%	2.62	**0.021**	0.019	0.077	5.79	0.023	0.407	**0.014**	5.90	0.023	0.510	**0.021**

heuristics. Results are shown in Figure 4. As we can see, the random instantiation of variables has almost no effect on the probability distribution of variable occurrences $f_v(k)$. However, heuristics tend to remove variables with high number of occurrences. As a consequence, after partially instantiating around 5% of the variables, the formula looses its scale-free property, and seems to follow an exponential distribution (see Table 2).

4 Formulas during SAT Solvers Search

We want to answer the question of what kind of formula a state-of-the-art SAT solver sees during the search. The question is important because, if we implement solvers specialized in industrial instances (assuming that they are scale-free) during the execution of the solver, when some variables are already instantiated, we can be dealing with a not scale-free formula anymore. This means that, when

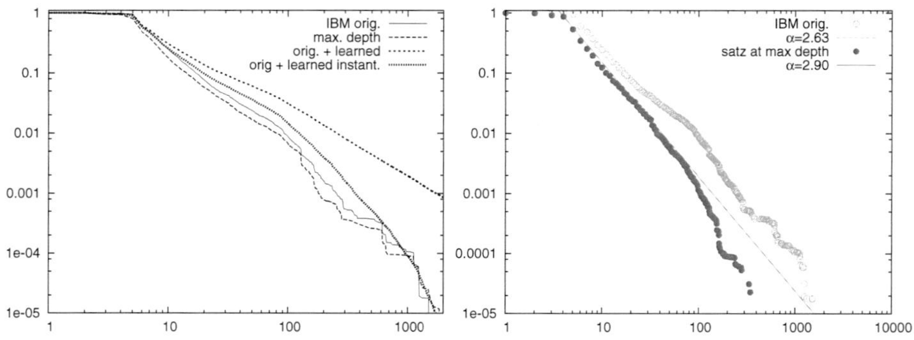

Fig. 5. Study of IBM formulas during the search. Left: with minisat, right: with satz.

a significant part of the variables are instantiated, the solver would do better changing its strategy.

Any complete SAT solver will backtrack immediately once it checks the current partial assignment is not consistent (in contrast to the setting in section 3), and second, state-of-the-art SAT solvers specialized on industrial instances augment the formula during search by adding new clauses, due to the learning mechanism they incorporate.

We have modified two very different SAT solvers, minisat [ES03]and satz [LA97]. Apart from the different heuristics and data structures these solvers incorporate, minisat applies a learning mechanism while satz does not.

We conducted experiments to answer the previous stated question, executing the solvers on each instance of the IBM family. We selected this family as the representative of scale-free formulas and random formulas as non scale-free formulas. The results reflect the average behavior of the family.

First, we study the formula under the longest partial assignment after 1000 seconds of search. Second we study, both the formula under the current partial assignment and the complete formula (original formula plus learned clauses) after 200000 decisions.

Figure 5 (left), shows the results of our experimentation on the IBM instances with minisat. As we can see, the scale-free structure is preserved in all cases. At maximal depth the distribution of frequencies is almost the same as in the original formulas. This seems to contradict the effect of partial assignments described in previous section but we have to remark that here the partial assignment is consistent. Moreover, it seems that the effect of the learned clauses makes the α exponent decrease.

We have repeated the same experiment with the same IBM formulas but after at most one hour of execution time of satz. Recall that here apart from applying a different heuristic we have not learned clauses. In Figure 5 (right), we can see that at the deepest assignments the formulas are still scale-free, although the exponent has been increased.

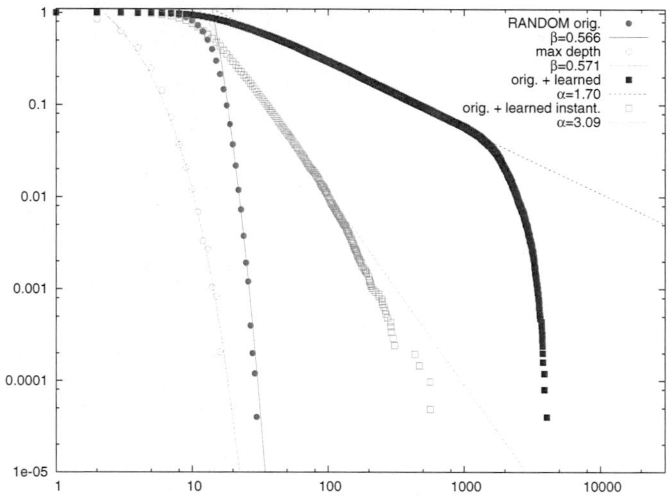

Fig. 6. Study of random formulas during the search

Therefore, very different SAT solvers seem to preserve the scale-free nature of formulas during their execution. Now the question is, what happens if we start with a random formula? For our experiment we have generated 50 random 3-CNF formulas of 500 variables at the phase transition point. Figure 6 shows the results. At the deepest decisions, after 1000 seconds, we see that the formulas still show an exponential decay with the same β as in the original formulas. However, after $2 \cdot 10^6$ decisions the formulas show a clear scale-free structure due to the addition of the learned clauses. As in the first experiment with the IBM family, the exponent α is smaller for the uninstantiated formula. To explain this phenomenon recall that the solvers like minisat, decide on the most active variables in learned clauses and learn clauses that contain decided variables. This creates an effect of *rich get richer* that has been proposed as a mechanism for creation of scale-free networks [BA99].

5 Conclusions

We have shown that most of the industrial formulas have a scale-free structure whereas random formulas have an Erdös-Rényi graph structure. This difference makes heuristics to perform better in industrial formulas than in random formulas.

We have observed that heuristically guided partial assignments (without guaranteeing consistency) make frequency distributions decay faster, destroying the power-law tail after instantiating 5% of the variables. However, if the assignments are consistent, as during the search in a SAT solver, we can instantiate up to 70% variables preserving the power-law tail (although increasing the exponent).

Finally, we have observed that the learning mechanism incorporated in modern SAT solvers tends to preserve the power-law distribution and even decrease its exponent.

References

[ABL09] Ansótegui, C., Bonet, M.L., Levy, J.: Towards industrial-like random SAT instances. In: Proc. of the 21st Int. Joint Conf. on Artificial Intelligence, IJCAI 2009 (2009)

[AJB99] Albert, R., Jeong, H., Barabási, A.-L.: The diameter of the www. Nature 401, 130–131 (1999)

[AJB00] Albert, R., Jeong, H., Barabási, A.-L.: Error and attack tolerance of complex networks. Nature 406, 378–482 (2000)

[BA99] Barabási, A.-L., Albert, R.: Emergence of scaling in random networks. Science 286, 509–512 (1999)

[BDIS05] Boufkhad, Y., Dubois, O., Interian, Y., Selman, B.: Regular random k-sat: Properties of balanced formulas. J. of Automated Reasoning 35, 181–200 (2005)

[CSN07] Clauset, A., Shalizi, C.R., Newman, M.E.J.: Power-law distributions in empirical data. Arxiv, 0706.1062 (2007)

[ER59] Erdős, P., Rényi, A.: On random graphs. Publicationes Mathematicae 6, 290–297 (1959)

[ES03] Eén, N., Sörensson, N.: An extensible SAT-solver. In: Giunchiglia, E., Tacchella, A. (eds.) SAT 2003. LNCS, vol. 2919, pp. 502–518. Springer, Heidelberg (2004)

[GFSB04] Gomes, C.P., Fernández, C., Selman, B., Bessière, C.: Statistical regimes across constrainedness regions. In: Wallace, M. (ed.) CP 2004. LNCS, vol. 3258, pp. 32–46. Springer, Heidelberg (2004)

[GHPW99] Gent, I.P., Hoos, H.H., Prosser, P., Walsh, T.: Morphing: Combining structure and randomness. In: Proc. of the 16th Nat. Conf. on Artificial Intelligence, AAAI 1999, pp. 654–660 (1999)

[JW90] Jeroslow, R.G., Wang, J.: Solving propositional satisfiability problems. Annals of Mathematics and Artificial Intelligence 1, 167–187 (1990)

[KS03] Kautz, H.A., Selman, B.: Ten challenges redux: Recent progress in propositional reasoning and search. In: Rossi, F. (ed.) CP 2003. LNCS, vol. 2833, pp. 1–18. Springer, Heidelberg (2003)

[KS07] Kautz, H.A., Selman, B.: The state of SAT. Discrete Applied Mathematics 155(12), 1514–1524 (2007)

[LA97] Li, C.M., Anbulagan: Look-ahead versus look-back for satisfiability problems. In: Smolka, G. (ed.) CP 1997. LNCS, vol. 1330, pp. 341–355. Springer, Heidelberg (1997)

[LADW05] Li, L., Alderson, D., Doyle, J.C., Willinger, W.: Towards a theory of scale-free graphs: Definition, properties, and implications. Internet Mathematics 2(4), 431–523 (2005)

[Sel00] Selman, B.: Satisfiability testing: Recent developments and challenge problems. In: Proc. of the 15th Annual IEEE Symposium on Logic in Computer Science, LICS 2000, p. 178 (2000)

[SKM97] Selman, B., Kautz, H.A., McAllester, D.A.: Ten challenges in propositional reasoning and search. In: Proc. of the 15th Int. Joint Conf. on Artificial Intelligence, IJCAI 1997, pp. 50–54 (1997)

[Wal01] Walsh, T.: Search on high degree graphs. In: Proc. of the 17th Int. Joint Conf. on Artificial Intelligence, IJCAI 2001, pp. 266–274 (2001)

[WS98] Watts, D.J., Strogatz, S.H.: Collective dynamics of 'small-world' networks. Nature 393, 440–442 (1998)

[XHHLB08] Xu, L., Hutter, F., Hoos, H., Leyton-Brown, K.: SATzilla: Portfolio-based algorithm selection for SAT. J. of Artificial Intelligence Research 32, 565–606 (2008)

A Gender-Based Genetic Algorithm for the Automatic Configuration of Algorithms*

Carlos Ansótegui[1], Meinolf Sellmann[2], and Kevin Tierney[2]

[1] Universitat de Lleida, Spain
carlos@diei.udl.cat
[2] Brown University, Department of Computer Science,
P.O. Box 1910, Providence, RI, 02912, USA
{sello,ktierney}@cs.brown.edu

Abstract. A problem that is inherent to the development and efficient use of solvers is that of tuning parameters. The CP community has a long history of addressing this task automatically. We propose a robust, inherently parallel genetic algorithm for the problem of configuring solvers automatically. In order to cope with the high costs of evaluating the fitness of individuals, we introduce a gender separation whereby we apply different selection pressure on both genders. Experimental results on a selection of SAT solvers show significant performance and robustness gains over the current state-of-the-art in automatic algorithm configuration.

1 Introduction

We consider the problem of automatic solver configuration. Practically all solvers have parameters that are partly fixed by the programmer and partly set by the user. In recent years, systems have been devised which automate the task of tuning parameters for a given set of training instances that are assumed to represent typical instances for the target algorithm.

There are several motivations for such an automation, the first being that it is of course time consuming to tune parameters and it may lead to better results when leaving the configuration of solvers to a computer rather than doing it by hand.

Moreover, it is conceivable that the existence of an effective tuning environment will cause algorithm developers to parameterize more aspects of their algorithms and thus leave more freedom for algorithmic solutions that are automatically tailored to the problems of individual users. In particular, many of the SAT solvers that are available today have parameters which cannot be set through the command line. These parameters have been fixed to values that the developers have found beneficial without knowledge about the particular instances a user may want to use the solver for. Automatic parameter tuning allows solvers to adapt to the final environment in which they need to perform. After being shipped, rather than relying on default parameters, an algorithm can be

* This work was partly supported by the projects TIN2007-68005-C04-04 and TIN2006-15662-C02-02 funded by the MEC, and by the the National Science Foundation through the Career: Cornflower Project (award number 0644113).

I.P. Gent (Ed.): CP 2009, LNCS 5732, pp. 142–157, 2009.

tuned automatically for the common tasks it is actually used for, and without requiring the user to learn about the algorithm parameters. For this very reason, Cplex 11 now comes with an automatic performance tuning tool.

Another argument for automatic solver configuration regards our own science: when we re-implement algorithms to conduct experimental comparisons with competing approaches, it is not unreasonable to assume that scientists spend much more time tuning their own algorithm than the algorithms of their competitors. A fair comparison could be achieved if all algorithms were tuned by an independent system. That way, we would come closer to understanding the true potential of algorithmic approaches rather than the ability of solver development teams to tune their solvers well. In this regard, it could be very interesting to add a new category to solver SAT/CSP/SMT competitions where solvers are first configured automatically on a training set and then evaluated on a related yet different test set.

1.1 Existing Approaches

Several approaches exist in the literature for the automatic tuning of algorithms. The first methods were created for tuning specific algorithms for a certain task. [13] devised a modular algorithm for solving constraint satisfaction problems (CSPs) and used a combination of exhaustive enumeration of all possible configurations and a parallel hill-climbing technique to automatically configure the system for a given CSP with an associated set of training instances. [4] classified local search (LS) approaches for SAT by means of context-free grammars and devised a genetic programming approach to select a good LS algorithm for a given set of SAT problems. [15] embedded a sequential parameter optimization approach in a wider framework for the design of evolutionary algorithms.

To tune the continuous parameters of general algorithms, [3] suggested an approach that determines good parameters for individual training instances. These parameters are found by trying configurations where parameters are at their extreme values and then fitting a regression function to the parameter/value tuples obtained in this way. The minimization of the resulting function yields a set of parameters for the given instance. A parameter set for the entire collection of instances was then obtained by averaging the parameter tuples for the individual instances.

Tuning problems with small sets of parameter configurations were considered in [2], a setting which is closely related to that in algorithm portfolios [6,7]. In this case, it is possible to race the different algorithms against each other, whereby a statistical test is used to eliminate inferior algorithms before the remaining algorithms are run on the next training instance.

In [14], Oltean used evolutionary algorithms by means of linear genetic programming. The genome of an individual is an encoding of an actual C-program for the problem to be solved, and crossover and mutation operators are problem dependent. The linear genetic program generates new individuals which replace the current worst individual in the population.

The CALIBRA system, proposed by [1], starts with a factorial design of the parameters. Once these initial parameter sets have been run and evaluated, an intensifying local search routine is started from a promising design, whereby the range of the parameters is limited according to the results of the initial factorial design experiments.

The only system we know of that can configure arbitrary algorithms with very large numbers of parameters was proposed by [9]. Their system, called ParamILS, conducts an iterated local search, whereby a special technique is used to limit the number of training instances that need to be run for each parameter set by focusing the test runs on promising parameter sets. In particular, a new set of parameters is not considered better than the current best until it has been evaluated on at least as many training instances as the current best. If a very large set of training instances is available, this approach allows quick movement through the search space while still avoiding an "over-tuning" effect which would be caused by considering few training instances only.

1.2 Our Approach

CALIBRA and ParamILS have shown that automatic configuration of algorithms is possible, and can, in fact, lead to massive improvements over hand-tuned parameter sets. Based on these successes, we aim to provide a configuration system which is very robust and provides high-quality parameter sets in an affordable amount of time, potentially by exploiting parallelism which is becoming more and more widely available given the current trends in hardware technology.

To this end, we propose a genetic algorithm for the problem of configuring solvers. There are two main reasons for this choice of approach. First of all, genetic algorithms are known to be very robust with respect to optimization problems that have undesirable objective landscapes [5]. Note that, in ordinary optimization, we usually have the freedom to adjust the objective in such a way that it is better suited for sequential local search which often yields good solutions faster than population-based approaches. In contrast, in our setting, where the target algorithm is given and the effect of changing parameters is a priori unknown, we must be able to cope with whatever objective landscape we encounter. The other reason is that genetic algorithms are inherently parallel. When trying to assess which individuals are competitive (the most time-intensive step in solver configuration), genetic algorithms allow us to race them against each other. Therefore, the time spent for the evaluation is determined by the good parameter sets, and this saves a lot of time in practice. In order to really exploit this last aspect, we introduce the concept of gender in the genetic algorithm. Before we apply this idea to solver configuration, we will explain the potential benefits of gender separation in genetic algorithms in the following section.

2 A Gender-Based Genetic Algorithm

The idea to exploit genders in genetic algorithms has been considered before in various publications, inspired by nature's example.

2.1 Related Work

In [10], Lis and Eiben use multiple genders for multi-objective optimization. Each gender is associated with one of the multiple objectives, and the fitness of each individual is evaluated according to the gender-specific fitness function. Cross-over is limited to mating individuals of different gender. This latter restriction used for single-objective problems is introduced in [16] which uses a gendered genetic algorithm to solve graph partitioning problems.

[12] discusses that, in nature, mate choice is more likely to guide evolution than natural selection. [18] introduces two genders where individuals of the first are evaluated according to standard fitness, while individuals of the second gender are associated with a second criterion, a so-called cooperative fitness. This second criterion depends on the potential mating partner. Mating couples are formed by selecting the fittest individuals from the first gender and mating them with good cooperative partners from the second.

Finally, Vrajitoru experiments with a population that consists of different gender types (self-fertilizing, sexual, and hermaphrodite) [19].

2.2 Gender-Specific Selection Pressure

Previously, gender separation has been realized only as a restriction of the way in which mating pairs are formed. We propose to take the gender separation beyond just the formation of parent couples. Instead, we propose to apply *different selection pressure* on the two gender populations. In particular, we apply intra-specific competition *only* in one part of the population. Individuals in this group must compete for the right of mating, and only the fittest in each generation win the right to mate with some of the individuals of the opposite gender. The individuals in this other group are not subjected to intra-specific selection.

With our application of tuning solver parameters in mind, we choose this setting for two reasons. First, the runtime of our algorithm will largely depend on the time that it takes to evaluate the fitness of individuals. Note that this requires running the given solver on some benchmark instances. By selecting only a small portion of the fittest individuals in one subgroup of the population, we can save a substantial number of fitness evaluations compared to the traditional way of applying a fitness ranking on all individuals.

Second, because of the expected very large cost to evaluate the fitness of an individual, in our application we will not be able to sustain a very large number of individuals in the population. Consequently, we cannot afford to lose partially good genes just because they first occurred in a less fit individual. By randomly assigning new offspring to one of the two groups, we increase the chance that good genes survive as they may occur in individuals which belong to the non-competitive part of the population. That is, the individuals in this group serve as a "variety store" of potentially beneficial gene-collections which are only indirectly subjected to the selection process. Note that recombining information from non-competitive solutions has been found beneficial in other contexts as well – for example when solving vehicle routing problems [17].

The individuals in the non-competitive part of the population can of course still be expected to improve in average fitness over time. However, this happens without the need to evaluate directly the fitness of these individuals: Less fit parents are likely to have less fit children and half of those can be expected to belong to the competitive part of the population. These children, if they are indeed not fit enough, will not be able to propagate their genes into the grandchild-generation. Consequently, the chances that the genetic line of unfit parents will survive are greatly diminished. Note that this

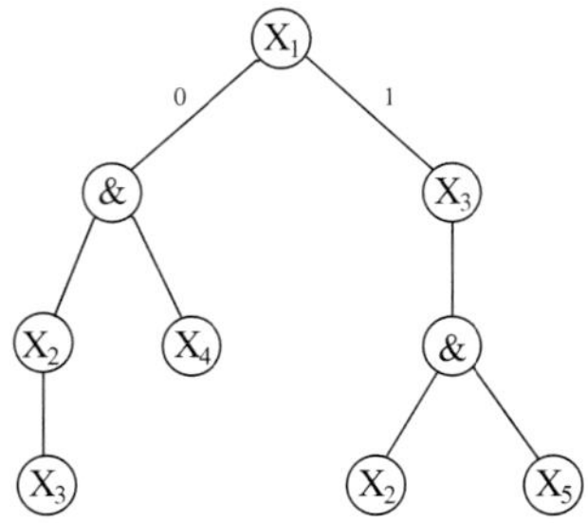

Fig. 1. Variable Tree

dynamic system agrees more closely with our modern understanding of evolution where the "survival of the fittest" is viewed more as the struggle for survival of genetic lines rather than that of individual beings.

2.3 Variable Trees

Apart from the fact that the introduction of gender will help reduce the number of fitness evaluations, in nature the invention of gender went hand-in-hand with the invention of recombination. Only sexual organisms have separate chromosomes which can be combined in arbitrary ways. This recombination process is realized by individuals having multiple sets of the entire genome – in humans for example there is one set of chromosomes from each parent. In our design we did not go quite as far as to introduce multiple copies of the genome. However, the idea that certain genes ought to be inherited as a set while others may be permuted and recombined arbitrarily aligns nicely with our expectation that certain groups of parameters will be linked more closely than others.

Therefore, we allow the user to define the design of the genome by passing a specific structure of the problem variables. With our application in mind, the variables will of course be the parameters of the solver to be tuned. In particular, to couple and decouple variables (parameters), the user passes a variable structure which is inspired by And/Or-trees (see, e.g., [11]). The idea is that And-nodes separate variables that can be optimized independently.

To specify the parameters of a solver and their relation, we distinguish three types of variables: continuous and integer variables, both associated with an upper and a lower bound, and categorical variables that come with an explicit list of feasible values. A *variable tree* is a tree where:

- Each node is labelled with a variable or the additional label "&", and each problem variable is associated with at least one node.
- Nodes associated with continuous or integer variables have at most one child, and And-nodes have at least two child-nodes.
- The children of categorical nodes partition the set of values that their parent variable can take. Branches leading to the children are labelled by the respective value(s) of the categorical variable.

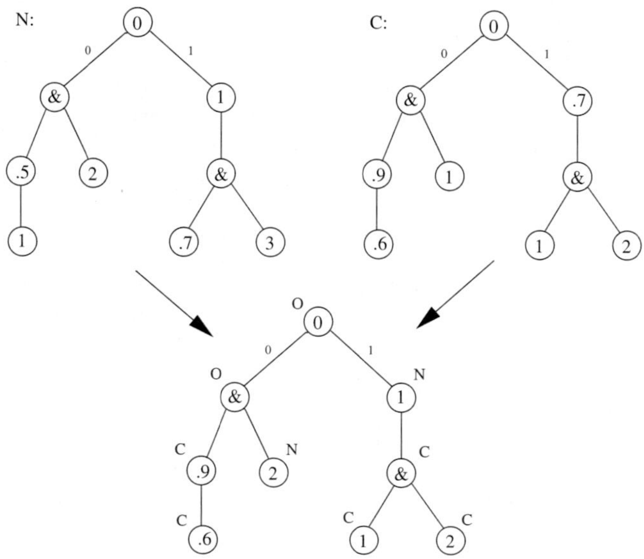

Fig. 2. Crossover Operator

In Figure 1, we illustrate a variable tree for the minimization of the function

$$(1-x_1)\left(\frac{x_2\sin(\pi(x_2-x_3))}{x_3}+(x_4-2)^2\right)+2x_1\left(|\frac{x_5}{x_3}-7|+(x_2x_3-1)^2\right)$$

whereby x_1 is categorical and takes values 0 or 1, x_2, x_3 are continuous and take values in $[\frac{1}{2}, 1]$, and x_4, x_5 are integer variables taking values in $\{1, 2, 3\}$. This function is used merely to show an interesting semantic tree for illustration purposes. We see that the structure reflects that, once x_1 is set to 0, the pair of variables x_2, x_3 can be optimized independently of x_4. Also, once x_1 is set to 1 and x_3 has been assigned a value, variables x_2 and x_5 can be optimized independently.

2.4 A Gender-Based Genetic Algorithm

We now have all concepts in place to describe our Gender-based Genetic Algorithm (GGA) for the automatic configuration of solvers. GGA uses parameters (X, P, M, A, S) which are used in the following way:

- **Initialization:** First, we randomly initialize the population and assign a gender C (for competitive) or N (for non-competitive) and an "age" of 1 to A years uniformly at random to each individual. In our experiments, we set A to 3.
- **Mating Rules:** Among the individuals with gender C, we select the top $X\%$ (in our experiments we set X to 10%). These have gained the right to mate in this season. $200/A\%$ of individuals of gender N are assigned uniformly at random to one of the mating individuals of gender C. The individuals of gender C then mate with all individuals of gender N which have been assigned to them.
- **Crossover:** Each mating of a couple results in one new individual with age 0 and random gender. The genome of the offspring is determined by traversing the variable

Crossover(ParameterTrees T_C, T_N, T_3)

```
curNode ← rootOf(T_3), nodeC ← rootOf(T_C), nodeN ← rootOf(T_N)
if (type[curNode]=And) OR (value[nodeC]=value[nodeN]) then
  label[curNode] ← O, value[curNode] ← value[nodeC]
else
  if rand() mod 2 = 0 then
    label[curNode] ← C, value[curNode] ← value[nodeC]
  else
    label[curNode] ← N, value[curNode] ← value[nodeN]
  end if
end if
S ← {curNode}
while (S ≠ ∅) do
  curNode ← pick(S), S ← S \ {curNode}
  for all childNodes of curNode do
    S ← S ∪ {childNode}
    nodeC ← correspondingNode(T_C, childNode)
    nodeN ← correspondingNode(T_N, childNode)
    if label[curNode]=O then
      if type[childNode]=OR then
        if value[nodeC]=value[nodeN] then
          label[childNode] ← O, value[childNode] ← value[nodeC]
        else
          if rand() mod 2 = 0 then
            label[childNode] ← C, value[childNode] ← value[nodeC]
          else
            label[childNode] ← N, value[childNode] ← value[nodeN]
          end if
        end if
      else
        label[childNode] ← O, value[childNode] ← value[nodeC]
      end if
      continue
    end if
    if label[curNode]=C then
      parLabl ← C, parVal ← value[nodeC]
      oParLabl ← N, oParVal ← value[nodeN]
    else
      parLabl ← N, parVal ← value[nodeN]
      oParLabl ← C, oParVal ← value[nodeC]
    end if
    if rand() mod 100 < δ then
      label[childNode] ← oParLabl, value[childNode] ← oParVal
    else
      label[childNode] ← parLabl, value[childNode] ← parVal
    end if
  end for
end while
```

Algorithm 1. Crossover Algorithm

tree top-down (compare with Algorithm 1 and Figure 2). A node can be labelled O ("open"), C, or N. If the root is an And-node, or if both parents agree on the value of the root-variable, we label it O. Otherwise, we randomly assign it label C or N (Lines 5 and 23). The algorithm continues by looking at the children of the root (and so on for each new node). If the label of the parent node is C (or N) then with probability $P\%$ we also label the child with C (N), otherwise with N (C) (Line 41). In our experiments we set P to 90%.

Finally, the variable assignment associated with the offspring is given by the values from the C (N) parent for all nodes labelled C (N). For variable-nodes labelled O both parents agree on its value, and we assign this value to the variable. Note that this procedure effectively combines a uniform crossover for child-variables of open And-nodes in the variable tree (thus exploiting the independence of different parts of the genome) and a randomized multiple-point crossover for variables that are more tightly connected.

- **Mutation:** As a final step to determine the offspring's genome, with probability $M\%$ we mutate the value of each variable (in our experiments, we set M to 10%). If we mutate a categorical variable, we choose a new value in its domain uniformly at random. For continuous and integer variables, we choose a new value according to a Gaussian distribution where the current value marks the expected value and the variance is set as $S\%$ of the variable's domain. In our experiments, we set S to 10%.
- **Death:** After the new offspring is created, all individuals' ages are increased by 1. Those with age greater than A are removed from the population. In combination with the mating rules that only $200/A\%$ of individuals of gender N mate in every season, this stabilizes the total population size.

Before we use this algorithm for the configuration of solvers, we first test it on some generic optimization functions of three different types: $f_1 = \sum_i (x_{2i}x_{2i+1} - p_i)^2$, $f_2 = \sum_i (x_{3i}x_{3i+1}x_{3i+2} - q_i)^2$, and $f_3 = \sum_i ((x_{3i}x_{3i+1} - v_i)^2 + (x_{3i}x_{3i+2} - w_i)^2)$, whereby all variables take integer values in $\{0, \ldots, 15\}$ and constants $p_i, v_i, w_i \in [0, 15^2]$, $q_i \in [0, 15^3]$ are chosen uniformly at random.

For each of these functions we compare three different variable trees that determine the genome structure: The completely independent structure (IND) with one And-node at the root and all variables as its children, the completely dependent structure (DEP) where all variables form one long chain, and the semantic structure (SEM) that results from the static analysis of each of the functions (with independent tuples (x_{2i}, x_{2i+1}) in f_1, $(x_{3i}, x_{3i+1}, x_{3i+2})$ in f_2 and f_3, whereby in the latter x_{3i+1} and x_{3i+2} are independent once x_{3i} has a value). Table 1 compares Gender-based GA (GGA) with the GA library GALib from [20]. The genome for the latter is determined by representing each variable as a four-bit string which tests showed to yield the best results. The cross-over and mutation probabilities were set to 1 and 0, respectively, which our own parameter tuner confirmed to be the best parameter settings for this optimization system. For our algorithm, we chose the parameter set $(10, 90, 10, 3, 10)$. That is, only the top 10% of competitive individuals are allowed to mate in each season; with probability 90% a child-variable inherits the value from the same parent as its parent-variable in the variable tree; the mutation-rate is 10%; individuals die after 3 mating seasons; and nodes in the child variable trees will randomly change their assignment from their parent's with probability 10%.

Table 1. Numerical Results for generic function minimization. We give the average solution value and, in parenthesis, the number of function evaluations (both in thousands) for 50 runs.

Prob-Pop-Gen	GGA IND	GGA DEP	GGA SEM	GA
f_1-1000-25	1.8(13.7)	29.8(13.7)	**1.44**(13.8)	7.19(26.0)
f_1-500-50	0.59(13.4)	28.9(12.8)	**0.4**(13.3)	4.56(25.5)
f_1-2000-25	1.54(27.1)	29.3(27.7)	**1.29**(27.6)	6.67(52.0)
f_1-1000-50	0.46(26.0)	28.7(26.1)	**0.31**(26.0)	4.41(51.0)
f_1-500-100	0.16(25.1)	28.3(25.0)	**0.13**(25.5)	2.69(50.5)
f_2-1000-25	1442(13.7)	6075(13.8)	**1229**(13.4)	4962(26.0)
f_2-500-50	392(13.3)	5273(13.0)	**340**(13.0)	3127(25.5)
f_2-2000-25	**1104**(27.4)	5121(25.3)	1119(27.6)	4405(52.0)
f_2-1000-50	307(25.6)	5886(27.5)	**304**(25.8)	3184(51.0)
f_2-500-100	141(25.3)	4830(25.4)	**124**(26.0)	2002(50.5)
f_3-1000-25	**16.2**(13.6)	18.5(13.9)	16.4(13.7)	43.2(26.0)
f_3-500-50	**5.75**(13.2)	8.01(13.5)	6.33(13.7)	31.9(25.5)
f_3-2000-25	14.7(27.5)	15.9(27.5)	**13.7**(27.2)	41.3(52.0)
f_3-1000-50	**4.86**(25.9)	6.40(26.5)	5.27(25.8)	31.8(51.0)
f_3-500-100	1.33(26.5)	1.7(27.1)	**1.25**(25.8)	22.5(50.5)

The results clearly show the benefits of introducing gender to reduce the number of function evaluations. Only half of the population is evaluated in each season. Moreover, we see that applying different selection pressure on both genders results in significantly improved solution quality. For example, when optimizing f_3 with a population size of 500 over 100 generations, the gender-based approach using the semantic genome structure (SEM) on average gives solution values around 1,250 at the cost of 25,800 fitness evaluations. The standard genetic algorithm performs 50,500 evaluations and still only achieves average solution values of 22,500.

In regard to the shape of parameter trees, we find that, in doubt, assuming independence or parameters appears better than assuming dependence, whereby the semantic hybridization of both can lead to improved performance.

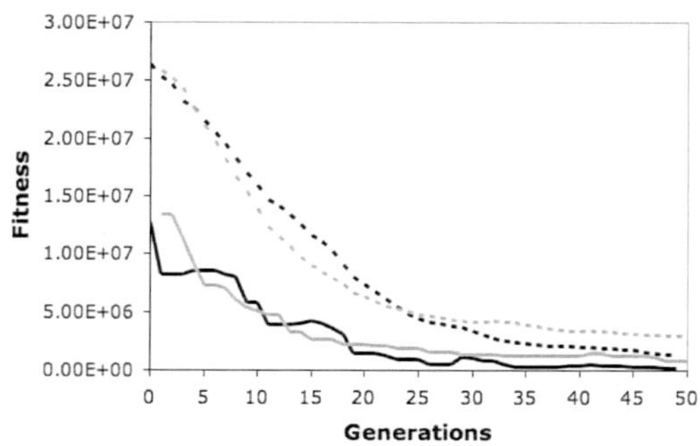

Fig. 3. Average (dashed) and best (solid) fitness for the competitive (black) and non-competitive (gray) populations when optimizing f_2 (averages over 25 runs)

In Figure 3 we show how the average and the best fitness in both parts of the population evolve over the course of several generations. Even though it may not have been expected, the average and best fitness in the non-competitive part of the population are not worse than in the competitive part. This is of course caused by the fact that gender is assigned randomly and the fitness of new individuals (no matter to which gender they belong) is determined by the fitness of both parents. Consequently, we found that the overall best solution was equally often found in both genders.

3 Automatic Solver Configuration

By representing the genome as a variable tree which is defined by the user of the configuration system, the outlined genetic algorithm can be applied directly for the task of tuning solver parameters. Note that our approach allows a separation of setting the structure of the parameters (which would typically be done by the developer of the solver) and selecting the training instances (which are best selected by the user of the solver). While using the same parameter categories as ParamILS, the variable tree representation does not require discretization of continuous parameters and offers a richer way of specifying parameter correlations.

To apply our genetic approach for solver configuration we only need to specify the function which selects the top $X\%$ of the competitive part of the population. We could of course run all individuals (i.e., parameter sets) on all training instances and then select the best ones. However, this costs a lot of time, especially when the task is to minimize the expected average runtime of the target algorithm, which is the most common evaluation criterion for solvers. In this case, we run the different parameter sets in parallel, and as soon as the best $X\%$ have finished, we interrupt all remaining runs. This way, the time spent is largely determined by the quality of the very good parameter sets.

For larger population sizes, creating too many parallel threads or processes is not appealing. In these instances, we partition the competitive population into smaller groups (typically between 5 and 10 members per available CPU) and select the best $X\%$ of this subgroup by running them in parallel. This is repeated until all groups have found their winners. In our experiments, we used groups of 8.

The size of the training set is also an important consideration. Many training instances are desirable so as to avoid an over-tuning effect where we find parameter sets which work particularly well for the training instance but do not generalize. On the other hand, evaluating each parameter set on all available instances costs a lot of time. [9] proposed a focused approach which results in more evaluations of better parameter sets. In our setting, we could employ the racing framework from [2] for selecting the top competitive individuals. However, we take the practical standpoint that we will normally not have access to as many training instances as we would like (which is necessary to obtain the theoretical guarantees that the probability of over-tuning converges to zero). Therefore, in each mating season we select a random subset of all available training instances and race the competitive individuals on those. In subsequent generations we linearly increase the relative size of the subset, so that the comparison of different parameter sets becomes more and more accurate the better the parameter sets become.

Table 2. Experimental results of a standard GA vs. GGA for configuring SAPS with a cutoff of 10 seconds. We report the average total CPU time over 10 runs of the tuners as well as the average solution quality. All times are given in seconds.

	GA		GGA		Improvement
	Avg.	Std. Dev	Avg.	Std. Dev.	[%]
Time	23.5K	4.48K	926	1.12K	96
Quality	1.79	1.57	0.07	0.01	96

4 Numerical Results

4.1 Gender Separation for Solver Configuration

We first compare our configurator with a "standard" genetic algorithm to demonstrate the advantages of the gender separation for configuration. We use a standard genetic algorithm along with our variable tree and crossover operations.

We compute 40 generations with a population of 30 individuals for both genetic algorithms. The standard GA does not perform a gender separation and uses a different mating scheme: Each individual in the population is evaluated in each generation. The oldest third of the population is replaced by new offspring. This is formed by mating two individuals from the population. The probability of an individual being chosen for mating is proportional to its fitness.

In Table 2 we report the configuration time as well as the runtime of the target SAT solver SAPS when using the final set of parameters for both configurators. The training was done on a set of 113 SAT instances, the quality of the configuration was evaluated on a set of 100 different SAT instances (see [9]). The tests were run on a AMD Athlon 64 3000+ CPU with 2 GB RAM.

We see that GGA works 20 times faster and returns configurations which are much better. Compared to running SAPS with the parameters found by the GA, SAPS requires only about 4% of the runtime using the parameter set returned by GGA. The great reduction in configuration time is mainly due to the fact that it is sufficient for GGA to find the top 10% of the competitive individuals. This can be done by racing parameter sets against each other, rather than evaluating them all. Since bad parameter settings can take a lot of time (the effect is softened by the cutoff of 10 seconds, but still significant), GA wastes a lot of time evaluating bad solver configurations.

This alone does not explain the greatly improved quality, though. As seen earlier in our preliminary experiments optimizing a generic function, the gender separation improves the solution quality by providing a store of genes which diversify the search and prevent us from getting stuck in local optima even though we aggressively select only the top 10% of the competitive individuals for mating.

4.2 Effect of Parameter Tree Structure

We evaluate our system on two of the same target algorithms that [9] used to show that their ParamILS approach outperforms the CALIBRA system from [1]. These are the previously considered SAPS solver and SAT4J, a systematic solver for SAT. We distinguish two different versions of SAT4J: SAT4J with a full parameter set some of which allow the solver to return unsatisfying SAT solutions, and SAT4J* with a limited parameter set all of which force the solver to return only feasible solutions.

The benchmark set for SAPS consists of 113 SAT instances for training and 100 different instances for testing. Since SAPS is a randomized solver, the instances are paired with 10 different seeds each. We used the SWGCP benchmark set from [9] for configuring SAT4J(*). It consists of 1000 SAT instances for training and 1000 different instances for testing. In addition, we compared our solver with ParamILS on the target algorithm SPEAR, another systematic SAT solver. SPEAR was tuned by ParamILS in [8]. Again, we used the SWGCP benchmark for training and testing.

We tuned 20 times for 5 CPU hours on a 64bit Linux Intel Xeon with 2.8GHz and 8GB RAM and then compared the average test performances of GGA when using the independent (IND), dependent (DEP), and semantic (SEM) parameter tree structures for tuning solvers SAPS, SAT4J*, and SPEAR. For SAPS, which only has four parameters, the semantic and the independent structure are the same, for the other two they are almost the same deviating only for a few parameters. We found that the test performance of IND and SEM were identical, 77ms for SAPS, 4.21s for SAT4J* and 2.03s for SPEAR. Using the fully dependent structure results in 88ms for SAPS, 4.88s for SAT4J*, and 5.25s for SPEAR. These results are in line with our finding in Section 2.4 that, in doubt, we will want to assume parameter independence. Tests on solvers with more complex parameter structures are needed to assess whether exploiting the semantic structure of parameters is beneficial for solver configuration.

4.3 GGA vs. ParamILS

We now compare GGA with ParamILS. We ran GGA for 100 generations with a population size of 200 for SAPS and 50 for SPEAR. For SAT4J(*), we ran GGA for 70 generations with a population of size 60. This resulted in total CPU times for configuration that never exceed the 10h cutoff which was used for ParamILS when configuring SAPS and SPEAR, and a 20h cutoff for SAT4J. Algorithms SAPS and SPEAR were run on a 32bit Linux Intel Core2 Quad CPU Q6600 with 2.4GHz and 3GB RAM, SAT4J(*) was run on a 64bit Linux Intel Xeon with 2.8GHz and 8GB RAM.

In Table 3 we show the results of 20 configuration runs of ParamILS[1] and our Gender-based genetic algorithm (GGA) when configuring solvers SAPS, SAT4J, and SAT4J*.

The table gives the final average computation time for the instances in the training as well as the test set. Comparing the training and test performances, we see that GGA very accurately assesses the expected performance. For SAPS and SAT4J*, ParamILS also assesses its performance quite accurately. However, for SAT4J, ParamILS returns parameters which perform quite badly on the training set and perform a lot better on the test set. For example, in configuration run 1 ParamILS finds a parameter set which actually works very well on the test set, although the performance on the training set is quite miserable. In configuration run 7, ParamILS also finds a set of parameters which actually works better on the test set than on the training set. However, the test performance is almost 100% worse than the average performance achieved by GGA.

---

[1] Thanks to Frank Hutter for providing support of ParamILS and the various solvers!.

Table 3. Experimental Results for Configuring SAPS (left), SAT4J (middle), and SAT4J* (right)

[ms] Run	ParamILS Train	ParamILS Test	GGA Train	GGA Test
1	51.55	52.12	37.25	35.44
2	53.91	51.45	46.15	41.69
3	55.31	50	55.85	49.52
4	55.11	51.8	36.07	33.4
5	53.72	50.91	46.15	40.1
6	54.96	52.4	35.68	33.56
7	54.67	52.79	34.69	32.2
8	55.11	49.91	37.54	32.76
9	56.24	51.09	38.24	35.42
10	56.26	51.31	34.7	33.53
11	55.02	52.26	35.99	34.52
12	54.29	51.61	36.1	33.99
13	54.31	51.42	36.35	33.59
14	56.58	52.31	37.42	34.58
15	57.38	54.15	38.79	36.66
16	57	54.09	52.98	52.25
17	56.31	52.6	35.78	32.68
18	58	53.73	38.59	35.06
19	54.52	51.83	35.83	33.9
20	61.47	55.85	36.57	34.56
⊘	55.6	52.2	39.3	36.5
σ	2.0	1.44	6.0	5.5

[s] Run	ParamILS Train	ParamILS Test	GGA Train	GGA Test
1	3.65	0.99	1.07	1.07
2	1.06	1.07	1.05	1.06
3	1.07	1.07	1.06	1.06
4	0.99	1.05	1.06	1.07
5	5.11	2.22	1.07	1.14
6	1.04	1.04	3.20	3.90
7	5.29	2.31	1.05	1.05
8	6.27	2.38	1.06	1.06
9	1.05	0.53	1.04	1.05
10	1.06	1.07	1.05	1.05
11	1.17	1.06	1.14	1.06
12	1.04	1.05	1.05	1.13
13	1.05	1.06	1.07	1.07
14	1.05	1.05	1.05	1.06
15	1.04	1.05	1.05	1.06
16	6.34	6.83	1.08	1.08
17	5.17	5.35	1.07	1.07
18	4.70	4.32	1.05	1.06
19	5.57	5.20	1.06	1.06
20	5.03	4.97	1.05	1.06
⊘	2.94	2.28	1.17	1.21
σ	2.2	1.92	0.48	0.63

[s] Run	ParamILS Train	ParamILS Test	GGA Train	GGA Test
1	2.20	2.86	3.19	2.92
2	5.06	5.41	3.99	3.66
3	4.70	5.78	3.01	2.92
4	2.44	2.58	2.90	2.99
5	2.56	2.57	3.06	3.14
6	2.48	2.76	2.92	2.93
7	2.52	2.77	3.00	2.95
8	2.55	2.66	2.90	2.92
9	2.80	2.96	3.01	3.00
10	2.57	2.92	2.96	2.93
11	2.45	3.92	2.87	2.93
12	4.90	5.05	2.87	2.79
13	2.49	3.93	3.01	2.81
14	2.41	2.92	2.96	2.81
15	5.25	5.93	3.01	2.70
16	4.21	4.36	3.29	3.12
17	4.89	5.39	4.91	4.88
18	2.54	2.63	3.34	3.24
19	2.48	2.55	2.92	2.92
20	2.45	2.42	3.04	3.06
⊘	3.2	3.62	3.16	3.08
σ	1.12	1.24	0.48	0.47

In terms of solution quality, GGA achieves significant improvements, outperforming ParamILS on all three configuration tasks. For SAPS, GGA's worst parameter set results in an average performance of 52.25 ms per test instance, which is still better than the average ParamILS parameter set which requires 52.4 ms. Comparing average test performances (see Table 4), GGA's parameters improve those of ParamILS when configuring SAPS by more than 30%. Note that the best parameter set found by ParamILS is significantly worse than the average performance achieved by GGA. Based on this observation, we conjecture that the poor performance of ParamILS for this solver is caused by an unfortunate discretization of continuous parameters for ParamILS – which GGA does not rely on as it handles continuous parameters directly.

For SAT4J and SAT4J*, ParamILS is able to find highly efficient parameter sets. However, for SAT4J GGA returns parameters which are on average 45% better than those provided by ParamILS, and for SAT4J* by over 14%. Considering that configuration is essentially a complex optimization problem, this is a big margin. For comparison, [9] reported a 19% improvement over CALIBRA when configuring SAPS so as to minimize median runtime.

Looking at the standard deviation over the different configuration runs, we see that the genetic approach performs much more robustly than iterated local search whose standard deviation is at times as large as its mean. Based on our experiments in Section 4.1, we attribute the solid performance of GGA mostly to the gender separation. However,

Table 4. Experimental results of ParamILS and GGA for various solvers. We give mean solver times on the training and test sets and, in brackets, the standard deviation over 20 configuration runs.

Problem	ParamILS		GGA		Welch's t-value		Improvement
	Train	Test	Train	Test	t(1-tail)	t(2-tail)	[%]
SAPS [ms]	55.8 (2.3)	52.4 (1.78)	38.8 (5.5)	36.0 (5.0)	< 0.01	< 0.01	31.30
SPEAR [s]	1.58 (0.088)	1.49 (0.087)	1.58 (0.086)	1.50 (0.077)	0.33	0.65	-0.67
SAT4J [s]	2.85 (2.12)	2.38 (1.97)	1.25 (0.67)	1.29 (0.76)	0.01	0.01	45.80
SAT4J* [s]	3.32 (1.11)	3.74 (1.28)	3.28 (0.89)	3.20 (0.81)	0.04	0.08	14.4

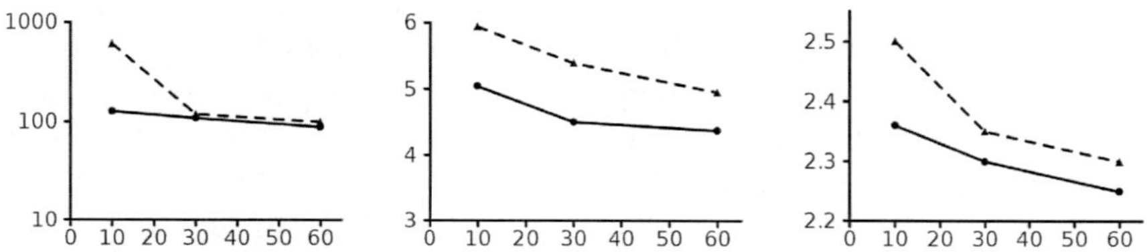

Fig. 4. Average test performance of 20 runs of GGA (solid) and ParamILS (dashed) after 10 minutes, 30 minutes, and 60 minutes tuning time for SAT solvers SAPS (left, msecs, log-scale), SAT4J* (middle, secs), and SPEAR (right, secs)

further tests are needed to find out which component of GGA is most important for its performance. Note that ParamILS was reported to find parameter sets which already improve on the manually set SAPS defaults by three to four orders of magnitude. The significant additional improvement by GGA is fairly surprising and shows just how hard it actually is to find near-optimal parameters.

In Table 4 we also report the results when configuring the SPEAR program. We find that both ParamILS and GGA perform equally well for this configuration problem. We have no actual lower bounds on the performance that can be achieved by SPEAR, but we conjecture that there is not much room for improving this algorithm any further.

In Figure 4, finally, we compare the test performance for very short tuning times, 10 minutes, 30 minutes, and 60 minutes, when tuning with ParamILS and GGA. We observe that GGA is able to identify good parameter sets early on. While this is an interesting insight which we added at the request of the reviewers, in practice we can expect that much more time will be spent for configuration as it is done off-line and pays off with every invocation of the tuned solver later.

5 Conclusion

We considered the problem of automatically configuring solvers. We proposed a genetic algorithm for this task and showed that it robustly provides high quality parameter sets which can significantly improve those of the pioneering system from [9]. To specify and exploit the dependencies of parameters, we introduced a special "variable tree" structure which indirectly defines the cross-over operator.

To improve performance, our approach introduced a gender separation which we believe to be of interest for genetic algorithms in general, especially when population sizes are small and the optimization costs are largely determined by the number of fitness evaluations. What makes a genetic approach appealing for solver configuration is that it offers the possibility to race parameter sets against each other. In combination with the gender separation, which allows us to focus on high quality parameter sets very aggressively, the evaluation of parameter sets is determined by the solver time needed with very good parameter settings. We conjecture that this is the most important advantage of GGA, but it is a subject of future research to identify the component (or combination of components) of GGA which is most important for its performance.

A practical advantage of using a population based approach is that it can be parallelized naturally. Our preliminary parallelization already resulted in substantial speedups. We are currently working on an efficient parallelization of our code which will provide the practical basis for configuring solvers that require more computation time. As future work, we are considering to locally improve good parameter sets which have been found, thus transforming our genetic algorithm into a memetic algorithm.

Acknowledgement

We would like to thank Warren Schudy for many helpful comments, especially in regard to the suggestion to model parameter structures as variable trees.

References

1. Adenso-Diaz, B., Laguna, M.: Fine-tuning of Algorithms using Fractional Experimental Design and Local Search. Operations Research 54(1), 99–114 (2006)
2. Birattari, M., Stuetzle, T., Paquete, L., Varrentrapp, K.: A Racing Algorithm for Configuring Metaheuristics. In: GECCO, pp. 11–18 (2002)
3. Coy, S.P., Golden, B.L., Runger, G.C., Wasil, E.A.: Using Experimental Design to Find Effective Parameter Settings for Heuristics. Journal of Heuristics 7(1), 77–97 (2001)
4. Fukunaga, A.: Automated discovery of local search heuristics for satisfiability testing. Evolutionary Computation 16(1), 31–61 (2008)
5. Goldberg, D.E.: Genetic algorithms in search, optimization, and machine learning. Addison-Wesley, Reading (1989)
6. Gomes, C., Selman, B.: Algorithm Portfolios. Artificial Intelligence 126(1-2), 43–62 (2001)
7. Huberman, B., Lukose, R., Hogg, T.: An Economics Approach to Hard Computational Problem. Science 265, 51–54 (2003)
8. Hutter, F., Babić, D., Hoos, H.H., Hu, A.J.: Boosting Verification by Automatic Tuning of Decision Procedures. FMCAD, 27–34 (2007)
9. Hutter, F., Hoos, H.H., Stützle, T.: Automatic Algorithm Configuration based on Local Search. In: AAAI, pp. 1152–1157 (2007)
10. Lis, J., Eiben, A.E.: A Multi-Sexual Genetic Algorithm for Multiobjective Optimization. In: IEEE International Conference on Evolutionary Computation, pp. 59–64 (1997)
11. Marinescu, R., Dechter, R.: And/Or Branch-and-Bound for Graphical Models. In: IJCAI, pp. 224–229 (2005)
12. Miller, G.F., Todd, P.M.: The Role of Mate Choice in Biocomputation. Evolution and Biocomputation, 169–204 (1995)

13. Minton, S.: Automatically Configuring Constraint Satisfaction Programs. Constraints 1(1), 1–40 (1996)
14. Oltean, M.: Evolving evolutionary algorithms using linear genetic programming. Evolutionary Computation 13(3), 387–410 (2005)
15. Preuss, M., Bartz-Beielstein, T.: Sequential Parameter Optimization Applied to Self-adaptation for Binary-coded Evolutionary Algorithms. Parameter Setting in Evolutionary Algorithms: Studies in Computational Intelligence, 91–119 (2007)
16. Rejeb, J., AbuElhaij, M.: New Gender Genetic Algorithm for Solving Graph Partitioning Problems. Circuits and Systems 1, 444–446 (2000)
17. Rochat, Y., Taillard, R.D.: Probabilistic Diversification and Intensification in Local Search for Vehicle Routing. Journal of Heuristics 1, 147–167 (1995)
18. Sanchez-Velazco, J., Bullinaria, J.A.: Gendered Selection Strategies in genetic Algorithms for Optimization. UKCI, 217–223 (2003)
19. Vrajitoru, D.: Simulating Gender Separation with Genetic Algorithms. In: GECCO, pp. 634–641 (2002)
20. Wall, M.: GAlib: A C++ Library of Genetic Algorithm Components. MIT, Cambridge (1996), `http://lancet.mit.edu/ga`

Filtering Numerical CSPs Using Well-Constrained Subsystems

Ignacio Araya, Gilles Trombettoni, and Bertrand Neveu

INRIA, Université de Nice-Sophia, CERTIS
{Ignacio.Araya,Gilles.Trombettoni,Bertrand.Neveu}@sophia.inria.fr

Abstract. When interval methods handle systems of equations over the reals, two main types of filtering/contraction algorithms are used to reduce the search space. When the system is well-constrained, interval Newton algorithms behave like a global constraint over the whole $n \times n$ system. Also, filtering algorithms issued from constraint programming perform an AC3-like propagation loop, where the constraints are iteratively handled one by one by a *revise* procedure. Applying a revise procedure amounts in contracting a 1×1 subsystem.

This paper investigates the possibility of defining contracting well-constrained subsystems of size k ($1 \leq k \leq n$). We theoretically define the *Box-k-consistency* as a generalization of the state-of-the-art Box-consistency. Well-constrained subsystems act as *global constraints* that can bring additional filtering w.r.t. interval Newton and 1×1 standard subsystems. Also, the filtering performed inside a subsystem allows the solving process to learn interesting multi-dimensional branching points, i.e., to bisect several variable domains simultaneously. Experiments highlight gains in CPU time w.r.t. state-of-the-art algorithms on decomposed and structured systems.

1 Introduction

When interval methods handle systems of equations over the reals, two main types of filtering/contraction algorithms are used to reduce the search space. When a system contains n unknowns/variables constrained by n equations, interval Newton algorithms behave like a global constraint over a linearization of the whole $n \times n$ system. Filtering algorithms issued from constraint programming handle 1×1 subsystems (one variable involved in one constraint) in an AC3-like propagation loop.

This paper investigates the possibility of filtering $k \times k$ subsystems, where the size $1 \leq k \leq n$. After introducing in Section 2 the necessary background about intervals, we define in Section 3 the Box-k-consistency achieved by our new algorithm. This partial consistency generalizes the well-known Box-consistency [2]. Due to the large amount of subsystems in a constraint system, we explain in Section 5 the criteria used to compute the Box-k-consistency in only certain subsystems that are made of equalities, connected and well-constrained. These subsystems are managed like *global constraints* [16,10] for enhancing the

I.P. Gent (Ed.): CP 2009, LNCS 5732, pp. 158–172, 2009.

filtering power. We detail in Section 4 the filtering (revise) procedure that filters one subsystem and makes it box-k consistent. The procedure expands a local search tree whose choice points are limited inside the subsystem, and uses a local interval Newton. This revise procedure has common points with the algorithm proposed in [5]. Their algorithm also performs a tree search where every node is filtered, before returning an outer approximation of the obtained sub-boxes. But it is applied to the whole system of equations and not to subsystems.

Section 5 details how the local search trees built inside subsystems allow a solving strategy to learn interesting multi-dimensional choice points in the global search tree, i.e., to bisect several variable domains simultaneously. These multi-dimensional branching points are called *multisplits*. Promising experiments highlight the benefits of our approach for *decomposed* and *structured* NCSPs.

2 Background

The algorithms presented in this paper aim at solving systems of equations or, more generally, numerical CSPs.

Definition 1. *A* **numerical CSP** *(NCSP) $P = (X, C, B)$ contains a set of constraints C and a set X of n variables. Every variable $x_i \in X$ can take a real value in the interval $[x_i]$ and B is the cartesian product (called a* **box***) $[x_1] \times ... \times [x_n]$. A solution of P is an assignment of the variables in X satisfying all the constraints in C.*

Since real numbers cannot be represented in computer architectures, note that the bounds of an interval $[x_i]$ should actually be defined as floating-point numbers. Most of the set operations can be achieved on boxes, such as inclusion and intersection. An operator `Hull` is often used to compute an outer approximation of the union of several boxes.

Definition 2. *Let $S = \{[b_1], ..., [b_n]\}$ be a set of boxes corresponding to a same n-set of variables.*

We call **hull** *of S, denoted by* `Hull`*(S), the minimal box including $[b_1], [b_2], ..., [b_n]$.*

To find all the solutions of an NCSP with interval-based techniques, the solving process starts from an initial box representing the search space and builds a search tree. The tree search **bisects** the current box, that is, **splits** on one dimension (variable) the box into two sub-boxes, thus generating one choice point. At every node of the search tree, filtering (also called *contraction*) algorithms reduce the bounds of the current box. These algorithms comprise **interval Newton** algorithms issued from the numerical analysis community [8,13] along with contraction algorithms issued from the constraint programming community. The process terminates with **atomic boxes** of size at most ϵ on every dimension.

The new contraction algorithm presented in this paper generalizes the famous `Box` algorithm that can enforce the *Box-consistency* property [2] defined as follows:

Definition 3. *An NCSP* (X, C, B) *is* **box-consistent** *if every pair* (c, x) *is box-consistent* ($c \in C$, $x \in X$ *and* x *is one of the variables involved in* c).

Consider a pair (c, x), *where* $c(x, y_1, ..., y_a) = 0$ *is an equation of arity* $a + 1$.[1] *Let* c' *be the equation* c *where the variables* y_i *are replaced by the current interval in* B: $c'(x) = c(x, [y_1], ..., [y_a]) = 0$. *The pair* (c, x) *is box-consistent if:*

- $0 \in c'([\underline{[x]}, +]) = c([\underline{[x]}, +], [y_1], ..., [y_a])$;
- $0 \in c'([-, \overline{[x]}]) = c([-, \overline{[x]}], [y_1], ..., [y_a])$.

$\underline{[x]}$, *resp.* $\overline{[x]}$, *denotes the lower bound, resp. the upper bound, of* $[x]$. $[\underline{[x]}, +]$ *denotes the tiny interval (of one u.l.p. large*[2]*) bounded by* $\underline{[x]}$ *and the following float.* $[-, \overline{[x]}]$ *denotes the tiny interval bounded by* $\overline{[x]}$ *and the float preceding* $\overline{[x]}$.

In practice, the `Box` algorithm performs an AC3-like propagation loop. For every pair (c, x), it reduces the bounds of $[x]$ such that the new left (resp. right) bound is the leftmost (resp. rightmost) solution of the univariate equation $c'(x) = 0$. Existing *revise* procedures use a *shaving* principle to narrow $[x]$: Slices $[s_i]$ inside $[x]$ with no solution are discarded by checking whether $c([s_i], [y_1], ..., [y_a])$ does not contain 0 and by using a univariate interval Newton.

Two other contraction algorithms are often used in solvers. `HC4` [2] whose revise procedure traverses twice the tree representing the mathematical expression of the constraint for narrowing all the involved variable intervals. `3B` [11] or a variant `3BCID` [17] uses a shaving refutation principle similar to SAC [6].

3 Box-k Partial Consistency

As explained above, the Box-consistency yields an outer approximation/box of 1×1 subsystems (c, x). The Box-k-consistency introduced in this paper generalizes Box-consistency by yielding an outer approximation of subsystems.

Definition 4. *Let* $P' = (X', C', B')$ *be a subsystem of a numerical CSP* $P = (X, C, B)$ ($|X'| = k$), *in which the (output) variables in* X' *are involved in at least one constraint in* C' *and the* **input variables** *(i.e., the variables involved in at least one constraint in* C' *which are not in* X'*) are replaced by their current interval in* B.

The subsystem P' *is* **box-k-consistent** *if there exists a k-box of size* 1 *u.l.p. on every face of the k-box* B' *for which all the constraints* c *in* C' *are "satisfied", i.e.,* $0 \in c(X')$.

If a box-k-consistent subsystem has an empty set of input variables, note that this subsystem is also global hull consistent [5]. Thus, like for the standard box-consistency, the presence of input variables makes the box-k-consistency weaker than global hull consistency.

[1] The definition of box-consistency can be straightforwardly extended to inequalities.

[2] One Unit in the Last Place is the gap between two very close floating-point numbers.

Fig. 1-left shows an example of a 2×2 subsystem. The outer box is box-consistent since it optimally approximates the solution set of constraints c_1 and c_2 individually. The inner box is box-2-consistent since it optimally approximates the set of six "thick" solutions to both constraints. Constraints are thick because the input variables (e.g., w_1, w_2, w_3) are replaced by intervals.

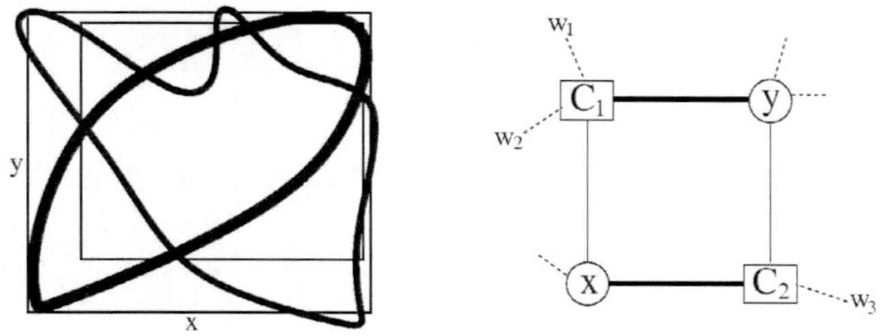

Fig. 1. Illustration of Box-2-consistency

Partial consistencies of NCSPs are generally defined modulo a precision ϵ that is used in practice by the corresponding algorithm to reach a fixpoint earlier. ϵ must then replace 1 u.l.p. in the previous definitions.

3.1 Benefits of Box-k-Consistency

The following example theoretically shows that a contraction obtained by a $k \times k$ subsystem may be stronger than contraction on 1×1 subsystems and on the whole $n \times n$ system performed by an interval Newton. Consider the NCSP $P = (\{x, y, z\},\ \{x - y = 0,\ x + y + z = 0,\ (z - 1)(z - 4)(2x + y + 2) = 0\},\ \{[-10^6, 10^6], [-10^6, 10^6][-10, 10]\})$.

Running `Box` and interval Newton on P does not filter the box. Achieving Box-2-consistency on the 2×2 subsystem ({x,y}, {x-y=0, x+y+z=0}) narrows

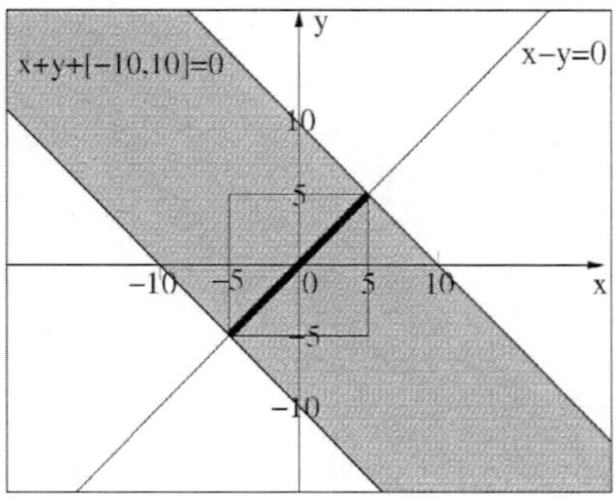

Fig. 2. Illustration of a subsystem of size 2, with $z = [-10, 10]$ as input variable. {[-5,5],[-5,5]} is box-2-consistent w.r.t. the 2 constraints $x - y = 0$ and $x + y + [z] = 0$.

the intervals of x and y to $[-5, 5]$ as shown in Fig. 2. Also, if branching was used to find solutions, only two bisections (choice points) would be necessary to isolate the 3 solutions $\{(\frac{-2}{3}, \frac{-2}{3}, \frac{4}{3}), (-0.5, -0.5, 1), (-2, -2, 4)\}$. We should highlight that Newton on the whole system does not contract the box because it contains several solutions, whereas Newton on the 2×2 subsystem does because it contains only one (thick) solution (segment in bold). Of course, this small example is didactic. Experiments described in Section 6 show larger and non-linear instances highlighting the benefits of structural partial consistencies over stronger partial consistencies like 3B-consistency [11].

3.2 Achieving Box-k-Consistency in Well-Constrained Subsystems of Equations

Enforcing Box-k-consistency in every subsystem of given size k is too time-consuming and counter-productive in practice. The number of subsets of k variables in a NCSP with n variables is high and one needs to consider only promising subsystems.

We have thus used several criteria to reduce the number of subsystems that are candidate. We first select subsystems with only equations (no inequalities) because equations bring a great reduction of the search space and have nice properties. To understand these properties, we have to pay attention to NCSPs that admit a finite number of solutions. These NCSPs contains n variables but also the same number n of independent equations (additional inequalities can reduce the number of solutions). Also, the corresponding *bipartite constraint graph* verifies the following structural/graph property [1].

Definition 5. *Let P be a system of n independent equations constraining n variables. The vertices of the* **bipartite constraint graph** *G corresponding to P are the n variables and the n equations, and edges connect one equation to its involved variables.*

The system of equations P is **(structurally) well-constrained** *if its constraint graph G has a perfect matching [7].*

For instance, Fig. 1-right shows the perfect matching (bold-faced edges) of the corresponding subgraph. This structural well-constriction can be viewed as a necessary condition to obtain a finite set of solutions. It appears that interval Newton also requires this condition (while it is of course not sufficient) for contracting a box. Indeed, if the system is not structurally well-constrained, the jacobian matrix will necessarily be singular [1]. Our subsystems fulfill this condition because Interval Newton is used by our new Box-k-Revise procedure (see Section 4) to achieve faster a box-k-consistent subsystem. (Also, the time complexity of interval Newton is cubic in the number of variables, so that it is sometimes intractable to apply it to very large NCSPs. Instead, we could use interval Newton only inside subsystems.)

We finally require our subsystems be connected for performance considerations. Indeed, if a given subsystem of size k contained several disconnected components of size at most k' ($k' < k$), we could make it box-k-consistent by achieving box-k'-consistency in every component.

To sum up, restricting the subsystems to well-constrained and connected subgraphs of equations has two virtues. First, it allows a strong filtering in specific subparts of the system, which is useful for sparse NCSPs or for (globally) under-constrained ones, e.g., systems mixing equalities and inequalities. Second, it allows the use of an interval Newton to faster contract the subsystem.

4 Contraction Algorithm Using Well-Constrained Subsystems as Global Constraints

Instead of contracting all the well-constrained subsystems of given size k, we have designed an AC3-like propagation that manages selected subsystems of different sizes: subsystems of size 1 but also well-constrained subsystems of larger size. Well-constrained subsystems are thus similar to *global constraints* [16,10] that can be defined by the user or automatically (see Section 6).

All the subsystems are first put into a propagation queue and revised in sequence. When a variable domain is reduced more than a ratio ρ_{propag}, all the subsystems involving this variable are pushed into the queue, if they are not already in it. This propagation process is just specialized by the revise procedure used for contracting the subsystems of size greater than 1 and detailed below.

4.1 The `Box-k` Revise Procedure

The revise procedure is based on a branch & prune method limiting the bisection to the k (output) variables X of the subsystem, and using a *breadth-first* search. At the end of this local tree search, the current box is replaced by the hull of the leaves of the local tree. The algorithm `Box-k-Revise` is a generic procedure that achieves a box-k-consistent subsystem. The procedure manages a list L of nodes that are leaves of the local tree. A leaf l in L has three significant components: $l.box$ designs the (n-dimensional) search space associated to the node; $l.precise$ is a boolean stating whether $l.box$ has reached the precision ϵ in all the dimensions (ϵ also yields the precision of the global solution); $l.certified$ is a boolean asserting whether $l.box$ contains a unique solution. The `box` parameter is the current global box (search space) when the revise procedure is called.

A combinatorial process (tree search) is performed by the `while` loop. At every iteration, one leaf in L, which is not *precise* and not *certified*, is selected, bisected and the two new sub-boxes are contracted. The search ends if all the leaves are tagged as *certified* or *precise* or if a limit τ_{leaves} in the number of leaves is reached. τ_{leaves} limits the memory storage requirement (see Section 4.5) and allows one to quickly propagate the obtained reductions to the other subsystems.

A leaf is simply selected in breadth-first order. We first tried a more sophisticated heuristic function for selecting a "large" box on the border of the hull of the different leaves. The idea was to maximize the gain in volume on the current

Algorithm 1. `Boxk-Revise` (**in-out** L, `box`; **in** X, C, ϵ, `subContractor`, τ_{leaves}, $\tau_{\rho_{io}}$)

```
UpdateLocalTree(L, box, X, C, ε, subContractor)
L' ← {l ∈ L s.t. ¬l.certified and ¬l.precise and ProcessLeaf?(l,X,C,τ_ρio)}
while 0 < L'.size and L.size < τ_leaves do
  l ← L'.front() /* Select a leaf in breadth-first order */
  (l1, l2) ←bisect(l, X)
  contract(l1, subContractor, X, C, ε)
  contract(l2, subContractor, X, C, ε)
  if l1.box ≠ ∅ then L.pushBack(l1) end if
  if l2.box ≠ ∅ then L.pushBack(l2) end if
  L.remove(l)
  L' ← {l ∈ L s.t. ¬l.certified and ¬l.precise and ProcessLeaf?(l,X,C,τ_ρio)}
end while
box ←hull(L) /* Outer approximation of the union of all the boxes l.box, l ∈ L */
```

global box in case the selected leaf would be eliminated by filtering. This multi-dimensional generalization of the `BoxNarrow` algorithm (that shaves the bounds of the handled interval in the `Box` algorithm) has been discarded because it did not bring a significant gain in performance.

Algorithm 2. `contract`(**in-out** l; **in** `subContractor`, X, C, ϵ)

```
if ¬l.precise then
  if ¬l.certified then subContractor(l.box) end if
  if l.box ≠ ∅ and I-Newton(l.box,X) then l.certified ← true end if
  if maxDiameter(l.box) < ε then l.precise ← true end if
end if
```

The procedure `contract` is mainly parameterized by the contraction procedure `subContractor` (`HC4` [2] or `3BCID` [17] in our experiments). The scope C of `subContractor` is the considered k-set of equations. After a call to `subContractor`, an interval Newton limited to the $k \times k$ subsystem is launched. If Newton certifies a unique solution in a leaf, `I-Newton` contracts *l.box* and returns *true* so that this leaf is tagged as *certified*.

4.2 The `S-kB-Revise` Variant

`S-kB-Revise` is the name of a variant of `Box-k-Revise` for which the entire system is used in the `contract` procedure. That is, the scope C of `subContractor` includes the whole n-set of constraints, instead of the k-set of constraints attached to the subsystem. With `S-kB-Revise`, the k-set of constraints in the subsystem is just used by interval Newton. This variant brings additional filtering, but at a higher cost.

4.3 Reuse of the Local Tree (Procedure `UpdateLocalTree`)

A simpler version of Algorithm 1 did not call the `UpdateLocalTree` procedure and simply initialized the list L with the current box. However, instead of performing an intensive search effort in only one subsystem, we preferred to quickly propagate the obtained reductions to the other subsystems. Therefore the `UpdateLocalTree` procedure reuses the local tree (i.e., its leaves) that has been saved in a previous call to Algorithm 1. Every leaf in the current list L is just updated by intersection with the current box and filtered with `subContractor`.

Algorithm 3. `UpdateLocalTree` (**in-out** L; **in** `box`, X, C, ϵ, `subContractor`)

```
if L = ∅ then
  L ← {Leaf(box)} /* Initialize the root of the local tree with the current box */
else
  for all l ∈ L do
    /* Update and contract every leaf of the stored local tree */
    if l.box ≠ (l.box ∩ box) then
      l.box ← l.box ∩ box
      contract(l, subContractor, C, ε))
      if l.box = ∅ then L.remove(l) end if
    end if
  end for
end if
```

In fact, the leaves of the local trees are also maintained in the global search tree. To do so, the list L is implemented as a *backtrackable* data-structure updated in case of backtracking. It avoids redoing the same job in the subsystems several times, in particular when the *multisplit* splitting heuristic is chosen (see Section 5).

4.4 Lazy Handling of a Leaf (Procedure `ProcessLeaf?`)

Our first experiments have shown us that handling a leaf in a local tree, i.e., bisecting it and contracting the two sub-boxes, was often counterproductive. We have then defined an input/output ratio ρ_{io} that decides whether a given leaf of box B must be handled in the local tree.

$$\rho_{io}(B, I, O, F) = \frac{Max_{x \in I}(smear(x))}{Max_{x \in O}(smear(x))}$$

The function `ProcessLeaf?` calculates ρ_{io} in a leaf. If this ratio is larger than a threshold $\tau_{\rho_{io}}$, the leaf will not be handled in the current revise procedure.

ρ_{io} is based on the well-known *smear* function [9] defined by:

$smear(x){:=}Max_{f \in F}(|\frac{\partial f}{\partial x}| \times Diam(x))$. This function is often used for selecting the next variable to be bisected in NCSPs (the variable with the largest smear evaluation).

The denominator of ρ_{io} can be directly explained by it: output variables (O) with a great smear evaluation (implying a small ratio ρ_{io}) often lead to a great contraction when they are bisected inside the local subsystem tree. Desiring a small impact of the input variables (I) is less intuitive. We understand that large input domains generally lead to large output domains (i.e., leaf boxes) in the subsystem and thus yields a poor reduction. The same argument holds in fact for the derivatives of functions. To illustrate this point, let us take a subsystem of size 1 like $0.001\,y + x^2 - 1 = 0$ (x is the output variable; $[x] = [y] = [-1, 1]$) having $\rho_{io} = \frac{0.002}{4} = 0.0005$. After one bisection on x, the subsystem contraction leads to a very small interval for x. A large interval would be obtained for x if the considered subsystem was $y + x^2 - 1 = 0$ with $\rho_{io} = \frac{2}{4} = 0.5$.

4.5 Properties of the Revise Procedure

The following proposition formalizes the correctness, the memory and time complexities of the procedure `Box-k-Revise`.

Proposition 1. *Let $P' = (X', C', B)$ be a subsystem of a CSP $P = (X, C, B)$, with $|X| = n$, $|C| = m$, $|X'| = |C'| = k$.*

The procedure `Box-k-Revise`*, called with $\tau_{leaves} = +\infty$ and $\tau_{\rho_{io}} = +\infty$, makes P' box-k-consistent.*

Let Diam be the largest interval diameter in B. Let d be $log_2(\frac{Diam}{\epsilon})$, the maximum number of times a given interval must be bisected to reach the precision ϵ. [3]

The memory complexity of `Box-k-Revise` *is $O(k\,\tau_{leaves})$.*

The number of calls to `subContractor` *is $O(k\,d\,\tau_{leaves})$.*

Proof. The correction is based on the combinatorial process performed by the procedure `Box-k-Revise`. Called with $\tau_{leaves} = +\infty$ and with the subsystem made of C', the procedure computes all the atomic boxes of precision ϵ in the subsystem before returning the hull of them, thus achieving roughly (i.e., assuming that the actual values of input variables are unknown) the global consistency of P'.

The memory complexity comes from the breadth-first search that must store the $O(\tau_{leaves})$ leaves of the local tree. The revise procedure works with n-dimensional boxes but, in order to save memory, stores at the end only k intervals of a $k \times k$ subsystem.

The number of calls to `subContractor` is bounded by the number of nodes in the local search tree. The number of leaves of this tree is τ_{leaves} (corresponding to *living* boxes that can contain solutions) plus the number of *dead* leaves eliminated by filtering. For any living leaf l, the number of nodes created in the tree to reach l is at most $2 \times d \times k$ since the root must be at most bisected d times in all its k dimensions. Although numerous such internal nodes are "shared" by several living leaves, this bounds the number of calls to a sub-filtering operator with $O(k\,d\,\tau_{leaves})$. □

[3] d generally falls between 20 and 60 in NCSPs occurring in practice.

Another property allows us to better understand the gain in contraction obtained by the `S-kB-Revise` variant (see Section 4.2).

Proposition 2. *Consider a propagation algorithm calling* `S-kB-Revise` *on all the subsystems of size k in a given NCSP P.*

This algorithm computes the $(k+2)B$-consistency of P.

The kB-consistency, introduced by Lhomme [11], is a strong partial consistency related to the k-consistency (in finite-domain CSPs) restricted to the bounds of intervals. $3B$-consistency is similar to SAC-consistency [6]. It is known to be stronger (i.e., to better contract) than box-consistency (i.e., box-1-consistency). It appears that this result can be generalized to any $k > 1$.

5 Multidimensional Splitting

It turns out that the `Box-k-Revise` procedure has not only a contraction effect, but also provides a new way to make choice points, that is, to build the (global) search tree. This new splitting strategy is called *multidimensional splitting* (in short *multisplit*).

Definition 6. *Consider a $k \times k$ subsystem P' defined inside an NCSP $P = (X, C, B)$. Consider a set S of m boxes associated to P' such that S contains all the solutions to P, and the m boxes obtained by projection on P' of the boxes in S are pairwise disjoint.*

A **multisplit** *of dimension k consists in splitting the search space B into the m boxes in S.*

In practice, the m boxes correspond to the leaves of a subsystem local tree. At the end of a `Box-k` propagation, our solving strategy makes a choice between a classical bisection and a multisplit. If all the subsystems have a ratio ρ_m larger than a user-defined threshold τ_m, then a standard bisection is performed. Otherwise, we multisplit the subsystem with the smallest ratio ρ_m, i.e., we replace the current box by the set L of m leaves associated to the local tree.

$$\rho_m = \frac{\sum_{l \in L} Volume(l)}{Volume(Hull(L))}$$

Multisplit generalizes a procedure used by `IBB` (see Section 6.1). `IBB` performs a multisplit once it finds the m solutions (i.e., atomic boxes) in a given block. The difference here is that a multisplit may occur with *non* atomic boxes whose size has not reached the required precision.

6 Experiments

The `Box-k` based propagation algorithm has been implemented in the `Ibex` open source interval-based solver in `C++` [4,3]. The variant with multisplit (`msplit`) performs a multisplit of a subsystem with the minimum ratio ρ_m, provided that $\rho_m < \tau_m$=0.99. All the competitors are also available in the same library, making the comparison fair.

6.1 Experiments on Decomposed Benchmarks

Ten *decomposed* benchmarks, described in [15,14], appear in Table 1. They have been previously decomposed by equational algorithms (*eq*) like maximum-matching, or by more sophisticated geometrical algorithms (*geo*). They are challenging for general-purpose interval methods, but can efficiently be solved by `IBB` [15,14].

Brief Description of `IBB`

`IBB` is dedicated to decomposed systems, i.e., sparse systems of equations that have been first decomposed into a sequence of *irreducible* [1] well-constrained blocks/subsystems. Inter-Block Backtracking handles every block in the order provided by the sequence. It interleaves contraction steps (performed by `HC4` and interval Newton) and bisections inside the block until atomic boxes (solutions) are obtained. Choice points are then made: the variables of the block are replaced by one of the atomic boxes, i.e., they are considered constant in subsequent blocks.

We understand that the `Box-k-Revise` procedure plus multisplit represents a generalization of `IBB` in that the input variables domains of a subsystem are not necessarily atomic and that a multisplit is not necessarily performed after a subsystem handling. In other terms, the `IBB` block handling is not a revise procedure, it is just an ad-hoc procedure embedded in a dedicated algorithm. Applied to decomposed systems, the only information that our new approach does not exploit is the order between blocks which provides to `IBB` a useful splitting heuristic.

Experimental Protocol

Every `Box-k` based strategy has been tuned with 6 different sets of parameter values: $\tau_{\rho_{io}}$ is 0.01, 0.2 or 0.8 (0.01 is always the best value on decomposed systems); the precision ρ_{propag} used in the `HC4` propagation is 1% or 10%; All the other parameters have been empirically fixed: the precision ρ_{propag} in the `Box-k` propagation is always 10%; the maximum number τ_{leaves} of leaves inside a subsystem tree is 10; the number of slices of `3BCID` in `Box-k(3BCID)` is 10. To be fair, the parameters of the competitor algorithms have been tuned so that 8 trials have been performed for `Box` and `HC4`, and 16 trials have been run for `3BCID`. For all the tests, the Newton ceil (size of maximum diameter under which interval Newton is run) is 10, and the same variable order is used in a round-robin strategy (except for `IBB` and for `Box-k` with multisplit).

The subsystems given to our `Box-k` propagation are defined automatically. The irreducible blocks produced by the `IBB` decomposition simply become the well-constrained subsystems handled by `Box-k-Revise`.

Results

Strategies based on `HC4`, `Box` and `3BCID` followed by interval Newton are not competitive at all with `Box-k` and `IBB` on the tested decomposed systems. The

Table 1. Experimental results on IBB benchmarks. The first 3 columns include the name of the system, its number n of variables and its number of solutions. The next three columns yield the CPU time (above) and the number of boxes, i.e., choice points (below), obtained on an Intel 6600 2.4 GHz by existing strategies based on HC4, Box or 3BCID followed by interval Newton (between two bisections selected in a round-robin way for the variable selection). The last four columns report the results obtained by our algorithms on the same computer: Box-k-Revise parameterized by subContractor=HC4 or subContractor=3BCID, with multisplit (msplit) or without. To be the closest to IBB, Box-k-Revise, and not the S-kB-Revise variant, is used by our constraint propagation algorithm.

Benchmark	n	#sols	**HC4**	**Box**	**3BCID**	**IBB**	**Box-k(HC4)**		**Box-k(3BCID)**	
								msplit		**msplit**
Chair(eq)	178	8	>3600	>3600	>3600	**0.27**	>3600	16.5	>3600	**0.52**
1x15,1x13,1x9,5x8,3x6,...								575		15
Latham(eq)	102	96	>3600	>3600	39.9	**0.17**	**0.94**	1.35	1.5	1.08
1x13,1x10,1x4,25x2,25x1					587		839	199	991	189
Ponts(eq)	30	128	33.4	33.4	1.89	**0.59**	6.85	8.19	0.79	**0.71**
1x14,6x2,4x1			20399	20399	357		783	231	307	231
Ponts(geo)	38	128	44.1	44.1	2.6	**0.16**	2.01	0.31	1.45	**0.39**
13x2,12x1			18363	18363	685		6711	767	6711	767
Sierp3(geo)	124	198	>3600	>3600	77.5	**0.62**	49.0	**1.38**	52.5	1.77
44x2,36x1					1727		84169	1513	84169	1513
Star(eq)	46	128	>3600	>3600	4.9	**0.05**	35.6	**0.12**	44.0	0.26
3x6,3x4,8x2					283		44195	263	44023	263
Tangent(eq)	28	128	77	77	2.1	**0.08**	1.74	**0.08**	1.87	0.14
1x4,10x2,4x1			390903	390903	753		12027	255	12235	255
Tangent(geo)	42	128	–	–	7.38	**0.08**	0.80	0.19	0.80	**0.19**
2x4,11x2,12x1					859		1415	251	1407	251
Tetra(eq)	30	256	1281	1281	12.3	**0.63**	33.6	1.06	13.57	**0.76**
1x9,4x3,1x2,7x1			607389	607389	1713		4619	483	2243	483
Sierp3(eq)	see Section 6.2					>5000	see Section 6.2			

comparison of Box-k against IBB is very positive because the CPU times reported for IBB are really the best that have never been obtained with any variant of this dedicated algorithm. Also, no timeout is reached by Box-k+multisplit and IBB is on average only twice faster than Box-k(3BCID) (at most 6 on Latham). As expected, the results confirm that multisplit is always relevant for decomposed benchmarks. For the benchmark Sierp3(eq) (the fractal Sierpinski at level 3 handled by an equational decomposition), an equational decomposition makes appear a large irreducible 50×50 block of distance constraints. This renders IBB unefficient on it (timeout).

6.2 Experiments on Structured Systems

Eight *structured* systems appear in Table 2. They are scalable chains of constraints of reasonable arity [12]. They are denoted *structured* because they are not sufficiently sparse to be decomposed by an equational decomposition, i.e., the system contains only one irreducible block, thus making IBB pointless. A brief and manual analysis of the constraint graph of every benchmark has led us to define a few well-constrained subsystems of reasonable size (between 2 and 10). In the same way, we have replaced the 50×50 block in Sierp3(eq) by 6×6 and 2×2 Box-k subsystems.

Table 2. Results on structured benchmarks. The same protocol as above has been followed, except that the solving strategy is more sophisticated. Between two bisections, the propagation with subsystems follows a `3BCID` contraction and an interval Newton. The four `Box-k` columns report the results obtained by the `S-kB-Revise` variant. The results obtained by `Box-k-Revise` are generally worse and appear, with multisplit only, in the last two columns.

Benchmark	n	#sols	**HC4**	**Box**	**3BCID**	**Box-k(HC4)**		**Box-k(3BCID)**		**Box-k-Revise**	
							msplit		**msplit**	HC4	3BCID
Bratu	60	2	58	626	48.7	47.0	**33.0**	135	126	86.4	96.2
29x3			15653	13707	79	39	17	43	25	125	129
Brent	10	1015	1383	127	**17.0**	28.5	20.2	44.9	31.0	20.8	34.9
2x5			7285095	42191	9849	2975	4444	4585	1309	5215	4969
BroydenBand	20	1	>3600	0.17	**0.11**	0.45	0.15	0.91	0.31	0.30	0.28
1x6,3x5				1	21	4	19	17	3	7	3
BroydenTri	30	2	1765	**0.16**	0.25	0.22	0.24	0.39	0.29	0.19	0.23
6x5			42860473	63	25	11	19	9	3	19	17
Reactors	30	120	>3600	>3600	288	340	315	81.4	**67.5**	250	194
3x10					39253	14576	10247	1038	788	35867	21465
Reactors2	10	24	>3600	>3600	28.8	**9.5**	12.3	10.4	12.2	9.93	11.9
2x5					128359	4908	10850	4344	5802	5597	5353
Sierp3Bis(eq)	83	6	>3600	>3600	4917	>3600	>3600	>3600	**389**	>3600	4503
1x14,6x6,15x2,3x1					44803				218		122409
Trigexp1	30	1	>3600	13	**0.08**	**0.08**	**0.08**	**0.08**	0.09	0.08	0.08
6x5				27	1	1	1	1	1	1	1
Trigexp2	11	0	1554	>3600	83.7	81.2	85.7	105	83.0	**80.6**	82.1
2x4,2x3			2116259		16687	15771	16755	3797	2379	15771	11795

Standard strategies based on `HC4` or `Box` followed by interval Newton are generally not competitive with `Bok-k` on the tested benchmarks. The solving strategy based on `S-kB-Revise` with `subContractor=3BCID` (column `Box-k(3BCID)`) appears to be a robust hybrid algorithm that is never far behind `3BCID` and is sometimes clearly better. The gain w.r.t. `3BCID` falls indeed between 0.7 and 12. The small number of boxes highlights the additional filtering power brought by well-constrained subsystems. Again, multisplit is often the best option.

The success of `Box-k` on `Sierp3Bis(eq)` has led us to try a particular version of `IBB` in which the inter-block filtering [15] is performed by `3BCID`. Although this variant seldom shows a good performance, it can solve `Sierp3(eq)` in 330 seconds.

6.3 Benefits of Sophisticated Features

Tables 3 has finally been added to show the individual benefits brought by two features: the user parameter $\tau_{\rho_{io}}$ driving the procedure `ProcessLeaf?` and the backtrackable list of leaves used to reuse the job achieved inside the subsystems.

Every cell reports the best result (CPU time in second) among both subcontractors. Multisplit is allowed in all the tests. The first line of results corresponds to the implemented and sophisticated revise procedure; the next ones correspond to simpler versions for which at least one of the two advanced features has been removed.

Three main observations can be drawn. First, when a significant gain is brought by the features on a given system, then this system is efficiently handled

Table 3. Benefits of the backtrackable data structure (BT) and of $\tau_{\rho_{io}}$ in the `Box-k`-based strategy. Setting $\tau_{\rho_{io}} = \infty$ means that subsystem leaves will be always processed in the revise procedure.

	Chair	Latham	Ponts(eq)	Ponts(geo)	Sierp3(geo)	Star	Tan(eq)	Tan(geo)	Tetra
BT, $\tau_{\rho_{io}}$	0.52	1.08	0.71	0.31	1.38	0.12	0.08	0.19	0.76
$\neg$ BT, $\tau_{\rho_{io}}$	10.8	4.61	1.51	1.27	23.9	2.34	0.71	1.58	2.13
BT, $\tau_{\rho_{io}} = \infty$	23.4	4.71	2.60	1.00	23.8	1.67	1.09	1.81	3.57
$\neg$ BT, $\tau_{\rho_{io}} = \infty$	24.2	6.60	2.80	1.11	23.9	2.40	1.15	1.82	3.54

	Bratu	Brent	BroyB.	BroyT.	Sierp3B(eq)	Reac.	Reac.2	Trigexp1	Trigexp2
BT, $\tau_{\rho_{io}}$	33.0	20.2	0.15	0.24	389	67	12.2	0.08	83
$\neg$ BT, $\tau_{\rho_{io}}$	33.2	21.0	0.14	0.23	411	97	12.0	0.07	85
BT, $\tau_{\rho_{io}} = \infty$	33.9	23.8	0.38	0.28	519	164	13.1	0.10	103
$\neg$ BT, $\tau_{\rho_{io}} = \infty$	33.0	28.7	0.40	0.38	533	401	18.7	0.07	148

against competitors in Tables 1 and 2. Second, $\tau_{\rho_{io}}$ seems to have a better impact on performance than the backtrackable list, but the difference is slight. Third, several systems are only slightly improved by one of both features, whereas the gain is significant when both are added together. This is true for most of the `IBB` benchmarks. On these systems, between 2 bisections in the search tree, it often occurs that a job inside *several* subsystems leads to identify atomic boxes (some others are not fully explored thanks to $\tau_{\rho_{io}}$). Although we multisplit only one of these subsystems, the job on the others is saved in the backtrackable list.

7 Conclusion

We have proposed a new type of filtering algorithms handling $k \times k$ well-constrained subsystems in an NCSP. $k \times k$ interval Newton calls and selected bisections inside such subsystems are useful to better contract decomposed and structured NCSPs. In addition, the local trees built inside subsystems allow a solving strategy to learn choice points bisecting several variable domains simultaneously.

Solving strategies based on `Box-k` propagations and multisplit have mainly three parameters: the choice between `Box-k-Revise` and `S-kB-Revise` (although `Box-k-Revise` seems better suited only for decomposed systems), the choice of sub-contractor (although `3BCID` seems to be often a good choice), and $\tau\rho_{io}$. This last parameter appears to be finally the most important one.

On decomposed and structured systems, our first experiments suggest that our new solving strategies are more efficient than standard general-purpose strategies based on `HC4`, `Box` or `3BCID` (with interval Newton). `Box-k`+multisplit can be viewed as a generalization of `IBB`. It can also solve large decomposed NCSPs with relatively small blocks in less than one second, but can also handle structured NCSPs that `IBB` cannot treat.

Subsystems have been automatically added in the decomposed systems, but have been manually added in the structured ones, as global constraints. In this paper, we have validated the fact that handling subsystems could bring additional contraction and relevant multi-dimensional choice points. The next step is to automatically select a relevant set of subsystems. We believe that an adaptation of maximum-matching machinery or other graph-based algorithms along with a criterion similar to ρ_{io} could lead to efficient heuristics.

Acknowledgments

We thank the referees for their helpful comments.

References

1. Ait-Aoudia, S., Jegou, R., Michelucci, D.: Reduction of Constraint Systems. In: Compugraphic (1993)
2. Benhamou, F., Goualard, F., Granvilliers, L., Puget, J.-F.: Revising Hull and Box Consistency. In: Proc. ICLP, pp. 230–244 (1999)
3. Chabert, G.: `http://www.ibex-lib.org` (2009)
4. Chabert, G., Jaulin, L.: Contractor Programming. Artificial Intelligence 173, 1079–1100 (2009)
5. Cruz, J., Barahona, P.: Global Hull Consistency with Local Search for Continuous Constraint Solving. In: Brazdil, P.B., Jorge, A.M. (eds.) EPIA 2001. LNCS (LNAI), vol. 2258, pp. 349–362. Springer, Heidelberg (2001)
6. Debruyne, R., Bessière, C.: Some Practicable Filtering Techniques for the Constraint Satisfaction Problem. In: Proc. IJCAI, pp. 412–417 (1997)
7. Dulmage, A.L., Mendelsohn, N.S.: Covering of Bipartite Graphs. Canadian Journal of Mathematics 10, 517–534 (1958)
8. Jaulin, L., Kieffer, M., Didrit, O., Walter, E.: Applied Interval Analysis. Springer, Heidelberg (2001)
9. Kearfott, R.B., Novoa III, M.: INTBIS, a portable interval Newton/Bisection package. ACM Trans. on Mathematical Software 16(2), 152–157 (1990)
10. Lebbah, Y., Michel, C., Rueher, M., Daney, D., Merlet, J.P.: Efficient and safe global constraints for handling numerical constraint systems. SIAM Journal on Numerical Analysis 42(5), 2076–2097 (2005)
11. Lhomme, O.: Consistency Tech. for Numeric CSPs. In: IJCAI, pp. 232–238 (1993)
12. Merlet, J.-P.: `http://www-sop.inria.fr/coprin/logiciels/ALIAS/Benches/benches.html` (2009)
13. Neumaier, A.: Int. Meth. for Systems of Equations. Cambridge Univ. Press, Cambridge (1990)
14. Neveu, B., Chabert, G., Trombettoni, G.: When Interval Analysis helps Interblock Backtracking. In: Benhamou, F. (ed.) CP 2006. LNCS, vol. 4204, pp. 390–405. Springer, Heidelberg (2006)
15. Neveu, B., Jermann, C., Trombettoni, G.: Inter-Block Backtracking: Exploiting the Structure in Continuous CSPs. In: Jermann, C., Neumaier, A., Sam, D. (eds.) COCOS 2003. LNCS, vol. 3478, pp. 15–30. Springer, Heidelberg (2005)
16. Régin, J.-C.: A Filtering Algorithm for Constraints of Difference in CSPs. In: Proc. AAAI 1994, pp. 362–367 (1994)
17. Trombettoni, G., Chabert, G.: Constructive Interval Disjunction. In: Bessière, C. (ed.) CP 2007. LNCS, vol. 4741, pp. 635–650. Springer, Heidelberg (2007)

Minimising Decision Tree Size as Combinatorial Optimisation*

Christian Bessiere[1], Emmanuel Hebrard[2], and Barry O'Sullivan[2]

[1] LIRMM, Montpelier, France
bessiere@lirmm.fr
[2] Cork Constraint Computation Centre
Department of Computer Science, University College Cork, Ireland
{e.hebrard,b.osullivan}@4c.ucc.ie

Abstract. Decision tree induction techniques attempt to find small trees that fit a training set of data. This preference for smaller trees, which provides a learning bias, is often justified as being consistent with the principle of Occam's Razor. Informally, this principle states that one should prefer the simpler hypothesis. In this paper we take this principle to the extreme. Specifically, we formulate decision tree induction as a combinatorial optimisation problem in which the objective is to minimise the number of nodes in the tree. We study alternative formulations based on satisfiability, constraint programming, and hybrids with integer linear programming. We empirically compare our approaches against standard induction algorithms, showing that the decision trees we obtain can sometimes be less than half the size of those found by other greedy methods. Furthermore, our decision trees are competitive in terms of accuracy on a variety of well-known benchmarks, often being the most accurate. Even when post-pruning of greedy trees is used, our constraint-based approach is never dominated by any of the existing techniques.

1 Introduction

Decision trees [5] are amongst the most commonly used classifiers in real-world machine learning applications. Part of the attraction of using a decision tree is that it is easy to use and interpret. For example, consider the data set for a simple classification task in Figure 1(a). Each training example is defined by a set of weather features (outlook, temperature, humidity, windy) and a class: $+$ ($-$) meaning I am happy (unhappy) to play outdoors under the given weather conditions. A decision tree for this training set is presented in Figure 1(b). The decision tree makes classifications by sorting the features of an instance through the tree from the root to some leaf node. At each internal node a test on a feature is performed, and each subtree corresponds to a possible outcome for that test. Classifications are made at the leaf nodes.

Traditional decision tree induction techniques attempt to find small trees that fit a training set of data. This preference for smaller trees is often justified as being consistent

* Bessiere is supported by the project CANAR (ANR-06-BLAN-0383-02). Hebrard and O'Sullivan are supported by Science Foundation Ireland (Grant number 05/IN/I886).

I.P. Gent (Ed.): CP 2009, LNCS 5732, pp. 173–187, 2009.

outlook	temp.	humidity	windy	play?
sunny	hot	high	false	-
sunny	hot	high	true	-
dull	hot	high	false	+
rain	mild	high	false	+
rain	cool	normal	false	+
rain	cool	normal	true	-
dull	cool	normal	true	+
sunny	mild	high	false	-
sunny	cool	normal	false	+
rain	mild	normal	false	+
sunny	mild	normal	true	+
dull	mild	high	true	+
dull	hot	normal	false	+
rain	mild	high	true	-

(a) An example data-set.

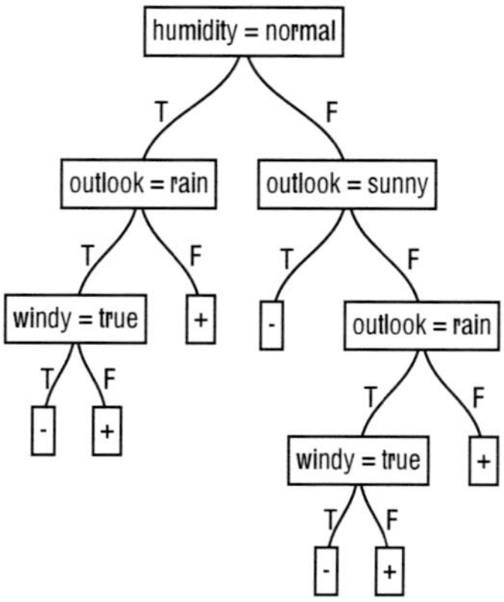

(b) A decision tree.

Fig. 1. An example decision tree learning problem

with the principle of Occam's Razor. Informally, this principle states that one should prefer simpler hypotheses. However, the majority of existing decision tree algorithms are greedy and rely on heuristics to find a small tree without search. While the decision trees found are usually small, there is no guarantee that much smaller, and possibly more accurate, decision trees exist. Finding small decision trees is often of great importance. Consider a medical diagnosis task in which each test required to diagnose a disease is intrusive or potentially risky to the well-being of the patient. In such an application minimising the number of such tests is of considerable benefit.

In this paper we study the problem of minimising decision tree size by regarding the learning task as a combinatorial optimisation problem in which the objective is to minimise the number of nodes in the tree. We refer to this as the *Smallest Decision Tree Problem*. We study formulations based on satisfiability, constraint programming, and hybrids with integer linear programming. We empirically compare our approaches against standard induction algorithms, showing that the decision trees we obtain can sometimes be less than half the size of those found by other greedy methods. Furthermore, our trees are competitive in terms of accuracy on a variety of well-known benchmarks, often being the most accurate. Even when post-pruning of greedy trees is used, our constraint-based approach is never dominated by any of the existing techniques.

The remainder of the paper is organised as follows. In Section 2 we present the technical background and define the problem we solve in this paper. We present a formulation of the problem using satisfiability (Section 3), constraint programming (Section 4), and a hybrid of constraint programming and linear programming (Section 5). Our experimental results are presented in Section 6. Finally, we position our approach with respect to the existing literature in Section 7, and conclude in Section 8 highlighting some directions for future work.

2 Background

SAT and Constraint Programming. A *propositional satisfiability* (SAT) formula consists of a set of Boolean variables and a set of clauses, where a clause is a disjunction of variables or their negation. The SAT problem is to find an assignment 0 (false) or 1 (true) to every variable, such that all clauses are satisfied. SAT is a very simple formalism, but it is extremely expressive, making the SAT problem NP-hard. In addition to its simplicity, SAT has the advantage that significant research effort has led to several extremely efficient SAT solvers being developed.

A *constraint network* is defined by a set of variables, each with a finite domain of values, and a set of constraints specifying allowed combinations of values for some subsets of variables. The constraint satisfaction problem is to find an assignment to each variable with a value from its domain such that all constraints are satisfied. *Constraint programming* (CP) involves expressing decision problems with constraint networks, called models of the problem to be solved.

The Smallest Decision Tree Problem. A standard approach to evaluating the quality of a machine learning technique is to first learn a hypothesis on a selected *training* set of examples, and then evaluate the *accuracy* of the learned hypothesis on a *test* set of examples. Although more sophisticated measurements can also be taken, we consider the problem of finding the smallest decision tree consistent with a training set of examples, which is known to be NP-Hard [2].

Let $\mathcal{E} = \{e_1, \dots, e_m\}$ be a set of *examples*, that is, Boolean valuations of a set $\mathcal{F}$ of *features*, and let $\mathcal{E}^+$, $\mathcal{E}^-$ be a partition of $\mathcal{E}$. We denote by $e[f]$ the valuation (0 or 1) of the feature $f \in \mathcal{F}$ in example $e \in \mathcal{E}$. Let $T = (X, U, r)$ be a binary tree rooted by $r \in X$, where $L \subseteq X$ denotes the set of leaves of T. A *decision tree* based on T is a labelled tree in which each internal node $x \in X \setminus L$ is labelled with an element of $\mathcal{F}$, denoted by $f(x)$. Each edge $(x, y) \in U$ is labelled with a Boolean $g(x, y)$, where $g(x, y) = 0$ if y is the left child of x and $g(x, y) = 1$ if y is the right child of x. The *size* of the decision tree is the number of nodes of T. Given $l \in L$, $p(l)$ denotes the path in T from the root r to leaf l. To each example $e \in \mathcal{E}$, we can associate the unique leaf $l(e) \in L$ such that every edge (x, y) in $p(l(e))$ is such that $e[f(x)] = g(x, y)$. A decision tree *classifies* a set of examples $\mathcal{E}$ iff for every pair $e_i \in \mathcal{E}^+, e_j \in \mathcal{E}^-$ we have $l(e_i) \neq l(e_j)$. Given a set of examples $\mathcal{E}$, we want to find a decision tree that classifies $\mathcal{E}$ with a minimum number of nodes. Alternatively, we can minimise the longest branch.

In the rest of the paper, we assume that all features admit a Boolean valuation. Categorical features can be encoded numerically, and a non-binary feature $f \in [1, \dots, s]$ can be represented by a set of binary features $f_1, \dots, f_s$. These Boolean features correspond to *equality* splits, that is where $f_v = 1$ stands for $f = v$ and $f_v = 0$ stands for $f \neq v$. It is standard in machine learning to split numerical data with a *disequality* split ($f[e] \leq v$ *or* $f[e] > v$). In order to allow these two types of split, we add to each example a second set of Boolean features $f'_1, \dots, f'_s$ standing for disequality splits. That is, where $f'_v = 1$ stands for $f \leq v$ and $f'_v = 0$ stands for $f > v$. For instance, let $f^1 \in [1..4], f^2 \in [1..4], f^3 \in [1..4]$ be three features and $e = \langle 2, 4, 1 \rangle$ be an example. The binary encoding would yield the following example: 0100 0001 1000 0111 0001 1111, on the set of Boolean features: $\{f^1_1, ..f^3_4, f'^1_1, ..f'^3_4\}$.

3 A SAT-Based Encoding

We first introduce a SAT model to find decision trees. This baseline approach requires a large number of clauses to represent the problem. Furthermore, guiding search with the generic heuristics of SAT solvers is not efficient (see Section 6). However, this approach underlines the critical aspects of this problem that need to be addressed in order to develop an efficient approach.

Given a binary tree $T = (X, U, r)$ and a training set $\mathcal{E}$, we present a SAT formula that is satisfiable iff there is a decision tree based on T that classifies $\mathcal{E}$.

Intuition. Given a set of features $\mathcal{F} = \{a, b, q, r\}$, suppose there are two examples e_i in $\mathcal{E}^+$ and e_j in $\mathcal{E}^-$ that have a similar value on the set of features $eq(e_i, e_j) = \{a, b\}$ (with $e_i[a] = e_j[a] = 0$, $e_i[b] = e_j[b] = 1$) and that differ on the set of features $\mathcal{F} \setminus eq(e_i, e_j) = \{q, r\}$. The SAT encoding has to ensure that e_i and e_j are not associated with the same leaf. Examples e_i and e_j are not both associated with a given leaf $l \in L$ iff there exists an edge $(x, y) \in p(l)$ such that:

$$f(x) \in \mathcal{F} \setminus eq(e_i, e_j) \ \vee \ (f(x) \in eq(e_i, e_j) \ \& \ g(x, y) \neq e_i[f(x)]).$$

The first case ensures that if $l(e_i)$ and $l(e_j)$ have x as a common ancestor, they appear in one of the two subtrees rooted in x; the second case ensures that none of $l(e_i), l(e_j)$ is equal to l since they will both branch on the opposite child of x.

Encoding. For every node $x \in X \setminus L$, for every feature $f \in \mathcal{F}$, we introduce a literal t_{xf}, whose value 1 will mean that node x is labelled with feature f. For each pair $e_i \in \mathcal{E}^+$ and $e_j \in \mathcal{E}^-$, for each leaf $l \in L$, we build a clause that forbids e_i and e_j to be classified at l. On the example above, suppose there is a path $p(l) = (x_1, x_2, l)$ in the tree such that x_2 is the left child of x_1 and l is the right child of x_2. We would add the clause: $t_{x_1q} \vee t_{x_1r} \vee t_{x_2q} \vee t_{x_2r} \vee t_{x_1b} \vee t_{x_2a}$. t_{x_1q} means x_1 is labelled with a feature that discriminates between e_i and e_j because $q \in \mathcal{F} \setminus eq(e_i, e_j)$. t_{x_1b} means the feature labelling x_1 will classify both e_i and e_j in the branch that does not lead to l because $p(l)$ uses the left child of x_1 whereas $e_i[b] = e_j[b] = 1$. Formally, we build the clauses:

$$\begin{aligned} &\Big(\bigvee_{(x,y)\in p(l), f \in eq(e_i,e_j) \ | \ g(x,y) \neq e_i[f]} t_{xf} \Big) \\ &\vee \Big(\bigvee_{(x,y)\in p(l), f \in \mathcal{F} \setminus eq(e_i,e_j)} t_{xf} \Big) \\ &\forall (e_i, e_j) \in \mathcal{E}^+ \times \mathcal{E}^-, \forall l \in L. \end{aligned} \tag{1}$$

The following clauses ensure that each node is labelled with at most one feature:

$$(\neg t_{xf} \ \vee \ \neg t_{xf'}), \quad \forall x \in X \setminus L, \forall f, f' \in \mathcal{F}. \tag{2}$$

By construction, a solution to the SAT formula defined above completely characterises a decision tree. Let M be such a solution. A node $x \in X \setminus L$ will be labelled with $f \in \mathcal{F}$ iff $M[t_{xf}] = 1$.

We add redundant clauses specifying that two nodes on a same path should not take the same feature as it speeds up the resolution process:

$$\bigwedge_{(x,y)\in p(l), (x',y') \in p(l), x \neq x'} (\neg t_{xf} \ \vee \ \neg t_{x'f}), \quad \forall l \in L, \forall f \in \mathcal{F}. \tag{3}$$

Complexity. Given $n = |X|, k = |\mathcal{F}|, m = |\mathcal{E}|$, the number of literals is in $O(nk)$. Observe that the number of literals is independent of the size of $\mathcal{E}$. However, the number of clauses strongly depends on the size of $\mathcal{E}$. We build at most $m^2 \cdot n/2$ clauses of type (1), each of length in $O(kn)$, and $n/2 \cdot k^2$ clauses of type (2), each of length in $O(1)$, which gives a space complexity in $O(kn^2m^2 + nk^2)$. There are at most $n/2 \cdot k$ conjunctions of n^2 binary clauses of type (3), which gives an extra space in $O(kn^3)$. Observe that we approximate the depth of T by n. This is a brute force approximation. If the tree is balanced, the depth will be in $O(log(n))$.

Observation. Our encoding has an interesting characteristic: it deals with 'useless' nodes for free. A node $x \in X \setminus L$ is useless if none of the examples in $\mathcal{E}$ will go through x to be classified, that is, $\forall e \in \mathcal{E}, x \notin p(l(e))$. In our encoding a node x is useless if it is not assigned any feature, that is, $\forall f \in \mathcal{F}, t_{xf} = 0$ in the solution. We then can add an extra type of redundant clauses to avoid decision trees going through useless nodes before reaching a real node:

$$\bigvee_{f \in \mathcal{F}} (t_{xf}) \vee \neg t_{yf'}, \quad \forall (x, y) \in U, y \in X \setminus L, \forall f' \in \mathcal{F}. \tag{4}$$

There are nk such clauses, each of size in $O(k)$, which gives a total size in $O(k^2 n)$.

4 A CP Model

The SAT encoding introduced in the previous section has several drawbacks. It does not scale well with the number of examples in the training set, and even less so with the depth of the decision tree. This latter problem is because the binary tree we encode is a 'superset' of the decision tree we find, and is fixed in advance. Moreover, when the number of examples m is large it would be too costly to maintain variables or clauses representing examples. We therefore introduce a special kind of set variable, where only the lower bound is stored and can be pruned. Usually, a set variable is represented using two reversible sets, one standing for the elements that *must* be in the set (lower bound) and one for the elements that *can* be in the set (upper bound). We implement these simplified set variables using a single reversible list of integers, representing the lower bound. The upper bound is implicit and the only possible operation is to shrink it to match the lower bound. This type of variable allows us to reason about large sets (sets of examples here), at a very low computational cost. Another observation from the SAT encoding is that starting from a complete tree is impractical, even for relatively small depths. We therefore use the expressivity of CP to get around this problem. We do not fix the binary tree on which to label. We simply assume an upper bound n on the number of nodes, and seek the smallest decision tree with n nodes or less. That is, both the tests to perform *and* the topology of the tree are decided within the model. We can therefore find potentially deep trees with a relatively small number of nodes.

4.1 Variables

We label nodes with integers from 1 to n, then for all $i \in [1..n]$ we introduce the following variables:

- $P_i \in [1..n]$: the index of the parent of node i.
- $L_i \in [1..n]$: the index of the left child of node i.
- $R_i \in [1..n]$: the index of the right child of node i.
- $N_i \in [0..2]$: the number of children of node i.
- $F_i \in [1..k]$: the index of the feature tested at node i.
- $D_{ij} \in \{0, 1\}$: "node j is a descendant of node i".
- $\emptyset \subseteq E_i \subseteq \{1, \ldots, m\}$: the subset of $\mathcal{E}$ such that the leaves associated with elements in E_i are all descendants of node i. We shall use the notation E_i^+ (resp. E_i^-) for $(E_i \cap \mathcal{E}^+)$ (resp. $(E_i \cap \mathcal{E}^-)$).
- $UB \in [0..n]$: an upper bound on the size of the decision tree (initialised to $n + 1$ and set to the size of the smallest decision tree found so far)

4.2 Constraint Program

The graph defined on nodes $\{1, \ldots, n\}$ with an edge (i, j) iff $P_j = i$ must form a tree. We use the TREE global constraint to enforce this requirement [4]. Notice that this constraint uses a data structure to store the set of descendants of every node. We make it explicit by using the Boolean variables D. We use a slightly modified version of the constraint that ensures the resulting graph is, in fact, composed of a single tree, and possibly a set of unconnected nodes; a node i is connected iff $P_i \neq i$ or $\exists j \neq i,\ P_j = i$. We, therefore, can add the constraint TREE(P, D) to the model.

Next, we channel the variables N, L, R and P with the following constraints, thus making sure that the tree is binary.

$$\forall i \neq j \in [1..n],\ P_j = i \Leftrightarrow ((L_i = j) \text{ xor } (R_i = j)). \tag{5}$$

$$\forall i,\ N_i = \sum_{j \neq i} P_j = i. \tag{6}$$

The next constraint ensures that no feature is tested twice along a branch.

$$D_{ij} = 1 \Rightarrow F_i \neq F_j. \tag{7}$$

Now we introduce some constraints to ensure that for all i, the variables E_i stand for the set of examples that shall be tested on node i.

$$L_i = j \Rightarrow E_j = \{k \mid k \in E_i \ \wedge\ e_k[F_i] = 0\}. \tag{8}$$

$$R_i = j \Rightarrow E_j = \{k \mid k \in E_i \ \wedge\ e_k[F_i] = 1\}. \tag{9}$$

For each node i, we ensure that unless all examples are classified, (i.e., there is no pair of examples with opposite polarity agreeing on the feature tested on this node) it cannot be a leaf.

$$\exists k \in E_i^+ \ \wedge\ \exists k' \in E_i^- \ \wedge\ e_k[F_i] = e_{k'}[F_i] \Rightarrow N_i > 0. \tag{10}$$

4.3 Inference

We introduce a number of implied constraints to improve this model. The first constraint ensures that the feature tested at a given node splits the examples in a non-trivial way.

$$\exists k \in E_i^+, \exists k' \in E_i^-, \text{ s.t. } e_k[F_i] \neq e_{k'}[F_i]. \tag{11}$$

This constraint does not improve the search, but it does help the subsequent constraint to work effectively. When, for a given node, every feature splits the examples so that both positive and negative examples are represented left and right, we know that this node will need not one, but two children. Let E be a set of examples and f be a feature, we denote by $l(f, E)$ (resp. $r(f, E)$) the cardinality of the subset of E that will be routed left (resp. right) when testing f.

$$l(F_i, E_i^+) \cdot l(F_i, E_i^-) \cdot r(F_i, E_i^+) \cdot r(F_i, E_i^-) \neq 0 \Rightarrow N_i = 2. \tag{12}$$

Due to the previous constraint, we can compute a lower bound on the number of past and future nodes that will be required. A simple sum constraint ensures we do not seek trees larger than one already found:

$$\sum_{i \in [1..n]} N_i < UB. \tag{13}$$

4.4 Symmetry Breaking

A search algorithm over the constraint model might repeatedly explore isomorphic trees where only node labellings change, significantly degrading performance. To avoid this, we ensure that the trees are ordered from root to leaves and from left to right by adding the following constraints:

$$\forall i \in [1..n-1],\ P_i \leq min(i, P_{i+1}). \tag{14}$$

$$\forall i \in [1..n],\ i \leq R_i \leq 2 * i + 2. \tag{15}$$

$$\forall i \in [1..n],\ i \leq L_i \leq min(2 * i + 1, R_i). \tag{16}$$

4.5 Search

Even when adding implied constraints and symmetry breaking constraints, the problem is often too large for the model above to explore a significant part of the search space in a reasonable amount of time. Thus, it is critical to use an efficient search heuristic in order to find good solutions quickly. We used the well known *information gain* heuristic, used in standard decision tree learning algorithms, as a search strategy. Therefore, the first branch explored by our constraint model is similar to that explored by C4.5 [6].

We also observed that diversifying the choices made by the search heuristic (via randomization) and using a *restart strategy* was beneficial. We used the following method: instead of branching on the feature offering the best information gain, we randomly picked among the three best choices. This small amount of randomization allowed us to restart search after unsuccessful dives. Each successive dive is bounded by the number of fails, initialised to 100 and then geometrically incremented by a factor of 1.5.

5 Hybrid CP and LP Model

Next we discuss a promising inference method to deduce a good lower bound on the number of nodes required to classify a set of examples. Consider a partial solution of the CP model, such that when a node i of the decision tree is assigned (that is, the parent of i and the feature tested on i are both known) then its parent is also assigned. It follows that the set of examples tested on an assigned node is perfectly known. We can compute a lower bound on the number of nodes required to classify this set of examples. By summing all these lower bounds for every assigned node without assigned children, we obtain a lower bound on the number of extra nodes that will be necessary to classify all yet unclassified examples. If this lower bound is larger than the number of available nodes we can backtrack, cutting the current branch in the search tree.

Consider a pair of examples (e_i, e_j) such that $e_i \in \mathcal{E}^+$ and $e_j \in \mathcal{E}^-$. We define $\delta(e_i, e_j)$ to be the set of *discrepancies* between examples e_i and e_j as follows: $\delta(e_i, e_j) = \{f \mid e_i[f] \neq e_i[f]\}$. Furthermore, we denote by C the corresponding collection of sets: $C(\mathcal{E}) = \{\delta(e_i, e_j) \mid e_i \in \mathcal{E}^+ \wedge e_j \in \mathcal{E}^-\}$. A *hitting set* for a collection $S_1, \ldots, S_n$ of sets is a set H such that $H \cap S_i \neq \emptyset, i \in 1..n$.

Theorem 1. *If a decision tree classifies a set of examples $\mathcal{E}$, the set of features tested in the tree is a hitting set of $C(\mathcal{E})$.*

Proof. Let F_T be the set of features tested in the decision tree and let $e_i \in \mathcal{E}^+$ and $e_j \in \mathcal{E}^-$. Clearly, in order to classify e_i and e_j, at least one of the features for which e_i and e_j disagree must be tested. That is, we have $F_T \cap \delta(e_i, e_j) \neq \emptyset$. Hence F_T is a hitting set for $C(\mathcal{E})$. □

Consequently, the size of the minimum hitting set on $C(\mathcal{E})$ is a lower bound on the number of distinct tests, and hence of nodes of a decision tree for classifying a set of examples $\mathcal{E}$. At the root node, this hitting set problem might be much too hard to solve, and moreover, it might not be a tight lower bound since tests can be repeated several times on different branches. However, during search on the CP model described above, there shall be numerous subtrees, each corresponding to a subset of $\mathcal{E}$, for which solving, or approximating the hitting set problem might give us a valuable bound. We can solve the hitting set problem using the following linear program on the set of Boolean variables $V = \{v_f \mid f \in \mathcal{F}\}$:

$$\text{minimise} \sum_{f \in \mathcal{F}} v_f \text{ subject to} : \forall c \in C(\mathcal{E}), \sum_{f \in c} v_f \geq 1.$$

This linear program can be solved efficiently by any LP solver. At each node of the *search* tree explored by the CP optimiser, let OP be the set of nodes of the *decision* tree whose parent is known (assigned) but children are unknown (not yet assigned). For each node $i \in OP$, we know the exact set of examples E_i to be tested on i. Therefore, a lower bound $lb(i)$, computed with the LP above, of the cardinality of the associated MINIMUM HITTING SET problem is also a valid lower bound on the number of descendants of i. Let I be the set of nodes (of the *decision* tree) already assigned, and

UB be the size of the smallest tree found so far. We can replace Constraint 13 with the following constraint:

$$|I| + \sum_{i \in OP} lb(i) < UB. \tag{17}$$

In order to avoid computing large linear relaxations too often, we use a *threshold* on the cardinality of $C(E_i)$ that is $|E_i^+| \times |E_i^-|$. Whenever this cardinality is larger than the threshold, we use $N_i + 1$ instead of $lb(i)$ in Constraint 17.

6 Experimental Results

We performed a series of experiments comparing our approach against the state-of-the-art in machine learning, as well as studying the scalability and practicality of our optimisation-based methods. An important distinction between our approach and standard greedy decision tree induction algorithms, is that we seek the *smallest* tree that has *perfect* classification accuracy on the training set. In this sense, our approach can be regarded as a form of knowledge compilation in which we seek the smallest compiled representation of the training data [3]. Standard decision tree induction algorithms can be forced to generate a tree that also has perfect classification accuracy on the data, but these trees tend to be large, since they *overfit* the training data. To overcome this overfitting, greedy methods are often post-pruned by identifying sub-branches of the tree that can be removed without having too significant an impact on its accuracy.

In our experiments, therefore, we compared the decision trees obtained from our optimisation approach against the decision trees obtained from standard tree induction methods, both unpruned and pruned. The results clearly show that the constraint programming approach produces very accurate trees, that are smaller than those found using standard greedy unpruned methods, and in a scalable manner. Even when compared against pruned trees, the accuracy of our decision trees is never the worst, and is often competitive with, or exceeds that of pruned trees built using standard methods.

All our experiments were run on a 2.6GHz Octal Core Intel Xeon with 12Gb of RAM running Fedora core 9. In Table 1, we report some characteristics (number of examples and features) of the benchmarks used in our experiments.

6.1 The Scalability of the SAT Encoding

In Table 1 we give the space complexity, in bytes, of the CNF encoding for each data set, assuming a maximum depth of 4 for the decision tree. In most cases, however, the depth of minimal trees is much larger. Recall that the space complexity of the SAT

Table 1. Characteristics of the data-sets, and the sizes of the corresponding SAT formulae

Benchmark	Weather	Mouse	Cancer	Car	Income	Chess	Hand w.	Magic	Shuttle	Yeast
#examples	14	70	569	1728	30162	28056	20000	19020	43500	1484
#features	10	45	3318	21	494	40	205	1887	506	175
CNF size (depth 4)	27K	3.5M	92G	842M	354G*	180G	248G	967G*	118G*	13G

formula increases exponentially with depth. Therefore, the reformulation is too large in almost all of our data sets (the results marked with an asterisk (*) were obtained using only 10% of the examples in the data set).

In order to study the behaviour of a SAT solver on this problem we ran SAT4J[1] on the SAT encoding of the two smallest data sets (`Weather` and `Mouse`). We used the depth of the smallest tree found by the CP model (see Section 6.2) to build the SAT encoding. To minimise the size of the decision tree, SAT4J features an ATLEASTK constraint ensuring that the number of 0's in a model is at least a given number K. The value of K is initialised to 0, and on each successful run, we set K to 1 plus the number of 0's in the previous model. We stop when either SAT4J returns false, or a time cutoff of 5 minutes has elapsed without improving the current model. We report the runtime for finding the best solution (sol.) and also the total elapsed time, including the time spent on proving or attempting to prove optimality (tot.). We also report the size (nodes) of the smallest tree found.

Benchmark	SAT model		
	time (sol.)	time (tot.)	tree size
`Weather`	0.14	0.37	9
`Mouse`	277.27	577.27	15

It is remarkable how well SAT4J can handle such large CNF files. For instance, the encoding of `Mouse` involves 74499 often very large clauses. However, it was clear the SAT method does not scale since these two tiny data sets produced large formulas. On the one hand, for the smaller data-set (`Weather`) SAT4J quickly found an optimal decision tree, but was slightly slower than the CP method (0.06s). On the other hand, for the larger data-set (`Mouse`), the CP model found a decision tree of 15 nodes in 0.05s, whilst the SAT required 277.27s to find a solution of equal quality but failed to prove its optimality within the 5 minute time limit.

6.2 The CP Model

In this experiment we compared the size of the decision trees produced by our CP classifier with respect to standard implementations of C4.5. We compare our results in terms of tree size and accuracy against WEKA [8] and ITI [7], using a number of data-sets from the UCI Machine Learning Repository[2]. For each data set, and for a range of ratios ($\frac{|training\ set|}{|test\ set|}$) we produced 100 random training sets of the given ratio by randomly sampling the whole data-set, using the remainder of the data for testing. Each classifier is trained on the same random sample. We report averages over the 100 runs for each classifier.

We first compare the size of the decision trees when they are complete (100% accurate classification) on the training data. Therefore, we switched off all *post-pruning* capabilities of WEKA and ITI. Moreover, WEKA also *pre-prunes* the tree, that is, it does not expand subtrees when the information gain measure becomes too small.

[1] `http://www.sat4j.org`
[2] `http://archive.ics.uci.edu/ml/`

We, therefore, modified WEKA to avoid this behaviour[3]. The following command lines were used for WEKA and ITI, respectively: `java weka.classifiers.trees.J48 -t train_set -T test_set -U -M 0` and `iti dir -ltrain_set -qtest_set -t`. The CP optimiser was stopped after spending five minutes without improving the current solution. We report the size of the tree found in the first descent of the CP optimiser and the size of the smallest tree found. We also report some search information – number of backtracks and runtime in seconds – to find the smallest tree.

The results for these unpruned decision trees are reported in the columns 'C4.5, no pruning' and 'cp' of Table 2. One could imagine that the information gain heuristic would be sufficient to find near-minimal trees with ITI or WEKA. The results show that this is not the case. The decision trees computed by ITI or WEKA without pruning are far from being minimal. Indeed C4.5 does not actively aim at minimising the tree size. Smaller decision trees can be found, and our CP model is effective in doing so.

It is somewhat surprising that even the first solution of the CP model is often better (in terms of tree size) than that of WEKA or ITI. This can be explained by the fact that we turn the data set into a numerical form, and then systematically branch using either equality or disequality splits. On the other hand, WEKA uses only equality splits on categorical features, and disequality splits on the numerical features. Our method allows tests with better information gain in certain cases.

In the rightmost columns of Table 2, we report the size of the pruned decision trees computed by ITI and WEKA. When compared against the pruned C4.5 trees, the CP tree is always dominated in terms of size. However, two points should be noted about this. Firstly, we have made no attempt to post-prune the CP trees. If we did so, we could expect a reduction in tree size, possibly comparable to that obtained for the trees generated using C4.5. Secondly, after pruning, the decision tree is no longer guaranteed to have 100% classification accuracy on the original training set.

Table 3 presents a detailed comparison of classification accuracy between the trees built using our CP approach and those built using WEKA and ITI, both pruned and unpruned. For each of the standard approaches we present the average classification accuracy of its trees based on 100 tests. In addition, we present the complement to 1 of the p-value, obtained from a paired t-test performed using the statistical computing system R[4]; this statistic is presented in the column labelled 'sig' (for significance). Suppose that for two methods, their average accuracy x and y over 100 runs are such that $x < y$, this value $(1 - p)$ can be interpreted as the probability that x is indeed less than y. For each reported average accuracy, we compute the significance of its relation to the CP accuracy. We regard a difference as statistically significant if the corresponding 'sig' value is at least 0.95. For example in the first line of the table the unpruned WEKA accuracy is 91.54, while the CP accuracy is 91.66. The significance of this difference is 0.42 which means that this is not a statistically significant difference.

The two right-most columns of Table 3 indicate the relative performance of the CP model. We assume that method A gives better trees than method B iff the accuracy is significantly better, and we define a dominance relation with respect to the CP model,

[3] This change was made in consultation with the authors of WEKA to ensure the system was not adversely affected.

[4] `http://www.r-project.org/`

Table 2. A comparison of the sizes of decision trees obtained from WEKA without pruning (WEKA), WEKA with pruning (WEKA (p)), ITI without pruning (ITI), ITI with pruning (ITI (p)), and CP. We also present statistics on the running time of the CP approach.

Benchmark	Prop.	C4.5, no pruning WEKA size	C4.5, no pruning ITI size	cp first size	cp best size	cp best time (s)	cp backtracks	C4.5, pruning WEKA (p) size	C4.5, pruning ITI (p) size
Cancer	0.2	11.66	11.94	10.92	9.22	47.12	17363.04	7.76	5.00
	0.3	16.12	16.74	14.30	12.48	27.22	7411.31	10.26	7.28
	0.5	23.48	25.98	20.40	18.48	45.05	8448.93	15.08	10.72
	0.7	31.10	36.18	26.22	24.26	38.00	6361.05	20.56	12.76
	0.9	37.98	41.56	32.40	30.08	57.92	7801.57	23.46	15.08
Car	0.05	30.09	23.38	24.82	18.52	8.48	250851.11	12.30	10.92
	0.1	46.67	37.20	40.16	30.12	26.32	1228232.51	18.98	17.04
	0.2	70.97	55.26	59.82	47.70	41.60	1840811.13	29.04	27.40
	0.3	87.67	68.96	74.16	60.06	29.66	1107230.15	37.14	34.40
	0.5	114.51	84.50	93.32	75.60	33.52	1025980.81	52.66	47.56
	0.7	139.22	96.12	105.54	86.30	32.05	839389.68	60.70	57.76
	0.9	161.73	100.76	115.24	92.02	44.81	1086473.89	66.45	64.62
Income	0.01	185.86	108.82	85.12	76.22	35.99	141668.94	34.08	26.68
	0.015	265.23	160.89	123.60	112.87	36.81	108710.27	46.37	38.95
	0.05	791.03	534.69	390.65	364.83	63.99	56624.72	124.21	122.39
Chess	0.01	126.84	89.46	81.54	66.58	49.13	1486914.05	1.00	18.46
	0.015	172.36	130.52	119.60	98.90	48.86	1376762.87	1.06	29.64
	0.05	434.54	372.54	317.20	274.66	41.57	360814.88	1.00	108.92
	0.1	735.26	644.22	525.48	458.80	66.11	112665.88	34.08	217.88
Hand writing (A)	0.01	11.54	14.24	10.66	8.78	10.86	22525.52	5.66	5.98
	0.015	14.52	17.92	13.66	10.66	21.24	38645.64	5.66	7.56
	0.05	33.80	42.24	31.50	24.16	28.60	28678.29	10.36	10.78
	0.1	51.84	67.68	48.22	39.58	40.62	51219.42	17.54	18.28
	0.2	76.16	109.92	75.40	62.98	41.37	51613.47	29.60	34.94
	0.3	94.22	144.36	95.26	80.28	40.93	53276.16	43.36	46.52
Hand writing (B)	0.01	18.28	19.68	16.30	12.82	19.27	84162.49	5.96	5.02
	0.015	24.58	27.66	22.04	17.06	20.19	95053.05	9.24	7.00
	0.05	64.60	70.30	55.92	47.10	40.20	211587.58	23.42	17.22
	0.1	106.56	119.58	95.78	84.76	33.72	171899.21	36.92	33.04
	0.2	180.36	199.72	160.20	145.04	42.05	212308.28	72.16	58.04
	0.3	240.18	268.92	214.48	196.42	33.64	146700.04	105.56	78.68
Hand writing (C)	0.01	14.12	15.64	13.08	10.34	17.10	58892.16	7.04	4.70
	0.015	19.38	21.00	17.80	14.00	20.54	79994.81	8.82	6.10
	0.05	48.64	52.62	43.16	35.52	30.89	119385.76	18.40	16.64
	0.1	80.86	88.42	73.76	61.96	44.43	150533.20	29.74	30.72
	0.2	132.36	146.52	121.50	104.74	39.99	113072.51	49.72	52.52
	0.3	175.44	193.56	160.64	139.36	49.49	135286.41	66.56	72.24
Magic	0.01	52.74	57.24	48.44	44.02	39.19	38038.14	32.70	17.10
	0.015	76.14	82.42	69.76	64.28	32.59	22821.48	45.54	24.16
	0.05	234.70	257.02	204.38	195.94	95.97	20816.56	132.36	70.48
Shuttle	0.05	19.02	25.20	12.96	8.06	48.58	7107.41	10.38	13.14
	0.1	24.18	31.54	16.14	10.88	69.49	5639.98	15.22	17.56
	0.2	28.46	35.38	20.76	13.62	14.94	3691.08	18.04	24.98
Yeast CYT	0.05	33.26	38.70	30.26	25.30	37.16	353351.68	21.72	9.84
	0.1	67.50	76.42	59.08	50.88	39.86	291145.95	36.52	19.14
	0.2	130.30	147.70	113.36	103.78	37.57	215424.60	64.64	37.50
	0.3	197.26	222.16	172.22	157.12	28.72	136221.02	96.18	53.58
	0.5	324.30	364.62	284.10	263.42	41.36	167958.64	147.66	87.82
	0.7	450.06	507.30	395.16	371.46	35.35	100452.08	199.52	122.10
Yeast MIT	0.05	22.00	24.46	18.62	15.70	11.55	65659.79	8.68	5.10
	0.1	42.60	47.38	37.12	31.62	22.06	118185.26	13.36	9.04
	0.2	81.70	92.26	70.80	63.30	21.04	79821.79	22.66	19.40
	0.3	121.48	138.72	103.28	94.88	28.96	117579.25	29.86	27.64
	0.5	200.92	227.88	169.32	156.60	33.17	128659.70	42.00	42.00
	0.7	279.12	310.26	232.72	217.30	36.27	117305.35	59.10	57.76

Table 3. A comparison of the classification accuracies of decision trees obtained from WEKA without pruning (WEKA), WEKA with pruning (WEKA (p)), ITI without pruning (ITI), ITI with pruning (ITI (p)), and CP

Benchmark	Prop.	WEKA		WEKA (p)		ITI		ITI (p)		CP	Relation	
		accur.	sig.	accur.	sig.	accur.	sig.	accur.	sig.	accur.	versus all	versus complete
Cancer	0.2	91.54	0.42	91.71	0.17	91.09	0.97	90.86	0.99	91.66	*among best*	*among best*
	0.3	91.74	0.59	91.96	0.09	91.76	0.47	91.96	0.07	91.93	*among best*	*among best*
	0.5	92.57	0.86	92.95	0.23	92.43	0.96	92.90	0.07	92.88	*among best*	*among best*
	0.7	92.85	0.96	93.36	0.07	93.33	0.17	93.90	0.96	93.39	*among best*	*among best*
	0.9	93.22	0.83	93.50	0.51	93.66	0.24	93.78	0.01	93.78	*among best*	*among best*
Car	0.05	88.34	0.54	87.22	0.99	88.69	0.18	87.26	0.99	88.61	*among best*	*among best*
	0.1	91.07	0.99	89.41	1.00	92.24	0.92	90.58	0.99	91.84	*among best*	*among best*
	0.2	94.13	0.99	92.88	1.00	94.89	0.26	94.23	0.99	94.83	*among best*	*among best*
	0.3	95.45	1.00	94.05	1.00	96.25	0.89	95.47	1.00	96.44	*among best*	*among best*
	0.5	96.87	1.00	96.14	1.00	97.67	0.99	97.13	1.00	97.92	*best*	*best*
	0.7	97.60	1.00	96.97	1.00	98.49	0.99	97.93	1.00	98.72	*best*	*best*
	0.9	97.92	1.00	97.67	1.00	99.18	0.24	98.61	0.99	99.22	*among best*	*among best*
income	0.01	78.00	0.96	80.46	1.00	76.61	1.00	78.79	0.87	78.47	*incomp.*	*best*
	0.015	78.58	0.99	81.30	1.00	77.11	1.00	79.26	0.47	79.12	*incomp.*	*best*
	0.05	79.76	1.00	82.79	1.00	77.76	1.00	80.41	0.34	80.45	*incomp.*	*best*
Chess	0.01	84.43	0.99	89.94	1.00	84.15	0.99	87.75	1.00	85.16	*incomp.*	*best*
	0.015	85.43	0.99	89.93	1.00	84.62	1.00	87.50	1.00	86.35	*incomp.*	*best*
	0.05	89.11	1.00	89.94	1.00	87.59	1.00	88.23	1.00	90.40	*best*	*best*
	0.1	91.53	1.00	90.22	1.00	89.84	1.00	89.48	1.00	92.91	*best*	*best*
Hand writing (A)	0.01	97.48	0.60	98.04	0.99	96.84	0.98	97.14	0.69	97.33	*incomp.*	*among best*
	0.015	97.90	0.97	98.48	1.00	97.39	0.97	97.87	0.91	97.66	*incomp.*	*incomp.*
	0.05	98.48	0.20	98.87	1.00	98.26	0.99	98.63	0.99	98.47	*incomp.*	*among best*
	0.1	98.83	0.81	99.05	1.00	98.60	1.00	98.85	0.89	98.80	*incomp.*	*among best*
	0.2	99.17	0.99	99.23	1.00	98.91	1.00	99.04	0.99	99.10	*incomp.*	*incomp.*
	0.3	99.35	1.00	99.36	1.00	99.06	1.00	99.14	1.00	99.26	*incomp.*	*incomp.*
Hand writing (B)	0.01	94.65	0.41	95.79	1.00	94.79	0.17	95.98	1.00	94.75	*among worst*	*among best*
	0.015	95.03	0.97	96.01	0.99	95.12	0.87	96.17	1.00	95.33	*incomp.*	*among best*
	0.05	96.29	0.98	97.00	1.00	96.42	0.19	97.15	1.00	96.43	*incomp.*	*among best*
	0.1	96.99	0.22	97.42	1.00	96.95	0.49	97.35	1.00	96.98	*among worst*	*among best*
	0.2	97.62	0.99	97.80	1.00	97.51	0.09	97.74	1.00	97.51	*among worst*	*among worst*
	0.3	97.94	0.99	98.03	1.00	97.79	0.39	97.93	1.00	97.80	*among worst*	*among worst*
Hand writing (C)	0.01	95.81	0.65	96.14	0.99	95.84	0.72	96.33	0.99	95.67	*among worst*	*among best*
	0.015	96.15	0.89	96.61	0.97	96.27	0.49	96.49	0.77	96.35	*among worst*	*among best*
	0.05	97.38	0.86	97.75	1.00	97.41	0.72	97.57	0.94	97.47	*among worst*	*among best*
	0.1	97.99	0.25	98.22	1.00	97.90	0.97	98.02	0.72	97.98	*incomp.*	*among best*
	0.2	98.43	0.57	98.63	1.00	98.37	0.87	98.42	0.18	98.41	*among worst*	*among best*
	0.3	98.63	0.63	98.76	1.00	98.58	0.99	98.64	0.31	98.65	*incomp.*	*among best*
Magic	0.01	75.00	0.93	75.84	0.40	75.04	0.96	76.61	0.99	75.65	*incomp.*	*among best*
	0.015	76.32	0.89	77.35	0.99	76.08	0.98	77.84	0.99	76.69	*incomp.*	*best*
	0.05	78.29	1.00	79.66	0.99	78.21	1.00	80.50	1.00	78.98	*incomp.*	*best*
Shuttle	0.05	99.77	1.00	99.72	1.00	99.76	1.00	99.71	1.00	99.85	*best*	*best*
	0.1	99.85	1.00	99.80	1.00	99.85	1.00	99.79	1.00	99.91	*best*	*best*
	0.2	99.93	1.00	99.90	1.00	99.92	0.99	99.90	1.00	99.95	*best*	*best*
Yeast CYT	0.05	65.21	0.74	65.34	0.83	63.46	0.99	65.11	0.55	64.79	*among best*	*among best*
	0.1	66.53	0.86	67.19	0.99	64.72	0.99	66.87	0.98	66.14	*incomp.*	*among best*
	0.2	66.93	0.15	68.04	0.99	65.94	0.99	68.04	0.99	66.98	*incomp.*	*among best*
	0.3	67.87	0.98	69.19	1.00	66.42	0.99	68.84	1.00	67.28	*incomp.*	*incomp.*
	0.5	68.53	0.40	70.19	1.00	67.79	0.99	69.73	1.00	68.41	*incomp.*	*among best*
	0.7	69.09	0.94	70.99	1.00	68.75	0.43	70.39	1.00	68.59	*among worst*	*among best*
Yeast MIT	0.05	80.21	0.52	83.32	1.00	80.27	0.42	83.51	1.00	80.50	*among worst*	*among best*
	0.1	81.14	0.05	85.24	1.00	80.53	0.92	84.46	1.00	81.12	*among worst*	*among best*
	0.2	81.92	0.13	85.83	1.00	81.46	0.95	84.98	1.00	81.89	*incomp.*	*among best*
	0.3	82.30	0.31	86.34	1.00	81.89	0.89	85.76	1.00	82.22	*among worst*	*among best*
	0.5	82.62	0.85	86.77	1.00	82.48	0.98	85.99	1.00	82.90	*incomp.*	*among best*
	0.7	82.50	0.99	86.37	1.00	82.65	0.99	86.11	1.00	83.17	*incomp.*	*among best*

based on this pairwise relation. We say that the CP model is the best, denoted "*best*" (resp. the worst, denoted "*worst*") iff it gives trees that are significantly better (resp. worse) than all other methods. We say that the CP model is among the best, denoted "*among best*" (resp. among the worst, denoted "*among worst*") if there is no other method giving better (resp. worse) trees, and if it is not the best (resp. worst). Finally, we

Table 4. Runtime, #Backtracks & Tree size (CP vs CP+LP)

Benchmark	CP model			CP+LP model		
	time	bts	tree size	time	bts	size
Cancer	44.37	16206	9.24	**31.64**	**10421**	**9.20**
	27.61	7098	12.48	**21.48**	**4569**	**12.44**
	42.82	7910	**18.50**	**42.54**	**4461**	18.62
Car	**8.75**	250449	**18.52**	12.30	**7733**	**18.52**
	29.13	1390076	30.10	51.15	**33182**	**29.58**
	48.21	2206249	47.64	87.05	**32979**	**46.46**
	29.69	1106445	60.06	48.97	**15829**	**59.18**
	33.53	1025011	**75.60**	53.40	**17715**	75.64
Income	**35.03**	140917	**76.22**	59.98	**11875**	76.68
	38.69	114533	**112.81**	89.08	**11074**	113.26
	63.3	54219.82	364.83	119.14	6802.40	368.87

say that it is incomparable, denoted "*incomp.*", iff there exists at least one method giving better trees and one giving worse trees. We report this comparison for each instance. In the penultimate column ("versus all") we compare against all methods. The CP model is the best in 13% of the cases, among the best in 20% of the cases, incomparable in 45% of the cases, among the worst in 22% of the cases, and is never the worst. In the last column ("versus complete"), we compare against only complete methods, that is, WEKA and ITI without post-pruning. The CP model is the best in 25% of the cases, among the best in 64% of the cases, incomparable in 7% of the cases, among the worst in 4% of the cases, and is never the worst.

It is clear that the CP generated trees are almost always better than those generated by standard decision tree methods that do not do pruning. Also, when compared against methods that use post-pruning, the CP approach is not dominated by either one. This is an encouraging result since it suggests that while the CP generated trees are competitive with advanced decision tree induction methods, they can only be improved further if they were also post-pruned.

6.3 Improving the CP Model Using LP

The aim of this experiment was to assess if the linear relaxation method introduced in Section 5 can improve the CP model. We run it using the same setting as described in Section 6.2, and on 3 benchmarks and with an arbitrary threshold[5] of 600 that was a good compromise. In Table 4, we report the runtime and number of backtracks to find the best solution, as well as the quality (tree size) of this solution.

We observe that the search space explored by the CP+LP model can be orders of magnitude smaller (see #backtracks), but the runtime can still be slightly worse because of the overhead of solving the linear relaxation. Thus, even if it is difficult to judge if overall the method is better than the basic CP model, we can expect it to scale well on harder problems.

7 Related Work

Decision trees are usually constructed using greedy algorithms [5,6] relying on a search bias that attempts to find smaller trees. Finding the minimum sized tree is NP-Hard [2].

[5] As defined in Section 5.

[1] have proposed a technique for improving the accuracy of a decision tree by selecting the next attribute to test as the one with the smallest expected consistent sub-tree size estimated using a sampling technique. The use of combinatorial optimisation to improve decision trees has also been reported [9], where the focus has been on determining linear-combination splits for the decision tree. These papers are not concerned with minimising overall tree-size. Our work contrasts with these approaches since we take an extreme view of Occam's Razor and seek to find the minimum sized decision tree using alternative approaches from the field of combinatorial optimisation. We have found that size can usually be reduced considerably, without negatively impacting accuracy.

8 Conclusion

We have presented a variety of alternative approaches to minimising the number of nodes in a decision tree. In particular, we have shown that while this problem can be formulated as either a satisfiability problem or a constraint program, the latter is more scalable. Our empirical results show the value of minimising decision tree size. We find smaller trees that are often more accurate than, but never dominated by, those found using standard greedy induction algorithms.

References

1. Esmeir, S., Markovitch, S.: Anytime learning of decision trees. Journal of Machine Learning Research 8, 891–933 (2007)
2. Hyafil, L., Rivest, R.L.: Constructing optimal binary decision trees is NP-complete. Inf. Process. Lett. 5(1), 15–17 (1976)
3. Katsirelos, G., Walsh, T.: A compression algorithm for large arity extensional constraints. In: Bessière, C. (ed.) CP 2007. LNCS, vol. 4741, pp. 379–393. Springer, Heidelberg (2007)
4. Prosser, P., Unsworth, C.: Rooted tree and spanning tree constraints. In: Workshop on Modelling and Solving Problems with Constraints, held at ECAI 2006 (2006)
5. Quinlan, J.R.: Induction of decision trees. Machine Learning 1(1), 81–106 (1986)
6. Quinlan, R.J.: C4.5: programs for machine learning. Morgan Kaufmann Publishers Inc., San Francisco (1993)
7. Utgoff, P.E., Berkman, N.C., Clouse, J.A.: Decision tree induction based on efficient tree restructuring. Machine Learning 29, 5–44 (1997)
8. Witten, I.H., Frank, E.: Data Mining: Practical machine learning tools and techniques. Morgan Kaufmann, San Francisco (2005)
9. Yüksektepe, F.Ü., Türkay, M.: A mixed-integer programming approach to multi-class data classification problem. European Journal of OR 173(3), 910–920 (2006)

Hull Consistency under Monotonicity

Gilles Chabert[1] and Luc Jaulin[2]

[1] Ecole des Mines de Nantes LINA CNRS UMR 6241,
4, rue Alfred Kastler 44300 Nantes, France
gilles.chabert@emn.fr

[2] ENSIETA, 2, rue François Verny 29806 Brest Cedex 9, France
luc.jaulin@ensieta.fr

Abstract. We prove that hull consistency for a system of equations or inequalities can be achieved in polynomial time providing that the underlying functions are monotone with respect to each variable. This result holds including when variables have multiple occurrences in the expressions of the functions, which is usually a pitfall for interval-based contractors. For a given constraint, an optimal contractor can thus be enforced quickly under monotonicity and the practical significance of this theoretical result is illustrated on a simple example.

1 Introduction

Solving constraint problems with real variables has been the subject of significant developments since the early 90's (see [3] for a comprehensive survey).

One of the key contribution is the concept of *hull consistency*, which is the counterpart of *bound consistency* in discrete constraint programming, as Definition 1 shows below.

Let us briefly trace the history. The underlying concepts of interval propagation appeared first in several pioneering papers [6,11,17,12] while consistency techniques for numerical CSP were formalized a few years later in [15,5]. A theoretical comparative study of consistencies was then conducted in [7,8]. Finally, hull consistency was made operational in [2,9] where the famous `HC4` algorithm is described.

Since hull consistency is based on the bound consistency of every isolated constraint, enforcing hull consistency in the general case (i.e., for arbitrary nonlinear equations) is a NP-hard problem [14]. On the practical side, this results into the inability to give a sharp enclosure when variables occur more than once in the expression of a constraint. This happens, in particular, with `HC4`.

We show that hull consistency can be enforced in polynomial time if the functions involved are all monotone.

Monotonicity follows the very intuitive idea that a function varies either in the same direction as a variable or in the opposite one (see Definition 2). It turns out that usual functions, i.e., built with arithmetic operators $(+,-,\times,/)$ and elementary functions (sin, exp, etc.) are analytic and therefore *most of the time* strictly monotone. In rigorous terms, *most of the time* means that, unless

I.P. Gent (Ed.): CP 2009, LNCS 5732, pp. 188–195, 2009.

the function is flat (or defined piecewise), the set of points that do not satisfy local strict monotonicity is of measure zero (in the sense of measure theory).

As a consequence, if a better contraction (or filtering) can be achieved under monotonicity, branch & prune algorithms should take advantage of it.

Monotonicity has been considered from the beginning of interval analysis [16], but with a motivation slightly different from ours. One of the most fundamental issues of interval analysis is the design of *inclusion functions* [13], i.e., methods for computing an enclosure of the range of a function on any given box. Of course, the sharper the better. Now, an optimal inclusion function (i.e., a method for computing the exact range on any box) can be built straightforwardly for monotone functions (see §2).

Hence, the main matter since that time has been to devise efficient way to *detect* monotonicity of a function f over a box $[x]$. One simple way to proceed is by checking that the gradient does not get null in $[x]$ which, in turn, requires an inclusion function for the gradient. The latter can then be based either on a direct interval evaluation, Taylor forms or the monotonicity test itself in a reentrant fashion (leading to the calculation of second derivatives, and so on).

Surprisingly enough, monotonicity has never been used so far in the design of *contractors*. Remember that, although related, computing a sharp enclosure for $\{f(x),\ x \in [x]\}$ and for $\{x \in [x] \mid f(x) = 0\}$ are quite different goals. As we already said, getting an optimal enclosure with a monotone function f is straightforward in the first case. But it is not so in the second case, especially when x is a vector of variables. Algorithm 1 below will provide an answer.

In the following, we first define properly the different concepts. The main result, a monotonicity-based polytime optimal contractor for a constraint, is then presented. Finally, we highlight the practical benefits of this contractor with a simple example.

1.1 Notations and Definitions

We consider throughout this paper a constraint satisfaction problem (CSP) with a vector of n real variables $x_1, \ldots, x_n$.

Domains of variables are represented by real intervals and a Cartesian product of intervals is called a *box*. Intervals and boxes will be surrounded by brackets, e.g., $[x]$. If $[x]$ is a box, x^- and x^+ will stand for the two opposite corners formed by the lower and upper bound respectively of each components (see Figure 2). Hence, x_i^- and x_i^+ will stand for the lower and upper bound respectively of the interval $[x]_i$. The width of an interval $[x]$ (width$[x]$) is $x^+ - x^-$.

Furthermore, given a mapping f on $\mathbb{R}^n$, we shall denote by $\{f = 0\}$ the constraint $f(x) = 0$ viewed as the set of all solution tuples, i.e.,

$$\{f = 0\} := \{x \in \mathbb{R}^n \mid f(x) = 0\}.$$

We can now give a definition of hull consistency.

Definition 1 (Hull consistency). *Let $\mathcal{P}$ be a constraint satisfaction problem involving a vector x of n variables and let $[x]$ be the domain of x.*

$\mathcal{P}$ is said to be hull consistent if for every constraint c and for all i ($1 \le i \le n$), there exists two points in $[x]$ which satisfy c and whose i^{th} coordinates are x_i^- and x_i^+ respectively.

The key property of hull consistency lies in the combination of local reasoning and interval representation of domains. This concept brought a decisive improvement to the traditional Newton-based numerical solvers that were basically only able to contract domains *globally*.

Definition 2 (Monotonicity). *A mapping $f : \mathbb{R} \to \mathbb{R}$ is* **increasing** *over an interval $[x]$ if $\forall a \in [x], \forall b \in [x] \quad a \le b \Longrightarrow f(a) \le f(b)$.*

A mapping $f : \mathbb{R} \to \mathbb{R}$ is **decreasing** *if $-f$ is increasing, and* **monotone** *if it is either decreasing or increasing.*

A mapping $f : \mathbb{R}^n \to \mathbb{R}$ is **increasing** *(resp.* **decreasing**, **monotone***) over a box $[x]$ if $\forall \tilde{x} \in [x]$ and $\forall i$, $1 \le i \le n$, $x_i \mapsto f(\tilde{x}_1, \ldots, \tilde{x}_{i-1}, x_i, \tilde{x}_{i+1}, \ldots, \tilde{x}_n)$ is increasing (resp. decreasing, monotone) over $[x]_i$.*

Strict monotonicity *is satisfied when formulas hold with strict inequalities.*

Let $f : [y] \subseteq \mathbb{R} \to \mathbb{R}$ be an increasing function. For any interval $[x] \subseteq [y]$, the infimum and the supremum of f on $[x]$ are $f(x^-)$ and $f(x^+)$. Hence, the following interval function:

$$[x] \to [f(x^-), f(x^+)]$$

is an optimal inclusion function for f. This result easily generalizes to monotone multivariate functions, by a componentwise repetition of the same argument.

2 Main Result

Enforcing hull consistency on a CSP boils down to enforcing bound consistency on every isolated constraint (cf. Definition 1). Giving an optimal contractor for a single constraint is thus the main issue, which we shall address below. We shall even focus on an equation $f(x) = 0$ (inequalities will be discussed further).

Consider first the univariate case (f has a single variable) and assume that f is differentiable. The set $\{f = 0\}$ can easily be bracketed by an interval Newton iteration (see, e.g., [16] for details on the operations involved):

$$[y] \leftarrow \Big(\tilde{y} - f(\tilde{y})/[f']([y])\Big) \cap [y]$$

where $[f']$ is a (convergent) inclusion function for f' and $\tilde{y}$ any point in $[y]$ such that $f(\tilde{y}) \neq 0$. Henceforth, we assume that a procedure $\texttt{univ\_newton}(f, [x], \varepsilon)$ is available. This procedure returns an interval $[y]$ such that both $[y^-, y^- + \varepsilon]$ and $[y^+ - \varepsilon, y^+]$ intersect $\{f = 0\}$. It can refer to any implementation of the univariate interval Newton iteration, such as the one given in [10].

Let us state the complexity. As noticed in the introduction, an analytic function is locally either strictly monotone or flat. Thus, it makes sense to assume strict monotonicity when dealing with complexity. The interval Newton iteration has a quadratic rate of convergence [1], i.e., the width of $[y]$ at every step is up to a constant factor less than the square of the width at the previous step. However, the quadratic rate is only achieved when the iteration is contracting, i.e., when $\tilde{y} - f(\tilde{y})/[f']([y]) \subseteq [y]$. While this condition is not fulfilled, the progression can be slow, as the following figure illustrates:

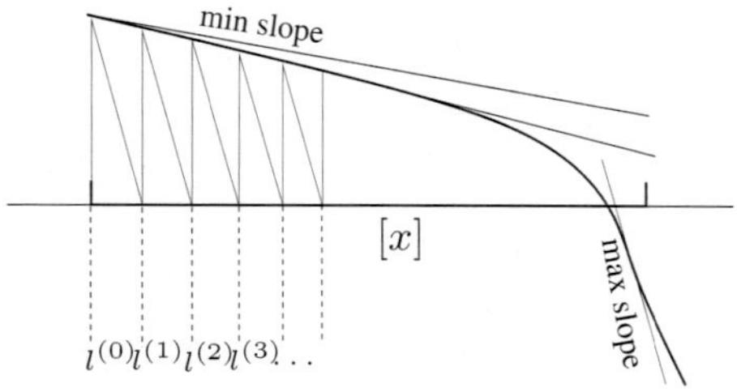

Fig. 1. Slow progression of the Newton iteration (with the left bound as point of expansion). The maximum slope on the interval $[x]$ (on the right side) is repeatedly encompassed in the interval computation of the derivative, which explains the slow progression. The successive lower bounds of the interval $[x]$ are $l^{(0)}, l^{(1)}, \ldots$.

When the point of expansion $\tilde{y}$ is the midpoint of $[y]$, the width of the interval is at least divided by two (this is somehow a way to interleave a dichotomy within the Newton iteration). Thus, the worst-case complexity of `univ_newton` with the midpoint heuristic is $O(log(w/\varepsilon))$, where w is the width of the initial domain. Finally, note that if f is not differentiable (or if no convergent inclusion function is available for f'), one can still resort to a simple dichotomy and achieve the same complexity.

The general algorithm (called `OCTUM`: *optimal contractor under monotonicity*) that works with a multivariate mapping $f : \mathbb{R}^n \to \mathbb{R}$ is given below. Note that `univ_newton` is called on the restriction of f to (axis-aligned) edges of the input box $[x]$. Since $(n-1)$ coordinates are fixed on an edge, the restriction is indeed a function from $\mathbb{R}$ to $\mathbb{R}$.

To ease the description of the algorithm, we will assume that the multivariate function f is increasing (according to Definition 2). Once the algorithm is understood, considering the other possible configurations makes no difficulty and just require a case-by-case adaptation. Lines 0 to 5 initializes the two vectors $x^{\ominus}$ and $x^{\oplus}$ that correspond to the vertices where f is minimized and maximized respectively. When f is increasing, $x^{\ominus}$ and $x^{\oplus}$ are just aliases for x^- and x^+. Line 6 checks that the box $[x]$ contains a solution (and otherwise, the algorithm returns the empty set). The main loop relies on the following fact (see Figure 2) that will be proven below. Remember that f is assumed to be increasing and that $[x]$ contains at least one solution. The minimum of x_i when x describes the solutions inside $[x]$ is then either reached

(1) on the edge where all the other variables are instantiated to their upper bound x_j^+ or
(2) on the face where $x_i = x_i^-$ (which means that no contraction can be made).

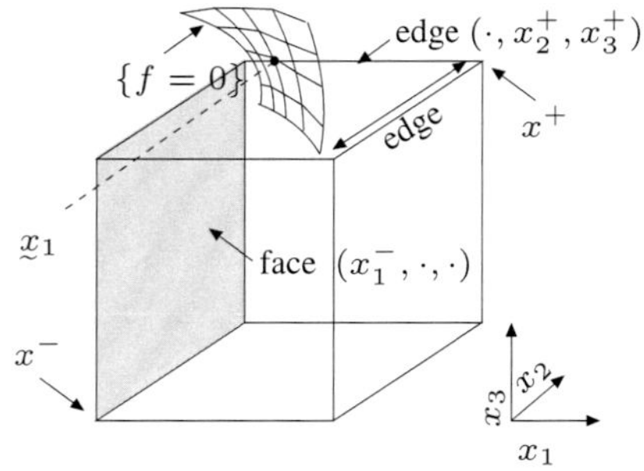

Fig. 2. The first component of the solutions inside the box either reaches its minimum on the face $(x_1^-, \cdot, \cdot)$ or on the edge $(\cdot, x_2^+, x_3^+)$

Furthermore, as soon as the minimum for a component x_i is met on an edge, the solution on this edge makes all x_j^+ $(j \neq i)$ consistent in the domain of the jth variable. To skip useless filtering operations for the upper bounds of the remaining variables, we use a flag named `sup` in the algorithm (see lines 17 and 18). Finally, when `univ_newton` is called, either the lower or upper bound of the resulting interval is considered, depending on which bound of $[x]_i$ is contracted (see lines 15 and 16 or 21 and 22). This ensures that no solution is lost.

Filtering the upper bound of x_i is entirely symmetrical. The complexity of `OCTUM` is $O(n \times log(\frac{\text{width}[x]}{\varepsilon}))$, where width$[x]$ stands for $\max_{1 \leq i \leq n} \text{width}[x]_i$. It can be qualified as a pseudo-linear complexity.

The `OCTUM` algorithm can be very easily extended to an inequality $f(x) \leq 0$ or $f(x) \geq 0$ by simply skipping narrowing operations on y_i^- or y_i^+ respectively.

The completeness and optimality of `OCTUM` relies on the following proposition.

Proposition 1. *Let $f : \mathbb{R}^n \to \mathbb{R}$ be a continuous increasing mapping and $[x]$ a box such that $\{f = 0\} \cap [x] \neq \emptyset$. Given i, $1 \leq i \leq n$ put:*

$$\underset{\sim}{x}_i := \inf\{x_i,\ x \in \{f = 0\} \cap [x]\}.$$

Then, one of the two options holds:

1. $\underset{\sim}{x}_i = x_i^-$,
2. *there exists $x \in [x]$ such that $x_i = \underset{\sim}{x}_i$ and for every $j \neq i$, $x_j = x_j^+$.*

Proof. Let x^* be a solution point in $[x]$ minimizing x_i, i.e., $f(x^*) = 0$ and $x_i^* = \underset{\sim}{x}_i$. If $x_i^* = x_i^-$, the first option holds and we are done. Assume $x_i^* > x_i^-$. If $n = 1$, the second option holds trivially. If $n > 1$, consider the following vector:

$$\overline{x} := (x_1^+, \ldots, x_{i-1}^+, x_i^*, x_{i+1}^+, \ldots, x_n^+).$$

Algorithm 1. OCTUM($f, [x], \varepsilon$)

Input: a monotone function f, a n-dimensional box $[x]$, $\varepsilon > 0$
Output: the smallest box $[y]$ enclosing $[x] \cap \{f = 0\}$, up to the precision ε

for $i = 1$ *to* n **do**
  if $(f \nearrow x_i)$ **then** $x_i^{\oplus} \leftarrow x_i^{+}$; // *$x^{\oplus}$ is the vertex where f is maximized*
  else $x_i^{\oplus} \leftarrow x_i^{-}$ // *$(f \nearrow x_i)$ means "f is increasing w.r.t. x_i"*
  if $(f \nearrow x_i)$ **then** $x_i^{\ominus} \leftarrow x_i^{-}$; // *$x^{\ominus}$ is the vertex where f is minimized*
  else $x_i^{\ominus} \leftarrow x_i^{+}$
if $f(x^{\ominus}) > 0$ *or* $f(x^{\oplus}) < 0$ **then return** $\emptyset$ // *check if a solution exists*
$[y] \leftarrow [x]$
sup $\leftarrow$ **false** // *true when $x_i^{\oplus}$ is consistent for all remaining i*
inf $\leftarrow$ **false** // *true when $x_i^{\ominus}$ is consistent for all remaining i*
for $i = 1$ *to* n **do**
  curr_sup $\leftarrow$ sup ; // *save the current value for the second block*
  if inf *is* **false then**
    $[s] \leftarrow$ univ_newton($t \mapsto f(x_1^{\oplus}, \ldots, x_{i-1}^{\oplus}, t, x_{i+1}^{\oplus}, x_n^{\oplus}), [y]_i, \varepsilon$)
    if $[s] \neq \emptyset$ **then**
      if $(f \nearrow x_i)$ **then** $y_i^{-} \leftarrow s^{-}$ // *update lower bound of x_i*
      else $y_i^{+} \leftarrow s^{+}$; // *update upper bound of x_i*
      sup $\leftarrow$ **true** // *the edge $x_j^{\oplus}$, $j \neq i$, contains a solution*
  if curr_sup *is* **false then**
    $[s] \leftarrow$ univ_newton($t \mapsto f(x_1^{\ominus}, \ldots, x_{i-1}^{\ominus}, t, x_{i+1}^{\ominus}, x_n^{\ominus}), [y]_i, \varepsilon$)
    if $[s] \neq \emptyset$ **then**
      if $(f \nearrow x_i)$ **then** $y_i^{+} \leftarrow s^{+}$
      else $y_i^{-} \leftarrow s^{-}$
      inf $\leftarrow$ **true**
return $[y]$

We will prove that $f(\overline{x}) = 0$. By contradiction, assume $f(\overline{x}) > 0$ (f being increasing). Since $\overline{x}_i = x_i^* > x_i^-$, there exists by continuity $\varepsilon > 0$ such that $x_i^* - \varepsilon > x_i^-$ and $f(\overline{x} - \varepsilon e_i) > 0$ (e_i being the ith unit vector). Now, x^* is the point in $[x]$ where f gets null with the smallest i^{th} coordinate. Hence, $f(x^* - \varepsilon e_i) < 0$.

Since f is continuous, f gets null somewhere on the segment joining $x^* - \varepsilon e_i$ and $\overline{x} - \varepsilon e_i$ because the sign of f is opposite at the two extremities. Since $[x]$ is convex, the corresponding point is inside $[x]$ and its ith component is $x_i^* - \varepsilon_i$, which contradicts the fact that x_i^* is the infimum among the solutions. ▲

3 A First Experiment

Consider the problem of characterizing the set of points (x, y) in $[-3, 0] \times [0, 3]$ satisfying $f(x, y) = 0$ with $f(x, y) = x^2y^2 - 9x^2y + 6xy^2 - 20xy - 1$.

Let us compare OCTUM with three other standard generic contractors (namely HC4 [2,9], BOX [4,18] and 3B [15]) as pruning steps of a classical branch & prune system. We have implemented a very naive method for detecting monotonicity, using an interval evaluation of the gradient that is systematically computed for every box (a better method would be to manage flags w.r.t. each variable in a *backtrackable* structure, each flag being set incrementally as soon as f is proven to be monotone). Even with this naive implementation, OCTUM yields better results, both in terms of *quality* (see Figure 3) and *quantity* (see Table below).

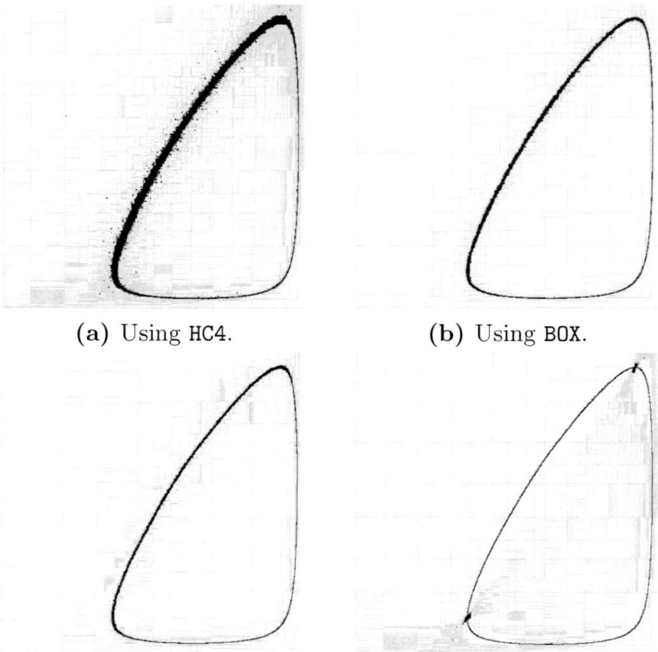

(a) Using HC4. (b) Using BOX.

(c) Using 3B. (d) Using OCTUM.

Fig. 3. Comparing the monotonicity-based contractor OCTUM with other standard operators. Black surfaces encompass the solutions while grey boxes represent the contracted parts. The thinnest black surface is obtained with OCTUM. Note however the two little marks in (d) that correspond to points where one component of the gradient gets null.

	Running time	Number of backtracks	Size of solution set
HC4	0.66s	28240	6928
BOX	1.37s	9632	3595
3B	1.89s	9171	2564
OCTUM	**0.40s**	**6047**	**1143**

4 Conclusion

We have proven that hull consistency can be achieved in polynomial time in the case of constraints involving monotone functions. Hull consistency amounts to bound consistency for each isolated constraint. We have given an algorithm called OCTUM that enforces bound consistency for an equation under monotonicity (and explained how to adapt it to inequalities). Hull consistency based on OCTUM can then be programed by simply embedding OCTUM in a classical AC3 propagation loop. A first experiment has illustrated the two nice properties of OCTUM: optimality and (pseudo-)linear complexity.

References

1. Alefeld, G., Mayer, G.: Interval Analysis: Theory and Applications. J. Comput. Appl. Math. 121(1-2), 421–464 (2000)
2. Benhamou, F., Goualard, F., Granvilliers, L., Puget, J.-F.: Revising Hull and Box Consistency. In: ICLP, pp. 230–244 (1999)
3. Benhamou, F., Granvilliers, L.: Continuous and interval constraints. In: Handbook of Constraint Programming, ch. 16, pp. 571–604. Elsevier, Amsterdam (2006)
4. Benhamou, F., McAllester, D., Van Hentenryck, P.: CLP(intervals) revisited. In: International Symposium on Logic programming, pp. 124–138. MIT Press, Cambridge (1994)
5. Benhamou, F., Older, W.J.: Applying Interval Arithmetic to Real, Integer and Boolean Constraints. Journal of Logic Programming 32, 1–24 (1997)
6. Cleary, J.G.: Logical Arithmetic. Future Computing Systems 2(2), 125–149 (1987)
7. Collavizza, H.: A Note on Partial Consistencies over Continuous Domains Solving Techniques. In: Maher, M.J., Puget, J.-F. (eds.) CP 1998. LNCS, vol. 1520, pp. 147–161. Springer, Heidelberg (1998)
8. Delobel, F., Collavizza, H., Rueher, M.: Comparing Partial Consistencies. Reliable Computing 5(3), 213–228 (1999)
9. Granvilliers, L., Benhamou, F.: Progress in the Solving of a Circuit Design Problem. Journal of Global Optimization 20(2), 155–168 (2001)
10. Hansen, E.R.: Global Optimization using Interval Analysis. Marcel Dekker, New York (1992)
11. Hyvönen, E.: Constraint Reasoning Based on Interval Arithmetic. In: IJCAI, pp. 1193–1198 (1989)
12. Hyvönen, E.: Constraint Reasoning Based on Interval Arithmetic—The Tolerance Propagation Approach. Artificial Intelligence 58, 71–112 (1992)
13. Jaulin, L., Kieffer, M., Didrit, O., Walter, E.: Applied Interval Analysis. Springer, Heidelberg (2001)
14. Kreinovich, V., Lakeyev, A., Rohn, J., Kahl, P.: Computational complexity and feasibility of data processing and interval computations. Kluwer, Dordrecht (1997)
15. Lhomme, O.: Consistency Techniques for Numeric CSPs. In: IJCAI, pp. 232–238 (1993)
16. Moore, R.: Interval Analysis. Prentice-Hall, Englewood Cliffs (1966)
17. Older, W.J., Vellino, A.: Extending Prolog with Constraint Arithmetic on Real Intervals. In: IEEE Canadian Conf. on Elec. and Comp. Engineering (1990)
18. Van Hentenryck, P., McAllester, D., Kapur, D.: Solving Polynomial Systems Using a Branch and Prune Approach. SIAM J. Numer. Anal. 34(2), 797–827 (1997)

A Constraint on the Number of Distinct Vectors with Application to Localization

Gilles Chabert[1], Luc Jaulin[2], and Xavier Lorca[1]

[1] Ecole des Mines de Nantes LINA CNRS UMR 6241,
4, rue Alfred Kastler 44300 Nantes, France
gilles.chabert@emn.fr, xavier.lorca@emn.fr
[2] ENSIETA, 2, rue François Verny 29806 Brest Cedex 9, France
luc.jaulin@ensieta.fr

Abstract. This paper introduces a generalization of the *nvalue* constraint that bounds the number of distinct values taken by a set of variables.The generalized constraint (called *nvector*) bounds the number of distinct (multi-dimensional) vectors. The first contribution of this paper is to show that this global constraint has a significant role to play with continuous domains, by taking the example of simultaneous localization and map building (SLAM). This type of problem arises in the context of mobile robotics. The second contribution is to prove that enforcing bound consistency on this constraint is NP-complete. A simple contractor (or propagator) is proposed and applied on a real application.

1 Introduction

This paper can be viewed as a follow-up of [7] on the application side and [1] on the theoretical side. It proposes a generalization of the `nvalue` global constraint in the context of a relevant application. The `nvalue` constraint is satisfied for a set of variables $x^{(1)}$,..., $x^{(k)}$ and an extra variable n if the cardinality of $\{x^{(1)}, \ldots, x^{(k)}\}$ (i.e., the number of distinct values) equals to n. This constraint appears in problems where the number of resources have to be restricted. The generalization is called `nvector` and matches exactly the same definition, except that the $x^{(i)}$ are vectors of variables instead of single variables. Of course, all the $x^{(i)}$ must have the same dimension (i.e., the same number of components) and the constraint is that the cardinality of $\{x^{(1)}, \ldots, x^{(k)}\}$ must be equal to n.

We first show that this new global constraint allows a much better modeling of the SLAM (*simultaneous localization and map building*) problem in mobile robotics. In fact, it allows to make automatic and thus robust a process part of which was performed by hand. Second, we classify the underlying theoretical complexity of the constraint. Finally, a simple algorithm is given and illustrated on a real example.

Since the application context involves continuous domains, we shall soon focus on the continuous case although the definition of `nvector` does not depend on the underlying type of domains.

I.P. Gent (Ed.): CP 2009, LNCS 5732, pp. 196–210, 2009.

The application is described in Section 2. We first give an informal description of what SLAM is all about. The model is then built step-by-step and its main limitation is discussed. Next, the constraint itself is studied from Section 3 to Section 5. After providing its semantic (Section 3), the complexity issue is analyzed (Section 4), and a simple contractor is introduced (Section 5). Finally, Section 6 shows how the *nvector* constraint is used in the SLAM problem modeling, and the improvements obtained are illustrated graphically.

In the rest of the paper, the reader is assumed to have basic knowledge on constraint programming over real domains and interval arithmetics [9]. Domains of real variables are represented by intervals. A Cartesian product of intervals is called a *box*. Intervals and boxes are surrounded by brackets, e.g., $[x]$. Vectors are in boldface letters. If $\mathbf{x}$ is a set of vectors, $\mathbf{x}^{(i)}$ stands for the ith vector and $x_j^{(i)}$ for the jth component of the ith vector. The same convention for indices carries over vector of boxes: $[\mathbf{x}^{(i)}]$ and $[x_j^{(i)}]$ are respectively the domains of $\mathbf{x}^{(i)}$ and $x_j^{(i)}$. Given a mapping f, $range(f, [\mathbf{x}])$ denotes the set-theoretical image of $[\mathbf{x}]$ by f and $f([\mathbf{x}])$ denotes the image of $[\mathbf{x}]$ by an interval extension of f.

2 Application Context

The nvector constraint can bring substantial improvements to constraint-based algorithms for solving the simultaneous localization and map building (**SLAM**) problem. This section describes the principles of the SLAM and provides a constraint model of this application.

2.1 Outline of the SLAM Problem

The SLAM problem can be described by an autonomous robot moving for a period of time in an unknown environment where the self-location cannot be performed accurately with the help of external equipments (typically, the GPS), and the observation of the environment can only be performed by the robot itself. Examples of such environments include under the sea and the surface of other planets (where the GPS is unavailable). Both limitations are also present indoor where the GPS is often considered unreliable.

In the SLAM problem, we have to compute as precisely as possible a map of the environment (i.e., the position of the detected objects), as well as the trajectory of the robot. The input data is the set of all measures recorded by the robot during the mission from its embedded sensors. These sensors can be divided into two categories: *proprioceptive*, those that allow to estimate the position of the robot itself (e.g.: a gyroscope) and *exteroceptive*, those that allows to detect objects of the environment (e.g.: a camera). Note that the positions of the robot at the beginning and the end of the mission are usually known with a good precision. Objects are called *landmarks* (or *seamarks* under the sea).

Since nothing can be observed from outside (should it be a landmark or the robot itself), uncertainties of measures get accumulated every time step. The

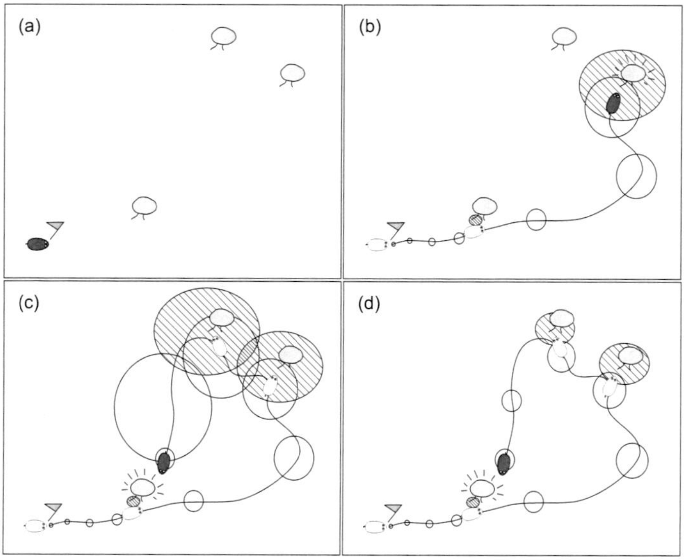

Fig. 1. A simple SLAM example. The robot is a mouse and landmarks are trees. Uncertainties on the robot position (resp. trees) are represented by blank (resp. hatched) ellipsis. (a) At initial time, the robot knows where it is. (b) The uncertainty on the robot position increases while it moves. When a second tree is detected, there is a significant uncertainty on its real position. (c) The first tree is met again and used to adjust the position of the robot. (d) With a backward computation of the trajectory, uncertainties on the previous positions and observations are reduced.

more the mission lasts, the more the robot gets lost. Likewise, detection of landmarks is achieved with less and less accuracy. However, when the robot detects again a landmark that was already placed with a good accuracy on the map, its position can be adjusted from that of the landmark and the whole process of estimation (trajectory and map building) can be refined (see Figure 1). Hence, localization and map building are two connected goals. This interdependence is one of the reason that makes traditional probabilistic approaches inadequate. In contrast, it makes no difficulty in the constraint programming framework, as shown below.

2.2 Basic Constraint Model

Let us now focus on a basic modeling of the SLAM. Many details on a real experiment (description of the robot, experimental setup, full constraint model, etc.) can be found in [7] and [8] that deal with SLAM in a submarine context.

The SLAM problem is cast into a CSP as follows. First, the motion of the autonomous robot obeys a differential equation: $\mathbf{p}'(t) = \mathbf{f}(\mathbf{u}(t))$, where $\mathbf{p}(t)$ is the position of the robot in space, $\mathbf{u}(t)$ a vector of m inputs (speed, rotation angles, etc.) and $\mathbf{f}$ a mapping from $\mathbb{R}^m$ to $\mathbb{R}^3$. This equation can be cast into a CSP using a classical interval variant of the Euler method. Details can be found in [7] and [8]. The discretization introduces a set of $(N+1)$ variables $\mathbf{p}^{(0)}, \ldots, \mathbf{p}^{(N)}$ where δ_t is the time lapse between to measures and $N\delta_t$ the total duration of the mission. These variables represent a discretization of the trajectory $\mathbf{p}$, i.e.,

$$\forall i,\ 0 \leq i \leq N, \quad \mathbf{p}^{(i)} = \mathbf{p}(t_0 + i\delta_t) \tag{1}$$

has to be fulfilled. The discretization also introduces N constraints.

Thus, the CSP provides a rigorous enclosure of the trajectory, i.e., for every possible input $\mathbf{u}(t)$ there exists a feasible tuple $(\mathbf{p}^{(0)}, \ldots, \mathbf{p}^{(N)})$ such that the trajectory $\mathbf{p}$ corresponding to the inputs satisfies (1).

2.3 Introducing Detections

Now that the motion of the vehicle has been cast into a CSP, let us take into account detections. In the mission, n landmarks have to be localized. Their coordinates will be denoted by $\mathbf{o}^{(1)}, \ldots, \mathbf{o}^{(n)}$. Once the mission is over, a human operator scans the waterfall of images provided by exteroceptive sensors. When a group of pixels are suspected to correspond to a landmark, a box encompassing the corresponding area is entered as a potential detection. The position of a pixel on the image can be directly translated into a distance between the landmark and the robot. First, the detection time $\tau(i)$ (i.e., the number of time steps since t_0) and the distance r_i are determined. Then, the landmark number $\sigma(i)$ is identified which amount to *match* detections with each others. Finally, a distance constraint $\texttt{dist}(\mathbf{p}^{(\tau(i))}, \mathbf{o}^{(\sigma(i))}) = r_i$ between the landmark and the robot is added into the model. Therefore, in [7], the model was augmented as follows:

$$(\mathcal{P}') \quad \left\{ \begin{array}{l} \text{additional variables:} \\ \quad \mathbf{o}^{(1)} \in [\mathbf{o}^{(1)}],\ \ldots,\ \mathbf{o}^{(n)} \in [\mathbf{o}^{(n)}] \\ \text{domains:} \\ \quad [\mathbf{o}^{(1)}] := (-\infty, +\infty),\ \ldots,\ [\mathbf{o}^{(n)}] := (-\infty, +\infty) \\ \text{additional constraints:} \\ \quad \texttt{dist}(\mathbf{p}^{(\tau(i))}, \mathbf{o}^{(\sigma(i))}) = r_i \quad (i = 1..k) \end{array} \right.$$

Humans are subject to three types of mistakes: (1) *Omission*, a landmark on the waterfall is missed by the operator; (2) *Illusion*, the operator adds a detection where there is no landmark; (3) *Mismatching*, the operator makes a wrong identification (do not match detections properly). On the one hand, the two first types of mistakes have been experimentally proven as irrelevant[1]. On the other hand, if identifying the *type of a landmark* is fairly easy, recognizing *one particular landmark* is very hard. In other words, the main difficulty in the operator's task is matching landmarks with each others. The bad news is that mismatching makes the model inconsistent. Hence, the third type of mistakes is critical and requires a lot of energy to be avoided. Up to now, in the experiments made in [7,8], matching was simply performed using a priori knowledge of seamark positions.

[1] The overall accuracy may suffer from a lack of detections but the consistency of the model is always maintained (at any time, many landmarks are anyway out of the scope of the sensors and somehow "missed"). Besides, perception is based on very specific visual patterns that rule out any confusion with elements of the environment.

2.4 Our Contribution

The ambition of our work is simply to skip the operator's matching phase. The idea is to use the knowledge on the number of landmarks to make this matching automatically by propagation. This is a realistic approach. In all the missions performed with submarine robots, a set of beacons is dropped into the sea to make the SLAM possible. The positions of the beacons at the bottom is not known because of currents inside water (consider also that in hostile areas they may be dropped by plane) but their cardinality is.

3 The Number of Distinct Vectors Constraint

Consider k vectors of variables $\mathbf{x}^{(1)}, \ldots, \mathbf{x}^{(k)}$ of equal dimension (which is 2 or 3 in practice) and an integer-valued variable n. The constraint $\texttt{nvector}(n, \{\mathbf{x}^{(1)}, \ldots, \mathbf{x}^{(k)}\})$ holds if there is n distinct vectors between $\mathbf{x}^{(1)}$ and $\mathbf{x}^{(k)}$ which can be written:

$$\texttt{nvector}(n, \{\mathbf{x}^{(1)}, \ldots, \mathbf{x}^{(k)}\}) \iff |\{\mathbf{x}^{(1)}, \ldots, \mathbf{x}^{(k)}\}| = n,$$

where $|\{\mathbf{x}^{(1)}, \ldots, \mathbf{x}^{(k)}\}|$ stands for the cardinality of the set of "values" taken by $\mathbf{x}^{(1)}, \ldots, \mathbf{x}^{(k)}$. An equivalent definition that better conform to the intuition behind is that the number of distinct "values" in $\{\mathbf{x}^{(1)}, \ldots, \mathbf{x}^{(k)}\}$ equals to n.

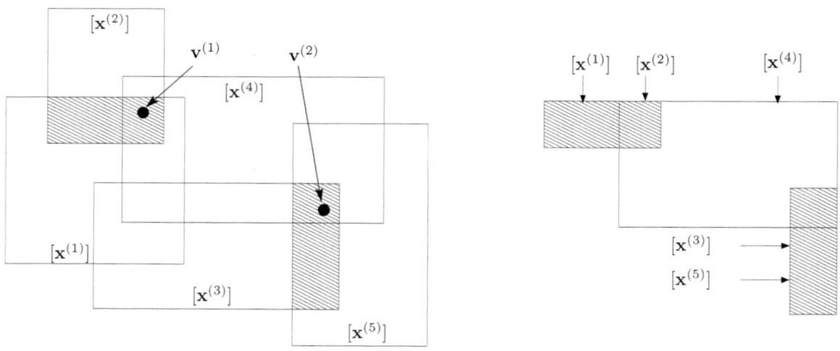

(a) Initial state (b) Final state

Fig. 2. (a) The atmost_nvector constraint with $n = 2$ and $k = 5$. The domains of $\mathbf{x}^{(i)}$ is the cross product of two intervals $[\mathbf{x}^{(i)}] = [x_1^{(i)}] \times [x_2^{(i)}]$. The constraint is satisfiable because there exists two vectors $\mathbf{v}^{(1)}$ and $\mathbf{v}^{(2)}$ such that $\forall i, 1 \leq i \leq n, [\mathbf{x}^{(i)}] \cap \{\mathbf{v}^{(1)}, \mathbf{v}^{(2)}\} \neq \emptyset$. The set of all such pairs $(\mathbf{v}^{(1)}, \mathbf{v}^{(2)})$ is represented by the two little dashed boxes. (b) Result of the bound consistency with respect to the atmost_nvector constraint. The domain of $[\mathbf{x}^{(4)}]$ encloses the two little rectangles.

A family of constraints can be derived from nvector in the same way as for nvalue [2]. The constraint more specifically considered in this paper is atmost_nvector that bounds the number of distinct vectors:

$$\texttt{atmost\_nvector}(n, \{\mathbf{x}^{(1)}, \ldots, \mathbf{x}^{(k)}\}) \iff |\{\mathbf{x}^{(1)}, \ldots, \mathbf{x}^{(k)}\}| \leq n.$$

It must be pointed out that, with continuous domains, the constraints `nvector` and `atmost_nvector` are operationally equivalent. Indeed, excepted in very particular situations where some domains are *degenerated* intervals (reduced to single points), variables can always take an infinity of possible values. This means that any set of k non-degenerated boxes always share *at least* n values. In the `nvector` constraint, only the upper bound ("*at most* n") can actually filters. This remark generalizes the fact that `all_diff` is always satisfied with non-degenerated intervals. Figure 2a shows an example with $n = 2$ (the variable is ground) and $k = 5$ in two dimensions. Figure 2b shows the corresponding result of a bound consistency filtering.

4 Operational Complexity

Enforcing generalized arc consistency (GAC) for the `atmost-nvalue` constraint is NP-hard [3]. Worse, simply computing the minimum number of distinct values in a set of domains is NP-hard [2]. However, when domains are intervals, these problems are polynomial [1,6]. The question under interest is to know if `atmost-nvector` is still a tractable problem when domains are boxes (vector of intervals), i.e., when the dimension is greater than 1. We will show that it is not.

Focusing on this type of domains is justified because continuous constraints are always handled with intervals. This also implies that bound consistency is the only acceptable form of filtering that can be applied with continuous domains.[2]

We shall restrict ourselves to the calculation of the minimum number of distinct values (also called the *minimum cardinality*) and to boxes of dimension 2, i.e., *rectangles*. As noticed in [2], the minimum number of distinct values shared by a set of rectangles is also the cardinality of the *minimum hitting set*. A hitting set is a collection of points that intersects each rectangle. In the two following subsections, we consider the equivalent problem of finding the cardinality of the *minimum clique partition* of a *rectangle graph*.

4.1 Rectangle Graphs and Clique Partitions

A *rectangle* graph is a n-vertices graph $G_r = (V_r, E_r)$ that can be extracted from n axis-aligned rectangles in the plane by (1) creating a vertex for each a rectangle, and (2) adding an edge when there is a non-empty intersection between two rectangles. Figure 3 depicts the two views of a rectangle graph.

A *clique partition* of a graph is a collection of complete subgraphs that partition all the vertices. A *minimal* clique partition is a clique partition with the smallest cardinality (i.e., with the smallest number of subgraphs). Since any set of pairwise intersecting rectangles intersect all mutually (by Helly's theorem), a k-clique in a rectangle graph represents the intersection of k rectangles in the

[2] Allowing gaps inside intervals and applying GAC filtering leads to unacceptable space complexity [4].

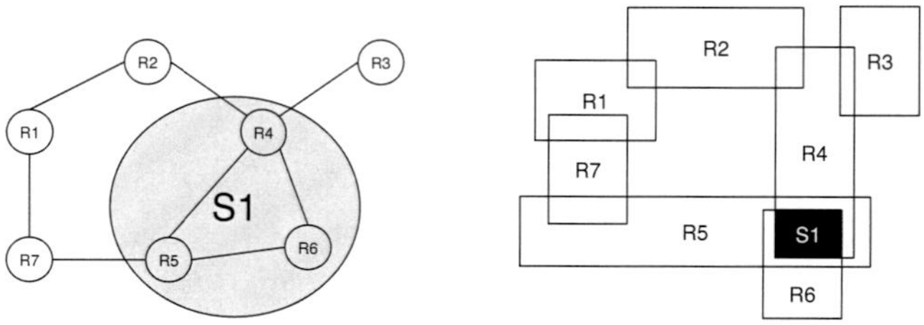

(a) Rectangle graph (b) Geometrical view

Fig. 3. A rectangle graph and its geometrical representation (axis-aligned rectangles)

geometrical view. For instance, the 3-clique $S1$ of G_r in Figure 3a represents the intersection of three rectangles $(R4, R5, R6)$ in Figure 3b, depicted by the black rectangle $S1$. As a consequence, looking for the minimum hitting set or the minimum clique partition are indeed equivalent problems for rectangle graphs. The final problem under consideration can then be formulated as follows:

Rectangle Clique Partition (RCP)
- *Instance*: A rectangle graph $G_r = (V_r, E_r)$ given in the form of $|V_r|$ axis-aligned rectangles in the plane and $k \leq |V_r|$.
- *Question*: Can V_r be partition into k disjoint sets $V_1, \ldots, V_k$ such that $\forall i,\ 1 \leq i \leq k$ the subgraph induced by V_i is a complete graph?

Proposition 1. *RCP is NP-complete.*

The fact that RCP belongs to NP is easy to prove: the given of a k–partition is a certificate that can be checked in polynomial time. The rest of the proof, i.e., the reduction of a NP-complete problem to RCP is given below. Note that the transformation below is inspired from that in [10] but contains fundamental differences.

4.2 Building a Rectangle Graph from a Cubic Planar Graph

The problem we will reduce to RCP involves *planar* graphs which require the introduction of extra vocabulary. An *embedding* of a graph G on the plane is a representation of G (see Figure 4) in which points are associated to vertices and arcs are associated to edges in such a way:

- the endpoints of the arc associated to an edge e are the points associated to the end vertices of e,
- no arcs include points associated to other vertices,
- two arcs never intersect at a point which is interior to either of the arcs.

A *planar* graph (Figure 4a) is a graph which admits an embedding on the plane (Figure 4b), and a *cubic* graph is a 3-regular graph, i.e., a graph in which every

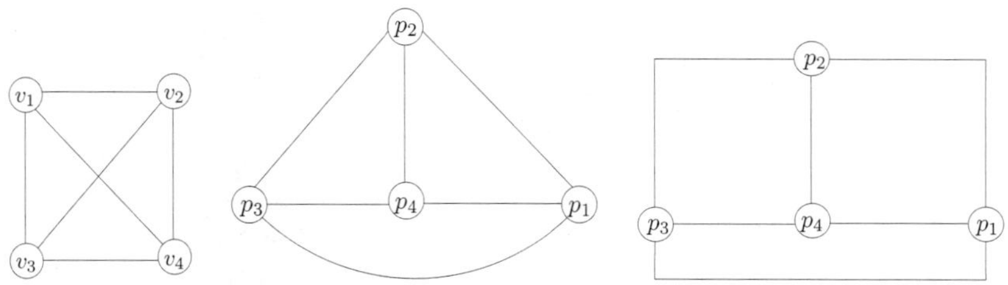

(a) Cubic Planar graph. (b) A possible embedding on the plane. (c) A rectilinear embedding on the plane.

Fig. 4. A (cubic) planar graph, one of its embedding on the plane, and one of its rectilinear embedding

vertex has 3 incident edges. A *rectilinear embedding* is an embedding where every arc is a broken line formed by horizontal and vertical segments (Figure 4c).

We are now in position to discuss about the transformation itself. Consider a planar cubic graph G_P. First, Tammassia *and al.* gave a polytime algorithm in [11] for computing a 5-rectilinear embedding of a planar graph, i.e., a rectilinear embedding where each broken line is formed by 5 segments. This is illustrated by Figure 5. Second, this embedding can be transformed into a rectangle graph using the patterns depicted on Figures 6a and 6b:

- Every segment of the 5-rectilinear embedding of G_p is replaced by a rectangle such that if two segments do not intersect in the 5-rectilinear embedding, the corresponding rectangles do not intersect (should they be flat).
- For segments having a vertex at a common endpoint, the corresponding rectangles intersect all mutually in the neighborhood of this vertex (Figure 6a). Remember that the case of a leaf node is irrelevant because G_p is cubic.
- For segments having a bend in common, the rectangles are disjoint.
- An extra rectangle is added in the neighborhood of each bend. This rectangle intersects the two rectangles associated to the segments (Figure 6b). The three rectangles cannot intersect all mutually due to the previous point.

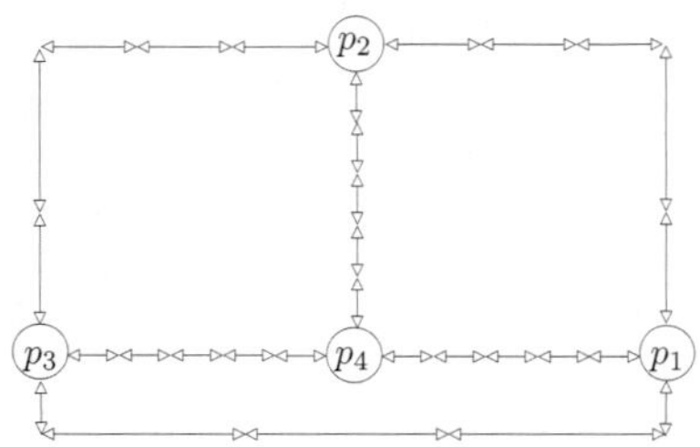

Fig. 5. A 5-rectilinear embedding of the cubic planar graph given in Figure 4 (segments are delineated by arrows)

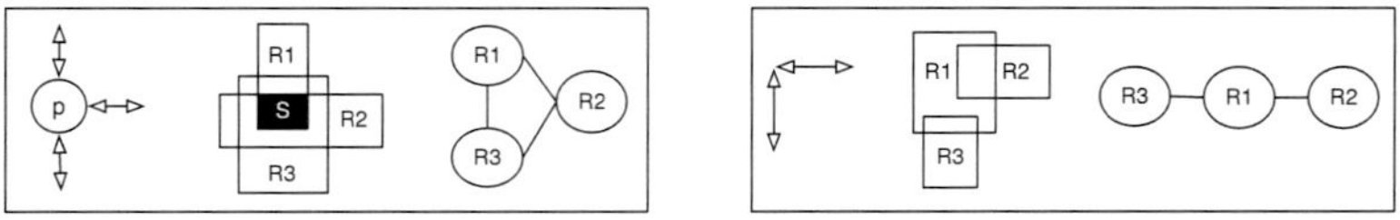

(a) Transformation of segments at a common endpoint.

(b) Transformation of segments having a bend in common.

Fig. 6. Atomic operations to transform the 5-rectilinear embedding of a cubic planar graph (left) into the geometrical view (middle) of a rectangle graph (right)

4.3 Reduction from Cubic Planar Vertex Cover

Consider the well-known *vertex cover* problem. This problem remains NP-Complete even for cubic planar graphs [12]. It is formally stated as follows:

Cubic Planar Vertex Cover (CPVC)

- *Instance*: A cubic planar graph $G_p = (V_p, E_p)$ and $k \leq |V_p|$.
- *Question*: Is there a subset $V' \subseteq V_p$ with $|V'| \leq k$ such each edge of E_p has at least one of its extremity in V'?

Lemma 1. *Let G_p be a n-vertex m-edges cubic planar graph and an integer value $k \leq n$. Let G_r be the rectangle graph obtained by the transformation of §4.2. The answer to CPVC with (G_p, k) is yes iff the answer to RCP with $(G_r, k + 4 \times m)$ is yes.*

Proof. This proof is illustrated in Figure 7.

Forward Implication: Assume a n-vertex m-edges cubic planar graph G_p has a vertex cover V' of cardinality k. We shall build a partition $\mathcal{P}$ of G_r, initially empty. To each edge $e = (v_i, v_j)$ of E_p corresponds a 2-degree chain c_e in G_r, with exactly nine vertices, from a vertex of the 3-clique p_i(associated to v_i) to a vertex of the 3-clique p_j (associated to v_j).

First, for every $v_i \in V'$, add the 3-clique p_i in $\mathcal{P}$. Since V' is a covering set, every chain has now one of its extreme vertex inside a clique of $\mathcal{P}$. The 8 other remaining vertices of the chain can then easily be partitioned into 4 additional cliques. Add these cliques to $\mathcal{P}$.

Once all these additional cliques are added to $\mathcal{P}$, the latter is a partition of G_r whose size is $k + 4 \times m$.

Backward Implication: Assume G_r can be partition into $k+4\times m$ cliques and let $\mathcal{P}$ be such a partition. We shall call a *free clique* a clique that only contain vertices of the same given chain. Similarly, a *shared clique* is a clique involving (extreme) vertices of at least two different chains.

Every chain contains 9 vertices. One can easily see that it requires at least 5 cliques to be partitioned, 4 of which are free. For every chain c, remove 4 free cliques of c from $\mathcal{P}$. Then there is still (at least) one clique left in $\mathcal{P}$ that involves a vertex of c. When this process is done, the number of remaining cliques in $\mathcal{P}$ is $k + 4 \times m - 4 \times m = k$.

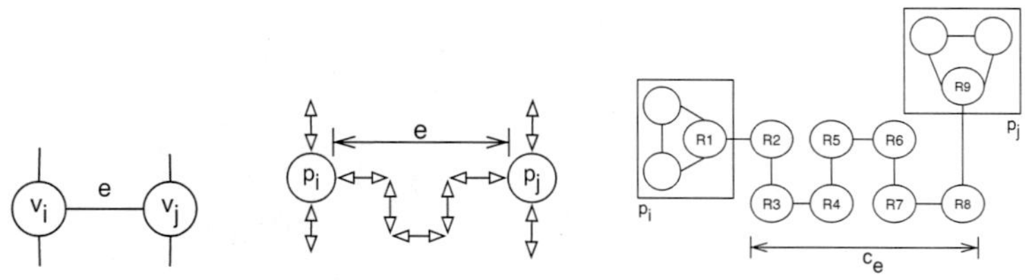

(a) Edge of G_p. (b) Rectilinear embedding view. (c) Corresponding *chain* (subgraph of G_r).

Fig. 7. Transforming an edge of the cubic planar graph G_P into, first, its rectilinear view, second, a rectangle graph G_r

Now, for every clique $\mathcal{C}$ of $\mathcal{P}$: if $\mathcal{C}$ is shared, put the corresponding vertex of G_P in V'. If $\mathcal{C}$ is free, consider the edge e of G_P associated to the chain and put anyone of the two endpoint vertices of e into V'. We have $|V'| = k$ and since every chain has a vertex in the remaining cliques of $\mathcal{P}$, every edge of G_P is covered by V'. □

5 A Very First Polytime Contractor

We shall now propose a simple algorithm for computing bound consistency with respect to `atmost_nvector`. Let us denote by *dim* the dimension of the vectors. The contractor is derived from the following implication:

$$\texttt{atmost\_nvector}(n, \{\mathbf{x}^{(1)}, \ldots, \mathbf{x}^{(k)}\}) \Longrightarrow$$
$$\texttt{atmost\_nvalue}(n, \{x_1^{(1)}, \ldots, x_1^{(k)}\}) \wedge \ldots \wedge \texttt{atmost\_nvalue}(n, \{x_{dim}^{(1)}, \ldots, x_{dim}^{(k)}\}).$$

Therefore, applying a contractor for `atmost_nvalue` with the projections of the boxes onto each dimension in turn gives a contractor for `atmost_nvector`. Since a contractor for `atmost_nvalue` enforcing GAC (hence, bound consistency) when domains are intervals already exists [1], we are done. This is achieved in $O(dim \times k\log(k))$. Although the purpose of this paper is not to describe an efficient contractor, we shall introduce right away a first (but significant) improvement of our naive algorithm.

The reader must simply admit that the contractor of $\texttt{atmost\_nvalue}(n, \{x_i^{(1)}, \ldots, x_i^{(k)}\})$ works in two steps (see Figure 8). First, it builds *groups* of variables. If the number of groups turns to be n, it is then proven that for all $x_i^{(j)}$ and $x_i^{(l)}$ that belong to the same group, $x_i^{(j)}$ and $x_i^{(l)}$ must satisfy $x_i^{(j)} = x_i^{(l)}$. Hence, the domains of all the variables of the same group G can be shrunk to their common intersection denoted by $[G]$. Furthermore, this process results in a set of n disjoint intervals $[G_1], \ldots, [G_n]$ that we will call *kernels*.

In the favorable case of n groups, the last step can be improved. If two variables $x_i^{(j)}$ and $x_i^{(l)}$ belong to the same group G, the whole vector variables $\mathbf{x}^{(j)}$ and $\mathbf{x}^{(l)}$

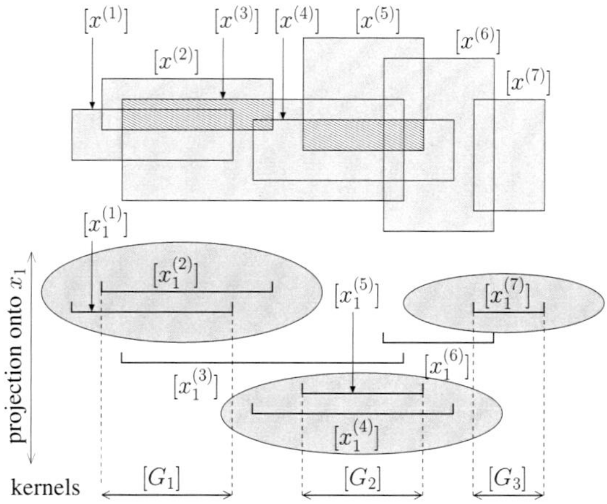

Fig. 8. Illustration of what the `at_most_nvalue` algorithm yields. Here, $k = 7$, $dim = 2$, $n = 3$ and the algorithm is run for the first projection (i.e., onto the horizontal axis). Three groups, represented by gray ellipsis, are identified, namely $G_1 = \{x_1^{(1)}, x_1^{(2)}\}$, $G_2 = \{x_1^{(4)}, x_1^{(5)}\}$ and $G_3 = \{x_1^{(7)}\}$. The variables of the same group are proven to be all equal (otherwise, the constraint is violated). E.g., $x_1^{(1)}$ and $x_1^{(2)}$ must satisfy $x_1^{(1)} = x_1^{(2)}$. Domains for the variables of a given group can be replaced by the corresponding kernel (e.g., $[x_1^{(1)}]$ and $[x_1^{(2)}]$ can be set to $[G_1]$). Notice that the kernels are all disjoint. In our suggested improvement, the whole boxes are intersected instead of the first components only. Hence the domains of $[\mathbf{x}^{(1)}]$ and $[\mathbf{x}^{(2)}]$ are intersected, which gives one of the hatched rectangles. The other hatched rectangle is $[\mathbf{x}^{(4)}] \cap [\mathbf{x}^{(5)}]$.

are actually constrained to be equal. Indeed, assume that $\mathbf{x}^{(j)}$ and $\mathbf{x}^{(l)}$ could take two different vectors. The $k - 2$ other variables necessarily share $n - 1$ different vectors because their i^{th} components must "hit" the $n - 1$ (disjoint) kernels $[G']$ with $G' \neq G$. This means that the overall number of distinct vectors is at least $(n - 1) + 2 > n$.

Hence, the algorithm of `atmost_nvalue` can be modified to intersect boxes (instead of just one component). This multiplies the complexity by dim (which is small in practice).

6 Experimental Evaluation: The SLAM Problem

This section shows how the `atmost_nvector` constraint allows to improve the modeling and resolution of the SLAM problem. We propose an extension of the original model given in Section 2, and provide a graphical validation.

We introduce k 3-dimensional variables $\mathbf{d}^{(1)}, \ldots, \mathbf{d}^{(k)}$ related to all the detections. A distance constraint involving each of the latters is added into the model, as well as a `nvector` constraint capturing the fact that only n landmarks exist.

$$(\mathcal{P}'') \quad \begin{cases} \text{additional variables:} \\ \quad \mathbf{d}^{(1)} \in [\mathbf{d}^{(1)}], \ \ldots, \ \mathbf{d}^{(k)} \in [\mathbf{d}^{(k)}] \\ \text{domains:} \\ \quad [\mathbf{d}^{(1)}] := (-\infty, +\infty), \ \ldots, \ [\mathbf{d}^{(k)}] := (-\infty, +\infty) \\ \text{additional constraints:} \\ \quad \mathtt{dist}(\mathbf{p}^{\tau(i)}, \mathbf{d}^{(i)}) = r_i \quad (i = 1..k) \\ \quad \mathtt{atmost} - \mathtt{nvector}(n, \{\mathbf{d}^{(1)}, \ldots, \mathbf{d}^{(k)}\}) \end{cases}$$

The introduction of the `atmost-nvector` constraint has provided the expected results. Let us first explain how the benefits of this constraint can be quantified.

In the context of differential equations, we are dealing with a very large number of variables and a sparse system with an identified structure. Moreover, the system is subject to many uncertainties. Under these conditions, the quality of the result is more relevant than the computation time, as we justify now.

No Choice Point. In applications, the number of variables can be huge (more than 1140000 in [7,8] in order to represent the three coordinates of the robot, the Euler angles, the speed, the altitude and the components of the rotation matrix at each time step). This prevents us for making choice points. Note that the situation is even worse since the solution set is not a thin trajectory (formed by isolated points). The numerous uncertainties make the best solution ever a thick beam of trajectories (as depicted in Figure 9). Continuums of solutions usually dissuade one from making choice points, even for much smaller problems.

Handcrafted Propagation. The size of our problem also prevents us for using a general-purpose AC3-like propagation algorithm, since building an adjacency matrix for instance would simply require too much memory. Furthermore, we have a precise knowledge of how the network of constraints is structured. Indeed, the discretization of the motion yields a single (long) chain of constraints and cycles in the network only appear with detection constraints. An appropriate strategy is then to base propagation on the detection constraints, which are much fewer. Every time a detection reduces significantly the domain of a position $\mathbf{p}^{(i)}$, the motion constraints are propagated forward (from $\mathbf{p}^{(i)}$ to $\mathbf{p}^{(N)}$) and backward (from $\mathbf{p}^{(i)}$ downto $\mathbf{p}^{(0)}$). In a nutshell, propagation in this kind of problems is guided by the physics.

Irrelevance of Computation Time. When the application is run, it takes a couple of seconds to load data of the sensors (which amount to initialize the domains of variables representing input vectors) and to precalculate expressions (e.g., rotation matrices). With or without the `nvector` constraint, propagation takes a negligible time (less than 1%) in comparison to this initialization. Therefore, focusing on computation time is not very informative.

Quality of the Result. Our contribution is to make propagation for the SLAM problem automatic whereas part of it was performed by a human operator so far. This is a result in itself. However, one may wonder which between the automatic matching and the operator's is most competitive. This, of course, is hard to evaluate a priori. In the experiment of [7,8], both have provided the optimal matching. The question that still remains is to know the extent to which our matching (the one provided by the algorithm in Section 5) improves the "quality" of the result, i.e., the accuracy of the trajectory.

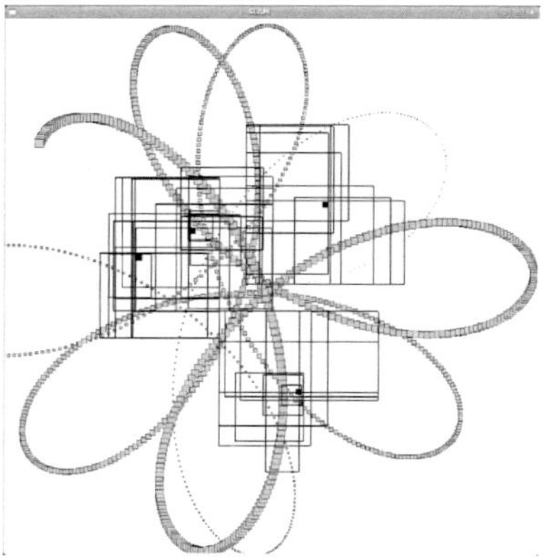

(a) Basic SLAM: trajectory and detections (1200 iterations).

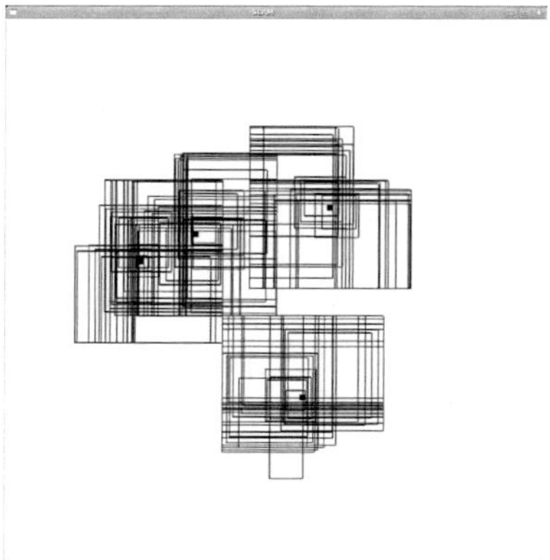

(b) Basic SLAM: Detections (4000 iterations).

(c) Basic SLAM: trajectory (4000 iterations).

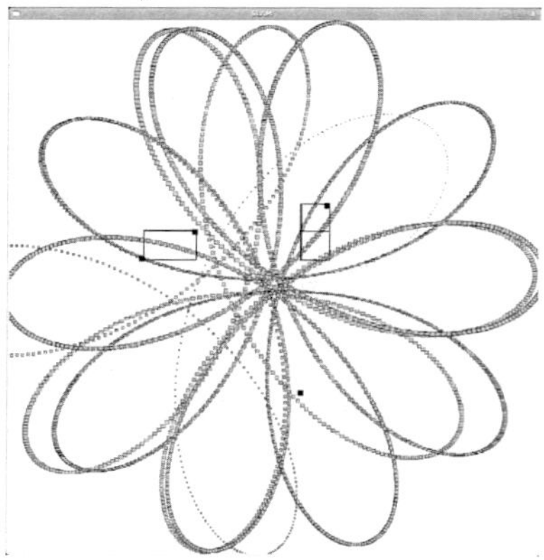

(d) Improved SLAM: trajectory and detections (4000 iterations).

Fig. 9. Comparing SLAM with the `atmost-nvector` contractor (*Improved SLAM*) and without(*Basic SLAM*). The trajectory is represented by gray filled boxes, detections by thick-border rectangles and landmarks by the little black squares. Figure 9(d) depicts the fixpoint of all the contractors: the four landmarks are very well localized and the trajectory is much thiner (after 4000 iterations, the largest diameter is 10 times less than in Figure 9(c)). Our algorithm has contracted many detections to optimal boxes around landmarks. We can also observe the weakness of our algorithm which has poorly reduced boxes whose projections on both axis encompass two projections of landmarks. This however exclude a detection which really encloses two landmarks: in this case, either there is a *real* ambiguity or it is the propagation to blame.

The idea was to make the robot looping around the same initial point so that many detections would intersect.[3] For this purpose, we have controlled the robot with a classical feedback loop which gives the expected cycloidal trajectory.

Four landmarks have been placed in the environment and we have basically considered that a landmark is detected everytime the distance between the robot and itself reaches a local minimum (if less than a reasonable threshold). The estimation of the landmark position is then calculated from this distance (with an additional noise) and a very rough initial approximation.

Figure 9 illustrates the effect of automatic matching on the estimation of the trajectory and the positions of the landmarks. All the results have been obtained in a couple of seconds by the Quimper system [5].

7 Conclusion

A somewhat natural generalization of the `nvalue` constraint, called `nvector` has been proposed. The `nvector` constraint can help modeling and solving many localization problems where a bound on the number of landmarks to localized is known. This has been illustrated on the SLAM problem and applied on a real experiment. We have also analyzed the complexity of this global constraint and given a simple contractor.

The benefit of this constraint in terms of modeling has a direct impact on the way data of the experiments have to be processed. Indeed, the constraint allows to avoid requiring someone that matches landmarks by hand. Hence, it reduces considerably the amount of work and the probability of mistake this operation entails.

The field of application is not restricted to the SLAM problem. Ongoing works show that the `nvector` is as crucial for the passive location of vehicles using TDOA (time difference of arrival) in signal processing. All these problems involve real variables. Hence, as a side contribution, this paper also offsets the lack of activity about global continuous constraints.

Future works include the design of more sophisticated contractors with benchmarking. The `nvector` constraint also leads up to the study of other global constraints. As soon as several estimations of the same landmark position are matched by `nvector`, this position satisfies indeed a global constraint (namely, the intersection of several spheres if estimations result from distance equations).

References

1. Beldiceanu, N.: Pruning for the minimum Constraint Family and for the number of distinct values Constraint Family. In: Walsh, T. (ed.) CP 2001. LNCS, vol. 2239, pp. 211–224. Springer, Heidelberg (2001)

[3] The experiment of [7,8] was not appropriate for this illustration because matching seamarks was actually too easy (6 seamarks and a rectangle graph of detections with 6 strongly connected components).

2. Bessière, C., Hébrard, E., Hnich, B., Kiziltan, Z., Walsh, T.: Filtering Algorithms for the NValue Constraint. In: Barták, R., Milano, M. (eds.) CPAIOR 2005. LNCS, vol. 3524, pp. 79–93. Springer, Heidelberg (2005)
3. Bessière, C., Hébrard, E., Hnich, B., Walsh, T.: The Complexity of Global Constraints. In: AAAI 2004, pp. 112–117 (2004)
4. Chabert, G.: Techniques d'Intervalles pour la Résolution de Systèmes d'Équations. PhD Thesis, Université de Nice-Sophia Antipolis (2007)
5. Chabert, G., Jaulin, L.: Contractor Programming. Artificial Intelligence 173, 1079–1100 (2009)
6. Gupta, U.I., Lee, D.T., Leung, Y.T.: Efficient Algorithms for Interval Graphs and Circular-Arc Graphs. Networks 12, 459–467 (1982)
7. Jaulin, L.: Localization of an Underwater Robot using Interval Constraint Propagation. In: Benhamou, F. (ed.) CP 2006. LNCS, vol. 4204, pp. 244–255. Springer, Heidelberg (2006)
8. Jaulin, L.: A Nonlinear Set-membership Approach for the Localization and Map Building of an Underwater Robot using Interval Constraint Propagation. IEEE Transaction on Robotics 25(1), 88–98 (2009)
9. Moore, R.: Interval Analysis. Prentice-Hall, Englewood Cliffs (1966)
10. Rim, C.S., Nakajima, K.: On Rectangle Intersection and Overlap Graphs. IEEE Transactions on Circuits and Systems 42(9), 549–553 (1995)
11. Tamassia, R., Tollis, I.G.: Planar Grid Embedding in Linear Time. IEEE Trans. Circuits Systems 36, 1230–1234 (1989)
12. Uehara, R.: NP-Complete Problems on a 3-connected Cubic Planar Graph and their Applications. Technical Report Technical Report TWCU-M-0004, Tokyo Woman's Christian University (1996)

Approximating Weighted Max-SAT Problems by Compensating for Relaxations

Arthur Choi, Trevor Standley, and Adnan Darwiche

Computer Science Department,
University of California, Los Angeles
Los Angeles, CA 90095
{aychoi,tstand,darwiche}@cs.ucla.edu

Abstract. We introduce a new approach to approximating weighted Max-SAT problems that is based on simplifying a given instance, and then tightening the approximation. First, we relax its structure until it is tractable for exact algorithms. Second, we compensate for the relaxation by introducing auxiliary weights. More specifically, we relax equivalence constraints from a given Max-SAT problem, which we compensate for by recovering a weaker notion of equivalence. We provide a simple algorithm for finding these approximations, that is based on iterating over relaxed constraints, compensating for them one-by-one. We show that the resulting Max-SAT instances have certain interesting properties, both theoretical and empirical.

1 Introduction

Relaxations are often used to tackle optimization problems, where a tractable relaxed problem is used to approximate the solution of an intractable one. Indeed, they are employed by a few recently proposed solvers for the maximum satisfiability (Max-SAT) problem [1,2], which have shown to be competitive for certain classes of benchmarks in recent Max-SAT evaluations. In these solvers, a given Max-SAT instance is relaxed enough until it is amenable to an exact solver. Upper bounds computed in the resulting relaxation are then used in a branch-and-bound search to find the Max-SAT solution of the original instance.

Whether a relaxation is used in a branch-and-bound search, or used as an approximation in and of itself, a trade-off must be made between the quality of a relaxation and its computational complexity. The perspective that we take in this paper, instead, is to take a given relaxation, infer from its weaknesses, and compensate for them. Since we assume that reasoning about the original problem is difficult, we can exploit instead what the relaxed problem is able to tell us, in order to find a tighter approximation.

In this paper, we propose a class of weighted Max-SAT approximations that are found by performing two steps. First, we *relax* a given weighted Max-SAT instance, which results in a simpler instance whose solution is an upper bound on that of the original. Second, we *compensate* for the relaxation by correcting for deficiencies that were introduced, which results in an approximation with

I.P. Gent (Ed.): CP 2009, LNCS 5732, pp. 211–225, 2009.

improved semantics and a tighter upper bound. These new approximations, can in turn be employed by other algorithms that rely on high quality relaxations.

More specifically, we relax a given weighted Max-SAT problem by removing from it certain equivalence constraints. To compensate for each equivalence constraint that we relax, we introduce a set of unit clauses, whose weights restore a weaker notion of equivalence, resulting in a tighter approximation. In the case a single equivalence constraint is relaxed, we can identify compensating weights by simply performing inferences in the relaxation. In the case multiple equivalence constraints are relaxed, we propose an algorithm that iterates over equivalence constraints, compensating for them one-by-one. Empirically, we observe that this iterative algorithm tends to provide monotonically decreasing upper bounds on the solution of a given Max-SAT instance.

Proofs are given in the Appendix, or in the full report [3], in the case of Theorem 4. For an introduction to modern approaches for solving and bounding Max-SAT problems, see [4].

2 Relaxing Max-SAT Problems

Max-SAT is an optimization variant of the classical Boolean satisfiability problem (SAT). Given a Boolean formula in clausal form, the goal is to find an assignment of variables X to truth values x or $\bar{x}$, that maximizes the number of satisfied clauses. In the *weighted* Max-SAT problem, each clause is associated with a non-negative weight and the goal is to find an assignment that maximizes the aggregate weight of satisfied clauses. The weighted *partial* Max-SAT problem further specifies clauses that are hard constraints, that must be satisfied by any solution.[1] In this paper, we focus on weighted Max-SAT problems, although we will formulate relaxations in terms of hard constraints, as we shall soon see.

Let $f = \{(C_1, w_1), \ldots, (C_m, w_m)\}$ be an instance of weighted Max-SAT over variables X, where w_j is the weight of a clause C_j. Let $\mathbf{X}$ denote the set of all variables in f, and let $\mathbf{x}$ denote an assignment of variables X to truth values x or $\bar{x}$ (we also refer to truth values as signs). An optimal Max-SAT assignment $\mathbf{x}^\star$ is an assignment that maximizes the aggregate weight of satisfied clauses: $\mathbf{x}^\star = \text{argmax}_\mathbf{x} \sum_{\mathbf{x} \models C_j} w_j$. We denote the corresponding value of a Max-SAT solution by $F^\star = \max_\mathbf{x} \sum_{\mathbf{x} \models C_j} w_j$. Note that a weighted Max-SAT instance f may have multiple assignments $\mathbf{x}^\star$ that are optimal, so we may refer to just the optimal value $F^\star$ when the particular optimal assignment $\mathbf{x}^\star$ is not relevant.

Let $\mathbf{Z} \subseteq \mathbf{X}$ denote a subset of the variables in instance f. We are also interested in Max-SAT solutions under partial assignments $\mathbf{z}$. More specifically, let $\mathbf{x} \sim \mathbf{z}$ denote that $\mathbf{x}$ and $\mathbf{z}$ are compatible instantiations, i.e., they set common variables to the same values. We then denote the value of a Max-SAT solution under a partial assignment $\mathbf{z}$ by

$$F(\mathbf{z}) = \max_{\mathbf{x} \sim \mathbf{z}} \sum_{\mathbf{x} \models C_j} w_j.$$

[1] In practice, hard constraints can be represented by clauses with large enough weights.

We will be particularly interested in the value of a Max-SAT solution when a single variable X is set to different signs, namely $F(x)$ when variable X is set positively, and $F(\bar{x})$ when variable X is set negatively. Note, for an optimal assignment $\mathbf{x}^\star$, we have $F^\star = F(\mathbf{x}^\star) = \max\{F(x), F(\bar{x})\}$, for any variable X.

Consider now the notion of an *equivalence* constraint:

$$(X \equiv Y, \infty) \stackrel{def}{=} \{(x \vee \bar{y}, \infty), (\bar{x} \vee y, \infty)\}$$

that is a hard constraint that asserts that X and Y should take the same sign. We consider in this paper the *relaxation* of Max-SAT instances that result from removing equivalence constraints. Clearly, when we remove an equivalence constraint from a Max-SAT instance f, more clauses can become satisfied and the resulting optimal value will be an upper bound on the original value $F^\star$.

It is straightforward to augment a Max-SAT instance, weighted or not, to an equivalent one where equivalence constraints can be relaxed. Consider, e.g.,

$$\{(a \vee b, w_1), (\bar{b} \vee c, w_2), (\bar{c} \vee d, w_3)\}.$$

We can replace the variable C appearing as a literal in the third clause with a clone variable C', and add an equivalence constraint $C \equiv C'$, giving us:

$$\{(a \vee b, w_1), (\bar{b} \vee c, w_2), (\bar{c}' \vee d, w_3), (c \vee \bar{c}', \infty), (\bar{c} \vee c', \infty)\}$$

which is equivalent to the original in that an assignment $\mathbf{x}$ of the original formula corresponds to an assignment $\mathbf{x}'$ in the augmented formula, and vice-versa, where the assignment $\mathbf{x}'$ sets the variable and its clone to the same sign. Moreover, the assignment $\mathbf{x}$ satisfies the same clauses in the original instance that assignment $\mathbf{x}'$ satisfies in the augmented instance (minus the equivalence constraint), and vice-versa. In this particular example, when we remove the equivalence constraint $C \equiv C'$, we have a relaxed formula composed of two independent subproblems: $\{(a \vee b, w_1), (\bar{b} \vee c, w_2)\}$ and $\{(\bar{c}' \vee d, w_3)\}$.

A number of structural relaxations can be reduced to the removal of equivalence constraints, including variable splitting [1,5,6], variable relabeling [2], and mini-bucket approximations [7,5]. In particular, these relaxations, which ignore variables shared among different clauses, can be restored by adding equivalence constraints.

3 Compensating for Relaxations

Suppose that we have simplified a Max-SAT instance f by relaxing equivalence constraints, resulting in a simpler instance h. Our goal now is to identify a Max-SAT instance g that is as tractable as the relaxed instance h, but is a tighter approximation of the original instance f.

We propose that we construct a new instance g by introducing auxiliary clauses into the relaxation. More specifically, for each equivalence constraint

$X \equiv Y$ relaxed from the original instance f, we introduce four unit clauses, with four auxiliary weights:

$$\{(x, w_x), (\bar{x}, w_{\bar{x}}), (y, w_y), (\bar{y}, w_{\bar{y}})\}.$$

Note that adding unit clauses to an instance does not impact significantly the complexity of solving it, in that the addition does not increase the treewidth of an instance. Thus, adding unit clauses does not increase the complexity of Max-SAT algorithms that are exponential only in treewidth [8]. Our goal now is to specify the auxiliary weights $w_x, w_{\bar{x}}, w_y, w_{\bar{y}}$ so that they compensate for the equivalence constraints relaxed, by restoring a weaker notion of equivalence.

3.1 Intuitions: A Simplified Case

In this section, we describe our proposal in the simpler case where a *single* equivalence constraint $X \equiv Y$ is removed. We shall make some claims without justification, as they follow as corollaries of more general results in later sections.

Say we remove a single equivalence constraint $X \equiv Y$ from f, resulting in a relaxation h. In any optimal assignment for instance f, variables X and Y are set to the same sign, because of the equivalence constraint $X \equiv Y$. If a Max-SAT assignment $\mathbf{x}^\star$ for the relaxation h happens to set variables X and Y to the same sign, then we know that $\mathbf{x}^\star$ is also a Max-SAT assignment for instance f. However, an optimal assignment for the relaxation h may set variables X and Y to different signs, thus the Max-SAT value $H^\star$ of the relaxation h is only an upper bound on the Max-SAT value $F^\star$ of the original instance f. The goal then is to set the weights $w_x, w_{\bar{x}}$ on X and $w_y, w_{\bar{y}}$ on Y to correct for this effect.

Consider first, in the relaxation, the values of $H(x)$ and $H(y)$, the Max-SAT values assuming that a variable X is set to a value x, and separately, that variable Y is set to a value y. If $H(x) \neq H(y)$ and $H(\bar{x}) \neq H(\bar{y})$, then we know that a Max-SAT assignment for the relaxation h sets X and Y to different signs: the Max-SAT value assuming X is set to x is not the same as the Max-SAT value assuming Y is set to y (and similarly for $\bar{x}$ and $\bar{y}$). Thus, we may want to set the weights $w_x, w_{\bar{x}}, w_y, w_{\bar{y}}$ so that $G(x) = G(y)$ and $G(\bar{x}) = G(\bar{y})$ in the compensation g, so that if there is a Max-SAT assignment that sets X to x, there is at least a Max-SAT assignment that also sets Y to y, even if there is no Max-SAT assignment setting both X and Y to the same sign at the same time.

We thus propose the following weaker notion of equivalence to be satisfied in a compensation g, to makeup for the loss of an equivalence constraint $X \equiv Y$:

$$\frac{1}{2}\begin{bmatrix} G(x) \\ G(\bar{x}) \end{bmatrix} = \frac{1}{2}\begin{bmatrix} G(y) \\ G(\bar{y}) \end{bmatrix} = \begin{bmatrix} w_x + w_y \\ w_{\bar{x}} + w_{\bar{y}} \end{bmatrix} \tag{1}$$

Before we discuss this particular choice of weights, consider the following properties of the resulting compensation g. First, we can identify when a compensation g satisfying Equation 1 yields exact results, just as we can with a relaxation h. In particular, if $\mathbf{x}^\star$ is an optimal assignment for g that sets the variables X and Y to the same sign, then: (1) assignment $\mathbf{x}^\star$ is also optimal for instance f; and

(2) $\frac{1}{2}G(\mathbf{x}^\star) = F(\mathbf{x}^\star)$. Moreover, a compensation g yields an upper bound that is *tighter* than the one given by the relaxation h:

$$F^\star \leq \frac{1}{2}G^\star \leq H^\star.$$

See Corollary 1 in the Appendix, for details.

To justify the weights we chose in Equation 1, consider first the following two properties, which lead to a notion of an ideal compensation. First, say that a compensation g has *valid configurations* if:

$$G(x) = G(y) = G(x, y) \qquad \text{and} \qquad G(\bar{x}) = G(\bar{y}) = G(\bar{x}, \bar{y}),$$

i.e., Max-SAT assignments that set X to a sign x also set Y to the same sign y, and vice versa; analogously if X is set to $\bar{x}$ or Y is set to $\bar{y}$. Second, say that a compensation g has *scaled values* if the optimal value of a valid configuration is proportional to its value in the original instance f, i.e., $G(x, y) = \kappa F(x, y)$ and $G(\bar{x}, \bar{y}) = \kappa F(\bar{x}, \bar{y})$ for some $\kappa > 1$. We then say that a compensation g is *ideal* if it has valid configurations and scaled values. At least for finding Max-SAT solutions, an ideal compensation g is just as good as actually having the equivalence constraint $X \equiv Y$. The following tells us that for any possible choice of weights, if the compensation is ideal then it must also satisfy Equation 1.

Proposition 1. *Let f be a weighted Max-SAT instance and let g be a compensation that results from relaxing a single equivalence constraints $X \equiv Y$ in f. If g has valid configurations and scaled values, with $\kappa = 2$, it also satisfies Eq. 1.*

Although a compensation satisfying Equation 1 may not always be ideal, it at least results in a meaningful approximation that is tighter than a relaxation. Note that we could have chosen a different value of κ, leading to equations slightly different from Equation 1, although the resulting approximation would be effectively the same. Moreover, the choice $\kappa = 2$ leads to simplified semantics, e.g., in the ideal case we can recover the exact values from the weights alone: $w_x + w_y = F(x, y)$ and $w_{\bar{x}} + w_{\bar{y}} = F(\bar{x}, \bar{y})$.

3.2 An Example

Consider the following weighted Max-SAT instance f with a single equivalence constraint $X \equiv Y$:

$$\begin{array}{llll} f: & (x \vee \bar{z}, 12) \ (y \vee \bar{z}, 6) \ (z, 30) \ (x \vee \bar{y}, \infty) \\ & (\bar{x} \vee \bar{z}, 3) \quad (\bar{y} \vee \bar{z}, 9) \qquad\quad (\bar{x} \vee y, \infty) \end{array}$$

which has a unique optimal Max-SAT assignment $\mathbf{x}^\star = \{X{=}x, Y{=}y, Z{=}z\}$, with Max-SAT value $F(\mathbf{x}^\star) = 12 + 6 + 30 = 48$. When we relax the equivalence constraint $X \equiv Y$, we arrive at a simpler instance h:

$$\begin{array}{ll} h: & (x \vee \bar{z}, 12) \ (y \vee \bar{z}, 6) \ (z, 30) \\ & (\bar{x} \vee \bar{z}, 3) \quad (\bar{y} \vee \bar{z}, 9) \end{array}$$

The relaxation h has a different optimal assignment $\mathbf{x}^\star = \{X{=}x, Y{=}\bar{y}, Z{=}z\}$, where variables X and Y are set to different signs. The optimal value is now $H(\mathbf{x}^\star) = 12 + 9 + 30 = 51$ which is greater than the value 48 for the original instance f. Now consider a compensation g with auxiliary unit clauses:

$$\begin{array}{llllll} g: & (x \vee \bar{z}, 12) & (y \vee \bar{z}, 6) & (z, 30) & (x, 27) & (y, 21) \\ & (\bar{x} \vee \bar{z}, 3) & (\bar{y} \vee \bar{z}, 9) & & (\bar{x}, 20) & (\bar{y}, 26) \end{array}$$

This compensation g satisfies Equation 1, as:

$$\frac{1}{2}\begin{bmatrix} G(x) \\ G(\bar{x}) \end{bmatrix} = \frac{1}{2}\begin{bmatrix} G(y) \\ G(\bar{y}) \end{bmatrix} = \begin{bmatrix} w_x + w_y \\ w_{\bar{x}} + w_{\bar{y}} \end{bmatrix} = \begin{bmatrix} 48 \\ 46 \end{bmatrix}$$

The compensation g has an optimal assignment $\mathbf{x}^\star = \{X{=}x, Y{=}y, Z{=}z\}$, the same as for the original instance f. It also has Max-SAT value $G(\mathbf{x}^\star) = 12 + 6 + 30 + 27 + 21 = 96$, where $\frac{1}{2}G(\mathbf{x}^\star) = F(\mathbf{x}^\star) = 48$.

Note that in this example, the weights happened to be integral, although in general, the weights of a compensation may be real-valued.

3.3 Compensations and Their Properties

In this section, we define compensations for the general case when multiple equivalence constraints are removed. Moreover, we formalize some of the properties we highlighted in the previous section.

Say then that we relax k equivalence constraints $X \equiv Y$. We seek a compensation g whose weights satisfy the condition:

$$\frac{1}{1+k}\begin{bmatrix} G(x) \\ G(\bar{x}) \end{bmatrix} = \frac{1}{1+k}\begin{bmatrix} G(y) \\ G(\bar{y}) \end{bmatrix} = \begin{bmatrix} w_x + w_y \\ w_{\bar{x}} + w_{\bar{y}} \end{bmatrix} \tag{2}$$

for each equivalence constraint $X \equiv Y$ relaxed. If a compensation g does indeed satisfy this condition, then it is possible to determine, in certain cases, when the optimal solution for a compensation is also optimal for the original instance f.

Theorem 1. *Let f be a weighted Max-SAT instance and let g be the compensation that results from relaxing k equivalence constraints $X \equiv Y$ in f. If the compensation g satisfies Equation 2, and if $\mathbf{x}^\star$ is an optimal assignment for g that assigns the same sign to variables X and Y, for each equivalence constraint $X \equiv Y$ relaxed, then:*

- *assignment $\mathbf{x}^\star$ is also optimal for instance f; and*
- $\frac{1}{1+k}G(\mathbf{x}^\star) = F(\mathbf{x}^\star)$.

Moreover, the Max-SAT value of a compensation g is an upper bound on the Max-SAT value of the original instance f.

Theorem 2. *Let f be a weighted Max-SAT instance and let g be the compensation that results from relaxing k equivalence constraints $X \equiv Y$ in f. If the compensation g satisfies Equation 2, then: $F^\star \leq \frac{1}{1+k}G^\star$*

We remark now that a relaxation alone has analogous properties. If an assignment $\mathbf{x}^\star$ is optimal for a relaxation h, and it is also a valid assignment for instance f (i.e., it does not violate the equivalence constraints $X \equiv Y$), then $\mathbf{x}^\star$ is also optimal for f, where $H(\mathbf{x}^\star) = F(\mathbf{x}^\star)$ (since they satisfy the same clauses). Otherwise, the Max-SAT value of a relaxation is an upper bound on the Max-SAT value of the original instance f. On the other hand, compensations are tighter approximations than the corresponding relaxation, at least in the case when a single equivalence constraint is relaxed: $F^\star \leq \frac{1}{2}G^\star \leq H^\star$. Although we leave this point open in the case where multiple equivalence constraints are relaxed, we have at least found empirically that compensations are never worse than relaxations. We discuss this point further in the following section.

The following theorem has implications for weighted Max-SAT solvers, such as [1,2], that rely on relaxations for upper bounds.

Theorem 3. *Let f be a weighted Max-SAT instance and let g be the compensation that results from relaxing k equivalence constraints $X \equiv Y$ in f. If compensation g satisfies Equation 2, and if $\tilde{\mathbf{z}}$ is a partial assignment that sets the same sign to variables X and Y, for any equivalence constraint $X \equiv Y$ relaxed, then: $F(\tilde{\mathbf{z}}) \leq \frac{1}{1+k}G(\tilde{\mathbf{z}})$*

Solvers, such as those in [1,2], perform a depth-first brand-and-bound search to find an optimal Max-SAT solution. They rely on upper bounds of a Max-SAT solution, under partial assignments, in order to prune the search space. Thus, any method capable of providing upper bounds tighter than those of a relaxation, can potentially have an impact in the performance of a branch-and-bound solver.

3.4 Searching for Weights

We now address the question: how do we actually find weights so that a compensation will satisfy Equation 2? Consider the simpler situation where we want weights for one particular equivalence constraint $X \equiv Y$. Ignoring the presence of other equivalence constraints that may have been relaxed, we can think of a compensation g as a compensation where only the single equivalence constraint $X \equiv Y$ being considered has been removed. The corresponding "relaxation" is found by simply removing the weights $w_x, w_{\bar{x}}, w_y, w_{\bar{y}}$ from g, for the single equivalence constraint $X \equiv Y$. More specifically, let $H_{x,y} = G(x,y) - [w_x + w_y]$ denote the Max-SAT value of the "relaxation," assuming that X and Y are set to x and y (and similarly for other configurations of X and Y). Given this "relaxation," we have a closed form solution for the weights, for a compensation g to satisfy Equation 2, at least for the one equivalence constraint $X \equiv Y$ being considered.

Theorem 4. *Let f be a weighted Max-SAT instance, let g be the compensation that results from relaxing k equivalence constraints in f, and let $X \equiv Y$ be one of k equivalence constraints relaxed. Suppose, w.l.o.g., that $H_{x,y} \geq H_{\bar{x},\bar{y}}$, and let:*

$$G^+ = \frac{1+k}{k} \max\left\{H_{x,y}, \frac{1}{2}[H_{x,\bar{y}} + H_{\bar{x},y}]\right\} \tag{3}$$

$$G^- = \frac{1+k}{k} \max\left\{H_{\bar{x},\bar{y}}, \frac{1}{1+2k}[H_{x,y} + kH_{x,\bar{y}} + kH_{\bar{x},y}], \frac{1}{2}[H_{x,\bar{y}} + H_{\bar{x},y}]\right\} \tag{4}$$

Algorithm 1. RelaxEq-and-Compensate (REC)

input: a weighted Max-SAT instance f with k equivalence constraints $X \equiv Y$
output: a compensation g satisfying Equation 2
main:

```
h ← result of relaxing all X ≡ Y in f
g ← result of adding to h weights w_x, w_x̄, w_y, w_ȳ for each X ≡ Y
initialize all weights w_x, w_x̄, w_y, w_ȳ, say to ½H⋆.
while weights have not converged do
    for each equivalence constraint X ≡ Y removed do
        update weights w_x, w_x̄, w_y, w_ȳ according to Equations 5 & 6
return g
```

If we set the weights for equivalence constraints $X \equiv Y$ to:

$$\begin{bmatrix} w_x \\ w_{\bar{x}} \end{bmatrix} = \frac{1}{2}\frac{1}{1+k}\begin{bmatrix} G^+ \\ G^- \end{bmatrix} + \frac{1}{4}\begin{bmatrix} H_{\bar{x},y} - H_{x,\bar{y}} \\ H_{x,\bar{y}} - H_{\bar{x},y} \end{bmatrix} \tag{5}$$

$$\begin{bmatrix} w_y \\ w_{\bar{y}} \end{bmatrix} = \frac{1}{2}\frac{1}{1+k}\begin{bmatrix} G^+ \\ G^- \end{bmatrix} + \frac{1}{4}\begin{bmatrix} H_{x,\bar{y}} - H_{\bar{x},y} \\ H_{\bar{x},y} - H_{x,\bar{y}} \end{bmatrix} \tag{6}$$

then equivalence constraint $X \equiv Y$ will satisfy Equation 2 in compensation g.

When the original instance f has been sufficiently relaxed, and enough equivalence constraints removed, then we will be able to compute the quantities $H_{x,y}$ efficiently (we discuss this point further in the following section). Theorem 4 then suggests an iterative algorithm for finding weights that satisfy Equation 2 for all k equivalence constraints relaxed, which is summarized in Algorithm 1.

This algorithm, which we call RelaxEq-and-Compensate (REC), initializes the weights of a given compensation g to some value, and iterates over equivalence constraints one-by-one. When it arrives at a particular equivalence constraint $X \equiv Y$, REC sets the weights according to Equations 5 & 6, which results in a compensation satisfying Equation 2, at least for that particular equivalence constraint. REC does the same for the next equivalence constraint, which may cause the previous equivalence constraint to no longer satisfy Equation 2. REC continues, however, until the weights of all equivalence constraints do not change with the application of Equations 5 & 6 (to some constant ϵ), at which point all equivalence constraints satisfy Equation 2.

We now make a few observations. First, if we set all weights $w_x, w_{\bar{x}}, w_y, w_{\bar{y}}$ of an initial compensation g_0 to $\frac{1}{2}H^\star$, then the initial approximation to the Max-SAT value is $\frac{1}{1+k}G_0^\star = H^\star$.[2] That is, the initial approximation is the same as the upper bound given by the relaxation h. We have observed empirically, interestingly enough, that when we start with these initial weights, every iteration of the REC algorithm results in a compensation g where the value of $\frac{1}{1+k}G^\star$ is no larger than that of the previous iteration. Theorem 2 tells us that a compensation g

[2] $G_0^\star = \max_{\mathbf{x}} G_0(\mathbf{x}) = \max_{\mathbf{x}}[H(\mathbf{x}) + \sum_{X\equiv Y} w_x + w_y]$
$= \max_{\mathbf{x}}[H(\mathbf{x}) + \sum_{X\equiv Y} \frac{1}{2}H^\star + \frac{1}{2}H^\star] = \max_{\mathbf{x}}[H(\mathbf{x})] + kH^\star = H^\star + kH^\star$.

satisfying Equation 2, which REC is searching for, yields a Max-SAT value $\frac{1}{1+k}G^\star$ that is an upper bound on the Max-SAT value $F^\star$ of the original instance f. This would imply that algorithm REC tends to provide monotonically decreasing upper bounds on $F^\star$, when starting with an initial compensation equivalent to the relaxation h. This would imply that, at least empirically, the value of $\frac{1}{1+k}G^\star$ is convergent in the REC algorithm. We have observed empirically that this is the case, and we discuss these points further in Section 4.

3.5 Knowledge Compilation

One point that we have yet to address is how to efficiently compute the values $H_{x,y}$ that are required by the iterative algorithm REC that we have proposed. In principle, any off-the-shelf Max-SAT solver could be used, where we repeatedly solve Max-SAT instances g where the variables X and Y are set to some values. However, when we remove k equivalence constraints, REC requires us to solve $4k$ Max-SAT instances in each iteration.

If, however, we relax enough equivalence constraints so that the treewidth is small enough, we can efficiently compile a given Max-SAT instance in CNF into decomposable negation normal form (DNNF) [9,10,11,12] (for details on how to solve weighted Max-SAT problems by compilation to DNNF, see [1,13]). Once our simplified instance g is in DNNF, many queries can be performed in time linear in the compilation, which includes computing at once all of the values $G(x), G(\bar{x})$ and $G(y), G(\bar{y})$, as well as the Max-SAT value $G^\star$. Computing each value $H_{x,y}$ can also be performed in time linear in the compilation, although lazy evaluation can be used to improve efficiency; see [1] for details.

We note that the required values can also be computed by message-passing algorithms such as belief propagation [14]. However, this would typically involve converting a weighted Max-SAT instance into an equivalent Bayesian network or factor graph, where general-purpose algorithms do not make use of the kinds of techniques that SAT solvers and compilers are able to. In contrast, knowledge compilation can be applied to solving Bayesian network tasks beyond the reach of traditional probabilistic inference algorithms [15].

4 Experiments

We evaluate here the REC algorithm on a selection of benchmarks. Our goals are: (1) to illustrate its empirical properties, (2) to demonstrate that compensations can improve, to varying extents, relaxations, and (3) that compensations are able to improve branch-and-bound solvers, such as CLONE, at least in some cases.

The relaxations that we consider are the same as those employed by the CLONE solver [1],[3] which in certain categories led, or was otherwise competitive with, the solvers evaluated in the 3rd Max-SAT evaluation.[4] CLONE relaxes a given Max-SAT instance by choosing, heuristically, a small set of variables to

[3] Available at `http://reasoning.cs.ucla.edu/clone/`

[4] Results of the evaluation are available at `http://www.maxsat.udl.cat/08/`

"split", where splitting a variable X simplifies a Max-SAT instance by replacing each occurrence of a variable X with a unique clone variable Y. This relaxation effectively ignores the dependence that different clauses have on each other due to the variable X being split. Note that such a relaxation is restored when we assert equivalence constraints $X \equiv Y$. For our purposes, we can then assume that equivalence constraints were instead relaxed.

Like CLONE, we relax Max-SAT instances until their treewidth is at most 8. Given this relaxation, we then constructed a compensation which was compiled into DNNF by the C2D compiler.[5] For each instance we selected, we ran the REC algorithm for at most 2000 iterations, and iterated over equivalence constraints in some fixed order, which we did not try to optimize. If the change in the weights from one iteration to the next was within 10^{-4}, we declared convergence, and stopped.

Our first set of experiments were performed on randomly parametrized grid models, which are related to the Ising and spin-glass models studied in statistical physics; see, e.g., [16]. This type of model is also commonly used in fields such as computer vision [17]. In these models, we typically seek a configuration of variables $\mathbf{x}$ minimizing a cost (energy) function of the form:

$$F(\mathbf{x}) = \sum_i \psi_i(x_i) + \sum_{ij} \psi_{ij}(x_i, x_j)$$

where variables X_i are arranged in an $n \times n$ grid, which interact via potentials ψ_{ij} over neighboring variables X_i and X_j. In our experiments, we left all $\psi_i(x_i) = 0$, and we generated each $\psi_{ij}(x_i, x_j) = -\log p$ with p drawn uniformly from $(0, 1)$. These types of models are easily reduced to weighted Max-SAT; see, e.g., [18]. Note, that the resulting weights will be floating-point values, which are not yet commonly supported by modern Max-SAT solvers. We thus focus our empirical evaluation with respect to a version of CLONE that was augmented for us to accommodate such weights.

We generated 10 randomly parametrized 16×16 grid instances and evaluated (1) the dynamics of the REC algorithm, and (2) the quality of the resulting approximation (although we restrict our attention here to 10 instances, for simplicity, the results we present are typical for this class of problems). Consider first Figure 1, where we plotted the quality of an approximation (y-axis) versus iterations of the REC algorithm (x-axis), for each of the 10 instances evaluated. We define the quality of an approximation as the error of the compensation $\frac{1}{1+k}G^\star - F^\star$, relative to the error of the relaxation $H^\star - F^\star$. That is, we measured the error

$$E = \frac{\frac{1}{1+k}G^\star - F^\star}{H^\star - F^\star}$$

which is zero when the compensation is exact, and one when the compensation is equivalent to the relaxation. Remember that we proposed to initialize the REC algorithm with weights that led to an initial compensation with an optimal

[5] Available at `http://reasoning.cs.ucla.edu/c2d/`

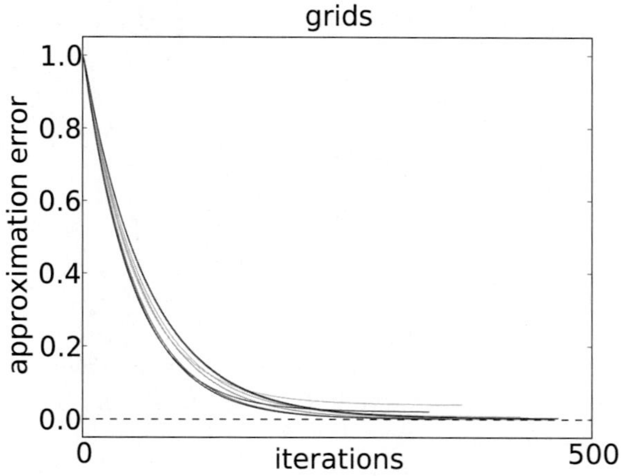

Fig. 1. Behavior of the REC algorithm random 16×16 grid instances. Note that color is used here to help differentiate plots, and is otherwise not meaningful.

value $\frac{1}{1+k} G_0^\star = H^\star$. Thus, we think of the error E as the degree to which the compensation is able to tighten the relaxation.

We make some observations about the instances depicted in Figure 1. First, all of the 10 instances converged before 500 iterations. Next, we see that the REC algorithm yields from iteration-to-iteration errors, and hence values $\frac{1}{1+k} G^\star$, that are monotonically decreasing. If this is the case in general, then this implies that these bounds $\frac{1}{1+k} G^\star$ are convergent in the REC algorithm, since a compensation satisfying Equation 2 is an upper bound on $F^\star$ (by Theorem 2). This implies, at least empirically, that the REC algorithm is indeed tightening a relaxation from iteration-to-iteration. Finally, we find that REC is capable of significantly improving the quality of an approximation, to exact or near-exact levels.

Given such improvement, we may well expect that a solver that relies on relaxations for upper bounds, such as CLONE, may benefit from an improved approximation that provides tighter bounds. In fact, using the relaxation alone, CLONE was unable to solve any of these instances, given a time limit of 10 minutes.[6] We thus augmented CLONE so that it can take advantage of the tighter REC approximation. In particular, we compensate for the relaxation that CLONE would normally use, and have CLONE use its tighter upper-bounds to prune nodes during branch-and-bound search.

This augmented CLONE algorithm now has one additional steps. Before we perform the branch-and-bound search (the CLONE step), we must first compensate for the relaxation (the REC step). The following table records the time, in seconds, to perform each step in the instances we considered:

[6] Experiments were performed on an Intel Xeon E5440 CPU, at 2.83GHz.

instance	1	2	3	4	5	6	7	8	9	10
REC	303	302	583	390	308	326	318	249	311	511
CLONE	528	1	45	2	253	121	47	14	12	10
total	831	303	628	392	561	447	365	263	323	521

Although CLONE was unable to solve any of these instances within 600 seconds with a relaxation alone, it was able to solve most instances within the same amount of time when enabled with a compensation. We further remark that there is ample room for improving the efficiency of our REC implementation, which is in Java and Jython (Jython is an implementation of the Python language in Java).

Finally, in Figure 2, we plot the performance of the REC algorithm on a subset of the AUC_PATHS benchmark from the weighted partial Max-SAT instances from the 2008 evaluation.[7] We find here that the REC algorithm is able to reduce the approximation error of a relaxation by roughly half in many instances, or otherwise appears to approach this level if allowed further iterations. We also see again that the REC has relatively stable dynamics. As CLONE was already able to efficiently solve all of the instances in the AUC_PATHS benchmark, CLONE did not benefit much from a compensation in this case.

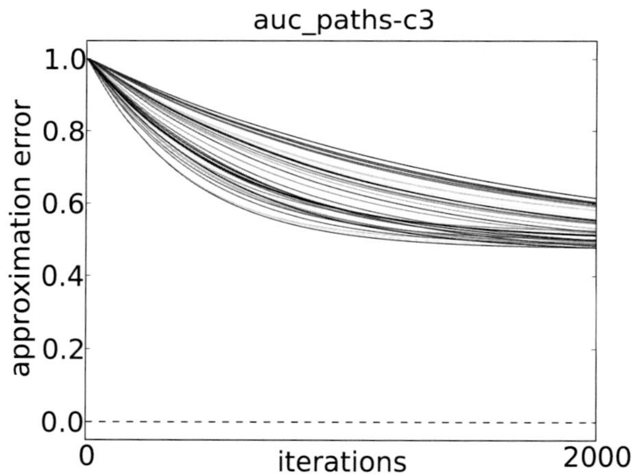

Fig. 2. Behavior of the REC algorithm in weighted partial Max-SAT instances (AUC_PATHS benchmarks)

5 Conclusion

In this paper, we proposed a new perspective on approximations of Max-SAT problems, that is based on relaxing a given instance, and then compensating for

[7] The instances selected were the 40 instances labeled `cat_paths_60_`p`_*.txt.wcnf` for $p \in \{100, 110, 120, 130, 140\}$.

the relaxation. When we relax equivalence constraints in a Max-SAT problem, we can perform inference on the simplified problem, identify some of its defects, and then recover a weaker notion of equivalence. We proposed a new algorithm, REC, that iterates over equivalence constraints, compensating for relaxations one-by-one. Our empirical results show that REC can tighten relaxations to the point of recovering exact or near-exact results, in some cases. We have also observed that, in some cases, these compensations can be used by branch-and-bound solvers to find optimal Max-SAT solutions, which they were unable to find with a relaxation alone.

Acknowledgments

This work has been partially supported by NSF grant #IIS-0713166.

References

1. Pipatsrisawat, K., Palyan, A., Chavira, M., Choi, A., Darwiche, A.: Solving weighted Max-SAT problems in a reduced search space: A performance analysis. Journal on Satisfiability, Boolean Modeling, and Computation 4, 191–217 (2008)
2. Ramírez, M., Geffner, H.: Structural relaxations by variable renaming and their compilation for solving MinCostSAT. In: Bessière, C. (ed.) CP 2007. LNCS, vol. 4741, pp. 605–619. Springer, Heidelberg (2007)
3. Choi, A., Standley, T., Darwiche, A.: Approximating weighted Max-SAT problems by compensating for relaxations. Technical report, CSD, UCLA (2009)
4. Li, C.M., Manyà, F.: MaxSAT, hard and soft constraints. In: Biere, A., Heule, M., van Maaren, H., Walsh, T. (eds.) Handbook of Satisfiability, pp. 613–631. IOS Press, Amsterdam (2009)
5. Choi, A., Chavira, M., Darwiche, A.: Node splitting: A scheme for generating upper bounds in Bayesian networks. In: UAI, pp. 57–66 (2007)
6. Siddiqi, S., Huang, J.: Variable and value ordering for MPE search. In: Proceedings of the 21st International Joint Conference on Artificial Intelligence (to appear, 2009)
7. Dechter, R., Rish, I.: Mini-buckets: A general scheme for bounded inference. J. ACM 50(2), 107–153 (2003)
8. Rish, I., Dechter, R.: Resolution versus search: Two strategies for SAT. J. Autom. Reasoning 24(1/2), 225–275 (2000)
9. Darwiche, A.: Decomposable negation normal form. Journal of the ACM 48(4), 608–647 (2001)
10. Darwiche, A.: On the tractability of counting theory models and its application to belief revision and truth maintenance. Journal of Applied Non-Classical Logics 11(1-2), 11–34 (2001)
11. Darwiche, A.: New advances in compiling CNF to decomposable negational normal form. In: Proceedings of European Conference on Artificial Intelligence, pp. 328–332 (2004)
12. Darwiche, A., Marquis, P.: A knowledge compilation map. Journal of Artificial Intelligence Research 17, 229–264 (2002)
13. Darwiche, A., Marquis, P.: Compiling propositional weighted bases. Artificial Intelligence 157(1-2), 81–113 (2004)

14. Pearl, J.: Probabilistic Reasoning in Intelligent Systems: Networks of Plausible Inference. Morgan Kaufmann Publishers, Inc., San Mateo (1988)
15. Chavira, M., Darwiche, A.: Compiling Bayesian networks with local structure. In: Proceedings of the 19th International Joint Conference on Artificial Intelligence (IJCAI), pp. 1306–1312 (2005)
16. Percus, A., Istrate, G., Moore, C.: Where statistical physics meets computation. In: Percus, A., Istrate, G., Moore, C. (eds.) Computational Complexity and Statistical Physics, pp. 3–24. Oxford University Press, Oxford (2006)
17. Szeliski, R., Zabih, R., Scharstein, D., Veksler, O., Kolmogorov, V., Agarwala, A., Tappen, M.F., Rother, C.: A comparative study of energy minimization methods for Markov random fields. In: Leonardis, A., Bischof, H., Pinz, A. (eds.) ECCV 2006. LNCS, vol. 3952, pp. 16–29. Springer, Heidelberg (2006)
18. Park, J.D.: Using weighted max-sat engines to solve MPE. In: AAAI/IAAI, pp. 682–687 (2002)

A Proofs

Proof (of Proposition 1). First, observe that:

$$G(x, y) = F(x, y) + [w_x + w_y] = \frac{1}{2}G(x, y) + [w_x + w_y]$$

since g has scaled values. Thus, $\frac{1}{2}G(x, y) = w_x + w_y$; similarly to show $\frac{1}{2}G(\bar{x}, \bar{y}) = w_{\bar{x}} + w_{\bar{y}}$. Since g has valid configurations, then g also satisfies Equation 1. □

In the remainder of this section, we call a complete assignment $\mathbf{x}$ *valid* iff $\mathbf{x}$ sets the same sign to variables X and Y, for every equivalence constraint $X \equiv Y$ removed from an instance f; analogously, for partial assignments $\mathbf{z}$. We will also use $\tilde{\mathbf{x}}$ and $\tilde{\mathbf{z}}$ to denote valid assignments, when appropriate.

Lemma 1. *Let f be a weighted Max-SAT instance and let g be the compensation that results from relaxing k equivalence constraints $X \equiv Y$ in f. If compensation g satisfies Equation 2, and if $\tilde{\mathbf{x}}$ is a complete assignment that is also valid, then $F(\tilde{\mathbf{x}}) \leq \frac{1}{1+k}G(\tilde{\mathbf{x}})$, with equality if $\tilde{\mathbf{x}}$ is also optimal for g.*

Proof. When we decompose $G(\tilde{\mathbf{x}})$ into the original weights, i.e., $F(\tilde{\mathbf{x}})$, and the auxiliary weights $w_x + w_y = \frac{1}{1+k}G(x)$ (by Equation 2) we have

$$G(\tilde{\mathbf{x}}) = F(\tilde{\mathbf{x}}) + \sum_{X \equiv Y} [w_x + w_y] = F(\tilde{\mathbf{x}}) + \sum_{X \equiv Y} \frac{1}{1+k} G(x).$$

Note that x is the value assumed by X in assignment $\tilde{\mathbf{x}}$. Since $G(x) \geq G(\tilde{\mathbf{x}})$,

$$G(\tilde{\mathbf{x}}) \geq F(\tilde{\mathbf{x}}) + \sum_{X \equiv Y} \frac{1}{1+k} G(\tilde{\mathbf{x}}) = F(\tilde{\mathbf{x}}) + \frac{k}{1+k} G(\tilde{\mathbf{x}})$$

and thus $\frac{1}{1+k}G(\tilde{\mathbf{x}}) \geq F(\tilde{\mathbf{x}})$. In the case where $G(\tilde{\mathbf{x}}) = G^{\star}$, we have $G(x) = G(\tilde{\mathbf{x}})$ for all $X \equiv Y$, so we have the equality $F(\tilde{\mathbf{x}}) = \frac{1}{1+k}G(\tilde{\mathbf{x}})$. □

Proof (of Theorem 1). Since $\mathbf{x}^\star$ is both optimal and valid, we know that $\frac{1}{1+k}G^\star = \frac{1}{1+k}G(\mathbf{x}^\star) = F(\mathbf{x}^\star)$, by Lemma 1. To show that $\mathbf{x}^\star$ is also optimal for the original instance f, note first that $G^\star = \max_{\mathbf{x}} G(\mathbf{x}) = \max_{\tilde{\mathbf{x}}} G(\tilde{\mathbf{x}})$. Then:

$$F(\mathbf{x}^\star) = \frac{1}{1+k}G^\star = \max_{\tilde{\mathbf{x}}} \frac{1}{1+k}G(\tilde{\mathbf{x}}) \geq \max_{\tilde{\mathbf{x}}} F(\tilde{\mathbf{x}}) = F^\star$$

using again Lemma 1. We can thus infer that $F(\mathbf{x}^\star) = F^\star$. □

Proof (of Theorem 2). Let $\mathbf{x}^\star$ be an optimal assignment for f. Since $\mathbf{x}^\star$ must also be valid, we have by Lemma 1 that

$$F^\star = F(\mathbf{x}^\star) \leq \frac{1}{1+k}G(\mathbf{x}^\star) \leq \frac{1}{1+k}G^\star$$

as desired. □

Proof (of Theorem 3). We have that

$$F(\tilde{\mathbf{z}}) = \max_{\tilde{\mathbf{x}} \sim \tilde{\mathbf{z}}} F(\tilde{\mathbf{x}}) \leq \max_{\tilde{\mathbf{x}} \sim \tilde{\mathbf{z}}} \frac{1}{1+k}G(\tilde{\mathbf{x}}) \leq \max_{\mathbf{x} \sim \tilde{\mathbf{z}}} \frac{1}{1+k}G(\mathbf{x}) = \frac{1}{1+k}G(\tilde{\mathbf{z}})$$

where the first inequality follows from Lemma 1. □

Proof of Theorem 4 appears in the Appendix of the full report [3].

Corollary 1. *Let f be a weighted Max-SAT instance and let g be the compensation that results from relaxing a single equivalence constraint $X \equiv Y$ in f. If compensation g satisfies Equation 1, then*

$$F^\star \leq \frac{1}{2}G^\star \leq H^\star.$$

Proof. The first inequality follows from Theorem 2. From the proof of Theorem 4 we know that either $\frac{1}{2}G^\star = F^\star \leq H^\star$ or

$$\frac{1}{2}G^\star = \frac{1}{2}[H(x,\bar{y}) + H(\bar{x},y)] \leq \max\{H(x,\bar{y}), H(\bar{x},y)\} \leq H^\star$$

thus we have the second inequality as well. □

Confidence-Based Work Stealing in Parallel Constraint Programming

Geoffrey Chu[1], Christian Schulte[2], and Peter J. Stuckey[1]

[1] National ICT Australia, Victoria Laboratory,
Department of Computer Science and Software Engineering,
University of Melbourne, Australia
{gchu,pjs}@csse.unimelb.edu.au
[2] KTH – Royal Institute of Technology, Sweden
cschulte@kth.se

Abstract. The most popular architecture for parallel search is work stealing: threads that have run out of work (nodes to be searched) steal from threads that still have work. Work stealing not only allows for dynamic load balancing, but also determines which parts of the search tree are searched next. Thus the place from where work is stolen has a dramatic effect on the efficiency of a parallel search algorithm.

This paper examines quantitatively how optimal work stealing can be performed given an estimate of the relative solution densities of the subtrees at each search tree node and relates it to the branching heuristic strength. An adaptive work stealing algorithm is presented that automatically performs different work stealing strategies based on the confidence of the branching heuristic at each node. Many parallel depth-first search patterns arise naturally from this algorithm. The algorithm produces near perfect or super linear algorithmic efficiencies on all problems tested. Real speedups using 8 threads range from 7 times to super linear.

1 Introduction

Architectures for parallel search in constraint programming typically use work stealing for distributing work (nodes to be searched) from running to idle threads, see for example [1,2,3,4]. Work stealing has often focused on keeping processors occupied. Its analysis often assumes that the amount of work to be done is fixed and independent of the work stealing scheme, for example [5].

While this is true for certain kinds of problems (finding all solutions, proving unsatisfiability), it is not true for others (finding a first solution, or finding an optimal solution). Such analyses fail to account for the fact that the place in the search tree from which work is stolen determines the search strategy and hence is bound to have a dramatic effect on efficiency.

Many approaches choose to steal from as close to the root of the search tree as possible, e.g. [3], as this tends to give the greatest granularity of work and minimizes the overhead of work stealing. However, this is not always the best strategy in terms of efficiency.

I.P. Gent (Ed.): CP 2009, LNCS 5732, pp. 226–241, 2009.

Effect of work stealing. Let us consider a relatively simple framework for parallel search. One thread begins with ownership of the entire search tree. When a thread finishes searching the subtree it was responsible for, it will steal an unexplored part of the search tree (a *job*) from its current owner. This continues until a solution is found or the entire search tree has been searched.

Example 1. The first problem we will consider is a simple model for small instances of the Traveling Salesman Problem, similar to [6]. In our experiments, with 8 threads, stealing left and low (as deep in the tree as possible) requires visiting a number of nodes equal to the sequential algorithm, while stealing high (near the root) requires visiting $\sim 30\%$ more nodes on average (see Table 1).

The explanation is simple. In one example instance, the sequential algorithm finds the optimal solution after 47 seconds of CPU time, after which it spends another ~ 300 seconds proving that no better solution exists. When the parallel algorithm is stealing left and low, all threads work towards finding that leftmost optimal solution, and the optimal solution is found in 47 seconds of CPU time as before (wall clock time $47/8 \approx 6$ seconds). After this, the search takes another 300 seconds of CPU time to conclude. Thus we have perfect linear speedup both in finding the optimal solution, and in proving that no better solution exist.

However, when the parallel algorithm is stealing high, only one thread actually explores the leftmost part of the search tree and works towards that leftmost optimal solution. The other seven threads explore other parts of the search tree, unfruitfully in this case. This time, the optimal solution is found in 47 seconds of *wall clock* time (376 seconds of CPU time!). The algorithm then spends another 200 seconds of CPU time proving optimality. That is, there is no speedup whatsoever for finding the optimal solution, but linear speedup for proving optimality. Since the optimal solution has been found so much later (376 seconds CPU time instead of 47 seconds), the threads search without the pruning benefits of the optimal solution. Hence, the number of nodes searched drastically increases, leading to a great loss of efficiency. Clearly, this effect gets worse as the number of threads increases. □

It may appear from this example that stealing left and low would be efficient for all problems. However, such a strategy can produce at best linear speedup.

Example 2. Consider the n-Queens problem. The search tree is very deep and a top level mistake in the branching will not be recovered from for hours.

Stealing low solves an instance within the time limit iff sequential depth first search solves it within the time limit. This is the case when a solution is in the very leftmost part of the search tree (only 4 instances out of 100, see Table 2).

Stealing high, in contrast, allows many areas of the search tree to be explored, so a poor choice at the root of the search tree is not as important. Stealing high results in solving 100 out of 100 instances tested. This is clearly far more robust than stealing low, producing greatly super-linear speedup. □

Veron *et al* [7] claim that linear and super linear speedups can be expected for branch and bound problems, but they fail to note that finding the optimal solution does not parallelize trivially as shown by Example 1. Rao and Kumar [8] (and others) show that super linear speedup ought to be consistently attainable for finding the first or the optimal solution for certain types of problems. Their analysis is valid if the search tree is random (i.e. the solutions are randomly distributed), but is not valid in systems where a branching heuristic orders the branches based on their likelihood of yielding a solution. The presence of such a branching heuristic makes linear speedup in finding solutions non-trivial. Gendron and Crainic [9] describe the issue and provide a description how the issue is handled in several systems. In general, the solutions utilise some kind of best-first criterion to guide how the problem is split up (see e.g. [10,11]).

Contributions. The paper contributes a quantitative analysis of how different work stealing strategies affect the total amount of work performed and explains the relationship between branching heuristic strength and optimal search strategy. It introduces *confidence-based work stealing* as an adaptive algorithm that, when provided with a user-defined *confidence*, will steal work in a near optimal manner. The confidence is the estimated ratio of solution densities between the subtrees at each node. The paper shows that confidence-based work stealing leads to very good algorithmic efficiencies, that is, not many more, and sometimes much less, nodes are explored than for sequential DFS (Depth First Search).

Although the analysis is done in the context of parallel search for constraint programming, the analysis is actually about the relationship between branching heuristic strength and the optimal search order created by that branching heuristic. Thus the analysis actually applies to all complete tree search algorithms whether sequential or parallel. The paper shows that when the assumptions about branching heuristic strength that lie behind standard sequential algorithms such as DFS, Interleaved Depth First Search (IDFS) [12], Limited Discrepancy Search (LDS) [13] or Depth-bounded Discrepancy Search (DDS) [14] are given to the algorithm as *confidence* estimates, the algorithm produces the exact same search patterns used in those algorithms. Thus the analysis and algorithm provides a framework which explains/unifies/produces all those standard search strategies. In contrast to the standard sequential algorithms which are based on rather simplistic assumptions about how branching heuristic strength varies in different parts of the search tree, our algorithm can adapt to branching heuristic strength on a node by node basis, potentially producing search patterns that are vastly superior to the standard ones. The algorithm is also fully parallel and thus the paper also presents parallel DFS, IDFS, LDS and DDS as well.

Confidence-based search shares the idea to explore more promising parts of the search tree first with other approaches such as impact-based search [15] and using solution counting [16] and constraint-level advice [17] for guiding search. However, there are two key differences. First, confidence-based search uses standard branchings (labelings) that define the shape of the search tree augmented by confidence. This makes the addition of confidence-based search to an existing constraint programming solver straightforward and allows us to reuse existing constraint models with user-defined branchings. Second, confidence-based search is designed to work in parallel.

2 Analysis of Work Allocation

In this section we show quantitatively that the strength of the branching heuristic determines the optimal place to steal work from. We will concentrate on the case of solving a satisfaction problem. The case for optimization is related since it is basically a series of satisfaction problems.

Preliminary definitions. A *constraint state* (C, D) consists of a system of constraints C over variables V with initial domain D assigning possible values $D(v)$ to each variable $v \in V$. The *propagation solver*, solv, repeatedly removes values from the domains of variables that cannot take part in the solution of some constraint $c \in C$, until no more values can be removed. It obtains a new domain $\text{solv}(C, D) = D'$. If D' assigns a variable the empty set the resulting state is a *failure state.* If D' assigns each variable a single value ($|D(v)| = 1, v \in V$) the resulting state is a *solution state.* Failure and solution states are *final states.*

Finite domain propagation interleaves propagation solving with search. Given a current (non-final) state (C, D) where $D = \text{solv}(C, D)$ the search process chooses a *search disjunction* $\vee_{i=1}^{n} c_i$ of *decision constraints* c_i $(1 \leq i \leq n)$ which is a consequence of the current state $C \wedge D$. The child states of (C, D) are calculated as $(C \wedge c_i, \text{solv}(C \wedge c_i, D))$, $1 \leq i \leq n$. Given a root state (C, D), this defines a *search tree* of states, where each non-final state is an internal node with children defined by the search disjunction and final states are leaves.

The *solution density* of a search tree T with k nodes and l solution state nodes is l/k.

Optimal split for binary nodes. For simplicity, assume that visiting each node in the search tree has roughly equal cost. Assuming an oracle that provides accurate information on solution density, work stealing from nodes whose subtrees have the highest solution densities will be optimal.

In practice however, the solution density estimates will not be perfect:

1. Any estimate of the solution density of a subtree will have a very high error, with a substantial chance that the solution density is actually zero.
2. The real solution densities, and hence the errors in the estimate, are highly correlated between nearby subtrees, as they share decision constraints from higher up in the tree.
3. The solution density estimate of a subtree should decrease as nodes in that subtree are examined without finding a solution. This is caused by:
   (a) As the most fruitful parts of the subtree are searched, the average solution density of the remaining nodes decrease.
   (b) The correlation between solution densities between nearby subtrees mean that the more nodes have failed in that subtree, the more likely the remaining nodes are to fail as well.

We have to take these issues into account when utilizing solutions densities to determine where to steal work. Given the actual solution density probability distribution for the two branches, we can calculate the expected number of nodes searched to find a solution. We derive the expression for a simple case. Suppose the solution density probability distribution is uniform, that is, it has equal probability of being any value between 0 and S where S is the solution density estimate. Let A and B be the solution density estimates for the left and right branch respectively, and assume a proportion p and $(1-p)$ of the processing power is sent down the left and right branch respectively. Then the expected number of nodes to be searched is given by the hybrid function:

$$f(A, B, p) = \begin{cases} \frac{1}{pA}(2 + \ln(\frac{pA}{(1-p)B})) & \text{for } pA > (1-p)B \\ \frac{1}{(1-p)B}(2 + \ln(\frac{(1-p)B}{pA})) & \text{otherwise} \end{cases} \qquad (1)$$

The shape of this function does not depend on the absolute values of A and B (which only serve to scale the function), but on their ratio, thus the shape is fixed for any fixed value of $r = A/(A+B)$.

The value of p which minimizes the function for given value of r is well approximated by the straight line $p = r$. In fact, the value of the f function at $p = r$ is no more than 2% higher than the true minimum for any r over the range of $0.1 \leq r < 0.9$. For simplicity, we will make this approximation from now on. This means that it is near optimal to divide the amount of processing power according to the ratio of the solution density estimate for the two branches. For example, if $r = 0.9$, which means that A is 9 times as high as B, then it is near optimal to send 0.9 of our processing power down the left branch and 0.1 of our processing power down the right.

Branching confidence. Define the *confidence* of a branching heuristic at each node as the ratio $r = A/(A+B)$. The branching heuristic can be considered *strong* when $r \to 1$, that is the solution density estimate of the left branch is far greater than for the right branch. In other words, the heuristic is really good at shaping the search tree so that solutions are near the left. In this case, our analysis shows that since r is close to 1, we should allocate almost all our processing power to the left branch every time. This is equivalent to stealing work for search as left and as low as possible. The branching heuristic is *weak* when $r \approx 0.5$, that is, the solution density estimate of the left branch and right branch are similar because the branching heuristic has no insight into where the solutions are. In this case, our analysis shows that since $r = 0.5$, the processing power should be distributed evenly between left and right branches at each node. This is equivalent to stealing work for search as high as possible.

3 Adaptive Work Stealing

Our analysis shows that the optimal work stealing strategy depends on the strength of the branching heuristic. Since we have a quantitative understanding of how optimal work stealing is related to branching heuristic strength, we can design a search algorithm that can automatically adapt and produce “optimal” search patterns when given some indication of the strength of the branching heuristic by the problem model. In this section, we flesh out the theory and discuss the implementation details of the algorithm.

3.1 Dynamically Updating Solution Density Estimates

Now we examine how solution density estimates should be updated during search as more information becomes available.

First we need to relate the solution density estimate of a subtree with root (C, D) with the solution density estimate of its child subtrees (the subtrees rooted at its child states $(C \wedge c_i, \text{solv}(C \wedge c_i, D))$). Consider an n-ary node. Let the subtree have solution density estimate S. Let the child subtree at the i-th branch have solution density estimate A_i and have size (number of nodes) k_i. If S and A_i are estimates of average solution density, then clearly: $S = \sum_{i=1}^{n} A_i k_i / \sum_{i=1}^{n} k_i$, i.e. the average solution density of the subtree is the weighted average of the solution densities of its child subtrees.

Uncorrelated subtrees. Assuming no correlation between the solution densities of subtrees, we have that if the first j child subtrees have been searched unsuccessfully, then the updated solution density estimate is $S = \sum_{i=j+1}^{n} A_i k_i / \sum_{i=j+1}^{n} k_i$. Assuming that k_i are all approximately equal, then the expression simplifies to:

$$S = \sum_{i=j+1}^{n} A_i/(n-j)$$

For example, suppose $A_1 = 0.3, A_2 = 0.2, A_3 = 0.1$, then initially, $S = (0.3 + 0.2 + 0.1)/3 = 0.2$. After branch 1 is searched, we have $S = (0.2 + 0.1)/2 = 0.15$, and after branch 2 is searched, we have $S = (0.1)/1 = 0.1$. This has the effect of reducing S as the branches with the highest values of A_i are searched, as the average of the remaining branches will decrease.

Correlated subtrees. Now we consider the case where there are correlations between the solution density estimates of the child subtrees. The correlation is likely since all of the nodes in a subtree share the constraint C of the parent state. Since the correlation is difficult to model we pick a simple generic model. Suppose the solution density estimates for each child subtree is given by $A_i = \rho A'_i$, where ρ represents the effect on the solution density due to the constraint added at the parent node, and A'_i represents the effect on the solution density due to constraints added within branch i. Then ρ is a common factor in the solution density estimates for each branch and represents the correlation between them. We have that:

$$S = \frac{\sum_{i=1}^{n} A_i k_i}{\sum_{i=1}^{n} k_i} = \rho \frac{\sum_{i=1}^{n} A'_i k_i}{\sum_{i=1}^{n} k_i}.$$

Suppose that when j out of n of the branches have been searched without finding a solution, the value of ρ is updated to $\rho \frac{n-j}{n}$. This models the idea that the more branches have failed, the more likely it is that the constraint C added at the parent node has already made solutions unlikely or impossible. Then, after j branches have been searched, we have: $S = \rho \frac{n-j}{n} \sum_{i=j+1}^{n} A'_i k_i / \sum_{i=j+1}^{n} k_i$. Assuming that all k_i are approximately equal again, the expression simplifies to:

$S = \rho \frac{n-j}{n} \sum_{i=j+1}^{n} A'_i/(n-j) = \frac{\rho}{n} \sum_{i=j+1}^{n} A'_i = \sum_{i=j+1}^{n} A_i/n$. Equivalently, we can write it as:

$$S = \frac{1}{n} \sum_{i=1}^{n} A_i \text{ and } A_i = 0 \text{ for } 1 \leq i \leq j \tag{2}$$

where we update A_i to 0 when branch i fails. The formula can be recursively applied to update the solution density estimates of any node in the tree given a change in solution density estimate in one of its subtrees.

Using confidence. In all of our results, the actual values of the solution densities are not required. We can formulate everything using *confidence*, the ratio between the solution densities of the different branches at each node. In terms of confidence, when a subtree is searched and fails the confidence values should be updated as follows:

Let r_i be the confidence value of the node i levels above the root of the failed subtree and r'_i be the updated confidence value. Let $\bar{r}_i = r_i, \bar{r}'_i = r'_i$ if the failed subtree is in the left branch of the node i levels above the root of the failed subtree and $\bar{r}_i = 1 - r_i, \bar{r}'_i = 1 - r'_i$ otherwise. Then:

$$\bar{r}'_i = (\bar{r}_i - \prod_{k=1}^{i} \bar{r}_i)/(1 - \prod_{k=1}^{i} \bar{r}_i) \tag{3}$$

3.2 Confidence Model

Given a confidence at each node, we now know how to steal work "optimally" and how to update confidence values as search proceeds. But how do we get an initial confidence at each node. Ideally, the problem modeler, with expert knowledge about the problem and the branching heuristic can develop a solution density heuristic that gives a confidence value at each node. However, this may not always happen, perhaps due to a lack of time or expertise. We can simplify things by using general confidence models. For example, we could assume that the confidence takes on an equal value *conf* for all nodes. This is sufficient to model general ideas like: the heuristic is strong or the heuristic is weak. Or we could have a confidence model that assigns $r = 0.5$ to the top d levels and $r = 0.99$ for the rest. This can model ideas like the heuristic is weak for the first d levels, but very strong after that, much like the assumptions used in DDS [14].

3.3 Algorithm

Given that we have a confidence value at each node, our confidence-based search algorithm will work as follows. The number of threads down each branch of a node is updated as the search progresses. When search for a subtree is finished, the confidence values of all nodes above the finished subtree are updated as described by Equation (3) above.

When work stealing is required, we start at the root of the tree. At each node we use the number of threads down each branch, the confidence value, to determine which branch to take. Given the number of threads down each branch is currently a and b respectively then if $|(a+1)/(a+b+1)-r| \leq |a/(a+b+1)-r|$ we go left, otherwise right (i.e., which move would split the work closer to the confidence value). We continue this process until we find an unexplored node, at which point we steal the subtree with that unexplored node as root.

There is an exception to this. Although we may sometimes want to steal as low as possible, we cannot steal too low, as then the granularity would become too small and communication costs will dominate the runtime. Thus we dynamically determine a granularity bound under which threads are not allowed to steal, e.g. 15 levels above the average fail depth. If the work stealing algorithm guides the work stealing to the level of the granularity bound, then the last unexplored node above the granularity bound is stolen instead. The granularity bound is dynamically adjusted to maintain a minimum average subtree size so that work stealing does not occur more often than a certain threshold.

Since the confidence values are constantly updated, the optimal places to search next changes as search progresses. In order for our algorithm to adapt quickly, we do not require a thread to finish the entire subtree it stole before stealing again. Instead, after a given *restart* time has passed, the thread returns the unexplored parts of its subtree to a master coordinating work stealing and steals work again from the top. The work frontier is stored at the master and the work is resumed when work stealing guides a thread to the area again (similar to the idea used in interleaving DFS [12]).

Example with reasonably high confidence. Suppose we know that the branching heuristic is reasonably strong, but not perfect. We may use $conf = 0.8$.

Suppose we have 8 threads. Initially, all confidence values are 0.8. When the 8 threads attempt to steal work at the root, thread 1 will go down the left hand side. Thread 2 will go down the left hand side as well. Thread 3 will go down the right hand side. Thread 4 will go down the left hand side, etc, until 6 threads are down left and 2 threads are down right. At node b, we will have 5 threads down the left and 1 thread down the right, and so on. The work stealing has strongly favored sending threads towards the left side of each node because of the reasonably high confidence values of 0.8. The result is shown in Figure 1(a).

Suppose as search progresses the subtree starting at node d finishes without producing a solution. Then we need to update the confidence values. Using Equation (3), the confidence value at node b becomes 0, and at node a 0.44. Now when the threads steal work from the root, the situation has changed. Since one of the most fruitful parts of the left branch has been completely searched without producing a solution, it has become much less likely that there is a solution down the left branch. The updated confidence value reflects this. Now the threads will be distributed such that 4 threads are down the left branch and 4 threads are down the right branch, as shown in Figure 1(b).

Next, perhaps the subtree starting at node j finishes. The confidence value at node e then becomes 0, the confidence value at node b remains 0 and the confidence value at node a becomes 0.14. The vast majority of the fruitful places in the left branch has been exhausted without finding a solution, and the

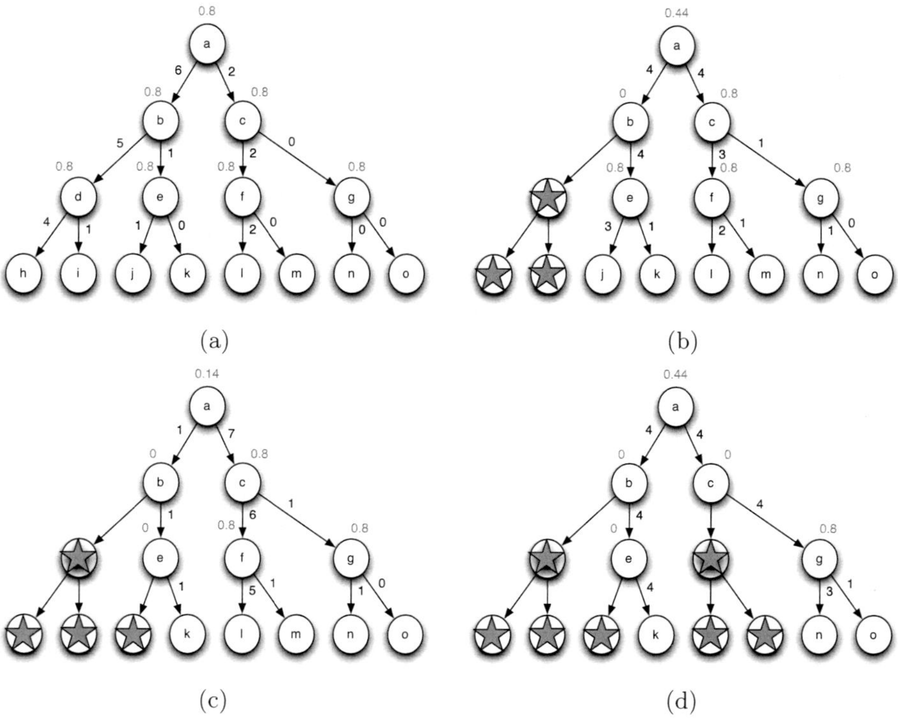

Fig. 1. Example with reasonably high confidence

confidence value at the root has been updated to strongly favor the right branch. The threads will now be distributed such that 7 threads go down the right and 1 thread goes down the left, as shown in Figure 1(c).

Next, suppose the subtree starting at node f finishes. The confidence value at node c becomes 0 and the confidence value at node a becomes 0.44. Since the most fruitful part of the right branch has also failed, the confidence value now swings back to favor the left branch more, as shown in Figure 1(d). This kind of confidence updating and redistribution of threads will continue, distributing the threads according to the current best solution density estimates. In our explanation here, for simplicity we only updated the confidence values very infrequently. In the actual implementation, confidence values are updated after every job is finished and thus occur much more frequently and in much smaller sized chunks.

Example with low confidence. For the second example, suppose we knew that the heuristic was very bad and was basically random. We may use $conf = 0.5$, i.e. the initial solution density estimates down the left and right branch are equal.

Suppose we have 4 threads. Initially, all the confidence values are 0.5. The threads will distribute as shown in Figure 2(a). This distributes the threads as far away from each other as possible which is exactly what we want. However, if the search tree is deep, and the first few decisions that the each threads makes are wrong, all threads may still get stuck and never find a solution.

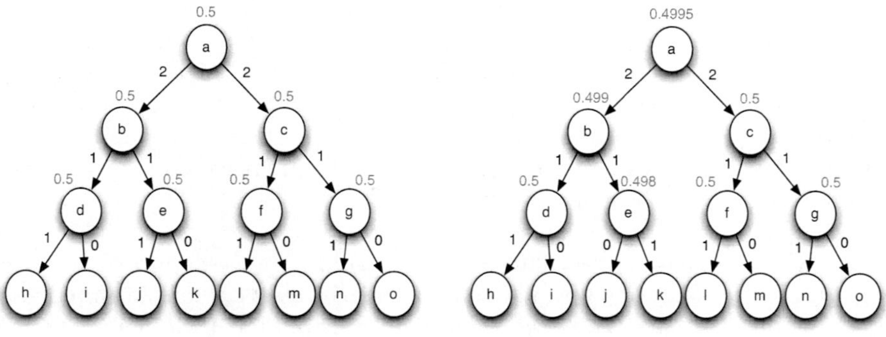

(a) (b)

Fig. 2. Example with low confidence

This is where the *restart* limit kicks in. After a certain time threshold is reached, the threads abandon their current search and begin work stealing from the root again. Since the confidence values are updated when they abandon their current job, they take a different path when they next steal work. For example, if the thread down node e abandons after having finished a subtree with root node at depth 10, then the confidence at node e becomes 0.498, the confidence at node b becomes 0.499, and the confidence at node a becomes 0.4995. Then when a thread steals work from the root, it will again go left, then right. When it gets to node e however, the confidence value is 0.498 and there are no threads down either branch, thus it will go right at this node instead of left like last time. The result is shown in Figure 2(b). The updated confidence value has guided the thread to an unexplored part of the search tree that is as different from those already searched as possible. This always happens because solution density estimates are decremented whenever a part of a subtree is found to have failed, so the confidence will always be updated to favour the unexplored parts of the search tree.

Emulating standard search patterns. As some other examples, we briefly mention what confidence models lead to some standard search patterns. DFS: *conf* = 1, *restart* = ∞. IDFS: *conf* = 1, *restart* = 1000. LDS: *conf* = 1-ϵ, *restart* = 1 node. DDS: *conf* = 0.5 if depth < d, 1-ϵ if depth $\geq$ d, *restart* = 1 node.

4 Experimental Evaluation

Confidence-based work stealing is implemented using Gecode 3.0.2 [18] with an additional patch to avoid memory management contention during parallel execution. The benchmarks are run on an Mac with 2× 2.8 GHz Xeon Quad Core E5462 processors with 4Gb of memory. However, due to memory limitations, we could not run the large instances of n-Queens or Knights on this machine. We run those two benchmarks on a Dell PowerEdge 6850 with 4× 3.0 GHz Xeon

Dual Core Pro 7120 CPUs with 32Gb of memory. 8 threads are used for the parallel search algorithm. We use a time limit of 20 min CPU time (2.5 min wall clock time for 8 threads), a restart time of 5 seconds, and a dynamic granularity bound that adjusts itself to try to steal no more than once every 0.5 seconds. We collected the following data: wall clock runtime, CPU utilization, communication overhead, number of steals, total number of nodes searched and number of nodes explored to find the optimal solution.

Optimization problems. In our first set of experiments we examine the efficiency of our algorithm for three optimization problems from Gecode's example problems. The problems are: Traveling Salesman Problem (TSP), Golomb-Ruler and Queens-Armies. A description of these problems can be found at [18]. We use the given search heuristic (in the Gecode example file) for each, except for TSP where we try both a strong heuristic based on maximising cost reduction and a weak heuristic that just picks variables and values in order. For TSP, we randomly generated many instances of an appropriate size for benchmarking. Only the size 12 and size 13 instances of Golomb Ruler, and only the size 9 and size 10 instances of Queen-Armies, are of an appropriate size for benchmarking. We use the simple confidence model with *conf* = 1, 0.66 and 0.5. The results are given in Table 1.

It is clear that in all of our problems, runtime is essentially proportional to the number of nodes searched, and it is highly correlated to the amount of time taken to find the optimal solution. The quicker the optimal solution is found, the fewer the nodes searched and the lower the total runtime. The communication cost, which includes all work stealing and synchronisation overheads, is less than 1% for all problems.

The strong heuristic in TSP is quite strong. Using *conf* = 1 achieves near perfect algorithmic efficiency, where algorithm efficiency is defined as the total number of nodes searched in the parallel algorithm vs the sequential algorithm. Other values of *conf* clearly cause an algorithmic slowdown. The optimal solution is found on average 3.2 and 3.3 times slower for *conf* = 0.66 and 0.5 respectively, resulting in an algorithmic efficiency of 0.81 and 0.80 respectively. The opposite is true when the weak heuristic is used. Using *conf* = 1 or 0.66 allows us to find the leftmost optimal solution in approximately the same number of nodes as the sequential algorithm, but using *conf* = 0.5 to reflect that the heuristic is weak allows the algorithm to find the optimal solution even faster, producing an algorithmic efficiency of 1.14.

The branching heuristic in Golomb Ruler is a greedy heuristic that selects the minimum possible value for the variable at each stage. This is a reasonable heuristic but by no means perfect. It turns out that for Golomb Ruler 12 and 13, the optimal solution does not lie directly in the left-most branch, and a certain degree of non-greediness leads to super-linear solution finding efficiencies.

The results for Queens-Armies show little difference depending on confidence. Clearly the heuristic is better than random at finding an optimal solution, and solution finding efficiency degrades slightly as we ignore the heuristic. But the overall nodes searched are almost identical for all confidence values, as the work

Table 1. Experimental results for optimization problems with simple confidence model. The results show: wall clock runtime in seconds (Runtime), speedup relative to the sequential version (Speedup), and runtime efficiency (RunE) which is Speedup/8, CPU utilization (CPU%), communication overhead (Comm%), number of steals (Steals), total number of nodes explored (Nodes), the algorithmic efficiency (AlgE) the total number of nodes explored in the parallel version versus the sequential version, the number of nodes explored to find the optimal solution (Onodes), and the solution finding efficiency (SFE) the total number of nodes explored in the parallel version to find the optimal versus the sequential version. Values for Runtime, CPU%, Comm%, Steals, Nodes, and Onodes are the geometric mean of the instances solved by all 4 versions.

TSP with strong heuristic, 100 instances (Mac)

conf	Runtime	Speedup	RunE	CPU%	Comm%	Steals	Nodes	AlgE	Onodes	SFE
Seq	313.3	—	—	100.0%	0.0%	—	5422k	—	1572k	—
1	38.2	7.25	0.91	96.5%	0.4%	708	5357k	1.01	1589k	0.99
0.66	47.2	5.88	0.74	93.7%	0.1%	319	6657k	0.81	5130k	0.31
0.5	48.0	5.77	0.72	92.9%	0.1%	467	6747k	0.80	5275k	0.30

TSP with weak heuristic, 100 instances (Mac)

conf	Runtime	Speedup	RunE	CPU%	Comm%	Steals	Nodes	AlgE	Onodes	SFE
Seq	347.8	—	—	99.0%	0.0%	—	7.22M	—	1.15M	—
1	46.7	7.45	0.93	99.4%	0.6%	1044	6.96M	1.04	1.09M	1.06
0.66	45.8	7.60	0.95	96.9%	0.1%	379	7.02M	1.03	1.10M	1.05
0.5	41.6	8.36	1.05	97.5%	0.1%	304	6.36M	1.14	0.96M	1.20

Golomb Ruler, 2 instances (n = 12, 13) (Mac)

conf	Runtime	Speedup	RunE	CPU%	Comm%	Steals	Nodes	AlgE	Onodes	SFE
Seq	562	—	—	100.0%	0.0%	—	9.71M	—	1.07M	—
1	69.0	8.15	1.02	99.3%	0.2%	572	8.96M	1.08	0.81M	1.33
0.66	59.0	9.54	1.19	99.3%	0.1%	346	7.58M	1.28	0.49M	2.21
0.5	65.2	8.63	1.08	99.3%	0.1%	259	8.42M	1.15	0.66M	1.63

Queen Armies, 2 instances (n = 9, 10) (Mac)

conf	Runtime	Speedup	RunE	CPU%	Comm%	Steals	Nodes	AlgE	Onodes	SFE
Seq	602	—	—	100.0%	0.0%	—	13.6M	—	845k	—
1	87.1	6.91	0.86	99.3%	0.7%	1521	14.5M	0.94	1878k	0.45
0.66	86.3	6.98	0.87	98.8%	0.2%	1143	14.5M	0.96	2687k	0.31
0.5	86.0	7.00	0.87	99.5%	0.2%	983	14.5M	0.95	2816k	0.30

required for the proof of optimality make up the bulk of the run time, and the proof of optimality parallelises trivially regardless of confidence.

Satisfaction problems. In our second set of experiments we examine the efficiency of our algorithm for three satisfaction problems from Gecode's examples [18]. The problems are: n-Queens, Knights, and Perfect-Square.

The sequential version solved very few instances of n-Queens and Knights. Furthermore, all those solves are extremely fast (< 3 sec) and are caused by

Table 2. Experimental results for satisfaction problems with simple confidence model

n-Queens, 100 instances (n = 1500, 1520, ..., 3480)

conf	Solved	Runtime	Speedup	RunE	CPU%	Comm%	Steals	Nodes	AlgE
Seq	4	2.9	—	—	99.9%	0.0%	—	1859	—
1	4	10.4	—	—	99.0%	86.6%	2	1845	—
0.66	29	18.0	—	—	81.6%	0.3%	9	15108	—
0.5	100	2.9	—	—	65.5%	1.6%	8	14484	—

Knights, 40 instances (n = 20, 22, ..., 98)

conf	Solved	Runtime	Speedup	RunE	CPU%	Comm%	Steals	Nodes	AlgE
Seq	7	0.22	—	—	99.9%	0.0%	—	1212	—
1	7	0.26	—	—	68.1%	59.7%	2	1150	—
0.66	13	0.50	—	—	48.0%	4.7%	8	8734	—
0.5	21	0.66	—	—	35.2%	6.0%	8	8549	—

Perfect-Square, 100 instances

conf	Solved	Runtime	Speedup	RunE	CPU%	Comm%	Steals	Nodes	AlgE
Seq	15	483.1	—	—	99.9%	0.0%	—	213k	—
1	13	72.3	6.68	0.83	98.0%	19.1%	419	216k	0.99
0.66	14	71.2	6.78	0.85	86.4%	2.9%	397	218k	0.98
0.5	82	8.9	54.02	6.75	89.0%	4.8%	21	32k	6.64

the search engine finding a solution at the very leftmost part of the search tree. Most of the time spent in those runs is from moving down to the leaf of the search tree rather than actual search and is not parallelisable, thus comparison of the statistics for the parallel vs sequential algorithms on those instances is not meaningful as there is very little work to parallelize. The number of instances solved is the more interesting statistic and is a better means of comparison. The parallel algorithm beats the sequential algorithm by an extremely large margin in terms of the number of instances solved.

n-Queens and Knights both have very deep subtrees and thus once the sequential algorithm fails to find a solution in the leftmost subtree, it will often end up stuck effectively forever. Modeling the fact that the branching heuristic is very weak at the top by using $conf = 0.5$ clearly produce a super linear speedup. The parallel algorithm solves 100 out of 100 instances of *n*-Queens compared to 4 out of 100 instances for the sequential algorithm or the parallel algorithm with $conf = 1$. The speedup cannot be measured as the sequential algorithm does not terminate for days when it fails to find a solution quickly. Similarly the parallel algorithm with $conf = 0.5$ solved 21 instances of Knights compared to 7 for the sequential and the parallel version with $conf = 1$.

Perfect Square's heuristic is better than random, but is still terribly weak. Using $conf = 0.5$ to model this once again produces super linear speedup, solving 82 instances out of 100 compared to 15 out of 100 for the sequential algorithm. We can compare run times for this problem as the sequential version solved a fair number of instances and those solves actually require some work (483 sec on average). The speedup in this case is 54 using 8 threads.

Table 3. Experimental results showing Nodes and algorithmic efficiency (AlgE) using accurate confidence values, where we follow the confidence value to degree α

	Seq		$\alpha = 1$		$\alpha = 0.5$		$\alpha = 0$		$\alpha = -0.5$		$\alpha = -1$	
Golomb-Ruler 12	5.31M	—	2.24M	2.37	3.48M	1.53	4.27M	1.24	10.8M	0.49	10.6M	0.50
Golomb-Ruler 13	71.0M	—	53.2M	1.34	57.6M	1.23	61.9M	1.15	74.8M	0.95	111M	0.64

Using accurate confidence values. So far, we have tested the efficiency of our algorithm using simple confidence models where the confidence value is the same for all nodes. This does not really illustrate the algorithm's full power. We expect that it should perform even better when node-specific confidence values are provided, so that we can actually encode and utilise information like, the heuristic is confident at this node but not confident at that node, etc. In our third set of experiments, we examine the efficiency of our algorithm when node specific confidence values are provided.

Due to our lack of domain knowledge, we will not attempt to write a highly accurate confidence heuristic. Rather, we will simulate one by first performing an initial full search of the search tree to find all solutions, then produce confidence estimates for the top few levels of the search tree using several strategies like, follow the measured solution density exactly, follow it approximately, ignore it, go against it, etc, to see what effect this has on runtime. Let α quantify how closely we follow the measured confidence value and let *conf* be the measured confidence value. Then we use the following formula for our confidence estimate: $conf' = \alpha \times conf + (1 - \alpha) \times 0.5$. If $\alpha = 1$, then we follow it exactly. If $\alpha = -1$, we go against it completely, etc. We use the Golomb-Ruler problem for our experiment as the full search tree is small enough to enumerate completely. The results are shown in Table 3.

The results show that using confidence values that are even a little biased towards the real value is sufficient to produce super linear speedup. And not surprisingly, going against the real value will result in substantial slowdowns.

5 Conclusion

By analysing work stealing schemes using a model based on solution density, we were able to quantitatively relate the strength of the branching heuristic with the optimal place to steal work from. This leads to an adaptive work stealing algorithm that can utilise confidence estimates to automatically produce "optimal" work stealing patterns. The algorithm produced near perfect or better than perfect algorithmic efficiency on all the problems we tested. In particular, by adapting to a steal high, interleaving search pattern, it is capable of producing super linear speedup on several problem classes. The real efficiency is lower than the algorithmic efficiency due to hardware effects, but is still quite good at a speedup of at least 7 at 8 threads. Communication costs are negligible on all problems even at 8 threads.

Acknowledgments. NICTA is funded by the Australian Government as represented by the Department of Broadband, Communications and the Digital Economy and the Australian Research Council.

References

1. Michel, L., See, A., Van Hentenryck, P.: Parallelizing constraint programs transparently. In: [20], pp. 514–528
2. Perron, L.: Search procedures and parallelism in constraint programming. In: Jaffar, J. (ed.) CP 1999. LNCS, vol. 1713, pp. 346–361. Springer, Heidelberg (1999)
3. Schulte, C.: Parallel search made simple. In: Beldiceanu, N., Harvey, W., Henz, M., Laburthe, F., Monfroy, E., Müller, T., Perron, L., Schulte, C. (eds.) Proceedings of TRICS: Techniques foR Implementing Constraint programming Systems, a post-conference workshop of CP 2000. Number TRA9/00, 55 Science Drive 2, Singapore 117599, pp. 41–57 (2000)
4. Rolf, C.C., Kuchcinski, K.: Load-balancing methods for parallel and distributed constraint solving. In: IEEE International Conference on Cluster Computing, pp. 304–309 (2008)
5. Kumar, V., Rao, V.N.: Parallel depth first search. Part II. Analysis. International Journal of Parallel Programming 16, 501–519 (1987)
6. Caseau, Y., Laburthe, F.: Solving small TSPs with constraints. In: Naish, L. (ed.) Proceedings of the Fourteenth International Conference on Logic Programming, Leuven, Belgium, pp. 316–330. The MIT Press, Cambridge (1997)
7. Véron, A., Schuerman, K., Reeve, M., Li, L.L.: Why and how in the ElipSys OR-parallel CLP system. In: Reeve, M., Bode, A., Wolf, G. (eds.) PARLE 1993. LNCS, vol. 694, pp. 291–303. Springer, Heidelberg (1993)
8. Rao, V.N., Kumar, V.: Superlinear speedup in parallel state-space search. In: Kumar, S., Nori, K.V. (eds.) FSTTCS 1988. LNCS, vol. 338, pp. 161–174. Springer, Heidelberg (1988)
9. Gendron, B., Crainic, T.G.: Parallel branch-and-bound algorithms: Survey and synthesis. Operations Research 42, 1042–1066 (1994)
10. Quinn, M.J.: Analysis and implementation of branch-and bound algorithms on a hypercube multicomputer. IEEE Trans. Computers 39, 384–387 (1990)
11. Mohan, J.: Performance of Parallel Programs: Model and Analyses. PhD thesis, Carnegie-Mellon University, Pittsburgh, PA, USA (1984)
12. Meseguer, P.: Interleaved depth-first search. In: [19], pp. 1382–1387
13. Harvey, W.D., Ginsberg, M.L.: Limited discrepancy search. In: Mellish, C.S. (ed.) Fourteenth International Joint Conference on Artificial Intelligence, Montréal, Québec, Canada, pp. 607–615. Morgan Kaufmann Publishers, San Francisco (1995)
14. Walsh, T.: Depth-bounded discrepancy search. In: [19], pp. 1388–1395
15. Refalo, P.: Impact-based search strategies for constraint programming. In: Wallace, M. (ed.) CP 2004. LNCS, vol. 3258, pp. 557–571. Springer, Heidelberg (2004)
16. Zanarini, A., Pesant, G.: Solution counting algorithms for constraint-centered search heuristics. In: [20], pp. 743–757

17. Szymanek, R., O'Sullivan, B.: Guiding search using constraint-level advice. In: Brewka, G., Coradeschi, S., Perini, A., Traverso, P. (eds.) ECAI, pp. 158–162. IOS Press, Amsterdam (2006)
18. Gecode Team: Gecode: Generic constraint development environment (2006), `http://www.gecode.org`
19. Pollack, M.E. (ed.): Fifteenth International Joint Conference on Artificial Intelligence. Morgan Kaufmann Publishers, Nagoya (1997)
20. Bessière, C. (ed.): CP 2007. LNCS, vol. 4741. Springer, Heidelberg (2007)

Minimizing the Maximum Number of Open Stacks by Customer Search

Geoffrey Chu and Peter J. Stuckey

NICTA Victoria Laboratory,
Department of Computer Science and Software Engineering,
University of Melbourne, Australia
{gchu,pjs}@csse.unimelb.edu.au

Abstract. We describe a new exact solver for the minimization of open stacks problem (MOSP). By combining nogood recording with a branch and bound strategy based on choosing which customer stack to close next, our solver is able to solve hard instances of MOSP some 5-6 orders of magnitude faster than the previous state of the art. We also derive several pruning schemes based on dominance relations which provide another 1-2 orders of magnitude improvement. One of these pruning schemes largely subsumes the effect of the nogood recording. This allows us to reduce the memory usage from an potentially exponential amount to a constant ~2Mb for even the largest solvable instances. We also show how relaxation techniques can be used to speed up the proof of optimality by up to another 3-4 orders of magnitude on the hardest instances.

1 Introduction

The *Minimization of Open Stacks Problem* (MOSP) [10] can be described as follows. A factory manufactures a number of different products in batches, i.e., all copies of a given product need to be finished before a different product is manufactured, so there are never two batches of the same product. Each customer of the factory places an order requiring one or more different products. Once one product in a customer's order starts being manufactured, a stack is opened for that customer to store all products in the order. Once all the products for a particular customer have been manufactured, the order can be sent and the stack is freed for use by another order. The aim is to determine the sequence in which products should be manufactured to minimize the maximum number of open stacks, i.e., the maximum number of customers whose orders are simultaneously active. The importance of this problem comes from the variety of real situations in which the problem (or an equivalent version of it) arises, such as cutting, packing, and manufacturing environments, or VLSI design. Indeed the problem appears in many different guises in the literature, including: graph path-width and gate matrix layout (see [3] for a list of 12 equivalent problems). The problem is known to be NP-hard [3].

We can formalize the problem as follows. Let P be a set of products, C a set of customers, and $c(p)$ a function that returns the set of customers who have ordered product $p \in P$. Since the products ordered by each customer $c \in C$ are

I.P. Gent (Ed.): CP 2009, LNCS 5732, pp. 242–257, 2009.

	p_1	p_2	p_3	p_4	p_5	p_6	p_7
c_1	X	.	.	.	X	.	X
c_2	X	.	.	X	.	.	.
c_3	.	X	.	X	.	X	.
c_4	.	.	X	X	.	X	X
c_5	.	.	X	.	X	.	.

(a)

	p_7	p_6	p_5	p_4	p_3	p_2	p_1
c_1	X	–	X	–	–	–	X
c_2	.	.	.	X	–	–	X
c_3	.	X	–	X	–	X	.
c_4	X	X	–	X	X	.	.
c_5	.	.	X	–	X	.	.

(b)

Fig. 1. (a) An example $c(p)$ function: $c_i \in c(p_j)$ if the row for c_i in column p_j has an X. (b) An example schedule: c_i is active when product p_j is scheduled if the row for c_i in column p_j has an X or a –.

placed in a stack different from that of any other customer, we use c to denote both a client and its associated stack. We say that customer c is active (or that stack c is open) at time k in the manufacturing sequence if there is a product required by c that is manufactured before or at time k, and also there is a product manufactured at time k or afterwards. In other words, c is active from the time the first product ordered by c is manufactured until the last product ordered by c is manufactured. The MOSP aims at finding a schedule for manufacturing the products in P (i.e., a permutation of the products) that minimizes the maximum number of customers active (or of open stacks) at any time. We call a problem with n customers and m products an $n \times m$ problem.

Example 1. Consider a 5×7 MOSP for the set of customers $C = \{c_1, c_2, c_3, c_4, c_5\}$, and set of products $P = \{p_1, p_2, p_3, p_4, p_5, p_6, p_7\}$, and a $c(p)$ function determined by the matrix M shown in Figure 1(a), where an X at position M_{ij} indicates that client c_i has ordered product p_j.

Consider the manufacturing schedule given by sequence $[p_7, p_6, p_5, p_4, p_3, p_2, p_1]$ and illustrated by the matrix M shown in Figure 1(b), where client c_i is active at position M_{ij} if the position contains either an X (p_j is in the stack) or an – (c_i has an open stack waiting for some product scheduled after p_j). Then, the active customers at time 1 are $\{c_1, c_4\}$, at time 2 $\{c_1, c_3, c_4\}$, at time 3 $\{c_1, c_3, c_4, c_5\}$, at time 4 $\{c_1, c_2, c_3, c_4, c_5\}$, at time 5 $\{c_1, c_2, c_3, c_4, c_5\}$, at time 6 $\{c_1, c_2, c_3\}$, and at time 7 $\{c_1, c_2\}$. The maximum number of open stacks for this particular schedule is thus 5. □

The MOSP was chosen as the subject of the first Constraint Modelling Challenge [6] posed in May 2005. Many different techniques were explored in the 13 entries to the challenge. The winning entry by Garcia de la Banda and Stuckey [2] concentrated on two properties of the MOSP problem. First, the permutative redundancy found in the MOSP problem leads naturally to a dynamic programming approach [2]. This approach is also largely equivalent to the constraint programming approach described in [5] where permutative redundancies are pruned using a table of no-goods. Second, by using a branch and bound method, the upper bound on the number of open stacks can be used to prune branches in various ways. These two techniques are very powerful and led to a solver that was an order of magnitude faster than any of the other entries in the 2005 MOSP challenge. The *Limited Open Stacks Problem, LOSP*(k), is the decision version of the problem, where we determine if for some fixed k there is an order of products that requires at most k stacks at any time. The best approach of [2] solves the MOSP problem by repeatedly solving $LOSP(k)$ and reducing k until this problem has no solution.

The search strategy used in this winning entry (branching on which product to produce next), is actually far from optimal. As was first discussed in [8] and shown in [9], branching on which customer stack to close next is never worse than branching on which product to produce next, and is usually much better, even when the number of customers is far greater than the number of products. This is the result of a simple dominance relation. In this paper we show that combining this search strategy with nogood recording produces a solver that is some 5-6 orders of magnitude faster than the winning entry to the Modelling Challenge on hard instances. We also derive several other dominance rules that provide a further 1-2 orders of magnitude improvement. One rule in particular largely subsumes the effect of the nogood recording. This allows us to reduce the memory usage from an potentially exponential amount to a constant ~2Mb for even the largest solvable instances. We also utilise relaxation techniques to speed up the proof of optimality for the hardest instances by a further 3-4 orders of magnitude. With all the improvements, our solver is able to solve all the open instances from the Modelling Challenge within 10 seconds!

2 Searching on Customers

Our solver employs a branch and bound strategy for finding the exact number of open stacks required for an MOSP instance. The MOSP instance is treated as a series of satisfaction problems LOSP(k), where at each stage, we ask whether there is a solution that uses no more than k stacks. If a solution is found, we decrease k and look for a better solution.

We briefly define what a dominance relation is. A dominance relation $\triangleright$ is a binary relation defined on the set of partial problems generated during a search algorithm. For a satisfaction problem, if P_i and P_j are partial problems corresponding to two subtrees in the search tree, then $P_i \triangleright P_j$ imply that if P_j has a solution, then P_i must have a solution. This means that if we are only interested in the satisfiability of the problem, and P_i dominates P_j, then as long as P_i is searched, P_j can be pruned.

The customer based search strategy is derived from following idea. Given a product order $U = [p_1, p_2, \ldots p_n]$, define a customer close order $T = [c_1, c_2, \ldots c_m]$ as the order in which customer stacks can close given U. Construct U' such that we first schedule all the products needed by c_1, then any products required by c_2, then those required by c_3, etc. It is easy to show that if U is a solution, then so is U'. Clearly this can be converted to a dominance relation. It is sufficient to search only product orderings where every product is required by the next stack to close. This can be achieved by branching on which customer stack to close next, and then scheduling exactly those products which are needed. The correctness of this search strategy is proved in [8] and [7]. In fact, it can be shown that the customer based search strategy never examines more nodes than the search strategy based on choosing which product to produce next (even when the strongest look ahead pruning of [2] is used).

	p_3	p_5	p_4	p_6	p_7	p_2	p_1
c_1	.	X	–	–	X	–	X|
c_2	.	.	X	–	–	–	X|
c_3	.	.	X	X	–	X|	.
c_4	X	–	X	X	X|	.	.
c_5	X	X|	.	.	.	.	.

(a)

	p_3	p_5	p_1	p_7	p_4	p_2	p_6
c_1	.	X	X	X|	.	.	.
c_2	.	.	X	–	X|	.	.
c_3	.	.	.	.	X	X	X|
c_4	X	–	–	X	X	–	X|
c_5	X	X|	.	.	.	.	.

(b)

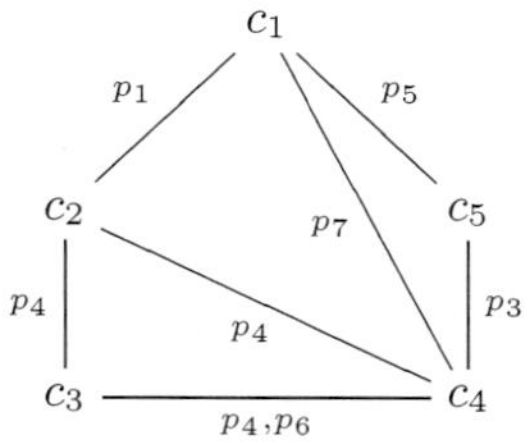

(c)

Fig. 2. (a) A schedule corresponding to customer order $[c_5, c_4, c_3, c_2, c_1]$. (b) A schedule corresponding to customer order $[c_5, c_1, c_2, c_3, c_4]$. (c) The customer graph (ignoring self edges) with edges labelled by the products that generate them.

Example 2. Consider the schedule U shown in Figure 1(b), the customers are closed in the order $\{c_4, c_5\}$ when p_3 is scheduled, then $\{c_3\}$ when p_2 is scheduled, then $\{c_2, c_1\}$ when p_1 is scheduled. Consider closing the customers in the order $T = [c_5, c_4, c_3, c_2, c_1]$ compatible with U. This leads to a product ordering, for example, of $U' = [p_3, p_5, p_4, p_6, p_7, p_2, p_1]$. The resulting scheduling is shown in Figure 2(a). It only requires 4 stacks (and all other schedules with this closing order will use the same maximum number of open stacks).

Define the *customer graph* $G = (V, E)$ for an open-stacks problem as: $V = C$ and $E = \{(c_1, c_2) \mid \exists p \in P, \{c_1, c_2\} \subseteq c(p)\}$, that is, a graph in which nodes represent customers, and two nodes are adjacent if they order the same product. Note that, by definition, each node is self-adjacent. Let $N(c)$ be the set of nodes adjacent to c in G. The customer graph for the problem of Example 1 is shown in Figure 2(c).

Rather than thinking in terms of products then, it is simpler to think of the MOSP problem entirely in terms of the customer graph. All functions $c(p)$ that produce the same customer graph have the same minimum number of stacks. Thus the products are essentially irrelevant. Their sole purpose is to create edges in the customer graph. Thus we can think of the MOSP this way. For each customer c, we have an interval during which their stack is open. If there is an edge between two nodes in the customer graph, then their intervals must overlap. If a customer close order that satisfies these constraints are found, an equivalent product ordering using the same number of stacks can always be found.

We define our terminology. At each node there is a set of customer stacks S that have been closed. The stacks which have been opened (not necessarily currently open) are $O(S) = \cup_{c \in S} N(c)$. The set of currently open stacks is given by $O(S) - \{c \in O(S) | N(c) \subseteq O(S)\}$. For each c not in S, define $o(c, S) = N(c) - O(S)$, i.e. the new stacks which will open if c is the next stack to close.

Define $open(c, S) = |o(c, S)|$ and $close(c, S) = |\{d | o(d, S) \subseteq o(c, S)\}|$, i.e. the new stacks that will open and the number of new stacks that will close respectively if we close c next.

Suppose that customers S are currently closed, then closing c requires opening $o(c, S)$, so the number of stacks required is $|O(S) - S \cup o(c, S)|$. If we are solving LOSP(k) and $|O(S) - S \cup o(c, S)| > k$ then it is not possible to close customer

c next, and we call the sequence $S ++ [c]$ *violating*. Note $|O(S) - S \cup o(c,S)| \geq open(c,S)$.

We define the *playable* sequences as follows: the empty sequence ϵ is playable; $S ++ [c]$ is playable if S is playable and $S ++ [c]$ is not *violating*.

A *solution* S of the LOSP(k) is a playable sequence of all the customers C. This leads to an algorithm for MOSP by simply solving LOSP(k) for k varying from $|C| - 1$ (there is definitely a solution with $|C|$) to 1, and returning the smallest k that succeeds.

```
MOSP(C, N)
   for (S ∈ 2^C) prob[S] := false
   for (k ∈ |C| − 1, ..., 1)
      if (¬ playable(∅,C,k,N) return k + 1

playable(S, R, k, N)
   if (prob[S]) return false
   if (R = ∅) return true
   for (c ∈ R)
      if (|O(S) − S ∪ o(c, S)| ≤ k)
         if (playable(S ∪ {c}, R − {c}, k, N)) return true
   prob[S] := true
   return false
```

This simple algorithm tries to schedule the customers in all possible orders using at most k stacks. The memoing of calls to playable just records which partial sets of customers, S, have been examined already by setting $prob[S] = true$. If a set has already been considered it either succeeded and we won't look further, or if it failed, we record the nogood and return $false$ when we revisit it.

The algorithm can be improved by noting that if $o(c_i, S) \subseteq o(c_j, S)$ and $i < j$, then clearly, we can always play i before j rather than playing j immediately, since closing j will close i in any case. Hence move j can be removed from the candidates R considered for the next customer to close.

Example 3. Reexamining the problem of Example 1 using the closing customer schedule of $[c_5, c_1, c_2, c_3, c_4]$ results in many possible schedules (all requiring the same maximum number of open stacks). One is shown in Figure 2(b). This uses 3 open stacks and is optimal since e.g. product p_4 always requires 3 open stacks.

3 Improving the Search on Customers

In this section we consider ways to improve the basic search approach by exploiting several other dominance relations.

3.1 Definite Moves

Suppose $S ++ [q]$ is playable and $close(q,S) \geq open(q,S)$, then if there is any solution extending S there is a solution extending $S ++ [q]$. This means that we can prune all branches other than q. Intuitively speaking, q is such a good move at this node that it is always optimal to play it immediately.

Theorem 1. *Suppose S ++ $[q]$ is playable and $close(q, S) \geq open(q, S)$, then if $U' = S$ ++ R is a solution, there exists a solution $U = S$ ++ $[q]$ ++ R'.*

Proof. Suppose there was a solution $U' = S$ ++ $[c_1, c_2, \ldots, c_m, q, c_{m+1}, \ldots, c_n]$. We claim that $U = S$ ++ $[q, c_1, \ldots, c_m, c_{m+1}, \ldots, c_n]$ is also a solution. The two sequences differ only in the placement of q. The number of stacks which are open at any time before, or any time after the set of customers $\{c_1, c_2, \ldots, c_m, q\}$ are played is identical for U and U', since it only depending on the set of customers closed and not the order. Thus if U' is a solution, then U has less than or equal to k open stacks at those times. Since S ++ $[q]$ is playable, the number of open stacks when q is played in U is also less than or equal to k. Finally, the number of open stacks when S ++ $[q, c_1, c_2, \ldots, c_i]$ has been played in U is always less than or equal to the number of open stacks when S ++ $[c_1, c_2, \ldots, c_i]$ has been played in U', because we have at most $open(q, S)$ extra stacks open, but at least $close(q, S)$ extra stacks closed. Thus the number of open stacks at these times are also less than or equal to k and U is a solution.

3.2 Better Moves

While definite moves are always worth playing we can find similarly that one move is always better than another. If both S ++ $[q]$ and S ++ $[r, q]$ are playable and $close(q, S \cup \{r\}) \geq open(q, S \cup \{r\})$ then if there is a solution extending S ++ $[r]$, there exists a solution extending S ++ $[q]$. This means that we do not need to consider move r at this node. Intuitively, q is so much better than r that rather than playing r now, it is always better to play q first.

Theorem 2. *Suppose S ++ $[q]$ and S ++ $[r, q]$ are playable and $close(q, S \cup \{r\}) \geq open(q, S \cup \{r\})$ then if $U' = S$ ++ $[r]$ ++ R is a solution there exists a solution $U = S$ ++ $[q]$ ++ R'.*

Proof. Suppose there was a solution $U' = S$ ++ $[r, c_1, c_2, .., c_m, q, c_{m+1}, \ldots, c_n]$. The conditions imply that if r is played now, q becomes a definite move. By the same argument as above, $U'' = S$ ++ $[r, q, c_1, \ldots, c_n]$ is also a solution. Now if we swap q with r, the number of new stacks opened before r increases by at most $open(q, S)$, but the number of new stacks closed before r increases by exactly $close(q, S)$. Also, playing q after S does not break the upperbound by our condition. Thus $U = S$ ++ $[q, r, c_1, \ldots, c_n]$ is also a solution. □

Although "better move" seems weaker than "definite move" as it prunes only one branch at a time rather than all branches but one, it is actually a generalisation, as by definition any "definite move" is "better" than all other moves. Our implementation of "better move" subsumes "definite move" so we will simply consider them as one improvement.

3.3 Old Move

Let $S = [s_1, s_2, \ldots, s_n]$. Suppose $U = [s_1, s_2, .., s_m, q, s_{m+1}, \ldots, s_n]$ is playable and we have previously examined the subtree corresponding to state $[s_1, s_2, .., s_m, q]$. Then we need not consider sequences starting with $U' = S +\!\!+ [q]$ because we will have already considered equivalent sequences earlier when searching from state $[s_1, s_2, .., s_m, q]$.

Theorem 3. *Let $S = [s_1, s_2, \ldots, s_n]$. Suppose that $S' = [s_1, s_2, .., s_m, q, s_{m+1}, \ldots, s_n]$ is playable, then if $U' = S +\!\!+ [q] +\!\!+ R$ is a solution then $U = S' +\!\!+ R$ is a solution.*

Proof. S' is playable by assumption so the number of open stacks at any time during S' is less than or equal to k. At any point after S', the number of open stacks are identical for U and U' since it only depends on the set of closed customers and not the order. Hence U is also a solution.

At any node, if it is found that at some ancestor node, the q branch has been searched and U is playable, then q can immediately be pruned. This pruning scheme was mentioned in [8], but it was incorrectly stated there. The author of [8] failed to note that the condition that U is playable is in fact crucial, because if U was not playable, then the set $S +\!\!+ [q]$ would have been pruned via breaking the upper bound and would not have in fact been previously explored, thus pruning it now would be incorrect.

Naively, it would appear to take $O(|C|^3)$ time to check the "old move" condition at each node. However, it is possible to do so in $O(|C|)$ time. At each node we keep a set $Q(S)$ of all the old moves. i.e. the set of moves q such that we can find $S' = [s_1, s_2, .., s_m, q, s_{m+1}, \ldots, s_n]$ which is playable, and such that move q has already been searched at the node $S'' = [s_1, .., s_m]$. Note that by definition, when a move r has been searched at the current node, r will be added to $Q(S)$. It is easy to calculate $Q(S +\!\!+ [s_{n+1}]))$ when we first reach that child node. First, $Q(S +\!\!+ [s_{n+1}]) \subseteq Q(S)$, since if $S' = [s_1, s_2, .., s_m, q, s_{m+1}, \ldots, s_n, s_{n+1}]$ is playable then by definition so is $S' = [s_1, s_2, .., s_m, q, s_{m+1}, \ldots, s_n]$. Second, to check if each $q \in Q(S)$ is also in $Q(S +\!\!+ [s_{n+1}])$, we simply have to check whether the last move in $S' = [s_1, s_2, .., s_m, q, s_{m+1}, \ldots, s_n, s_{n+1}]$ is playable, as all the previous moves are already known to be playable since $q \in Q(S)$. Checking the last move takes constant time so the total complexity is $O(|C|)$. There are some synergies between the "better move" improvement and the "old move" improvement. If $q \in Q(S)$ and q is better than move r, then we can add r to $Q(S)$ as well. This allows "old move" to prune sets that we have never even seen before.

3.4 Upperbound Heuristic

In this section, we describe an upperbound heuristic which was found to be very effective on our instances. A good heuristic for finding an optimal solution is useful from a practical point of view if no proof of optimality is required. It is also a crucial component for the relaxation techniques described in the next

Table 1. Comparison of upperbound heuristic, versus complete search on some difficult problems. Times in milliseconds.

Instance	Orig. time	Heur. time	Speedup
100-100-2	4136.3	39.8	104.0
100-100-4	4715.5	43.3	108.8
100-100-6	8.2	12.8	0.6
100-100-8	9.2	6.3	1.5
100-100-10	1.6	1.3	1.3
125-125-2	1159397.3	385.1	3010.3
125-125-4	2593105.1	398.9	6500.1
125-125-6	8975.9	424.9	21.1
125-125-8	187.8	146.1	1.3
125-125-10	22.2	8.3	2.7

subsection which can give several orders of magnitude speedup on the proof of optimality for hard problems.

In [2] the authors tried multiple branching heuristics in order to compute and upper bound, but only applied them in a greedy fashion, effectively searching only 1 leaf node for each. We can do much better by performing an incomplete search where we are willing to explore a larger number of nodes, but still much fewer than a complete search. Simple ways of doing this using our complete search engine include, sorting the choices according to some criteria, and only trying the first m moves for some m. Or trying all the moves which are no worse than the best by some amount e, etc.

One heuristic that is extremely effective is to only consider the moves where we close a customer stack that is currently open, the intuition being that if a stack is not even open yet, there is no point trying to close it now. Although this seems intuitively reasonable, it is in fact not always optimal. In practice however, an incomplete search using this criteria is very fast, and finds the optimal solution almost all the time, and several orders of magnitude faster than the complete search for some hard instances. The reason for its strength comes from its ability to exploit a not quite perfect dominance relation. Almost all the time, subtrees where we close a stack that is not yet open is dominated by one where we close a currently open stack, and thus we can exploit this to prune branches similarly to what we did in Section 3. The dominance is not always true however, so using such a pruning rule makes it an incomplete, heuristic search. The procedure ub_MOSP is identical to that for MOSP except that the line **for** $(c \in R)$ is replaced by **for** $(c \in R \cap O(S))$.

See Table 1 for a brief comparison of the times required to find the optimal solution.

3.5 Relaxation

Relaxation has been used in [4] in the context of a local search method. The idea there was to try to relax the problem in such a way that solution density is increased and thus better solutions can be found quicker. However, those methods are of no help for proving optimality. In this section we show how relaxation can be used to speed up the proof of optimality.

As was seen in the experimental results in [2], the sparser instances of MOSP are substantially harder than denser instances of MOSP given the same number of customers and products. This can be explained by the fact that in sparser instances, each customer has far fewer neighbours in the customer graph, thus many more moves would fall under the upper bound limit at each node and both the depth and the branching factor of the search tree are dramatically increased compared to a dense instance of the same size.

However, the sparsity of these instances also leads to a potential optimization. Since the instance is sparse and the optimum is low (e.g. 20-50 for a 125×125 problem) it is possible that not all of the constraints are actually required to force the lower bound. It is possible that there is some small "unsatisfiable core" of customers which are producing the lower bound. If such an unsatisfiable core exists and can be identified, we can potentially remove a large number of customers from the problem and make the proof of optimality much quicker. It turns out that this is often possible.

First, we will show how we can relax the MOSP instance. Naively, we can simply delete an entire node in the customer graph and remove all edges containing that node. This represents the wholesale deletion of some constraints and of course is a valid relaxation. However, we can do much better using the following result from [1] (although only informal arguments are given for correctness)

Lemma 1. *If G' is some contraction of G, where G represents the customer graph of an MOSP instance, then G' is a relaxation of G.*

So by using this lemma, we can get some compensation by retaining some of the edges when we remove a node. Next we need to identify the nodes which can be removed/merged without loosening the lower bound on the problem.

The main idea is that the longer a customer's stack is open in the optimal solutions, the more likely it is that that customer is contributing to the lower bound, since removal of such a customer would mean that there is a high chance that one of the optimal solutions can reduce to one needing one fewer stack. Thus we want to avoid removing such customers. Instead we want to remove or merge customers whose stacks are usually open for a very short time. One naïve heuristic is to greedily remove nodes in the customer graph with the lowest degree. Fewer edges coming out of a node presumably means that the stack is open for a shorter period of time on average.

A much better heuristic comes from the following idea. Suppose there exist a node c such that any neighbour of c is also connected to most of the neighbours of c, then when c is forced open by the closure of one of those neighbours, that neighbour would also have forced most of the neighbours of c to open, and thus c will be able to close soon afterwards and will only be open for a short time. The condition that most neighbours of c are connected to most other neighbours of c is in fact quite common for sparse instances due to the way that the customer graph is generated from the products (each product produces a clique in the graph). To be more precise, in our implementation, the customers are ranked according to:

$$F(c) = \sum_{c' \in N(c)} |N(c) - N(c')|/|N(c)| \tag{1}$$

This is a weighted average of how many neighbours c' of c are not connected to each neighbour of c. The weights represents the fact that neighbours with fewer neighbours are more likely to close early and be the one that forces c to open. We merge the node c with the highest value of $F(c)$ with the neighbouring node c' with the highest value of $|N(c) - N(c')|$, as that node stands to gain the highest number of edges.

Although we have a good heuristic for finding nodes to merge, it is quite possible to relax too much to the point that the relaxed problem has a solution lower than the true lower bound of the original problem, in which case it will be impossible to prove the true lower bound using this relaxed problem. Thus it is important that we have a quick way of finding out when we have relaxed too much. This is where the very fast and strong upperbound heuristic of the previous subsection is needed. The overall relaxation algorithm is as follows:

```
relax_MOSP(C, N)
  ub := ub_MOSP(C, N)                          % ub is an upper bound
  (C', N') := (C, N)
  while (|C'| > ub)
    (C', N') := merge_one_pair(C', N')         % relax problem
  while ((C, N) ≠ (C', N'))
    relax_ub := ub_MOSP(C', N')
    if (relax_ub < ub)                         % too relaxed to prove lb
      (C', N') := unmerge_one_pair(C', N')     % unrelax problem
    else
      lb := MOSP(C', N')                       % compute lowerbound
      if (lb < ub)                             % too relaxed to prove lb
        (C', N') := unmerge_one_pair(C', N')   % unrelax problem
      else return ub                           % lb = ub
  return MOSP(C, N)                            % relaxation failed!
```

As can be seen, the upperbound heuristic is necessary to find a good (optimal) solution quickly. It is also used to detect when we are too relaxed as quickly as possible so that we can unrelax. If the upperbound heuristic is sufficiently good, we will quickly be able to find a relaxation that removes as many customers as possible without being too relaxed. If the upperbound heuristic is weak however, we could waste a lot of time searching in a problem that is in fact too relaxed to ever give us the true lowerbound. In practice, we have found that our upperbound heuristic is quite sufficient for the instances we tested it on.

There are a few optimisations we can make to this basic algorithm. Firstly, when an unmerge is performed, we can attempt to extend the last solution found to a solution of this less relaxed problem. If the solution extends, then it is still too relaxed and we need to unmerge again. This saves us having to actually look for a solution to this problem. Secondly, naively, when we perform an unrelax, we can simply unmerge the last pair of nodes that were merged. However, we can do better. One of the weaknesses of the current algorithm is that the nodes to be merged are chosen greedily using equation (1). If this happens to choose a bad relaxation that lowers the lowerbound early on, then we will not be able to remove any more customers beyond that point. We can fix this to some extent by choosing which pair of nodes to unmerge when we unrelax. We do this by considering each of the problems that we get by unmerging each pair of the current merges. If the last solution found does not extend to a solution for one of these, then we choose that unmerge, as this unrelaxation gives us a chance to prove the true lowerbound. If the last solution extends to a solution for all of them, we unmerge the last pair as per usual. This helps to get rid of early mistakes in merging and is useful on several of our instances.

4 Experimental Evaluation

In this section we demonstrate the performance of our algorithm, and the effect of the improvements. The experiments were performed on a Xeon Pro 2.4GHz processor with 2Gb of memory. The code implementing the approaches were compiled using g++ with -O3 optimisation.

4.1 Modelling Challenge Instances

Very stringent correctness tests were performed in view of the large speedups achieved. All versions of our solver were run on the 5000+ instances used in the 2005 model challenge [6], as well as another 100,000 randomly generated instances of size 10×10 to 30×30 and various densities. The answers were identical with the solver of [2].

We compare our solver with the previous state of the art MOSP solver, on which our solver is based. The results clearly show that our solver is orders of magnitude faster than the original version. Getting an exact speedup is difficult as almost all of the instances that the original version can solve are solved trivially by our solver in a few milliseconds, whereas instances that our solver finds somewhat challenging are completely unsolvable by the original version.

Our solver was able to solve all the open problems from the Modelling Challenge: SP2, SP3, and SP4. Table 2 compares these problems with the best results

Table 2. Results on the open problems from the Constraint Modelling Challenge 2005, comparing versus the the winner of the challenge [2]. Times in milliseconds.

	[2]			This paper		
	Best	Nodes	Time	Optimal	Nodes	Time
SP2	19	25785	1650	19	1236	7
SP3	36	949523	∼3600000	34	84796	410
SP4	56	3447816	∼14400000	53	1494860	9087

from the Challenge by [2]. The nodes and times (in milliseconds) for [2] are for finding the best solution they can. The times for our method are for the full solve including proof of optimality (using all improvements).

4.2 Harder Random Instances

Of the 5000+ instances used in the 2005 challenge, only SP2, SP3 and SP4 take longer than a few milliseconds for our solver to solve. Thus we generate some difficult random instances for this experiment. First we specify the number of customers, number of products and the average number of customers per product. We then calculate a density that will achieve the specified average number of customers per product. The customer vs product table is then randomly generated using the calculated density to determine when to put 1's and 0's. As a post condition, we throw away any instance where the customer graph can be decomposed into separate components. This is done because we want to compare on instances of a certain size, but if the customer graph decomposes, then the instance degenerates into a number of smaller and relatively trivial instances.

We generate 5 instances for each of the sizes 30×30, 40×40, 50×50, 75×75, 100×100, 125×125, 100×50, 50×100, and average number of customer per product values of 2, 4, 6, 8, 10, for a total of 200 instances.

Ideally, we want to measure speedup by comparing total solve time. However, as mentioned before, the instances that our solver finds challenging are totally unsolvable by the original. Table 3 is split into two parts. Above the horizontal line are the instances where the original managed to prove optimality. Here, nodes, time (in milliseconds) and speedup are for the total solve. Below the line the original cannot prove optimality. Here, nodes, time and speedup are for the finding a solution that is at least as good as the solver of [2] could find. The column δOpt shows the average distance this solution is from the optimal. Note that our approach finds and proves the optimal in all cases although the statistics for this are not shown in the table. Time to find an equally good solution is not necessarily a good indication of the speedup achievable for the full solve, as other factors like branching heuristics come into play. However, the trend is quite clear. The original solver is run with its best options. Our MOSP algorithm is run with "better move", "old move" and nogood recording turned on (but no relaxation). Both solvers have a node limit of 2^{25} iterations.

Note that because a single move can close multiple stacks, it is possible to completely solve an instance using fewer nodes than there are customers. This occurs frequently in the high density instances. Thus the extremely low node counts shown here are not errors. The speedup is around 2-3 orders of magnitude for the smallest problems (30×30), and around 5-6 orders of magnitude for the hardest problems that the original version can solve (40×40). The speedup appears to grow exponentially with problem size. We cannot get any speedup numbers for the harder instances since the original cannot solve them. However, given the trend in the speedup, it would not be surprising if the speedup for a full solve on the hardest instances solvable by our solver (100×100) was in the realms of 10^{10} or more.

Table 3. Comparing customer search versus [2] on harder random instances. Search is to find the best solution found by [2] with node limit 2^{25}.

Instance	δOpt	[2] Nodes	[2] Time(ms)	This paper Nodes	This paper Time(ms)	Speedup
30-30-2	0	14318	480	408	4.4	109
30-30-4	0	48232	1981	158	2.0	979
30-30-6	0	89084	2750	56	0.9	3136
30-30-8	0	83558	2010	18	0.4	5322
30-30-10	0	18662	506	8	0.2	2335
40-40-2	0	669384	192917	1472	14.9	12942
40-40-4	0	3819542	227087	556	6.1	36959
40-40-6	0	11343235	625892	217	2.9	218062
40-40-8	0	8379706	334392	49	0.8	403272
40-40-10	0	3040040	98194	20	0.4	229305
50-50-2	0	12356205	1300311	6344	65.8	19758
50-50-4	0.2	5612259	446409	219	2.7	164728
50-50-6	0.2	7949026	510831	45	0.8	636409
50-50-8	0.2	525741	28337	15	0.4	75274
50-50-10	0	16809	784	7	0.2	3411
75-75-2	0.8	2485310	420935	7030	76.8	5484
75-75-4	2.6	3507703	666784	63	1.4	486669
75-75-6	1.2	4412286	756032	59	1.4	548132
75-75-8	1.2	4121046	519778	19	0.6	841336
75-75-10	0.6	1198282	120087	15	0.5	244128
100-100-2	2.2	3008009	765131	481	9.4	81653
100-100-4	4.8	6777140	2017286	145	3.3	619257
100-100-6	4	1269071	347970	39	1.4	241145
100-100-8	4.4	1686045	414456	31	1.1	363468
100-100-10	1.6	4195494	789039	15	0.7	1097494
125-125-2	1.8	7418402	3276210	36672	436.8	7500
125-125-4	3.8	3412379	1559691	916	20.4	76286
125-125-6	6	6076996	2643707	57	2.3	1144180
125-125-8	6.2	942290	321050	28	1.5	217007
125-125-10	3.4	170852	45798	24	1.3	35647
50-100-2	0.2	90076	9971	97	1.6	6290
50-100-4	1	1973322	139300	23	0.6	220776
50-100-6	0.6	1784	116	13	0.4	301
50-100-8	0	97	9	5	0.2	47
50-100-10	0	99	8	3	0.2	39
100-50-2	0	2393401	438220	11117	133.7	3279
100-50-4	0.4	14211006	3499389	183260	1592.3	2198
100-50-6	1.2	5326088	1395417	1569	21.4	65163
100-50-8	0.6	1522796	408908	3506	45.8	8932
100-50-10	1	3594743	906559	524	10.6	85710

Next we examine the effect of each of our improvements individually by disabling them one at a time. The three improvements we test here are "better move", "old move" and nogood recording. We use only the instances which are

Table 4. (a) Comparing the effects of each optimisation in turn, and (b) comparing the effects of relaxation

	(a)			(b)			
Instance	Better	Old	Nogood	No relax(ms)	Relax(ms)	Removed	Speedup
100-100-2	25.2	63.7	1.21	603120	370	51.2	1630.6
100-100-4	8.94	5.95	1.01	266205	4798	20.8	55.5
100-100-6	1.90	1.71	0.91	10344	3909	8	2.6
100-100-8	1.54	1.30	0.66	551	712	2.4	0.8
100-100-10	2.96	2.75	1.00	46	94	0.6	0.5
125-125-2	—	—	—	59642993	3284	62.2	18161.8
125-125-4	—	—	—	29768563	251634	24.8	118.3
125-125-6	11.9	3.10	0.98	810678	167384	9	4.8
125-125-8	1.65	1.17	0.98	18781	11978	5	1.6
125-125-10	1.27	0.98	0.64	768	1041	3	0.7

solvable without the improvements and non-trivial, i.e. the 100×100 instances and the easier 125×125 instances. For each improvement, we show the relative slowdown compared to the version with all three optimisations on.

As Table 4(a) shows, both "better move" and "old move" can produce up to 1 to 2 orders of magnitude speedup on the harder instances. The lower speedups are from instances that are already fairly easy and solvable in seconds. The results from disabling the nogood recording are very interesting. It is known from previous work, e.g. the DP approach of [2] and the CP approach of [5] that nogood recording or equivalent techniques produce several orders of magnitude speedup. However, these approaches require (in the worst case) exponential memory usage for the nogood table. It appears however that once we have the "old move" improvement, we can actually turn off nogood recording without a significant loss of performance. In fact, some instances run faster. Thus our "old move" improvement largely subsumes the effect of the huge nogood tables used in the DP [2] or CP [5] approaches and reduces the memory usage from an exponential to a linear amount. The solver of [2] uses up all 2Gb of main memory in ~5 min with nogood recording. However, our new solver using "old move" pruning uses a constant amount of memory < 2Mb even for $125{\times}125$ problems.

4.3 Relaxation

In the following set of experiments, we demonstrate the effectiveness of our relaxation technique. For each of our largest instances, we show in Table 4(b) the total runtime (in milliseconds) without relaxation, with relaxation, and the number of customers that was successfully removed without changing the lower bound, as well as the speedup for relaxation. Both versions are run with the customer search strategy, "better move" and "old move" improvements.

As can be seen from the results in Table 4(b), relaxation is most effective for sparse instances where we can get up to 3-4 orders of magnitude improvement. There is a slight slowdown for several dense instances but that is because they are trivial to begin with (take < 1s). The sparser the instance, the more customers can be removed without changing the lower bound and the greater the speedup from the relaxation technique. For the hardest instances, 125-125-2, it is often possible to remove some 60-70 of the 125 customers without changing the bound. This reduces the proof of optimality that normally takes 10+ hours into mere seconds. The 125-125-4 instances are now comparatively harder, since we are only able to remove around 25 customers and get a speedup of ~100. Relaxation is largely ineffective for the denser instances like 125-125-8,10. However, dense instances are naturally much easier to solve anyway, so we have speedup where it is needed the most.

Our relaxation techniques are also useful if we only wish to prove a good lower bound rather than the true lower bound. For example, if we only insist on proving a lower bound that is 5 less than the true optimum, then ~45 customers can be removed from the 125-125-4 instances and the bound can be proved in seconds. This is again several orders of magnitude speedup compared to using a normal complete search to prove such a bound. In comparison, although the HAC lower bound heuristic of [1] uses virtually no time, it gives extremely weak lower bounds for the 125-125-4 instances, which are some 30 stacks below the optimum and are of little use.

5 Conclusion

In this paper we show how combining nogood recording with the customer based search strategy of [8] yields a solver that is 5-6 orders of magnitude faster than the previous state of the art MOSP solver. We show how exploiting several dominance relations leads to the the "definite move", "better move" and "old move" improvements. These produce a further 1-2 orders of magnitude improvement. The "old move" improvement in particular is able to subsume the effect of pruning using the extremely large nogood table. This allows us to reduce the memory usage of our solver from an amount exponential in the size of the problem to a constant ~2Mb. Finally we show how relaxation techniques can be used to speed up the proof of optimality of the hardest instances by another 3-4 orders of magnitude.

Acknowledgments. NICTA is funded by the Australian Government as represented by the Department of Broadband, Communications and the Digital Economy and the Australian Research Council.

References

1. Becceneri, J.C., Yannasse, H.H., Soma, N.Y.: A method for solving the minimization of the maximum number of open stacks problem within a cutting process. Computers & Operations Research 31, 2315–2332 (2004)
2. Garcia de la Banda, M., Stuckey, P.J.: Dynamic programming to minimize the maximum number of open stacks. INFORMS JOC 19(4), 607–617 (2007)
3. Linhares, A., Yanasse, H.H.: Connections between cutting-pattern sequencing, VLSI design, and flexible machines. Computers & Operations Research 29, 1759–1772 (2002)

4. Prestwich, S.: Increasing solution density by dominated relaxation. In: 4th Int. Workshop on Modelling and Reformulating Constraint Satisfaction Problems (2005)
5. Shaw, P., Laborie, P.: A constraint programming approach to the min-stack problem. Constraint Modelling Challenge 2005 [6]
6. Smith, B., Gent, I.: Constraint modelling challenge report (2005), `http://www.cs.st-andrews.ac.uk/~ipg/challenge/ModelChallenge05.pdf`
7. Wilson, N., Petrie, K.: Using customer elimination orderings to minimise the maximum number of open stacks. In: Constraint Modelling Challenge 2005 [6]
8. Yannasse, H.H.: On a pattern sequencing problem to minimize the maximum number of open stacks. EJOR 100, 454–463 (1997)
9. Yannasse, H.H.: A note on generating solutions of a pattern sequencing problem to minimize the maximum number of open orders. Technical Report LAC-002/98, INPE, São José dos Campos, SP, Brazil (1998)
10. Yuen, B.J., Richardson, K.V.: Establishing the optimality of sequencing heuristics for cutting stock problems. EJOR 84, 590–598 (1995)

Using Relaxations in Maximum Density Still Life

Geoffrey Chu[1], Peter J. Stuckey[1], and Maria Garcia de la Banda[2]

[1] NICTA Victoria Laboratory,
Department of Computer Science and Software Engineering,
University of Melbourne, Australia
{gchu,pjs}@csse.unimelb.edu.au
[2] Faculty of Information Technology,
Monash University, Australia
mbanda@infotech.monash.edu.au

Abstract. The Maximum Density Sill-Life Problem is to fill an $n \times n$ board of cells with the maximum number of live cells so that the board is stable under the rules of Conway's Game of Life. We reformulate the problem into one of minimising "wastage" rather than maximising the number of live cells. This reformulation allows us to compute strong upper bounds on the number of live cells. By combining this reformulation with several relaxation techniques, as well as exploiting symmetries via caching, we are able to find close to optimal solutions up to size $n = 100$, and optimal solutions for instances as large as $n = 69$. The best previous method could only find optimal solutions up to $n = 20$.

1 Introduction

The *Game of Life* was invented by John Horton Conway and is played on an infinite board. Each cell c in the board is either alive or dead at time t. The live/dead state at time $t+1$ of cell c, denoted as $state(c, t+1)$, can be obtained from the number l of live neighbours of c at time t and from $state(c, t)$ as follows:

$$state(c,t+1) = \begin{cases} l < 2 & \text{dead} & \text{[Death by isolation]} \\ l = 2 & state(c,t) & \text{[Stable condition]} \\ l = 3 & \text{alive} & \text{[Birth condition]} \\ l > 3 & \text{dead} & \text{[Death by overcrowding]} \end{cases}$$

The board is said to be a *still-life* at time t if it is unchanged by these rules, i.e., it is the same at $t+1$. For example, an empty board is a still-life. Given a finite $n \times n$ region where all other cells are dead, the *Maximum Density Still-life Problem* aims at computing the highest number of live cells that can appear in a still life for the region. The density is thus expressed as the number of live cells over the $n \times n$ region.

The raw search space of the Maximum Density Still-life Problem has size 2^{n^2}. Thus, it is extremely difficult even for "small" values of n. Previous search methods using IP [1] and CP [2] could only solve up to $n = 9$, while a CP/IP hybrid method with symmetry breaking [2] could solve up to $n = 15$. An attempt using

I.P. Gent (Ed.): CP 2009, LNCS 5732, pp. 258–273, 2009.

bucket elimination [6] reduced the time complexity to $O(n^2 2^{3n})$ but increased the space complexity to $O(n2^{2n})$. This method could solve up to $n = 14$ before it ran out of memory. A subsequent improvement that combined bucket elimination with search [7], used less memory and was able to solve up to $n = 20$. In this paper we combine various techniques to allow us to solve instances up to $n = 50$ completely or almost completely, i.e. we prove upper bounds and find solutions that either achieve that upper bound or only have 1 less live cell. We also obtain solutions that are no more than 4 live cells away from our proven upper bound all the way to $n = 100$. The largest completely solved instance is $n = 69$.

The contributions of this paper are as follows:

- We give a new insightful proof that the maximum density of live cells in an infinite still life is $\frac{1}{2}$. This proof allows us to reformulate the maximum density still-life problem in terms of minimising "wastage" rather than maximising the number of live cells.
- We derive tight lower bounds on wastage (which translate into upper bounds on the number of live cells) that can be used for pruning.
- We define a static relaxation of the original problem that allows us to calculate closed form equations for an upper bound on live cells for all n. And we do this in constant time by using CP with caching. We conjecture that this upper bound is either equal to the optimum, or only 1 higher for all n.
- We identify a subset of cases for which we can improve the upper bound by 1 by performing a complete search on center perfect solutions. This completes the proof of optimality for the instances where our heuristic search was able to find a solution with 1 cell less than the initial upper bound.
- We define a heuristic incomplete search using dynamic relaxation as a lookahead method that can find optimal or near optimal solutions all the way up to $n = 100$.
- We find optimal solutions for n as large as 69, more than 3 times larger than any previous methods, and with a raw search space of 2^{4761}.

2 Wastage Reformulation

The maximum density of live cells in an infinite still life is known to be $\frac{1}{2}$ [5,4]. However, that proof is quite complex and only applies to the infinite plane. In this section we provide a much simpler proof that can easily be extended to the bounded case and gives much better insight into the possible sub-patterns that can occur in an optimal solution. The proof is as follows.

First we assign an area of 2 to each cell (assume the side length is $\sqrt{2}$). We will partition the area of each dead cell into two 1 area pieces and distribute them among its live neighbours according to the local pattern found around the dead cell. It is clear that if we can prove that all live cells on the board end up with an area ≥ 4, then it follows that the density of live cells is $\leq \frac{1}{2}$.

The area of a dead cell is assigned only to orthogonal neighbouring live cells, i.e., those that share an edge with the dead cell. We describe as "wastage" the

Table 1. Possible patterns around dead cells, showing where they donate their area and any wastage of area

Pattern:

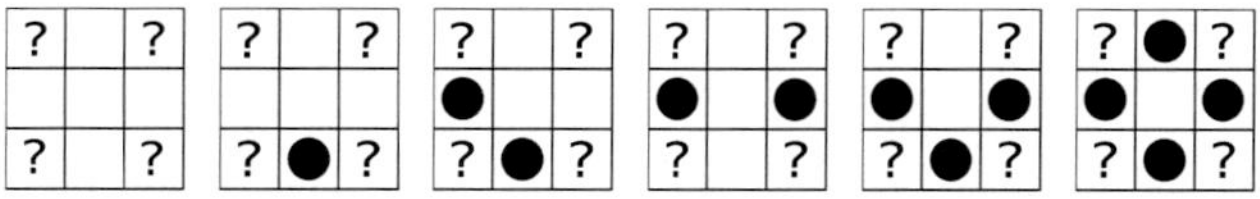

Beneficiaries	{}	{ S }	{S, W}	{E, W}	{E, W}	{}
Wastage	2	1	0	0	0	2

Table 2. Contributions to the area of a live cell from its South neighbour

Pattern:

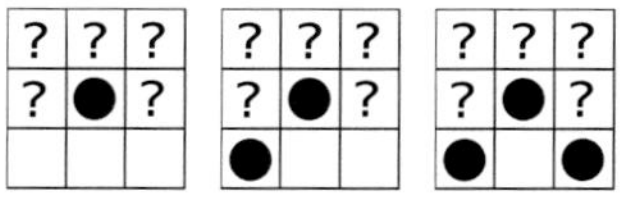

Area received	1	1	0

area from a dead cell that does not need to be assigned to any live cell for the proof to work (i.e., it is not needed to reach our 4 target for live cells). Table 1 shows all possible patterns of orthogonal neighbours (up to symmetries) around a dead cell. Live cells are marked with a black dot, dead cells are unmarked, and cells whose state is irrelevant for our purposes are marked with a "?".

Each pattern indicates the beneficiaries, i.e., the North, East, South or West neighbours that receive 1 area from the center dead cell, and the resulting amount of wastage. As it can be seen from the table, a dead cell gives 1 area to each of its live orthogonal neighbours if it has ≤ 2 live orthogonal neighbours, 1 area to the two opposing live orthogonal neighbours if it has 3, and no area if it has 4. Note that, in the table, wastage occurs whenever the dead cell has ≤ 1 or 4 live orthogonal neighbours. As a result, we only need to examine the 3 bordering cells on each side of a live cell, to determine how much area is obtained from the orthogonal neighbour on that side. For example, the area obtained by the central live cell from its South neighbour is illustrated in Table 2.

The area obtained by a live cell can therefore be computed by simply adding up the area obtained from its four orthogonal neighbours. Since each live cell starts off with an area of 2, it must receive at least 2 extra area to end up with an area that is ≥ 4. Let us then look at all possible patterns around a live cell and see where the cell will receive area from. Table 3 shows all possible neighbourhoods of a live cell (up to symmetries). For each pattern, it shows the benefactors, i.e., the North, East, South or West neighbours that give 1 area to the live cell, and the resulting amount of wastage, which occurs whenever a live cell receives more than 2.

Note that the last pattern does not receive sufficient extra area, just 1 from the South neighbour. However, the last two patterns always occur together in unique pairs due to the still-life constraints (each of the central live cells has 3 neighbours so the row above the last pattern, and the row below the second last

Table 3. Possible patterns around a live cell showing area benefactors and any wastage

Pattern: 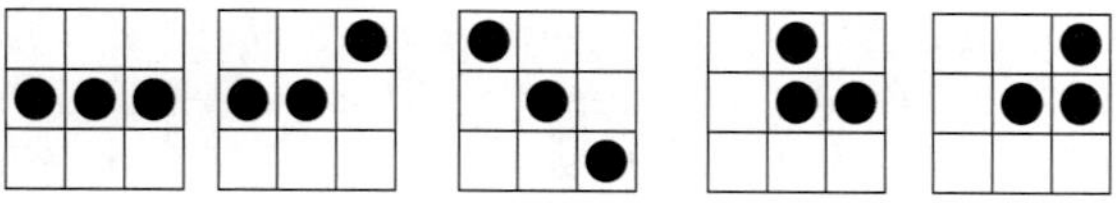

Benefactors:	{N,S}	{N,E,S}	{N,E,S,W}	{S,W}	{N,S,W}
Wastage:	0	1	2	0	1

Pattern: 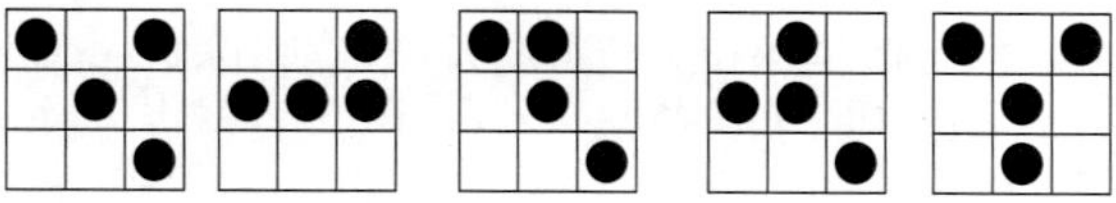

Benefactors:	{S,W}	{N,S}	{E,S,W}	{E,S}	{E,W}
Wastage:	0	0	1	0	0

Pattern: 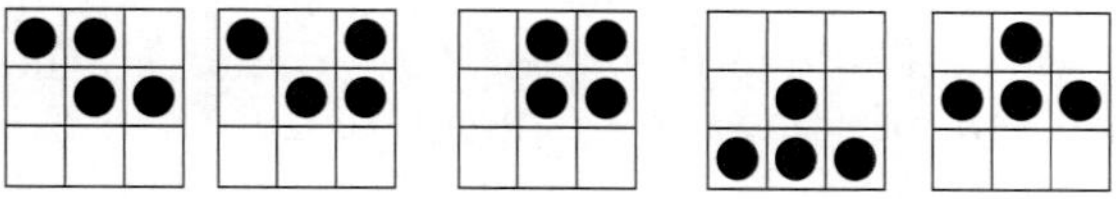

Benefactors:	{S,W}	{S,W}	{S,W}	{N,E,W}	{S}
Wastage:	0	0	0	1	−1

pattern must only consist of dead cells). Hence, we can transfer the extra 1 area from the second last pattern to the last.

Clearly, all live cells end up with an area ≥ 4, and this completes our proof that the maximum density on an infinite board is $\frac{1}{2}$.

The above proof is not only much simpler than that of [5,4], it also provides us with good insight into how to compute useful bounds for the case in which the board is finite. In particular, it allows us to know exactly how much we have lost from the theoretical maximum density by looking at the amount of "wastage" produced by the patterns in the currently labeled cells.

To achieve this, we reformulate the objective function in the Bounded Maximum Density Still Life Problem as follows. For each cell c, let $P(c)$ be the 3×3 pattern around that cell. Note that if c is on the edge of the $n \times n$ region, the dead cells beyond the edge are also included in this pattern. Let $w(P)$ be the wastage for each 3×3 pattern as listed in Tables 1 and 3. Define $w(c)$ for each cell c as follows. If c is within the $n \times n$ region, then $w(c) = w(P(c))$. If c is in the row immediately beyond the $n \times n$ region and shares an edge with it (there are $4n$ such cells), then $w(c) = 1$ if the cell in the $n \times n$ region with which it shares an edge is dead, and $w(c) = 0$ otherwise. For all other c, let $w(c) = 0$. Let $W = \sum w(c)$ over all cells.

Theorem 1. *Wastage and live cells are related by*

$$live\_cells = \frac{n^2}{2} + n - \frac{W}{4} \quad (1)$$

Proof. We adapt the proof for the infinite board to the $n \times n$ region. Let us assign 2 area to each cell within the $n \times n$ region, and 1 area to each of the $4n$ cell in the row immediately beyond the edge of the $n \times n$ region. Now, for each dead cell within the $n \times n$ region, partition the area among its live orthogonal neighbours as before. For each dead cell in the row immediately beyond the $n \times n$ region, give its 1 area to the cell in the $n \times n$ region with which it shares an edge. Again, since the last two 3×3 patterns listed above must occur in pairs, we transfer an extra 1 area from one to the other. Note also that the second last pattern of Table 3 cannot appear on the South border (which would mean that the last pattern appeared outside the shape) since it is not stable in this position. Clearly, after the transfers, all live cells once again have ≥ 4 area, and wastage for the 3×3 patterns centered around cells within the $n \times n$ region remain the same. However, since we are in the finite case, we also have wastage for the cells which are in the row immediately beyond the edge of the $n \times n$ region. These dead cells always give 1 area to the neighbouring cell which is in the $n \times n$ region. If that cell is live, the area is received. If that cell is dead, that 1 area is wasted. The reformulation above counts all these wastage as follows. The total amount of area that was used was $2n^2$ from the cells within the $n \times n$ region and $4n$ from the $4n$ cells in the row immediately beyond the edge, for a total of $2n^2 + 4n$. Now, 4 times live cells will be equal to the total area minus all the area wasted, and hence we end up with Equation (1). □

We can trivially derive some upper bounds on the number of live cells using this equation. Clearly $W \geq 0$ and, thus, we have live cells $\leq \lfloor \frac{n^2}{2} + n \rfloor$. Also, by the still life constraints, there cannot be three consecutive live cells along the edge of the $n \times n$ region. Hence, there is always at least 1 wastage per 3 cells along the edge and we can improve the bound to live cells $\leq \lfloor \frac{n^2}{2} + n - \lfloor \frac{1}{3} n \rfloor \rfloor$. While this bound is very close to the optimal value for a small n, it differs from the true optimum by $O(n)$ and will diverge from the optimum for a large n. We provide a better bound in the next section.

3 Closed form Upper Bound

Although in the infinite case there are many patterns that can achieve exactly $\frac{1}{2}$ density, it turns out that in the bounded case, the boundary constraints force significant extra wastage. As explained in the previous section, the still life constraints on the edge of the $n \times n$ region trivially force at least 1 wastage per 3 edge cells, however, it can be shown by exhaustive search that even this theoretical minimal wastage of $1/3$ per edge cell is unattainable for an infinitely long edge due to the still life constraints on the inside of the $n \times n$ region.

There is also no way to label the corner without producing some extra wastage. For example, for a 6×6 corner, naively we expect to have $(6+6)/3 = 4$ wastage forced by the still life constraints on the boundary. However, due to the still life constraints within the corner, there is actually no way to label a 6×6 corner without at least 6 wastage.

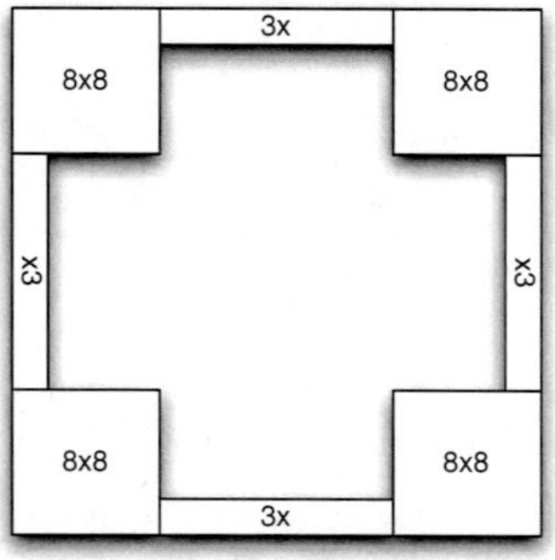

Fig. 1. The relaxed version of the problem, only filling in the 8×8 corners and the 3 rows around the edge

Examination of the optimal solutions found by other authors show that, in all instances, all or almost all wastage is found either in the corners or within 3 rows from the edge. This leads to the following conjecture:

Conjecture 1. All wastage forced by the boundary constraints of an $n \times n$ region must appear either in the 8×8 corners, or within 3 rows from the edge.

If this conjecture is true, then it should be possible to find a good lower bound on the amount of forced wastage simply by examining the corners and the first few rows from the edge.

We thus perform the following relaxation of the bounded still life problem. We keep only the variables representing the four 8×8 corners of the board, as well as the variables representing cells within 3 rows of the edge (see Figure 1). All variables in the middle are removed. The still life constraints for fully surrounded cells, including cells on the edge of the $n \times n$ region, remain the same. The still life constraints for cells neighbouring on removed cells are relaxed as follows: dead cells are always considered to be consistent with the problem constraints, and live cells are considered to be consistent as long as they do not have more than 3 live neighbours (otherwise any extension to the removed cells would violate the original constraints). The objective function is modified as follows: fully surrounded cells have their wastage counted as before, live cells with neighbouring removed cells have no wastage counted, and dead cells with neighbouring removed cells have 1 wastage counted if and only if it is surrounded by ≥ 3 dead unremoved cells (since any extension to the removed cells will result in a pattern with ≥ 1 wastage). Clearly, since we have relaxed the constraints, and also potentially ignored some wastage in our count, any lower bound we get on wastage in this relaxed problem is a valid lower bound on the wastage in the original problem.

Since the constraint graph of this relaxed problem has bounded, very low path-width, it can easily be solved using CP with caching in $O(n)$ time (see [3] Theorem 13.2). In practice, solving the 8×8 corners is the hardest and takes $8s$. An example corner solution is shown in Figure 2(a). Solving the width 3

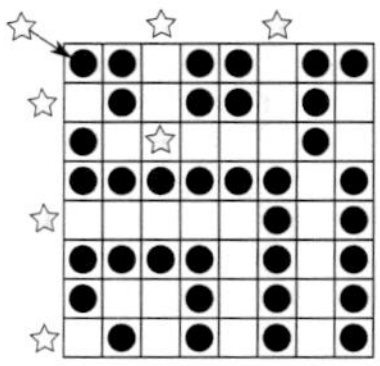

(a)

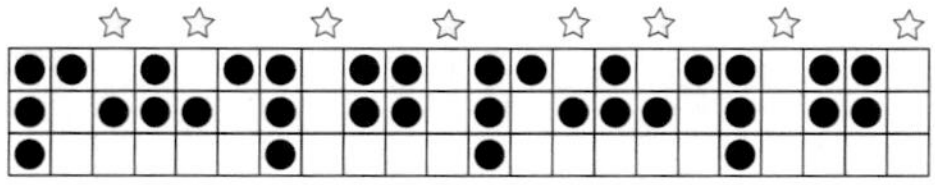

(b)

Fig. 2. An optimal 8×8 North West corner pattern with 7 wastage highlighted, and (b) the periodic pattern for an optimal North edge with 4 wastage per period 11 highlighted

Table 4. Upper bound by relaxation for small n, shown in bold if it is equal to the optimal solution

n	optimal	upper bound
8	36	**36**
9	43	44
10	54	55
11	64	65
12	76	77
13	90	91
14	104	**104**
15	119	120
16	136	**136**
17	152	**152**
18	171	172
19	190	**190**
20	210	**210**

edge takes milliseconds. The results of calculating the bound from the relaxed problem for small n is shown in Table 4.

Because of the high symmetry of the edge subproblem (full translational symmetry), the edge bounds starts to take on a periodic pattern for n sufficiently large, at which point we can derive a closed form equation for their values for all n. The periodicity comes from the fact that it can be shown that the optimal periodic edge pattern (see Figure 2(b)) has period 11, and any sufficiently long optimal edge pattern will have a series of these in the middle.

Since the set of optimal solutions for the 8× 8 corners remain the same for any $n > 16$, and we can derive a closed form equation for the edge bounds for large n, we can calculate a closed form equation for the lower bound on wastage for the whole relaxed problem for any n sufficiently large. For $n \geq 50$ it is:

$$min\_wastage = \begin{cases} 8 + 16 \times \lfloor n/11 \rfloor, n \equiv 0, 1, 2 & \mod 11 \\ 12 + 16 \times \lfloor n/11 \rfloor, n \equiv 3, 4 & \mod 11 \\ 14 + 16 \times \lfloor n/11 \rfloor, n \equiv 5 & \mod 11 \\ 16 + 16 \times \lfloor n/11 \rfloor, n \equiv 6, 7 & \mod 11 \\ 18 + 16 \times \lfloor n/11 \rfloor, n \equiv 8 & \mod 11 \\ 20 + 16 \times \lfloor n/11 \rfloor, n \equiv 9, 10 & \mod 11 \end{cases} \quad (2)$$

A lower bound on wastage can be converted into an upper bound on live cells using Equation (1):

$$live\_cells \leq \left\lfloor \frac{2n^2 + 4n - min\_wastage}{4} \right\rfloor \tag{3}$$

We will call the live cell upper bound calculated from our closed form wastage lower bounds the "closed form upper bound". If Conjecture 1 is true and all forced wastage must appear within a few rows of the edge, then this closed form upper bound should be extremely close to the real optimal value.

Conjecture 2. The maximum number of live cells in an $n \times n$ region is

$$\left\lfloor \frac{2n^2 + 4n - min\_wastage}{4} \right\rfloor$$

or one less than this, where *min_wastage* is the optimal solution to the relaxed problem of Figure 1 (given by Equation (2) for $n \geq 50$). □

Conjecture 2 is true for previously solved n. As it can be seen from Table 4, for $n \leq 20$, our upper bound is never off the true optimum by more than 1, and is often equal to it.

As our new results in Table 5 show, our conjecture is also true for at least up to $n = 50$. The bound is also achievable exactly for n as high as 69. For larger n our solver is too weak to guarantee finding the optimal solution and thus we cannot verify the conjecture. We believe the conjecture is true because although not all 3×3 patterns are perfect (waste free), there are a large number of perfect ones and there easily appears to be enough different combinations of them to label the center of an $n \times n$ region perfectly. Indeed, since we already know that there are many ways to label an infinite board perfectly, the boundary constraints are the only constraints that can force wastage.

4 Finding Optimal Solutions

In the previous section, we found a relaxation that allows us to find very good upper bounds in constant time. However, in order to find an optimal solution it is still necessary to tackle the size 2^{n^2} search space. The previous best search based methods could only find optimal solutions up to $n = 15$, with a search space of 2^{225}. Here, we attempt to find solutions up to $n = 100$, which has a search space of 2^{10000}, a matter of 3000 orders of magnitude difference. Furthermore, optimal solutions are extremely rare: 1682 out of 2^{169} for $n = 13$, 11 out of 2^{196} for $n = 14$, and so on [7]. Clearly, we are going to need some extremely powerful pruning techniques.

For some values of n, the closed form upper bound calculated in the previous section is already the true upper bound. For such instances we simply need to find a solution that achieves the upper bound and, therefore, we can use an incomplete search. We take advantage of this by looking only for solutions of a particular form: those where all wastage lies in the 8×8 corners or within 3 rows of the edge, and the "center" is labeled perfectly with no wastage whatsoever. We call such solutions "center perfect". Our choice is motivated by the fact that, while there are many ways to label the center perfectly, there are very few ways to label the edge with the minimum amount of wastage. Since we are only allowed a very limited number of wastage on the entire board, they should not be used to deal with the center cells. Searching only for center perfect solutions, allows us to implement our solver more simply and to considerably reduce the search space.

4.1 Dynamic Relaxation as Lookahead

The main technique whereby we make solution finding feasible is through the use of dynamic relaxation as a lookahead technique. We label the board k rows at a time (call each set of k rows a "super-row"), and column by column within each super-row. Our implementation can set the width of the super row anywhere between 4 and 8, although our experimental evaluation has not shown any obvious difference in performance. At each node in our search tree, we perform a relaxation of the current subproblem, which we solve exactly as a lookahead. If we cannot find a (good enough) solution to the lookahead problem we fail and try another labelling.

The relaxation consists of all the unlabeled variables within k rows from the currently labeled variables, a 16×8 "thick edge" chunk of variables on each side, a width 3 edge down each side, the bottom two 8×8 corners, and the bottom width 3 edge (see Figure 3(a)). It also includes the 2 rows of already labeled variables immediately bordering on the unlabeled region. The values of these labeled variables form boundary conditions for the relaxed problem at this particular node. We will discuss the choice of this shape later on. We relax the still life constraints and objective function as before.

The relaxed subproblem is an optimisation problem. If by solving the relaxed subproblem, we find that the minimum amount of wastage among the unlabeled variables, plus the amount of wastage already found in the labeled variables, exceeds the wastage upper bound for the original problem, then clearly there are no solutions in the current branch and the node can be pruned. Since the constraint graph of the relaxed problem has low path-width, it can be solved in linear time using CP with caching. In practice though, we exploit symmetries and caching to solve it incrementally in constant time (see below). Lower bounds calculated for particular groups of variables (such as corners, edges, and super-rows) are cached. These can be reused for other subproblems. We also use these cached results as a table propagator, as they can immediately tell us which particular value choices will lead to wastage lower bounds that violate the current upper bound.

The relaxed problem can be solved incrementally in constant time as follows. When we travel to a child node, the new relaxed problem at that node is not that

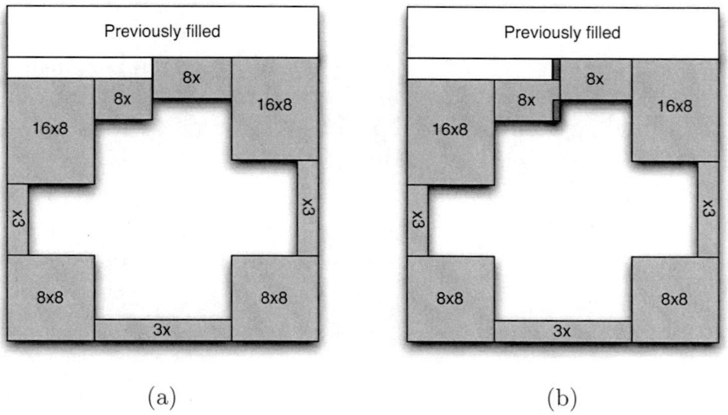

Fig. 3. Dynamic relaxation lookahead for still-life search: (a) shows darkened the area of lookahead, (b) shows the change in lookahead on labelling an additional column when super-row width is 4 and lookahead width is 8

different from the one at the previous node (see Figure 3(a) and (b), respectively). A column of k variables from the previous relaxation has now been removed from the relaxation and labeled (new top dark column in Figure 3(b)), and a new column of k variables is added to the relaxation (new bottom dark column in Figure 3(b)). By caching previous results appropriately, it is possible to lookup all the solutions to the common part of the two relaxed problems (cached when solving the previous relaxation). Now we simply need to extend those solutions to the new column of k variables, which takes $O(2^k)$ time but is constant for a fixed k. There is sufficient time/memory to choose a k of up to 12. The larger k is, the greater the pruning achieved but the more expensive the lookahead is. In practice, $k = 8$ seems to work well. With $k = 8$, we are able to search around 1000 nodes per second and obtain a complete depth 400+ lookahead in constant time.

The choice of the shape of the lookahead is important, and is based on our conjectures as well as on extensive experimentation. The most important variables to perform lookahead on are the variables where wastage is most likely to be forced by the current boundary constraints. Conjecture 1 can be applied to our relaxed subproblems as well. It is very likely that the wastage forced by the boundary constraints can be found within a fixed number of rows from the current boundary. Thus, the variables near the boundary are the ones we should perform lookahead on. The reason for the 16×8 thick edge chunk of variables comes from our experimentation. It was found that the lookahead is by far the weakest around the "corners" of the subproblem, where we have boundary constraints on two sides. A lookahead using a thinner edge (initially width 3) often failed to see the forced wastage further in and, thus, allowed the solver to get permanently stuck around those "corners". By using a thicker edge, we increase the ability of the lookahead to predict forced wastage and thus we get stuck less often. A 16×8 thick edge is still not sufficiently large to catch all forced wastage, but we are unable to make it any larger due to memory constraints, as the memory requirement is exponential in the width of the lookahead.

Interestingly, if Conjecture 1 is indeed true for the subproblems, and we can get around the memory problems and perform a lookahead with a sufficient width to catch all forced wastage, then the lookahead should approach the strength of an oracle, and it may be possible to find optimal solutions in roughly polynomial time! Indeed, from our results (see Table 5), the run time required to find optimal solutions does not appear to be growing anywhere near exponentially in n^2.

4.2 Search Strategy

The search strategy is also very important. A traditional depth first branch and bound strategy is doomed to failure, as it will tend to greedily use up the wastage allowance to get around any problem it encounters. If it maxes out on the wastage bound early on, then it is extremely unlikely that it will be able to label the rest of the board without breaking the bound. However, we may not be able to tell this until much later on, since the center can often be labeled perfectly. Therefore, the search will reach extremely deep parts of the search tree, even though there is no chance of success and we will essentially be stuck forever. Instead, we need a smarter search strategy that has more foresight.

Intuitively, although labeling the top parts of the board optimally is difficult, it is much easier than labeling the final part of the board, since the final part of the board will have boundary constraints on every side, making it extremely difficult to get a perfect labeling with minimal wastage. Thus, we would like to save all our wastage allowance until the end. We accomplish this by splitting the search into two phases. The first phase consists of labeling all parts of the board other than the last 8 rows, and the second phase consists of the last 8 rows. In the first phase, we use a variation of limited discrepancy search (LDS), designed to preserve our wastage allowance as much as possible. We describe this in the next paragraph. If we manage to get to the last 8 rows, then we switch to a normal depth first search where the wastage allowance is used as necessary to finish off labeling the last few rows.

Our LDS-like algorithm is as follows. We define the discrepancy as the amount by which a value choice causes our wastage lower bound to increase. Note that this is totally different from defining it as the amount of wastage caused by a value choice. For example, if our lookahead tells us that a wastage of 10 is unavoidable among the unlabeled variables, and we choose a value with 1 wastage, after which our lower bound for the rest of the unlabeled variables is 9, then there is no discrepancy. This is because even though we chose a value that caused wastage, the wastage was forced. We are thus still labeling optimally (as far as our lookahead can tell). On the other hand, if the lookahead tells us that 10 wastage is unavoidable, and we choose a value with 0 wastage, after which our lower bound becomes 11, then there is a discrepancy. This is because, even though we did not cause immediate wastage, we have in fact labeled sub-optimally as this greedy choice causes more wastage later on. Discrepancy based on increases in wastage lower bounds is far better than one based on immediate wastage, as greedily avoiding wastage often leads to substantially more wastage later on! Note that by definition, there can be multiple values at each node with the same discrepancy, all of which should be searched at that discrepancy level.

Our search also differs from traditional LDS in the way that discrepancies are searched. Traditional LDS assumes that the branching heuristic is weak at the top of the tree, and therefore tries discrepancies at the top of the tree first. This involves a lot of backtracking to the top of the tree and is extremely inefficient for our case, since the branching heuristic for still life is not weak at the top of the tree and our trees are very deep. Hence, we order the search to try discrepancies at the leaves first, which makes the search much closer to depth-first search. This is substantially more efficient and is in fact crucial for solving our relaxations incrementally. We also modify LDS by adding local restarts. This is based on the fact that, for problems with as large a search space as this, complete search methods are doomed to failure, as mistakes require an exponential time to fix. Instead, we do incomplete LDS by performing local restarts at random intervals dependant on problem size, during which we backtrack by a randomised amount which averages to two super-rows. Value choices with the same number of discrepancies are randomly reordered each time a node is generated, so when a node is re-examined, the solver can take another path. Given the size of the search space, we do not wait until all nodes with the current discrepancy are searched before we try a higher discrepancy. Instead, after a reasonable amount of time (around 20 min) we give up and increase the discrepancy. Essentially, what we are doing is to try to use as little of our wastage at the top as possible and save more for the end. But if it feels improbable that we can label the top with so little wastage, then we use more of our allowance for the top.

5 Improving the Upper Bound

The incomplete search described in the previous section does not always find a solution equal to the closed form upper bound in the allocated time. For some of the cases in which this happens, we can perform a different kind of simplified search that also allows us to prove optimality. This simplified search is based on Equation (3) and, in particular, on the fact that wastage lower bounds are converted into live cell upper bounds by being rounded down. For instance, if our wastage lower bound gives us $live\_cells \leq \lfloor 100.75 \rfloor$, then we actually have $live\_cells \leq 100$. This means that the live cell upper bound we get is actually slightly stronger whenever the numerator is not divisible by 4. Since the upper bound is strengthened, it means that we do not always have to achieve the minimum amount of forced wastage in order to achieve the live cell upper bound. Let us define $spare = (2n^2 + 4n - min\_wastage) \mod 4$. The *spare* value (0, 1, 2 or 3) tells you how many "unforced" (by the boundary constraints) wastage we can afford to have and still achieve the closed form upper bound. In other words, although the boundary constraints force a certain number of wastage, we can afford to have *spare* wastage anywhere in the $n \times n$ region and still achieve the closed form upper bound.

It is interesting to consider the instances of n for which $spare = 0$. We believe that these are the instances where our closed form upper bound is most likely to be off by one. This is because when $spare = 0$, the closed form upper bound can only be achieved if there are no "unforced" wastage, and this can make the problem unsolvable. In other words, when $spare = 0$, any optimal solution that achieves the closed form upper bound must also be center perfect, since the edge and corner variables alone are sufficient to force all the wastage allowed. Thus, a complete search on center perfect solutions is sufficient to prove unsatisfiability of this value and improve the upper bound by 1. If we have already found a solution with 1 less live cell than the closed form upper bound, this will constitute a full proof of optimality. A complete search on center perfect solutions with minimum wastage is in fact quite feasible. This is because the corner, edges and center all have to be labeled absolutely perfectly with no unforced wastage, and there are very few ways to do this. For $spare > 0$ however, none of this is possible, as the spare "unforced" wastage can occur in an arbitrary position in the board, and proving that no such solution exists requires staggeringly more search.

6 Results

Our solver is written in C++ and compiled in g++ with O3 optimisation. It is run on a Xeon Pro 2.4GHz processor with 2Gb of memory. We run it for all n between 20 and 100. Smaller values of n have already been solved and are trivial for our solver. In Table 5, we list for each instance n, the lower bound (best solution found), the upper bound (marked with an asterisk if improved by 1 through complete search), the time in seconds taken to find the best solution, and the time in seconds taken to prove the upper bound. Each instance was run for at most 12 hours.

There are two precomputed tables which are used for all instances and are read from disk. These include tables for the 8×8 corner which takes $8s$ to compute and 16k memory to store, and a "thick edge" table consisting of a width 8 edge which takes 17 minutes to compute and ∼400Mb of memory. Since these are calculated only once and used for all instances the time used is not reflected in the table.

As can be seen, "small" instances like $20 \leq n \leq 30$ which previously took days or were unsolvable can now be solved in a matter of seconds (after paying the above fixed costs). The problem gets substantially harder for larger n. Beyond $n \sim 40$, we can no longer reliably find the optimal solution. However, we are still able to find solutions which are no more than 3-4 cells off the true optimum all the way up to $n = 100$. The run times for the harder instances have extremely high variance due to the large search space and the scarcity of the solutions. If the solver gets lucky, it can find a solution in mere seconds. Otherwise, it can take hours. Thus, the run time numbers are more useful as an indication of what is feasible, rather than as a precise measure of how long it takes.

Our memory requirements are also quite modest compared to previous methods. The previous best method used an amount of memory exponential in n

Table 5. Results on large max-density still life problems. Optimal answers are shown in bold. Upper bounds which are improved by complete search are shown starred. Times in seconds to one decimal place (in reality 0.1 usually represents a few milliseconds).

n	lower	upper	lb. time	ub. time
20	**210**	210	0.1	0.1
21	**232**	232	0.4	0.1
22	**253**	253	3.4	0.1
23	**276**	276	0.1	0.1
24	**301**	301*	0.5	0.1
25	**326**	326	0.6	0.1
26	**352**	352*	2.1	6.2
27	**379**	379	51	0.1
28	406	407	1.4	0.1
29	**437**	437	2.5	0.1
30	**466**	466*	45	0.3
31	**497**	497	0.6	0.1
32	**530**	530*	1815	0.1
33	**563**	563	60	0.1
34	**598**	598	207	0.1
35	**632**	632*	1459	1.9
36	**668**	668	0.1	0.1
37	**706**	706	1.1	0.1
38	743	744	43	0.1
39	**782**	782*	3.4	3.3
40	**823**	823*	3.3	10.0
41	**864**	864*	553	2.1
42	906	907	1.6	0.1
43	949	950	2176	0.1
44	**993**	993*	285	3.4
45	**1039**	1039	3807	0.1
46	1084	1085*	101	3.4
47	1131	1132	244	0.1
48	1180	1181	265	0.1
49	**1229**	1229*	563	10
50	1279	1280	9.2	0.1
51	1330	1331	105	0.1
52	1381	1383	354	0.1
53	**1436**	1436	4326	0.1
54	1489	1490*	25219	3.3
55	1543	1545	296	0.1
56	1601	1602	484	0.1
57	1657	1659	816	0.1
58	1716	1717	1950	0.1
59	1774	1776	992	0.1
60	1834	1836	4.3	0.1
61	**1897**	1897	15648	0.1
62	1957	1959*	62	0.1
63	2021	2023	7594	0.1
64	2085	2087	389	0.1
65	2150	2152	137	0.1
66	2217	2218	2296	0.1
67	2284	2285	26137	0.1
68	2351	2354	3149	0.1
69	**2422**	2422	14755	0.1
70	2490	2493	641	0.1
71	2562	2564	3077	0.1
72	2634	2636	866	0.1
73	2706	2709	433	0.1
74	2781	2783	3575	0.1
75	2856	2858	5440	0.1
76	2932	2934*	5879	3.4
77	3009	3011	5298	0.1
78	3088	3090	20865	0.1
79	3166	3169	2768	0.1
80	3247	3249	8328	0.1
81	3327	3330	113	0.1
82	3410	3412	2	0.1
83	3492	3495	10849	0.1
84	3576	3579*	1083	3.6
85	3661	3664*	3666	0.6
86	3748	3751	7628	0.1
87	3835	3838	957	0.1
88	3923	3926	5047	0.1
89	4012	4015	1837	0.1
90	4102	4105*	7047	3.2
91	4193	4196	605	0.1
92	4286	4289	8843	0.1
93	4379	4382	2254	0.1
94	4473	4476	3669	0.1
95	4568	4571	10871	0.1
96	4664	4667	16801	0.1
97	4761	4764	36205	0.1
98	4859	4862*	3462	3.4
99	4958	4961	7660	0.1
100	5058	5062	15458	0.1

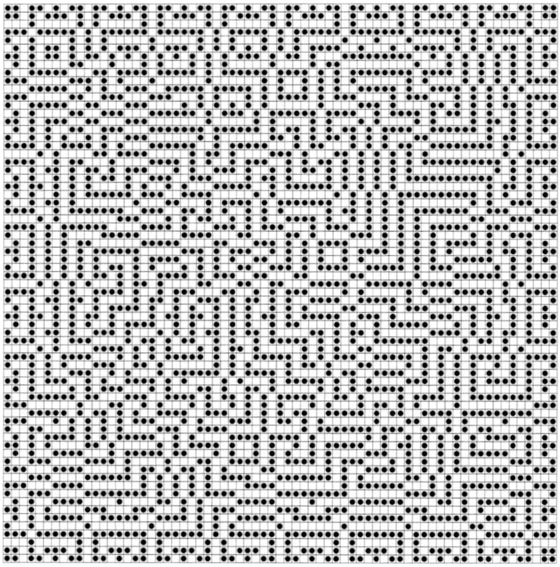

Fig. 4. An optimal solution to 69×69

and could not be run for $n > 22$. Our solver, on the other hand, uses a polynomial amount of memory. For $n = 100$ for example, we use $\sim$ 400 Mb for the precomputed tables and $\sim$ 120 Mb for the actual search.

7 Conclusion

We reformulate the Maximum Density Still Life problem into one of minimising wastage. This allows us to calculate very tight upper bounds on the number of live cells and also gives insight into the patterns that can yield optimal solutions. Using a boundary based relaxation, we are able to prove in constant time an upper bound on the number of live cells for all n which is never off the true optimum by more than one for all $n \leq 50$. We further conjecture that this holds true for all n and thus we may have found the optimum value for all n up to a margin of error of 1. By using dynamic relaxation as a lookahead learning/pruning technique, we are able to produce a complete depth 400+ lookahead that can be calculated in constant time. This prunes the search so powerfully that we can find optimal (or near optimal) solutions for a problem where the search space grows as 2^{n^2}. The largest n for which the problem is completely solved is $n = 69$ (shown in Figure 4). Further, we have proved upper and lower bounds that differ by no more than 4 for all n up to 100. All of our solutions can be found at `www.csse.unimelb.edu.au/~pjs/still-life/`

Acknowledgments. We would like to thank Michael Wybrow for helping us generate the still life pictures used in this paper. NICTA is funded by the Australian

Government as represented by the Department of Broadband, Communications and the Digital Economy and the Australian Research Council.

References

1. Bosch, R.: Integer programming and conway's game of life. SIAM Review 41(3), 596–604 (1999)
2. Bosch, R., Trick, M.: Constraint programming and hybrid formulations for three life designs. In: Proceedings of the International Workshop on Integration of AI and OR Techniques in Constraint Programming for Combinatorial Optimization Problems, CP-AI-OR 2002, pp. 77–91 (2002)
3. Dechter, R.: Constraint Processing. Morgan Kaufmann, San Francisco (2003)
4. Elkies, N.: The still-life density problem and its generalizations. arXiv:math/9905194v1
5. Elkies, N.: The still-life density problem and its generalizations. Voronoi's Impact on Modern Science: Book I, 228–253 (1998), arXiv:math/9905194v1
6. Larrosa, J., Dechter, R.: Boosting search with variable elimination in constraint optimization and constraint satisfaction problems. Constraints 8(3), 303–326 (2003)
7. Larrosa, J., Morancho, E., Niso, D.: On the practical use of variable elimination in constraint optimization problems: 'still-life' as a case study. Journal of Artificial Intelligence Research (2005)

Constraint-Based Graph Matching

Vianney le Clément[1], Yves Deville[1], and Christine Solnon[2,3]

[1] Université catholique de Louvain, Department of Computing Science and Engineering, Place Sainte-Barbe 2, 1348 Louvain-la-Neuve, Belgium
vianney.leclement@student.uclouvain.be,
Yves.Deville@uclouvain.be
[2] Université de Lyon
[3] Université Lyon 1, LIRIS, CNRS UMR5205, 69622 Villeurbanne Cedex, France
christine.solnon@liris.cnrs.fr

Abstract. Measuring graph similarity is a key issue in many applications. We propose a new constraint-based modeling language for defining graph similarity measures by means of constraints. It covers measures based on univalent matchings, such that each node is matched with at most one node, as well as multivalent matchings, such that a node may be matched with a set of nodes. This language is designed on top of Comet, a programming language supporting both Constraint Programming (CP) and Constraint-Based Local Search (CBLS). Starting from the constraint-based description of the measure, we automatically generate a Comet program for computing the measure. Depending on the measure characteristics, this program either uses CP —which is better suited for computing exact measures such as (sub)graph isomorphism— or CBLS —which is better suited for computing error-tolerant measures such as graph edit distances. First experimental results show the feasibility of our approach.

1 Introduction

In many applications graphs are used to model structured objects such as, e.g., images, design objects, molecules or proteins. In these applications, measuring graph similarity is a key issue for classification, pattern recognition or information retrieval. Measuring the similarity of two graphs involves finding a best matching between their nodes. Hence, graph similarity measures are closely related to graph matching problems. There exist many different kinds of graph matchings, ranging from exact matchings such as (sub)graph isomorphism to error-tolerant matchings such as (extended) edit distances.

Exact matchings may be solved by complete approaches such as, e.g., Nauty [1] for graph isomorphism and Vflib [2] for subgraph isomorphism. These approaches exploit invariant properties such as node degrees to prune the search space and they are rather efficient on this kind of problems. However, in many real world applications one looks for (sub)graph isomorphisms which satisfy additional constraints such as, e.g., label compatibility between the matched nodes. Dedicated approaches such as Vflib can only handle equality constraints between matched labels; other constraints cannot be used during the search process to reduce the search space.

Error-tolerant matchings involve finding a best matching, that optimizes some given objective function which evaluates the similarity induced by the matching. They are usually solved by numerical methods [3,4,5], or by heuristic approaches which explore

I.P. Gent (Ed.): CP 2009, LNCS 5732, pp. 274–288, 2009.

the search space of all possible matchings in an incomplete way, using heuristics to guide the search such as, e.g., genetic algorithms [6], greedy algorithms [7], reactive tabu search [8], and Ant Colony Optimization [9]. These algorithms do not guarantee to find optimal solutions; as a counterpart, they usually find rather good solutions within reasonable CPU times. Algorithms dedicated to error-tolerant matchings may be used to solve exact matching problems by defining appropriate edit costs. However, they may be less efficient than dedicated approaches such as Nauty or Vflib.

Contribution. Graph matching problems may be defined by means of constraints on the cardinality of the matching, and on edges and labels. Hence, we introduce a modeling language for defining graph matchings by means of constraints. This language allows one to define new graph matching problems in a declarative way, by a simple enumeration of constraints. It covers both exact and error-tolerant matchings. We show that this language can be used to define existing matching thus giving a uniform framework to these different matching problems.

Graph matching problems defined by means of constraints are solved by constraint solvers that are embedded into the language. We more particularly consider two different kinds of solving approaches: a complete approach, which combines a tree search with filtering techniques, and an incomplete approach, based on local search. Our system is designed on top of Comet, a constraint-based modeling language supporting both tree search and local search. Starting from the constraint-based description of the matching, we automatically generate a Comet program for computing it. Depending on the constraints, this program either uses tree search or local search, thus choosing the most efficient approach for solving the considered matching problem. A similar approach is taken in [10] for solving scheduling problems.

Outline of the Paper. Section 2 gives an overview of existing graph matching problems. Section 3 briefly describes the constraint programming paradigm on which our approach is based. Section 4 introduces a set of constraints that may be used to define graph matching problems. Section 5 shows how to use these constraints to define existing graph matching problems. Section 6 discusses implementation issues and Section 7 gives some preliminary experimental results.

2 Graph Matching Problems

We consider labeled directed graphs defined by $G = (N, E, L)$ such that N is a set of nodes, $E \subseteq N \times N$ is a set of directed edges, and $L : N \cup E \to \mathbb{N}$ is a function that associates labels to nodes and edges. Throughout this paper, we assume that the labeled graphs to be matched are $G_1 = (N_1, E_1, L_1)$ and $G_2 = (N_2, E_2, L_2)$ such that $N_1 \cap N_2 = \varnothing$.

Functional Matchings. A (total) functional matching between G_1 and G_2 is a function $f : N_1 \to N_2$ which matches each node of G_1 with a node of G_2. When the matching preserves all edges of G_1, i.e., $\forall (u_1, v_1) \in E_1, (f(u_1), f(v_1)) \in E_2$, it is called a *graph homomorphism* [11].

In many cases, f is injective so that each node of G_2 is matched to at most one node of G_1, i.e., $\forall u_1, v_1 \in N_1, u_1 \neq v_1 \Rightarrow f(u_1) \neq f(v_1)$. In this case the matching is said to be *univalent*. Particular cases are *subgraph isomorphism*, when f is a homomorphism, and *graph isomorphism*, when f is bijective and f^{-1} is also a homomorphism. Note that while subgraph isomorphism is NP-complete, graph isomorphism is not known to be nor NP-complete nor in P.

These different problems can be extended to the case where f is a partial function such that some nodes of N_1 are not matched to a node of N_2. In particular, the *maximum common subgraph* corresponds to the partial injective matching which preserves edges and maximizes the number of matched nodes or edges.

These different problems can also be extended to take into account node and edge labels, thus leading to the *graph edit distance* [12] or the *graph matching problem* of [13]. For example, the graph edit distance involves finding a partial injective matching which minimizes the sum of deletion (resp. addition) costs associated with the labels of the nodes and edges of G_1 (resp. G_2) that are not matched, and substitution costs associated with labels of nodes and edges that are matched but that have different labels.

Relational Matchings. Many real-world applications involve comparing objects described at different granularity levels and, therefore, require *multivalent* matchings, such that each node may be matched with a (possibly empty) set of nodes. In particular, in the field of image analysis, some images may be over-segmented whereas some others may be under-segmented so that several regions of one image correspond to a single region of another image. In this case, the matching is no longer a function, but becomes a *relation* $M \subseteq N_1 \times N_2$, and the goal is to find the best matching, i.e., the matching which maximizes node, edge and label matchings while minimizing the number of split nodes (that are matched with more than one node). Graph similarity measures based on multivalent matchings have been proposed, e.g., in [7,14].

All these problems, ranging from maximum common subgraph to similarity measures based on multivalent matchings, are NP-hard.

3 Constraint-Based Modeling

Constraint Programming (CP) is an attractive alternative to dedicated approaches: it provides high level languages to declaratively model Constraint Satisfaction Problems (CSPs); these CSPs are solved in a generic way by embedded constraint solvers [15].

Many embedded constraint solvers are based on a complete tree search combined with filtering techniques which reduce the search space. We have proposed in [16] and [17] filtering algorithms that are respectively dedicated to graph and subgraph isomorphism problems. These filtering algorithms exploit the global structure of the graphs to drastically reduce the search space. We have experimentally shown that they allow CP to be competitive, and in some cases to outperform, dedicated approaches such as Nauty or Vflib.

Embedded constraint solvers may also be based on local search. In this case, the search space is explored in an incomplete way by iteratively performing local modifications, using some metaheuristics such as tabu search or simulated annealing to escape from locally optimal solutions. We have introduced in [8] a reactive tabu search approach for solving multivalent graph matching problems.

Comet [18] is a constraint-based modeling language which supports both complete tree search and local search. A Comet program is composed of two parts: (1) a high-level model describing the problem by means of constraints, constraint combinators, and objective functions; (2) a search procedure expressed at a high abstraction level.

```
1  include "matching";
2
3  bool[,] adj1 = ...
4  bool[,] adj2 = ...
5  SimpleGraph<Mod> g1(adj1);
6  SimpleGraph<Mod> g2(adj2);
7
8  Matching<Mod> m(g1,g2);
9  m.post(cardMatch(g1.getAllNodes(), 1, 1));
10 m.post(injective(g1.getAllNodes()));
11 m.post(matchedToSomeEdges(g1.getAllEdges()));
12 m.close();
13
14 DefaultGMSynthesizer synth();
15 GMSolution<Mod> sol = synth.solveMatching(m);
16 print(sol);
```

Fig. 1. Example of $\mathcal{SI}$ matching problem solved with our Comet prototype

Our system for modeling and solving graph matching problems is designed on top of Comet. An example of program for modeling and solving a subgraph isomorphism problem is given in Fig. 1.

As for every Comet program, this program consists in two parts. In the first part (lines 9–13), the problem is modeled by means of high-level constraints. The first two constraints specify the cardinality of the matching to search, which must match each node of G_1 with exactly one node of G_2 (line 10), and which must be injective (line 11). The last constraint specifies that the matching must preserve edges (line 12). Note that problem-dependent constraints may be very easily added. We introduce in section 4 the different constraints that may be used to model graph matching problems, and we show in section 5 how to use these constraints to define existing graph matching problems.

In the second part (lines 15–16), a synthesizer is called to automatically generate a Comet program for computing the solution. Depending on the constraints, this program either uses tree search or local search, thus choosing the most appropriate approach. This synthesizer is described in Section 6.

4 Constraints for Modeling Graph Matching Problems

A graph matching problem between two directed graphs $G_1 = (N_1, E_1, L_1)$ and $G_2 = (N_2, E_2, L_2)$ involves finding a matching $M \subseteq N_1 \times N_2$ which satisfies some given constraints. These constraints actually specify the considered matching problem. In this section, we introduce different constraints that may be used to define graph matching problems. To make the following easier to read, we denote by $M(u)$ the set of nodes that are matched to a node u by M, i.e., $\forall u \in N_1, M(u) = \{v \in N_2 | (u,v) \in M\}$ and $\forall u \in N_2, M(u) = \{v \in N_1 | (v,u) \in M\}$.

Fig. 2 describes the basic constraints for modeling graph matching problems.

The first set of constraints enables to specify the minimum and maximum number of nodes a node is matched to. For ease of use, these constraints are also defined for a set U of nodes, to constrain the number of nodes matched to every node in U.

$$
\begin{array}{rl}
\multicolumn{2}{c}{\text{Let } M \subseteq N_1 \times N_2, u, v \in N_1 \cup N_2, U \subseteq N_1 \cup N_2, ub, lb \in \mathbb{N}, L = L_1 \cup L_2,} \\
\multicolumn{2}{c}{D \subseteq (N_1 \cup N_2) \times (N_1 \cup N_2).} \\
\hline
\text{MinMatch}(M,u,lb) & \equiv lb \leq \#M(u) \\
\text{MinMatch}(M,U,lb) & \equiv \forall u \in U : \text{MinMatch}(M,u,lb) \\
\text{MaxMatch}(M,u,ub) & \equiv \#M(u) \leq ub \\
\text{MaxMatch}(M,U,ub) & \equiv \forall u \in U : \text{MaxMatch}(M,u,ub) \\
\text{CardMatch}(M,u,lb,ub) & \equiv \text{MinMatch}(M,u,lb) \wedge \text{MaxMatch}(M,u,ub) \\
\text{CardMatch}(M,U,lb,ub) & \equiv \forall u \in U : \text{CardMatch}(M,u,lb,ub) \\
\hline
\text{Injective}(M,U) & \equiv \forall u,v \in U, u \neq v : M(u) \cap M(v) = \varnothing \\
\hline
\text{MatchedToSomeEdges}(M,u,v) & \equiv \exists u' \in M(u), \exists v' \in M(v) : (u',v') \in E_1 \cup E_2 \\
\text{MatchedToSomeEdges}(M,D) & \equiv \forall (u,v) \in D : \text{MatchedToSomeEdges}(M,u,v) \\
\text{MatchedToAllEdges}(M,u,v) & \equiv \forall u' \in M(u), \forall v' \in M(v) : (u',v') \in E_1 \cup E_2 \\
\text{MatchedToAllEdges}(M,D) & \equiv \forall (u,v) \in D : \text{MatchedToAllEdges}(M,u,v) \\
\hline
\text{MatchSomeNodeLabels}(M,u) & \equiv \exists v \in M(u) : L(u) = L(v) \\
\text{MatchSomeEdgeLabels}(M,u,v) & \equiv \exists u' \in M(u), \exists v' \in M(v) : (u',v') \in E_1 \cup E_2 \\
 & \quad \wedge L(u,v) = L(u',v') \\
\text{MatchAllNodeLabels}(M,u) & \equiv \forall v \in M(u) : L(u) = L(v) \\
\text{MatchAllEdgeLabels}(M,u,v) & \equiv \forall u' \in M(u), \forall v' \in M(v) : (u',v') \in E_1 \cup E_2 \\
 & \quad \wedge L(u,v) = L(u',v') \\
\hline
\end{array}
$$

Fig. 2. Basic constraints for modeling graph matching problems

The second set of constraints enables to state that a set U of nodes is *injective*, i.e., that the nodes of this set are matched to different nodes. There is a simple relationship between the maximum numbers of matched nodes and injective nodes when U is equal to N_1 (resp. N_2): in this case, every node of N_2 (resp. N_1) must be matched to at most one node of N_1 (resp. N_2), i.e.,

$$\text{Injective}(M,N_1) \Leftrightarrow \text{MaxMatch}(M,N_2,1)$$
$$\text{Injective}(M,N_2) \Leftrightarrow \text{MaxMatch}(M,N_1,1)\ .$$

The third set of constraints allows one to specify that a couple of nodes must be matched to a couple of nodes related by an edge. MatchedToSomeEdges ensures that there exists at least one couple of matched nodes which is related by an edge whereas MatchedToAllEdges ensures that all couples of matched nodes are related by edges. Note that, when $M(u) = \varnothing$ or $M(v) = \varnothing$, the MatchedToSomeEdges(M, u,v) constraint is violated whereas MatchedToAllEdges(M,u,v) is satisfied. Note also that when $\#M(u) = \#M(v) = 1$, the two constraints MatchedToSomeEdges(M,u,v) and MatchedToAllEdges(M,u,v) are equivalent. Note finally that these constraints are meaningful only when u and v belong to the same graph. For ease of use, these constraints are also defined for a set D of couples of nodes to constrain every couple of D to be matched to edges.

The last set of constraints enables to specify that labels of matched nodes or edges must be equal. On the one hand MatchSomeNodeLabels (resp. MatchSomeEdgeLabels) ensures that there is at least one matched node (resp. edge) with the same label. On the other hand MatchAllNodeLabels (resp. MatchAllEdgeLabels) ensures that all matched nodes (resp. edges) have the same label.

All these constraints may either be posted as hard ones, so that they cannot be violated, or as soft ones, so that they may be violated at some given cost. Soft constraints are posted by using a specific method which has two arguments: the constraint and the cost associated with its violation.

5 Modeling Graph Matching Problems by Means of Constraints

We now show how to model classical graph matching problems with the constraints introduced in the previous section. Note that different (equivalent) formulations of these problems are possible.

Exact matching problems are modeled with hard constraints. For graph homomorphism ($\mathcal{GH}$), the matching must be a total function which preserves edges, i.e.,

$$\mathcal{GH}(M,G_1,G_2) \equiv \text{CardMatch}(M,N_1,1,1) \wedge \text{MatchedToSomeEdges}(M,E_1).$$

For subgraph isomorphism ($\mathcal{SI}$), we add an injective constraint to $\mathcal{GH}$, i.e.,

$$\mathcal{SI}(M,G_1,G_2) \equiv \mathcal{GH}(M,G_1,G_2) \wedge \text{Injective}(M,N_1).$$

For induced subgraph isomorphism ($\mathcal{ISI}$), we add a MatchedToAllEdges constraint to $\mathcal{SI}$ in order to ensure that when the two nodes of an edge of G_2 are matched to nodes of G_1, these matched nodes are related by an edge in G_1, i.e.,

$$\mathcal{ISI}(M,G_1,G_2) \equiv \mathcal{SI}(M,G_1,G_2) \wedge \text{MatchedToAllEdges}(M,E_2).$$

For graph isomorphism ($\mathcal{GI}$), we check that the matching is a bijective total function which preserves edges, i.e.,

$$\mathcal{GI}(M,G_1,G_2) \equiv \text{CardMatch}(M,N_1 \cup N_2,1,1) \wedge \text{MatchedToSomeEdges}(M,E_1 \cup E_2).$$

Constraints can also be used for modeling approximate matching problems such as the maximum common subgraph ($\mathcal{MCS}$). In this case, one has to combine hard constraints (for ensuring that the matching is a partial function) with soft constraints (for maximizing the number of edges of G_1 which are matched), i.e.,

$$\begin{aligned}\mathcal{MCS}(M,G_1,G_2) \equiv\ & \text{MaxMatch}(M,N_1 \cup N_2,1) \\ & \wedge\ \forall(u,v) \in E_1, \text{soft}(\text{MatchedToSomeEdges}(M,u,v),1)\ .\end{aligned}$$

By defining the cost of violation of each MatchedToSomeEdges soft constraint to 1, we ensure that the optimal solution will have a cost equal to the number of edges of G_1 which are not in the common subgraph. Hence, the number of edges in the common subgraph is equal to the number of edges of G_1 minus the cost of the optimal solution.

For maximum common induced subgraph ($\mathcal{MCIS}$), one has to replace the soft MatchedToSomeEdges constraint by a hard MatchedToAllEdges constraint as edges between matched nodes must be preserved. To maximize the number of nodes that are matched, we add a soft MinMatch constraint. More precisely,

$$\begin{aligned}\mathcal{MCIS}(M,G_1,G_2) \equiv\ & \text{MaxMatch}(M,N_1 \cup N_2,1) \\ & \wedge\ \text{MatchedToAllEdges}(M,E_1 \cup E_2) \\ & \wedge\ \forall u \in N_1, \text{soft}(\text{MinMatch}(M,u,1),1)\ .\end{aligned}$$

By defining the cost of violation of each MinMatch soft constraint to 1, we ensure that the optimal solution will have a cost equal to the number of nodes of G_1 which are not in the common subgraph. Hence, the number of nodes in the common subgraph is equal to the number of nodes of G_1 minus the cost of the optimal solution.

The graph edit distance ($\mathcal{GED}$) generalizes $\mathcal{MCS}$ by taking into account edge and node labels. This distance is computed with respect to some edit costs which are given by the user. Let us note $c_d(l)$ (resp. $c_a(l)$) the edit cost associated with the deletion (resp. addition) of the label l, and $c_s(l_1,l_2)$ the edit cost associated with the substitution of label l_1 by label l_2. The graph edit distance may be defined by $\mathcal{GED}(M,G_1,G_2) \equiv$

$$\begin{aligned}
&\text{MaxMatch}(M,N_1 \cup N_2,1)\\
\wedge\ &\forall u \in N_1, \text{soft}(\text{MinMatch}(M,u,1), c_d(L_1(u)))\\
\wedge\ &\forall u \in N_2, \text{soft}(\text{MinMatch}(M,u,1), c_a(L_2(u)))\\
\wedge\ &\forall u \in N_1, \text{soft}(\text{MatchAllNodeLabels}(M,u), c_s(L_1(u), L_2(M(u))))\\
\wedge\ &\forall (u,v) \in E_1, \text{soft}(\text{MatchedToSomeEdges}(M,u,v), c_d(L_1(u,v)))\\
\wedge\ &\forall (u,v) \in E_2, \text{soft}(\text{MatchedToSomeEdges}(M,u,v), c_a(L_2(u,v)))\\
\wedge\ &\forall (u,v) \in E_1, \text{soft}(\text{MatchAllEdgeLabels}(M,u,v), c_s(L_1(u,v), L_2(M(u),M(v))))\ .
\end{aligned}$$

In this case, violation costs of soft constraints are defined by edit costs. For the nodes of G_1 (resp. G_2), the cost of violation of the MinMatch constraint is equal to the edit cost of the deletion (resp. addition) of node labels as this constraint is violated when a node of G_1 (resp. G_2) is not matched to a node of G_2 (resp. G_1), thus indicating that this node must be deleted (resp. added). The cost of the soft MatchAllNodeLabels constraint is equal to the edit cost of substituting the label of u by the label of the node it is matched to in G_2. Similar soft constraints are posted on edges to define deletion, addition and substitution costs.

Finally, one may also define multivalent graph matching problems, such that one node may be matched to a set of nodes. Let us consider for example the extended graph edit distance ($\mathcal{EGED}$) defined in [14]. This distance extends $\mathcal{GED}$ by adding two edit operations for splitting and merging nodes. Let us note $c_p(u,U)$ (resp. $c_m(U,u)$) the edit cost associated with the splitting of the node $u \in N_1$ into the set of nodes $U \subseteq N_2$ (resp. the merging of the set of nodes $U \subseteq N_1$ into the node $u \in N_2$). $\mathcal{EGED}$ may be defined by replacing the hard MaxMatch constraint of $\mathcal{GED}$ by two soft MaxMatch constraints which respectively evaluate split and merged nodes, i.e.,

$$\begin{aligned}
&\forall u \in N_1, \text{soft}(\text{MaxMatch}(M,u,1), c_p(u,M(u))\\
\wedge\ &\forall u \in N_2, \text{soft}(\text{MaxMatch}(M,u,1), c_m(u,M(u)))\ .
\end{aligned}$$

6 Comet Prototype

As illustrated in Fig. 1, constraint-based graph matching in our Comet prototype is done in two parts. First, high level constraints modeling the problem are posted. Second, a synthesizer is called to solve the problem by means of CP and/or CBLS techniques.

Canonical form of Modeling Constraints. High-level constraints, called modeling constraints, implement the `GMConstraint<Mod>` interface. Such constraints, like those described in section 4, are stated on the nodes, edges and labels of the graph and are posted by the model, implemented by Matching<Mod>, as hard or soft constraints. For soft constraints, an additional parameter specifies the cost of a violation.

To easily state characteristics of the matching, we introduce *canonical* constraints aggregating all the modeling constraints of a type. For example, all MinMatch and MaxMatch constraints will be aggregated into a single cardinality canonical constraint knowing the lower and upper bounds of the matchings of each node. The model maintains two constraint stores, for hard and soft canonical constraints respectively. When a hard (resp. soft) modeling constraint is posted to the model, its postCanonical (resp. postSoftCanonical) method is called with the hard (resp. soft) constraint store. This method can add or modify canonical constraints within the store. Typically a modeling constraint will only modify its associated canonical constraint. The key concept is that there must not exist more than one canonical constraint of each type, identified by a unique string, in a constraint store. Apart for stating characteristics, this has another benefit: global constraints can be generated by the synthesizer.

Once the model is closed, i.e. no more constraints may be added, each canonical constraint gets a chance to modify itself or other constraints, in order to make the whole model canonical, through the canonify method. In a canonical model, we cannot add any modeling constraint without altering the described matching problem. For example, the Injective constraint adjusts the cardinality constraint if every node of a graph belongs to an injective set. The canonify method should report the canonical constraints it has modified, so that the model can iterate the procedure, using a static dependency graph of the canonical constraints, untill it reaches a fix-point. Methods of the different interfaces are depicted below.

```
class Matching<Mod> { ...
  void post(GMConstraint<Mod> constraint);
  void postSoft(GMConstraint<Mod> constraint, int cost);
}
interface GMConstraint<Mod> {
  void postCanonical(GMConstraintStore<Mod> store);
  void postSoftCanonical(GMConstraintStore<Mod> store, int cost);
}
interface CanonicalGMConstraint<Mod> {
  string getId();
  set{string} canonify(GMConstraintStore<Mod> store);
  void postCP(Solver<CP> cp, GMVarStore<CP> vars);
  void postLS(SetConstraintSystem<LS> S, GMVarStore<LS> vars);
}
```

Synthesizer. Once the model is closed, a synthesizer is called to effectively solve the problem. The canonical representation of the model allows to compute various kinds of characteristics such as whether the matching is functional, univalent or multivalent, or even the class of the problem (such as those defined in section 4). The GMCharacteristic<Mod> interface allows to define such characteristics.

This synthesizer has three tasks, i.e., create variables, post the CP and/or CBLS constraints, and perform the search.

Creating Variables. To represent the matching, a variable x_u is associated to each node $u \in N$; the value of x_u denotes the matching of node u, that is $M(u)$. We assume that nodes are represented by positive integers. The type and domain of these variables depend on the MinMatch and MaxMatch constraints, as described below:

MinMatch	MaxMatch	Type	Domain
1	1	int	N
0	1	int	$N \cup \{\perp\}$
Otherwise		set{int}	2^N

The $\perp$ value denotes that $M(u) = \varnothing$. It is implemented as the negative node number, i.e., $-u$, ensuring a unique $\perp$ value for each node of a graph. Of course, for nodes in G_1, the domain is restricted to N_2, and similarly for nodes in G_2.

Depending on the constraint solver chosen by the synthesizer (complete incremental search (CP) or incomplete local search (LS)), the variables are declared as CP variables or LS variables in the Comet language. As set variables do not yet exist in Comet CP, we have implemented these with a simple boolean array. A similar limitation led us to reimplement the constraint interface in Comet LS.

Since a variable is declared for nodes in G_1 as well as for nodes in G_2, the matching M is redundantly represented. Depending on the chosen solver, channeling constraints (CP) or Comet invariants (LS) are added to relate these two sets of variables. This redundant representation allows the solver to choose the best variables to construct a solution and to perform the search.

The association between a node and its variable is available to the various constraints through the GMVarStore<CP> and GMVarStore<LS> interfaces. As the creation of variables is likely to be the same for every synthesizer, the default implementations of these interfaces handle the variable creation in their constructor.

Posting the Constraints and Performing the Search. Once the variables are created, the synthesizer asks the canonical (hard and soft) constraints to post themselves with the postCP or postLS methods. These methods take two arguments: the solver and the GMVarStore containing the associations between nodes and variables. In CP, soft constraints are implemented by an objective function. In LS, hard constraints are either handled by a neighborhood ensuring that they cannot be violated, or they are posted as soft constraints with much higher violation costs.

A synthesizer solves a problem either in CP or in LS. The default choice of CP or LS, as implemented by DefaultGMSynthesizer, depends on the constraints: if the maximum number of matched nodes is 1 for every node of one graph and if all constraints are hard ones, then the synthesizer chooses CP, and the variables associated with these nodes are used as choice variables in the tree search; otherwise, the synthesizer chooses LS and the variables associated with the graph with the fewest set variables are used for defining neighborhoods.

Implementing the Constraints. We now focus on how constraints for modeling matching problems are implemented in our framework.

Node Cardinality. For univalent matchings, the cardinality constraints (MinMatch, MaxMatch, and CardMatch) are already implemented by variable domains. For multivalent matchings, nodes are associated with set variables and we post inequality constraints on the cardinality of these sets.

If the matching is a surjective function from N_1 to N_2, i.e. MinMatch$(M,N_2,1)$ and MaxMatch$(M,N_1,1)$, a redundant constraint is added to the CP solver. This is a particular case of the global cardinality constraint `cardinality`$(N_2,1,[x_u|u \in N_1],\#N_1)$ [19] which here holds when at least 1 variable (and at most $\#N_1$) is assigned to each value of N_2. A similar constraint is posted if the matching is surjective from N_2 to N_1.

Injective Set. For univalent matchings, Injective(M,U) constraints are implemented by `alldifferent`$([x_u|u \in U])$ constraints. Note that we associate a different $\perp$ value to every different node u (defined by $-u$) so that `alldifferent` is not violated when several nodes are matched to $\perp$.

For multivalent matchings such that some variables in the set U are implemented as `set{int}` variables, additional constraints of the form $x_{u1} \cap x_{u2} = \varnothing$ (with x_{u1} and x_{u2} set variables) and $x_v \notin x_u$ (with x_u a set variable and x_v an integer variable) are posted.

Edges. If we consider univalent matchings, MatchedToSomeEdges(M,u,v) (resp. MatchedToAllEdges(M,u,v)) is easily implemented as $(x_u,x_v) \in E_1 \cup E_2$ (resp. $x_u \neq \perp \wedge x_v \neq \perp \Rightarrow (x_u,x_v) \in E_1 \cup E_2$). When the matching is multivalent, LS constraints are generated.

Additional Constraints. The system is open and modular so that new constraints may be defined. For instance, we introduced the constraint CommonNeighbor(u,v) which holds if $M(u)$ and $M(v)$ share at least one neighbor.

Current Limitations of the System. The prototype is about 4,200 lines of Comet code. In the current implementation, the cost of soft constraints must be a fixed value; hence $\mathcal{GED}$ and $\mathcal{EGED}$ are not yet supported. The CP part of the current prototype does not support soft constraints and MatchedToEdges constraints for multivalent matching. This is not very limitative as CP is not really adapted for these matching problems.

In the current implementation, a matching problem is solved either with CP or with LS. In the future, we plan to allow the solver to combine CP with LS.

The analysis of the characteristics of the matching problem is rather limited. It can however detect the standard matching problems. We plan to add additional global constraints in the CP part in order to speed up the search process. In particular, we plan to integrate the redundant constraints of [20], the filtering algorithm of [17] for the subgraph isomorphism problem, and the filtering algorithm of [16] for the graph isomorphism problem.

The metaheuristics in the LS part are still basic (tabu search); this will be extended and adapted to matching problems. In particular, we plan to implement the reactive tabu search algorithm of [8].

7 Experimental Results

We report experiments on the subgraph isomorphism problem and on a pattern recognition problem using CP, and on the maximum common subgraph using LS.

Table 1. Comparison of synthesizer/CP and Vflib on $\mathcal{SI}$

	Vflib					Synthesizer/CP				
class	solv.%	mean	std	min	max	solv.%	mean	std	min	max
P200	75	61.8	93.0	0.5	309.2	100	6.1	1.9	2.6	9.5
P600	0	-	-	-	-	100	234.3	70.5	105.8	402.6
P1000	0	-	-	-	-	15	522.8	107.8	370.3	599.1
V200	82	63.2	112.9	0.0	574.7	62	84.3	132.2	0.8	530.9
V200+1	82	63.8	114.2	0.0	583.3	71	69.5	108.4	0.8	524.7
V200+2	82	64.4	115.1	0.0	582.9	77	63.5	117.1	0.7	540.2
V200+3	82	64.9	115.8	0.0	582.7	79	50.9	119.0	0.7	531.9
V200+4	82	65.5	116.8	0.0	584.8	85	40.0	93.8	0.7	510.3
V200+5	81	59.4	102.8	0.0	459.6	87	31.5	84.5	0.7	507.2

7.1 Subgraph Isomorphism Using a CP Solver

We evaluate our system on the subgraph isomorphism problem as modeled in Fig. 1. Given the characteristics of the problem, the default synthesizer uses CP. Our model is compared with the state of the art Vflib C++ library [2].

The benchmark contains two families of randomly generated graphs. In the first family (P*), graphs are randomly generated using a power law distribution of degrees $P(d = k) = k^{-\lambda}$: this distribution corresponds to scale-free networks which model a wide range of real networks, such as social, Internet, or neural networks [21]. We only report experiments on graphs generated with the standard value $\lambda = 2.5$. Each class contains 20 different instances. For each instance, we first generate a connected target graph which node degrees are bounded between 5 and 8. Then, a connected pattern graph is extracted from the target graph by randomly selecting 90% of nodes and edges. All instances of classes P200, P600 and P1000 are feasible instances that respectively have 200, 600, and 1000 nodes.

The second family (V*) is taken from the Vflib benchmarks [22]. Class V200 contains the 100 instances from the class called si2_r001_m200 in Vflib. These instances were randomly generated using a uniform distribution, the target graph has 200 nodes, and the source graph has 40 nodes, i.e. 20% of the target (see [22] for details). In order to assess the modularity of our approach, we generated the classes V200+k (with $k \in \{1,2,3,4,5\}$) by adding k additional CommonNeighbor(u,v) constraints in the Subgraph Isomorphism model. In order to ensure the existence of a solution (instances without solution are all solved by the CP solver in about a second), we have randomly chosen (u,v) from the pairs of nodes satisfying the additional constraint from the solutions found. Such side constraints cannot be directly handled by Vflib; so we added to Vflib an algorithm to filter the solutions satisfying these additional constraints.

Table 1 gives for every class the percentage of solved instances within a CPU time limit of 600s on a Core 2 Quad (only one core used) 2,4 Ghz with 2Go of RAM. It also gives the execution time of the solved instances (mean, standard deviation, minimum and maximum times in seconds).

On the P* instances, the synthesized CP algorithm solves much more instances and is also much more efficient than the standard Vflib. On the V200 class, Vflib solves more instances. However, as Vflib is not able to actively exploit additional constraints to prune the tree during the search, results of Vflib on V200+k classes are almost identical than for V200; there is a slight overhead for the postprocess filtering the solutions. On the contrary, the synthesized CP solver can exploit these additional constraints during the search, increasing the number of solved instances and reducing the computation time. The more additional constraints, the more solved instances and the less computation time. From V200+4, the CP solver outperforms Vflib. This clearly shows the interest of a constraint-based approach compared to specialized algorithms.

It should be noticed that the CP approach could be made more efficient by a full implementation of the iterative filtering described in [17].

7.2 Pattern Recognition Using a CP Solver

We now illustrate our solver on a pattern recognition problem which involves finding patterns in images. Graphs are generated from images by extracting interest points (corresponding to salient points) and computing a Delaunay triangulation on them. Finding patterns in images amounts to finding connected sets of faces in graphs modeling images. This problem lies part-way between subgraph isomorphism and induced subgraph isomorphism as some edges in the target graph are mandatory and some others are optional. This problem may be solved with subgraph isomorphism, assuming post-processing is done on the found solutions to check that all mandatory edges are matched. In a pattern recognition context, it may be meaningful to introduce an additional constraint stating that the distance between two nodes in the pattern should be similar to the distance of the corresponding target nodes (up to a small delta value). We have implemented this constraint as a CP propagator with a forward-checking consistency level. Note that such constraints cannot be handled with Vflib as it can only handle equality constraints between matched labels.

Table 2 gives the results for target graphs with 100, 500 and 1000 nodes. Five pattern graphs are extracted from each target graphs by selecting a connected subset of faces which respectively contains 5%, 10%, 20%, 33%, and 50% of the target faces. Table 2 shows us that Vflib is better on small instances, but is outperformed by our system on larger ones. Using the additional constraint globally improves the performances, even though only forward checking has been used. This shows the interest of the more flexible constraint-based approach in real-world applications.

7.3 Maximum Common Subgraph Using an LS Solver

We now evaluate the feasibility of our approach on the maximum common subgraph ($\mathcal{MCS}$) problem as described in Section 5, using an LS solver. Existing solvers like vflib cannot handle the MCS problem. Also, [3,4,5,7] do not handle the MCS. Different complete approaches have been compared on the MCS in [23] but experimental results are limited to graphs up to 30 nodes so that we have not compared our approach on these benchmarks.

The benchmarks are also taken from the Vflib benchmarks [2]. The classes M25, M50 and M100 contains 20 instances from the classes called mcs50_r02 in Vflib. The graphs have respectively 25, 50 and 100 nodes and have been randomly generated (see

Table 2. Execution time in seconds of subgraph isomorphism for pattern recognition without ($\mathcal{SI}$) and with ($\mathcal{SI}+$) additional constraint on distances. Rows represent the number of vertices of the target graphs, columns show the sizes of the pattern graphs as a percentage of the target.

		Synthesizer/CP					Vflib				
		5%	10%	20%	33%	50%	5%	10%	20%	33%	50%
$\mathcal{SI}$	100	0.8	0.5	0.7	**0.1**	0.2	**0.0**	**0.0**	**0.0**	2.0	**0.0**
	500	19.3	4.7	**10.5**	**15.8**	**30.7**	**0.1**	**0.1**	246.7	192.3	–
	1000	**30.6**	**595.8**	**119.0**	**152.3**	–	86.7	–	–	–	–
$\mathcal{SI}+$	100	0.3	0.1	0.1	0.1	0.2					
	500	3.0	4.4	9.5	16.9	28.9					
	1000	16.1	47.8	82.5	148.0	–					

Table 3. Maximum common subgraph with LS. Average (standard deviation) of execution time, number of iterations, and percentage of edges in the best found solution. The model has been executed 10 times per instance.

class	time	iterations	edges%
M25	8.5 (2.5)	7768.1 (2301.3)	48.3 (1.1)
M50	33.9 (10.7)	8023.8 (2543.3)	40.2 (0.5)
M100	141.5 (46.4)	8398.4 (2755.0)	34.5 (0.2)

[22] for details). It is known that these graphs have a common *induced* subgraph with 50% of the nodes, but this lower bound does not provide any information on the size of maximum common partial subgraph in terms of edges.

A basic tabu search is generated by the synthesizer. The MaxMatch(M,N_2) hard constraint is maintained through the neighborhood, by either swapping the value of two matched nodes in G_1, by matching a node in G_1 to an unmatched node of G_2, or by removing a matching in G_1. The time limit is fixed at 20,000 iterations, but the search stops after 5,000 iterations without global improvement. A random new solution is also generated after 1,000 iterations without global improvement. The results are reported in Table 3.

It is difficult to compare an LS approach with complete algorithms for maximum common subgraph as these algorithms cannot handle graphs with 100 nodes [24]. This also justifies the choice of an LS solver by our default synthesizer. The generated LS solver could be improved in many ways. These benchmarks are presented to assess the feasibility of generating an LS solver by the synthesizer.

8 Conclusion

Measuring graph similarity is a key issue in many applications. We proposed a new constraint-based modeling language for defining graph matching problems by means of constraints. It covers both univalent and multivalent matchings, and we have shown that it may be used to define many different existing matching problems in a very declarative way. Such a constraint-based formulation of the different matching problems actually stressed out their shared features and differences in a very concise way.

We have built a synthesizer which is able to automatically generate a Comet program to solve matching problems modeled with our language. Depending on the characteristics of the matching, the synthesizer either uses a branch and propagate approach —which is better suited for computing exact matchings such as (sub)graph isomorphism— or a local search approach —which is better suited for computing error-tolerant matching such as graph edit distances. First experimental results showed the feasibility of our approach on subgraph isomorphism and maximum common subgraph problems.

As future work, we will extend the prototype to lift the current limitations : handling soft constraints in CP, integrating new filtering algorithms in CP, improving the analysis of the matching characteristics, integrating several LS metaheuristics, and combining CP and LS solvers.

Acknowledgments. The authors want to thank the anonymous reviewers for their helpful comments. Christine Solnon acknowledges an ANR grant BLANC 07-1_184534: this work was done in the context of project SATTIC. This research is also partially supported by the Interuniversity Attraction Poles Programme (Belgian State, Belgian Science Policy).

References

1. McKay, B.D.: Practical graph isomorphism. Congressus Numerantium 30, 45–87 (1981)
2. Cordella, L.P., Foggia, P., Sansone, C., Vento, M.: Performance evaluation of the vf graph matching algorithm. In: ICIAP, pp. 1172–1177 (1999)
3. Umeyama, S.: An eigendecomposition approach to weighted graph matching problems. IEEE Transactions on Pattern Analysis and Machine Intelligence 10(5), 695–703 (1988)
4. Almohamad, H., Duffuaa, S.: A linear programming approach for the weighted graph matching problem. IEEE Transactions on Pattern Analysis and Machine Intelligence 15(5), 522–525 (1993)
5. Zaslavskiy, M., Bach, F., Vert, J.: A path following algorithm for the graph matching problem. In: Elmoataz, A., Lezoray, O., Nouboud, F., Mammass, D. (eds.) ICISP 2008. LNCS, vol. 5099, pp. 329–337. Springer, Heidelberg (2008)
6. Cross, A., Wilson, R., Hancock, E.: Inexact graph matching using genetic search. Pattern Recognition 30, 953–970 (1997)
7. Champin, P.A., Solnon, C.: Measuring the similarity of labeled graphs. In: Ashley, K.D., Bridge, D.G. (eds.) ICCBR 2003. LNCS, vol. 2689, pp. 80–95. Springer, Heidelberg (2003)
8. Sorlin, S., Solnon, C.: Reactive tabu search for measuring graph similarity. In: Brun, L., Vento, M. (eds.) GbRPR 2005. LNCS, vol. 3434, pp. 172–182. Springer, Heidelberg (2005)
9. Sammoud, O., Solnon, C., Ghedira, K.: Ant Algorithm for the Graph Matching Problem. In: Raidl, G.R., Gottlieb, J. (eds.) EvoCOP 2005. LNCS, vol. 3448, pp. 213–223. Springer, Heidelberg (2005)
10. Monette, J.N., Deville, Y., Van Hentenryck, P.: AEON: Synthesizing scheduling algorithms from high-level models. In: Proceedings of 2009 INFORMS Computing Society Conference (2009)
11. Vosselman, G.: Relational Matching. LNCS, vol. 628. Springer, Heidelberg (1992)
12. Bunke, H.: On a relation between graph edit distance and maximum common subgraph. Pattern Recognition Letters 18, 689–694 (1997)

13. Zaslavskiy, M., Bach, F., Vert, J.P.: A path following algorithm for graph matching. In: Elmoataz, A., Lezoray, O., Nouboud, F., Mammass, D. (eds.) ICISP 2008 2008. LNCS, vol. 5099, pp. 329–337. Springer, Heidelberg (2008)
14. Ambauen, R., Fischer, S., Bunke, H.: Graph Edit Distance with Node Splitting and Merging. In: Hancock, E.R., Vento, M. (eds.) GbRPR 2003. LNCS, vol. 2726, pp. 95–106. Springer, Heidelberg (2003)
15. Tsang, E.: Foundations of Constraint Satisfaction. Academic Press, London (1993)
16. Sorlin, S., Solnon, C.: A parametric filtering algorithm for the graph isomorphism problem. Constraints 13(4), 518–537 (2008)
17. Zampelli, S., Deville, Y., Solnon, C., Sorlin, S., Dupont, P.: Filtering for Subgraph Isomorphism. In: Bessière, C. (ed.) CP 2007. LNCS, vol. 4741, pp. 728–742. Springer, Heidelberg (2007)
18. Van Hentenryck, P., Michel, L.: Constraint-Based Local Search. The MIT Press, Cambridge (2005)
19. Quimper, C., Golynski, A., Lopez-Ortiz, A., van Beek, P.: An efficient bounds consistency algorithm for the global cardinality constraint. Constraints 10(1), 115–135 (2005)
20. Larrosa, J., Valiente, G.: Constraint satisfaction algorithms for graph pattern matching. Mathematical. Structures in Comp. Sci. 12(4), 403–422 (2002)
21. Barabasi, A.L.: Linked: How Everything Is Connected to Everything Else and What It Means. Plume (2003)
22. De Santo, M., Foggia, P., Sansone, C., Vento, M.: A large database of graphs and its use for benchmarking graph isomorphism algorithms. Pattern Recogn. Lett. 24(8), 1067–1079 (2003)
23. Bunke, H., Foggia, P., Guidobaldi, C., Sansone, C., Vento, M.: A comparison of algorithms for maximum common subgraph on randomly connected graphs. In: Caelli, T.M., Amin, A., Duin, R.P.W., Kamel, M.S., de Ridder, D. (eds.) SPR 2002 and SSPR 2002. LNCS, vol. 2396, pp. 123–132. Springer, Heidelberg (2002)
24. Sorlin, S.: Mesurer la similarité de graphes. PhD thesis, Université Claude Bernard, Lyon I, France (2006)

Constraint Representations and Structural Tractability

David A. Cohen, Martin J. Green*, and Chris Houghton

Department of Computer Science,
Royal Holloway, University of London, UK

Abstract. The intractability of the general CSP has motivated the search for restrictions which lead to tractable fragments. One way to achieve tractability is to restrict the structure of the instances. As much of the work in this area arises from similar work on databases it has been a natural assumption that all constraint relations are explicitly represented. If this is the case then all instances with an acyclic hypergraph structure are tractable. Unfortunately this result does not hold if we are allowed to represent constraint relations implicitly: the class of SAT instances with acyclic hypergraph structure is NP-hard.

Continuing the work of Chen and Grohe on the succinct GDNF representation we develop the theory of structural tractability for an extension to the table constraint that has a succinct representation of SAT clauses. This mixed representation is less succinct than the GDNF representation but more succinct than the table representation.

We prove a strict hierarchy of structural tractability for the GDNF, the mixed, and the explicit representations of constraint relations. Using this proof we are able to show that the mixed representation provides novel tractable structural classes. Since the mixed representation naturally extends SAT, this provides a useful result, extending known structural tractability results for SAT.

Under a natural restriction we are able precisely to capture the tractable structural classes for this mixed representation. This gives us an extension of Grohe's dichotomy theorem for the tractability of classes of relational structures with a fixed signature. In particular it captures the tractability of some classes of unbounded arity, specifically the class of CSPs with precisely one constraint.

1 Introduction

A *CSP* is a collection of variables, some subsets of which are constrained. The CSP paradigm has proved to be very useful in a variety of contexts, for example planning [1] and scheduling [2].

In a CSP each variable has a *domain* of possible values. Each *constraint* enforces that the assignment of values to the variables in its constraint *scope* lie in its constraint *relation*.

* This work was supported by EPSRC Grant EP/C525949/1.

I.P. Gent (Ed.): CP 2009, LNCS 5732, pp. 289–303, 2009.

We define a class of CSPs to be *tractable* if there is a solution algorithm which solves any member of the class in polynomial time. There has been considerable success in identifying tractable CSP classes [3,4].

A *relational structure* permitted by a CSP is a labelled ordered hypergraph of the constraint scopes: hyperedges with the same label having the same relation. A class of CSPs is called *structural* if membership of the class is determined purely by consideration of the *relational structures* of the CSPs. This paper is concerned with tractable structural classes.

Since many structural tractability results are derived from similar results for relational databases [5] it has been the standard assumption that constraint relations are listed explicitly in the *representation* of a CSP. The assumption is at odds with the way that many modern solvers process constraints. During backtrack search propagators are used that, when the search state changes, efficiently test for membership and propagate consequences. The use of propagators makes the use of implicit representations natural for some types of constraint.

For example, *SAT instances* are encoded using *clauses*, which it would be senseless to represent explicity. Using clauses allows CSP solvers to efficiently propagate constraints using so-called watched literals [6]. In fact, there are many other types of *global constraint* [7], with natural implicit representations and effective propagators.

When both the domain size and the constraint arity are bounded tractability is unaffected by representation of constraint relations. The following example shows that tractability can critically depend on representation.

Example 1. The hypergraph structure of a CSP is simply a hypergraph whose vertices are the variables and whose hyperedges are the constraint scopes.

Acyclicity is the natural generalisation to hypergraphs of tree structure for graphs and has equivalent desirable properties: the class of CSPs with acyclic hypergraph structure, with extensionally represented constraints, is tractable [5].

Acyclicity of a hypergraph can be determined by repeatedly: removing hyperedges contained in other hyperedges and deleting isolated vertices. A hypergraph is acyclic if this algorithm deletes all vertices [5,8]. For graphs it is easy to see that the algorithm characterises forests, hence acyclicity.

The standard SAT encoding represents a clause with r variables by listing the scope and sign of each literal in the single disallowed assignment. The class of SAT instances is not tractable [9].

We now show that the class of SAT instances with acyclic hypergraph structure, with standard SAT representation, is not tractable since any SAT instance can be reduced in polynomial time to a pair of acyclic SAT instances.

Let P be any SAT instance. We construct two SAT instances.

- Obtain P_F by adding the clause "not every variable is False" to P.
- Obtain P_T by adding the clause "not every variable is True" to P.

Since these instances have scopes containing all variables the acyclicity identification algorithm trivially shows them to be acyclic.

If P has a solution, then at least one of the two instances, P_T and P_F, will have a solution. Conversely, if at least one of these two acyclic instances has a solution, then this will be a solution to P.

This has serious implications for the theory of tractability. We have to understand which structures are tractable when we use more compact representations. Some work has begun which addresses this deficit.

1.1 Our Contribution

In this paper we continue the study of succinct representations in the theory of CSP tractability.

Recently, Chen and Grohe [10] described the GDNF representation and gave an exact characterisation of the tractable structural classes. Continuing their work we develop the theory of structural tractability for an extension to the table constraint that has a succinct representation for SAT clauses. This mixed representation is less succinct than the GDNF representation but more succinct than the table representation.

We show that the set of tractable structural classes for the mixed representation is distinct both from the set of tractable structural classes for the (more succinct) GDNF representation and from the set of tractable structural classes for the (less succinct) table representation. As part of this proof we describe a class of structures that is not tractable for the GDNF representation yet is tractable for the mixed representation. This class shows that the mixed representation provides novel tractable structural classes. Since the mixed representation naturally extends SAT this extends known structural tractability results for SAT [11,12].

We have been unable to find a dichotomy for general structural classes for the mixed CSP representation. However, under a natural restriction we are able precisely to capture the tractable structural classes for this mixed representation. This result extends slightly the dichotomy result for structural classes with explicitly represented relations proved by Grohe [13].

2 Background and Definitions

In this section we define basic notions and describe some of the key results in the area.

Definition 1. *A* ***CSP*** *is a triple $\langle V, D, C\rangle$, where V is a set of* ***variables****, D is a function which maps each variable $v \in V$ to a set of values, $D(v)$, called the* ***domain*** *of v, and C is a set of* ***constraints****.*

Each element of C is a pair $\langle \sigma, \rho\rangle$ where $\sigma \in V^*$ is a list of variables called the ***scope****. The length r of σ is called the* ***arity*** *of the constraint. The* ***relation****, ρ, of the constraint is a subset of the product $D(\sigma[1]) \times \cdots \times D(\sigma[r])$ of the domains of the variables in the scope.*

A ***solution*** *to $\langle V, D, C\rangle$ is a mapping s from V into $\bigcup_{v\in V} D(v)$ which* ***satisfies*** *each constraint: that is, for each $\langle \sigma, \rho\rangle \in C$ we have $s(\sigma) \in \rho$.*

In order to discuss the complexity of solving classes of CSPs we need to define a *representation*. We will define several such representations that differ only in the way they represent constraints.

Definition 2. *Let $c = \langle \sigma, \rho \rangle$ be a constraint of arity r.*

The ***positive*** *(or* ***extensional*** *or* ***table***) ***representation*** *of c is a list of the variables in σ followed by the tuples in ρ.*

The ***negative representation*** *of c is a list of the variables in σ followed by the tuples in $D(\sigma[1]) \times \cdots \times D(\sigma[r])$ that are not in ρ.*

The ***mixed representation*** *of c is a Boolean flag, which if T is followed by the positive representation of c, and if F by the negative representation.*

The ***GDNF representation*** *[10,14,15,16] of c is a list of the variables in σ followed by a list of expressions of the form $A[1] \times \cdots \times A[r]$ where ρ is the union of these set products.*

Extensional representation of constraint relations is unnecessarily verbose and so leads to anomalies in the theory of tractability. The same is the case for explicitly listed domains and so instead we want implicitly to infer the domain elements that can occur in solutions from the constraint relations. In the positive and GDNF representations this is sufficient.

In the negative (and hence the mixed) representations domain values which do not occur as literals in any forbidden tuple could appear in solutions. All such (missing) literals in any domain are equivalent in the sense that they are interchangeable in any solution.

So, as a general method, we represent any domain containing such elements with the symbol $+$ and the remaining domains, all of whose elements are inferred by the representations of constraints, using the symbol $-$.

Definition 3. *Let Θ be any representation. A Θ* ***instance*** *is a list of variables, each followed by the symbol $+$ or $-$ representing its domain, followed by a list of Θ representations of constraints.*

We say that a class $\mathcal{T}$ of Θ instances is ***tractable*** *if there exists a solution algorithm that runs in time polynomial in the size of the input Θ instance.*

Example 2. Consider the SAT instance whose implicit logical clauses are:

$v_1 \vee v_2$, $\overline{v_2} \vee \overline{v_3} \vee v_4$ and $v_1 \vee v_4$.

A mixed representation of this CSP is the following:

$$\begin{array}{l} v_1, -, v_2, -, v_3, +, v_4, -, \\ \qquad T, (v_1, v_2), (F, T), (T, F), (TT) \\ \quad F, (v_2, v_3, v_4), (T, T, F), \\ \qquad T, (v_1, v_4), (T, F), (F, T), (T, T) \end{array}$$

Here only the domain of v_3 cannot be completely determined from the constraint tuples.

When we are comparing different representations a key notion will be the relative size of the two different representations.

Definition 4. *We say that a representation Φ is* **as succinct** *as a representation Θ if there is a polynomial p for which, given any CSP P represented in Θ with size $|P|_\Theta$ there exists a representation of P in Φ of size at most $p(|P|_\Theta)$.*

2.1 Key Results

Much of the theoretical work on CSP structure is now couched in terms of relational structures rather than simple hypergraphs. This is because relational structures are rich enough to capture tractable classes that are not definable just using the hypergraph structure. Example 4 in Section 2.2 demonstrates this using simple structures.

We can motivate relational structures in a less theoretical way by observing that we sometimes build CSPs from others in specified ways. We may take the dual, or a hidden variable representation. We may add auxiliary variables or use channelling constraints. In each such case, certain of the constraints in our instance may be limited to a different CSP language from others. For example, some channelling constraints or constraints to auxiliary variables may be functional. Such extra information about the structure of an instance is simply not available once we have abstracted to a hypergraph.

On the other hand the relational structure of an instance captures precisely the fact that scopes may be of different types. Instead of just having one hyperedge relation we have several: one for each type of constraint. Here theory and practice meet. A CSP *may* not get harder when it is reformulated even though the hypergraph structure may be more complex. Structural tractability using a relational structure begins to capture this idea.

Definition 5. *A* ***relational structure*** *$\langle V, R_1, \ldots, R_m\rangle$ is a set V and a list of* ***relations*** *over V.*

The CSP $\langle V, D, C\rangle$ **permits** *structure $\langle V, R_1, \ldots, R_m\rangle$ when there is a partition of the constraints $C = C_1 \cup \cdots \cup C_m$ and, for $i = 1, \ldots, m$, each $c \in C_i$ has the same relation and $R_i = \{\sigma \mid \langle \sigma, \rho\rangle \in C_i\}$.*

Example 3. Consider again the SAT instance of Example 2 whose implicit logical clauses are:

$v_1 \vee v_2$, $\overline{v_2} \vee \overline{v_3} \vee v_4$ and $v_1 \vee v_4$.

This instance has the same relation on two scopes and so permits the two distinct structures:

$\langle V, \{\langle v_1, v_2\rangle, \langle v_1, v_4\rangle\}, \{\langle v_2, v_3, v_4\rangle\}\rangle$ and
$\langle V, \{\langle v_1, v_2\rangle\}, \{\langle v_1, v_4\rangle\}, \{\langle v_2, v_3, v_4\rangle\}\rangle$.

Definition 6. *Let $\mathcal{H}$ be a class of relational structures and Θ be any representation. We define $\Theta(\mathcal{H})$ to be the class of Θ representations of CSPs permitting a structure in $\mathcal{H}$.*

Specifically:
- Positive($\mathcal{H}$) *is the class of Positive representations of CSPs permitting a structure in* $\mathcal{H}$.
- Negative($\mathcal{H}$) *is the class of Negative representations of CSPs permitting a structure in* $\mathcal{H}$.
- Mixed($\mathcal{H}$) *is the class of Mixed representations of CSPs permitting a structure in* $\mathcal{H}$.
- GDNF($\mathcal{H}$) *is the class of GDNF representations of CSPs permitting a structure in* $\mathcal{H}$.

For bounded arity and the positive representation, the structural classes are precisely determined by Theorem 1. The statement of this theorem refers to cores and tree width of relational structures so we define these notions here[1].

Definition 7. *A relational structure* $\langle V, R_1, \ldots, R_m\rangle$ *is a* ***substructure*** *of a relational structure* $\langle V', R'_1, \ldots, R'_m\rangle$ *if* $V \subseteq V'$ *and, for each* i, $R_i \subseteq R'_i$.

A ***homomorphism*** *from a relational structure* $\langle V, R_1, \ldots, R_m\rangle$ *to a relational structure* $\langle V', R'_1, \ldots, R'_m\rangle$ *is a mapping* $h : V \to V'$ *such that for all* i *and all tuples* $t \in R_i$ *we have* $h(t) \in R'_i$.

A relational structure $\mathbb{S}$ *is a* ***core*** *if there is no homomorphism from* $\mathbb{S}$ *to a proper substructure of* $\mathbb{S}$. *A core of a relational structure* $\mathbb{S}$ *is a substructure* $\mathbb{S}'$ *of* $\mathbb{S}$ *such that there is a homomorphism from* $\mathbb{S}$ *to* $\mathbb{S}'$ *and* $\mathbb{S}'$ *is a core. It is well known that all cores of a relational structure* $\mathbb{S}$ *are isomorphic. Therefore, we often speak of the core,* Core($\mathbb{S}$)*, of* $\mathbb{S}$. *For a class,* $\mathcal{H}$*, of relational structures, we denote by* Core($\mathcal{H}$) *the class of relational structures* $\{\text{Core}(\mathbb{S}) \mid \mathbb{S} \in \mathcal{H}\}$.

Definition 8. *Let* $\mathbb{S} = \langle V, R_1, \ldots, R_m\rangle$ *be any relational structure.*

The ***Gaifman graph****,* $G(\mathbb{S})$*, of* $\mathbb{S}$ *has vertex set* V. *A pair of vertices* $\{v, w\}$ *is an edge of* $G(\mathbb{S})$ *when there is a tuple* t *of some* R_i *containing both* v *and* w.

Given an ordering $v_1 < \ldots, < v_n$ *of* V *the* ***induced graph*** $I(S, <)$ *is obtained from the Gaifman graph by adding edges. Process the vertices, in order, from* v_n *to* v_1. *When* v_r *is processed, and for some* $i, j < r$ *for which* $\{v_i, v_r\}$ *and* $\{v_j, v_r\}$ *are both edges of* $I(S, <)$*, we add the edge* $\{v_i, v_j\}$ *to* $I(S, <)$.

After this process, the ***width*** *of any* v_r *is the number of its earlier neighbours:* $|\{v_i \mid i < r, \{v_i, v_r\} \in I(S, <)\}|$. *The width of* $\mathbb{S}$*, for this ordering, is the maximum width of any* $v \in V$.

The ***tree width****,* tw($\mathbb{S}$)*, of* $\mathbb{S}$ *is its minimal width over all orderings. For a class,* $\mathcal{H}$*, of relational structures, we denote by* tw($\mathcal{H}$) *the maximum tree width of any structure in* $\mathcal{H}$. *We say* tw($\mathcal{H}$) $= \infty$ *if the tree width is unbounded.*

Theorem 1 (Grohe [13], Corollary 19). *Assuming that W[1] is not FPT*[2]*, for every recursively enumerable*[3] *class* $\mathcal{H}$ *of relational structures of bounded arity,* Positive($\mathcal{H}$) *is tractable if and only if* tw(Core($\mathcal{H}$)) $< \infty$.

---

1 We have simplified quoted theorems slightly for the context of this paper.

2 Many complexity results rely on a standard complexity theoretical assumption, that the class W[1] is not equal to the class FPT. This is the parameterised complexity analogue of the assumption that NP is not equal to P.

3 Recursively enumerable just means recognisable by a Turing machine. Non-recursively enumerable are of no practical interest.

The following straightforward result about how tractability is preserved when a representation is replaced with a less succinct representation is stated without proof.

Theorem 2. *Let Φ be a representation that is as succinct as a representation Θ. Any class which is structurally tractable for Φ is also structurally tractable for Θ.*

In other words, as the representation becomes more succinct, the tractable classes get smaller. This allows us to make a very important observation.

Corollary 1. *Assuming that W[1] is not FPT. Let $\mathcal{H}$ be a recursively enumerable class of relational structures with bounded arity.* $\text{Negative}(\mathcal{H})$, $\text{Mixed}(\mathcal{H})$, *and* $\text{GDNF}(\mathcal{H})$ *are tractable for any bounded domain size if and only if* $\text{tw}(\text{Core}(\mathcal{H})) < \infty$.

Proof. For bounded arity and domain size the positive representation is as succinct as each of these three other representations. □

For any succinct representation, it is thus only classes of structures with unbounded arity, or classes of CSPs with unbounded domain size, whose tractability must still be characterised.

2.2 Unbounded Arity and Succinct Representations

Unfortunately, as we have shown in Example 1, acyclicity is not a sufficient condition to guarantee tractability for succinct representations.

Recently, Chen and Grohe [10] described the GDNF representation. They identified precisely the tractable structural classes. Their characterisation of tractable classes relies on the so-called incidence graph of a relational structure.

Definition 9. *Let $\mathbb{S} = \langle V, R_1, \ldots, R_m \rangle$ be any relational structure and let:*

$$L(\mathbb{S}) = \{\langle i, t \rangle \mid i \in \{1, \ldots, m\}, t \in R_i\}.$$

*The **incidence graph**,* $\text{IncG}(\mathbb{S})$*, of $\mathbb{S}$ is the bipartite graph, $\langle V', E \rangle$, where*

- $V' = V \cup L(\mathbb{S})$, *and*
- $E = \{\langle v, \langle i, t \rangle \rangle \mid v \in V, \langle i, t \rangle \in L(\mathbb{S}), \exists j,\ t[j] = v\}$.

*The **incidence width** of relational structure $\mathbb{S}$, denoted* $\text{iw}(\mathbb{S})$*, is the tree width of its incidence graph, that is,* $\text{tw}(\text{IncG}(\mathbb{S}))$*. We say* $\text{iw}(\mathcal{H}) = \infty$ *if the incidence width is unbounded.*

Theorem 3 (Theorem 14 of Chen and Grohe [10]). *Assuming that W[1] is not FPT. Let $\mathcal{H}$ be a recursively enumerable class of relational structures. Then* $\text{GDNF}(\mathcal{H})$ *is tractable if and only if* $\text{iw}(\text{Core}(\mathcal{H})) < \infty$.

The following simple example shows that, for structural tractability, it is not enough to consider just hypergraph structure.

Example 4. Consider the class of relational structures $\mathcal{H} = \{\mathbb{A}_\rangle \mid \rangle = \infty, \in, \ldots\}$, where $\mathbb{A}_n = \langle \{1, \ldots, n\}, R_1, \ldots, R_n \rangle$, and each $R_i = \{\langle 1, \ldots, n \rangle\}$. It is clear that $\mathbb{A}_n$ cannot be be homorphically equivalent to any of its proper substructures: each relation has only one tuple. So $\mathbb{A}_n$ is a core. Furthermore its incidence graph is the complete n, n bipartite graph which has tree width n. Choose any infinite subset $\mathcal{H}'$ of $\mathcal{H}$. By Theorem 3, we know that GDNF($\mathcal{H}'$) is intractable.

Recall that the hypergraph structure of a CSP is simply a hypergraph whose vertices are the variables of the instance and whose edges are the constraint scopes.

Now let $\mathcal{M}$ be any class of hypergraphs of unbounded arity. It is no harder to solve the CSPs whose hypergraph structures are substructures of those in $\mathcal{M}$ than it to solve the CSPs whose hypergraph structures are precisely the hypergraphs of $\mathcal{M}$. Certainly the hypergraphs in $\mathcal{M}$ have substructures of unbounded arity, each of which is a hypergraph containing a single hyperedge. For the hypergraph with a single hyperedge with k vertices there is a CSP with this hypergraph structure that permits the relational structure $\mathbb{A}_k$. So the set of CSPs with hypergraph structure in $\mathcal{M}$ permit an infinite number of relational structures of $\mathcal{H}$. So the class of CSPs whose hypergraph structure lies in $\mathcal{M}$ is intractable for the GDNF representation.

On the other hand, consider the class of relational structures $\mathcal{H}' = \{\mathbb{B}_i \mid i = 1, 2, \ldots\}$, where $\mathbb{B}_n$ has universe $\{1, \ldots, n\}$ and just one n-ary relation containing just one tuple. The incidence width of each such relational structure is one. This class of relational structures has unbounded arity but GDNF($\mathcal{H}'$) is tractable.

So, just considering hypergraph structure there are **no** tractable structural classes of unbounded arity for the GDNF representation. By considering relational structure we can find many such tractable structural classes.

3 The Mixed Representation

In this section we prove a dichotomy theorem for structural tractability for the mixed representation. We first need to define the *interaction width* of a relational structure. Intuitively, if variables are in a very strong sense equivalent then they can be merged before solving a CSP instance. The interaction width captures the resulting maximum arity after this reduction and allows us to extend tractability results to classes with unbounded arity. As a simple example, any tractable bounded arity class of SAT instances remains tractable after arbitrary duplication of any subset of the variables. However, without the reduction using interaction width even this class is not captured by current theory.

Definition 10. *For any relational structure $\mathbb{S} = \langle V, R_1, \ldots, R_m \rangle$ we define, for any $v \in V$, $\tau(v)$ to be $\{\langle i, t \rangle \mid t \in R_i$ and $\exists j,\ t[j] = v\}$.*

We say that $\{v', w'\} \subseteq V$ is $\mathbb{S}$-similar to $\{v, w\} \subseteq V$ if there exists some E_i and $t, t' \in E_i$ where for some j, k, $t[j,k] = \langle v, w \rangle$ and $t'[j,k] = \langle v', w' \rangle$. The equivalence relation $\mathbb{S}$-equivalence is the transitive closure of this relation.

We say that $v \in V$ and $w \in V$ are τ-equivalent if either $v = w$ or, for every set $\{v', w'\} \subseteq V$ which is $\mathbb{S}$-equivalent to $\{v, w\}$ we have $\tau(v') = \tau(w')$. This is an equivalence relation for V and we denote by $\mathrm{Reg}(v)$ the τ-equivalence class of v.

The ***interaction width*** *of $\mathbb{S}$ is:*

$$\mathrm{intw}(\mathbb{S}) = \max_{\langle v_1,\ldots,v_k\rangle \in E_i}^{i=1,\ldots,n} |\{\mathrm{Reg}(v_1), \ldots, \mathrm{Reg}(v_k)\}| \, .$$

For a class, $\mathcal{H}$, of relational structures, we denote by $\mathrm{intw}(\mathcal{H})$ the maximum interaction width of any structure in $\mathcal{H}$. We say $\mathrm{intw}(\mathcal{H}) = \infty$ if the interaction width is unbounded.

The proof of the dichotomy theorem requires the definition of appropriate merged structures which have trivial τ-equivalence classes. By construction τ-equivalence induces an equivalence on the columns of the relations of any relational structure. Hence the following are well-defined.

Definition 11. *Let $\mathbb{S} = \langle V, R_1, \ldots, R_m\rangle$ be a relational structure. For any R_i choose $\mathrm{mergeCol}(R_i)$ to be any tuple of indices of representatives of the distinct τ-equivalence classes of the columns of R_i.*

For any $t \in R_i$ we now define $\mathrm{mergeCol}(t)$ to be the tuple of τ-equivalence classes of the variables in t, which occur at the indices of the tuple $\mathrm{mergeCol}(R_i)$. Then $\mathrm{Mrg}(R_i) = \{\mathrm{mergeCol}(t) \mid t \in R_i\}$.

We can then define the ***merged structure*** *of $\mathbb{S}$ to be*

$$\mathrm{Mrg}(\mathbb{S}) = \langle\{\mathrm{Reg}(v) \mid v \in V\}, \mathrm{Mrg}(R_1), \ldots, \mathrm{Mrg}(R_m)\rangle \quad .$$

For a class, $\mathcal{H}$, of relational structures, we denote by $\mathrm{Mrg}(\mathcal{H})$ the set of merged structures of the structures in $\mathcal{H}$.

Definition 12. *Let $\mathbb{S} = \langle V, R_1, \ldots, R_m\rangle$ be a relational structure and $\mathrm{Reg}(v_1), \ldots, \mathrm{Reg}(v_q)$ be an enumeration of the distinct τ-equivalence classes. For $i = 1, \ldots, q$ define the binary relation $\mathrm{Reg}_i = \{\langle \mathrm{Reg}(v_i), y\rangle \mid y \in \mathrm{Reg}(v_i)\}$.*

We define the ***extended merged structure*** *of $\mathbb{S}$ to be*

$$\mathrm{ExtMrg}(\mathbb{S}) = \left\langle V \cup \{\mathrm{Reg}(x) \mid x \in V\}, \mathrm{Mrg}(R_1), \ldots, \mathrm{Mrg}(R_m), \mathrm{Reg}_1, \ldots, \mathrm{Reg}_q \right\rangle .$$

We can now state the main result of this section.

Theorem 4. *Assuming that W[1] is not FPT. Let $\mathcal{H}$ be a recursively enumerable class of relational structures with bounded interaction width. The class $\mathrm{Mixed}(\mathcal{H})$ is tractable if and only if $\mathrm{tw}(\mathrm{Core}(\mathrm{Mrg}(\mathcal{H}))) < \infty$.*

This result is a natural extension of Grohe's result stated as Theorem 1 which characterises some additional tractable classes of unbounded arity and in particular includes the class of CSPs with precisely one constraint.

The proof of this theorem is quite involved, so in order to save space we will only sketch it here.

Proof. Let $\mathcal{H}$ be any class of relational structures with bounded interaction width.

Suppose $\mathrm{tw}(\mathrm{Core}(\mathrm{Mrg}(\mathcal{H}))) = \infty$.

In this case we will show that $\mathrm{Mixed}(\mathcal{H})$ is intractable.
Consider the class of relational structures $\mathrm{ExtMrg}(\mathcal{H})$.
Since for all $\mathbb{S} \in \mathcal{H}$ we have that $\mathrm{Mrg}(\mathbb{S})$ is a substructure of $\mathrm{ExtMrg}(\mathbb{S})$ we have that $\mathrm{tw}(\mathrm{Core}(\mathrm{ExtMrg}(\mathcal{H}))) \geq \mathrm{tw}(\mathrm{Core}(\mathrm{Mrg}(\mathcal{H})))$.
So $\mathrm{tw}(\mathrm{Core}(\mathrm{ExtMrg}(\mathcal{H}))) = \infty$.
However, $\mathrm{ExtMrg}(\mathcal{H})$ has bounded arity, since the arity of $\mathrm{Mrg}(\mathcal{H})$ is equal to $\mathrm{intw}(\mathcal{H})$ and the extended merged structure adds only binary relations.
By Grohe's result, Theorem 1, $\mathrm{Positive}(\mathrm{ExtMrg}(\mathcal{H}))$ is intractable.
Given any instance P in $\mathrm{Positive}(\mathrm{ExtMrg}(\mathcal{H}))$, for each assignment to an interaction region variable in the extended merged structure, we choose a representative assignment for each of the variables in that interaction region, allowed by P. We use these extensions to generate constraint tuples in an "unmerged" version of P.
This "unmerges" P, in polynomial time, to an instance of $\mathrm{Positive}(\mathcal{H})$, which is thereby shown to be an intractable subset of $\mathrm{Mixed}(\mathcal{H})$.

Suppose $\mathrm{tw}(\mathrm{Core}(\mathrm{Mrg}(\mathcal{H}))) < \infty$.

In this case we will show that $\mathrm{Mixed}(\mathcal{H})$ is tractable.
We solve an instance $P \in \mathrm{Mixed}(\mathcal{H})$ in polynomial time as follows.
Choose an order for the variables of each interaction region. Now merge assignments to these variables into tuples which will be the domain values for the interaction region variable in the merged structure. We now build the tuples for the constraint relations of the merged structure. These are lists of tuples, one for each interaction region in the scope.
We then build the domain for each interaction region variable. If the variable is in the scope of any positive constraint then the domain is $-$. If it is only in the scope of negative constraints then it may have domain $+$.
This reduces the class $\mathrm{Mixed}(\mathcal{H})$ to a mixed class with bounded arity where the tree width of the core is bounded.
We now have to convert negative constraints to positive constraints in a way that preserves solutions.
Suppose that every constraint with some particular region in its scope is negative. If the domain of this variable is $+$ then remove each such constraint and add a new unary constraint on this variable which just allows the "special" value, which we will denote by $*$.
We can just enumerate all positive acceptable tuples for our remaining negative constraints in polynomial time, since the domains of each interaction region are inferred.
We have now reduced our original instances to the tractable class of positive instances with bounded arity and cores of bounded tree width. We lastly have to explain how to convert a solution to one of these derived instances to an original solution. The only difficulty here arises for variables which have been assigned the value $*$. All other variables have been assigned tuple values and these convert directly to assignments of the original variables.

However the $*$ value can only be given if each original constraint on that interaction region was negative. In this case we can enumerate all possible tuples, for the original region, until we find one not inferred by the negative constraints. This value can then be assigned. This is polynomial time as it requires listing each inferred region tuple (at most) once and finally the chosen acceptable tuple. □

4 Place in the Hierarchy

In this section we prove that there is a *strict* structural tractability hierarchy for the mixed, GDNF and positive representations. In doing so, we describe a class of structures, $\mathcal{H}$, which is not tractable for the GDNF representation yet is tractable for the mixed representation. This shows that the mixed representation provides novel tractable structural classes. Since the mixed representation naturally extends SAT we have extended the known structural tractability results for SAT [11,12].

Proposition 1. *Let* $\mathcal{H}$ *by any class of relational structures. If* Mixed($\mathcal{H}$) *is tractable then so is* Positive($\mathcal{H}$). *If* GDNF($\mathcal{H}$) *is tractable then so is* Mixed($\mathcal{H}$).

Proof. Since the positive representation is only linearly larger than the mixed representation it follows immediately that the mixed is as succinct as the positive.

Given a positively represented constraint we can use the straightforward construction of Chen and Grohe [10] which generates a product of unary sets for each allowed assignment. This may be done in linear time, and the resulting GDNF representation of the constraint is (approximately) the same size.

Given a negatively represented constraint we can use a result by Katsirelos and Walsh [16]. They show that any negatively represented constraint may be converted to an equivalent GDNF representation of the constraint, with a polynomial number of set products in polynomial time using a simple algorithm which descends a decision tree with a polynomial number of leaves.

Using Theorem 2 we are done. □

The following theorem then makes the hierarchy strict.

Theorem 5. *There is a class* $\mathcal{H}$ *of relational structures for which* Positive($\mathcal{H}$) *is tractable but* Mixed($\mathcal{H}$) *is intractable.*

There is a class $\mathcal{H}$ *of relational structures for which* Mixed($\mathcal{H}$) *is tractable but* GDNF($\mathcal{H}$) *is intractable.*

Proof. Recall Example 1 where we showed that the class of acyclic structures is not tractable for the standard encoding of SAT. This implies that this class is not tractable for the mixed representation.

Consider the relational structure $\mathbb{H}_n = \langle V_n, E_1, E_2, \ldots, E_n \rangle$ where:

$$V_n = \{v_1, v_2, \ldots v_n, w_1^1, w_2^1, w_2^2, \ldots, w_n^1, \ldots, w_n^n\}$$

and the relations are:

$$E_1 = \{e_1 = \langle v_1, v_2, \ldots, v_n, w_1^1 \rangle\}$$
$$\vdots$$
$$E_n = \{e_n = \langle v_1, v_2, \ldots, v_n, w_n^1, \ldots, w_n^n \rangle\}$$

We depict the structural hypergraph, $H(\mathbb{H}_n)$, of $\mathbb{H}_n$ in Fig. 1 (left).

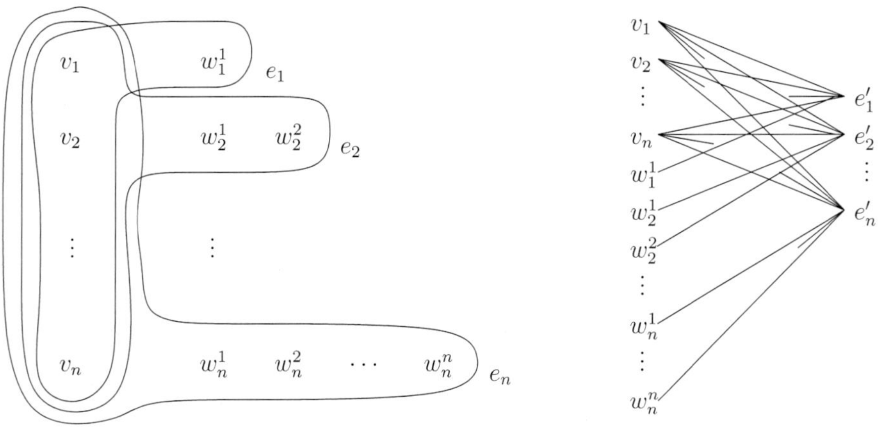

Fig. 1. (left) $H(\mathbb{H}_n)$ (right) $\mathrm{IncG}(\mathbb{H}_n)$

The interaction width of $\mathbb{H}_n$ is 2, since for each i, the relational tuple e_i has only two interaction regions, $\{v_1, v_2, \ldots, v_n\}$ and $\{w_i^1, \ldots, w_i^i\}$.

What is more, it is straightforward to show that the merged structure of $\mathbb{H}_n$ is tree-structured, so the tree width of the merged structure of $\mathbb{H}_n$ is 1.

By consideration of the arities of the relations we can see that each of these relational structures is a core. So to determine the tractability of these structures for the GDNF representation we have only to consider their incidence width.

The incidence graph, $\mathrm{IncG}(\mathbb{H}_n)$, of $\mathbb{H}_n$ is the bipartite graph $\langle V_n', E_n' \rangle$ such that $V_n' = V_n \cup L(\mathbb{H}_n)$, where

$$L(\mathbb{H}_n) = \{e_1' = \langle 1, e_1 \rangle, e_2' = \langle 2, e_2 \rangle, \ldots, e_n' = \langle n, e_n \rangle\}, \text{ and}$$

$$\begin{aligned} E_n' = &\{\langle v_1, e_1' \rangle, \ldots, \langle v_n, e_1' \rangle, \langle w_1^1, e_1' \rangle\} \\ &\cup \{\langle v_1, e_2' \rangle, \ldots, \langle v_n, e_2' \rangle, \langle w_2^1, e_2' \rangle, \langle w_2^2, e_2' \rangle\} \\ &\vdots \\ &\cup \{\langle v_1, e_n' \rangle, \ldots, \langle v_n, e_n' \rangle, \langle w_n^1, e_n' \rangle, \ldots, \langle w_n^n, e_n' \rangle\} . \end{aligned}$$

We depict the incidence graph, $\mathrm{IncG}(\mathbb{H}_n)$, of $\mathbb{H}_n$ in Fig. 1 (right). The tree width of $\mathrm{IncG}(\mathbb{H}_n)$ is at least n, since the vertices $\{v_1, v_2, \ldots, v_n\}$ and $\{e_1', e_2', \ldots, e_n'\}$ of $\mathrm{IncG}(\mathbb{H}_n)$ form a complete bipartite subgraph.

Let $\mathcal{H}$ be the class of relational structures $\{\mathbb{H}_i \mid i = 1, 2, \ldots\}$. The infinite class $\mathcal{H}$ has bounded interaction width together with bounded tree width of the merged structures, but has unbounded tree width of the incidence graphs. It follows that GDNF($\mathcal{H}$) is intractable and Mixed($\mathcal{H}$) is tractable. □

5 Bounded Interaction Width and Structural Tractability

In this section we describe a class relational structures which is tractable for the GDNF representation and so for the mixed representation, but has unbounded interaction width. This demonstrates that bounded interaction width is not a necessary condition for structural tractability for the mixed representation.

Consider the relational structure $\mathbb{J}_n = \langle V_n, F_n, G_1, \ldots, G_n \rangle$ where

$$V_n = \{v_1, v_2, \ldots v_n, w_1^1, w_2^1, w_2^2, \ldots, w_n^1, \ldots, w_n^n\}$$

and the relations are:

$$\begin{aligned} F_n &= \{f_n = \langle v_1, v_2, \ldots, v_n \rangle\} \\ G_1 &= \{g_1 = \langle v_1, w_1^1 \rangle\} \\ &\vdots \\ G_n &= \{g_n = \langle v_n, w_n^1, \ldots, w_n^n \rangle\} \end{aligned}$$

We depict the structural hypergraph, $H(\mathbb{J}_n)$, of $\mathbb{J}_n$ in Fig. 2 (left).

The interaction width of $\mathbb{J}_n$ is n, since the relational tuple f_n has a separate interaction with each of the n hyperedges $g_1, g_2, \ldots, g_n$.

Consideration of the arities of the relations is enough to show that these relational structures are cores, so to determine if this class of structures is tractable for the GDNF representation we need only consider their incidence width.

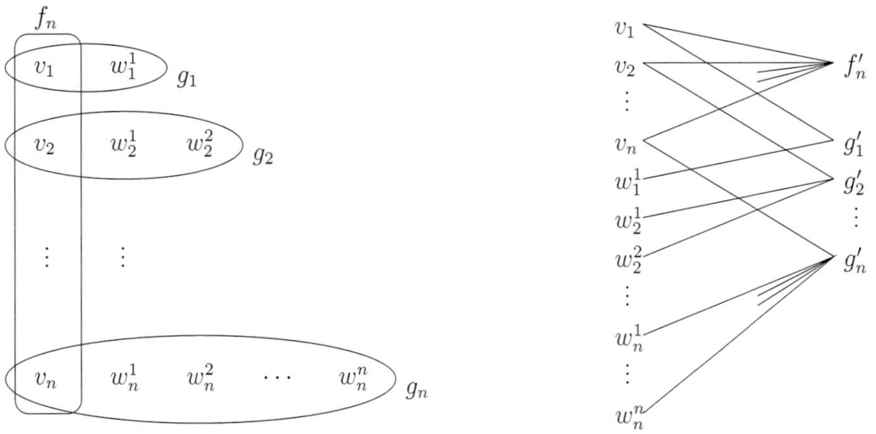

Fig. 2. (left) $H(\mathbb{J}_n)$ (right) IncG($\mathbb{J}_n$)

Consider the incidence graph, $\mathrm{IncG}(\mathbb{J}_n)$, of $\mathbb{J}_n$ which is the bipartite graph $\langle W, F'_n \rangle$ such that $W = V_n \cup L(\mathbb{J}_n)$, where

$$L(\mathbb{J}_n) = \{f'_n = \langle 1, f_n \rangle, g'_1 = \langle 2, g_1 \rangle, g'_2 = \langle 3, g_2 \rangle, \ldots, g'_n = \langle n+1, g_n \rangle\}, \text{ and}$$

$$\begin{aligned} F'_n = & \{\langle v_1, f'_n \rangle, \ldots, \langle v_n, f'_n \rangle\} \\ & \cup \{\langle v_1, g'_1 \rangle, \langle w_1^1, g'_1 \rangle\} \\ & \vdots \\ & \cup \{\langle v_n, g'_n \rangle, \langle w_n^1, g'_n \rangle \ldots, \langle w_n^n, g'_n \rangle\} . \end{aligned}$$

We depict the incidence graph, $\mathrm{IncG}(\mathbb{J}_n)$, of $\mathbb{J}_n$ in Fig. 2 (right). It is straightforward to show that the tree width of $\mathrm{IncG}(\mathbb{J}_n)$ is 1. An ordering of the vertices of $\mathrm{IncG}(\mathbb{J}_n)$ that witnesses this fact is

$$[f'_n, v_1, g'_1, w_1^1, v_2, g'_2, w_2^1, w_2^2, \ldots, v_n, g'_n, w_n^1, \ldots, w_n^n] .$$

Let $\mathcal{J}$ be the class of relational structures $\{\mathbb{J}_i \mid i = 1, 2, \ldots\}$.

Since $\mathrm{iw}(\mathrm{Core}(\mathcal{J})) < \infty$ we know, by Theorem 3 that $\mathrm{GDNF}(\mathcal{J})$ is tractable. Then, by Proposition 1 we know that $\mathrm{Mixed}(\mathcal{J})$ is also tractable. However, $\mathrm{tw}(\mathrm{Core}(\mathrm{Mrg}(\mathcal{H}))) = \infty$ so Theorem 4 does not characterise all tractable structural classes for the mixed representation.

6 Conclusion

Acyclicity is not a sufficient condition for tractability when we consider more succinct representations: in particular, those that allow us succinctly to specify SAT instances.

However, we have representations GDNF and mixed with simple width based characterisations of structurally tractable classes. For the GDNF representation this characterisation is a complete dichotomy: classes which do not have this bounded width are W[1] hard. We do not yet have a dichotomy for structural classes of mixed representations of CSPs, except in the special case of bounded interaction width, and this motivates further research.

We have also shown that the hierarchy of structural tractability is strict for positive, mixed and GDNF representations. We used this characterisation to construct a novel structurally tractable class for SAT.

References

1. Kautz, H.A., Selman, B.: Planning as satisfiability. In: Proceedings of the Tenth European Conference on Artificial Intelligence (ECAI 1992), pp. 359–363 (1992)
2. van Beek, P.: Reasoning about qualitative temporal information. Artificial Intelligence 58, 297–326 (1992)

3. Gottlob, G., Leone, N., Scarcello, F.: A comparison of structural CSP decomposition methods. Artif. Intell. 124(2), 243–282 (2000)
4. Bulatov, A., Jeavons, P., Krokhin, A.: Classifying the complexity of constraints using finite algebras. SIAM J. Comput. 34(3), 720–742 (2005)
5. Beeri, C., Fagin, R., Maier, D., Yannakakis, M.: On the desirability of acyclic database schemes. Journal of the ACM 30, 479–513 (1983)
6. Moskewicz, M., Madigan, C., Zhao, Y., Zhang, L., Malik, S.: Chaff: engineering an efficient sat solver. In: DAC 2001: Proceedings of the 38th conference on Design automation, pp. 530–535. ACM, New York (2001)
7. van Hoeve, W.J., Katriel, I.: Global constraints. In: Rossi, F., van Beek, P., Walsh, T. (eds.) Handbook of Constraint Programming. Elsevier, Amsterdam (2006)
8. Graham, M.: On the universal relation. Technical report, University of Toronto (1979)
9. Garey, M., Johnson, D.: Computers and Intractability: A Guide to the Theory of NP-Completeness. Freeman, San Francisco (1979)
10. Chen, H., Grohe, M.: Constraint satisfaction with succinctly specified relations. In: Creignou, N., Kolaitis, P., Vollmer, H. (eds.) Complexity of Constraints. Number 06401 in Dagstuhl Seminar Proceedings, Dagstuhl, Germany, Internationales Begegnungs- und Forschungszentrum für Informatik (IBFI), Schloss Dagstuhl, Germany (2006)
11. Szeider, S.: On fixed-parameter tractable parameterizations of SAT. In: Giunchiglia, E., Tacchella, A. (eds.) SAT 2003. LNCS, vol. 2919, pp. 188–202. Springer, Heidelberg (2004)
12. Houghton, C., Cohen, D., Green, M.J.: The effect of constraint representation on structural tractability. In: Benhamou, F. (ed.) CP 2006. LNCS, vol. 4204, pp. 726–730. Springer, Heidelberg (2006)
13. Grohe, M.: The complexity of homomorphism and constraint satisfaction problems seen from the other side. J. ACM 54(1), 1 (2007)
14. Focacci, F., Milano, M.: Global cut framework for removing symmetries. In: Walsh, T. (ed.) CP 2001. LNCS, vol. 2239, pp. 77–92. Springer, Heidelberg (2001)
15. Katsirelos, G., Bacchus, F.: Generalized NoGoods in CSPs. In: AAAI, pp. 390–396 (2005)
16. Katsirelos, G., Walsh, T.: A compression algorithm for large arity extensional constraints. In: Bessière, C. (ed.) CP 2007. LNCS, vol. 4741, pp. 379–393. Springer, Heidelberg (2007)

Asynchronous Inter-Level Forward-Checking for DisCSPs

Redouane Ezzahir[1,2], Christian Bessiere[1], Mohamed Wahbi[1,2], Imade Benelallam[2], and El Houssine Bouyakhf[2]

[1] LIRMM/CNRS, U. of Montpellier 2, France
bessiere@lirmm.fr, ezzahir@lirmm.fr, wahbi@lirmm.fr
[2] LIMIARF/FSR, U. of Mohammed V Agdal, Morroco
imade.benelallam@ieee.org, bouyakhf@fsr.ac.ma

Abstract. We propose two new asynchronous algorithms for solving Distributed Constraint Satisfaction Problems (DisCSPs). The first algorithm, AFC-ng, is a nogood-based version of Asynchronous Forward Checking (AFC). The second algorithm, Asynchronous Inter-Level Forward-Checking (AILFC), is based on the AFC-ng algorithm and is performed on a pseudo-tree ordering of the constraint graph. AFC-ng and AILFC only need polynomial space. We compare the performance of these algorithms with other DisCSP algorithms on random DisCSPs in two kinds of communication environments: Fast communication and slow communication. Our experiments show that AFC-ng improves on AFC and that AILFC outperforms all compared algorithms in communication load.

1 Introduction

Distributed Constraint Satisfaction Problems (*DisCSPs*) is a general framework for solving distributed problems. DisCSPs have a wide range of applications in multi-agent coordination, such as distributed resource allocation problems [1], distributed scheduling problems [2], sensor networks [3], and log-based reconciliation [4].

DisCSPs are composed of agents, each holding its local constraint network. Variables in different agents are connected by constraints. Agents assign values to their variables, attempting to generate a locally consistent assignment that is also consistent with all constraints between agents [5,6]. To achieve this goal, agents check the value assignments to their variables for local consistency and exchange messages among them to check consistency of their proposed assignments against constraints among variables that belong to different agents.

Several efficient distributed algorithms for solving DisCSPs have been developed in the last decade. Synchronous Backtrack (SBT) is the simplest DisCSP search algorithm that performs assignments sequentially and synchronously. Only the agent holding a Current Partial Assignment (*CPA*) performs an assignment or backtrack [7]. The first complete asynchronous search algorithm for DisCSPs is the Asynchronous Backtracking ABT [8,5,9]. In ABT, agents perform assignments asynchronously and send out messages to constraining agents, informing them about their assignments. Due to the asynchronous nature of agents operations, the global assignment state at any particular time during the run of asynchronous backtracking is in general inconsistent. Nogoods

I.P. Gent (Ed.): CP 2009, LNCS 5732, pp. 304–318, 2009.

are used to prevent the construction of globally inconsistent solutions. Another promising algorithm for DisCSPs is the Asynchronous Forward-Checking (AFC) algorithm [10,11]. This algorithm is based on the forward checking (FC) algorithm for CSPs, but performs forward checking asynchronously.

In this paper, we present two new asynchronous algorithms for solving DisCSPs. The first one is based on Asynchronous Forward Checking (AFC) and uses nogood recording. We call it Nogood-Based AFC (AFC-ng). The second one is based on AFC-ng and is performed on a pseudo-tree ordering of the constraint graph. We call it Asynchronous Inter-Level Forward-Checking (AILFC).

This paper is organized as follows. Section 2 gives the necessary background on DisCSPs. Sections 3 and 4 describe the algorithms AFC-ng and AILFC. Correctness proofs are given in Section 5. Section 6 presents an experimental evaluation of our proposed algorithms against three other well-known distributed algorithms. Section 7 summarizes several related works and we conclude the paper in Section 8.

2 Background

2.1 Distributed Constraint Satisfaction Problems

The Distributed Constraint Satisfaction Problem *DisCSP* has been formalized in [6] as a tuple $(\mathcal{A}, \mathcal{X}, \mathcal{D}, \mathcal{C})$, where $\mathcal{A}$ is a set of agents $\{A_1, \ldots, A_k\}$, $\mathcal{X}$ is a set of variables $\{x_1, \ldots, x_n\}$, where each x_i is controlled by one agent in $\mathcal{A}$. $\mathcal{D}$=$\{D(x_1), \ldots, D(x_n)\}$ is a set of domains, where $D(x_i)$ is a finite set of values to which variable x_i may be assigned. Only the agent who is assigned a variable has control on its value and knowledge of its domain. $\mathcal{C}$ is a set of binary constraints that specify the combinations of values allowed for the two variables they involve. A constraint $c_{ij} \in \mathcal{C}$ between two variables x_i and x_j is a subset of the Cartesian product $D(x_i) \times D(x_j)$.

For simplicity purposes, we consider a restricted version of DisCSP where each agent owns exactly one variable. We identify the agent number with its variable index. We also consider that the total order among agents that is used by a search algorithm is the lexicographic ordering $x_i \prec x_j$ if $i < j$.

We assume that communication between two agents is not necessarily generalized FIFO (aka causal order) channels [12]. Thus, all agents maintain their own counter, called Ctr, and increment it whenever they change their value. The current value of the counter *tags* each generated assignment. An *assignment* for an agent $A_i \in \mathcal{A}$ is a tuple (x_i, v_i, Ctr_i) where v_i is a value from the domain of x_i and Ctr_i is the tag value.

A *nogood* ng for value c for variable x_k is a clause of the form $x_i = a \wedge x_j = b \wedge \ldots \Rightarrow x_k \neq c$, meaning that the assignment $x_k = c$ (i.e., the right hand side $Rhs(ng)$ of ng) is inconsistent with the assignments $x_i = a, x_j = b, \ldots$ (i.e., the left hand side $Lhs(ng)$ of ng). When every value of a variable x_k is ruled out by a nogood, these nogoods are resolved computing a new nogood $newNg$. Let x_j be the lowest variable in the left-hand side of the nogoods, with $x_j = b$. $Lhs(newNg)$ is the conjunction of the left-hand sides of all nogoods except $x_j = b$. $Rhs(newNg)$ is $x_j \neq b$.

Definition 1 (Current Partial Assignment (CPA)). *Given an agent $A_i \in \mathcal{A}$, a CPA is an ordered set of assignments* $\{(x_1, v_1, Ctr_1), \ldots, (x_{i-1}, v_{i-1}, Ctr_{i-1}) \mid x_1 \prec \ldots \prec x_{i-1} \prec x_i\}$.

Definition 2 (AgentView). *The agent view of an agent $A_i \in \mathcal{A}$ stores the newest assignments received from agents that precede A_i in the ordering $\prec$. It has a form similar to a* CPA *and is initialized to the set of empty assignments* $\{(x_j, \emptyset, 0) \mid i \neq j\}$.

Definition 3 (Time-stamp). *A time-stamp is an ordered list of counters $\langle Ctr_1, Ctr_2, \ldots, Ctr_k \rangle$. When comparing (lexicographically) two time-stamps, the most up to date is one which is lexicographically greater, that is, the one with greatest value on the first counter on which they differ, if any, otherwise the longest one.*

2.2 Asynchronous Forward-Checking (AFC)

AFC is based on the Forward-Checking (FC) algorithm for CSPs but it performs the forward checking phase asynchronously [10,11]. As in synchronous backtracking, agents assign their variables only when they hold the current partial assignment (CPA). The CPA is a unique message that is passed from one agent to the next one in the ordering. The CPA carries the partial assignment that agents attempt to extend into a complete solution by assigning their variables on it. Forward checking is performed as follows. Every agent that sends the CPA to its successor also sends copies of the CPA to all agents whose assignments are not yet on the CPA. Agents that received CPAs update domains of their variables, removing all values that are in conflict with assignments on the received CPA.

An agent that generates an empty domain as a result of a forward-checking operation initiates a backtrack by sending *Not_OK* messages which carry the inconsistent partial assignment which caused the empty domain. *Not_OK* messages are sent to all agents with unassigned variables on the (inconsistent) CPA. When an agent holding a *Not_OK* receives a CPA, it sends this CPA back in a backtrack message. When multiple agents reject a given assignment by sending *Not_OK* messages, only the first agent that will receive a CPA and is holding a relevant *Not_OK* message will eventually backtrack. After receiving a new CPA, the *Not_OK* message becomes obsolete when the CPA it carries is no longer a subset of the received CPA.

An improved backtrack method for AFC was described in Section 6 of [11]. Instead of just sending *Not_OK* messages to all agents unassigned in the CPA, the agent who detects the empty domain can itself initiate a backtrack operation. It sends a backtrack message to the last agent assigned in the inconsistent CPA in addition to the *Not_OK* messages to all agents not instantiated in the inconsistent CPA. The agent who receives a backtrack message generates (if it is possible) a new CPA that will dominate older ones thanks to the time-stamp mechanism (see Definition 3).

3 Nogood-Based AFC

The nogood-based Asynchronous Forward-Checking (AFC-ng) is based on AFC but it tries to enhance the asynchronism of the forward phase. The two main features of AFC-ng are the following. First, an agent finding an empty domain no longer sends

```
procedure Start()
InitMyAgentView();
end ← false; myAgentView.Consistent ← true;
if(self = IA) then Assign();
while(¬end)
    msg ← getMsg();
    switch(msg.type)
        CPA       : ProcessCPA(msg);
        BackCPA   : ProcessBackCPA(msg);
        Terminate : ProcessTerminate(msg);

procedure InitMyAgentView()
myAgentView ← {(x_j, ∅, 0) | x_j ≺ self};

procedure Assign()
if(∃v ∈ myInitialDomain, ∄ng ∈ myNogoodStore | Rhs(ng) = v) then
    myValue ← ChooseValue(); /*not eliminated by myNogoodStore*/
    myCtr ← myCtr+1; CPA ← myAgentView ∪ {(self, myValue, myCtr)};
    SendCPA(CPA);
else Backtrack();

procedure SendCPA(CPA)
next ← getNextAgent();
if(next=nil) then BroadcastMsg:Terminate(myAgentView); end ← true;
else for each x_j ≻ self do sendMsg:CPA(CPA,next) to x_j;
```

Fig. 1. Nogood-based AFC algorithm running by agent self (Part 1)

Not_OK messages. It resolves the nogoods attached to its values and sends the backtrack message to the lower agent in the resolved nogood. Hence, multiple backtracks may be performed at the same time coming from different agents having an empty domain. These backtracks are sent concurrently by these different agents to different destinations. The re-assignments of the destination agents then happen simultaneously and generate several CPAs. However, the CPA coming from the highest level in the search tree will eventually dominate all others. Interestingly, the search process with the new CPA of highest level can use nogoods reported by the (killed) lower level processes, so that it benefits from their computational effort. Second, each time an agent performs a forward-check, it revises its *initial* domain, (including values already removed by a stored nogood) in order to store the best nogoods for removed values (one nogood per value). When comparing two nogoods eliminating the same value, the nogood with the *highest possible lowest variable* involved is selected (HPLV heuristic) [13]. As a result, when an empty domain is found, the resolvent nogood contains variables as high as possible in the ordering, so that the backtrack message is sent as high as possible, thus saving unnecessary search effort [9].

Description of the Algorithm

We call $self$ the variable that points to the agent itself. An AFC-ng agent $self$ executes the code shown in Figures 1 and 2. The data structure. $myInitialDomain$ contains all values of the initial domain of $self$. $self$ stores a nogood per removed value in

```
ProcessCPA(msg)
19: if(¬myAgentView.Consistent ∧ myAgentView ⊂ msg.CPA) then return;
20: splitlevel ← CompareTimeStamp(myAgentView, msg.CPA);
21: if(splitlevel > 0) then
22:    UpdateMyAgentView(msg.CPA, splitlevel); myAgentView.Consistent ← true;
23:    FC_ReviseInitialDomain();
24:    if(∀v ∈ myInitialDomain, ∃ng ∈ myNogoodStore | Rhs(ng) = v) then Backtrack();
25:    else CheckAssign(msg.Next)

procedure CheckAssign(next)
26: if(next = self) then Assign();

procedure Backtrack()
27: newNg ← solve(myNogoodStore);
28: if(newNg = empty) then BroadcastMsg:Terminate(∅); end ← true; return;
29: for each x_j ≻ Rhs(newNg) do
30:    myAgentView.Value[x_j] ← unknown ;
31:    for each ng ∈ myNogoodStore do
32:       if(x_j ∈ Lhs(ng)) then remove(ng, myNogoodStore);
33: myAgentView.Consistent ← false; myValue ← empty; CPA ← myAgentView;
34: SendMsg:BackCPA(CPA, newNg) to Rhs(newNg);

ProcessBackCPA(msg)
35: if(¬myAgentView.Consistent ∧ myAgentView ⊂ msg.CPA) then return;
36: splitlevel ← CompareTimeStamp(myAgentView, msg.CPA);
37: if(splitlevel = 0 ∧ myValue = RhsValue(msg.Nogood)) then
38:    add(msg.Nogood, myNogoodStore); myValue ← empty; Assign();

ProcessTerminate(msg)
39: end ← true ; myValue ← empty;
40: if(msg.CPA ≠ ∅) then myValue ← msg.CPA.Value[self];

procedure UpdateMyAgentView(CPA, splitlevel)
41: for each j ≥ splitlevel do myAgentView[j] ← CPA[j]; /* update value and Ctr */
42: for each ng ∈ myNogoodStore do
43:    if Lhs(ng) is inconsistent with myAgentView then remove(ng, myNogoodStore);

procedure FC_ReviseInitialDomain()
44: for each v ∈ myInitialDomain do
45:    if(¬Consistent(v, myAgentView)) then
46:       store the best nogood for v; /* according to the HPLV heuristic*/

function CompareTimeStamp(view, CPA)
47: from j ← 1 to size(CPA) do
48:    if (Ctr(CPA[j]) > Ctr(view[j])) then return j;
49:    else if (Ctr(CPA[j]) < Ctr(view[j])) then return −1;
50: return 0;
```

Fig. 2. Nogood-based AFC algorithm running by agent self (Part 2)

$myNogoodStore$. $self$ calls the procedure Start() in which $self$ initiates its agent view (line 1) by setting counters to zeros (line 10). The agent view contains a consistency flag that represents whether the partial assignment it holds is consistent. If $self$ is the initializing agent (IA), it initiates the search by calling procedure assign() (line 3). All agents performing the main loop wait for messages, and process received messages according to their types (line 4-9).

When calling assign() $self$ tries to find an assignment, which is consistent with its agent view. If $self$ succeeds, it increments its counter Ctr, generates a CPA from its agent view augmented by $self$ assignment (line 13), and then sends forward the CPA to every agent whose assignments are not yet on the CPA, or reports a solution, when the CPA includes all agents assignments (line 17). Before sending any CPA, $self$ attaches to every CPA message the ID of his successor (line 18). Only if the receiver ID equals that attached to the CPA message, the receiver performs an assignment (line 26). When $self$ fails to find a consistent assignment, it calls procedure Backtrack() (line 15).

Agents use time-stamps to detect and discard obsolete CPAs. Function CompareTimeStamp($view$, CPA) returns the index $splitlevel$ of the first counter on which $view$ and CPA differ if CPA is newest (see Definition 3) or contains $view$ (line 48). If $view$ is newest, it returns -1. When $view$ and CPA are identical or when CPA is included in $view$ CompareTimeStamp returns 0.

Whenever $self$ receives a CPA, procedure ProcessCPA() is called. $self$ checks its agent view status. If it is not consistent and the agent view is a subset of the received CPA, this means that $self$ has already backtracked, then $self$ does nothing (line 19). Otherwise, $self$ compares the time-stamp of its agent view with the one of the received CPA by calling CompareTimeStamp (line 20). If the received CPA is newest, $self$ updates its agent view and marks it consistent (lines 21-22). Procedure UpdateMyAgentView (lines 41-43) sets the agent view and the nogood store to be consistent with the received CPA. Each nogood in the nogood store containing a value for a variable different from that received in the CPA will be deleted (line 43). Next, $self$ calls procedure FC_ReviseInitialDomain() (in line 23) to store nogoods for values inconsistent with the new agent view or to try to find a better nogood for values already having one in the nogood store (line 46). A nogood is better according to the *HPLV* heuristic if the lowest variable in the body of the nogood is higher.

When every value of $self$'s variable is ruled out by a nogood (line 24), the procedure Backtrack is called. These nogoods are resolved by computing a new nogood $newNg$ (line 27). If the new nogood is empty, $self$ terminates execution after sending a $Terminate$ message to all agents in the system meaning that problem is unsolvable (line 28). Otherwise, $self$ updates its agent view by removing assignments of every agent that is strictly greater than the last agent ($Rhs(newNg)$) in the $newNg$. $self$ also updates its nogood store by removing obsolete nogoods. Finally it marks its agent view as inconsistent and it initiates a backtrack procedure by sending one $BackCPA$ message to the lower priority agent ($Rhs(newNg)$) involved in the $newNg$ (line 34).

The $BackCPA$ message carries the $newNg$ and the inconsistent CPA containing assignments of all agents smaller than or equal to $Rhs(newNg)$ in the agent ordering (lines 29-30). $self$ remains in an inconsistent state until receiving a new CPA holding at least one agent assignment with counter higher than that in the agent view of $self$ (lines 21-22).

When a $BackCPA$ message is received, $self$ checks the validity of received $BackCPA$ using agent view status and time-stamp (lines 35-36). If $BackCPA$ is accepted (line 37), $self$ removes its last assignment, adds attached nogood to its nogood store, and calls the procedure assign() (line 38).

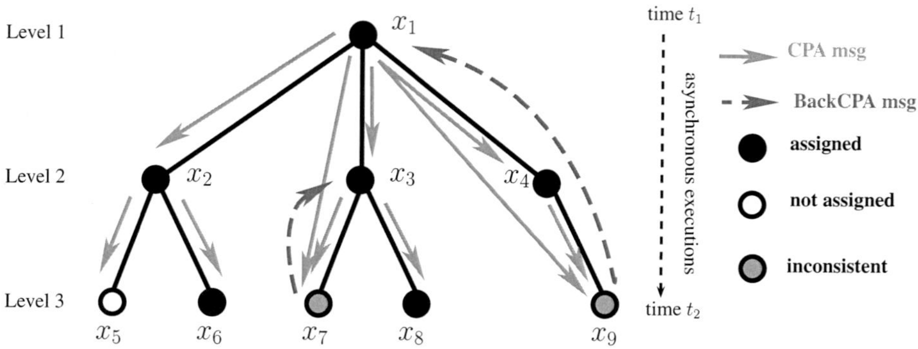

Fig. 3. An example of the AILFC execution

ProcessTerminate procedure is called when an agent receives a $Terminate$ message. It marks end flag $true$ to stop the main loop (line 39). If attached CPA is empty then there is no solution. Otherwise, agent solution is retrieved from the CPA (line 40).

4 Asynchronous Inter Level Forward-Checking

A DisCSP can be represented by a constraint graph $G = (X, E)$, whose nodes represent the variables and edges represent the constraints (that is, $X = \mathcal{X}$ and $\{x_i, x_j\} \in E \Leftrightarrow c_{ij} \in \mathcal{C}$). The graph can be re-arranged to form a pseudo-tree [14]. A *pseudo-tree* $G_{PT} = (X, r, E, U)$ for the graph G is defined by a root node $r \in X$ and a directed tree $T = (X, U)$ rooted in r such that for any edge $\{x_i, x_j\} \in E$, x_i and x_j are not in different branches of T. For any arc $(x_i, x_j) \in U$, the node x_i is the parent of the node x_j. If x_i is the parent of x_j, then x_j is a child of x_i. A node x_i is an ancestor of a node x_j if x_i is the parent of x_j or an ancestor of the parent of x_j. A node x_j is a descendant of a node x_i if x_i is an ancestor of x_j. A leaf is a node that has no child. In our implementation, the pseudo-tree is built by a DFS traversal of the graph. Thus, we have $U \subseteq E$.

The AILFC algorithm is based on AFC-ng performed on a pseudo-tree ordering of the constraint graph (built in a preprocessing step). Agents are prioritized according to the pseudo-tree ordering in which each agent has a single parent and various children. Using this priority ordering, AILFC performs multiple AFC-ng processes on the paths from the root to the leaves. The root initiates the search by generating a CPA, assigning its value on it, and sending CPA messages to its linked descendants (including its children) that share a constraint with it. Each child that receives a copy of the CPA performs AFC-ng on the sub-problem restricted to its ancestors (agents that are assigned in the CPA) and the set of its descendants. Therefore, instead of giving the privilege of assigning to only one agent, all agents who are in disjoint subtrees may assign their variables simultaneously. So, the Inter-Level Forward Checking is performed asynchronously on each path from the root to any leaf. AILFC thus exploits the potential speed-up of a parallel exploration in the processing of distributed problems.

An execution of AILFC on a sample DisCSP problem is shown in Figure 3. At time t_1, the root x_1 sends copies of the CPA on messages to its linked descendants (including its children). Children x_2, x_3 and x_4 assign their values simultaneously in the received CPAs and then perform concurrently the AILFC algorithm. Agents x_7, and x_9 only perform a forward- checking. At time t_2, x_9 finds an empty domain and sends a $BackCPA$ message to x_1. At the same time, other CPAs propagate down through the other paths. For instance, a CPA has propagated down from x_3 to x_7 and x_8. x_7 detects an empty domain and sends a nogood to x_3 attached on a $BackCPA$ message. For the CPA that propagates on the path (x_1, x_2, x_5) (resp. (x_1, x_2, x_6)), x_5 (resp. x_6) successfully assigned its value and initiated a solution detection. However, when x_1 receives the $BackCPA$ from x_9, it initiates a new search process by sending a new copy of the CPA which will kill any CPA where x_1 is assigned its old value.

In AFC-ng, a solution is reached when the last agent receives the CPA and succeeds in assigning its variable. In AILFC, the situation is different because a CPA can reach a leaf without being complete. When all agents are assigned and no constraint is violated, this state is a global solution and the network has reached quiescence, meaning that no message is traveling through it. Such a state can be detected using specialized snapshot algorithms [15], but AILFC uses a different mechanism that allows to detect solutions before quiescence. AILFC uses an additional type of message called $Accepted$ that inform parents of the acceptance of their CPA. Termination can be inferred earlier and the number of messages required for termination detection can be reduced. A similar technique of solution detection was used in the AAS algorithm [16].

The mechanism of solution detection is as follows: whenever a leaf node succeeds in assigning its value, it sends an $Accepted$ message to its parent. This message contains the CPA that was received from the parent incremented by the value-assignment of the leaf node. When a non-leaf agent $self$ receives $Accepted$ messages from all its children that are all compatible with each other, all compatible with $self$'s agent view and with $self$'s value, $self$ builds an $Accepted$ message being the conjunction of all received $Accepted$ messages plus $self$'s value-assignment. If $self$ is the root a solution is found, and $self$ broadcasts this solution to all agents. Otherwise, $self$ sends the built $Accepted$ to its parent.

Description of the Algorithm

A preprocessing step before starting the AILFC algorithm is performed to convert the constraint graph into a pseudo-tree. $Children(self) \subset \mathcal{A}$ is the set of children of agent $self$ in the pseudo-tree, $Desc(self)$ is the set of its descendants and $linkedDesc(self) \subset Desc(self)$ is the set of its descendants (including its children) that are constrained with $self$. $Parent(self) \in \mathcal{A}$ is the parent of agent $self$ and $Ancestors(self) \subset \mathcal{A}$ is the set of its ancestors (including its parent).

In Figure 4, we present only the procedures that are new or different from those of AFC-ng in Figures 1 and 2. In InitMyAgentView(), the agent view of $self$ is initialized to the set $Ancestors(self)$. Ctr is set to 0 for each agent in $Ancestors(self)$ (line 11). The new data structure storing the received $Accepted$ messages is initialized to the empty set (line 12). In SendCPA(CPA), instead of sending copies of the CPA to all agents not yet instantiated on it, $self$ sends copies of the CPA only to its linked

```
procedure Start()
..:
10:      Accepted  : ProcessAccepted(msg);

procedure InitMyAgentView()
11: myAgentView ← {(x_j, ∅, 0) | x_j ∈ Ancestors(self)};
12: for each child ∈ children(self) accepted[child] ← ∅; /*For Solution Detection*/

procedure SendCPA(CPA)
13: if(children(self) = ∅) then
14:    SolutionDetection();
15: else for each desc ∈ linkedDesc(self) do sendMsg:CPA(CPA, self) to desc;

procedure CheckAssign(ancestor)
16: if(Parent(self) = ancestor) then Assign();

procedure SolutionDetection()
17: if(children(self) = ∅) then
18:    SendAccepted(myAgentView ∪ {(self, myValue, myCtr)}, self) to Parent(self);
19: else PA ← BuildAccepted();
20: if(PA ≠ ∅) then
21:    if(self = root) then Broadcast(Terminate, PA); end ← true;
22:    else SendAccepted(PA, self) to Parent(self);

ProcessAccepted(msg)
23: if(accepted[msg.Sender]=∅∨CompareTimeStamp(msg.CPA,accepted[msg.Sender])>0)then
24:    accepted[msg.Sender] ← msg.CPA;
25:    SolutionDetection();

function BuildAccepted()
26: PA ← myAgentView ∪ {(self, myValue, myCtr)};
27: for each child ∈ children(self) do
28:    if(accepted[child]= ∅ ∨ ¬Compatible(PA,accepted[child])) return ∅;
29:    else PA ← PA ∪ accepted[child];
30: return PA
```

Fig. 4. New lines/procedures of AILFC with respect to AFC-ng

descendants ($linkedDesc(self)$) (line 15). When the set $linkedDesc(self)$ is empty (i.e., $self$ is a leaf), $self$ calls the procedure SolutionDetection to build and send an $Accepted$ message. In CheckAssign($ancestor$), $self$ assigns its value if the CPA was received from its parent (line 16) (i.e., if $ancestor$ is the parent of $self$).

In SolutionDetection(), if $self$ is a leaf ($Children(self)$ is empty), it sends an $Accepted$ message to its parent. The $Accepted$ message sent by $self$ contains its agent view incremented by its assignment (lines 17-18). If $self$ is not a leaf, it calls the BuildAccepted() procedure to build an accepted partial solution PA (line 19). If the returned partial solution PA is not empty and $self$ is the root, PA is a solution of the problem. Then, $self$ broadcasts it to other agents including the system agent and sets the end flag to $true$ (line 21). Otherwise, $self$ sends an $Accepted$ message containing PA to its parent (line 22).

In ProcessAccepted(msg), when $self$ receives an $Accepted$ message from its $child$ for the first time, or when msg is newer than that received before (lines 23-24), $self$ stores the content of this message and calls the SolutionDetection procedure (line 25).

In BuildAccepted(), if an accepted partial solution is reached. $self$ generates a partial solution PA incrementing its agent view with its assignment (line 26). Next, $self$ loops over the set of $Accepted$ messages received from its children. If at least one $child$ has never sent an *Accepted* message or the *Accepted* message is incompatible with PA, then the partial solution has not yet been reached and the function returns empty (lines 27-28). Otherwise, the partial solution PA is incremented by the *Accepted* message of *child* (line 29). Finally, the accepted partial solution is returned (line 30).

5 Correctness Proofs

Theorem 1. *AFC-ng is sound, complete, and terminates.*

The argument for soundness is close to the one given in [11,17]. The fact that agents only forward consistent partial solution on the CPAs messages at only one place in function assign() (line 14), implies that the agents receive only consistent assignments. A solution is reported by the last agent only in function SendCPA(CPA) at line 17. At this point, all agents have assigned their variables, and their assignments are consistent. Thus the AFC-ng algorithm is sound.

For completeness, we need to show that AFC-ng is able to terminate and does not report inconsistency if a solution exists.

Lemma 1. *AFC-ng is guaranteed to terminate.*

For sake of clarity, we assume that the order in which AFC-ng assigns the variables is the lexicographic ordering $X_1, X_2, \dots, X_n$. We define the total order o on CPAs as follows. Let I_1 be an assignment on $X_1, \dots, X_{k_1}$, I_2 be an assignment on $X_1, \dots, X_{k_2}$, and s be the smallest index on which I_1 and I_2 differ. $I_1 \prec_o I_2$ if and only if $s = k_1 + 1$ or the value $I_1[s]$ is chosen before the value $I_2[s]$ by the value ordering heuristics on variable X_s given the CPA $I_1[1..s-1]$.

To prove the lemma we prove that AFC-ng performs a finite number of backtrack steps. In AFC-ng, several backtracks can be performed simultaneously as they are generated concurrently by different agents to different destinations. The re-assignments of destination agents then happen simultaneously, generating several CPAs. However, the CPA at the highest level in the search hierarchy tree will eventually dominate all others thanks to its greater time-stamp (see line 21 in Figure 2). Thus, every backtrack step may be represented by the backtrack at the highest level. The agent X_i who has received that backtrack of highest level has to replace its previous assignment v_i in the CPA by a new one v'_i because the backtrack message contains a nogood rejecting value v_i. If v_i was not the first value chosen by X_i since it has received the current CPA from X_{i-1} then we know that all other values v_j preferred to v_i were ruled out by a nogood at the time v_i was chosen. Now, the CPA on $X_1, \dots, X_{i-1}$ has not changed since then, otherwise this would not be the highest backtrack. As a result, the nogoods rejecting values v_j preferred to v_i are still valid and v'_i is necessarily the *next* preferred value in the heuristic order. By definition of the order o, the new CPA obtained is greater than the previous one according to o because it has not changed on $X_1, \dots, X_{i-1}$ and v'_i is less preferred than v_i. Since o is a total order and since there are a finite number of variables and a finite number of values per variable, there will be a finite number of new CPAs generated. Now, each backtrack of highest level generates a new CPA. Thus, AFC-ng performs a finite number of backtracks.

Lemma 2. *AFC-ng cannot infer inconsistency if a solution exists.*

Whenever a newer CPA or a $BackCPA$ message is received, AFC-ng agent updates its nogood store. Hence, for every CPA that may potentially lead to a solution, agents only store valid nogoods. In addition, every nogood resulting from a CPA is redundant with regard to the DisCSP to solve. Since all additional nogoods are generated by logical inference when a domain wipe-out occurs, the empty nogood cannot be inferred if the network is satisfiable. This mean that AFC-ng is able to produce all solutions.

Theorem 2. *AILFC algorithm is sound, complete, and terminates.*

AILFC agents only forward consistent partial assignments (CPAs). Hence, leaf agents receive only consistent CPAs. Thus, leaf agents send Accepted message only holding consistent assignments to their parent. Since a parent builds an $Accepted$ message only when the $Accepted$ messages received from all its children are compatible with each other and all compatible with its own value, the $Accepted$ message it sends contains a consistent partial solution. The root broadcasts a solution only when it can build itself such an $Accepted$ message. Therefore, the solution is correct and AILFC is sound.

AILFC performs multiple AFC-ng processes on the paths of the pseudo-tree from the root to the leaves. Thus, it inherits the completeness property of AFC-ng (empty nogood cannot be inferred if the network is satisfiable (see Lemma 2). It also appears that the agent of high priority cannot fall into an infinite loop. By induction on the level of the pseudo-tree no agent can fall in such a loop, which ensures the termination of AILFC.

6 Experimental Evaluation

In this section we compare experimentally AFC-ng and AILFC to three other algorithms: AFC, ABT, and ABT-Hyb [18]. Algorithms are tested on the same static agents ordering using *max-degree* heuristic and the same nogood selection heuristic (*HPLV*). For ABT and ABT-Hyb we implemented an improved version of Silaghi's solution detection [12] and counters for tagging assignments. This allows to better treat non-causal order channels [12]. All experiments were performed on the DisChoco platform [19] in which agents are simulated by Java threads that communicate only through message passing.

The algorithms are tested on uniform binary random DisCSPs which are characterized by $\langle n,\ d,\ p_1,\ p_2\rangle$, where n is the number of agents/variables, d the number of values per variable, p_1 the network connectivity defined as the ratio of existing binary constraints, and p_2 the constraint tightness defined as the ratio of forbidden value pairs. We solved 100 instances of two classes of constraints graph: sparse graph $\langle 20, 10, 0.25, p_2\rangle$ and dense graph $\langle 20, 10, 0.75, p_2\rangle$. We vary the tightness from 0.10 to 0.90 by steps of 0.10.

We evaluate the algorithms performance by the average of total messages sent [20] (including system messages) and the average of Equivalent Non-Concurrent Constraint Checks (ENCCCs) [21]. ENCCCs are a weighted sum of processing and communication time. We simulate two scenarios of communication: fast communication (where message delay is null and ENCCCs reduce to standard NCCCs), and slow communication with uniform random message delay (where the cost of the delay is between 500 and 1000 constraint checks.)

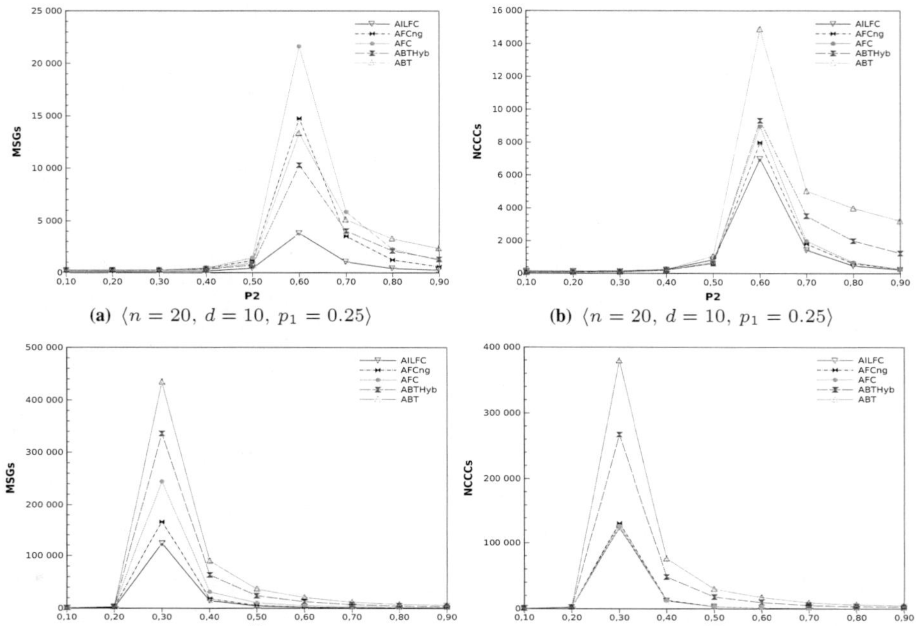

(a) $\langle n = 20,\ d = 10,\ p_1 = 0.25\rangle$ (b) $\langle n = 20,\ d = 10,\ p_1 = 0.25\rangle$

(c) $\langle n = 20, d = 10, p_1 = 0.75\rangle$ (d) $\langle n = 20,\ d = 10,\ p_1 = 0.75\rangle$

Fig. 5. Total number of messages sent and NCCCs on fast communication

Fast Communication

Figure 5 presents performance of AILFC, AFC-ng, AFC, ABT and ABT-hyb running on a fast communication environment. The figure shows that in both types of constraint graphs (sparse and dense), AILFC has the lowest communication load (#MSGs). Concerning NCCCs, AILFC is the fastest algorithm on sparse graphs (Fig. 5(b)). On dense graphs AILFC behaves like AFC and AFC-ng. Comparing AFC-ng with AFC, Fig. 5 shows that they perform the same number of NCCCs but AFC-ng exchanges less messages than AFC. Comparing AFC-ng with ABT and ABT-hyb, Fig. 5 shows that in both types of constraint graphs AFC-ng is faster than ABT-hyb and ABT. However, on sparse graphs, Fig 5(a) shows that AFC-ng sends more messages than ABT-hyb and ABT.

Slow Communication

In Figure 6 we report experimental results with slow communication. The figure shows that AILFC is again the best algorithm in terms of number of messages. Concerning ENCCCs, as in the fast communication environment, AILFC is faster than or equal to AFC and AFC-ng depending on whether the graph is sparse or dense. The comparison of AFC and AFC-ng shows a pattern close to the one observed with fast communication: AFC-ng is better, or slightly better, both in terms of messages and ENCCCs. The main difference between fast and slow communication is the performance of ABT and ABT-hyb. Whereas they remain expensive in terms of messages, they become the best algorithms in terms

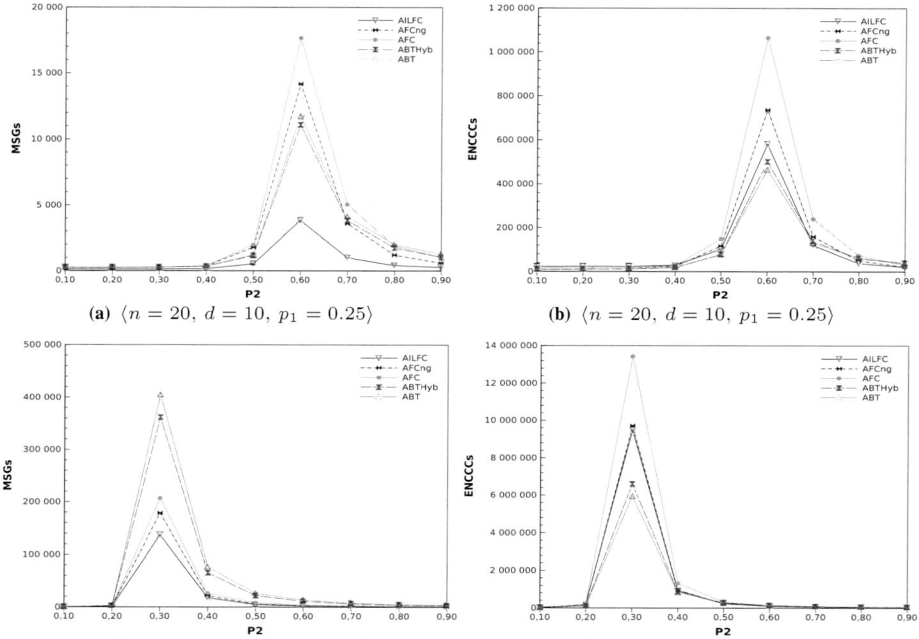

(a) $\langle n = 20, \ d = 10, \ p_1 = 0.25 \rangle$ (b) $\langle n = 20, \ d = 10, \ p_1 = 0.25 \rangle$

(c) $\langle n = 20, \ d = 10, \ p_1 = 0.75 \rangle$ (d) $\langle n = 20, \ d = 10, \ p_1 = 0.75 \rangle$

Fig. 6. Total number of messages sent and ENCCCs on slow communication

of ENCCCs, with a slight advantage to ABT. This confirms that in slow communication environment, the more the algorithm is asynchronous, the better it is.

Discussion

A first observation on these experiments is that ABT, ABT-hyb on one side, and AFC, AFC-ng on the other side, show quite opposite patterns. If message passing is not an issue, ABT and ABT-hyb are good choices with slow communication whereas AFC and AFC-ng are good when communication is fast. A second observation is that AILFC is always better than or equivalent to AFC-ng, which is better than or equivalent to AFC, both in terms of messages and amount of processing (ENCCCs). If limiting the communication load is important, AILFC is the best among all both for fast and slow communication. AILFC benefits both from running separate search processes in disjoint problem subtrees, which pays off when a graph is sparse, and from using the same mechanism as AFC-ng, which pays off when agents are highly connected (dense graphs).

7 Other Related Work

In [18,7] the performance of asynchronous (ABT), synchronous (Synchronous Conflict BackJumping (SCBJ)), and hybrid approaches (ABT-Hyb) was studied. It is shown that ABT-Hyb improves over ABT and that SCBJ requires less communication effort than

ABT-Hyb. In Interleaved Asynchronous Backtracking (IDIBT) [22], agents participate in multiple processes of asynchronous backtracking. Each agent keeps a separate *AgentView* for each search process in IDIBT. The number of search processes is fixed by the first agent in the ordering. The performance of concurrent asynchronous backtracking [22] was tested and found to be ineffective for more than two concurrent search processes [22]. Dynamic Distributed BackJumping (DDBJ) was presented in [17]. It is an improved version of the basic AFC. It combines the concurrency of an asynchronous dynamic backjumping algorithm, and the computational efficiency of the AFC algorithm, coupled with the *possible conflict heuristics* of dynamic value and variable ordering. As in DDBJ, AFC-ng performs several backtracks simultaneously. However, AFC-ng should not be confused with DDBJ. DDBJ is based on dynamic ordering and requires additional messages to compute ordering heuristics. In AFC-ng, all agents that received a $BackCPA$ message continue search concurrently. Once a more up to date CPA is received by an agent, all nogoods already stored can be kept if consistent with that CPA.

8 Conclusion

Two new complete, asynchronous algorithms are presented. The first algorithm, Nogood-Based Asynchronous Forward Checking (AFC-ng), is an improvement on AFC. The second, Asynchronous Inter-Level Forward-Checking (AILFC), is based on AFC-ng and is performed on a pseudo-tree re-arrangement of the constraint graph. Experiments ran on random DisCSPs show that AFC-ng improves AFC both in fast and slow communication environments. Experiments show that AILFC is the more robust algorithm in both communication types. In particular, it is the best in terms of messages sent. In slow communication environments, the performance of algorithms that perform variable assignments sequentially deteriorates. This is observed for AFC and AFC-ng, and, less significantly for ABT-Hyb, when compared to ABT.

References

1. Petcu, A., Faltings, B.: A value ordering heuristic for distributed resource allocation. In: Faltings, B.V., Petcu, A., Fages, F., Rossi, F. (eds.) CSCLP 2004. LNCS (LNAI), vol. 3419, pp. 86–97. Springer, Heidelberg (2005)
2. Wallace, R.J., Freuder, E.: Constraint-based multi-agent meeting scheduling: effects of agent heterogeneity on performance and privacy loss. In: Proceeding of the 3rd workshop on distributed constrait reasoning, DCR 2002, Bologna, pp. 176–182 (2002)
3. Fernández, C., Béjar, R., Krishnamachari, B., Gomes, K.: Communication and computation in distributed CSP algorithms. In: Van Hentenryck, P. (ed.) CP 2002. LNCS, vol. 2470, pp. 664–679. Springer, Heidelberg (2002)
4. Chong, Y.L., Hamadi, Y.: Distributed log-based reconciliation. In: ECAI, pp. 108–112 (2006)
5. Yokoo, M.: Algorithms for distributed constraint satisfaction problems: A review. Autonomous Agents & Multi-Agent Systems 3, 198–212 (2000)
6. Yokoo, M., Durfee, E.H., Ishida, T., Kuwabara, K.: Distributed constraint satisfaction problem: Formalization and algorithms. IEEE Trans. on Data and Kn. Eng. 10, 673–685 (1998)

7. Zivan, R., Meisels, A.: Synchronous vs asynchronous search on DisCSPs. In: Proceeding of 1st European Workshop on Multi Agent System, EUMAS, Oxford (2003)
8. Yokoo, M., Durfee, E.H., Ishida, T., Kuwabara, K.: Distributed constraint satisfaction for formalizing distributed problem solving. In: IEEE Intern. Conf. Distrb. Comp. Sys., pp. 614–621 (1992)
9. Bessiere, C., Maestre, A., Brito, I., Meseguer, P.: Asynchronous backtracking without adding links: a new member in the ABT family. Artificial Intelligence 161(1-2), 7–24 (2005)
10. Meisels, A., Zivan, R.: Asynchronous forward-checking for distributed CSPs. In: Zhang, W. (ed.) Frontiers in Artificial Intelligence and Applications. IOS Press, Amsterdam (2003)
11. Meisels, A., Zivan, R.: Asynchronous forward-checking for DisCSPs. Constraints 12(1), 131–150 (2007)
12. Silaghi, M.C.: Generalized dynamic ordering for asynchronous backtracking on DisCSPs. In: DCR workshop, AAMAS 2006, Hakodate, Japan (2006)
13. Maestre, A., Bessiere, C.: Improving asynchronous backtracking for dealing with complex local problems. In: ECAI, pp. 206–210 (2004)
14. Freuder, E.C., Quinn, M.J.: Taking advantage of stable sets of variables in constraint satisfaction problems. In: IJCAI 1985, pp. 1076–1078 (1985)
15. Mani, K.C., Lamport, L.: Distributed snapshots: determining global states of distributed systems. ACM Trans. Comput. Syst. 3(1), 63–75 (1985)
16. Silaghi, M.C., Faltings, B.: Asynchronous aggregation and consistency in distributed constraint satisfaction. Artificial Intelligence 161(1-2), 25–54 (2005)
17. Nguyen, T., Sam-Hroud, D., Faltings, B.: Dynamic distributed backjumping. In: Proceeding of 5th workshop on DCR 2004, Toronto (2004)
18. Brito, I., Meseguer, P.: Synchronous, asynchronous and hybrid algorithms for DisCSP. In: Workshop on DCR 2004, CP 2004, Toronto (2004)
19. Ezzahir, R., Bessiere, C., Belaissaoui, M., Bouyakhf, E.H.: Dischoco: a platform for distributed constraint programming. In: Proceeding of Workshop on Distributed Constraint Reasoning of IJCAI 2007, pp. 16–21 (2007)
20. Lynch, N.A.: Distributed Algorithms. Morgan Kaufmann Series (1997)
21. Chechetka, A., Sycara, K.: No-commitment branch and bound search for distributed constraint optimization. In: Proc. of AAMAS 2006, pp. 1427–1429. ACM, New York (2006)
22. Hamadi, Y.: Interleaved backtracking in distributed constraint networks. Int. J. of AI Tools 11, 167–188 (2002)

From Model-Checking to Temporal Logic Constraint Solving

François Fages and Aurélien Rizk

EPI Contraintes, INRIA Paris-Rocquencourt,
BP105, 78153 Le Chesnay Cedex, France
Francois.Fages@inria.fr, Aurelien.Rizk@inria.fr
http://contraintes.inria.fr

Abstract. In this paper, we show how model-checking can be generalized to temporal logic constraint solving, by considering temporal logic formulae with free variables over some domain $\mathcal{D}$, and by computing a validity domain for the variables rather than a truth value for the formula. This allows us to define a continuous degree of satisfaction for a temporal logic formula in a given structure, opening up the field of model-checking to optimization. We illustrate this approach with reverse-engineering problems coming from systems biology, and provide some performance figures on parameter optimization problems with respect to temporal logic specifications.

1 Introduction

Temporal logics were introduced for program verification by Pnueli [29] as specification languages for expressing the behavior of sequential as well as concurrent programs. Temporal logics essentially extend classical logic, used for describing states, with temporal operators (**X** "at next state", **F** "finally at some future state", **G** "globally at all future states", **U** "until") and computation path quantifiers (**E** "on some path", **A** "on all paths"). Temporal logics have proven useful for formalizing and verifying the behavior of a broad variety of systems ranging from electronic circuits to software programs, physical systems and more recently biological systems, in either boolean [16,10], discrete [7], stochastic [8,22] or continuous [32,17,9,3,10,5] settings.

There has been a large body of work for generalizing model-checking techniques, initially developed for discrete systems [28,13], to quantitative transition systems [1,34,35,23,20,26,14,15]. One common approach is to represent infinite sets of states with finite structures, e.g. by symbolic or numerical constraints, and develop decision or semi-decision procedures for checking the validity of closed temporal logic formulae over such structures.

In computational systems biology however, one core problem is of a reverse-engineering nature: it consists in inferring a mechanistic model of a cell process from knowledge of the elementary interactions and from the observation of the system's global behavior under various conditions (milieu, stress, mutations etc.) [9].

I.P. Gent (Ed.): CP 2009, LNCS 5732, pp. 319–334, 2009.

Beyond verifying whether an already built model does reproduce some dynamical properties specified in temporal logic, model-checking techniques need thus be generalized to model-synthesis techniques, including parameter optimization with respect to temporal specifications. Indeed, most kinetic parameters cannot be measured and must be inferred from the global behavior of the biological system. However, the boolean valuation of temporal logic formulae is not very helpful to guide the search for kinetic parameter values since it does not quantify how far a system is from satisfying its specification.

To answer this question, we generalize the model-checking problem, i.e. deciding whether a closed temporal logic formula is true in a given structure, to a constraint solving problem, i.e. determining the validity domain of the free variables of a given temporal logic formula that make it true in a given structure (see example 2). A continuous degree of satisfaction for a closed temporal logic formula ϕ can then be defined for a given structure, as the distance between the validity domain of a pattern formula ψ obtained by replacing the constants in ϕ by variables, and the actual constant values in ϕ.

In this paper, we present a temporal logic constraint solving algorithm, in a very general first-order setting of Quantifier-Free Computation Tree Logic (QFCTL) formulae with constraints over some arbitrary computational domain $\mathcal{D}$. Then we describe our particular implementation in the Biochemical Abstract Machine BIOCHAM[1] [19] using linear constraint solving over the reals and numerical integration of (non-linear) parametric ordinary differential equations (ODE), and provide some performance figures on a benchmark of kinetic parameter optimization problems with respect to temporal specifications, coming from molecular systems biology.

Such ODE models were considered by Janssen, Van Hentenryck and Deville in [24] for enclosing their solutions and finding their stable states by constraint consistency methods. Our approach consists in using temporal logic to formalize the dynamical properties of the solutions, in a much more flexible way than by specifying their stable states, or than by curve fitting, allowing us to express concentration or time thresholds or oscillation constraints for instance. In our previous work in [9], we introduced the idea of searching parameter values by model-checking with a generate and test algorithm that was limited in practice to 2-3 parameters. In [17] we introduced a constraint solving algorithm for Linear Time Logic (LTL) queries with interval constraints over the reals, interpreted in a single finite trace. In [32] we used it for parameter search in higher dimensions over a single trace and in [33] for robustness analysis using a solver for linear constraints over the reals. Here we generalize this approach with a new constraint solving algorithm for branching time logic (QFCTL), presented in an abstract setting of constraints over some arbitrary domain D. The generalization to branching time logic is not trivial since the labeling procedure must be generalized to a fixpoint algorithm. We believe that this generality is important

[1] BIOCHAM, and the examples of this paper, are available at
`http://contraintes.inria.fr/BIOCHAM`

for relating model-checking to constraint solving independently of a particular application.

The idea of allowing free variables in temporal logic formulae to extract information from a model is however not new. It was introduced by William Chan in [11] in the propositional case with the notion of temporal logic queries, where one free variable stands as a place holder for a propositional formula that makes the query true. Following Chan's seminal paper, temporal queries have been investigated by many authors, but to our knowledge, always in a propositional setting and not in a first-order setting with constraints over some computation domain allowing to compute validity domains for variables. In [30,14], model-checking procedures for various infinite-state structures have been presented as constraint-solving procedures, however the question of generalizing model-checking to constraint solving for temporal logic formulae containing free variables was not mentioned. Furthermore, the procedures inspired from logic programming were semi-decision procedures, whereas we shall present here decision procedures.

The rest of the paper is organized as follows. Section 2 presents the quantifier free fragment of first-order computation tree logic with constraints over some domain $\mathcal{D}$, noted QFCTL($\mathcal{D}$), and transpose the propositional CTL model-checking algorithm to this setting. Section 3 describes a QFCTL constraint solving algorithm which computes validity domains for variables by fixpoint iteration in quadratic time in the size of the structure. Section 4 studies the case where the underlying constraint domain $\mathcal{D}$ is a metric space, and defines in this case a real-valued degree of satisfaction of a QFCTL formula in a given structure. This continuous valuation in $[0, 1]$ of temporal logic formulae is defined using the validity domain of a QFCTL formula with free variables. This opens up the field of model-checking to continuous optimization with respect to temporal logic specifications over the reals. Section 5 describes our implementation in the Biochemical Abstract Machine BIOCHAM [19] of a QFCTL constraint solver over the reals, restricted to non branching traces, and evaluate its performance on a benchmark of systems biology parameter optimization problems.

2 Quantifier-Free Computation Tree Logic QFCTL

2.1 Syntax

In this paper, we consider a general setting of first-order temporal logic formulae without quantifiers, interpreted in some fixed structure. Let Σ be a signature of constant, function and predicate symbols interpreted over some fixed computation domain $\mathcal{D}$. For the sake of simplicity, we assume that the predicate symbols are closed under negation, i.e. each predicate p comes with its dual $\overline{p}$ such that for all $e_1, \ldots, e_n \in \mathcal{D}$, $\models_{\mathcal{D}} \neg p(e_1, \ldots, e_n)$ if and only if $\models_{\mathcal{D}} \overline{p}(e_1, \ldots, e_n)$.

Let $\mathcal{V}$ be an infinite set of variables, among which a finite set $V \subseteq \mathcal{V}$ of *state variables* (also called rigid variables) is distinguished. An *atomic constraint* is an atomic formula formed over Σ and $\mathcal{V}$ and a *constraint*, noted $c, \ldots$, is a conjunction of atomic constraints.

Quantifier-free first-order computation tree logic (QFCTL*) formulae are formed using the atomic constraints, the logical connectives $\neg$, $\vee$, $\wedge$, $\Rightarrow$, the path quantifiers: **E** (exists a path), **A** (forall paths) and the temporal operators: **X** (next), **F** (finally, at some time point), **G** (globally, at all time points), **U** (until), **W** (weak until). QFCTL denotes the fragment of QFCTL* in which the temporal operators are immediately preceded by a path quantifier.

The set of variables occurring in a formula ϕ is denoted by $V(\phi)$. We say that a formula ϕ is *closed* if $V(\phi) \subseteq V$, i.e. if it contains only state variables.

Example 1. Let us consider inequality constraints over the reals and the QFCTL* formula $\mathbf{EF}\ (x = v_1 \wedge \mathbf{X}\ (x = v_2 \wedge v_1 < v_2 \wedge \mathbf{X}\ (x = v_3 \wedge v_3 < v_2)))$, where x is a state variable and v_1, v_2, v_3 are free variables. This formula expresses that on some path (**E**), at some time point (**F**), the value of x is v_1, and at next time point (**X**), x gets a greater value v_2, and at next time point (**X**), x gets a lesser value, i.e. v_2 is local maximum for x.

2.2 Semantics

QFCTL* formulae are interpreted in branching structures, called Kripke structures, over some computation domain $\mathcal{D}$. We assume that the constraint satisfiability problem in $\mathcal{D}$, i.e. whether $\models_{\mathcal{D}} \exists X\ c$ for a constraint c where $X = V(c)$, is decidable. In the following, the notation $\exists(c)$ stands for the existencial closure of constraint c, i.e; $\exists X\ c$ where $X = V(c) \setminus V$.

A *state*, $s : V \to \mathcal{D}$, is a snapshot of values of state variables at a given time. A state is thus represented by a $\mathcal{D}$-valuation of the variables in V. We will write $s(v)$ for the value of state variable v in state s, and by extension, $s(e)$ for the value of a closed expression e (made up of state variables, function and constraint symbols) in state s. The set of states over V and $\mathcal{D}$ is denoted by $\mathcal{S}(V, \mathcal{D})$, or simply $\mathcal{S}$ when V and $\mathcal{D}$ are implicit.

A *Kripke structure over a set of variables V and a domain $\mathcal{D}$* is a quadruple $K = (S, R, V, \mathcal{D})$ where

- $S \subset \mathcal{S}(V, \mathcal{D})$ is a set of states over V and $\mathcal{D}$,
- $R \subset S \times S$ is a left-total relation over S (i.e. $\forall s \in S\ \exists t \in S\ (s, t) \in R$) called the state transition relation.

We say that a Kripke structure $K = (S, R, V, \mathcal{D})$ is *finite* if S is finite, and *finite state* if $\mathcal{S}(V, \mathcal{D})$ is finite (which implies that K and $\mathcal{D}$ are finite),

A path in K is an infinite sequence of states, noted $\pi = (s_0, s_1, ...)$, such that $(s_i, s_{i+1}) \in R$ for all $i \geq 0$. For such a path π and an integer k, π^k denotes the kth suffix path $(s_k, s_{k+1}, ...)$.

Definition 1. *A closed QFCTL* $(\Sigma, V, \mathcal{D})$ *formula* ϕ *is* true in a state s in a Kripke structure $K(S, R, V, \mathcal{D})$, *if the relation* $K, s \models_{\mathcal{D}} \phi$ *holds following the inductive definition given in Table 1.*

It is worth noting that the only difference between the inductive definition of Table 1 and the usual definition for propositional CTL* formulae [13] is in the

Table 1. Inductive definition of the truth values of closed QFCTL* formulae in a Kripke structure

$K, s \models_{\mathcal{D}} c$	if c is an atomic constraint and $\models_{\mathcal{D}} s(c)$,
$K, s \models_{\mathcal{D}} \mathbf{E}\ \phi$	if for some path π starting from s, $K, \pi \models_{\mathcal{D}} \phi$,
$K, s \models_{\mathcal{D}} \mathbf{A}\ \phi$	if for all paths π starting from s, $K, \pi \models_{\mathcal{D}} \phi$,
$K, \pi \models_{\mathcal{D}} \phi$	if $K, s \models_{\mathcal{D}} \phi$ where s is the first state of π,
$K, \pi \models_{\mathcal{D}} \mathbf{X}\ \phi$	if $K, \pi^1 \models_{\mathcal{D}} \phi$,
$K, \pi \models_{\mathcal{D}} \mathbf{F}\ \phi$	if there exists $k \geq 0$ s.t. $K, \pi^k \models_{\mathcal{D}} \phi$,
$K, \pi \models_{\mathcal{D}} \mathbf{G}\ \phi$	if for all $k \geq 0$, $K, \pi^k \models_{\mathcal{D}} \phi$,
$K, \pi \models_{\mathcal{D}} \phi \mathbf{U}\ \phi'$	if there exists $k \geq 0$ s.t. $K, \pi^k \models_{\mathcal{D}} \phi'$ and $K, \pi^j \models_{\mathcal{D}} \phi$ for all $0 \leq j < k$.
$K, \pi \models_{\mathcal{D}} \phi \mathbf{W}\ \phi'$	if either for all $k \geq 0$, $K, \pi^k \models_{\mathcal{D}} \phi$ or there exists $k \geq 0$ s.t. $K, \pi^k \models_{\mathcal{D}} \phi \wedge \phi'$ and for all $0 \leq j < k$, $K, \pi^j \models_{\mathcal{D}} \phi$.
$K, \pi \models_{\mathcal{D}} \neg\phi$	if $K, \pi \not\models_{\mathcal{D}} \phi$,
$K, \pi \models_{\mathcal{D}} \phi\ \wedge\ \phi'$	if $K, \pi \models_{\mathcal{D}} \phi$ and $K, \pi \models_{\mathcal{D}} \phi'$,
$K, \pi \models_{\mathcal{D}} \phi\ \vee\ \phi'$	if $K, \pi \models_{\mathcal{D}} \phi$ or $K, \pi \models_{\mathcal{D}} \phi'$,
$K, \pi \models_{\mathcal{D}} \phi \Rightarrow \phi'$	if $K, \pi \models_{\mathcal{D}} \phi'$ or $K, \pi \not\models_{\mathcal{D}} \phi$,

base case of an atomic constraint c. Such an atomic constraint is closed and is evaluated in $\mathcal{D}$ by checking its validity: $\models_{\mathcal{D}} s(c)$.

Definition 2. *A QFCTL* $(\Sigma, \mathcal{V}, \mathcal{D})$ formula ϕ is* satisfiable *in a Kripke structure $K = (S, R, V, \mathcal{D})$, noted (by a slight abuse of notation) $K \models_{\mathcal{D}} \exists Y\ \phi$, where $Y = \mathcal{V}(\phi) \setminus V$, if there exists a state $s \in S$ and a valuation $\rho : Y \to \mathcal{D}$ such that $K, s \models_{\mathcal{D}} \rho(\phi)$.*

Example 2. Let us consider the following finite Kripke structure over the reals, composed of five states, where state variable x takes values 1 to 5 respectively:

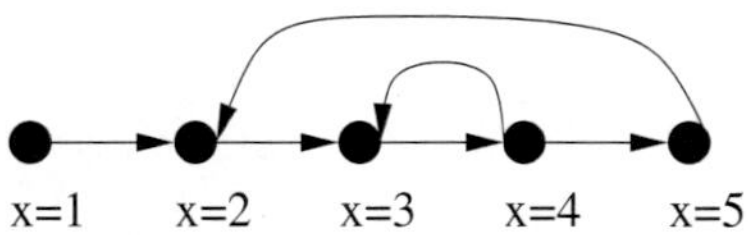

The QFCTL formula $\mathbf{EG}\ (x \leq V)$ has one free variable V. This formula is satisfiable in all states. It is true for all valuations of $V \geq 5$ in the last state, and for all valuations of $V \geq 4$ in the other states. The QFCTL constraint solving problem consists in computing these validity domains for the free variables of the formula in each state.

QFCTL* formulae enjoy the classical duality properties: $\neg\mathbf{E}\ \phi = \mathbf{A}\ \neg\phi$, $\neg\mathbf{X}\ \phi = \mathbf{X}\ \neg\phi$, $\neg\mathbf{F}\ \phi = \mathbf{G}\ \neg\phi$, $\neg(\phi_1 \mathbf{U}\ \phi_2) = (\neg\phi_2 \mathbf{W}\ \neg\phi_1)$ Moreover, **F** and **G** can be defined as abbreviations: $\mathbf{F}\ \phi = (true\mathbf{U}\ \phi)$, $\mathbf{G}\ \phi = (\phi\mathbf{W}\ false)$. Similarly, **W** can be defined from **G** and **U** by $\phi_1 \mathbf{W}\ \phi_2 = \mathbf{G}\ \phi_1 \vee (\phi_1 \mathbf{U}\ \phi_1 \wedge \phi_2)$. By assuming that the constraint language is closed under negation, negations (and implications) can be eliminated from QFCTL formulae by pushing them down to the constraints. Without loss of generality, we will thus sometimes restrict the QFCTL* formulae to *negation-free normal forms*, formed using $\vee$, $\wedge$, **E** , **A** , **X** , **U** , and **G** only.

2.3 QFCTL Model-Checking Algorithm

The (global) model-checking problem is the problem of determining the set of states in which a given temporal logic formula is true in a given finite Kripke structure. In particular, it solves the (local) decision problem of determining whether a given formula is true in a given initial state. The *QFCTL($\Sigma, V, \mathcal{D}$) model-checking problem* is the following:

Input: a finite Kripke structure $K = (S, R, V, \mathcal{D})$, a closed QFCTL($\Sigma$,V,$\mathcal{D}$) formula ϕ,

Output: the set of states $s \in S$ such that $K, s \models_{\mathcal{D}} \phi$.
By assuming that closed constraints can be evaluated in $\mathcal{D}$, the usual propositional CTL model-checking algorithm for finite Kripke structures [13] can be generalized to QFCTL as follows:

Algorithm 1

1. *Construct the graph (S, R) of K*
2. *for each subformula ψ of ϕ, taken in increasing order, add ψ to the labels of*
   (a) *the states s s.t. $\models_{\mathcal{D}} s(\psi)$ if ψ is an atomic constraint,*
   (b) *the states not labelled by ψ_1 if $\psi = \neg\psi_1$,*
   (c) *the states labelled by ψ_1 and ψ_2 if $\psi = \psi_1 \wedge \psi_2$ (similarly for $\vee, \Rightarrow$),*
   (d) *the immediate predecessors of states labeled by ψ_1 if $\psi =$ **EX** ψ_1*
   (e) i. *the states labelled by ψ_2*
       ii. *and their predecessors labelled by ψ_1*
       *if $\psi =$ **E** $(\psi_1$**U** $\psi_2)$*
   (f) i. *the states of non trivial strongly connected components labelled by ψ_1,*
       ii. *and their predecessors labelled by ψ_1*
       *if $\psi =$ **EG** ψ_1,*
3. *return the states labeled by ϕ.*

Example 3. On the Kripke structure of example 2, the model-checking algorithm evaluates the formula **EG** $(x \leq 4)$ by labeling the states with the subformulas as follows:

state	$x = 1$	$x = 2$	$x = 3$	$x = 4$	$x = 5$
step *(a)*	$x \leq 4$	$x \leq 4$	$x \leq 4$	$x \leq 4$	
step *(f)i.*			**EG** $(x \leq 4)$	**EG** $(x \leq 4)$	
step *(f)ii.*	**EG** $(x \leq 4)$	**EG** $(x \leq 4)$	**EG** $(x \leq 4)$	**EG** $(x \leq 4)$	

By using Tarjan's linear time algorithm for computing strongly connected components as usual [13], we get

Proposition 3. *Algorithm 1 solves the model-checking problem for QFCTL formulae over a finite Kripke structure K and a computation domain $\mathcal{D}$ in time $O(|K| * |\phi| * f(|\phi|))$, where ϕ is the formula to verify, $|\phi|$ its size (number of subformulae) and $f(n)$ is the time complexity for checking the $\mathcal{D}$-validity of a closed constraint of size n.*

3 QFCTL Constraint Solving Algorithm

Definition 4. *The QFCTL($\Sigma, \mathcal{V}, \mathcal{D}$)* satisfiability problem *is the following:*

Input: *a finite Kripke structure* $K = (S, R, V, \mathcal{D})$, *a QFCTL($\Sigma$,$\mathcal{D}$) formula* ϕ *where* $Y = V(\phi) \setminus V$,

Output: *set of states* s *such that* $K, s \models_{\mathcal{D}} \exists Y\ \phi$.

The constraint solving problem *is distinguished from the satisfiability problem by asking moreover to compute the validity domains of the variables:*

Output: *set of pairs* (s, ρ) *where* s *is a state and* $\rho : Y \to \mathcal{D}$ *is a valuation s.t.* $K, s \models_{\mathcal{D}} \rho(\phi)$, *assuming a finite representation of an infinite set of valuations, e.g. by a finite set of constraints.*

3.1 Fixpoint Computation of Validity Domains

It is well known that a propositional CTL formula ϕ can be identified with the set of states which satisfy it, i.e. $\{s \in \mathcal{S} \mid K, s \models_{\mathcal{D}} \phi\}$, and that for finite Kripke structures, the basic CTL operators can be characterized as the least or greatest fixpoints of certain monotonic operators in $2^{\mathcal{S}} \to 2^{\mathcal{S}}$, called predicate transformers [13,2].

Interestingly, this fixpoint characterization extends to the solutions of QFCTL formulae containing free variables, by associating, to a QFCTL formula ϕ, a set of states given with constraints for the free variables of ϕ describing the solutions of the satisfiability problem for ϕ.

Table 2. Fixpoint characterization of the constrained states satisfying a QFCTL formula

$[c] = \{(s|c) \ : \ s \in S \text{ and } \models_{\mathcal{D}} \exists(s(c))\}$ for a constraint c,

$[\mathbf{EX}\ \phi] = ex([\phi])$	$[\mathbf{AX}\ \phi] = ax([\phi])$
$[\mathbf{EF}\ \phi] = \mu Z.[\phi] \cup ex(Z)$	$[\mathbf{AF}\ \phi] = \mu Z.[\phi] \cup ax(Z)$
$[\mathbf{EG}\ \phi] = \nu Z.[\phi] \sqcap ex(Z)$	$[\mathbf{AG}\ \phi] = \nu Z.[\phi] \sqcap ax(Z)$
$[\mathbf{E}\ (\phi_1 \mathbf{U}\ \phi_2)] = \mu Z.[\phi_2] \cup ([\phi_1] \sqcap ex(Z))$	$[\mathbf{A}\ (\phi_1 \mathbf{U}\ \phi_2)] = \mu Z.[\phi_2] \cup ([\phi_1] \sqcap ax(Z))$
$[\mathbf{E}\ (\phi_1 \mathbf{W}\ \phi_2)] = \nu Z.[\phi_1] \cup ([\phi_2] \sqcap ex(Z))$	$[\mathbf{A}\ (\phi_1 \mathbf{W}\ \phi_2)] = \nu Z.[\phi_1] \cup ([\phi_2] \sqcap ax(Z))$

Let $\mathcal{S}_C$ be the set of state-constraint pairs, noted $s|c$, where s is a state and c is a $\mathcal{D}$-satisfiable constraint. Let us consider the set lattice $(2^{\mathcal{S}_C}, \emptyset, \mathcal{S}_C, \cup, \sqcap)$ with the operations of set union $\cup$ and constrained intersection $\sqcap$, i.e. intersection of states with a satisfiable constraint conjunction:

$$A \sqcap B = \{(s|c \wedge c') \ : \ s|c \in A,\ s|c' \in B,\ \models_{\mathcal{D}} \exists(c \wedge c')\}.$$

We shall not describe subsumption checks in this presentation, however it is worth noticing that the lattice top element $\mathcal{S}_C$ of all constrained states is logically equivalent to the element of all states given with the constraint true $\{(s|true) : s \in \mathcal{S}\}$. This element will be used as top element in computations.

Let $ex, ax : 2^{\mathcal{S}_\mathcal{C}} \to 2^{\mathcal{S}_\mathcal{C}}$ be the two constrained predicate transformers (associated to CTL operators **EX** and **AX**) defined by:

$$ex(Z) = \{(s|c) \in \mathcal{S}_\mathcal{C} \ : \ \exists (s,t) \in R \ s.t. \ t|c \in Z\},$$

$$ax(Z) = \{(s|c_1 \wedge \ldots \wedge c_n) \in \mathcal{S}_\mathcal{C}) \ : \{t \ : \ (s,t) \in R\} = \{t_1, \ldots, t_n\}, \\ \forall i, \ 1 \leq i \leq n, \ \ t_i|c_i \in Z, \ \models_\mathcal{D} \exists (c_1 \wedge \ldots \wedge c_n)\}.$$

Let us consider the fixpoint equations between QFCTL formulae and sets of state-constraint pairs given in Table 2. These equations can be used to compute the validity domains of the free variables of a QFCTL formula in each state, by finite fixpoint iteration.

Example 4. For the formula **EG** $(x \leq V)$ and the Kripke structure of Example 2, the fixpoint iteration provides the following result:

state	$x=1$	$x=2$	$x=3$	$x=4$	$x=5$
$x \leq V$	$1 \leq V$	$2 \leq V$	$3 \leq V$	$4 \leq V$	$5 \leq V$
$\tau^0_{\mathbf{EG}\ (x\leq V)}$	true	true	true	true	true
$\tau^1_{\mathbf{EG}\ (x\leq V)}$	$1 \leq V$	$2 \leq V$	$3 \leq V$	$4 \leq V$	$5 \leq V$
$\tau^2_{\mathbf{EG}\ (x\leq V)}$	$2 \leq V$	$3 \leq V$	$4 \leq V$	$4 \leq V$	$5 \leq V$
$\tau^3_{\mathbf{EG}\ (x\leq V)}$	$3 \leq V$	$4 \leq V$	$4 \leq V$	$4 \leq V$	$5 \leq V$
$\tau^4_{\mathbf{EG}\ (x\leq V)}$	$4 \leq V$	$4 \leq V$	$4 \leq V$	$4 \leq V$	$5 \leq V$
$\tau^5_{\mathbf{EG}\ (x\leq V)} = [\mathbf{EG}\ (x \leq V)]$	$4 \leq V$	$4 \leq V$	$4 \leq V$	$4 \leq V$	$5 \leq V$

While for the formula **AG** $(x \leq V)$ involving greatest fixpoint computation, starting from all states labelled by the constraint true, we get the fixpoint

state	$x=1$	$x=2$	$x=3$	$x=4$	$x=5$
$[\mathbf{AG}\ (x \leq V)]$	$5 \leq V$	$5 \leq V$	$5 \leq V$	$5 \leq V$	$5 \leq V$

Let us say that an operator τ over a set S is *bounded* if $\forall s \in S\ \exists i \in N\ \tau^{i+1}(s) = \tau^i(s)$. It is clear from the proof of Knaster-Tarski-Kleene theorem that:

Proposition 1. *A bounded monotonic operator τ over a lattice $(L, \bot, \top, \sqcup, \sqcap)$ admits a least fixpoint equal to $\tau^i(\bot)$ for some $i \geq 0$, and a greatest fixpoint equal to $\tau^j(\top)$ for some $j \geq 0$.*

Lemma 1. *The constrained predicate transformers ex and ax, and the constrained predicate transformers associated to QFCTL operators, are monotonic and bounded in finite Kripke structures.*

Proof. As for monotonicity, it is clear that if $Z_1 \subset Z_2$ then $ex(Z_1) \subseteq ex(Z_2)$, $ax(Z_1) \subseteq ax(Z_2)$. The same goes for the predicate transformers of QFCTL operators since they proceed by intersection or union of monotonic operators.

Predicate transformers ex and ax are also bounded since otherwise one could exhibit an infinite chain of states such that $(s_i, s_{i+1}) \in R$ with $\forall i, j,\ 1 \leq i < j,\ s_i \neq s_j$, a contradiction in a finite Kripke structure. The same goes for the other predicate transformers since they are built by union or intersection of bounded operators.

Proposition 2 (soundness). *If $s|c \in [\phi]$ then $K, s \models_{\mathcal{D}} \rho(\phi)$ for every valuation ρ such that $\models_{\mathcal{D}} \rho(c)$.*

Proof. By structural induction on ϕ.

Proposition 3 (completeness). *If $K, s \models_{\mathcal{D}} \rho(\phi)$ then there exists $s|c \in [\phi]_K$ such that $\models_{\mathcal{D}} \rho(c)$.*

Proof. By structural induction on ϕ.

We thus get:

Theorem 1. *The satisfiability problem of QFCTL formulae in a finite Kripke structure over a domain $\mathcal{D}$ with a decidable language of constraints, is decidable.*

Proposition 4. *The number of fixpoint iteration steps in the QFCTL constraint solving algorithm 2 is in $O(n * k^2)$ where n is the size of the formula and k is the number of states.*

Proof. The algorithm proceeds by iteratively computing constrained states for the subformulae of the formula, hence in at most n steps. For each subformula, the algorithm computes a fixpoint of constrained states by iteration on constrained states, hence in at most k^2 steps. Each elementary step involves constraint satisfiability checking operations whose time complexity is not counted here.

This quadratic complexity in the number of states must be contrasted with the linear complexity of the QFCTL model-checking algorithm (Prop. 3). The linear complexity of model-checking relies on Tarjan's algorithm for computing strongly connected components of the structure for **EG** formulae. For computing validity domains however, one would need to consider the different circuits of the structure in order to label the states with appropriate domains for **EG** formulae (see example 2). The fixpoint computation shows that this labeling can be done in quadratic time, whereas a naive algorithm considering all the circuits of the finite Kripke structure would require an exponential time.

4 QFCTL Formulae over a Metric Space $\mathcal{D}$

In this section, we consider the case where the computation domain $\mathcal{D}$ is a metric space, i.e. a domain given with a distance function $d : \mathcal{D}^2 \to \mathbb{R}$.

4.1 Continuous Valuation of QFCTL Formulae

In this general setting of metric spaces as computation domain, the QFCTL constraint solving algorithm provides a mean to evaluate closed QFCTL formula continuously in the interval $[0, 1]$, instead of by a Boolean value.

For this, given a QFCTL formula ϕ, a QFCTL *pattern formula* $\psi(x_1, ..., x_k)$ is introduced by replacing some constants in ϕ by new variables $\{x_1, ..., x_k\}$:

we have $\phi = \psi(v_1, ..., v_k)$ for some instantiation of the variables by domain values $v_1, ..., v_k$. The satisfaction degree of ϕ is then defined using the distance between the validity domain of the variables $x_1, ..., x_k$ in ψ and the objective values $(v_1, ..., v_k)$ in $\mathbb{R}^k$.

Definition 5. *The* violation degree $vd(\phi, \psi)$ *of a QFCTL formula* ϕ *in a Kripke structure* K *with respect to a pattern formula* $\psi(\boldsymbol{y})$ *such that* $\phi = \psi(\boldsymbol{v})$ *for some real values* $\boldsymbol{v}$*, is the euclidean distance* d *between* $\boldsymbol{v}$ *and the initial state validity domain* $D_\psi^{\boldsymbol{y}}$ *of the free variables of* ψ*, or* $+\infty$ *if* ψ *is not satisfiable:*

$$vd(\phi, \psi) = min_{\boldsymbol{v'} \in D_\psi^{\boldsymbol{y}}} d(\boldsymbol{v'}, \boldsymbol{v})$$

The satisfaction degree, $sd(\phi, \psi) \in [0, 1]$, *of* ϕ *with respect to* ψ *is obtained by normalization:*

$$sd(\phi, \psi) = \frac{1}{1 + vd(\phi, \psi)}$$

Example 5. For instance, in example 3, the formulae **EG** $(x \leq 2)$ and **AG** $(x \leq 2)$ can now be given a continuous degree of satisfaction with respect to the pattern formulae **EG** $(x \leq V)$ and **AG** $(x \leq V)$ respectively. This is obtained simply by computing the distance between the validity domain of V and the objective value 2. This gives the following satisfaction degrees:

state	$x = 1$	$x = 2$	$x = 3$	$x = 4$	$x = 5$
$sd($**EG** $(x \leq 2)$	1/3	1/3	1/3	1/3	1/4
$sd($**AG** $(x \leq 2))$	1/4	1/4	1/4	1/4	1/4

Interestingly, this notion of satisfaction degree gives also rise naturally to a pseudometric between Kripke structures with respect to a temporal specification ϕ, by considering the difference in the degrees of satisfaction of ϕ between both structures.

4.2 Parameter Optimization with Respect to a QFCTL Formula

Because it quantifies how far a Kripke structure K is from a given specification, the satisfaction degree can be used to guide the search when one tries to modify a Kripke structure to make it satisfy a specification. When $\mathcal{D} = \mathbb{R}$, this enables the use of (non-linear) continuous optimization methods, where state valuations are the variables being optimized and where the satisfaction degree provides a "black box" fitness function [32].

An optimization problem is thus defined for the satisfaction degree of a QFCTL formula ϕ^* w.r;t. a valuation ρ_ϕ of some of its variables. This has an intuitive interpretation : ϕ^* is the kind of property considered while ρ_ϕ defines the objective valuation of the actual property.

In the implementation described below, we use the covariance matrix adaptive evolution strategy of Hansen and Ostermeier [21], as non-linear optimization method for maximizing the satisfaction degree of QFCTL specifications.

4.3 Robustness Estimation with Respect to QFCTL($\mathbb{R}_{\text{lin}}$) Specifications

The notion of satisfaction degree can also be used to define a degree of robustness of a behavior described in temporal logic with respect to a set of perturbations, and estimate it computationally [33]. This robustness degree is defined as the mean value of the satisfaction degree of the property of interest over all admissible perturbations, possibly weighted by probabilities. This definition is an adaptation of the general definition given by Kitano [25] to the temporal logic setting:

Definition 6. *[33] Let P be a set of perturbations, $prob(p)$ be the probability of perturbation p, s be the initial state of the numerical trace of the system under perturbation $p \in P$. The* robustness degree $R_{\phi,P}$ *of a property ϕ with respect to P is the real value* $R_{\phi,P} = \int_{p \in P} sd(\phi, s) prob(p) dp$

In the case of an infinite perturbation set, this robustness degree can be estimated by sampling. The robustness degree can be used to compare models and can even be integrated as a criterion in the parameter optimization procedure [32].

4.4 Implementation

Our current implementation of QFCTL constraint solving is restricted to linear constraints over the reals and to linear Kripke structures, i.e. numerical traces without branching. The constraint solving problem of a QFCTL formula on a numerical trace, the computation of the satisfaction degree of a formula and its use as a fitness function for parameter optimization are implemented in version 2.8 of the freely available tool BIOCHAM, a modeling environment for the analysis of biological systems [19]. The computation of validity domains is handled by a simplified version of the QFCTL constraint solving algorithm dealing only with a single numerical trace, i.e. a finite linear structure [33]. The atomic constraints are linear constraints over the reals. Their satisfiability is checked using the Parma Polyhedra Library [4].

5 Applications in Systems Biology

5.1 Context of Molecular Systems Biology

The use of temporal logics in systems biology relies on a logical paradigm which consists in making the following identifications:

biological model = (quantitative) transition system
biological properties = temporal logic formulae
automatic validation = model-checking

In this domain, temporal logics have been used in recent years in many applications, either as query languages of large interaction maps [10] or gene regulatory networks [5], or as specification languages of biological properties known or inferred [17] from experiments, and used for validating models, discriminating between models and proposing new biological experiments [7], finding parameter values [9], or estimating robustness [32,6].

The difficulty inherent in using quantitative, but still incomplete, uncertain and imprecise biological knowledge makes the modeling problem a challenging task. Temporal logics help cope with this difficulty by providing a powerful specification language of the behavior of the system. The advantage of temporal logics is particularly explicit in comparison with the essentially qualitative properties considered in dynamical systems theory (e.g. multistability, existence of oscillations) or with the exact quantitative properties considered in optimization theory (e.g. curve fitting) as it allows us to express both qualitative (e.g. some protein is eventually produced) and quantitative (e.g. a concentration exceeds 10) properties.

Numerical traces representing evolution over a given time span of biological species can be obtained either from measurements in biological experiments, or from simulations by numerical integration of (non-linear) ODE models. QFCTL formulae are interpreted on these finite traces by adding a loop on the last state which only changes the meaning of $\mathbf{X}$ [32]. Non-linear continuous optimization methods can then be used to optimize a biological model with respect to a QFCTL specification. Note that in this case, the Kripke structure is optimized indirectly, by optimizing the biological model producing it.

5.2 Parameter Optimization with Respect to QFCTL($\mathbb{R}_{\mathbf{lin}}$) Specifications

We examine here the problem of finding parameter values of biological models such that the numerical trace obtained by simulation satisfies a given specification. We consider several examples of parameter optimization problems [32] in three ordinary differential equation models. We provide the CPU time (in seconds) required for optimization and for a single validity domain computation.

The first model is the budding yeast cell cycle ODE model of Chen et al. [12] which displays how some proteins interact to form an heterodimer known as maturation promoting factor (*MPF*) playing a key role in the control of mitotic cycles. For given sets of kinetic parameter values, *MPF* exhibits periodic activity peaks. We use formulae ϕ_1^*, ϕ_2^* and ϕ_3^* to see if it is possible to respectively find values in order to have higher *MPF* peaks, greater *MPF* amplitude, or shorter oscillation periods.

The second model is a model of the MAPK signal transduction cascade in the cell [27]. In [31], oscillations have been found in this model of the cascade of enzymatic reactions is directional and does not contain any negative feedback reaction. In [18] we analyzed this phenomenon in terms of the negative circuits of the influence graph associated to a reaction graph. Here, we use ϕ_5 to find parameter values and initial conditions that exhibit sustained oscillations with some amplitude constraint on the protein complex *MAPKp1p2*. Formula ϕ_4 illustrates a curve fitting problem on two time points and the protein complex *MEKRAFp1*.

The last example is a design problem. Given a specification of a synthetic gene transcriptional cascade system, whose input is *EYFP* we search for transcription rate parameter values with well-timed and fast-switching constraints (formula ϕ_6).

The following QFCTL formulae are considered with the objective valuations given in column ρ_ϕ of Table 3:

$$\phi_1^* = EF([MPF] > max)$$

$$\begin{aligned}\phi_2^* = \; & EF([MPF] > x1 \wedge EF([MPF] < x2 \\ & \wedge EF([MPF] > x1 \wedge EF([MPF] < x2)))) \\ & \wedge x1 - x2 > a\end{aligned}$$

$$\begin{aligned}\phi_3^* = \; & EF(d([Cdc2])/dt < 0 \wedge EX(d([Cdc2])/dt > 0 \wedge Time = t1 \\ & \wedge EX(F(d([Cdc2])/dt > 0 \wedge EX(d([Cdc2])/dt < 0 \wedge Time = t2)) \\ & \wedge t2 - t1 > p\end{aligned}$$

$$\begin{aligned}\phi_4^* = \; & EG(Time = 30 \rightarrow [MEKRAFp1] = u \\ & \wedge Time = 60 \rightarrow [MEKRAFp1] = v) \wedge \cdots\end{aligned}$$

$$\begin{aligned}\phi_5^* = \; & EF([MAPKp1p2] > x1 \wedge EF([MAPKp1p2] < x2)) \\ & \wedge x1 - x2 > a\end{aligned}$$

$$\begin{aligned}\phi_6^* = \; & EG(Time < t_1 \rightarrow [EYFP] < 10^3) \\ & \wedge EG(Time > t_2 \rightarrow [EYFP] > 10^5) \\ & \wedge t_1 > b1 \wedge t_2 < b_2 \wedge t_2 - t_1 < b_3\end{aligned}$$

Table 3 summarizes our performance evaluation on these benchmarks. The dimension of the search space (i.e. number of parameters) does not determine alone the complexity of an optimization problem : problems involving ϕ_4^* and ϕ_5^* have

Table 3. Parameter optimization benchmark where #parameters is the number of parameters of the biological model being optimized, $|V|$ the number of species in that model (i.e the number of variables in the produced numerical trace), last $|S|$ the number of states in the last produced numerical trace, $|Y|$ the number of formula variables, CPU time the time in seconds required to complete the optimization on a Core 2 Duo 2GHz, #iterations the number of calls to the fitness function (i.e validity domain computations and numerical simulations of the model) and CPU time for val. domain in one iteration the time to compute the validity domain on the last simulated trace.

Model	#parameters	$\|V\|$	last $\|S\|$	ϕ^*	ρ_ϕ	$\|Y\|$	CPU time (s)	#iterations	CPU time (s) for val. domain in one iteration
Cell cycle	2	6	186	ϕ_1^*	max=0.3	1	27	132	0.02
Cell cycle	2	6	204	ϕ_2^*	a=0.3	3	131	138	1.2
Cell cycle	8	6	267	ϕ_3^*	p=20	3	23	40	0.39
MAPK	22	22	35	ϕ_4^*	$u, v.. = 0.03, 0.04..$	6	259	611	0.001
MAPK	37	22	234	ϕ_5^*	a=0.5	3	453	868	0.17
Cascade	15	3	346	ϕ_6^*	b_1, b_2, b_3	5	70	96	0.48
Cascade					$= 150, 150, 450$				

high dimensions and are the longest to solve but ϕ_3^* and ϕ_6^* are faster than ϕ_2^* despite their much higher dimensions. But even with search space dimensions as high as 37, by guiding the search with the continuous valuation of QFCTL formulae, it is possible to find solutions with only less than a thousand calls to the fitness function. Notice that the time required to compute the validity domains is in general only a small fraction of the total CPU time. This can be explained by the optimization method overhead and more importantly by the time required to generate traces by numerical integration.

6 Conclusion

We have shown that the QFCTL constraint satisfiability problem is decidable in finite Kripke structures over an arbitrary computation domain with a decidable language of constraints, i.e. that any constraint solver can be lifted to a temporal logic constraint solver over finite Kripke structures. We have presented a generic QFCTL constraint solver which computes validity domains for the free variables of a formula, in quadratic time in the number of states, and linear time in the size of the formula, apart from the basic constraint satisfiability checks.

We have shown that the computation of validity domains for QFCTL constraints over metric spaces makes it possible to define a continuous measure of satisfaction of QFCTL formulae, opening up the field of model-checking to optimization. This has been illustrated with central computational issues in systems biology, from which our present work originates, for inferring kinetic parameter values in structural models from the observed behaviors of the system formalized in temporal logic with numerical constraints. It should be clear however that these methods for parameter optimization with respect to temporal logic specifications and robustness analyses, should be of a wider application spectrum in dynamical systems, reverse-engineering and synthesis of hybrid systems.

Acknowledgements. We gracefully acknowledge discussions with Sylvain Soliman and Grégory Batt on this topic, partial support from the European FP6 Strep project TEMPO, and the reviewer's comments for improving the presentation.

References

1. Abdulla, P.A., Cerans, K., Jonsson, B., Tsay, Y.-K.: General decidability theorems for infinite-state systems. Logic in Computer Science, pp. 313–321 (1996)
2. Emerson, E.A., Clarke, E.M.: Characterizing correctness properties of parallel programs using fixpoints. In: de Bakker, J.W., van Leeuwen, J. (eds.) ICALP 1980. LNCS, vol. 85, pp. 169–181. Springer, Heidelberg (1980)
3. Antoniotti, M., Policriti, A., Ugel, N., Mishra, B.: Model building and model checking for biochemical processes. Cell Biochemistry and Biophysics 38, 271–286 (2003)
4. Bagnara, R., Hill, P.M., Zaffanella, E.: The Parma Polyhedra Library: Toward a complete set of numerical abstractions for the analysis and verification of hardware and software systems. Science of Computer Programming 72(1-2), 3–21 (2008)

5. Batt, G., Ropers, D., de Jong, H., Geiselmann, J., Mateescu, R., Page, M., Schneider, D.: Validation of qualitative models of genetic regulatory networks by model checking: Analysis of the nutritional stress response in *Escherichia coli*. Bioinformatics 21(suppl. 1), 19–28 (2005)
6. Batt, G., Yordanov, B., Weiss, R., Belta, C.: Robustness analysis and tuning of synthetic gene networks. Bioinformatics 23(18), 2415–2422 (2007)
7. Bernot, G., Comet, J.-P., Richard, A., Guespin, J.: A fruitful application of formal methods to biological regulatory networks: Extending thomas' asynchronous logical approach with temporal logic. Journal of Theoretical Biology 229(3), 339–347 (2004)
8. Calder, M., Vyshemirsky, V., Gilbert, D., Orton, R.: Analysis of signalling pathways using the continuous time markow chains. In: Plotkin, G. (ed.) Transactions on Computational Systems Biology VI. LNCS (LNBI), vol. 4220, pp. 44–67. Springer, Heidelberg (2006); CMSB 2005 Special Issue
9. Calzone, L., Chabrier-Rivier, N., Fages, F., Soliman, S.: Machine learning biochemical networks from temporal logic properties. In: Priami, C., Plotkin, G. (eds.) Transactions on Computational Systems Biology VI. LNCS (LNBI), vol. 4220, pp. 68–94. Springer, Heidelberg (2006); CMSB 2005 Special Issue
10. Chabrier, N., Fages, F.: Symbolic model checking of biochemical networks. In: Priami, C. (ed.) CMSB 2003. LNCS, vol. 2602, pp. 149–162. Springer, Heidelberg (2003)
11. Chan, W.: Temporal-logic queries. In: Emerson, E.A., Sistla, A.P. (eds.) CAV 2000. LNCS, vol. 1855, pp. 450–463. Springer, Heidelberg (2000)
12. Chen, K.C., Csikász-Nagy, A., Györffy, B., Val, J., Novàk, B., Tyson, J.J.: Kinetic analysis of a molecular model of the budding yeast cell cycle. Molecular Biology of the Cell 11, 396–391 (2000)
13. Clarke, E.M., Grumberg, O., Peled, D.A.: Model Checking. MIT Press, Cambridge (1999)
14. Delzanno, G., Podelski, A.: Constraint-based deductive model checking. STTT 3(3), 250–270 (2001)
15. Egerstedt, M., Mishra, B.: (2008)
16. Eker, S., Knapp, M., Laderoute, K., Lincoln, P., Meseguer, J., Kemal Sönmez, M.: Pathway logic: Symbolic analysis of biological signaling. In: Proceedings of the seventh Pacific Symposium on Biocomputing, January 2002, pp. 400–412 (2002)
17. Fages, F., Rizk, A.: On temporal logic constraint solving for the analysis of numerical data time series. Theoretical Computer Science 408(1), 55–65 (2008)
18. Fages, F., Soliman, S.: From reaction models to influence graphs and back: a theorem. In: Fisher, J. (ed.) FMSB 2008. LNCS (LNBI), vol. 5054, pp. 90–102. Springer, Heidelberg (2008)
19. Fages, F., Soliman, S., Rizk, A.: BIOCHAM v2.8 user's manual. In: INRIA (2009), `http://contraintes.inria.fr/BIOCHAM`
20. Halbwachs, N., Proy, Y.e., Roumanoff, P.: Verification of real-time systems using linear relation analysis. In: Formal Methods in System Design, pp. 157–185 (1997)
21. Hansen, N., Ostermeier, A.: Completely derandomized self-adaptation in evolution strategies. Evolutionary Computation 9(2), 159–195 (2001)
22. Heath, J., Kwiatkowska, M., Norman, G., Parker, D., Tymchyshyn, O.: Probabilistic model checking of complex biological pathways. In: Priami, C. (ed.) CMSB 2006. LNCS (LNBI), vol. 4210, pp. 32–47. Springer, Heidelberg (2006)
23. Henzinger, T.A., Ho, P.-H., Wong-Toi, H.: HYTECH: A model checker for hybrid systems. In: Grumberg, O. (ed.) CAV 1997. LNCS, vol. 1254, pp. 460–463. Springer, Heidelberg (1997)

24. Janssen, M., Van Hentenryck, P., Deville, Y.: A constraint satisfaction approach for enclosing solutions to parametric ordinary differential equations. SIAM Journal on Numerical Analysis 40(5) (2002)
25. Kitano, H.: Towards a theory of biological robustness. Molecular Systems Biology 3, 137 (2007)
26. Larsen, K.G., Pettersson, P., Yi, W.: Uppaal in a nutshell. Int. Journal on Software Tools for Technology Transfer 1, 134–152 (1997)
27. Levchenko, A., Bruck, J., Sternberg, P.W.: Scaffold proteins may biphasically affect the levels of mitogen-activated protein kinase signaling and reduce its threshold properties. PNAS 97(11), 5818–5823 (2000)
28. Clarke, E.M., Emerson, E.A.: Design and synthesis of synchronization skeletons using branching-time temporal logic. In: Kozen, D. (ed.) Logic of Programs 1981. LNCS, vol. 131, pp. 52–71. Springer, Heidelberg (1982)
29. Pnueli, A.: The temporal logic of programs. In: FOCS, pp. 46–57 (1977)
30. Podelski, A.: Model checking as constraint solving. In: Palsberg, J. (ed.) SAS 2000. LNCS, vol. 1824, pp. 22–37. Springer, Heidelberg (2000)
31. Qiao, L., Nachbar, R.B., Kevrekidis, I.G., Shvartsman, S.Y.: Bistability and oscillations in the huang-ferrell model of mapk signaling. PLoS Computational Biology 3(9), 1819–1826 (2007)
32. Rizk, A., Batt, G., Fages, F., Soliman, S.: On a continuous degree of satisfaction of temporal logic formulae with applications to systems biology. In: Heiner, M., Uhrmacher, A.M. (eds.) CMSB 2008. LNCS (LNBI), vol. 5307, pp. 251–268. Springer, Heidelberg (2008)
33. Rizk, A., Batt, G., Fages, F., Soliman, S.: A general computational method for robustness analysis with applications to synthetic gene networks. BioInformatics 25(12), 169–178 (2009)
34. Tevfik, B., Richard, G., William, P.: Symbolic model checking of infinite state systems using presburger arithmetic. In: Grumberg, O. (ed.) CAV 1997. LNCS, vol. 1254, pp. 400–411. Springer, Heidelberg (1997)
35. Wolper, P., Boigelot, B.: Verifying systems with infinite but regular state space. In: Y. Vardi, M. (ed.) CAV 1998. LNCS, vol. 1427, pp. 88–97. Springer, Heidelberg (1998)

Exploiting Problem Structure for Solution Counting

Aurélie Favier[1], Simon de Givry[1], and Philippe Jégou[2]

[1] INRA MIA Toulouse, France
{afavier,degivry}@toulouse.inra.fr
[2] Université Paul Cézanne, Marseille, France
philippe.jegou@univ-cezanne.fr

Abstract. This paper deals with the challenging problem of counting the number of solutions of a CSP, denoted #CSP. Recent progress have been made using search methods, such as BTD [15], which exploit the constraint graph structure in order to solve CSPs. We propose to adapt BTD for solving the #CSP problem. The resulting exact counting method has a worst-case time complexity exponential in a specific graph parameter, called *tree-width*. For problems with sparse constraint graphs but large tree-width, we propose an iterative method which approximates the number of solutions by solving a partition of the set of constraints into a collection of partial chordal subgraphs. Its time complexity is exponential in the maximum clique size - the *clique number* - of the original problem, which can be much smaller than its tree-width. Experiments on CSP and SAT benchmarks shows the practical efficiency of our proposed approaches.

1 Introduction

The Constraint Satisfaction Problem (CSP) formalism offers a powerful framework for representing and solving efficiently many problems. Finding a solution is NP-complete. A more difficult problem consists in counting the number of solutions. This problem, denoted #CSP, is known to be #P-complete [27]. This problem has numerous applications in computer science, particularly in AI, e.g. in approximate reasoning [23], in diagnosis [18], in belief revision [5], *etc.*

In the literature, two principal classes of approaches have been proposed. The first class, methods find exactly the number of solutions. The second class, methods propose approximations. For the first class, a natural approach consists in extending classical search algorithms such as FC or MAC in order to enumerate all solutions. But the more solutions there are, the longer it takes to enumerate them.

Here, we are interested in search methods that exploit the problem structure, providing time and space complexity bounds. This is the case for the d-DNNF compiler [6] and AND/OR graph search [8,9] for counting. We propose to adapt Backtracking with Tree-Decomposition (BTD) [15] to #CSP. This method was initially proposed for solving structured CSPs. Our modifications to BTD are similar to what has been done in the AND/OR context [8,9], except that BTD is based on a cluster tree-decomposition instead of a pseudo-tree, which naturally enables BTD to exploit dynamic variable orderings inside clusters whereas AND/OR search uses a static ordering.

Most of the recent work on counting has been realized on #SAT, the model counting problem associated with SAT [27]. Exact methods for #SAT extend systematic SAT

I.P. Gent (Ed.): CP 2009, LNCS 5732, pp. 335–343, 2009.

solvers, adding component analysis [3] and caching [26] for efficiency. Approaches using approximations estimate the number of solutions.

They propose poly-time or exponential time algorithms which must offer reasonably good approximations of the number of solutions, with theoretical guarantees about the quality of the approximation, or not. Again, most of the work has been done on #SAT by sampling either the original OR search space [28,12,10,17], or the original AND/OR search space [11]. All these methods except that in [28] provide a lower bound on the number of solutions with a high-confidence interval obtained by randomly assigning variables until solutions are found. A possible drawback of these approaches is that they might find no solution within a given time limit due to inconsistent partial assignments. For large and complex problems, this results in zero lower bounds or it requires time-consuming parameter (e.g sample size) tuning in order to avoid this problem. Another approach involves reducing the search space by adding streamlining XOR constraints [13,14]. However, it does not guarantee that the resulting problem is easier to solve.

In this paper, we propose to relax the problem, by partitioning the set of constraints into a collection of structured chordal subproblems. Each subproblem is then solved using our modified BTD. This task should be relatively easy if the original instance has a sparse graph[1]. Finally, an approximate number of solutions on the whole problem is obtained by combining the results of each subproblem. The resulting approximate method called `ApproxBTD` gives also a trivial upper bound on the number of solutions. Other relaxation-based counting methods have been tried in the literature such as mini-bucket elimination and iterative join-graph propagation [16], or in the related context of Bayesian inference, iterative belief propagation and the edge deletion framework [4][2]. These approaches do not exploit the local structure of instances as it is done by search methods such as BTD, thanks to local consistency and dynamic variable ordering.

In the next section, we introduce notation and tree-decomposition. Section 3 describes BTD for counting and Section 4 presents ApproxBTD for approximate counting. Experimental results are given in Section 5, then we conclude.

2 Preliminaries

A CSP is a quadruplet $\mathcal{P} = (X,D,C,R)$. X and D are sets of n variables and finite domains. The domain of variable x_i is denoted d_{x_i}. The maximum domain size is d. C is a set of m constraints. Each constraint $c \in C$ is a set $\{x_{c_1},\dots,x_{c_k}\}$ of variables. A relation $r_c \in R$ is associated with each constraint c such that r_c represents (in intension) the set of allowed tuples over $d_{x_{c_1}} \times \cdots \times d_{x_{c_k}}$. An *assignment* $Y = \{x_1,\dots,x_k\} \subseteq X$ is a tuple $\mathcal{A} = (v_1,\dots,v_k)$ from $d_{x_1} \times \cdots \times d_{x_k}$. A constraint c is said *satisfied* by $\mathcal{A}$ if $c \subseteq Y, \mathcal{A}[c] \in r_c$, *violated* otherwise. A solution is a complete assignment satisfying all the constraints. The structure of a CSP can be represented by the graph (X,C), called

[1] In fact, it depends on the tree-width of the subproblems, which is bounded by the maximum clique size of the original instance. In the case of a sparse graph, we expect this size to be small. This forbids using our approach for solution counting in CSPs with global constraints.

[2] It starts by solving an initial polytree-structured subproblem, further augmented by progressively recovering some edges, until the whole problem is solved. ApproxBTD starts directly with a possibly larger chordal subproblem.

the *constraint graph*, whose vertices are the variables of X and with an edge between two vertices if the corresponding variables share a constraint.

A *tree-decomposition* [22] of a CSP $\mathcal{P}$ is a pair $(\mathcal{C},\mathcal{T})$ with $\mathcal{T}=(I,F)$ a tree with vertices I and edges F and $\mathcal{C}=\{C_i : i \in I\}$ a family of subsets of X, such that each cluster C_i is a node of $\mathcal{T}$ and satisfies: (1) $\cup_{i\in I} C_i = X$, (2) for each constraint $c \in C$, there exists $i \in I$ with $c \subseteq C_i$, (3) for all $i,j,k \in I$, if k is on a path from i to j in $\mathcal{T}$, then $C_i \cap C_j \subseteq C_k$. The width of a tree-decomposition $(\mathcal{C},\mathcal{T})$ is equal to $max_{i\in I}|C_i|-1$. The tree-width of $\mathcal{P}$ is the minimum width over all its tree-decompositions. Finding an optimal tree-decomposition is NP-Hard [2]. In the following, from a tree-decomposition, we consider a rooted tree (I,F) with root C_1 and we note $Sons(C_i)$ the set of son clusters of C_i and $Desc(C_j)$ the set of variables which belong to C_j or to any descendant C_k of C_j in the subtree rooted in C_j.

3 Exact Solution Counting with BTD

The essential property of tree decomposition is that assigning $C_i \cap C_j$ (C_j is a son of C_i) separates the initial problem into two subproblems, which can then be solved independently. The first subproblem rooted in C_j is defined by the variables in $Desc(C_j)$ and by all the constraints involving *at least* one variable in $Desc(C_j) \setminus C_i$. The remaining constraints, together with the variables they involve, define the remaining subproblem.

A tree search algorithm can exploit this property by using a suitable variable ordering : the variables of any cluster C_i must be assigned before the variables that remain in its son clusters. In this case, for any cluster $C_j \in Sons(C_i)$, once $C_i \cap C_j$ is assigned, the subproblem rooted in C_j conditioned by the current assignment $\mathcal{A}$ of $C_i \cap C_j$ can be solved independently of the rest of the problem. The exact number of solutions nb of this subproblem may then be recorded, called a #good and represented by a pair $(\mathcal{A}[C_i \cap C_j], nb)$, which means it will never be computed again for the same assignment of $C_i \cap C_j$. This is why algorithms such as BTD or AND / OR graph search are able to keep the complexity exponential in the size of the largest cluster only.

BTD is described in Algorithm 1. Given an assignment $\mathcal{A}$ and a cluster C_i, BTD looks for the number of extensions $\mathcal{B}$ of $\mathcal{A}$ on $Desc(C_i)$ such that $\mathcal{A}[C_i\text{-}V_{C_i}] = \mathcal{B}[C_i\text{-}V_{C_i}]$. V_{C_i} denotes the set of unassigned variables of C_i. The first call is to BTD($\emptyset, C_1, C_1$) and it returns the number of solutions. Inside a cluster C_i, it proceeds classically by assigning a value to a variable and by backtracking if any constraint is violated. When every variable in C_i is assigned, BTD computes the number of solutions of the subproblem induced by the first son of C_i, if there is one. More generally, let us consider C_j, a son of C_i. Given a current assignment $\mathcal{A}$ on C_i, BTD checks whether the assignment $\mathcal{A}[C_i \cap C_j]$ corresponds to a #good. If so, BTD multiplies the recorded number of solutions with the number of solutions of C_i with $\mathcal{A}$ as its assignment. Otherwise, it extends $\mathcal{A}$ on $Desc(C_i)$ in order to compute its number of consistent extensions nb. Then, it records the #good $(\mathcal{A}[C_i \cap C_j], nb)$. BTD computes the number of solutions of the subproblem induced by the next son of C_i. Finally, when each son of C_i has been examined, BTD tries to modify the current assignment of C_i. The number of solutions of C_i is the sum of solution counts for every assignment of C_i. The time (resp. space) complexity of BTD for #CSP is the same as for CSP: $O(n.m.d^{w+1})$ (resp. $O(n.s.d^s)$) with $w+1$ the size of the largest C_k and s the size of the largest intersection $C_i \cap C_j$ (C_j is a son of C_i). In practice, for problems with large tree-width, BTD runs out of time and memory, as shown in Section 5. In this case, we are interested in an approximate method.

Algorithm 1. BTD($\mathcal{A}$, C_i, V_{C_i}) : integer

```
    if V_{C_i} = ∅ then
        if Sons(C_i) = ∅ then return 1;
        else
            S ← Sons(C_i) ; NbSol ← 1 ;
            while S ≠ ∅ and NbSol ≠ 0 do
                choose C_j in S ; S ← S − {C_j};
                if (A[C_i ∩ C_j], nb) is a #good in P then NbSol ← NbSol × nb; else
                    nb ← BTD(A, C_j, V_{C_j} − (C_i ∩ C_j));
                    record #good (A[C_i ∩ C_j], nb) of C_i/C_j in P;
                    NbSol ← NbSol × nb;
            return NbSol;
    else
        choose x ∈ V_{C_i} ; d ← d_x ; NbSol ← 0 ;
        while d ≠ ∅ do
            choose v in d ; d ← d-{v};
1           if A ∪ {x ← v} does not violate any c ∈ C then
                NbSol ← NbSol+BTD(A ∪ {x ← v}, C_i, V_{C_i}-{x});
        return NbSol;
```

4 Approximate Solution Counting with ApproxBTD

We consider here CSPs that are not necessarily structured. We can define a collection of subproblems of a CSP by partitioning the set of constraints, that is the set of edges in the graph. We remark that each graph (X,C) can be partitioned into k subgraphs (X_1,E_1), ..., (X_k,E_k), such that $\cup X_i = X$, $\cup E_i = C$ and $\cap E_i = \emptyset$, and such that each (X_i,E_i) is chordal[3]. So, each (X_i,E_i) can be associated to a structured subproblem $\mathcal{P}_i$ (with corresponding sets of variables X_i and constraints E_i), which can be efficiently solved using BTD. Assume that $\mathcal{S}_{\mathcal{P}_i}$ is the number of solutions for each subproblem $\mathcal{P}_i$, $1 \leqslant i \leqslant k$. We will estimate the number of solutions of $\mathcal{P}$ exploiting the following property. We denote $\mathbb{P}_{\mathcal{P}}(\mathcal{A})$ the probability of "$\mathcal{A}$ *is a solution of* $\mathcal{P}$". $\mathbb{P}_{\mathcal{P}}(\mathcal{A}) = \frac{\mathcal{S}_{\mathcal{P}}}{\prod_{x\in X} d_x}$.

Property 1. *Given a CSP $\mathcal{P} = (X,D,C,R)$ and a partition $\{\mathcal{P}_1,...,\mathcal{P}_k\}$ of $\mathcal{P}$ induced by a partition of C in k elements.*

$$\mathcal{S}_{\mathcal{P}} \approx \left\lceil \left(\prod_{i=1}^{k} \frac{\mathcal{S}_{\mathcal{P}_i}}{\prod_{x\in X_i} d_x} \right) \times \prod_{x\in X} d_x \right\rceil$$

Notice that the approximation returns an exact answer if all the subproblems are independent ($\cap X_i = \emptyset$) or $k = 1$ ($\mathcal{P}$ is already chordal) or if there exists an inconsistent

[3] A graph is chordal if every cycle of length at least four has a chord, i.e. an edge joining two non-consecutive vertices along the cycle.

subproblem $\mathcal{P}_i$. Moreover, we can provide a trivial upper bound on the number of solutions due to the fact that each subproblem $\mathcal{P}_i$ is a relaxation of $\mathcal{P}$ (the same argument is used in [21] to construct an upper bound).

$$\mathcal{S}_{\mathcal{P}} \leq \min_{i\in[1,k]} \left\lceil \frac{\mathcal{S}_{\mathcal{P}_i}}{\prod_{x\in X_i} d_x} \times \prod_{x\in X} d_x \right\rceil$$

Algorithm 2. ApproxBTD($\mathcal{P}$) : integer

Let $G' = (X', C')$ be the constraint graph associated with $\mathcal{P}$;
$i \leftarrow 0$;
while $G' \neq \emptyset$ **do**
    $i \leftarrow i+1$;
    Compute a partial chordal subgraph (X_i, E_i) of G' ;
    Let $\mathcal{P}_i$ be the subproblem associated with (X_i, E_i) ;
    $\mathcal{S}_{\mathcal{P}_i} \leftarrow$BTD($\emptyset, C_1^i, C_1^i$) with C_1^i the root cluster of the tree-decomposition of $\mathcal{P}_i$;
    $G' \leftarrow (X', C' - E_i)$ with X' be the set of variables induced by C'-E_i ;
$k \leftarrow i$;
return $\left\lceil \prod_{i=1}^{k} \frac{\mathcal{S}_{\mathcal{P}_i}}{\prod_{x\in X_i} d_x} \times \prod_{x\in X} d_x \right\rceil$;

Our proposed method called ApproxBTD is described in Algorithm 2. Applied to a problem $\mathcal{P}$ with constraint graph (X, C), the method builds a partition $\{E_1, ..., E_k\}$ of C such that the constraint graph (X_i, E_i) is chordal for all $1 \leqslant i \leqslant k$. Subproblems associated to (X_i, E_i) are solved with BTD. The method returns an approximation to the number of solutions of $\mathcal{P}$ using Property 1.

The number of iterations is less than n (at least we have $n-1$ edges (a tree) at each iteration or vertices have been deleted). Each chordal subgraph and its associated optimal tree-decomposition can be computed in $O(nm)$ [7][4]. Moreover, we guarantee that the tree-width w (plus one) of each computed chordal subgraph is at most equal to K, the maximum clique size (the *clique number*) of $\mathcal{P}$. Let w^* be the tree-width of $\mathcal{P}$, we have $w+1 \leq K \leq w^*+1$. Finally, the time complexity of ApproxBTD is $O(n^2md^K)$.

5 Experimental Results

We implemented BTD and ApproxBTD counting methods on top of `toulbar2` C++ solver[5]. The experimentations were performed on a 2.6 GHz Intel Xeon computer with 4GB running Linux 2.6.27-11-server. Reported times are in seconds. We limit to one hour the time spent for solving a given instance ('-' time out, 'mem' memory out). In BTD (line 1), we use generalized arc consistency (only for constraints with 2 or 3 unassigned variables) instead of backward checking, for efficiency reasons. The *min domain / max degree* dynamic variable ordering, modified by a conflict back-jumping heuristic [19], is used inside clusters. Our methods exploit a binary branching scheme. The variable is assigned to its first value or this value is removed from the domain.

[4] It returns a maximal subgraph for binary CSP. For non-binary CSP, we do not guarantee subgraph maximality and add to the subproblem all constraints totally included in the subgraph.

[5] http://mulcyber.toulouse.inra.fr/projects/toulbar2 version 0.8.

We performed experiments on SAT and CSP benchmarks[6]. We selected academic (random k-SAT *wff*, All-Interval Series *ais*, Towers of Hanoi *hanoi*) and industrial (circuit fault analysis *ssa* and *bit*, logistics planning *logistics*) satisfiable instances. CSP benchmarks are graph coloring instances (counting the number of optimal solutions) and genotype analysis in complex pedigrees [25]. This problem involves counting the number of consistent genotype configurations satisfying genotyping partial observations and Mendelian laws of inheritance. The corresponding decision problem is NP-complete [1]. We selected instances from [25], removing genotyping errors beforehand.

We compared BTD with state-of-the-art #SAT solvers `Relsat` [3] v2.02 and `Cachet` [26] v1.22, and also `c2d` [6] v2.20 which also exploits the problem structure. Both methods uses *MinFill* variable elimination ordering heuristic (except for *hanoi* where we used the default file order) to construct a tree-decomposition / d-DNNF. We also compared ApproxBTD with approximation method `SampleCount` [12]. With parameters $(s = 20, t = 7, \alpha = 1)$, `SampleCount-LB` provides an estimated lower bound on the number of solutions with a high-confidence interval (99% confidence), after seven runs. With parameters $(s = 20, t = 1, \alpha = 0)$, called `SampleCount-A` in the following table, it gives only an approximation without any guarantee, after the first run of `SampleCount-LB`. CSP instances are translated into SAT by using the direct encoding (one Boolean variable per domain value, one clause per domain to enforce at least a domain value is selected, and a set of binary clauses to forbid multiple value selection).

The following table summarizes our results. The columns are : instance name, number of variables, number of constraints / clauses, width of the tree-decomposition, exact number of solutions if known, time for `c2d`, `Cachet`, `Relsat`, and BTD; for ApproxBTD : maximum tree-width for all chordal subproblems, approximate number of solutions, and time; and for `SampleCount-A` and `SampleCount-LB` : approximate number of solutions, corresponding upper bound, and time. We reported total CPU times as given by c2d, cachet, relsat (precision in seconds). For BTD and ApproxBTD, the total time does not include the task of finding a variable elimination ordering. For SampleCount, we reported total CPU times with the bash command "time". We noticed that BTD can solve instances with relatively small tree-width (except for *le450* which has few solutions). Exact #SAT solvers generally perform better than BTD on SAT instances (except for *hanoi5*) but have difficulties on translated CSP instances. Here, BTD maintaining arc consistency performed better than #SAT solvers using unit propagation. Our approximate method ApproxBTD exploits a partition of the constraint graph in such a way that the resulting subproblems to solve have a small tree-width ($w \leq 11$) on these benchmarks. It has the practical effect that the method is relatively fast whatever the original tree-width. The quality of the approximation found by ApproxBTD is relatively good and it is comparable (except for ssa and logistics benchmarks) to `SampleCount`, which takes more time. For graph coloring, BTD and ApproxBTD outperform also a dedicated CSP approach (*2_Insertion_3* $\geq$ 2.3e12, *mug100_1* $\geq$ 1.0e28 and *games120* $\geq$ 4.5e42 in 1 minute each; *myciel5* $\geq$ 4.1e17 in 12 minutes, times were measured on a 3.8GHz Xeon as reported in [14]).

[6] From `www.satlib.org`, `www.satcompetition.org` and `mat.gsia.cmu.edu/COLOR02/`

Instances	Vars	w	Solutions	c2d	cachet	relsat	BTD	ApproxBTD				SampleCount-A		SampleCount-LB	
	(Bool vars)			Time	Time	Time	Time	w	Solutions		Time	Solutions	Time	Solutions	Time
SAT															
wff.3.100.150	100	39	1.8e21	-	-	-	mem	2	$\approx$ **2.21e21**	$\leq$ 1.95e27	0.02	$\approx$ 1.37e21	959.8	-	-
wff.3.150.525	150	92	1.4e14	-	-	**2509**	mem	2	$\approx$ 9.93e14	$\leq$ 7.80e40	0.2	$\approx$ **3.80e14**	0.68	$\geq$ 2.53e12	4.65
ssa7552-038	1501	25	2.84e40	**0.15**	0.22	67	0.72	5	$\approx$ 2.47e37	$\leq$ 1.51e138	1.05	$\approx$ **1.11e40**	134.2	$\geq$ 3.54e38	1162.06
ssa7552-160	1391	12	7.47e32	0.12	**0.08**	5	0.29	5	$\approx$ 1.56e34	$\leq$ 2.23e113	0.82	$\approx$ **5.08e32**	144.6	$\geq$ 2.31e31	1293
2bitcomp_5	125	36	9.84e15	0.47	**0.15**	1	11.53	5	$\approx$ 9.50e16	$\leq$ 1.75e31	0.03	$\approx$ **4.37e15**	0.184	$\geq$ 3.26e15	1.23
2bitmax_6	252	58	2.10e29	18.71	**1.57**	20	mem	5	$\approx$ 2.69e30	$\leq$ 4.27e65	0.14	$\approx$ **1.62e29**	1.676	$\geq$ 2.36e26	10.34
ais10	181	116	296	17.14	29.31	**6**	390.32	9	$\approx$ 1	$\leq$ 2.86e22	1.05	$\approx$ **124**	45.93	$\geq$ 20	312.74
ais12	265	181	1328	1162.75	2169	**229**	-	11	$\approx$ **1**	$\leq$ 1.64e40	3.78	$\approx$ 0	9.156	$\geq$ 0	9.17
logistics.a	828	116	3.8e14	-	**3.82**	10	mem	10	$\approx$ 1	$\leq$ 2.33e147	13.30	$\approx$ **7.25e11**	170.9	$\geq$ 0	605.33
logistics.b	843	107	2.3e23	-	**12.4**	433	mem	13	$\approx$ 1	$\leq$ 2.28e143	13.72	$\approx$ **2.13e23**	198.7	$\geq$ 0	229.86
hanoi4	718	46	1	3.41	32.69	3	**1.72**	6	$\approx$ **1**	$\leq$ 8.65e107	1.57	$\approx$ 0	5.24	$\geq$ 0	5.25
hanoi5	1931	58	1	-	-	-	**25.46**	7	$\approx$ **1**	$\leq$ 2.62e298	15.69	$\approx$ 0	6.14	$\geq$ 0	6.16
coloring															
2-Insertions_3	37 (148)	9	6.84e13	235	-	-	**7.9**	1	$\approx$ **1.91e13**	$\leq$ 6.00e17	0.01	$\approx$ 4.73e12	1.00	$\geq$ 4.73e12	7.42
2-Insertions_4	149 (596)	38	-	-	-	-	-	1	$\approx$ 1.30e22	$\leq$ 1.64e71	0.07	$\approx$ 0	3.76	$\geq$ 0	3.78
DSJC125.1	125 (625)	65	-	-	-	-	-	3	$\approx$ 1.23e13	$\leq$ 2.27e70	0.13	$\approx$ 0	73.14	$\geq$ 0	73.17
games120	120 (1080)	41	-	-	-	-	-	8	$\approx$ 1.12e78	$\leq$ 1.92e99	9.83	$\approx$ 0	13.76	$\geq$ 1.35e61	91.07
GEOM30a	30 (180)	6	4.98e14	0.86	-	-	**0.08**	5	$\approx$ **7.29e14**	$\leq$ 1.81e15	0.03	$\approx$ 1.23e13	0.432	$\geq$ 3.28e12	3.70
GEOM40	40 (240)	5	4.1e23	1	-	-	**0.09**	5	$\approx$ **4.42e23**	$\leq$ 1.10e24	0.02	$\approx$ 2.14e20	1.552	$\geq$ 6.50e19	9.31
le450_5a	450 (2250)	315	3840	-	343.68	**326**	953.24	4	$\approx$ **1**	$\leq$ 2.41e216	2.87	$\approx$ 0	8.56	$\geq$ 0	8.58
le450_5b	450 (2250)	318	120	-	242.30	**187**	1167.78	4	$\approx$ **1**	$\leq$ 5.71e216	2.90	$\approx$ 0	8.59	$\geq$ 0	8.61
le450_5c	450 (2250)	315	120	-	**20.79**	57	43.66	4	$\approx$ **1**	$\leq$ 1.49e201	6.65	$\approx$ 0	110.5	$\geq$ 0	110.56
le450_5d	450 (2250)	299	960	-	**16.07**	36	85.03	4	$\approx$ **1**	$\leq$ 8.58e200	6.64	$\approx$ 0	54.57	$\geq$ 0	54.58
mug100_1	100 (400)	3	1.3e37	0.19	-	-	**0.02**	2	$\approx$ **5.33e37**	$\leq$ 7.08e41	0.01	$\approx$ 4.2e34	2.08	$\geq$ 4.20e34	15.59
myciel5	47 (282)	21	-	-	-	-	mem	1	$\approx$ 7.70e17	$\leq$ 8.53e32	0.02	$\approx$ 7.29e17	0.86	$\geq$ 7.29e17	6.37
pedigree															
parkinson	34 (340)	4	3.56e10	31.34	5.33	344	**0.12**	2	$\approx$ 2.33e10	$\leq$ 6.3e11	0.02	$\approx$ **4.08e10**	3.58	$\geq$ 9.17e8	25.62
moissac3	72 (1350)	2	4.89e35	0.98	0.09	$<$ 1	**0.02**	2	$\approx$ **3.40e35**	$\leq$ 6.11e35	0.02	$\approx$ 1.18e34	16.93	$\geq$ 6.75e31	118.41
langladeM7	427 (8818)	8	6.73e196	-	-	-	**31.79**	2	$\approx$ **3.81e193**	$\leq$ 8.07e203	0.14	$\approx$ 3.03e179	816.9	-	-

6 Conclusion

In this paper, we have proposed two methods for counting solutions of CSPs. These methods are based on a structural decomposition of CSPs. We have presented an exact method, which is adapted to problems with small tree-width. For problems with large tree-width and sparse constraint graph, we have presented a new approximate method whose quality is comparable with existing methods and which is much faster than other approaches and which requires no parameter tuning (except the choice of a tree decomposition heuristic). Other structural parameters [20,24] should deserve future work.

References

1. Aceto, L., Hansen, J.A., Ingólfsdóttir, A., Johnsen, J., Knudsen, J.: The complexity of checking consistency of pedigree information and related problems. Journal of Computer Science Technology 19(1), 42–59 (2004)
2. Arnborg, S., Corneil, D., Proskurowski, A.: Complexity of finding embeddings in a k-tree. SIAM Journal of Discrete Mathematics 8, 277–284 (1987)
3. Bayardo, R., Pehoushek, J.: Counting models using connected components. In: AAAI 2000, pp. 157–162 (2000)
4. Choi, A., Darwiche, A.: An edge deletion semantics for belief propagation and its practical impact on approximation quality. In: Proc. of AAAI, pp. 1107–1114 (2006)
5. Darwiche, A.: On the tractable counting of theory models and its applications to truth maintenance and belief revision. Journal of Applied Non-classical Logic 11, 11–34 (2001)
6. Darwiche, A.: New advances in compiling cnf to decomposable negation normal form. In: Proc. of ECAI, pp. 328–332 (2004)
7. Dearing, P.M., Shier, D.R., Warner, D.D.: Maximal chordal subgraphs. Discrete Applied Mathematics 20(3), 181–190 (1988)
8. Dechter, R., Mateescu, R.: The impact of and/or search spaces on constraint satisfaction and counting. In: Proc. of CP, Toronto, CA, pp. 731–736 (2004)
9. Dechter, R., Mateescu, R.: And/or search spaces for graphical models. Artif. Intell. 171(2-3), 73–106 (2007)
10. Gogate, V., Dechter, R.: Approximate counting by sampling the backtrack-free search space. In: Proc. of AAAI 2007, Vancouver, CA, pp. 198–203 (2007)
11. Gogate, V., Dechter, R.: Approximate solution sampling (and counting) on and/or search spaces. In: Proc. of CP 2008, Sydney, AU, pp. 534–538 (2008)
12. Gomes, C.P., Hoffmann, J., Sabharwal, A., Selman, B.: From sampling to model counting. In: Proc. of IJCAI, pp. 2293–2299 (2007)
13. Gomes, C.P., Sabharwal, A., Selman, B.: Model counting: A new strategy for obtaining good bounds. In: Proc. of AAAI-06, Boston, MA (2006)
14. Gomes, C.P., van Hoeve, W.-J., Sabharwal, A., Selman, B.: Counting CSP solutions using generalized XOR constraints. In: Proc. of AAAI 2007, Vancouver, BC, pp. 204–209 (2007)
15. Jégou, P., Terrioux, C.: Hybrid backtracking bounded by tree-decomposition of constraint networks. Artificial Intelligence 146, 43–75 (2003)
16. Kask, K., Dechter, R., Gogate, V.: New look-ahead schemes for constraint satisfaction. In: Proc. of AI&M (2004)
17. Kroc, L., Sabharwal, A., Selman, B.: Leveraging belief propagation, backtrack search, and statistics for model counting. In: Perron, L., Trick, M.A. (eds.) CPAIOR 2008. LNCS, vol. 5015, pp. 127–141. Springer, Heidelberg (2008)

18. Satish Kumar, T.K.: A model counting characterization of diagnoses. In: Proc. of the 13th International Workshop on Principles of Diagnosis (2002)
19. Lecoutre, C., Sais, L., Tabary, S., Vidal, V.: Last conflict based reasoning. In: Proc. of ECAI 2006, Trento, Italy, pp. 133–137 (2006)
20. Nishimura, N., Ragde, P., Szeider, S.: Solving #sat using vertex covers. Acta Inf. 44(7), 509–523 (2007)
21. Pesant, G.: Counting solutions of CSPs: A structural approach. In: Proc. of IJCAI, pp. 260–265 (2005)
22. Robertson, N., Seymour, P.D.: Graph minors II: Algorithmic aspects of tree-width. Algorithms 7, 309–322 (1986)
23. Roth, D.: On the hardness of approximate reasonning. Artificial Intelligence 82(1-2), 273–302 (1996)
24. Samer, M., Szeider, S.: A fixed-parameter algorithm for #sat with parameter incidence treewidth (2006)
25. Sanchez, M., de Givry, S., Schiex, T.: Mendelian error detection in complex pedigrees using weighted constraint satisfaction techniques. Constraints 13(1), 130–154 (2008)
26. Sang, T., Bacchus, F., Beame, P., Kautz, H., Pitassi, T.: Combining component caching and clause learning for effective model counting. In: SAT 2004, Vancouver, Canada (2004)
27. Valiant, L.G.: The complexity of computing the permanent. Theoretical Computer Sciences 8, 189–201 (1979)
28. Wei, W., Selman, B.: A new approach to model counting. In: Bacchus, F., Walsh, T. (eds.) SAT 2005. LNCS, vol. 3569, pp. 324–339. Springer, Heidelberg (2005)

Solving a Location-Allocation Problem with Logic-Based Benders' Decomposition

Mohammad M. Fazel-Zarandi and J. Christopher Beck

Department of Mechanical and Industrial Engineering
University of Toronto
Toronto, Ontario M5S 3G8, Canada
{fazel,jcb}@mie.utoronto.ca

Abstract. We address a location-allocation problem that requires deciding the location of a set of facilities, the allocation of customers to those facilities under facility capacity constraints, and the allocation of the customers to trucks at those facilities under per truck travel-distance constraints. We present a hybrid approach that combines integer programming and constraint programming using logic-based Benders' decomposition. Computational performance against an existing integer programming model and a tabu search approach demonstrates that the Benders' model is able to find and prove optimal solutions an order of magnitude faster than an integer programming model while also finding better feasible solutions in less time for the majority of problem instances when compared to the tabu search.

1 Introduction

Location-routing problems are well-studied, challenging problems in the area of logistics and fleet management [1]. The goal is to find the minimum cost solutions that decides on a set of facilities to open, the allocation of clients and vehicles to each facility, and finally the creation of a set of routes for each vehicle. Given the difficulty of this problem, Albareda-Sambola et al. [2] recently introduced a *location-allocation* problem which simplifies the routing aspect by assuming a full truckload per client. Multiple clients can be served by the same vehicle if the sum of the return trips is less than the maximum travel distance of the truck.

In this paper, we develop a logic-based Benders' decomposition [3] for the location-allocation problem. We compare our approach empirically to an integer programming (IP) model and to a sophisticated tabu search [2]. Our experimental results demonstrate an order of magnitude improvement over the IP model in terms of time required to find and prove optimality and significant improvement over the tabu search approach in terms of finding high-quality feasible solutions with small CPU time. To our knowledge, this is a first attempt to solve a location-allocation problem using logic-based Benders' decomposition.

I.P. Gent (Ed.): CP 2009, LNCS 5732, pp. 344–351, 2009.

2 Problem Definition and Existing Approaches

The capacity and distance constrained plant location problem (CDCPLP) [2] considers a set of capacitated facilities, each housing a number of identical vehicles for serving clients. Clients are served by full return trips from the facility. The same vehicle can be used to serve several clients as long as its daily workload does not exceed a given total driving distance. The goal is to select the set of facilities to open, determine the number of vehicles required at each opened site, and assign clients to facilities and vehicles in the most cost-efficient manner. The assignments must be feasible with respect to the facilities' capacities and the maximum distance a vehicle can travel.

Formally, let J be the set of potential facilities (or sites) and I be the set of clients. Each facility, $j \in J$, is associated with a fixed opening cost, f_j, and a capacity, b_j (e.g., a measure of the volume of material that a facility can process). Clients are served by open facilities with a homogeneous set of vehicles. Each vehicle has a corresponding fixed utilization cost, u, and a maximum total daily driving distance, l. Serving client i from site j generates a driving distance, t_{ij}, for the vehicle performing the service, consumes a quantity, d_i, of the capacity of the site, and has an associated cost, c_{ij}. The available vehicles at a site are indexed in set K with parameter $\overline{k} \geq |K|$ being the maximum number of vehicles at any site. Albareda-Sambola et al. formulate an integer programming (IP) model of the problem as shown in Figure 1, where the decision variables are:

$$p_j = \begin{cases} 1, \text{if facility } j \text{ is open} \\ 0, \text{otherwise} \end{cases}$$

$$z_{jk} = \begin{cases} 1, \text{if a } k\text{th vehicle is assigned to site } j \\ 0, \text{otherwise} \end{cases}$$

$$x_{ijk} = \begin{cases} 1, \text{if client } i \text{ is served by the } k\text{th vehicle of site } j \\ 0, \text{otherwise} \end{cases}$$

The objective function minimizes the sum of the costs of opening the facilities, using the vehicles, and the travel. Constraint (1) ensures that each client is served by exactly one facility. The driving distance limits are defined by constraint (2). Constraint (3) limits the demand allocated to facility j. Constraints (4) and (5) ensure that a client cannot be served from a site that has not been opened nor by a vehicle that has not been allocated. Constraint (6) states that at a site, vehicle k will not be used before vehicle $k-1$.

Albareda-Sambola et al. compare the IP performance to that of a three-level nested tabu search. The outermost level decides the open facilities, the middle level, the assignment of clients to facilities, and the innermost level, the assignment of clients to trucks. Tabu search is done on each level in a nested fashion: first neighborhoods that open, close, and exchange facilities are used to find a feasible facility configuration, then, using that configuration, the client assignment neighborhoods are explored, and finally the truck assignment is searched

$$\min \sum_{j \in J} f_j p_j + u \sum_{j \in J} \sum_{k \in K} z_{jk} + \sum_{i \in I} \sum_{j \in J} c_{ij} \sum_{k \in k} x_{ijk}$$

$$
\begin{array}{lll}
\text{s.t.} \displaystyle\sum_{j \in J} \sum_{k \in K} x_{ijk} = 1 & i \in I & (1) \\
\displaystyle\sum_{i \in I} t_{ij} x_{ijk} \leq l \cdot z_{jk} & j \in J, k \in K & (2) \\
\displaystyle\sum_{i \in I} \sum_{k \in K} d_i x_{ijk} \leq b_j p_j & j \in J & (3) \\
z_{jk} \leq p_j & j \in J, k \in K & (4) \\
x_{ijk} \leq z_{jk} & i \in I, j \in J, k \in K & (5) \\
z_{jk} \leq z_{jk-1} & j \in J, k \in K \backslash \{1\} & (6) \\
x_{ijk}, p_j, z_{jk} \in \{0, 1\} & i \in I, j \in J, k \in K & (7)
\end{array}
$$

Fig. 1. An IP model of the CDCPLP [2]

over. Search then returns (i.e., as in a nested-loop) to the client assignments and eventually back to the facility openings. Computational results showed strong performance for the tabu search: it was able to find close-to-optimal solutions within a few minutes of CPU time.

3 A Logic-Based Benders' Decomposition Approach

In Benders' decomposition [3], a problem is partitioned into a master problem and a subproblem, which are solved iteratively until the optimal solution is found. When the subproblem is infeasible subject to current master solution, a cut that eliminates at least the current master solution is added to the master problem. The cut ensures that all future solutions are closer to being feasible.

The CDCPLP can be decomposed into a location-allocation master problem (LAMP) and a set of truck assignment subproblems (TASPs). The LAMP is concerned with choosing the open facilities, allocating clients to these sites, and deciding on the number of trucks at each site. The TASP assigns clients to specific vehicles and can be modeled as a set of independent bin-packing problems: clients are allocated to the trucks so that the total-distance constraint on each truck is satisfied. We use IP for the master problem and CP for the subproblems.

The Location-Allocation Master Problem. An IP formulation of LAMP is shown in Figure 2 where p_j is as defined above and:

$$x_{ij} = \begin{cases} 1, \text{ if client } i \text{ is served by site } j \\ 0, \text{ otherwise} \end{cases}$$

$$numVeh_j : \text{number of vehicles assigned to facility } j$$

$$\min \sum_{j \in J} f_j p_j + \sum_{i \in I} \sum_{j \in J} c_{ij} x_{ij} + u \sum_{j \in J} numVeh_j$$

$$\text{s.t.} \sum_{j \in J} x_{ij} = 1 \qquad i \in I \qquad (8)$$

$$\sum_{i \in I} t_{ij} x_{ij} \leq l \cdot \overline{k} \qquad j \in J \qquad (9)$$

$$t_{ij} x_{ij} \leq l \qquad i \in I, j \in J \ (10)$$

$$\sum_{i \in I} d_i x_{ij} \leq b_j p_j \qquad j \in J \qquad (11)$$

$$numVeh_j \geq \left\lceil \frac{\sum_{i \in I} t_{ij} x_{ij}}{l} \right\rceil \quad j \in J \qquad (12)$$

$$cuts \qquad (13)$$

$$x_{ij} \leq p_j \qquad i \in I, j \in J \ (14)$$

$$x_{ij}, p_j \in \{0, 1\} \qquad i \in I, j \in J \ (15)$$

Fig. 2. An IP model of the LAMP

Constraint (8) ensures that all clients are served by exactly one facility. The distance limitations are defined by constraints (9) and (10). Constraint (11) limits the demand assigned to facility j. Constraint (12) defines the minimum number of vehicles assigned to each site. *cuts* are constraints that are added to the master problem each time one of the subproblems is not able to find a feasible solution. Initially, *cuts* is empty.

The cut for a given TASP j after iteration h is:

$$numVeh_j \geq numVeh^*_{jh} - \sum_{i \in I_{jh}} (1 - x_{ij}), \quad j \in J_h$$

where, $I_{jh} = \{i \mid x^h_{ij} = 1\}$ is the set of clients assigned to facility j in iteration h, J_h is the set of sites for which the subproblem is infeasible in iteration h, and $numVeh^*_{jh}$ is the minimum number of vehicles needed at site j to serve the clients that were assigned. Informally, the summation is the maximal decrease in the minimal number of trucks needed given the clients reassigned to other facilities: the largest possible reduction in reassigning one client is one truck. The form of this cut is directly inspired by the Benders' cut for scheduling with makespan minimization formulated by Hooker [4].

The Truck Assignment Subproblem. Given the set of clients allocated (I_j) and the number of vehicles assigned to an open facility ($numVeh_j$), the goal of the TASP is to assign clients to the vehicles of each site such that the vehicle travel-distance constraints are satisfied. The TASP for each facility can be modeled as a bin-packing problem. A CP formulation of TASP is shown in Figure 3 where: *load* is an array of variables such that $load[k] \in \{0, ..., l\}$ is the total distance assigned to vehicle $k \in \{1, ..., numVehBinPacking_j\}$, *truck* is an array of decision variables, one for each client $i \in I_j$, such that $truck[i] \in \{1, ..., numVeh_j\}$ is the index of the truck assigned to client i, and *dist* is the vector of distances

$$\min \ numVehBinPacking_j$$

$$\text{s.t. } pack(load, truck, dist) \quad (16)$$
$$numVeh_j \leq numVehBinPacking_j < numVehFFD_j \quad (17)$$

Fig. 3. A CP model of the TASP

between site j and client $i \in I_j$. The pack global constraint (16) maintains the load of the vehicles given the distances and assignments of clients to vehicles [5]. The upper and lower bounds on the number of vehicles is represented by constraint (17).

Algorithm 1 shows how we solve the sub-problems in practice. We first use the first-fit decreasing (FFD) heuristic (line 3) to find $numVehFFD_j$, a heuristic solution to the sub-problem. If this value is equal to the value assigned by the LAMP solution, $numVeh_j$, then the sub-problem has been solved. Otherwise, in line 5 we solve a series of satisfaction problems using the CP formulation, setting $numVehBinPacking_j$ to each value in the interval $[numVeh_j..numVehFFD_j - 1]$ in increasing order.

4 Computational Results

We compare our Benders' approach to the IP and tabu search models in turn. Unless otherwise noted, the tests were performed on a Duo Core AMD 270 CPU with 1 MB cache, 4 GB of main memory, running Red Hat Enterprise Linux 4. The IP model was implemented in ILOG CPLEX 11.0. The Benders' IP/CP approach was implemented in ILOG CPLEX 11.0 and ILOG Solver 6.5.

IP vs. Benders'. We generated problems following exactly the same method as Albareda-Sambola et al [2]. We start with the 25 instances of Barceló et al. [6] in three sizes: 6 instances of size 20×10 (i.e., 20 clients, 10 possible facility sites), 11

Algorithm 1: Algorithm for solving the TASP

SolveTASP():

1 cuts = $\emptyset$
2 **for** *each facility* **do**
3     $numVehFFD$ = runFFD()
4   **if** $numVehFFD > numVeh_j$ **then**
5         $numVehBinPacking$ = runCPBinPacking()
6   **if** $numVehBinPacking > numVeh_j$ **then**
7   cuts ← cuts + new cut
8 return cuts

Table 1. The mean CPU time (seconds) and percentage of unsolved problem instances (% Uns.) for the IP and Benders' approaches and for the Benders' approach, the mean number of iterations. Overall indicates the mean results over all problem instances–recall that each subset has a different number of instances.

Problem	Uncorrelated						Correlated					
	IP		Benders'			Time	IP		Benders'			Time
Set	Time	% Uns.	Time	% Uns.	Iter	Ratio	Time	% Uns.	Time	% Uns.	Iter	Ratio
20 × 10	252	0	33	0	3.8	9.6	65	0	24	0	4.1	6.3
30 × 15	55303	23	17593	6	16	13.6	29514	12	8065	2	13.1	20.5
40 × 20	144247	75	72598	35	26.2	2.4	79517	38	28221	10	22.8	6.5
Overall	70553	34	30980	14	16.3	9.1	38447	17	12585	4	14.0	12.6

instances of size 30×15, and 8 instances of size 40×20. The fixed facility opening cost, f_j, demands for each client, d_i, assignment costs, c_{ij}, and facility capacities, b_j, are extracted from Barceló et al. with the exception that the facility capacities are multiplied by 1.5 as they are very tight. Six different pairs of truck distance limit, l, and truck usage cost, u, values are then used to create different problem sets: $(40, 50), (40, 100), (50, 80), (50, 150), (100, 150), (100, 300)$. Finally, the travel distances, t_{ij}, are randomly generated based on the costs, c_{ij} in two different conditions. In the *correlated* condition: $t_{ij} = scale(c_{ij}, [15, 45]) + rand[-5, 5]$. The first term is a uniform scaling of c_{ij} to the integer interval $[15, 45]$ while the second term is a random integer uniformly generated on the interval $[-5, 5]$. In the *uncorrelated* condition, $t_{ij} = rand[10, 50]$. Overall, therefore, there are 300 problem instances: 25 original instances times 6 (l, u) conditions times 2 correlated/uncorrelated conditions.

Table 1 compares the mean CPU time in seconds required to solve each problem instance for each set. In all cases, 48 hours was used as a maximum time. The "Time Ratio" for a given instance is calculated as the IP run-time divided by the Benders' run-time. The mean over each instance in each subset was then calculated. For unsolved instances, the 48-hour time limit was used. As can be seen Benders' is able to solve substantially more problems than IP and, on average, has a run-time about an order-of-magnitude faster.

Figure 4 shows a scatter-plot of the run-times of each problem instances for both IP and the Benders' approach. Both axes are log-scale and the points below the $x = y$ line indicate lower run-time for the Benders' approach. On all but 26 of the 300 instances, the Benders' achieves equivalent or better run-time.

Tabu search vs. Benders'. One of the weaknesses of a Benders' decomposition approach is that usually the first globally feasible solution found is the optimal solution. This means that cutting off runs due to a time-limit will result in no feasible solutions. For problems too large for a Benders' approach to find optimality, another algorithm is needed to find a good but not necessarily optimal solution. Metaheuristic techniques, such as tabu search, are widely used for location and routing problems for this purpose [7].

With our Benders' formulation, however, we have a globally feasible, suboptimal solution at each iteration. In generating a cut, we find the minimum number of trucks needed at each facility. This number of trucks constitutes a

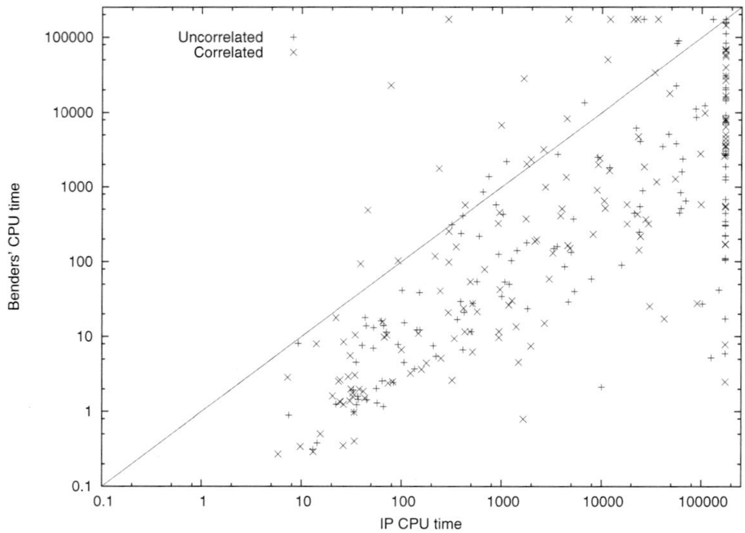

Fig. 4. Run-time of IP model (x-axis, log-scale) vs. Benders' IP/CP model (y-axis, log-scale) of the 300 problem instances. Points below the $x = y$ line indicate lower run-time for the Benders' model.

feasible solution even though fewer trucks were assigned in the master solution.[1] Thus, at the end of each iteration, we have a globally feasible solution.

Albareda-Sambola et al. used 19 medium and large problem instances to evaluate their tabu search approach, reporting run-times and bounds on the optimality gap. All instances are correlated and have (l, u) values of $(50, 80)$. We received these exact instances form the authors.[2] We believe that the run-time for our Benders' model to find the *first* feasible solution and the gap from optimality provide some basis for comparison.

Table 2 presents the mean and median time for the first iteration and mean percentage gap from optimal for the Benders' approach. This is compared to the run-time on a 2.4GHz Pentium IV and mean percentage gap from optimal reported by Albareda-Sambola et al.[3] The columns labeled "# Dom." indicate the number of problems in each set for which one approach was clearly dominant with respect to both lower CPU time and lower % gap. As can be seen, on average the Benders' approach is 13.6% worse than Tabu search with respect to the mean CPU time to find a feasible solution but finds solutions with substantially smaller optimality gaps. However, Benders' exhibits three run-time outliers that obscure

[1] This is not true when the minimal number of trucks required at a facility is greater than $\overline{k}$. This did not occur in any of our experiments.

[2] We would like to thank Maria Albareda-Sambola for providing these instances.

[3] Albareda-Sambola et al. presented the cost of their best solution and bounds on the percentage gap. As we found the optimal solutions we were able to calculated the exact gap from optimality for the tabu search.

Table 2. The mean and median CPU time (seconds), the mean percentage gap from optimal and the bounds of that gap for Tabu, and the number of instances for which each approach dominated the other

Problem Set	Benders'				Tabu			
	Time		% Gap	# Dom.	Time		% Gap	# Dom.
	Mean	Median	Mean		Mean	Median	Mean	
30×15	60.4	13.5	2.07	4	60.5	66.1	4.12	0
40×20	185.4	86.2	1.82	6	148.7	163.0	10.41	0
Overall	113.0	39.5	1.96	10	97.6	78.8	6.77	0

the results. Out of 19 problem instances, Benders' finds a better solution faster than tabu on 10 instances while tabu search was not able to find a better solution faster than Benders' for any instance.

5 Conclusion

In this paper, we presented a novel logic-based Benders' decomposition approach to a location-allocation problem. Our approach was able to substantially outperform an existing IP model by finding and proving optimality, on average, more than ten times faster. Our approach also performed better than an existing tabu search in finding good, feasible solutions in a short time.

Acknowledgments. This research was supported in part by the Natural Sciences and Engineering Research Council, Canadian Foundation for Innovation, Ontario Ministry for Research and Innovation, Alcatel-Lucent, and ILOG, S.A.

References

1. Drezner, Z.: Facility Location. A Survey of Applications and Methods. Springer Series in Operations Research. Springer, New York (1995)
2. Albareda-Sambola, M., Fernández, E., Laporte, G.: The capacity and distance constrained plant location problem. Computers & Operation Research 36(2), 597–611 (2009)
3. Hooker, J.N., Ottosson, G.: Logic-based benders decomposition. Mathematical Programming 96 (2003)
4. Hooker, J.N.: A hybrid method for planning and scheduling. Constraints 10, 385–401 (2005)
5. Shaw, P.: A constraint for bin packing. In: Wallace, M. (ed.) CP 2004. LNCS, vol. 3258, pp. 648–662. Springer, Heidelberg (2004)
6. Barceló, J., Fernández, E., Jornsten, K.: Computational results from a new lagrangean relaxation algorithm for the capacitated plant locating problem. European Journal of Operational Research 53, 38–45 (1991)
7. Laporte, G.: Location-routing problems. In: Golden, B.L., Assad, A.A. (eds.) Vehicle routing: methods and studies, pp. 163–197. North-Holland, Amsterdam (1988)

Lazy Clause Generation Reengineered

Thibaut Feydy and Peter J. Stuckey

National ICT Australia, Victoria Laboratory
Department of Computer Science and Software Engineering
The University of Melbourne, Australia
{tfeydy,pjs}@csse.unimelb.edu.au

Abstract. Lazy clause generation is a powerful hybrid approach to combinatorial optimization that combines features from SAT solving and finite domain (FD) propagation. In lazy clause generation finite domain propagators are considered as clause generators that create a SAT description of their behaviour for a SAT solver. The ability of the SAT solver to explain and record failure and perform conflict directed backjumping are then applicable to FD problems. The original implementation of lazy clause generation was constructed as a cut down finite domain propagation engine inside a SAT solver. In this paper we show how to engineer a lazy clause generation solver by embedding a SAT solver inside an FD solver. The resulting solver is flexible, efficient and easy to use. We give experiments illustrating the effect of different design choices in engineering the solver.

1 Introduction

Lazy clause generation [1] is a hybrid of finite domain (FD) propagation solving and SAT solving that combines some of the strengths of both. Essentially in lazy clause generation, a FD propagator is treated as a clause generator, that feeds a SAT solver with a growing clausal description of the problem. The advantages of the hybrid are: it retains the concise modelling of the problem of an FD system, but it gains the SAT abilities to record nogoods and backjump, as well as use activities to drive search. The result is a powerful hybrid that is able to solve some problems much faster than either SAT or FD solvers.

The original lazy clause generation solver was implemented as a summer student project, where a limited finite domain propagation engine was constructed inside a SAT solver.

In this paper we discuss how we built a robust generic lazy clause generation solver in the G12 system. The crucial difference of the reengineered solver is that the SAT solver is treated as a propagator in an FD solver (hence reversing the treatment of which solver is master). This approach is far more flexible than the original design, more efficient, and available as a backend to the Zinc compiler.

We discuss the design decisions that go into building a robust lazy clause generation solver, and present experiments showing the effect of these decisions. The new lazy clause generation solver is a powerful solver with the following features:

I.P. Gent (Ed.): CP 2009, LNCS 5732, pp. 352–366, 2009.

- Powerful modelling: any Zinc (or MiniZinc) model executable by the G12 FD solver can be run using the lazy clause generation solver.
- Excellent default search: if no search strategy is specified then the default VSIDS search is usually very good.
- Programmed search with nogoods: on examples with substantial search the solver usually requires orders of magnitude less search than the FD solver using the same search strategy.
- Flexible global constraints: since decomposed globals are highly effective we can easily experiment with different decompositions.

The resulting system is a powerful combination of easy modelling and highly efficient search. It competes against the best FD solutions, often on much simpler models, and against translation to SAT. In many cases using the lazy clause generation solver with default settings to solve a simple FD statement of the problem gives very good results

2 Background

2.1 Propagation-Based Constraint Solvers

Propagation-based constraint solving models constraints c as propagators, that map the set of possible values of variables (a domain) to a smaller domain by removing values that cannot take part in any solution. The key advantage of this approach is that propagation is "composable", propagators for each constraint can be constructed independently, and used in conjunction.

More formally. A *domain* D is a mapping from a fixed (finite) set of variables $\mathcal{V}$ to finite sets of integers. A *false domain* D is a domain with $D(x) = \emptyset$ for some $x \in \mathcal{V}$. A domain D_1 is *stronger* than a domain D_2, written $D_1 \sqsubseteq D_2$, if $D_1(x) \subseteq D_2(x)$ for all $x \in \mathcal{V}$. A range is a contiguous set of integers, we use *range* notation $[\,l \mathinner{..} u\,]$ to denote the range $\{d \in \mathcal{Z} \mid l \leqslant d \leqslant u\}$ when l and u are integers. We shall be interested in the notion of a *starting domain*, which we denote D_{init}. The starting domain gives the initial values possible for each variable. It allows us to restrict attention to domains D such that $D \sqsubseteq D_{init}$.

An *integer valuation* θ is a mapping of variables to integer values, written $\{x_1 \mapsto d_1, \ldots, x_n \mapsto d_n\}$. We extend the valuation θ to map expressions and constraints involving the variables in the natural way.

Let *vars* be the function that returns the set of variables appearing in a valuation. We define a valuation θ to be an element of a domain D, written $\theta \in D$, if $\theta(x_i) \in D(x_i)$ for all $x_i \in vars(\theta)$.

A *constraint* c over variables $x_1, \ldots, x_n$ is a set of valuations θ such that $vars(\theta) = \{x_1, \ldots, x_n\}$. We also define $vars(c) = \{x_1, \ldots, x_n\}$. We will *implement* a constraint c by a set of propagators that map domains to domains. A *propagator* f is a monotonically decreasing function from domains to domains: $f(D) \sqsubseteq D$, and $f(D_1) \sqsubseteq f(D_2)$ whenever $D_1 \sqsubseteq D_2$. A propagator f is *correct* for a constraint c iff for all domains D

$$\{\theta \mid \theta \in D\} \cap c = \{\theta \mid \theta \in f(D)\} \cap c$$

This is a very weak restriction, for example the identity propagator is correct for all constraints c.

Example 1. For the constraint $c \equiv x_0 \Leftrightarrow x_1 \leqslant x_2$ the function f defined by

$$f(D)(x_0) = D(x_0) \cap (\{0 \mid \max D(x_1) > \min D(x_2)\} \cup \{1 \mid \min D(x_1) \leqslant \max D(x_2)\})$$

$$f(D)(x_1) = \begin{cases} D(x_1), & \{0,1\} \subseteq D(x_0) \\ \{d \in D(x_1) \mid d \leqslant \max D(x_2)\}, & D(x_0) = \{1\} \\ \{d \in D(x_1) \mid d > \min D(x_2)\}, & D(x_0) = \{0\} \end{cases}$$

$$f(D)(x_2) = \begin{cases} D(x_2), & \{0,1\} \subseteq D(x_0) \\ \{d \in D(x_2) \mid d \geqslant \min D(x_1)\}, & D(x_0) = \{1\} \\ \{d \in D(x_1) \mid d < \max D(x_2)\}, & D(x_0) = \{0\} \end{cases}$$

is a correct propagator for c. Let $D(x_0) = [\,0\,..\,1\,]$, $D(x_1) = [\,7\,..\,9\,]$, $D(x_2) = [\,-3\,..\,5\,]$ then $f(D)(x_0) = \{0\}$.

A *propagation solver* $solv(F, D)$ for a set of propagators F and a domain D finds the greatest mutual fixpoint of all the propagators $f \in F$.

In practice the propagation solver $solv(F, D)$ is carefully engineered to take into account which propagators must be at fixed point and do not need to be reconsidered. It also will include priorities on propagators so that cheap propagators are executed before expensive ones. See [2] for more details.

2.2 SAT Solvers

Propagation based SAT solvers [3] are specialized propagation solvers with only Booleans variables, a built-in conflict based search and clausal constraints of the form $l_1 \vee l_2 \vee \cdots \vee l_n$ where l_i is a *literal* (a Boolean variable or its negation).

Unit propagation consists of detecting a conflict or fixing a literal once all other literals in a clause have been fixed to false. SAT solvers can perform unit propagation very efficiently using watch literals.

Conflict analysis is triggered each time a conflict is detected. By traversing a reverse implication graph (ie. remembering which clause fixed a literal), SAT solvers build a *nogood*, or conflict clause, which is added to the constraint store.

Conflict analysis allows SAT solvers to find the last satisfiable decision level, to which they can backjump, i.e. backtrack to a point before the last choicepoint.

SAT solvers maintain *activities* of the variables seen during conflict analysis. The heuristic used prioritizes variables that are the most involved in recent conflicts. This allow them to use a *conflict driven* or *activity based* search [3].

2.3 Original Lazy Clause Generation

The original lazy clause generation hybrid solver [1] works as follows. Propagators are considered as clause generators for the SAT solver. Instead of applying propagator f to domain D to obtain $f(D)$, whenever $f(D) \neq D$ we build a clause that encodes the change in domains. In order to do so we must link the integer variables of the finite domain problem to a Boolean representation.

We represent an integer variable x with domain $D_{init}(x) = [\, l \,..\, u \,]$ using the Boolean variables $[\![x = l]\!], \ldots, [\![x = u]\!]$ and $[\![x \leqslant l]\!], \ldots, [\![x \leqslant u - 1]\!]$. The variable $[\![x = d]\!]$ is true if x takes the value d, and false if x takes a value different from d. Similarly the variable $[\![x \leqslant d]\!]$ is true if x takes a value less than or equal to d and false if x takes a value greater than d. For integer variables with $D_{init}(x) = [\, 0 \,..\, 1 \,]$ we simply treat x as a Boolean variable.

Not every assignment of Boolean variables is consistent with the integer variable x, for example $\{[\![x = 5]\!], [\![x \leqslant 1]\!]\}$ requires that x is both 5 and $\leqslant 1$. In order to ensure that assignments represent a consistent set of possibilities for the integer variable x we add to the SAT solver clauses $DOM(x)$ that encode $[\![x \leqslant d]\!] \rightarrow [\![x \leqslant d + 1]\!]$ and $[\![x = d]\!] \leftrightarrow ([\![x \leqslant d]\!] \wedge \neg[\![x \leqslant d - 1]\!])$. We let $DOM = \cup\{DOM(v) \mid v \in \mathcal{V}\}$.

Any set of literals A on these Boolean variables can be converted to a domain: $domain(A)(x) = \{d \in D_{init}(x) \mid \forall [\![c]\!] \in A. vars(l) = \{x\} \Rightarrow x = d \models c\}$, that is the domain of all values for x that are consistent with all the Boolean variables related to x. Note that it may be a false domain.

Example 2. For example the assignment $A = \{[\![x_1 \leqslant 8]\!], \neg[\![x_1 \leqslant 2]\!], \neg[\![x_1 = 4]\!], \neg[\![x_1 = 5]\!], \neg[\![x_1 = 7]\!], [\![x_2 \leqslant 6]\!], \neg[\![x_2 \leqslant -1]\!], [\![x_3 \leqslant 4]\!], \neg[\![x_3 \leqslant -2]\!]\}$ is consistent with $x_1 = 3, x_1 = 6$ and $x_1 = 8$. hence $domain(A)(x_1) = \{3, 6, 8\}$. For the remaining variables $domain(A)(x_2) = [\, 0 \,..\, 6 \,]$ and $domain(A)(x_3) = [\, -1 \,..\, 4 \,]$. □

In the lazy clause generation solver, search is controlled by the SAT engine. After making a decision, unit propagation is performed to reach a unit propagation fixpoint with assignment A. Every fixed literal is then translated into a domain change, creating a new domain $D = domain(A)$, and the appropriate propagators are woken up. When we find a propagator f where $f(D) \neq D$ the propagator does not directly modify the domain D but instead generates a set of clauses C which explain the domain changes. Each clause is added to the SAT solver, starting a new round of unit propagation. This continues until fixpoint when the next SAT decision is made. See Figure 1(a). Adding an explanation of failure will force the SAT solver to fail and begin its process of nogood construction. It then backjumps to where the nogood would first propagate, and on untrailing the domain D must be reset back to its previous state.

Example 3. Suppose the SAT solver decides to set $[\![y \leqslant 1]\!]$ and unit propagation determines that $\neg[\![x_1 \leqslant 6]\!]$. Assuming the current domain $D(x_0) = [\, 0 \,..\, 1 \,]$, $D(x_1) = [\, 1 \,..\, 9 \,]$, $D(x_2) = [\, -3 \,..\, 5 \,]$ then the domain changes to $D'(x_1) = [\, 7 \,..\, 9 \,]$ and propagators dependent on the lower bound of x_1 are scheduled, including for example the propagator f for $x_0 \Leftrightarrow x_1 \leqslant x_2$ from Example 1. When applied to domain D' it obtains $f(D')(x_0) = \{0\}$. The clausal explanation of the change in domain of x_1 is $\neg[\![x_1 \leqslant 6]\!] \wedge [\![x_2 \leqslant 5]\!] \rightarrow \neg x_0$. This becomes the clause $[\![x_1 \leqslant 6]\!] \vee \neg[\![x_2 \leqslant 5]\!] \vee \neg x_0$. This is added to the SAT solver. Unit propagation sets the literal $\neg x_0$. This creates domain $D''(x_0) = \{0\}$ which causes the propagator f to be re-examined but no further propagation occurs.

Assuming $domain(A) \sqsubseteq D$, then when clauses C that explain the propagation of f are added to the SAT database containing DOM and unit propagation is

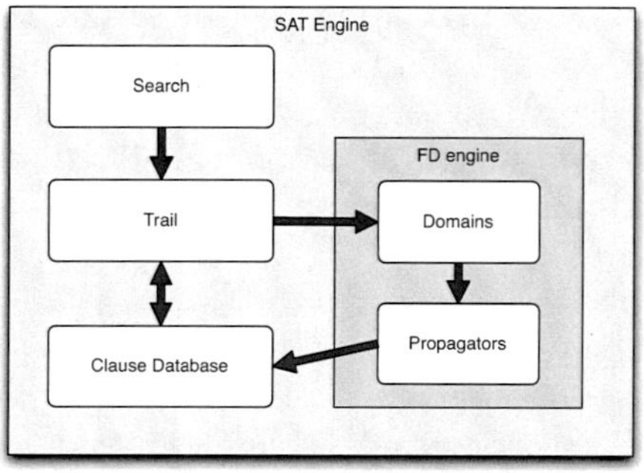

(a)

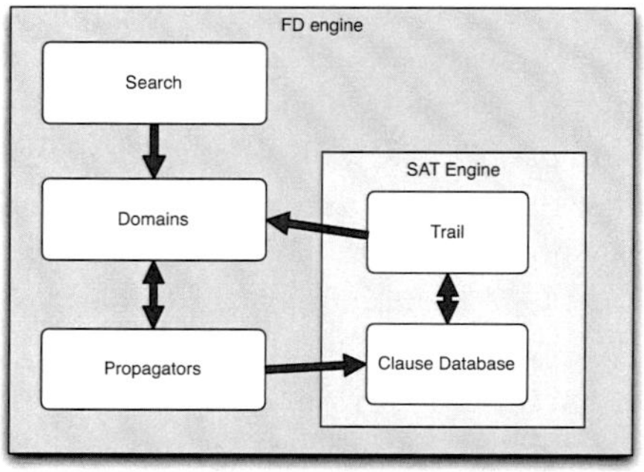

(b)

Fig. 1. (a) The original architecture for lazy clause generation, and (b) the new architecture

performed, then the resulting assignment A' will be such that $domain(A') \sqsubseteq f(D)$. Using lazy clause generation we can show that the SAT solver maintains an assignment which is at least as strong the domains of an FD solver [1].

The advantages over a normal FD solver are that we automatically have the nogood recording and backjumping ability of the SAT solver applied to our FD problem, as well as its activity based search.

3 Lazy Clause Generation as a Finite Domain Solver

The original lazy clause generation solver used a SAT solver as a master solver and had a cut down finite domain propagation engine inside. This approach meant that the search was not programmable, but built into the SAT solver and minimization was available only as dichotomic search, on top of SAT search.

3.1 The New Solver Architecture

The new lazy clause generation solver is designed as an extension of the existing G12 finite domain solver. It is a backend for the Zinc compiler which can be used wherever the finite domain solver is used. The new solver architecture is illustrated in Figure 1(b).

Search is controlled by the finite domain solver. When a variables domain changes propagators are woken as usual, and placed in priority queue. The SAT solver unit propagation engine, acts as a global propagator. Whenever a literal is set or clauses are posted to the SAT solver, then this propagator is scheduled for execution at the highest priority.

When a propagator f is executed that updates a variable domain ($f(D) \neq D$) or causes failure ($f(D)$ is a false domain) it posts an *explanation* clause to the SAT solver that explains the domain reduction or failure. This will schedule the SAT propagator for execution.

The SAT propagator when executed computes a unit fixpoint. Then each of the literals fixed by unit propagation causes the corresponding domain changes to be made in the domain, which may wake up other propagators. The cycle of propagation continues until a fixpoint is reached. Note that other work (e.g. [4]) has suggested using a SAT solver to implement a global propagator inside a CP solver.

3.2 Encoding of Finite Domain Variables

In the new architecture integer variables are implemented as usual with a representation of bounds, and domains with holes, and queues of events for bounds changes, fixing a variable and removing an interior value. Concrete variables are restricted to be *zero based*, that is have initial domains that range over values $[0..n]$, views [5] are used to encode non-zero based integer variables.

The lazy clause generation solver associates each integer variable with a set of Boolean variables. Changes in these Boolean variables will be reflected in the domains of the integer variables. There are two possible ways of doing this:

The Array Encoding. The array encoding of integer variables is an encoding with two arrays :

- An array of inequality literals $[\![x \leqslant d]\!]$, $d \in [0..n-1]$
- An array of equality literals $[\![x = d]\!]$, $d \in [0..n]$

inequality literals are generated eagerly whereas equality literals are generated lazily. When a literal $[\![x = d]\!]$ has to be generated we post the *domain clauses*: $[\![x = d]\!] \rightarrow [\![x \leqslant d]\!]$, $[\![x = d]\!] \rightarrow \neg[\![x \leqslant d-1]\!]$, and $\neg[\![x \leqslant d-1]\!] \wedge [\![x \leqslant d]\!] \rightarrow [\![x = d]\!]$. The array encoding is linear in the size of the initial integer domain, while bound updates are linear in the size of the domain reduction.

The List Encoding. The list encoding generates inequality and equality literals lazily when they are required for propagation or explanation. As such the size of the encoding is linear in the number of generated literals, and a bound update is linear in the number of generated literals that will be fixed by the update.

When a literal $[\![x \leqslant d]\!]$ has to be generated :

- we determine the closest existing bounds: $l = \max\{d' \mid [\![x \leqslant d']\!] \text{ exists}, d' < d\}$, $u = \min\{d' \mid [\![x \leqslant d']\!] \text{ exists}, d < d'\}$
- we post the new *domain clauses* : $[\![x \leqslant l]\!] \rightarrow [\![x \leqslant d]\!], [\![x \leqslant d]\!] \rightarrow [\![x \leqslant u]\!]$

When a literal $[\![x = d]\!]$ has to be generated we first generate the literals $[\![x \leqslant d]\!]$ and $[\![x \leqslant d-1]\!]$ if required, then proceed as for the array encoding.

Caching the positions of the largest $[\![x \leqslant d]\!]$ which is *false* (lower bound $d+1$) and smallest $[\![x \leqslant d]\!]$ which is *true* (upper bound d) allows access performances similar to the array encoding for the following reasons. An inequality literal is required either to explain another variable update or to reduce the domain of the current variable. In the first case, the inequality literal is most likely the one corresponding to the current bound, in which case it is cached. In the latter case, where we reduce a variable bound by an amount δ, a sequence of δ clauses will have to be propagated in the array encoding.

When a new literal is generated, previous sequence literals becomes redundant. When a literal $[\![x \leqslant d]\!]$ is inserted where $l < d < u$ then the binary clause $[\![x \leqslant l]\!] \rightarrow [\![x \leqslant u]\!]$ becomes redundant. At most n redundant constraints exists after n literals have been generated. However these redundant constraints alter the propagation order and can have a negative impact on the nogoods generated during conflict analysis.

By default the solver uses array encoding for variables with "small" (< 10000) initial domains and list encoding for larger domains.

Use of Views. We use views [5] to avoid creating additional variables. A *view* is a monotonic, injective function from a variable to a domain. In practice, we use affine views, and each variable is an affine view over a zero-based variable.

Given a variable x where $D_{init}(x) = [\, l \,..\, u \,]$ where $l \neq 0$ we represent x as a view $x = x_c + l$ where $D_{init}(x_c) = [\, 0 \,..\, u - l \,]$. Given a variable x and a new variable y defined as $y = ax + b$, let x_c be the concrete variable of x, ie. $\exists a', \exists b'$ such that $x = a'x_c + b'$ then y is defined as $y = a'ax_c + (a'b + b')$.

Using views rather than creating fresh variables has the following advantages over creating new concrete variables :

- *space savings.* This is especially true with the array encoding which is always linear in the domain size.
- *literal reuse.* Reusing literals means stronger and shorter nogoods.

3.3 Propagator Implementation

For use in a lazy clause generation each propagator should be extended to explain its propagations and failures. Note that the new lazy clause generation system can handle propagators that do not explain themselves by treating their propagations as decisions, but this significantly weakens the benefits of lazy clause generation.

When a propagator is run, then all its propagations are reflected by changing the domains of integer variables, as well as adding *explanation clauses* to the SAT solver that explains the propagation made. Note that it must also explain failure.

Once we are using lazy clause generation we need to reassess the best possible way to implement each constraint.

Linear Constraints. Linear constraints $\sum_{i \in 1 \ldots n} a_i x_i \leqslant a_0$ and $\sum_{i \in 1 \ldots n} a_i x_i = a_0$ are among the most common constraints. But long linear constraints do not generate very reusable explanations, since they may involve many variables. It is worth considering breaking up a long linear constraint into smaller units. So e.g. $\sum_{i \in 1 \ldots n} a_i x_i = a_0$ becomes $s_1 = a_1 x_1 + a_2 x_2$, ... $s_{i+1} = s_i + a_{i+1} x_{i+1}$, ..., $s_n = s_{n-1} + a_n x_n$, $a_0 = s_n$. Note that for finite domain propagation alone this is guaranteed to result in the same domains (on the original variables) [6]. The decomposition adds many new variables and slows propagation considerably, but means explanations are more likely to be reused. We shall see that the *size* of the intermediate sums s_i will be crucial in determining the worth of this translation.

Reified Constraints. The lazy clause generation solver does not have regular reified constraints but only implications constraints of the form $l \Rightarrow c$ where l is a literal and c is a constraint. This has the advantage that the events of this implication constraint are the events of the non reified version plus an event on l being asserted. Similarly the explanations for $l \Rightarrow c$ are the same as for c with $\neg l$ disjoined. The main advantage is that often we do not need both directions.

Example 4. Consider the constraint $x_1 + 2 \leqslant x_2 \vee x_2 + 5 \leqslant x_1$. The usual decomposition is $b_1 \Leftrightarrow x_1 + 2 \leqslant x_2$, $b_2 \Leftrightarrow x_2 + 5 \leqslant x_1$, $(b_1 \vee b_2)$. A better decomposition is $b_1 \Rightarrow x_1 + 2 \leqslant x_2$, $b_2 \Rightarrow x_2 + 5 \leqslant x_1$, $(b_1 \vee b_2)$ since the only propagation possible if the falsity of one of the inequalities forcing a Boolean to be false, which forces the other Boolean to be true and the other inequality to hold. Note that e.g. $x_0 \Leftrightarrow x_1 \leqslant x_2$ is implemented as $x_0 \Rightarrow x_1 \leqslant x_2$, $\neg x_0 \Rightarrow x_1 > x_2$ illustrating the need for the lhs to be a literal rather than a variable.

3.4 Global Propagators

Rather than create complex explanations for global constraints it is usually easier to build decompositions. Learning for decomposed globals is stronger, and can regain the benefits of the global view that are lost by decomposition. If we are using decomposition to define global propagators then we can easily experiment with different definitions. It is certainly worth reconsidering which decomposition to use in particular for lazy clause generation.

Element Constraints. An element constraint $\texttt{element}(x, a, y)$ which enforces that $y = a[x]$ where a is a fixed array of integers indexed on the range $[\,0\,..\,n\,]$ can be implemented simply as the binary clauses $\wedge_{k=0}^{n} [\![x = k]\!] \rightarrow [\![y = a[k]]\!]$ which enforces domain consistency.

GCC. We propose a new decomposition of the global cardinality constraint (and by specialisation also the alldifferent constraint) which exploits the property of our solver that maintaining the state of the literals $[\![x = k]\!]$ and $[\![x \leqslant k]\!]$ is cheap as it is part of the integer variable encoding. $\texttt{gcc}([x_1, \ldots, x_n], [c_1, \ldots, c_m])$ enforces that the value i occurs c_i times in $x_1, \ldots, x_n$. We introduce $m+1$ sum variables $s_0, \ldots, s_m$ defined by $s_i = \sum_{j \in 1 \ldots n} [\![x_j \leqslant i]\!]$ and post the following constraints $s_m - s_0 = \sum_{i \in 1..m} c_i$ and $\forall i \in 1 \ldots m,\ s_i - s_{i-1} = c_i$. To generate holes in the domains we add the constraints $\forall i \in 1 \ldots m, c_i = \sum_{j \in 1 \ldots n} [\![x_j = i]\!]$.

3.5 Extending the SAT Solver

SAT solvers need to be slightly extended to be usable with lazy clause generation.[1] The first extension is to communicate domain information back to the

[1] Although we manage this by building code outside the SAT solver code, leaving it untouched, but accessing its data structures.

propagation solver, e.g. when $[\![x \leqslant d]\!]$ is set true we remove from $D(x)$ the values greater than d, when is set false we remove values less than or equal to d, similarly for $[\![x = d]\!]$.

Lazy clause generation adds new clauses as search progresses of three kinds: domain clauses, explanation clauses, and nogood clauses. Usually a SAT solver only posts nogood clauses. On posting a nogood it immediately backjumps to the first place the nogood clause could unit propagate. We don't have such a luxury in lazy clause generation, since the SAT solver is not in charge of search, and indeed it may be unaware of choices that did not affect any of its variables.

When the SAT solver can backjump a great distance because a failure is found to not depend on the last choice, we have to mimic this. This is managed by checking the SAT solver first in each propagation loop, before applying any search decision. If unit propagation in SAT still detects failure, then we can immediately fail, and continue backtracking upward to the first satisfiable ancestor state.

A feature of the dual modelling inherent in lazy clause generation is that explanation clauses are redundant information, since they can be regenerated by the propagators whenever they could unit propagate.[2] Hence we can choose to delete these clauses from the SAT solver on backtracking. This reduces the number of clauses in the database, but means that more expensive propagators need to be called more often. We can select whether to delete explanations or not, by default they are deleted.

3.6 Search

Search is controlled by the FD solver, but we can make use of information from the SAT solver. We can perform:

VSIDS search. The SAT solver search heuristic VSIDS [3], based on activity, can be used to drive the search. At each choice point we retrieve the highest activity literal from the SAT solver and try setting it true or false. This is the default search for the lazy solver. Because of lazy encodings, it may be necessary to interleave search with the generation of new literals for unfixed variables, as not all literals encoding the variable domain exist initially, and in the end we need to fix all the finite domain variables. As in SAT solvers, we restart the search from time to time.

Finite Domain Search. One of the main advantages of the solver presented here compared to the solver presented in [1] is the ability to use programmed specialized finite domain searches if they are specified in the model.

Branch and bound Search. We use incremental branch and bound rather than dichotomic branch and bounds with restart due to the incrementality of our SAT solver. This differs with other SAT solver based approach such as [7] and [1].

[2] Except in cases where that the clause is stronger than the propagator. (See [1]).

Hybrid Search. We can of course build new hybrids of finite domain programmed search that make use of the activity values from the SAT solver as part of the search. We give an example in Section 4.3

4 Experiments

The experiments were run on Core 2 T8300 (2.40 GHz), except the experiments from 4.4 and 4.6 which were run respectively on a Pentium D 3.0 GHz and a Xeon 3.0 GHz for comparison with cited experiments. All experiments were run on one core. We use the following scheme for expressing variants of our approach: l = G12 lazy clause generation solver, f = G12 normal finite domain solver; v = VSIDS search, s = problem specific programmed search, h = hybrid search (see Section 4.3). When we turn off optimizations we place them after a minus: d = no deletion, a = list encoding (no arrays), r = normal reified constraints rather than single implication ones from Section 3.3, and w = no views.

4.1 Arithmetic Puzzles

The Grocery Puzzle [8] is a tiny problem but its intermediate variables have bounds up to 2^{38}. It cannot be solved using the array encoding. SEND-MORE-MONEY is another trivial problem, but here if we break the linear constraint (which has coefficients up to 9000) into ternary constraints the array encoding requires a second to solve because of the size of intermediate sum variables. Applying the list encoding on the decomposed problem, and either encoding on the original form require only a few milliseconds. These simple examples illustrate why the lazy list encoding is necessary for a lazy clause generation solver.

4.2 Constrained Path Covering Problem

The constrained path covering problem is a problem which arises in transportation planning and consists of finding a covering of minimum cardinality of a directed network. Each node $n \in Nodes$ except the start and end nodes have a positive cost $cost[n]$, and the total cost of a path cannot exceed a fixed bound. A CP model for this problem associates predecessor ($prev[n]$)/successor($next[n]$) variables to each node, as well as a cumulative cost $cumul[n]$, related by the following constraints :

$$\forall n \in Nodes.cumul[n] - cost[n] = cumul[prev[n]]$$
$$\forall n \in Nodes.\forall p \in Nodes.prev[n] = p \Leftrightarrow next[p] = n$$

In Table 1, we compare the lazy clause generation solver with default search (lv) and a specialized finite domain search (ls), as well as the G12 FD solver (fs) with the same search. We also compare creating fresh variables (lv-w,ls-w) for the result of the element constraints generated above ($cumul[prev[n]]$), as opposed to using views. The benchmark CPCP-n-m has n nodes and m edges.

The specialized finite domain search clearly outperforms VSIDS on these problems. Avoiding creating variables by using views improves search as well. This

Table 1. Constrained Path Covering Problem

	Times(sec)					Choicepoints				
	lv	ls	lv-w	lv-w	fs	lv	ls	lv-w	lv-w	fs
CPCP-17-89	0.40	0.17	0.76	0.27	**0.08**	572	**63**	563	**63**	1905
CPCP-23-181	9.02	**0.25**	36.74	0.38	0.28	31521	**449**	44423	1198	9149
CPCP-30-321	>600	**0.53**	>600	0.80	0.64	?	**804**	?	1595	9666
CPCP-37-261	>600	**1.59**	>600	2.70	>600	?	**1689**	?	2067	?
CPCP-37-495	>600	**0.99**	>600	1.44	>600	?	**1745**	?	3348	?
Average	>361.89	**0.71**	367.5	1.12	>240.2	?	**950**	1654	1231	?

is especially true with VSIDS which can be explained by the addition of useless literals, which just confuse its discovery of the "hard parts" of the problem.

The lazy clause generation solver, while slower than the finite domain solver, scales a lot better due to huge search reductions and wins for all but the easiest instance.

4.3 Radiation

Radiation scheduling [9] builds a plan for delivering a specific pattern of radiation by multiple exposures. The best search for this problem first fixes the variables shared by subproblems then fixes the subproblem variables, for each subproblem independently. Then if any subproblem is unsatisfiable we can use cuts to backtrack directly to search again the shared variables. For these experiments since we are restricted to Zinc search which does not support cuts, we simply search first on the shared variables and then on the subproblem variables in turn.

We use square matrices of size 6 to 8 with maximum intensity ranging from 8 to 10 constructed as in [9]. We ran these instances using VSIDS (lv), as well as the specialized finite domain search (without cuts) (ls), as well as a hybrid search (lh) where we use the specialized search on the shared variables, and then VSIDS on the remaining variables. We also run the FD solver (fs) with specialized search.

Each instance was run with the original linear inequalities, as well as with a decomposition into ternary inequalities, introducing intermediate sums. These linear sums are short (6−8 Boolean variables) and have small coefficients (1−10).

In this case, introducing intermediate sums definitely improved nogood generation as the choice point count is systematically reduced. On average, the specialized search outperforms VSIDS search, although the difference is reduced by the constraint decomposition, which strengthens the reusability of the explanations and nogoods generated. The hybrid search outperforms both the finite domain search and VSIDS on most instances. The FD solver is not competitive on any but the smallest instances because of the lack of explanation.

4.4 Open Shop Scheduling Problem

An open shop scheduling problem n-m-k is defined by n jobs and m machines, where each job consist of m tasks each requiring a different machine. The

	Time(sec)							Choicepoints(x1000)						
	Long linear				Ternary			Long linear				Ternary		
	fs	lv	ls	lh	lv	ls	lh	fs	lv	ls	lh	lv	ls	lh
6-08-1	2.25	**0.40**	0.61	**0.40**	0.60	0.78	0.58	72.8	1.46	1.40	**1.27**	1.29	1.33	1.31
6-08-2	**0.16**	0.37	0.53	0.41	0.54	0.72	0.53	**0.90**	1.40	1.07	2.00	1.35	1.06	1.03
6-08-3	1.12	**0.36**	0.80	0.48	0.61	1.06	0.65	48.9	1.60	2.25	1.59	**1.44**	2.09	1.56
6-09-1	2.11	0.35	0.39	**0.26**	0.52	0.46	0.36	39.2	1.19	0.61	0.46	1.43	0.52	**0.45**
6-09-2	6.68	**0.70**	1.46	0.74	0.86	1.66	1.00	271.7	3.02	4.56	2.99	**2.32**	3.98	2.78
6-09-3	432.4	0.84	1.62	**0.77**	0.96	1.86	1.06	10497	2.96	5.01	2.98	**2.66**	4.68	2.79
7-08-1	>1800	0.69	1.18	**0.56**	0.82	1.42	0.88	?	2.19	2.47	**1.11**	1.32	2.48	1.12
7-08-2	1378	**0.42**	0.89	0.54	0.76	1.09	0.78	60310	**0.78**	1.49	0.90	1.31	1.46	0.89
7-08-3	299.2	1.00	1.67	**0.82**	1.44	1.89	1.18	13767	4.49	3.95	2.26	3.49	3.57	**2.19**
7-09-1	>1800	1.17	1.63	**0.79**	1.49	2.05	1.20	?	3.48	3.59	**1.71**	3.02	3.53	1.79
7-09-2	>1800	4.05	8.62	**3.10**	3.82	9.14	4.08	?	12.4	27.9	10.7	**7.88**	23.5	9.30
7-09-3	5.60	1.11	2.04	**1.00**	1.44	2.30	1.40	199.2	4.27	4.69	**2.90**	3.47	4.22	2.58
8-09-1	950.3	3.42	4.36	**1.99**	2.94	5.11	3.14	27814	8.29	7.52	**3.99**	4.39	7.31	3.92
8-09-2	14.8	2.01	2.58	**1.50**	1.86	3.21	2.26	424.4	5.24	3.72	3.10	**2.72**	3.49	3.21
8-09-3	31.70	5.94	7.39	**3.02**	7.02	7.56	4.30	1345	14.9	18.3	10.9	**10.7**	15.3	8.23
8-10-1	1033	45.72	34.50	**18.76**	35.28	30.07	24.78	39494	51.7	64.0	41.5	**36.1**	40.1	38.1
8-10-2	>1800	26.16	21.47	**8.49**	11.41	20.74	12.47	?	39.4	47.1	18.9	**15.7**	33.0	18.6
8-10-3	>1800	93.68	37.11	**17.80**	54.41	31.11	20.62	?	88.1	96.8	52.8	55.8	63.5	**41.3**
Av.	>706	10.47	7.16	**3.41**	7.04	6.79	4.51	?	13.7	16.5	9.0	8.7	12.0	**7.8**

Fig. 2. Radiation problem : time and choice points

objective is to find a minimal schedule such that each pair of tasks (i, j) from the same job or machine are not overlapping, which is represented by the constraint $s_i + d_i \leqslant s_j \vee s_j + d_j \leqslant s_i$, where s_i and s_j are the start time of the tasks and d_i and d_j are the (fixed) durations of the tasks. An open-shop problem of size $n \times m$ has nm variables and $(nm)(nm + 1)/2$ non-overlapping constraints.

The benchmarks used are from [7]. In Table 2(a) we compare: our default lazy clause generation solver (lv) and without deletion (lv-d); the solver presented in [1] (cutsat); and the static translation approach of [7] (csp2sat) using MiniSAT version 2.0. All solvers use VSIDS search. We do not compare against f for this and subsequent problems since they all use VSIDS search. The table shows that our approach, using branch and bound rather than dichotomic search and with a slightly different propagation, vastly outperforms cutsat which beats cps2sat. Deleting previously generated explanations also substantially improves the results.

In Table 2(b) we compare results on smaller instances for different variations of lv. We can see that the overhead of the list representation is substantial, while the use of one directional reification also has significant benefits, although this is lessened by deletion.

4.5 Hoist Scheduling

We tested our lazy clause generation solver on the hoist scheduling problem presented in [10]. Example *j*-*h*-*p* has j jobs, h hoists and parallel tracks if $p = y$. We compare a simple Zinc model, run with default settings (lv) and without deletion (lv-d) as well as by static translation to SAT using [11] (fzntini), all of these using VSIDS search, against the carefully crafted Eclipse model [10] using its specialized finite domain search, run either with the Eclipse finite domain solver ic or with a finite domain and linear programing hybrid iclin using COIN-OR [12] as the linear solver. The results in Table 3 shows that our approach using

Table 2. Open shop scheduling: (a) comparing with previous approaches on hard instances, and (b) comparing the two variable representations on easier instances

(a)

tai	lv-d	lv	cutsat	csp2sat
20-20-1	55.27	**31.69**	283.1	1380.3
20-20-2	341.6	**47.54**	497.8	1520.1
20-20-3	56.63	**37.80**	270.7	1367.6
20-20-4	93.47	**38.14**	269.9	1361.3
20-20-5	50.74	**47.94**	278.8	1397.0
20-20-6	57.62	**35.26**	324.2	1405.6
20-20-7	79.20	**38.44**	455.3	1439.9
20-20-8	130.40	**41.42**	424.8	1420.8
20-20-9	44.54	**32.41**	246.1	1377.8
20-20-10	49.27	**38.84**	242.2	1346.8
Average	95.88	**39.96**	329.8	1401.7

(b)

tai	lv-d	lv	lv-ad	lfd-a	lv-dr	lv-r
15-15-1	18.63	11.73	37.47	45.28	22.94	**11.24**
15-15-2	12.68	**11.68**	70.77	76.50	24.80	14.11
15-15-3	18.87	**13.44**	49.78	96.58	15.27	15.10
15-15-4	17.61	**8.47**	55.06	54.12	14.83	9.21
15-15-5	21.02	12.34	77.30	74.45	35.36	**12.16**
15-15-6	32.44	**12.75**	26.12	39.85	20.67	16.97
15-15-7	22.66	**15.80**	33.25	45.25	23.27	16.54
15-15-8	17.12	**13.55**	29.84	21.54	12.97	13.87
15-15-9	29.68	23.23	99.44	71.24	24.51	**17.11**
15-15-10	15.1	**10.85**	124.21	41.79	47.61	17.32
Average	20.58	**13.38**	60.32	56.66	24.22	14.36

Table 3. Hoist scheduling results

Example	fzntini	lv-d	lv	ic	iclin	Example	fzntini	lv-d	lv	ic	iclin
4-1-n	1153.3	3.46	2.81	**2.18**	8.57	6-1-n	>1800	1.23	**1.18**	4.09	13.9
4-2-n	458.4	0.72	0.66	0.3	**0.1**	6-2-n	>1800	1.80	1.78	0.7	**0.15**
4-2-y	358.5	0.35	0.34	**0.3**	2.04	6-2-y	827.7	**1.01**	2.31	4.81	28.8
4-3-n	493.5	0.55	0.55	0.39	**0.09**	6-3-n	1524.8	0.63	0.68	0.6	**0.14**
4-3-y	272.6	**0.32**	0.33	0.4	1.0	6-3-y	780.0	0.86	0.72	**0.56**	4.43
5-1-n	>1800	3.98	**3.68**	10.8	52.9	7-1-n	>1800	**1.39**	1.46	3.65	9.59
5-2-n	1090.2	0.40	0.77	0.5	**0.1**	7-2-n	>1800	6.02	4.82	0.70	**0.18**
5-2-y	594.0	**0.53**	0.57	8.55	50.2	7-2-y	927.9	19.5	18.0	**17.2**	100.1
5-3-n	983.9	0.40	0.42	0.5	**0.12**	7-3-n	>1800	0.53	0.78	0.80	**0.17**
5-3-y	484.3	0.44	0.72	0.7	**0.12**	7-3-y	912.8	**0.66**	0.92	0.80	4.71
						Average	>1083	2.26	**2.17**	2.93	13.95

a simple model is competitive with the specialized models with hand-written search, especially on the hardest instances. We see that lazy clause generation is competitive whereas static translation (fzntini) struggles because of the size of the resulting SAT model (in results not shown). Clause deletion does not seem as advantageous as in the open-shop benchmarks.

4.6 Quasi-Group Completion

A $n \times n$ *latin square* is a square of values $x_{ij}, 1 \leqslant i, j \leqslant n$ where each number $[1..n]$ appears exactly once in each row and column. It is represented by constraints

$$\texttt{alldifferent}([x_{i1}, \ldots, x_{in}]),\ 1 \leqslant i \leqslant n$$
$$\texttt{alldifferent}([x_{1j}, \ldots, x_{nj}]),\ 1 \leqslant j \leqslant n$$

Table 4. Comparison of all different decomposition on quasi group completion problems

	Time (seconds)				Choicepoints			
	gcc	bnd	bnd+	diseq	gcc	bnd	bnd+	diseq
qcp-25-264-0-ext	**6.44**	1164.34	166.19	31.48	1759	15171	**1635**	77110
qcp-25-264-1-ext	**44.80**	>1800	1577.96	>1800	**13773**	?	17099	?
qcp-25-264-2-ext	**2.53**	730.13	116.25	68.74	**421**	10016	971	153128
qcp-25-264-3-ext	**157.58**	>1800	>1800	1473.21	**46063**	?	?	702897
qcp-25-264-4-ext	**22.30**	1334.34	712.71	>1800	**6697**	17322	7648	?
qcp-25-264-5-ext	**12.58**	1449.90	459.54	537.30	**3785**	18254	4679	380392
qcp-25-264-6-ext	**341.62**	>1800	>1800	170.99	**83871**	?	?	216433
qcp-25-264-7-ext	**6.08**	1265.34	159.93	178.15	1423	14289	**1342**	200123
qcp-25-264-8-ext	**3.01**	546.69	75.73	23.08	**553**	6051	586	76995
qcp-25-264-9-ext	**12.66**	>1800	638.59	36.18	**3303**	?	5484	94814
qcp-25-264-10-ext	**5.30**	914.16	121.65	123.26	981	11128	**979**	200110
qcp-25-264-11-ext	0.81	15.76	14.46	**0.35**	**0**	**0**	**0**	399
qcp-25-264-12-ext	0.80	37.53	14.65	**0.55**	**0**	590	**0**	3960
qcp-25-264-13-ext	**0.80**	337.77	14.62	1.03	**0**	4259	**0**	11408
qcp-25-264-14-ext	**4.77**	1106.12	146.86	347.84	**1183**	13412	1251	323175
Average	**41.47**	>1047.35	>522.13	>439.48	**10920.8**	?	?	?

The quasigroup completion problem (QCP) is a latin square problem where some of the x_{ij} are given. These are challenging problems which exhibit phase transition behaviour. We use instances from the 2008 CSP Solver Competition [13].

We compare several decompositions of the `alldifferent` constraint all using our default solver lv. The diseq decomposition is the usual decomposition into disequalities $\neq$. The gcc decomposition explained in Section 3.4, strengthens propagation by doing some additional bounds propagation. The bnd decomposition is a decomposition that maintains bounds-consistency [14], while bnd+ is a modification of bnd where we replace each expression $[\![x_i \leqslant d]\!] \wedge \neg[\![x_i \leqslant d-1]\!]$ with $[\![x_i = d]\!]$ to obtain a decomposition which combines the propagation of bnd and diseq. The different variations only require changing the definition of `alldifferent` included in the Zinc model.

The results are shown in Table 4. While the bnd+ decomposition is the strongest its size is prohibitive. The gcc decomposition is comprehensively best hitting the right tradeoff of strength of propagation versus size of decomposition. Comparing with results from the CSP Solver competition 2008, only two solvers could solve more than 2 of these problems (using the diseq model) in 1800s (on a 3GHz Xeon): choco2_dwdeg, requiring an average $>$ 608.8s (2 timeouts), and choco2_impwdeg, requiring $>$ 776.8s (3 timeouts)

5 Conclusion

The reengineered lazy clause generation solver is highly flexible hybrid constraint programming solver that combines the modelling and search flexibility of finite domain solving with the learning and adaptive search capabilities of SAT solvers. It forces us to reconsider many design choices for finite domain propagation. The resulting solver is highly competitive and able to tackle problems that are beyond the scope of either finite domain or SAT solvers alone. It also illustrates that the combination of specialized finite domain search with nogoods can be extremely powerful.

Acknowledgments. NICTA is funded by the Australian Government as represented by the Department of Broadband, Communications and the Digital Economy and the Australian Research Council.

References

1. Ohrimenko, O., Stuckey, P.J., Codish, M.: Propagation = lazy clause generation. In: Bessière, C. (ed.) CP 2007. LNCS, vol. 4741, pp. 544–558. Springer, Heidelberg (2007)
2. Schulte, C., Stuckey, P.J.: Efficient constraint propagation engines. ACM Trans. Program. Lang. Syst. 31 (2008)
3. Moskewicz, M.W., Madigan, C.F., Zhao, Y., Zhang, L., Malik, S.: Chaff: Engineering an efficient sat solver. In: Procs. DAC 2001, pp. 530–535 (2001)
4. Bacchus, F.: GAC via unit propagation. In: Bessière, C. (ed.) CP 2007. LNCS, vol. 4741, pp. 133–147. Springer, Heidelberg (2007)
5. Schulte, C., Tack, G.: Views and iterators for generic constraint implementations. In: van Beek, P. (ed.) CP 2005. LNCS, vol. 3709, pp. 817–821. Springer, Heidelberg (2005)
6. Harvey, W., Stuckey, P.: Improving linear constraint propagation by changing constraint representation. Constraints 8(2), 173–207 (2003)
7. Tamura, N., Taga, A., Kitagawa, S., Banbara, M.: Compiling finite linear CSP into SAT. In: Benhamou, F. (ed.) CP 2006. LNCS, vol. 4204, pp. 590–603. Springer, Heidelberg (2006)
8. Schulte, C., Smolka, G.: Finite Domain Constraint Programming in Oz. A Tutorial, `http://www.mozart-oz.org/documentation/fdt/`
9. Baatar, D., Boland, N., Brand, S., Stuckey, P.J.: Minimum cardinality matrix decomposition into consecutive-ones matrices: CP and IP approaches. In: Van Hentenryck, P., Wolsey, L.A. (eds.) CPAIOR 2007. LNCS, vol. 4510, pp. 1–15. Springer, Heidelberg (2007)
10. Rodosek, R., Wallace, M.: A generic model and hybrid algorithm for hoist scheduling problems. In: Maher, M.J., Puget, J.-F. (eds.) CP 1998. LNCS, vol. 1520, pp. 385–399. Springer, Heidelberg (1998)
11. Huang, J.: Universal booleanization of constraint models. In: Stuckey, P.J. (ed.) CP 2008. LNCS, vol. 5202, pp. 144–158. Springer, Heidelberg (2008)
12. Lougee-Heimer, R.: The Common Optimization INterface for operations research: Promoting open-source software in the operations research community. IBM Journal of Research and Development 47, 57–66 (2003)
13. International CSP Solver Competition, `http://www.cril.univ-artois.fr/CPAI08/`
14. Bessiere, C., Katsirelos, G., Narodytska, N., Quimper, C.G., Walsh, T.: Decompositions of all different, global cardianlity and related constraints. In: IJCAI (2009)

The Proper Treatment of Undefinedness in Constraint Languages

Alan M. Frisch[1] and Peter J. Stuckey[2,⋆]

[1] Artificial Intelligence Group, Dept. of Computer Science, Univ. of York, UK
frisch@cs.york.ac.uk

[2] National ICT Australia. Dept. of Computer Science and Software Engineering, Univ. of Melbourne, Australia
pjs@csse.unimelb.edu.au

Abstract. Any sufficiently complex finite-domain constraint modelling language has the ability to express undefined values, for example division by zero, or array index out of bounds. This paper gives the first systematic treatment of undefinedness for finite-domain constraint languages. We present three alternative semantics for undefinedness, and for each of the semantics show how to map models that contain undefined expressions into equivalent models that do not. The resulting models can be implemented using existing constraint solving technology.

1 Introduction

Finite-domain constraint modelling languages enable us to express complicated satisfaction and optimization problems succinctly in a data independent way. Undefinedness arises in any reasonable constraint modelling language because, for convenience, modellers wish to use *functional syntax* to express their problems and in particular they want to be able to use *partial functions*. The two most common partial functions used in constraint models are division, which is undefined if the denominator is zero, and array lookup, $a[i]$, which is undefined if the value of i is outside the range of index values of a. Other partial functions available in some constraint modelling systems are square root, which is undefined on negative values, and exponentiation, which is undefined if the exponent is negative.

A survey of some existing constraint languages and solvers shows a bewildering pattern of behaviour in response to undefined expressions. Fig. 1 shows the results of solving five problems in which undefinedness arises with five finite-domain solvers: ECLiPSe 6.0 #42 [1], SWI Prolog 5.6.64 [2], SICStus Prolog 4.0.2 [3], OPL 6.2 [4] and MiniZinc 1.0 [5,6]. The three rightmost columns, which are explained later in the paper, show the correct answers according to the three

⋆ Much of this research was conducted while Alan Frisch was a visitor at the Univ. of Melbourne. His visit was supported by the Royal Society and the Univ. Melbourne. For many useful discussions we thank the members of the Essence and Zinc research teams, especially Chris Jefferson and Kim Marriott. We thank David Mitchell for advise about complexity. NICTA is funded by the Australian Government as represented by the Dept. of Broadband, Communications and the Digital Economy and the Australian Research Council.

I.P. Gent (Ed.): CP 2009, LNCS 5732, pp. 367–382, 2009.

Problem	ECLiPSe	SWI	SICStus	OPL	MiniZinc	Relational	Kleene	Strict
(1) $y \in \{0,1\}$ $1/y = 2 \lor y < 1$	$y \mapsto 0$	$y \mapsto 0$	none	none	$y \mapsto 0$	$y \mapsto 0$	$y \mapsto 0$	none
(2) $y \in \{-1,0\}$ $1 = \sqrt{y} \lor y < 0$	none					$y \mapsto -1$	$y \mapsto -1$	none
(3) $y \in \{3,4\}$ $a[y] = 1 \lor y > 3$ where a is $[1,4,9]$ and indexed 1..3				$y \mapsto 4$	$y \mapsto 4$	$y \mapsto 4$	$y \mapsto 4$	none
(4) $y \in \{0,1,2\}$ $\mathtt{T} \lor 1/y = 1$	$y \mapsto 0$ $y \mapsto 1$	$y \mapsto 0$ $y \mapsto 1$ $y \mapsto 2$	$y \mapsto 1$ $y \mapsto 2$	$y \mapsto 1$ $y \mapsto 2$	$y \mapsto 0$ $y \mapsto 1$ $y \mapsto 2$	$y \mapsto 0$ $y \mapsto 1$ $y \mapsto 2$	$y \mapsto 0$ $y \mapsto 1$ $y \mapsto 2$	$y \mapsto 1$ $y \mapsto 2$
(5) $y \in \{0,1\}$ $\neg(1/y = 1)$	$y \mapsto 0$	$y \mapsto 0$	none	none	$y \mapsto 0$	$y \mapsto 0$	none	none

Fig. 1. Examples of how undefinedness is handled

semantics we introduce. Note that T is notation for *true*, and empty cells indicate that either the solver does not provide a square root function or that it does not allow an array to be accessed with a decision variable using functional notation. All five example problems involve Boolean operators ($\neg$ or $\lor$) because it is on such constraints that differences arise between and among implementations, our intuitions, and the three semantics introduced in this paper.

The first thing to notice is the disagreement among the solvers. The only compatible pairs of solvers are SICStus and OPL, and MiniZinc and SWI. The second observation is that some of the solvers behave irregularly. Problems (1), (2) and (3) are analogous—they just involve different partial functions—yet ECLiPSe finds a solution to (1) but not (2) and OPL finds a solution to (3) but not (1).

The issue of undefinedness in constraint languages and solvers has been the attention of almost no systematic thought. Consequently, as these examples show, implementations treat undefinedness in a rather haphazard manner and users do not know what behaviour to expect when undefinedness arises.

This paper directly confronts the two fundamental questions about undefinedness in constraint languages: What is the intended meaning of a model containing partial functions and how can those models be implemented? We address these questions by considering a simple modelling language, $\mathcal{E}$, that has two partial functions: division and array lookup. We first present three alternative truth-conditional semantics for the language: the relational semantics, the Kleene semantics, and the strict semantics. Each is obtained by starting with a simple intuition and pushing it systematically through the language. Following the standard convention that "$f(a) = \bot$" means that f is a partial function that is undefined on a, all three semantics use the value $\bot$ to represent the result of division by zero and out-of-bounds array lookups. The semantics differ in how other operators, including logical connectives and quantifiers, behave when applied to expressions that denote $\bot$. On models in which undefinedness does not arise, the semantics agree with each other, with existing implementations, and with our intuitions.

After presenting the three semantics for $\mathcal{E}$ we show how each can be implemented. Existing constraint modelling languages are implemented by mapping a constraint with nested operations into an existentially quantified conjunction of un-nested, or flat, constraints. For example, $b \vee (x/y \geq z)$ gets mapped to

$$\exists\, b', t.\, (t = x/y) \wedge (b' = t \geq z) \wedge (\mathtt{T} = b \vee b').$$

Solvers then use libraries that provide procedures for propagating each of the flat constraints. Two difficulties confront attempts to use this approach when expressions can denote $\perp$. Firstly, existing propagation procedures do not handle $\perp$. For example, the propagator that handles $t = x/y$ can bind an integer value to t when the values of x and y are known and y is non-zero, but cannot bind $\perp$ to t if y is known to be zero. Secondly, the transformations that flatten nested constraints are equivalence preserving in classical logic, but some are not equivalence preserving in a non-classical semantics that uses $\perp$. For example, rewriting $\mathtt{T} \vee \mathit{exp}$ to $\mathtt{T}$ in not equivalence preserving for the strict semantics.

This paper employs a novel approach for implementing the three semantics for $\mathcal{E}$. Rather than transform constraints in $\mathcal{E}$ to flattened constraints, we transform the $\mathcal{E}$-model to another one that has the same solutions but in which undefinedness cannot arise. A different transformation is used for each of the semantics. Since undefinedness cannot arise in the resulting $\mathcal{E}$ models, they can be implemented using the well-understood techniques that are standardly used in the field.

2 A Simple Constraint Language

We use a simplified form of Essence [7], called $\mathcal{E}$, as our language for modelling decision problems, not just problem instances. Every model in $\mathcal{E}$ has exactly three statements, signalled by the keywords `given`, `find` and `such that`. As an example consider the following model of the graph colouring problem.

`given`	k:`int`, n:`int`, Edge: `array`$[1..n, 1..n]$ `of bool`
`find`	Colour: `array`$[1..n]$ `of int`$(1..k)$
`such that`	$\forall v{:}1..n-1.\ \forall v'{:}v..n.\ \mathit{Edge}[v, v'] \rightarrow \mathit{Colour}[v] \neq \mathit{Colour}[v']$

The `given` statement specifies three parameters: k, the number of colours; n, the number of vertices in the graph to be coloured; and Edge, an incidence matrix specifying the graph to be coloured. The integers $1..n$ represent the vertices of the graph. The `find` statement says that the goal of the problem is to find Colour, an array that has an integer $1..k$ for each vertex. Finally the `such that` statement requires that a solution must satisfy the constraint that for any two nodes, if there is an edge between them then they must have different colours.

The language has three main syntactic categories: statements, expressions and domains. Each expression of the language has a unique type that can be determined independently of where the expression appears. Where τ is a type we write $\mathit{exp}{:}\tau$ to denote an arbitrary expression of type τ. The types of the language are `int`, `bool`, and `array` $[IR]$ `of` τ, where τ is any type and IR is an integer range specifying the index values of the array. Throughout the language, an integer range, always denoted IR, is of the form exp_1:`int`..exp_2:`int` and never contains a decision variable. Notice that the array constructor can be nested; for example `array`$[1..10]$ `of array` $[0..5]$ `of int` is a type. We often abbreviate "`array` $[l_1..u_1]$ `of` $\cdots$ `of array` $[l_n..u_n]$" as "`array` $[l_1..l_n, \ldots, l_n..u_n]$."

Domains are used to associate a set of values with a parameter or decision variable. A domain is either (1) `bool`, (2) `int`, (3) of the form `int` (IR) or (4) of the form `array` $[IR]$ `of` *Dom*, where *Dom* is a domain. A domain is finite if it is constructed without using case (2) of the definition. The non-terminal *FDom* is used for finite domains.

The syntax of the three statements is as follows (where $n \geq 0$):

`given` $NewId_1{:}Dom_1, \ldots, NewId_n{:}Dom_n$
  where Dom_i can contain an occurrence of $NewId_j$ only if $i \leq j$.
`find` $NewId_1{:}FDom_1, \ldots, NewId_n{:}FDom_n$
`such that` exp_1:`bool`$, \ldots,$ exp_n:`bool`

Finally, let's consider the syntax of expressions, starting with the atomic expressions. The integer constants are written in the usual way and the constants of type `bool` are `T` and `F`. Each identifier that has been declared as a parameter or decision variable is an expression whose type is determined by the domain given in the declaration. A quantified variables can appear within the scope of its quantifier. As will be seen, quantified variables are always of type `int`.

The following are non-atomic expressions of type `int`:

- exp_1:`int` *intop* exp_2:`int`, where *intop* is one of $+$, $-$, $*$, or $/$,
- $-exp_1$:`int`,
- `boolToInt`(*exp*:`bool`), and
- $\sum NewId{:}IR.\ exp$:`int`

The symbol "/" is for integer division. An example of an integer expression using these constructs is $\sum i{:}0..n{-}1.$ `boolToInt`$(a[i] = 0)$, which counts up the number of 0 entries in a .

The following are non-atomic expressions of type `bool`:

- exp_1:`bool` *boolop* exp_2:`bool`, where *boolop* is one of $\wedge$, $\vee$, $\rightarrow$ or $\leftrightarrow$.
- $\neg exp$:`bool`.
- exp_1:`int` *compop* exp_2:`int`, where *compop* is one of $=$, $\neq$, $\leq$ or $<$.
- $\mathcal{Q}\, NewId{:}IR.\ exp$:`bool`, where $\mathcal{Q}$ is a logical quantifier, $\exists$ or $\forall$.

Finally, the following expression is of type τ:

- $AR[exp$:`int`$]$, where AR is of type "`array` $[IR]$ `of` τ", for some τ. Notice that *exp*:`int` may contain free variables.

We often abbreviate "$AR[i_1]\cdots[i_n]$" as "$AR[i_1,\ldots,i_n]$".

For simplicity we assume that each identifier *NewId* occurring in $\forall NewId{:}IR.\,exp$, $\exists NewId{:}IR.\,exp$ or $\sum NewId{:}IR.\,exp$ is a new identifier that appears nowhere else in the model except in *exp*.

3 The Semantics of $\mathcal{E}$

This section presents three alternative semantic accounts of $\mathcal{E}$. In each undefinedness arises in only two ways: dividing by zero and indexing into an array with a value that is out of bounds. The three accounts differ only in how they determine whether an expression is undefined if it contains an undefined subexpression. For models that are *safe*—those in which division by zero and out-of-bounds indices do not arise—the three semantics agree with each other and, we believe, with the intuitions of constraint modellers and the behaviour of constraint solvers. For safe models, solvers do not exhibit a haphazard pattern of behaviour.

This section first presents the part of the semantics that the three have in common. Then three subsections describe the distinctive parts of the three semantics.

The purpose of these semantics is to identify the solutions of an instance of an $\mathcal{E}$ model—that is, what assignments to decision variables satisfy what instances. To be clear, our focus is defining the truth conditions of the language, not on defining the behaviour of a decision procedure for satisfiability or any other program.

As the semantics defines solutions of instances, we start by defining the instances of a model: a pair $\langle M, I\rangle$ is a *problem instance* if M is an $\mathcal{E}$ model and I is an *instantiation* for M. An instantiation for M maps each parameter of M to a value that is appropriate as determined by the `given` statement of the model. If the `given` statement of M is "`given` $NewId_1{:}Dom_1, \ldots, NewId_n{:}Dom_n$", then an instantiation I of M maps each parameter $NewId_i$ to a member of the set denoted by Dom_i. The denotation of Dom_i, written $[\![Dom_i]\!]^I$, must be taken relative to I since Dom_i may itself contain parameters; for example in "$\texttt{array}[a * b..c * b]\ \texttt{of int}$" the symbols a, b and c may be parameters. The following rules define the semantics of domains.

- $[\![exp_l..exp_u]\!]^I = \bot$ if $[\![exp_l]\!]^I = \bot$ or $[\![exp_u]\!]^I = \bot$ (1)
  $= \{i \in \mathbb{Z} \mid [\![exp_l]\!]^I \le i \le [\![exp_u]\!]^I\}$ otherwise
- $[\![\texttt{int}(IR)]\!]^I = \emptyset$ if $[\![IR]\!]^I = \bot$
  $= [\![IR]\!]^I$ otherwise
- $[\![\texttt{bool}]\!]^I = \{\texttt{T}, \texttt{F}\}$
- $[\![\texttt{int}]\!]^I = \mathbb{Z}$ (That is, the set of all integers.)
- $[\![\texttt{array}\ [IR]\ \texttt{of}\ DOM]\!]^I = \emptyset$ if $[\![IR]\!]^I = \bot$ or $[\![IR]\!]^I = \emptyset$
  $= [\![IR]\!]^I \longrightarrow [\![DOM]\!]^I$ otherwise.

Notice that an array denotes a total function over the set of index values that are within bounds. This set may be empty. Also notice that some models have no instantiations because a parameter may have a domain denoting the empty set.

The semantics must dictate whether an instance $\langle M, I\rangle$ of a model is satisfied by an assignment A to the decision variables of M. An assignment must map each decision variable to an appropriate value. If the `find` statement of M is "`find` $NewId_1{:}FDom_1, \ldots, NewId_n{:}FDom_n$", then an assignment A for $\langle M, I\rangle$ maps each decision variable $NewId_i$ to a member of $[\![FDom_i]\!]^I$. Notice that $[\![FDom_i]\!]^I$ may be the empty set, in which case the instance has no assignments and hence no solutions.

Finally, quantified variables are handled in the same way as in first-order logic—that is, denotations are taken relative to an assignment g that maps each quantified variable to an appropriate value.

We write $[\![M]\!]^{I,A,g}$ to mean the denotation of instance $\langle M, I\rangle$ with respect to A and g. As the denotation function is defined compositionally, we extend the notation and write $[\![exp]\!]^{I,A,g}$ where exp is any expression of $\mathcal{E}$.

Our primary intuition regarding undefinedness is that an assignment A is a solution to an instance if the constraints of the instance all denote T with respect to the assignment, and this is the case even if the same constraints denote undefined with respect to other assignments. Thus we say that an instance I of model M is satisfied by assignment A if $[\![c]\!]^{I,A} = \mathtt{T}$ for every constraint c in M. This intuition would be violated by a solver that aborts if one assignment generates an error condition such as division by zero even though other assignments are solutions.

It remains to define the semantics of the expressions of $\mathcal{E}$. This section defines the semantics for expressions where they agree for all three semantics.

Let us start with the atomic expressions. In all assignments, "T", "F", "1", "2", "3", etc. denote T, F, 1, 2, 3, etc. For other atomic expressions we have:

- $[\![\alpha]\!]^{I,A,g} = I(\alpha)$ if α is a parameter
  $= A(\alpha)$ if α is a decision variable
  $= g(\alpha)$ if α is a quantified variable.

Now consider the operators that are used to build up integer expressions. For every binary integer operator *intop* we have

- $[\![exp_1 \; intop \; exp_2]\!]^{I,A,g} = \bot$ if $[\![exp_1]\!]^{I,A,g} = \bot$ or $[\![exp_2]\!]^{I,A,g} = \bot$
  $= [\![exp_1]\!]^{I,A,g} \; [\![intop]\!]^{I,A,g} \; [\![exp_2]\!]^{I,A,g}$ otherwise

where $[\![intop]\!]^{I,A,g}$ is the obvious operation. For division $[\![exp_1/exp_2]\!]^{I,A,g} = \bot$ if $[\![exp_2]\!]^{I,A,g} = 0$. Unary operators are handled in a similar manner.

Now consider summation expressions. If g is a variable assignment, then $\sigma[x \mapsto d]$ is the assignment that is identical to σ with the possible exception that it maps x to d.

- $[\![\sum x{:}IR.\; exp]\!]^{I,A,g} =$ if $[\![IR]\!]^{I,A,g} = \emptyset$ then 0
  else if $[\![IR]\!]^{I,A,g} = \bot$ then $\bot$
  else if $[\![exp]\!]^{I,A,g[x \mapsto d]} = \bot$ for some $d \in [\![IR]\!]^{I,A,g}$ then $\bot$
  else the sum of $[\![exp]\!]^{I,A,g[x \mapsto d]}$ for all $d \in [\![IR]\!]^{I,A,g}$

Notice that $\sum x{:}1.. -1.\; x/0$ denotes 0 with respect to any assignments since $1.. -1$ denotes the empty set.

The integer range associated with the summation quantifier and, indeed, all quantifiers, may contain free occurrences of quantified variables. So semantic rule (1) must be generalised to take assignments to quantified variables.

- $[\![exp_l..exp_u]\!]^{I,A,g} = \bot$ if $[\![exp_l]\!]^{I,A,g} = \bot$ or $[\![exp_u]\!]^{I,A,g} = \bot$
  $= \{i \in \mathbb{Z} \mid [\![exp_l]\!]^{I,A,g} \leq i \leq [\![exp_u]\!]^{I,A,g}\}$ otherwise

3.1 Semantics 1: A Three-Valued Kleene Semantics

This semantics follows the approach used by Frisch et. al. [8] in giving a semantics to ESSENCE. Three truth values are used — T, F and $\bot$ — where the intuition is that $\bot$ indicates a lack of information. Thus, $\mathtt{T} \vee \bot$ is T because it is T regardless of whether the "unknown" value $\bot$ is T or F. Similarly, $\mathtt{T} \wedge \bot$ is $\bot$ because it could be T or F depending on the "unknown" value of the second argument. This results in the Boolean connectives of the three-valued propositional logic of Kleene [9, §64].

$\wedge$	T	F	$\bot$
T	T	F	$\bot$
F	F	F	F
$\bot$	$\bot$	F	$\bot$

$\vee$	T	F	$\bot$
T	T	T	T
F	T	F	$\bot$
$\bot$	T	$\bot$	$\bot$

$\rightarrow$	T	F	$\bot$
T	T	F	$\bot$
F	T	T	T
$\bot$	T	$\bot$	$\bot$

$\leftrightarrow$	T	F	$\bot$
T	T	F	$\bot$
F	F	T	$\bot$
$\bot$	$\bot$	$\bot$	$\bot$

$\neg$	
T	F
F	T
$\bot$	$\bot$

boolToInt	
T	1
F	0
$\bot$	$\bot$

Existential quantification should behave like disjunction, which yields:

- $[\![\exists x{:}IR.\ exp{:}\mathtt{bool}]\!]^{I,A,g}$
  $= \mathtt{T}$ if $[\![IR]\!]^{I,A,g} \neq \bot$ and $[\![exp{:}\mathtt{bool}]\!]^{I,A,g[x\mapsto d]} = \mathtt{T}$ for some $d \in [\![IR]\!]^{I,A,g}$
  $= \mathtt{F}$ if $[\![IR]\!]^{I,A,g} \neq \bot$ and $[\![exp{:}\mathtt{bool}]\!]^{I,A,g[x\mapsto d]} = \mathtt{F}$ for all $d \in [\![IR]\!]^{I,A,g}$
  $= \bot$ otherwise

The rule for universal quantification is obtained from this by interchanging "T" and "F." Notice that $\exists x{:}1..{-1}.\ 1/0 = 7$ denotes F with respect to any assignments since $1..-1$ denotes the empty set.

For integer comparison and array lookup we have:

- $[\![exp_1{:}\mathtt{int}\ compop\ exp_2{:}\mathtt{int}]\!]^{I,A,g} = \bot$ if $[\![exp_1]\!]^{I,A,g} = \bot$ or $[\![exp_2]\!]^{I,A,g} = \bot$
  $= [\![exp_1]\!]^{I,A,g}\ [\![compop]\!]^{I,A,g}\ [\![exp_2]\!]^{I,A,g}$ otherwise
- $[\![AR[exp{:}\mathtt{int}]]\!]^{I,A,g} =$ if $[\![AR]\!]^{I,A,g} = \bot$ or $[\![exp{:}\mathtt{int}]\!]^{I,A,g} = \bot$ then $\bot$
  else if the function $[\![AR]\!]^{I,A,g}$ is not defined on $[\![exp{:}\mathtt{int}]\!]^{I,A,g}$ then $\bot$
  else $[\![AR]\!]^{I,A,g}([\![exp{:}\mathtt{int}]\!]^{I,A,g})$

Again $[\![compop]\!]^{I,A,g}$ is the obvious operation.

3.2 Semantics 2: A Three-Valued Strict Semantics

The second semantic account of $\mathcal{E}$ is strict in that any compound expression is undefined whenever one of its sub-expressions is undefined. It is straightforward to specify the semantics based on this principle.

Here is the rule for the existential quantification, the remainder are similar:

- $[\![\exists x{:}IR.\ exp]\!]^{I,A,g} =$ if $[\![IR]\!]^{I,A,g} = \bot$ or $[\![exp]\!]^{I,A,g[x\mapsto d]} = \bot$ for some $d \in [\![IR]\!]^{I,A,g}$ then $\bot$
  else if $[\![exp]\!]^{I,A,g[x\mapsto d]} = \mathtt{T}$ for some $d \in [\![IR]\!]^{I,A,g}$ then T
  else F

As in the Kleene semantics, a consequence of the rule for existential quantification is that $\exists x{:}1..-1.\ 1/0 = 7$ denotes F with respect to any assignment. Finally, the semantic rules for the comparison operators, the boolToInt operator and for indexing into arrays are the same as for the Kleene semantics.

3.3 Semantics 3: A Two-Valued Relational Semantics

The third semantic account for $\mathcal{E}$ is based on the observation that undefinedness results from the application of partial functions and the view that functional notation is a shorthand for relational notation. So, instead of thinking of division as a function, one could think of it as a relation, $\mathtt{div}(x, y, z)$, which holds if and only if the result of dividing x by y is z (equivalently $y \times z = x \wedge y \neq 0$). Hence $\mathtt{div}(5, 0, 3)$ denotes F not $\bot$.

In this semantics there can be undefined integer expressions, but all Boolean expressions are either T or F. Thus the Boolean operators $\neg$, $\wedge$, $\vee$, $\rightarrow$ and $\leftrightarrow$ as well as `boolToInt` have their usual classical interpretation. The rules for the logical quantifiers are:

- $[\![\exists x{:}IR.\ exp{:}\texttt{bool}]\!]^{I,A,g}$
  $= \texttt{T}$ if $[\![IR]\!]^{I,A,g} \neq \bot$ and $[\![exp{:}\texttt{bool}]\!]^{I,A,g[x\mapsto d]} = \texttt{T}$ for some $d \in [\![IR]\!]^{I,A,g}$
  $= \texttt{F}$ otherwise

The rule for indexing into an array AR of type "`array`[IR] of `bool`" is:

- $[\![AR[exp{:}\texttt{int}]]\!]^{I,A,g} =$ if $[\![AR]\!]^{I,A,g} = \bot$ or $[\![exp{:}\texttt{int}]\!]^{I,A,g} = \bot$ then F
  else if the function $[\![AR]\!]^{I,A,g}$ is not defined on $[\![exp{:}\texttt{int}]\!]^{I,A,g}$ then F
  else $[\![AR]\!]^{I,A,g}([\![exp{:}\texttt{int}]\!]^{I,A,g})$

If AR is an array of any other type, then the rule is the same as for the other two semantics.

3.4 Comparison of the Semantics

Here we briefly state some of the properties of and relationships among the three semantics. Space limitations preclude the presentation of examples or proofs. We invite the reader to revisit Fig. 1 and confirm that the three semantics produce the results shown in the last three columns.

Theorem 1. *Let e be any expression, I be any instantiation, A be any assignment and g be any variable assignment. Let s, k and r be the value of $[\![e]\!]^{I,A,g}$ in the strict, Kleene and relational semantics, respectively. If $s \neq \bot$ then $s = k$. If $k \neq \bot$ then $k = r$.* □

Theorem 2. *Let M be any $\mathcal{E}$ model. If in the strict semantics I is an instance of M and A is a solution to $\langle M, I\rangle$ then in the Kleene semantics I is an instance of M and A is a solution to $\langle M, I\rangle$. If in the Kleene semantics I is an instance of M and A is a solution to $\langle M, I\rangle$ then in the relational semantics I is an instance of M and A is a solution to $\langle M, I\rangle$.* □

In both the Kleene and relational semantics, a decision variable is the same as a prenexed existential quantifier. This is not the case in the strict semantics.

If e_1 and e_2 are integer expressions, then in both the Kleene and strict semantics $e_1 \neq e_2$ and $\neg(e_1 = e_2)$ are logically equivalent. Similarly, $e_1 < e_2$ and $\neg(e_2 \leq e_1)$ are logically equivalent in these two semantics. However, neither logical equivalence holds in general in the relational semantics.

In the Kleene and relational semantics, for every Boolean expression e, F and $\texttt{F} \wedge e$ are logically equivalent and T and $\texttt{T} \vee e$ are logically equivalent. Neither of these equivalences holds in general in the strict semantics.

In the Kleene and strict semantics, for every integer range IR and every Boolean expression ϕ, $\neg\forall x{:}IR.\ \phi$ and $\exists x{:}IR.\ \neg\phi$ are logically equivalent and $\forall x{:}IR.\ \neg\phi$ and $\neg\exists x{:}IR.\ \phi$ are logically equivalent. Neither equivalence generally holds in the relational semantics. For example, $\forall x{:}1/0..1/0.\ \neg(1 = 1)$ denotes F in all assignments but $\neg\exists x{:}1/0..1/0.\ 1 = 1$ denotes T in all assignments.

4 Transforming Constraints in $\mathcal{E}$

This section shows how, for each of the three semantics, a model can be transformed into one that has the same instances and solutions but whose constraints are *safe*. Space limitations require us to make the simplifying assumption that all expressions in `given` and `find` statements and in integer ranges associated with quantifiers are safe. More rigorously we say that an occurrence of an expression e is unsafe if $[\![e]\!]^{I,A,g} = \bot$ for some I, A and g. Furthermore an occurrence of an expression is unsafe if it contains any unsafe occurrence. Otherwise, an occurrence is said to be safe.

Let us first introduce the idea behind the transformations. Here we write $\phi[e]$ to denote an expression containing an occurrence of e. A subsequent reference to $\phi[e']$ denotes the same expression but with e replaced by e'. As an example to illustrate the transformations, consider transforming an atomic Boolean expression $A[e'/e]$ that is unsafe because e could denote 0. The expression $A[e'/e]$ may occur within a complex constraint.

In the relational semantics the basic idea is to transform $A[e'/e]$ to $\exists a'{:}\texttt{nz}.\, a' = e \wedge A[e'/a']$. Here `nz` is the domain of all non-zero integers, so the resulting expression is safe. Notice that the resulting expression is false if e denotes 0.

In the strict semantics the basic idea is to transform $A[e'/e]$ in the same way as the relational semantics but also to add to the `such that` statement the constraint $e \neq 0$. This is a simplification because e may contain free variables.

The transformation for the Kleene semantics depends on the polarity of the occurrence of $A[e'/e]$. If it occurs in a positive context, then it is transformed in the same way as the relational semantics. However, if $A[e'/e]$ occurs in a negative context then it is transformed to $(\exists a'{:}\texttt{nz}.\, a' = e \wedge A[e'/a']) \vee e = 0$. This expression is true if e denotes 0, which has the same effect as making the expression false in a positive context.

These basic transformations are logically correct except for the case `boolToInt` in the Kleene semantics, which is difficult to handle and requires special treatment. Unfortunately, these basic transformations add existential variables to the interior of a constraint and do not propagate efficiently. The actual transformations avoid these problems but, consequently, are much more complicated.

Now let's proceed to consider the actual transformations. In performing transformations to render a model safe, we need to pass through a language called $\mathcal{E}^+$, which is the same as $\mathcal{E}$ but with four additional features, each of which has the same denotation in each of the three semantics.

- A new kind of *IR*, called `nz`, which denotes the set of all non-zero integers.
- An additional *boolop*, $\Leftrightarrow$, where $x \Leftrightarrow y$ denotes `T` if x and y take the same value (including $\bot$) and `F` otherwise.
- The operator $\texttt{boolToInt}_0$, which denotes the function that maps `T` to 1, `F` to 0, and $\bot$ to 0.
- A domain can contain "`array` $[l..u]$" where l and u can contain integer expressions of the form $\texttt{MIN}\, x{:}IR.exp$ and $Qmax\, x{:}IR.exp$, respectively. If the *IR* associated with either form of expression denotes $\emptyset$ then the "`array`

$[l..u]$" in which it occurs denotes an array with an empty set of indices. Otherwise, $[\![\texttt{MIN}\, x{:}IR.exp]\!]^{I,A,g}$ and $[\![\texttt{MIN}\, x{:}IR.exp]\!]^{I,A,g}$) are the minimum and maximum values of $\{[\![exp]\!]^{I,A,g[x\mapsto d]} | d \in [\![IR]\!]^{I,A,g}\}$.

Theorem 3. *The set of safe $\mathcal{E}^+$ models are the same in all three semantics. If M is a safe model, then its instantiations are the same in all three semantics and every instance of M has the same solutions in all three semantics.* □

By reductions from Diophantine problems it is straightforward to show that neither the safe nor the unsafe expressions of $\mathcal{E}$ are recursively enumerable. Therefore, for the transformations we assume the existence of a procedure that can determine that some expressions are safe, though it can not detect all safe expressions. We place three requirements on such a procedure: (1) if the procedure says that an expression is safe, then it is safe; (2) the procedure identifies as safe every expression of the form exp/i, where i is an integer variable with domain `nz`; and (3) the procedure identifies as safe every expression of the form and $AR[i]$, where i is an integer variable and the domain of i and the index range of AR are defined with syntactically identical expressions. If this procedure identifies an expression as safe, then we say that the expression is *provably safe*, otherwise we say that it is *possibly unsafe*.

Our transforms work by eliminating all possibly unsafe expressions and are correct even when the eliminated expression happens to be safe. However, a more accurate estimate of safeness results in the transformations making fewer changes to the model, thus producing a simpler model.

Example 1. Consider the model in $\mathcal{E}$ of the form:

`given`	m: `array`$[1..10]$ `of` `int`$(1..5)$
`find`	x:`int`$(1..20)$, y:`int`$(-3..3)$
`such that`	$(\forall j{:}1..4.\ \sum i{:}j..9.\ \texttt{boolToInt}(m[i] \geq m[x]) \leq j/(y^2-5))$

then the expression $m[i]$ is provably safe, and we may assume expressions $m[x]$ and $j/(y^2-5)$ are possibly unsafe, although a more sophisticated analysis could determine that $y^2 - 5$ cannot take the value 0. □

The transforms for all three semantics use the common sub-procedure TransformExtract, given in Fig. 2, to transform a model M.

Example 2. Consider applying TransformExtract to the following model:

`find`	x:`int`$-1..10$
`such that`	$1 \leq 1/\texttt{boolToInt}(7/x \leq 1)$

First A must be chosen to be $7/x \leq 1$. (The transform cannot choose A to be the entire constraint as this contains a possibly unsafe Boolean expression.) Steps (a), (b) and (c) are executed, resulting in

`find`	x:`int`$-1..10, b_1$:`bool`
`such that`	$1 \leq 1/\texttt{boolToInt}(b_1)$,
	$b_1 \Leftrightarrow 7/x \leq 1$

TransformExtract(M)

While there is an occurrence A of a possibly unsafe Boolean expression in a constraint of M such that every occurrence of a Boolean expression within A is provably safe and the constraint does not contain an occurrence of "$\Leftrightarrow$" do:

Let b be an identifier that does not occur in M.
If no free quantified variables occur in A then
(a) Replace occurrence A in M with b.
(b) Add b:`bool` to the decision variables of M.
(c) Add to M the constraint $b \Leftrightarrow A$.

Otherwise
(d) Let $y_1, \ldots, y_m$ be, in order, the variables attached to the quantifiers whose scope includes A.
Let $y_{i_1}, \ldots, y_{i_n}$ be, in order, those variables in $y_1, \ldots, y_m$ that occur in A.
Let $l_i..u_i$ be the integer range expression over which each y_i is quantified.
Let MIN_j be the expression $\texttt{MIN}\ y_1{:}l_1..u_1.\ \cdots\ \texttt{MIN}\ y_{j-1}{:}l_{j-1}..u_{j-1}.\ l_j$.
Let MAX_j be the expression $\texttt{MAX}\ y_1{:}l_1..u_1.\ \cdots\ \texttt{MAX}\ y_{j-1}{:}l_{j-1}..u_{j-1}.\ u_j$.
(e) Replace occurrence A in M with $b[y_{i_1}, \ldots, y_{i_n}]$.
(f) Add "b:`array`[$MIN_{i_1}..MAX_{i_1}, \ldots, MIN_{i_n}..MAX_{i_n}$]`of bool`" to the decision variables of M.
(g) Add to M the constraint $\forall y_1{:}l_1..u_1.\ \cdots\ \forall y_n{:}l_n..u_n.\ b[y_{i_1}, \ldots, y_{i_n}] \Leftrightarrow A$

Fig. 2. TransformExtract

Next A is chosen to be $1 \leq 1/\texttt{boolToInt}(b_1)$ and steps (a), (b) and (c) are executed, resulting in

$$\begin{array}{ll} \texttt{find} & x{:}\texttt{int}{-1..10}, b_1{:}\texttt{bool}, b_2{:}\texttt{bool} \\ \texttt{such that} & b_2, \\ & b_2 \Leftrightarrow 1 \leq 1/\texttt{boolToInt}(b_1), \\ & b_1 \Leftrightarrow 7/x \leq 1 \end{array}$$

□

Let us consider the relationship between a model M and the model M' that results from applying TransformExtract to M. We say that an assignment α' is an extension of an assignment α if every variable assigned by α is also assigned by α' and the two assign the same values to the variables of α.

Theorem 4. *Let M be a model and M' be the result of applying TransformExtract to M. Let I be any instantiation for M (and hence for M'). Then in any of the three semantics, assignment α is a solution to $\langle M, I\rangle$ if and only if some extension of α is a solution to $\langle M', I\rangle$.* □

After performing TransformExtract a model consists of two disjoint sets of constraints: M', the original constraints modified by steps (a) and (e) of TransformExtract, and B, the set of constraints added by steps (c) and (g) of TransformExtract. Notice that every constraint in M' is provably safe and every constraint in B is possibly unsafe. The transformation needed to make B provably safe is different for each of the three semantics. We consider each in turn.

4.1 Transformations for the Relational Semantics

To obtain a safe model for the relational semantics, Transform2Rel, as shown in Fig. 3 is performed. The transformations introduce a new variable a' to take

Transform2Rel(M)

(a) Perform TransformExtract(M).
(b) While M contains a possibly unsafe occurrence of $b \Leftrightarrow C$ perform Transform2Pos($M, b \Leftrightarrow C$).

Transform2Pos($M, b \Leftrightarrow C$)

(c) If C contains an possibly unsafe expression of the form exp'/exp where exp is a provably safe expression then replace $b \Leftrightarrow C$ with

$$\begin{aligned} \exists\, a'{:}\texttt{nz}.\ \exists\, b'{:}\texttt{bool}.\ & b' \Leftrightarrow (a' = exp) \wedge \\ & b \Leftrightarrow (b' \wedge C\{exp \mapsto a'\}) \wedge \\ & exp \neq 0 \rightarrow b' \end{aligned}$$

(d) If C contains a possibly unsafe expression of the form $AR[exp]$, where AR is an expression of type `array` [l..u] of τ, and exp is a provably safe expression then replace $b \Leftrightarrow C$ with

$$\begin{aligned} \exists\, a'{:}l..u.\ \exists\, b'{:}\texttt{bool}.\ & b' \Leftrightarrow (a' = exp) \wedge \\ & b \Leftrightarrow (b' \wedge C\{exp \mapsto a'\}) \wedge \\ & (l \leq exp \wedge exp \leq u) \rightarrow b' \end{aligned}$$

Fig. 3. Transform2Rel

the place of exp, and a new Boolean b' to capture whether $a' = exp$. a' has a domain that forces the resulting expression to be provably safe. The complexity arises in capturing the cases where a' and exp differ in value. If $a' \neq exp$ then the Boolean b is forced to be `F`. The third conjunct is required since otherwise we could choose $a' \neq exp$ and make b `F` when indeed the expression will not lead to undefined. The third conjunct forces $a' = exp$ if this will not result in an undefined expression.

Theorem 5. *Let M be a model resulting from the application of TransformExtract and let M' be the result of applying Transform2Rel to M. Let I be any instantiation for M (and hence for M'). Then M' is safe and in the relational semantics $\langle M', I\rangle$ and $\langle M, I\rangle$ have the same solutions.* □

4.2 Transformations for the Strict Semantics

The strict semantics is the simplest to implement. The full transformation is given in Fig. 4.

Transform2Strict(M)

(a) Perform TransformExtract(M).
(b) While M contains a possibly unsafe occurrence of $b \Leftrightarrow C$ do:
 (c) If C contains a possibly unsafe expression of the form exp'/exp, where exp is provably safe then replace $b \Leftrightarrow C$ with

$$\exists\, a'{:}\texttt{nz}.\ a' = exp \wedge (b \Leftrightarrow C\{exp \mapsto a'\})$$

 (d) If C contains a possibly unsafe expression of the form $AR[exp]$, where AR is an expression of type `array` [l..u] of τ, then replace $b \Leftrightarrow C$ with

$$\exists\, a'{:}l..u\ .\ a' = exp \wedge (b \Leftrightarrow C\{exp \mapsto a'\})$$

Fig. 4. Transform2Strict

Theorem 6. *Let M be a model resulting from the application of TransformExtract and let M' be the result of applying Transform2Strict to M. Let I be any instantiation for M (and hence for M'). Then M' is safe and in the strict semantics $\langle M', I\rangle$ and $\langle M, I\rangle$ have the same solutions.* □

4.3 Transformations for the Kleene Semantics

The Kleene semantics is the most difficult to make safe. This is the only semantics where Boolean expressions can really take the value ⊥ (in the strict semantics if this occurs then there can be no solution). In order to transform this correctly to an effectively two valued semantics that is supported by the underlying constraint solvers we need to take into account whether a Boolean expression occurs in a positive context, where undefined will be equivalent to `F` for satisfiability, or a negative context where undefined will be equivalent to `T` for satisfiability.

The Transform2Kleene transformation, shown in Fig. 5, converts an $\mathcal{E}$ model into a safe $\mathcal{E}^+$ model. Step (a) replaces each occurrence of ↔ with two implications. The effect, as will be seen, is that every expression appears in either a positive or negative context, but not both. Step (b) replaces each occurrence of `boolToInt` with an equivalent expression containing two occurrences of $\texttt{boolToInt}_0$. This is done because the remainder of the transformation correctly deals with $\texttt{boolToInt}_o$ without any special provisions. Though each of these first two steps can make the model exponentially larger, a more sophisticated version of these steps could avoid this by introducing new Boolean variables. Step (c) performs TransformExtract, just as in the other two transforms. Finally Step (d) replaces all possibly unsafe expressions with ones that are provably safe. Positive occurrences are handled as in the relational transformation; negative occurrences employ a new transformation, Transform2Neg, also shown in Fig. 5.

Given an $\mathcal{E}^+$ expression which does not include or `boolToInt` or ↔ we can define the context of each Boolean expression appearing in a model as follows.

- For `such that` $exp_1, \ldots, exp_n$ each of $exp_i, 1 \leq i \leq n$ appear positively.
- For $\texttt{boolToInt}_0(exp)$ then exp appears positively.
- If $\neg exp$ appears positively then exp appears negatively, and if $\neg exp$ appears negatively then exp appears positively.
- If $exp \vee exp'$ or $exp \wedge exp'$ appear in manner H (positively or negatively) then exp and exp' appear in manner H.
- If b appears in manner H (positively or negatively) and $b \Leftrightarrow exp$ occurs in M then exp appears in manner H.

Note that since the model results from TransformExtract the rules for expressions of the form $b \Leftrightarrow C$ are unambiguous since there is exactly one such expression for each introduced b.

Theorem 7. *Let M be a model resulting from the application of TransformExtract. Let M' be the result of applying Transform2Kleene to M. Then Let I be any instantiation for M (and hence for M'). Then M' is safe and in the Kleene semantics $\langle M', I\rangle$ and $\langle M, I\rangle$ have the same solutions.* □

Transform2Kleene(M)

(a) Replace every occurrence in M of an expression of the form $exp_1 \leftrightarrow exp_2$ with $(exp_1 \rightarrow exp_2) \wedge (exp_2 \rightarrow exp_1)$.
(b) Replace every occurrence in M of an expression of the form $\texttt{boolToInt}(exp)$ with $\texttt{boolToInt}_0(exp)/(\texttt{boolToInt}_0(exp) + \texttt{boolToInt}_0(\neg exp))$.
(c) Perform TransformExtract(M)
(d) While M contains a possibly unsafe occurrence of $b \Leftrightarrow C$ do:
    (e) if b occurs positively in M then perform Transform2Pos($M, b \Leftrightarrow C$)
    (f) otherwise perform Transform2Neg($M, b \Leftrightarrow C$).

Transform2Neg($M, b \Leftrightarrow C$)

(g) If C contains a possibly unsafe expression of the form exp'/exp where exp is provably safe then replace $b \Leftrightarrow C$ with

$$\exists a'{:}\texttt{nz}.\ \exists b'{:}\texttt{bool}.\ \exists b''{:}\texttt{bool}.\ \begin{array}{l} b' \Leftrightarrow (a' = exp) \wedge \\ b'' \Leftrightarrow (exp \neq 0) \wedge \\ b \Leftrightarrow (\neg b'' \vee (b' \wedge C\{exp \mapsto a'\})) \wedge \\ b'' \rightarrow b' \end{array}$$

(h) If C contains a possibly unsafe expression of the form $AR[exp]$, where AR is an expression of type $\texttt{array}$ [l..u] of τ and exp is provably safe, then replace $b \Leftrightarrow C$ with

$$\exists a'{:}l..u.\ \exists b'{:}\texttt{bool}.\ \exists b''{:}\texttt{bool}.\ \begin{array}{l} b' \Leftrightarrow (a' = exp) \wedge \\ b'' \Leftrightarrow (l \leq exp \wedge exp \leq u) \\ b \Leftrightarrow (\neg b'' \vee (b' \wedge C\{exp \mapsto a'\})) \wedge \\ b'' \rightarrow b' \end{array}$$

Fig. 5. Transform2Kleene

5 Solving the Transformed Models

The transformations defined in the previous section create models that are safe; undefinedness cannot occur. we can now replace $\Leftrightarrow$ by $\leftrightarrow$ and $\texttt{boolToInt}_0$ by $\texttt{boolToInt}$ since these operators are identical in two-valued logic. The models are still not directly executable in a constraint solver which takes a set of finite-domain variables and conjunction of constraints on these variables. In order to create such a final form we need to map the resulting model in $\mathcal{E}$ further, principally unrolling loops and flattening. Since the models are safe the existing mappings should respect the semantics of the model.

The reader may be concerned that the transforms introduce new variables with the infinite domain nz. This is unproblematic since it can be shown that if a search assigns values to all the finite-domain variables, then the value of each infinite-domain variable either becomes irrelevant to determining satisfiability or is fixed by propagation (even using simple propagators). As an example, consider the variable a' introduced by line (c) of Transform2Pos. If search fixes the value of exp to 0 and then propagation on $b' \Leftrightarrow a' = exp$ fixes b' to F and propagation on $b' \Leftrightarrow a' = exp$ can fix b to F without knowing the value of a'. On the other hand, if search fixes exp to a value other than 0, then propagation on $exp \neq 0 \rightarrow b'$ fixes b' to T and propagation on $b \Leftrightarrow (a' = exp)$ forces a' to be fixed to the same value as exp.

It can be shown that with only weak assumptions about propagators, none of the transformations weaken propagation. As an example, consider the model:

```
find        x:int(1..6)
such that   ¬(12/x ≥ 4)
```

Assuming this model is implemented by the constraint $div(12, x, t) \wedge (b \leftrightarrow t \geq 4) \wedge \neg b$, then enforcing domain consistency results in the domains: $t : 2..3$, $x : 4..6$.

The model is safe so there is no need to transform it. However if we did transform it for the relational semantics then the result, after converting existential variables to decision variables, would be

```
find        x:int(1..6), a':nz, b:bool, b':bool
such that   b' ⇔ (a' = x) ∧ b ⇔ (b' ∧ (12/a' ≥ 4)) ∧ (x ≠ 0 → b') ∧ ¬b
```

Assuming this model is implemented by the constraint $(b \leftrightarrow b' \wedge b_2) \wedge (b' \leftrightarrow a' = x) \wedge (b_2 \leftrightarrow t \geq 4) \wedge div(12, a', t) \wedge (b_3 \rightarrow b') \wedge (b_3 \leftrightarrow x \neq 0) \wedge \neg b$ then enforcing domain consistency results in the domains: $b = \mathtt{F}$, $b_3 = \mathtt{T}$, $b' = \mathtt{T}$, $b_2 = \mathtt{F}$, $t : 2..3$, $a' : 4..6$, and $x : 4..6$.

6 Conclusion

As modelling languages become more expressive, it becomes more likely that a modeller creates models where undefinedness occurs. A clear understanding of how undefinedness is treated by a modelling language is vital to both the modeller and the system's implementer. For the modeller misunderstanding may result in modelling errors giving the wrong results (including incorrect optimal values) when undefinedness is silently translated to failure. For the systems builder a great deal of care must be taken to ensure that transformations and optimizations of the systems do not change the meaning of the model.

Fig. 1 shows that without a clear understanding of undefinedness implementers have struggled to implement a proper treatment of undefinedness. Our work shows how undefinedness can be properly treated in a simple constraint language and we believe that this approach can be extended easily to handle richer languages. Already our work has informed the development of MiniZinc and generated a bug report for ECLiPSe. We expect our work to result in bug reports in additional languages and to alter and inform the development of other languages such as ESSENCE′. Finally, other subtle issues are likely to arise in the development of highly-expressive modelling languages; our work suggests that the use of semantics could be valuable in resolving those issues.

References

1. Apt, K.R., Wallace, M.: Constraint Logic Programming Using ECLiPSe. Cambridge University Press, Cambridge (2006)
2. SWI Prolog (2009), `http://www.swi-prolog.org/`
3. Intelligent Systems Laboratory: SICStus Prolog. Swedish Institute of Computer Science (2009), `http://www.sics.se/isl/sicstuswww/site/index.html`
4. Van Hentenrcyk, P.: The OPL Optimization Programming Language. MIT Press, Cambridge (1999)

5. Nethercote, N., Stuckey, P., Becket, R., Brand, S., Duck, G., Tack, G.: Minizinc: Towards a standard CP modelling language. In: Bessière, C. (ed.) CP 2007. LNCS, vol. 4741, pp. 529–543. Springer, Heidelberg (2007)
6. Marriott, K., Nethercote, N., Rafeh, R., Stuckey, P.J., Garcia de la Banda, M., Wallace, M.: The design of the Zinc modelling language. Constraints 13, 229–267 (2008)
7. Frisch, A.M., Harvey, W., Jefferson, C., Hernández, B.M., Miguel, I.: Essence: A constraint language for specifying combinatorial problems. Constraints 13, 268–306 (2008)
8. Frisch, A.M., Grum, M., Jefferson, C., Martínez Hernández, B., Miguel, I.: The essence of ESSENCE: A language for specifying combinatorial problems. In: Proc. of the 4th Int. Workshop on Modelling and Reformulating Constraint Satisfaction Problems, pp. 73–88 (2005)
9. Kleene, S.C.: Introduction to Metamathematics. Van Nostrand, New York (1952)

Search Spaces for Min-Perturbation Repair

Alex S. Fukunaga*

Global Edge Institute, Tokyo Institute of Technology, Meguro, Tokyo, Japan
fukunaga@is.titech.ac.jp

Abstract. Many problems require minimally perturbing an initial state in order to repair some violated constraints. We consider two search spaces for exactly solving this minimal perturbation repair problem: a standard, difference-based search space, and a new, commitment-based search space. Empirical results with exact search algorithms for a min-cost virtual machine reassignment problem, a minimal perturbation repair problem related to server consolidation in data centers, show that the commitment-based search space can be significantly more efficient.

1 Introduction

Most research on constraint satisfaction and optimization focus on generating solutions from scratch – given a set of variables and constraints, generate an (optimal) assignment that satisfies the constraints. In practice, there are many situations where it is necessary to find solutions to constraint satisfaction problems that are as close as possible to a given, initial state. One example is related to *server consolidation*, the use of virtual machines to consolidate multiple servers onto fewer servers [1]. There is currently great interest in server consolidation due to opportunities for improved energy efficiency and cost reduction. Server consolidation can be modeled as a bin packing problem [2]. Consider a set of n virtual machines (VMs), where each VM has a *weight* (representing resource demand). Given m physical servers, each with a some *capacity* (representing aggregate CPU/RAM resources) the server consolidation problem is the problem of assigning the n VMs to the m physical servers, such that each VM is assigned to exactly one physical server, and for every physical server, the sum of the assigned VM demands is within the capacity of the physical server.

Suppose that after the initial server assignments are made, the resource requirements of the VMs deviate from the original forecasts, resulting in some servers being overloaded. VMs can be reassigned among the servers in order to rebalance the loads. However, migration of a VM between servers incurs costs (e.g., system administration costs, possible downtime for a service). The *Min-Cost Virtual Machine Reassignment Problem* (VMRP) seeks a new assignment of VMs to servers such that no server is overloaded, and the number of jobs that are moved from their initial assignment is minimized. Heuristics for this problem were investigated in [3]. An approximation for a similar problem was considered in [4]. Similar problems arise for process migration in distributed systems.

* This research was supported by JSPS, JST, and the Okawa Foundation.

I.P. Gent (Ed.): CP 2009, LNCS 5732, pp. 383–390, 2009.

Another scenario where a solution similar to a given initial state is desired occurs in staff scheduling. Employees express preferences regarding when they want to work, but their preferences must be balanced against the staffing demands and constraints of the business, requiring a schedule that satisfies staffing requirements while deviating minimally from employee preferences.

This general class of *Min-Perturbation Repair Problems* (MPRP) seeks to *repair* an initial assignment of values to variables (i.e., find a conflict-free solution), with minimal *cost*, where cost is the number of differences between a candidate solution and the initial assignment. While a standard CSP seeks a solution which does not violate any constraints, the MPRP imposes the additional goal of minimizing the distance from an initial assignment of values to variables.

This paper considers search algorithms for the MPRP, focusing on alternative search spaces. In Sect. 2, we consider a standard, difference based search space which has been used in previous work on the MPRP, as well as a new, commitment-based search space. Using the VMRP as a case study, we experimentally evaluate these search spaces (Sect. 3). We discuss related work in Sect. 4, and conclude in Sect. 5

2 Search Spaces for Minimal Perturbation Repair

Given an initial variable assignment $I = \{x_1 = v_1, ..., x_n = v_n\}$, let D_i be the set of states which have exactly i variables whose value are different from that of the initial state I. We call the set $D = D_1 \cup ...D_n$ the *difference space*, or *D-space*. The root node of this search space is I. Nodes at depth d of the search tree contain variable assignments which differ by d assignments from I. Each edge in the tree changes the value of one variable which has not yet been changed by any ancestor. A standard depth-first branch-and-bound (DFBNB) can be applied to explore this search space. Problem-specific pruning techniques, such as those described in Sec 3, are applied at each node.

Instead of a depth-first search strategy, we can also use a best-first search strategy, such as Iterative-Deepening A* (IDA*) [5], which expands nodes in a best-first order using linear space (at the cost of reopening some nodes). The admissible heuristic function, h, used by IDA* is the same as the lower bounding function used for DFBNB, and the d-th iteration of IDA* explores the subset of the DFBNB D-space search tree where at each node, the sum $f = g + h \leq d$, where g is the number of differences from the initial state in the current solution, and h is the lower bound on the additional number of differences required to find a conflict-free solution. Ran et al [6] applied iterative-deepening in D-space to solve a minimal perturbation problem for binary CSPs.

We now introduce *commitment*, a useful concept for MPRP search algorithms. A variable x is *committed* to value v at node N if x is assigned to v at N and every descendant of N, and *uncommitted* otherwise. For variables $x_1, ..., x_n$, we denote a search state as $S = \{x_1 = v_1, ..., x_n = v_n\}$, or more concisely, $\{v_1, ..., v_n\}$. Furthermore, the values are underlined if the variable is committed to that value. In a 2-variable MPRP where the current assignments are $x_1 = 1, x_2 = 2$, and we have committed $x_1 = 1$, we can denote this state as $\{\underline{x_1 = 1}, x_2 = 2\}$, or more concisely, $\{\underline{1}, 2\}$. At the root node of this search space, the variables are assigned the values of the initial assignment I, and all variables are uncommitted. Furthermore, we annotate a value to be different from the initial state I with an asterisk (*). Thus, $\{\underline{1}, \underline{3}^*, 2\}$ denotes a state where x_1 is committed to value 1, x_2 to 3, and x_3 is uncommitted (but assigned to 2), and where the value of x_2 differs from the value of x_2 in the initial state. Fig. 1 shows a search tree for the VMRP. The sibling nodes are ordered according to a variable ordering implemented in the search algorithm (lex order in Fig. 1).

One issue with the D-space search tree is symmetry, e.g., given two variables x_1 and x_2, assigning $x_1 = 2$ first, followed by $x_2 = 1$, is symmetric to assigning $x_2 = 1$, then $x_1 = 2$. One approach to eliminating such symmetries is a standard nogood-based approach [6]. We use a different approach, which eliminates symmetries by asserting commitments on the siblings of a node. For example, in Fig. 1, node B is the result of starting with the initial state A and moving the VM with weight 20 from $S_{c=10}$, the server with capacity 10, to $S_{c=25}$, the server with capacity 25. For all of the siblings of node B, we commit this VM to $S_{c=10}$ (i.e., forbid moving it to $S_{c=25}$). This is equivalent to saying that assigning the VM with weight 20 to $S_{c=10}$ is nogood for all siblings and descendants of B. Thus, a separate representation for nogoods is unnecessary. This method generalizes straightforwardly when there are more than 2 possible values.[1]

An explicit notion of commitment (as opposed to just assignment) in a MPRP is very useful for purposes other than symmetry elimination. For example, in a VMRP, if we have committed a VM with weight 8 to a server S with capacity 10, then all nodes which assign more than 2 additional units of demand to S can be pruned. Note that merely *assigning* a VM with weight 8 (for example, in the initial assignment I) does not allow the same pruning, because it is possible to move that VM out of S, allowing another VM with demand greater than 2 to be assigned to S. It is the *commitment* which allows us to prune. Similarly, note that this type of pruning is not captured by the nogoods used in [6]. Commitments also constrain the feasible domains for a variable, which has a significant impact on the effectiveness of variable ordering (we use most-constrained ordering).

A new, alternative search space for searching the space of commitments, rather than differences, is a *commitment-based search space* (C-space), where each node in the search tree represents a partially committed assignment of variables to values, and edges represent a commitment of a variable to some value. For each variable, we represent its current value, as well as whether a commitment has been made to the value. The root node of this search space is I. Nodes at depth d of the search tree contain variable assignments with d commitments.[2] A sample C-space search tree is shown in Fig. 1. Each edge represents a single commitment. While all edges in D-space have cost 1, some edges in C-space have cost 0 (e.g,. the edges $a \rightarrow b, c \rightarrow d, d \rightarrow f, f \rightarrow h$ in Fig. 1). As with D-space, C-space can be searched using either the DFBNB or IDA*.

---

1 Another class of symmetries, not handled by nogoods or commitments, arises with low-precision instances, e.g., multiple VMs with the same weight; this is future work.

2 In D-space, nodes at depth d can contain more than d commitments because of the commitments asserted for symmetry elimination.

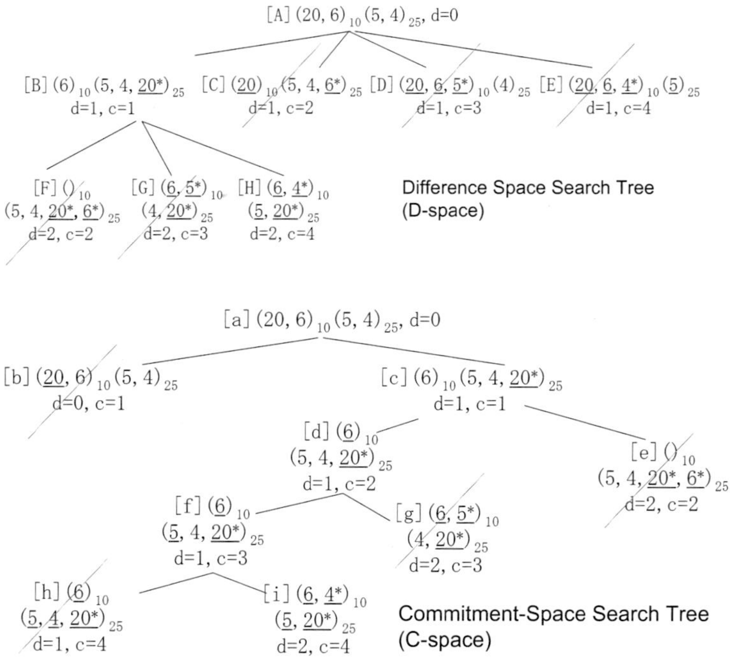

Fig. 1. Difference-Space and Commitment-Space search trees for VMRP with 2 servers ($c_1 = 10, c_2 = 25$), and 4 VMs ($w_1 = 20, w_2 = 6, w_3 = 5, w_r = 4$). The initial assignment is $I = \{1, 1, 2, 2\}$. For each node, $d = \#$ of perturbations from the initial assignment I, and $c = \#$ of commitments. Underlined values are committed, and an asterisk (*) indicates that the committed value is different from I. For example, node H in D-space has 4 committed variables, where 2 have values different from the initial assignment. Infeasible nodes are pruned (slash through node).

D-space can be seen as the subset of C-space where all committed decision variables are assigned a value different from the initial assignment. For example, given variables x_1, x_2 and initial assignment $I = \{1, 2\}$, the assignment $\{\underline{3}, 2\}$ is in D-space because v_1 is committed to a value that is different than in I, but $\{\underline{1}, 2\}$ is not in D-space because x_1 is committed to the same value as in I. Each node in D-space corresponds to a unique assignment of variables to values. In contrast, C-space has multiple nodes representing the same assignment of values to variables, except that the nodes have different commitments. For example, the search state $\{1, 2\}$ and $\{\underline{1}, 2\}$ represent the same variable assignments, but in the latter, a commitment has been made to assign x_1 to 1 for all of its descendants, whereas the former has not made any commitments.

In a problem with n variables with a domain of m possible values, the search tree for D-space contains $T_D = m^n$ nodes, because there is a 1-to-1 correspondence between the unique assignments of variables to values and the nodes in the D-space search tree (after elimination of symmetric nodes), and there are m^n unique assignments. This tree does not have a regular branching factor; the first level below the root node consists of the $n(m-1)$ assignments which differ from the initial assignment by exactly 1.

The root node of the C-space search tree is the initial assignment, and at each tree depth, we commit a variable to one of m values (one of the m values is the initial value in I). Thus, the C-space search tree has size $T_C = 1+m+m^2+...m^n$, and by straightforward manipulations, $T_C = (m^{n+1}-1)/(m-1)$. Comparing T_D and T_C, T_C/T_D approaches 2 in the worst case (when $m=2$).

Despite the redundancy in C-space compared to D-space, C-space has a lower branching factor (m) compared to D-space (depends on node; $n(m-1)$ at first level of the search tree). Given a comparable number of nodes, and the same set of pruning techniques, a narrower, deeper tree is easier to search than a wider tree, because a successful pruning in the narrower tree tends to prune more nodes than a successful pruning at the same depth in the wider tree. Thus, the main, potential advantage of C-space is in reorganizing the search tree structure from the relatively "top-heavy" D-space tree to a relatively narrow tree with branching factor m, at the cost of some redundancy.

3 Experimental Comparison of MPRP Algorithms for the VMRP

We evaluated search algorithms for the VMRP based on the search strategies described above, enhanced with the following VMRP-specific bounds.

A server is *oversubscribed* if the sum of the VM weights assigned to it exceeds its capacity. A lower bound LB_O (for pruning in DFBNB and for the admissible heuristic h in IDA*) is computed as follows: For each oversubscribed server S, sort the uncommitted VMs assigned to S in non-decreasing order, and count the number of VMs that must be removed from S in this order such that usage no longer exceeds capacity. For example, for the VMRP state $\{(5,6)(4,3)(\underline{10},1,2)\}$ where server capacity is 10, $LB_O = 3$. This is because either the VM with weight 5 or 6 must move from the first server, and the 1 and 2 must move from the third server (the VM with weight 10 is excluded from consideration because it is uncommitted). In addition, there are some bounds based on the wasted space in the servers, which we detailed in a workshop paper [7]. However, since those other VMRP-specific techniques affect all search strategies equally, and have less than a factor of 2 impact on runtime, we omit the details here due to space.

A common scenario for server consolidation in practice involves consolidating tens of services into fewer servers, where the target ratio of VMs to physical servers is commonly around 3-5 [8]. We generated solvable, random benchmarks based on this scenario as follows. For each server s_j, $1 \le j \le m$ (all servers with a capacity of 1000), VMs were uniformly generated in the range $[200, 400]$ and assigned to s_j until the remaining capacity was under 100. At that point, one 'filler' VM was generated, whose weight was constrained such that the slack (remaining capacity) in s_j was between 0 and 20. Minimizing the slack in this way increases the instance difficulty. Then all the VMs were removed from the servers, shuffled and reassigned to the servers in a round-robin manner, resulting in a solvable VMRP instance where each server has a balanced number of VMs, but some

Table 1. Virtual Machine Reassignment Problem results with varying # of servers (m) The *fail* column indicates # of instances (out of 30) not solved within the time limit (600 seconds/instance). The *time* and *nodes* columns show average runtime and nodes generated on the successful runs, excluding failed runs.

	Commitment Space (C-space)									D-space					
	DFBNB			IDA*			IDA*/B			IDA*			IDA*/NG		
m	**fail**	time	nodes	**fail**	time	nodes	**fail**	time	nodes	**fail**	time	nodes	**fail**	time	nodes
5	0	0.1	34481	0	0.05	10379	0	0.23	81431	0	0.17	42783	0	0.18	25770
6	0	1.9	5.0e5	0	0.4	7.2e4	0	3.3	1.1e6	0	3.0	7.6e5	0	4.54	5.6e5
8	4	119.9	2.0e7	0	32.4	5.3e6	**8**	157.0	5.4e7	**8**	104.7	2.9e7	**6**	52.6	6.6e6
10	**28**	112.5	1.5e7	**14**	150.6	2.1e7	**29**	251.3	1.1e8	**29**	156.3	5.2e7	**23**	145.8	13.5e7
12	**30**	-	-	**26**	267	2.8e7	**30**	-	-	**30**	-	-	**30**	-	-

servers are overloaded. We tested each of the search algorithm configurations on 30 instances with m varying from 5 to 12, with a time limit of 600 seconds per instance on a 3.0GHz Intel Core2 processor. We used a most-constrained variable ordering and a lexicographic value ordering for all algorithms (results using different variable and value ordering strategies were similar). In addition to combinations of C-space and D-space with DFBNB and IDA*, we also ran two additional configurations: (1) The C-space+IDA*/B configuration is similar to C-space+IDA*, except that all pruning is disabled for nodes which have the same variable *assignment* as their parents (i.e., the "extra" nodes resulting from cost 0 edges in C-space which are not present in D-space, such as nodes b, d, f, h in Fig. 1). At all other nodes, pruning is enabled as in C-space+IDA*. This tests how pruning at these "extra" nodes impacts C-Space+IDA*. (2) The D-space+IDA*/NG configuration is similar to D-space+IDA*, except that this algorithm is based on the iterative deepening algorithm in [6], which, on the d-th iteration, enumerates variable assignments with at most d differences compared to the initial assignment. However, it does not assert commitments, and nogoods are used for symmetry pruning. While this searches the same space of variable assignments as D-space, the lack of commitments means that variable ordering and bound-based pruning techniques are less effective.

Table 1 shows the results. The *fail* column indicates the number of instances (out of 30) that were not solved within the time limit. The *time* and *nodes* columns show average time spent and nodes generated on the successful runs, excluding the failed runs.

As shown in Table 1, C-space+IDA* significantly outperformed the other algorithms. D-space+DFBNB is not shown due to space, but performed significantly worse than D-space+IDA* and C-space+DFBNB. C-space+IDA*/B performs significantly worse than C-space+IDA*, and is comparable with D-space+IDA*, indicating that in fact, pruning at the "extra", inner nodes in C-space plays a significant role in enhancing the performance relative to D-space. The poor performance of D-space+IDA*/NG, which does not assert any commitments, shows the importance of exploiting commitments for pruning and variable ordering. Similar results were obtained with VM sizes in the range [100, 300].

4 Related Work

Several techniques for solving CSPs and combinatorial optimization problems which apply some backtracking strategy to a nonempty variable assignment (similar to our algorithms) have been previously proposed, including [9,10]. Although these techniques use a partial or complete variable assignment to guide the search for a solution, they do not have an explicit goal or mechanism for ensuring a *minimal* count of perturbations from a particular initial assignment.

Previous work has explicitly addressed minimal perturbation. Ran, et al. proposed iterative-deepening in D-space for binary CSPs [6], with symmetry pruning based on nogoods. El Sakkout and Wallace [11] considered a minimal cost repair problem for scheduling. They consider difference functions that can be expressed linearly - the MPRP objective of minimizing difference count is excluded ([11],p.368). Their probe backtracking algorithm does not explicitly consider the initial schedule, and reschedules from scratch. Barták et al. investigated over-constrained CSPs for which there is likely to be no feasible solution without violated constraints [12], and studied methods to seek a maximal assignment of consistent variables which also differs minimally from an initial state. They also studied an iterative repair (local search) algorithm biased to seek minimal perturbation solutions for course timetabling [13].

IDA* in C-space is related to Limited Discrepancy search (LDS) [14] and its variants, as both algorithms search a space which is characterized by some notion of "discrepancy". LDS can be viewed as a best-first search, where the cost of a node is the number of discrepancies (from the first value returned by a value ordering heuristic) [15]. Let δ be the class of all value ordering heuristics where the first value returned by the ordering is the value in the initial state. Using some value ordering from δ, we can implement LDS in C-space which is similar to IDA* in C-space. The differences are: (1) On the d-th iteration, LDS explores nodes with up to d discrepancies, while IDA* searches nodes with cost estimate (Sect. 2) $f \leq d$. The reason for this difference is that the goal of LDS to find a solution for a standard CSP – it is not explicitly trying to find a min-perturbation solution. The lack of a nontrivial bound/heuristic means that LDS performs even less pruning than C-space+IDA*/B (which at least prunes at all non-redundant nodes) (2) LDS specifies a specific *value ordering strategy*, i.e., a policy from the class δ, whereas IDA* (as well as DFBNB) does not specify a particular value ordering among sibling nodes (our VMRP experiments used lexical ordering). Thus, LDS, when applied directly to the MPRP, is a special case of C-space IDA* with a trivial lower bound ($h = 0$) and a value ordering heuristic from δ. On the VMRP, we found that all the C-space algorithms in Table 1 significantly outperforms LDS (more than an order of magnitude).

5 Conclusions

We investigated exact search algorithms for repairing constraint violations with a minimal number of perturbations from an initial state. Starting with a straightforward difference-based search space which enumerates differences from the initial assignment, we showed that introducing an explicit notion of commitments to values allows elimination of symmetries (subsuming nogoods), as well as domain-specific pruning. We then propose a commitment space, which reorganizes the search tree into a regular, narrow structure at the cost of some redundancy. Experimental results for the Virtual Machine Reassignment Problem, a min-perturbation variant of bin packing, show that search in C-space is significantly more efficient than search in D-space, and that the combination of C-space with IDA* result in the best performance for the VMRP. Although we used the VMRP as an example, C-space and D-space are general notions for MPRP problems, and future work will investigate additional MPRP domains. Some preliminary results for an antenna-scheduling related domain are in [7].

References

1. Vogels, W.: Beyond server consolidation. ACM Queue 6(1) (2008)
2. Gupta, R., Bose, S., Sundarrajan, S., Chebiyam, M., Chakrabarti, A.: A two-stage heuristic algorithm for solving the server consolidation problem. In: IEEE Int. Conf. on Services Computing (2008)
3. Ajiro, Y.: Recombining virtual machines to autonomically adapt to load changes. In: Proc. 22nd Conf. of the Japanese Society for Artificial Intelligence (2008)
4. Aggarwal, G., Motwani, R., Zhu, A.: The load rebalancing problem. In: Proc. 15th ACM Symp. on parallel algorithms and architectures, pp. 258–265 (2003)
5. Korf, R.: Depth-first iterative-deepening: an optimal admissible tree search. Artificial Intelligence 27(1), 97–109 (1985)
6. Ran, Y., Roos, N., van den Herik, H.: Approaches to find a near-minimal change solution for dynamic CSPs. In: Proc. CPAIOR, pp. 378–387 (2002)
7. Fukunaga, A.: Search algorithms for minimal cost repair problems. In: Proc. CP/ICAPS 2008 Workshop on Constraint Satisfaction Techniques for Planning and Scheduling Problems (2008)
8. Ajiro, Y.: NEC System Platform Research Labs. Personal Communication (2009)
9. Beck, C.: Solution-guided multi-point constructive search for job shop scheduling. Journal of Artificial Intelligence Research 29, 49–77 (2007)
10. Verfaillie, G., Schiex, T.: Solution reuse in dynamic constraint satisfaction problems. In: Proc. AAAI, Seattle, Washington, pp. 307–312 (1994)
11. El-Sakkout, H., Wallace, M.: Probe backtrack search for minimal perturbation in dynamic scheduling. Constraints 5, 359–388 (2000)
12. Barták, R., Müller, T., Rudová, H.: A new approach to modeling and solving minimal perturbation problems. In: Apt, K.R., Fages, F., Rossi, F., Szeredi, P., Váncza, J. (eds.) CSCLP 2003. LNCS (LNAI), vol. 3010, pp. 233–249. Springer, Heidelberg (2004)
13. Müller, T., Rudová, H., Barták, R.: Minimal perturbation in course timetabling. In: Burke, E.K., Trick, M.A. (eds.) PATAT 2004. LNCS, vol. 3616, pp. 126–146. Springer, Heidelberg (2005)
14. Harvey, W., Ginsberg, M.: Limited discrepancy search. In: Proc. IJCAI, pp. 607–615 (1995)
15. Korf, R.: Improved limited discrepancy search. In: Proc. AAAI, pp. 286–291 (1996)

Snake Lex: An Alternative to Double Lex

Andrew Grayland[1], Ian Miguel[1], and Colva M. Roney-Dougal[2]

[1] School of Comp. Sci., St Andrews, UK
{andyg,ianm}@cs.st-and.ac.uk
[2] School of Maths and Stats, St Andrews, UK
colva@mcs.st-and.ac.uk

Abstract. Complete row and column symmetry breaking in constraint models using the lex leader method is generally prohibitively costly. Double lex, which is derived from lex leader, is commonly used in practice as an incomplete symmetry-breaking method for row and column symmetries. Double lex is based on a *row-wise* canonical variable ordering. However, this choice is arbitrary. We investigate other canonical orderings and focus on one in particular: *snake ordering*. From this we derive a corresponding incomplete set of symmetry breaking constraints, *snake lex*. Experimental data comparing double lex and snake lex shows that snake lex is substantially better than double lex in many cases.

1 Introduction

A *variable symmetry* in a constraint model is a bijective mapping from the set of variables to itself that maps (non-)solutions to (non-)solutions. The set of (non-)solutions reachable by applying all symmetry mappings to one (non-)solution forms an *equivalence class*. Restricting search to one member (or a reduced set of members) of each equivalence class can dramatically reduce systematic search: *symmetry breaking*. The *lex leader* method for breaking variable symmetries adds a constraint per symmetry so as to allow only one member of each equivalence class [3]. The lex leader method can produce a huge number of constraints and sometimes adding them to a model can prove counterproductive. One commonly-occurring case is when trying to break symmetries in a matrix, where any row (or column) can be permuted with any other row (or column): *row and column symmetries* [4].

It is often possible to achieve good results by breaking a smaller set of symmetries: *incomplete symmetry breaking*. One method to do this for row and column symmetries is *double lex* [4], which constrains adjacent pairs of rows and adjacent pairs of columns. Although this method is incomplete, it can reduce search dramatically. Double lex is derived from a reduction of a complete set of lex constraints created with a *row-wise* ordering as the canonical member of each equivalence class. The use of row-wise ordering is, however, arbitrary. The possible benefits of varying canonical orderings of lex constraints in graceful graph models was investigated in [11]. This paper investigates other canonical orderings and selects one (*snake ordering*) that looks promising to investigate further. From it, we create a new set of incomplete symmetry breaking constraints (*snake lex*). An empirical analysis demonstrates that snake lex can often deliver substantially better results than double lex.

I.P. Gent (Ed.): CP 2009, LNCS 5732, pp. 391–399, 2009.

2 Background

A constraint satisfaction problem is a triple $(\mathcal{X}, D, \mathcal{C})$, where $\mathcal{X}$ is a finite set of variables. Each $x \in \mathcal{X}$ has a finite set of values $D(x)$ (its *domain*). The finite set $\mathcal{C}$ consists of constraints on the variables. Each $c \in \mathcal{C}$ is defined over a sequence, $\mathcal{X}'$, of variables in $\mathcal{X}$. A subset of the Cartesian product of the domains of the members of $\mathcal{X}'$ gives the set of allowed value combinations. A *complete assignment* maps every variable to a member of its domain. A complete assignment satisfying all constraints is a *solution*. A *variable symmetry* of a CSP is a bijection $f : \mathcal{X} \to \mathcal{X}$ such that $\{\langle x_i, a_i\rangle \ : x_i \in \mathcal{X}, a_i \in D(x_i)\}$ is a solution if and only if $\{\langle f(x_i), a_i\rangle \ : \ x_i \in \mathcal{X}, a_i \in D(f(x_i))\}$ is a solution. The set of all variable symmetries of a CSP is closed under composition of functions and inversion, and so forms a group, the *variable symmetry group* of the CSP.

Row and column symmetries appear commonly in CSP models that contain matrices [2,7,8,9]. When it is possible to map any ordered list of distinct rows to any other such list of the same length, with the same being true for columns, then there is *complete* row and column symmetry. For an n by m matrix there are $n! \times m!$ symmetries.

For a symmetry group of size s the lex leader method produces a set of $s - 1$ lex constraints to provide complete symmetry breaking. We first decide on a canonical order for the variables in $\mathcal{X}$, then post constraints such that this ordering is less than or equal to the permutation of the ordering by each of the symmetries. Consider the following 2×2 matrix with complete row and column symmetry, where $x_i \in \mathcal{X}$:

$$\begin{matrix} x_{11} & x_{12} \\ x_{21} & x_{22} \end{matrix}$$

If we choose a row-wise canonical variable ordering, in this case $x_{11}x_{12}x_{21}x_{22}$, then we can generate the following 3 lex constraints to break all the symmetries.

row swap:	$x_{11}x_{12}x_{21}x_{22} \leq_{\text{lex}} x_{21}x_{22}x_{11}x_{12}$
column swap:	$x_{11}x_{12}x_{21}x_{22} \leq_{\text{lex}} x_{12}x_{11}x_{22}x_{21}$
both swapped:	$x_{11}x_{12}x_{21}x_{22} \leq_{\text{lex}} x_{22}x_{21}x_{12}x_{11}$

Although this example is trivial, breaking all row and column symmetries by adding lex constraints is generally counter-productive since we have to add $(n! \times m!) - 1$ symmetry breaking constraints to the model. Double lex [4] is a commonly used incomplete symmetry breaking method for row and column symmetries. This method involves ordering the rows of a matrix and (independently) ordering the columns. This produces only $n + m - 2$ symmetry breaking constraints, shown below for the 2×2 example:

$$x_{11}x_{12} \leq_{\text{lex}} x_{21}x_{22}$$
$$x_{11}x_{21} \leq_{\text{lex}} x_{12}x_{22}$$

The double lex constraints can be derived from the lex leader generated based upon a row-wise canonical variable ordering. One method of doing so is to use reduction rules 1, 2 and 3′ as given in [5] and [6]. Let α, β, γ, and δ be strings of variables, and x and y be individual variables. The reduction rules are:

1 If $\alpha = \gamma$ entails $x = y$ then a constraint c of the form $\alpha x \beta \leq_{\text{lex}} \gamma y \delta$ may be replaced with $\alpha\beta \leq_{\text{lex}} \gamma\delta$.
2 Let $C = C' \cup \{\alpha x \leq_{\text{lex}} \gamma y\}$ be a set of constraints. If $C' \cup \{\alpha = \gamma\}$ entails $x \leq y$, then C can be replaced with $C' \cup \{\alpha \leq_{\text{lex}} \gamma\}$.

3′ Let $C = C' \cup \{\alpha x \beta \leq_{\text{lex}} \gamma y \delta\}$ be a set of constraints. If $C' \cup \{\alpha = \gamma\}$ entails $x = y$, then C can be replaced with $C' \cup \{\alpha\beta \leq_{\text{lex}} \gamma\delta\}$.

Rule 1 is subsumed by Rule 3′, but is often useful as a preprocess, as it reasons over an individual constraint. Algorithms to implement these rules are described in [6,10].

3 In Search of an Alternative Canonical Ordering

Double lex performs well [4], but we are not aware of work exploring similar-sized sets of incomplete symmetry-breaking constraints derived from other canonical variable orderings. In order to identify other useful canonical orderings we examine a large number of small cases and test interesting candidates on benchmark problems. A lex constraint of the form $x_1 x_2 \dots x_m \leq_{\text{lex}} y_1 y_2 \dots y_m$ consists of m *pairs* of variables. We compare canonical orderings by counting the pairs remaining after reducing the entire set of corresponding lex leader constraints by Rules 1, 2 and 3'. We hypothesize that fewer remaining pairs will promote reduced search through more effective propagation.

We began by examining all 6! = 720 canonical variable orderings of a 2×3 matrix. The algorithms described in [10] and [6] were used to reduce the lex constraints, and the distribution of pairs remaining is shown in Fig. 1. The smallest number of pairs remaining after reduction was 15, which was obtained for 108 of the canonical orderings. The ordering $x_{11}x_{21}x_{22}x_{12}x_{13}x_{23}$ reduces to 15 pairs and is interesting because it has a regular form, and might therefore be expected to produce a regular set of incomplete symmetry-breaking constraints equivalent to double lex. Standard row-wise ordering, $x_{11}x_{12}x_{13}x_{21}x_{22}x_{23}$, resulted in 23 pairs, so by this measure is a poor ordering.

Fig. 1 also shows the results of a similar experiment for all 8! canonical orderings of the 2×4 matrix. The ordering $x_{11}x_{21}x_{22}x_{12}x_{13}x_{23}x_{24}x_{14}$, the natural extension to that identified in the 2×3 results, reduces to just 30 pairs (row-wise reduces to 109).

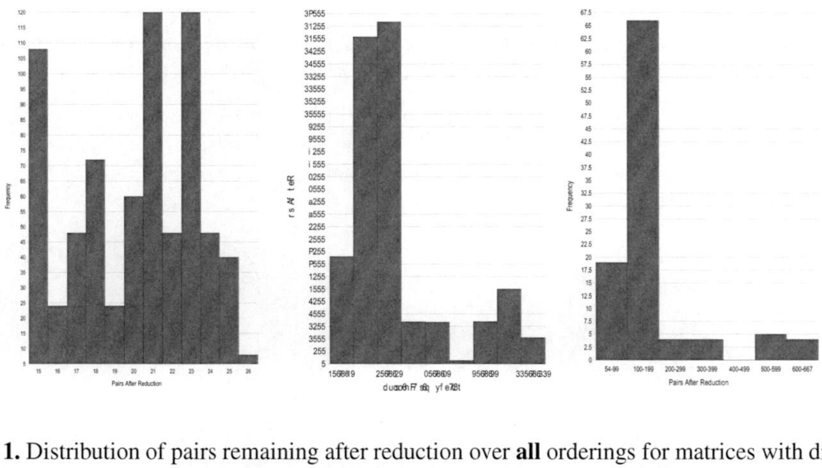

Fig. 1. Distribution of pairs remaining after reduction over **all** orderings for matrices with dimensions 2×3 (left) , 2×4 (centre). Distribution (right) of pairs remaining after reductions for a 2×5 matrix on 100 random, columnwise snake and row-wise orderings.

In general this ordering is simple to describe: columnwise the ordering starts at the top-left corner then moves down the column. It then moves to the neighbouring element in the next column and then upwards. This pattern continues until all variables have been listed. The row-wise snake variant is described similarly. We call this *snake ordering*.

We sampled 100 orderings of the 2×5 matrix, and also tested columnwise snake and row-wise orderings. Figure 1 summarises the results, in which snake ordering again proves to be one of the best measured by remaining pairs (reducing to 54 vs. 655 for row-wise). To gain an indication of the importance of matrix dimension we compared row-wise against snake ordering on a 3×3 matrix. In this case, the two orderings produced much more similar results (reduction to 88 pairs for snake, 92 for row-wise), suggesting that matrix dimension *is* significant. Nonetheless, our results suggest that snake ordering is worth investigating further.

4 Snake Lex

Recall that double lex is derived by reducing a complete set of lex constraints with a row-wise ordering. In this section, we derive a corresponding small, easily-described set of constraints for the snake ordering, called *snake lex*. First we give a formal definition of columnwise snake ordering. Row-wise snake ordering is defined similarly.

Definition 1. *Let* $\mathcal{X} = (x_{ij})_{n \times m}$ *be a matrix of variables. The* columnwise snake ordering *on variables is* $x_{11}, x_{21}, \ldots, x_{n1}, x_{n2}, \ldots, x_{12}, \ldots, x_{1m}, \ldots x_{nm}$, *if* m *is odd, and* $x_{11}, x_{21}, \ldots, x_{n1}, x_{n2}, \ldots, x_{12}, \ldots, x_{nm}, \ldots x_{1m}$ *if* m *is even. That is, snake order on variables starts at row 1, column 1. It goes down the first column to the bottom, then back* up *along the second column to the first row. It continues, alternating along the columns until all variables have been ordered.*

From the columnwise snake ordering, columnwise snake lex can be derived:

Definition 2. *Let* $\mathcal{X}(x_{ij})_{n \times m}$ *be a matrix of variables. The* columnwise snake lex *set of constraints,* $\mathcal{C}$*, is defined as follows.* $\mathcal{C}$ *contains* $2m-1$ *column constraints, beginning*

$$\begin{array}{ll} c_1 & x_{11}x_{21}\cdots x_{n1} \leq_{\text{lex}} x_{12}x_{22}\cdots x_{n2} \\ c_2 & x_{11}x_{21}\cdots x_{n1} \leq_{\text{lex}} x_{13}x_{23}\cdots x_{n3} \\ c_3 & x_{n2}x_{(n-1)2}\cdots x_{12} \leq_{\text{lex}} x_{n3}x_{(n-1)3}\cdots x_{13} \\ c_4 & x_{n2}x_{(n-1)2}\cdots x_{12} \leq_{\text{lex}} x_{n4}x_{(n-1)4}\cdots x_{14} \\ & \vdots \end{array}$$

and finishing with

$$c_{2m-1} \quad x_{1(m-1)}\cdots x_{n(m-1)} \leq_{\text{lex}} x_{1m}\cdots x_{nm}$$

if m *is odd and*

$$c_{2m-1} \quad x_{n(m-1)}\cdots x_{1(m-1)} \leq_{\text{lex}} x_{nm}\cdots x_{nm}$$

if m is even. $\mathcal{C}$ contains $n-1$ row constraints. If m is odd then these are

$$\begin{array}{ll} r_1 & x_{11}x_{22}x_{13}\dots x_{1m} \leq_{\text{lex}} x_{21}x_{12}x_{23}\dots x_{2m} \\ r_2 & x_{21}x_{32}x_{23}\dots x_{2m} \leq_{\text{lex}} x_{31}x_{22}x_{33}\dots x_{3m} \\ & \vdots \\ r_{n-1} & x_{(n-1)1}\dots x_{(n-1)m} \leq_{\text{lex}} x_{n1}\dots x_{nm}. \end{array}$$

If m is even then these are:

$$\begin{array}{ll} r_1 & x_{11}x_{22}x_{13}\dots x_{2m} \leq_{\text{lex}} x_{21}x_{12}x_{23}\dots x_{1m} \\ r_2 & x_{21}x_{32}x_{23}\dots x_{3m} \leq_{\text{lex}} x_{31}x_{22}x_{33}\dots x_{2m} \\ & \vdots \\ r_{n-1} & x_{(n-1)1}\dots x_{nm} \leq_{\text{lex}} x_{n1}\dots x_{(n-1)m}. \end{array}$$

The following theorem shows that columnwise snake lex is derived from the columnwise snake lex leader, and is therefore sound.

Theorem 1. *The columnwise snake lex constraints are sound.*

PROOF. We show that each constraint can be derived from a constraint in the full set of lex leader constraints by applying Rule 1 and then using only a prefix.

In each case the left hand side of the unreduced lex leader constraint is

$$x_{11}x_{21}\dots x_{n1}x_{n2}x_{(n-1)2}\dots x_{12}x_{13}\dots x_{n3}x_{n4}\dots$$

We first consider the column constraints, c_k. First let $k \equiv 1 \bmod 4$ and $a = (k+1)/2$. The symmetry which swaps columns a and $a+1$ and fixes everything else gives

$$Ax_{1a}x_{2a}\dots x_{na}B \leq_{\text{lex}} Cx_{1(a+1)}x_{2(a+1)}\dots x_{n(a+1)}D,$$

where A, B, C and D are strings of variables and $A = C$. Rule 1 removes A and C, and then considering only the first n pairs gives constraint c_k. If $k \equiv 2 \bmod 4$ then let $a = k/2$, replace $a+1$ by $a+2$ on the right hand side, and apply the same argument.

Next let $k \equiv 3 \bmod 4$ and $a = (k+1)/2$. The symmetry which swaps columns a and $a+1$ and fixes everything else gives a constraint of the form

$$\begin{aligned} Ax_{na}x_{(n-1)a}\dots x_{1a}B \leq \\ Cx_{n(a+1)}x_{(n-1)(a+1)}\dots x_{1(a+1)}D, \end{aligned}$$

where A, B, C and D are strings of variables and $A = C$. Again, using Rule 1 on A and C, and then taking only a prefix gives constraint c_k. If $k \equiv 0 \bmod 4$ then let $a = k/2$, replace $a+1$ by $a+2$ on the right hand side, and apply the same argument. Consider now the rows. There is a symmetry that interchanges rows a and $a+1$ and fixes everything else. The unreduced lex leader constraint for this symmetry is:

$$\begin{aligned} x_{11}\dots x_{a1}x_{(a+1)1}\dots x_{n1}x_{n2}\dots x_{(a+1)2}x_{a2}\dots \leq_{\text{lex}} \\ x_{11}\dots x_{(a+1)1}x_{a1}\dots x_{n1}x_{n2}\dots x_{a2}x_{(a+1)2}\dots \end{aligned}$$

Rule 1 deletes all pairs of the form (x_{ij}, x_{ij}) to obtain:

$$\begin{aligned} x_{a1}x_{(a+1)1}x_{(a+1)2}x_{a2}x_{a3}x_{(a+1)3}x_{(a+1)4}x_{a4}\dots \leq_{\text{lex}} \\ x_{(a+1)1}x_{a1}x_{a2}x_{(a+1)2}x_{(a+1)3}x_{a3}x_{a4}x_{(a+1)4}\dots \end{aligned}$$

Rule 1 simplifies $x_{ai}x_{(a+1)i} \leq_{\text{lex}} x_{(a+1)i}x_{ai}$ to $x_{ai} \leq x_{(a+1)i}$, and similarly for $x_{(a+1)i}x_{ai} \leq x_{ai}x_{(a+1)i}$, resulting in constraint r_k:

$$x_{a1}x_{(a+1)2}x_{a3}x_{(a+1)4} \cdots \leq_{\text{lex}} x_{(a+1)1}x_{a2}x_{(a+1)3}x_{a4} \cdots$$

Since each of the snake lex constraints is derived by first applying Rule 1 to a lex leader constraint and then taking only a prefix, the snake lex constraints are sound. □

There are similarities between the columnwise snake lex and double lex constraints on columns. Columnwise snake lex constrains the first column to be less than or equal to both the second and the third columns. It also constrains the *reverse* of the second column to be less than or equal to the *reverse* of the third and fourth columns. This pattern continues until the penultimate column is compared with the last. As an example, consider a 4×3 matrix. Double lex constrains adjacent columns (left), while snake lex produces the set of constraints on the columns on the right:

$$\begin{aligned} x_{11}x_{21}x_{31} &\leq_{\text{lex}} x_{12}x_{22}x_{32}, \\ x_{12}x_{22}x_{32} &\leq_{\text{lex}} x_{13}x_{23}x_{33}, \\ x_{13}x_{23}x_{33} &\leq_{\text{lex}} x_{14}x_{24}x_{34}. \end{aligned} \qquad \begin{aligned} x_{11}x_{21}x_{31} &\leq_{\text{lex}} x_{12}x_{22}x_{32}, \\ x_{11}x_{21}x_{31} &\leq_{\text{lex}} x_{13}x_{23}x_{33}, \\ x_{32}x_{22}x_{12} &\leq_{\text{lex}} x_{33}x_{23}x_{13}, \\ x_{32}x_{22}x_{12} &\leq_{\text{lex}} x_{34}x_{24}x_{14}, \\ x_{13}x_{23}x_{33} &\leq_{\text{lex}} x_{14}x_{24}x_{34}. \end{aligned}$$

Generally, given m columns and n rows, double lex adds $m-1$ constraints on columns, each with n pairs. Snake lex adds $2m-1$ constraints on columns, each with n pairs. We could increase the number of double lex constraints to the same number as snake lex by allowing each column to be less than or equal to the column two to its right, however *lex-chain*, which considers an entire set of rows or columns globally, has been shown to perform no better than double lex in practice[1].

We next consider the rows. Double lex gives the following for our sample matrix:

$$\begin{aligned} x_{11}x_{12}x_{13}x_{14} &\leq_{\text{lex}} x_{21}x_{22}x_{23}x_{24}, \\ x_{21}x_{22}x_{23}x_{24} &\leq_{\text{lex}} x_{31}x_{32}x_{33}x_{34}. \end{aligned}$$

The snake lex method is slightly more complicated, but gives the same number of constraints. We take the first two rows and zig zag between them to produce a string of variables starting at row 1, column 1, and a second string starting at row 2, column 1. We then constrain the first of these strings to be lexicographically less than or equal to the second one. Next we produce a similar constraint between rows i and $i+1$ for all i. The set of constraints for our 3×4 matrix are:

$$\begin{aligned} x_{11}x_{22}x_{13}x_{24} &\leq_{\text{lex}} x_{21}x_{12}x_{23}x_{14}, \\ x_{21}x_{32}x_{23}x_{34} &\leq_{\text{lex}} x_{31}x_{22}x_{33}x_{24}. \end{aligned}$$

Generally, double lex and snake lex both add $n-1$ row constraints, each with m pairs.

Thus far, we have considered columnwise snake ordering. We can also consider rowwise snake ordering, which may be useful if, for example, the rows are more heavily constrained (by the problem constraints) than the columns. To do so we simply *transpose* the matrix and then order as before. The transpose of our example 3×4 matrix is shown below (left), along with the corresponding constraints (right):

$$
\begin{array}{ccc}
 & & x_{11}x_{12}x_{13}x_{14} \leq_{\text{lex}} x_{21}x_{22}x_{23}x_{24}, \\
x_{11} & x_{21} & x_{31} \quad\quad x_{11}x_{12}x_{13}x_{14} \leq_{\text{lex}} x_{31}x_{32}x_{33}x_{34}, \\
x_{12} & x_{22} & x_{32} \quad\quad x_{24}x_{23}x_{22}x_{21} \leq_{\text{lex}} x_{34}x_{33}x_{32}x_{31}, \\
x_{13} & x_{23} & x_{33} \quad\quad x_{11}x_{22}x_{31} \leq_{\text{lex}} x_{12}x_{21}x_{32}, \\
x_{14} & x_{24} & x_{34} \quad\quad x_{12}x_{23}x_{32} \leq_{\text{lex}} x_{13}x_{22}x_{33}, \\
 & & x_{13}x_{24}x_{33} \leq_{\text{lex}} x_{14}x_{23}x_{34}.
\end{array}
$$

Note that the double lex constraints do not change for this transposition, hence double lex is insensitive to switching between a row-wise and columnwise search ordering.

5 Experimental Results

We used four benchmark problem classes to compare snake and double lex empirically. All models used exhibit row and column symmetry. Preliminary experimentation revealed the superiority of row-wise snake lex on the tested instances, which we therefore used throughout. This is correlated with the rows being significantly longer than the columns in 3 of 4 classes. For each class we carried out four experiments per instance. We tested double lex and snake lex, each with row-wise and then snake static variable heuristics, separating the effects of the search order from those of symmetry breaking. Each time given is the mean of five trials and all results are presented in Fig. 2. The solver used was MINION.

The first problem class studied is *balanced incomplete block design* (BIBD) [9], in which b blocks and v objects must be related such that: each block contains k objects, each object appears in r blocks, and each pair of objects must appear together in a block λ times. We use the standard model, a $v \times b$ 0/1 matrix, with columns summing to k, rows summing to r, and the scalar product of every pair of rows equal to λ. Note that the parameters v, k and λ of the BIBD fix the values of b and r. Results show that snake lex outperforms double lex in every tested case. Where it was possible to find all solutions, snake lex both reduces search and breaks more symmetry (finds fewer symmetrically-equivalent solutions). The single solution cases show a speed up over double lex of several orders of magnitude, possibly due to interaction with the λ constraint.

The second problem class is the *equidistant frequency permutation array* problem (EFPD) [8], which is to find a set of c codewords drawn from q symbols such that each symbol occurs λ times in each codeword and the hamming distance between every pair of codewords is d. Our model [8] is a $c \times q\lambda$ matrix of variables with domain $\{1, ..., d\}$. In order to give a fair comparison between the four combinations of search order and lex constraints, we picked six instances with no solutions. This means that we are *not* examining the case where we did best in the previous experiment, that of finding the first solution, but instead exhausting the search space. Solve time decreases by around 30% when the lex method is changed from double to snake lex, and then by a further 30% when the search order is changed to snake. Notice also that the search time *increases* when double lex is used in conjunction with the snake order, suggesting both that it is important for the search order to mirror the constraints, and that it is the snake lex constraints that are causing the improved solve times.

The third problem class is the *fixed length error correcting codes* (FLECC) [7] problem, which is to find d length-q codewords drawn from 4 symbols such that the *Lee*

BIBD:

Lex:	**Double**		**Snake**	
Search:	Row	Snake	Row	Snake
(v, k, λ)				
(7, 3, 5)	8.02,600557	8.38,606002	6.79,490784	6.78,455984
(7, 3, 6)	70.47,4979664	75.86,4321932	55.87,3984264	50.47,3448478
(7, 3, 20)	0.52,17235	0.47,15010	0.04,577	0.02,321
(7, 3, 30)	2.80,67040	2.53,60600	0.05,754	0.02,481
(7, 3, 40)	9.14,182970	8.58,168915	0.05,1063	0.03,641
(7, 3, 45)	16.09,278310	14.94,258860	0.06,1526	0.03,721
(7, 3, 50)	25.11,406525	23.70,380455	0.08,1671	0.05,801

EFPD:

Lex:	**Double**		**Snake**	
Search:	Row	Snake	Row	Snake
(q, λ, d, c)				
(3, 3, 5, 9)	4.1,164755	4.3,174677	3.6,120106	2.9,110666
(3, 4, 5, 9)	26.1,941061	27.1,1127011	15.9,537270	11.5,448329
(3, 5, 5, 10)	45.9,1556198	53.5,1841688	29.6,855950	20.0,678540
(3, 6, 5, 10)	68.9,2064850	76.8,2439205	39.7,1046091	27.4,811734
(3, 7, 5, 11)	94.5,2496190	103.0,2890581	54.5,1194583	35.0,910269
(3, 8, 5, 12)	124.6,2756291	123.4,3187617	71.0,1337286	43.1,1003119

FLECC:

Lex:	**Double**		**Snake**	
Search:	Row	Snake	Row	Snake
(q, d, c)				
(9, 5, 5)	23.34,662995	8.23,199720	13.09,293735	0.83,19817
(8, 6, 4)	0.73,5607	0.74,6122	0.53,6850	0.52,6359
(15, 5, 22)	0.77,15100	0.72,15136	0.41,8235	0.39,8205
(20, 5, 30)	3.28,51216	3.28,51223	1.31,19601	1.30,19603
(25, 5, 40)	3.88,49030	4.00,49002	1.44,17558	1.42,17600
(30, 5, 50)	4.56,49175	4.83,49112	1.86,17668	1.70,17750

Howell:

Lex:	**Double**		**Snake**		**Nodes**
Search:	Row	Snake	Row	Snake	All
(s, n)					
(4, 9)	0.98	0.97	1.19	1.23	60,799
(5, 11)	28.18	27.22	29.65	30.27	3,557,740
(6, 13)	1148.3	1153.8	1179.6	1196.2	231,731,046

Fig. 2. Double vs. row-wise snake lex, with both row and snake search ordering. BIBD & FLECC: searches above double line are for all solutions, remainder for one solution. Format: (time(s), nodes), except Howell, where nodes uniform throughout.

distance (see cited paper for definition, or CSPLib 36) between every pair of codewords is c. Our model [7] uses a $d \times q$ matrix, with each of q symbols appearing once in each row, and Lee distance c between each pair of rows. As with the BIBD test case both all solution and single solution problems were tested. Results show a speedup of almost two orders of magnitude in the all solution case. In the all solutions case, snake lex (with either order) requires on average less than half the time that the double lex with row-wise order uses, and the speedup seems to be increasing with instance size.

The final experiment involved solving the *Howell design* [2] problem. A Howell design is an s by s matrix M, whose coefficients are unordered pairs of symbols from a set $\mathcal{S}$ of size n. Each of the $n(n-1)/2$ pairs occurs at most once. Every row and every column of M contain at least one copy of every symbol from $\mathcal{S}$. There exists an *empty* symbol which can be used as many times as necessary, provided all other constraints are met. Three unsolvable instances were tested. Results show that, in this case, snake lex is slightly slower than double lex. One reason for this could be that Howell Designs are square (cf. Section 3). Note, however, that the nodes taken is the same throughout.

The extra time taken could therefore simply be due to the overhead incurred by snake lex adding a slightly larger number of constraints than double lex.

6 Discussion

This paper highlights the need for further investigation into incomplete row and column symmetry breaking. We have demonstrated that double lex can be outperformed by snake lex in many instances and indeed, there may be another variable ordering that can create a small set of constraints capable of beating both. Future research will investigate the production on the fly of tailored sets of lex constraints, depending on a variable ordering specified by a modeler. The initial direction of future work will be to examine the other lex leaders found to have a small number of pairs upon reduction comparing them to both double lex and the snake lex and to investigate whether the shape of the matrix affects solve times, particularly with square matrices.

Acknowledgements. A. Grayland is supported by a Microsoft Research EPSRC CASE studentship. I. Miguel is supported by a Royal Academy of Engineering/EPSRC Research Fellowship. C. M. Roney-Dougal is partially supported by EPSRC grant number EP/C523229/1.

References

1. Carlsson, M., Beldiceanu, N.: Arc-consistency for a chain of lexicographic ordering constraints. Technical Report T2002-18, SICS (2002)
2. Colbourn, C.J., Dinitz, J.H.: The CRC Handbook of Combinatorial Designs. CRC Press, Inc., Florida (1996)
3. Crawford, J., Ginsberg, M., Luks, E., Roy, A.: Symmetry-breaking predicates for search problems. In: Proc. KR, pp. 148–159 (1996)
4. Flener, P., Frisch, A.M., Hnich, B., Kiziltan, Z., Miguel, I., Pearson, J., Walsh, T.: Breaking row and column symmetries in matrix models. In: Van Hentenryck, P. (ed.) CP 2002. LNCS, vol. 2470, pp. 462–476. Springer, Heidelberg (2002)
5. Frisch, A.M., Harvey, W.: Constraints for breaking all row and column symmetries in a three-by-two-matrix. In: Proc. Symcon (2003)
6. Grayland, A., Jefferson, C., Miguel, I., Roney-Dougal, C.: Minimal ordering constraints for some families of variable symmetries. Annals Math. A. I. (to appear, 2009)
7. Hammons, A.R., Vijay Kumar, P., Calderbank, A.R., Sloane, N.J.A., Sol, P.: The z 4 -linearity of kerdock, preparata, goethals, and related codes. IEEE Trans. Inform. Theory 40(2), 301–319 (1994)
8. Huczynska, S.: Equidistant frequency permutation arrays and related constant composition codes. Circa Preprint 2009/2 (2009)
9. Meseguer, P., Torras, C.: Solving strategies for highly symmetric csps. In: IJCAI, pp. 400–405 (1999)
10. Öhrman, H.: Breaking symmetries in matrix models. MSc Thesis, Dept. Information Technology, Uppsala University (2005)
11. Smith, B.M.: Sets of Symmetry Breaking Constraints. In: Proceedings of SymCon 2005, the 5th International Workshop on Symmetry in Constraints (2005)

Closing the Open Shop: Contradicting Conventional Wisdom

Diarmuid Grimes[1], Emmanuel Hebrard[1], and Arnaud Malapert[2]

[1] Cork Constraint Computation Centre & University College Cork, Ireland
{d.grimes,e.hebrard}@4c.ucc.ie
[2] EMN/LINA UMR CNRS 6241, Cirrelt
arnaud.malapert@emn.fr

Abstract. This paper describes a new approach for solving disjunctive temporal problems such as the open shop and job shop scheduling domains. Much previous research in systematic search approaches for these problems has focused on developing problem specific constraint propagators and ordering heuristics. Indeed, the common belief is that many of these problems are too difficult to solve without such domain specific models. We introduce a simple constraint model that combines a generic adaptive heuristic with naive propagation, and show that it often outperforms state-of-the-art solvers for both open shop and job shop problems.

1 Introduction

It is usually accepted that the most efficient methods for solving *Open shop* and *Job shop* scheduling problems are local search algorithms, such as tabu search for job shop [16,17] and particle swarm optimization for open shop [20]. However, constraint programming often remains the solution of choice. It is indeed relatively competitive [4,23] whilst providing a more flexible approach. For instance, one can add domain-specific constraints without entailing major design revisions. Moreover, it is commonly believed that the most efficient CP models are those based on strong inference methods, such as *Edge Finding* [7,18], and specific search strategies, such as *Texture* [9].

In this paper we build on the results of [1] which showed that for open shop problems a constraint programming approach with strong inference and domain specific search strategies outperforms the best local search algorithms. Here, we introduce a constraint model combining simple propagation methods with the generic *weighted degree* heuristic [5] and empirically show that the complex inference methods and search strategies can, surprisingly, be advantageously replaced by this naive model.

An $n \times m$ open shop problem (OSP) involves n jobs and m machines. A job is composed of m tasks, each associated with a duration and a machine. Each machine maps to exactly one task in each job, and can only run a single task at a time. The objective is to minimise the *makespan* M, that is, the total duration to run all tasks. A job shop problem (JSP) is identical except the order of the tasks within each job is fixed.

Our approach relies on a standard model and generic variable ordering and restart policy. First, each pair of conflicting tasks, whether because they belong to the same job, or share a machine, are associated through a *disjunctive* constraint, with a Boolean

I.P. Gent (Ed.): CP 2009, LNCS 5732, pp. 400–408, 2009.

variable standing for their relative ordering. Following the standard search procedure for this class of problems, the search space can thus be restricted to the partial orders on tasks. Second, the choice of the next (Boolean) variable to branch on is made by combining the current domain sizes of the associated two tasks with the weighted degree of the corresponding ternary disjunctive constraint. Third, we use restarts, together with a certain amount of randomization and nogood recording from restarts [13].

We demonstrate that this simple approach outperforms more sophisticated constraint models using state of the art heuristics and filtering algorithms implemented in Choco [8] and Ilog Solver, both on OSPs and JSPs. We believe that the weighted degree heuristic, within this model, is extremely effective at identifying the contentious pairs of tasks. In Section 2 we first describe state of the art constraint models and strategies for open shop and job shop scheduling problems. Next, we describe a lighter model as well as the search strategies that we use to empirically support our claims. In Section 3, we present an experimental comparison of our model with state-of-the-art solvers for open shop and job shop scheduling problems.

2 Constraint Models

Here, we describe the two models that we compared in our experiments on the open shop and job shop benchmarks. The first model is the accepted state of the art, using strong inference (global unary resource constraints with the Edge Finding filtering algorithm) as well as specific search heuristics. The second is our lighter and simpler approach, relying on ternary reified disjunctive constraints, and a variable heuristic largely based on the generic weighted degree heuristic [5]. We shall refer to the former as the "heavy" model, and to the latter as the "light" model throughout.

Edge Finding + Profile, "Heavy" Model: In this model, a global filtering algorithm is used for each unary resource. Let T denote a set of tasks sharing an unary resource and Ω denote a subset of T . We consider the three following propagation rules:

Not First/Not Last: This rule determines if the task t_i cannot be scheduled after or before a set of tasks Ω. In that case, at least one task from the set must be scheduled after (resp. before). The domain of task t_i can be updated accordingly.

Detectable Precedence: If a precedence $t_i \prec t_j$ can be discovered by comparing t_i and t_j's time bounds, then their domains can be updated with respect to all the predecessors or successors.

Edge Finding: This filtering technique determines that some tasks must be executed first or last in a set $\Omega \subseteq T$. It is the counterpart of the first rule.

Search Strategy: The branching scheme is that proposed in [3] and denoted *Profile*. We select a critical pair of tasks sharing the same unary resource and impose an ordering. This heuristic, based on the probabilistic profile of the tasks, determines the most constrained resources and tasks. At each node, the resource and the time point with the maximum contention are identified, then a pair of tasks that rely most on this resource at this time point are selected (it is also ensured that the two tasks are not already connected by a path of temporal constraints).

Once the pair of tasks has been chosen, the order of the precedence has to be decided. For that purpose, we retain one of the three randomized value ordering heuristics from the same paper: centroid. The centroid is a real deterministic function of the domain and is computed for the two critical tasks. The centroid of a task is the point that divides its probabilistic profile equally. We commit the sequence which preserves the ordering of the centroids of the two tasks. If the centroids are at the same position, a random ordering is chosen. (For a more detailed discussion on filtering techniques for disjunctive scheduling problems we would point the reader to [2].)

Simple Disjunction + Weighted Degree, "Light" Model: The starting time of each task t_i is represented by a variable $t_i \in [0..M - d_i]$. Next, for every pair of unordered tasks t_i, t_j sharing a job or a machine we introduce a Boolean variable b_{ij} standing for the ordering between t_i and t_j. A value of 0 for b_{ij} means that task t_i should precede task t_j, whilst a value of 1 stands for the opposite ordering. The variables t_i, t_j and b_{ij} are linked by the following constraint, on which Bounds Consistency (BC) is maintained:

$$b_{ij} = \begin{cases} 0 \Leftrightarrow t_i + d_i \leq t_j \\ 1 \Leftrightarrow t_j + d_j \leq t_i \end{cases}$$

For n jobs and m machines, this model therefore involves $nm(m + n - 2)/2$ Boolean variables for OSPs, $nm(m + n - 2)/4$ for JSPs, and as many disjunctive constraints.

Search Strategy: Instead of searching by assigning a starting time to a single value on the left branches, and forbidding this value on the right branches, it is common to branch on *precedences*. In the heavy model, an unresolved pair of tasks t_i, t_j is selected and the constraint $t_i + d_i \leq t_j$ is posted on the left branch whilst $t_j + d_j \leq t_i$ is posted on the right branch. In our model, branching on the Boolean variables precisely simulates this strategy and thus significantly reduces the search space. Indeed, it has been observed (for instance in [15]) that the existence of a partial ordering of the tasks (compatible with start times and durations, and such that its projection on any job or machine is a total order) is equivalent to the existence of a solution. In other words, if we successfully assign all Boolean variables, the existence of a solution is guaranteed.

We use the weighted degree heuristic [5], which chooses the variable maximising the total weight of neighbouring constraints, initialised to its degree. A constraint's weight is incremented by one each time the constraint causes a failure during search. We show in Section 3, that the weighted degree heuristic is very efficient in this context. It is important to stress that the behaviour of this heuristic is dependent on the modelling choices. Indeed, two different, yet logically equivalent, sets of constraints may distribute the weights differently. In this model, every constraint involves one and only one search variable. Moreover, the relative light weight of the model allows the search engine to explore much more nodes, thus learning weights quicker.

However, at the start of the search, this heuristic is completely uninformed since every Boolean variable has the same degree (i.e. 1). We use the domain size of the two tasks t_i, t_j associated to every disjunct b_{ij} to inform the variable selection method until the weighted degrees effectively kick in. We denote $w(ij)$ the number of times the search failed while propagating the constraint on t_i, t_j and b_{ij}. We pick the variable minimising the sum of the residual time windows of the two tasks' starting times, divided by the weighted degree: $(max(t_i) + max(t_j) - min(t_i) - min(t_j) + 2)/w(ij)$

3 Experimental Section

All experiments reported in this paper were run on an Intel Xeon 2.66GHz machine with 12GB of ram on Fedora 9. Each algorithm run on a problem had an overall time limit of 3600s. Unless otherwise stated, values were chosen lexically, ties were broken randomly and nogoods were recorded from restarts.

We first provide evidence that, contrary to popular belief, the domain-specific model and heuristics are unnecessary for solving these problems. For OSPs, we compare our model with the heavy model which was recently shown to be the state-of-the-art on these benchmarks, matching the best metaheuristics on the Taillard benchmarks, and outperforming both exact and approximate methods on the other two benchmarks [1]. We implemented our algorithm in Mistral [11] and, for better comparison on the OSPs, in Choco. The code and parameters used for the heavy model are those used in [1].

Next, on JSPs, we compare the same light model implemented in Mistral, with the heavy model implemented in Ilog Scheduler (algorithm denoted "randomized restart" in [4]). Once again, we got the code for our comparison from the author and used the same parameters as were used in that paper.

3.1 Open Shop Scheduling

We used three widely studied sets of instances: Brucker [6], Gueret & Prins [10], and Taillard [21]. The problems range from size 3x3 to 20x20, with 192 instances overall. For all experiments, a sample of 20 runs with different seeds was performed.

The Choco models use a simple randomised constructive heuristic detailed in [1], and referred to as "CROSH", to calculate a good initial upper bound, followed by branch and bound search. The restarting strategy used for the light model (both Choco and Mistral) was geometric [22], with a base of 256 failures, and a multiplicative factor of 1.3. The heavy model uses the Luby restarting sequence [14] order 3 with a scale factor tuned to the problem dimensions. We also ran the most effective Mistral model which only differs from the Choco light model in that a dichotomic search was used instead of CROSH to get an initial upper bound. The lower bound lb is set to the duration of the longest job/machine, whilst the upper bound ub in the dichotomic search is initialised by a greedy algorithm. We repeatedly solve the decision problem with a makespan fixed to $\frac{ub+lb}{2}$, updating lb and ub accordingly, until they have collapsed.

Figure 1 is a log-log plot of the average (a) time and (b) number of nodes taken by the two models to solve each of the 192 open shop instances. Next, in Table 1, we selected a subset of 11 of the hardest problems based on overall results, choosing problems for which at least one of the light and heavy models took over 20 seconds to solve. This subset consisted of one Gueret-Prins, 4 Taillard and 6 Brucker instances, respectively of order 10, 20 and 7-8. The results are presented in terms of average time and average nodes (where a model failed to prove optimality on a problem its time is taken as 3600s, so in these cases the average time is a lower bound).

Both Choco models proved optimality on all 11 instances. However, the light model is generally much faster (only slower on 1 of the 11 instances), even though it explores many more nodes. The Mistral light model was fastest on all problems (nodes for Mistral include those explored in dichotomic search, hence the difference). It is also of

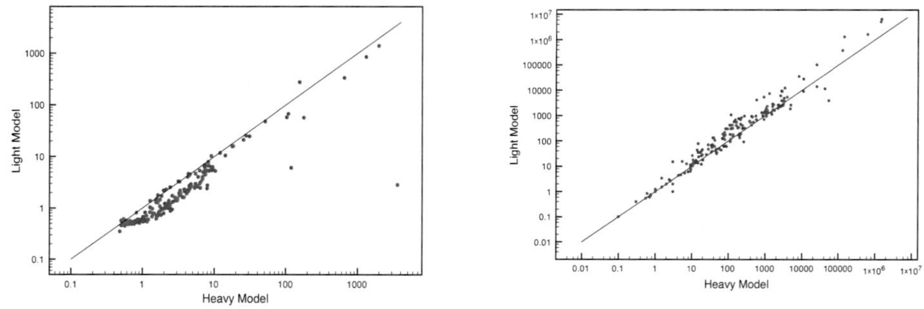

(a) Time Comparison (b) Node Comparison

Fig. 1. Log-log plots of Choco Light vs Heavy models on open shop problems

interest to note that on the hardest problems (6 Brucker instances), the heavy model was slightly quicker on average than the (Choco) light model at finding the optimal solution (310s vs 360s), but was 3 times slower at proving optimality once the solution had been found (385s vs 125s). This reinforces our belief that the weighted degree heuristic is adept at identifying the most contentious variables.

In order to better understand the importance of the heuristics to the two models we ran the same experiments but with the heuristics swapped, "Light-profile" and "Heavy-domwdeg". As can be seen in the table, swapping heuristics resulted in worse performance in both cases (failing to prove optimality on some problems for both), albeit more noticeably for combining profile with the light model.

Table 1. Results (Time) For Hard Open Shop Scheduling Problems

Instances	Light		Heavy		Light-profile		Heavy-domwdeg		Mistral	
	Time (s)	Nodes	Time (s)	Nodes	Time (s)	Nodes	Time (s)	Nodes	Time (s)	Nodes
GP10-01	6.1	4K	118.4	53K	2523.2	6131K	9.6	3K	**0.3**	3K
j7-per0-0	854.5	5.1M	1310.9	1.4M	979.1	4.3M	3326.7	2.6M	**327.0**	8.7M
j7-per10-2	57.6	0.4M	102.7	0.1M	89.5	0.4M	109.6	0.1M	**33.5**	1.2M
j8-per0-1	1397.1	6.4M	1973.8	1.5M	1729.3.	6.0M	3600.0	2.4M	**427.1**	10M
j8-per10-0	24.6	0.10M	30.9	0.02M	19.7	0.05M	68.0	0.06M	**17.6**	0.47M
j8-per10-1	275.2	1.3M	154.8	0.1M	92.8	0.3M	796.7	0.7M	**89.3**	2.3M
j8-per10-2	335.3	1.6M	651.4	0.6M	754.0	2.7M	697.1	0.6M	**93.1**	2.3M
tai-20-1	25.4	2K	27.5	3K	2524.7	1458K	34.4	3K	**3.9**	21K
tai-20-2	56.5	11K	178	42K	3600.0	2261K	81.9	10K	**14.1**	61K
tai-20-7	47.8	9K	60.9	11K	3600.0	2083K	63.7	8K	**9.5**	47K
tai-20-8	66.8	14K	108.3	25K	3600.0	2127K	84.8	11K	**8.2**	39K
Total Avg	286.1	1.3M	428.9	0.4M	1774.3	2.5M	806.6	0.6M	**93.0**	2.3M

Finally, we investigated whether the difference in restarting strategies might account for the improvement with the light model. We ran the Luby strategy on the light model and the geometric strategy on the heavy model. Again, in both cases, the change lead to a degradation in performance.

Search Heuristics. We next assess the impact of the two aspects of the domwdeg heuristic, task size (dom) and weighted degree ($wdeg$), on the light model.

The following experiments were run using the Mistral light model with the same settings as before. However, due to the nature of wdeg and dom, all variable heuristics were further randomized by randomly choosing from their top three choices when no tie existed. Furthermore the cutoff for dichotomic search was 30 seconds for each (lb ub) to allow all heuristics to achieve a good initial upper bound. We present our results in terms of average makespan found (mks), average percent proven optimal and the average time over the sample of 20 runs for each problem.

Table 2. Variable Heuristic Comparison: Open Shop Scheduling Problems

Instances	dom			wdeg			domwdeg		
	Mks	Opt(%)	Time(s)	Mks	Opt(%)	Time (s)	Mks	Opt(%)	Time (s)
Brucker	1019.7	89.2	1529.2	1019.5	100.0	381.4	1019.5	100.0	249.7
Taillard	1253.7	0.0	3600.0	1215.2	93.7	1423.9	1214.7	100.0	8.8
Total Avg	1113.4	53.0	2358.5	1097.7	97.0	798.4	1097.6	100.0	153.3

Table 2 presents our findings on the Brucker and Taillard instances of Table 1. The results clearly show the effectiveness of wdeg on these problems. It proved optimality on average 97% of the time, and in the cases where it failed to prove optimality the average makespan found was within one of the optimal value. However, as previously mentioned, it suffers from a lack of discrimination at the start of search. For example, in the tai-20-* problems there are 7600 variables all with an initial weighted degree of 1. Discrimination will only occur after the first failure which, given the size of the problems and the looseness of constraints, can occur deep in search, especially with a heuristic that is random up until at least one failure occurs.

The domain heuristic is relatively poor on these problems (only proving optimality 53% of the time and finding poor makespans for the tai-20-* problems). However, the addition of this information to the wdeg heuristic (*domwdeg*) results in 100% optimality. The domain factor has two benefits, it provides discrimination at the start of search, and it improves the quality of the initial weights learnt due to its fail firstness.

3.2 Job Shop Scheduling

We now compare the light model, implemented in Mistral, with a randomized restart algorithm used in [4], which is implemented in Ilog scheduler. It should be noted that we are not comparing with Beck's SGMPCS algorithm, but with the randomized restarts approach that was used as an experimental comparison by Beck. This algorithm is nearly identical to the heavy model introduced in Section 2.

All parameters for the algorithm are taken from [4], so the restart strategy is Luby with an initial failure limit of 1 (i.e there is no scale factor). The variable ordering heuristic is the profile heuristic described earlier. Randomization is added by randomly selecting with uniform probability from the top 10% most critical pairs of (machine, time point). Finally, the standard constraint propagation techniques for scheduling are used, such as time-table [19], edge-finding [18], and balance constraints [12]. However nogood recording from restarts isn't part of the algorithm. We ran experiments on the same cluster described earlier, using Ilog scheduler 6.2.

The Mistral parameters are the same as for the open shop problems, with the exception that we used a 300 second cutoff for the dichotomic search in order to achieve a good initial upper bound. Furthermore, we used a static value ordering heuristic, where each pair of operations on a machine were ordered based on their relative position in their job (i.e. if precedences on two jobs state that t_i is second and t_j is fourth on their respective jobs, then b_{ij} will first branch on 0 during search, i.e. t_i precedes t_j). (We experimented with several value heuristics and this proved to be the best.)

Table 3 describes our results on 4 sets of 10 JSP instances proposed by Taillard [21]. The different sets have problems of different sizes (#jobs x #machines). The four sets are of size 20x15, 20x20, 30x15, 30x20, respectively. For each instance of each set, 10 randomized runs were performed, since problems were rarely solved to optimality within the cutoff. (This set of experiments took roughly 33 days of CPU time.)

We present our results in terms of averages over each set of problems. In particular, the average of the mean makespan found per problem in each set, the average of the best makespan found for each problem in each set, and average standard deviation.

Table 3. Job Shop Scheduling Problems

Instances	Scheduler			Mistral		
	Mean	Best	Std Dev	Mean	Best	Std Dev
tai11-20	1411.1	1409.9	11.12	**1407.3**	**1392.9**	24.35
tai21-30	**1666.0**	1659.0	13.51	1666.8	**1655.1**	22.91
tai31-40	1936.1	1927.1	18.67	**1921.6**	**1899.2**	37.79
tai41-50	2163.1	2153.1	17.85	**2143.3**	**2119.5**	42.71

As expected, since we used faster hardware, the results we obtained with Ilog Scheduler match, or improve slightly on the values reported in [4]. Both approaches perform similarly on the first two sets, although best solutions found by Mistral are consistently better than Ilog scheduler. Mistral scales up better on the next two (larger) sets, indeed its average mean makespan for each set is better than Scheduler's average best makespan in both sets. It is interesting to note that the standard deviation is much larger for the lighter model. The down side is that it means our approach is less robust. However, it also means that using parallel computing should improve the lighter model more than it would improve Scheduler's results.

4 Conclusion

In this paper we have shown that, contrary to popular belief, disjunctive temporal problems (such as in the scheduling domain) can be efficiently solved by combining naive propagation with the generic weighted degree heuristic. Important additional factors in such an approach are restarting, nogood recording from restarts, a good method for finding an initial upper bound, and an element of randomization. We have shown that such an approach can often outperform the state of the art solvers for open shop and job shop scheduling problems.

However, our approach was not able to match the results of solution guided multi point constructive search (SGMPCS) for job shop scheduling problems [4]. It is our belief that a similar solution guided approach can be incorporated into our model to improve its performance, and this is the direction we intend to take our future work.

Acknowledgements

The authors would like to thank Hadrien Cambazard for his valuable comments, and Chris Beck for supplying the code for the job shop scheduling comparison. This work was supported by Science Foundation Ireland under Grant 05/IN/I886.

References

1. Malapert, A., Cambazard, H., Guéret, C., Jussien, N., Langevin, A., Rousseau, L.-M.: An Optimal Constraint Programming Approach to the Open-Shop problem. Submitted to INFORMS, Journal of Computing (2008)
2. Baptiste, P., Le Pape, C., Nuijten, W.: Constraint-Based Scheduling: Applying Constraint Programming Techniques to Scheduling Problems. Kluwer Academic Publishers, Dordrecht (2001)
3. Beck, J.C., Davenport, A.J., Sitarski, E.M., Fox, M.S.: Texture-Based Heuristics for Scheduling Revisited. In: AAAI/IAAI, pp. 241–248 (1997)
4. Christopher Beck, J.: Solution-Guided Multi-Point Constructive Search for Job Shop Scheduling. Journal of Artificial Intelligence Research 29, 49–77 (2007)
5. Boussemart, F., Hemery, F., Lecoutre, C., Sais, L.: Boosting Systematic Search by Weighting Constraints. In: Proceedings of the 16th European Conference on Artificial Intelligence (ECAI 2004), Valencia, Spain, August 2004, pp. 482–486 (2004)
6. Brucker, P., Hurink, J., Jurisch, B., Wöstmann, B.: A Branch & Bound Algorithm for the Open-shop Problem. In: GO-II Meeting: Proceedings of the Second International Colloquium on Graphs and Optimization, pp. 43–59. Elsevier Science Publishers B. V., Amsterdam (1997)
7. Carlier, J., Pinson, E.: An Algorithm for Solving the Job-shop Problem. Management Science 35(2), 164–176 (1989)
8. The choco team. choco: an Open Source Java Constraint Programming Library. In: The Third International CSP Solver Competition, pp. 31–40 (2008)
9. Fox, M.S., Sadeh, N.M., Baykan, C.A.: Constrained heuristic search. In: IJCAI, pp. 309–315 (1989)
10. Guéret, C., Prins, C.: A new Lower Bound for the Open Shop Problem. Annals of Operations Research 92, 165–183 (1999)
11. Hebrard, E.: Mistral, a Constraint Satisfaction Library. In: The Third International CSP Solver Competition, pp. 31–40 (2008)
12. Laborie, P.: Algorithms for Propagating Resource Constraints in AI Planning and Scheduling: Existing Approaches and new Results. Artificial Intelligence 143(2), 151–188 (2003)
13. Lecoutre, C., Sais, L., Tabary, S., Vidal, V.: Nogood Recording from Restarts. In: IJCAI, pp. 131–136 (2007)
14. Luby, M., Sinclair, A., Zuckerman, D.: Optimal Speedup of Las Vegas Algorithms. In: ISTCS, pp. 128–133 (1993)
15. Meiri, I.: Combining Qualitative and Quantitative Constraints in Temporal Reasoning. In: AAAI, pp. 260–267 (1991)
16. Nowicki, E., Smutnicki, C.: A Fast Taboo Search Algorithm for the Job Shop Problem. Manage. Sci. 42(6), 797–813 (1996)

17. Nowicki, E., Smutnicki, C.: An Advanced Tabu Search Algorithm for the Job Shop Problem. Journal of Scheduling 8(2), 145–159 (2005)
18. Nuijten, W.: Time and Resource Constraint Scheduling: A Constraint Satisfaction Approach. PhD thesis, Eindhoven University of Technology (1994)
19. Le Pape, C.: Implementation of Resource Constraints in ILOG SCHEDULE: A Library for the Development of Constraint-Based Scheduling Systems. Intelligent Systems Engineering 3, 55–66 (1994)
20. Sha, D.Y., Hsu, C.-Y.: A new Particle Swarm Optimization for the Open Shop Scheduling Problem. Computers & Operation Research 35(10), 3243–3261 (2008)
21. Taillard, E.: Benchmarks for Basic Scheduling Problems. European Journal of Operations Research 64, 278–285 (1993)
22. Walsh, T.: Search in a Small World. In: IJCAI, pp. 1172–1177 (1999)
23. Watson, J.-P., Beck, J.C.: A Hybrid Constraint Programming / Local Search Approach to the Job-Shop Scheduling Problem. In: Perron, L., Trick, M.A. (eds.) CPAIOR 2008. LNCS, vol. 5015, pp. 263–277. Springer, Heidelberg (2008)

Reasoning about Optimal Collections of Solutions

Tarik Hadžić, Alan Holland, and Barry O'Sullivan

Cork Constraint Computation Centre
Department of Computer Science, University College Cork, Ireland
{t.hadzic,a.holland,b.osullivan}@4c.ucc.ie

Abstract. The problem of finding a collection of solutions to a combinatorial problem that is optimal in terms of an inter-solution objective function exists in many application settings. For example, maximizing diversity amongst a set of solutions in a product configuration setting is desirable so that a wide range of different options is offered to a customer. Given the computationally challenging nature of these multi-solution queries, existing algorithmic approaches either apply heuristics or combinatorial search, which does not scale to large solution spaces. However, in many domains compiling the original problem into a compact representation can support computationally efficient query answering. In this paper we present a new approach to find optimal collections of solutions when the problem is compiled into a multi-valued decision diagram. We demonstrate empirically that for real-world configuration problems, both exact and approximate versions of our methods are effective and are capable of significantly outperforming state-of-the-art search-based techniques.

1 Introduction

Knowledge compilation provides a basis for efficiently processing queries that would otherwise be intractable. Compact representations, such as automata, decision diagrams, and negation normal forms, are of critical importance to solve combinatorial problems in a number of application settings such as product configuration, diagnosis and planning. In a product configuration domain, for instance, one needs to find solutions in an interactive manner, so fast response times are required. Since this is typically an $\mathcal{NP}$-complete task, configuration problems have been compiled into automata [1] and binary decision diagrams [2,3] to guarantee fast response times. In other domains, such as diagnosis and planning, a much harder class of probabilistic inference tasks needs to be solved. They usually rely on counting the number of solutions, which is $\#\mathcal{P}$-complete in most non-trivial settings. Therefore, compiled representations such as decomposable negation normal forms (DNNFs) [4] have been proven to be critical for enhancing inference. In each of these domains, compilation effort is traded in favour of faster online inference but at the cost of a potentially exponential increase in memory requirements. Fortunately, in practice, many compiled representations of real world problems are compact.

In this paper we investigate whether we can exploit knowledge compilation to enhance inference for another class of challenging queries that require computing optimal *collections* of solutions. In many settings we not only want to find a solution that satisfies a set of constraints, but we also want to find solutions that ensure that *other* solutions

I.P. Gent (Ed.): CP 2009, LNCS 5732, pp. 409–423, 2009.

exist that satisfy particular constraints, or optimize a particular objective function. For example, in a configuration setting we may wish to find a set of solutions that are as diverse as possible, given a specific measure of diversity such as Hamming distance. For inexact database query answering, we might wish to find a set of tuples from a database that are as close as possible to a user-specified query but as mutually diverse as possible.

This paper offers the following contributions. Firstly, we present a novel approach to answering multi-solution queries over a multi-valued decision diagram (MDD) when we wish to find a collection of solutions that are optimal with respect to an arbitrary additive objective function. Secondly, we show how to exploit the compactness of MDDs by transforming the problem of multi-solution optimization into a problem of single-solution optimization over a *product MDD*. Thirdly, we reduce the memory requirement of the method by using an implicit optimization scheme that avoids explicitly generating the expanded MDD, along with an approximate optimization scheme that limits the number of nodes that can be processed during optimization. Fourthly, we show how our approach can be instantiated to exactly solve various important inter-solution distance queries. Finally, we empirically evaluate the practicability of our approach on real-world product configuration problems and demonstrate its effectiveness in comparison to current state-of-the-art search-based approaches. We conclude with a discussion of the significance of our results and possible future work.

2 Preliminaries

Constraint Satisfaction. We assume that the original problem is given in the form of a *constraint satisfaction problem*.

Definition 1 (Constraint Satisfaction Problem). *A* constraint satisfaction problem *(CSP), $\mathcal{P}(X, D, F)$, is defined by a set of variables $X = \{x_1, \ldots, x_n\}$, a set of domains $D_1, \ldots, D_n$ such that variable x_i takes a value from D_i, and a set of constraints $F =_{\text{def}} \{f_1, \ldots, f_m\}$ that restrict the set of consistent assignments to the variables in X.*

The set of solutions Sol of a CSP $\mathcal{P}$ is a subset of $D_1 \times \ldots \times D_n$ such that every element $s \in Sol$ satisfies all constraints in F. The CSP is a versatile modeling framework, capable of expressing problems from various application domains. Consider the following pedagogical example from a product configuration domain.

Example 1 (T-Shirt CSP). We are interested in selecting a T-shirt which is defined by three attributes: the color (black, white, red, or blue), the size (small, medium, or large) and the print ("Men In Black" - MIB or "Save The Whales" - STW). There are two rules that define what are valid combinations: if we choose the MIB print then the color black has to be chosen as well, and if we choose the small size then the STW print (including a big picture of a whale) cannot be selected as the picture of a whale does not fit on the small shirt. The implicit representation (X, D, F) of the T-shirt example consists of variables $X = \{x_1, x_2, x_3\}$ representing color, size and print. Variable domains are $D_1 = \{0, 1, 2, 3\}$ $(black, white, red, blue)$, $D_2 = \{0, 1, 2\}$ $(small, medium, large)$, and $D_3 = \{0, 1\}$ (MIB, STW). The two rules translate to $F = \{f_1, f_2\}$, where f_1 is $x_3 = 0 \Rightarrow x_1 = 0$ $(MIB \Rightarrow black)$ and f_2 is $(x_2 = 0 \Rightarrow x_3 \neq 1)$ $(small \Rightarrow not\ STW)$. ♢

The CSP model is summarized in Fig. 1(a). There are $|D_1| \cdot |D_2| \cdot |D_3| = 24$ possible assignments. Eleven of these assignments are valid configurations and they form the solution space shown in Fig. 1(b).

Variables:
$x_1 \in \{0, 1, 2, 3\}$
$x_2 \in \{0, 1, 2\}$
$x_3 \in \{0, 1\}$

Constraints:
$\mathbf{f_1}: x_3 = 0 \Rightarrow x_1 = 0$
$\mathbf{f_2}: x_2 = 0 \Rightarrow x_3 \neq 1$

(a) T-Shirt Constraint Satisfaction Problem

color	size	print
black	*small*	*MIB*
black	*medium*	*MIB*
black	*medium*	*STW*
black	*large*	*MIB*
black	*large*	*STW*
white	*medium*	*STW*
white	*large*	*STW*
red	*medium*	*STW*
red	*large*	*STW*
blue	*medium*	*STW*
blue	*large*	*STW*

(b) Solution space of the T-Shirt CSP

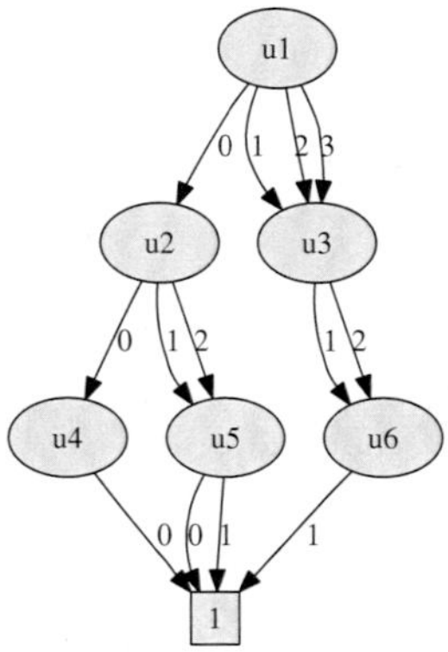

(c) MDD representation of the T-Shirt CSP

Fig. 1. T-Shirt example. The CSP formulation is presented in Fig. 1(a), the solution space in Fig. 1(b) and the corresponding MDD in Fig. 1(c). Every row in a solution space table corresponds to a unique path in the MDD.

Multivalued Decision Diagrams. Multivalued Decision diagrams (MDDs) are compressed representations of solution sets $Sol \subseteq D_1 \times \ldots \times D_n$. They are rooted directed acyclic graphs (DAGs) where each node u corresponds to a variable x_i and each of its outgoing edges e corresponds to a value $a \in D_i$. Paths in the MDD correspond to solutions in Sol.

Definition 2 (Multi-valued Decision Diagram). *An MDD M is a rooted directed acyclic graph (V, E), where V is a set of vertices containing the special terminal vertex* $\mathbf{1}$ *and a root $r \in V$. Furthermore, var : $V \rightarrow \{1, \ldots, n+1\}$ is a labeling of all nodes with a variable index such that var$(\mathbf{1}) = n + 1$. Each edge $e \in E$ is defined by a triple (u, u', a) of its start node u, its end node u' and an associated value a.*

We work only with *ordered* MDDs. A total ordering $<$ of the variables is assumed such that for all edges (u, u', a) $var(u) < var(u')$. For convenience we assume that the variables in X are ordered according to their indices. On each path from the root to the terminal, every variable labels exactly one node. An MDD encodes a CSP solution set $Sol \subseteq D_1 \times \ldots \times D_n$, defined over variables $\{x_1, \ldots, x_n\}$. To check whether an assignment $\mathbf{a} = (a_1, \ldots, a_n) \in D_1 \times \ldots \times D_n$ is in Sol we traverse M from the root, and at every node u labeled with variable x_i, we follow an edge labeled with a_i. If there is no such edge then $\mathbf{a}$ is not a solution. Otherwise, if the traversal eventually ends in terminal $\mathbf{1}$ then $\mathbf{a} \in Sol$. In Fig. 1(c) we show an MDD that encodes the solution set from Fig. 1(b). For every node we show its unique identifier. The variable ordering used is $x_1 < x_2 < x_3$, i.e. $var(u_1) = 1$, $var(u_2) = var(u_3) = 2$, $var(u_4) = var(u_5) = var(u_6) = 3$.

Ordered MDDs can be considered as being arranged in *layers* of vertices, each layer being labeled with the same variable index. We will denote by V_i the set of all nodes labeled with x_i, $V_i = \{u \in V \mid var(u) = i\}$. Similarly, we will denote by E_i the set of all edges originating in V_i, i.e. $E_i = \{e(u, u', a) \in E \mid var(u) = i\}$. In our T-Shirt MDD for example, $V_2 = \{u_2, u_3\}$ and $E_2 = \{(u_2, u_4, 0), (u_2, u_5, 1), (u_2, u_5, 2), (u_3, u_6, 1), (u_3, u_6, 2)\}$. We will denote by $p : u_1 \rightsquigarrow u_2$ any path in an MDD from u_1 to u_2. Also, edges between u and u' will be sometimes denoted as $e : u \rightarrow u'$. A value a of an edge $e(u, u', a)$ will sometimes be denoted as $v(e)$, while a partial assignment associated with path p will be denoted as $v(p)$. Every path corresponds to a unique assignment. Hence, the set of all solutions represented by the MDD is $Sol = \{v(p) \mid p : r \rightsquigarrow \mathbf{1}\}$. In fact, every node $u \in V_i$ can be associated with a subset of solutions $Sol(u) = \{v(p) \mid p : u \rightsquigarrow \mathbf{1}\} \subseteq D_i \times \ldots \times D_n$. For example, in the T-Shirt MDD, $Sol(u_2) = \{(0,0), (1,0), (1,1), (2,0), (2,1)\}$.

A distinct advantage of decision diagrams is that they can represent the size of the solution set they encode in an exponentially smaller data structure by *merging isomorphic subgraphs*. Two nodes u_1, u_2 are isomorphic if they encode the same solution set $Sol(u_1) = Sol(u_2)$. The T-Shirt MDD from Fig. 1(c) is merged. In addition to merging isomorphic nodes, one can *remove redundant nodes* which reduces the size by at most a linear factor. This complicates MDD-based algorithms since the removal of nodes introduces *long edges*. Therefore, throughout the paper we always assume an ordered *merged* MDD. Given a variable ordering there is a unique merged MDD for a given CSP $\mathcal{P}(X, D, F)$ and its solution set Sol. The size of an MDD depends critically on the ordering, and could vary exponentially. It can grow exponentially with the number of variables, but in practice, for many interesting constraints the size is surprisingly small. An MDD for a given CSP $\mathcal{P}(X, D, F)$ is typically constructed by first creating an MDD M_i for each constraint f_i, and then using pairwise conjunctions to construct $M_1 \wedge \ldots \wedge M_m$. A conjunction of two MDDs $M_i \wedge M_j$ can be performed in worst-case quadratic time and space $O(|M_i| \cdot |M_j|)$, where $|M|$ denotes the size of the MDD $|V| + |E|$. Hence the final MDD can be exponentially large $O(|M_1| \cdot \ldots \cdot |M_n|)$.

3 Optimal Collections of Solutions

In many settings we do not want to find just a single solution satisfying a set of constraints $\mathcal{P}(X, D, F)$, but we want to retrieve *collections* of such solutions that should be *optimal* with respect to some cost function. Particularly challenging are those notions of optimality that require evaluating inter-solution relationships within a collection in order to evaluate the cost of entire collection. We will refer to such notions of optimality as *multi-solution costs*. For example, computing a collection of k most expensive solutions with respect to some cost function $c(x_1, \ldots, x_n) : D \rightarrow \mathbb{R}$ is *not* a multi-cost query since the cost of an entire collection depends only on the costs of individual members, without considering their inter-relationships. However, retrieving k solutions that are as *diverse* as possible *is* a multi-solution query since the diversity of a collection can be evaluated only by looking at distances between all the members in the collection.

Another example of a multi-solution query can be found in risk management and reasoning about robust solutions. We may wish to find a high-quality solution for which a contingency solution exists that is robust to specific types of failure. While the quality of both original and contingent solution depends on each solution individually, there is a *penalty* associated with switching from one solution to another. Hence, for a given failure probability, the expected cost of such a pair can be estimated only by considering both solutions simultaneously.

Formally, we are given a CSP $\mathcal{P}(X, D, F)$ representing its solution set Sol. We wish to retrieve a combination of k solutions $s^1, \ldots, s^k$ where $s^j = (s^j_1, \ldots, s^j_n) \in Sol$ and $j = 1, \ldots, k$ that is *optimal* with respect to a cost function C:

$$C : \overbrace{(D_1 \times \cdots \times D_n) \times \cdots \times (D_1 \times \cdots \times D_n)}^{k} \to \mathbb{R}.$$

In this paper we restrict our attention to a subclass of *additive* cost functions C, i.e. the cost functions which can be represented as a sum of variable-level cost functions $C_i : (D_i)^k \to \mathbb{R}$:

$$C(s^1_1, \ldots, s^1_n, \ldots, s^k_1, \ldots, s^k_n) = \sum_{i=1}^{n} C_i(s^1_i, \ldots, s^k_i).$$

Many interesting real-world notions of optimality can be expressed as additive costs, such as notions of diversity, and similarity based on *Hamming distance* and other forms of distances between categorical and numerical variables in recommender systems research. When retrieving only single solutions (for $k = 1$) restriction to additive cost function still leads to a very general class of cost functions that subsume objective functions of entire areas of research such as *integer linear programming*.

Recall that, without loss of generality, we focus on retrieving k solutions that *maximize* the value of the cost function C; finding minimal tuples is a symmetric problem. Formally, the problem we are addressing in this paper is as follows.

Problem 1 (Optimal k-ary Collection of Solutions). Find k solutions $(s^1, \ldots, s^k)$ from Sol that are optimal with respect to the following objective:

$$maximize \sum_{i=1}^{n} C_i(s^1_i, \ldots, s^k_i),\ s^j \in Sol,\ j = 1, \ldots, k.$$

The state-of-the-art approach in the constraint programming community to reasoning about multiple solutions of constraint satisfaction problems is presented in [5,6] where the authors investigated reasoning about diversity and similarity of solutions. The core of their approach is based on introducing k-copies of variables, $X^j = \{x^j_1, \ldots, x^j_n\}$, $x^j_i \in D_i$, where $j = 1, \ldots, k$ and $i = 1, \ldots, n$. Corresponding copies of the original constraints $F^j = \{f^j_1, \ldots, f^j_m\}$ where $j = 1, \ldots, k$, are posted over the new sets of the variables. The resulting CSP has $n \cdot k$ variables and $m \cdot k$ constraints. By setting an objective function

$$C(x_1^1,\ldots,x_n^1,\ldots,x_1^k,\ldots,x_n^k) = \sum_{i=1}^{n} C_i(x_i^1,\ldots,x_i^k)$$

the problem of finding an optimal collection of solutions becomes a problem of finding a single optimal solution in such an expanded CSP model. The work in [5,6] focused not just on additive cost functions, but considered non-additive definitions of distances as well, for which a number of complexity results were demonstrated and practical solution approaches were suggested.

4 An MDD-Based Approach to Computing Optimal Collections

In many application domains, such as product configuration, the set of solutions Sol can be compactly represented as a multi-valued decision diagram (MDD). In cases when MDDs are compact, they can be effectively exploited to enhance answering a number of tasks related to feasibility and optimality of individual solutions. A number of otherwise $\mathcal{NP}$-hard tasks become linear in the size of the MDD. In particular, for an additive cost function, finding optimal solutions reduces to efficient shortest and longest path computations over an acyclic graph (see e.g. [1,7]). We investigate here whether we can, in an analogous manner, exploit the compressed MDD representation of the original solution set Sol and the additive nature of a multi-solution cost function to enhance search for optimal collections of solutions.

4.1 An MDD-Friendly CSP Encoding

The CSP encoding from the previous section is particularly well suited in a search setting using a standard CSP solver, as additional variables and constraints can be easily posted based on the original problem formulation. However, it is not clear how this encoding would help us to exploit an existing MDD representation of the solution space Sol. Even if we are able to choose a variable ordering that leads to a compact MDD representation, we would still have to optimize the sum of *non-unary* cost functions $C_i(x_i^1,\ldots,x_i^k)$ over an MDD which cannot be reduced to efficient shortest-path-based computations.

We therefore suggest an alternative encoding of the original CSP. A core aspect of our construction is to introduce k-dimensional variables $\mathbf{x_i} \in \mathbf{D_i}$, $i = 1,\ldots,n$ where

$$\mathbf{D_i} =_{\text{def}} \overbrace{D_i \times \cdots \times D_i}^{k} = (D_i)^k \quad i = 1,\ldots,n.$$

An assignment to variables $\mathbf{x_1},\ldots,\mathbf{x_n}$ represents k solutions. Every value $\mathbf{v_i}$ in the domain of $\mathbf{x_i}$ corresponds to a vector of values $(v_i^1,\ldots,v_i^k) \in (D_i)^k$ from domain D_i in the original CSP. An assignment to the variables $\{\mathbf{x}_1 = \mathbf{v}_1,\ldots,\mathbf{x}_n = \mathbf{v}_n\}$, is a *solution* if for each coordinate $j = 1,\ldots,k$, vectors $(v_1^j,\ldots,v_n^j)$ are solutions in the original CSP. We will use $\mathbf{Sol}$ to denote the set of all such solutions.

4.2 Constructing a k-MDD

We will now show that a corresponding MDD representation of the solution set **Sol**, which we call a k-MDD or a *product* MDD and denote as M^k, with respect to a variable ordering $\mathbf{x_1} \prec \ldots \prec \mathbf{x_n}$, can be generated directly from M. We will denote such an MDD as $M^k(V^k, E^k)$ where V^k and E^k denote vertices and edges. Let V_i^k and E_i^k denote the vertices and edges in the i-th layer, respectively. To construct the MDD it suffices to notice that for every *product* of vertices from the i-th layer V_i of the original MDD there is a unique vertex in V_i^k, i.e. $V_i^k \equiv (V_i)^k$. Analogously, for every combination of edges from E_i there is a unique edge in E_i^k, i.e. $E_i^k \equiv (E_i)^k$. In other words, we introduce a vertex $\mathbf{u} \in V_i^k$ for every k-tuple of vertices $(u_1, \ldots, u_k) \in (V_i)^k$. We will denote this relationship as $\mathbf{u} \equiv (u_1, \ldots, u_k)$ or write simply $\mathbf{u}(u_1, \ldots, u_k)$. We introduce an edge $\mathbf{e}(e_1, \ldots, e_k) \in E_i^k$ between the nodes $\mathbf{u}(u_1, \ldots, u_k) \in V_i^k$, $\mathbf{u}'(u_1', \ldots, u_k') \in V_{i+1}^k$ whenever each pair of nodes (u_j, u_j'), $j = 1, \ldots, k$, is connected with an edge e_j in M, i.e. if $e_j : u_j \rightarrow u_j'$, $\forall j = 1, \ldots, k$.

An example of such a construction is provided in Figure 2. An MDD M for the initial CSP is presented in Figure 2(a), encoding solutions $\{(1,1,1),(1,2,2),(2,3,2)\}$. Each node is labeled with a unique identifier rather than a variable label. Root node r corresponds to variable x_1, nodes a, b to variable x_2, and nodes c, d to variable x_3. The k-MDD M^2 (for $k = 2$), is shown in Figure 2(b). For the sake of clarity, we label every node (except root r and terminal $\mathbf{1}$) with combinations of the corresponding nodes in M, to illustrate the construction process. For the same reason, we label every edge in M^2 with a combination of values of the corresponding edges in M. For example, by combining a node a with a node b we get a product node (a, b). Furthermore, since an edge $r \rightarrow a$ has a label 1, and an edge $r \rightarrow b$ has a label 2 in M, an edge $r \rightarrow (a, b)$ has a label $(1, 2)$ in M^2.

From the definition of the product MDD M^k, its size can be computed exactly: $|V^k| = \sum_{i=1}^n |V_i|^k$, and $|E^k| = \sum_{i=1}^n |E_i|^k$. The time to construct M^k is linear in the number of operations $\Theta(|V^k| + |E^k|)$.

Furthermore, the label of each edge $\mathbf{e}(e_1, \ldots, e_k)$ can be identified with a k-tuple of values, $v(\mathbf{e}) = (v(e_1), \ldots, v(e_k))$. This can then, in a natural way, be extended to the valuation of paths, $v(\mathbf{p})$, so that every solution in **Sol** corresponds to a tuple of solutions in $Sol \times \cdots \times Sol$. From the definition of M^k, it then easily follows that for every product node $\mathbf{u} \equiv (u_1, \ldots, u_k)$, it holds that $Sol(\mathbf{u}) = Sol(u_1) \times \cdots \times Sol(u_k)$. Based on these observations, we can now prove the following property.

Lemma 1. *If an MDD M is merged then the k-MDD M^k is also merged.*

Proof (Proof by contradiction). Assume that an MDD M is merged. Since for every $\mathbf{u} \in V_i^k$ such that $\mathbf{u} \equiv (u_1, \ldots, u_k)$, it holds that $Sol(\mathbf{u}) = Sol(u_1) \times \cdots \times Sol(u_k)$, if two different nodes $\mathbf{u_1}(u_1^1, \ldots, u_1^k), \mathbf{u_2}(u_2^1, \ldots, u_2^k) \in V_i^k$ are isomorphic, then it must also hold that $Sol(\mathbf{u_1}) = Sol(\mathbf{u_2})$. This would imply that $Sol(u_1^j) = Sol(u_2^j)$, $j = 1, \ldots, k$. Since $\mathbf{u_1} \neq \mathbf{u_2}$, for at least one j, $u_1^j \neq u_2^j$. This indicates that the original MDD M was not merged, which is a contradiction. □

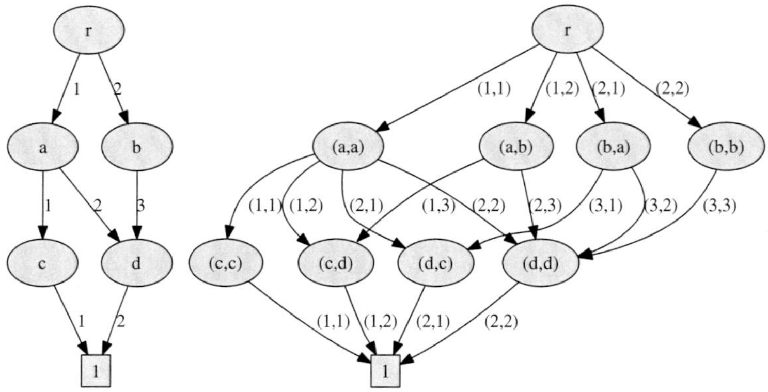

(a) An MDD for the initial problem.

(b) A corresponding product MDD.

Fig. 2. An example of the construction of a product MDD M^2 from the original MDD M. In order to illustrate the construction process, nodes in M^2 are labeled with the corresponding combinations of nodes from M. Analogously, labels on edges in M^2 are combinations of labels of the corresponding edges in M.

4.3 Optimization over k-MDDs

A critical advantage of a k-MDD M^k in comparison to MDDs obtained by the original CSP encoding, is that it allows optimization of a cost function C in *linear time* in its size. Namely, every edge $\mathbf{e}(e_1, \ldots, e_k) \in E_i^k$ corresponds to an assignment to variables $(x_i^1, \ldots, x_i^k) = (v(e_1), \ldots, v(e_k))$ and can be therefore labeled with a cost $C_i(\mathbf{e}) =_{\text{def}} C_i(v(e_1), \ldots, v(e_k))$. In an analogous manner we define the cost of a path $\mathbf{p}$ in a k-MDD as $C(\mathbf{p}) =_{\text{def}} \sum_{\mathbf{e_i} \in \mathbf{p}} C_i(\mathbf{e}_i)$. A k-MDD from Figure 2(b) labeled with respect to costs induced by *Hamming distance* is shown in Figure 3(a). Recall that a Hamming distance $\delta_i(v, v')$ for two assignments to a single variable x_i is 0 if $v = v'$, and 1 otherwise. A Hamming distance between two assignments to all n variables, $\mathbf{v^1}(v_1^1, \ldots, v_n^1)$, $\mathbf{v^2}(v_1^2, \ldots, v_n^2)$ is $\delta(\mathbf{v^1}, \mathbf{v^2}) = \sum_{i=1}^n \delta_i(v_i^1, v_i^2)$.

The cost of the longest path in M^k with respect to a labeling with respect to C as described above corresponds to the maximum value of a k-collection of solutions from Sol, i.e. to a solution to our Problem 1. In our example the longest path has length 3, i.e. there is a pair of solutions that differs on all three variable assignments. One such path is $r \rightarrow (a, b) \rightarrow (c, d) \rightarrow \mathbf{1}$, leading to a pair of solutions $((1, 1, 1), (2, 3, 2))$. The longest path is computed in a standard fashion, using Algorithm 1, computing for each node $\mathbf{u}$, in a bottom-to-top breadth-first search traversal, $MAX[\mathbf{u}] := max\{C(\mathbf{p}) \mid \mathbf{p} : \mathbf{u} \leadsto \mathbf{1}\}$.

The space requirements of our scheme are dominated by representing the MDD M^k, which requires $\Theta(\sum_{i=1}^n |V_i|^k + \sum_{i=1}^n |E_i|^k)$ space. The time complexity is determined by updates to $MAX[\mathbf{u}]$ on line 1 of Algorithm 1. Assuming that each update takes constant time, the time complexity is $\Theta(\sum_{i=1}^n |E_i|^k)$, since there is an update for each edge in M^k.

Algorithm 1. `Longest-Path(`M^k, C`)`

```
Data: MDD M^k(V^k, E^k), cost function C
MAX[1] = 0;
foreach i = n, ..., 1 do
    foreach u ∈ V_i^k do
        MAX[u] = −∞;
        foreach e : u → u' do
            MAX[u] = max{MAX[u], C_i(e) + MAX[u']};
return MAX[r];
```

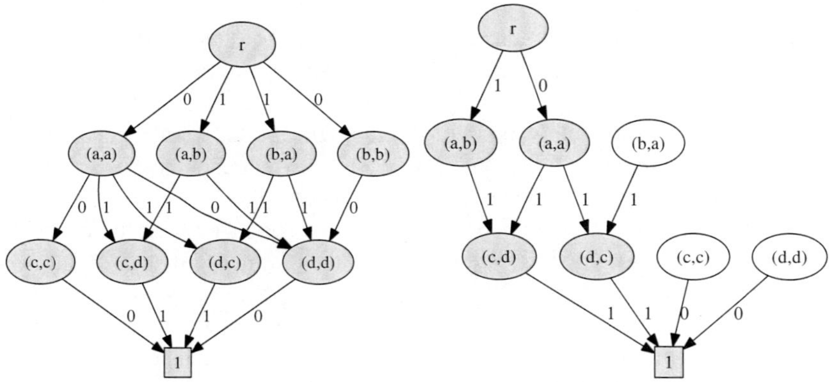

(a) Cost labeling of a k-MDD. (b) Approximate optimization for $T = 2$.

Fig. 3. Exact and approximate optimization over the k-MDD from Figure 2(b). Edges are labeled with respect to Hamming distance costs instead of values. In approximate optimization we process at most T of the most expensive nodes in each layer. Nodes that are processed are indicated as shaded in the figure.

5 Improvements over k-MDD Optimization

In the previous section we developed an approach to multi-solution queries that exploits the compactness of the initial MDD M, and the additivity of the cost function C. The complexity of our approach grows exponentially with k. It is guaranteed to succeed whenever the size of the MDD is small and whenever k is small. We now consider situations when this is not the case and present several extensions of the basic optimization scheme to reduce time and space requirements.

5.1 Implicit Optimization

Due to a one-to-one correspondence between the vertices and edges of M^k to k-tuples of the vertices and edges in M, we can execute Algorithm 1 *without constructing* M^k. The statement of the algorithm is identical – with the only difference being that instead of iterating over nodes $\mathbf{u} \in V_i^k$ we iterate over tuples $(u_1, \ldots, u_k) \in (V_i)^k$. Instead of updating $MAX[\mathbf{u}]$ with respect to $\mathbf{e} \in E_i^k$ we update $MAX[u_1, \ldots, u_k]$ with respect to tuples $(e_1, \ldots, e_k) \in (E_i)^k$. The space complexity depends only on maintaining and updating cost labels $MAX[u_1, \ldots, u_k]$. Since we have to store such labels for every k-tuple of nodes in each layer, and since we have to perform an update for every k-tuple of edges in each layer, the space complexity is $\Theta(\sum_{i=1}^{n} |V_i|^k)$ while time complexity is still $\Theta(\sum_{i=1}^{n} |E_i|^k)$. A significant reduction in space requirements can be further achieved by maintaining MAX labels only for tuples of nodes in two neighboring layers at a time. The space complexity then reduces to $\Theta(max_{i=2}^{n}\{|V_i|^k + |V_{i-1}|^k\})$.

5.2 Approximate Node-Limited Optimization

The implicit optimization scheme helps us meet memory restrictions but only in cases when there are no layers that are excessively large. However, in some domains this is not true and certain layers may be very large compared to others. Furthermore, even if we do satisfy memory restrictions, we still perform the same number of operations $\Theta(\sum_{i=1}^{n} |E_i|^k)$ as the original scheme. In settings where response times are important, this might be unacceptable.

We suggest an approximate version of the implicit optimization scheme. The main idea is to limit the number of nodes that can be processed at each layer, to a predefined threshold T. In each layer we traverse and process tuples $(u_1, \ldots, u_k)$ in *decreasing* order with respect to the maximal cost $MAX[u_1, \ldots, u_k]$. For each tuple, we update all the *parent* tuples. An approximate version of the Hamming distance optimization for $T = 2$ is illustrated in Figure 3(b). Algorithm 2 illustrates this approach. For each layer i we maintain the set of nodes S_i that have been updated in previous iterations. Furthermore, we maintain a *sorted* set with respect to the MAX labels, so that efficient access to the most expensive tuples $\mathbf{u}$ is possible.

Algorithm 2. `Approximate-Longest-Path(`M, C`)`

Data: MDD $M(V, E)$, cost function C, node threshold T
$S_i \leftarrow \emptyset, i = 1, \ldots, n$;
$S_{n+1} \leftarrow \{\mathbf{1}\}$;
$MAX[\mathbf{1}] \leftarrow 0$;
foreach $i = n+1, \ldots, 2$ **do**
    nodeCount $= 0$;
    while nodeCount $\leq T$ **do**
        $\mathbf{u}' =$ next most expensive tuple from S_i;
        nodeCount $++$;
        foreach $\mathbf{e} : \mathbf{u} \rightarrow \mathbf{u}'$ **do**
            if $\mathbf{u} \notin S_{i-1}$ **then**
                $MAX[\mathbf{u}] = -\infty$;
                $S_{i-1} \leftarrow S_{i-1} \cup \{\mathbf{u}\}$;
            $MAX[\mathbf{u}] = max\{MAX[\mathbf{u}], C_i(\mathbf{e}) + MAX[\mathbf{u}']\}$;
return $MAX[\mathbf{r}]$;

Complexity analysis. At the i-th layer we traverse and expand at most T nodes. If d_i is the largest number of incoming edges to any node in V_i, then the largest number of updated nodes from $(V_{i-1})^k$ is $T\cdot(d_i)^k$, i.e. $|S_{i-1}| \leq T\cdot(d_i)^k$. Hence, assuming that we maintain two neighboring layers during the traversal, the maximum space requirements are $max_{i=2}^n\{|S_i| + |S_{i-1}|\} \in O(max_{i=2}^n T \cdot ((d_{i+1})^k + (d_i)^k))$.

Since we perform at most $(d_i)^k$ updates for T nodes in each layer, we execute at most $\sum_{i=2}^{n+1} T \cdot (d_i)^k$ updates. However, in each update with respect to edge $\mathbf{e} : \mathbf{u} \to \mathbf{u}'$, we have to check whether parent node $\mathbf{u} \in S_{i-1}$. For typical implementations of the set structure (binary search tree) each membership check takes a logarithmic number of operations, i.e. $log_2(|S_{i-1}|) \leq log_2(T \cdot (d_i)^k)$. Hence, the time requirements of our scheme are $O(\sum_{i=2}^{n+1} T \cdot d_i \cdot log_2(T \cdot (d_i)^k))$.

6 Applications and Extensions

In general, we are interested in applications where it is desirable to return a collection of solutions in which there is a relationship between those solutions in terms of a *distance* [8,5]. Ensuring solution diversity is the most natural setting where such queries are desirable, but diversity can be defined in a number of interesting ways that depend on the application domain. We present some examples of useful distance measures and the associated settings that can benefit from our approach.

Maximum Hamming Distances. In this case the distance between two solutions is measured in terms of the number of variables with different assignments. We can use Algorithm 1 to solve the $\mathcal{NP}$-hard *Maximum Hamming Distance* problem [8,9,5] using quadratic space $O(max_{i=2}^n\{|V_i|^2 + |V_{i-1}|^2\})$ and quadratic time $O(\sum_{i=1}^n |E_i|^2)$ with respect to the input MDD M (which can be exponentially large in the input problem specification). The problem of finding solutions that maximize the Hamming Distance has applications in error correcting codes and electronics.

Optimally Diverse Collections. Given a solution set Sol and a desired size of subset k, we wish to compute a subset $S \subseteq Sol$ with maximum diversity among all k-subsets of Sol. For example, we might want to determine the most diverse collection of products that are initially shown to a new user before his preferences are elicited. For a tuple of paths $\mathbf{p}(p_1,\ldots,p_k)$, we define its cost $C(\mathbf{p})$ as the sum of all pairwise distances: $C(\mathbf{p}) = sum(\delta,(p_1,\ldots,p_k)) = \sum_{i,j=1}^k \delta(p_i,p_j)$. This is an additive objective function since

$$C(\mathbf{p}) = \sum_{i=1}^n sum(\delta,(e_i^1,\ldots,e_i^k)) = \sum_{i=1}^n C_i(\mathbf{e_i})$$

and, therefore, the problem can be solved by instantiating Algorithm 1. Note, however, that evaluating the cost of an edge, $C_i(\mathbf{e}) = sum(\delta,(e^1,\ldots,e^k))$, no longer takes constant time, but requires in worst case $k\cdot(k-1)/2$ summations. Therefore, the time complexity of the algorithm is $\Theta(\sum_{i=1}^n |E_i|^k \cdot k^2)$, while the space complexity remains $\Theta(max_{i=2}^n\{|V_i|^k + |V_{i-1}|^k\})$.

Subset Diversity. Several web search engines offer the user an option to view “Similar Pages”. The concept of finding a set of diverse solutions with some common attributes is powerful because it assists a user in an interactive setting to search a possibly large set of solutions more efficiently. We can support such solution discovery in a combinatorial problem setting using our approach. A trivial extension to Algorithm 1 involves fixing certain assignments in M^k and then finding a diverse set of solutions with this common assignment. One could envisage many potential settings, such as product configuration, where this concept may be valuable. For example, the user may wish to receive a diverse set of options for a product with certain attributes fixed.

Risk Management and Contingency Planning. When a solution faces a risk of failure, *e.g.* because a particular assignment becomes inconsistent in a dynamic setting, it is important that an alternative solution can be formed without making alterations to the initial solution that are too costly [10]. For queries seeking a solution that minimizes expected losses, it is necessary to determine the cost of a repair for all possible failures in a tractable manner. In such a setting, the robustness of any solution is inextricably linked to its neighboring solutions thereby presenting an application domain for distance queries. For example, in queries related to finding *robust solutions*, for every possible solution x there is a probability of its failure $P(x) \in [0, 1]$, in which case a *contingent* solution x' should be provided instead of x. Such a contingent solution x' should be of high value $c(x')$ but also as close as possible to the original solution x.

7 Empirical Evaluation

Given an MDD $M(V, E)$, let $V_{max} = max_{i=1}^{n}|V_i|$ denote the largest *node-width*, and let $E_{max} = max_{i=1}^{n}|E_i|$ denote the largest *edge-width* in the MDD M. The space and time complexities of our approach are determined by $(V_{max})^k$ and $(E_{max})^k$, respectively. While this is exponential in k, in many practical applications k is often small. It is argued, for example, that for recommender systems the optimal number of solutions to present to a user is $k = 3$ [11].

7.1 Scalability for Real-World Configuration Problems

We generated MDDs for a number of real-world configuration instances from CLib.[1] We present their relevant properties in Table 1. The first five columns indicate instance name, number of variables, number of solutions, size of compiled representation on disk, number of MDD nodes $|V|$, number of MDD edges $|E|$, maximal node-width V_{max} and edge-width E_{max}.

Table 1 indicates that even small MDDs can successfully represent huge solution spaces, thus highlighting the computational advantage of our methodology. We use node-widths V_{max} and edge-widths E_{max} as indicators of the worst-case values of dynamic programming data entries that must be stored simultaneously in our scheme, and the number of operations that have to be performed. The first five instances are quite manageable even for the exact optimization when k is small. However, the *Big-PC* instance and widely investigated *Renault* instance [1] might be out of reach, particularly with respect to space requirements.

[1] `http://www.itu.dk/research/cla/externals/clib`

Table 1. MDD properties for some real-world problems. (*) The original version of the Renault instance has 101 variables, two of which are unconstrained and therefore removed from the problem.

Instance	Variables	Solutions	Size (KB)	MDD Nodes	MDD Edges	V_{max}	E_{max}
ESVS	26	2^{31}	5	96	220	24	66
FS	23	2^{24}	41	767	2017	348	648
Bike2	34	2^{26}	56	933	1886	194	457
PC2	41	2^{20}	237	3907	6136	534	1211
PC	45	2^{20}	298	4875	7989	640	2240
Big-PC	124	2^{83}	7,945	100272	132889	11537	11961
Renault	99*	2^{41}	9,891	329135	426212	8609	17043

7.2 Comparison between Exact and Approximate Optimization

We evaluated the performance of the implicit optimization version of Algorithm 1 and the approximate node-limited Algorithm 2 using *Hamming distance* as a cost function when looking for two most distant solutions. The results are presented in Table 2. For the approximate scheme we indicate the smallest node threshold for the first five instances that is sufficient to reach optimality. For the largest two instances we indicate the smallest multiple of 1000 for which the threshold is sufficiently large to allow the discovery of the optimal solution. We can see that we can easily handle the first five instances. Furthermore, our exact scheme is able to handle both the largest instances *Big-PC* and *Renault*. The heuristic scheme that was used for instance *Renault* in [5] leads to an estimate of maximum Hamming distance 60. We computed the exact value, which is 71. Furthermore, for a threshold $T = 30000$ our approximate scheme is able to compute the maximum Hamming distance using an order-of-magnitude less time. It requires processing only a tiny fraction of all tuples that an exact scheme would process.

Table 2. Empirical results for exact and approximate version of our algorithmic scheme

Instance	Exact		Approximate		
	Longest Path	Time (s)	Longest Path	Time (s)	Threshold
ESVS	26	0.09	26	0.09	3
FS	22	0.1	22	0.09	1
Bike2	32	0.11	32	0.1	2
PC2	34	0.19	34	0.09	2
PC	37	0.29	37	0.1	39
Big-PC	118	27.81	118	1.92	30000
Renault	71	98.4	71	11.79	30000

7.3 Comparison against CSP Search Approach

Hebrard et al. evaluated their approach using a Renault instance [5]. We re-implemented a variant of that approach in Mistral,[2] based on the original paper and consultations with the authors. We used the model and objective function as described in Section 3 (based on introducing k copies of variables and constraints). We tried several search schemes, but the most effective one was to first branch on all X^i variables before branching on X^{i+1} variables, and to use restarts with a random value selection heuristic. The results for Hamming distance significantly outperformed the scheme from [5] in terms of both time and the Hamming distance found: in that paper a distance of 60 was achieved in 10 seconds. The variant we implemented found a pair of solutions at distance 69 within 0.39 seconds. However, finding a pair of solutions with distance 70 took 16min 1.55s, while finding a pair with distance 71 took $3h$, $32min$, 8s. The algorithm was unable to prove optimality even after more than 24 hours of computation.

Hence, while the search-based approach is very competitive for solution-rich instances with large number of distant pairs, finding the exact optimal solution seems to be very challenging if the optimal collection is tightly constrained, i.e. if there are not many collections satisfying the optimality criteria. In comparison, the exact MDD-based approach is able to find the optimal solution within 98 seconds. The results clearly demonstrate that MDD representations of the solution set Sol can be exploited to significantly enhance finding optimal collections of solutions. This result provides compelling evidence that the k-MDD approach significantly improves upon the state-of-the-art for finding optimal collections of solutions with an inter-solution cost function.

8 Related Work

A number of researchers have studied the problem of finding sets of solutions to combinatorial problems. For example, work in constraint satisfaction ensures that the Hamming distance between the two solutions is maximized [8,9] or that the solutions are within some distance of each other [12]. Dahllof [13] studied the problem of finding two XSAT models at maximum Hamming distance. Diversity and similarity are also important concepts in case-based reasoning and recommender systems [11,14]. Typically, we want chose a diverse set of cases from the case-base, but retrieve cases based on similarity. Hebrard *et al.* propose a number of practical solution methods in constraint programming for enforcing diversity (or similarity) between solutions [5], as well as the problem of satisfying complex distance queries in a constraint-based setting [6]. However, all of these approaches use search-based methods and are incapable of exploiting compiled representations. Our approach is the first one that utilizes compiled representation in this setting and is able to retrieve tightly-constrained optimal collections of solutions even in cases when it is prohibitively expensive for search-based methods.

9 Conclusion

We presented the first approach to retrieving optimal collections of solutions when the solution space has been compiled into a multi-valued decision diagram. We exploited

[2] `http://4c.ucc.ie/~ehebrard/Software.html`

the compactness of a compiled representation using our k-MDD approach, and then reduced the memory requirements and introduced an approximation scheme for larger problems. We discussed some important real-world queries that can now be handled using our approach. We empirically demonstrated the practical usability of our schemes on real-world configuration benchmarks, and demonstrated its superiority over a state-of-the-art search-based approach when the computation of exact optimal collections is required. In conclusion, we affirmed the veracity of our hypothesis; namely that knowledge compilation can be successfully exploited for reasoning about optimal collections of solutions.

In future work we plan to extend the algorithmic schemes presented in this paper and to evaluate their applicability in domains that include risk management and network reliability.

Acknowledgments. Tarik Hadzic is supported by an IRCSET/Embark Initiative Postdoctoral Fellowship Scheme. Barry O'Sullivan is supported by Science Foundation Ireland (Grant Number 05/IN/I886). We would like to thank Emmanuel Hebrard for the help with adopting the algorithms from [5].

References

1. Amilhastre, J., Fargier, H., Marquis, P.: Consistency restoration and explanations in dynamic CSPs-application to configuration. Artificial Intelligence (2002)
2. Hadzic, T., Subbarayan, S., Jensen, R.M., Andersen, H.R., Møller, J., Hulgaard, H.: Fast backtrack-free product configuration using a precompiled solution space representation. In: PETO Conference, DTU-tryk, June 2004, pp. 131–138 (2004)
3. Bryant, R.E.: Graph-based algorithms for boolean function manipulation. IEEE Transactions on Computers (1986)
4. Darwiche, A., Marquis, P.: A knowledge compilation map. Journal of Artificial Intelligence Research 17, 229–264 (2002)
5. Hebrard, E., Hnich, B., O'Sullivan, B., Walsh, T.: Finding diverse and similar solutions in constraint programming. In: Proceedings of AAAI 2005(July 2005)
6. Hebrard, E., O'Sullivan, B., Walsh, T.: Distance constraints in constraint satisfaction. In: IJCAI, pp. 106–111 (2007)
7. Hadzic, T., Hooker, J.N.: Cost-bounded binary decision diagrams for 0-1 programming. In: Van Hentenryck, P., Wolsey, L.A. (eds.) CPAIOR 2007. LNCS, vol. 4510, pp. 84–98. Springer, Heidelberg (2007)
8. Angelsmark, O., Thapper, J.: Algorithms for the maximum hamming distance problem. In: Faltings, B.V., Petcu, A., Fages, F., Rossi, F. (eds.) CSCLP 2004. LNCS (LNAI), vol. 3419, pp. 128–141. Springer, Heidelberg (2005)
9. Crescenzi, P., Rossi, G.: On the hamming distance of constraint satisfaction problems. Theor. Comput. Sci. 288(1), 85–100 (2002)
10. Holland, A., O'Sullivan, B.: Weighted super solutions for constraint programs. In: Proceedings of AAAI 2005, Pittsburgh, Pennsylvania (July 2005)
11. Shimazu, H.: Expertclerk: Navigating shoppers buying process with the combination of asking and proposing. In: IJCAI, pp. 1443–1450 (2001)
12. Bailleux, O., Marquis, P.: Some computational aspects of distance-sat. Journal of Automated Reasoning 37(4), 231–260 (2006)
13. Dahllöf, V.: Algorithms for max hamming exact satisfiability. In: Deng, X., Du, D.-Z. (eds.) ISAAC 2005. LNCS, vol. 3827, pp. 829–838. Springer, Heidelberg (2005)
14. Smyth, B., McClave, P.: Similarity vs. diversity. In: Aha, D.W., Watson, I. (eds.) ICCBR 2001. LNCS (LNAI), vol. 2080, pp. 347–361. Springer, Heidelberg (2001)

Constraints of Difference and Equality: A Complete Taxonomic Characterisation*

Emmanuel Hebrard[1], Dániel Marx[2], Barry O'Sullivan[1], and Igor Razgon[1]

[1] Cork Constraint Computation Centre
Department of Computer Science, University College Cork, Ireland
{e.hebrard,b.osullivan,i.razgon}@4c.ucc.ie
[2] Budapest University of Technology and Economics
Budapest, Hungary
dmarx@cs.bme.hu

Abstract. Many combinatorial problems encountered in practice involve constraints that require that a set of variables take distinct or equal values. The AllDifferent constraint, in particular, ensures that all variables take distinct values. Two soft variants of this constraint were proposed in [4], defined either with respect to a so-called *variable* or *graph*-based cost function. When requiring similarity, as opposed to diversity, one can consider the dual definition either for the cost or for the basic constraint itself, that is, AllEqual in our case. Six cost functions can be defined by exploring every combination of these definitions. It is therefore natural to study the complexity of achieving arc consistency and bounds consistency on them. From our earlier work on this topic an open problem remained, namely achieving bounds consistency on the maximisation of the SoftAllDiff constraint when considering the graph-based cost. In this paper we resolve this problem. Therefore, we give a complete taxonomy of constraints of equality and difference, based on the alternative objective functions used for the soft variants.

1 Introduction

Constraints for reasoning about the diversity or similarity of a set of variables are ubiquitous in constraint programming. For example, in a university timetabling problem we will want to ensure that all courses taken by a particular student are held at different times. Similarly, in meeting scheduling we will want to ensure that the participants of a meeting are scheduled to meet at the same time and in the same place. Sometimes, when the problem is over-constrained, we may wish to maximise the extent to which these constraints are satisfied. Consider again our timetabling example: we might wish to maximise the number of courses that are scheduled at different times when a student's preferences cannot all be met.

* Hebrard, O'Sullivan and Razgon are supported by Science Foundation Ireland (Grant Number 05/IN/I886). Marx is supported by the Magyary Zoltán Felsőoktatási Közalapítvány and the Hungarian National Research Fund (OTKA grant 67651).

I.P. Gent (Ed.): CP 2009, LNCS 5732, pp. 424–438, 2009.

In a constraint programming setting, these requirements are normally specified using global constraints. One of the most commonly used global constraints is the AllDifferent [6], which enforces that *all* variables take pair-wise different values. A soft version of the AllDifferent constraint, the SoftAllDiff, has been proposed by the authors of [4]. They proposed two cost metrics for measuring the degree of satisfaction of the constraint, which are to be minimised or maximised: *graph-* and *variable*-based cost. The former counts the number of equalities, whilst the latter counts the number of variables, violating an AllDifferent constraint. When we wish to enforce that a set of variables take equal values, we can use the AllEqual, or its soft variant, the SoftAllEqual constraint, which we recently introduced [3].

When considering these two constraints (AllDifferent and AllEqual), these two costs (graph-based and variable-based) and objectives (minimisation and maximisation) we can define eight algorithmic problems related to constraints of difference and equality. In fact, because the graph-based costs of AllDifferent and AllEqual are dual, only six distinct problems are defined.

When we introduced the SoftAllEqual constraint one open problem remained: namely, the design of an algorithm for achieving bounds consistency on the SoftAllEqual constraint when the objective is to maximise the number of equalities achieved in the decomposition graph of the constraint, i.e. the SoftAllEqual constraint defined by the graph-based cost. In this paper we resolve this open question, and propose an efficient bounds consistency algorithm for this case. This result enables us to fully characterise the complexity of achieving arc consistency and bounds consistency on each of the eight constraints in this class. This paper, therefore, provides a complete taxonomy of constraints of difference and equality.

The remainder of this paper is organised as follows. In Section 2 we introduce the necessary technical background. A complete taxonomy of constraints of equality and difference is presented in Section 3. In Section 4 we present the main technical contribution of the paper, namely the complexity of achieving bounds consistency on the SoftAllEqual when the objective is to optimise the graph-based cost. A filtering algorithm is proposed in Section 5. Concluding remarks are made in Section 6.

2 Background

Constraint Satisfaction. A constraint satisfaction problem (CSP) is a triplet $\mathcal{P} = (\mathcal{X}, \mathcal{D}, \mathcal{C})$ where $\mathcal{X}$ is a set of variables, $\mathcal{D}$ a mapping of variables to sets of values (without loss of generality, we assume $\mathcal{D}(X) \subset \mathbb{Z}$ for all $X \in \mathcal{X}$, and we denote by $min(X)$ and $max(X)$ the minimum and maximum values in $\mathcal{D}(X)$, respectively) and $\mathcal{C}$ a set of constraints that specify allowed combinations of values for subsets of variables. An assignment of a set of variables $\mathcal{X}$ is a set of pairs S such that $|\mathcal{X}| = |S|$ and for each $(X, v) \in S$, we have $X \in \mathcal{X}$ and $v \in \mathcal{D}(X)$. A constraint $C \in \mathcal{C}$ is *arc consistent* (AC) iff, when a variable in the scope of C is assigned any value, there exists an assignment of the other variables in C such that C is satisfied. This satisfying assignment is called a *domain support* for the value. Similarly, we call a *range support* an assignment satisfying C, but where values, instead of being taken from the domain of each variable ($v \in \mathcal{D}(X)$), can be any integer between the minimum and maximum of this domain following the natural order on $\mathbb{Z}$, that is, $v \in [min(X), \ldots, max(X)]$. A constraint $C \in \mathcal{C}$ is *range consistent* (RC) iff, every value of every variable in the scope of C has a range support. A constraint $C \in \mathcal{C}$ is *bounds consistent* (BC) iff, for every variable X in the scope of C, $min(X)$ and $max(X)$ have a range support. Given a CSP $\mathcal{P} = (\mathcal{X}, \mathcal{D}, \mathcal{C})$, we shall use the following notation throughout the paper: n shall denote the number of variables, i.e., $n = |\mathcal{X}|$; m shall denote the number of distinct unary assignments, i.e., $m = \sum_{X \in \mathcal{X}} |\mathcal{D}(X)|$; Λ shall denote the total set of values, i.e., $\Lambda = \bigcup_{X \in \mathcal{X}} \mathcal{D}(X)$; finally, λ shall denote the total number of distinct values, i.e., $\lambda = |\Lambda|$;

Soft Global Constraints. Adding a cost variable to a constraint to represent its degree of violation is now common practice in constraint programming. This model was introduced in [7]. It offers the advantage of unifying hard and soft constraints since arc consistency, along with other types of consistencies, can be applied to such constraints with no extra effort. As a consequence, classical constraint solvers can solve over-constrained problems modelled in this way without modification. This approach was applied to a number of other constraints, for instance in [9].

Two natural cost measures have been explored for the AllDifferent and for a number of other constraints. The *variable-based cost* counts how many variables need to change in order to obtain a valid assignment for the hard constraint. The *graph-based cost* counts how many times a component of a decomposition of the constraint is violated. Typically these components correspond to edges of a decomposition graph, e.g. for an AllDifferent constraint, the decomposition graph is a clique and an edge is violated if and only if both variables connected by this edge share the same value. For instance, still for the AllDifferent constraint, the following example shows two solutions involving four variables $X_1, \ldots, X_4$ each with domain $\{a, b\}$:

$$S_1 = \{(X_1, a), (X_2, b), (X_3, a), (X_4, b)\}$$

$$S_2 = \{(X_1, a), (X_2, b), (X_3, b), (X_4, b)\}$$

In both solutions, at least two variables must change (e.g., X_3 and X_4) to obtain a valid solution. Therefore, the variable-based cost is 2 for S_1 and S_2. However, in S_1 only two edges are violated, (X_1, X_3) and (X_2, X_4), whilst in S_2, three edges are violated, (X_2, X_3), (X_2, X_4) and (X_3, X_4). Thus, the graph-based cost of S_1 is 2 whereas it is 3 for S_2.

3 Taxonomy

In this section we introduce the taxonomy of the soft constraints related to AllDifferent and AllEqual. We consider the eight algorithmic problems

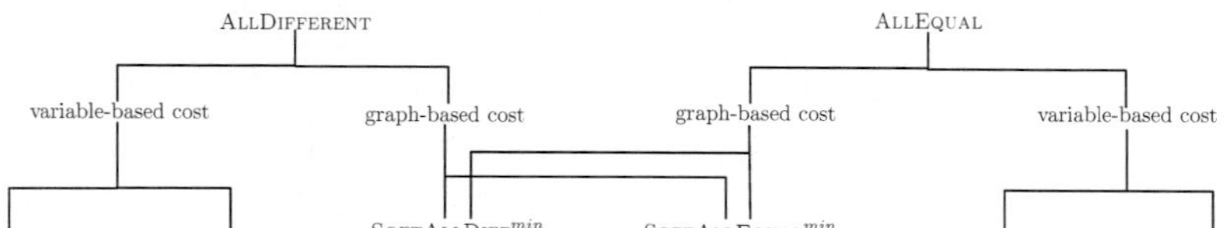

	SoftAllDiff$_V^{min}$	SoftAllDiff$_V^{max}$	SoftAllDiff$_G^{min}$ SoftAllEqual$_G^{max}$	SoftAllEqual$_G^{min}$ SoftAllDiff$_G^{max}$	SoftAllEqual$_V^{min}$	SoftAllEqual$_V^{max}$
AC	$\mathcal{O}(n\sqrt{m})$	NP-hard	$\mathcal{O}(nm)$	NP-hard	$\mathcal{O}(m)$	$\mathcal{O}(\sqrt{n}m)$
BC	$\mathcal{O}(n\sqrt{m})$	$\mathcal{O}(nlog(n))$	$\mathcal{O}(nm)$	$\mathcal{O}(min(\lambda^2, n^2)nm)$	$\mathcal{O}(m)$	$\mathcal{O}(nlog(n))$

Fig. 1. Complexity of optimising difference and equality

related to constraints of difference and equality defined by combining these two constraints, two costs (graph-based and variable-based), and two objectives (minimisation and maximisation). In fact, because the graph-based costs of AllDifferent and AllEqual are dual, only six different problems are thus defined. We close the last remaining cases: the complexity of achieving AC and BC SoftAllEqual$_V^{min}$ in this section, and that of achieving BC on SoftAllDiff$_G^{max}$ in Sections 4 and 5. Based on these results, Figure 1 can now be completed.

The next six paragraphs correspond to the six columns of Figure 1, i.e., to the twelve elements of the taxonomy. For each of them, we briefly outline the current state of the art, using the following assignment as a running example to illustrate the various costs: $S_3 = \{(X_1, a), (X_2, a), (X_3, a), (X_4, a), (X_5, b), (X_6, b), (X_7, c)\}$.

SoftAllDiff*: Variable-based cost, Minimisation.*

Definition 1 (SoftAllDiff$_V^{min}$)

$$\text{SoftAllDiff}_V^{min}(\{X_1, ..X_n\}, N) \Leftrightarrow N \geq n - |\{v \mid X_i = v\}|.$$

Here the cost to minimise is the number of variables that need to be changed in order to obtain a solution satisfying an AllDifferent constraint. For instance, the cost of S_3 is 4 since three of the four variables assigned to a as well as one of the variables assigned to b must change. This objective function was first studied in [4] where the authors give an algorithm for achieving AC in $\mathcal{O}(n\sqrt{m})$. To our knowledge, no algorithm with better time complexity for the special case of bounds consistency has been proposed for this constraint.

SoftAllDiff*: Variable-based cost, Maximisation.*

Definition 2 (SoftAllDiff$_V^{max}$)

$$\text{SoftAllDiff}_V^{max}(\{X_1, ..X_n\}, N) \Leftrightarrow N \leq n - |\{v \mid X_i = v\}|.$$

Here the same cost is to be maximised. In other words, we want to minimise the number of distinct values assigned to the given set of variables, since the complement of this number to n is exactly the number of variables to modify in order to obtain a solution satisfying an AllDifferent constraint. For instance, the cost of S_3 is 4 and the number of distinct values is $7-4=3$. This constraint was studied under the name AtMostNValues in [1] where the authors introduce an algorithm in $\mathcal{O}(nlog(n))$ to achieve BC, and in [2] where the authors show that achieving AC is NP-hard since the problem is isomorphic to Min Hitting Set.

SoftAllDiff: *Graph-based cost, Minimisation &* SoftAllEqual: *Graph-based cost, Maximisation.*

Definition 3 (SoftAllDiff$_G^{min}$ $\simeq$ SoftAllEqual$_G^{max}$)

$$\text{SoftAllDiff}_G^{min}(\{X_1,..X_n\},N) \Leftrightarrow N \geq |\{\{i,j\} \mid X_i = X_j \ \&\ i \neq j\}|.$$

Here the cost to minimise is the number of violated constraints when decomposing AllDifferent into a clique of binary NotEqual constraints. For instance, the cost of S_3 is 7 since four variables share the value a (six violations) and two share the value b (one violation). Clearly, it is equivalent to maximising the number of violated binary Equal constraints in a decomposition of a global AllEqual. Indeed, these two costs are complementary to $\binom{n}{2}$ of each other (on S_3: $7+14=21$). An algorithm in $\mathcal{O}(nm)$ for achieving AC on this constraint was introduced in [8]. Again, to our knowledge there is no algorithm improving this complexity for the special case of BC.

SoftAllEqual: *Graph-based cost, Minimisation &* SoftAllDiff: *Graph-based cost Maximisation.*

Definition 4 (SoftAllEqual$_G^{min}$ $\simeq$ SoftAllDiff$_G^{max}$)

$$\text{SoftAllEqual}_G^{min}(\{X_1,..X_n\},N) \Leftrightarrow N \geq |\{\{i,j\} \mid X_i \neq X_j \ \&\ i \neq j\}|.$$

Here we consider the same two complementary costs, however we aim at optimising in the opposite way. In [3] the authors show that achieving AC on this constraint is NP-hard, however the complexity of achieving BC is left as an open question. In this paper we show that computing the optimal cost can be done in $\mathcal{O}(min(n\lambda^2, n^3))$ thus demonstrating that BC can be achieved in polynomial time.

SoftAllEqual: *Variable-based cost, Minimisation.*

Definition 5 (SoftAllEqual$_V^{min}$)

$$\text{SoftAllEqual}_V^{min}(\{X_1,..X_n\},N) \Leftrightarrow N \geq n - max_{v \in \Lambda}(|\{i \mid X_i = v\}|)$$

Here the cost to minimise is the number of variables that need to be changed in order to obtain a solution satisfying an AllEqual constraint. For instance, the cost of S_3 is 3 since four variables already share the same value. This is equivalent to maximising the number of variables sharing a given value. Therefore this bound can be computed trivially by counting the occurrences of every value in the domains, that is, in $\mathcal{O}(m)$. On the other hand, pruning the domains according to this bound without degrading the time complexity is not as trivial, so we show how it can be done.

Theorem 1. *AC on* SoftAllEqual$_V^{min}$ *can be achieved in* $\mathcal{O}(m)$ *steps.*

Proof. We suppose, without loss of generality, that the current upper bound on the cost is k. We first compute the number of occurrences $occ(v)$ for each value $v \in \Lambda$, which can be done in $\mathcal{O}(m)$. There are three cases to consider:

1. First, consider the case where no value appears in $n - k$ domains or more ($\forall v \in \Lambda,\ occ(v) < n - k$). In this case the constraint is violated, hence every value is inconsistent.
2. Second, consider the case where at least one value v appears in the domains of at least $n - k + 1$ variables ($\exists v \in \Lambda,\ occ(v) > n - k$). In this case we can build a support for every value $w \in \mathcal{D}(X)$ by assigning all variables in $\mathcal{X} \setminus X$ with v if possible. The resulting assignment has a cost of k, hence every value is consistent.
3. Otherwise, if neither of the two cases above hold, we know that no value appears in more than $n - k$ domains, and that at least one appears $n - k$ times, let W denote the set of such values. In this case, the pair (X, v) is inconsistent iff $v \notin W$ & $W \subset \mathcal{D}(X)$.

   We first suppose that this condition does not hold and show that we can build a support. If $v \in W$ then clearly we can assign every possible variable to v and achieve a cost of k. If $W \not\subset \mathcal{D}(X)$, then we consider w such that $w \in W$ and $w \notin \mathcal{D}(X)$. By assigning every variable with w when possible we achieve a cost of k.

   Now we suppose that this condition holds and show that (X, v) does not have an AC support. Indeed once X is assigned to v the domains are such that no value appear in $n - k$ domains or more, since *every* value in W has now one less occurrence, hence we are back to Case 1.

Computing values satisfying the condition above can be done easily once the number of occurrences have been computed. In Case 3, the domain can be pruned down to the set W of values whose number of occurrences is $n - k$. □

SoftAllEqual: *Variable-based cost, Maximisation.*

Definition 6 (SoftAllEqual$_V^{max}$)

$$\text{SoftAllEqual}_V^{max}(\{X_1, ..X_n\}, N) \Leftrightarrow N \leq n - max_{v \in \Lambda}(|\{i \mid X_i = v\}|)$$

Here the same cost has to be maximised. In other words we want to minimise the maximum cardinality of a value. For instance, the cost of S_3 is 3, that is the complement to n of the maximum cardinality of a value ($3 = 7 - 4$). This is exactly equivalent to applying a Global Cardinality constraint (considering only the upper bounds on the cardinalities). In [5] the authors introduce an algorithm in $\mathcal{O}(\sqrt{n}m)$ and in $\mathcal{O}(nlog(n))$ for achieving AC and BC, respectively, on this constraint.

4 The Complexity of Bounds Consistency on SoftAllEqual$_G^{min}$

In this section we introduce an efficient algorithm that, assuming the domains are discrete intervals, computes the maximum possible number of pairs of equal values in an assignment. This algorithm allows us to close the last remaining open complexity question in Figure 1: BC on the SoftAllEqual$_G^{min}$ constraint. We then improve this algorithm, first by reducing the time complexity thanks to a preprocessing step, before turning it into a filtering method in Section 5.

We start by introducing additional terminology. Given two integers a and b, $a \leq b$, we say that the set of all integers x, $a \leq x \leq b$ is an *interval* and denote it by $[a, b]$. Let $\mathcal{X}$ be the set of variables of the considered CSP and assume that the domains of all the variables of $\mathcal{X}$ are sub-intervals of $[1, \lambda]$. We denote by $\mathbf{ME}(\mathcal{X})$ the set of all assignments P to the variables of $\mathcal{X}$ such that the number of pairs of equal values of P is the maximum possible. The subset of $\mathcal{X}$ containing all the variables whose domains are subsets of $[a, b]$ is denoted by $\mathcal{X}_{a,b}$. The subset of $\mathcal{X}_{a,b}$ including all the variables containing the given value c in their domains is denoted by $\mathcal{X}_{a,b,c}$. Finally the number of pairs of equal values in an element of $\mathbf{ME}(\mathcal{X}_{a,b})$ is denoted by $C_{a,b}(\mathcal{X})$ or just $C_{a,b}$ if the considered set of variables is clear from context. For notational convenience, if $b < a$, then we set $\mathcal{X}_{a,b} = \emptyset$ and $C_{a,b} = 0$. The value $C_{1,\lambda}(\mathcal{X})$ is the number of equal pairs of values in an element of $\mathbf{ME}(\mathcal{X})$.

Theorem 2. *$C_{1,\lambda}(\mathcal{X})$ can be computed in $\mathcal{O}((n + \lambda)\lambda^2)$ steps.*

Proof. The problem is solved by a dynamic programming approach: for every a, b such that $1 \leq a \leq b \leq \lambda$, we compute $C_{a,b}$. The main observation that makes it possible to use dynamic programming is the following: in every $P \in \mathbf{ME}(\mathcal{X}_{a,b})$, there is a value c ($a \leq c \leq b$) such that every variable $X \in \mathcal{X}_{a,b,c}$ is assigned value c. To see this, let value c be a value that is assigned by P to a maximum number of variables. Suppose that there is a variable X with $c \in \mathcal{D}(X)$ that is assigned by P to a different value, say c'. Suppose that c and c' appear on x and y variables, respectively. By changing the value of X from c' to c, we increase the number of equalities by $x - (y - 1) \geq 1$ (since $x \geq y$), contradicting the optimality of P.

Notice that $\mathcal{X}_{a,b} \setminus \mathcal{X}_{a,b,c}$ is the disjoint union of $\mathcal{X}_{a,c-1}$ and $\mathcal{X}_{c+1,b}$ (if $c-1 < a$ or $c+1 > b$, then the corresponding set is empty). These two sets are independent in the sense that there is no value that can appear on variables from both sets. Thus it can be assumed that $P \in \mathbf{ME}(\mathcal{X}_{a,b})$ restricted to $\mathcal{X}_{a,c-1}$ and $\mathcal{X}_{c+1,b}$ are elements of $\mathbf{ME}(\mathcal{X}_{a,c-1})$ and $\mathbf{ME}(\mathcal{X}c + 1, b)$, respectively. Taking into consideration all possible values c, we get

$$C_{a,b} = \max_{c, a \leq c \leq b} \left(\binom{|\mathcal{X}_{a,b,c}|}{2} + C_{a,c-1} + C_{c+1,b} \right). \qquad (1)$$

In the first step of the algorithm, we compute $|\mathcal{X}_{a,b,c}|$ for all values of a, b, c. For each triple a, b, c, it is easy to compute $|\mathcal{X}_{a,b,c}|$ in time $\mathcal{O}(n)$, hence all these values

Algorithm 1. Computing $C_{1,\lambda}(\mathcal{X})$

```
  ∀ 1 ≤ a,b,c ≤ λ, δ_{a,b,c} ← |X_{a,b,c}| ← C_{a,b} ← 0;
  foreach k ∈ [0, λ − 1] do
      foreach a ∈ [1, λ] do
          b ← a + k;
          foreach X ∈ X_{a,b} do
1             δ_{a,b,min(X)} ← δ_{a,b,min(X)} + 1;
2             δ_{a,b,max(X)+1} ← δ_{a,b,max(X)+1} − 1;
          foreach c ∈ [a, b] do
3             |X_{a,b,c}| ← |X_{a,b,c−1}| + δ_{a,b,c};
4             C_{a,b} ← max(C_{a,b}, (binom(|X_{a,b,c}|, 2) + C_{a,c−1} + C_{c+1,b}));
  return C_{1,λ};
```

can be computed in time $\mathcal{O}(n\lambda^3)$. However, the running time can be reduced to $\mathcal{O}((n+\lambda)\lambda^2)$ as follows. For each pair a, b, we determine all the values $|\mathcal{X}_{a,b,c}|$, $a \le c \le b$ in time $\mathcal{O}(n+\lambda)$. More precisely, we define $\delta_{a,b,c} = |\mathcal{X}_{a,b,c}| - |\mathcal{X}_{a,b,c-1}|$ and compute $\delta_{a,b,c}$ for every $a < c \le b$ (Alg. 1, Line 1-2). Observe that if $\mathcal{D}(X) = [a', b']$ for some $X \in \mathcal{X}_{a,b}$, then X contributes only to two of the $\delta_{a,b,c}$ values: it increases $\delta_{a,b,a'}$ by 1 and decreases $\delta_{a,b,b'+1}$ by 1. Thus by going through all variables, we can compute the $\delta_{a,b,c}$ values for a fixed a, b and for all $a \le c \le b$ in time $\mathcal{O}(n)$ and we can also compute $|\mathcal{X}_{a,b,a}|$ in the same time bound. Now we can compute the values $|\mathcal{X}_{a,b,c}|$, $a < c \le b$ in time $\mathcal{O}(\lambda)$ by using the equality $|\mathcal{X}_{a,b,c}| = |\mathcal{X}_{a,b,c-1}| + \delta_{a,b,c}$ iteratively (Alg. 1, Line 3).

In the second step of the algorithm, we compute all the values $C_{a,b}$. We compute these values in increasing order of $b-a$. If $a = b$, then $C_{a,b} = \binom{|\mathcal{X}_{a,a,a}|}{2}$. Otherwise, values $C_{a,c-1}$ and $C_{c+1,b}$ are already available for every $a \le c \le b$, hence $C_{a,b}$ can be determined in time $\mathcal{O}(\lambda)$ using Eq. (1) (Alg. 1, Line 4). Thus all the values $C_{a,b}$ can be computed in time $\mathcal{O}(\lambda^3)$, including $C_{1,\lambda}$, which is the value of the optimum solution of the problem. Using standard techniques (storing for each $C_{a,b}$ a value c that minimises (1)), a third step of the algorithm can actually produce a variable assignment that obtains the maximum value. □

Algorithm 1 computes the largest number of equalities one can achieve by assigning a set of variables with interval domains. It can therefore be used to find an optimal solution to either SoftAllDiff$_G^{max}$ or SoftAllEqual$_G^{min}$. Notice that for the latter one needs to take the complement to $\binom{n}{2}$ in order to get the value of the violation cost. Clearly, it follows that achieving range or bounds consistency on these two constraints can be done in polynomial time, since Algorithm 1 can be used as an oracle for testing the existence of a range support. We give an example of the execution of Algorithm 1 in Figure 2. A set of ten variables, from X_1 to X_{10} are represented. Then we give the table $C_{a,b}$ for all pairs $a, b \in [1, \lambda]$.

The complexity can be further reduced if $\lambda >> n$. Let $\mathcal{X}$ be a set of variables with interval domains on $[1, \lambda]$. Consider the function $occ : Q \mapsto [0..n]$, where

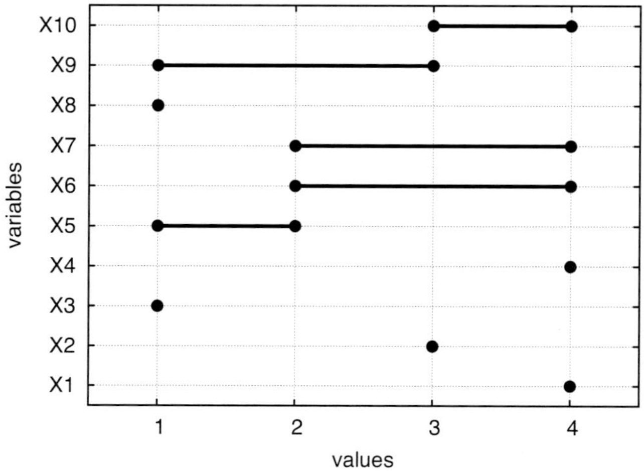

$C_{a,b}$	$a=1$	$a=2$	$a=3$	$a=4$
$b=1$	1			
$b=2$	$\mathcal{X}_{1,2,1}+C_{2,2}=3$	0		
$b=3$	$\mathcal{X}_{1,3,1}+C_{2,3}=6$	0	0	
$b=4$	$\mathcal{X}_{1,4,1}+C_{2,4}=16$	$\mathcal{X}_{2,4,4}+C_{2,3}=6$	$\mathcal{X}_{3,4,4}+C_{3,3}=3$	1

Fig. 2. A set of intervals, and the corresponding dynamic programming table ($C_{a,b}$)

$Q \subset \mathbb{Q}$ is a set of values of the form $a/2$ for some $a \in \mathbb{Z}$, such that $min(Q) = 1$ and $max(Q) = \lambda$. Intuitively, the value of $occ(a)$ is the number of variables whose domain interval encloses the value a, more formally:

$$\forall a \in Q,\ occ(a) = |\{X \mid X \in \mathcal{X}, min(X) \leq a \leq max(X)\}|.$$

Such an occurrence function, along with the corresponding set of intervals, is depicted in Figure 3. The *crest* of the function *occ* is an interval $[a,b] \in Q$ such that for some $c \in [a,b]$, *occ* is monotonically increasing on $[a,c]$ and monotonically decreasing on $[c,b]$. For instance, on the set intervals represented in Figure 3, $[1,15]$ is a crest since it is monotonically increasing on $[1,12]$ and monotonically decreasing on $[12,15]$.

Let $\mathcal{I}$ be a partition of $[1,\lambda]$ into a set of intervals such that every element of $\mathcal{I}$ is a crest. For instance, $\mathcal{I} = \{[1,15],[16,20],[21,29],[30,42]\}$ is such a partition for the set of intervals shown in Figure 3. We denote by $R_{\mathcal{I}}(\mathcal{X})$ the reduction of $\mathcal{X}$ by the partition $\mathcal{I}$. The reduction has as many variables as $\mathcal{X}$ (2) but the domains are replaced with the set of intervals in $\mathcal{I}$ that overlap with the corresponding variable in $\mathcal{X}$ (3).

$$R_{\mathcal{I}}(\mathcal{X}) = \{X' \mid X \in \mathcal{X}\} \tag{2}$$

$$\forall X \in \mathcal{X},\ \mathcal{D}(X') = \{I \mid I \in \mathcal{I}\ \&\ \mathcal{D}(X) \cap I \neq \emptyset\} \tag{3}$$

For instance, the set of intervals depicted in Figure 3 can be reduced to the set shown in Figure 2, where each element in $\mathcal{I}$ is mapped to an integer in $[1,4]$.

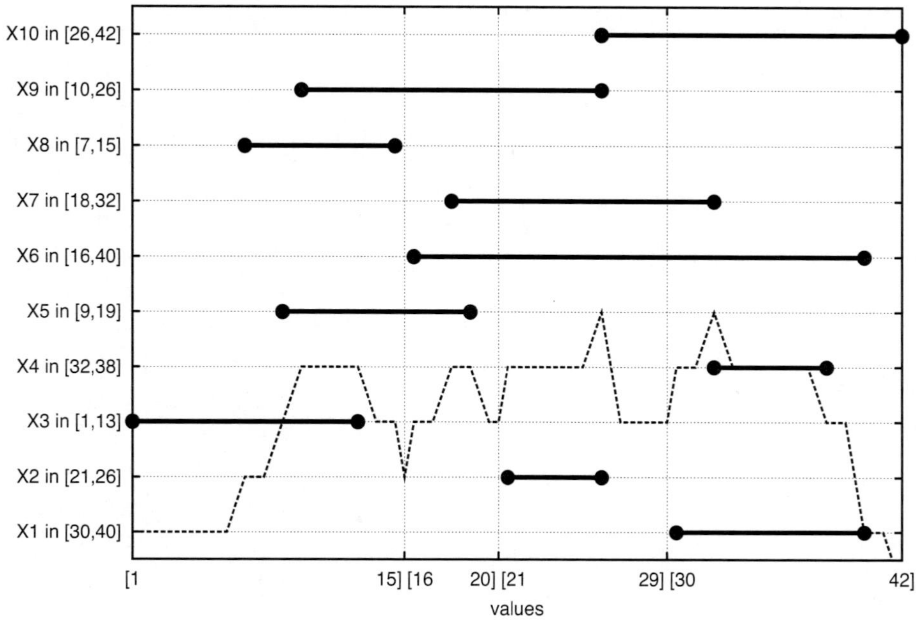

Fig. 3. Some intervals and the corresponding *occ* function

Theorem 3. *If $\mathcal{I}$ is a partition of $[1, \lambda]$ such that every element of $\mathcal{I}$ is a crest of occ, then* $\mathbf{ME}(\mathcal{X}) = \mathbf{ME}(R_{\mathcal{I}}(\mathcal{X}))$.

Proof. First, we show that for any optimal solution $s \in \mathbf{ME}(\mathcal{X})$, we can produce a solution $s' \in \mathbf{ME}(R_{\mathcal{I}}(\mathcal{X}))$ that has the same number of equalities as s. Indeed, for any value a, consider every variable X assigned to this value, that is, such that $s[X] = a$. Let $I \in \mathcal{I}$ be the crest containing a; by definition we have $I \in \mathcal{D}(X')$. Therefore we can assign all these variables to the same value I.

Now we show the opposite, that is, given a solution to the reduced problem, one can build a solution to the original problem with the same cost. The key observation is that, for a given crest $[a, b]$, all intervals overlapping with $[a, b]$ have a common value. Indeed, suppose that this is not the case, that is, there exists $[c_1, d_1]$ and $[c_2, d_2]$ both overlapping with $[a, b]$ such that $d_1 < c_2$. Then $occ(d_1) > occ(d_1 + \frac{1}{2})$ and similarly $occ(c_2 - \frac{1}{2}) < occ(c_2)$. However, since $a \leq d_1 < c_2 \leq b$, $[a, b]$ would not satisfy the conditions for being a crest, hence a contradiction. Therefore, for a given crest I, and for every variable X' such that $s'[X'] = I$, we can assign X to this common value. □

We show that this transformation can be achieved in $\mathcal{O}(nlog(n))$ steps. Observe that $\delta_{1,a,\lambda}$, if it was defined on Q rather than $[1, \lambda]$, would in fact be the derivative of *occ*. Moreover, we can compute it in $\mathcal{O}(nlog(n))$ steps as shown in Algorithm 2. We first compute the non-null values of $\delta_{1,a,\lambda}$ by looping through each variable $X \in \mathcal{X}$ (Line 1). Then we sort them, and finally we create the partition into

Algorithm 2. Computing a partition into crests $\mathcal{I}$

$\delta \leftarrow \emptyset$;
1 **foreach** $X \in \mathcal{X}$ **do**
  if $\exists(min(X), k) \in \delta$ **then**
    replace $(min(X), k)$ with $(min(X), k+1)$ in δ;
  else
    add $(min(X), 1)$ to δ;
  if $\exists(max(X)+1, k) \in \delta$ **then**
    replace $(max(X)+\frac{1}{2}, k)$ with $(max(X)+\frac{1}{2}, k-1)$ in δ;
  else
    add $(max(X)+\frac{1}{2}, -1)$ to δ;
sort δ by increasing first element;
$\mathcal{I} \leftarrow \emptyset$;
$min \leftarrow max \leftarrow 1$;
2 **while** $\delta \neq \emptyset$ **do**
  $polarity \leftarrow pos$;
  $k = 1$;
  repeat
    pick and remove the first element (a, k) of δ;
    $max \leftarrow round(a) - 1$;
    if $polarity = pos$ & $k < 0$ **then** $polarity \leftarrow neg$;
  until $polarity = pos$ *or* $k < 0$;
  add $[min, max]$ to $\mathcal{I}$;
  $min \leftarrow max + 1$;

crests by going through the derivative once and identifying the inflection points. Clearly, the number of elements in δ is bounded by $2n$. Therefore, the complexity of Algorithm 2 is dominated by the sorting of its elements, hence the $\mathcal{O}(nlog(n))$ worst-case time complexity.

Therefore, we can replace every crest by a single value as a preprocessing step and then run Algorithm 1. Moreover, since the number of crests is bounded by n, we obtain the following theorem, where n stands for the number of variables, λ for the number of distinct values, and m for the sum of all domain sizes.

Theorem 4. *RC on* SoftAllEqual$_G^{min}$ *can be achieved in* $\mathcal{O}(min(\lambda^2, n^2)nm)$ *steps.*

Proof. If $\lambda \leq n$ then one can achieve range consistency by iteratively calling Algorithm 1 after assigning each of the $\mathcal{O}(m)$ unit assignments $((X, v)\ \forall X \in \mathcal{X}, v \in \mathcal{D}(X))$. The resulting complexity is $\mathcal{O}(n\lambda^2)m$ (the term λ^3 is absorbed by $n\lambda^2$ due to $\lambda \leq n$). Otherwise, we apply the above $\mathcal{O}(nlog(n))$ procedure and similarly achieve range consistency after that. Since after the reformulation $\lambda = \mathcal{O}(n)$, the resulting complexity is $\mathcal{O}(n^3m)$. □

5 A Filtering Method for SoftAllEqual$_G^{min}$

In this section, we show that in the particular case where the cost variable N to minimise is such that $max(N) = (\binom{n}{2} - C_{1,\lambda})$, RC can be achieved in the same time complexity as that for computing $C_{1,\lambda}$. In other words, once the current lower bound, computed by Algorithm 1, exactly matches the required value $max(N)$, we can compute and prune range inconsistent values at no extra computational cost. This is an important particular case. Indeed, on the one hand, if $max(N)$ is not as high as this lower bound, the constraint should fail hence no pruning is required. On the other hand, if $max(N)$ is strictly larger than this lower bound, it is less likely that pruning should occur since the constraint is less tight.

We show how one can compute all the values participating in an optimal solution in $\mathcal{O}(min(n\lambda^2, \lambda^3))$ steps. When the cost $(\binom{n}{2} - C_{1,\lambda})$ of an optimal solution matches $max(N)$ all values either participate in an optimal solution or in no solution at all.

In the first step we run Algorithm 1, hence in the rest of the section we assume that all values $C_{a,b}(\mathcal{X})$ and $|X_{a,b,c}|$ are known (i.e. can be computed in a constant time). Moreover, we introduce the following additinal notation:

- Let $V_{a,b}(\mathcal{X})$ be the set of all values c such that $C_{a,b}(\mathcal{X}) = \binom{|\mathcal{X}_{a,b,c}|}{2} + C_{a,c-1}(\mathcal{X}) + C_{c+1,b}(\mathcal{X})$. In other words $c \in V_{a,b}(\mathcal{X})$ if there is an optimal assignment to the set of variables whose domains are subintervals of $[a,b]$, where each variable containing c in its domain is assigned with c. By the above assumption, given an interval $[a,b]$ and $c \in [a,b]$, it is possible to check in $\mathcal{O}(1)$ whether $c \in V_{a,b}(\mathcal{X})$.
- Let $H(\mathcal{X})$ be the *descendance* graph naturally defined by V, that is, whose set of vertices are all sub-intervals $[a,b]$ of $[1,n]$ and there is an arc $([a,b],[c,d])$ if and only if $a = c$ & $(d+1) \in V_{a,b}(\mathcal{X})$ or $d = b$ & $(c-1) \in V_{a,b}(\mathcal{X})$. In other words, given an interval $[a,b]$, any value $c \in V_{a,b}(\mathcal{X})$ such that $a < c < b$ defines two 'children' of $[a,b]$: one is $[a,c-1]$, the other is $[c+1,d]$. If $c = a$ or $c = d$ then there is only one child, namely $[c+1,d]$ and $[a,c-1]$, respectively (of course, if $a = b$ then no other interval is a child of $[a,b]$).
- We say that $[c,d]$ is a *descendant* of $[a,b]$ if $[c,d] = [a,b]$ or $H(\mathcal{X})$ has a path from $[a,b]$ to $[c,d]$. Since computing whether $c \in V_{a,b}(\mathcal{X})$ can be done in $\mathcal{O}(1)$, it is possible to compute in $\mathcal{O}(\lambda)$ the set of arcs leaving $[a,b]$. Therefore, by applying a DFS- or BFS-like method, it is possible to compute in $\mathcal{O}(\lambda^3)$ the set of all descendants of $[1,\lambda]$.
- We say that the interval $[a,b]$ is a *witness* of (X,c) if and only if $a \leq min(X) \leq c \leq max(X) \leq b$, $[a,b]$ is a descendant of $[1,\lambda]$ and $c \in V_{a,b}(\mathcal{X})$.

The intuition behind the following proof is that for an assignment (X,c) to belong to an optimal solution $P \in \mathbf{ME}(\mathcal{X})$, two conditions need to hold. First, the value c must be involved in an optimal solution, that is, it must belong to some set $V_{a,b}(\mathcal{X})$ such that $[a,b]$ is a descendant of $[1,\lambda]$. Second, there must exist at least one variable $X \in \mathcal{X}$ whose domain is included in $[a,b]$ and such that $c \in \mathcal{D}(X)$. The notion of witness defined above encapsulates these two conditions, we therefore look for a witness $[a,b]$ for (X,c).

We shall proceed by induction: if c belongs to $V_{1,\lambda}(\mathcal{X})$, then $[1,\lambda]$ is a witness. Otherwise, there is some value $d \in V_{1,\lambda}(\mathcal{X})$ such that either $c \in [a,d-1]$ or in $c \in [d+1,b]$. Moreover, (X,c) belongs to the optimal assignment of the variables whose domains are subsets of one of these intervals. We show in the proof that, by proceeding with such an inductive descent, we will eventually encounter a descendant $[a', b']$ of $[1, \lambda]$ such that $c \in V_{a',b'}(\mathcal{X})$ and this interval $[a', b']$ will be the desired witness of (X, c).

The proof of the opposite direction will require a more careful consideration of the descendance relation defined in the list above. In particular, we will notice that if $[a', b']$ is a descendant of $[1, \lambda]$ then any optimal assignment P to the variables whose domains are subsets of $[a', b']$ is a subset of an optimal assignment of $\mathcal{X}$. It will follow that if (X, c) participates in such an assignment P then it participates in a globally optimal assignment as well.

Lemma 1. *(X, c) belongs to an assignment $P \in \mathbf{ME}(\mathcal{X})$ if and only if (X, c) has a witness.*

Proof. Let $[a, b]$ be the domain of X and let $[a', b']$ be a witness of $[a, b]$. Observe first that since $[a', b']$ is a descendant of $[1, \lambda]$, any element $P' \in \mathbf{ME}(\mathcal{X}_{a,b})$ is a subset of an element of $\mathbf{ME}(\mathcal{X})$. This is clear if $[a', b'] = [1, \lambda]$. Assume that $([1, \lambda], [a', b'])$ is an arc of $H(\mathcal{X})$ and assume without loss of generality that $a' = 1$ and $(b' + 1) \in V_{1,\lambda}(\mathcal{X})$. That is, $C_{1,\lambda}(\mathcal{X}) = \binom{|\mathcal{X}_{1,\lambda,b'+1}|}{2} + C_{a',b'}(\mathcal{X}) + C_{b'+2,\lambda}(\mathcal{X})$. That is, any assignment to $\mathcal{X}_{a',b'}$ that results in $C_{a',b'}(\mathcal{X})$ of equal pairs of values (in other words, any assignment of $\mathbf{ME}(\mathcal{X}_{a',b'})$) can be extended to an assignment of $\mathbf{ME}(\mathcal{X})$. If the distance in $H(\mathcal{X})$ between $[1, \lambda]$ and $[a', b']$ is greater than 1 the observation can be proved by induction applying the argument for distance 1 along the shortest path from $[1, \mathcal{X}]$ to $[a, b]$.

Since $c \in V_{a',b'}(\mathcal{X})$ there is $P' \in \mathbf{ME}(\mathcal{X}_{a',b'})$ where all the variables having c in their domains are assigned with c. Since $[a, b] \subseteq [a', b']$, $(X, c) \in P'$. According to the previous paragraph P' is a subset of assignment $P \in \mathbf{ME}(\mathcal{X})$. Consequently, $(X, c) \in P$ as required.

Conversely, assume that $(X, c) \in P \in \mathbf{ME}(\mathcal{X})$. Observe that in this case either $c \in V_{1,\lambda}(\mathcal{X})$ or there is an interval $[a^*, b^*]$ such that $([1, \lambda], [a^*, b^*])$ is an arc in $H(\mathcal{X})$ and $[a, b] \subseteq [a^*, b^*]$. Indeed assume that $c \notin V_{1,\lambda}(\mathcal{X})$ and let $P \in \mathbf{ME}(\mathcal{X})$ such that $(X, c) \in P$. As has been observed in Theorem 2, there is a value d such that for any variable X' having d in its domain, $(X', d) \in P$. It follows that $d \in V_{1,\lambda}(\mathcal{X})$ and hence $c \neq d$. Then either $[a, b] \subseteq [1, d - 1]$ (i.e. $[a^*, b^*] = [1, d - 1]$) or $[a, b] \subseteq [d + 1, \lambda]$ (i.e. $[a^*, b^* = [d + 1, \lambda]$) because otherwise X contains d in its domain and hence $(X, d) \in P$ in contradiction to our assumption that $(X, c) \in P$.

We proceed by induction on the difference between $\lambda - 1$ and $b - a$. If the difference is 0 then, since $[a, b] \subseteq [1, \lambda]$, it follows that $[a, b] = [1, \lambda]$. Thus $[a, b]$ cannot be a proper sub-interval of $[1, \lambda]$ and hence, according to the previous paragraph, $c \in V_{1,\lambda}(\mathcal{X})$. Clearly, $[1, \lambda]$ is a witness of $[a, b]$. Assume now that $(\lambda - 1) - (b - a) > 0$. If $c \in V_{1,\lambda}(\mathcal{X})$ then $[1, \lambda]$ is a witness of $[a, b]$. Otherwise, let $[a^*, b^*]$ be as in the previous paragraph. By the induction assumption, (X, c) has a witness $[a', b']$ with respect to $\mathcal{X}_{a^*,b^*}$. In other words, $[a, b] \subseteq [a', b']$, $c \in V_{a',b'}(\mathcal{X}_{a^*,b^*})$ and $[a', b']$ is a descendant of $[a^*, b^*]$ in $H(\mathcal{X}_{a^*,b^*})$. Since $\mathcal{X}_{a',b'} \subseteq \mathcal{X}_{a^*,b^*}, c \in V_{a',b'}(\mathcal{X})$. It is also not hard to observe that $H(\mathcal{X}_{a^*,b^*})$ is a sub-graph of $H(\mathcal{X})$. It follows that $[a', b']$ is a witness of (X, c) with respect to $\mathcal{X}$. □

Lemma 1 allows us to achieve RC in the following way. For each value (X, c) such that $\mathcal{D}(X) = [a, b]$, check for all super-intervals $[a', b']$ whether $[a', b']$ is a witness of $[a, b]$. Since there are $\mathcal{O}(n\lambda)$ possible values, $\mathcal{O}(\lambda^2)$ super-intervals of the given interval and the witness relation can be checked in $\mathcal{O}(1)$, bounds consistency can be achieved in $\mathcal{O}(n\lambda^3)$. This runtime can be further reduced if before exploring the values we compute an auxiliary Boolean three-dimensional array $MarkedValues$ using the following procedure. Order the sub-intervals $[a, b]$ of $[1, \lambda]$ by decreasing difference $b - a$ and explore them according to this order. For the given interval $[a, b]$, explore all values $c \in [a, b]$ and set $MarkedValues[a][b][c]$ to 1 if and only if the one of the following conditions is true:

1. $[a, b]$ is a descendant of $[1, \lambda]$ in $H(\mathcal{X})$ and $c \in V_{a,b}(\mathcal{X})$;
2. $MarkedValues[a-1][b][c] = 1$ (only if $a > 1$);
3. $MarkedValues[a][b+1][c] = 1$ (only if $b < \lambda$).

If none of the above conditions are satisfied then $MarkedValues[a][b][c]$ is set to 0. Clearly, computing the $MarkedValues$ array takes $\mathcal{O}(\lambda^3)$. Having completed the above procedure, the following lemma holds.

Lemma 2. *Let $X \in \mathcal{X}$ be a variable with domain $[a, b]$. Then for any $c \in [a, b]$, (X, c) belongs to an assignment $P \in \mathbf{ME}(\mathcal{X})$ iff $MarkedValues[a][b][c] = 1$.*

Proof. Assume that $MarkedValues[a][b][c] = 1$. If the first condition among the above three is satisfied then $[a, b]$ is a witness of (X, c). Otherwise, let $[a', b']$ be the super-interval of $[a, b]$ that caused $MarkedValues[a][b][c]$ to be set to 1. If the first condition is satisfied with respect to $[a', b']$ then $[a, b]$ is a witness of (X, c). Otherwise there is a super-interval $[a'', b'']$ that caused $MarkedValues[a'][b'][c]$ to be set to 1. Proceeding to argue in this way we explore a sequence of intervals such that every next element in this sequence is a strict super-interval of its predecessor. Clearly this sequence is of finite length and $MarkedValues[a^*][b^*][c]$ for the last element $[a^*, b^*]$ of this sequence is set according to Condition 1. Hence, $[a^*, b^*]$ is a witness of (X, c). Thus if $MarkedValues[a][b][c] = 1$ then (X, c) has a witness and (X, c) belongs to an assignment $P \in \mathbf{ME}(\mathcal{X})$ according to Lemma 1.

Conversely assume that (X, c) belongs to an assignment $P \in \mathbf{ME}(\mathcal{X})$. Let $[a', b']$ be a witness of (X, c) existing according to Lemma 1. Then, by Condition 1, $MarkedValues[a'][b'][c] = 1$. It is not hard to show that $MarkedValues[a][b][c]$ is set to 1 by the inductive application of Conditions 2 and 3. □

Theorem 5. *If the minimum number of allowed equal pairs of values is $C_{a,b}(\mathcal{X})$, RC on $\text{SoftAllEqual}_G^{min}$ can be achieved in $\mathcal{O}((n + \lambda)\lambda^2)$ steps.*

Proof. Achieving BC in this case can be done by filtering all values that do not belong to an assignment $P \in \mathbf{ME}(\mathcal{X})$. By Lemma 2, this can be done by exploring all values (X, c) and for each of them checking in $\mathcal{O}(1)$ whether $MarkedValues[a][b][c] = 1$ where $[a, b]$ is the domain of X. Since the $MarkedValues$ array can be computed in $\mathcal{O}((n + \lambda)\lambda^2)$ (the computation includes all the previously discussed computational steps) and there are $\mathcal{O}(n\lambda)$ values, the theorem follows. □

6 Conclusion

Constraints for reasoning about the number of different assignments to a set of variables are ubiquitous in constraint programming and artificial intelligence. In this paper we considered the global constraints AllDifferent and AllEqual, and their optimisation variants, SoftAllDiff and SoftAllEqual, respectively. A major technical contribution of the paper is an efficient algorithm for optimising the cost of the SoftAllEqual constraint when the objective is to maximise the number of equalities achieved in the decomposition graph of the constraint. Therefore, we give a complete characterisation of these constraints. This paper can be regarded as providing a complete taxonomy of constraints of difference and equality.

References

1. Beldiceanu, N.: Pruning for the minimum constraint family and for the number of distinct values constraint family. In: Walsh, T. (ed.) CP 2001. LNCS, vol. 2239, pp. 211–224. Springer, Heidelberg (2001)
2. Bessière, C., Hebrard, E., Hnich, B., Kiziltan, Z., Walsh, T.: Filtering algorithms for the nvalueconstraint. Constraints 11(4), 271–293 (2006)
3. Hebrard, E., O'Sullivan, B., Razgon, I.: A soft constraint of equality: Complexity and approximability. In: Stuckey, P.J. (ed.) CP 2008. LNCS, vol. 5202, pp. 358–371. Springer, Heidelberg (2008)
4. Petit, T., Régin, J.-C., Bessière, C.: Specific filtering algorithms for over-constrained problems. In: Walsh, T. (ed.) CP 2001. LNCS, vol. 2239, pp. 451–463. Springer, Heidelberg (2001)
5. Quimper, C.-G., López-Ortiz, A., van Beek, P., Golynski, A.: Improved algorithms for the global cardinality constraint. In: Wallace, M. (ed.) CP 2004. LNCS, vol. 3258, pp. 542–556. Springer, Heidelberg (2004)
6. Régin, J.-C.: A filtering algorithm for constraints of difference in csps. In: AAAI, pp. 362–367 (1994)
7. Régin, J.-C., Petit, T., Bessière, C., Puget, J.-F.: An original constraint based approach for solving over constrained problems. In: Dechter, R. (ed.) CP 2000. LNCS, vol. 1894, pp. 543–548. Springer, Heidelberg (2000)
8. van Hoeve, W.J.: A hyper-arc consistency algorithm for the soft alldifferent constraint. In: Wallace, M. (ed.) CP 2004. LNCS, vol. 3258, pp. 679–689. Springer, Heidelberg (2004)
9. van Hoeve, W.J., Pesant, G., Rousseau, L.-M.: On global warming: Flow-based soft global constraints. J. Heuristics 12(4-5), 347–373 (2006)

Synthesizing Filtering Algorithms for Global Chance-Constraints*

Brahim Hnich[1], Roberto Rossi[2], S. Armagan Tarim[3], and Steven Prestwich[4]

[1] Faculty of Computer Science, Izmir University of Economics, Turkey
brahim.hnich@ieu.edu.tr
[2] Logistics, Decision and Information Sciences, Wageningen UR, The Netherlands
roberto.rossi@wur.nl
[3] Operations Management Division, Nottingham University Business School, UK
armtar@yahoo.com
[4] Cork Constraint Computation Centre, University College Cork, Ireland
s.prestwich@4c.ucc.ie

Abstract. Stochastic Constraint Satisfaction Problems (SCSPs) are a powerful modeling framework for problems under uncertainty. To solve them is a P-Space task. The only solution approach to date compiles down SCSPs into classical CSPs. This allows the reuse of classical constraint solvers to solve SCSPs, but at the cost of increased space requirements and weak constraint propagation. This paper tries to overcome some of these drawbacks by automatically synthesizing filtering algorithms for global chance-constraints. These filtering algorithms are parameterized by propagators for the deterministic version of the chance-constraints. This approach allows the reuse of existing propagators in current constraint solvers and it enhances constraint propagation. Experiments show the benefits of this novel approach.

1 Introduction

Stochastic Constraint Satisfaction Problems (SCSPs) are a powerful modeling framework for problems under uncertainty. SCSPs were first introduced in [10] and further extended in [9] to permit multiple chance-constraints and a range of different objectives in order to model combinatorial problems under uncertainty. SCSP is a PSPACE-complete problem [10]. The approach in [9] compiles down SCSPs into deterministic equivalent CSPs. This makes it possible to reuse existing solvers, but at the cost of increased space requirements and of hindering constraint propagation. In this paper we overcome some of these drawbacks by automatically synthesizing filtering algorithms for global chance-constraints. These filtering algorithms are built around propagators for the deterministic version of the chance-constraints. Like the approach in [9], our approach reuses the propagators already available for classical CSPs. But, unlike [9], our approach uses fewer decision variables and strengthens constraint propagation. Our results

* Brahim Hnich is supported by the Scientific and Technological Research Council of Turkey (TUBITAK) under Grant No. SOBAG-108K027.

I.P. Gent (Ed.): CP 2009, LNCS 5732, pp. 439–453, 2009.

show that our approach is superior to the one in [9], since it achieves stronger pruning and therefore it proves to be more efficient in terms of run time and explored nodes.

The paper is structured as follows: in Section 2 we provide the relevant formal background; in Section 3 we discuss the structure of a SCSP solution; in Section 4 we describe the state-of-the-art approach to SCSPs; in Section 5 we discuss our novel approach; in Section 6 we present our computational experience; in Section 7 we provide a brief literature review; finally, in Section 8 we draw conclusions.

2 Formal Background

A Constraint Satisfaction Problem (CSP) consists of a set of variables, each with a finite domain of values, and a set of constraints specifying allowed combinations of values for some variables. A *solution* to a CSP is an assignment of variables to values in their respective domains such that all of the constraints are satisfied. Constraint solvers typically explore partial assignments enforcing a local consistency property. A constraint c is *generalized arc consistent* (*GAC*) iff when a variable is assigned any of the values in its domain, there exist compatible values in the domains of all the other variables of c. In order to enforce a local consistency property on a constraint c during search, we employ filtering algorithms that remove inconsistent values from the domains of the variables of c. These filtering algorithms are repeatedly called until no more values are pruned. This process is called *constraint propagation.*

An m-stage SCSP is defined as a 7-tuple $\langle V, S, D, P, C, \theta, L\rangle$, where V is a set of decision variables and S is a set of stochastic variables, D is a function mapping each element of V and each element of S to a domain of potential values. In what follows, we assume that both decision and stochastic variable domains are finite. P is a function mapping each element of S to a probability distribution for its associated domain. C is a set of chance-constraints over a non-empty subset of decision variables and a subset of stochastic variables. θ is a function mapping each chance-constraint $h \in C$ to θ_h which is a threshold value in the interval $(0, 1]$. $L = [\langle V_1, S_1\rangle, \ldots, \langle V_i, S_i\rangle, \ldots, \langle V_m, S_m\rangle]$ is a list of *decision stages* such that each $V_i \subseteq V$, each $S_i \subseteq S$, the V_i form a partition of V, and the S_i form a partition of S.

The solution of an m-stage SCSP is, in general, represented by means of a *policy tree* [9]. The arcs in such a policy tree represent values observed for stochastic variables whereas nodes at each level represent the decisions associated with the different stages. We call the policy tree of an m-stage SCSP that is a solution a *satisfying policy tree.*

3 Satisfying Policy Trees

In order to simplify the presentation, we assume without loss of generality, that each $V_i = \{x_i\}$ and each $S_i = \{s_i\}$ are singleton sets. All the results can be easily extended in order to consider $|V_i| > 1$ and $|S_i| > 1$. In fact, if S_i comprises more than one random variable, it is always possible to aggregate these variables into a single multivariate random variable [5] by convoluting them. If V_i comprises more than one decision variable, the following discussion still holds, provided that the term $DecVar$, which we will introduce in the next paragraph, is interpreted as a set of decision variables.

Let $S = \{s_1, s_2, \ldots, s_m\}$ be the set of all stochastic variables and $V = \{x_1, x_2, \ldots, x_m\}$ be the set of all decision variables. In an m-stage SCSP, the policy tree has

$$\mathcal{N} = 1 + |s_1| + |s_1| \cdot |s_2| + \ldots + |s_1| \cdot |s_2| \cdot \ldots \cdot |s_{m-1}|$$

nodes, where $|s_j|$ denotes the cardinality of $D(s_j)$. We adopt the following node and arc labeling schemes for the policy tree of an m-stage SCSP. The depth of a node can be uniquely associated with its respective decision stage, more specifically V_i is associated with nodes at depth $i-1$. We label each node with $\langle DecVar, DecVal, Index\rangle$ where $DecVar$ is a decision variable that must be assigned at the decision stage associated with the node, $DecVal \in D(DecVar)$ is the value that this decision variable takes at this node, and $Index \in \{0, \ldots, \mathcal{N}-1\}$ is a unique index for this node. Each arc will be labeled with $\langle StochVar, StochVal\rangle$ where $StochVar \in S$ and $StochVal \in D(StochVar)$. According to our labeling scheme, the root node has label $\langle x_1, \bar{x}_1, 0\rangle$ where $\bar{x}_1$ is the value assigned to the variable x_1 associated with the root node and the index of the root node is 0. The root node is at depth 0. For each value $\bar{s}_1 \in D(s_1)$, we have an arc leaving the root node labeled with $\langle s_1, \bar{s}_1\rangle$. The $|s_1|$ nodes connected to the root node are labeled from 1 to $|s_1|$. For each node at depth 1, we label each of $|s_2|$ arcs with $\langle s_2, \bar{s}_2\rangle$ for each $\bar{s}_2 \in D(s_2)$. For the nodes at depth 2, we label them from $\langle x_2, \bar{x}_2, |s_1|+1\rangle$ to $\langle x_2, \bar{x}_2, |s_1| + |s_1|.|s_2|\rangle$, and so on until we label all arcs and all nodes of the policy tree. A path p from the root node to the last arc can be represented by the sequence of the node and arc labelings, i.e. $p = [\langle x_1, \bar{x}_1, 0\rangle, \langle s_1, \bar{s}_1\rangle, \ldots, \langle x_m, \bar{x}_m, k\rangle, \langle s_m, \bar{s}_m\rangle]$. Let Ψ denote the set of all distinct paths of a policy tree. For each $p \in \Psi$, we denote by $arcs(p)$ the sequence of all the arc labelings in p whereas $nodes(p)$ denotes the sequence of all node labelings in p. That is $arcs(p) = [\langle s_1, \bar{s}_1\rangle, \ldots, \langle s_m, \bar{s}_m\rangle]$ whereas $nodes(p) = [\langle x_1, \bar{x}_1, 0\rangle, \ldots, \langle x_m, \bar{x}_m, j\rangle]$. We denote by $\Omega = \{arcs(p) | p \in \Psi\}$ the set of all scenarios of the policy tree. The probability of $\omega \in \Omega$ is given by $\Pr\{\omega\} = \prod_{i=1}^{m} \Pr\{s_i = \bar{s}_i\}$, where $\Pr\{s_i = \bar{s}_i\}$ is the probability that stochastic variable s_i takes value $\bar{s}_i$.

Now consider a chance-constraint $h \in C$ with a specified threshold level θ_h. Consider a policy tree $\mathcal{T}$ for the SCSP and a path $p \in \mathcal{T}$. Let $h_{\downarrow p}$ be the deterministic constraint obtained by substituting the stochastic variables in h with the corresponding values ($\bar{s}_i$) assigned to these stochastic variables in $arcs(p)$. Let $\bar{h}_{\downarrow p}$ be the resulting tuple obtained by substituting the decision variables in $h_{\downarrow p}$ by the values ($\bar{x}_i$) assigned to the corresponding decision variables in $nodes(p)$. We say that h is *satisfied wrt to a given policy tree* $\mathcal{T}$ iff

$$\sum_{p \in \Psi : \bar{h}_{\downarrow p} \in h_{\downarrow p}} \Pr\{arcs(p)\} \geq \theta_h.$$

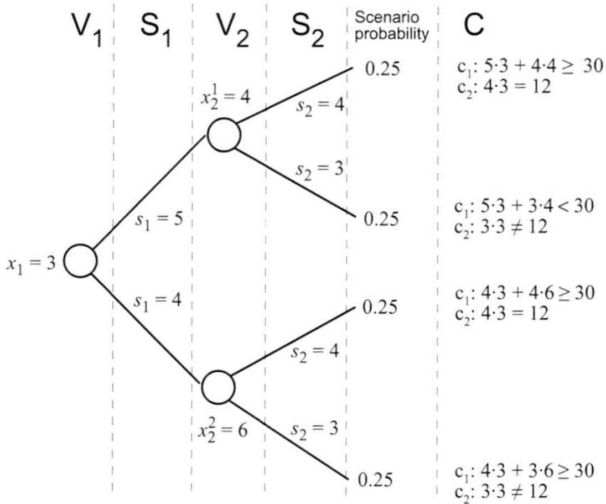

Fig. 1. Policy tree for the SCSP in Example 1

Definition 1. *Given an m-stage SCSP $\mathcal{P}$ and a policy tree $\mathcal{T}$, $\mathcal{T}$ is a satisfying policy tree to $\mathcal{P}$ iff every chance-constraint of $\mathcal{P}$ is satisfied wrt $\mathcal{T}$.*

Example 1. Let us consider a two-stage SCSP in which $V_1 = \{x_1\}$ and $S_1 = \{s_1\}$, $V_2 = \{x_2\}$ and $S_2 = \{s_2\}$. Stochastic variable s_1 may take two possible values, 5 and 4, each with probability 0.5; stochastic variable s_2 may also take two possible values, 3 and 4, each with probability 0.5. The domain of x_1 is $\{1, \ldots, 4\}$, the domain of x_2 is $\{3, \ldots, 6\}$. There are two chance-constraints[1] in C, $c_1 : \Pr\{s_1x_1 + s_2x_2 \geq 30\} \geq 0.75$ and $c_2 : \Pr\{s_2x_1 = 12\} \geq 0.5$. In this case, the decision variable x_1 must be set to a unique value before random variables are observed, while decision variable x_2 takes a value that depends on the observed value of the random variable s_1. A possible solution to this SCSP is the satisfying policy tree shown in Fig. 1 in which $x_1 = 3, x_2^1 = 4$ and $x_2^2 = 6$, where x_2^1 is the value assigned to decision variable x_2, if random variable s_1 takes value 5, and x_2^2 is the value assigned to decision variable x_2, if random variable s_1 takes value 4. The four labeled paths of the above policy tree are as follows:

$$p_1 = [\langle x_1, 3, 0\rangle, \langle s_1, 5\rangle, \langle x_2, 4, 1\rangle, \langle s_2, 4\rangle], \quad p_2 = [\langle x_1, 3, 0\rangle, \langle s_1, 5\rangle, \langle x_2, 4, 1\rangle, \langle s_2, 3\rangle],$$
$$p_3 = [\langle x_1, 3, 0\rangle, \langle s_1, 4\rangle, \langle x_2, 6, 2\rangle, \langle s_2, 4\rangle], \quad p_4 = [\langle x_1, 3, 0\rangle, \langle s_1, 4\rangle, \langle x_2, 6, 2\rangle, \langle s_2, 3\rangle].$$

As the example shows, a solution to a SCSP is not simply an assignment of the decision variables in V to values, but it is instead a satisfying policy tree.

4 Scenario-Based Approach to Solve SCSPs

In [9], the authors discuss an equivalent scenario-based reformulation for SCSPs. This reformulation makes it possible to compile SCSPs down into conventional

[1] In what follows, for convenience, we will denote a chance-constraint by using the notation "$\Pr\{\langle cons\rangle\} \geq \theta$", meaning that constraint $\langle cons\rangle$, constraining decision and random variables, should be satisfied with probability greater or equal to θ.

Constraints:

(1)	$(5x_1^1 + 4x_2^1 \geq 30) \leftrightarrow (Z_{c_1}^1 = 1)$	(6)	$(4x_1^1 = 12) \leftrightarrow (Z_{c_2}^1 = 1)$
(2)	$(5x_1^1 + 3x_2^1 \geq 30) \leftrightarrow (Z_{c_1}^2 = 1)$	(7)	$(3x_1^1 = 12) \leftrightarrow (Z_{c_2}^2 = 1)$
(3)	$(4x_1^1 + 4x_2^2 \geq 30) \leftrightarrow (Z_{c_1}^3 = 1)$	(8)	$(4x_1^1 = 12) \leftrightarrow (Z_{c_2}^3 = 1)$
(4)	$(4x_1^1 + 3x_2^2 \geq 30) \leftrightarrow (Z_{c_1}^4 = 1)$	(9)	$(3x_1^1 = 12) \leftrightarrow (Z_{c_2}^4 = 1)$
(5)	$\sum_{\omega=1}^4 0.25 Z_{c_1}^\omega \geq \theta_{c_1}$	(10)	$\sum_{\omega=1}^4 0.25 Z_{c_2}^\omega \geq \theta_{c_2}$

Decision variables:

$x_1 \in \{1, 2, 3, 4\}$, $x_2^1 \in \{3, 4, 5, 6\}$,

$x_2^2 \in \{3, 4, 5, 6\}$, $Z_h^\omega \in \{0, 1\}$ $\forall \omega = 1, \ldots, 4;$ $\forall h \in \{c_1, c_2\}$.

Fig. 2. Deterministic equivalent CSP for Example 1

(non-stochastic) CSPs. For example, the multi-stage SCSP described in Example 1 is compiled down to its deterministic equivalent CSP shown in Fig. 2. The decision variables x_1^1, x_2^1, and x_2^2 represent the nodes of the policy tree. The variable x_1 is decided at stage 1 so we have one copy of it (x_1^1) whereas since x_2 is to be decided at stage 2 and since s_1 has two values, we need two copies for x_2, namely x_2^1 and x_2^2. Chance-constraint c_1 is compiled down into constraints (1),...,(5), whilst chance-constraint c_2 is compiled down into constraints (6),...,(10). Constraints (1),...,(4) are reification constraints in which every binary decision variable $Z_{c_1}^\omega$ is 1 iff in scenario $\omega \in \{1, \ldots, 4\}$ constraint $\bar{s}_1 x_1^1 + \bar{s}_2 x_2^i \geq 30$ — where $i \in \{1, 2\}$ identifies the copy of decision variable x_2 associated with scenario ω — is satisfied. Finally, constraint (5) enforces that the satisfaction probability achieved must be greater or equal to the required threshold $\theta_{c_1} = 0.75$. A similar reasoning applies to constraints (6),...,(10).

The scenario-based reformulation approach allows us to exploit the full power of existing constraint solvers. However, it has a number of serious drawbacks that might prevent it from being applied in practice. These drawbacks are:

Increased Space Requirements: For each chance-constraint, $|\Omega|$ extra Boolean variables and $|\Omega|+1$ extra constraints are introduced. This requires more space and might increase the solution time;

Hindering Constraint Propagation: the holistic CSP heavily depends on reification constraints for constraint propagation, which is a very weak form of propagation. Also, if the chance-constraint involves a global constraint (e.g., $\Pr\{\text{alldiff}(x_1, s_1, x_2)\} \geq \theta$), then the corresponding reification constraints (e.g., $\text{alldiff}(x_1^1, \bar{s}_1, x_2^1) \leftrightarrow Z_w$) cannot simply be supported in an effective way in terms of propagation by any of the current constraint solvers.

5 Generic Filtering Algorithms

In this section we show how to overcome the drawbacks discussed above.

5.1 Theoretical Properties

Like the approach in [10], in order to solve an m-stage SCSP, we introduce a decision variable for each node of the policy tree. Given an SCSP $\langle V, S, D, P, C, \theta, L\rangle$, we let $\mathcal{PT}$ be an array of decision variables indexed from 0 to $\mathcal{N}-1$ representing the space of all possible policy trees. The domains of these variables are defined as follows:

- $D(\mathcal{PT}[i]) = D(x_1)$, $i \in M_1 = \{0\}$,
- $D(\mathcal{PT}[i]) = D(x_2)$, $i \in M_2 = \{1, \ldots, |s_1|\}$,
- $D(\mathcal{PT}[i]) = D(x_3)$, $i \in M_3 = \{(1 + |s_1|), \ldots, (|s_1| \cdot |s_2|)\}$,
- ...
- $D(\mathcal{PT}[i]) = D(x_m)$, $i \in M_m = \{(1 + |s_1| \cdot |s_2| \cdot \ldots \cdot |s_{m-2}|), \ldots, (|s_1| \cdot |s_2| \cdot \ldots \cdot |s_{m-1}|)\}$.

This array of decision variables is shared among the constraints in the model similarly to what happens with decision variables in classic CSPs. In what follows, we will discuss how to propagate chance-constraints on the policy tree decision variable array.

Definition 2. *Given a chance-constraint $h \in C$ and a policy tree decision variable array $\mathcal{PT}$, a value v in the domain of $\mathcal{PT}[i]$ is* **consistent wrt** *h iff there exists an assignment of values to variables in $\mathcal{PT}$ that is a satisfying policy wrt h, in which $\mathcal{PT}[i] = v$.*

Definition 3. *A chance-constraint $h \in C$ is* **generalized arc-consistent** *iff every value in the domain of every variable in $\mathcal{PT}$ is consistent wrt h.*

Definition 4. *A SCSP is* **generalized arc-consistent** *iff every chance-constraint is generalized arc-consistent.*

For convenience, given a chance-constraint $h \in C$, we now redefine $h_{\downarrow p}$ as the resulting deterministic constraint in which we substitute every decision variable x_i in h with decision variable $\mathcal{PT}[k]$, where $\langle x_i, -, k\rangle$ is an element in $nodes(p)$, and — according to our former definition — in which we substitute every stochastic variable s_i with the corresponding values $(\bar{s}_i)$ assigned to s_i in $arcs(p)$. Note that the deterministic constraint $h_{\downarrow p}$ is a classical constraint, so a value v in the domain of any decision variable *is consistent* iff there exist compatible values for all other variables such that $h_{\downarrow p}$ is satisfied, otherwise v is inconsistent. We denote by $h^{i,v}_{\downarrow p}$ the constraint $h_{\downarrow p}$ in which decision variable $\mathcal{PT}[i]$ is set to v. $h^{i,v}_{\downarrow p}$ *is consistent* if value v in $D(\mathcal{PT}[i])$ is consistent w.r.t. $h_{\downarrow p}$. Let $\Psi_i = \{p \in \Psi | h_{\downarrow p}$ *constrains* $\mathcal{PT}[i]\}$. We introduce $f(i, v)$ as follows:

$$f(i, v) = \sum_{p \in \Psi_i : h^{i,v}_{\downarrow p} \text{ is consistent}} \Pr\{arcs(p)\},$$

where $f(i, v)$ is the sum of the probabilities of the scenarios in which value v in the domain of $\mathcal{PT}[i]$ is consistent. As the next proposition shows, one can exploit this to identify a subset of the inconsistent values.

Proposition 1. *For any* $i \in M_k$ *and value* $v \in D(\mathcal{PT}[i])$, *if*

$$f(i,v) + \sum_{j \in M_k, j \neq i} \max(j) < \theta_h,$$

then v *is inconsistent wrt* h*; where* $\max(j) = \max\{f(j,v) | v \in D(\mathcal{PT}[j])\}$.

Proof (Sketch)**.** The assignment $\mathcal{PT}[i] = v$ is consistent w.r.t. h iff the satisfaction probability of h is greater or equal to θ_h. From the definition of $f(i,v)$ and of $\max(j)$ it follows that, if $f(i,v) + \sum_{j \in M_k, j \neq i} \max(j) < \theta_h$, when $\mathcal{PT}[i] = v$, the satisfaction probability of h is less than θ_h even if we choose the best possible value for all the other variables. □

5.2 Filtering Algorithms

We now describe our generic filtering strategy for chance-constraints. We distinguish between two cases: the case when $\theta_h < 1$ and the case where $\theta_h = 1$. In the first case, we design a specialized filtering algorithm whereas for the second case we provide a reformulation approach that is more efficient. Both methods, however, are parameterized with a filtering algorithm $\mathcal{A}$ for the deterministic constraints $h_{\downarrow p}$ for all $p \in \Psi$ that maintains GAC (or any other level of consistency). This allows us to reuse existing filtering algorithms in current constraint solvers and makes our filtering algorithms generic and suitable for any global chance-constraint.

Case 1: Algorithm 1 takes as input chance-constraint h, $\mathcal{PT}$, and a propagator $\mathcal{A}$. It filters from $\mathcal{PT}$ inconsistent values wrt h. For each decision variable and each value in its domain, we initialize $f[i,v]$ to 0 in line 2. In line 5, we iterate through the scenarios in Ψ. For each scenario, we create a copy c of constraint $h_{\downarrow p}$ and of the decision variables it constrains. Then we enforce GAC on c using $\mathcal{A}$. We iterate through the domain of each copy of the decision variable at index i and, if a given value v has support, we add the probability associated with the current scenario to the respective $f[i,v]$ (line 10). It should be noted that, for each scenario, constraint c is dynamically generated every time the filtering algorithm runs, and also that these constraints are never posted into the model. They are only used to reduce the domains of the copies of the associated decision variables. In line 12, for each variable $i \in \{0, \ldots, \mathcal{N}-1\}$ we compute the maximum support probability $f[i,v]$ provided by a value v in the domain of $\mathcal{PT}[i]$, and we store it at $max[i]$. In line 16, for each stage $k \in \{1, \ldots, m\}$, we store in $g[k]$ the sum of the $max[i]$ of all variables $i \in M_k$. Finally, (line 20) at stage k we prune from $D(\mathcal{PT}[i])$ any value v that makes $g[k]$ strictly smaller than θ_h when we replace $max[i]$ in $g[k]$ with $f[i,v]$.

Theorem 1. *Algorithm 1 is a sound filtering algorithm.*

Proof (Sketch) Soundness. When a value v is pruned by Algorithm 1, Proposition 1 is true. Thus, any pruned value v is inconsistent. □

Algorithm 1 fails to prune some inconsistent values because such values are supported by values that may become inconsistent at a later stage of the algorithm.

Algorithm 1. Filtering Algorithm

input : h; $\mathcal{PT}$; $\mathcal{A}$.
output: Filtered $\mathcal{PT}$ wrt h.

begin
  for *each* $i \in \{0, \ldots, \mathcal{N}-1\}$ **do**
    for *each* $v \in D(\mathcal{PT}[i])$ **do**
      $f[i, v] \leftarrow 0$;
  for *each* $p \in \Psi$ **do**
    Create a copy c of $h_{\downarrow p}$ and of the decision variables it constrains;
    Enforce GAC on c using $\mathcal{A}$;
    for *each index i of the variables in c* **do**
      for *each v in domain of the copy of* $\mathcal{PT}[i]$ **do**
        $f[i, v] \leftarrow f[i, v] + \Pr\{arcs(p)\}$;
    delete c and the respective copies of the decision variables;
  for *each* $i \in \{0, \ldots, \mathcal{N}-1\}$ **do**
    $max[i] \leftarrow 0$;
    for *each* $v \in D(\mathcal{PT}[i])$ **do**
      $max[i] \leftarrow \max(max[i], f[i, v])$;
  for *each* $k \in \{1, \ldots, m\}$ **do**
    $g[k] \leftarrow 0$;
    for *each* $i \in M_k$ **do**
      $g[k] \leftarrow g[k] + max[i]$
  for *each* $k \in \{1, \ldots, m\}$ **do**
    for *each* $i \in M_k$ **do**
      for *each* $v \in \mathcal{PT}[i]$ **do**
        if $g[k] - max[i] + f[i, v] < \theta_h$ **then**
          prune value v from $D(\mathcal{PT}[i])$;
end

We illustrate these situations with an example. Consider a 2-stage SCSP in which $V_1 = \{x_1\}$, where $x_1 \in \{1, 2\}$, $S_1 = \{s_1\}$, where $s_1 \in \{a, b\}$, $V_2 = \{x_2\}$, where $x_2 \in \{1, 2, 3\}$, and $S_2 = \{s_2\}$, where $s_2 \in \{a, b\}$. Let $Pr\{s_i = j\} = 0.5$ for all $i \in \{1, 2\}$ and $j \in \{a, b\}$. Let h be the chance-constraint with $\theta_h = 0.75$. In this constraint, for the first scenario ($s_1 = a$ and $s_2 = a$) the only consistent values for $\mathcal{PT}[0]$ and $\mathcal{PT}[1]$ are 1 and 2 respectively. For the second scenario ($s_1 = a$ and $s_2 = b$) the only consistent values for $\mathcal{PT}[0]$ and $\mathcal{PT}[1]$ are 2 and 1 respectively. For the third scenario ($s_1 = b$ and $s_2 = a$) the only consistent values for $\mathcal{PT}[0]$ and $\mathcal{PT}[2]$ are 1 and 3 respectively. For the fourth scenario ($s_1 = b$ and $s_2 = b$) the only consistent values for $\mathcal{PT}[0]$ and $\mathcal{PT}[2]$ are 1 and 3 respectively. Our algorithm originally introduces three decision variables $\mathcal{PT}[0] \in \{1, 2\}$, $\mathcal{PT}[1] \in \{1, 2, 3\}$, and $\mathcal{PT}[2] \in \{1, 2, 3\}$. Assume that at some

Table 1. Example of inconsistent values gone undetected

$\mathcal{PT}[0]$	$f[0,v]$	$\mathcal{PT}[1]$	$f[1,v]$	$\mathcal{PT}[2]$	$f[2,v]$
1	0.75	1	0.25	3	0.5
2	0.25	2	0.25		

stage during search, the domains become $\mathcal{PT}[0] \in \{1,2\}$, $\mathcal{PT}[1] \in \{1,2\}$, and $\mathcal{PT}[2] \in \{3\}$. In Table 1, the values that are not pruned by Algorithm 1 when $\theta = 0.75$ are underlined. Only value 2 in the domain of $\mathcal{PT}[0]$ is pruned. But, value 2 was providing support to value 1 in the domain of $\mathcal{PT}[1]$. This goes undetected by the algorithm and value 1 for $\mathcal{PT}[1]$ no longer provides $f[1,v] = 0.25$ satisfaction, but 0. Thus, there exists no satisfying policy in which $\mathcal{PT}[1] = 1$! We can easily remedy this problem by repeatedly calling Algorithm 1 until we reach a fixed-point and no further pruning is done. We denote as $\mathcal{H}$ this modified algorithm.

Theorem 2. *Algorithm $\mathcal{H}$ runs in $O(|\Omega| \cdot a \cdot \mathcal{N}^2 \cdot d^2)$ time and in $O(\mathcal{N} \cdot d + p)$ space where a is the time complexity of $\mathcal{A}$, p is its space complexity, and d is the maximum domain size.*

Proof (Sketch): Time complexity. In the worst case, Algorithm 1 needs to be called $\mathcal{N} \cdot d$ times in order to prune at each iteration just one inconsistent value. At each of these iterations, the time complexity is dominated by complexity of line 7 assuming that $|\Omega| \gg |V|$. Enforcing GAC on each of the $|\Omega|$ constraints runs in a time using algorithm $\mathcal{A}$. In the worst case, we need to repeat this whole process $\mathcal{N} \cdot d$ times in order to prune at each iteration just one inconsistent value. Thus the time complexity of this step is in $|\Omega| \cdot a \cdot \mathcal{N} \cdot d$. The overall time complexity is therefore in $O(|\Omega| \cdot a \cdot \mathcal{N}^2 \cdot d^2)$ time.

(Sketch) Space complexity. The space complexity is dominated by the size of $\mathcal{PT}$ and by the space complexity of $\mathcal{A}$. $\mathcal{PT}$ requires $\mathcal{N} \cdot d$ space whereas $\mathcal{A}$ requires p space. Therefore, the modified algorithm runs in $O(\mathcal{N} \cdot d + p)$ space. □

In Table 2 we report the pruned values for Example 1 achieved by $\mathcal{H}$. The values that are not pruned when $\theta = 0.75$ are underlined. Note that if we propagate the constraints in the model generated according to the approach described in [9] and shown in Fig. 2, no value is pruned.

Even though algorithm $\mathcal{H}$ is a sound filtering algorithm, it is unfortunately still incomplete.

Theorem 3. *The level of consistency achieved by algorithm $\mathcal{H}$ on global chance-constraint h is weaker than GAC on h.*

Proof. Consider a 2-stage SCSP where $V_1 = \{x_1\}$ where $x_1 \in \{1,2\}$, $S_1 = \{s_1\}$ where $s_1 \in \{a,b\}$, $V_2 = \{x_2\}$ where $x_2 \in \{1,2\}$, and $S_2 = \{s_2\}$ where $s_2 \in \{a,b\}$. Let $Pr\{s_i = j\} = 0.5$ for all $i \in \{1,2\}$ and $j \in \{a,b\}$. Let h be the

Table 2. Pruning for Example 2 after calling Algorithm $\mathcal{H}$

$\mathcal{PT}[0]$	$f[0,v]$	$\mathcal{PT}[1]$	$f[1,v]$	$\mathcal{PT}[2]$	$f[2,v]$
1	0.0	3	0.25	3	0.0
2	0.5	4	0.5	4	0.25
3	1.0	5	0.5	5	0.5
4	1.0	6	0.5	6	0.5

Table 3. Filtered domains

$\mathcal{PT}[0]$	$f[0,v]$	$\mathcal{PT}[1]$	$f[1,v]$	$\mathcal{PT}[2]$	$f[2,v]$
1	1	1	0.5	1	0.5
2	1	2	0.5	2	0.5

chance-constraint with $\theta_h = 0.75$. Furthermore, for the first scenario ($s_1 = a$ and $s_2 = a$) the consistent tuples for x_1 and x_2 are in $\{\langle 1,1\rangle\ \langle 2,1\rangle\ \langle 2,2\rangle\}$. For the second scenario ($s_1 = a$ and $s_2 = b$) the consistent tuples for x_1 and x_2 are in $\{\langle 1,2\rangle\ \langle 2,1\rangle\ \langle 2,2\rangle\}$. For the third scenario ($s_1 = b$ and $s_2 = a$) the consistent tuples for x_1 and x_2 are in $\{\langle 1,1\rangle\ \langle 2,1\rangle\ \langle 2,2\rangle\}$. For the fourth scenario ($s_1 = b$ and $s_2 = b$) the consistent tuples for x_1 and x_2 are in $\{\langle 1,2\rangle\ \langle 2,1\rangle\ \langle 2,2\rangle\}$. Algorithm $\mathcal{H}$ introduces three decision variables $\mathcal{PT}[i] \in \{1,2\}$ for all $i \in \{0,1,2\}$. Table 3 shows the result of algorithm $\mathcal{H}$. None of the values is pruned, but there exists no satisfying policy in which $\mathcal{PT}[0] = 1$. □

Indeed, we conjecture that maintaining GAC on a global chance-constraint is intractable in general even if maintaining GAC on its deterministic version is polynomial.

Case 2: When $\theta_h = 1$ the global chance-constraint h can be reformulated as $h_{\downarrow p}, \forall p \in \Psi$. If all deterministic constraints are simultaneously GAC, then this reformulation is equivalent to algorithm $\mathcal{H}$. Nevertheless, even in this special case, we still lose in terms of pruning.

Theorem 4. *GAC on h is stronger than GAC on the reformulation.*

Proof. We consider the same example as in the previous proof but with $\theta_h = 1$ instead. All deterministic constraints are simultaneously GAC, but $\mathcal{PT}[i] = 1$ cannot be extended to any satisfying policy. □

6 Computational Experience

In this section, we present our computational experience, which shows that our approach outperforms the state-of-the-art approach in [9] both in terms of run time and explored nodes, and that it is also able to achieve stronger pruning.

In our experiments we considered a number of randomly generated SCSPs. The SCSPs considered feature five chance-constraints over 4 integer decision variables, $x_1, \ldots, x_4$ and 8 stochastic variables, $s_1, \ldots, s_8$. The decision variable domains are: $D(x_1) = \{5, \ldots, 10\}$, $D(x_2) = \{4, \ldots, 10\}$, $D(x_3) = \{3, \ldots, 10\}$, and $D(x_4) = \{6, \ldots, 10\}$. The domains of stochastic variables s_1, s_3, s_5, s_7 comprise 2 integer values each. The domains of stochastic variables s_2, s_4, s_6, s_8 comprise 3 integer values each. The values in these domains have been randomly generated as uniformly distributed in $\{1, \ldots, 5\}$. Each value appearing in the domains of random variables s_1, s_3, s_5, s_7 is assigned a realization probability of $\frac{1}{2}$. Each value appearing in the domains of random variables s_2, s_4, s_6, s_8 is assigned a realization probability of $\frac{1}{3}$. There are five chance-constraints in the model, the first embeds an equality, $c_1 : \Pr\{x_1 s_1 + x_2 s_2 + x_3 s_3 + x_4 s_4 = 80\} \geq \alpha$, the second and the third embed inequalities, $c_2 : \Pr\{x_1 s_5 + x_2 s_6 + x_3 s_7 + x_4 s_8 \leq 100\} \geq \beta$ and $c_3 : \Pr\{x_1 s_5 + x_2 s_6 + x_3 s_7 + x_4 s_8 \geq 60\} \geq \beta$. Parameters α and β take values in $\{0.005, 0.01, 0.03, 0.05, 0.07, 0.1\}$ and $\{0.6, 0.7, 0.8\}$, respectively. The fourth chance-constraint embeds again an inequality, but in this case the constraint is defined over a subset of all the decision and random variables in the model: $c_4 : \Pr\{x_1 s_2 + x_3 s_6 \geq 30\} \geq 0.7$. Finally, the fifth chance-constraint embeds an equality also defined over a subset of all the decision and random variables in the model: $c_5 : \Pr\{x_2 s_4 + x_4 s_8 = 20\} \geq 0.05$.

We considered 3 possible stage structures. In the first stage structure we have only one stage, $\langle V_1, S_1 \rangle$, where $V_1 = \{x_1, \ldots, x_4\}$ and $S_1 = \{s_1, \ldots, s_8\}$. In the second stage structure we have two stages, $\langle V_1, S_1 \rangle$ and $\langle V_2, S_2 \rangle$, where $V_1 = \{x_1, x_2\}$, $S_1 = \{s_1, s_2, s_5, s_6\}$, $V_2 = \{x_3, x_4\}$, and $S_1 = \{s_3, s_4, s_7, s_8\}$. In the third stage structure we have four stages, $\langle V_1, S_1 \rangle$, $\langle V_2, S_2 \rangle$, $\langle V_3, S_3 \rangle$, and $\langle V_4, S_4 \rangle$, where $V_1 = \{x_1\}$, $S_1 = \{s_1, s_5\}$, $V_2 = \{x_2\}$, $S_1 = \{s_2, s_6\}$, $V_3 = \{x_3\}$, $S_3 = \{s_3, s_7\}$, and $V_4 = \{x_4\}$, $S_4 = \{s_4, s_8\}$.

The propagation strategy discussed in Section 5 requires an existing propagator $\mathcal{A}$ for the deterministic constraints. Since the only constraints appearing in the SCSPs described above are linear (in)equalities, we borrowed a simple bound-propagation procedure for linear (in)equalities implemented in Choco 1.2 [6], a JAVA open source CP solver. The variable selection heuristic used during the search is the *domain over dynamic degree* strategy, while the value selection heuristic selects values from decision variable domains in *increasing* order.

In order to assess efficiency and effectiveness, we compared our approach (GCC) — which models the discussed SCSPs using five global chance-constraints, one for each chance-constraint in the model — against the deterministic equivalent CSPs generated using the state-of-the-art scenario-based approach in [9] (SBA).

Firstly, we wish to underline that SBA, the approach discussed in [9], requires a much larger number of constraints and decision variables to model the problems above. Specifically, the single-stage problem is modeled, in [9], using 6484 decision variables and 6485 constraints, while GCC — our approach — requires only 4 decision variables and 5 constraints; this is mainly due to the fact that, in addition to the 4 decision variables required to build the policy tree, SBA introduces 1296 binary decision variables for each of the 5 chance-constraints in the model; furthermore, SBA also introduces 1297 reification constraints for each chance-constraint in the model, similarly to what shown in Example 1 (Fig. 2). The two-stage problem is modeled by SBA using 6554 decision variables (74 for the policy tree and 6480 binary decision variables) and 6485 constraints, while GCC requires only 74 decision variables and 5 constraints; finally, the four-stage problem is modeled by SBA using 6739 decision variables and 6485 constraints, while GCC requires only 259 decision variables and 5 constraints.

As discussed above, in our comparative study we considered 18 different possible configurations for the parameters α and β. For each of these configurations, we generated 15 different probability distributions — i.e. sets of values in the domains — for the random variables in our model. These probability distributions were divided in three groups and employed to generate 5 single-stage problems, 5 two-stage problems and 5 four-stage problems. Therefore the test bed comprised, in total, 270 instances. To each instance we assigned a time limit of 240 seconds for running the search. The computational performances of GCC and SBA are compared in Fig. 3. All the experiments were performed on an Intel(R) Centrino(TM) CPU 1.50GHz with 2Gb RAM. The solver used for our test is Choco 1.2 [6].

In the test bed considered, GCC solved, in the given time limit of 240 seconds, all the instances that SBA could solve within this time limit. In contrast, SBA was often not able to solve — within the given time limit of 240 secs — instances that GCC could solve in a few seconds. More specifically, both GCC and SBA could solve 90 over 90 1-stage instances; on average GCC explored roughly 5 times less nodes and was about 2.5 times faster than SBA for these instances. GCC could solve 45 over 90 2-stage instances, while SBA could only solve 18 of them; on average GCC explored roughly 400 times less nodes and was about 13 times faster than SBA for these instances. Finally, GCC could solve 31 over 90 4-stage instances, while SBA could only solve 10 of them; on average GCC explored roughly 300 times less nodes and was about 15 times faster than SBA for these instances.

In our computational experience, we also compared the effectiveness of the filtering performed by SBA and GCC. In order to do so, we considered 90 two-stage feasible instances randomly generated according to the strategy discussed above (5 different probability distributions for the random variables and 18 different configurations for the parameters α and β). We considered a solution for each of these instances, we randomly picked subsets of the decision variables in the problem, we assigned them to the value they take in this solution, we propagated according to SBA and GCC, respectively, and we compared the percentage of values pruned by each of these two approaches. In Fig. 4 we show the results of this comparison, which is performed for a number of decision variables assigned that ranges from 0% — this corresponds to a root node propagation — to 90% of the decision variables that appear in the policy tree. In the graph, for each percentage of decision variables assigned, we report — in percentage on the total amount of values in the initial decision variable domains — the minimum, the maximum, and the average number of values pruned from the domains. As it appears from the graph, if we consider the minimum percentage of values pruned

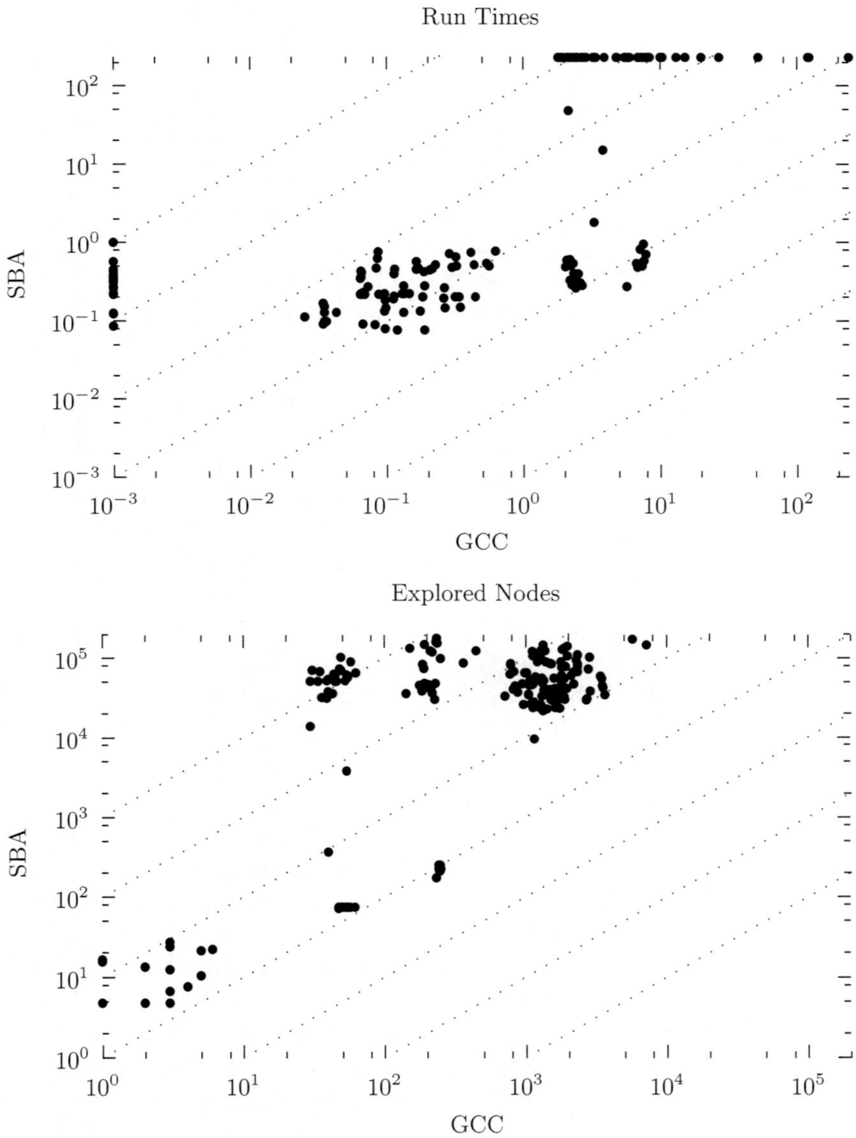

Fig. 3. Scatter graphs for our computational experience. The top graph compares the run time performance of SBA and GCC for the 270 instances in our test bed. The bottom graph shows, instead, a comparison in terms of explored nodes.

by the two approaches, GCC always achieves a stronger pruning than SBA in the worst case. Furthermore, as the maximum percentage of values pruned reported in the graph witnesses, GCC is able to achieve a much stronger pruning than SBA in the best case. On average, GCC always outperforms SBA, by filtering

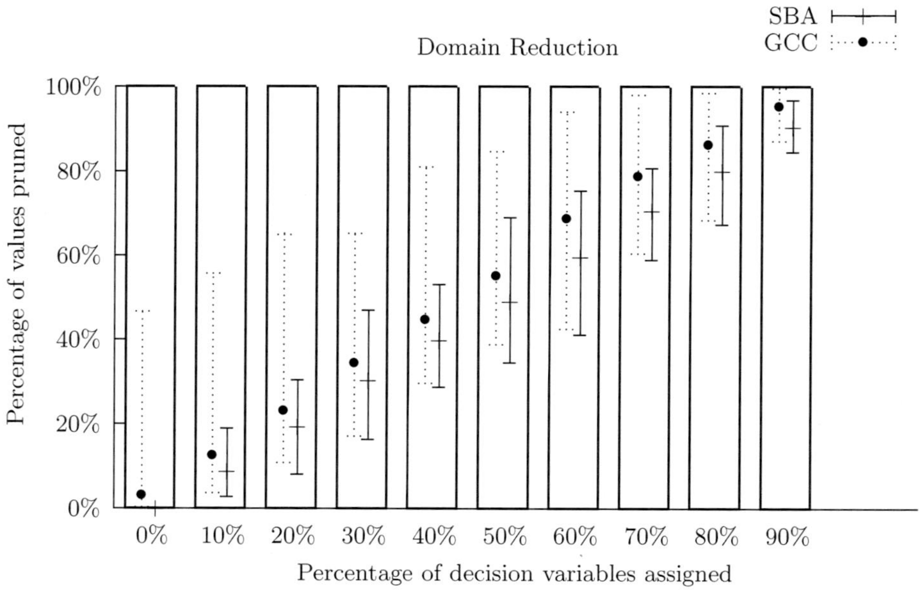

Fig. 4. Effectiveness of the filtering performed by SBA and GCC

up to 8.64% more values when 60% of the decision variables are assigned and at least 3.11% more values at the root node.

7 Related Works

Closely related to our approach are [7,8]. In these works ad-hoc filtering strategies for handling specific chance-constraints are proposed. However, the filtering algorithms presented in both these works are special purpose, incomplete, and do not reuse classical propagators for conventional constraints. Other search and consistency strategies, namely a backtracking algorithm, a forward checking procedure [10] and an arc-consistency [1] algorithm have been proposed for SCSPs. But these present several limitations and cannot be directly employed to solve multi-stage SCSPs as they do not explicitly feature a policy tree representation for the solution of a SCSP. Finally, efforts that try to extend classical CSP framework to incorporate uncertainty have been influenced by works that originated in different fields, namely *chance-constrained programming* [4] and *stochastic programming* [3]. To the best of our knowledge the first work that tries to create a bridge between Stochastic Programming and Constraint Programming is by Benoist et al. [2]. The idea of employing a scenario-based approach for building up constraint programming models of SCSPs is not novel, since Tarim et al. [9] have already used this technique to develop a fully featured language — Stochastic OPL — for modeling SCSPs. Our work proposes an orthogonal approach to solving SCSPs that could easily be integrated with the compilation approach of [9] to make it more efficient.

8 Conclusions

We proposed generic filtering algorithms for global chance-constraints. Our filtering algorithms are parameterized with conventional propagators for the corresponding deterministic version of the global chance-constraint. Our experimental results show that our approach outperforms the approach in [9], both in terms of run time and explored nodes. We also showed, experimentally, that our approach produces stronger pruning than the approach in [9]. An interesting open question is to determine if it is tractable to maintain GAC on global chance-constraints for which GAC on the corresponding deterministic constraints is tractable. Future works may investigate the tractability of GAC for classes of global chance-constraints and ways of making algorithm $\mathcal{H}$ incremental.

References

1. Balafoutis, T., Stergiou, K.: Algorithms for stochastic csps. In: Benhamou, F. (ed.) CP 2006. LNCS, vol. 4204, pp. 44–58. Springer, Heidelberg (2006)
2. Benoist, T., Bourreau, E., Caseau, Y., Rottembourg, B.: Towards stochastic constraint programming: A study of online multi-choice knapsack with deadlines. In: Walsh, T. (ed.) CP 2001. LNCS, vol. 2239, pp. 61–76. Springer, Heidelberg (2001)
3. Birge, J.R., Louveaux, F.: Introduction to Stochastic Programming. Springer, New York (1997)
4. Charnes, A., Cooper, W.W.: Deterministic equivalents for optimizing and satisficing under chance constraints. Operations Research 11(1), 18–39 (1963)
5. Jeffreys, H.: Theory of Probability. Clarendon Press, Oxford (1961)
6. Laburthe, F., The OCRE project team: Choco: Implementing a cp kernel. Technical report, Bouygues e-Lab, France (1994)
7. Rossi, R., Tarim, S.A., Hnich, B., Prestwich, S.: A global chance-constraint for stochastic inventory systems under service level constraints. Constraints 13(4), 490–517 (2008)
8. Rossi, R., Tarim, S.A., Hnich, B., Prestwich, S.D.: Cost-based domain filtering for stochastic constraint programming. In: Stuckey, P.J. (ed.) CP 2008. LNCS, vol. 5202, pp. 235–250. Springer, Heidelberg (2008)
9. Tarim, S.A., Manandhar, S., Walsh, T.: Stochastic constraint programming: A scenario-based approach. Constraints 11(1), 53–80 (2006)
10. Walsh, T.: Stochastic constraint programming. In: van Harmelen, F. (ed.) European Conference on Artificial Intelligence, ECAI 2002, Proceedings, pp. 111–115. IOS Press, Amsterdam (2002)

An Interpolation Method for CLP Traversal

Joxan Jaffar, Andrew E. Santosa, and Răzvan Voicu

School of Computing, National University of Singapore
{joxan,andrews,razvan}comp.nus.edu.sg

Abstract. We consider the problem of exploring the search tree of a CLP goal in pursuit of a target property. Essential to such a process is a method of tabling to prevent duplicate exploration. Typically, only actually traversed goals are memoed in the table. In this paper we present a method where, upon the successful traversal of a subgoal, a *generalization* of the subgoal is memoed. This enlarges the record of already traversed goals, thus providing more pruning in the subsequent search process. The key feature is that the abstraction computed is guaranteed not to give rise to a spurious path that might violate the target property.

A driving application area is the use of CLP to model the behavior of other programs. We demonstrate the performance of our method on a benchmark of program verfication problems.

1 Introduction

In this paper we present a general method for optimizing the traversal of general search trees. The gist of the method is backward-learning: proceeding in a depth-first manner, it discovers an *interpolant* from the completed exploration of a subtree. The interpolant describes properties of a more general subtree which, importantly, preserves the essence of the original subtree with respect to a *target property*. We show via experiments that often, the generalized tree is considerably more general than the original, and therefore its representation is considerably smaller.

Our method was originally crafted as a means to optimize the exploration of states in computation trees, which are used as a representation of program behaviour in program analysis and verification. Such a representation can be symbolic in that a single node represents not one but a possibly infinite set of concrete program states or traces. The importance of a computation tree stems from the fact that it can represent a *proof* of some property of the program. Building such a tree in fact is an instance of a search problem in the sense of Constraint Programming, see eg. [1], and viewed as such, the problem of state-space exploration essentially becomes the problem of traversing a search tree. In this circumstance, the target property can simply be a predicate, corresponding to a *safety property*. Or it can be something more general, like the projection onto a set of distinguished variables; in this example, preserving the target property would mean that the values of these variables remain unchanged.

More concretely, consider a CLP derivation tree as a decision tree where a node has a conjunction of formulas symbolically representing a set of states. Its successor node has an incrementally larger conjunction representing a new decision. Suppose the target nodes are the terminal nodes. During a depth-first traversal, whenever a path in the tree

I.P. Gent (Ed.): CP 2009, LNCS 5732, pp. 454–469, 2009.

is traversed completely, we compute an *interpolant* at the target node. Where F denotes the formula in this node and T denotes the target property, this interpolant is a formula F' such that $F \models F'$ and $F' \models T$. (Failure is reported if no such F' can be found, ie: that $F \not\models T$.) Any such F' not only establishes that this node satisfies the target property, but also establishes that a generalization of F will also suffice. This interpolant can now be propagated back along the same path to ancestor states resulting in their possible generalizations. The final generalization of a state is then the conjunction of the possible generalizations of derivation paths that emanate from this state.

One view of the general method is that it provides an enhancement to the general method of *tabling* which is used in order to avoid duplicate or redundant search. In our case, what is tabled is not the encountered state itself, but a *generalization* of it.

The second part of the paper presents a specific algorithm for both computing and propagating interpolants throughout the search tree. The essential idea here is to consider the formulas describing subgoals as syntactic entities, and then to use serial constraint replacement successively on the individual formulas, starting in chronological order of the introduction of the formulas. In this way, we achieve efficiency and can still obtaining good interpolants.

1.1 Related Work

Tabling for logic programming is well known, with notable manifestation in the SLG resolution [2,3] which is implemented in the XSB logic programming system [4]. As mentioned, we differ by tabling a generalization of an encountered call.

Though we focus on examples of CLP representing other programs, we mentioned that we have employed an early version of the present ideas for different problems. In [5], we enhanced the dynamic programming solving of resource-constrained shortest path (RCSP) problems. This kind of example is similar to a large class of combinatorial problems.

Our interpolation method is related to various no-good learning techniques in CSP [6] and conflict-driven and clause learning techniques in SAT solving [7,8,9,10]. These techniques identify subsets of the minimal *conflict set* or *unsatisfiable core* of the problem at hand w.r.t. a subtree. This is similar to our use of interpolation, where we generalize a precondition "just enough" to continue to maintain the verified property.

An important alternative method for proving safety of programs is translating the verification problem into a Boolean formula that can then be subjected to SAT or SMT solvers [11,12,13,14,15,16]. In particular, [8] introduces bounded model checking, which translates k-bounded reachability into a SAT problem. While practically efficient in case when the property of interest is violated, this approach is in fact incomplete, in the sense that the state space may never be fully explored. An improvement is presented in [17], which achieves unbounded model checking by using interpolation to successively refine an abstract transition relation that is then subjected to an external bounded model checking procedure. Techniques for generating interpolants, for use in state-of-the-art SMT solvers, are presented in [18]. The use of interpolants can also be seen in the area of theorem-proving [19].

In the area of program analysis, our work is related to various techniques of abstract interpretation, most notably *counterexample-guided abstraction refinement* (*CEGAR*)

(a)	(b)
$\langle 0\rangle$ `if` $(*)$ `then` $x := x+1$ $\langle 1\rangle$ `if` $(y \geq 1)$ `then` $x := x+2$ $\langle 2\rangle$ `if` $(y < 1)$ `then` $x := x+4\langle 3\rangle$	$p_0(X,Y)$:- $p_1(X,Y)$. $p_0(X,Y)$:- $p_1(X',Y), X' = X+1$. $p_1(X,Y)$:- $p_2(X,Y), Y < 1$. $p_1(X,Y)$:- $p_2(X',Y), X' = X+2, Y \geq 1$. $p_2(X,Y)$:- $p_3(X,Y), Y \geq 1$. $p_2(X,Y)$:- $p_3(X',Y), X' = X+4, Y < 1$. $p_3(X,Y,X,Y)$.

Fig. 1. A Program and Its CLP Model

(a)	(b)
$p_0(X)$:- $p_1(X'), X' = X+1$. $p_0(X)$:- $p_1(X)$. $p_1(X)$:- $p_2(X'), X' = X+2$. $p_1(X)$:- $p_2(X)$. $p_2(X)$:- $p_3(X'), X' = X+4$. $p_2(X)$:- $p_3(X)$. $p_3(X)$.	$p_0(I,N,X,Y)$:- $p_1(I,N,X,Y), \varphi(X)$. $p_0(I,N,X,Y)$:- $p_1(I,N,X,Y), \neg\varphi(X)$. $p_1(I,N,X,Y)$:- $p_2(I,N,X,Y), I \leq N$. $p_1(I,N,X,Y)$:- $p_4(I,N,X,Y), I > N$. $p_2(I,N,X,Y)$:- $p_3(I,N,X,Y'), Y' = Y \times (-Y)$. $p_3(I,N,X,Y)$:- $p_1(I',N,X,Y), I' = I+1$. $p_4(I,N,X,Y)$.

Fig. 2. CLP Programs

[20,21,22], which perform successive refinements of the abstract domain to discover the right abstraction to establish a safety property of a program. An approach that uses interpolation to improve the refinement mechanism of CEGAR is presented in [22,23]. Here, interpolation is used to improve over the method given in [21], by achieving better locality of abstraction refinement. Ours differ from CEGAR formulations in some important ways: first, instead of *refining*, we *abstract* a set of (concrete) states. In fact, our algorithm abstracts a state after the computation subtree emanating from the state has been completely traversed, and the abstraction is constructed from the interpolations of the constraints along the paths in the subtree. Thus a second difference: our algorithm interpolates a *tree* as opposed to a *path*. More importantly, our algorithm does not traverse *spurious paths*, unlike abstract interpretation. We shall exemplify this difference in a comparison with BLAST [24] in Section 6.

2 The Basic Idea

Our main application area is the state-space traversal of imperative programs. For this, we model imperative programs in CLP. Such modeling has been presented in various works [25,26,27] and is informally exemplified here. Consider the imperative program of Fig. 1 (a). Here the $*$ denotes a condition of nondeterministic truth value. We also augment the program with program points enclosed in angle brackets. The CLP model of the same program is shown in Fig. 1 (b).

To exemplify our idea, let us first consider a simpler CLP program of Fig. 2 (a), which is a model of an imperative program. The execution of the CLP program results in the derivation tree shown in Fig. 3 (top). The derivation tree is obtained by reduction using p_i predicates in the CLP model. In Fig. 3 (top), we write a CLP goal $p_k(\tilde{X}), \varphi$ as $\langle k \rangle, \varphi'$ where φ' is a simplification of φ by projecting on the variables $\tilde{X}$, and an arrow denotes a derivation step.

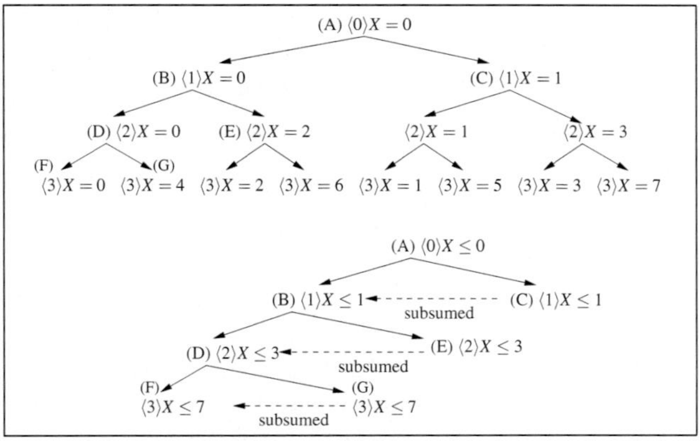

Fig. 3. Interpolation

Starting in a goal satisfying $X = 0$, all derivation sequences are depicted in Fig. 3 (top). Suppose the target property is that $X \leq 7$ at program point $\langle 3 \rangle$. The algorithm starts reducing from (A) to (F). (F) is labelled with $X = 0$ which satisfies $X \leq 7$. However, a more general formula, say $X \leq 5$, would also satisfy $X \leq 7$. The constraint $X \leq 5$ is therefore an interpolant, since $X = 0$ implies $X \leq 5$, and $X \leq 5$ in turn implies $X \leq 7$. We could use such an interpolant to generalize the label of node (F). However, we would like to use as general an interpolant as possible, and clearly in this case, it is $X \leq 7$ itself. Hence, in Fig. 3 (bottom) we replace the label of (F) with $\langle 3 \rangle$ $X \leq 7$. In this way, node (G) with label $\langle 3 \rangle$ $X = 4$ (which has not yet been traversed) is now *subsumed* by node (F) with the new label (since $X = 4$ satisfies $X \leq 7$) and so (G) need not be traversed anymore. Here we can again generate an interpolant for (G) such that it remains subsumed by (F). The most general of these is $X \leq 7$ itself, which we use to label (G) in Fig. 3 (bottom).

We next use the interpolants of (F) and (G) to produce a generalization of (D). Our technique is to first compute candidate interpolants from the interpolants of (F) and (G), w.r.t. the reductions from (D) to (F) and from (D) to (G). The final interpolant of (D) is the conjunction of these candidate interpolants. In this process, we first rename the variables of (F) and (G) with their primed versions, such that (F) and (G) both have the label $X' \leq 7$. First consider the reduction from (D) to (F), which is in fact equivalent to a `skip` statement, and hence it can be represented as the constraint $X' = X$. It can be easily seen that the label $X = 0$ of (D) entails $X' = X \models X' \leq 7$. Here again we compute an interpolant. The interpolant here would be entailed by $X = 0$ and entails $X' = X \models X' \leq 7$. As interpolant, we choose $X \leq 7$[1].

Similarly, considering the goal reduction (D) to (G) as the augmentation of the constraint $X' = X + 4$, we obtain a candidate interpolant $X \leq 3$ for (D). The final interpolant for (D) is the conjunction of all candidates, which is $X \leq 7 \wedge X \leq 3 \equiv X \leq 3$. We label

[1] This interpolant corresponds to the *weakest precondition* [28] w.r.t. the statement $X := X$ and the target property $X \leq 7$, however, in general the obtained precondition need not be the weakest, as long as it is an interpolant.

(D) with this interpolant in Fig. 3 (bottom). In this way, (E) is now subsumed by (D), and its traversal for the verification of target property is not necessary.

We then generate an interpolant for (E) in the same way we did for (G). By repeating the process described above for other nodes in the tree, we obtain the smaller tree of Fig. 3 (bottom), which is linear in the size of the program. This tree represents the part of the symbolic computation tree that would *actually be traversed* by the algorithm. Hence, while the tree's size is exponential in the number of `if` statements, our algorithm prunes significant parts of the tree, speeding up the process.

2.1 With Infeasible Sequences

Now consider Fig. 1 (a) where there are *infeasible* sequences ending in goals with unsatisfiable constraint, which are depicted in Fig. 4 (top). Let the target property be $X \leq 5$ at point $\langle 3 \rangle$. A key principle of our re-labeling process is that it *preserves the infeasibility* of every derivation sequence. Thus, we must avoid re-labeling a node too generally, since that may turn infeasible paths into feasible ones. To understand why this is necessary, let us assume, for instance, re-labelling (E), whose original label is $X = 2, Y \geq 1$, into $X = 2$. This would yield $X = 6$ at point $\langle 3 \rangle$ (a previously unreachable goal), which no longer entails the target property $X \leq 5$.

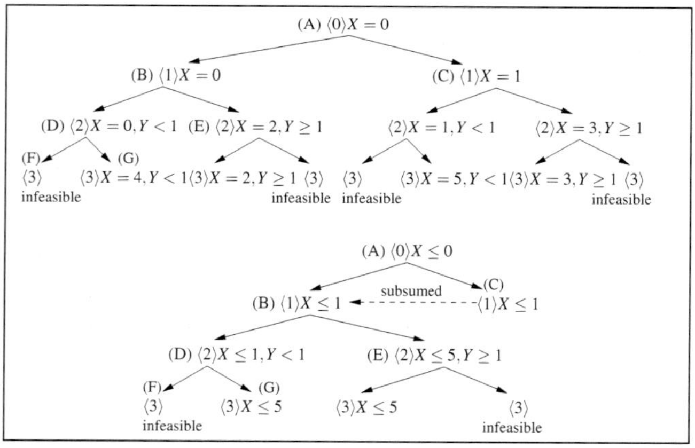

Fig. 4. Interpolation of Infeasible Sequences

Next note that the path ending at (F) is also infeasible. Applying our infeasibility preservation principle, we keep (F) labeled with *false*, and therefore the only possible interpolant for (F) is *false* itself. This would produce the interpolant $Y < 1$ at (D) since this is the most general condition that preserves the infeasibility of (F). Note that here, $Y < 1$ is implied by the original label $X = 0, Y < 1$ of (D) and implies $Y \geq 1 \models$ *false*, which is the weakest precondition of *false* w.r.t. the negation of the `if` condition on the transition from (D) to (F).

Now consider (G) with $X = 4, Y < 1$ and note that it satisfies $X \leq 5$. (G) can be interpolated to $X \leq 5$. As before, this would produce the precondition $X \leq 1$ at (D). The final interpolant for (D) is the *conjunction* of $X \leq 1$ (produced from (G)) and $Y < 1$ (produced from (F)). In this way, (E) cannot be subsumed by (D) since its label does not satisfy both the old and the new labels of (D). Fortunately, however, after producing the interpolant for (B), the node (C) can still be subsumed.

In Section 5 we detail an efficient technique to generate interpolants called *serial constraint replacement*, which is based on constraint deletion and slackening. This technique is briefly explained next.

2.2 Loops

Our method assumes that the derivation tree is finite in the sense that there is no infinite sequence of distinct goals in a single path, although the tree may contain cyclic derivations. Here we discuss how we may compute an interpolant from cyclic derivations.

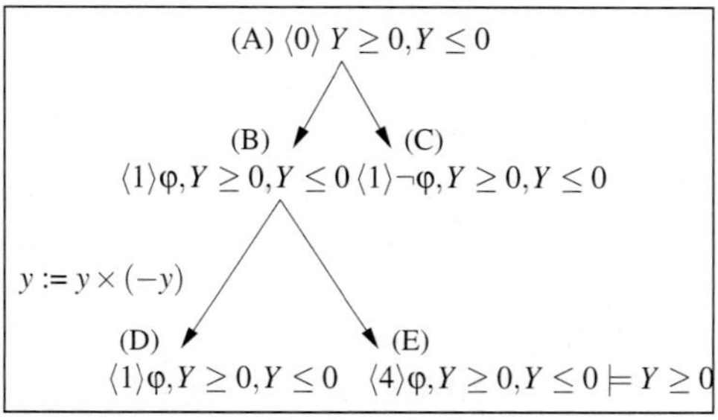

Fig. 5. Loop Interpolation

Consider Fig. 2 (b) which is a program with cyclic derivation. The program contains some constraint φ. The derivation tree, where the initial goal is $Y \geq 0, Y \leq 0$, is shown in Fig. 5. The tree is finite because in (D), the second occurrence of point $\langle 1\rangle$, is *subsumed* by the ancestor (B). This subsumption is enabled by a *loop invariant* made available by some external means at node (B), and which renders unnecessary the expansion of node (D) (the computation tree is"closed" at (D)).

In the spirit of the example, we now attempt to generalize node (B) in order to avoid having to traverse node (C) (which must be traversed if (B) were not generalized).

Let us first examine the path (A), (B), (D). The constraint φ can be removed from (B) so that the resulting goal remains a loop invariant. This is because φ is not related to the other variables. More importantly, φ is itself a loop invariant, and we come back to this later.

Next we attempt to remove the constraint $Y \leq 0$. The resulting goal at (B) now has the constraint $Y \geq 0$. But this goal is no longer invariant. That is, the computation tree at this new node (B) is such that the corresponding node (D) is *not* subsumed by (B). A similar situation would arise if we kept $Y \leq 0$ and deleted $Y \geq 0$.

The only possibility we have left is to check if we can remove *both* of $Y \leq 0$ and $Y \geq 0$. Indeed, that is possible, since the sequence (A), (B), (D), from which all constraints $\varphi, Y \leq 0, Y \geq 0$ are removed, is such that (B) subsumes (D). Indeed, in this case, the generalized (B) subsumes all the goals at point $\langle 1\rangle$.

So far, we have shown that for the sequence (A), (B), (D), at node (B), we could perform the following kinds of deletions: (1) delete nothing, (2) delete φ alone, (3) delete both of $Y \leq 0$ and $Y \geq 0$, and (4) delete all of $\varphi, Y \leq 0$ and $Y \geq 0$. (That is, we exclude the case where we delete just one of $Y \leq 0$ and $Y \geq 0$.) We would then have a new sequence where (B) continues to subsume (D).

Let us now examine the sequence (A), (B), (E), which is the second derivation sequence emanating from (B). Note that (E) is a target goal, and we require that $Y \geq 0$ here. Thus the choices (3) or (4) above made for the sequence (A), (B), (D) would not be suitable for this path, because these deletions would remove all constraints on Y.

It thus becomes clear that the best choice is (2). That is, we end up generalizing (B) by deleting only the constraint φ. With this as an interpolant, it is now no longer necessary to traverse the subtree of goal (C).

In summary, a final goal that is subsumed by one of its ancestors is already labelled with a *path invariant*, that is, a invariant that holds only for the particular derivation sequence. However, it is still possible to generalize the final goal by deleting any individual constraint that also appears at the ancestor goal, and it is *itself invariant*. Note that in this example, φ was invariant through the cycle, unlike $Y \leq 0$ or $Y \geq 0$. A rather straightforward idea then is to consider only invariant constraints as candidates for deletion within a sequence. We detail this in sections 4 and 5.2.

3 CLP Preliminaries

We first briefly overview CLP [29]. The *universe of discourse* is a set of terms, integers, and arrays of integers. A *constraint* is written using a language of functions and relations. In this paper, we will not define the constraint language explicitly, but invent them on demand in accordance with our examples. Thus the terms of our CLP programs include the function symbols of the constraint language.

An *atom*, is as usual, of the form $p(\tilde{t})$ where p is a user-defined predicate symbol and the $\tilde{t}$ a tuple of terms. A *rule* is of the form A :- $\tilde{B}, \phi$ where the atom A is the *head* of the rule, and the sequence of atoms $\tilde{B}$ and the constraint ϕ constitute the *body* of the rule. A *goal* G has exactly the same format as the body of a rule. We say that a rule is a (constrained) *fact* if $\tilde{B}$ is the empty sequence. A *ground instance* of a constraint, atom and rule is defined in the obvious way.

A *substitution* simultaneously replaces each variable in a term or constraint into some expression. We specify a substitution by the notation $[\tilde{E}/\tilde{X}]$, where $\tilde{X}$ is a sequence $X_1, \ldots, X_n$ of variables and $\tilde{E}$ a list $E_1, \ldots, E_n$ of expressions, such that X_i is replaced by E_i for all $1 \leq i \leq n$. Given a substitution θ, we write as $E\theta$ the application of the substitution to an expression E. A *renaming* is a substitution which maps variables variables. A *grounding* is a substitution which maps each variable into a value in its domain.

In this paper we deal with goals of the form $p_k(\tilde{X}), \Psi(\tilde{X})$, where p_k is the predicate defined in a CLP model of an imperative program and $\Psi(\tilde{X})$ is a constraint on $\tilde{X}$. Given a goal $\mathcal{G} \equiv p_k\tilde{X}), \Psi(\tilde{X})$, $[\![\mathcal{G}]\!]$ is the set of the groundings θ of the primary variables $\tilde{X}$ such that $\tilde{\exists}\Psi(\tilde{X})\theta$ holds. We say that a goal $\overline{\mathcal{G}} \equiv p_k(\tilde{X}), \overline{\Psi}(\tilde{X})$ *subsumes* another goal $\mathcal{G} \equiv p_{k'}(\tilde{X}'), \Psi(\tilde{X}')$ if $k = k'$ and $[\![\overline{\mathcal{G}}]\!] \supseteq [\![\mathcal{G}]\!]$. Equivalently, we say that $\overline{\mathcal{G}}$ is a *generalization* of $\mathcal{G}$. We write $\mathcal{G}_1 \equiv \mathcal{G}_2$ if $\mathcal{G}_1$ and $\mathcal{G}_2$ are generalizations of each other. We say that a sequence is *subsumed* if its last goal is subsumed by another goal in the sequence.

Given two goals $\mathcal{G}_1 \equiv p_k(\tilde{X}_1), \Psi_1$ and $\mathcal{G}_2 \equiv p_k(\tilde{X}_2), \Psi_2$ sharing a common program point k, and having disjoint sets of variables, we write $\mathcal{G}_1 \wedge \mathcal{G}_2$ to denote the goal $p_k(\tilde{X}_1), (\tilde{X}_1 = \tilde{X}_2, \Psi_1, \Psi_2)$.

Let $G \equiv (B_1, \cdots, B_n, \phi)$ and P denote a goal and program respectively. Let $R \equiv A$:- $C_1, \cdots, C_m, \phi_1$ denote a rule in P, written so that none of its variables appear in G. Let $A = B$, where A and B are atoms, be shorthand for equations between their corresponding arguments. A *reduct* of G using R is of the form

$$(B_1, \cdots, B_{i-1}, C_1, \cdots, C_m, B_{i+1}, \cdots, B_n, B_i = A \wedge \phi \wedge \phi_1)$$

provided $B_i = A \wedge \phi \wedge \phi_1$ is satisfiable.

A *derivation sequence* is a possibly infinite sequence of goals $G_0, G_1, \cdots$ where $G_i, i > 0$ is a reduct of G_{i-1}. If there is a last goal G_n with no atoms called *terminal goal*, we say that the derivation is *successful*. In order to prove safety, we test that the goal implies the safety condition. A derivation is ground if every reduction therein is ground. Given a sequence τ defined to be $\mathcal{G}_0, \mathcal{G}_1, \ldots, \mathcal{G}_n$, then $cons(\tau)$ is all the constraints of the goal $\mathcal{G}_n$. We say that a sequence is *feasible* if $cons(\tau)$ is satisfiable, and *infeasible* otherwise. Moreover, we say that a derivation sequence τ is *successful*, when it is feasible and k is the final point.

The *derivation tree* of a CLP has as branches its derivation sequences. In this tree, the *ancestor-descendant* relation between nodes is defined in the usual way. A leaf of this tree is *cyclic* if its program point appears at one of its ancestors, which will be called the *cyclic ancestor* of the leaf in question. A derivation tree is *closed* if all its branches are either successful, infeasible, or subsumed. Given a CLP with a derivation tree T, whose root is a goal $\mathcal{G}$, we denote by $T[\mathcal{G}'/\mathcal{G}]$ the tree obtained by replacing the root $\mathcal{G}$ by a new goal $\mathcal{G}'$, and relabeling the nodes of T to reflect the rules represented by the edges of T. In other words, $T[\mathcal{G}'/\mathcal{G}]$ represents the symbolic computation tree of the same program, started at a different goal.

Informally, we say that two closed trees T and T' have the *same shape* if their sequences can be uniquely paired up such that, for every pair of sequences (τ, τ'), we have: (a) τ is a sequence in T, and τ' is a sequence in T'; (b) τ and τ' have the same sequence of predicates; and (c) τ and τ' are both simultaneously either successful, infeasible, or subsumed.

Given a target property represented as a condition $\Psi(\tilde{X})$ on system variables, we say that a final goal $\mathcal{G}$ is *safe* if $[\![\mathcal{G}]\!] \subseteq [\![p_k(\tilde{X}), \Psi(\tilde{X})]\!]$.

We end this section with a definition of the notion of interpolant.

Definition 1 (Interpolant). *A goal $\mathcal{G}_I$ is an* interpolant *for closed tree T with root $\mathcal{G}$ if:*

- *all its successful sequences end in safe goals;*
- *$\mathcal{G}_I$ subsumes $\mathcal{G}$,*
- *T and $T[\mathcal{G}_I/\mathcal{G}]$ have the same shape.* □

4 The Algorithm

In this section, we describe an idealized algorithm to traverse a computation tree of a given goal $\mathcal{G}$. The recursive procedure computes for each such $\mathcal{G}$ a possibly infinite set of interpolants. These interpolants are then propagated to the parent goal $\mathcal{G}_p$ of $\mathcal{G}$. The eventual completion of the traversal of $\mathcal{G}_p$ is likewise augmented by a computation of its interpolant. Clearly this process is most naturally implemented recursively by a depth-first traversal. Our main technical result is that all interpolants are *safe*, that is, all computation trees resulting from a goal subsumed by an interpolant are safe.

The algorithm is presented in Fig. 6. Its input is a goal $\mathcal{G}$. We assume that there is a target property that $\Psi_f(\tilde{X}_f)$ must hold at a target point k_f. Without loss of generality, we also assume that $\mathcal{G}$ and Ψ_f do not share variables, and before the execution, the memo table containing computed interpolants is empty. The idea is to compute interpolants that are as general as possible. The function `solve` is initially called with the initial goal. We first explain several subprocedures that are used in Fig. 6:

- `memoed(`$\mathcal{G}$`)` tests if $\mathcal{G}'$ is in the memo table such that $\mathcal{G}'$ subsumes $\mathcal{G}$. If this is the case, it returns the set of all such $\mathcal{G}'$.
- `memo(`I`)` records the set I of interpolants in the memo table.
- $\text{WP}(\mathcal{G},\rho)$ is a shorthand for the condition $\rho \models \Psi'$ where ρ is the constraint in the rule used to produce the reduct $\mathcal{G} \equiv p_k(\tilde{X}'), \Psi'$; $\tilde{X}$ and $\tilde{X}'$ are the variables appearing in this rule.

In what follows, we discuss each of the five cases of our algorithm. Each case is characterized by a proposition that contributes to the proof of the correctness theorem stated at the end of this section.

First, the algorithm tests whether the input goal $\mathcal{G}$ is already memoed, in which case, the return value of `memoed(`$\mathcal{G}$`)` is returned. The following proposition holds:

Proposition 1. *If $\mathcal{G}' \in \texttt{memoed}(\mathcal{G})$ then $\mathcal{G}'$ is an interpolant of $\mathcal{G}$.*

Next consider the case where current goal $\mathcal{G}$ is *false*. Here the algorithm simply returns a singleton set containing the goal $p_k(\tilde{X})$,*false*. The following proposition is relevant to the correctness of this action.

Proposition 2. *The goal $p_k(\tilde{X})$, false is an interpolant of a false goal.*

Next consider the case where current goal $\mathcal{G}$ is terminal. If $\mathcal{G}$ is unsafe, that is $[\![\mathcal{G}]\!] \not\subseteq [\![p_k(\tilde{X}_f), \Psi_f(\tilde{X}_f)]\!]$, the entire function aborts, and we are done. Otherwise, the algorithm returns *all* generalizations $\overline{\mathcal{G}}$ of $\mathcal{G}$ such that $[\![\overline{\mathcal{G}}]\!] \subseteq [\![p_k(\tilde{X}_f), \Psi_f(\tilde{X}_f)]\!]$. The following proposition is relevant to the correctness of this action.

Proposition 3. *When the target property is specified by $p_{k_f}(\tilde{X}_f), \Psi_f$ where k_f is a final program point, and the goal $\mathcal{G}$ is safe, then its generalization $\overline{\mathcal{G}}$ is an interpolant of $\mathcal{G}$, where $[\![\overline{\mathcal{G}}]\!] \subseteq [\![p_{k_f}(\tilde{X}_f), \Psi_f]\!]$.*

Next consider the case where current goal $\mathcal{G}$ is a looping goal, that is, $\mathcal{G}$ is subsumed by an ancestor goal. Here we compute a set of generalizations of the ancestor goal such that the same execution from the ancestor goal to the current input goal still results in a cycle. In other words, we return the set of all possible generalizations of the ancestor goal such that when the same reduction sequence (with constraint Φ along the sequence) is traversed to the current goal, the goal remains subsumed by the ancestor goal. The following proposition is relevant to the correctness of this action.

Proposition 4. *If $\mathcal{G} \equiv p_k(\tilde{X}), \Psi$ is a goal with ancestor $p_k(\tilde{X}'), \Psi'$ such that $\Psi \equiv \Psi' \wedge \Phi$, then $p_k(\tilde{X}), \overline{\Psi}[\tilde{X}/\tilde{X}']$ is an interpolant of $\mathcal{G}$ where $\overline{\Psi} \wedge \Phi \models \overline{\Psi}[\tilde{X}/\tilde{X}']$ if all successful goals in the tree are safe.*

Finally we consider the recursive case. The algorithm represented by the recursive procedure `solve`, given in Figure 6, applies all applicable CLP rules to create new goals from $\mathcal{G}$. It does this by reducing $\mathcal{G}$. It then performs recursive calls using the reducts. Given the return values of the recursive calls, the algorithm computes the interpolants of the goal $\mathcal{G}$ by an operation akin to weakest precondition propagation. The final set of interpolants for $\mathcal{G}$ is then simply the intersection of the sets of interpolants returned by recursive calls.

```
solve(𝒢 ≡ p_k(X̃),Ψ) returns  a set of interpolants
    case (I = memoed(𝒢)): return I
    case 𝒢 is false (Ψ ≡ false): return {p_k(X̃),false}
    case 𝒢 is target (k = k_f):
        if (Ψ[X̃_f/X̃] ⊭ Ψ_f) abort else return {𝒢̄ : [[𝒢̄]] ⊆ [[p_{k_f}(X̃_f),Ψ_f]]}
    case 𝒢 is cyclic:
        let cyclic ancestor of 𝒢 be p_k(X̃'),Ψ' and Ψ ≡ Ψ' ∧ Φ
        return {p_k(X̃),Ψ̄[X̃/X̃'] : Ψ̄ ∧ Φ ⊨ Ψ̄[X̃/X̃']}
    case otherwise:
        foreach rule p_k(X̃) :- p_k'(X̃'),ρ(X̃,X̃'):
            I := I ∩ {p_k(X̃),WP(𝒢̄',ρ) : 𝒢̄' ∈ solve(p_k'(X̃'),Ψ ∧ ρ)}
        endfor
        memo(I) and return I
```

Fig. 6. The Interpolation Algorithm

Proposition 5. *Let $\mathcal{G}$ have the reducts $\mathcal{G}_i$ where $1 \leq i \leq n$. Let $\overline{\mathcal{G}}_i \in$ solve$(\mathcal{G}_i)$, $1 \leq i \leq n$. Then $\overline{\mathcal{G}}$ is an interpolant of $\mathcal{G}$ if for all $1 \leq i \leq n$, $\overline{\mathcal{G}} \equiv \cap_{i=1}^{n}\{p_k(\tilde{X}), \textsc{wp}(\overline{\mathcal{G}}_i, \rho_i)\}$.*

The following theorem follows from Propositions 1 through 5:

Theorem 1 (Safety). *The algorithm in Fig. 6 correctly returns interpolants.*

We note that the generation of interpolants here employs a notion of *weakest precondition* in the literature [28,30]. Given a goal transition induced by a statement s and represented as input-output constraint $\rho(\tilde{X}, \tilde{X}')$, the weakest precondition of a condition Ψ' is $\rho(\tilde{X}, \tilde{X}') \models \Psi'$. By our use of interpolation, we do not directly use the weakest precondition to generalize a goal, a technique which is notoriously inefficient [31], but we instead use interpolants, which approximate the weakest precondition, are efficient to compute, and yet still generalize the input goal. That is, instead of WP, we use another function INTP$(\mathcal{G}, \rho)$ such that when $\mathcal{G}$ is $p_{k'}(\tilde{X}'), \Psi'$, then INTP$(\mathcal{G}, \rho)$ is a constraint $\overline{\Psi}$ such that $\overline{\Psi}$ entails $(\rho \models \Psi')$. In the next section, we demonstrate an algorithm that implements INTP based upon an efficient implementation of constraint deletions and "slackening."

5 Serial Constraint Replacement

We now present a general practical approach for computing an interpolant. Recall the major challenges arising from the idealized algorithm in Fig. 6:

- not one but a *set* of interpolants is computed for each goal traversed;
- even for a single interpolant, there needs to be efficient way to compute it;
- interpolants for the descendants of a goal need to be combined by a process akin to weakest-precondition propagation, and then these results need to be conjoined.

Given a set of constraints, we would like to generalize the maximal number of constraints that would preserve some "interpolation condition". Recall that this condition is (a) being unsatisfiable in the case we are dealing with a terminal (*false*) goal, (b) implying a target property in case we are dealing with a target goal, or (c) implying that the subsumed terminal node remains subsumed.

Choosing a subset of constraints is clearly an inefficient process because there are an exponential number of subsets to consider. Instead, we order the constraints according to the execution order, and process the constraints *serially*. While not guaranteed to find the smallest subset, this process is efficient and more importantly, attempts to generalize the constraints *in the right order* because the earliest constraints, those that appear in the most goals along a path, are generalized first.

The computation of interpolants is different across the three kinds of paths considered. Case (a) and (b) are similar and will be discussed together, and we discuss separately case (c).

5.1 Sequences Ending in a False or Target Goal

Consider each constraint Ψ along the path to the terminal goal in turn. If the terminal goal were *false*, we replace Ψ with a more general constraint if the goal remains *false*, In case the terminal goal were a target, we replace Ψ with a more general constraint if the goal remains safe. We end up concretely with a subset of the constraints in the terminal goal, and this defines that the interpolant is the goal with the replacements realized. We next exemplify.

Note that a program statement gives rise to a constraint relating the states before and after execution of the statement Consider the following imperative program and its CLP model:

$\langle 0 \rangle$ $x := x+1$	$p_0(X,Y,Z)$:- $p_1(X',Y,Z), X' = X+1.$
$\langle 1 \rangle$ `if` $(z \geq 0)$ `then`	$p_1(X,Y,Z)$:- $p_2(X,Y,Z), Z \geq 0.$
$\langle 2 \rangle$ $\quad y := x$	$p_1(X,Y,Z)$:- $p_3(X,Y,Z), Z < 0.$
$\langle 3 \rangle$	$p_2(X,Y,Z)$:- $p_3(X,Y',Z), Y' = X.$
	$p_3(X,Y,Z).$

The sequence of constraints obtained from the derivation sequence which starts from the goal $p_0(X_0,Y_0,Z_0), X_0 = 0, Y_0 = 2$ and goes along $\langle 0 \rangle$, $\langle 1 \rangle$, $\langle 2 \rangle$, to $\langle 3 \rangle$ is $X_0 = 0, Y_0 = 2, X_1 = X_0 + 1, Z_0 \geq 0, Y_1 = X_1$ for some indices 0 and 1 denoting versions of variables. At this point, we need to *project* the constraints onto the *current variables*, those that represent the current values of the original program variables. At $\langle 3 \rangle$, the projection of interest is $X = 1, Y = 1, Z \geq 0$.

If the target property were $Y \geq 0$ at point $\langle 3 \rangle$, then it holds because the projection implies it. In the case of an infeasible sequence (not the case here), our objective would be to preserve the infeasibility, which means that we test that the constraints imply the target condition *false*.

In general, then, we seek to generalize a projection of a list of constraints in order to obtain an interpolant. Here, we simply replace a constraint with a more general one as long as the result satisfies the target property. For the example above, we could delete (replace with *true*) the constraints $Y_0 = 2$ and $Z_0 \geq 0$ and still prove the target property $Y \geq 0$ at $\langle 3 \rangle$.

In Table 1 we exemplify both the deletion and the slackening techniques using our running example. The first column of Table 1 is the executed statements column (we

Table 1. Interpolation Techniques

Statement	No Interpolation		Deletion		Deletion and Slackening (1)	
	Constraint	Projection	Constraint	Projection	Constraint	Projection
$\{x=0,y=2\}$ $\langle 0\rangle$	$X_0=0,Y_0=2$	$X=0,Y=2$	$X_0=0$	$X=0$	$X_0\geq 0$	$X\geq 0$
$x:=x+1$ $\langle 1\rangle$	$X_1=X_0+1$	$X=1,Y=2$	$X_1=X_0+1$	$X=1$	$X_1=X_0+1$	$X\geq 1$
if $(z\geq 0)$ $\langle 2\rangle$	$Z_0\geq 0$	$X=1,Y=2,Z\geq 0$	(none)	$X=1$	(none)	$X\geq 1$
$y:=x$ $\langle 3\rangle$	$Y_1=X_1$	$X=1,Y=1,Z\geq 0$	$Y_1=X_1$	$X=1,Y=1$	$Y_1=X_1$	$X\geq 1,Y=X$

represent the initial goal in curly braces). During the first traversal without abstraction, the constraints in the second column is accumulated. The projection of the accumulated constraints into the primary variables is shown in the third column. As mentioned, this execution path satisfies the target property $Y_1 \geq 0$. We generalize the goals along the path using one of two techniques:

- **Constraint deletion.** Here we replace a constraint with *true*, effectively deleting it. This is demonstrated in the fourth column of Table 1. Since the constraint $Z_0 \geq 0$ and $Y_0 = 2$ do not affect the satisfaction of the target property, they can be deleted. The resulting projections onto the original variables is shown in the fifth column, effectively leaving Y unconstrained up to $\langle 2\rangle$, while removing all constraints on Z.
- **Slackening.** Another replacement techniqe is by replacing equalities with non-strict inequalities, which we call *slackening*. For example, replacing the constraint $X_0 = 0$ with $X_0 \geq 0$ in the fifth column of Table 1 would not alter the unsatisfiability of the constraint system. (We would repeat this exercise with $X_0 \leq 0$.) The actual replacement in the sixth column results in the more general interpolants in the seventh column. Recall the demonstration of slackening in Section 2.

5.2 Sequences Ending in a Subsumed Goal

Consider now the case of a sequence τ ending in a goal which is subsumed by an ancestor goal. Say τ is $\tau_1\tau_2$ where τ_1 depicts the prefix sequence up to the subsuming ancestor goal. The subsumption property can be expressed as

$$cons(\tau) \models cons(\tau_1)[\tilde{X}_i/\tilde{X}]$$

Following the spirit of the previous subsection, we now seek to replace any individual constraint in τ_1 as long as this subsumption holds. (Note that replacing a constraint in τ_1 automatically replaces a constraint in τ, because τ_1 is a prefix of τ.)

However, there is one crucial difference with the previous subsection. Here we shall only be replacing an individual constraint Ψ that is *itself invariant* (for point k) in the sequence. The reason for this is based on the fact that in order to propagate an interpolant (now represented as a single goal, and not a family), the interpolants for descendant nodes need to be simply conjoined in order to form the interpolant for the parent goal (explained in the next section). This may result in re-introduction of a replaced constraint Ψ in τ_1. The condition that Ψ is itself invariant guarantees that even if Ψ is re-introduced, we still have an interpolant that is invariant for the cycle.

5.3 Propagating Interpolants

One key property of serial constraint replacement is the ease with which interpolants are generated from various derivation paths that share some prefix. Recall in the previous sections that we need to produce a common interpolant for the intermediate nodes in the tree from the interpolants of their children. Here, we compute candidate interpolants of a parent node from the interpolants of the children. Note that each interpolant is now simply a conjunction of constraints. Then, the interpolant of the parent is *simply the conjunction* of the candidate interpolants (cf. the intersection of interpolants for the recursive case of the algorithm in Section 4).

Table 2. Propagating Interpolants

Statement	Deletion and Slackening (2)		Combination (1), (2)	
	Constraint	Projection	Constraint	Projection
$\{x=0, y=2\}$ $\langle 0\rangle$	$Y_0 \geq 1$	$Y \geq 1$	$X_0 \geq 0, Y_0 \geq 1$	$X \geq 0, Y \geq 1$
$x := x+1$ $\langle 1\rangle$	(none)	$Y \geq 1$	$X_1 = X_0 + 1$	$X \geq 1, Y \geq 1$
if $(z \geq 0)$ $\langle 3\rangle$	(none)	$Y \geq 1$		

Let us now consider another path through the sample program, one that visits $\langle 0\rangle$, $\langle 1\rangle$, and $\langle 3\rangle$ without visiting $\langle 2\rangle$. The statements and the interpolation along this path by deletion and slackening are shown in the first to third columns of Table 2. Note that this path, and the path $\langle 0\rangle$, $\langle 1\rangle$, $\langle 2\rangle$, and $\langle 3\rangle$ considered before, share the initial goal $p_0(X_0, Y_0, Z_0), X_0 = 0, Y_0 = 2$ and the first statement of the program. Now to compute the actual interpolant for $\langle 0\rangle$ and $\langle 1\rangle$, we need to consider the interpolants generated from both paths. Using our technique, we can simply conjoin the constraints at each level to obtain the common interpolants. This is exemplified in the fourth column of Table 2. As can be seen, the resulting interpolants are simple because they do not contain disjunction. The resulting projection is shown in the fifth column. The execution of the program, starting from the goal represented by the projection, along either of the possible paths, is guaranteed to satisfy the target property.

6 Experimental Evaluation

We implemented a prototype verifier using CLP($\mathcal{R}$) constraint logic programming system [32]. This allows us to take advantage of built-in Fourier-Motzkin algorithm [33] and meta-level facilities for the manipulation of constraints. We performed the experiments on a 1.83GHz Macbook Pro system.

6.1 Array Bounds Verification by Constraint Deletion

We verify that at each array access point, the possible index values are legal. The programs "FFT" and "LU" are from the Scimark benchmark suite, and "linpack" from Toy [34]. Here we manually provide an invariant for each loop to finitize the traversal. The specific interpolation method used is the constraint deletion.

In Table 3, "Goals" indicates the cost of traversal, and the time is in seconds. The fairly large "linpack" program (907 LOC) is run in four cases.

Table 3. Array Bounds Verification

Problem	LOC	No Interpolation States	Time	Interpolation States	Time
FFT	165	2755	10.62	120	0.10
LU	102	298	0.39	138	0.17
$linpack^{200}$	907	6385	19.70	332	0.53
$linpack^{400}$	907	8995	151.65	867	2.47
$linpack^{600}$	907	16126	773.63	867	2.47
linpack	907	∞	∞	867	2.47

For the first three, we apply a depth bound for the search tree, progressively at 200, 400, and 600 (denoted in the table as $linpack^b$, where b is the depth bound) to demonstrate the change in the size of the search tree, which is practically infinite when there is no depth bound. Using interpolation, on the other hand, we can completely explore the computation tree. In all cases, the number of goals visited is significantly reduced as a result of interpolation.

6.2 Resource Bound Verification by Constraint Deletion and Slackening

Here we seek to verify an upper bound for a certain variable, representing resource usage. In our examples, the resource of interest is the execution time of a program, modeled as a monotonically increasing instrumented variable.

Table 4. Resource Bound Verification

Problem	LOC	No Interpolation States	Time	Interpolation States	Time
decoder	27	344	1.42	160	0.49
sqrt	16	923	27.13	253	7.46
qurt	40	1104	38.65	290	11.39
janne_complex	15	1410	48.64	439	7.87
$statemate^{20}$	1298	21	0.05	21	0.08
$statemate^{30}$	"	1581	2.93	48	0.16
$statemate^{40}$	"	∞	∞	71	0.24
statemate	"	∞	∞	1240	17.09

We consider the "decoder", "sqrt", "qurt", "janne_complex" and "statemate" programs from the Mälardalen benchmark [35] used for testing WCET (worst-case execution time) analysis. One challenge to accurate WCET is that program fragments can behave significantly different when invoked in different contexts. For the "statemate" problem, we limit the depth bound of the goal-space search without interpolation into cases 20, 30, and 40, and we also store the summarization of the maximum resource usage of a computation subtree in the memo table. In Table 4, "$statemate^n$" denotes verification runs of "statemate" with n as the the depth bound. For "statemate," we limit the number of iteration of a loop in the program to two (actual number of iterations when all variables are initialized to 0). The "statemate" program displays a significant amount of dependencies between Boolean conditions in a path. The more dependency between statements there is, the less the reduction that can be obtained by interpolation. For example, in the experiment "$statemate^{30}$" the reduction in goal space from 1581 to 48. More notably, the interpolation based goal exploration can in fact completely explore the goal space for the "statemate" experiment, traversing just 1240 goals.

Finally, we compare with the CEGAR tool BLAST, on a fragment[2], of "statemate." As in BLAST, we combine statements in a block into single transition to facilitate proper comparison of the number of search tree nodes. Our prototype completed the verification in traversing 142 search tree nodes. With default options (breadth-first, no heuristics), BLAST traverses 1410 nodes. The difference is essentially due to spurious paths and constraint slackening.

[2] BLAST ran out of memory when run with the full program.

Acknowledgement

We thank Dirk Beyer for his help on BLAST.

References

1. Marriott, K., Stuckey, P.J.: Programming with Constraints. MIT Press, Cambridge (1998)
2. Chen, W., Warren, D.S.: Tabled evaluation with delaying for general logic programs. J. ACM 43(1), 20–74 (1996)
3. Swift, T.: A new formulation of tabled resolution with delay. In: Barahona, P., Alferes, J.J. (eds.) EPIA 1999. LNCS (LNAI), vol. 1695, pp. 163–177. Springer, Heidelberg (1999)
4. Sagonas, K., Swift, T., Warren, D.S., Freire, J., Rao, P., Cui, B., Johnson, E., de Castro, L., Dawson, S., Kifer, M.: The XSB System Version 2.5 Volume 1: Programmer's Manual (June 2003)
5. Jaffar, J., Santosa, A.E., Voicu, R.: Efficient memoization for dynamic programming with ad-hoc constraints. In: 23rd AAAI, pp. 297–303. AAAI Press, Menlo Park (2008)
6. Frost, D., Dechter, R.: Dead-end driven learning. In: 12th AAAI, pp. 294–300. AAAI Press, Menlo Park (1994)
7. Bayardo Jr., R.J., Schrag, R.: Using csp look-back techniques to solve real-world sat instances. In: 14th AAAI/9th IAAI, pp. 203–208. AAAI Press, Menlo Park (1997)
8. Biere, A., Cimatti, A., Clarke, E.M., Zhu, Y.: Symbolic model checking without BDDs. In: Cleaveland, W.R. (ed.) TACAS 1999. LNCS, vol. 1579, pp. 193–207. Springer, Heidelberg (1999)
9. Moskewicz, M.W., Madigan, C.F., Zhao, Y., Zhang, L., Malik, S.: Chaff: Engineering an efficient SAT solver. In: 38th DAC, pp. 530–535. ACM Press, New York (2001)
10. Silva, J.P.M., Sakallah, K.A.: GRASP—a new search algorithm for satisfiability. In: ICCAD 1996, pp. 220–227. ACM and IEEE Computer Society (1996)
11. McMillan, K.L.: Applying SAT methods in unbounded symbolic model checking. In: Brinksma, E., Larsen, K.G. (eds.) CAV 2002. LNCS, vol. 2404, p. 250. Springer, Heidelberg (2002)
12. Lahiri, S.K., Nieuwenhuis, R., Oliveras, A.: SMT techniques for fast predicate abstraction. In: [36], pp. 424–437
13. Jones, R.B., Dill, D.L., Burch, J.R.: Efficient validity checking for processor verification. In: Rudell, R.L. (ed.) ICCAD 1995, pp. 2–6. IEEE Computer Society Press, Los Alamitos (1995)
14. Barrett, C., Dill, D.L., Levitt, J.R.: Validity checking for combinations of theories with equality. In: Srivas, M., Camilleri, A. (eds.) FMCAD 1996. LNCS, vol. 1166, pp. 187–201. Springer, Heidelberg (1996)
15. Stump, A., Barrett, C., Dill, D.L.: CVC: A cooperating validity checker. In: Brinksma, E., Larsen, K.G. (eds.) CAV 2002. LNCS, vol. 2404, pp. 500–504. Springer, Heidelberg (2002)
16. Barrett, C., Berezin, S.: CVC Lite: A new implementation of the cooperating validity checker. In: Alur, R., Peled, D.A. (eds.) CAV 2004. LNCS, vol. 3114, pp. 515–518. Springer, Heidelberg (2004)
17. McMillan, K.L.: Interpolation and SAT-based model checking. In: Hunt Jr., W.A., Somenzi, F. (eds.) CAV 2003. LNCS, vol. 2725, pp. 1–13. Springer, Heidelberg (2003)
18. Cimatti, A., Griggio, A., Sebastiani, R.: Efficient interpolant generation in satisfiability modulo theories. In: Ramakrishnan, C.R., Rehof, J. (eds.) TACAS 2008. LNCS, vol. 4963, pp. 397–412. Springer, Heidelberg (2008)
19. McMillan, K.L.: An interpolating theorem prover. TCS 345(1), 101–121 (2005)

20. Clarke, E.M., Grumberg, O., Jha, S., Lu, Y., Veith, H.: Counterexample-guided abstraction refinement for symbolic model checking. J. ACM 50(5), 752–794 (2003)
21. Henzinger, T.A., Jhala, R., Majumdar, R., Sutre, G.: Lazy abstraction. In: 29th POPL, pp. 58–70. ACM Press, New York (2002); SIGPLAN Notices 37(1)
22. Henzinger, T.A., Jhala, R., Majumdar, R., McMillan, K.L.: Abstractions from proofs. In: 31st POPL, pp. 232–244. ACM Press, New York (2004)
23. McMillan, K.L.: Lazy abstraction with interpolants. In: [36], pp. 123–136
24. Beyer, D., Henzinger, T., Jhala, R., Majumdar, R.: The software model checker Blast. Int. J. STTT 9, 505–525 (2007)
25. Flanagan, C.: Automatic software model checking using CLP. In: Degano, P. (ed.) ESOP 2003. LNCS, vol. 2618, pp. 189–203. Springer, Heidelberg (2003)
26. Jaffar, J., Santosa, A.E., Voicu, R.: Modeling systems in CLP. In: Gabbrielli, M., Gupta, G. (eds.) ICLP 2005. LNCS, vol. 3668, pp. 412–413. Springer, Heidelberg (2005)
27. Delzanno, G., Podelski, A.: Constraint-based deductive model checking. Int. J. STTT 3(3), 250–270 (2001)
28. Dijkstra, E.W.: A Discipline of Programming. Prentice-Hall Series in Automatic Computation. Prentice-Hall, Englewood Cliffs (1976)
29. Jaffar, J., Maher, M.J.: Constraint logic programming: A survey. J. LP 19/20, 503–581 (1994)
30. Bjørner, N., Browne, A., Manna, Z.: Automatic generation of invariants and intermediate assertions. TCS 173(1), 49–87 (1997)
31. Flanagan, C., Saxe, J.B.: Avoiding exponential explosion: Generating compact verificatio n conditions. In: 28th POPL, pp. 193–205. ACM Press, New York (2001)
32. Jaffar, J., Michaylov, S., Stuckey, P.J., Yap, R.H.C.: The CLP($\mathcal{R}$) language and system. ACM TOPLAS 14(3), 339–395 (1992)
33. Jaffar, J., Maher, M.J., Stuckey, P.J., Yap, R.H.C.: Output in CLP($\mathcal{R}$). In: Proc. Int. Conf. on Fifth Generation Computer Systems, Tokyo, Japan, vol. 2, pp. 987–995 (1992)
34. Toy, B.: Linpack.c (1988), `http://www.netlib.org/benchmark/linpackc`
35. Mälardalen WCET research group benchmarks (2006), `http://www.mrtc.mdh.se/projects/wcet/benchmarks.html`
36. Ball, T., Jones, R.B. (eds.): CAV 2006. LNCS, vol. 4144, pp. 532–546. Springer, Heidelberg (2006)

Same-Relation Constraints

Christopher Jefferson[1], Serdar Kadioglu[2,⋆], Karen E. Petrie[3], Meinolf Sellmann[2,⋆], and Stanislav Živný[3]

[1] University of St Andrews, School of Computer Science, St Andrews, UK
[2] Brown University, Department of Computer Science, Providence, USA
[3] Oxford University, Computing Laboratory, Oxford, UK

Abstract. The ALLDIFFERENT constraint was one of the first global constraints [17] and it enforces the conjunction of one binary constraint, the not-equal constraint, for every pair of variables. By looking at the set of all pairwise not-equal relations at the same time, AllDifferent offers greater filtering power. The natural question arises whether we can generally leverage the knowledge that sets of pairs of variables all share the same relation. This paper studies exactly this question. We study in particular special constraint graphs like cliques, complete bipartite graphs, and directed acyclic graphs, whereby we always assume that the *same* constraint is enforced on all edges in the graph. In particular, we study whether there exists a tractable GAC propagator for these global Same-Relation constraints and show that AllDifferent is a huge exception: most Same-Relation Constraints pose NP-hard filtering problems. We present algorithms, based on AC-4 and AC-6, for one family of Same-Relation Constraints, which do not achieve GAC propagation but outperform propagating each constraint individually in both theory and practice.

1 Motivation

The ALLDIFFERENT constraint was one of the first global constraints [17] and it enforces the conjunction of one binary constraint, the not-equal constraint, for every pair of variables. By looking at the set of all pairwise not-equal relations at the same time, AllDifferent offers greater filtering power while incurring the same worst-case complexity as filtering and propagating the effects of not-equal constraints for each pair of variables individually. The natural question arises whether we can leverage the knowledge that sets of pairs of variables all share the same relation in other cases as well. We investigate in particular binary constraint satisfaction problems (BCSPs) with special associated constraint graphs like cliques (as in AllDifferent), complete bipartite graphs (important when a relation holds between all variables X in a subset of I and Y in J), and directed acyclic graphs (apart from bounded tree width graphs the simplest generalization of trees), whereby we always assume that the *same* constraint is enforced on all edges in the graph. We refer to the conjunction of the same binary relation over any set of pairs in a BCSP as a *Same-Relation Constraint*.

⋆ This work was supported by the National Science Foundation through the Career: Cornflower Project (award number 0644113).

I.P. Gent (Ed.): CP 2009, LNCS 5732, pp. 470–485, 2009.

2 Theory Background

We study the complexity of achieving GAC on binary CSPs with one same-relation constraint. The classes of structures considered are all constraint graphs as defined in [12]. The ultimate goal is to classify, for a given constraint graph, which same-relation constraints admit a polynomial-time GAC, and which do not. It is well known that the CSP is equivalent to the HOMOMORPHISM problem between relation structures [8]. Most theoretical research has been done on the case where either the domain size or the arities of the constraints are bounded.

We deal with the problem where both the domain size and the arities of the constraints are unbounded (so-called *global constraints*). Since we are interested in the complexity of achieving GAC, we allow all unary constraints. In mathematical terms, we study the LIST HOMOMORPHISM problem. For a CSP instance $\mathcal{P}$, achieving GAC on $\mathcal{P}$ is equivalent to checking solvability of $\mathcal{P}$ with additional unary constraints [3]. Note that showing that a problem does not have a polynomial-time decision algorithm implies that this problem also does not have a polynomial-time algorithm for achieving GAC, which is a more general problem.

Generalizing the result of Freuder [10], Dalmau et al. showed that CSPs with bounded tree-width modulo homomorphic equivalence are solvable in polynomial time [5]. Grohe showed[1] that this is the only tractable class of bounded arity, defined by structure [13]. In other words, a class of structures with unbounded tree-width modulo homomorphic equivalence is not solvable in polynomial time.

In the case of binary CSPs, we are interested in classes of constraint graphs with a same-relation constraint. The first observation is that classes of graphs with bounded tree-width are not only tractable, but also permit achieving GAC in polynomial time (even with separate domains, and even with different constraints: different domains are just unary constraints which do not increase the tree-width). The idea is that such graphs have bounded-size separating sets, and the domains on these separating sets can be explicitly recorded in polynomial space (dynamic programming approach) [9,6,13]. Therefore, we are interested in classes of graphs with unbounded tree-width.

3 Clique Same-Relation

First we look at cliques. The famous ALLDIFFERENT constraint is an example of a same-relation constraint which, if put on a clique, is tractable – in case of AllDifferent, because the microstructure is perfect [18] and also has a polynomial-time GAC [17].

Definition 1. *Given a set of values D and a set of variables $\{X_1, \ldots, X_n\}$, each associated with its own domain $D_i \subseteq D$, and a binary relation $R \subseteq D \times D$, an assignment $\sigma : \{X_1, \ldots, X_n\} \to D$ satisfies the* Clique Same-Relation Constraint *CSR on the relation R if and only if for all i and j such that $1 \leq i, j \leq n, i \neq j$, it holds that $(\sigma(X_i), \sigma(X_j)) \in R$.*

[1] Assuming a standard assumption from parameterized complexity theory FPT $\neq$ W[1], see [7] for more details.

3.1 Complexity of Achieving GAC

Despite the tractability of AllDifferent, in general enforcing the same binary relation over all pairs of variables of a CSP yields a hard filtering problem.

Theorem 1. *Deciding whether CSR is satisfiable for an arbitrary binary relation is NP-hard.*

Proof. We reduce from the CLIQUE problem. Assume we are given an undirected graph $G = (V, E)$ and a value $k \in \mathbb{N}$ and need to decide whether G contains a clique of size k. We construct a CSP with just one CSR constraint in the following way. We introduce k variables $X_1, \ldots, X_k$, each associated with domain V. The relation R is defined as $R \leftarrow \{(a, b) \in V^2 \mid a \neq b, \{a, b\} \in E\}$. We claim that CSR on the relation $(X_1, \ldots, X_k, R)$ is satisfiable if and only if G contains a clique of size k.

"⇒" Assume there is an assignment $\sigma : \{X_1, \ldots, X_k\} \rightarrow V$ that satisfies CSR. Then, $C \leftarrow \{\sigma(X_1), \ldots, \sigma(X_k)\} \subseteq V$ is a clique because CSR enforces R for each pair of variables, and thus that there exists an edge between all pairs of nodes in C. Furthermore, $|C| = k$ since R forbids that the same node is assigned to two different variables.

"⇐" Now assume there exists a clique $C = \{v_1, \ldots, v_k\} \subseteq V$ with $|C| = k$. Setting $\sigma(X_i) \leftarrow v_i$ gives a satisfying assignment to CSR because for all $i \neq j$ we have that $(\sigma(X_i), \sigma(X_j)) = (v_i, v_j) \in E$ with $v_i \neq v_j$, and thus $(\sigma(X_i), \sigma(X_j)) \in R$. □

Corollary 1. *Achieving GAC for the CSR is NP-hard.*

In fact, we can show more: Even when we limit ourselves to binary symmetric relations which, for each value, forbid *only one other value*, deciding the satisfiability of the CSR is already intractable. This shows what a great exception AllDifferent really is. Even the slightest generalization already leads to intractable filtering problems.

Theorem 2. *Deciding CSR is NP-hard even for relations where each value appears in at most one forbidden tuple.*

Proof. We reduce from SAT. Given a SAT instance with k clauses over n variables, we consider an instance of CSR with k variables, each corresponding to one clause. Let D be $\{\langle 1, T\rangle, \langle 1, F\rangle, \ldots, \langle n, T\rangle, \langle n, F\rangle\}$. We define $R \subseteq D \times D$ to be the binary symmetric relation which forbids, for every $1 \leq i \leq n$, the set of tuples $\{(\langle i, T\rangle, \langle i, F\rangle), (\langle i, F\rangle, \langle i, T\rangle)\}$. Note that R is independent of the clauses in the SAT instance.

Each clause in the SAT instance is encoded into the domain restriction on the corresponding variable. For instance, the clause $(x_1 \vee \neg x_2 \vee x_3)$ encodes as the domain $\{\langle 1, T\rangle, \langle 2, F\rangle, \langle 3, T\rangle\}$.

Any solution to this CSR instance, which can contain at most one of $\langle i, T\rangle$ and $\langle i, F\rangle$ for any $1 \leq i \leq n$, gives a solution to the SAT instance (as each variable must be assigned a literal in its clause). SAT variables which are not assigned a value can be given any value without compromising satisfiability. Analogously, a feasible assignment to the SAT formula maps back to a satisfying assignment to CSR in the same way: in any clause, take any of the literals in the solution which satisfy that clause and assign the variable that value. □

3.2 Restriction on the Size of Domain

Our proof that CSR is intractable required both an increasing number of variables and increasing domain size. The question arises whether the problem becomes tractable when the domain size is limited. The following shows that the CSR is indeed tractable when the domain size is bounded.

Lemma 1. *For a constraint CSR for a symmetric relation R, it is possible to check if an assignment satisfies the constraint given only:*

- *a promise that any domain value d such that $\langle d, d \rangle \notin R$, is used at most once, and*
- *the set S of domain values assigned.*

By ensuring that there are no two distinct values $s_1, s_2 \in S$ such that $\langle s_1, s_2 \rangle \notin R$.

Proof. If CSR is violated, there must be two variables which do not satisfy R. This could occur either because two variables are assigned the same domain value d such that the assignment $\langle d, d \rangle$ is forbidden by R, or two variables are assigned different values d_1, d_2 such that the tuple $\langle d_1, d_2 \rangle$ which do not satisfy R.

Lemma 1 provides a useful tool for characterizing the satisfying assignments to CSR, which we will use to devise a general filtering algorithm.

Theorem 3. *Achieving GAC for the CSR is tractable for bounded domains.*

Proof. Since the definition of CSR requires that the relation holds in both directions, it is sufficient to consider symmetric relations only. Then, Lemma 1 shows that satisfying assignments can be expressed by the set of allowed values and only using values d such that $\langle d, d \rangle$ is forbidden by R at most once. We shall show how given a CSR constraint, given a set of values S which satisfies Lemma 1 and a list of sub-domains for the variables in the scope of the constraint, we can find if an assignment with values only in S exists in polynomial time.

Given such a set of domain values S, we call values d such that $\langle d, d \rangle \in R$ *sink values*. Note that in any assignment which satisfies the CSR constraint and contains assignments only in S, changing the assignment of any variable to a sink value will produce another satisfying assignment. Therefore, without loss of generality, we can assume every variable which could be assigned any sink value in S is assigned such a value.

This leaves only variables whose domains contain only values which can occur at most once. This is exactly equivalent to an ALLDIFFERENT constraint, and can be solved as such.

Finally, note that for a domain of size d, there are 2^d subsets of the domain, and this places a very weak bound on the subsets of the domain which will satisfy the conditions of Lemma 1. Therefore, for any domain size there is a fixed bound on how many subsets have to be checked.

Theorem 3 shows that achieving GAC for the CSR is tractable for bounded domains, although the algorithm presented here is not practical. There are a number of simple ways its performance could be improved which we will not consider here as we are merely interested in theoretical tractability.

An interesting implication of our result is the following. Consider a symmetric relation with at most k allowed tuples for each domain value; that is, given $R \subseteq D \times D$, we require that for each $d \in D$, $|\{\langle d, .\rangle \in R\}| \leq k$ for some k.

Corollary 2. *Let k be a bound on the number of allowed tuples involving each domain value in R. If k is bounded, then achieving GAC for the CSR is tractable. If k is unbounded, then solving CSR is NP-hard.*

Proof. If k is bounded, then after assigning a value to an arbitrary variable achieving GAC for the CSR reduces to the bounded domain case (see Theorem 3) . The unbounded case follows from Theorem 1. □

4 Bipartite Same-Relation

After studying complete constraint graphs in the previous section, let us now consider the complete bipartite case. This is relevant for CSPs where a set of variables is partitioned into two sets A and B and the same binary constraint is enforced between all pairs of variables possible between A and B.

Definition 2. *Given a set of values D and two sets of variables $A = \{X_1, \ldots, X_n\}$ and $B = \{X_{n+1} \ldots, X_m\}$, each associated with its own domain $D_i \subseteq D$, and a binary relation $R \subseteq D \times D$, an assignment $\sigma : \{X_1, \ldots, X_n\} \rightarrow D$ satisfies the* Bipartite Same-Relation Constraint *BSR on relation R if and only if $\forall X_i \in A, X_i \in B$ it holds that $(\sigma(X_i), \sigma(X_j)) \in R$.*

4.1 Complexity of Achieving GAC

At first, the BSR appears trivially tractable because once an allowed assignment is found between any pair of variables in both parts of the bipartite graph, these values can be assigned to all variables. Indeed, as any bipartite graph (with at least one edge) is homomorphically equivalent to a single edge, such CSPs instance are easy to solve [5,13] (using the fact that CSPs are equivalent to the HOMOMORPHISM problem).

However, we have to take into account that the domains of the variables may be different; in other words, unary constraints are present. This fact causes the problem of achieving GAC for the BSR to become intractable. In mathematical terms, instead of facing a HOMOMORPHISM problem (which is trivial on bipartite graphs), we must deal with the LIST-HOMOMORPHISM problem.

Theorem 4. *Deciding whether BSR is satisfiable is NP-hard.*

Proof. We reduce from the CSR satisfaction problem which we showed previously is NP-hard. Assume we are given the CSR constraint on relation R over variables $\{X_1, \ldots, X_n\}$ with associated domains $D_1, \ldots, D_n$. We introduce variables $Y_1, \ldots, Y_n$, $Z_1, \ldots, Z_n$ and set $A \leftarrow \{Y_1, \ldots, Y_n\}$, $B \leftarrow \{Z_1, \ldots, Z_n\}$. The domain of variables Y_i and Z_i is $\{(i, k) \mid k \in D_i\}$. Finally we define the relation P over the tuples $((i, k), (j, l))$ where $1 \leq i, j \leq n$ and either $i = j \wedge k = l$ or $i \neq j \wedge (k, l) \in R$. We claim that BSR on A, B and P is satisfiable if and only if CSR on R and the X_i is.

"⇒" Let σ denote a solution to BSR on A,B and P. For all i the initial domains and the definition of P imply that $\sigma(Y_i) = \sigma(Z_i) = (i, k_i)$ for some $k_i \in D_i$. Define $\tau : \{X_1, \ldots, X_n\} \to D$ by setting $\tau(X_i) \leftarrow k_i$. Let $1 \leq i, j \leq n$ with $i \neq j$. Then, since $((i, k_i), (j, k_j)) \in P$, $(\tau(X_i), \tau(X_j)) = (k_i, k_j) \in R$. And therefore, τ satisfies CSR for the relation R.

"⇐" Let τ denote a solution to CSR on R and the X_i. Then, σ with $\sigma(Y_i) \leftarrow \sigma(Z_i) \leftarrow (i, \tau(X_i))$ satisfies BSR on A, B and P. □

Corollary 3. *Achieving GAC for the BSR is NP-hard.*

5 DAG Same-Relation

In the previous sections we showed that achieving GAC for cliques and complete bipartite graphs is hard. Now we go on to show that a simple generalization of trees to directed graphs is intractable. When the binary relation that we consider is not symmetric, each edge in the constraint graph is directed. The generalization of trees (which we know are tractable) to the directed case then results in directed acyclic graphs (DAGs).

Definition 3. *Let D be a set of values, X be a set of variables $X = \{X_1, \ldots, X_n\}$ and G be a directed acyclic graph (DAG) $G = \langle X, A \rangle$. Each variable is associated with its own domain $D_i \subseteq D$. Given a binary relation $R \subseteq D \times D$, an assignment $\sigma : \{X_1, \ldots, X_n\} \to D$ satisfies the* DAG Same-Relation Constraint *DSR on relation R if and only if $\forall\, 1 \leq i, j \leq n$ such that $(i, j) \in A$, it holds that $(\sigma(X_i), \sigma(X_j)) \in R$.*

5.1 Complexity of Achieving GAC

Somewhat surprisingly, we find that even the simple graph structure of DAGs yields intractable filtering problems: bipartite graphs, with the orientation of all edges from one partition to the other, form a DAG. Therefore, Theorem 4 proves that solving DSR is NP-hard.

The question arises whether DAGs become tractable when we know that the direction on every arc is truly enforced by the constraints. Let us consider anti-symmetric relations. A relation R is anti-symmetric if for all a and b, $(a, b) \in R$ and $(b, a) \in R$ implies $a = b$. First we show that irreflexive antisymmetric relations can be NP-hard on DAGs.

Lemma 2. *Deciding satisfiability if DSR is NP-hard even for irreflexive antisymmetric relations.*

Proof. We use the equivalence between the CSP and the HOMOMORPHISM problem [8]. Solving an instance of DSR on relation R is equivalent to the question of whether there is a homomorphism[2] between the digraph A and digraph R. This problem is known as ORIENTED GRAPH COLORING [19]. The complexity of this problem

[2] A homomorphism between two directed graphs (digraphs) $G = \langle V(G), A(G) \rangle$ and $H = \langle V(H), A(H) \rangle$ is a mapping $f : V(G) \to V(H)$ which preserves arcs, that is, $(u, v) \in A(G)$ implies $(f(u), f(v)) \in A(H)$.

was studied in [15], and Swart showed that the ORIENTED GRAPH COLORING problem is polynomial-time solvable for R on at most 3 vertices, and NP-complete otherwise, even when restricted to DAGs [20]. Note this proves more that almost all asymmetric relations are NP-hard on DAGs. □

Note that it follows[3] from a recent result of Hell et al. that solving DSR is NP-hard also for reflexive antisymmetric relations [14].

6 Grid Same-Relation

Another interesting class of graphs are *grids*. For $m, n \geq 1$, the $(m \times n)$-grid is the graph with vertex set $\{(i, j) | 1 \leq i \leq m, 1 \leq j \leq n\}$ and an edge between two different vertices (i, j) and (i', j') if $|i - i'| + |j - j'| = 1$.

Definition 4. *Given a set of values D and a set of variables*

$$\{X_{1,1}, \ldots, X_{1,m}, \ldots, X_{m,1}, \ldots, X_{m,n}\},$$

each associated with its own domain $D_i \subseteq D$, and a binary relation $R \subseteq D \times D$, an assignment $\sigma : \{X_{1,1}, \ldots, X_{m,n}\} \rightarrow D$ satisfies the Grid Same-Relation Constraint *GSR on relation R if and only if for all distinct pairs of points (i, j) and (i', j') such that $|i - i'| + |j - j'| = 1$, it holds that $\sigma(X_{i,j}, X_{i',j'}) \in R$.*

Once more, we find:

Lemma 3. *Deciding satisfiability of GSR is NP-hard.*

Proof. We reduce from the CLIQUE problem. Let $\langle G, k\rangle$ be an instance of CLIQUE, where $G = \langle V, E\rangle$ is an undirected graph without loops and $V = \{1, \ldots, n\}$. The goal is to determine whether there is a clique of size k in G. We define an instance of GSR with variables $X_{1,1}, \ldots, X_{k,k}$ and a relation R. For every $1 \leq i \leq k$, the domain of $X_{i,i}$ is $\{\langle i, i, u, u\rangle \mid 1 \leq u \leq n\}$. For every $1 \leq i \neq j \leq k$, the domain of $X_{i,j}$ is $\{\langle i, j, u, v\rangle \mid \{u, v\} \in E\}$. We define the relation R as follows: $\{\langle i, j, u, v\rangle, \langle i', j', u', v'\rangle\}$ belongs to R if and only if the following two conditions are satisfied:

1. $i = i' \Rightarrow [(u = u') \,\&\, (v \neq v')]$
2. $j = j' \Rightarrow [(v = v') \,\&\, (u \neq u')]$

We claim that G contains a clique of size k if and only if GSR on the $X_{i,j}$ and R is satisfiable.

"⇒" Assume there exists a clique $C = \{v_1, \ldots, v_k\} \subseteq V$ in G with $|C| = k$. We claim that setting $\sigma(X_{i,i}) \leftarrow \langle i, i, v_i, v_i\rangle$ for all $1 \leq i \leq k$, and $\sigma(X_{i,j}) \leftarrow \langle i, j, v_i, v_j\rangle$ for all $1 \leq i \neq j \leq k$ gives a satisfying assignment to GSR. Let $X_{i,j}$ and $X_{i',j'}$ be two variables such that $|i - i'| + |j - j'| = 1$. Let $\sigma(X_{i,j}) = \langle i, j, u, v\rangle$ and $\sigma(X_{i',j'}) = \langle i', j', u', v'\rangle$. If $i = i'$, then $u = u'$ and $v \neq v'$ from the definition of σ. If $j = j'$, then $v = v'$ and $u \neq u'$ from the definition of σ. Hence in both cases, $\{\langle i, j, u, v\rangle, \langle i', j', u', v'\rangle\} \in R$.

[3] Private communication with A. Rafiey.

"$\Leftarrow$" Assume there is a solution σ to GSR on relation $(X_{1,1},\ldots,X_{k,k},R)$. From the definition of R, observe that for every fixed i there exists u_i such that $\sigma(X_{i,j}) = \langle i,j,u_i,.\rangle$ for every j. (In other words, the third argument of every row is the same.) Similarly, for every fixed j there exists v_j such that $\sigma(X_{i,j}) = \langle i,j,.,v_j\rangle$ for every i. (In other words, the fourth argument of every column is the same.) By these two simple observations, for every $1 \le i \le k$, there is no v and $j \neq j'$ such that $\sigma(X_{i,j}) = \langle i,j,u_i,v\rangle$ and $\sigma(X_{i,j'}) = \langle i,j',u_i,v\rangle$. (In other words, the fourth arguments of every row are all different.) Assume, for contradiction, that $\sigma(X_{i,j}) = \langle i,j,u_i,v\rangle$ and $\sigma(X_{i,j'}) = \langle i,j',u_i,v\rangle$ for some $v \neq u_i$ and $j' \neq j$; that is, the value v occurs more than once in the i-th row. But then $\sigma(X_{j',j'}) = \langle j',j',v,v\rangle$ as $X_{i,j'}$ and $X_{j',j'}$ are in the same column, and $\sigma(X_{j',j}) = \langle j',j,v,x\rangle$ as $X_{j',j}$ and $X_{j',j'}$ are in the same row.. But as $X_{i,j}$ and $X_{j',j}$ are in the same column, and $\sigma(X_{i,j}) = \langle i,j,u_i,v\rangle$, we get $x = v$. In other words, $\sigma(X_{j',j}) = \langle j',j,v,v\rangle$. But this is a contradiction as σ would not be a solution to GSR. Similarly, we can prove the same results for columns. Moreover, the same argument shows that if $\sigma(X_{i,j}) = \langle i,j,u,v\rangle$, then $\sigma(X_{j,i}) = \langle j,i,v,u\rangle$. Using this kind of reasoning repeatedly shows that there is a set of k different values $C = \{u_1,\ldots,u_k\}$ such that $\delta(X_{i,j}) = \langle i,j,u_i,u_j\rangle$. From the definition of R, C forms a clique in G. □

Corollary 4. *Achieving GAC for the GSR is intractable.*

7 Decomposing Same Relation Constraints

We proved a series of negative results for Same Relation Constraints (SRCs). Even for simple constraint graphs like DAGs and grids, SRCs pose intractable filtering problems. In this section, we investigate whether we can exploit the fact that the same relation is enforced on all edges of a constraint graph to achieve GAC on the corresponding binary CSP, where we consider the *collection* of individual binary constraints. This will achieve the same propagation as propagating each constraint in isolation, unlike for example the AllDifferent constraint [17], which propagates the conjunction constraint. However, by making use of the added structure of SRC, we will show how both theoretical and practical performance gains can be achieved. We begin with the clique same-relation constraint.

7.1 Decomposing CSR

Using AC-4 [16] or any of its successors to achieve GAC, we require time $O(n^2d^2)$ for a network of n^2 binary constraints over n variables with domain size $|D| = d$. By exploiting the fact that the *same* relation holds for all pairs of variables, we can speed up this computation.

AC-4 Approach. We follow in principle the approach from AC-4 to count the number of supports for each value. The core observation is that a value l has the same number of supports $k \in D_i$ *no matter to which* D_j l *belongs*. Therefore, it is sufficient to introduce counters $supCount[i,l]$ in which we store the number of values in D_i which support

1: **Init-CSR** $(X_1, \ldots, X_n, R)$
2: **for all** $l \in D$ **do**
3: $\quad S_l \leftarrow \{k \in D \mid (k, l) \in R\}$
4: **end for**
5: **for all** $l \in D$ **do**
6: $\quad$ **for all** $1 \leq i \leq n$ **do**
7: $\quad\quad supCount[i, l] \leftarrow |D_i \cap S_l|$
8: $\quad$ **end for**
9: **end for**

Algorithm 1. Root-Node Initialization for the CSR Constraint

$l \in D_j$ for any $j \neq i$. In Algorithm 1 we show how these counters are initialized at the root by counting the number of values in the domain of each variable that supports any given value l.

In Algorithm 2, we show how to filter the collection of binary constraints represented by the CSR so that we achieve GAC on the collection. The algorithm proceeds in two phases. In the first phase, lines 2-12, we update the counters based on the values that have been removed from variable domains since the last call to this routine. We assume that this incremental data is given in $\Delta_1, \ldots, \Delta_n$. For each value l that has lost all its support in some domain D_i as indicated by the corresponding counter $supCount[i, l]$ becoming 0, we add the tuple (i, l) to the set Q. The tuple means that l has to be removed from all D_j where $j \neq i$. In the second phase, lines 13-26, we iterate through Q to perform these removals and to update the counters accordingly. If new values must be removed as a consequence, they are added to Q.

Lemma 4. *Algorithm 2 achieves GAC on the collection of binary constraints represented by the CSR constraint in time $O(nd^2)$ and space $O(nd)$ where n is the number of variables and d is the size of the value universe D.*

Proof.

- **GAC:** The method is sound and complete as it initially counts all supports for a value and then and only then removes this value when all support for it is lost.
- **Complexity:** The space needed to store the counters is obviously in $\Theta(nd)$, which is linear in the input when the initial domains for all variables are stored explicitly. Regarding time complexity, the dominating steps are step 6 and step 19. Step 6 is carried out $O(nd^2)$ times. The number of times Step 19 is carried out is $O(nd)$ times the number of times that step 13 is carried out. However, step 13 can be called no more than $2d$ times as the same tuple (i, l) cannot enter Q more than twice: after a value $l \in D$ has appeared in two tuples in Q it is removed from all variable domains. □

AC-6 Approach. As it is based on the same idea as AC-4, the previous algorithm is practically inefficient in that it always computes all supports for all values. We can improve the algorithm by basing it on AC-6 [1,2] rather than AC-4. Algorithms 3-5 realize this idea. In Algorithm 3 we show how to modify the initialization part. First,

```
Filter-CSR (X_1, ..., X_n, R, Δ_1, ..., Δ_n)
Q ← ∅
for all i = 1 ... n do
    for all k ∈ Δ_i do
        for all l ∈ S_k do
            supCount[i, l] ← supCount[i, l] − 1
            if supCount[i, l] == 0 then
                Q ← Q ∪ {(i, l)}
            end if
        end for
    end for
end for
while Q ≠ ∅ do
    (i, l) ∈ Q, Q ← Q \ {(i, l)}
    for all j ≠ i do
        if l ∈ D_j then
            D_j ← D_j \ {l}
            for all k ∈ S_l do
                supCount[j, k] ← supCount[j, k] − 1
                if supCount[j, k] == 0 then
                    Q ← Q ∪ {(j, k)}
                end if
            end for
        end if
    end for
end while
```

Algorithm 2. AC-4-based Filtering Algorithm for the CSR Constraint

the set of potential supports S_l for a value $l \in D$ is now an ordered tuple rather than a set. Second, support counters are replaced with support indices. The new variable $supIndex[i, l]$ tells us the index of the support in S_l which is currently supporting l in D_i. Algorithm 4 shows how to find a new support for a given value l from the domain of the variable i. The algorithm iterates through the ordered support list until it reaches to the end of it, line 6, in which case it returns a failure or until it finds a new support value k. Set-variable $T_{v,k}$ is used to store the set of values that are currently being supported by value k that is in the domain of variable v. In case a new support value l is found, $T_{v,k}$ is extended by the value l, line 10. These three new data structures, S_l, $supIndex[i, l]$ and $T_{v,k}$, allow us to quickly find values for which a new support needs to be computed, which can be done incrementally as a replacement support may only be found after the current support in the chosen ordering. This way, the algorithm never needs to traverse more than all potential supports for all values.

Finally, Algorithm 5 provides a general outline to our AC-6 based propagator which again works in two phases, similar to its AC-4 based counter-part. In the first phase, lines 2-12, we scan the values that have been removed from variable domains since the last call to this routine. We assume that this incremental data is given in $\Delta_1, \ldots, \Delta_n$. In line 6, we look for a new support for each value l that was previously supported by a

```
initSup (X_1, ..., X_n, R)
for all l ∈ D do
    S_l ← (k ∈ D | (k, l) ∈ R)
end for
T ← ∅
for all l ∈ D do
    for all 1 ≤ i ≤ n do
        supIndex[i, l] ← 0
        newSup(X_1, ..., X_n, R, i, l)
    end for
end for
```

Algorithm 3. Support Initialization for the CSR Constraint

value k from the domain of variable i, which is now lost. If the value l is not supported anymore, it is removed from the set $T_{i,k}$ and the tuple (i,l) is added to the queue which means that l has to be removed from all D_j where $j \neq i$. In the second phase, lines 13-26, we iterate through Q to perform these removals and to update the set variables $T_{j,l}$ accordingly. If new values must be removed as a consequence, they are added to Q.

7.2 Decomposing BSR

Analogously to our results on the CSR Constraint, we can exploit again the knowledge that there is the same relation on all edges in the complete bipartite constraint graph. Again, the time that AC-4 or AC-6 would need to achieve GAC on the collection of binary constraints that is represented by the BSR is in $O(n^2d^2)$. Following the same idea as for CSR, we can reduce this time to $O(nd^2)$.

We can devise an AC-6 based propagator for BSR in line with Algorithms 3-5. We can still use the set of potential supports S_l for a value which stores ordered tuples and the support indices $supIndex[i,l]$ which tell us the index of the support in S_l which is currently supporting l in D_i. The only required modification is that the set variable $T_{v,k}$ now has to be replaced with $T^A_{v,k}$ and $T^B_{v,k}$ to distinguish between the partitions of the constraint graph.

```
bool newSup (X_1, ..., X_n, R, i, l)
supIndex[i, l] ← supIndex[i, l] + 1
while supIndex[i, l] ≤ |S_l| and S_l[supIndex[i, l]] ∉ D_i do
    supIndex[i, l] ← supIndex[i, l] + 1
end while
if supIndex[i, l] > |S_l| then
    return false
else
    k ← S_l[supIndex[i, l]]
    T_{v,k} ← T_{v,k} ∪ {l}
    return true
end if
```

Algorithm 4. Support Replacement for the CSR Constraint

```
Filter-CSR (X_1, ..., X_n, R, Δ_1, ..., Δ_n)
Q ← ∅
for all i = 1...n do
    for all k ∈ Δ_i do
        for all l ∈ T_{i,k} do
            if !newSup(X_1, ..., X_n, R, i, l) then
                Q ← Q ∪ {(i, l)}
                T_{i,k} ← T_{i,k} \ {l}
            end if
        end for
    end for
end for
while Q ≠ ∅ do
    (i, l) ∈ Q, Q ← Q \ {(i, l)}
    for all j ≠ i do
        if l ∈ D_j then
            D_j ← D_j \ {l}
            for all k ∈ T_{j,l} do
                if !newSup(X_1, ..., X_n, R, j, k) then
                    Q ← Q ∪ {(j, k)}
                    T_{j,l} ← T_{j,l} \ {k}
                end if
            end for
        end if
    end for
end while
```

Algorithm 5. AC-6-based Filtering Algorithm for the CSR Constraint

Lemma 5. *Achieving GAC on the collection of binary constraints represented by the CSR constraint in time $O(nd^2)$ and space $O(nd)$ where n is the number of variables and d is the size of the value universe D.*

8 Experiments

The purpose of our experiments is to demonstrate the hypothesis that our CSR propagator brings substantial practical benefits, and that therefore this area of research has both theoretical as well as practical merit. To this end, we study two problems that can be modeled by the CSR. We show that filtering can be sped up significantly by exploiting the knowledge that all pairs of variables are constrained in the same way. Note that the traditional AC6 and the improved version for CSR achieve the exact same consistency, thus causing identical search trees to be explored. We therefore compare the time needed per choice point, without comparing how the overall model compares with the state-of-the-art for each problem as this is beyond the scope of this paper.

We performed a number of comparisons between our AC-4 and AC-6 based algorithms, and found that the AC-6 algorithm always outperformed the AC-4 algorithm.

Table 1. Average Speed-up for the Stable Marriage Problem

		Couples								
		10	15	20	25	30	35	40	45	50
Probability	0.6	2.3	4.8	6.3	9.1	11.0	11.4	12.4	12.2	12.3
of attraction	0.9	3.8	4.9	6.5	6.9	8.6	9.2	10.1	11.4	12.9

This is not surprising, based both on our theoretical analysis, and previous work showing AC-6 outperforms AC-4 [1]. Because of this and space limitations we only present experiments using our AC-6 based algorithm.

All experiments are implemented using the Minion constraint solver on a Macbook with 4GB RAM and a 2.4GHz processor. These experiments show that CSR is a very robust and efficient propagator, never resulting in a slow-down in any of our experiments and producing over a hundred times speedup for larger instances.

Stable Marriage. The first problem we consider is the Stable Marriage Problem. It consists in pairing n men and n women into couples, so that no husband and wife in different couples prefer each other to their partner. We refer the reader to [11] for a complete definition and discussion of previous work on this problem. We use the hardest variant of this problem, where the men and women are allowed ties in their preference lists, and to give a list of partners they refuse to be married to under any circumstances. These two extensions make the problem NP-hard.

The standard model used in the CP literature keeps a decision variable for each man and woman where each domain consists of the indices of the corresponding preference list. The model then posts a constraint for each man-woman pair consisting of a set of no good pairs of values. We use the following, alternative model. Our model has one decision variable for each couple, whose domain is all possible pairings. We post between every pair of variables that the couples are 'stable', and also do not include any person more than once.

This model is inferior to using a specialized n-ary constraint for the stable marriage problem [21], the intention is not to provide a new efficient model for the stable marriage problem. The reason we consider this model here is to show the benefits of using a CSR constraint in place of a clique of binary constraints.

The instances we consider are generated at random, with a fixed probability that any person will refuse to be married to another. This allows us to vary the number of tuples in the constraints. As the search space is the same size regardless of if we use our specialized CSR propagator or a clique of constraints (they achieve the same level of consistency after all), we show only the speed-up that our algorithm provides per search node. So, 2.0 means that our algorithm solved the problem twice as fast, or searched twice as many search nodes per second. The CSR propagator was never slower in any experiment we ran. For each instance we generated and solved 20 problems and took the average speed-up per choice point.

Table 1 shows the results. We note that CSR always provides a sizable improvement in performance, that only increases as problem instances get larger and harder, increasing up to over 10 times faster for larger instances.

Table 2. Average Speed-up for the Table Planning Problem

		People Per Table							
		30	40	50	60	70	80	90	100
Probability of edge	0.4	15	37	58	76	90	105	117	136
	0.5	11	33	51	66	80	81	83	91
	0.6	13	31	49	63	77	76	77	78
	0.8	7	18	21	27	33	34	36	38
	0.9	5	6	8	14	15	19	20	22

The reason that the gain begins to reach a limit is that the size of the domains of the variable increases as the square of the number of people, meaning the cliques of size 50 have variables of domain 2500. Book-keeping work in the solver and algorithm for such large domains begins to dominate. Nevertheless, our algorithm is still over 10 times faster for these large problems.

Table Planning. The second problem we consider is the *Table Planning Problem*. The Table Planning Problem is the problem of sitting a group of people at tables, so that constraints about who will sit with each other are satisfied. Problems like this one often occur in the planning of events.

In this paper, an instance of the Table Planning Problem (TPP) is a triple $\langle T, S, R \rangle$ where T is the number of tables and S is the size of each table. This implies there are $S \times T$ people to sit. R is a symmetric relation on the set $\{1, \ldots, S \times T\}$, which i is related to j if people i and j are willing to sit on the same table. A solution to the TPP therefore is a partition of people, denoted by the set $\{1, \ldots, S \times T\}$, where each part of the partition represents a table. Therefore in any solution each member of this partition must be of size S and all pairs of numbers within it must satisfy R.

We consider instances of TPP with three tables and where R is generated randomly, with some fixed probability of each edge, and its symmetric image, being added. The model we use is an $S \times T$ matrix, with each variable having domain $\{1, \ldots, S \times T\}$. The constraints are that each row (representing a table) has a clique of the constraint R and a single AllDifferent constraint on all the variables. We consider representing the cliques of the constraint R either as separate constraints, or using our propagator.

As we know that the size of search will be identical, regardless of how the cliques of R are implemented, we show only the speed-up achieved by our improved propagator. We run each of the following experiments for ten different randomly generated relations, and take an average of the speed-ups achieved. We measure speed-up based on the number of nodes per second searched in ten minutes, or how long it takes to find the first solution or prove no solution exists, whichever is fastest.

Our results are presented in Table 2. We observe large, scalable gains for larger problems, with over 20 times speed-up for the densest problems. For sparser constraints, we even achieve over 100 times speed-up using our CSR propagator instead of a clique of constraints. This shows again how well the CSR propagator scales for larger problems, achieving immense practical improvements.

9 Conclusions and Future Work

We have looked at generalizing the famous AllDifferent to cliques of other constraints, and also other standard patterns such as bipartite, directed acyclic, and grid graphs. Unlike with the AllDifferent case, these constraints pose intractable filtering problems. By making use of the structure however, we can still provide substantial improvements in both theoretical and practical performance using new, generic algorithms. We have performed benchmarking across two problems using an AC-6 based decomposition algorithm on the CSR constraint. The experimental results show substantial gains in performance, proving that is worth exploiting the structure of same-relation constraints.

In the future, now we have laid down a theoretical framework, we will consider further benchmarks. In particular, we are interested to study how same-relation constraints interact with other global constraints.

References

1. Bessière, C., Cordier, M.O.: Arc-consistency and arc-consistency again. In: AAAI, pp. 108–113 (1993)
2. Bessière, C., Freuder, E., Régin, J.C.: Using Constraint Metaknowledge to Reduce Arc Consistency Computation. Artificial Intelligence 107(1), 125–148 (1999)
3. Bessière, C., Hebrard, E., Hnich, B., Walsh, T.: The Complexity of Reasoning with Global Constraints. Constraints 12(2), 239–259 (2007)
4. Bulatov, A.: Tractable Conservative Constraint Satisfaction Problems. LICS, 321–330 (2003)
5. Dalmau, V., Kolaitis, P.G., Vardi, M.Y.: Constraint Satisfaction, Bounded Treewidth, and Finite-Variable Logics. In: Van Hentenryck, P. (ed.) CP 2002. LNCS, vol. 2470, pp. 310–326. Springer, Heidelberg (2002)
6. Dechter, R., Pearl, J.: Tree Clustering for Constraint Networks. Artificial Intelligence 38(3), 353–366 (1989)
7. Downey, R., Fellows, M.: Parametrized Complexity. Springer, Heidelberg (1999)
8. Feder, T., Vardi, M.: The Computational Structure of Monotone Monadic SNP and Constraint Satisfaction. SIAM Journal on Computing 28(1), 57–104 (1998)
9. Freuder, E.: A Sufficient Condition for Backtrack-bounded Search. Journal of the ACM 32(4), 755–761 (1985)
10. Freuder, E.: Complexity of K-Tree Structured Constraint Satisfaction Problems. In: AAAI, pp. 4–9 (1990)
11. Gent, I.P., Irving, R.W., Manlove, D.F., Prosser, P., Smith, B.M.: A Constraint Programming Approach to the Stable Marriage Problem. In: Walsh, T. (ed.) CP 2001. LNCS, vol. 2239, pp. 225–239. Springer, Heidelberg (2001)
12. Gottlob, G., Scarcello, F.: A comparison of structural CSP decomposition methods. Artificial Intelligence 124(2), 243–282 (1999)
13. Grohe, M.: The complexity of homomorphism and constraint satisfaction problems seen from the other side. Journal of the ACM 54(1) (2007)
14. Hell, P., Huang, J., Rafiey, A.: List Homomorphism to Reflexive Digraphs: Dichotomy Classification (submitted, 2009)
15. Klostermeyer, W., MacGillivray, G.: Homomorphisms and oriented colorings of equivalence classes of oriented graphs. Discrete Mathematics 274(1-3), 161–172 (2004)
16. Mohr, R., Henderson, T.: Arc and path consistency revisited. Artificial Intelligence 28(2), 225–233 (1986)

17. Régin, J.C.: A filtering algorithm for constraints of difference in CSPs. AAAI 1, 362–367 (1994)
18. Salamon, A.Z., Jeavons, P.G.: Perfect Constraints Are Tractable. In: Stuckey, P.J. (ed.) CP 2008. LNCS, vol. 5202, pp. 524–528. Springer, Heidelberg (2008)
19. Sopena, E.: Oriented graph coloring. Discrete Mathematics 229(1-3), 359–369 (2001)
20. Swarts, J.S.: The Complexity of Digraph Homomorphisms: Local Tournaments, Injective Homomorphisms and Polymorphisms. PhD thesis, University of Victoria (2008)
21. Unsworth, C., Prosser, P.: An n-ary constraint for the stable marriage problem. In: IJCAI, pp. 32–38 (2005)

Dialectic Search*

Serdar Kadioglu and Meinolf Sellmann

Brown University, Department of Computer Science,
P.O. Box 1910, Providence, RI 02912, U.S.A.
{serdark,sello}@cs.brown.edu

Abstract. We introduce Hegel and Fichte's dialectic as a search meta-heuristic for constraint satisfaction and optimization. Dialectic is an appealing mental concept for local search as it tightly integrates and yet clearly marks off of one another the two most important aspects of local search algorithms, search space exploration and exploitation. We believe that this makes dialectic search easy to use for general computer scientists and non-experts in optimization. We illustrate dialectic search, its simplicity and great efficiency on four problems from three different problem domains: constraint satisfaction, continuous optimization, and combinatorial optimization.

1 Introduction

Local search (LS) is a powerful algorithmic concept which is frequently used to tackle combinatorial problems. While originally developed for constrained optimization, beginning with the seminal work of Selman et al. [23] in the early 90ies local search algorithms have become extremely popular to solve also constraint satisfaction problems. Today, many highly efficient SAT solvers are based on local search. Recently there have also been developed general purpose constraint solvers that are based on local search [33].

The general idea of local search is easy to understand and often used by non-experts in optimization to tackle their combinatorial problems. There exists a wealth of modern hybrid LS paradigms like iterated local search (ILS) [29], very large scale neighborhood search [1,31], or variable neighborhood search [17]. By far the most prevalent LS methods used by non-experts are simulated annealing [22,6,26] and tabu search [11,12].

Simulated annealing (SA) is inspired by the physical annealing process in metallurgy. The method starts out by performing a random walk as almost all randomly generated neighbors are accepted in the beginning. It then smoothly transitions more and more into a hill-climbing heuristic when neighbors are more and more unlikely to be accepted the more they degrade the solution quality. In tabu search (TS) we move to the best solution in the neighborhood of the current solution, no matter whether that neighbor improves the current solution or not. To avoid cycling, a tabu list is maintained that dynamically excludes neighbors which we may have visited already in the near past. Typically, the latter is achieved by excluding neighbors that have certain problem-specific properties which were observed recently in the search.

* This work was supported by the National Science Foundation through the Career: Cornflower Project (award number 0644113).

I.P. Gent (Ed.): CP 2009, LNCS 5732, pp. 486–500, 2009.

Both concepts are very popular with non-experts because they are easy to understand and to implement. However, to achieve a good heuristic performance for a given problem, the vanilla methods rarely work well without significant tuning and experimentation. In particular, it has often been observed that SA is able to find high-quality solutions only when the temperature is lowered very slowly or more sophisticated neighborhoods and techniques like reheats are used. TS, on the other hand, often finds good solutions much earlier in the search than SA. However, the vague definition of the tabu-concept is difficult to handle for non-experts. If the criteria that define which neighbors are currently tabu are too broad, then many neighbors which have actually not been visited earlier are tabu. Then, so-called aspiration criteria need to be introduced to override the tabu list. Moreover, the tabu tenure is of great practical importance and difficult to tune. There exist sophisticated methods to handle this problem like reactive TS [3] which dynamically adapts the length of the tabu list and other techniques such as strategic oscillation or ejection chaining.

We argue that these techniques outside the core methods are too involved for non-experts and that there is a need for a simple method that is easy to handle for anyone with a general background in constraints. The objective of this work is to provide such a meta-heuristic which, by design, draws the user's attention to the most important aspects of any efficient local search procedure. To this end, in the next section we introduce *dialectic search*. In the sections thereafter, we provide empirical evidence that demonstrates the effectiveness of the general approach on four different problems from highly different problem domains: constraint satisfaction, continuous optimization, and discrete optimization.

2 Dialectic Search

Without being able to make any assumptions about the search landscape, there is no way to extrapolate search experience and any unexplored search point is as good as any other. Only when we observe statistical features of the landscape which are common to many problem instances we may be able to use our search experience as leverage to predict where we may find improving solutions. The most commonly observed and exploited statistical feature is the correlation of fitness and distance [21]. It gives us a justification for intensifying the search around previously observed high quality solutions.

While the introduction of a search bias based on predictions where improving solutions may be found is the basis of any improvement over random search, it raises the problem that we need to introduce a second force which prevents us from investigating only a very small portion of the search space. This is an inherent problem of local search as the method does not allow us to memorize, in a compact way, all previously visited parts of the search space. In artificial intelligence, the dilemma of having to balance the wish for improving solutions with the need to diversify the search is known as the exploitation-exploration trade-off (EET). It has been the subject of many practical experiments as well as theoretical studies, for example on bandit problems [24].

SA and TS address the EET in very different ways. SA explores a lot in the beginning and then shifts more and more towards exploitation by lowering the acceptance rate of worsening neighbors. TS, on the other hand, mixes exploitation and exploration in every step by moving to the best neighbor which is not tabu. The often extremely good performance of TS indicates that binding exploration and exploitation steps more tightly together is beneficial. However, the idea to mix exploration and exploitation in the same local search step is arguably what makes TS so opaque to the non-expert and what causes the practical problems with defining the tabu criteria, tabu tenure, aspiration criteria, etc.

2.1 A Meta-heuristic Inspired by Philosophy

We find an LS paradigm where exploration and exploitation are tightly connected yet clearly separated from each other in philosophy: Hegel and Fichte's Dialectic [18,9]. Their concept of intellectual discovery works as follows: The current model is called the *thesis*. Based on it, we formulate an *antithesis* which negates (parts of) the thesis. Finally, we merge thesis and antithesis to achieve the *synthesis*. The merge is guided by the principle of *Aufhebung*. The latter is German and has a threefold meaning: First, that parts of the thesis and the antithesis are preserved ("aufheben" in the sense of "bewahren"). Second, that certain parts of thesis and antithesis are annihilated ("aufheben" in the sense of "ausloeschen"). And third, that the synthesis is better than thesis and antithesis ("aufheben" in the sense of "aufwerten"). The synthesis then becomes the new thesis and the process is iterated.

Analyzing Hegel and Fichte's dialectic, we find that it strikes an appealing balance between exploration and exploitation. In essence, the formulation of an antithesis enforces search space exploration, while the optimization of thesis and antithesis allows us to exploit and improve. Furthermore, while in each step both exploration and exploitation play their part, they are clearly marked off of one another and can be addressed separately. We argue that this last aspect is what makes dialectic search very easy to handle.

2.2 Dialectic Search

We outline the dialectic search meta-heuristic in Algorithm 1. After initializing the search with a first solution, we first improve it by running a greedy improvement heuristic. We initialize a global counter which we use to terminate the search after a fixed number (GLOBALLIMIT) of global iterations.

In each global iteration, we perform local dialectic steps, whereby the quality of the resulting solution of each such local step is guaranteed not to degrade. Again, a counter (local) is initialized which counts the number of steps in which we did not improve the objective.

In each dialectic step, we first derive an antithesis from the thesis, which is immediately improved greedily. We assume that the way how the antithesis is generated is randomized. The synthesis is then generated by merging thesis and antithesis in a profitable way, very much like a cross-over operator in genetic algorithms. Here we assume that 'Merge' returns a solution which is different from the thesis, but may coincide with the antithesis. In case that the greedily improved synthesis is actually worse than the thesis, we return to the beginning of the loop and try improving the (old) thesis again by trying a new antithesis. In case that the synthesis improves the best solution (bestSolution) ever seen, we update the latter. If the synthesis at least improves the thesis, the no-improvement counter 'local' is reset to zero. Then, the synthesis becomes the new thesis.

```
1: Dialectic Search
2: thesis ← InitSolution()
3: thesis ← GreedyImprovement(thesis)
4: global ← 0
5: bestSolution ← thesis
6: bestValue ← Objective(bestSolution)
7: while global++<GLOBALLIMIT do
8:    local ← 0
9:    while local++<LOCALLIMIT do
10:       antithesis ← GreedyImprovement(Modify(thesis))
11:       synthesis ← Merge(thesis,antithesis)
12:       synthesis ← GreedyImprovement(synthesis)
13:       thesisValue ← Objective(thesis)
14:       synthesisValue ← Objective(synthesis)
15:       if thesisValue<synthesisValue then
16:          goto Line 9
17:       end if
18:       if synthesisValue<bestValue then
19:          bestSolution ← synthesis
20:          bestValue ← synthesisValue
21:       end if
22:       if synthesisValue<thesisValue then
23:          local ← 0
24:       end if
25:       thesis ← synthesis
26:    end while
27:    thesis ← antithesis
28: end while
29: return bestSolution
```

Algorithm 1. Dialectic Search

Finally, when the number of non-improving local improvement steps is exceeded, we make the last antithesis the new thesis and start over with the next global step.

So what dialectic search does is this: For a given assignment (the thesis), it greedily improves it. Then it tries to improve the solution further by generating randomized modifications (an antithesis) of the current assignment, greedily improving it, and then combining the two assignments to form a new assignment, which is also greedily improved (the synthesis). If this new assignment is at least as good, it is considered the new current assignment. If this process does not result in improvements for a while, then the search moves to the modified assignment and continues searching from there.

As any meta-heuristic, the general outline of dialectic search that we gave above leaves certain steps open. In genetic algorithms, for example, we need to define mutation and cross-over operators. In dialectic search, we need to specify how the thesis is transformed into an antithesis, how an assignment is greedily improved, and how thesis and antithesis are combined to form the synthesis. These functions must be defined for each problem individually. The contribution of dialectic search is that it manages

```
Merge (thesis, antithesis)
bestValue ← INFINITY
S ← {i | thesis[i] ≠ antithesis[i]}
while S ≠ ∅ do
   bestMoveValue ← INFINITY
   for all i ∈ S do
      margin ← SwitchMargin(thesis,antithesis,i)
      if margin < bestMoveValue then
         bestMoveValue ← margin, bestMove ← i
      end if
   end for
   S ← S \ {bestMove}
   thesis[bestMove] ← antithesis[bestMove]
   thesisValue ← thesisValue−bestMoveValue
   if thesisValue ≤ bestValue then
      synthesis ← thesis
      bestValue ← thesisValue
   end if
end while
return synthesis
```

Algorithm 2. A Procedure to Compute the Synthesis

the balance between exploitation and exploration, which is arguably the hardest part when devising a new local search procedure. With dialectic search, the user can focus on both tasks separately. When defining how antitheses are formed (function 'Modify'), the task is pure search space exploration. When improving a solution greedily (function 'GreedyImprovement'), the task is pure exploitation. The rule of thumb is that the antithesis is a randomized perturbation of parts of the thesis and the greedy improvement consists in moving to the best neighbor until a local minimum is reached.

Only when the synthesis is computed ('Merge'), both exploration and exploitation play a role as we would obviously like to find a very good combination of thesis and antithesis. In Algorithm 2 we give a function for computing the synthesis from two given assignments, the thesis and the antithesis. The procedure works iteratively. In each step we consider the variables on which thesis and antithesis differ and by what margin the objective changes when a variable in the thesis is re-assigned to the corresponding value in the antithesis. We perform the best change and iterate until we reach the antithesis. Like this, we generate a path from thesis to antithesis, and we return as synthesis the best solution on the path.

The idea to merge thesis and antithesis is well-founded by the empirical finding that optimization problems often exhibit a correlation between the fitness of local optima and their average distance to each other, i.e., a "big valley" structure [2]. The particular Algorithm 2 is inspired by the path relinking technique [13] and represents of course only one possible way of merging thesis and antithesis. Depending on the background of the reader, the function presented may also be viewed as a kind of tabu search as variables which have already been assigned their target value are no longer allowed to change their value. Another way to look at the problem of generating the synthesis is to view it as an optimization problem itself, where the task is to find the best combination of thesis and antithesis. Thus, dialectic search is implicitly related to iterated local search [29], variable neighborhood search [17], and very large scale neighborhood search [1,31].

3 Constraint Satisfaction

We first test dialectic search on problems from the constraint satisfaction domain, the costas arrays problem (CAP) and the magic squares problem (MSP).

3.1 Costas Arrays

A costas array [16] is a pattern of n marks on an $n \times n$ grid, one mark per row and one per column, in which the $n(n-1)/2$ vectors between the marks are all different. Such patterns are important as they provide a template for generating radar and sonar signals with ideal ambiguity functions [8,10]. A model for CAP is to define an array of variables $X_1, \ldots, X_n$ which form a permutation. For each length $l \in \{1, \ldots, n-1\}$, we add $n-l$ more variables $X_1^l, \ldots, X_{n-l}^l$, whereby each of these variables is assigned the difference of $X_i - X_{i+l}$ for $i \in \{1, \ldots, n-l\}$. These additional variables form a difference triangle as shown in Figure 1. Each line of this difference triangle must not contain any value twice. That is, the CAP is simply a collection of AllDifferent constraints on $X_1, \ldots, X_n$ and $X_1^l \ldots, X_{n-l}^l$ for all $l \in \{1, \ldots, n-1\}$.

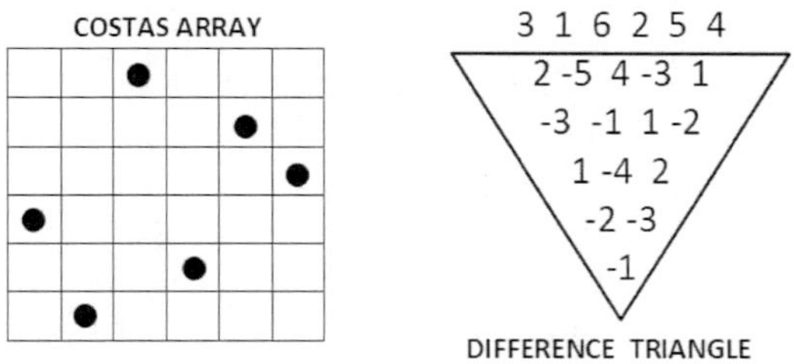

Fig. 1. 6x6 Costas Array 316254

Costas arrays can be constructed using the generation methods based on the theory of finite fields for infinitely many n. However, there is no construction method for all n and it is, e.g., unknown whether there exists a costas array of order 32. We devise a simple dialectic search for the problem and compare with tabu search.

Objective, Initialization and Greedy Improvement. As objective, we use the sum of the square of the violations of all AllDifferent constraints in the difference triangle. Our initial costas array is a random permutation of the numbers from 1 to n. As greedy improvement heuristic, we consider pairs of variables X_i and X_j and compute the cost-delta that would result from flipping the values of the two cells. We commit the pair that would decrease the violations the most and iterate until no possible flip results in a cost improvement anymore, i.e, when we are stuck in a local minimum.

Table 1. Numerical Results for the Costas Array Problem. We compare tabu search and dialectic search in terms of minimum, maximum, standard deviation and average solution time in seconds over 100 runs.

	Minimum		Maximum		Std. Deviation		Average	
Order	TS	Dialectic	TS	Dialectic	TS	Dialectic	TS	Dialectic
13	0.03	**0.00**	1.24	**0.27**	0.24	**0.05**	0.25	**0.05**
14	0.03	**0.00**	4.9	**2.07**	0.82	**0.31**	0.96	**0.26**
15	0.04	0.04	22.9	**6.84**	3.45	**1.33**	3.59	**1.31**
16	0.13	**0.1**	95.8	**32.6**	19.5	**7.11**	21.83	**7.74**
17	1.03	**0.65**	741	**250**	126	**49.4**	114	**53.4**
18	5.49	**4.43**	2568	**1936**	613	**370**	696	**370**

Antithesis and Synthesis. Hegel defined the antithesis as the negation of the thesis. For non-binary variables it is not uniquely defined what the negation of a variable assignment is. We interpret the negation of an assignment to mean that the variable is assigned a different value. For the CAP, we define an antithesis as follows. First, we determine randomly the fraction of variables that must change their value. Then, we compute an antithesis by iteratively switching the values of two cells, whereby in each step we choose the pair of cells which yields the best solution. Note that this procedure is closely related to the greedy improvement heuristic. The difference is that, in the antithesis computation, cells which have already switched values are not allowed to change their values anymore. As synthesis, we return the best solution found while moving from thesis to antithesis in this iterative way.

Numerical Results. In Table 1 we compare this simple approach with the tabu search algorithm using the quadratic neighborhood which is implemented in COMET. This algorithm was shown to be highly competitive compared to specialized procedures for constraint satisfaction in [34]. Unless otherwise stated, all tests in this paper were run on a Pentium III 733MHz machine with 512Mb RAM. Our algorithms are implemented in C++ and compiled using GCC 4.3, with the -O3 flag. COMET models are run using the just-in-time (-j2) compiler flag.

Even though the tabu search approach incorporates sophisticated techniques like an adaptive tabu tenure procedure, we see that the simple dialectic search algorithm is superior and outperforms TS in terms of average solution time and the minimal and maximal time needed in 100 trials. Moreover, the standard deviation shows that dialectic search performs far more robustly and predictably than TS.

3.2 Magic Squares

Our next problem from constraint satisfaction domain is the magic squares problem [27]. A magic square of order n is an $n \times n$ square that contains all numbers from 1 to n^2 such that the sum of each row, each column, and both main diagonals equals the "magic sum" $n(n^2-1)/2$. Although the problem of constructing a magic square is easy (there exist polynomial-time construction methods), magic squares are notoriously hard for systematic constraint programming approaches. The best systematic approach was

Table 2. Numerical Results for the Magic Square Problem. We present minimum, maximum and average solution time in seconds for tabu, adaptive search, and dialectic search. Tabu search and dialectic search results are averaged over 100 runs. The adaptive search results are taken from [7] who ran their algorithms 10 times on each problem.

		Minimum			Maximum			Average		
Order	Vars	TS	Adaptive	Dialectic	TS	Adaptive	Dialectic	TS	Adaptive	Dialectic
20	400	13	0.1	0.12	56.8	7.35	14.6	18	3.41	**2.95**
30	900	98.8	0.67	0.38	800	52.5	54.2	135	18.1	**15.2**
40	1600	458	10.1	1.1	1.31K	166	360	541	58.1	**53.3**
50	2500	1.45K	44.5	3.23	36K	648	584	2.51K	203	**150**
60	3600	4.04K	-	5.94	5.23K	-	1.45K	4.48K	-	**361**
70	4900	9.56K	-	11.8	12.5K	-	3.01K	10.5K	-	**711**
80	6400	18.6K	-	17.8	23.8K	-	6.28K	20.3K	-	**1.80K**
90	8100	33.7K	-	30.4	45.7K	-	13.2K	37.3K	-	**3.46K**
100	10000	60.4K	-	46.3	67.1K	-	22.7K	64K	-	**5.02K**

presented in [15] and can only construct magic squares of orders up to 18 efficiently. [7] have proposed an adaptive local search for the problem which is able to construct magic squares of order 50 within 200 seconds on a Pentium III 733MHz machine.

Objective, Initialization and Greedy Improvement. We propose a simple dialectic search algorithm for the problem. As objective, we sum the squares of the deviations from the magic sum for all rows, columns, and the two main diagonals. We start with a random permutation of the numbers from 1 to n^2 and assign them to the cells in the square row by row. As greedy improvement heuristic, we consider pairs of variables and compute the cost-delta that would result from flipping the values of the two cells. We commit the best such flip and iterate until no more cost improvement is possible anymore.

Antithesis and Synthesis. For the magic squares problem we interpret antithesis as permuting the values of a fraction of the variables. The fraction of variables that must change their value is determined randomly. We compute an antithesis by iteratively switching the values of two cells, whereby in each step we choose the pair of cells that would yield the most improvement in the objective value. Again, cells that have already been altered are not allowed to change anymore. As synthesis we return the best solution observed while moving from thesis to antithesis.

Numerical Results. In Table 2 we compare this simple approach with tabu search algorithm using the quadratic neighborhood which is implemented in COMET and the adaptive local search algorithm from [7]. The adaptive search results presented in [7] were carried out on a Pentium III 733MHz, the same specs as our own machine. We see that the dialectic search algorithm, despite its great simplicity, clearly outperforms the existing approaches. Adaptive search appears more robust than dialectic search in the comparison. Note, however, that dialectic search was run 100 times while the results reported in [7] are based on a very small set of 10 runs only.

We attribute the superior performance of dialectic search on this problem to its ability to be greedy without running into the problem of being stuck in local optima. In fact, we found that magic squares is actually a very easy problem and reacts very well to aggressive exploitation strategies, allowing us to efficiently search very large spaces of $(n^2)!$ potential solutions (for magic squares of order 100, that's $10,000! > 10^{35,000}$). Using greedy improvements and a couple of synthesis moves alone, it is extremely easy to find solutions with objectives 1 or 2, meaning that only one or two rows, columns, or main diagonals have a deviation from the magic sum of 1. To get from here to a real magic square is then result of a search where we consider a sequence of solutions of the same quality (note the $<$ instead of $\leq$ in line 15 of Algorithm 1 which ensures that we move to the synthesis even if it has the same cost as the thesis) and usually very few (mostly zero and occasionally one or two) global steps where the search actually moves to the antithesis instead of the synthesis and continues from there.

4 Continuous Optimization

We next apply our dialectic search algorithm to continuous optimization, the problem of finding the minimum of an n-dimensional, real-valued function over a box polytope (i.e., the only constraints are lower and upper bounds on the continuous variables). Continuous optimization problems arise in many practical application areas, like VLSI design, chemical engineering, and trajectory planning. The problem is relatively simple for functions that are differentiable and for which zero points of the derivatives can be computed. However, for higher-dimensional functions with many local minima, continuous optimization can become a challenging task. We present a simple dialectic search algorithm for the problem and compare it with simulated annealing.

Initial Solution. An initial solution is obtained by assigning to each variable a value chosen uniformly at random from the variable's domain interval.

Antithesis and Synthesis. The antithesis is determined by selecting a random variable with value x^0 from the current solution and changing it to a new random value x^1. To compute the synthesis, we conduct an equi-distant walk from thesis to antithesis. At each step of the walk, we move $|x^0 - x^1|/K$ towards the antithesis. The best solution encountered during this walk is returned as the synthesis.

Numerical Results. The performance of dialectic search is examined on three well-known functions; Rastrigin, De Jong's noiseless function #4, and Alpine, with dimensions 20 and 50. DeJong's function is convex and unimodal whereas Rastrigin and Alpine functions are highly multimodal and exhibit many local minima. In Figure 2 we give the definition, boundary values and visualization of each function. The minimum objective value in all cases is zero. We compare our results with the SA implementation from [30] which is known to be robust, easy to use and applicable to complex continuous problems. Table 3 shows that dialectic search robustly provides very good solutions at little cost also on this problem domain.

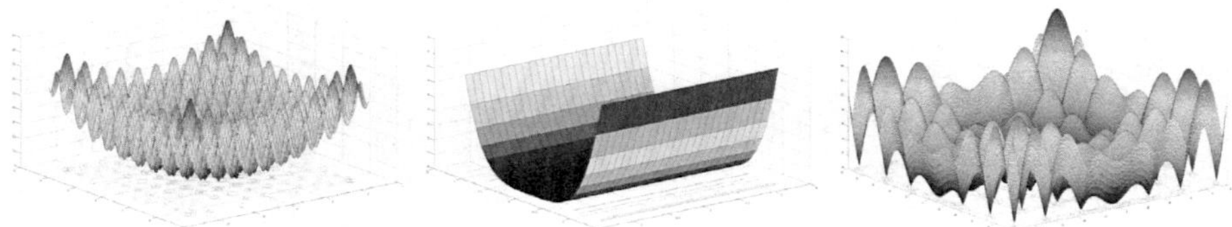

(a) Rastrigin (b) DeJong (c) Alpine

Fig. 2. $Rastrigin(x) = 10n + \sum_{i=1}^{n}(x_i^2 - 10cos(2\pi x_i))$ where $-5.12 \leq x_i \leq 5.12$, $DeJong(x) = \sum_{i=1}^{n} ix_i^4$ where $-1.28 \leq x_i \leq 1.28$ and $Alpine(x) = \sum_{i=1}^{n} |x_i sin(x_i) + 0.1x_i|$ where $-10 \leq x_i \leq 10$

Table 3. Continuous Optimization. We give average minimum value and average number of function evaluations over 250 runs for continuous function minimization with dimensions 20 and 50. SA cooling factors are set to 0.98 and 0.99.

	Dialectic		SA-0.98		SA-0.99	
Function	Value	Eval.	Value	Eval.	Value	Eval.
Rastr.-20	**$< 10^{-3}$**	**208K**	24.4	3.4M	22.4	6.8M
Rastr.-50	**$< 10^{-3}$**	**818K**	87.3	8.3M	86.8	9.9M
DeJong-20	$< 10^{-3}$	**848**	$< 10^{-3}$	946	$< 10^{-3}$	946
DeJong-50	$< 10^{-3}$	3.7K	$< 10^{-3}$	**2.5K**	$< 10^{-3}$	**2.5K**
Alpine-20	$< 10^{-3}$	**86K**	$< 10^{-3}$	1M	$< 10^{-3}$	2M
Alpine-50	$< 10^{-3}$	**458K**	$< 10^{-3}$	2.9M	$< 10^{-3}$	5.8M

5 Constrained Optimization – Set Covering

Our final evaluation of the dialectic search paradigm is on one of the most studied NP-hard combinatorial optimization problems, the set cover problem (SCP): Given a finite set $S := \{1, \ldots, m\}$ of items, and a family $F := \{S_1, \ldots, S_n \subseteq S\}$ of subsets of S, and a cost function $c : F \rightarrow \mathbb{R}^+$, the objective is to find a subset $C \subseteq F$ such that $S \subseteq \bigcup_{S_i \in C} S_i$ and $\sum_{S_i \in C} c(S_i)$ is minimized. The SCP has numerous practical applications such as crew scheduling for airlines or railway companies [19,20,5], location of emergency facilities [32], and production planning in various industries [36].

Initial Solution and Greedy Improvement. A simple greedy construction for SCP is to pick sets one by one until a cover is found. [35] compare 7 different criteria how the next set is chosen (like the set which covers the most uncovered items, the set with least costs, the set with best cost over newly covered items ratio, and several variations of the latter). It was suggested to choose one of the criteria at random in each step of the greedy construction. Run around 30 times, this randomized approach was reported to yield good solutions, and we use this method to initialize our search. As greedy improvement heuristic, we simply remove redundant sets, if any. If there are several, we first pick a set which leaves the fewest items uncovered.

Antithesis and Synthesis. As antithesis, we pick a randomized subset T of the current selection C, whereby we choose the size of this subset randomly between one half,

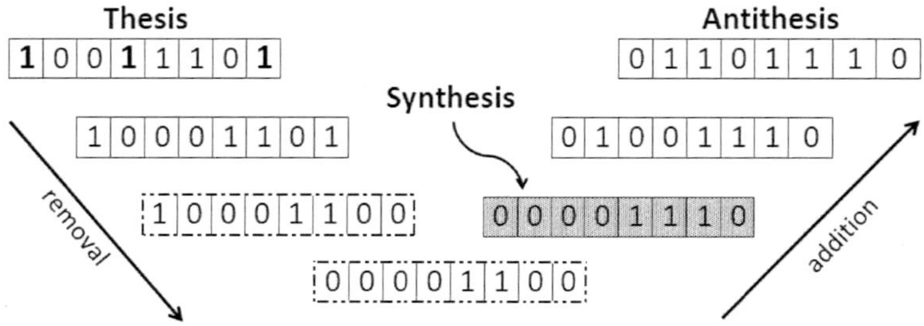

Fig. 3. Function 'Merge' for the SCP. The decision to select a bag or not is represented as a binary variable. The bold variables in the thesis correspond to a randomized subset T of the current selection C. Dashed boxes are used to indicate solutions that do not form a cover. The synthesis is the lowest-cost cover found on the walk from thesis to antithesis.

one third, and one quarter of the cardinality of C. T is empty first and then augmented iteratively by selecting two sets whose removal would leave the fewest items uncovered which are still covered by $C \setminus T$. One of the two sets is chosen uniformly at random and added to T. We repeat this until T has the desired size. If $A \leftarrow F \setminus T$ does not cover all items, we greedily add sets in T to A until it is a cover. A becomes our antithesis.

To obtain a synthesis, we conduct a greedy walk from the thesis to the antithesis. This walk consists of two phases. In the first phase, we remove all sets in C that are not part of A. In the second phase, we greedily select a set in A which minimizes the cost over newly covered items and repeat until we obtain a cover which is returned as the synthesis. Figure 3 illustrates such a greedy walk from thesis to antithesis.

Numerical Results. We compare this simple dialectic search with the iterative greedy algorithm, ITEG, from [25] and the tabu search, TS, from [28]. We consider 70 well-known benchmark instances that are available from the OR library [4]. These instances involve up to 400 items and 4000 sets. In order to compare with ITEG and TS which were developed for the uni-cost SCP, the costs of all sets are set to one. ITEG was run on a multi-user Silicon Graphics IRIX Release 6.2 IP25, 194MHz MIPS R10000 processor and TS was run on a Pentium 4 with 2.4GHz. When comparing with ITEG, we use again our Pentium III 733MHz machine and we divide the cutoff times reported for ITEG by a factor of 4 which corresponds to the SPECint95 ratio of the two machines used. For the comparison with TS, we use an AMD Athlon 64 X2 Dual Core Processor 3800 2.0 GHz machine which is slightly slower than the machine used in [28].

Tables 4 and 5 summarize the results. Due to space restrictions we cannot show all results on individual instances.We therefore report aggregate results for each of the different benchmark classes. It should be noted that the developers of the TS approach tuned the tabu tenure on and for each of these sets individually. Similarly, the developers of ITEG set the algorithm parameters to a suitable value for each benchmark class. In contrast, Hegel was run with one set of parameters on all instances in all classes. As we can see, Hegel provides high quality solutions very quickly. With the exception of classes '4' and '5' where it performs slightly worse on average, Hegel produces equally good or better results than ITEG in sometimes substantially less time. In terms of the

Table 4. Numerical Results for the Set Cover Problem. We present the average solution (standard deviation), best solution (standard deviation), average time to find the best solution, and the time limit used. The results are averaged for each benchmark class in the OR library. Hegel was run 50 times on each instance, ITEG data were taken from [25] who ran their algorithms 10 times on each instance.

	AvgSol		BestSol		AvgTime		TimeLimit		
Class	ITEG	Dialectic	ITEG	Dialectic	ITEG	Dialectic	ITEG	Dialectic	Speedup
a	38.78	**38.77** (0.16)	38.6	38.6	-	1.59 (1.38)	7.5	7.5	1
b	22.04	**22.00** (0.04)	22.0	22	-	0.47 (0.23)	15	2.5	6
c	43.44	43.44 (0.42)	43.0	43	-	3.00 (2.48)	10	10	1
d	25.00	**24.86** (0.15)	25.0	**24.4**	-	0.74 (0.49)	27.5	5	5.5
e	5.00	5.00 (0.00)	5.0	5.0	-	0.00 (0.00)	2.5	0.1	25
4	**38.07**	38.43 (0.28)	37.8	37.8	-	0.56 (0.49)	2.5	2.5	1
5	**34.47**	34.51 (0.35)	34.1	34.1	-	0.76 (0.56)	2.5	2.5	1
6	20.86	**20.76** (0.11)	20.8	**20.6**	-	0.24 (0.24)	15	2.5	6
nre	17.04	**17.00** (0.00)	17.0	17.0	-	0.42 (0.09)	8.5	1	8.5
nrf	10.50	**10.44** (0.49)	10.0	10	-	0.58 (0.21)	16.5	1	16.5
nrg	62.82	**62.56** (0.47)	62.0	**61.6**	-	2.85 (0.98)	6.5	5	1.3
nrh	34.78	**34.49** (0.5)	34.0	34.0	-	1.62 (0.54)	15	2.5	6

Table 5. Numerical Results for the Set Cover Problem. We present the average runtime (standard deviation) in seconds for finding the best solution in each run, as well as the average solution quality and the best solution quality. Results are averaged for all instances in each benchmark class in the OR library. Hegel was run 50 times on each instance and TS data were taken from [28] who ran their algorithms 10 times on each instance.

	AvgSol		BestSol		AvgTime		
Class	TS	Dialectic	TS	Dialectic	TS	Dialectic	Speedup
a	**38.66** (0.24)	38.74 (0.16)	**38.4**	38.6	4.3 (3.78)	**1.78** (1.63)	2.4
b	22.02 (0.06)	**22.00** (0)	22	22	7.02 (6.98)	**0.49** (0.25)	14
c	43.5 (0.44)	**43.45** (0.41)	43	43	7.86 (7.16)	**2.97** (2.45)	2.6
d	25 (5.04)	**24.81** (0.12)	24.8	**24.4**	14.4 (14.4)	**1.07** (0.77)	13.4
e	5 (0)	5 (0)	5	5	0 (0)	0 (0)	0
4	**37.92** 0.27)	38.20 (0.30)	**37.7**	37.8	**0.67** (0.83)	1.63 (1.80)	0.4
5	34.36 (0.35)	**34.28** (0.15)	34.1	34.1	1.87 (2.35)	**1.85** (1.77)	1
6	20.78 (0.06)	**20.66** (0.09)	20.6	20.6	**0.26** (0.54)	0.72 (0.69)	0.3
nre	17.14 (0.3)	**16.98** (0.06)	17	**16.6**	5.94 (11.3)	**0.50** (0.46)	11.7
nrf	10.62 (0.5)	**10** (0)	10	10	31.4 (61.96)	**1.31** (0.90)	23.8
nrg	62.7 (0.6)	**62.25** (0.47)	61.8	**61.2**	32.0 (32.3)	**4.33** (2.28)	7.3
nrh	34.88 (0.44)	**34.03** (0.19)	34	**33.8**	22.4 (57.5)	**3.49** (2.20)	6.4

best solutions found over the different runs, when computing the average for each class, Hegel always performs as good or better than ITEG.

Comparing with TS, Hegel is performing slightly worse on classes '4' and 'a' and outperforms TS in terms of solution quality otherwise, at times quite substantially (see classes 'nrf' and 'nrh'). Moreover, Hegel always finds the best solution earlier, leading to speed-ups of up to a factor of 23.

Finally, in terms of individual instances, Hegel found formerly unknown improving solutions on four instances (d4(24), nre1(16), nrg3(61), nrg5(61)), that is over 5% of all instances in one of the best studied benchmark sets in OR.

It took about 20 man-days to develop and test this algorithm which was by far the most time we spent on any algorithm presented in this paper. Despite this short development time, our algorithm outperforms the state-of-the-art approaches on set covering. This shows that the dialectic search meta-heuristic effectively leads to simple and highly efficient local search approaches.

6 Conclusion

We proposed to use Hegel and Fichte's dialectic as a meta-heuristic search paradigm and demonstrated its power and effectiveness by solving four problems from three greatly different problem domains: constraint satisfaction, continuous optimization, and combinatorial optimization.

With very little effort the dialectic search paradigm allowed us to devise a local search algorithm for the costas arrays problem and the magic squares problem. Moreover, with very little effort we devised a local search algorithm for the set covering problem, one of the most intensively studied problems in the operations research literature which has been the subject of many research projects and on which entire Ph.D. theses have been written. Even though we spent only about 20 man-days on this algorithm, it outperforms the fastest algorithms from the very rich literature which were individually tuned on and for each class of benchmark problems. In contrast, our algorithm is the same for all problems from all classes, it did not undergo any sophisticated tuning, and it still provides solutions of the same or better quality in less time.

We conclude that Hegel and Fichte's dialectic provides an appealing framework for devising highly efficient local search algorithms for anyone working on constraints. We believe that the reason for the simplicity of use is primarily caused by the fact that dialectic search allows us to develop functions for exploitation and exploration in separation. We outlined a close relation with existing techniques, especially tabu search, iterated local search, variable neighborhood search, and very large scale neighborhood search. We welcome the view that dialectic search represents a special case of all of these methods as it may help the research community to further improve dialectic search. Our own objective was to devise a meta-heuristic which is, on one hand, general enough to be applied to a great variety of problems and which, on the other hand, is specific enough to guide the user to develop effective problem-specific methods for search space exploration and exploitation. We believe we found a search paradigm which strikes a good balance between being specific and being general in Hegel and Fichte's philosophy of dialectic.

References

1. Ahuja, R.K., Ergun, O., Orlin, J.B., Punnen, A.P.: A survey of Very Large Scale Neighborhood Search Techniques. Discrete Applied Mathematics 123, 75–102 (2002)
2. Boese, K.D.: Cost versus distance in the traveling salesman problem. TR, CSD-950018, UCLA (1995)

3. Battiti, R., Tecchiolli, G.: The reactive tabu search. ORSA Journal on Computing 6(2), 126–140 (1994)
4. Beasley, J.E.: OR-Library: Distributing test problems by electronic mail. Operations Research Society 41, 1069–1072 (1990)
5. Caprara, A., Fischetti, M., Toth, P., Vigo, D., Guida, P.L.: Algorithms for Railway Crew Management. Mathematical Programming 79, 125–141 (1997)
6. C̃erny, V.: Thermodynamics Approach to the Travelling Salesman Problem: An Efficient Simulation Algorithm. Optimization Theory Appl. 45, 41–51 (1985)
7. Codognet, P., Diaz, D.: Yet Another Local Search Method for Constraint Solving. In: AAAI (2001)
8. Costas, J.P.: A study of a class of detection waveforms having nearly ideal range-Doppler ambiguity properties. Proceedings of the IEEE 72(8), 996–1009 (1984)
9. Fichte, J.G.: Wissenschaftslehre, The Science of Knowledge. Cambridge University Press, Cambridge (1982); translated by P. Heath and J. Lachs (1794)
10. Freedman, A., Levanon, N.: Staggered Costas signals. IEEE Trans. Aerosp. Electron Syst. AES-22(6), 695–701 (1986)
11. Glover, F.: Tabu Search, Part I. ORSA Journal on Computing 1(3), 190–206 (1989)
12. Glover, F.: Tabu Search, Part II. ORSA Journal on Computing 2(1), 4–32 (1990)
13. Glover, F., Laguna, M., Marti, R.: Fundamentals of scatter search and path relinking. TR, 80309-0419, University of Colorado (2000)
14. Goffe, W.L., Ferrier, G.D., Rogers, J.: Global Optimization and Statistical Functions with Simulated Annealing. Journal of Econometrics 60(1-2), 65–99 (1994)
15. Gomes, C., Sellmann, M.: Streamlined constraint reasoning. In: Wallace, M. (ed.) CP 2004. LNCS, vol. 3258, pp. 274–289. Springer, Heidelberg (2004)
16. Golomb, S.W., Taylor, H.: Two Dimensional Synchronization Patterns for Minimum Ambiguity. IEEE Trans. Informat. Theory IT-28(4), 600–604 (1982)
17. Hansen, P., Mladenovic, N.: Variable Neighbourhood Search: Principles and Applications. European Journal of Operational Research 130, 449–467 (2001)
18. Hegel, G.W.F.: Phänomenologie des Geistes, Phenomenology of Spirit, translated by Miller, A.V. Oxford University Press, Oxford (1807) (1977)
19. Hoffmann, K.L., Padberg, M.W.: Solving Airline Crew Scheduling Problems by Branch-and-Cut. Management Science 39(6), 657–682 (1993)
20. Housos, E., Elmoth, T.: Automatic Optimization of Subproblems in Scheduling Airline Crews. Interfaces 27(5), 68–77 (1997)
21. Jones, T., Forrest, S.: Fitness Distance Correlation as a Measure of Problem Difficulty for Genetic Algorithms. In: Genetic Algorithms, pp. 184–192 (1995)
22. Kirkpatrick, S., Gelatt, C.D., Vecchi, M.P.: Optimization by Simulated Annealing. Science 220(4598), 671–680 (1983)
23. Selman, B., Levesque, H.J., Mitchell, D.G.: A New Method for Solving Hard Satisfiability Problems. In: AAAI, pp. 440–446 (1992)
24. Macready, W.G., Wolpert, D.H.: Bandit Problems and the Exploration/Exploitation Tradeoff. IEEE Transactions on Evolutionary Computation 2(2), 2–22 (1998)
25. Marchiori, E., Steenbeek, A.: An Iterated Heuristic Algorithm for the Set Covering Problem. In: WEA (1998)
26. Metropolis, N., Rosenbluth, A.W., Rosenbluth, M.N., Teller, A.H., Teller, E.: Equations of State Calculations by Fast Computing Machines. Chemical Physics 21(6), 1087–1092 (1953)
27. Moran, J.: The Wonders of Magic Squares. Vintage, New York (1982)
28. Musliu, N.: Local Search Algorithm for Unicost Set Covering Problem. In: Ali, M., Dapoigny, R. (eds.) IEA/AIE 2006. LNCS (LNAI), vol. 4031, pp. 302–311. Springer, Heidelberg (2006)

29. Stuetzle, T.: Iterated local search for the quadratic assignment problem. TR, AIDA-99-03, Darmstadt University of Technology (1999)
30. SIMANN. Fortran Simulated Annealing code (2004), `http://wueconb.wustl.edu/~goffe`
31. Thompson, P.M., Orlin, J.B.: The theory of cyclic transfers Working Paper No. OR, 200-89, Operations Research Center. MIT, Cambridge (1989)
32. Toregas, C., Swain, R., ReVelle, C., Bergman, L.: The Location of Emergency Service Facilities. Operational Research 19(6), 1363–1373 (1971)
33. Van Hentenryck, P., Michel, L.: Constraint-Based Local Search. The MIT Press, Cambridge (2005)
34. Van Hentenryck, P., Michel, L.: Synthesis of constraint-based local search algorithms from high-level models. In: AAAI (2007)
35. Vasko, F.J., Wilson, G.R.: An Efficient Heuristic for Large Set Covering Problems. Naval Research Logistics Quarterly 31, 163–171 (1984)
36. Vasko, F.J., Wolf, F.E.: Optimal Selection of Ingot Sizes via Set Covering. Operations Research 35, 115–121 (1987)

Restricted Global Grammar Constraints*

George Katsirelos[1], Sebastian Maneth[2], Nina Narodytska[2], and Toby Walsh[2]

[1] NICTA, Sydney, Australia
george.katsirelos@nicta.com.au
[2] NICTA and University of NSW, Sydney, Australia
sebastian.maneth@nicta.com.au, ninan@cse.unsw.edu.au,
toby.walsh@nicta.com.au

Abstract. We investigate the global GRAMMAR constraint over restricted classes of context free grammars like deterministic and unambiguous context-free grammars. We show that detecting disentailment for the GRAMMAR constraint in these cases is as hard as parsing an unrestricted context free grammar. We also consider the class of linear grammars and give a propagator that runs in quadratic time. Finally, to demonstrate the use of linear grammars, we show that a weighted linear GRAMMAR constraint can efficiently encode the EDITDISTANCE constraint, and a conjunction of the EDITDISTANCE constraint and the REGULAR constraint.

1 Introduction

In domains like staff scheduling, regulations can often be naturally expressed using formal grammars. Pesant [9] introduced the global REGULAR constraint to express problems using finite automaton. Sellmann [14] and Quimper and Walsh [11] then introduced the global GRAMMAR constraint for context-free grammars. Unlike parsing which only finds a single solution, propagating such a constraint essentially considers all solutions. Nevertheless, a propagator for the REGULAR constraint runs in linear time and a propagator for the GRAMMAR constraint runs in cubic time just like the corresponding parsers. Subsequently, there has been research on more efficient propagators for these global constraints [12,10,4,7,6]. Whilst research has focused on regular and unrestricted context-free languages, a large body of work in formal language theory considers grammars between regular and context-free. Several restricted forms of context-free grammars have been proposed that permit linear parsing algorithms whilst being more expressive than regular grammars. Examples of such grammars are LL(k), LR(1), and LALR. Such grammars play an important role in compiler theory. For instance, yacc generates parsers that accept LALR languages.

In this paper we explore the gap between the second and third levels of the Chomsky hierarchy for classes of grammars which can be propagated more efficiently than context-free grammars. These classes of grammar are attractive because either they have a linear or quadratic time membership test (e.g., LR(k) and linear grammars, respectively) or they permit counting of strings of given length in polynomial time (e.g.,

* NICTA is funded by the Australian Government's Department of Broadband, Communications, and the Digital Economy and the Australian Research Council.

I.P. Gent (Ed.): CP 2009, LNCS 5732, pp. 501–508, 2009.

unambiguous grammars). The latter may be useful in branching heuristics. One of our main contributions is a lower bound on the time complexity for propagating grammar constraints using deterministic or unambiguous grammars. We prove that detecting disentailment for such constraints has the same time complexity as the best parsing algorithm for an arbitrary context-free grammar. Using LL(k) languages or unambiguous grammars does not therefore improve the efficiency of propagation. Another contribution is to show that linearity of the grammar permits quadratic time propagation. We show that we can encode an EDITDISTANCE constraint and a combination of an EDITDISTANCE constraint and a REGULAR constraint using such a linear grammar. Experimental results show that this encoding is very efficient in practice.

2 Background

A context-free grammar is a tuple $G = \langle N, T, P, S \rangle$, where N is a finite set of *non-terminal* symbols, T is a finite set of *terminal* symbols, P is a set of *productions*, and $S \in N$ is the *start symbol*. A production is of the form $A \to \alpha$ where $A \in N$ and $\alpha \in (N \cup T)^+$. The derivation relation $\Rightarrow_G$ induced by G is defined as follows: for any $u, v \in (N \cup T)^*$, $uAv \Rightarrow_G u\alpha v$ if there exists a production $A \to \alpha$ in P. Sometimes we additionally index $\Rightarrow_G$ by the production that is applied. The transitive, reflexive closure of $\Rightarrow_G$ is denoted by $\Rightarrow_G^*$. A string in $s \in T^*$ is *generated* by G if $S \Rightarrow_G^* s$. The set of all strings generated by G is denoted $L(G)$. Note that this does not allow the *empty string* ε as right-hand side of a production. Hence, $\varepsilon \notin L(G)$. This is not a restriction: we can add a new start symbol Z with productions $Z \to \varepsilon \mid S$ to our grammars. Our results can easily be generalized to such ε-enriched grammars. We denote the length of a string s by $|s|$. The *size* of G, denoted by $|G|$, is $\sum_{A \to \alpha \in P} |A\alpha|$.

A context-free grammar is in *Chomsky form* if all productions are of the form $A \to BC$ where B and C are non-terminals or $A \to a$ where a is a terminal. Any ε-free context-free grammar G can be converted to an equivalent grammar G' in Chomsky form with at most a linear increase in its size; in fact, $|G'| \leq 3|G|$, see Section 4.5 in [3]. A context-free grammar is in *Greibach form* if all productions are of the form $A \to a\alpha$ where a is a terminal and α is a (possibly empty) sequence of non-terminals. Any context-free grammar G can be converted to an equivalent grammar G' in Greibach form with at most a polynomial increase in its size; in fact, the size of G' is in $O(|G|^4)$ in general, and is in $O(|G|^3)$ if G has no chain productions of the form $A \to B$ for nonterminals A, B, see [1]. A context-free grammar is *regular* if all productions are of the forms $A \to w$ or $A \to wB$ for non-terminals A, B and $w \in T^+$.

3 Simple Context-Free Grammars

In this section we show that propagating a *simple* context-free grammar constraint is at least as hard as parsing an (unrestricted) context-free grammar. A grammar G is *simple* if it is in Greibach form, and for every non-terminal A and terminal a there is at most one production of the form $A \to a\alpha$. Hence, restricting ourselves to languages recognized by simple context-free grammars does not improve the complexity of propagating a global grammar constraint. Simple context-free languages are included in the deterministic context-free languages (characterized by deterministic push-down automata), and also in the $LL(1)$ languages [13], so this result also holds for propagating these classes of languages. Given finite sets $D_1, \dots, D_n$, their *Cartesian product language* $L(R_{D_1,\dots,D_n})$ is the cross product of the domains $\{a_1 a_2 \cdots a_n \mid a_1 \in D_1, \dots, a_n \in D_n\}$. Following [14], we define the global GRAMMAR constraint:

Definition 1. *The* GRAMMAR$([X_1, \dots, X_n], G)$ *constraint is true for an assignment variables X iff a string formed by this assignment belongs to $L(G)$.*

From Definition 1, we observe that finding a support for the grammar constraint is equivalent to intersecting the context-free language with the Cartesian product language of the domains.

Proposition 1. *Let G be a context-free grammar, $X_1, \dots, X_n$ be a sequence of variables with domains $D(X_1), \dots, D(X_n)$. Then $L(G) \cap L(R_{D(X_1),\dots,D(X_n)}) \neq 0$ iff* GRAMMAR$([X_1, \dots, X_n], G)$ *has a support.*

Context-free grammars are effectively closed under intersection with regular grammars. To see this, consider a context-free grammar G in Chomsky form and a regular grammar R. Following the "triple construction", the intersection grammar has non-terminals of the form $\langle F, A, F' \rangle$ where F, F' are non-terminals of R and A is a non-terminal of G. Intuitively, $\langle F, A, F' \rangle$ generates strings w that are generated by A and also by F, through a derivation from F to wF'. If $A \rightarrow BC$ is a production of G, then we add, for all non-terminals F, F', F'' of R, the production $\langle F, A, F'' \rangle \rightarrow \langle F, B, F' \rangle \langle F', C, F'' \rangle$. The resulting grammar is $O(|G|n^3)$ in size where n is the number of non-terminals of R. This is similar to the construction of Theorem 6.5 in [3] which uses push-down automata instead of grammars. Since emptiness of context-free grammars takes linear time (cf. [3]) we obtain through Proposition 1 a cubic time algorithm to check whether a global constraint GRAMMAR$([X_1, \dots, X_n], G)$ has support. In fact, this shows that we can efficiently propagate more complex constraints, such as the conjunction of a context-free with a regular constraint. Note that if R is a Cartesian product language then the triple construction generates the same result as the CYK based propagator for the GRAMMAR constraint [14,11].

We now show that for *simple* context-free grammars G, detecting disentailment of the constraint GRAMMAR$([X_1, \dots, X_n], G)$, i.e. testing whether it has a solution, is at least as hard as parsing an arbitrary context-free grammar.

Theorem 1. *Let G be a context-free grammar in Greibach form and s a string of length n. One can construct in $O(|G|)$ time a simple context-free grammar G' and in $O(|G|n)$ time a Cartesian product language $L(R_{D(X_1),\dots,D(X_n)})$ such that $L(G') \cap L(R_{D(X_1),\dots,D(X_n)}) \neq \emptyset$ iff $s \in L(G)$.*

Proof. The idea behind the proof is to determinize an unrestricted context free grammar G by mapping each terminal in G to a set of pairs – the terminal and a production that can consume this terminal. This allows us to carry information about the derivation inside a string in G'. Then, we construct a Cartesian product language $L(R_{D(X_1),\dots,D(X_n)})$ over these pairs so that all strings from this language map only to the string s. Let $G = \langle N, T, P, S\rangle$ and fix an arbitrary order of the productions in P. We now construct the grammar $G' = \langle N, T', P', S\rangle$. For every $1 \leq j \leq |P|$, if the j-th production of P is $A \to a\alpha$ then let (a, j) be a new symbol in T' and let the production $A \to (a, j)\alpha$ be in P'. Next, we construct the Cartesian product language. We define $D(X_i) = \{(a, j) | (s_i = a) \wedge (a, j) \in T'\}$, $i = 1, \ldots, n$ and s_i is the i-th letter of s. Clearly, G' is constructed in $O(|G|)$ time and $L(R_{D(X_1),\ldots,D(X_n)})$ in $O(|P|n)$ time.

($\Rightarrow$) Let $L(G') \cap L(R_{D(X_1),\ldots,D(X_n)})$ be non empty. Then there exits a string s' that belongs to the intersection. Let $s' = (a_1, i_1)\cdots(a_n, i_n)$. By the definition of $L(R_{D(X_1),\ldots,D(X_n)})$, the string $a_1a_2\cdots a_n$ must equal s. Since $s' \in L(G')$, there must be a derivation by G of the form

$$S \Rightarrow_{G,p_1} a_1\alpha \Rightarrow_{G,p_2} a_1a_2\alpha' \cdots \Rightarrow_{G,p_n} a_1 \cdots a_n$$

where p_j is the j-th production in P. Hence, $s \in L(G)$.

($\Leftarrow$) Let $s \in L(G)$. Consider a derivation sequence of the string s. We replace every symbol a in s that was derived by the i-th production of G by (a, i). By the construction of G', the string s' is in $L(G')$. Moreover, s' is also in $L(R_{D(X_1),\ldots,D(X_n)})$. □

Note that context-free parsing has a quadratic time lower bound, due to its connection to matrix multiplication [8]. Given this lower bound and the fact that the construction of Theorem 1 requires only linear time, we can deduce the following.

Corollary 1. *Let G be a context-free grammar. If G is simple (or deterministic or $LL(1)$) then detecting disentailment of* GRAMMAR$([X_1, \ldots, X_n], G)$ *is at least as hard as context-free parsing of a string of length n.*

We now show the converse to Theorem 1 which reduces intersection emptiness of a context-free with a regular grammar, to the membership problem of context-free languages. This shows that the time complexity of detecting disentailment for the GRAMMAR constraint is the same as the time complexity of the best parsing algorithm for an arbitrary context free grammar. Therefore, our result shows that detecting disentailment takes $O(n^{2.4})$ time [2], as in the best known algorithm for Boolean matrix multiplication. It does not, however, improve the asymptotic complexity of a domain consistency propagator for the GRAMMAR constraint [14,11].

Theorem 2. *Let $G = \langle N, T, P, S\rangle$ be a context-free grammar and $L(R_{D(X_1),\ldots,D(X_n)})$ be Cartesian product language. One can construct in time $O(|G| + |T|^2)$ a context-free grammar G' and in time $O(n|T|)$ a string s such that $s \in L(G')$ iff $L(G) \cap L(R_{D(X_1),\ldots,D(X_n)}) \neq \emptyset$.*

Proof (Sketch). We assign an index to each terminal in T. For each position i of the strings of R, we create a bitmap of the alphabet that describes the terminals that may appear in that position. The j-th bit of the bitmap is 1 iff the symbol with index j may appear at position i. The string s is the concatenation of the bitmaps for each position and has size $n|T|$. First, we add $B \to 0$ and $B \to 1$ to G'. For each terminal in T with index j, we introduce $T_j \to B^{j-1}1B^{|T|-j}$ into G' to accept any bitmap with 1 at the j-th position. Then, for each production in G of the form $A \to a\alpha$ such that the index of a is j, we add $A \to T_j\alpha$ to G'. In this construction, every production in G' except for those with T_i on the left hand side can be uniquely mapped to a production in G. It can be shown that $s \in L(G')$ iff $L(G) \cap L(R_{D(X_1),\dots,D(X_n)}) \neq \emptyset$. □

4 Linear Context-Free Grammars

A context-free grammar is *linear* if every production contains at most one non-terminal in its right-hand side. The linear languages are a proper superset of the regular languages and are a strict subset of the context-free languages. Linear context-free grammars possess two important properties: (1) membership of a given string of length n can be checked in time $O(n^2)$ (see Theorem 12.3 in [15]), and (2) the class is closed under intersection with regular grammars (to see this, apply the "triple construction" as explained after Proposition 1). The second property opens the possibility of constructing a polynomial time propagator for a conjunction of the the linear GRAMMAR and the REGULAR constraints. Interestingly, we can show that a CYK-based propagator for this type of grammars runs in quadratic time. This is then the third example of a grammar, besides regular and context-free grammars, where the asymptotic time complexity of the parsing algorithm and that of the corresponding propagator are equal.

Theorem 3. *Let G be a linear grammar and* GRAMMAR$([X_1,\dots,X_n],G)$ *be the corresponding global constraint. There exists a domain consistency propagator for this constraint that runs in $O(n^2|G|)$ time.*

Proof. We convert $G = \langle N, T, P, S\rangle$ into CNF. Every linear grammar can be converted into the form $A \to aB$, $A \to Ba$ and $A \to a$, where $a, b \in T$ and $A, B \in N$ (see Theorem 12.3 of [15]) in $O(|G|)$ time. To obtain CNF we replace every terminal $a \in T$ that occurs in a production on the right hand side with a new non-terminal Y_a and introduce a production $Y_a \to a$.

Consider the CYK-based domain consistency propagator for an arbitrary context-free grammar constraint [11,14]. The algorithm runs in two stages. In the first stage, it constructs in a bottom-up fashion a dynamic programing table $V_{n\times n}$, where an element A of $V_{i,j}$ is a potential non-terminal that generates a substring from the domains of variables $[X_i,\dots,X_{i+j}]$. In the second stage, it performs a top-down traversal of V and marks an element A of $V_{i,j}$ iff it is reachable from the starting non-terminal S using productions of the grammar and elements of V. It then removes unmarked elements, including terminals. If it removes a terminal at column i of the table, it prunes the corresponding value of variable X_i.

The complexity of this algorithm is bounded by the number of possible 1-step derivations from each non-terminal in the table. Let $G' = \langle N', T', P', S'\rangle$ be an arbitrary context free grammar. There are $O(|N'|n^2)$ non-terminals in the table and each non-terminal can be expanded in $O(F'(A)n)$ possible ways, where $F'(A)$ is the number of productions in G' with non-terminal A on the left-hand side. Therefore, the total time complexity of the propagator for unrestricted context-free grammars is $n^2\sum_{A\in N'} nF'(A) = O(n^3|G'|)$. In contrast, the number of possible 1-step derivations from each non-terminal in linear grammars is bounded by $O(F(A))$. Therefore, the propagator runs in $O(n^2|G|)$ for a linear grammar G. □

Theorem 3 can be extended to the weighted form of the linear GRAMMAR constraint, WEIGHTEDCFG [5]. A weighted grammar is annotated with a weight for each production and the weight of a derivation is the sum of all weights used in it. The linear WEIGHTEDCFG$(G, Z, [X_1, \ldots, X_n])$ constraint holds iff an assignment X forms a string belonging to the weighted linear grammar G and the minimal weight derivation of X is less than or equal to Z. The domain consistency propagator for the WEIGHTEDCFG constraint is an extension of the propagator for GRAMMAR that computes additional information for each non-terminal $A \in V_{i,j}$—the minimum and the maximum weight derivations from A. Therefore, this algorithm has the same time and space asymptotic complexity as the propagator for GRAMMAR, so the complexity analysis for the linear WEIGHTEDCFG constraint is identical to the non-weighted case.

It is possible to restrict linear grammars further, so that the resulting global constraint problem is solvable in *linear time*. As an example, consider "fixed-growth" grammars in which there exists l and r with $l + r \geq 1$ such that every production is of the form either $A \rightarrow w \in T^+$ or $A \rightarrow uBw$ where the length of $u \in T^*$ equals l and the length of $w \in T^*$ equals r. In this case, the triple construction (explained below Proposition 1) generates $O(|G|n)$ new non-terminals implying linear time propagation (similarly, CYK runs in linear time as it only generates non-terminals on the diagonal of the dynamic program). A special case of fixed-growth grammars are regular grammars which have $l = 1$ and $r = 0$ (or vice versa).

5 The EDITDISTANCE Constraint

To illustrate linear context-free grammars, we show how to encode an edit distance constraint into such a grammar. EDITDISTANCE$([X_1, \ldots, X_n, Y_1, \ldots, Y_m], N)$ holds iff the edit distance between assignments of two sequences of variables **X** and **Y** is less than or equal to N. The edit distance is the minimum number of deletion, insertion and substitution operations required to convert one string into another. Each of these operations can change one symbol in a string. W.L.O.G. we assume that $n = m$. We will show that the EDITDISTANCE constraint can be encoded as a linear WEIGHTEDCFG constraint. The idea of the encoding is to parse matching substrings using productions of weight 0 and to parse edits using productions of weight 1.

We convert EDITDISTANCE$([\mathbf{X}, \mathbf{Y}], N)$ into a linear WEIGHTEDCFG$([\mathbf{Z}_{2n+1}, N, G_{ed})$ constraint. The first n variables in the sequence **Z** are equal to the sequence **X**, the variable Z_{n+1} is ground to the sentinel symbol $\#$ so that the grammar can distinguish the sequences **X** and **Y**, and the last n variables of the sequence **Z** are equal to the reverse of the sequence **Y**. We define the linear weighted grammar G_{ed} as follows. Rules $S \rightarrow dSd$ with weight $w = 0, \forall d \in D(X) \cup D(Y)$, capture matching terminals, rules $S \rightarrow d_1 S d_2$ with $w = 1, \forall d_1 \in D(X), d_2 \in D(Y), d_1 \neq d_2$, capture replacement, rules $S \rightarrow dS|Sd$ with $w = 1, \forall d \in D(X)$, capture insertions and deletions. Finally, the rule $S \rightarrow \#$ with weight $w = 0$ generates the sentinel symbol. As discussed in the previous section, the propagator for the linear WEIGHTEDCFG constraint takes $O(n^2|G|)$ time. Down a branch of the search tree, the time complexity is $O(n^2|G|ub(N))$.

We can use this encoding of the EDITDISTANCE constraint into a linear WEIGHTEDCFG constraint to construct propagators for more complex constraints. For instance, we can exploit the fact that linear grammars are closed under intersection

Table 1. Performance of the encoding into WEIGHTEDCFG constraints shown in: number of instances solved in 60 sec / average number of choice points / average time to solve

n	N	ED_{Dec}			$ED_{\wedge}$		
		#solved	#choice points	time	#solved	#choice points	time
15	2	**100**	**29**	**0.025**	**100**	6	0.048
20	2	**100**	661	0.337	**100**	**6**	**0.104**
25	3	93	2892	2.013	**100**	**10**	**0.226**
30	3	71	6001	4.987	**100**	**12**	**0.377**
35	4	58	5654	6.300	**100**	**17**	**0.667**
40	4	40	3140	4.690	**100**	**17**	**0.985**
45	5	36	1040	2.313	**100**	**19**	**1.460**
50	5	26	1180	4.848	**100**	**24**	**1.989**
TOTALS							
solved/total			524 /800			**800** /800	
avg time for solved			2.557			**0.732**	
avg choice points for solved			2454			**14**	

with regular grammars to propagate efficiently the conjunction of an EDITDISTANCE constraint and REGULAR constraints on each of the sequences **X**,**Y**. More formally, let **X** and **Y** be two sequences of variables of length n subject to the constraints REGULAR$(\mathbf{X}, R_1)$, REGULAR$(\mathbf{Y}, R_2)$ and EDITDISTANCE$(\mathbf{X}, \mathbf{Y}, N)$. We construct a domain consistency propagator for the conjunction of these three constraints, by computing a grammar that generates strings of length $2n + 1$ which satisfy the conjunction. First, we construct an automaton that accepts $\mathcal{L}(R_1)\#\mathcal{L}(R_2)^R$. This language is regular and requires an automaton of size $O(|R_1| + |R_2|)$. Second, we intersect this with the linear weighted grammar that encodes the EDITDISTANCE constraint using the "triple construction". The size of the obtained grammar is $G_\wedge = |G_{ed}|(|R_1| + |R_2|)^2$ and this grammar is a weighted linear grammar. Therefore, we can use the linear WEIGHTEDCFG$(\mathbf{Z}, N, G_\wedge)$ constraint to encode the conjunction. Note that the size of $G_\wedge$ is only quadratic in $|R_1|+|R_2|$, because G_{ed} is a linear grammar. The time complexity to enforce domain consistency on this conjunction of constraints is $O(n^2|G_\wedge|) = O(n^2d^2(|R_1| + |R_2|)^2)$ for each invocation and $O(n^2d^2(|R_1| + |R_2|)^2ub(N))$ down a branch of the search tree.

To evaluate the performance of the WEIGHTEDCFG$(\mathbf{Z}, N, G_\wedge)$ constraint we carried out a series of experiments on random problems. In out first model the conjunction of the EDITDISTANCE constraint and two REGULAR constraints was encoded with a single WEIGHTEDCFG$(\mathbf{Z}, N, G_\wedge)$ constraint. We call this model $ED_\wedge$. The second model contains the EDITDISTANCE constraint, encoded as WEIGHTEDCFG$(\mathbf{Z}, N, G_{ed})$, and two REGULAR constraints. The REGULAR constraint for the model ED_{Dec} is implemented using a decomposition into ternary table constraints [11]. The WEIGHTEDCFG constraint is implemented with an incremental monolithic propagator [5]. The first REGULAR constraint ensures that there are at most two consecutive values one in the sequence. The second encodes a randomly generated string of 0s and 1s. To make problems harder, we enforced the EDITDISTANCE constraint and the REGULAR constraints on two sequences $\mathbf{X}\#(\mathbf{Y})^R$ and $\mathbf{X}'\#(\mathbf{Y}')^R$ of the same length $2n+1$. The EDITDISTANCE constraint and the first REGULAR constraint are identical for these two sequences, while **Y** and **Y**′ correspond to different randomly generated strings of 0s and 1s. Moreover, **X** and **X**′ overlap on 15% of randomly chosen variables. For each possible value of $n \in \{15, 20, 25, 30, 35, 40, 45, 50\}$, we generated 100 instances. Note that n is the length of each sequence **X**, **Y**, **X**′ and **Y**′. N is the maximum edit distance between **X** and **Y** and between **X**′ and **Y**′. We used a random value and variable ordering and a time out of 60 sec. Results for different values of n are presented in Table 1. As can be seen from the table, the model $ED_{\wedge}$ significantly outperforms the model ED_{Dec} for larger problems, but it is slightly slower for smaller problems. Note that the model $ED_{\wedge}$ solves many more instances compared to ED_{Dec}.

6 Conclusions

Unlike parsing, restrictions on context free grammars such as determinism do not improve the efficiency of propagation of the corresponding global GRAMMAR constraint. On the other hand, one specific syntactic restriction, that of linearity, allows propagation in quadratic time. We demonstrated an application of such a restricted grammar in encoding the EDITDISTANCE constraint and more complex constraints.

References

1. Blum, N., Koch, R.: Greibach normal form transformation revisited. Inf. Comput. 150, 112–118 (1999)
2. Coppersmith, D., Winograd, S.: Matrix multiplication via arithmetic progressions. J. Symbolic Comput. 9, 251–280 (1990)
3. Hopcroft, J.W., Ullman, J.D.: Introduction to automata theory, languages, and computation. Addison-Wesley, Reading (1979)
4. Kadioglu, S., Sellmann, M.: Efficient context-free grammar constraints. In: AAAI, pp. 310–316 (2008)
5. Katsirelos, G., Narodytska, N., Walsh, T.: The weighted CFG constraint. In: Perron, L., Trick, M.A. (eds.) CPAIOR 2008. LNCS, vol. 5015, pp. 323–327. Springer, Heidelberg (2008)
6. Katsirelos, G., Narodytska, N., Walsh, T.: Reformulating global grammar constraints. In: van Hoeve, W.J., Hooker, J.N. (eds.) CPAIOR 2009. LNCS, vol. 5547, pp. 132–147. Springer, Heidelberg (2009)
7. Lagerkvist, M.: Techniques for Efficient Constraint Propagation. PhD thesis, KTH, Sweden (2008)
8. Lee, L.: Fast context-free grammar parsing requires fast boolean matrix multiplication. J. ACM 49, 1–15 (2002)
9. Pesant, G.: A regular language membership constraint for finite sequences of variables. In: Wallace, M. (ed.) CP 2004. LNCS, vol. 3258, pp. 482–495. Springer, Heidelberg (2004)
10. Quimper, C., Walsh, T.: Decompositions of grammar constraints. In: AAAI, pp. 1567–1570 (2008)
11. Quimper, C.G., Walsh, T.: Global grammar constraints. In: Benhamou, F. (ed.) CP 2006. LNCS, vol. 4204, pp. 751–755. Springer, Heidelberg (2006)
12. Quimper, C.G., Walsh, T.: Decomposing global grammar constraints. In: Bessière, C. (ed.) CP 2007. LNCS, vol. 4741, pp. 590–604. Springer, Heidelberg (2007)
13. Rozenberg, G., Salomaa, A.: Handbook of Formal Languages, vol. 1. Springer, Heidelberg (2004)
14. Sellmann, M.: The theory of grammar constraints. In: Benhamou, F. (ed.) CP 2006. LNCS, vol. 4204, pp. 530–544. Springer, Heidelberg (2006)
15. Wagner, K., Wechsung, G.: Computational Complexity. Springer, Heidelberg (1986)

Conflict Resolution

Konstantin Korovin, Nestan Tsiskaridze, and Andrei Voronkov

The University of Manchester
{korovin,tsiskarn,voronkov}@cs.man.ac.uk

Abstract. We introduce a new method for solving systems of linear inequalities over the rationals—the *conflict resolution method*. The method successively refines an initial assignment with the help of newly derived constraints until either the assignment becomes a solution of the system or a trivially unsatisfiable constraint is derived. We show that this method is correct and terminating. Our experimental results show that conflict resolution outperforms the Fourier-Motzkin method and the Chernikov algorithm, in some cases by orders of magnitude.

1 Introduction

In this paper we introduce a new algorithm for checking solvability of systems of linear inequalities, called *conflict resolution*. The method works with such a system S and an assignment to variables σ (initially arbitrary) and refines the assignment trying to make it into a solution of S. If such a refinement is impossible, it is due to a pair of inequalities in S of the forms $x + p \geq 0$ and $-x + q \geq 0$ with some properties. In this case we *resolve* the conflict by adding a new equation $p + q \geq 0$. The use of this rule makes the method similar to the Fourier-Motzkin variable elimination.

The first (rather naive) implementation of the method shows that in practice it normally behaves better, and sometimes orders of magnitude better than the Fourier-Motzkin method and the Chernikov algorithm. These experiments are confirmed by some properties of our method proved in this paper, namely, that it never derives redundant (in some natural sense) inequalities. The conflict resolution algorithm is well-suited both for proving inconsistency of systems of linear constraints and for finding satisfying assignments.

It is a bit too early to extensively compare this method to the existing implementations of the simplex or interior point methods since not much is known about the best strategies, optimisations and modifications of the method but we hope it can eventually become competitive also with the best previously known methods. In this paper we only present some initial experimental results.

This paper is structured as follows. Section 2 defines main notions. Section 3 overviews the Fourier-Motzkin variable elimination method. In Section 4 we

I.P. Gent (Ed.): CP 2009, LNCS 5732, pp. 509–523, 2009.

introduce the conflict resolution and the assignment refinement rules used in our method and prove some of their properties. Section 5 presents the conflict resolution algorithm CRA. We prove that the algorithm is correct and terminating. In Section 6 we compare our method with the Fourier-Motzkin variable elimination: we show that our method does not derive redundant inequalities and give an example of a sequence of systems on which the Fourier-Motzkin method is exponential while our method is linear independently of the strategy used and the order of rule applications. In Section 7 we briefly discuss how the method can be modified to handle strict inequalities and linear programming. Section 8 is dedicated to the implementation and Section 9 to experiments with our method. Finally, in Section 10 we mention related work on avoiding redundancy in the Fourier-Motzkin method.

2 Preliminaries

Let $\mathbb{Q}$ denote the set of rationals. Throughout the paper we denote by n a positive integer and by X a finite set of variables $\{x_1, \ldots, x_n\}$. We call a *rational linear constraint over* X either a formula $a_n x_n + \ldots + a_1 x_1 + b \diamond 0$, where $\diamond \in \{\geq, >, =, \neq\}$ and $a_i \in \mathbb{Q}$ for $1 \leq i \leq n$, or one of the formulas $\bot, \top$. The formula $\bot$ is always false and $\top$ is always true. The constraints $\bot$ and $\top$ are called *trivial.* For brevity, in the sequel we will call such rational linear constraints over X simply *linear constraints.* We call a *system of linear constraints* any finite set of linear constraints.

In this paper we will describe several algorithms for solving finite sets of rational linear constraints. Let $\succ$ be a total order on X. Without loss of generality we assume $x_n \succ x_{n-1} \succ \ldots \succ x_1$. A constraint is called *normalised* if it is of the form $\bot$, $\top$, $x_k + q \diamond 0$ or $-x_k + q \diamond 0$, where $\diamond$ is as defined above, x_k is the maximal variable in the respective constraint, and q does not contain x_k. Evidently, every constraint can be effectively transformed into an equivalent normalised constraint. In the sequel, we assume that all constraints are normalised.

We define an *assignment* σ over the set of variables X as a mapping from X to $\mathbb{Q}$, i.e. $\sigma : X \to \mathbb{Q}$. Given an assignment σ, a variable $x \in X$ and a value $v \in \mathbb{Q}$, we call the *update of* σ *at* x *by* v, denoted by σ_x^v, the assignment obtained from σ by changing the value of x by v and leaving the values of the other variables unchanged.

For a linear polynomial q over X, denote by $q\sigma$ the value of q after replacing all variables $x \in X$ by the corresponding values $\sigma(x)$. An assignment σ is called a *solution of a linear constraint* $q \diamond 0$ if $q\sigma \diamond 0$ is true; it is a *solution of a system* of linear constraints if it is a solution of every constraint in the system. If σ is a solution of a linear constraint c (or a system S of such constraints), we also say that σ *satisfies* c (respectively, S), denoted by $\sigma \models c$ (respectively, $\sigma \models S$), otherwise we say that σ *violates* c (respectively, S). A system of linear constraints is said to be *satisfiable* if it has a solution.

For simplicity, we consider only algorithms for solving systems of linear constraints of the form $q \geq 0$, $\perp$ and $\top$ and discuss the general case later.

3 Fourier-Motzkin Elimination

In this section we briefly describe the Fourier-Motzkin elimination method. Consider a system S of linear constraints. The method either determines that S has no solution, or finds at least one. The method is based on an iterative algorithm changing S by eliminating a variable at each step. We assume that the variables are eliminated according to the order $\succ$, that is, x_n is eliminated first. At each step, if the maximal variable in the current system of linear constraints is x_k, we denote the current system by S_k, thus $S_n = S$. When the algorithm terminates, we obtain a system containing only trivial constraints, we denote this system by S_0.

Let $k > 0$. The system S_{k-1} is obtained from S_k by (i) adding new linear constraints as follows: for every pair of linear constraints $x_k + p \geq 0$ and $-x_k + q \geq 0$ in S_k we add to S_{k-1} a new constraint $p + q \geq 0$ and (ii) removing all linear constraints containing x_k.

One can show that the original system S is unsatisfiable if and only if S_0 contains $\perp$. If S_0 does not contain $\perp$, we can build a solution σ of S using the following observation. An assignment σ satisfies S_k if and only if σ satisfies S_{k-1} and

$$x_k\sigma \in [\max\{-p\sigma \mid (x_k + p \geq 0) \in S_k\}, \min\{q\sigma \mid (-x_k + q \geq 0) \in S_k\}]\,. \quad (1)$$

As usual, we assume that the minimum of the empty set is $+\infty$ and the maximum of it is $-\infty$. Condition (1) essentially says that the value of x_k lies in a certain interval determined by the values of variables $x_1, \ldots, x_{k-1}$. One can prove that this interval is non-empty whenever σ satisfies S_{k-1}. Thus, we can change any solution σ of S_{k-1} into a solution of S_k by updating σ at x_k by an arbitrary value in this interval. In this way we can build a solution to $S = S_n$ as follows. We start with an arbitrary assignment σ (which obviously satisfies S_0) and update it at $x_1, \ldots, x_n$ as described above. In fact, all solutions of the initial system can be derived this way.

Note that the Fourier-Motzkin algorithm applied to a set of linear constraints always terminates and generates only a finite number of linear constraints. However, the algorithm is in general exponential.[1] In general, the number of linear constraints in S_{k-1} is in the worst case quadratic in the number of constraints in S_k.

Unlike the Fourier-Motzkin method, our conflict resolution method does not eliminate variables. It uses the rule deriving $p + q \geq 0$ from $x_k + p \geq 0$ and $-x_k + q \geq 0$ but derives new constraints in a more restrictive way.

[1] Some papers claim it is double-exponential but we could not find any paper proving this. Schrijver [11] defines a sequence of systems of the size $O(n^3)$ on which the method generates $O(2^n)$ constraints. Some papers refer to [3] as giving an example of double-exponential behaviour but [3] only repeats the example from [11] verbatim.

4 Conflict Resolution

In this section we introduce our *conflict resolution method* for solving systems of linear rational constraints.

Let c be a linear constraint. If the maximal variable in c is x_k, then we say that k is the *level* of c. If c contains no variables, then we define the level of c to be 0. Note that, since all constraints are assumed to be normalised, a constraint written in the form $x_k + p \geq 0$ or $-x_k + q \geq 0$ is of the level k. The notion of level induces a partial order on linear constraints, which we will denote also by $\succ$, as follows. For two linear constraints c_1 and c_2, we have $c_1 \succ c_2$ if and only if the level of c_1 is strictly greater than the level of c_2.

We call a *state* a pair (S, σ), where S is a system of linear constraints and σ an assignment. Let $\mathbb{S} = (S, \sigma)$ be a state and k a positive integer. We say that $\mathbb{S}$ *contains a k-conflict* $(x_k + p \geq 0, -x_k + q \geq 0)$ if (i) both $x_k + p \geq 0$ and $-x_k + q \geq 0$ are linear constraints in S and (ii) $p\sigma + q\sigma < 0$. Instead of "k-conflict" we will sometimes simple say "conflict". Note that if σ is a solution of S, then $\mathbb{S}$ contains no conflicts.

We will now formulate our method. Given a system S of linear constraints, it starts with an initial state (S, σ), where σ is an arbitrary assignment and repeatedly transforms the current state either by either adding a new linear constraint to S or updating the assignment. We will formulate these rules below as transformation rules on states $\mathbb{S} \Rightarrow \mathbb{S}'$, meaning that $\mathbb{S}$ can be transformed into $\mathbb{S}'$. Let k be an integer such that $1 \leq k \leq n$.

The conflict resolution rule (CR) (at the level k) is the following rule:

$$(S, \sigma) \Rightarrow (S \cup \{p + q \geq 0\}, \sigma),$$

where (S, σ) contains a k-conflict $(x_k + p \geq 0, -x_k + q \geq 0)$.

The assignment refinement rule (AR) (at the level k) is the following rule:

$$(S, \sigma) \Rightarrow (S, \sigma^v_{x_k}),$$

where

1. σ satisfies all constraints in S of the levels $0, \ldots, k-1$.
2. σ violates at least one constraint in S of the level k.
3. $\sigma^v_{x_k}$ satisfies all constraints in S of the level k.

We will call any instance of an inference rule an *inference*. Thus, our algorithm will perform CR-inferences and AR-inferences.

Note that the conflict resolution rule derives a linear constraint violated by σ:

Lemma 1. *Let* (S, σ) *contain a k-conflict* $(x_k + p \geq 0, -x_k + q \geq 0)$. *Then* $\sigma \not\models p + q \geq 0$. ❑

Let us introduce a new notation. For any system S of linear constraints, a non-negative integer k and an assignment σ denote

$$L(S,\sigma,k) \stackrel{\text{def}}{=} \max\{-p\sigma \mid (x_k + p \geq 0) \in S\};$$
$$U(S,\sigma,k) \stackrel{\text{def}}{=} \min\{q\sigma \mid (-x_k + q \geq 0) \in S\};$$
$$I(S,\sigma,k) \stackrel{\text{def}}{=} [L(S,\sigma,k), U(S,\sigma,k)].$$

For every set S of linear constraints and a positive integer k, denote by $S_{=k}$ (respectively, $S_{<k}$) the subset of S consisting of all constraints of the level k (respectively, of all levels strictly less than k).

Lemma 2. *(i) Condition (2) of the assignment refinement rule implies $x_k\sigma \notin I(S,\sigma,k)$. (ii) Condition (3) of the assignment refinement rule is equivalent to $v \in I(S,\sigma,k)$. (iii) The interval $I(S,\sigma,k)$ is non-empty if and only if $\mathbb{S}$ contains no k-conflicts.*

Proof. (i) We assume that $x_k\sigma \in I(S,\sigma,k)$ and prove that σ satisfies $S_{=k}$. Take any constraint in $S_{=k}$. Without loss of generality assume that it has the form $x_k + p \geq 0$. Since $x_k\sigma \in I(S,\sigma,k)$, we have $x_k\sigma \geq L(S,\sigma,k)$, that is, $x_k\sigma \geq \max\{-p\sigma \mid (x_k + p \geq 0) \in S\}$. This implies $x_k\sigma \geq -p\sigma$, hence σ is a solution of $x_k + p \geq 0$.

(ii) In one direction, assume $v \in I(S,\sigma,k)$. Note that $x_k\sigma^v_{x_k} = v$, so $x_k\sigma^v_{x_k} \in I(S,\sigma,k)$ Using the same arguments as in (i) but with σ replaced by $\sigma^v_{x_k}$ we can prove $\sigma^v_{x_k} \models S_{=k}$. In the other direction, assume $\sigma^v_{x_k} \models S_{=k}$. We have to prove $v \in I(S,\sigma,k)$, that is, $v \geq L(S,\sigma,k)$ and $v \leq U(S,\sigma,k)$. We will only prove the former condition, the latter one is similar. The former condition means $v \geq \max\{-p\sigma \mid (x_k + p \geq 0) \in S\}$. To prove it, we have to show that for all constraints of the form $x_k + p \geq 0$ in S (and hence in $S_{=k}$) we have $v \geq -p\sigma$. Since p may only contain variables in $\{x_1, \ldots, x_{k-1}\}$ and σ agrees with $\sigma^v_{x_k}$ on all such variables, we have $-p\sigma = -p\sigma^v_{x_k}$, so $v \geq -p\sigma^v_{x_k}$. Using $x_k\sigma^v_{x_k} = v$, we obtain $x_k\sigma^v_{x_k} \geq -p\sigma^v_{x_k}$, hence $\sigma^v_{x_k}$ is a solution of $x_k + p \geq 0$, and we are done.

(iii) We will prove that $I(S,\sigma,k)$ is empty if and only if $\mathbb{S}$ contains a k-conflict. In one direction, assume $I(S,\sigma,k)$ is empty. Then $L(S,\sigma,k) > U(S,\sigma,k)$. Note that this implies that both $L(S,\sigma,k)$ and $U(S,\sigma,k)$ are finite. Since they are finite, $S_{=k}$ contains two constraints of the form $x_k + p \geq 0$ and $-x_k + q \geq 0$ such that $-p\sigma = L(S,\sigma,k)$ and $q\sigma = U(S,\sigma,k)$. This and $L(S,\sigma,k) > U(S,\sigma,k)$ implies $-p\sigma > q\sigma$, and so $0 > p\sigma + q\sigma$. Therefore, $(x_k + p \geq 0, -x_k + q \geq 0)$ is a k-conflict. The proof in other direction is similar. ❑

The following is a key lemma for our method.

Lemma 3. *Let (S,σ) be a state and $1 \leq k \leq n$. Let σ satisfy all constraints in S of the levels $0,\ldots,k-1$ and violate at least one constraint of the level k. If $I(S,\sigma,k)$ is empty, then the conflict resolution rule at the level k is applicable and the assignment refinement rule at this level is not applicable. If $I(S,\sigma,k)$ is non-empty, then the assignment refinement rule at the level k is applicable and the conflict resolution rule at this level is not applicable.*

Proof. Suppose $I(S, \sigma, k)$ is empty. By Lemma 2 (iii) $\mathbb{S}$ contains a k-conflict, so the conflict resolution rule is applicable at the level k. Since $I(S, \sigma, k)$ is empty, by Lemma 2 (ii) condition (3) of the assignment refinement rule is violated, so the assignment refinement rule at this level is not applicable.

Suppose that $I(S, \sigma, k)$ is non-empty. Then by Lemma 2 (iii) $\mathbb{S}$ contains no conflict, so the conflict resolution rule at the level k is not applicable. Take an arbitrary value $v \in I(S, \sigma, k)$. By Lemma 2 (ii) condition (3) of the assignment refinement holds. Conditions (1) and (2) of this rule hold by the assumptions of this lemma, so the assignment refinement rule is applicable. ❑

5 The Conflict Resolution Algorithm

The *Conflict Resolution Algorithm CRA* is given as Algorithm 1.

Let us note that the algorithm is well-defined, that is, the interval $I(S, \sigma, k)$ at line 11 is non-empty. Indeed, the algorithm reaches this line if (S, σ) contains no conflict at the level k (by line 6). Then $I(S, \sigma, k)$ is non-empty by Lemma 2 (iii).

Example 1. This example illustrates the algorithm. Let S_0 be the following set of constraints.

$$
\begin{array}{rrrrl}
x_4 - & 2x_3 & & + \ \ x_1 + 5 \geq 0 & (1) \\
-x_4 - & x_3 & - 3x_2 & - 3x_1 + 1 \geq 0 & (2) \\
-x_4 + & 2x_3 & + 2x_2 & + \ \ x_1 + 6 \geq 0 & (3) \\
 & -x_3 & + \ \ x_2 & - 2x_1 + 5 \geq 0 & (4) \\
 & x_3 & & + 3x_1 - 1 \geq 0 & (5)
\end{array}
$$

Algorithm 1. The Conflict Resolution Algorithm CRA

Input: A set S of linear constraints.
Output: A solution of S or "unsatisfiable".

1: **if** $\bot \in S$ **then return** "unsatisfiable"
2: $\sigma :=$ arbitrary assignment;
3: $k := 1$
4: **while** $k \leq n$ **do**
5:     **if** $\sigma \not\models S_{=k}$ **then**
6:         **while** (S, σ) contains a k-conflict $(x_k + p \geq 0, -x_k + q \geq 0)$ **do**
7:             $S := S \cup \{p + q \geq 0\}$; ▷ application of **CR**
8:             $k :=$ the level of $(p + q \geq 0)$;
9:             **if** $k = 0$ **then return** "unsatisfiable"
10:         **end while**
11:         $\sigma := \sigma^v_{x_k}$, where v is an arbitrary value in $I(S, \sigma, k)$ ▷ application of **AR**
12:     **end if**
13:     $k := k + 1$
14: **end while**
15: **return** σ

Assume that the initial assignment σ maps all variables to 0. The algorithm starts at the level 0. The sets $S_{=0}, S_{=1}, S_{=2}$ are empty, so the assignment σ trivially satisfies them. However, it violates constraint (5) and so violates $S_{=3}$. The interval $I(S, \sigma, 3)$ is $[1, 5]$. It is non-empty, so by Lemma 3 we can apply the assignment refinement rule at level 3 by updating σ at x_3 by any value in $[1, 5]$. Let us choose, for example, the value 4. Let σ_1 denote the newly obtained assignment $\{x_4 \mapsto 0, x_3 \mapsto 4, x_2 \mapsto 0, x_1 \mapsto 0\}$. Now we move to the next level 4. There is a 4-conflict between constraints (1) and (2) (line 6). We make a CR-inference between these two clauses deriving a new constraint

$$- x_3 - x_2 - \tfrac{2}{3}x_1 + 2 \geq 0 \tag{6}$$

added to the set S at line 7. According to line 8 of the algorithm we set the level k to the level of the new constraint, that is, to 3. Now there are no more conflicts on level 3 and we have $I(S, \sigma, 3) = [1, 2]$. We should update the assignment at x_3 by an arbitrary value in this interval. Suppose, for example, that we have chosen 1 as the value for x_3 obtaining $\{x_4 \mapsto 0, x_3 \mapsto 1, x_2 \mapsto 0, x_1 \mapsto 0\}$ and increase k by 1 proceeding to level 4. At this moment all constraints at level 4 are satisfied and the algorithm terminates returning σ. ❏

Our aim is to prove that the algorithm is correct and terminating.

Theorem 1. *The conflict resolution algorithm CRA always terminates. Given an input set of constraints S_0, if CRA outputs "unsatisfiable", then S_0 is unsatisfiable. If CRA outputs an assignment σ, then σ is a solution of S_0.*

This theorem will be proved after a series of lemmas establishing properties of the algorithm. In these lemmas we always denote the input set of constraints by S_0.

Lemma 4. *At any step of the algorithm the set S is equivalent to S_0, that is, S and S_0 have the same set of solutions.*

Proof. Observe that line 7 is the only line that changes S. It is easy to see that the application of this line does not change the set of solutions of S since the constraint $p + q \geq 0$ added to S is implied by S. ❏

The following lemma is obvious.

Lemma 5. *Every constraint occurring in S at any step of the CRA algorithm belongs to the set of constraints derived by the Fourier-Motzkin algorithm applied to S_0.* ❏

Lemma 6. *The assignment σ at lines 4 and 6 satisfies $S_{<k}$.*

Proof. By induction on the number of iterations of the outermost while-loop. Before the first iteration the property is obvious since $k = 1$ and $\perp \notin S$. So we assume that the property holds before an iteration of the loop and show it holds after this iteration. If $\sigma \models S_{=k}$ at line 5, then by $\sigma \models S_{<k}$ we have $\sigma \models S_{<k+1}$. It

remains to consider the case when $\sigma \not\models S_{=k}$ at line 5. In this case the algorithm may enter the internal while-loop starting at line 6. It is easy to see that at the exit of this loop the property is satisfied as well, since k only decreases in the loop and the new constraint $p + q \geq 0$ is at the level k. So it remains to show that after line 11 we have $\sigma \models S_{=k}$. But this is guaranteed by Lemma 2 (ii), so we are done. ❑

Lemma 7. *Let* (S, σ) *contain a conflict* $(x_k + p \geq 0, -x_k + q \geq 0)$ *at line 6. Then we have* $(p + q \geq 0) \notin S$.

Proof. By Lemma 6 at line 6 we have $\sigma \models S_{<k}$. But we have $\sigma \not\models (p + q \geq 0)$, hence $(p + q \geq 0) \notin S_{<k}$. Since the level of $(p + q \geq 0)$ is strictly less than k this implies $(p + q \geq 0) \notin S$. ❑

This lemma means that the same constraint will never be added again to S. In fact, the algorithm has a much stronger property formulated below in Lemma 8.

Let us now give the proof of Theorem 1.

Proof. We start with proving termination. By Lemma 7 the algorithm never adds the same constraint twice. By Lemma 5 we can add only a finite number of different constraints. Therefore, condition on line 6 can hold only a finite number of times. From the moment this condition becomes permanently false, k will always increase by 1, so the outermost while-loop will terminate.

Suppose now that the algorithm returns "unsatisfiable". If this happens at line 1, then $\bot \in S_0$, so S_0 is unsatisfiable. Otherwise, this happens at line 9. Then $\sigma \not\models p + q \geq 0$ by Lemma 1. Since $k = 0$, then the constraint $p + q \geq 0$ contains no variables, so this constraint is trivial and unsatisfiable. By Lemma 4, this constraint is implied by S_0, hence S_0 is unsatisfiable too.

It remains to consider the case when the algorithm return an assignment σ. This only can happen at the last line of the algorithm. At this line, $k = n + 1$. By Lemma 6, σ satisfies $S_{<n+1}$. Note that $S_{<n+1} = S$, so σ also satisfies S. By Lemma 4, S is equivalent to S_0, hence σ also satisfies S_0. ❑

6 Conflict Resolution and the Fourier-Motzkin Method

We say, that a CR-inference at a level k is *redundant* w.r.t. a state (S, σ) if the conclusion of this inference is a consequence of constraints in $S_{<k}$. Let us prove a key property that distinguishes our algorithm from the Fourier-Motzkin method.

Lemma 8. *Every CR-inference performed by the CRA algorithm is non-redundant.*

Proof. Suppose that the algorithm performs a redundant inference adding $p+q \geq 0$ at line 7. Then by the definition of redundancy $p+q \geq 0$ is implied by $S_{<k}$. By Lemma 6 we have $\sigma \models S_{<k}$, then σ must also satisfy $p + q \geq 0$. This contradicts to the definition of a conflict. ❑

To illustrate this lemma, let us come back to Example 1. Note that in this example we have not applied the conflict resolution inference between constraints (1) and (3). It is easy to see that the conclusion of this inference is implied by constraints (4) and (5) at smaller levels, therefore *this inference would not be applied independently of the choices of assignments made by the algorithm.*

Let us now show that the Fourier-Motzkin algorithm cannot polynomially simulate our algorithm in a very strong sense. This example is taken from [11]. It contains all inequalities of the form $\pm x_k \pm x_l \pm x_m \geq 0$, where $n \geq k > l > m \geq 1$. Evidently, the size of the system is $O(n^3)$ and there exists only a single solution assigning 0 to all variables. It is shown in [11] that the Fourier-Motzkin method generates exponentially many (in n) inequalities for this example. Let σ be an arbitrary assignment. Our method will start generating conflicts from level 3 containing 8 inequalities until it updates σ so that $x_1\sigma = x_2\sigma = x_3\sigma = 0$. After that it will proceed to level 4, where the interval $I(S, \sigma, 4)$ will consist of a single point 0. The assignment refinement will set $x_4\sigma$ to 0 and no conflicts will be generated. The same will happen with all levels greater than 4, so the algorithm will terminate in a linear number of steps. Essentially, apart from the initial work on level 3, the conflict resolution algorithm will only evaluate every inequality once and so work in time linear in the size of the system, that is $O(n^3)$. Note that this running time does not depend on either the choice of the initial assignment or the choice of values in the assignment refinement inferences.

7 Extensions

In this section we briefly mention two extensions of the method: one is for working with strict inequalities and another one for linear programming.

The modification of the algorithm for working with *strict inequalities* $p > 0$ is straightforward. First, when we consider the interval

$$I(S, \sigma, k) = [L(S, \sigma, k), U(S, \sigma, k)]$$

if any endpoint of this interval corresponds to a strict inequality, we use a semi-open or an open interval instead. For example, if there is a strict inequality $(x_k + p > 0) \in S$ such that $-p\sigma = L(S, \sigma, k)$ but no strict inequality $(-x_k + q > 0) \in S$ such that $q\sigma = U(S, \sigma, k)$, then we use the semi-open interval

$$I(S, \sigma, k) = (L(S, \sigma, k), U(S, \sigma, k)].$$

Second, the result of the conflict resolution rule is a strict inequality if at least one of the premises is strict. It is not hard to generalise our method to deal with disequalities $p \neq 0$ as well.

To use our algorithm for linear programming, we can use the following trick. Suppose, for example, that we want to find a maximum of a linear function p. To this end we assume that the constraint do not contain the variable x_1 and add the equality $p - x_1 = 0$. After that we use our algorithm with the only modification that we always select the maximal possible value for x_1 in the assignment refinement rule. A special care should be taken when we have no a priory upper bound on x_1. However, using the method for linear programming is beyond the scope of this paper.

8 Implementation

In this section we briefly describe details of our implementation of the conflict resolution algorithm. Our implementation works with linear constraints of the form $q \diamond 0$, for $\diamond \in \{\geq, >, =\}$. In order to compare conflict resolution with other methods for solving systems of linear constraints we also implemented the standard Fourier-Motzkin algorithm and the Chernikov algorithm [4] using the same data structures as used in the implementation of CRA.

Informally, the Chernikov algorithm extends the Fourier-Motzkin algorithm, (see Section 3) with the following restriction on added linear constraints. Let $S = S_n$ be the set of linear constraints. With each linear constraint we associate a set of initial constraints used in the derivation of this constraint, called the *index set*. Define the index set of an initial constraint $c \in S_n$ to be $\{c\}$. Let $k > 0$. The system S_{k-1} is obtained from S_k by removing all linear constraints containing x_k and adding new linear constraints as follows. For every pair of linear constraints $x_k + p \geq 0$ and $-x_k + q \geq 0$ in S_k, with index sets I, J respectively, we add to S_{k-1} a new constraint $p + q \geq 0$ with the index set $I \cup J$, if the following conditions (i-ii) hold. Let l be the level of $p + q \geq 0$ (see Section 4), then (i) the cardinality of $I \cup J$ is less than or equal to $n - l + 1$, and (ii) there is no constraint c in S_{k-1} of the level l with the index set $U \subseteq (I \cup J)$. It is shown in [4] that the original system S is unsatisfiable if and only if S_0 contains $\bot$.

Our implementation of conflict resolution follows Algorithm 1. There are a number of key parameters that can be used to fine-tune the CRA algorithm, namely

1. strategies for selecting conflicts,
2. strategies for selecting values in the assignment refinement rule,
3. order on variables.

Let us briefly describe possible choices for these parameters in our current implementation.

The strategy for selecting conflicts in the current implementation is based on *maximal overlaps*, as described below. At the line 7 of Algorithm 1 we select a k-conflict $x_k + p \geq 0$ and $-x_k + q \geq 0$ in S (i.e. $p\sigma + q\sigma < 0$), such that $-p\sigma = L(S, \sigma, k)$ and $q\sigma = U(S, \sigma, k)$. To explain the rationale behind this strategy let us extend our notion of redundancy to constraints. We call a constraint c at a level k *redundant* if this constraint is implied by $S_{<k+1} - \{c\}$. One can modify our algorithm and show that any redundant constraint can be removed[2].

It is not hard to prove that constraints $x_k + p \geq 0$ such that $-p\sigma = L(S, \sigma, k)$ are "almost" non-redundant in the following sense.

[2] We cannot remove redundant constraints simultaneously since removal of a redundant constraint can make another previously redundant constraint non-redundant.

Lemma 9. *Consider the set* S^+ *of all constraints at a level* k *having the form* $x_k + p \geq 0$*. Consider its subset* S' *consisting of all constraints* $x_k + p \geq 0$ *such that* $-p\sigma = L(S, \sigma, k)$*. Then* S' *is not implied by* $S_{<k} \cup (S^+ - S')$*.* ❏

One can formulate a symmetric property for constraints $-x_k + q \geq 0$ such that $q\sigma = U(S, \sigma, k)$.

Although our algorithm does not perform redundant inferences, the system may contain redundant constraints at a level k for two reasons: (i) it may contain redundant constraints initially; and (ii) addition of new constraints to a level k may make other constraints at this and higher levels redundant. Choosing a k-conflict $x_k + p \geq 0$ and $-x_k + q \geq 0$ in S (i.e. $p\sigma + q\sigma < 0$), such that $-p\sigma = L(S, \sigma, k)$ and $q\sigma = U(S, \sigma, k)$ does not, in general, guarantee, that the constraints forming the conflict are non-redundant but it guarantees that they are "almost" non-redundant in the sense of Lemma 9.

We tried several strategies for selecting values in the assignment refinement rule. One of the strategies is just selection of the middle point in the interval $I(S, \sigma, k)$ (line 11). Our experiments show that using this strategy frequently results in a rapid growth of the sizes of numerators and denominators of rational values in the assignment. In order to avoid this problem we used the following strategy for selecting the update values from the interval $I(S, \sigma, k)$. First, if the endpoints of $I(S, \sigma, k)$ coincide, then we select one of them. Otherwise, we select a rational number n/m in $I(S, \sigma, k)$ such that (i) m is the least power of 2 among denominators of all rationals in $I(S, \sigma, k)$, and (ii) n is such that, n/m is the closest rational to the middle point of the interval, among all rationals satisfying (i). It is possible to show that a rational satisfying both (i) and (ii) always exists. In particular, if $I(S, \sigma, k)$ contains integer points, then our strategy will select an integer in $I(S, \sigma, k)$ closest to the middle point. As our experiments show, such choice of values considerably simplifies the assignment values and constraint evaluation.

The last parameter of the CRA algorithm we consider here is the order on variables. In the current implementation the order on variables is selected randomly before the run of the CRA algorithm. We believe that a careful selection of the order on variables based on the properties of the input problem can considerably improve the performance of our implementation and we will make experiments with the order selection in the nearest future.

The CRA algorithm is implemented in C++ using the GMP library for arbitrary precision arithmetic[3]. Thus, all computations with rational numbers are done with arbitrary precision. We implemented two different representations of constraints, one using arrays of the size n to store vectors and one storing only non-zero coefficients. Not surprisingly, on randomly generated problems the first implementation is slightly better in both time and space while the second one can be much faster (and consume much less space) on non-random problems, where vectors are normally sparse.

Finally, let us note that in the context of satisfiability modulo theories (SMT), it is desirable for solvers to be incremental and be able to generate explanations

[3] `http://gmplib.org/`

for the unsatisfiability. The CRA algorithm and our implementation can easily be made incremental: after adding/removing constraints we can always continue with the current assignment, moreover the CRA never performs redundant inferences and in particular, never performs the same inference twice (unless the conclusion was removed). Explanations can be generated from the proofs of unsatisfiability which are easily extractable from runs of the CRA algorithm.

9 Experimental Results

In this section we experimentally evaluate our implementation of the conflict resolution algorithm, the Fourier-Motzkin algorithm and the Chernikov algorithm. We implemented the algorithm in C++ using the GNU Multiple Precision Arithmetic Library (GMP) for handling arbitrary-precision rationals.

We compare our implementation with CVC3 [1] and Barcelogic [10], which are well-developed solvers for satisfiability modulo theories (SMT). CVC3 incorporates a variant of the Fourier-Motzkin algorithm and Barcelogic incorporates the simplex algorithm for reasoning with linear arithmetic. Let us note that our implementation is at a very early stage, no preprocessing was used and crucial for efficiency heuristics such as selection of suitable variable order are yet to be implemented. Already for this implementation, our experimental results are very encouraging, showing that the conflict resolution algorithm is considerably more efficient in solving linear constraints than the standard Fourier-Motzkin algorithm. For example, an order of magnitude difference occurs already on small problems.

We evaluated the solvers on two sets of benchmarks[4]. The first set of benchmarks consists of randomly generated systems of linear constraints. The second set of benchmarks consists of systems of linear constraints extracted from real-life SMT problems [2], using our tool called Hard Reality (HRT) [9]. All experiments were run on a Linux laptop with CPU 2.8GHz and memory 4Gb.

Table 1. Randomly Generated Problems

4000 problems vars 3-12 (unsat/sat)				
	CRA	CVC3	FM	Ch
timeout (20s)	0/0	11/9	790/329	149/10
av. time (s)	0/0	0/0	0.4/0.1	0.6/0.1

400 problems vars 13-22 (unsat/sat)				
	CRA	CVC3	FM	Ch
timeout (20s)	5/2	21/33	183/144	155/65
av. time (s)	0.2/0.3	0/0	0.1/0.5	1.9/0.6

Results for randomly generated problems are shown in Table 1. The conflict resolution algorithm can solve all 4000 randomly generated problems with the number of variables ranging from 3 to 12 (within the total time of 7 seconds) and on the problems with the number of variables ranging from 13 to 22 fails only on 7. The CVC3 implementation of the Fourier-Motzkin algorithm fails to solve 20 problems and 54 problems respectively. Our implementation of the

[4] http://www.cs.man.ac.uk/~korovink/cra_bench

Table 2. Hard Reality Problems

304 problems (unsat)				
	CRA	CVC3	FM	Ch
timeout (60s)	1	4	44	42
av. time	0.2	0.13	0.1	0.12

$$\begin{array}{rrrrrl}
2x_5 & - 3x_4 & + x_3 & - 3x_2 & - 2x_1 & + 3 \geq 0 \\
2x_5 & + x_4 & - 2x_3 & & - 2x_1 & + 2 \geq 0 \\
-x_5 & & & + 3x_2 & + x_1 & + 2 \geq 0 \\
-3x_5 & & + 2x_3 & & - 3x_1 & - 2 \geq 0 \\
x_5 & - 2x_4 & & - 2x_2 & + 3x_1 & - 2 \geq 0 \\
-2x_5 & + 2x_4 & - 3x_3 & - x_2 & + 2x_1 & + 3 > 0 \\
3x_5 & - 2x_4 & + 2x_3 & + 3x_2 & + 2x_1 & + 1 > 0 \\
x_5 & & & & + 2x_1 & + 2 > 0 \\
& 2x_4 & - x_3 & - 3x_2 & - x_1 & + 3 = 0
\end{array}$$

Fig. 1. A randomly generated problem

Fourier-Motzkin algorithm solves considerably fewer problems than CRA. The Chernikov algorithm improves over the Fourier-Motzkin but solves considerably fewer problems than CRA.

Table 2 compares the solvers on the problems extracted from SMT benchmarks using the Hard Reality Tool. These problems have different structure than the randomly generated problems, in particular the number of variables and constraints are considerably higher, most of problems contain several hundred of different variables and constraints.

The CRA also solves more problems in the Hard Reality benchmarks than any of CVC3, Fourier-Motzkin, and Chernikov algorithms. The average time of the CRA is a bit higher than of CVC3 due to extra solved problems. Indeed, in a pairwise comparison on all solved problems in these benchmarks the CRA is faster than CVC3.

One of the most striking examples showing the difference in performance is shown in Figure 1. The problem on this figure which was randomly generated and contains 5 variables and 10 linear constraints.

The standard Fourier-Motzkin algorithm run on this problem generated over 280 million linear constraints, while the conflict resolution algorithm generated only 21 constraints. However, this example is not outstanding as compared to our other experiments described above.

Compared to the simplex algorithm, the conflict resolution shows promising potential. In Table 3 the CRA is compared to Barcelogic. Already our non-optimized implementation is faster than Barcelogic on a number of problems, although Barcelogic can solve more problems than CRA within 20 seconds.

To summarise, our experiments show that a naive implementation of the conflict resolution algorithm outperforms the Fourier-Motzkin and Chernikov algorithms in solving systems of linear constraints and has promising potential

Table 3. CRA vs Barcelogic

400 problems vars 13-22 (unsat/sat)				
	faster	same	av. time	timeout (20s)
Barcelogic	28/29	146/167	0.04/0	0/0
CRA	23/7	146/167	0.2/0.3	5/2
400 problems vars 23-32 (unsat/sat)				
Barcelogic	110/67	31/88	0.25/1.0	0/0
CRA	63/41	31/88	0.7/1.6	60/37

compared to the simplex algorithm. For the future work we are planning to extend the CRA algorithm with various heuristics for choosing conflicts, order on variables, values in the assignment update rule and methods for avoiding unnecessary re-evaluation of constraints.

10 Related Work

In this section we compare various modifications of the Fourier-Motzkin method with the conflict resolution method.

Most of modifications of the Fourier-Motzkin method aim at identifying potentially redundant constraints by providing some easy-to-check sufficient conditions for redundancy. One of the most prominent methods for restricting generation of redundant constraints was suggested by Chernikov [4]. His idea is to associate with each constraint some bookkeeping information on how this constraint was derived. Under certain conditions a newly derived constraint can be shown to be redundant based on this information (see Section 8). There are a number of extensions and modifications of this and other ideas developed over the past decades (e.g., [5,8,6,7]). Our notion of redundancy seems to be orthogonal to that of Chernikov and the others, in particular it is based on the ordering on constraints and semantic entailment from the smaller constraints. One of important properties of the conflict resolution algorithm is that it never performs redundant inferences as defined in this paper. As a future work, we will investigate whether it is possible to combine our notion of redundancy with restrictions used by other methods.

11 Conclusions

We presented a new algorithm for solving systems of linear constraints, called conflict resolution. The method successively refines an initial assignment with the help of newly derived constraints until either the assignment becomes a solution of the system or the inconsistency of the initial system is proved. We have shown that this method is correct and terminating. The conflict resolution method has a number of attractive properties such as blocking of redundant inferences. We implemented our method and experimental results show that on the majority of problems we tried conflict resolution considerably outperforms well-developed methods such Fourier-Motzkin and Chernikov algorithms. We are currently working on improving our implementation and integration of crucial for efficiency heuristics such as various strategies for conflict selection, assignment refinement and variable order.

References

1. Barrett, C., Cesare Tinelli, C.: CVC3. In: Damm, W., Hermanns, H. (eds.) CAV 2007. LNCS, vol. 4590, pp. 298–302. Springer, Heidelberg (2007)
2. Barrett, C., Ranise, S., Stump, A., Tinelli, C.: The Satisfiability Modulo Theories Library, SMT-LIB (2008), `http://www.SMT-LIB.org`
3. Chandru, V.: Variable elimination in linear constraints. Comput. J. 36(5), 463–472 (1993)
4. Chernikov, S.N.: Linejnye Neravenstva. Nauka, Moscow (1968) (in Russian)
5. Duffin, R.J.: On Fourier's analyse of linear inequality systems. Mathematical Programming Study 1, 71–95 (1974)
6. Imbert, J., Van Hentenryck, P.: A note on redundant linear constraints. Technical Report CS-92-11, CS Department, Brown University (1992)
7. Jaffar, J., Maher, M.J., Roland, P.S., Yap, R.H.C.: Projecting CLP($\mathcal{R}$) constraints. New Generation Computing 11 (1993)
8. Kohler, D.A.: Projection of Convex Polyhedral Sets. PhD thesis, University of California, Barkaley (1967)
9. Korovin, K., Voronkov, A.: Hard Reality Tool (submitted, 2009), `http://www.cs.man.ac.uk/~korovink/hr`
10. Nieuwenhuis, R., Oliveras, A.: Decision Procedures for SAT, SAT Modulo Theories and Beyond. The BarcelogicTools (Invited Paper). In: Sutcliffe, G., Voronkov, A. (eds.) LPAR 2005. LNCS (LNAI), vol. 3835, pp. 23–46. Springer, Heidelberg (2005)
11. Schrijver, A.: Theory of Linear and Integer Programming. John Wiley and Sons, Chichester (1998)

Propagator Groups

Mikael Z. Lagerkvist and Christian Schulte

KTH – Royal Institute of Technology, Sweden
{zayenz,cschulte}@kth.se

Abstract. This paper introduces propagator groups as an abstraction for controlling the execution of propagators as implementations of constraints. Propagator groups enable users of a constraint programming system to program how propagators within a group are executed.

The paper exemplifies propagator groups for controlling both *propagation order* and *propagator interaction.* Controlling propagation order is applied to debugging constraint propagation and optimal constraint propagation for Berge-acyclic propagator graphs. Controlling propagator interaction by encapsulating failure and entailment is applied to general reification and constructive disjunction. The paper describes an implementation of propagator groups (based on Gecode) that is applicable to any propagator-centered constraint programming system. Experiments show that groups incur little to no overhead and that the applications of groups are practically usable and efficient.

1 Introduction

Over the last two decades, an array of techniques to control the execution of groups of propagators have been developed:

- S-boxes [10] support the debugging of constraint propagation by grouping several propagators into a conceptually bigger propagator that typically captures some problem structure. Then, the drastically fewer executions of the bigger propagator can be understood by a modeler during debugging. While it is generally acknowledged that the lack of support for debugging is one of the key obstacles to learning constraint programming, none of today's mainstream systems support it.
- The optimal propagation of Berge-acyclic propagator graphs requires to order the execution of a group of propagators and is generally acknowledged to be significant for efficient propagation [16, 3]. However, no mainstream constraint programming system implements it.
- General reification (reifying arbitrary constraints and groups of constraints) and constructive disjunction are well known concepts that can improve propagation considerably [13]. Both require to control the execution and interaction of groups of propagators. However, both concepts are only supported to some extent by Propia [15] and general reification by Mozart [18, 19].

I.P. Gent (Ed.): CP 2009, LNCS 5732, pp. 524–538, 2009.

In summary, these powerful and even essential techniques are known but either not widely available or not even implemented. All these techniques have in common that they are based on controlling the execution of groups of propagators. This paper introduces *propagator groups* as a new abstraction that supports the user-level implementation of the above mentioned techniques. A group collects a set of propagators and defers their scheduling and execution to a user-supplied routine. This makes scheduling and execution programmable, while the implementation of propagator groups requires only very small and local changes to the underlying constraint programming system.

Similar to groups, spaces in Mozart [19] also encapsulate failure and entailment. However, the lack of control over propagation order and access to local variables makes them unable to implement propagator ordering and constructive disjunction. The Propia system [15] supports constructive disjunction and to some extent generalized reification through a nested search procedure. However, it can not express propagator ordering.

Contributions. This paper contributes *propagator groups* as a new abstraction for constraint programming systems that allows many interesting techniques to be implemented by a user without modifying the underlying system. The paper contributes a simple yet expressive architecture for groups and techniques for their efficient implementation. Moreover, the paper shows how propagator groups can be applied to debugging, optimal propagator ordering, general reification, and constructive disjunction. These applications are shown to be efficient with propagator groups. In particular, the paper contributes the first implementation of optimal propagation for Berge-acyclic propagator graphs.

Plan of the paper. In the following section the necessary background on constraint programming is given. In Sect. 3 a model and implementation of propagator groups is presented. Sections 4 and 5 apply groups to debugging, optimal propagation ordering, general reification, and constructive disjunction. The final section concludes.

2 Constraint Programming

Constraint programming is based on two important concepts, variables (together with their associated domains) and constraints.

Variables and domains. There is a finite set of variables *Var* and a finite set of values *Val*. A *domain* $d \in \mathit{Dom}$ is a set of values a variable can take, $\mathit{Dom} = \mathcal{P}(\mathit{Val})$. A *store* $s \in \mathit{Store}$ is a complete mapping from variables to domains, $\mathit{Store} = \mathit{Var} \rightarrow \mathit{Dom}$. An *assignment* is a store where the range of the function is restricted to singleton sets $(\{\{v\} \mid v \in \mathit{Val}\})$.

Set relations $\sim$ are lifted to a pair of stores S_1 and S_2 in the natural, point-wise way $(S_1 \sim S_2 = \forall x \in \mathit{Var}.\, S_1(x) \sim S_2(x))$. A store S_1 is *stronger* than a store S_2, written $S_1 \leq S_2$, if $S_1 \subseteq S_2$. A store S_1 is *strictly stronger* than a store

S_2, written $S_1 < S_2$, if $S_1 \leq S_2$ and $S_1 \neq S_2$. The *disagreement set* $\text{dis}(S_1, S_2)$ of stores S_1 and S_2 is defined as $\{x \in \mathit{Var} \mid S_1(x) \neq S_2(x)\}$.

A tuple of values over variables $x_1, \ldots, x_n$ can be turned into a store in the following way: for a variable $x \in \mathit{Var}$, $\text{store}(\langle v_1, \ldots, v_n\rangle)(x)$ is defined as $\{v_l\}$ if $x = x_l$ for some $l \in \{1, \ldots, n\}$ and as Val otherwise.

Constraints. A *constraint* $c \in \mathit{Con}$ over the set of variables $\text{var}(c) = \{x_1, \ldots, x_n\}$ is defined by the set of assignments that are solutions to the constraint, $\mathit{Con} = \mathcal{P}(\{\langle v_1, \ldots, v_n\rangle \mid v_i \in \mathit{Val}\})$. A store can be turned into a constraint over variables $x = \{x_1, \ldots, x_n\}$ using $\text{cons}(S, x) = \{\langle v_1, \ldots, v_n\rangle \mid \forall i. v_i \in S(x_i)\}$.

A *constraint satisfaction problem* (CSP) is a pair of a set of constraints C and a store S, $\langle C, S\rangle \in \mathcal{P}(\mathit{Con}) \times \mathit{Store}$. A tuple $\langle v_1, \ldots, v_n\rangle$ in a constraint over variables $x_1, \ldots, x_n$ is *valid* under the store S iff $\forall i. v_i \in S(x_i)$. An assignment a is a *solution* to the CSP $\langle C, S\rangle$, both over variables $\{x_1, \ldots, x_n\}$, iff the assignment is a solution for each constraint, $\forall c \in C. \langle a(x_1), \ldots, a(x_n)\rangle \in c$, and the assignment is valid for the store, $a \subseteq S$. The solutions to a CSP, $\text{sol}(C, S)$, is the set of all assignments that are solutions.

A constraint c is *entailed* by the store S iff $\text{cons}(S, \text{var}(c)) \subseteq c$. Entailed constraints can safely be removed from the CSP since they no longer restrict the set of solutions.

Propagators. To solve a CSP, a constraint programming system uses propagators as implementations of constraints.

A *propagator* is a function p that takes a store S as input and returns a tuple $\langle \mathit{stat}, S'\rangle$, where stat is a status message and S' is a new store. The status message will be ignored when only the resulting store is interesting. A propagator p must be *contracting* ($p(S) \leq S$ for all stores S) and *monotonic* (if $S_1 \leq S_2$ then $p(S_1) \leq p(S_2)$ for all stores S_1, S_2). A store S is a *fix-point* of a propagator p iff $p(S) = S$. The status-message indicates if the propagator is entailed (*entailed*) or if it has detected failure ($\bot$). A propagator is *entailed* for a store S iff for all stores $S' \leq S$ it holds that $p(S') = S'$. Entailed propagators can safely be removed from the pool of propagators, since they will do no more pruning. A propagator that reports failure indicates that there are no solutions left, and propagation can be aborted.

A propagator p references variables $\text{var}(p) = \{x_1, \ldots, x_n\}$ and is said to *implement* its induced constraint c_p. The induced constraint is defined as the set of assignments that the propagator identifies as solutions:

$$c_p = \{\ \langle v_1, \ldots, v_n\rangle \mid p(\text{store}(\langle v_1, \ldots, v_n\rangle)) = \text{store}(\langle v_1, \ldots, v_n\rangle)\ \}$$

For a given constraint c, any propagator p such that $c_p = c$ can be used. Note that many different propagators exist for the very same constraint, typically differing in propagation strength and efficiency.

Constraint programs. Analogously to constraint satisfaction problems, a store and a set of propagators can be combined to form a *constraint program* (CP) $\langle P, S\rangle$. The set of solutions $\text{sol}(P, S)$ to a CP is defined as $\text{sol}(\{c_p \mid p \in P\}, S)$.

```
Propagate(S)
begin
    Q ← P;
    while Q ≠ ∅ do
        choose and remove p from Q;
        ⟨stat, S'⟩ ← p(S);
        if stat = ⊥ then
            return failure;
        if stat = entailed then
            remove p from P;
        foreach p ∈ ⋃_{x∈dis(S,S')} prop(x) ∩ P do
            Schedule(p);
        S ← S';
    return S;
end

Schedule(p)
begin
    Q ← Q ∪ {p};
end
```

Algorithm 1. Propagator-centered propagation

Variable dependencies. To manage propagation efficiently, a constraint programming system needs to know which propagators may affect which variables, and for which variables a domain change might make a propagator not be at a fix-point. For the purposes of this paper, the set of referenced variables for the propagator can be used as the dependencies.

Dependencies are used in propagation as follows: if a store S is a fix-point of a propagator p, then any store $S' \leq S$ with $\text{var}(p) \cap \text{dis}(S, S') = \emptyset$ is also a fix-point of p. To better characterize how propagators and variables are organized in an implementation, the set of propagators $\text{prop}(x)$ depending on a variable x is defined as $p \in \text{prop}(x)$ if and only if $x \in \text{var}(p)$.

Propagation. Constraint propagation refers to the process of finding the largest mutual fix-point (equivalently, the weakest mutual fix-point with respect to the strength of stores) of the set of propagators from an initial store S that propagation starts from. Since propagators are defined to be monotonic and contracting functions, it is guaranteed that there exists a unique largest mutual fix-point. The cornerstone of a propagation algorithm is to maintain some representation of which propagators might not be at fix-point. Propagator-centered propagation is controlled by the set of propagators still to be propagated (as opposed to variable-centered that maintains a set of modified variables).

Propagator-centered propagation is shown in Algorithm 1. It is assumed that all propagators are contained in the global propagator set P. The global queue Q contains propagators not known to be at fix-point. The choose operation to get the next propagator from Q is left unspecified, but a realistic implementation bases the decision on priority or cost, see for example [21].

Algorithm 1 does not spell out some details. A real system will most probably use events for variable modifications instead of a simple list of dependencies. Furthermore, whether a propagator is at a fix-point or not after it has been propagated should be taken into account to avoid needless re-execution. For a complete discussion of constraint propagation algorithms, see [2], and see [20] for the implementation of these algorithms in constraint programming systems.

3 Groups

A group is an execution manager for a set of propagators. It controls the order of propagation as well as the handling of failure and entailment. The only change to the system is that propagators that belong to a group must be scheduled in their group rather than globally. Running a group is done through a propagator.

Section 3.1 presents a model of groups. Section 3.2 details the implementation of the model. Section 3.3 evaluates the overhead of groups.

3.1 Model

A *group* is an abstraction that supports the operations *schedule* and *propagate*. The first is used for scheduling a propagator that belongs to the group. The second operation is used to run the propagators scheduled in the group.

A relation group (p) is defined for all propagators p. This relation maps a propagator to the group it should be scheduled in. If the propagator is not a member of any group, the relation is empty. To keep the number of concepts small, only propagators can be scheduled and executed by the main propagation loop. Each group g has an associated propagator that is called its *controller propagator*. The controller propagator is given by controller (g). The set of propagators in P that should be scheduled in a group g is given by $\mathrm{prop}(g) = \{p \in P \mid \mathrm{group}(p) = g\}$ (the inverse of the group $(\cdot)$ relation).

A group together with its controller propagator needs to maintain the requirements of a propagator: it should be contracting and monotonic. This must be ensured by all group implementations.

Propagate(S)
    ▷ As in Algorithm 1

Schedule(p)
begin
    if $\mathrm{group}(p) \neq \emptyset$ **then**
        $\mathrm{group}(p).\mathrm{schedule}(p)$;
        $Q \leftarrow Q \cup \mathrm{controller}(\mathrm{group}(p))$;
    else
        $Q \leftarrow Q \cup \{p\}$;
end

Algorithm 2. Propagator-centered propagation with groups

Algorithm 2 presents the propagation algorithm that supports groups. The only difference from Algorithm 1 is that when scheduling a propagator p, a check is made to see if the propagator belongs to a group. If so, p is scheduled in that group $g = \text{group}(p)$ and controller (g) is added to the global queue.

```
Basic group g
begin
    q : Queue;
    schedule(p) begin
        q.push(p);
    end
    propagate(S) : ⟨Status, Store⟩ begin
        while ¬q.empty do
            p ← q.pop;
            ⟨stat, S'⟩ ← p(S);
            foreach p ∈ ⋃_{x∈dis(S,S')} prop(x) ∩ P do
                Schedule(p);
            if stat = ⊥ then
                return ⟨⊥, S'⟩;
            if stat = entailed then
                remove p from P;
                if prop(g) = ∅ then
                    return ⟨entailed, S'⟩;
            S ← S';
        return ⟨∅, S⟩;
    end
end
```

Algorithm 3. Basic group g with a single flat queue of propagators

Algorithm 3 shows a basic group that maintains a queue of propagators to be executed. This group implements no special behavior except grouping a set of propagators together. Failure of any of the propagators is reported directly. The group is only entailed if all its propagators are entailed (checked by $\text{prop}(g) = \emptyset$). The basic controller propagator runs the group g by executing $g.\mathbf{propagate}(\cdot)$ and reports entailment and failure accordingly. As will be seen in Sect. 5, this basic group can be used as a building block for more advanced controller propagators.

3.2 Implementation

A group is a simple interface that specifies one function for scheduling a propagator that belongs to that group, `schedule`, and one function for running the propagators scheduled in the group, `propagate`. How scheduling is done is left to the group implementation, as is the method for running the propagators.

To implement the group $(\cdot)$ relation each propagator has a pointer to a group. This means that each propagator needs one extra word of memory. The standard group is the null group, meaning that the scheduling is done in the normal system queue. If there is a group different from the null group, then the scheduling for that propagator is delegated to the group. This incurs an overhead of one test against null per propagator scheduling. Additionally, one function call per propagator scheduled in a group instead of the global queue needs to be done (the call to g.`schedule`$(\cdot)$).

Execution of a group is managed by the controller propagator. This means that the system does not need to be aware of groups except when scheduling a propagator. A difference from the model is that the way in which the controller propagator for a group is added to the set of propagators to run is programmable; it is done by the `schedule` function of the group. The benefit is that sometimes the controller propagator is guaranteed to be scheduled anyway, and in those cases a back-link to the propagator (represented in the model by $\text{controller}(g)$ for a group g) does not need to be maintained. If such a link is desired, the user can add it to the group that needs it with a pointer

The basic group from Algorithm 3 uses the computed set $\text{prop}(\cdot)$ to check if there are any propagators left to be scheduled in the group. In a real implementation, computing the set on the fly is not feasible. Instead the cardinality of the set is maintained as a member of a group, and is updated by propagators as they are created and removed.

Optimized implementation of $\text{group}(\cdot)$. Implementing the $\text{group}(\cdot)$ connection with a pointer in each propagator wastes memory for propagators not belonging to a group. By adding group-scheduled versions of all propagators, only those propagators that belong to a group contains the pointer. Checking if a propagator has a group pointer can be done cheaply using a tag-bit in pointers to it.

The main overhead in this design is the work in adding an extra optional group-scheduled version of each propagator. Additionally, it would preclude moving a propagator dynamically from the general pool into a group.

The implementation of the model has been done in the Gecode system [22] version 3.0.2. The general description of the implementation here is applicable to any propagator-centered system and is by no means specific to Gecode.

3.3 Evaluation

The implementation of groups touches few parts of the core system, and should therefore incur a small overhead. To evaluate this, the queens problem is tried in two variants. Problems `queens-n-*` use the naive model with a quadratic number of binary not-equals constraints. Problems `queens-s-*` use three large alldifferent constraints instead (albeit with naive propagation only to guarantee that both variants have the same search space). The two different versions test the behavior of the system using many small and few but large propagators.

The experiments have been run on an Athlon 64 3500+ with 2GB of RAM running Ubuntu Linux with gcc version 4.2.4. Times are computed as the average of at least 20 runs with a coefficient of deviation of less than 2%. The `queens-*-200` instances were limited to searching 100 000 nodes.

Table 1. Basic overhead of Groups. Systems compared are without groups (`plain`), with groups added but not used (`groups`), and with groups added and used for scheduling (`scheduling`). Time is given in milliseconds and memory in kilobytes allocated.

	plain		groups		scheduling	
Problem	time	memory	time	memory	time	memory
`queens-n-10`	0.16	63	0.16	63	0.17	63
`queens-n-100`	30.38	7 245	28.10	7 885	30.55	7 885
`queens-n-200`	1 876.30	45 779	1 913.25	50 323	2 061.33	50 323
`queens-s-10`	0.05	19	0.05	19	0.05	19
`queens-s-100`	1.21	355	1.22	356	1.14	356
`queens-s-200`	726.10	1 958	715.05	1 958	708.35	1 958

The results are presented in Table 1. The difference in time between `plain` (the base system) and `groups` (groups added but not used) is not significant; it is less than 2% in the worst case. The inevitable overhead associated with scheduling through groups instead of in the normal queue is reasonably low at around 7% in the worst case. The slightly larger memory-overhead for programs with many propagators is due to the fact that each propagator has an additional field for the group it belongs to. If the memory overhead is prohibitive, the memory-optimized design where each propagator pointer is tagged can be used.

4 Controlling Propagation Order

In a modern constraint programming system, the execution order is defined by the system and works on the granularity of single propagators. Propagator groups can be used for debugging by giving a high-level view of the propagation process (Sect. 4.1). For some sets of propagators, the optimal ordering of propagators is known statically. By following this order instead of the generic order chosen by the system, the optimal number of propagation steps can be achieved (Sect. 4.2).

4.1 Debugging

In many constraint models, the high-level constraints that the model is expressed in can each correspond to many smaller concrete propagators in the system used. As demonstrated in [10], grouping these smaller concrete constraints into larger entities that represent the high-level conceptual constraints is beneficial for understanding the propagation-process. If a high-level view of the propagation process is presented, then stepping through propagation is meaningful.

Implementation. Using the basic group presented in Algorithm 3, it is easy to group propagators into hierarchical groups. As long as the propagators in a group are of roughly the same complexity level, the single-queue group works well. If larger groups of different types of propagators is used, an implementation using multiple queues such as presented in [21] could be used.

Table 2. Grouped propagator execution. Model compared are without groups (`plain`) and with groups (`groups`). Time is given in milliseconds and steps are average propagation steps per node.

	plain		groups	
Problem	time	steps	time	steps
`sudoku-val-5`	5.45	7.55	5.41	4.15
`sudoku-dom-5`	3.27	22.86	3.68	9.75
`sudoku-val-66`	89.38	10.27	99.37	4.68
`sudoku-dom-66`	0.61	314.00	0.96	47.00
`sudoku-val-87`	1 433.37	8.53	1 484.20	4.20
`sudoku-dom-87`	5.00	59.35	6.93	16.57

Evaluation. In the basic constraint programming model for a $n \times n$ Sudoku, there are $3n$ alldifferent constraints. These constraints are composed of three sets that work on rows, columns, and boxes respectively. This division has the interesting property that no two constraints from the same set share a variable.

To illustrate the constraint propagation for a Sudoku problem, it is natural to divide the propagators into the three groups for rows, columns, and boxes. To test this, three instances were run using naive propagation and domain propagation. The numbers refer to the instance-number in the Gecode example. As seen in Table 2, the efficiency of solving a Sudoku is roughly the same, but the number of propagation steps per node is reduced. Thanks to the reduced number of steps, going through the changes between each step becomes feasible.

Using the ability to group the propagators into larger groups, a visualization-system such as the one developed in [14] can be used without getting an overwhelming detail of information during debugging.

4.2 Optimal Propagation Ordering

Decompositions of global constraints that do not sacrifice propagation is an interesting research topic [16, 3, 4, 5]. In some cases (such as the decomposition of the regular constraint into extensional constraints [16]) the decomposition uses a Berge-acyclic propagator graph, which ensures that the local propagation achieves global domain consistency [1]. One benefit of such a decomposition is that the optimal ordering of the propagation is known, with one forward and one backward pass being sufficient to reach a fix-point.

Unfortunately, no constraint programming system supports the specification of propagation order on such a fine-grained level. This means that while the complexity of propagating the decomposition in the optimal order can be calculated, the actual complexity of the decomposition depends on the propagation order that a particular system implements.

Implementation. An *ordered propagation group* g contains a list with the propagators in prop(g). This list is sorted according to the order in which the propagation should be done. For a Berge-acyclic propagator graph any topological sorting of the propagators works. The controller propagator is the same as for the basic group: it simply runs the group.

On activation, the group does one forward and one backward pass through the list of propagators. When inspected, each propagator is executed until it reaches a fix-point if it has been scheduled. After the two passes, the group is at a fix-point if the propagators form a Berge-acyclic propagator graph.

The implementation does not try to find the set of propagators that should be executed without inspecting all of the propagators since that would complicate the scheduling operation. If the overhead of running through all the propagators is too high for some applications, advisors [12] could be used to record the first and last position that needs to be inspected, delimiting the range of propagators that need to be executed. This is a good compromise, since the scheduling operation can be kept at constant time complexity.

Evaluation. Consider a simple ordering problem on n variables x_i with domain $\{1, \ldots, n\}$. Each pair of consecutive variables is ordered using $x_i < x_{i+1}$. For a system that uses queues for scheduling, as most system do, $O(n^2)$ runs through the propagators are needed to ensure that the variables are assigned their respective values through propagation. The bad behavior is because the scheduling will ensure that only forward-passes are made through the list of propagators. Using a group to schedule the propagators in the optimal order, two passes through the propagators are sufficient to ensure that the variables are assigned.

In Table 3, the simple example is tried for varying number of variables. At small sizes ($n = 10$) there is no measurable overhead to using groups even though the relative benefits in number of steps is smaller. As the sizes grow larger, the complexity difference becomes apparent, with orders of magnitude difference in the time. This example is conservative in that the individual propagators are among the cheapest propagators that exist. The costlier the individual propagators are, the more important it is to use a good propagation ordering.

Table 3. Ordered propagator execution. Model compared are without groups (`plain`) and with groups (`groups`). Time is given in milliseconds and steps are propagation steps. For `group`, the steps are the number of steps inside the controlling group.

	`plain`		`groups`	
Problem	time	steps	time	steps
`order-10`	0.04	36	0.03	18
`order-100`	2.95	4 851	0.39	198
`order-1000`	304.90	498 501	7.31	1 998
`order-5000`	7 743.35	12 492 501	103.35	9 998

5 Controlling Propagator Interaction

Since a group is responsible for executing a propagator, it is possible to control how failure of the propagator is handled. Furthermore, a group also encapsulates entailment, which is dual to failure (it represents failure of the negated constraint). Using these facilities, it is possible to implement general reification (Sect. 5.1) and constructive disjunction (Sect. 5.2) using groups.

5.1 General Reification

A reified constraint $c \leftrightarrow (b = 1)$ reflects if the constraint c holds into a Boolean variable b. Reification is commonly available in constraint programming systems for simple constraints using specialized propagators. For larger and more complex constraints such as alldifferent, the effort of implementation often outweighs the benefit of having a reified version of the constraint.

The basic pattern of propagation for a reified constraint looks as follows.

$$
\begin{aligned}
c \leftrightarrow (b = 1) \ :=& \\
c \text{ holds} &\Rightarrow \text{propagate } b = 1 \\
\neg c \text{ holds} &\Rightarrow \text{propagate } b = 0 \\
b = 1 \text{ holds} &\Rightarrow \text{propagate } c \\
b = 0 \text{ holds} &\Rightarrow \text{propagate } \neg c
\end{aligned}
$$

Implementation. Given a constraint c and a Boolean variable b a simple implementation of the reified constraint $c \leftrightarrow (b = 1)$ can be done by posting the constraint in a basic group g. Instead of using the global store-variables $x = \text{var}(c)$ copies of the variables, x', are made and used for the constraint, giving a modified store S'. The equality relation for the stores S and S' is extended in the obvious way. The controlling propagator proceeds through the following steps:

- Add new domain reductions from x to copied variables x' and run the group.
- If g is failed, set $b = 0$ and report entailment.
- If g is entailed, set $b = 1$. If $S = S'$, report entailment.
- If b is set to one, add local domain reductions to the global store.

Unfortunately, the above implementation of generic reification lacks the possibility to propagate $\neg c$, instead it has to wait for failure or entailment of c. This is not a problem with the approach, but a consequence of the problem of propagating negated constraints and is something that is handled in the same way in general reification in Mozart [18, 19]. To implement reification in Propia [15], both the constraint and the negated constraint must be expressed as CLP clauses.

Evaluation. As a base evaluation of using groups for reification, the standard no-overlaps constraints for perfect square packing are tried. For each pair of squares i and j with coordinates x and y and size s, they do not overlap iff $(x_i + s_i \leq x_j) \vee (x_j + s_j \leq x_i) \vee (y_i + s_i \leq y_j) \vee (y_j + s_j \leq y_i)$. Three versions are tested: using normal reification; using normal reification on copied variables; and using groups to implement reification. The numbers refer to the instance numbers in the Gecode example. As seen in Table 4 the overhead from using groups for reification is not unreasonable, especially compared with using standard reified propagators on copied variables which is always worse than groups.

Table 4. Reified no-overlap constraint for the perfect square problem. Models compared use normal reification (`reified`), reification on copied variables (`copied`), and an implementation using groups (`group`). Columns v and c indicate the number of variables and the number of reified constraints for no-overlap. Time is given in milliseconds and memory in kilobytes allocated.

			`reified`		`copied`		`group`	
Problem	c	v	time	memory	time	memory	time	memory
`perfsq-0`	840	3360	280.83	4 998	618.35	8 646	493.80	7 366
`perfsq-1`	924	3696	1 530.05	5 254	3 848.30	8 902	3 164.80	7 750
`perfsq-2`	924	3696	388.57	8 326	650.08	12 166	565.12	10 886
`perfsq-3`	1012	4048	461.20	5 510	1 034.44	9 541	865.02	8 134
`perfsq-4`	1012	4048	2 281.11	15 175	3 314.45	19 783	2 960.56	18 436

The Equidistant Frequency Permutation Array (EFPA) problem [9] is a combinatorial design problem. One of the proposed improvements in [9] to the model includes a reified version of alldifferent. Unfortunately, this is not available in most constraint programming systems. Two versions are tested: using a decomposition representing a reified alldifferent; and using a group to implement it.

Table 5. Reified alldifferent for the EFPA problem. Models compared use a decomposition (`decomposed`) and an implementation using groups (`group`) for the reified alldifferent constraint. Time is given in milliseconds.

	`decomposed`		`group`	
$\langle d, \lambda, q, v\rangle$	time	nodes	time	nodes
$\langle 4, 3, 4, 6\rangle$	0.03	2 192	0.06	1 045
$\langle 4, 4, 4, 8\rangle$	3.23	20 564	2.32	10 104

As shown in Table 5, the additional pruning from the true reified alldifferent propagator using groups pays of in the number of nodes that need to be explored. The time is slightly improved, although not as much. Using groups, it is possible to try a reified version of alldifferent in a reasonably efficient manner.

5.2 Constructive Disjunction

Given a constraint $c \vee c'$ where c and c' share common variables x, the use of the reified decomposition $c \leftrightarrow (b = 1) \wedge c' \leftrightarrow (b' = 1) \wedge b + b' \geq 1$ sacrifices propagation. Consider a disjunctive resource constraint between two tasks 1 and 2 with start-times $s_1 \in \{1..10\}$ and $s_2 \in \{1..10\}$ and durations $d_1 = 6$ and $d_2 = 7$. The reified construction $s_1 + d_1 \leq s_2 \leftrightarrow (b_1 = 1) \wedge s_2 + d_2 \leq s_1 \leftrightarrow (b_2 = 1) \wedge b_1 + b_2 \geq 1$ does not propagate any new information. It is not hard to see that the domains could be reduced, giving $s_1 \in \{1..4, 8..10\}$ and $s_2 \in \{1..3, 7..10\}$.

While it is possible to write a specialized propagator to handle disjunctive resources, it may not be cost-efficient to do so. Furthermore, it does not handle the general case of propagating disjunctive constraints. The technique of constructive disjunction [11, 13, 7, 6, 23] can be used to get full propagation. The basic scheme is to add renamed copies of the variables, as in the previous section, and to run the disjunctive constraints on each of the copies. For any given variable x that is shared among the disjuncts, the union of the domains of the variable-copies for the disjuncts is the new domain.

Implementation. For each disjunct c_i, a new copy of the constraint variables $\mathrm{var}(c_i)$ in the store S is made, giving a new store S_i. The propagators for each disjunct c_i are put in a basic group g_i. By putting the disjuncts in separate groups, the status of each disjunct (failure, entailment) can be checked. The controlling propagator proceeds through the following steps:

- For each variable x, add the new domain-reductions to each copied store S_i.
- Run each group g_i that is not yet failed.
- If all groups are failed, report failure.
- If there exists an entailed group g_i where $S_i = S$, report entailment.
- For each variable x shared among all non-failed groups G, set $x = \cup_{g_j \in G} x_j$.

Implementing a constructive disjunction propagator is mostly straightforward. One interesting aspect is that the largest part of the code and logic is related to handling failed and entailed groups so that the (relatively) expensive propagator is not run needlessly and so that no propagation is missed.

Implementing constructive disjunction using groups is similar to the hard-coded implementation from the cc(FD) system [11]. The difference is that with groups the system is kept unaware of constructive disjunction. Programming constructive disjunction in Propia [15] is similar to groups in that the system is not hard-coded for constructive disjunction, while it is different in that it uses nested search to evaluate each disjunct. This is an approach that saves memory and trades it for computation time.

Evaluation. The value of constructive disjunction as a general technique has been investigated previously [23]. To give a basic evaluation of the implementation, the disjunctive resource example from above is tried with normal reification and with constructive disjunction. The branching used is to try the median value. The example is scaled with a factor k representing time granularity on both variables and durations. While this is a small artificial example, it is based on the common serialization constraint. The results in Table 6 show that the use of constructive disjunction gives an important speed-up. For the case where the use of constructive disjunction is desired, groups enable the use without incurring overhead for the system implementer.

Table 6. Simple use of constructive disjunction. Time is given in milliseconds.

	`reified`		`constructive`	
k	time	nodes	time	nodes
100	0.00	402	0.00	18
1 000	0.02	4 002	0.01	182
10 000	0.20	40 002	0.04	1 802
100 000	1.97	400 002	0.18	18 002

6 Conclusions

The addition of groups to a propagator-oriented constraint programming system is a small, simple, and minimal extension that allows several interesting and useful techniques to be implemented at the user-level without any additional support from the system. In particular, groups enable the first implementation of optimal propagation ordering for Berge-acyclic constraint graphs. The implementation is simple, and the overhead of the system is kept low.

The advantage of being able to experiment with defining the order of propagation and to control the execution of propagators opens up for many new interesting topics. For example, combining the analysis of constraint graphs from [8] with groups for optimal propagator ordering would allow optimal propagator ordering for large subsets of the constraints without requiring the user to specify these patterns.

The results show that groups can implement reification for complex constraints without reification-support and constructive disjunction with a moderate overhead.

Acknowledgements. The authors would like to thank Guido Tack and the anonymous reviewers for comments that considerably improved this paper.

References

[1] Beeri, C., Fagin, R., Maier, D., Yannakakis, M.: On the desirability of acyclic database schemes. J. ACM 30(3), 479–513 (1983)
[2] Bessiere, C.: Constraint propagation. In: Rossi, et al. (eds.) [17], ch. 3, pp. 29–84
[3] Bessiere, C., Hebrard, E., Hnich, B., Kiziltan, Z., Walsh, T.: SLIDE: A useful special case of the CARDPATH constraint. In: Ghallab, M., Spyropoulos, C.D., Fakotakis, N., Avouris, N.M. (eds.) ECAI, pp. 475–479 (2008)
[4] Bessiere, C., Hebrard, E., Hnich, B., Kiziltan, Z., Walsh, T.: Range and roots: Two common patterns for specifying and propagating counting and occurrence constraints. Artificial Intelligence (2009)
[5] Bessiere, C., Katsirelos, G., Narodytska, N., Quimper, C.-G., Walsh, T.: Decompositions of all different, global cardinality and related constraints. In: IJCAI (2009)
[6] Carlson, B., Carlsson, M.: Compiling and executing disjunctions of finite domain constraints. In: ICLP, pp. 117–131 (1995)

[7] Codognet, C., Codognet, P.: Guarded constructive disjunction: Angel or demon? In: Montanari, U., Rossi, F. (eds.) CP 1995. LNCS, vol. 976, pp. 345–361. Springer, Heidelberg (1995)

[8] Francis, K., Stuckey, P.J.: Constraint propagation for loose constraint graphs. In: Cho, Y., Wainwright, R.L., Haddad, H., Shin, S.Y., Koo, Y.W. (eds.) SAC, pp. 334–335. ACM, New York (2007)

[9] Gent, I.P., McKay, P., Miguel, I., Nightingale, P., Huczynska, S.: Modelling equidistant frequency permutation arrays in constraints. In: SARA (2009)

[10] Goualard, F., Benhamou, F.: Debugging constraint programs by store inspection. In: Deransart, P., Małuszyński, J. (eds.) DiSCiPl 1999. LNCS, vol. 1870, pp. 273–297. Springer, Heidelberg (2000)

[11] Hentenryck, P.V., Saraswat, V.A., Deville, Y.: Design, implementation, and evaluation of the constraint language cc(FD). In: Podelski, A. (ed.) Constraint Programming: Basics and Trends. LNCS, vol. 910, pp. 293–316. Springer, Heidelberg (1995)

[12] Lagerkvist, M.Z., Schulte, C.: Advisors for incremental propagation. In: Bessière, C. (ed.) CP 2007. LNCS, vol. 4741, pp. 409–422. Springer, Heidelberg (2007)

[13] Müller, T., Würtz, J.: Constructive Disjunction in Oz. In: Krall, A., Geske, U. (eds.) 11. Workshop Logische Programmierung, Technische Universität Wien, September 27-29 (1995)

[14] Paltzer, N.: Debugging constraint propagation. Master's thesis, Saarland University, Germany (March 2008)

[15] Provost, T.L., Wallace, M.: Domain independent propagation. In: FGCS, pp. 1004–1011 (1992)

[16] Quimper, C.-G., Walsh, T.: Global grammar constraints. In: Benhamou, F. (ed.) CP 2006. LNCS, vol. 4204, pp. 751–755. Springer, Heidelberg (2006)

[17] Rossi, F., van Beek, P., Walsh, T.: Handbook of constraint programming (2006)

[18] Schulte, C.: Programming deep concurrent constraint combinators. In: Pontelli, E., Santos Costa, V. (eds.) PADL 2000. LNCS, vol. 1753, pp. 215–229. Springer, Heidelberg (2000)

[19] Schulte, C.: Programming Constraint Services. LNCS (LNAI), vol. 2302. Springer, Heidelberg (2002)

[20] Schulte, C., Carlsson, M.: Finite domain constraint programming systems. In: Rossi, et al. (eds.) [17], ch. 14, pp. 495–526

[21] Schulte, C., Stuckey, P.J.: Efficient constraint propagation engines. Transactions on Programming Languages and Systems 31(1), 2:1–2:43 (2008)

[22] The Gecode team. Generic constraint development environment (2006), `http://www.gecode.org`

[23] Würtz, J., Müller, T.: Constructive disjunction revisited. In: Görz, G., Hölldobler, S. (eds.) KI 1996. LNCS, vol. 1137, pp. 377–386. Springer, Heidelberg (1996)

Efficient Generic Search Heuristics within the EMBP Framework

Ronan Le Bras[1,2], Alessandro Zanarini[1,2], and Gilles Pesant[1,2]

[1] École Polytechnique de Montréal, Montreal, Canada
[2] CIRRELT, Université de Montréal, Montreal, Canada
{Ronan.LeBras,Alessandro.Zanarini,Gilles.Pesant}@cirrelt.ca

Abstract. Accurately estimating the distribution of solutions to a problem, should such solutions exist, provides efficient search heuristics. The purpose of this paper is to propose new ways of computing such estimates, with different degrees of accuracy and complexity. We build on the Expectation-Maximization Belief-Propagation (EMPB) framework proposed by Hsu et al. to solve Constraint Satisfaction Problems (CSPs). We propose two general approaches within the EMBP framework: we firstly derive update rules at the constraint level while enforcing domain consistency and then derive update rules globally, at the problem level. The contribution of this paper is two-fold: first, we derive new generic update rules suited to tackle any CSP; second, we propose an efficient EMBP-inspired approach, thereby improving this method and making it competitive with the state of the art. We evaluate these approaches experimentally and demonstrate their effectiveness.

Keywords: Constraint Satisfaction, Search Heuristics, Probabilistic Reasoning, Expectation Maximization, Belief Propagation.

1 Introduction

In this paper we address Constraint Satisfaction Problems (CSPs). To guide the backtrack search, we estimate the percentages of solutions that have a given variable assigned a particular value. Accurately estimating this distribution of the solutions, should such solutions exist, provide efficient search heuristics. Previous work [4] already shows a positive correlation between accuracy of the estimates and efficiency of the heuristic. The more accurate the estimations are, the more efficient the heuristic is. On one side of this accuracy spectrum, a distribution in which every variable-value assignment is equally probable represents a totally random heuristic. On the other side lie exact marginal probabilities for a randomly sampled solution of the problem. The first approach is trivial whereas the second remains intractable for hard problems. Indeed, finding the proportion of solutions that assign a variable a particular value is a generalization of the problem of detecting backbone variables. (A *backbone variable* is defined as a variable that takes the same value in every solution of a given problem; thus, the marginal distribution of such a variable corresponds to a probability of 1

I.P. Gent (Ed.): CP 2009, LNCS 5732, pp. 539–553, 2009.

for this particular value, and 0 for the other values in its domain.) Since even an approximation of a backbone is intractable in general [6], the problem of estimating the distribution of solutions is *a fortiori* also intractable in general for hard problems. The purpose of this paper is therefore not to present exact estimation methods for these distributions but to specify approximate methods, with different degrees of accuracy and of complexity.

In this probabilistic environment, inference methods such as message-passing algorithms were proven to be very effective. Even if they are more traditionally applied in probabilistic inference, their aptitude to estimate marginal probability makes them particularly suitable for a backtrack search framework. For instance, Belief Propagation (BP) [10,8] (and later on Generalized Belief Propagation [13]) was developed to tackle inference problems such as problems arising in computer vision or error-correcting coding theory. Kask et al. [5] then demonstrated that BP was especially efficient when applied to CSPs and used as a value-ordering heuristic. Mezard et al. [9] invented Survey Propagation, which eventually turned out to be nothing less than the state of the art to solve large random Boolean satisfiability (SAT) problems [1,7]. More recently, Hsu et al. [3,4] suggested Expectation Maximization (EM) variants of these two major message-passing algorithms.

Here we exploit Expectation-Maximization Belief Propagation (EMBP) which was proposed by [3] to address the Quasigroup with Holes (QWH) problem. The contribution of this paper is two-fold: first, we derive new generic update rules suited to tackle any CSP; second, we propose an efficient EMBP-inspired approach, thereby improving this method and making it competitive with the state of the art. We evaluate these approaches experimentally and demonstrate their effectiveness.

The following section presents the EMBP algorithm. Section 3 then introduces the local consistency extensions of this algorithm, while Section 4 describes the global consistency approach. Section 5 reports the experimental results on three problems. Section 6 discusses connections between the EMBP framework and search heuristics from the literature. Final comments are given in Section 7.

2 Expectation-Maximization Belief Propagation

A *Constraint Satisfaction Problem* (CSP) consists of a finite set of variables $\mathcal{X} = \{x_1, x_2, \ldots, x_n\}$ with finite domains $\mathcal{D} = \{D_1, D_2, \ldots, D_n\}$ such that $x_i \in D_i$ for all i, together with a finite set of constraints $\mathcal{C}$, each on a subset of $\mathcal{X}$. A constraint $C_i \in \mathcal{C}$ is a subset $T(C_i)$ of the Cartesian product of the domains of the variables that are in C_i. We write $X(C_i)$ to denote the set of variables involved in C_i and we call tuple $\tau \in T(C_i)$ an allowed combination of values of $X(C_i)$.

This problem can be modeled indifferently as a Factor Graph, a Pairwise Markov random field or a Bayesian Network [13]. These models then define a joint probability function $P(x_1, x_2, \ldots, x_n)$. Thus, estimating the distribution of solutions boils down to approximating marginal probabilities $P(x_i), \forall i \in 1..n$.

Here lies the purpose of inference methods, and particularly of message-passing algorithms.

2.1 EMBP Framework

Like other message-passing algorithms, EMBP [3] iteratively adjusts the probability for a variable x_i to be assigned a particular value v in a randomly chosen solution. We will denote $\theta_{x_i}(v)$ and call *bias* such a probability. Hence, θ_{x_i} represents a probability distribution over the values in $D(x_i)$ which approximates the exact marginal distribution $P(x_i)$. The EMBP framework also introduces a binary-vector random variable y that indicates whether each constraint is satisfied. EMBP proceeds by sending two kind of messages (thus falling into the "message-passing" category), as illustrated on Figure 1. First, a variable sends its probability distribution to the constraints (Figure 1(a)) so that the latter hypothetically compute the probability of satisfying tuples. Then, every variable retrieves this information from its relevant constraints (Figure 1(b)) in order to update its probability distribution.

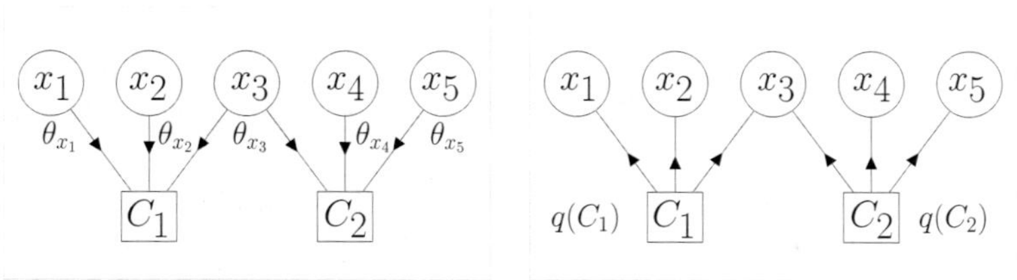

(a) E-Step (b) M-Step

Fig. 1. Illustrating Expectation-Maximization messages for the following example: $\mathcal{X} = \{x_1, x_2, x_3, x_4, x_5\}$, $\mathcal{C} = \{C_1, C_2\}$, with $X(C_1) = \{x_1, x_2, x_3\}$ and $X(C_2) = \{x_3, x_4, x_5\}\}$

2.2 EMBP General Update Rule

Let Θ be the vector of variable biases θ_{x_i}. In essence, the basic goal of the EM algorithm consists of finding the variable biases Θ that maximize the probability $P(y|\Theta)$ that the constraints are satisfied. However, the EMBP methodology assumes that the vector of variables y was originally generated by using not only the parameters Θ, but also hidden variables that we did not observe. In a CSP environment, these latent variables (that we will call z) are actually the satisfying configurations, i.e. tuples, of the constraints. In other words, the variable z ranges over all valid solutions of the problem. We denote by S_z the support of the distribution of z.

Hence, we now want to maximize $P(y, z|\Theta)$. The difficulty lies in the fact that we cannot marginalize on z since it would assume that we can observe the solutions of the problem. Instead, we divide this computation into two iterative steps. In the first step of the EM algorithm (the E-Step), we hypothesize the distribution $Q(z) = P(z|y, \Theta)$. The distribution function $Q(z)$ represents each solution probability given the biases Θ and given the observation y that the constraints are satisfied. Figure 1(a) illustrates this step, in which every constraint C receives the probability distribution θ_{x_i} (where $x_i \in X(C)$) and computes the probability of the satisfying configurations given these distributions. In the second step (the M-Step), we revise the variable biases Θ. As illustrated in Figure 1(b), the variables adjust their distribution taking into account the probability of the valid tuples of the constraints.

Within the Expectation-Maximization (EM) methodology, the E-step firstly assumes the independence between constraints to compute $Q(z)$ as we hypothetically consider only satisfying configurations. Thus, $Q(z)$ becomes the product of the probabilities of the constraints, given a particular set of biases Θ. Therefore, $Q(z) = \prod_{i=1}^{m}(q(C_i))$, where $q(C_i)$ is the probability of a given configuration for the constraint C_i.

The M-step then enforces the dependence between constraints, and yields the following general formula:

$$\theta_{x_i}(v) = \frac{1}{\eta} \sum_{C_k \in \mathcal{C}: x_i \in X(C_k)} \left(\sum_{z \in S_z : x_i = v} Q(z) \right) \tag{1}$$

Here η is a normalizing constant and equals the summation of the numerator over all values of v. For more details about how to derive Equation (1), please refer to [3].

As a result, EMBP provides a general algorithm to compute variable biases. Within this framework, the definition of $Q(z)$ remains however unspecified. The methods presented in the following section exploit this degree of freedom.

2.3 Computing EMBP: A Tradeoff between Accuracy and Complexity

One might consider determining S_z, the solution space of the problem, and then computing $Q(z)$ exactly. This approach is however intractable since it implies expressing each solution of the problem to estimate the biases in order to find a solution. Hence, an approximation of $Q(z)$ is required. The methods described in this paper (summarized in Table 1) gradually improve the accuracy of the estimation of $Q(z)$ and, by doing so, increase the complexity of its computation. In that regard, the methods need to find the supports either locally (i.e. at the

Table 1. EMBP-based search heuristics

Heuristics	Consistency
EMBP-a	local Arc-Consistency
EMBP-Lsup	local X-Consistency
EMBP-Gsup	global X-Consistency

constraint level - *Lsup*) or globally (i.e. after propagating the whole constraint network - *Gsup*) enforcing a level of consistency denoted by X.[1]

3 EMBP and Local Consistency

Whereas Equation (1) sums over the probabilities of all satisfying tuples, the following local X-consistency methods (EMBP-a and EMBP-Lsup) are designed to approximate this summation by simplifying it.

3.1 EMBP-a for the `alldifferent` Constraint

Originally, Hsu et al. [3] suggest such an approximation for the `alldifferent` constraint. The probability that variable x_i is assigned the value v can be approximated by the probability that no other variable in the constraint takes the value v. The approach is therefore ensuring pairwise consistency between x_i and the other variables in the `alldifferent` constraint. Thus, the authors obtain the following update rule for a set $\mathcal{C}_a$ of alldifferent constraints:

$$\begin{aligned}\theta_{x_i}(v) &= \frac{1}{\eta} \sum_{C_k \in \mathcal{C}_a : x_i \in X(C_k)} \left(\prod_{x_j \in X(C_k) \backslash x_i} \sum_{v' \in D(x_j) \backslash v} \theta_{x_j}(v') \right) \\ &= \frac{1}{\eta} \sum_{C_k \in \mathcal{C}_a : x_i \in X(C_k)} \left(\prod_{x_j \in X(C_k) \backslash x_i} (1 - \theta_{x_j}(v)) \right)\end{aligned} \tag{2}$$

where η is again a normalizing constant.

3.2 EMBP-Lsup

We propose to derive local X-consistency EMBP methods, which are a fairly natural extension of EMBP-a. The reasoning behind these methods to compute $\theta_{x_i}(v)$ for a given constraint is to consider all assignments $x_j = v'$ that are X-consistent with the assignment $x_i = v$ within this constraint. In other words, these methods take into account domain reductions at the constraint level when enforcing X-consistency. Essentially, the method processes one constraint at a time, and for each variable-value assignment looks for the supports that are X-consistent and updates the biases using the following formula:

$$\theta_{x_i}(v) = \frac{1}{\eta} \sum_{C_k \in \mathcal{C} : x_i \in X(C_k)} \left(\prod_{x_j \in X(C_k) \backslash x_i} \sum_{v' \in \tilde{D}_{x_i = v}(x_j)} \theta_{x_j}(v') \right) \tag{3}$$

where $\tilde{D}_{x_i=v}(x_j)$ represents the reduced domain of the variable x_j after assigning $x_i = v$ and enforcing X-consistency on C_i.

[1] Note that the constraints involved can even implement different levels of consistency, as is common in practice.

Considering that ensuring pairwise consistency between one variable and the others is reminiscent of arc consistency, EMBP-a indeed falls into the set of local X-consistency EMBP methods, with X standing for arc consistency. Nonetheless if we consider stronger consistency, such as domain consistency, the improvement is twofold: it provides better accuracy since the variable biases rely only on supports that are domain consistent (rather than arc consistent for EMBP-a) and it is easily implementable for any constraint for which we can enforce domain consistency. The method can be easily extended to consider any form of consistency. As in the general EMBP update rule, EMBP-Lsup still assumes independent constraints and only the M-Step enforces the dependence between the constraints.

Algorithm 1 shows the pseudo-code to update $\theta_{x_i}(v)$ at iteration t. P_{C_i} represents the contribution of C_i to $\theta_{x_i}(v)$ and S_{x_j} the summation for the variable x_j. Firstly, the algorithm picks up a constraint C_i whose variable set contains x_i, then propagates the assignment $x_i = v$ (line 3). For every variable x_j in the constraint's scope, we add up the value biases of the reduced domain (line 8) and multiply the result with P_{C_i} (line 9). Once the algorithm is done with processing constraint C_i, it adds the contribution of this constraint to $\theta_{x_i}(v)$, rolls back the effect of the propagation (lines 10–11), and continues to iterate over the remaining constraints. Note that when all the $\theta_{x_i}(\cdot)$ have been computed, they have to be normalized.

1 $\theta_{x_i}^{t+1}(v) = 0$;
2 **for** *each constraint* $C_i \in C : x_j \in X(C_i)$ **do**
3   enforce X-consistency on C_i with $x_i = v$;
4     $P_{C_i} = 1$;
5   **for** *each variable* $x_j \in X(C_i) \setminus x_i$ **do**
6         $S_{x_j} = 0$;
7   **for** *each value* $v' \in \tilde{D}_{x_i=v}(x_j)$ **do**
8             $S_{x_j} = S_{x_j} + \theta_{x_j}^{t}(v')$;
9         $P_{C_i} = P_{C_i} \times S_{x_j}$;
10     $\theta_{x_i}^{t+1}(v) = \theta_{x_i}^{t+1}(v) + P_{C_i}$;
11   rollback propagation C_i;
12 normalize $\theta_{x_i}^{t+1}(v)$;

Algorithm 1. EMBP-Lsup update algorithm for $\theta_{x_i}(v)$ at iteration t

Given k the number of constraints in which x_i is involved, n their maximum arity, d the maximum cardinality of the variable domains and P the worst-case complexity of the constraints' propagation algorithms, the worst-case time complexity of Algorithm 1 is $\mathcal{O}(kP+knd)$. To reach convergence, the code shown in Algorithm 1 should be run for every variable-value pair at each iteration of the EMBP. In order to improve the efficiency of the procedure, the propagation of the constraints is performed once and the information related to $\tilde{D}_{x_i=v}$ is cached since the reduced domains remain constant over the iterations.

This method aims to be a tradeoff between accuracy and performance: at any given time no propagation other than the one of the processed constraint is performed, which is less expensive than propagating the whole model, but also less accurate.

4 EMBP and Global Consistency

Introducing a new method that we call EMBP-Gsup, we suggest to go one step further in terms of accuracy for the computation of $Q(z)$. Indeed, in EMBP-Gsup, the problem is considered as a whole and the method directly exploits the dependence between constraints when computing $Q(z)$. EMBP-Lsup considers supports that are X-consistent for one constraint at a time whereas EMBP-Gsup will improve the quality of the approximation by taking into account supports that are X-consistent after propagating each problem's constraint. In the new representation underlying the approximation of $Q(z)$, instead of having m constraints (see Figure 1), we now consider only one, in which every variable in $\mathcal{X}$ is involved.

The update formula is then:

$$\begin{aligned} \theta_{x_i}(v) &= \frac{1}{\eta} \prod_{x_j \in \mathcal{X} \backslash x_i} \sum_{v' \in \hat{D}_{x_i=v}(x_j)} \theta_{x_j}(v') \\ &= \frac{1}{\eta} \prod_{x_j \in \mathcal{X} \backslash x_i} \left(1 - \sum_{v' \in D(x_j) \backslash \hat{D}_{x_i=v}(x_j)} \theta_{x_j}(v') \right) \end{aligned} \tag{4}$$

where $D(x_j)$ is the domain of x_j before assigning $x_i = v$ and $\hat{D}_{x_i=v}(x_j)$ stands for the reduced domain of the variable x_j after assigning $x_i = v$ and enforcing X-consistency on each problem constraint.

The method indirectly depends on the problem modelling since it depends on the overall inference power underlying the specific modelling. However, this approach also offers the possibility of using different levels of consistency during the probing phase (when we compute $\hat{D}_{x_i=v}$) and during the effective propagation phase of the search.

Algorithm 2 shows the pseudo-code to update $\theta_{x_i}(v)$ at iteration t. The main difference with Algorithm 1 lies in the fact that there is no summation over the constraints since the contribution of each constraint is implicitly reflected in $\hat{D}_{x_i=v}(x_j)$ due to the initial propagation over the whole problem (line 2). In practice the experiments revealed that it is faster to iterate over the delta domain of the variables (which is the complementary set of $\hat{D}_{x_i=v}(x_j)$) rather than over $\hat{D}_{x_i=v}(x_j)$ itself (line 5).

Given K the total number of constraints, N the total number of variables, d the maximum cardinality of the variable domains and P the worst-case complexity of the constraints' propagation algorithms, the worst-case time complexity of Algorithm 2 is $\mathcal{O}(KP + Nd)$. Even though not obvious from the worst-case

$\theta^{t+1}_{x_i}(v) = 1;$
enforce X-consistency on the whole problem with $x_i = v$;
for *each variable* $x_j \in \mathcal{X} \setminus x_i$ **do**
    $S_{x_j} = 1;$
    for *each value* $v' \in D(x_j) \setminus \hat{D}_{x_i=v}(x_j)$ **do**
        $S_{x_j} = S_{x_j} - \theta^{t}_{x_j}(v');$
    $\theta^{t+1}_{x_i}(v) = \theta^{t+1}_{x_i}(v) \times S_{x_j};$
rollback propagation of $x_i = v$;
normalize $\theta^{t+1}_{x_i}(v)$;

Algorithm 2. EMBP-Gsup update algorithm for $\theta_{x_i}(v)$ at iteration t

complexities, Algorithm 2 happens to be more time consuming than Algorithm 1 since we usually have $K \gg k$ and $N \gg n$. In order to avoid propagation at each iteration of the EMBP, the information related to $D(x_j) \setminus \hat{D}_{x_i=v}$ is also computed once and then cached, as in Algorithm 1.

5 Experiments

In this section we evaluate search heuristics based on the methods we propose. Fundamentally, at each node of the search tree, after computing the variable biases according to EMBP-a, EMBP-Lsup, or EMBP-Gsup, the search heuristic branches on the variable-value pair that presents the highest bias. We evaluate the proposed search heuristics on three benchmark problems - the Nonogram problem, the Quasigroup With Holes problem and the Magic Square problem. We then compare our results with six other heuristics, rndMinDom, MaxSD, IlogIBS, IlogAdvIBS, RSC-LA, and RSC2-LA. Before presenting the results of the experiments, we detail these heuristics and explain the rationale behind the selection of these six other search heuristics:

- rndMinDom randomly picks up a variable with the smallest domain size and then randomly selects a value from its domain. We present the results for this heuristic as a benchmark for fairly uninformed yet very common heuristics.

- MaxSD stands for maximum Solution Density and belongs to the family of solution counting based heuristics. This heuristic exploits counting information provided by the constraints in order to branch on the part of the search tree where it is likely to find a higher number of solutions [14]. MaxSD is considered the state-of-the-art for solving the hard QWH problems.

- Impact Based Search [11] methods first choose the variable whose instantiation triggers the largest search space reduction (highest impact) that is approximated as the reduction of the Cartesian product of the domains of the variables. The impact is either approximated as the average reduction observed during the search or computed exactly at a given node of the search (the exact computation provides better information but is more time consuming). IlogIBS represents

Impact Based Search where the impacts are approximated. IlogAdvIBS chooses a subset of 5 variables with the best approximated impacts and then it breaks ties based on the exact impacts while further ties are broken randomly [11].

- RSC-LA stands for restricted singleton consistency look-ahead heuristic [2]. RSC-LA maintains restricted singleton consistency during the search while RSC2-LA maintains this level of consistency for a subset of variables whose domain size equals 2. While enforcing singleton consistency, the method collects look-ahead information under the form of impacts and uses it in a manner similar to [11]. RSC-LA is similar to EMBP-Gsup since they both perform a complete look-ahead procedure at every choice point. Hence, it is definitely worth comparing the results for these two approaches, as they only differ on how to aggregate the look-ahead information.

All tests were performed with Ilog Solver 6.5 on a AMD Opteron 2.4GHz with 1GB of RAM; for the heuristics that have some sort of randomization we present the arithmetic averages over 10 runs. In the heuristics based on EMBP, the bias computation is performed at every node. At each node, we randomly initialize the biases and iterate until convergence or until a fixed number of iterations is reached. Even though EMBP methods do guarantee convergence [3], they converge to a *local* optimum, which might differ from the exact marginalizations. Nonetheless, convergence is relatively fast and every iteration is quite time consuming so we decided to limit to 5 the number of iterations during these experiments.

Nonogram. A Nonogram (problem 12 of CSPLib) is built on a rectangular $n \times m$ grid and requires filling in some of the squares in the unique feasible way according to some clues given on each row and column. As a reward, one gets a pretty monochromatic picture. Each individual clue indicates how many sequences of consecutive filled-in squares there are in the row (column), with their respective size in order of appearance. Each sequence is separated from the others by at least one blank square but we know little about their actual position in the row (column). Such clues can be modeled with `regular` constraints (the actual automata $\mathcal{A}_i^r, \mathcal{A}_j^c$ are not difficult to derive but lie outside the scope of this paper):

$$\begin{array}{ll} \texttt{regular}((x_{ij})_{1 \le j \le m}, \mathcal{A}_i^r) & 1 \le i \le n \\ \texttt{regular}((x_{ij})_{1 \le i \le n}, \mathcal{A}_j^c) & 1 \le j \le m \\ x_{ij} \in \{0, 1\} & 1 \le i \le n,\ 1 \le j \le m \end{array}$$

We experimented with 180 instances[2] of sizes ranging from 16×16 to 32×32. We enforced domain consistency on the constraints and set a timeout of 10 minutes for each run.

Figure 2 shows the percentage of instances solved within a given time. In Nonograms the additional constraint inference performed by EMBP does not bring a clear advantage over simply using information such as impacts. In fact, as shown in the plot, this problem is a fairly easy one, most of the instances are

[2] Instances taken from http://www.blindchicken.com/~ali/games/puzzles.html

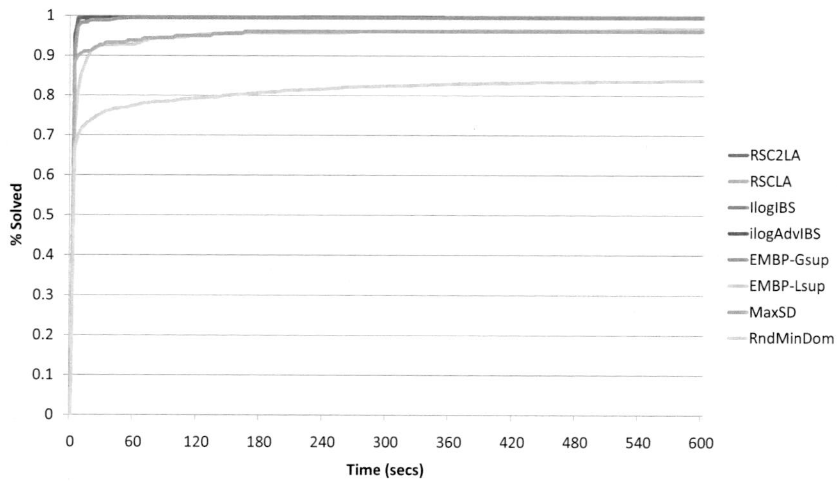

Fig. 2. Percentage of solved instances vs time for 180 Nonogram instances

solved within few seconds; yet EMBP-Gsup, despite its inherent overhead, performs basically the same as RSC-based and IBS-based heuristics. MaxSD and EMBP-Lsup are slightly behind in this test, being able to solve fewer instances. Finally, our baseline RndMinDom is significantly slower compared to the rest of the heuristics. What the figure doesn't show is that our proposed heuristics dramatically improve the total number of backtracks (by three orders of magnitude) over any other heuristic tested except *RSC-LA* methods, even if it does not translate to the best overall total running time (mainly due to a single instance that timed out).

Quasigroup with Holes. A Latin Square of order n is defined on a $n \times n$ grid whose cells each contain an integer from 1 to n such that each integer appears exactly once per row and column. The Quasigroup with Holes (QWH) problem gives a partially-filled Latin Square instance and asks to complete it. It can be modeled easily as follows:

$$
\begin{array}{ll}
\texttt{alldifferent}((x_{ij})_{1\le j\le n}) & 1 \le i \le n \\
\texttt{alldifferent}((x_{ij})_{1\le i\le n}) & 1 \le j \le n \\
x_{ij} = d & (i,j,d) \in S \\
x_{ij} \in \{1,2,\ldots,n\} & 1 \le i,j \le n
\end{array}
$$

where S represents the set of pre-assigned cells.

The set of instances is the same as in [14]: 40 instances with $n = 30$ and a percentage of holes around 42% (near the phase transition). We enforced domain consistency on the constraints and set a timeout of 20 minutes for each run.

For this test we also added the heuristic dom/ddeg; min conflicts that chooses the variable with the lowest ratio of domain size over dynamic degree and then selects the value with the minimum number of conflicts (this has been considered

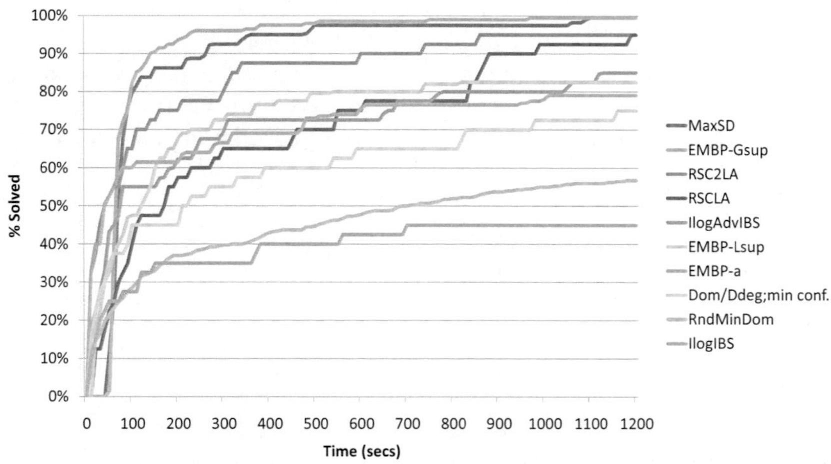

Fig. 3. Percentage of solved instances vs time for 40 hard QWH instances of order 30

for years one of the best custom heuristics for QWH). For the heuristic MaxSD, the counting algorithm has been set as in [14].

Figure 3 shows the results (presented as in the Nonogram problem). Despite the fact that EMBP-Lsup is significantly more lightweight than EMBP-Gsup, its performance does not match that of EMBP-Gsup because of poorer heuristic guidance. EMBP-Gsup performs very well in terms of number of backtracks (at least one order of magnitude better than any other heuristic) and scores the best total time. Again most of the computation time is spent at each node of the search tree in propagating the whole problem for each variable-value pair but the accuracy of the variables' biases definitely pays off. This overhead can be seen in Figure 3: the heuristic takes some time to get close to the leaves of the search tree to find solutions. Therefore the heuristic hardly solves any instance within the first minute. However it is able to solve about 75% of the instances in 90 seconds. MaxSD presents the same behavior, due to the sampling algorithm used to count the number of solutions of the `alldifferent` constraints which causes a significant overhead. Nonetheless, the total running time of EMBP-Gsup is significantly lower. The heuristics that were scoring the best running times on the Nonogram problem (RSC-based and IBS-based heuristics) struggle more here and their running time goes from almost two to seven times the running time of EMBP-Gsup to solve the same number of instances. The total number of instances solved is also significantly lower compared to EMBP-Gsup. Finally, it is interesting to see how Hsu et al.'s approach behaves with respect to EMBP-Lsup: the two methods are similar (although Hsu et al.'s only applies to `alldifferent` constraints), the only difference being that ours enforces domain consistency on the constraints whereas EMBP-a keeps a weak form of arc consistency. Hence, EMBP-a is able to perform more backtracks w.r.t. EMBP-Lsup in the same amount of time but has poorer heuristic guidance.

Magic Square. This very old puzzle is built on a $n \times n$ grid. The task is to place the first n^2 integers in the grid so that each row, column and main diagonal sums up to the same value. A partially filled Magic Square Problem asks for a solution, if one exists. It can be made harder to solve than the traditional version starting from a blank grid. More formally, here is a model for the Magic Square Problem:

$$\begin{array}{ll} \texttt{alldifferent}((x_{i,j})_{1\le i,j\le n}) & \\ \sum_{1\le j\le n} x_{i,j} = sum & 1 \le i \le n \\ \sum_{1\le i\le n} x_{i,j} = sum & 1 \le j \le n \\ \sum_{1\le i\le n} x_{i,i} = sum & \\ \sum_{1\le i\le n} x_{(n+1-i),i} = sum & \\ x_{i,j} \in \{1, n^2\} & 1 \le i \le n,\ 1 \le j \le n \end{array}$$

where $sum = \frac{n(n^2+1)}{2}$. On the `alldifferent` constraint, domain consistency is enforced, while for the equality `knapsack` constraint, bound consistency is enforced. The set of instances is the same as in [12]: 40 instances with prefilled cells (in order to avoid trivial solutions) — half of the instances have 10 preset variables and the other half, 50. A timeout of one hour was set for each run.

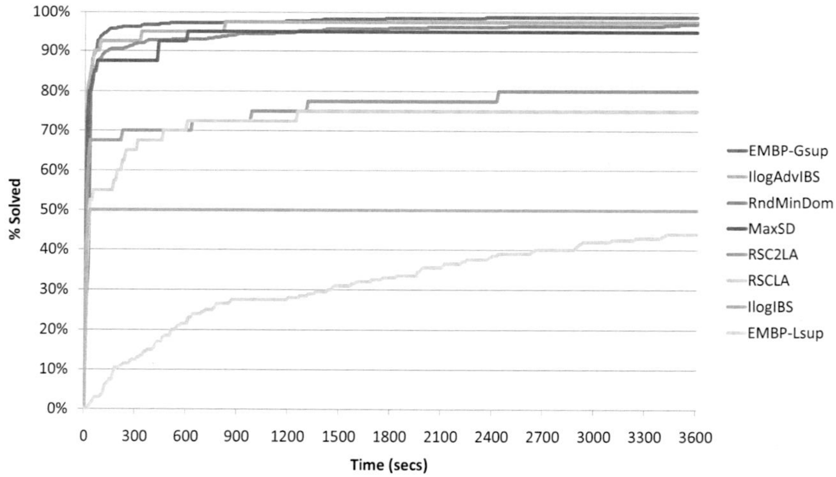

Fig. 4. Percentage of solved instances vs time for 40 Magic Square instances

Figure 4 shows the results as presented on the previous two problems. EMBP-Gsup is the best performing heuristic, outperforming by about 35% the second best heuristic (IlogAdvIBS) in terms of total time (including timeouts). As shown in the plot, in this problem the heuristic is well suited both for hard and easy instances. Again, the number of backtracks is the lowest among the group of heuristics by at least one order of magnitude. Despite being fairly good on the QWH problem, MaxSD falls far behind in this test taking more than double the time of EMBP-Gsup.

As before, EMBP-Lsup was not able to provide an interesting performance with respect to EMBP-Gsup. However, it is worth mentioning that the EMBP-Gsup underlying algorithm provides at no extra cost a reduced form of singleton consistency whereas EMBP-Lsup does not enforce implicitely such form of consistency.

Overall, compared to some state-of-the-art heuristics, EMBP-Gsup turns out to be the most consistent heuristic on the problems considered. RSC2-LA performs better on Nonogram instances but not on QWH and it fails to solve 20% of the Magic Square instances. Impact-Based Search heuristics achieve comparable performance on Nonogram but finish far behind on QWH instances. Finally, MaxSD performs very well on QWH but not as well as EMBP-Gsup when considering Nonogram and Magic Square.

6 Discussion

In this section we draw connections between some of the heuristics used in this paper.

Solution counting algorithms [14] propose an approach which is closely related to the EMBP method. Indeed, constraint-centered solution counting also offers marginals of the variables for a specific constraint. This method however considers constraints separately. In the EMBP framework, this would be equivalent to considering one independent factor graph for each constraint. Hence, constraint-level solution counting estimates marginal distributions of *a priori* independent constraints. Within a backtrack search tree, this approach then suggests to consider a basic aggregation of these marginals but misses a more global reasoning that considers the dependence between the constraints. Also, the first iteration of the local X-consistency EMBP method would give the exact computation of the solution densities as defined in [14] (again assuming variable biases are uniformly initialized) if we were able to exactly compute $Q(z)$.

The RSC-LA methods proposed by Correia and Barahona in [2] also present strong similarities with EMBP-Gsup. In our case, we are building search heuristics using look-ahead information which in return provides restricted singleton consistency. Conversely, in [2], the authors ensure restricted singleton consistencies and take advantage of the information provided during look-ahead procedures to derive search heuristics. Compared to [2], when computing EMBP-Gsup, we thus also benefit from restricted singleton consistency that allows us to shave the search tree with every variable-value pair that is inconsistent. As a result, both methods present a similar overhead of computation time at each node. The difference in the search heuristics lies in the fact that [2] exploits impact information whereas we are performing inference reasoning.

There are also some interesting connections between IBS and EMBP methods. Were we to uniformly initialize the variable biases, the first iteration of EMBP-Gsup would compute the impact of every variable-value pair as the reduction of the Cartesian product of the domain, as [11,2] would do. Subsequent iterations further refine this impact, thereby generating a sort of "weighted" impact.

7 Conclusion and Open Issues

This paper provided generic and efficient heuristics built upon the EMBP framework. Whereas previous EMBP proposals in [3] addressed problems involving only `alldifferent` constraints, we lifted that restriction with our contribution. Furthermore, we provided a more efficient formulation that achieves very promising performance and is competitive with the state of the art — it was the most consistent on the problems considered. The number of backtracks for our proposed heuristics was also consistently much lower, thus indicating excellent heuristic guidance and some potential for runtime improvement if parts of the computation are optimized or approximated (e.g. see the next paragraph). An important step ahead has been achieved compared to the approaches presented in [3] and [14]. While these previous approaches respectively require constraint-specific update rules and constraint-specific solution counting algorithms, EMBP-Gsup and EMBP-Lsup are completely general and easily pluggable into any model.

Several important open issues still remain. First, even though accurate biases certainly provide useful information for an underlying search framework, the question remains of determining the most efficient way to use it. Indeed, when used within a variable-ordering heuristic, a method providing a set of biases still needs to define a branching strategy. Here the possibilities are numerous. For example, we can choose to branch on the highest bias as we have done during these experiments, but we could also choose the highest strength (difference between a bias and the reciprocal of the domain size) or even remove the value with the lowest bias from a variable domain. Second, instead of computing a survey at each node of the search tree, we could use previous information. For instance, we could set more than one variable at once, or also compute partial surveys and keep track of the domain reduction information, as an IBS heuristic would do. Finally, in future work, we would like to derive equivalent update rules for the Expectation-Maximization Survey Propagation (EMSP) [4] and exploit the promising potential of Survey Propagation.

Acknowledgements

The authors would like to thank Eric Hsu for useful comments about the EMBP framework and the anonymous referees for their constructive comments. The authors also thank the FCI-Relève program for financing the equipment on which the experiments were partially conducted. This research was supported in part by an NSERC discovery grant and an FQRNT doctoral scholarship.

References

1. Braunstein, A., Mézard, M., Zecchina, R.: Survey propagation: An algorithm for satisfiability. Random Structures Algorithms 27(2), 201–226 (2005)
2. Correia, M., Barahona, P.: On the integration of singleton consistencies and look-ahead heuristics. In: CSCLP, pp. 62–75 (2007)

3. Hsu, E.I., Kitching, M., Bacchus, F., McIlraith, S.A.: Using expectation maximization to find likely assignments for solving csp's. In: AAAI, pp. 224–230 (2007)
4. Hsu, E.I., Muise, C.J., Beck, J.C., McIlraith, S.A.: Applying probabilistic inference to heuristic search by estimating variable bias. In: Proceedings of the 1st International Symposium on Search Techniques in Artificial Intelligence and Robotics (at AAAI 2008), Chicago, IL, USA, July 13–14, vol. WS-08-10, pp. 68–75 (2008)
5. Kask, K., Dechter, R., Gogate, V.: Counting-based look-ahead schemes for constraint satisfaction. In: Wallace, M. (ed.) CP 2004. LNCS, vol. 3258, pp. 317–331. Springer, Heidelberg (2004)
6. Kilby, P., Slaney, J.K., Thibaux, S., Walsh, T.: Backbones and backdoors in satisfiability. In: Veloso, M.M., Kambhampati, S. (eds.) AAAI, pp. 1368–1373. AAAI Press / The MIT Press (2005)
7. Kroc, L., Sabharwal, A., Selman, B.: Survey propagation revisited. In: 23rd UAI, Vancouver, BC, July 2007, pp. 217–226 (2007)
8. Kschischang, F.R., Frey, B.J., Loeliger, H.A.: Factor graphs and the sum-product algorithm. IEEE Transactions on Information Theory 47(2), 498–519 (2001)
9. Mezard, M., Parisi, G., Zecchina, R.: Analytic and algorithmic solution of random satisfiability problems. Science 297(5582), 812–815 (2002)
10. Pearl, J.: Probabilistic Reasoning in Intelligent Systems: Networks of Plausible Inference. Morgan Kaufmann, San Francisco (1988)
11. Refalo, P.: Impact-based search strategies for constraint programming. In: Wallace, M. (ed.) CP 2004. LNCS, vol. 3258, pp. 557–571. Springer, Heidelberg (2004)
12. Trick, M.A.: A dynamic programming approach for consistency and propagation for knapsack constraints. In: Annals of Operations Research, vol. 118, pp. 73–84 (2003)
13. Yedidia, J.S., Freeman, W.T., Weiss, Y.: Understanding belief propagation and its generalizations. Morgan Kaufmann Publishers Inc., San Francisco (2003)
14. Zanarini, A., Pesant, G.: Solution counting algorithms for constraint-centered search heuristics. Constraints 14(3), 392–413 (2009)

Failed Value Consistencies for Constraint Satisfaction

Christophe Lecoutre and Olivier Roussel

Univ Lille Nord de France, F-59000 Lille, France
UArtois, CRIL, F-62307 Lens, France
CNRS, UMR 8188, F-62307 Lens, France
{lecoutre,roussel}@cril.fr

Abstract. In constraint satisfaction, basic inferences rely on some properties of constraint networks, called consistencies, that allow the identification of inconsistent instantiations (also called nogoods). Two main families of consistencies have been introduced so far: those that permit us to reason from variables such as (i, j)-consistency and those that permit us to reason from constraints such as relational (i, j)-consistency. This paper introduces a new family of consistencies based on the concept of failed value (a value pruned during search). This family is orthogonal to previous ones.

1 Introduction

Any user of a constraint solver ideally expects the system to be robust and clever enough to automatically identify all relevant properties of a problem instance. These properties typically depend on the structure of the instance and can help solving it. Some of the approaches to reach that goal are inferences from strong consistencies, adaptive heuristics, nogood recording and automatic symmetry breaking. They permit an efficient exploration of the search space, learning much useful information before or during search so as to avoid exploring fruitless combinations of values of variables.

Usually, backtrack search is used for solving instances of the Constraint Satisfaction Problem (CSP). Backtrack search combines a depth-first exploration to instantiate variables and a backtracking mechanism to deal with encountered dead-ends. During search, some values are proved to be inconsistent, i.e. not to participate to any solution - we call them failed values. Interestingly, it is known [13] that failed values "convey" some information: given a satisfiable binary CSP instance P, for any pair (x, a) where x is a variable of P and a a value in the domain of x, if there is no solution containing a for x, then there is necessarily a variable $y \neq x$ which is assigned a value b such that (y, b) is not compatible with (x, a). It is then possible to dynamically and iteratively decompose problem instances [13,2].

In this paper, we propose to exploit failed values in a different manner. By locally reasoning from failed values, we show that some inferences can be performed within reasonable complexities. In particular, we develop a new family of domain-filtering consistencies based on failed values and show that they are complementary to (incomparable with) usual ones. They contribute to prune the search space and also offer a lazy detection of a generalized form of the substitutability relation. Algorithms checking or enforcing consistencies based on failed values can be naturally grafted to any constraint

I.P. Gent (Ed.): CP 2009, LNCS 5732, pp. 554–568, 2009.

propagation engine in order to reinforce the filtering strength of the search algorithm. This approach may represent an interesting contribution to the quest for robust solvers.

After recalling some technical background, two basic consistencies, called FVC and AFVC and based on failed values, are presented. Next, algorithms for checking FVC and enforcing AFVC are introduced, and AFVC is compared to substitutability and usual consistencies. Finally, before presenting some preliminary experimental results, we show that an entire class of consistencies based on failed values can be naturally defined.

2 Technical Background

A constraint network (CN) P is composed of a finite set of n variables, denoted by $vars(P)$, and a finite set of e constraints, denoted by $cons(P)$. Each variable x has an associated domain, denoted by $dom(x)$, that contains the finite set of values that can be assigned to x. Each constraint c involves a set of variables, called the *scope* of c and denoted by $scp(c)$. It is defined by a relation, denoted by $rel(c)$, which contains the set of tuples allowed for the variables involved in c. A *binary* constraint involves exactly 2 variables, and a *non-binary constraint* strictly more than 2 variables. For a binary constraint c_{xy} such that $scp(c_{xy}) = \{x, y\}$, if $(a, b) \in rel(c_{xy})$, we say that (x, a) and (y, b) are compatible. We also say that (x, a) and (y, b) are compatible if there is no binary constraint between x and y.

In this paper, we shall consider given an initial CN P^{init} and a current CN P derived from P^{init} by potentially reducing variable domains. The initial domain of a variable x is denoted by $dom^{init}(x)$ whereas the current domain is denoted by $dom^P(x)$ or more simply by $dom(x)$. A (current) value of P is a pair (x, a) with $x \in vars(P)$ and $a \in dom(x)$. For any variable x, we always have $dom(x) \subseteq dom^{init}(x)$ and denote this fact by $P \preceq P^{init}$. More generally, given two CNs P and P', we note $P' \preceq P$ iff P and P' are defined on the same set of variables and the same set of constraints, and for every variable x in $vars(P) = vars(P')$, we have $dom^{P'}(x) \subseteq dom^P(x)$. $P' \prec P$ iff $P' \preceq P$ and there exists a variable x such that $dom^{P'}(x) \subset dom^P(x)$.

An instantiation I of a set $X = \{x_1, \ldots, x_k\}$ of variables is a set $\{(x_1, a_1), \ldots, (x_k, a_k)\}$ such that $\forall i, a_i \in dom^{init}(x_i)$; the set X of variables occurring in I is denoted by $vars(I)$ and each value a_i is denoted by $I[x_i]$. An instantiation I on a CN P is an instantiation of a set $X \subseteq vars(P)$; it is complete if $vars(I) = vars(P)$, partial otherwise. I is valid on P iff $\forall (x, a) \in I, a \in dom(x)$ $(= dom^P(x))$. $I[x/a]$ is the instantiation obtained from I by replacing the value assigned to x in I by a. An instantiation I covers a constraint c iff $scp(c) \subseteq vars(I)$, and satisfies a constraint c with $scp(c) = \{x_1, \ldots, x_r\}$ iff a) I covers c and b) the tuple $(a_1, \ldots, a_r)$, such that $\forall i, a_i = I[x_i]$, is allowed by c, i.e. $(a_1, \ldots, a_r) \in rel(c)$. An instantiation I on a CN P is locally consistent iff a) I is valid on P and b) every constraint of P covered by I is satisfied by I. It is locally inconsistent otherwise. A solution of P is a complete instantiation on P that is locally consistent. An instantiation I on a CN P is globally inconsistent, or a nogood, iff it cannot be extended to a solution of P. It is globally consistent otherwise.

A CN is said to be *satisfiable* iff it admits at least one solution. The Constraint Satisfaction Problem (CSP) is the NP-complete task of determining whether a given CN is satisfiable or not. A CSP instance is defined by a CN which is solved either by finding a solution or by proving unsatisfiability. To solve a CSP instance, a depth-first search algorithm with backtracking can be applied, where at each step of the search, a variable assignment is performed followed by a filtering process called constraint propagation. Typically, constraint propagation algorithms are based on some properties of CNs that allow us to identify and remove some values which cannot occur in any solution. Such properties are called domain-filtering consistencies [6,4].

Let us introduce some classical consistencies. An instantiation I on P is a support (resp. a conflict) for a value (x, a) on a constraint c involving x iff I is valid, $I[x] = a$ and I satisfies (resp. does not satisfy) c. A value (x, a) of P is GAC-consistent (GAC stands for Generalized Arc Consistency) iff there exists a support for (x, a) on every constraint of P involving x. P is GAC-consistent iff every value of P is GAC-consistent. For binary CNs, GAC is referred to as AC (Arc Consistency). For any CN P, we know that there exists a greatest GAC-consistent network, denoted by $GAC(P)$ and called the GAC-closure of P, which is equivalent to P and such that $GAC(P) \preceq P$. When the domain of a variable of P is empty, P is clearly unsatisfiable, which is denoted by $P = \bot$. The CN $P|_{x=a}$ is obtained from P by removing every value $b \neq a$ from $dom(x)$. A value (x, a) of P is SAC-consistent (SAC stands for Singleton Arc Consistent) iff $GAC(P|_{x=a}) \neq \bot$. P is SAC-consistent iff every value of P is SAC-consistent.

Any domain-filtering consistency allows us to identify and remove inconsistent values. In order to compare the pruning capability of different consistencies, we can introduce a preorder [6]. Let ϕ and ψ be two consistencies. ϕ is stronger than ψ, denoted by $\phi \trianglerighteq \psi$, iff whenever ϕ holds on a CN P, ψ also holds on P. ϕ is strictly stronger than ψ, denoted by $\phi \triangleright \psi$ iff $\phi \trianglerighteq \psi$ and there exists at least one CN P such that ψ holds on P but not ϕ. When some consistencies cannot be ordered (none is stronger that the other), we say that they are *incomparable*. For classical domain-filtering consistencies defined on binary CNs, we have: SAC $\triangleright$ MaxRPC $\triangleright$ PIC $\triangleright$ AC (MaxRPC and PIC are respectively Max-Restricted Path Consistency and Path Inverse Consistency, see [3]).

3 Consistencies Based on Failed Values

In this section, we present two new basic consistencies. The first identifies nogoods of any size whereas the second identifies inconsistent values (i.e. nogoods of size 1). These two consistencies are based on *failed values* which are simply values proved to be inconsistent (e.g. during search). Failed values "convey" some information:

Lemma 1 (directly derived from [13]). *If a value (x, a) of a CN P is globally inconsistent then every solution S of P is such that $S[x/a]$ violates at least one constraint of P involving x.*

Proof. $S[x/a]$ is not a solution of P since (x, a) is globally inconsistent. This means that at least one constraint of P is not satisfied by $S[x/a]$. But we know that every constraint c of P that does not involve x is satisfied by $S[x/a]$ because the restriction of $S[x/a]$ over $scp(c)$ is exactly the restriction of S over $scp(c)$. Consequently, at least one constraint of P involving x is not satisfied by $S[x/a]$. □

If P is a binary CN, then every solution of P contains a value for a variable $y \neq x$ which is not compatible with (x, a). A failed value is defined as follows:

Definition 1. *Let P and P' be two CNs such that $P' \prec P$. A* failed value *of P' with respect to P is a value (x, a) of P such that $P|_{x=a}$ is unsatisfiable and $a \notin dom^{P'}(x)$.*

In practice, a failed value is a value pruned from a CN because it has been proved to be inconsistent. At any time during search, a failed value can be identified by inference and/or search methods [7,16]. For example, if $P^{init}|_{x=a}$ is shown to be unsatisfiable, clearly, a can be removed from $dom^{init}(x)$. We then obtain a smaller CN P with (x, a) being a failed value of P (with respect to P^{init}). However, note that failed values can be defined with respect to any intermediate CN reached during search. We need now to introduce *conflict sets*.

Definition 2. *Let P be a CN, x be a variable of P and $a \in dom^{init}(x)$.*

- *The* conflict set *of (x, a) on a constraint c of P involving x, denoted by $\chi_P(c, x, a)$, is the set of valid instantiations I of $scp(c) \setminus \{x\}$ on P such that $I \cup \{(x, a)\}$ does not satisfy c.*
- *The* conflict set *of (x, a) on P is $\chi_P(x, a) = \cup_{c \in cons(P) | x \in scp(c)} \chi_P(c, x, a)$.*

For every conflict set χ, $vars(\chi) = \cup_{I \in \chi} vars(I)$. When possible, we simplify $\chi_P(x, a)$ into $\chi(x, a)$. Figure 1 shows[1] a simple CN P with a binary constraint between w and x and a ternary constraint between w, y and z. Here $\chi(w, a) = \{\{(x, b)\}, \{(y, a), (z, a)\}\}$ and $\chi(w, c) = \{\{(x, b)\}, \{(x, c)\}, \{(y, c), (z, c)\}\}$; note that $\{(x, b)\}$ and $\{(y, a), (z, a)\}$ are two instantiations in $\chi(w, a)$ of size 1 and 2.

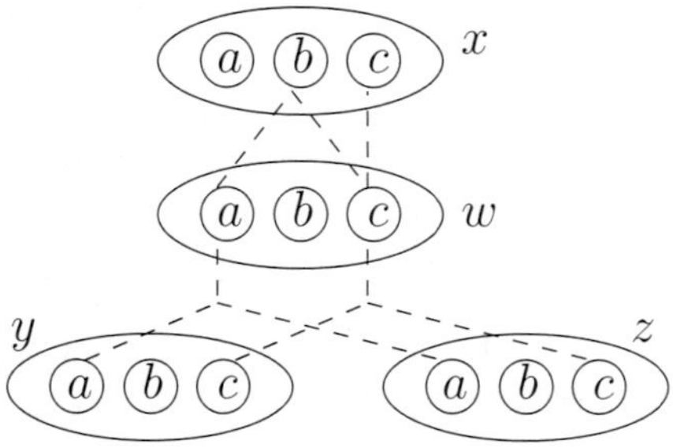

Fig. 1. Illustration of conflict sets

Failed values and instantiations can be connected as follows:

Definition 3. *Let (x, a) be a failed value of a CN P and I a valid instantiation on P.*

- *(x, a) is* covered *by I iff $vars(\chi_P(x, a)) \subseteq vars(I)$.*
- *(x, a) is* verified *by I iff $\exists J \in \chi_P(x, a) \mid J \subseteq I$.*

[1] In each figure of this paper, solid (resp. dashed) edges represent allowed (resp. forbidden) tuples. The absence of edges between two variables means that there is no binary constraint involving them.

Note that a failed value verified by an instantiation is not necessarily covered by it. However, it is shown below that when a failed value is covered by an instantiation but not verified, a nogood is identified. FVC (Failed Value Consistency) is a general nogood-identifying consistency.

Definition 4. *Let P be a constraint network.*

- *A valid instantiation I on P is* FVC-consistent *for a failed value (x, a) of P iff either (x, a) is not covered by I or (x, a) is verified by I.*
- *A valid instantiation I on P is* FVC-consistent *iff it is FVC-consistent for every failed value of P; otherwise, I is said to be* FVC-inconsistent.

Assume that (w, c) in Figure 1 is a failed value. $I = \{(x, a), (y, c), (z, c)\}$ is an instantiation that verifies (w, c) because I contains $\{(y, c), (z, c)\} \in \chi(w, c)$. $I' = \{(x, a)\}$ does not verify (w, c) but is FVC-consistent for (w, c) because (w, c) is not covered by I'. $I'' = \{(x, a), (y, a), (z, a)\}$ is FVC-inconsistent because (w, c) is both covered by I'' and not verified by I''.

Proposition 1. *Any FVC-inconsistent instantiation is globally inconsistent.*

Proof. Without any loss of generality, we consider here that I is a valid instantiation on a constraint network P that is FVC-inconsistent for a failed value (x, a) of P with respect to P^{init}; we have $P \prec P^{init}$. We know that there is no solution of P^{init} involving (x, a) because (x, a) is a failed value of P wrt P^{init}. We can even say more: for every solution S of P^{init}, the complete instantiation $S[x/a]$ is not a solution because at least one constraint involving x is violated (see Lemma 1). Because I is FVC-inconsistent for (x, a), we know that I covers $vars(\chi_P(x, a))$ and (x, a) is not verified by I. This means that it is not possible to extend I into a complete instantiation I' on P such that $I'[x/a]$ violates at least one constraint involving x. Every solution of P is a solution of P^{init} (since $P \prec P^{init}$) and every solution S of P^{init} is such that $S[x/a]$ violates at least one constraint involving x. We can deduce that I is globally inconsistent. □

Otherwise stated, some nogoods can be identified via deleted values (that are themselves nogoods). These nogoods are not necessarily of size 1. For example, in Figure 2 there is a failed value (w, a) and three binary constraints involving w. Any valid instantiation

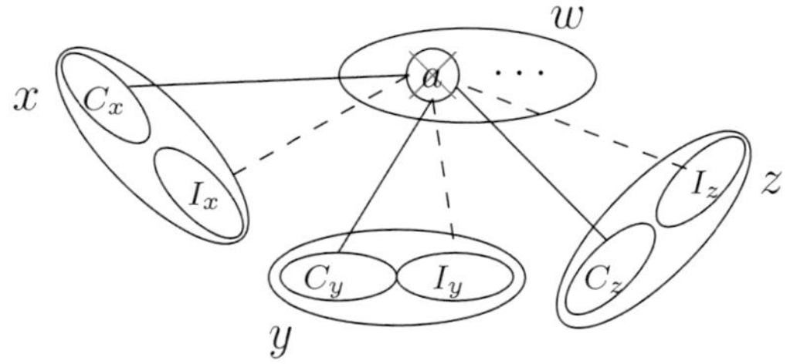

Fig. 2. A failed value (w, a), its compatible values in C_x, C_y and C_z and its incompatibles values in I_x, I_y and I_z

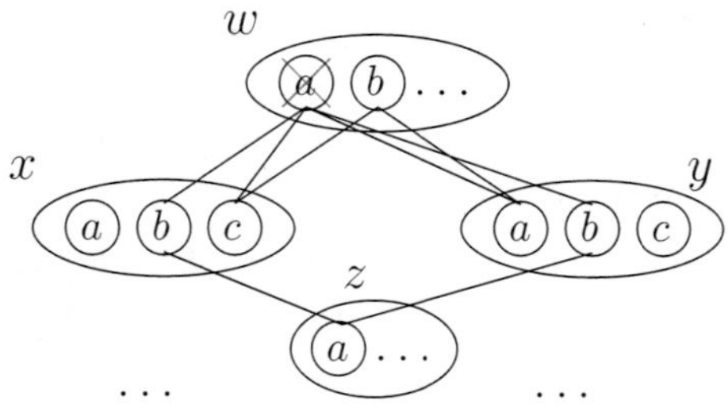

Fig. 3. Illustration of AFVC

of $\{x, y, z\}$ is globally inconsistent if it only contains values compatible with (w, a). In other words, every tuple in $C_x \times C_y \times C_z$ is a nogood (of size 3).

Interestingly, inference can be conducted differently by reasoning between each value and each failed value (through its conflict set). More precisely, we can define a related domain-filtering consistency, called Arc Failed Value Consistency (AFVC).

Definition 5. *Let P be a constraint network.*

- *A value (x, a) of P is AFVC-consistent for a failed value (y, b) of P iff (x, a) can be extended to a locally consistent instantiation verifying (y, b).*
- *A value (x, a) of P is AFVC-consistent iff (x, a) is AFVC-consistent for every failed value of P; otherwise, (x, a) is said to be AFVC-inconsistent.*
- *P is AFVC-consistent iff every value of P is AFVC-consistent.*

For the first item above, note that we may have $x = y$. In this case, necessarily, we have $a \neq b$ since (x, a) is a current value whereas (y, b) is a pruned one. Note that AFVC can be regarded as a local consistency since it suffices to reason from the conflict set of each failed value. In particular for binary constraints, a value (x, a) is AFVC-consistent for a failed value (y, b) iff (x, a) is compatible with a valid value in $\chi(y, b)$. The AFVC algorithm given in Section 4 is based on this simple observation.

Proposition 2. *Any AFVC-inconsistent value is globally inconsistent.*

Proof. Let (x, a) be a current value of P that is AFVC-inconsistent for a failed value (y, b) of P. Suppose that there exists a solution S of P such that $S[x] = a$. Necessarily, as (x, a) is AFVC-inconsistent for (y, b), $S[y/b]$ cannot violate any constraint involving y, and consequently is a solution of P. This contradicts the fact that (y, b) is globally inconsistent (since it is a failed value), and consequently our hypothesis. We can deduce that (x, a) is globally inconsistent. □

As an illustration, let us consider the CN (partially) depicted in Figure 3. This CN can be completed (see dots) so that some classical consistencies hold (e.g. arc consistency). We assume here that (w, a) is a failed value and $\chi(w, a) = \{\{(x, a)\}, \{(y, c)\}\}$. Observe that (w, b) and (z, a) are AFVC-inconsistent. Indeed, (w, b) (resp. (z, a)) is not compatible with any value in $\chi(w, a)$.

Restricted to nogoods of size 1, FVC is strictly weaker than AFVC (the proof is omitted). We have:

Proposition 3. *Let (x, a) be a value of P. If $I = \{(x, a)\}$ is FVC-inconsistent then (x, a) is AFVC-inconsistent.*

Finally, one can show that AFVC verifies certain properties (e.g. see [1,3,16]) that permits us to define the AFVC-closure.

Proposition 4. *For any CN P, there exists a greatest AFVC-consistent CN equivalent to P, called the AFVC-closure of P and denoted by $AFVC(P)$, such that $AFVC(P) \preceq P$.*

$AFVC(P)$ can be obtained by iteratively removing, in any order, values that are not AFVC-consistent.

4 Algorithms for FVC and AFVC

We propose to embed filtering algorithms based on failed values within MAC (Maintaining Arc Consistency) [23]. MAC is a backtrack search algorithm that develops a binary search tree: at each node, an uninstantiated variable x is selected, a value a in $dom(x)$ is selected, a left subtree starting with a branch labelled with the positive decision $x = a$ (variable assignment) is first explored and a right subtree starting with a branch labelled with the negative decision $x \neq a$ (value refutation) is later explored. The consistency enforced at each node is GAC.

```
function checkFailedValue(P: CN, (x, a): failed value):Boolean
begin
    if isValid(res[x, a]) then
        return true
    foreach constraint c of P such that x ∈ scp(c) do
        τ ← seekConflict(c, x, a)
        if τ ≠ nil then
            res[x, a] ← τ
            return true
    return false
end
function checkFVC(P: CN, F: set of failed values):Boolean
begin
    foreach failed value (x, a) ∈ F do
        if ¬checkFailedValue(P, (x, a)) then
            return false
    return true
end
```

Algorithm 1. Checking FVC

For binary CNs, an immediate solution to make use of FVC is to post, for each failed value (x, a), a non-binary constraint whose scope is $vars(\chi(x, a))$: its associated relation forbids any instantiation FVC-inconsistent for (x, a). For our example in Figure 2, we would obtain a ternary constraint c_{xyz} such that $rel(c_{xyz}) = dom(x) \times dom(y) \times dom(z) \setminus C_x \times C_y \times C_z$. Interestingly, one may conceive efficient filtering algorithms (propagators) to enforce GAC on such constraints. However, this approach is intrusive and its generalization to non-binary CNs is rather complex. This is why we propose a weakened approach for the general case: checking FVC at each node of the search tree by checking that no identified failed value is currently covered and unverified. To identify failed values during search, it suffices to keep the set F of negative decisions labelling the current branch (i.e. the branch leading from the root of the search tree to the current node). Note that failed values that correspond to values removed when enforcing GAC can be discarded because these failed values are always verified (as long as no domain wipe-out occurs): for each of these values, there exists a constraint with only conflicts.

To check FVC, Algorithm 1 is called. For each failed value, we seek a conflict; seekConflict(c, x, a) seeks a conflict for (x, a) on c. When a conflict is found for a failed value (x, a), it is stored in a backtrack-stable data structure called res; this plays the role of residues as in [17] when establishing GAC. The validity of the residual conflict is first tested at line 3. When no conflict exists for a failed value, $false$ is returned, which forces the search algorithm to backtrack. The worst-case space complexity of this algorithm is $O(nd)$ whereas its worst-case time complexity is $O(|F|ed^{r-1})$ where $|F|$ denotes the number of failed values in F, e the number of constraints, d the greatest domain size and r the greatest constraint arity. For binary constraint networks, the worst-case time complexity is only $O(|F|ed)$.

We now propose an algorithm to enforce the domain-filtering consistency AFVC on binary constraint networks. Given a binary CN P with a set of failed values F, the procedure enforceAFVC (see Algorithm 2) computes $AFVC(P)$ and returns $false$ when a domain wipe-out occurs. The data structures are as follows. For each failed value (x, a), $\chi(x, a)$ is considered to be an array of values indexed from 1 to $length(\chi(x, a))$. These values correspond to the instantiations (of size 1 since P is binary) in the conflict set of (x, a). The 2-dimensional array $last$ maps each pair composed of a failed value (x, a) in F and a value (y, b) of P to an an integer corresponding to the index of the most recent value found in $\chi(x, a)$ that is present in P and compatible with (y, b): $last[(x, a)][(y, b)]$ indicates the position of the last found (so-called) AFVC-support for (y, b) on (x, a). For each value (z, c) of P, $S(z, c)$ is a list storing the pairs (failed value,value) for which (z, c) is the last found AFVC-support. Structures S and $last$ are inspired from those used in AC6 and AC2001 (see e.g. [3]).

Certainly, dynamically computing (or updating) conflict sets at each node would be prohibitive. This is why we consider that conflict sets are computed for the initial problem instance by the call initialize(P^{init}). The function enforceAFVC tries to identify an AFVC-support for each pair (failed value,value). If no support can be found for a value (y, b) on a failed value (x, a), b is removed from $dom(y)$ and (y, b) is added to the propagation queue Q. Each removed value $(z, c) \in Q$ is "propagated": a new support must be found for each pair stored in $S(z, c)$ (this is done from the last recorded position).

The worst-case space and time complexities of enforceAFVC are $O(nd(M + |F|))$ and $O(M|F|nd)$ respectively, where n is the number of variables, d the greatest

```
procedure initialize(P: binary CN)
begin
    foreach value (x, a) of P do
        χ(x, a) ← ∅
        S(x, a) ← ∅
    foreach constraint c_xy of P do
        foreach tuple (a, b) ∈ dom(x) × dom(y) | (a, b) ∉ rel(c_xy) do
            add (x, a) to χ(y, b)
            add (y, b) to χ(x, a)
end
function seekAFVCSupport((x, a): failed value, (y, b): value): Boolean
begin
    position ← last[(x, a)][(y, b)] + 1
    while position ≤ length(χ(x, a)) do
        (z, c) ← χ(x, a)[position]
        if c ∈ dom(z) ∧ (y, b) and (z, c) are compatible then
            last[(x, a)][(y, b)] ← position
            add ((x, a), (y, b)) to S(z, c)
            return true
        position ← position + 1
    return false
end
function enforceAFVC(P: binary CN, F: set of failed values): Boolean
begin
    Q ← ∅
    foreach failed value (x, a) ∈ F do
        foreach value (y, b) of P do
            last[(x, a)][(y, b)] ← 0
            if ¬seekAFVCSupport((x, a), (y, b)) then
                remove b from dom(y)
                if dom(y) = ∅ then
                    return false
                add (y, b) to Q
    while Q ≠ ∅ do
        pick and delete (z, c) from Q
        foreach ((x, a), (y, b)) ∈ S(z, c) do
            if b ∈ dom(y) ∧ ¬seekAFVCSupport((x, a), (y, b)) then
                remove b from dom(y)
                if dom(y) = ∅ then
                    return false
                add (y, b) to Q
        S(z, c) ← ∅
end
```

Algorithm 2. Enforcing AFVC

domain size, $|F|$ the number of failed values and M the maximum size of a conflict set ($M = max_{(x,a) \in P^{init}} |\chi(x,a)|$). Indeed, the space required by χ arrays is $O(ndM)$, and that required by both $last$ and S is $O(|F|nd)$. On the other hand, the cumulated complexity of seekAFVCSupport for each pair (failed value,value) is $O(M)$, and there are $O(|F|nd)$ different pairs. Note that initialize has a $O(ed^2)$ time complexity but this is amortized since initialize is only called initially (for P^{init}). By definition, $|F| < nd$ and $M < nd$ hence $M|F|nd < n^3d^3$. In practice, it may be worthwhile to bound the number of failed values and/or to bound the maximum size of conflict sets by a constant in order to concentrate only on promising failed values. If both are bound, the complexity becomes $O(nd)$. The algorithm presented above can be easily adapted to be used at each node of the search tree developed by MAC (the complexity remains the same for a branch of the search tree).

5 Substitutability and Usual Consistencies

Neighborhood substitutability is a weak form of substitutability [11] that can be related to consistencies based on failed values. A value $a \in dom(x)$ is neighborhood substitutable for a value $b \in dom(x)$ iff for every constraint c involving x and every support I for (x,b) on c, $I[x/a]$ is a support for (x,a) on c. For example, it can be exploited as a reduction operator by applying a convergent sequence of neighborhood substitution deletions [5]. We have the following interesting proposition.

Proposition 5. *If a value (x,a) is neighborhood substitutable for a value (x,b) on a CN $P' \succ P$, if (x,a) is a failed value of P and if (x,b) is a value of P then (x,b) is AFVC-inconsistent.*

Proof. The definition of neighborhood substitutability can be reformulated as: (x,a) is neighborhood substitutable for (x,b) iff $\chi(x,a) \subseteq \chi(x,b)$. If (x,a) is a failed value of P, then it is not possible to extend (x,b) into a consistent instantiation that verifies (x,a). (x,b) is then AFVC-inconsistent. □

AFVC can be seen as a lazy dynamic mechanism to detect values that can be substituted (and are globally inconsistent). Importantly, it allows us to identify inconsistent values for which no neighborhood substitutable value exists. Indeed, a value (x,b) is AFVC-inconsistent if the conflict set of (x,b) is included in the conflict set of a failed value. However, whereas only values for the same variable are considered for neighborhood substitutability, AFVC is more general. An illustration is given in Figure 3: (w,a) is substitutable for (w,b) but not for (z,a) since $w \neq z$.

Interestingly, AFVC is incomparable with most of the domain-filtering consistencies. More precisely, it is incomparable with "usual" consistencies, i.e. local consistencies ϕ that do not rely on failed values and that verify the four basic properties: a) ϕ holds on any CN only involving entailed constraints (a constraint is entailed on P iff it is satisfied by every valid instantiation on its scope), b) ϕ holds on any CN iff it holds on each of its connected sub-networks, c) there exist unsatisfiable CNs where ϕ holds and d) there exist some CNs where ϕ does not hold. For example, (G)AC, SAC, PIC, . . . are "usual" but global consistency (defined as: any locally consistent instantiation can be extended to a solution) is not usual.

Proposition 6. *AFVC is incomparable with usual consistencies.*

Proof. Let us consider a "usual" local consistency ϕ. Let us consider a (satisfiable) CN P_1 that only contains entailed constraints, a CN P_2, unsatisfiable but ϕ-consistent, on a separate set of variables, and the problem $P = P_1 \cup P_2$. Since P_2 is unsatisfiable, any value (x, a) of P_1 can be identified as a failed value (e.g. after search). We now assume that (x, a) is a failed value of P_1. Since P_1 contains only entailed constraints, $\chi(x, a) = \varnothing$ and therefore, no instantiation can verify (x, a). Thus, P is not AFVC-consistent. In contrast, ϕ holds on P (by hypothesis), hence $\phi \ntrianglerighteq$ AFVC. Besides, AFVC $\ntrianglerighteq \phi$: it suffices to choose a CN P with no failed value such that ϕ does not hold. Consequently, AFVC and ϕ are incomparable. □

6 A Hierarchy of Consistencies

In [10], Freuder introduced the general class of (i, j)-consistencies. Informally, a constraint network is (i, j)-consistent iff every locally consistent instantiation of a set of i variables can be extended to a locally consistent instantiation involving any j additional variables. Arc consistency, path consistency [20,19] and path inverse consistency (PIC) [12] all belong to this class since they correspond to $(1, 1)$-consistency, $(2, 1)$-consistency and $(1, 2)$-consistency, respectively. Another important class of consistencies defined in terms of (existing) constraints is that of relational (i, m)-consistencies [8]. Informally, a constraint network is relational (i, m)-consistent iff for every set C of m constraints and every set $X \subseteq Y$ of i variables, where $Y = \cup_{c \in C} scp(c)$, every locally consistent instantiation of X can be extended to a valid instantiation of Y satisfying each constraint of C. Generalized arc consistency and relational path-inverse-consistency [4] respectively correspond to relational $(1, 1)$-consistency and relational $(1, 2)$-consistency.

Here, we propose a new general class of original consistencies, based on the concept of failed value.

Definition 6 (Failed Value (i, f)-consistency). *P is FV(i, f)-consistent iff for every set X of i variables of P and every set Y of f failed values of P, every locally consistent instantiation of X can be extended to a locally consistent instantiation verifying each failed value in Y.*

From this general definition, many consistencies can be derived: Arc Failed Value Consistency (AFVC) is FV$(1, 1)$-consistency; Path Failed Value Consistency (PFVC) is FV$(2, 1)$-consistency; Path-Inverse Failed Value Consistency (PIFVC) is FV$(1, 2)$-consistency. Inspired from their variable-based counterparts, MaxRPC and SAC, two additional natural consistencies are introduced.

Definition 7 (MaxFVC). *A value (x, a) of P is MaxFVC-consistent iff for every failed value (y, b) of P, (x, a) can be extended to a locally consistent instantiation I verifying (y, b) such that for every additional failed value (z, c) of P, I can be extended to a locally consistent instantiation verifying (z, c).*

Definition 8 (SAFVC). *A value (x, a) of P is SAFVC-consistent iff $AFVC(P|_{x=a}) \neq \bot$.*

Proposition 7. *PIFVC $\rhd$ AFVC and SAFVC $\rhd$ AFVC.*

Proof. From definitions, we directly deduce that PIFVC $\trianglerighteq$ AFVC and SAFVC $\trianglerighteq$ AFVC. To show strictness, let us consider Figure 4 where a CN P (partially depicted) admits two failed values (w, a), and (x, a). Here, we have $\chi(w, a) = \{\{(y, b)\}, \{(z, b)\}\}$ and $\chi(x, a) = \{\{(y, a)\}, \{(z, b)\}\}$. P is AFVC-consistent but neither SAFVC-consistent nor PIFVC-consistent. Indeed, (z, a) cannot be extended to an instantiation verifying both failed values and $AFVC(P|_{z=a}) = \bot$. □

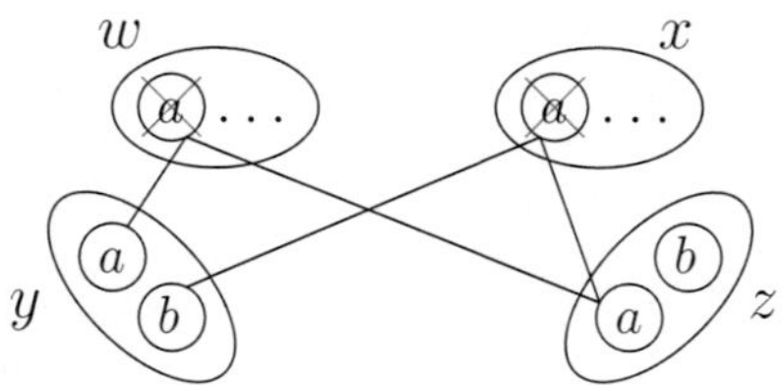

Fig. 4. AFVC holds but neither SAFVC nor PIFVC holds

Figure 5 summarizes the relations between domain-filtering consistencies based on failed values. Due to lack of space, proofs are omitted.

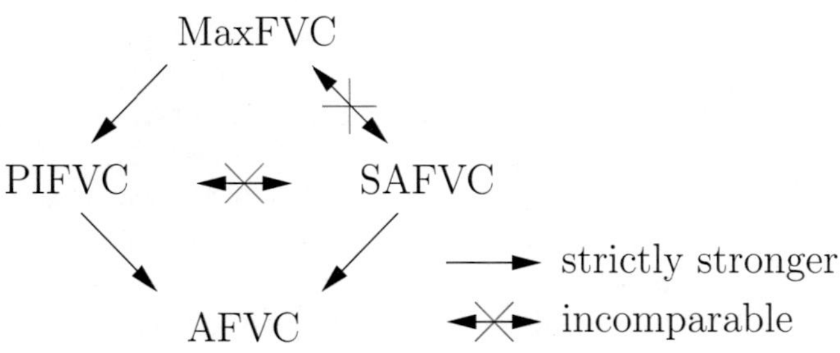

Fig. 5. Relationships between domain-filtering consistencies based on failed values

7 Preliminary Experimental Results

In order to show the practical interest of consistencies based on failed values, we have performed experiments with our constraint solver Abscon. On a computer equipped with a 3GHz processor and 2GB of RAM under Linux, MAC was employed with $dom/ddeg$ and $lexico$ as variable[2] and value ordering heuristics, as our baseline. We first compared, on binary instances, MAC with MAC embedding the procedure (see

[2] Using $dom/wdeg$ does not guarantee exploring the same search tree when additional filtering is performed. This is why we have chosen $dom/ddeg$.

Table 1. Impact of checking FVC and enforcing AFVC on binary instances

	MAC		MAC+FVC		MAC+AFVC	
Instance	CPU	nodes	CPU	nodes	CPU	nodes
Graph coloring						
1-fullins-4-4	106	$7M934$	4.8	$215,812$	5.6	$110,636$
2-fullins-4-4	10.7	$177,424$	4.8	$67,924$	2.3	$12,764$
2-insertions-4-3	10.5	$455,533$	3.2	$112,886$	5.4	$62,753$
2-insertions-5-3	3.1	$7,767$	1.8	$3,494$	2.4	$2,941$
Composed						
composed-25-5	178	$10M864$	85	$4M838$	148	$2M333$
composed-25-7	106	$7M934$	4.78	$215,812$	5.6	$110,636$
Job-shop						
os-taillard-4-100-1	84	$1M870$	30.2	$247,383$	83	$206,845$
os-taillard-4-100-3	7.6	$147,270$	2.3	$29,698$	3.0	$17,818$
os-taillard-4-95-2	7.5	$195,665$	3.3	$51,957$	19.7	$41,226$
Queen Attacking						
qa-5	7.5	$318,601$	6.4	$240,495$	10.0	$238,940$
qa-6	311	$7M703$	259	$5M883$	443	$5M576$

Algorithm 1) that checks FVC at each search step (denoted by MAC+FVC) and also with MAC embedding a function that enforces AFVC at each search step (denoted by MAC+AFVC). For our preliminary tests, we have implemented a less sophisticated version of Algorithm 2. This algorithm does enforce AFVC but is quite simpler (and theoretically less efficient) as it does not integrate $last$ and S structures.

Table 1 shows the results obtained on some binary instances[3] with these three backtrack search variants in terms of CPU time and number of visited nodes. Because $dom/ddeg$ is a non-adaptive heuristic, we have the guarantee that MAC+FVC visits less nodes than MAC and that MAC+AFVC visits less nodes than MAC+FVC. In Table 1, for some instances, the number of visited nodes is significantly reduced when consistencies based on failed values are used. MAC+FVC is clearly the most efficient algorithm since it is usually better than MAC+AFVC and sometimes one order of magnitude faster than MAC alone. Interestingly, reasoning from failed values allows us to benefit from the structure (related to substitutability for graph coloring and job-shop instances) contained in these instances. On some other series of instances (not presented here) including random ones, MAC+FVC does not prune the search tree. However, we have observed that this is never penalizing in terms of CPU time. This is because the worst-case time complexity of checking FVC is limited and the number of failed values is usually small. Although one may be disappointed by the relative inefficiency of MAC+AFVC, one should consider that a non optimized AFVC algorithm has been used here and that many developments to control the complexity of AFVC (or even stronger consistencies) remain to be studied.

[3] Available at `http://www.cril.fr/~lecoutre/research/benchmarks/`

Table 2. Impact of checking FVC on non-binary instances

	MAC		MAC+FVC	
Instance	CPU	nodes	CPU	nodes
Dimacs				
hole-08	5.8	$0M699$	1.9	$0M177$
hole-09	80	$9M062$	16	$1M753$
aim-100-1-6-sat-3	21.4	$3M173$	5.01	$0M505$
aim-100-1-6-sat-4	160	$22M400$	47.1	$5M292$
Renault-mod				
renault-mod-15	505	$1M969$	118	$0M511$
renault-mod-18	$1,193$	$6M812$	253	$1M419$
renault-mod-43	796	$4M221$	23	$0M127$
renault-mod-45	85.5	$0M591$	15.5	$89,494$

Table 2 shows the results obtained on some series of non-binary instances. Clearly, MAC+FVC outperforms MAC alone. Finally, note that MAC+FVC was used in Abscon during the third constraint solver competition[4] with good results.

8 Conclusion

In this paper, we have shown how values detected as globally inconsistent during search, and called failed values, can be useful to prune the search space through the introduction of a new class of consistencies that are orthogonal to the usual ones. Whereas FVC is a consistency that is cheap to check and that makes MAC more robust, AFVC and its direct extensions require further developments to determine the best way of controlling the enforcement of these new domain-filtering consistencies. We have also noticed that AFVC allows us to detect in a lazy manner a generalized form of neighborhood substitutability.

Although it is related to approaches that eliminate redundancies by posting constraints (for SAT) [21,14] or decomposing problems [13], this way of reasoning has been developed so as to yield a hierarchy of increasingly stronger consistencies. For binary constraint networks, failed values basically permit us to identify and exploit a form of nogood in the same spirit as generalized nogood in [15], global cut seed in [9], kernel [22] and partial state [18]. We believe that identifying common properties to these different approaches is an exciting perspective.

References

1. Apt, K.R.: Principles of Constraint Programming. Cambridge University Press, Cambridge (2003)
2. Bennaceur, H., Lecoutre, C., Roussel, O.: A decomposition technique for solving Max-CSP. In: Proceedings of ECAI 2008, pp. 500–504 (2008)

[4] See `http://www.cril.univ-artois.fr/CPAI08/`

3. Bessiere, C.: Constraint propagation. In: Handbook of Constraint Programming, ch. 3. Elsevier, Amsterdam (2006)
4. Bessiere, C., Stergiou, K., Walsh, T.: Domain filtering consistencies for non-binary constraints. Artificial Intelligence 72(6-7), 800–822 (2008)
5. Cooper, M.C.: Fundamental properties of neighbourhood substitution in constraint satisfaction problems. Artificial Intelligence 90, 1–24 (1997)
6. Debruyne, R., Bessiere, C.: Domain filtering consistencies. Journal of Artificial Intelligence Research 14, 205–230 (2001)
7. Dechter, R.: Constraint processing. Morgan Kaufmann, San Francisco (2003)
8. Dechter, R., van Beek, P.: Local and global relational consistency. Theoretical Computer Science 173(1), 283–308 (1997)
9. Focacci, F., Milano, M.: Global cut framework for removing symmetries. In: Walsh, T. (ed.) CP 2001. LNCS, vol. 2239, pp. 77–92. Springer, Heidelberg (2001)
10. Freuder, E.C.: A sufficient condition for backtrack-bounded search. Journal of the ACM 32(4), 755–761 (1985)
11. Freuder, E.C.: Eliminating interchangeable values in constraint satisfaction problems. In: Proceedings of AAAI 1991, pp. 227–233 (1991)
12. Freuder, E.C., Elfe, C.: Neighborhood inverse consistency preprocessing. In: Proceedings of AAAI 1996, pp. 202–208 (1996)
13. Freuder, E.C., Hubbe, P.D.: Using inferred disjunctive constraints to decompose constraint satisfaction problems. In: Proceedings of IJCAI 1993, pp. 254–261 (1993)
14. Gallo, G., Urbani, G.: Algorithms for testing the satisfiability of propositional formulae. Journal of Logic Programming 7(1), 45–61 (1989)
15. Katsirelos, G., Bacchus, F.: Generalized nogoods in CSPs. In: Proceedings of AAAI 2005, pp. 390–396 (2005)
16. Lecoutre, C.: Constraint networks: techniques and algorithms. ISTE/Wiley (2009)
17. Lecoutre, C., Hemery, F.: A study of residual supports in arc consistency. In: Proceedings of IJCAI 2007, pp. 125–130 (2007)
18. Lecoutre, C., Sais, L., Tabary, S., Vidal, V.: Exploiting past and future: Pruning by inconsistent partial state dominance. In: Bessière, C. (ed.) CP 2007. LNCS, vol. 4741, pp. 453–467. Springer, Heidelberg (2007)
19. Mackworth, A.K.: Consistency in networks of relations. Artificial Intelligence 8(1), 99–118 (1977)
20. Montanari, U.: Network of constraints: Fundamental properties and applications to picture processing. Information Science 7, 95–132 (1974)
21. Purdom, P.W.: Solving satisfiability with less searching. IEEE transactions on pattern analysis and machine intelligence 6(4), 510–513 (1984)
22. Razgon, I., Meisels, A.: A CSP search algorithm with responsibility sets and kernels. Constraints 12(2), 151–177 (2007)
23. Sabin, D., Freuder, E.C.: Contradicting conventional wisdom in constraint satisfaction. In: Proceedings of CP 1994, pp. 10–20 (1994)

A Precedence Constraint Posting Approach for the RCPSP with Time Lags and Variable Durations

Michele Lombardi and Michela Milano

DEIS, Università di Bologna V.le Risorgimento 2, 40136, Bologna, Italy

Abstract. Resource Constrained Project Scheduling Problem is a very important problem in project management, manufacturing and resource optimization. We focus on a variant of RCPSP with time lags and variable activity durations. The solving approach is based on Precedence Constraint Posting that adds new precedence constraints to the original project graph so that all resource conflicts are solved and a consistent assignment of start times can be computed for whatever combination of activity durations. We propose a novel method for computing resource conflicts based on the minimum flow on the resource graph and we use it in an efficient complete search strategy. We experiment the approach on instances coming from the scheduling of parallel applications on multi processor systems on chip.

1 Introduction

In this work we tackle the Resource Constrained Project Scheduling Problem (RCPSP) with minimum and maximum time lags and variable durations. RCPSP aims to schedule a set of activities subject to precedence constraints and the limited availability of resources.

The classical RCPSP formulation is based on a directed acyclic graph where nodes are activities and arcs are precedence relations. Activities have release time and deadlines and use a certain amount of finite capacity resources. The problem consists in finding an assignment of start times to activities, such that no resource capacity is exceed at any point of time and the *makespan* is minimized. Here we introduce two variants to the classical RCPSP formulation: time lags (i.e., minimum and maximum time interval) between activities and variable duration. We assume activity durations are not known a priori, but range at run time between a worst and best case execution time. We want to schedule all activities such that, whatever their duration is, all temporal and resource constraints are satisfied.

This problem variant finds a broad spectrum of industrial and design applications, whenever uncertainty about durations is an issue and there are deadlines to be mandatory met. As a case study, we consider predictable scheduling of parallel computer programs on multiprocessor systems, subject to hard real time constraints. It is common in the embedded system research community to

I.P. Gent (Ed.): CP 2009, LNCS 5732, pp. 569–583, 2009.

cast a system design problem into that of scheduling a so called Task Graph (indeed equivalent to a project graph), annotated with durations and resource requirements, on a target hardware platform (set of processors, memories, communication channels...). The objective is generally to maximize the application performance (i.e. minimize the makespan) or to meet a fixed deadline. An acknowledged issue is that task durations are data dependent and therefore not a-priori known, despite lower and upper bounds can safely be identified; this is especially troublesome when dealing with real time applications, when a specified performance must be guaranteed for whatever execution scenario. Minimum and maximum time lags are useful as they enable the representation of setup times for specific sequences of tasks: for example, if dynamic voltage scaling is supported, a task t_1 executing on a processors at clock frequency f_1 cannot be scheduled *immediately* after a task t_2 with frequency f_2, as requires the insertion of some setup time to enable frequency switching.

We adopt a Precedence Constraint Posting approach (see [14]): the solution we provide is an augmented project graph; this is the original project graph plus a *fixed* set of new precedence constraints, such that all possible resource conflicts are cleared and a consistent assignment of start times can be computed for whatever combination of activity durations at run time. In practice, once a solution (i.e. an augmented graph) is given, a feasible assignment of start times to activities can be computed in polynomial time.

The main contribution of the paper concerns the detection of minimal conflict sets based on the minimal flow problem and its integration in an efficient complete search strategy. Moreover, to the best of the authors knowledge, this is the first paper dealing with both variable durations and minimum/maximum time lags in RCPSP. We perform computational experiments on instances taken from the field of system design, in particular for scheduling parallel multi-task applications on multiprocessor systems. We compare our approach with a complete approach implemented in [9] and we show order of magnitude speed ups due mainly to the search strategy based on the min flow conflict detection.

2 Problem Definition

The most classical RCPSP is defined on a directed acyclic graph $\langle A, E \rangle$ (referred to as *project graph* in the following), where A is a set of n activities a_i having fixed duration d_i, and E is a set of directed edges (t_i, t_j), defining precedence relations over the set A. Time window constraints exist on the start and end time of each activity; specifically, we assume every activity has to start after a specified *release time* (rs_i) and to end before a specified *deadline* (dl_i). Without loss of generality, we assume there is a single source activity (a_0) with no ingoing arcs and a single sink activity (a_{n-1}) with no outgoing arcs. Each activity requires a certain amount $req(a_i, r_k)$ of one or more renewable resources r_k within a set R; all resources have finite capacity $cap(r_k)$. The problem consists in finding a *schedule* (that is, an assignment of start times to activities), such that no resource capacity is exceed at any point of time and the overall completion time (*makespan*) is minimized. In the RCPSP with minimum and maximum time lags each arc (a_i, a_j) is labeled with a minimum and maximum value (d_{ij}, D_{ij}), such that the time distance between a_i and a_j cannot be lower than d_{ij} nor higher than D_{ij}. Finally, unlike in classical RCPSP definition, we assume activity durations are not know a priori, but ranging at execution time between a lower and an upper bound d_i and D_i.

We require the time window constraints to be met at run time for every possible scenario, that is for every possible combination of duration values. Therefore, all possible resource conflicts should be cleared so that a consistent assignment of start times can be computed for whatever combination of activity durations. Finally, dealing with variable activity durations requires to reformulate the notiton of optimality: in particular we are interested here in finding the best possible global deadline which can be met by a given project.

3 PCP: Background and Related Work

We adopt a Precedence Constraint Posting approach (PCP, see [14]); in PCP possible resource conflicts are resolved off-line by adding a fixed set of precedence constraints between the involved activities. The resulting augmented graph defines a set of possible schedules, rather than a schedule in particular. More precisely, the graph can be used to devise an on-line linear time policy to schedule an activity a_i depending on the execution history so far; in particular when the project is executed the start time of each activity a_i: (1) must be within the time window (2) must be greater than the end time of all activity predecessors. The provided solution graph must be such that if these rules are followed, a feasible assignment of start times is guaranteed to be found for every possible combination of task durations. This amounts to enforcing dynamic controllability, as defined in [20]: a project is controllable if we can guarantee all deadline constraints are met; in particular, dynamic controllability requires that the activity starting times should be decided knowing only already executed activities.

Central to any PCP approach is the definition of Minimal Conflict Set (MCS), making its first appearance in [4] (1983), where a branching scheme based on resolution of the so called "minimal forbidden sets" is first proposed. A MCS is a set of activities collectively overusing one of the problem resources and such that the removal of a single activity from the set wipes out the conflict as well; additionally, the activities must have the possibility to overlap in time. Following [8], we formally define a MCS for a resource r_k as a set of activities such that:

1. $\sum_{a_i \in MCS} req(a_i, r_k) > cap(r_k)$
2. $\forall a_i \in MCS \; : \; \sum_{a_j \in MCS \setminus \{a_i\}} req(a_j, r_k) \leq cap(r_k)$
3. $\forall a_i, a_j \in MCS$ with $i < j \; : \; start(a_i) \preceq end(a_j) \wedge end(a_j) \preceq start(a_i)$ is consistent with current state of the model, where $t_i \preceq t_j$ denotes that a time instant t_i can occur before t_j.

where 1. requires the set to be a conflict, 2. is the minimality condition and 3. requires activities to be possibly overlapping. A MCS can be resolved by posting a single precedence constraint between any pair of activities in the set; complete search can thus be performed by using MCS as choice points and posting on each branch a precedence constraint (also referred to as *resolver*). This is the case of many PCP based works: for example [8] makes use of complete search to detect MCS and proposes a heuristic to rank possible resolvers. Other branch and bound approaches based on posting precedence constraints to resolve MCS are reported in [18], where minimum and maximum time lags are also considered; variable activity durations are not taken into account in any of these works.

One of the key difficulties with complete PCP approaches is that detecting MCS by complete enumeration can be time consuming, as their number is in general exponential in the size of the graph. A possible way to overcome this issue is to only consider MCS which can occur given for a specified "execution policy": this is the case for example of the already cited work [18] where an earliest start policy (start every activity as soon as possible) is implicitly assumed.

Having an earliest start policy [16,17] is also a basic assumption for many stochastic RCPSP approaches; here variable task durations are explicitly considered and usually modeled as stochastic variables with known distribution. Most works in this area assume the objective function is to minimize the makespan and focus on computing restrictions of earliest start policies to resolve resource conflicts on-line. This is the case of preselective policies, introduced in [4,5], and a priori selecting for each minimal conflict set an activity to be delayed in case the conflict occurs. A major drawback with this approach is the need to enumerate all possible MCS [3,2]. To overcome this limitation, [19] introduces so called linear preselective policies: in this case a fixed priority is assigned to each activity such that when a conflict is about to occur at run time, the lowest priority activity is always delayed. To the best of the authors knowledge, no stochastic RCPSP approach has considered minimum and maximum time lags so far. Another way to fix the issue of efficient conflict detection is to drop completeness and resort to heuristic methods; see for example the method described in [14], which also incorporates time reasoning by representing the target project as a simple temporal network; many feature of this efficient and expressive model will be leveraged by our approach.

Our answer to the conflict detection issue is to cast the detection of a conflict to a (polynomial complexity) minimum flow problem by exploiting the transitivity of the solution graph. This is a first major contribution of this work. A related technique is outlined in [13], where a maximum (instead of minimum) flow algorithm is used to extract usage envelopes from a particular activity-resource graph; this latter technique is also employed in [15]. Finally, for excellent overviews about methods for the RCPSP problem and dealing with uncertainty in scheduling see [1,7,6].

4 Description of the Approach

We propose a PCP based approach for the RCPSP with minimum and maximum time lags and variable, bounded activity durations. The method performs complete search branching on MCS. The first major contribution w.r.t. similar approaches like [8] is the use of an efficient, polynomial time, MCS detection procedure based on the solution of a minimum flow problem.

Similarly to [14], our approach incorporates an expressive and efficient time model, taking care of consistency checking of time windows constraints and enabling the detection of possibly overlapping activities. Our second main contribution is the extension of such a time model in order to provide efficient time reasoning with variable durations and enable constant time consistency/overlapping check.

4.1 The Time Model

The time model we devised relates to Simple Temporal Networks with Uncertainty (STNU, [20]), for which dynamic controllability has been proved to be achievable in polynomial time (see [12]).

The model itself consists of a set T of temporal events or *time points* t_i with associated time windows, connected by directional binary constraints so as to form a directed graph. Such constraints are in either of the two forms:

1. $t_i \xrightarrow{[d_{ij},D_{ij}]} t_j$ (*free* constraint, with STNU terminology), meaning that a value δ in the interval $[d_{ij}, D_{ij}]$ must exist such that $t_j = t_i + \delta$; equivalently we can state that $d_{ij} \leq t_j - t_i \leq D_{ij}$.
2. $t_i \xrightarrow{[d_{ij}:D_{ij}]} t_j$ (*contingent* constraint, with STNU terminology), meaning that t_i and t_j must have enough flexibility to let $t_j = t_i + \delta$ hold for each value $\delta \in [d_{ij}, D_{ij}]$.

Both d_{ij} and D_{ij} are assumed to be non-negative; therefore, unlike in STN, a constraint cannot be reverted. The above elements are sufficient to represent a variety of time constraints; in particular, an instance of the problem at hand, can be modeled by:

1. introducing two time points s_i, e_i for the start and the end of each activity, respectively with time windows $[st_i, \infty]$ and $[0, dl_i]$;
2. adding a precedence constraint $s_i \xrightarrow{[d_i:D_i]} e_i$ for each activity;
3. adding a precedence constraint $e_i \xrightarrow{[d_{ij},D_{ij}]} s_j$ for each arc in the project graph.

We rely on CP and constraint propagation for dynamic controllability checking, rather than using a specialized algorithm like in [12,11]. In particular, we are interested in maintaining for each time point t_i four different bounds, namely: the start (s_p) and the end (e_p) of the time region where the associated event can occur (*possible region*); and the start (s_o) and the end (e_o) of a time region (*obligatory region*) such that, if t_i is forced to occur out of there, dynamic controllability is compromised. An alternative view is that the obligatory region keeps track of the amount of flexibility left to a time point.

Figure 1 shows as example of such bounds: s_p and e_p delimit the region where each t_i *can* occur at run: for example t_1 can first occur at time 10, if t_0 occurs

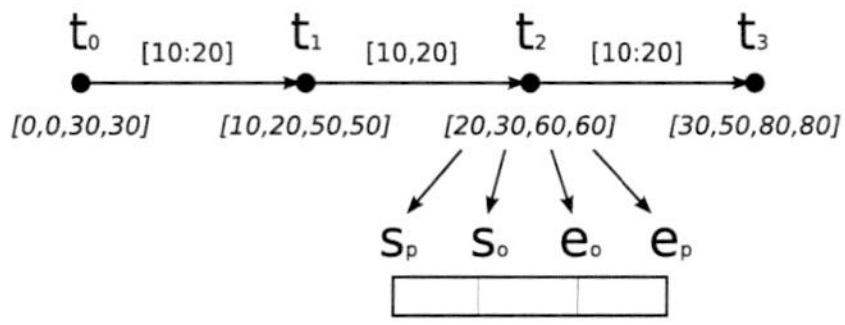

Fig. 1. Time bounds for a simple time point network

at 0 and has run time duration 10; similarly t_2 can first occur at 20 as at least 10 time units must pass between t_1 and t_2 due to the precedence constraint. As for the upper bounds, note that t_2 cannot occur after time 60, or there would be a value $\delta \in [10, 20]$ with no support in the time window of t_3; conversely, t_1 can occur as late as time 50, since there is *at least* a value $\delta \in [10, 20]$ with a support in the time window of t_2. Consider now bounds on the obligatory region: note that if (for instance) t_1 is *forced* to occur before time 20 the network is no longer dynamic controllable, as in that case there would not be sufficient time span between t_0 and t_1. Similarly, t_2 cannot be forced to occur later than time 60 or there would be a value $\delta \in [10, 20]$ such that the precedence constraint between t_2 and t_3 cannot be satisfied.

In general, bounds on the possible region (s_p, e_p) are needed to enable efficient dynamic controllability checking (if the time window of every time point is not empty) and constant time overlapping checking; bounds on the obligatory region (s_o, e_o) are a novel contribution and let one check in constant time whether a new precedence constraint can be added without compromising dynamic controllability. With this trick it is possible for example to remove inconsistent ordering relations between tasks that cannot be trivially inferred starting from the possible region.

We maintain the described bounds by introducing two CP integer variables $\mathtt{T}^{\mathtt{m}}, \mathtt{T}^{\mathtt{M}}$ for each time point, such that $\min(\mathtt{T}^{\mathtt{m}})$ marks the start of the possible region and $\max(\mathtt{T}^{\mathtt{m}})$ tells how far this start can be pushed forward; similarly, $\max(\mathtt{T}^{\mathtt{M}})$ marks the end of the possible region and $\min(\mathtt{T}^{\mathtt{M}})$ tells how far this end can be pulled back; consequently, we map s_p to $\min(\mathtt{T}^{\mathtt{m}})$, e_o to $\max(\mathtt{T}^{\mathtt{m}})$, s_o to $\min(\mathtt{T}^{\mathtt{M}})$, e_p to $\max(\mathtt{T}^{\mathtt{M}})$; then we post for each t_i: $\mathtt{T}^{\mathtt{m}}_{\mathtt{i}} \leq \mathtt{T}^{\mathtt{M}}_{\mathtt{i}}$. This ensures $s_p \leq s_o$, $e_o \leq e_p$ and triggers a failure whenever s_p is pushed beyond e_o (the time point is forced to occur after the obligatory region) or e_p is pulled before s_o (the time point is forced to occur before the obligatory region). Note that if dynamic controllability has to be kept, the case $s_p > e_o$ never occurs. For each constraint $t_i \xrightarrow{[d,D]} t_j$, we perform the following filtering:

$$d \leq \max(\mathtt{T}^{\mathtt{m}}_{\mathtt{j}}) - \max(\mathtt{T}^{\mathtt{m}}_{\mathtt{i}}) \leq D \qquad d \leq \max(\mathtt{T}^{\mathtt{M}}_{\mathtt{j}}) \max(\mathtt{T}^{\mathtt{M}}_{\mathtt{i}}) \leq D$$

$$d \leq \min(\mathtt{T}^{\mathtt{m}}_{\mathtt{j}}) - \min(\mathtt{T}^{\mathtt{m}}_{\mathtt{i}}) \leq D \qquad d \leq \min(\mathtt{T}^{\mathtt{M}}_{\mathtt{j}}) \min(\mathtt{T}^{\mathtt{M}}_{\mathtt{i}}) \leq D$$

which can be done by posting $d \leq \mathtt{T}^{\mathtt{m}}_{\mathtt{j}} - \mathtt{T}^{\mathtt{m}}_{\mathtt{i}} \leq D$ and $d \leq \mathtt{T}^{\mathtt{M}}_{\mathtt{j}} - \mathtt{T}^{\mathtt{M}}_{\mathtt{i}} \leq D$. The rationale behind the filtering rules can be explained by looking at Figure 2A.

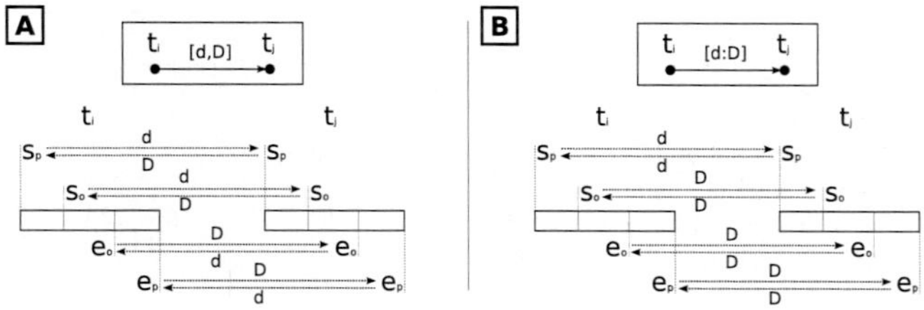

Fig. 2. (A) Filtering rules for free constraints; (B) Filtering rules for contingent constraints

For example $s_p(t_j)$ cannot be less than d time units away from $s_p(t_i)$, or no δ exists in $[d, D]$ such that $t_j = t_i + \delta$; so $\min(\mathtt{T}_j^m)$ can be updated to $\min(\mathtt{T}_i^m) + d$, in case it is less than that value: this is depicted in the figure with an arrow going from $s_p(t_i)$ to $s_p(t_j)$. By reasoning in the same fashion one can derive all other filtering rules. Similarly, for each contingent constraint $t_i \xrightarrow{[d:D]} t_j$ we perform the following filtering:

$$\max(\mathtt{T}_i^m) + D = \max(\mathtt{T}_j^m) \qquad \max(\mathtt{T}_i^M) + D = \max(\mathtt{T}_j^M)$$
$$\min(\mathtt{T}_i^m) + d = \min(\mathtt{T}_j^m) \qquad \min(\mathtt{T}_i^M) + D = \min(\mathtt{T}_j^M)$$

As in the previous case, figure 2B gives a pictorial intuition of the rationale behind the rules. Now, neither $s_p(t_j)$ can be closer than d time units to $s_p(t_i)$, nor $s_p(t_i)$ can be farther than d units from $s_p(t_j)$; otherwise there would exist a δ value in $[d, D]$ such that t_j could not be equal to $t_i + \delta$. This explains the second filtering rule on the leftmost column. By reasoning similarly one can derive all other rules. The described filtering runs in polynomial time and is *sufficient* to enforce dynamic controllability on the network. A *necessary and sufficient* filtering for dynamic controllability can be obtained by simply removing all rules involving $\max(\mathtt{T}_j^m)$ and $\min(\mathtt{T}_j^M)$.

Additionally, we keep track of all precedence constraints in the time model as a directed graph $\langle T, C \rangle$, referred to as *time graph*. Here T is the set of time points and $(t_i, t_j) \in C$ iff there is a chain of precedence constraints between t_i and t_j. The reason for having a graph is that it is not possible in general to detect whether a time point has to come before another by just looking at the time bounds: for example in Figure 1 we have $s_p(t_0) \leq e_p(t_1)$; nevertheless the precedence constraints force t_0 to execute before t_1. In the following, we will write $t_i \preceq t_j$ if an edge (t_i, t_j) exists in C (we write $t_i \prec t_j$ if the precedence relation is strict). Transitive closure is kept on the time graph by dynamic updates with cubic complexity along a branch on the search tree.

4.2 Search Strategy

One of the key difficulties with complete search based on MCS branching is how to detect and choose conflict sets to branch on; this stems from the fact that the number of MCS is in general exponential in the size of the project graph, hence complete enumeration incurs the risk of combinatorial explosion. We propose

Algorithm 1. Overview of the search strategy

```
set best MCS so far (best) to ∅
for r_k ∈ R do
   find a conflict set S by solving a minimum flow problem
   if weight of S is higher than cap(r_k) then
      refine S to a minimal conflict set S'
      if S' is better than the best MCS so far then
         best = S'
      end if
   end if
end for
if best = ∅ then
   the problem is solved
else
   open a choice point branching on possible resolvers of best
end if
```

to detect possible conflict set by solving a minimum flow problem on a specific resource r_k, as described in [10]; the method has the major advantage of having polynomial complexity. Note however the conflict set found is not guaranteed to be minimal, nor to be well suited to open a choice point. We coped with this issue by adding a conflict improvement step. An overview of the adopted search strategy is shown in Algorithm 1. In the next section each of the steps will be described in deeper detail; the adopted criterion to evaluate the quality of a conflict set will be given as well.

Conflict Set Detection: The starting observation for the minimum flow based conflict detection is that, if the problem contains a *minimal* conflict set, it also contains a non necessarily minimal conflict set, i.e. a conflict set not necessarily satisfying the minimality condition in the definition of section 2; let us refer to this as a CS. Therefore we can check the existence of an MCS on a given resource r_k by checking the existence of any CS. Moreover, as the activities in a CS must have the possibility to overlap, they always form a stable set (or independent set) on the augmented project graph, further annotated with all precedence constraint which can be detected by time reasoning; if we assign to each activity the requirement $req(t_i, r_k)$ as a weight, a stable set S is a CS iff $\sum_{a_i \in S} req(a_i, r_k) > cap(r_k)$. We refer to such weighted graph as *resource graph* $\langle A, E_R \rangle$, where $(a_i, a_j) \in E_R$ iff $a_i \preceq a_j$ or $e_p(e_i) \leq s_p(s_j)$. We can therefore check the existence of a MCS on a resource r_k by finding the maximum weight independent set on the resource graph and checking its total weight; this amounts to solve the following ILP model P':

$$
\begin{aligned}
\mathbf{P}' : \max \; & \sum_{a_i \in A} req(a_i, r_t) x_i \\
\text{s.t.} \; & \sum_{a_i \in \pi_j} x_i \leq 1 \quad \forall \pi_j \in \Pi \\
& x_i \in \{0, 1\}
\end{aligned} \tag{1}
$$

$$
\begin{aligned}
\mathbf{P}'' : \min \; & \sum_{\pi_j \in \Pi} y_j \\
\text{s.t.} \; & \sum_{a_i \in \pi_j} y_j \geq req(a_i, r_t) \quad \forall a_i \in A \\
& y_j \in \{0, 1\}
\end{aligned} \tag{2}
$$

where x_i are the decision variables and $x_i = 1$ iff activity a_i is in the selected set; Π is the set of all path in the graph (in exponential number) and π_j is a path in Π. As for the constraints (1) consider that, due to the transitivity of temporal relations, a clique on the resource graph is always a path from source to sink. In any independent set no two nodes can be selected from the same clique, therefore, no more than one activity can be selected from each path π_j in the set Π of all graph paths.

The corresponding dual problem is P'', where variable y_j is path π_j is selected; that is, finding the maximum weight stable set on a transitive graph amounts to find the minimum set of source-to-sink paths such that all nodes are covered by a number of paths at least equal to their requirement (constraints (2)). Note that, while the primal problem features an exponential number of constraints, its dual has an exponential number of variables. One can however see that the described dual is equivalent to route the least possible amount of flow from source to sink, such that a number of minimum flow constraints are satisfied; therefore, by introducing a real variable f_{ij} for each edge in E_R, we get:

$$\begin{aligned} \min \ & \textstyle\sum_{a_j \in E^+(a_0)} f_{0j} & \\ \text{s.t.} \ & \textstyle\sum_{a_j \in E^-(a_i)} f_{ji} \geq req(a_i, r_t) \quad \forall a_i \in A & (3) \\ & \textstyle\sum_{a_j \in E^-(a_i)} f_{ji} = \sum_{a_j \in E^+(a_i)} f_{ij} \quad \forall a_i \in A \setminus \{a_0, a_{n-1}\} & (4) \\ & f_{ij} \geq 0 & \end{aligned}$$

where $E^+(a_i)$ denotes the set of direct successors of a_i and $E^-(a_i)$ denotes the set of direct predecessors. One can note this is a flow minimization problem. Constraints (3) are the same as constraints (2), while the flow balance constraints (4) for all intermediate activities are implicit in the previous model. The problem can be solved starting for an initial feasible solution by iteratively reducing the flow with the any embodiment of the inverse Ford-Fulkerson's method, with complexity $O(|E_R| \cdot \mathcal{F})$ (where $\mathcal{F}$ is the value of the initial flow). Once the final flow is known, activities in the source-sink cut form the maximum weight independent set.

In our approach we solve the minimum flow problem by means of the Edmond-Karp's algorithm. On this purpose each activity a_i has to be split into two subnodes a'_i, a''_i; the connecting arc (a'_i, a''_i) is then given minimum flow requirement $req(a_i, r_k)$; every arc $(a_i, a_j) \in E_R$ is converted into an arc (a''_i, a'_j) and assigned minimum flow requirement 0. An initial solution is computed by:

1. selecting each arc (a'_i, a''_i) in the new graph with minimum flow requirement $req(a_i, r_k) > 0$
2. routing $req(a_i, r_k)$ units of flow along a backward path from a'_i to a'_0
3. routing $req(a_i, r_k)$ units of flow along a forward path from a''_i to a''_{n-1}

minor optimizations are performed in order to reduce the value of the initial flow. If at the end of the process the total weight of the independent set is higher than $cap(r_k)$, then a CS has been identified.

Reduction to MCS: Once a conflict set has been identified, a number of issues still have to be coped with; namely (1) the detected CS is not necessarily minimal and (2) the detected CS does not necessarily yield a good choice point. Branching on non-minimal CS can result in exploring unnecessary search paths; extracting a MCS from a given CS rises however no practical trouble, as it can be done very easily in polynomial time. The second issue is instead more complex, as it requires to devise a good MCS evaluation heuristic. We propose to tackle both problems by performing local search based intensification.

As evaluation criterion for a given conflict set S we use the lexicographic composition of (1) the number of precedence constraints which *can* be posted between pairs of activities (a_i, a_j) with $a_i, a_j \in S$, and of (2) the size of the set itself (i.e. $|S|$); for both criteria, lower values are better. Note a precedence constraint *cannot* be posted on (a_i, a_j) iff neither of the followings holds:

1. $a_j \prec a_i$ in the time graph (where $\prec$ denotes a strict precedence relation)
2. $s_o(e_i) > e_o(s_j)$ in the time model, as briefly discussed in section 4.1 (where e_i and s_j are time points representing the end of a_i and the start of a_j)

As a consequence, local search first naturally moves towards CS yielding choice points with a small number of branches: this goes in the same direction of the "minimum size domain" variable selection criterion. Once the number of branches in the resulting choice point cannot be further reduced, nodes are removed from the CS turning it into a minimal conflict set. Note the total weight must always exceed the capacity $cap(r_k)$.

Given a conflict set S, we consider the following pool of local search moves:

1. $\texttt{add}(\texttt{S}, \texttt{a}_\texttt{i})$: a node $a_i \notin S$ is added to S, all nodes $a_j \in S$ such that $(a_i, a_j) \in E_R$ or $(a_j, a_i) \in E_R$ are removed from the set. The number of precedence constraints that can be posted is updated accordingly. The move has quadratic complexity.
2. $\texttt{del}(\texttt{S}, \texttt{a}_\texttt{i})$: a node $a_i \in S$ is removed from S; the number of precedence constraints that can be posted is updated accordingly. The move has linear complexity.

At each local search step all `del` moves for $a_i \in S$ are evaluated, together with all `add` moves for every immediate predecessor or successor of activities in S; the best move strictly improving the current set is then chosen. The process stops when a local optimum is reached.

Opening a choice point: As a MCS is selected, the next stage of the search step is to open a choice point. Let $RS = (a_{i_0}, a_{j_0}), \dots (a_{i_{m-1}}, a_{j_{m-1}})$ be the list of pairs of nodes in the set such that a precedence constraints can be posted. Then the choice point can be recursively expressed as:

$$CP(RS) = \begin{cases} \texttt{post}(a_{i_0}, a_{j_0}) \text{ if } |RS = 1| \\ \texttt{post}(a_{i_0}, a_{j_0}) \vee [\texttt{forbid}(a_{i_0}, a_{j_0}) \wedge CP(RS \setminus (a_{i_0}, a_{j_0}))] \end{cases}$$

where (a_{i_0}, a_{j_0}) always denotes the first pair in the sequence being processed and shrunk. The operation $\texttt{post}(a_{i_0}, a_{j_0})$ amounts to add the constraint $e_i \xrightarrow{[0,\infty]} s_j$ in the time model, and $\texttt{forbid}(a_{i_0}, a_{j_0})$ consists in adding $s_j \xrightarrow{[1,\infty]} e_i$ (strict precedence relation). Prior to actually building the choice point, all precedence constraints could in principle be sorted according to some heuristic criterion; note however this is not done in the current implementation: introducing some score to rank precedence constraints (e.g. the one proposed in [8], based on preserved search space) is part of planned future work.

Detecting unsolvable conflict sets: At search time, it is in principle possible for a conflict set to become unsolvable, that is: no resolver can be posted at all. Despite being not too frequent in practice and having no impact on the method convergence, this situation can have a substantial impact on the solver performance if not promptly detected. On this purpose an additional step is added at the beginning of each search step where an attempt to identify unsolvable conflict sets is performed. Observe that a conflict set S is unsolvable iff for each pair $a_i, a_j \in S$ neither $e_i \xrightarrow{[0,\infty]} s_j$ nor $e_j \xrightarrow{[0,\infty]} s_i$ can be posted. In practice, if we build an *undirected* graph where an edge connecting a_i and a_j is present if such a situation holds, an unsolvable CS for a resource r_k always forms a clique with weight higher than $cap(r_k)$.

As neither the special graph we have just described nor its complement are transitive, the minimum flow based method cannot be used to detect an unsolvable CS. We therefore resorted to complete search, taking advantage of the very sparse structure quite often exhibited by the special graph. During search, nodes are selected according to their degree (number of adjacent edges) in the graph, deprived of the currently selected set; such degree is dynamically updated as new nodes are selected. In order to limit the search effort we finally run the process with a fail limit, which was empirically set to $10\times$ the size of the project graph.

5 Experimental Results

The described approach was implemented on top of ILOG Solver 6.3. The PCP method with minimum flow based MCS detection was compared to a solver exploiting a CP model in order to find MCS (see [9] for details). In the following, we refer to the first approach as MF solver (Minimum Flow), and to the latter as CS solver (Complete Search). In particular, we performed tests on RCPSP instances derived from a system design problem. Given a computer application described as a graph and a target platform, the problem consists in finding a schedule guaranteed to meet a global deadline constraint; this is a very relevant issue in the design of real time systems. Extensive experimentation on different type of scheduling problems is planned for future work, as well as a comparison with related methods such as Complete MCS Search [8]. Nodes in the application graph denote tasks/processes or data communication activities; each node has a priori unknown duration, bounded by a worst case and a best case execution time. The considered platform features 16 processors (resources with capacity 1) and 32 communication devices (resources with capacity 10); this is indeed representative of an advanced platform. Each task requires a processors and each data communication activity requires up to 9 units of two communication devices.

Table 1. Results on the first group of instances (growing number of nodes) for (A) the CS solver and (B) the MF solver

	N	FL	TT	T	F	N_{MCS}	T_{MCS}	T_O	MEM	FEAS T	FEAS T_{MCS}	INF T	INF T_{MCS}
A	41-49	0.59	—	7.03	22	164	2.82	0	11M	0.49	23.20	0.13	1.10
	56-66	0.56	0.96	90.12	1832	1140	67.34	1	28M	1.37	30.56	8.84	116.56
	75-82	0.55	—	49.24	30	287	18.07	0	55M	2.93	39.20	0.71	2.90
	93-103	0.56	0.97	130.34	81	533	55.79	3	110M	8.42	66.14	1.66	6.57
	111-116	0.56	0.89	251.08	71	720	129.22	4	172M	18.84	89.00	2.14	5.67
	121-128	0.59	0.92	365.05	96	666	169.58	6	222M	22.11	75.75	5.82	11.25
B	41-49	0.60	—	0.39	6	155	0.37	0	0.22M	0.05	0.05	0.00	0.01
	56-66	0.56	0.91	1.07	6	212	1.04	1	0.28M	0.15	0.15	0.00	0.01
	75-82	0.56	—	3.32	18	309	3.25	0	0.32M	0.41	0.39	0.09	0.09
	93-103	0.57	0.98	11.29	51	537	11.18	2	0.42M	1.29	1.27	0.23	0.23
	111-116	0.55	0.92	27.52	1175	1549	26.86	4	0.49M	3.28	3.22	0.09	0.08
	121-128	0.59	0.89	36.64	82	713	36.36	6	0.50M	3.19	3.19	1.80	1.78

Table 2. Results on the second group of instances (growing branching factor) for (A) the CS solver and (B) the MF solver

	BF	FL	TT	T	F	N_{MCS}	T_{MCS}	TO	MEM	FEAS T	FEAS T_{MCS}	INF T	INF T_{MCS}
A	2-4	0.58	1.00	51.21	13	305	21.09	1	107M	3.05	38.22	0.62	0.00
	3-5	0.56	0.99	67.51	14	405	31.21	1	125M	4.97	55.44	0.66	0.11
	4-6	0.55	1.00	84.40	48	565	42.63	1	134M	6.72	75.00	0.83	2.78
	5-7	0.54	0.98	71.15	15	476	33.13	2	133M	5.36	66.50	0.62	0.00
	6-8	0.52	0.99	102.02	20	667	54.95	4	160M	8.40	88.00	0.67	0.33
B	2-4	0.58	—	3.74	48	385	3.67	0	0.33M	0.34	0.33	0.15	0.15
	3-5	0.56	—	4.17	7	423	4.09	0	0.36M	0.54	0.53	0.01	0.01
	4-6	0.56	0.80	41.00	5418	6230	39.65	1	0.39M	5.78	5.58	0.03	0.03
	5-7	0.54	0.49	4.88	6	455	4.78	2	0.37M	0.67	0.63	0.00	0.01
	6-8	0.51	0.93	6.16	6	576	6.06	4	0.42M	0.87	0.86	0.00	0.00

The following testing process has been devised: for each available instance a very loose deadline requirement is first computed (namely, the sum of all worst case durations); next binary search is performed iteratively tightening a feasible upper bound and an infeasible lower bound on the achievable deadline; the process converges to the best possible deadline and provides information about the performance of the solvers, as well as an indication of the tightness to the best deadline constraint which can be reached by both solvers. A time limit of 900 seconds was set on the whole test process; all experiments were run on an Intel Core2 Duo with 2GB of RAM. Being real world instances for the problem at hand too scarce in number for a meaningful evaluation, we resorted to using a random instance generator, specially devised to mimic the structure of real world applications[1]: i.e. with quite many relations, fork (with many successors) and join (with many predecessors) tasks almost fully nested, some added arcs between these nested hierarchical structures. In particular, we generated two groups of instances by scaling (1) the number of nodes in the graph and (2) the branching factor (i.e. number of outgoing arcs of certain nodes in the graph).

[1] Both the generator and the instances are available at http://www.lia.deis.unibo.it/Staff/MicheleLombardi/

Table 3. Performance of the MF solver when some features are turned off

	NO UCS FINDER						NO LOCAL SEARCH					
N	T	N_{MCS}	N_{UCS}	TO	N_{MF}	N_{LS}	T	N_{MCS}	N_{UCS}	TO	N_{MF}	N_{UF}
41-49	0.37	155	0	0	914	562	0.43	198	36	0	973	242
56-66	1.04	212	0	1	1417	1042	1.07	243	2	1	1383	253
75-82	3.52	325	0	0	3355	2933	3.96	393	41	0	3533	442
93-103	11.96	538	0	2	7660	7168	14.02	727	159	2	8685	894
111-116	18.51	631	0	5	8566	8133	48.65	1318	279	4	23744	1606
121-128	40.98	713	0	6	15499	14983	133.13	2049	1088	5	46073	3145

Table 1 summarizes results on the first group of instance, respectively for the CS (A) and MF solver (B). Here the branching factor was fixed to the range 3-5; the worst case duration for each task is around twice the best case duration; those value are definitely reasonable for real world problems. Each row refers to a group of 10 instances, and shows the minimum and maximum number of nodes (**N**), the average solution time (**T**) and number of fails (**F**) for the entire test process, the number of MCS used to branch ($\mathbf{N_{MCS}}$) and the time spent to detect them ($\mathbf{T_{MCS}}$); the number of timed out instances and the total memory usage are respectively in columns **TO** and **MEM**. The solution time and the time to detect MCS is reported for single feasible and unfeasible runs as well (**FEAS**, **INF**). Finally, columns **FL** and **TT** deserve special attention; whenever a feasible solution is found a flexibility indicator can be given by the ratio between the best case completion time and the worst case completion time for the produced graph; the average of such indicator is reported in **FL**. When a timeout occurs, the ratio between the current lower bound and the current upper bound on the best achievable deadline is computed and used as a tightness indicator; the average of such indicator is reported in **TT**. As one can see, the MF solver reports improvements of around one order of magnitude both in the total solution time and in the time required to detect MCS; in particular, the latter has to be mainly ascribed to the flow based conflict detection method, while the former also benefits from the much more efficient time model used in the MF solver; the compactness of the time model also has a leading role in the drastic improvement in memory usage. The flexibility and the tightness achieved by the two solvers are comparable; note in particular the very high values of the tightness indicator reached by both the approaches. Table 2 shows the same results for the second group of instances; here the number of nodes in the graph is always between 74 and 94, while the branching factor spans the interval reported for each row in the column **BF**. As one can observe, the trend of all results is around the same as the previous case, with the addition of the flexibility indicator getting higher as the branching factor increases. Note however the MF solver sometimes achieves considerably worse values for the tightness indicator value; this indicates the MF solver tends to stop earlier (i.e. for more loose bounds) in case of timeouts.

A last set of experiments was finally performed to test the impact of the local search (LS) and the unsolvable conflict set (UCS) detector; in fact, as those feature are not necessary for the method to converge, it is reasonable to question their actual effect on the solver efficiency. Table 3 reports results for those tests, performed on the first group of instances previously discussed. Both for the case when the UCS finder and the local search are turned off, the table shows the solution time (**T**), the number of processed MCS ($\mathbf{N_{MCS}}$) and the number of detected UCS ($\mathbf{N_{UCS}}$); the number of timed out instances (**TO**) is indicated as well. Columns $\mathbf{N_{MF}}$, $\mathbf{N_{LS}}$, $\mathbf{N_{UF}}$ respectively report the number of times the minimum flow, local search and UCS finder algorithm are run. The results show that the actual advantage of incorporating a UCS finder is shadowed by the efficacy of the local search CS improver, to the point that sometimes a better run time is achieved without the feature. On the other hand, no question arise about the effective utility of the local search method during MCS detection. Quite surprisingly however, turning LS off helps reducing the number of timeouts: this has to be further investigated.

6 Conclusion

An efficient complete solver for facing Resource Constraint Project Scheduling with minimum and maximum time lags and variable durations is proposed. The main contributions are: an effective time model inherited by STNU, an efficient algorithm for conflict set detection and its encapsulation in a sophisticated search strategy. Current research is aimed at introducing objective functions, and in taking into account run time policies.

Acknowledgement

Many thanks to Valentina Cacchiani for the useful advices about algorithms for finding stable sets. The work described in this publication was supported by the PREDATOR Project funded by the European Community's 7th Framework Programme, Contract FP7-ICT-216008.

References

1. Brucker, P., Drexl, A., Möhring, R., Neumann, K., Pesch, E.: Resource-constrained project scheduling: Notation, classification, models, and methods. European Journal of Operational Research 112(1), 3–41 (1999)
2. Stork, F.: Branch-and-bound algorithms for stochastic resource-constrained project scheduling. Technical Report Research Report No. 702/2000, Technische Universität Berlin (2000)
3. Stork, F.: Stochastic resource-constrained project scheduling. PhD thesis, Technische Universität Berlin (2001)
4. Igelmund, G., Radermacher, F.J.: Algorithmic approaches to preselective strategies for stochastic scheduling problems. Networks 13(1), 29–48 (1983)
5. Igelmund, G., Radermacher, F.J.: Preselective strategies for the optimization of stochastic project networks under resource constraints. Networks 13(1), 1–28 (1983)

6. Herroelen, W., Leus, R.: Project scheduling under uncertainty: Survey and research potentials. European Journal of Operational Research 165(2), 289–306 (2005); Project Management and Scheduling
7. Beck, J.C., Davenport, A.J.: A survey of techniques for scheduling with uncertainty (2002), `http://www.eil.utoronto.ca/profiles/chris/gz/uncertainty-survey.ps`
8. Laborie, P.: Complete MCS-Based Search: Application to Resource Constrained Project Scheduling. In: IJCAI, pp. 181–186 (2005)
9. Benini, L., Lombardi, M., Milano, M.: Robust non-preemptive hard real-time scheduling for clustered multicore platforms. In: Proc. of DATE 2009 (2009)
10. Golumbic, M.: Algorithmic Graph Theory And Perfect Graphs, 2nd edn. Elsevier, Amsterdam (2004)
11. Morris, P.H., Muscettola, N.: Temporal dynamic controllability revisited. In: AAAI, pp. 1193–1198 (2005)
12. Morris, P.H., Muscettola, N., Vidal, T.: Dynamic control of plans with temporal uncertainty. In: IJCAI, pp. 494–502 (2001)
13. Muscettola, N.: Computing the envelope for stepwise-constant resource allocations. In: Van Hentenryck, P. (ed.) CP 2002. LNCS, vol. 2470, pp. 139–154. Springer, Heidelberg (2002)
14. Policella, N., Cesta, A., Oddi, A., Smith, S.F.: From precedence constraint posting to partial order schedules: a csp approach to robust scheduling. AI Commun. 20(3), 163–180 (2007)
15. Policella, N., Oddi, A., Smith, S.F., Cesta, A.: Generating robust partial order schedules. In: Wallace, M. (ed.) CP 2004. LNCS, vol. 3258, pp. 496–511. Springer, Heidelberg (2004)
16. Möhring, R.H., Radermacher, F.J., Weiss, G.: Stochastic scheduling problems I - General strategies. Mathematical Methods of Operations Research 28(7), 193–260 (1984)
17. Möhring, R.H., Radermacher, F.J., Weiss, G.: Stochastic scheduling problems II - set strategies. Mathematical Methods of Operations Research 29(3), 65–104 (1985)
18. De Reyck, B., Herroelen, W.: A branch-and-bound procedure for the resource-constrained project scheduling problem with generalized precedence relations. European Journal of Operational Research 111(1), 152–174 (1998)
19. Möhring, R.H., Stork, F.: Linear preselective policies for stochastic project scheduling. Mathematical Methods of Operations Research 52(3), 501–515 (2000)
20. Vidal, T., Fargier, H.: Handling contingency in temporal constraint networks: from consistency to controllabilities. J. Exp. Theor. Artif. Intell. 11(1), 23–45 (1999)

SOGgy Constraints: Soft Open Global Constraints

Michael J. Maher

NICTA* and University of NSW
Sydney, Australia
Michael.Maher@nicta.com.au

Abstract. We investigate soft open constraints. We generalize and unify classes of soft constraints and adapt them to the open setting. We give sufficient conditions for generalized classes of decomposition-based and edit-based soft constraints to be contractible. Finally, we outline a propagator for an open generalized edit-based soft REGULAR constraint.

1 Introduction

Open global constraints [1,8,11] allow variables to be added to the global constraint during execution, which is vital when we want to interleave problem construction and problem solving. Soft constraints are useful for addressing problems that might be overconstrained. In this paper we investigate soft open constraints, that is, soft constraints in the sense of Petit *et al* [14] that are dynamic, or open, in the sense of Barták [1]. Until now, such constraints have not been investigated.

The main focus is on the ability to recognise when soft constraints are contractible – a property that ensures that propagators for closed constraints can be re-used for the corresponding open constraint. We give sufficient conditions for generalized classes of decomposition-based and edit-based soft constraints to be contractible. We also outline a propagator for an open generalized soft REGULAR constraint to demonstrate the ease with which a propagator for a closed constraint can be adapted for an open constraint.

The reader is assumed to have a basic knowledge of constraint programming, CSPs, global constraints, and filtering, as might be found in [6,15,3]. In particular, global constraints not defined here can be found in [3]. Background on open constraints is given in Section 2, while Section 3 is the main section, discussing when soft constraints are contractible, and Section 4 outlines a propagator for a soft open REGULAR constraint.

2 Open Constraints

In this paper, we view a global constraint as a relation over a sequence of variables $[X_1, X_2, \ldots, X_n]$ (also denoted by $\boldsymbol{X}$). Other arguments of a constraint are considered parameters and are assumed to be fixed before execution.

* NICTA is funded by the Australian Government as represented by the Department of Broadband, Communications and the Digital Economy and the Australian Research Council through the ICT Centre of Excellence program.

I.P. Gent (Ed.): CP 2009, LNCS 5732, pp. 584–591, 2009.

We formalize the semantics of a constraint C as a formal language L_C. A word $d_1 d_2 \ldots d_n$ appears in L_C iff the constraint $C([X_1, X_2, \ldots, X_n])$ has a solution $X_1 = d_1, \ldots X_n = d_n$. When it is convenient, we will describe languages with Kleene regular expressions. We will need the following definitions later. Let $P(L) = \{w \mid \exists u \; wu \in L\}$ denote the set of prefixes of a language L, called the prefix-closure of L. We say L is *prefix-closed* if $P(L) = L$.

In an open global constraint, variables may be appended to the end of the sequence of variables during execution. A constraint may also be closed during execution, at which point no further variables may be appended. The scope of such constraints changes during the execution, and we refer to the state of the constraint at some point in the execution as an *occurrence* of the constraint.

There are several models of open constraint that have been proposed. Barták's model [1] is straightforward: the constraint involves a sequence of variables to which variables may be added. Thus the type of the constraint is unchanged, whether the constraint is open or closed. The model of van Hoeve and Régin [8] uses a set variable S describing a set of object variables, rather than a sequence, to represent the collection of variables in the constraint. A third model [11] uses a sequence of variables, with an additional integer variable N denoting the length of the sequence. In this paper we focus on Barták's model.

In general, a propagator for a closed constraint may be unsound for the corresponding open constraint. That is, the propagator may make an inference that turns out to be unjustified once the sequence of variables $\boldsymbol{X}$ is extended to $\boldsymbol{X}Y$. The property of *contractibility* was introduced in [10] to characterize those constraints for which a closed propagator is sound for the open constraint, that is, all domain reductions remain sound if the sequence of variables is extended. It was shown there that a constraint C is contractible iff L_C is prefix-closed. For contractible constraints it appears to be straightforward to extend closed implementations to open constraints [10]. Propagators also have been designed for some incontractible constraints [8,11].

3 Contractibility of Soft Constraints

We consider "soft" constraints in the style of [14][1]. In such constraints there is a violation measure, which measures the degree to which an assignment to the variables violates the associated "hard" constraint, and solutions are assignments that satisfy an upper bound on the violation measure. Thus such soft constraints have the form $m(\boldsymbol{X}) \leq Z$, where m is the violation measure[2]. We refer to the hard constraint as $C(\boldsymbol{X})$ and the corresponding soft constraint as $C_s(\boldsymbol{X}, Z)$.

Assessing the contractability of such constraints is made easier by a result of [10]. We say a function f on a sequence is *non-decreasing* if $f(\boldsymbol{a}) \leq f(\boldsymbol{b})$ for every $\boldsymbol{a}$ and $\boldsymbol{b}$ where $\boldsymbol{a}$ is a prefix of $\boldsymbol{b}$. (Here the notation $\boldsymbol{a}$ denotes a sequence $a_1 \ldots a_n$ or $[a_1, \ldots, a_n]$.)

[1] Consideration of other forms of soft constraint are left for future research.

[2] Also called violation cost [14]. We assume that violation measures return non-negative values, and return 0 exactly when the hard constraint holds.

Proposition 1. *Let C be a global constraint and suppose C can be expressed as $f(\boldsymbol{X}) \leq Z$. Then C is contractible iff f is a non-decreasing function.*

Thus, to evaluate whether or not soft constraints are contractible we must consider the form of the violation measure, and whether it forms a non-decreasing function.

We can recognise four broad classes of violation measures: those based on constraint decomposition, edit distance, graph properties, and the semantics of a specific constraint. We address these classes in the following subsections.

3.1 Decomposition-Based Violation Measures

In the *decomposition-based* violation measures[3], a decomposition of C into an equivalent conjunction of more elementary constraints is identified, and the violation measure of C_s for a particular assignment is the number of elementary constraints in the decomposition that are violated (i.e. not satisfied) by the assignment. Some work prefers the decomposition into binary constraints [14,7], but there seems no reason to be limited by this restriction, and any canonical decomposition should serve.

When the decomposition of $C(\boldsymbol{X})$ is a subset of the decomposition of $C(\boldsymbol{X}Y)$ then it is clear that the number of violations for any assignment cannot decrease when the sequence of variables is extended. Hence, the violation measure is non-decreasing and the soft constraint C_s is contractible.

We can extend this result to decompositions that introduce new variables. Let $C(\boldsymbol{X}) \leftrightarrow \exists \boldsymbol{U} \bigwedge_{c \in \mathcal{C}} c(\boldsymbol{X}, \boldsymbol{U})$ be the decomposition of $C(\boldsymbol{X})$, where $\mathcal{C}$ is the collection of elementary constraints in the decomposition of $C(\boldsymbol{X})$. We define the violation measure m as follows: for every assignment v to the variables $\boldsymbol{X}$, $m(\boldsymbol{X}) = \min num\_viol_{v'}(\mathcal{C})$ where we minimize over all extensions v' of v to $\boldsymbol{U}$, and $num\_viol_{v'}(S)$ is the number of unsatisfied constraints in S under v'.

We assume that the decomposition can be expressed as a formula of the form $\exists \boldsymbol{U} \bigwedge_i p_i(\boldsymbol{X}, \boldsymbol{U})$, where each $p_i(\boldsymbol{X}, \boldsymbol{U})$ denotes an occurrence of an elementary constraint over a subset of the variables $\boldsymbol{X}, \boldsymbol{U}$, and constants. We permit duplicate occurrences of a constraint, which allows a form of weighting within a decomposition. We denote such a formula by $(\boldsymbol{X}, \boldsymbol{U}, S)$, where S is the collection of elementary constraints. We say that one formula $(\boldsymbol{X}, \boldsymbol{U}, S)$ is *covered* by another formula $(\boldsymbol{W}, \boldsymbol{V}, T)$ if there is a variable renaming θ that maps $\boldsymbol{X}$ into $\boldsymbol{W}$ and $\boldsymbol{U}$ into $\boldsymbol{V} \cup \boldsymbol{W}$ such that the multiset $S\theta$ is a submultiset of T[4] .

We can now provide a sufficient condition for a soft constraint with a decomposition-based violation measure to be contractible.

Proposition 2. *Let C_s be a soft constraint with a decomposition-based violation measure. Let $(\boldsymbol{X}, \boldsymbol{U}, S)$ be the decomposition of $C(\boldsymbol{X})$ and $(\boldsymbol{X}Y, \boldsymbol{V}, T)$ be the decomposition of $C(\boldsymbol{X}Y)$. If $(\boldsymbol{X}, \boldsymbol{U}, S)$ is covered by $(\boldsymbol{X}Y, \boldsymbol{V}, T)$ via a renaming that is the identity on $\boldsymbol{X}$ then C_s is contractible.*

[3] This includes primal graph based violation costs [14].

[4] A multiset M_1 is a *submultiset* of multiset M_2 if, for every element $a \in M_1$, the multiplicity of a in M_1 is less than or equal to its multiplicity in M_2.

For example, CONTIGUITY is implemented in [9] essentially by the decomposition

$$\text{CONTIGUITY}(\boldsymbol{X}) \leftrightarrow \exists \boldsymbol{L}, \boldsymbol{R}, X_0, X_{n+1} \bigwedge_{i=1}^{n} C'(X_{i-1}, R_{i-1}, L_i, X_i, R_i, L_{i+1}, X_{i+1})$$

for a certain constraint C'. The decomposition of CONTIGUITY($\boldsymbol{XY}$) covers that of CONTIGUITY($\boldsymbol{X}$), and hence a soft CONTIGUITY constraint under the decomposition-based violation measure is contractible.

In general, any constraint definition based on composition of smaller constraints can be viewed as a decomposition. Thus, as a consequence of the previous Proposition, we have a counterpart of results of [10] for decomposition-based soft constraints.

Corollary 1. *If C is a global constraint definable using only the meta-constraints* SLIDE *and* SPLASH*, existential quantification, conjunction, and constraints on a fixed finite prefix of $\boldsymbol{X}$, then the soft constraint C_s based on that decomposition is contractible.*

As a result, decomposition-based soft ALLDIFFERENT, INTERDISTANCE, SEQUENCE and $\leq_{lex}$ constraints are contractible, using the decompositions mentioned in [10]. Different decompositions for the same constraint may lead to different violation measures, with possibly different contractibility of the corresponding soft constraints. Thus it is necessary to specify the decomposition involved.

Rather than using $num\_viol_{v'}$ to count the *number* of violations in a decomposition, we can instead sum the *amount* of violation (by v') of the constraints in the decomposition. This gives rise to the value-based violation measure for GCC [14,7] and the measure used for the soft SEQUENCE constraint [12]. In the latter case, the decomposition of $C(\boldsymbol{X})$ is a subset of the decomposition of $C(\boldsymbol{XY})$ and it follows that the soft SEQUENCE constraint is contractible. More generally

Proposition 3. *If $C(\boldsymbol{X})$ is covered by $C(\boldsymbol{XY})$ via a substitution θ that leaves $\boldsymbol{X}$ unchanged then C_s is contractible under such measures.*

This proposition does not apply to the value-based soft GCC because the decomposition of $C(\boldsymbol{X})$ involves constraints over $\boldsymbol{X}$. These constraints $c(\boldsymbol{X})$ cannot be covered by the corresponding $c(\boldsymbol{XY})$. In fact, the value-based soft GCC is not contractible, for essentially the same reason that the hard GCC is not contractible: lower bounds on the number of occurrences of a value can lead to the violation measure decreasing as the sequence gets longer.

Covering is a purely syntactic relationship; it is independent of the meaning of the constraints involved. Semantically equivalent (conjunctions of) constraints do not, in general, cover each other. To take a trivial example, constraints $X+0 = Y$ and $X*1 = Y$ are semantically equivalent, but do not cover each other because they are syntactically different. As a result, Propositions 2 and 3 may fail to recognise cases of contractibility.

3.2 Edit-Based Violation Measures

The *edit-based* violation measures use a notion of edit distance, which is the minimum number of edit operations required to transform a word into a word of L_C. There are many possible edit operations but the common ones are: to substitute one letter for another, to insert a letter, to delete a letter, and to transpose two adjacent letters. This class includes the *variable-based* violation measures [14,7], since such measures are simply edit distances where substitution is the only edit operation. The *object-based* measures of [4] are edit distances where deletion is the only edit operation.

The edit-based violation measures for closed constraints are not appropriate for open constraints, because they fail to take into account that the current sequence of variables may be extended with more variables.

For example, consider an open constraint C where $L_C = abc + defghi$ and an occurrence of the constraint $C([X_1, X_2, X_3])$. If $X_1 = d$, $X_2 = e$ and $X_3 = f$ then the edit distance of this instance to L_C is 3, even though this instance is completely accurate if the sequence of variables is extended. Similarly, if $L_C = abc$ and we have an occurrence $C([X_1, X_2])$ with $X_1 = a$ and $X_2 = b$ then the edit distance is 1, even though there is no violation.

To take account of the possibility that a sequence of variables may be extended, we employ the edit distance to $P(L_C)$, the prefix-closure of L_C. With this choice, the violation measure in any occurrence is a lower bound on the violation measure of the final occurrence of the constraint.

In [10] the prefix-closure was used to approximate a constraint so that constraint propagation is sound when the constraint is open. In [11] the approximation was used to design propagators for specific constraints. The use of the prefix closure in this paper is somewhat different from its use in [10,11]: rather than using $P(L_C)$ as an approximation to L_C, $P(L_C)$ is used to formulate what it means to be an open soft constraint.

To address a wide range of edit-based measures, we generalize the measures. We allow non-negative weights $\alpha, \beta, \gamma, \delta$ for the edit operations substitution, insertion, deletion and transposition, respectively, and let n_s, n_i, n_d, n_t be the number of the respective operations used in an edit. Then we define $m(w) = \min_{edits} \alpha n_s + \beta n_i + \gamma n_d + \delta n_t$ to be the minimum, over all edits that transform w to an element of $P(L_C)$, of the weighted sum of the edit operations.

Proposition 4. *Let C_s be a soft constraint with a weighted edit-based violation measure. Suppose $\alpha \leq \delta$ or $\beta \leq \delta$.*

Then C_s is contractible.

It follows that edit-based measures that only involve substitutions, insertions and deletions provide contractible constraints. Thus the variable-based measures [14,7], the object-based measures [4], and the edit-based measures of [7] induce contractible soft constraints. However, when transpositions are allowed and have a comparatively low cost, an edit-based violation measure can lead to a soft constraint that is not contractible.

Example 1. Suppose $\delta < \alpha$, $\delta < \beta$, and $\delta < \gamma$. Consider a REGULAR constraint for $(ab)^* + (ab)^*a$, which is a prefix-closed language. The word *abba* has edit distance δ, by transposing the last two letters, but its prefix *abb* has edit distance $\min\{\alpha, \beta, \gamma\}$, since we could either substitute a for b, insert a before the second b, or delete a b. Thus the weighted edit-distance violation measure is not non-decreasing and hence, by Proposition 1, the corresponding soft open REGULAR constraint is not contractible.

This example reinforces a point made earlier: the introduction of $P(L_C)$ to the definition of edit-based violation measure plays a different role than its use in [10,11]; in this case, its use does not ensure contractibility.

3.3 Graph Property-Based Violation Measures

The *graph property-based* violation measures [4] are based on representation of global constraints by graph properties [2,3]. A global constraint C is represented by an initial directed graph G_i, where each vertex corresponds to a variable and each edge has an attached constraint on the variables, and a graph property[5]. An assignment v to variables is a solution to C iff the graph G_f, obtained by deleting all edges whose constraints are not solved by v and all isolated vertices, satisfies the graph property. A graph property is a simple constraint on a characteristic of G_f such as the number of vertices, edges, sources or sinks, or the number or size of connected components or strongly connected components, etc. The violation measure of C under v is the violation measure of the graph property on G_f.

This framework has great flexibility; [4] presents three different violation measures for ALLDIFFERENT in this framework. Unfortunately, this makes it difficult to identify classes of constraints that are contractible. We need to narrow our focus to get a result.

Proposition 5. *Let C be a global constraint, defined in the graph property framework, and C_s be the corresponding graph property-based soft constraint.*

Suppose $G_i^{\mathbf{X}}$ is a subgraph of $G_i^{\mathbf{X}Y}$ for every sequence $\mathbf{X}$[6]; the only graph characteristics used to define C are the number of vertices , number of arcs, and the size of the largest (strongly) connected component in G_f; and the graph properties have the form of constant upper bounds on these graph characteristics.

Then C_s is contractible.

This result is applicable to CONTIGUITY as defined in [3] and the third form of ALLDIFFERENT in [4] (slightly altered to use $\leq$ in place of $=$ in the graph property, which is necessary in an open setting). It also applies to ATMOST, DISJOINT, and GOLOMB.

The assumption that $G_i^{\mathbf{X}}$ is a subgraph of $G_i^{\mathbf{X}Y}$ is not a strong assumption, given that most arc generators in [3] have this property. However, the other assumptions are very limiting.

[5] In general, multiple graphs and/or graph properties may be used.

[6] We use the sequence of variables as a superscript, to distinguish graphs.

3.4 Other Violation Measures

The last class of violation measures are *ad hoc* measures derived from the semantics of C. Because formulations of the measures are related to the semantics of the specific constraint and, perhaps, to a specific application, it is difficult to make broad statements about the nature of these measures and, hence, the contractibility of the corresponding soft constraints.

4 Propagator

The focus of this paper is not propagators, but we give a brief outline of a propagator for an open soft REGULAR under a weighted edit-based measure, building on the minimum cost network flow propagator designed in [7]. As discussed in Section 3.2, we use the prefix-closure of the given automaton. The propagator of [7] addressed only substitutions, insertions and deletions, but it can be extended to also handle transpositions. Transposition violation arcs have the form (q_k^i, q_l^{i+2}) where $\delta(q_k, d_{i+1}) = q$ and $\delta(q, d_i) = q_l$ for some state q and some $d_i \in D(X_i)$, $d_{i+1} \in D(X_{i+1})$. Violation arcs of the different types have capacity 1 and cost α, β, γ or δ, depending on type. When a new variable is appended to the sequence of variables, an additional copy of the states of the automaton must be added, with additional arcs representing both automata transitions and violations, somewhat like the open REGULAR constraint in [11]. The propagation algorithm is essentially the same as the one for edit-based SOFTREGULAR in [7]. The only difference is that reductions in variable domains may require transposition violation arcs to be deleted.

Because some of the variables in an open constraint will be unspecified during part of the execution, we need to adapt the definition of consistency. The following is an appropriate form of domain consistency for Barták's model [11].

Definition 1. *Given a domain D, an occurrence of a constraint $C(\boldsymbol{X})$ is* open D-consistent *if*

for every $X_i \in \boldsymbol{X}$ and every $d \in D(X_i)$ there is a word $d_1 \dots d_n$ in L_C such that $d_i = d$, $|\boldsymbol{X}| \leq n$, and $d_j \in D(X_j)$ for $j = 1, \dots, |\boldsymbol{X}|$.

Proposition 6. *Suppose $\alpha \leq \delta$ or $\beta \leq \delta$, and $\beta + \gamma \leq 2\delta$.*

The propagator described above achieves open D-consistency for the soft open global REGULAR *constraint under the weighted edit-distance violation measure.*

The proof relies on a lemma that states that if $\beta + \gamma \leq 2\delta$ then there is an edit of minimal cost where any letter subject to transposition is not subject to any other edit operation. It also relies on Proposition 4 above, Proposition 8 of [11] and Corollary 6 of [7].

5 Conclusions

We have broadened some classes of violation measures, and used these classes to establish that a wide range of soft constraints are contractible. Notice that the contractibility of a soft constraint is independent of the contractibility of the underlying hard constraint. The edit-based soft REGULAR constraint is contractible if all edit operations have weight 1, even when the underlying REGULAR constraint is not contractible. Conversely, a contractible hard constraint can give rise to an uncontractible soft constraint if the violation measure is chosen badly, as we saw in Example 1. Indeed, it appears that contractibility depends more on the violation measure than on the hard constraint. For example, soft GCC is contractible under many edit-based measures, such as the variable-based measure, but not under the decomposition-oriented value-based measure.

Contractibility only supports the adaptation of closed constraint propagators to open propagators. It seems orthogonal to issues like the usefulness of the violation measure for an application, the complexity of propagation, and the design of efficient propagators. These issues require further research.

Acknowledgements. Thanks to the referees for their insightful comments.

References

1. Barták, R.: Dynamic Global Constraints in Backtracking Based Environments. Annals of Operations Research 118, 101–119 (2003)
2. Beldiceanu, N.: Global Constraints as Graph Properties on a Structured Network of Elementary Constraints of the Same Type. In: Dechter, R. (ed.) CP 2000. LNCS, vol. 1894, pp. 52–66. Springer, Heidelberg (2000)
3. Beldiceanu, N., Carlsson, M., Rampon, J.-X.: Global Constraint Catalog, SICS Technical Report T2005:08. Current version, `http://www.emn.fr/x-info/sdemasse/gccat/`
4. Beldiceanu, N., Petit, T.: Cost Evaluation of Soft Global Constraints. In: Régin, J.-C., Rueher, M. (eds.) CPAIOR 2004. LNCS, vol. 3011, pp. 80–95. Springer, Heidelberg (2004)
5. Bessière, C., Hebrard, E., Hnich, B., Kiziltan, Z., Walsh, T.: SLIDE: a useful special case of the CardPath constraint. In: ECAI 2008, pp. 475–479 (2008)
6. Dechter, R.: Constraint Processing. Morgan Kaufmann, San Francisco (2003)
7. van Hoeve, W.-J., Pesant, G., Rousseau, L.-M.: On Global Warming: Flow-Based Soft Global Constraints. Journal of Heuristics 12(4-5), 347–373 (2006)
8. van Hoeve, W.-J., Régin, J.-C.: Open Constraints in a Closed World. In: Beck, J.C., Smith, B.M. (eds.) CPAIOR 2006. LNCS, vol. 3990, pp. 244–257. Springer, Heidelberg (2006)
9. Maher, M.J.: Analysis of a Global Contiguity Constraint. In: Proc. Workshop on Rule-Based Constraint Reasoning and Programming (2002)
10. Maher, M.J.: Open Contractible Global Constraints. In: IJCAI 2009, pp. 578–583 (2009)
11. Maher, M.J.: Open Constraints in a Boundable World. In: van Hoeve, W.J., Hooker, J.N. (eds.) CPAIOR 2009. LNCS, vol. 5547, pp. 163–177. Springer, Heidelberg (2009)
12. Maher, M., Narodytska, N., Quimper, C.-G., Walsh, T.: Flow-based propagators for the SEQUENCE and related global constraints. In: Stuckey, P.J. (ed.) CP 2008. LNCS, vol. 5202, pp. 159–174. Springer, Heidelberg (2008)
13. Pesant, G.: A Regular Language Membership Constraint for Finite Sequences of Variables. In: Wallace, M. (ed.) CP 2004. LNCS, vol. 3258, pp. 482–495. Springer, Heidelberg (2004)
14. Petit, T., Régin, J.-C., Bessière, C.: Specific filtering algorithms for over-constrained problems. In: Walsh, T. (ed.) CP 2001. LNCS, vol. 2239, pp. 451–463. Springer, Heidelberg (2001)
15. Rossi, F., van Beek, P., Walsh, T. (eds.): Handbook of Constraint Programming. Elsevier, Amsterdam (2006)

Exploiting Problem Decomposition in Multi-objective Constraint Optimization

Radu Marinescu

Cork Constraint Computation Centre
University College Cork, Ireland
r.marinescu@4c.ucc.ie

Abstract. Multi-objective optimization is concerned with problems involving multiple measures of performance which should be optimized simultaneously. In this paper, we extend AND/OR Branch-and-Bound (AOBB), a well known search algorithm, from mono-objective to multi-objective optimization. The new algorithm MO-AOBB exploits efficiently the problem structure by traversing an AND/OR search tree and uses static and dynamic mini-bucket heuristics to guide the search. We show that MO-AOBB improves dramatically over the traditional OR search approach, on various benchmarks for multi-objective optimization.

1 Introduction

A *Constraint Optimization Problem* (COP) is the minimization (or maximization) of an objective function subject to a set of constraints on the possible values of a set of independent decision variables. Many real-world problems involve multiple measures of performance or objectives which should be optimized simultaneously. In contrast with single function optimization, the simultaneous optimization of multiple, possibly conflicting, objective functions does not in general admit a single, perfect solution. Instead, *Multi-objective Constraint Optimization Problems* (MO-COP) tend to be characterized by a set of alternatives which must be considered equivalent in the absence of information concerning the relevance of each objective relative to the others. Therefore, solving a MO-COP is to find the *efficient frontier*, namely the set of equivalent or non-dominated costs of the set of feasible solutions.

Most complete algorithms for solving MO-COPs typically fall within one of the following two categories: *search* and *inference*. Search-based algorithms (*e.g.* depth-first Branch-and-Bound) transform a problem into a set of subproblems by selecting a variable and considering the assignment of each of its domain values. The subproblems are solved in sequence applying recursively the same transformation rule. These algorithms are not sensitive to the problem structure, have a time complexity which is exponential in the number of variables, but can operate in polynomial space. Inference-based algorithms (*e.g.*, variable elimination) are known to be good at exploiting the structural information encoded in the problem. These methods apply a sequence of transformations that reduce the problem size, while preserving the solution space of the problem. Their time and space complexity is exponential in a topological parameter called *treewidth* (always less than or equal to the number of variables). Due to their high space requirements, when the treewidth is large, the latter methods are often impractical.

I.P. Gent (Ed.): CP 2009, LNCS 5732, pp. 592–607, 2009.

The AND/OR search space for COPs [1] is a relatively new framework for search that is sensitive to the problem structure, often resulting in substantially improved performance. It is based on a pseudo tree that captures conditional independencies in the problem, resulting in a search tree exponential in the depth of the pseudo tree, rather than in the number of variables. The AND/OR Branch-and-Bound (AOBB) is a recent search algorithm introduced in [2] that traverses depth-first an AND/OR search tree associated with a single objective COP instance.

In this paper, we extend the AND/OR search tree from mono-objective to multi-objective optimization. We then present a multi-objective depth-first AND/OR Branch-and-Bound algorithm (MO-AOBB) for finding the efficient frontier of a MO-COP instance. The efficiency of the proposed algorithm also depends on the accuracy of its guiding heuristic function. We investigate a class of partitioning-based heuristics which is based on the *multi-objective mini-bucket elimination* introduced recently in [3]. The mini-bucket heuristics can be either pre-compiled or generated dynamically at each node in the search tree. We evaluate the impact of our advancement on several benchmarks for MO-COPs, including risk-conscious combinatorial auctions, MAX-SAT-ONE problem instances, bi-objective weighted vertex cover problems, as well as a set of MO-COP instances derived from real-world mono-objective COP instances. Our results show conclusively that the new multi-objective AND/OR Branch-and-Bound algorithm improves dramatically over state-of-the-art search algorithms traversing the traditional OR space, in many cases by several orders of magnitude.

The paper is organized as follows. Section 2 provides background on COPs, AND/OR search trees for COPs as well as on MO-COPs. In Section 3 we extend the AND/OR search tree from mono-objective to multi-objective optimization. Section 4 presents the multi-objective AND/OR Branch-and-Bound search algorithm for finding the efficient frontier of a MO-COP instance. Section 5 is dedicated to our empirical evaluation, while Section 6 provides concluding remarks and directions of future research.

2 Background

2.1 Mono-objective Constraint Optimization Problems

A *Constraint Optimization Problem* (COP) is a triple $\mathcal{P} = \langle \mathbf{X}, \mathbf{D}, \mathbf{F} \rangle$, where $\mathbf{X} = \{X_1, ..., X_n\}$ is a set of variables, $\mathbf{D} = \{D_1, ..., D_n\}$ is a set of finite domains and $\mathbf{F} = \{f_1, ..., f_r\}$ is a set of cost functions. Cost functions can be either *soft* or *hard* (constraints). Without loss of generality we assume that hard constraints are represented as (bi-valued) cost functions. Allowed and forbidden tuples have cost 0 and ∞, respectively. The scope of function f_i, denoted $scope(f_i) \subseteq \mathbf{X}$, is the set of arguments of f_i. The goal is to find a complete value assignment to the variables, $\boldsymbol{x} = (x_1, ..., x_n)$, that minimizes the objective function defined by $\mathcal{F}(\mathbf{X}) = \sum_{i=1}^{r} f_i(Y_i)$, where $Y_i \subseteq \mathbf{X}$.

Given a COP instance, its *primal graph* G has a node for each variable and an edge connects any two nodes whose variables appear in the scope of the same function. The *induced graph* of G relative to an ordering d of its variables is obtained by processing the nodes in reverse order of d. For each node all its earlier neighbors are connected, including neighbors connected by previously added edges. The *width* of a node is the number of edges connecting it to nodes lower in the ordering. The *treewidth* of a graph along d, denoted $w^*(d)$, is the maximum width of nodes in the induced graph.

2.2 AND/OR Search Trees for COP

The AND/OR search space [1] is a unifying framework for advanced algorithmic schemes for graphical models, including constraint networks and cost networks. Its main virtue consists in exploiting independencies between variables during search, which can provide exponential speedups over traditional structure-blind search methods (called here OR search). The search space is defined using a backbone *pseudo tree* [4].

Definition 1 (pseudo tree). *Given an undirected graph $G = (V, E)$, a directed rooted tree $\mathcal{T} = (V, E')$ defined on all its nodes is called a* pseudo tree *if any edge of G that is not included in E' is a back-arc in $\mathcal{T}$, namely it connects a node to an ancestor in $\mathcal{T}$.*

Given a COP instance $\mathcal{P} = \langle \mathbf{X}, \mathbf{D}, \mathbf{F} \rangle$, its primal graph G and a pseudo tree $\mathcal{T}$ of G, the associated AND/OR search tree $S_{\mathcal{T}}(\mathcal{P})$ has alternating levels of OR and AND nodes. The OR nodes are labeled X_i and correspond to the variables. The AND nodes are labeled $\langle X_i, x_i \rangle$ (or just x_i) and correspond to value assignments of the variables. The structure of the AND/OR search tree is based on the underlying pseudo tree $\mathcal{T}$. The root of the AND/OR search tree is an OR node labeled with the root of $\mathcal{T}$. The children of an OR node X_i are AND nodes labeled with assignments $\langle X_i, x_i \rangle$ that are consistent with the assignments along the path from the root. The children of an AND node $\langle X_i, x_i \rangle$ are OR nodes labeled with the children of variable X_i in $\mathcal{T}$.

A *solution tree* T of an AND/OR search tree $S_{\mathcal{T}}(\mathcal{P})$ is an AND/OR subtree such that: (1) it contains the root of $S_{\mathcal{T}}(\mathcal{P})$, s; (2) if a non-terminal AND node $n \in S_{\mathcal{T}}(\mathcal{P})$ is in T then all of its children are in T; (3) if a non-terminal OR node $n \in S_{\mathcal{T}}(\mathcal{P})$ is in T then exactly one of its children is in T;

It was shown [4,5,1] that given a COP instance and a pseudo tree $\mathcal{T}$ of depth m, the size of the AND/OR search tree based on $\mathcal{T}$ is $O(n \cdot k^m)$, where k bounds the domains of variables. Moreover, a COP having treewidth w^* has a pseudo tree of depth at most $w^* \cdot \log n$, and therefore it has an AND/OR search tree of size $O(n \cdot k^{w^* \cdot \log n})$.

2.3 Multi-objective Constraint Optimization Problems

Let us consider now problems with p objectives. A *cost vector* (also called a *p-vector*) $\boldsymbol{v} = (v_1, ..., v_p)$ is a vector of p components where each $v_j \in \mathbb{R}$ represents the cost with respect to objective j. A cost vector which has all components equal to 0 is denoted by $\{\mathbf{0}\}$. A *multi-objective function* f is a function that associates a cost vector to each assignment of its scope. Let $\boldsymbol{v}$ and $\boldsymbol{u}$ be two cost vectors. We say that $\boldsymbol{v}$ dominates $\boldsymbol{u}$, denoted by $\boldsymbol{v} \leq \boldsymbol{u}$, if $\forall j, v_j \leq u_j$. We say that $\boldsymbol{v} < \boldsymbol{u}$ if $\boldsymbol{v} \leq \boldsymbol{u}$ and $\boldsymbol{v} \neq \boldsymbol{u}$. The sum of cost vectors is the usual point-wise sum.

Let α be a set of cost vectors. We define its *non-dominated closure* as $\parallel \alpha \parallel = \{\boldsymbol{u} \in \alpha \mid \forall \boldsymbol{v} \in \alpha, \boldsymbol{v} \not< \boldsymbol{u}\}$. A set of cost vectors closed under non-domination is called *frontier*. Let α and β be two frontiers. We say that α *dominates* β, denoted by $\alpha \leq \beta$, if $\forall \boldsymbol{v} \in \beta$, $\exists \boldsymbol{u} \in \alpha$ s.t. $\boldsymbol{u} \leq \boldsymbol{v}$. We say that $\alpha < \beta$, if $\alpha \leq \beta$ and $\alpha \neq \beta$. The sum of frontiers is $\alpha + \beta = \parallel \{\boldsymbol{w} = \boldsymbol{u} + \boldsymbol{v} \mid \boldsymbol{u} \in \alpha, \boldsymbol{v} \in \beta\} \parallel$.

X_0	X_1	X_2	f_1
0	0	0	(5,0)
0	0	1	(2,0)
0	1	0	(3,0)
0	1	1	(2,0)
1	0	0	(2,0)
1	0	1	(3,0)
1	1	0	(0,0)
1	1	1	(2,0)

X_0	X_1	X_3	f_2
0	0	0	(1,0)
0	0	1	(4,0)
0	1	0	(0,0)
0	1	1	(2,0)
1	0	0	(6,0)
1	0	1	(5,0)
1	1	0	(6,0)
1	1	1	(5,0)

X_1	X_3	X_4	f_3
0	0	0	(1,0)
0	0	1	(3,0)
0	1	0	(5,0)
0	1	1	(4,0)
1	0	0	(1,0)
1	0	1	(3,0)
1	1	0	(5,0)
1	1	1	(4,0)

X_0	g_0
0	(0,0)
1	(0,1)

X_1	g_1
0	(0,0)
1	(0,2)

X_2	g_2
0	(0,0)
1	(0,3)

X_3	g_3
0	(0,0)
1	(0,4)

X_4	g_4
0	(0,0)
1	(0,5)

(a) Bi-objective functions

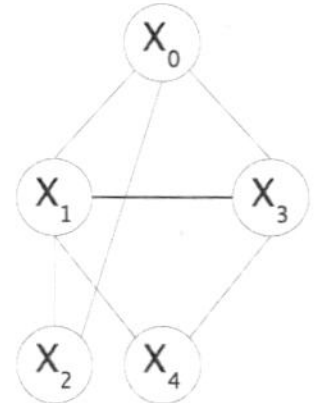

(b) Primal graph

Fig. 1. An example of a MO-COP instance with 2 objectives

A *Multi-objective Constraint Optimization Problem* (MO-COP) is a triple $\mathcal{M} = \langle \mathbf{X}, \mathbf{D}, \mathbf{F} \rangle$, where $\mathbf{X} = \{X_1, ..., X_n\}$, $\mathbf{D} = \{D_1, ..., D_n\}$ and $\mathbf{F} = \{f_1, ..., f_r\}$ are variables, domains and multi-objective functions, respectively. The sum of functions in $\mathbf{F}$ defines the objective function $\mathcal{F}$. The only difference with respect to mono-objective COP is that multi-objective functions are used. A *solution* is a complete assignment $\boldsymbol{x} = (x_1, ..., x_n)$. A solution $\boldsymbol{x}$ is *efficient* or *Pareto optimal* if $\forall \boldsymbol{x}'$, $\mathcal{F}(\boldsymbol{x}') \nless \mathcal{F}(\boldsymbol{x})$. The *efficient frontier* is the set $\mathcal{E} = \{\mathcal{F}(\boldsymbol{x}) \mid \forall \boldsymbol{x}', \mathcal{F}(\boldsymbol{x}') \nless \mathcal{F}(\boldsymbol{x})\}$. The optimization task is to compute $\mathcal{E}$, and for each $\boldsymbol{v} \in \mathcal{E}$, one (of the possibly many) Pareto optimal assignment with cost $\boldsymbol{v}$[1].

As in the mono-objective case, each MO-COP instance has an associated *primal graph*, which is computed in the same way: nodes correspond to the variables and an arc connects any pair of nodes whose variables belong to the scope of the same function.

Example 1. Figure 1 shows a MO-COP instance with 2 objectives, 5 bi-valued variables and 8 cost functions. The first objective is defined by functions f_1, f_2 and f_3, while functions g_0, g_1, g_2, g_3 and g_4 form the second objective, respectively. The corresponding bi-objective functions are given in Figure 1(a). For example, the cost vector associated with $f_1(X_0 = 0, X_1 = 0, X_2 = 0)$ has two components, 5 and 0, which represent the costs with respect to the first and second objective, respectively. The primal graph is shown in Figure 1(b). The problem has three solutions: (00000), (01000) and (01100) with non-dominated costs $(7, 0)$, $(4, 2)$ and $(3, 5)$, respectively.

3 Weighted AND/OR Search Trees for MO-COP

In this section we extend the AND/OR search tree from mono-objective to multi-objective optimization. Given a MO-COP instance $\mathcal{M} = \langle \mathbf{X}, \mathbf{D}, \mathbf{F} \rangle$ and its primal graph G, the AND/OR search tree $S_{\mathcal{T}}(\mathcal{M})$ of $\mathcal{M}$ is driven by a pseudo tree $\mathcal{T}$ of G, as in the mono-objective case. The arcs from nodes OR nodes X_i to AND nodes $\langle X_i, x_i \rangle$ in $S_{\mathcal{T}}(\mathcal{M})$ are annotated by *weights* derived from the multi-objective cost functions in $\mathbf{F}$. Each node n in the weighted search tree is associated with a *value* $v(n)$ which stands for the answer to the optimization query restricted to the conditioned subproblem below n.

[1] In some cases, $\mathcal{E}$ may be too large and must be approximated. In this paper we do not consider this situation.

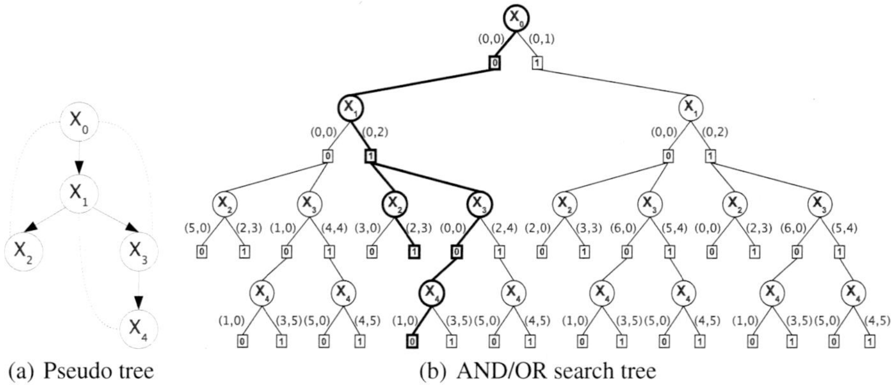

Fig. 2. AND/OR search tree for the MO-COP instance from Fig. 1

Definition 2 (weight). *The* weight $w(n, n')$ *of the arc from the OR node* n *labeled* X_i *to the AND node* n' *labeled* $\langle X_i, x_i \rangle$ *is a cost vector defined as the sum of all the multi-objective cost functions whose scope includes* X_i *and is fully assigned along the path from the root to* $\langle X_i, x_i \rangle$*, evaluated at the values along the path.*

Definition 3 (value). *The* value $v(n)$ *of a node* n *in a weighted AND/OR search tree of a MO-COP instance* $\mathcal{M} = \langle \mathbf{X}, \mathbf{D}, \mathbf{F} \rangle$ *with* p *objectives is defined recursively as follows (where* $succ(n)$ *are the children of* n*):*

(1) $v(n) = \{\mathbf{0}\}$*, if* $n = \langle X_i, x_i \rangle$ *is a terminal AND node;*
(2) $v(n) = \sum_{n' \in succ(n)} v(n')$*, if* $n = \langle X_i, x_i \rangle$ *is an internal AND node;*
(3) $v(n) = \| \{(w(n, n') + v(n')) \mid n' \in succ(n)\} \|$*, if* $n = X_i$ *is an OR node.*

It is easy to see that the value $v(n)$ of a node in the AND/OR search tree $S_{\mathcal{T}}(\mathcal{M})$ is the set of cost vectors representing the efficient frontier of the subproblem rooted at n, subject to the current variable instantiation along the path from the root to n. If n is the root of $S_{\mathcal{T}}(\mathcal{M})$, then $v(n)$ is the efficient frontier of the initial problem.

Example 2. Figure 2(b) shows the AND/OR search tree for the MO-COP instance from Figure 1, relative to the pseudo tree from Figure 2(a). The numbers on the OR-to-AND arcs are the weights corresponding to the function values. For example, the weight of the arc from the OR node X_2 to its AND child $\langle X_2, 0 \rangle$, along the path $(X_0, \langle X_0, 0 \rangle, X_1, \langle X_1, 0 \rangle, X_2, \langle X_2, 0 \rangle)$, is $(5, 0)$ and is obtained by summing the values of the functions $f_1(X_0, X_1, X2)$ and $g_2(X_2)$, whose scopes contain X_2 and are fully instantiated along that path. A solution tree that corresponds to the assignment $(X_0 = 0, X_1 = 1, X_2 = 1, X_3 = 0, X_4 = 0)$ with cost $(3, 5)$ is highlighted in Figure 2(b). The OR node X_2 along the path $(X_0, \langle X_0, 1 \rangle, X_1, \langle X_1, 0 \rangle, X_2)$ has value $(2, 0)$ which is the non-dominated closure of the frontier below it, namely $\{(2, 0), (3, 3)\}$.

4 Multi-objective AND/OR Branch-and-Bound Search

Multi-objective problems can be solved using a depth-first Branch-and-Bound schema, which traverses a traditional OR search tree of possible assignments in a depth-first manner and outputs the efficient frontier $\mathcal{E}$ of the problem [6,7]. The algorithm, called MO-BB, handles two frontiers, called *lower* and *upper bound frontiers*. The upper bound frontier is the set of cost vectors of the best solutions found so far and is an overestimation of the efficient frontier $\mathcal{E}$. During search, MO-BB computes at each visited node a lower bound frontier [8] of the current subproblem using a heuristic evaluation function. MO-BB backtracks if the upper bound dominates the lower bound, because it implies that the current partial assignment cannot lead to any new efficient cost vector.

In this section, we apply the general principles of AND/OR search and extend MO-BB into a Branch-and-Bound algorithm that explores an AND/OR rather than a regular OR search tree for finding the efficient frontier of a MO-COP instance. The proposed algorithm is an extension of the recently introduced single objective AND/OR Branch-and-Bound (AOBB) algorithm [2] to the multi-objective optimization case. We start by revisiting the notion of partial solution trees [9] to represent sets of solution trees.

Definition 4 (partial solution tree). *A partial solution tree T' of an AND/OR search tree $S_{\mathcal{T}}(\mathcal{M})$ is a subtree which: (1) contains the root node s of $S_{\mathcal{T}}(\mathcal{M})$; (2) if n is an OR node in T' then it contains one of its AND child nodes in $S_{\mathcal{T}}(\mathcal{M})$, and if n is an AND node it contains all its OR children in $S_{\mathcal{T}}(\mathcal{M})$. A node of T' is a* tip *node if it has no children in T'. A tip node of T' is either a* terminal *node (if it has no children in $S_{\mathcal{T}}(\mathcal{M})$), or a* non-terminal *node (if it has children in $S_{\mathcal{T}}(\mathcal{M})$).*

A partial solution tree may be extended (possibly in several ways) to a full solution tree. It represents $extension(T')$, the set of all full solution trees which can extend it. Clearly, a partial solution tree all of whose tip nodes are terminal is a solution tree.

4.1 Heuristic Lower Bounds on Partial Solution Trees

We next define the notion of heuristic evaluation function of a partial solution tree, which will be used to guide the AND/OR Branch-and-Bound search. Like in OR search, we assume a given heuristic evaluation function $h(n)$ associated with each node in the search tree such that $h(n)$ is an underestimation of the efficient frontier $v(n)$ of the conditioned subproblem below n.

Definition 5 (heuristic evaluation function). *Given a partial solution tree T'_n rooted at node n in the AND/OR tree $S_{\mathcal{T}}(\mathcal{M})$, the* heuristic evaluation function *$f(T'_n)$, is defined recursively by: (1) if T'_n consists of a single node n, then $f(T'_n) = h(n)$; (2) if n is an OR node having the AND child m in T'_n, then $f(T'_n) = w(n,m) + f(T'_m)$; (3) if n is an AND node having OR children $m_1, ..., m_k$ in T'_n, then $f(T'_n) = \sum_{i=1}^{k} f(T'_{m_i})$.*

Clearly, by definition, $f(T'_n)$ is a lower bound frontier that underestimates the efficient frontier of the subproblem represented by T'_n. During search, the algorithm maintains also an upper bound frontier $ub(s)$ of the efficient frontier $v(s)$, where s is the root of the search tree. Given the current partial solution tree T'_s and if the upper bound frontier $ub(s)$ dominates the lower bound frontier $f(T'_s)$ (*i.e.*, $ub(s) \leq f(T'_s)$), then searching below the current tip node t of T'_s is guaranteed not to lead to any new efficient cost vector and, therefore, search below t can be safely halted.

Algorithm 1. MO-AOBB

```
   Data: A MO-COP instance M = ⟨X, D, F⟩ with p objectives, pseudo tree T.
   Result: Efficient frontier E of M.
1  create an OR node s labeled X_1                  // Create and initialize the root node
2  v(s) ← ∅; OPEN ← {s}
3  while OPEN ≠ ∅ do
4      n ← top(OPEN); remove n from OPEN
5      if n is an OR node, labeled X_i then                                    // EXPAND
6          for x_i ∈ D_{X_i} do
7              create an AND node n' labeled ⟨X_i, x_i⟩
8              v(n') ← {0}; h(n') ← heuristic(n')
9              w(n, n') ← Σ_{f∈F, X_i∈scope(f)} f(asgn(π_n))
10             succ(n) ← succ(n) ∪ {n'}
11     else if n is an AND node, labeled ⟨X_i, x_i⟩ then
12         let T'_s be the current partial solution tree
13         if v(s) ⋠ f(T'_s) then
14             for X_j ∈ children_T(X_i) do
15                 create an OR node n' labeled X_j
16                 v(n') ← ∅; h(n') ← heuristic(n')
17                 succ(n) ← succ(n) ∪ {n'}
18     Add succ(n) on top of OPEN
19     while succ(n) == ∅ do                                                   // PROPAGATE
20         let par be the parent of n
21         if n is an OR node, labeled X_i then
22             if X_i == X_1 then                                   // Search is complete
23                 return v(n)
24             v(par) ← v(par) + v(n)
25         if n is an AND node, labeled ⟨X_i, x_i⟩ then
26             v(par) ←|| v(par) ∪ {(w(par, n) + v(n))} ||
27         remove n from succ(par)
28         n ← par
```

4.2 The Branch-and-Bound Algorithm

The depth-first AND/OR Branch-and-Bound search algorithm, MO-AOBB, that traverses an AND/OR search tree is described by Algorithm 1. It takes as input a MO-COP instance $\mathcal{M} = \langle \mathbf{X}, \mathbf{D}, \mathbf{F} \rangle$ with p objectives as well as a pseudo tree $\mathcal{T}$ of its primal graph and returns the efficient frontier of $\mathcal{M}$. The fringe of the search is maintained by a stack called OPEN. The current node is denoted by n, its parent by par, the current path by π_n and its assignment by $asgn(\pi_n)$. The successors of the current node are denoted by $succ(n)$, while the children of a variable X_i in $\mathcal{T}$ are denoted by $children_{\mathcal{T}}(X_i)$.

Each node n in the search tree maintains its current value $v(n)$, which is updated based on the values of its children. For OR nodes, the current $v(n)$ is an upper bound frontier on the efficient frontier below n. Initially, $v(n)$ is set to $\emptyset$ if n is OR, and $\{\mathbf{0}\}$ if n is AND, respectively. Procedure heuristic(n) computes the lower bound $h(n)$ on the efficient frontier of the subproblem below n, conditioned on $asgn(\pi_n)$.

In the EXPAND step, the current node n is expanded in the usual way, depending on whether it is an AND or an OR node (lines 5–17). The successors of an OR node X_i are AND nodes labeled by the domain values x_i of X_i. The successors of an AND node $\langle X_i, x_i \rangle$ are OR nodes labeled by the children X_j of X_i in the pseudo tree $\mathcal{T}$. The algorithm also computes the heuristic evaluation function $f(T'_s)$ of the current partial solution tree T'_s being explored, based on Definition 5. The search below n is terminated if the current upper bound frontier $v(s)$ dominates $f(T'_s)$ (line 13).

The node values are updated by the PROPAGATE step (lines 19–28). It is triggered when a node has an empty set of descendants (note that as each successor is evaluated, it is removed from the set of successors in line 27). This means that all its children have been evaluated, and their values are already determined. If the current node is the root, then the search terminates with its value, namely the efficient frontier (line 23). If n is an OR node, then its parent par is an AND node, and par updates its current value $v(par)$ by summation with the value of n (line 24). An AND node n propagates its value to its parent par in a similar way, by the non-dominated closure (lines 25–26). Finally, the current node n is set to its parent par (line 28), because n was completely evaluated. Search continues either with a *propagation* step (if conditions are met) or with an *expansion* step. The algorithm can be easily instrumented to also record the solution trees that correspond to the cost vectors in the efficient frontier. Clearly,

Theorem 1. *MO-AOBB is sound and complete for MO-COP. In a problem with a single objective function (i.e., $p = 1$), the algorithm MO-AOBB is equivalent to AOBB.*

While the time complexity of MO-AOBB is bounded by $O(n \cdot exp(m))$, the size of the AND/OR search tree, in practice, the running time of MO-AOBB is expected to be much better than the worst-case bound because the pruning mechanism via the heuristic evaluation function will prune unpromising portions of the search space.

4.3 Partitioning-Based Lower Bound Heuristics

We now describe briefly two general schemes for generating the heuristic estimates $h(n)$, based on the multi-objective Mini-Bucket approximation. These schemes are parameterized by the Mini-Bucket i-bound, thus allowing for a controllable trade-off between heuristic strength and its computational overhead.

Mini-Bucket Elimination. (MBE) [10] is an approximation algorithm designed to avoid the high time and space complexity of *Bucket Elimination* (BE) [11], by partitioning large buckets into smaller subsets, called *mini-buckets*, each containing at most i (called i-bound) distinct variables. The mini-buckets are then processed separately. The algorithm computes a lower bound (assuming minimization) on the optimal solution to a COP. The complexity is time and space $O(exp(i))$. More recently, [3] extended MBE(i) from mono-objective to multi-objective optimization problems, yielding *Multi-objective Mini-Bucket Elimination* (MO-MBE). Given a MO-COP instance, MO-MBE(i) outputs not only a *lower bound frontier* of the efficient frontier, but also the collection of augmented buckets, which form the basis for the heuristics generated.

Static Mini-Bucket Heuristics. In the past, [12,2] showed that the intermediate functions generated by the MBE(i) can be used to compute a heuristic function that underestimates the minimal extension of the current assignment in both OR and AND/OR search trees for COP. Here, we extend the idea to multi-objective optimization. Consider a MO-COP instance $\mathcal{M}$ and the ordered set of augmented buckets $\{B(X_1), ..., B(X_n)\}$ generated by MO-MBE(i) along a DFS traversal of the pseudo tree $\mathcal{T}$ of $\mathcal{M}$. Given a node n in the AND/OR search tree relative to $\mathcal{T}$, the *static mini-bucket heuristic* function $h(n)$ is computed as follows: (1) if n is an AND node labeled $\langle X_j, x_j \rangle$, then $h(n)$ is the sum of all intermediate functions that were generated in buckets corresponding to descendants of X_j in $\mathcal{T}$ and reside in bucket $B(X_j)$ or the buckets corresponding to the ancestors of X_j in $\mathcal{T}$; (2) if n is an OR node labeled X_j, then $h(n) = \| \{(w(n,m) + h(m)) \mid m \in succ(n)\} \|$.

Dynamic Mini-Bucket Heuristics. Rather than pre-compiling the mini-bucket heuristic information, it is possible to generate it dynamically, during search [2,7]. Specifically, given the set of buckets $\{B(X_1), ..., B(X_n)\}$ ordered along a DFS traversal of $\mathcal{T}$, a node n in the AND/OR search tree and given the current partial assignment $asgn(\pi_n)$ along the path to n, the *dynamic mini-bucket heuristic* function $h(n)$ is computed as follows: (1) if n is an AND node labeled $\langle X_j, x_j \rangle$, then $h(n)$ is the sum of the intermediate functions that reside in bucket $B(X_j)$ and were generated by MO-MBE(i), conditioned on $asgn(\pi_n)$, in the buckets corresponding to the descendants of X_j in $\mathcal{T}$; (2) if n is an OR node labeled X_j, then $h(n) = \| \{(w(n,m) + h(m)) \mid m \in succ(n)\} \|$. Given an i-bound, the dynamic mini-bucket heuristic implies a much higher computational overhead compared with the static version. However, the bounds generated dynamically may be far more accurate since some of the variables are assigned and will therefore yield smaller functions and less partitioning.

5 Experiments

To evaluate our multi-objective AND/OR Branch-and-Bound approach we have conducted a number of experiments on several MO-COP problem classes such as bi-objective combinatorial auction, bi-objective weighted vertex cover problems, MAX-SAT-ONE problem instances as well as bi-objective Weighted CSPs derived from standard mono-objective instances. We implemented our algorithms in C++ and carried out all experiments on a 2.4GHz dual quad-core with 8GB of RAM running Linux Ubuntu 8.10.

We considered two classes of depth-first AND/OR Branch-and-Bound search algorithms guided by static and dynamic mini-bucket heuristics. They are denoted by MO-AOBB+SMB(i) and MO-AOBB+DMB(i), respectively. We compare these algorithms against traditional depth-first OR Branch-and-Bound algorithms with static and dynamic mini-bucket heuristics, denoted by MO-BB+SMB(i) and MO-BB+DMB(i), respectively. The parameter i represents the mini-bucket i-bound and controls the accuracy of the heuristic. For reference, we also ran the multi-objective Russian Doll Search algorithm, denoted by MO-RDS, which was recently introduced in [6]. MO-RDS explores a traditional OR search tree and extends the well known Russian Doll Search [13] from mono-objective to multi-objective optimization. Algorithms MO-BB+DMB(i) and MO-RDS were evaluated extensively in [6,7] and shown to outperform dramatically state-of-the-art algorithms for multi-objective optimization such as ϵ-constraint based search [14], multi-objective iterative deepening search [3] and multi-objective bucket elimination [3]. The pseudo trees that guide the AND/OR search algorithms were generated using the *min-fill* heuristic [2]. All competing OR search algorithms used a static variable ordering which was derived from the pseudo tree as well.

In all our experiments we report the CPU time in seconds and the number of nodes visited for solving the problems. We also specify the problems' parameters such as the number of variables (n), maximum domain size (k), the number of cost functions (c), the depth of the pseudo tree (h) and the treewidth of the graph (w^*). In addition, we also record the size of the efficient frontier found ($|\mathcal{E}|$). The best performance points are highlighted in each table. A "-" stands for exceeding the time limit.

Bi-objective Weighted Vertex Cover. Given a graph $G = (V, E)$, a *vertex cover* is a subset of vertices $S \subseteq V$ such that $\forall(u, v) \in E$, either $u \in S$ or $v \in S$. The *minimum vertex cover* is a vertex cover of minimum size. In the weighted version every vertex v has an associated weight $w(v)$ and the *weighted minimum vertex cover* is a vertex cover S with minimum $F(S) = \sum_{v \in S} w(v)$. In the bi-objective version, each vertex v has two weights $w_1(v)$ and $w_2(v)$ and the task is to minimize the two associated objective functions simultaneously [3].

We generated random graphs with parameters (V, E, C), where V is the number of vertices, E is the number of edges and C is the maximum weight, as suggested in [3]. Instances were generated by randomly selecting E edges. The two weights associated with each vertex were generated uniformly at random between 0 and C. Table 1 reports the results obtained on two classes of random graphs: `sparse` (with $V \in \{60, 70, 80, 90\}$ and $E \in \{100, 120, 140, 160\}$), and `medium` (with $V \in \{30, 40, 50, 60\}$ and $C \in \{90, 120, 150, 180\}$), respectively. In both cases we set $C = 30$. The columns are indexed by the mini-bucket i-bound and the table entries represent an average over 10 random instances generated for each parameter configuration.

When looking at the algorithms guided by mini-bucket heuristics, we observe that both MO-AOBB+SMB(i) and MO-AOBB+DMB(i) outperformed dramatically the corresponding OR Branch-and-Bound algorithms MO-BB+SMB(i) and MO-BB+DMB(i), across all reported i-bounds. For example, on `sparse` graphs of size (60,100), MO-AOBB+SMB(10) solved all instances in about 3 seconds, while MO-BB+SMB(10) exceeded the 1 hour time limit. In terms of the quality of the heuristic, we also see that static mini-buckets with a relatively large i-bound represent the best choice. However, if larger i-bounds are not possible, dynamic mini-buckets with small i-bounds are preferred, especially on `sparse` problems. For problems from the `medium` class, MO-RDS proved to be cost effective with respect to the mini-bucket based algorithms.

Bi-objective Combinatorial Auctions. In combinatorial auctions, an auctioneer has a set of goods to sell and the buyers submit a set of bids on indivisible subsets of goods [15]. In *risk-conscious* auctions, the auctioneer wants also to control the risk of not being paid after a bid has been accepted, because it may cause large losses in revenue [16]. Let $M = \{1, ..., n\}$ be the set of goods to be auctioned and let $\mathcal{B} = \{B_1, ..., B_m\}$ be the set of bids. A bid B_j is defined by a triple (S_j, p_j, r_j), where $S_j \subseteq M$, p_j is the bid price and r_j is the probability of failure, respectively. The auctioneer must decide which bids to accept under the constraint that each good is allocated to at most one bid. The first objective is to maximize the auctioneer profit. The second objective is to minimize risk. Assuming independence, after a logarithmic transformation of probabilities, this objective can also be expressed as an additive function [7].

Table 1. Result for bi-objective weighted vertex cover problems using static and dynamic mini-bucket heuristics. Time limit 1 hour. Overall best performance points are boxed.

(V, E) (w*, h) |𝓔|	MO-RDS		MO-BB+SMB(i) MO-AOBB+SMB(i) MO-BB+DMB(i) MO-AOBB+DMB(i) i=2		MO-BB+SMB(i) MO-AOBB+SMB(i) MO-BB+DMB(i) MO-AOBB+DMB(i) i=4		MO-BB+SMB(i) MO-AOBB+SMB(i) MO-BB+DMB(i) MO-AOBB+DMB(i) i=6		MO-BB+SMB(i) MO-AOBB+SMB(i) MO-BB+DMB(i) MO-AOBB+DMB(i) i=8		MO-BB+SMB(i) MO-AOBB+SMB(i) MO-BB+DMB(i) MO-AOBB+DMB(i) i=10	
	time	nodes	time	nodes	time	nodes	time	nodes	time	nodes	time	nodes
sparse												
(60, 100)			-	-	-	-	-	-	-	-	-	-
(10, 19)			13.27	109,182	14.10	53,051	3.39	49,877	3.28	49,317	**3.27**	49,104
14	108.74	12,774,381	-	-	-	-	-	-	-	-	-	-
			6.25	52,984	5.78	49,149	6.23	49,081	6.80	49,076	7.13	49,076
(70, 120)			-	-	-	-	-	-	-	-	-	-
(12, 22)			41.74	253,516	10.87	119,060	7.69	106,561	7.15	104,145	**6.98**	103,529
16	1560.06	224,998,226	-	-	-	-	-	-	-	-	-	-
			14.80	112,891	12.60	103,527	13.44	103,275	14.83	103,246	16.21	103,242
(80, 140)			-	-	-	-	-	-	-	-	-	-
(14, 26)			444.19	2,466,147	74.70	747,408	57.01	689,606	52.33	675,044	**51.40**	672,451
17	-	-	-	-	-	-	-	-	-	-	-	-
			103.91	735,773	83.71	671,259	85.34	669,980	91.02	698,885	96.33	698,860
(90, 160)			-	-	-	-	-	-	-	-	-	-
(16, 27)			1815.14	6,047,711	112.88	731,474	58.93	568,307	51.77	551,728	**46.09**	536,089
29	-	-	-	-	-	-	-	-	-	-	-	-
			169.56	704,192	84.37	533,688	90.17	532,051	100.80	531,590	118.75	531,519
medium												
(30, 90)			-	-	-	-	-	-	-	-	-	-
(11, 19)			18.56	356,178	17.76	353,469	17.43	351,828	**17.27**	351,304	**17.27**	351,098
6	**0.05**	8,314	-	-	-	-	-	-	-	-	-	-
			21.89	351,295	24.18	351,081	25.39	351,039	26.17	351,027	26.76	351,026
(40, 120)			-	-	-	-	-	-	-	-	-	-
(14, 23)			139.99	2,312,847	133.06	2,291,297	130.34	2,277,972	129.44	2,274,436	**128.74**	2,271,194
10	**0.97**	113,751	-	-	-	-	-	-	-	-	-	-
			158.00	2,271,245	173.75	2,269,975	179.67	2,269,844	184.42	2,269,771	189.31	2,269,739
(50, 150)			-	-	-	-	-	-	-	-	-	-
(18, 27)			1356.05	19,902,899	1386.05	20,737,676	1306.88	19,962,033	1305.10	19,906,542	**1303.90**	19,874,904
8	**21.56**	2,252,573	-	-	-	-	-	-	-	-	-	-
			1415.91	18,386,161	1455.62	17,863,968	1469.87	17,559,896	1485.19	17,484,973	1499.06	17,414,634
(60, 180)			-	-	-	-	-	-	-	-	-	-
(21, 32)			3116.71	42,000,631	3054.23	41,731,445	3037.54	41,831,396	3031.55	41,834,376	**3028.49**	41,689,599
12	**296.14**	21,624,533	-	-	-	-	-	-	-	-	-	-
			3112.38	36,825,713	3122.18	34,713,221	3129.76	34,158,428	3138.68	33,822,995	3148.01	33,801,126

For our purpose, we generated mono-objective auctions from the `paths` distribution of the CATS suite [15] and randomly added failure probabilities to the bids in the range 0 to 1. Figure 3 summarizes the results obtained on randomly generated auctions with 30 goods and increasing number of bids (each data point represents an average over 10 random instances). These problem instances tend to become highly connected as the number of bids increases. The average treewidth ranged between 8 (for 30 bids) and 53 (for 150 bids), while the average pseudo tree depth was between 16 (for 30 bids) and 88 (for 150 bids), respectively. We report on three values of the mini-bucket i-bound, namely $i = 2$, $i = 8$ and $i = 14$, respectively.

We see that MO-AOBB+SMB(i) is better than MO-BB+SMB(i) at relatively small i-bounds when the heuristic is weak. This demonstrates the benefit of AND/OR versus classical OR search when the heuristic estimates are relatively weak and the algorithms rely primarily on search rather than pruning via the heuristic evaluation function. For example, on auctions with 60 bids, the average running time of MO-AOBB+SMB(2) was about 17 seconds, while MO-BB+SMB(2) took on average 1,951 seconds. As the i-bound increases and the corresponding heuristics are strong enough to prune the search space substantially, the difference between AND/OR and OR search decreases. When focusing on dynamic mini-bucket heuristics, we see that MO-AOBB+DMB(i) is better than MO-BB+DMB(i) at relatively small i-bounds, but the difference is not that

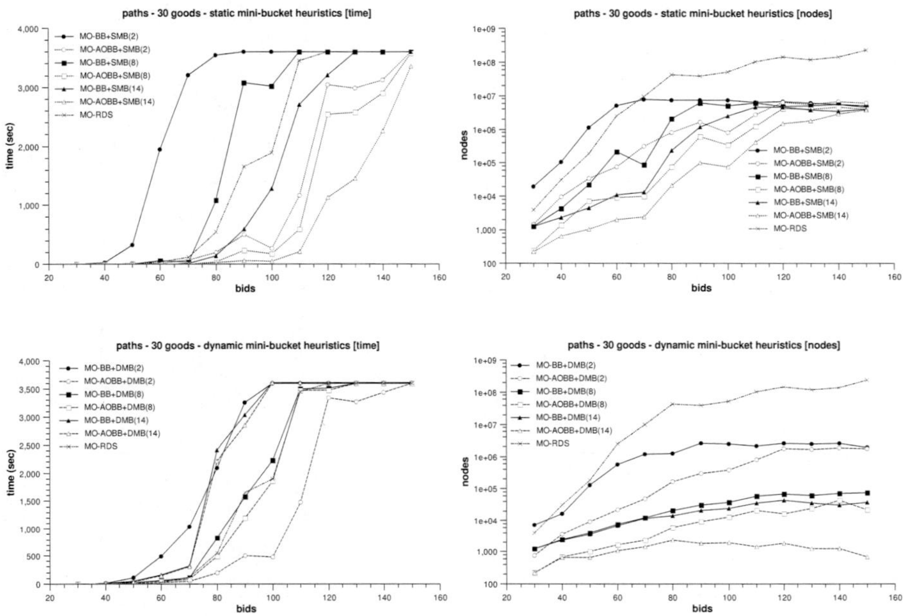

Fig. 3. Results for bi-objective combinatorial auctions with 30 goods and increasing number of bids. CPU time and nodes visited using static (top) and dynamic (bottom) mini-bucket heuristics. Time limit 1 hour. Average treewidth $w^* \in [8, 53]$, average pseudo tree depth $h \in [16, 88]$.

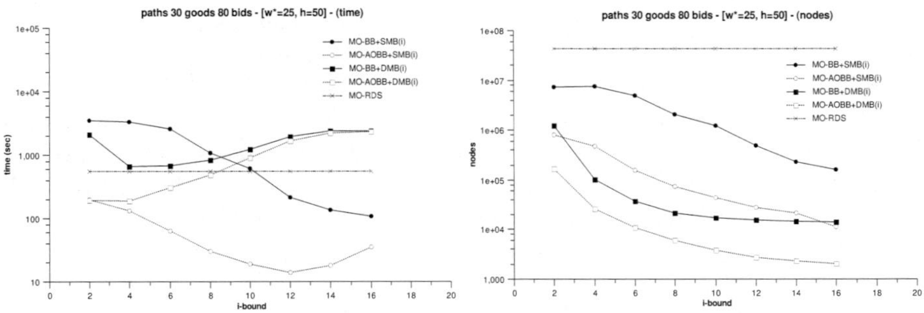

Fig. 4. Impact of the i-bound on bi-objective combinatorial auctions with 30 goods and 80 bids using static and dynamic mini-bucket heuristics. CPU time (left) and nodes visited (right). Average treewidth $w^* = 25$, average pseudo tree depth $h = 50$.

prominent as in the static case. This is because these heuristics are far more accurate compared with the pre-compiled ones and the savings in number of nodes caused by traversing the AND/OR search tree do not translate into additional time savings. On this domain, we see that MO-AOBB+SMB(i) with relatively large i-bounds outperformed MO-RDS. However, MO-RDS was superior to MO-AOBB+DMB(i), except for the smallest reported i-bound, in which case MO-AOBB+DMB(2) was competitive.

Table 2. Results for MAX-SAT-ONE instances using static and dynamic mini-bucket heuristics. Time limit 30 minutes. Overall best performance points are highlighted.

instance (n, c, k) (w*, h) $\vert\mathcal{E}\vert$	MO-RDS		MO-BB+SMB(i) MO-AOBB+SMB(i) MO-BB+DMB(i) MO-AOBB+DMB(i) i=6		MO-BB+SMB(i) MO-AOBB+SMB(i) MO-BB+DMB(i) MO-AOBB+DMB(i) i=8		MO-BB+SMB(i) MO-AOBB+SMB(i) MO-BB+DMB(i) MO-AOBB+DMB(i) i=10		MO-BB+SMB(i) MO-AOBB+SMB(i) MO-BB+DMB(i) MO-AOBB+DMB(i) i=12	
	time	nodes	time	nodes	time	nodes	time	nodes	time	nodes
aim-50-1-6no-1			-	-	-	-	14.20	43,072	12.46	36,837
(50, 119, 2)			-	-	-	-	7.98	27,929	**6.49**	22,054
(15, 31)	-	-	-	-	237.72	27,739	152.70	24,457	180.40	24,244
8			-	-	220.02	13,184	134.92	9,903	164.30	9,690
aim-50-1-6no-2					-	-	-	-	54.78	148,563
(50, 127, 2)			-	-	-	-	899.74	2,425,636	**35.15**	106,091
(19, 31)	-	-	271.66	48,118	295.76	36,233	314.72	33,226	535.52	32,712
11			218.24	27,115	237.42	16,524	251.69	13,926	450.64	13,504
aim-50-1-6no-3			-	-	-	-	-	-	32.03	79,194
(50, 120, 2)			-	-	1125.50	3,820,213	379.20	1,127,093	**3.84**	13,015
(17, 27)	-	-	335.04	92,522	182.50	77,540	176.86	76,623	223.34	76,528
10			233.46	26,465	84.97	11,500	78.85	10,583	125.51	10,488
aim-50-1-6no-4			-	-	423.19	900,024	312.10	660,848	433.63	778,334
(50, 126, 2)			-	-	311.94	657,833	**269.36**	592,225	379.00	730,513
(20, 30)	-	-	1102.11	94,624	587.73	44,434	721.36	38,521	1135.44	38,127
11			1040.83	72,361	531.54	23,047	667.06	17,323	1068.47	16,931
aim-50-1-6yes1-1			-	-	-	-	619.58	1,417,018	68.76	153,541
(50, 127, 2)			80.02	213,427	50.02	140,081	**25.14**	78,158	26.28	68,358
(18, 29)	155.33	177,981,953	187.92	115,004	208.41	113,821	269.87	113,774	305.64	113,680
10			67.23	30,086	85.67	28,911	148.03	28,865	183.73	28,771
aim-50-1-6yes1-2			-	-	74.05	183,418	75.17	198,489	30.43	75,738
(50, 126, 2)			596.16	1,757,792	38.33	106,507	61.72	175,580	**20.11**	57,089
(16, 34)	-	-	394.17	65,879	277.48	36,430	342.56	34,197	380.54	33,463
9			356.38	48,982	244.55	20,111	310.87	17,880	348.11	17,146
aim-50-1-6yes1-3			-	-	-	-	653.68	1,027,972	525.84	846,067
(50, 128, 2)			-	-	150.80	397,432	61.28	146,520	**42.28**	100,060
(20, 27)	-	-	169.36	88,156	230.36	87,401	226.42	85,373	264.30	85,144
10			87.62	34,017	147.73	33,262	144.41	31,256	181.32	31,028
aim-50-1-6yes1-4			159.71	362,140	66.09	142,424	24.08	62,518	18.81	51,603
(50, 127, 2)			43.91	136,779	28.46	76,451	17.46	48,822	**13.15**	38,478
(19, 29)	146.56	22,063,228	55.05	24,839	100.66	24,434	132.36	23,524	280.12	23,317
9			43.19	11,842	88.56	11,435	121.40	10,525	267.98	10,318

The better performance of MO-RDS relative to the AND/OR algorithms can be explained by a reduced computational overhead for generating its guiding heuristic.

Figure 4 plots the running time and number of nodes visited by MO-AOBB+SMB(i) and MO-AOBB+DMB(i) (resp. MO-BB+SMB(i) and MO-BB+DMB(i)), on auctions with 30 goods and 80 bids. Focusing on MO-AOBB+SMB(i), for example, we see that its running time, as a function of i, forms a U-shaped curve. At first ($i = 2$) it is high, then as the i-bound increases the total time decreases (when $i = 12$ the time is 14.01), but then as i increases further the time starts to increase again. The same behavior can be observed in the case of MO-AOBB+DMB(i) as well.

MAX-SAT-ONE Instances. Let $\mathcal{F}$ be a Boolean formula in conjunctive normal form (CNF). MAX-SAT is the problem of finding a truth assignment such that the number of satisfied clauses in $\mathcal{F}$ is maximized. MAX-ONE is the problem of finding a model for $\mathcal{F}$ with a maximum number of variables assigned to true. MAX-SAT-ONE is the problem of maximizing both the number of satisfied clauses and variables assigned to true, as described in [3].

Table 2 shows the results obtained for experiments with 8 CNF instances from the DIMACS benchmark[2]. As before, we observe that MO-AOBB+SMB(i) offers the overall best performance, especially for relatively large i-bounds. For example, on the *aim-50-1-6yes1-1*, MO-AOBB+SMB(10) is about 25 times faster than MO-BB+SMB(10),

[2] Available online at: ftp://dimacs.rutgers.edu/pub/challenge/satisfiability/benchmarks/cnf/

Table 3. Results for bi-objective WCSPs derived from standard mono-objective WCSPs using static mini-bucket heuristics. Time limit 10 hours. Overall best performance points are boxed.

instance (n, c, k)	(w*, h) $\|\mathcal{E}\|$	MO-RDS		MO-BB+SMB(i) MO-AOBB+SMB(i) i=14		MO-BB+SMB(i) MO-AOBB+SMB(i) i=16		MO-BB+SMB(i) MO-AOBB+SMB(i) i=18		MO-BB+SMB(i) MO-AOBB+SMB(i) i=20	
		time	nodes	time	nodes	time	nodes	time	nodes	time	nodes
c432	(27, 45)	-	-	-	-	-	-	-	-	-	-
(432, 648, 2)	31			38.99	96,805	**25.15**	59,330	31.21	62,597	39.95	50,753
c499	(23, 55)	-	-	25040.22	17,482,361	27633.93	20,140,468	5275.72	5,263,053	1611.60	1,266,736
(499, 748, 2)	24			21967.54	16,840,017	24407.10	19,498,060	4290.26	4,620,932	**668.43**	625,078
c880	(27, 67)	-	-	-	-	-	-	-	-	-	-
(881, 1323, 2)	53			-	-	-	-	9088.16	9,723,797	**3181.03**	3,959,806
s386	(19, 44)	-	-	-	-	-	-	-	-	-	-
(172, 258, 2)	13			-	-	-	-	**14.30**	73,572	20.85	73,372
s1423	(24, 54)	-	-	-	-	-	-	-	-	-	-
(749, 1125, 2)	119			-	-	-	-	**3990.74**	1,193,420	4284.95	1,195,986
s1488	(47, 67)	-	-	-	-	-	-	-	-	-	-
(667, 1000, 2)	10			-	-	-	-	2510.72	11,708,708	**1751.86**	9,392,631
s1494	(48, 69)	-	-	-	-	-	-	-	-	-	-
(661, 991, 2)	14			-	-	-	-	**2794.69**	17,273,392	2924.80	16,089,333
90-14-1	(22, 66)	-	-	-	-	-	-	-	-	-	-
(196, 392, 2)	12			19.45	130,651	**7.96**	23,559	13.26	17,677	24.92	16,761
90-16-1	(24, 82)	-	-	-	-	-	-	-	-	-	-
(256, 512, 2)	8			328.61	1,073,957	122.99	446,091	139.18	441,316	**84.88**	127,279
cpcs360b	(20, 27)	-	-	-	-	-	-	-	-	-	-
(360, 720, 2)	76			**85.17**	119,216	113.62	119,216	181.77	119,200	524.64	119,200
blockmap-05-01	(18, 48)	-	-	-	-	-	-	-	-	-	-
(700, 1400, 2)	1			**3.51**	3,483	11.61	3,417	39.74	2,269	52.32	2,110
blockmap-05-02	(20, 53)	-	-	-	-	-	-	-	-	-	-
(855, 1710, 2)	2			**5.72**	5,579	14.52	4,296	54.78	3,484	119.87	3,463
blockmap-05-03	(22, 59)	-	-	-	-	-	-	-	-	-	-
(1005, 2010, 2)	2			**7.59**	11,325	15.78	30,324	29.03	26,454	65.96	20,459
mm-03-08-03	(20, 57)	-	-	-	-	-	-	-	-	-	-
(1220, 2440, 2)	1			**10.50**	47,783	20.06	7,669	57.27	6,426	105.90	6,257

and about 6 times faster than MO-RDS, respectively. When using dynamic mini-bucket heuristics, we see again that MO-AOBB+DMB(i) is competitive only for relatively small i-bounds, due to computational overhead issues. Note also that MO-RDS could solve only two instances, namely *aim-50-1-6yes1-1* and *aim-50-1-6yes1-4*. The relatively worse performance of MO-RDS on this domain can be explained by the quality of its guiding heuristic function which was far inferior to the mini-bucket based ones.

Bi-objective Weighted CSPs. We considered a set of standard mono-objective Weighted CSP [17] benchmarks from the UCI Graphical Models Repository (available online at: *http://graphmod.ics.uci.edu/group/Repository*). For our purpose, we converted each of these instances into a bi-objective optimization problem by adding a secondary additive objective function defined by $F_2 = \sum_{i=1}^{n} g(X_i)$, where $g(X_i)$ is a unary cost function defined on variable X_i with costs generated uniformly at random between 0 and 10.

Table 3 shows the results obtained for experiments with 14 WCSP instances using static mini-bucket heuristics. We did not report results with dynamic mini-bucket heuristics because of the prohibitively large computational overhead associated with relatively large i-bounds. We see that MO-AOBB+SMB(i) offers the overall best performance on this domain, outperforming its competitors by several orders of magnitude. For example, MO-AOBB+SMB(16) solves the *90-14-1* instance in about 8 seconds, whereas MO-BB+SMB(16) exceeds the 10 hour time limit. We also note that MO-RDS was not able to solve any of the test instance within the time limit, while MO-BB+SMB(i) could solve only one instance, namely *c499*.

Summary of the Empirical Evaluation. Our empirical evaluation on several classes of multi-objective optimization problems demonstrated conclusively that the AND/OR Branch-and-Bound tree search algorithms guided by static mini-bucket heuristics were the best performing algorithms overall. The difference between MO-AOBB+SMB(i) and the OR tree search counterpart MO-BB+SMB(i) was more pronounced at relatively small i-bounds (corresponding to relatively weak heuristic estimates) and amounted to several orders of magnitude in terms of both running time and size of the search space explored. For larger i-bounds, when the heuristic estimates are strong enough to prune the search space substantially, the difference between MO-AOBB+SMB(i) and MO-BB+SMB(i) decreased. We also showed that MO-AOBB+SMB(i) was able in many cases to outperform dramatically recent state-of-the-art solvers for multi-objective optimization such as MO-RDS. With dynamic mini-bucket heuristics, MO-AOBB+DMB(i) proved competitive only for relatively small i-bounds due to computational overhead issues. This suggests that these heuristics can be considered when space is restricted. When comparing the AND/OR algorithms guided by mini-bucket heuristics with MO-RDS, we observed that the former were better on relatively sparse networks while on highly connected networks, the latter provided sometimes a better alternative.

6 Conclusion

The paper investigates the impact of AND/OR search spaces for graphical models on multi-objective optimization. We introduced a general multi-objective AND/OR Branch-and-Bound algorithm and specialized it with two schemes for generating heuristic estimates that can guide the search. Our empirical evaluation on several benchmarks for MO-COP showed that the new AND/OR algorithms improved dramatically over traditional OR ones, in many cases by several orders of magnitude. Future work includes the extension of the algorithm to explore a search *graph* rather than a tree, via caching, as well as to investigate alternative control strategies such as best-first or frontier search in the context of MO-COPs. We also plan to apply the AND/OR search principle to the MO-RDS algorithm. Finally, we can exploit *point quad-trees* for the representation of the efficient frontier in order to access the data structure more efficiently.

References

1. Dechter, R., Mateescu, R.: AND/OR search spaces for graphical models. Artificial Intelligence 171(2-3), 73–106 (2007)
2. Marinescu, R., Dechter, R.: AND/OR branch-and-bound for graphical models. In: International Joint Conference on Artificial intelligence (IJCAI), pp. 224–229 (2005)
3. Rollon, E., Larrosa, J.: Bucket elimination for multi-objective optimization problems. Journal of Heuristics 12, 307–328 (2006)
4. Freuder, E.C., Quinn, M.J.: Taking advantage of stable sets of variables in constraint satisfaction problems. In: IJCAI, pp. 1076–1078 (1985)
5. Bayardo, R., Miranker, D.: A complexity analysis of space-bound learning algorithms for the constraint satisfaction problem. In: AAAI, pp. 298–304 (1996)

6. Rollon, E., Larrosa, J.: Multi-objective Russian doll search. In: National Conference on Artificial Intelligence (AAAI), pp. 249–254 (2007)
7. Rollon, E., Larrosa, J.: Constraint optimization techniques for multiobjective branch and bound search. In: International Conference on Logic Programming, ICLP (2008)
8. Ehrgott, M., Gandibleux, X.: Bounds and bound sets for biobjective combinatorial optimization problems. Notes in Economics and Mathematical Systems 507, 241–253 (2001)
9. Nilsson, N.J.: Principles of Artificial Intelligence. Tioga, Palo Alto (1980)
10. Dechter, R., Rish, I.: Mini-buckets: A general scheme of approximating inference. Journal of ACM 50(2), 107–153 (2003)
11. Dechter, R.: Bucket elimination: A unifying framework for reasoning. Artificial Intelligence 113, 41–85 (1999)
12. Kask, K., Dechter, R.: A general scheme for automatic generation of search heuristics from specification dependencies. Artificial Intelligence 129(1-2), 91–131 (2001)
13. Verfaillie, G., Lematre, M., Schiex, T.: Russian doll search for solving constraint optimization problems. In: AAAI, pp. 181–187 (1996)
14. Ehrgott, M., Gandibleux, X.: Multiple criteria optimization. State of the art. Annotated bibliographic surveys. Kluwer Academic Publishers, Dordrecht (2002)
15. Leyton-Brown, K., Pearson, M., Shoham, Y.: Towards a universal test suite for combinatorial auction algorithms. In: ACM Electronic Commerce, pp. 66–76 (2000)
16. Holland, A.: Risk Management in Combinatorial Auctions. PhD thesis, University College Cork (2005)
17. Bistarelli, S., Montanari, U., Rossi, F.: Semiring based constraint solving and optimization. Journal of ACM 44(2), 309–315 (1997)

Search Space Extraction

Deepak Mehta, Barry O'Sullivan, Luis Quesada, and Nic Wilson

Cork Constraint Computation Centre
Department of Computer Science, University College Cork, Ireland
{d.mehta,b.osullivan,l.quesada,n.wilson}@4c.ucc.ie

Abstract. Systematic tree search is often used in conjunction with inference and *restarts* when solving challenging Constraint Satisfaction Problems (CSPs). In order to improve the efficiency of constraint solving, techniques that learn during search, such as constraint weighting and nogood learning, have been proposed. Constraint weights can be used to guide heuristic choices. Nogood assignments can be avoided by adding additional constraints. Both of these techniques can be used in either one-shot systematic search, or in a setting in which we frequently restart the search procedure. In this paper we propose a third way of learning during search, generalising previous work by Freuder and Hubbe. Specifically, we show how, in a restart context, we can guarantee that we avoid revisiting a previously visited region of the search space by extracting it from the problem. Likewise, we can avoid revisiting inconsistent regions of the search space by extracting inconsistent subproblems, based on a significant improvement upon Freuder and Hubbe's approach. A major empirical result of this paper is that our approach significantly outperforms MAC combined with weighted degree heuristics and restarts on challenging constraint problems. Our approach can be regarded as an efficient form of learning that does not rely on constraint propagation. Instead, we rely on a reformulation of a CSP into an equivalent set of CSPs, none of which contain any of the search space we wish to avoid.

1 Introduction

Systematic tree search is often used in conjunction with inference and restarts when solving challenging Constraint Satisfaction Problems (CSPs). Maintaining Arc Consistency (MAC) is considered to be one of the best generic systematic algorithms for solving constraint satisfaction problems [1]. However, if search branches into an insoluble subtree, as with all backtrack search algorithms, recovering from this mistake can take an extremely long time due to thrashing.

Good search heuristics significantly help the efficiency of systematic search algorithms. Recently, a class of variable ordering heuristics that attempt to branch on the most critically constrained variables have been proposed. The most effective of these involve maintaining weights for each constraint based on the number of times inconsistency was caused by propagating it; these are the weighted degree class of heuristics [2].

Combining MAC-based systematic search with a random restarts strategy can also significantly improve performance over a one-shot approach using systematic search alone [3]. The insight here is that by restarting search once we have failed to solve a problem within a pre-specified effort threshold, it can pay off to simply start again.

I.P. Gent (Ed.): CP 2009, LNCS 5732, pp. 608–622, 2009.

However, in order to ensure that search remains complete, in subsequent runs we need to increase our maximum effort threshold.

A complementary approach to improving search is to learn to avoid regions of the search space that we know will only lead to failure. The general approach here is to use nogood recording [4], where each nogood represents an assignment to a subset of the variables which does not lead to a solution. Nogoods and restarts have also been combined with MAC-based search with positive results [5].

In this paper we take a different approach. Freuder and Hubbe [6] proposed to remove the search space of a failed subproblem discovered at some point in the search tree from the forward phase of the forward checking search algorithm. However, in their approach when the search algorithm backtracks, the search space associated with this failed subproblem is restored. Therefore, the algorithm may revisit the same failed search space in a different branch of the search tree.

Extending the work of [6] we avoid regions of the search space by extracting them from the unvisited search space of the problem. In other words, once the failed subproblem is extracted it is never revisited. Specifically, in a restart context, we want to guarantee that we avoid revisiting a previously visited region of the search space. Likewise, we want to avoid revisiting inconsistent regions of the search. We propose an approach to making such guarantees based on a significant improvement upon Freuder and Hubbe's earlier work. We show how previously visited search space can be extracted so that it is not revisited upon restarting as well as showing that extracting inconsistent subproblems can be achieved, in a much larger-scale way than reported previously. A major empirical result of this paper is that our approach significantly outperforms MAC combined with weighted degree heuristics and restarts on challenging constraint problems. Our approach can be regarded as an efficient form of learning that does not rely on constraint propagation. Instead, we rely on a reformulation of a CSP into an equivalent set of CSPs, none of which contain any of the search space we wish to avoid.

The remainder of this paper is organised as follows. Section 2 presents the necessary background required for this paper. We describe the principle of search space extraction in Section 3. In Section 4 we present an algorithm called MACER for avoiding visited search space when using a restart-based search algorithm. In Section 5 we present an algorithm called MACER$^+$ that can be used to extract unsatisfiable cores, providing an alternative approach to nogood recording. We present and discuss our empirical results in Section 6, clearly highlighting the strength of our approach. A number of conclusions and directions for future work are discussed in Section 7.

2 Background

A constraint network (CN) P is a triple $(\mathcal{X}, \mathcal{C}, D)$ where $\mathcal{X}$ is a set of n variables and $\mathcal{C}$ is a set of e constraints. Each variable $X \in \mathcal{X}$ is associated with a domain (a set of values), which is denoted by $D(X)$. We use d to denote the maximum domain size. Each constraint $C \in \mathcal{C}$ is associated with a set of variables on which the constraint C is defined, which is denoted by $scope(C)$. For simplicity, we restrict our attention to binary CNs, where the constraints involve two variables. A *possibility* is an assignment of values to all the variables. A *solution* of a CN is a possibility that satisfies all the constraints. A CN is said to be *satisfiable* if and only if it admits at least one solution. In general, determining the satisfiability of a CN is NP-complete. Solving a CN involves either finding one (or more) solution or determining that the CN has no solution.

The basic procedure to solve a CN is systematic backtracking, which involves repeated selection of an unassigned variable and assigning a value from its domain. If a variable is selected such that none of its possible values are consistent with the previously assigned variables, then the algorithm backtracks and attempts to assign an alternative untried value. The search stops when either a single solution is found, or all the combinations of instantiations to variables have been tried. This basic procedure for solving a CN has been augmented by using variable ordering and value ordering heuristics, local consistency techniques, branching strategies, randomisation and restarts, etc.

A local consistency technique is used to remove values from the domains that do not participate in any solution. Usually after each variable assignment during search, a domain consistency (e.g., arc consistency) is enforced (Chapter 4 of [7]). A constraint $C \in \mathcal{C}$ is arc consistent if $\forall X \in scope(C)$ and $\forall a \in D(X)$, C is satisfied. A constraint network is arc consistent if $\forall C \in \mathcal{C}$ is arc consistent. Maintaining Arc Consistency (MAC) during search is considered to be one of the most efficient and generic approach to solve constraint networks.

A branching strategy defines a search tree. The two most well-known branching strategies are k-way branching and binary branching. In k-way branching, when a variable X with k values in its domain is selected for instantiation, k branches are formed. Here each branch corresponds to an assignment of a value to the selected variable. In binary branching, when a variable X is selected for instantiation, its values are assigned via a sequence of binary choices. If the values are assigned in the order $v_1, v_2, \ldots, v_k$, then two branches are formed for the value v_1, associated with $X = v_1$ and $X \neq v_1$ respectively. Crucially, the constraint $X \neq v_1$ is propagated, before selecting another variable-value pair. In this paper, we only focus on k-way branching. However, all the ideas that are presented can be extended to binary-branching.

Definition 1 (Decision). *Let $P = (\mathcal{X}, \mathcal{C}, D)$ be a CN and (X, v) be a pair such that $X \in \mathcal{X}$ and $v \in D(X)$. The assignment $X = v$ is called a decision.*

Definition 2 (Restriction of a network). *Let P be a CN and let $\triangle$ be an ordered set (sequence) of decisions taken during search, $P|_\triangle$ is the CN derived from P such that, for any decision $(X = v) \in \triangle$, $D(X)$ is set to $\{v\}$ in $P|_\triangle$, and the other domains are the same as in P.*

Definition 3 (Filtering of a network). *Let ϕ be an inference operator that enforces a local consistency. $\phi(P)$ is the CN derived from P obtained after applying the inference operator ϕ.*

In this paper, ϕ is equivalent to arc consistency, unless stated otherwise.

Definition 4 (Unsatisfiable network). *Let P be a CN. If there exists a variable with an empty domain in P then P is unsatisfiable, which is denoted by $P = \perp$.*

Definition 5 (Nogood). *Let P be a CN and $\triangle$ be a set of decisions. $\triangle$ is a nogood of P if and only if $\phi(P|_\triangle)$ is unsatisfiable.*

Definition 6 (Subnetwork). *Let $P = (\mathcal{X}^P, \mathcal{C}^P, D^P)$ be a CN. $Q = (\mathcal{X}^Q, \mathcal{C}^Q, D^Q)$ is a subnetwork of P if and only if $\mathcal{X}^Q \subseteq \mathcal{X}^P$, $\mathcal{C}^Q \subseteq \mathcal{C}^P$ and $D^Q \subseteq D^P$.*

Definition 7 (Unsatisfiable core). *Let P be an inconsistent CN. Q is an unsatisfiable core of P if Q is an unsatisfiable subnetwork of P.*

Although MAC is considered to be one of the best approaches for solving constraint networks, it can get stuck in an unpromising search and can thrash too many times if wrong decisions are taken earlier in the search tree. In order to avoid this thrashing, randomisation and restart strategies are used [8].

3 Principles of Search Space Extraction

The techniques presented in this paper rely on the extraction of subnetworks from a set of networks. In this section we formally define the principles of the approach.

Definition 8 (Size of a network). *Let $P = (\mathcal{X}, \mathcal{C}, D^P)$ be a constraint network. The size of P is the product of the domain sizes of its variables, i.e., $\prod_{X \in \mathcal{X}} |D^P(X)|$.*

Definition 9 (Intersection of networks). *Let $P = (\mathcal{X}, \mathcal{C}, D^P)$ and $P' = (\mathcal{X}, \mathcal{C}, D^{P'})$ be constraint networks. We define $P \sqcap P'$ to be the constraint network $(\mathcal{X}, \mathcal{C}, D^{P \sqcap P'})$, where $D^{P \sqcap P'}$ is defined by $\forall X \in \mathcal{X}$, $D^{P \sqcap P'}(X) = D^P(X) \cap D^{P'}(X)$.*

Notice that $P \sqcap P' = \perp$ if there exists $X \in \mathcal{X}$ such that $D^P(X) \cap D^{P'}(X) = \emptyset$

Definition 10 (Subsumption of networks). *Let $P = (\mathcal{X}, \mathcal{C}, D^P)$ and $P' = (\mathcal{X}, \mathcal{C}, D^{P'})$ be constraint networks. We say that P subsumes P', $P' \sqsubset P$ (resp. $P' \sqsubseteq P$), if for each variable $X \in \mathcal{X}$, $\mathcal{D}^{P'}(X) \subset \mathcal{D}^P(X)$ (resp. $\mathcal{D}^{P'}(X) \subseteq \mathcal{D}^P(X)$).*

Let $P = (\mathcal{X}, \mathcal{C}, D^P)$ and $E = (\mathcal{X}, \mathcal{C}, D^E)$ be constraint networks, where $\mathcal{X} = \{X, Y, Z\}$, $D^E = \{X \to \{a\}, Y \to \{a, b\}, Z \to \{a, b, c\}\}$, and $D^P = \{X \to \{a, b, c, d\}, Y \to \{a, b, c, d\}, Z \to \{a, b, c, d\}\}$. Notice that $E \sqsubset P$. Figure 1 shows a tree describing the extraction of E from P, which results in the set of networks $\{R_1, R_2, R_3\}$. At every level of the tree a variable V is selected such that $D^P(V) \neq D^P(V) \cap D^E(V)$, and two branches (or subproblems) are created by splitting the domain of V. The left hand side subproblem has those values in $D^P(V)$ that are also in $D^E(V)$. The right hand side subproblem has those values of $D^P(V)$ that are not in $D^E(V)$. This is repeatedly done until the left hand side is equivalent to $P \sqcap E$. In this case $P \sqcap E$ is equivalent to E since $E \sqsubseteq P$.

Definition 11 (Extraction of networks). *Let $P = (\mathcal{X}, \mathcal{C}, D^P)$ and $E = (\mathcal{X}, \mathcal{C}, D^E)$ be constraint networks. The extraction of E from P under the variable ordering r, denoted as $P \ominus_r E$, is defined as the singleton set $\{P\}$ if $P \sqcap E = \perp$. $P \ominus_r E$ is $\emptyset$ if $P \sqsubseteq E$. Otherwise, $P \ominus_r E$ is the set of networks $\{R\} \cup (M \ominus_r E)$, where $R = (\mathcal{X}, \mathcal{C}, D^R)$ and $M = (\mathcal{X}, \mathcal{C}, D^M)$, such that for the first variable X in $\mathcal{X}$, under r, satisfying $D^P(X) \neq D^P(X) \cap D^E(X)$, $D^R(X) = D^P(X) - D^E(X)$, $D^M(X) = D^P(X) \cap D^E(X)$, and $\forall Y \neq X$, $D^R(Y) = D^M(Y) = D^P(Y)$.*

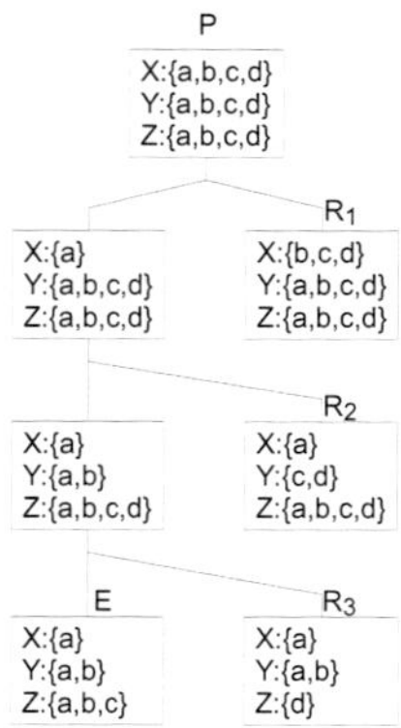

Fig. 1. The extraction of E from P results in the set of networks $\{R_1, R_2, R_3\}$

Definition 11 follows the algorithm described in [6]. In those cases where no confusion arises, we simplify the notation by omitting the variable ordering. If $E \sqcap P \neq \bot$, then the extraction of E from P, i.e. $P \ominus E = \{R_1, \ldots, R_k\}$, is a minimum set of k constraint networks such that the union of their possibilities is the set of possibilities that are in P but not in E. The sets of possibilities of these networks are mutually disjoint. Notice that $\{R_1, \ldots, R_k, E\}$ can also be seen as a partition of P. The cardinality of $P \ominus E$ is linear with respect to the number of the domains of the variables on which $P \sqcap E$ and P are different. The worst-case time complexity of extracting a network from another is $\mathcal{O}(n^2\, d)$, where n is the number of variables in the network and d is the maximum domain size. Any solution of $R_i \in P \ominus E$ is a solution of P. If E is inconsistent, the set of solutions of $P \ominus E$ is equal to the set of solutions of P.

Let $\mathcal{P} = \{P_1, \ldots, P_p\}$ and $\mathcal{E} = \{E_1, \ldots, E_e\}$ be two sets of networks. $\ominus$ can be lifted to operate on sets of networks in the natural way. $\mathcal{P} \ominus \mathcal{E}$ is the set of networks obtained after extracting E_j from P_i, for all $1 \leq i \leq p$ and for all $1 \leq j \leq e$, i.e., $\bigcup_{1 \leq i \leq p, 1 \leq j \leq e} P_i \ominus E_j$.

4 Extraction of Visited Search Space

When a search algorithm is used in conjunction with a restart strategy, whenever a cutoff (in terms of the number of nodes or the number of failures) is reached, the search restarts. Let $\triangle_l$ be the last branch (nogood) before restarting search. If we are only interested in finding a single solution, then there is no solution in any part of the search space explored before $\triangle_l$. Therefore, ideally, when we restart we do not want to revisit any part of the visited insoluble region. One way of doing this is to extract the visited search space until $\triangle_l$ from the original problem. In this way whenever we restart we can guarantee that the same possibility will never be explored.

4.1 The Unvisited Search Space

Let P be a constraint network. Each dead-end in the search tree can be seen as a nogood. Each nogood $\triangle_i$ can be associated with an inconsistent problem $P|_{\triangle_i}$. Let

$\triangle_l$ be the last nogood, and let $vars(\triangle_l)$ be the set of variables occurring in $\triangle_l$. Let $\mathcal{E} = \{P|_{\triangle_1}, P|_{\triangle_2}, \ldots, P|_{\triangle_l}\}$ be the set of inconsistent networks that represent the visited search space until $\triangle_l$. In order to ensure that no possibility is revisited, we need to search in the set of networks $\{P\} \ominus \mathcal{E}$. The search space represented by $\{P\} \ominus \mathcal{E}$ can be represented with a set of networks $\mathcal{R}$ defined as follows: for each variable $X_i \in vars(\triangle_l)$, a constraint network $R_i \in \mathcal{R}$ is created such that for all variables Y instantiated before X_i in $\triangle_l$, $D^{R_i}(Y) = \{\triangle_l(Y)\}$, for the variable X_i, $D^{R_i}(X_i) = \{v \in D^P(X_i) | v > \triangle_l(X_i)\}$ and for the remaining variables Z that are not instantiated $D^{R_i}(Z) = D^P(Z)$. It follows from the definition of $\mathcal{R}$ that $|\mathcal{R}| \leq |\triangle_l|$.

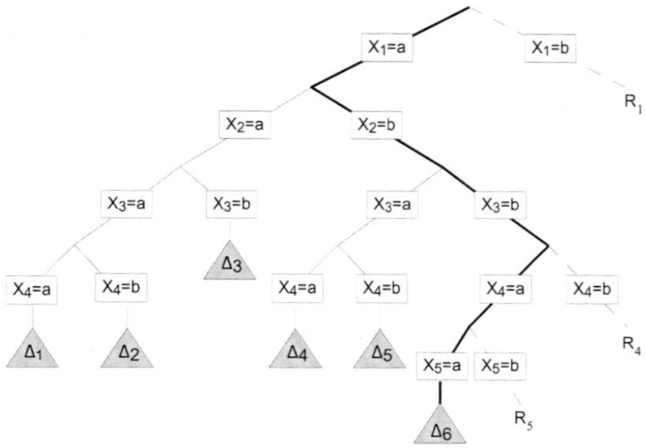

Fig. 2. A search tree consisting of visited and unvisited search spaces

Figure 2 depicts a typical search tree of some problem P. Here each triangle represents a dead-end, which is associated with a nogood. The branch of the search tree that is shown in bold corresponds to the last failed nogood before restarting search. So, $\triangle_l$ is equal to $\triangle_6$. At each node of the last failed branch, we create a subnetwork by restricting the domain of the variable associated with that node with those values that are not tried yet. The set of these networks is $\mathcal{R} = \{R_1, R_4, R_5\}$. Notice that at nodes $X_2 = b$ and $X_3 = b$ no such subnetworks are created since there are no more values left to try. In order to avoid revisiting a possibility in the already visited search space we need to search in $\{P\} \ominus \{P|_{\triangle_1}, P|_{\triangle_2}, \ldots, P|_{\triangle_6}\}$, which is $\mathcal{R} = \{R_1, R_4, R_5\}$.

4.2 Maintaining the Unvisited Search Space

In this section we present the algorithm that **m**aintains **a**rc **c**onsistency on the input problem during search and **e**xtracts visited search space before **r**estarting search (MACER). The result of this extraction is a set of networks associated with unexplored possibilities.

The pseudo-code for MACER is presented in Algorithm 1. In this algorithm $\mathcal{R}$ is used to denote the set of networks associated with unexplored possibilities and `sf` is a Boolean variable that is used to denote whether the solution is found or not. Initially $\mathcal{R}$ is a singleton set containing the input network P, and `sf` is set to `false`. MACER repeatedly selects and removes any network Q from $\mathcal{R}$ until a solution is found or

Algorithm 1. MACER(P)

```
sf ← false
R ← {P}
while (sf = false ∧ R ≠ ∅ ) do
    f ← 0 // the number of failures is set to 0
    mf is set to a value depending on the restart strategy
    select and remove any Q from R
    MACE (Q, R, f, mf, sf)
```

$\mathcal{R}$ becomes empty and invokes MACE to determine the satisfiability of Q. MACE may generate more subnetworks while exploring Q. These subnetworks are added to $\mathcal{R}$. Before invoking MACE the number of failures f is initialised to 0 and mf is set to a cutoff value in terms of the number of failures depending upon the restart strategy. P is proved to be inconsistent when $\mathcal{R}$ becomes empty and sf = false.

Algorithm 2. MACE($P, \mathcal{R}$, f, mf, sf)

Require:
- P : input CSP $(\mathcal{X}, \mathcal{C}, D)$
- $\mathcal{R}$: set of networks representing the unvisited search space.
- f : number of failures
- mf: threshold on the number of failures
- sf: solution found

```
if X = ∅ then
    sf ← true
else
    select and remove any variable X from X
    V ← D^P(X)
    repeat
        select and remove any value v from V
        P' ← AC(P|_{X=v})
        if P' ≠⊥ then
            MACE(P', R, f, mf, sf)
        else
            f ← f + 1
        restart_search ← (f > mf)
    until V = ∅ ∨ sf ∨ restart_search
    if restart_search ∧ V ≠ ∅ then
        CREATESUBNETWORK(P, R, X, V)
```

The pseudo-code for MACE is presented in Algorithm 2. If $\mathcal{X}$ is empty then a solution is found. Otherwise, a variable X is selected from $\mathcal{X}$ for the instantiation. The algorithm continues to branch until all the values of the variable X are tried, or a solution is found or restart_search is set to True. After each assignment of the variable X, the problem is made arc-consistent. If there is no domain wipeout after enforcing arc consistency then a recursive call is made, otherwise the number of failures is incremented. If the number of failures reaches the maximum number of allowed failures then `restart_search` is set to `True`. If `restart_search` is set to `True` and if there are still some values in V that have not been considered for the instantiation of the variable X then the function CREATESUBNETWORK is invoked. The algorithm MACE terminates when either of the following is true: all the values of the first variable of the input network are tried, that means the problem is insoluble, or a solution is found, or a restart search is triggered.

The pseudo-code for CREATESUBNETWORK is shown in Algorithm 3. This algorithm creates a subnetwork by copying the domains of the future and past variables as they are at that level and considering only those values of the current variable X that are untried, which is basically the set V. When $V = \emptyset$ no subnetwork is created. Notice that saving the subnetwork as described before also preserves the propagation carried out due to the initialisation of variables before the current variable X.

Algorithm 3. CREATESUBNETWORK$((\mathcal{X}, \mathcal{C}, D^P), \mathcal{R}, X, V)$

$P' \leftarrow (\mathcal{X}, \mathcal{C}, D^{P'})$
$D^{P'}(X) \leftarrow V$
for all $Y \in \mathcal{X}$ such $X \neq Y$ **do**
    $D^{P'}(Y) \leftarrow D^P(Y)$
$\mathcal{R} \leftarrow \mathcal{R} \cup \{P'\}$

If $\triangle_l$ is the last branch explored before restarting search then at most $|\triangle_l|$ networks are created. Therefore, the worst-case time and space complexities of creating these networks at each restart is $\mathcal{O}(|\triangle_l| n\, d)$, where n is the number of variables and d is the maximum domain size of the variables. If r is the number of restarts required to solve a constraint network P, and if m is the maximum number of variables involved in the last branches of all those restarts, then the space complexity of $\mathcal{R}$ is $\mathcal{O}(r\, m\, n\, d)$. Notice that the number of networks in $\mathcal{R}$ of MACER grows linearly with respect to the number of restarts.

5 Extraction of Unsatisfiable Cores

In this section we present an extension of MACER that not only extracts the visited search space but also extracts an unsatisfiable core of the selected subnetwork Q from the set of networks of $\mathcal{R}$ after proving the inconsistency of Q. We call this approach MACER$^+$, the pseudo-code for which is presented in Algorithm 4. Before invoking MACE, the algorithm saves the number of networks of $\mathcal{R}$ in `nr` (Line 7). If MACE does not find any solution in Q and if the remaining networks in $\mathcal{R}$ is equal to `nr`, MACER$^+$ infers that Q is unsatisfiable. If Q is unsatisfiable, then an unsatisfiable core Q' is obtained from Q. An unsatisfiable core of Q can be computed by determining those variables which were involved in those constraints that were needed in proving the unsatisfiability of the network Q. Effectively this means that constraints on those variables were activated by the underlying arc consistency algorithm of MACE, which led to domain updates.

In MACER$^+$, each variable $X \in \mathcal{X}$ is associated with a Boolean variable `active`, which is used to distinguish the variables that are activated from those that are not. Before searching in Q, `active` is set to `false` for all the variables in $\mathcal{X}$ (Line 8). During search while enforcing arc consistency whenever a revision is effective, `active` is set to `true` for the variables involved in the corresponding constraint. This approach for computing an unsatisfiable core is similar to the `pcore` approach presented in [9].

Given $Q = (\mathcal{X}, \mathcal{C}, D^Q)$, an unsatisfiable core of Q that is extracted from the remaining networks in $\mathcal{R}$ can be defined as $(\mathcal{X}, \mathcal{C}, D^{Q'})$, where $\forall X \in \mathcal{X}$, if `active`$[X] =$ `true`, $D^{Q'}(X) = D^Q(X)$; otherwise $D^{Q'}(X) = D^P(X)$ (Line 12 – 16). Notice that the domains of the variables that are not involved in the unsatisfiable core are set to the domains that the variables had in the input problem P. The complexity of constructing this unsatisfiable core Q' is $\mathcal{O}(n\, d)$ because it basically involves in traversing the domains of the variables. As the worst-case time complexity of extracting a network from another network is $\mathcal{O}(n^2\, d)$, the worst-case time complexity of extracting the unsatisfiable core from the networks in $\mathcal{R}$ is $\mathcal{O}(n^2\, d\, |\mathcal{R}|)$, where n is the number of variables in the network and d is the maximum domain size.

Algorithm 4. MACER$^+(P)$

1: `sf` $\leftarrow$ `false`
2: $\mathcal{R} \leftarrow \{P\}$
3: **while** (`sf=false` $\wedge\ \mathcal{R} \neq \emptyset$) **do**
4:   $f \leftarrow 0$ // the number of failures is set to 0
5:   `mf` is set to a value depending on the restart strategy
6:   select and remove any $Q = (\mathcal{X}, \mathcal{C}, D^Q)$ from $\mathcal{R}$
7:   `nr` $\leftarrow |\mathcal{R}|$
8:   $\forall X \in \mathcal{X}$, active$[X]$ = `false`
9:   MACE$(Q, \mathcal{R}$, `f`, `mf`, `sf`$)$
10:   **if** `nr` $= |\mathcal{R}| \wedge$ `sf` $=$ `false` **then**
11:    $Q' \leftarrow (\mathcal{X}, \mathcal{C}, D^{Q'})$
12:    **for all** $X \in \mathcal{X}$ **do**
13:     **if** active[X] **then**
14:      $D^{Q'}(X) \leftarrow D^Q(X)$
15:     **else**
16:      $D^{Q'}(X) \leftarrow D^P(X)$
17:    $\mathcal{R} \leftarrow \mathcal{R} \ominus \{Q'\}$

Each time the extraction of an unsatisfiable core from a network in $\mathcal{R}$ is performed, the number of networks in $\mathcal{R}$ can increase by n in the worst-case. Therefore, the cardinality of $\mathcal{R}$ can increase exponentially with respect to the number of extracted unsatisfiable cores. If an unsatisfiable core is extracted each time a restart takes place, the cardinality of $\mathcal{R}$ can also increase exponentially with respect to the number of restarts.

6 Experimental Results

In this section we present some results that demonstrate the effectiveness of extracting inconsistent search space in the context of restart search.

We perform experiments on a variety of problems that were used as benchmarks in the CP solver competition (http://cpai.ucc.ie/05/Benchmarks.html). The instances of the following problems were used: forced random binary problem, quasigroup completion problem, quasigroup with holes, queen-attacking problem, job-shop scheduling problem and modified radio link frequency assignment problems. The details of these problems are described in [10]. All the problems have only binary constraints.

All the search algorithms were tested using the conflict-directed variable ordering heuristic *dom/wdeg* [2]. As the domain over weighted degree heuristic has been shown to be the most efficient generic variable ordering heuristic, it was a natural choice to use it with all the approaches. Values were chosen lexically for all approaches. All the search algorithms used AC-3 as an underlying arc consistency algorithm. We used the geometric restart strategy. The initial number of allowed failures was set to 30 and at each restart the number of failures was increased by a factor of 1.5. All the algorithms were implemented in C. All the experiments were performed on a Linux machine with Pentium M (CPU 2 GHz and 1GB of RAM) processor. The performances of all the approaches are measured in terms of search nodes (#nodes), failures (#failures), consistency checks (#checks), and runtime in milliseconds (time). We used the time limit of 60 minutes to stop the search.

Table 1 presents the comparison between the performances of MAC, MACR (MAC with restarts) and MACER (MAC with restarts and extraction of visited search space) on a variety of problems. The first column denotes the name of the problem and the number in the brackets indicates the number of instances; their average results are presented in the corresponding row. The column #us denotes the number of instances not solved by MAC for a given problem. Out of 81 instances MAC could not solve 6 instances within the time limit while both MACR and MACER solved all the instances. We observe that MACER, that does not revisit the same search space, outperforms MACR significantly. Of course, because of the differences in the exploration of the search tree, it is not always possible to expect these improvements. It is recalled that MACER is also flexible enough to choose between different networks. In other words, it can jump out of an unpromising search tree and can choose a different branch. However, for these results, MACER simply uses a stack to save the networks and selects the network from the top of the stack.

When MAC is used in conjunction with a restart strategy and the dom/wdeg variable ordering heuristic, weights associated with the constraints are preserved from one run to the other. This may help MACR to select the hard variables higher up in the search tree and may increase the chances of detecting unsatisfiability of the network early in the subsequent search [3] [9]. Before restarting search, MACER decomposes the problem into many networks. In the subsequent search it may identify the unsatisfiable core, which may be present in the other subnetworks as well. Therefore, it may need to prove the unsatisfiability of the same core in more than one network, which could be expensive. In order to avoid this, we proposed MACER^+. This phenomenon was observed for the unsatisfiable instances of the modified RLFAP [11]. The results of these experiments are summarised in Table 2. Here scen11_fk corresponds to scen11 with the k highest frequencies removed from the domains of the variables [12]. MACER could solve only

Table 1. Average Results for MAC, MACR and MACER with *dom/wdeg* variable ordering heuristic

		MAC		MACR		MACER	
prob	#us	#nodes	time	#nodes	time	#nodes	time
FRB-30-15 (5)	0	3,295	231	6,105	402	3,492	231
FRB-35-17 (5)	0	37,380	1,847	73,477	3,606	18,417	928
FRB-40-19 (5)	0	148,869	8,153	677,241	37,243	266,219	15,028
FRB-45-21 (5)	0	1,744,149	138,638	6,852,925	696,081	1,760,702	191,296
QCP-15 (21)	0	118,830	5,862	162,178	16,867	91,184	13,158
QWH-20 (10)	0	431,149	27,149	653,910	232,737	202,050	83,188
QWH-25 (5)	0	885,218	104,150	1,138,775	842,187	695,072	570,987
QA-5 (1)	0	82,602	3,508	671	30	133	8
QA-6 (1)	0	595,069	68,617	58,603	5,021	39,585	2,691
QA-7 (1)	1	-	-	534,823	62,376	24,625	2,446
JS-E0DDR1 (10)	1	13,166,763	311,769	379	57	1,231	86
JS-E0DDR2 (10)	4	177	25	275	50	331	57
PIGEON-10 (1)	0	623,529	1,419	2,359,038	4,100	619,662	1,702
PIGEON-11 (1)	0	6,235,300	5,772	19,414,596	20,336	6,180,739	6,981
PIGEON-12 (1)	0	68,588,311	63,921	218,708,760	205,964	69,621,744	72,858

Table 2. Results for some hard instances of Modified RLFAP

	algorithm	#checks	time	#nodes	#failures
scen11_f4	MACER$^+$	2,328,392,085	52,577	995,105	797,616
	MACER	28,091,626,592	1,507,003	12,125,814	9,648,834
	MACR	3,538,062,253	80,202	1,553,784	1,241,818
	MAC	-	-	-	-
scen11_f3	MACER$^+$	6,735,996,236	149,429	2,824,173	2,257,039
	MACER	-	-	-	-
	MACR	9,484,306,821	196,341	4,164,335	3,328,121
	MAC	-	-	-	-
scen11_f2	MACER$^+$	17,812,913,144	470,833	7,833,259	6,242,426
	MACER	-	-	-	-
	MACR	46,476,404,826	1,169,450	19,882,045	15,727,705
	MAC	-	-	-	-
scen11_f1	MACER$^+$	67,212,745,483	1,791,228	29,094,864	23,389,498
	MACER	-	-	-	-
	MACR	101,349,033,658	2,904,688	45,423,629	36,428,007
	MAC	-	-	-	-

one instance for which it took two orders of magnitude more time than MACR. When MACER$^+$ identifies an unsatisfiable core it tries to extract it from the remaining networks, and thus it may avoid proving the unsatisfiability of the same core many times. We remark that MACER$^+$ extracts an unsatisfiable core from a given network only when the ratio of the size of the network and the number of restarts is less than or equal to the size of the core being extracted. Results depicted in the table clearly show that the extraction of unsatisfiable cores can make a significant difference in terms of checks, time, nodes and failures. As the problem becomes harder, MACER$^+$ starts performing better. MAC could not solve any of the four instances within a time limit of 60 minutes.

A heuristic used for selecting a network Q from the set of remaining networks $\mathcal{R}$ can have a substantial impact on the cardinality of $\mathcal{R}$ in MACER$^+$. For example, if a heuristic always selects a network Q having maximum size and if MACE could not prove its unsatisfiability or could not find a solution within a given cutoff value, then more subnetworks would be added to $\mathcal{R}$. Although the cardinality of $\mathcal{R}$ increases linearly with respect to the number of restarts, the extraction of unsatisfiable cores can increase it

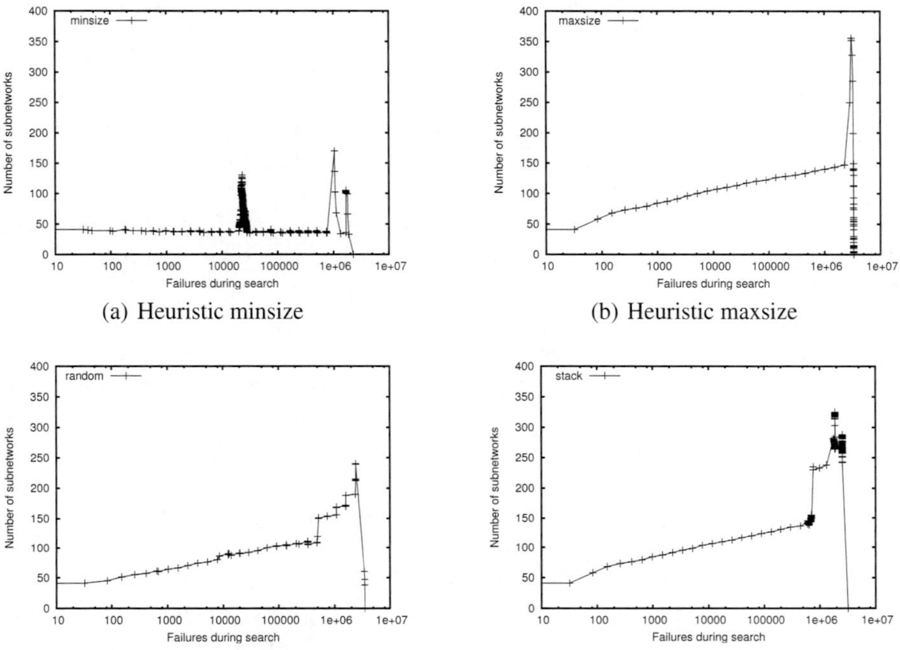

(a) Heuristic minsize

(b) Heuristic maxsize

(c) Heuristic random

(d) Heuristic stack

Fig. 3. The number of networks in $\mathcal{R}$ during search with the heuristics (a) minsize, (b) maxsize, (c) random and (d) stack

exponentially. If Q' is an unsatisfiable core of the selected network Q and if its extraction from each network in $\mathcal{R}$ generates at least k networks then $\mathcal{R} \ominus \{Q'\}$ would result in at least $k \times |\mathcal{R}|$ networks.

We investigated the impact of using various simple heuristics like minimum size of the network (minsize), maximum size of the network (maxsize), random selection of the network (random), and stack-wise selection of the network (stack). The results obtained by using these heuristics for the instance scen11_f3 are presented in Figure 3. The graphs plot the relation between the cardinality of $\mathcal{R}$ and the total number of failures encountered so far during search. Although the difference between the performances (in terms of the number of failures) of different heuristics when used with MACER$^+$ is not much, the cardinality of $\mathcal{R}$ at any given point is the lowest for the heuristic minsize. The maximum number of the unexplored networks in $\mathcal{R}$ at any given point during the search is 170 with the minsize heuristic, while with the heuristics maxsize, random and stack, it is 356, 240 and 325 respectively. Keeping the cardinality of $\mathcal{R}$ low improves the efficiency of the algorithm in terms of time. The results for MACER$^+$ in Table 2 are obtained by using the heuristic minsize. We remark that the extraction of an unsatisfiable core from the remaining networks can also decrease the cardinality of $\mathcal{R}$, since the extraction can also generate empty networks, which are discarded. This can be observed in the descending slopes of the graphs.

In MACER$^+$ the extraction of an unsatisfiable core of Q from the set of networks $\mathcal{R}$ is done after determining the inconsistency of Q. This can also be seen as extracting the subnetwork resulting from the root node of Q from the remaining networks when the algorithm backtracks to that node without finding a solution. This approach can be extended further by extracting the inconsistent subnetwork resulting from any node of the search tree of Q from the set of networks associated with the unexplored possibilities. This approach is denoted by MACER*. In the following we show that $dom/wdeg$ heuristic may not always perform efficiently on a certain class of problem. We identify and present that problem class and show that MACER*can solve such problem more efficiently.

Let P_1 denote a Model B random problem [13]. In this model a random CSP instance is characterised by (n_1, d, e, t), where n_1 is the number of variables, d is the uniform domain size of each variable, e is the number of constraints, t is the number of incompatible pairs in each constraint. Let P_2 denotes a slightly modified version of the Domino problem. An instance of P_2 is characterised by (n_2), where n_2 is the number of variables, the domain of each variable is $\{1, 2, \ldots, n_2 - 1\}$, the set of constraints is $\{c_{i(i+1)} | i < n_2\} \cup \{c_{1n_2}\}$, and the set of pairs of allowed values for each constraint is $\{(i, i+1) | i < n_2 - 1\} \cup \{(n_2 - 1, 1)\}$. Informally it is an undirected constraint graph with a cycle constraint. We generate problem $P_1 \oplus P_2$ by merging P_1 and P_2. In particular we merge a satisfiable instance of a dense loosely constrained Model B random problem with an unsatisfiable instance of sparse tightly constrained modified version of the Domino problem. Table 3 presents the comparison of MAC, MACR, and MACER* on some instances of the described problem.

Table 3. For each class of the composed problem, 20 instances were generated and their mean performances are reported. Time-out was set to 60 minutes.

problem parameters	algorithm	#checks	time	#nodes	#failures
	MACER*	197,131,624	48,204	190,322	102,165
$(40, 10, 732, 11) \oplus (10)$	MACR	2,152,566,608	570,989	2,146,392	1,143,770
	MAC	7,822,074,968	1,820,608	7,793,099	4,139,248
	MACER*	765,651,246	192,186	668,703	351,096
$(50, 10, 1000, 10) \oplus (5)$	MACR	-	-	-	-
	MAC	-	-	-	-

The *dom/wdeg* variable ordering heuristic associates a counter with each constraint. Whenever a domain is wiped out while enforcing arc consistency, the weight of the corresponding constraint is incremented by 1. The weighted degree of a variable is the sum of the weights of the constraints in which the variable is involved. In the beginning of the search the dom/wdeg heuristic behaves like dom/deg heuristic, and therefore for an instance of the $P_1 \oplus P_2$ problem that are used here, it selects variables from P_1. If finding a solution for P_1 is not easy then it fails and the result is that it increments the weights of the variables of P_1. The weights of P_1 are incremented so much that both MAC and MACR spend a lot of time in solving P_1. To illustrate this point we present the final weights of one of the instances of $(40, 10, 732, 11) \oplus (10)$ in Figure 4. The variables 1 to 40 and 41 to 50 on the x-axis correspond to P_1 and P_2 respectively. For the $(50, 10, 1000, 10) \oplus (5)$ problem class, neither MAC nor MACR solved a single

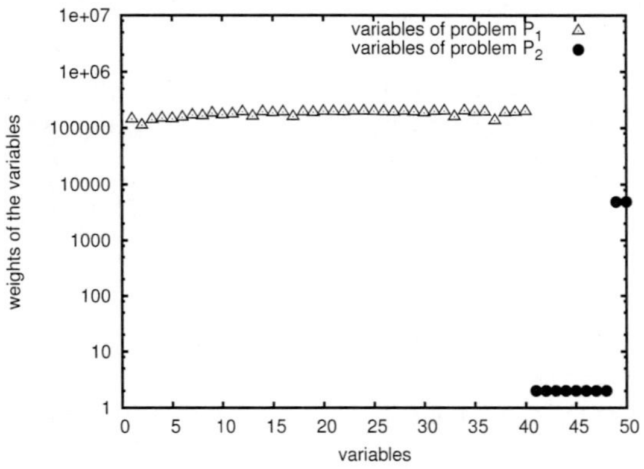

Fig. 4. Weight profiles for an instance of $(40, 10, 732, 11) \oplus (10)$

instance within the time limit. MACER* is very efficient. As soon as it finds a solution of P_1 it extracts an unsatisfiable core associated with P_2 from the complete search space, which results in proving the inconsistency of the problem.

In our preliminary investigation MACER* proved to be expensive for the other problems. The reason is that in our current implementation it is necessary to traverse the entire set $\mathcal{R}$ and perform the following operations: check whether the inconsistent sub-network can be extracted from each network in $\mathcal{R}$, and if the latter is true, it performs the extraction. Although the search space could be reduced drastically, this comes at the expense of huge space complexity and the overhead is such that it does not compensate for the reduction in the number of failures and visited nodes. Nevertheless, we believe that this overhead can be reduced significantly and may prove to be useful for many problems if we store the networks in an efficient data structure such as multi-valued decision diagrams, which is also one of the future directions for this work.

7 Conclusions and Future Work

In this paper we presented a new approach for avoiding revisiting the search space in a restart context. First we proposed the algorithm MACER that extracts the inconsistent visited search space by maintaining unvisited search space. Second we proposed the algorithm MACER$^+$ that not only extracts the visited search space but also extracts unsatisfiable cores from the remaining search space. We also proposed MACER*, which can be seen as an extension of MACER$^+$. Empirical results confirm the effectiveness of extracting inconsistent search space.

There are many potential future directions of this work. The foremost is to use smart data structures like multi-valued decision diagrams to store the networks associated with the unexplored possibilities in order to extract networks efficiently. We would also like to establish the relationship between nogood recording and extraction of inconsistent search space in order to compare the level of pruning achieved by these two different techniques. For some problem, it may be possible to use symmetry properties to learn more inconsistent networks and extract them also from the remaining search space.

Acknowledgements

This material is based upon works supported by the Science Foundation Ireland under Grants No. 05/IN/I886 and No. 08/PI/I1912, and Embark Post Doctoral Fellowships No. CT1080049908 and No. CT1080049909.

References

1. Sabin, D., Freuder, E.C.: Contradicting conventional wisdom in constraint satisfaction. In: ECAI, pp. 125–129 (1994)
2. Boussemart, F., Hemery, F., Lecoutre, C., Saïs, L.: Boosting systematic search by weighting constraints. In: Proceedings of the Thirteenth European Conference on Artificial Intelligence (2004)
3. Grimes, D., Wallace, R.J.: Learning to identify global bottlenecks in constraint satisfaction search. In: FLAIRS Conference, pp. 592–597 (2007)
4. Schiex, T., Verfaillie, G.: Nogood recording for static and dynamic constraint satisfaction problems. In: ICTAI, pp. 48–55 (1993)
5. Lecoutre, C., Sais, L., Tabary, S., Vidal, V.: Nogood recording from restarts. In: IJCAI, pp. 131–136 (2007)
6. Freuder, E.C., Hubbe, P.D.: Extracting constraint satisfaction subproblems. In: IJCAI, pp. 548–557 (1995)
7. Rossi, F., van Beek, P., Walsh, T.: Handbook of Constraint Programming. Elsevier Science Inc., Amsterdam (2006)
8. Gomes, C.P., Selman, B., Kautz, H.A.: Boosting combinatorial search through randomization. In: AAAI/IAAI, pp. 431–437 (1998)
9. Hemery, F., Lecoutre, C., Sais, L., Boussemart, F.: Extracting mucs from constraint networks. In: ECAI, pp. 113–117 (2006)
10. Boussemart, F., Hemery, F., Lecoutre, C.: Description and representation of the problems selected for the first international constraint satisfaction solver competition. In: van Dongen, M. (ed.) Proceedings of the Second International Workshop on Constraint Propagation and Implementation. Solver Competition, vol. 2, pp. 7–26 (2005)
11. Cabon, B., De Givry, S., Lobjois, L., Schiex, T., Warners, J.: Radio link frequency assignment. Journal of Constraints 4, 79–89 (1999)
12. Bessière, C., Chmeiss, A., Saïs, L.: Neighborhood-based variable ordering heuristics for the constraint satisfaction problem. In: Walsh, T. (ed.) CP 2001. LNCS, vol. 2239, pp. 565–569. Springer, Heidelberg (2001)
13. Gent, I., MacIntyre, E., Prosser, P., Smith, B., Walsh, T.: Random constraint satisfaction: Flaws and structure. Journal of Constraints 6(4), 345–372 (2001)

Coalition Structure Generation Utilizing Compact Characteristic Function Representations

Naoki Ohta[1], Vincent Conitzer[2], Ryo Ichimura[1], Yuko Sakurai[1], Atsushi Iwasaki[1], and Makoto Yokoo[1]

[1] Department of ISEE, Kyushu University, Fukuoka 819-0395, Japan
{ohta@agent,ichimura@agent,sakurai@agent,iwasaki@, yokoo@}is.kyushu-u.ac.jp

[2] Department of Computer Science, Duke University, Durham, NC 27708, USA
conitzer@cs.duke.edu

Abstract. This paper presents a new way of formalizing the Coalition Structure Generation problem (CSG), so that we can apply constraint optimization techniques to it. Forming effective coalitions is a major research challenge in AI and multi-agent systems. CSG involves partitioning a set of agents into coalitions so that social surplus is maximized. Traditionally, the input of the CSG problem is a black-box function called a *characteristic function*, which takes a coalition as an input and returns the value of the coalition. As a result, applying constraint optimization techniques to this problem has been infeasible. However, characteristic functions that appear in practice often can be represented concisely by a set of rules, rather than a single black-box function. Then, we can solve the CSG problem more efficiently by applying constraint optimization techniques to the compact representation directly.

We present new formalizations of the CSG problem by utilizing recently developed compact representation schemes for characteristic functions. We first characterize the complexity of the CSG under these representation schemes. In this context, the complexity is driven more by the number of rules rather than by the number of agents. Furthermore, as an initial step towards developing efficient constraint optimization algorithms for solving the CSG problem, we develop mixed integer programming formulations and show that an off-the-shelf optimization package can perform reasonably well, i.e., it can solve instances with a few hundred agents, while the state-of-the-art algorithm (which does not make use of compact representations) can solve instances with up to 27 agents.

Keywords: Multiagent systems, coalition structure generation, constraint optimization.

1 Introduction

Coalition formation is an important capability in automated negotiation among self-interested agents. Coalition structure generation (CSG) involves partitioning

I.P. Gent (Ed.): CP 2009, LNCS 5732, pp. 623–638, 2009.

a set of agents into coalitions so that social surplus is maximized. This problem has become a popular research topic in AI and multi-agent systems. Possible applications of CSG include distributed vehicle routing (Sandholm and Lesser, 1997), multi-sensor networks (Dang et al., 2006), etc. The CSG problem is equivalent to a *complete set partition problem* (Yeh, 1986), and various algorithms for solving the CSG problem have been developed. Sandholm et al. (1999) propose an anytime algorithm with worst-case guarantees. However, to obtain an optimal coalition structure, this algorithm must check all coalition structures. Thus, the worst-case time complexity is $O(n^n)$, where n is the number of agents. On the other hand, Dynamic Programming (DP) based algorithms (Yeh, 1986; Rothkopf et al., 1998; Rahwan and Jennings, 2008b) are guaranteed to find an optimal solution in $O(3^n)$. Shehory and Kraus (1998) propose a greedy algorithm that puts constraints on the possible size of the coalitions.

Arguably, the state-of-the-art algorithm is the IP (integer partition) algorithm (Rahwan et al., 2007). This is an anytime algorithm, which divides the search space into partitions based on integer partition, and performs branch & bound search. Although the worst-case time complexity for obtaining an optimal solution is $O(n^n)$, in practice, IP is much faster than DP based algorithms. Furthermore, Rahwan and Jennings (2008a) introduce an extension of the IP algorithm that utilizes DP for preprocessing.

As far as we are aware, all existing works on CSG assume that the characteristic function is represented implicitly, and we have oracle access to the function—that is, the value of a coalition (or a coalition structure as a whole) can be obtained using some procedure. This is because representing an *arbitrary* characteristic function explicitly requires $\Theta(2^n)$ numbers, which is prohibitive for large n. When a characteristic function is represented by a black-box function, there is no room for applying constraint optimization techniques. Thus, this problem has been irrelevant to the CP community.

However, characteristic functions that appear in practice often display significant structure, and it is likely that such characteristic functions can be represented much more concisely. Indeed, recently, several new methods for representing characteristic functions have been developed (Ieong and Shoham, 2005; Conitzer and Sandholm, 2004, 2006). These representation schemes capture characteristics of interactions among agents in a natural and concise manner, and can reduce the representation size significantly. Surprisingly, to our knowledge, these representation schemes have not yet been used for CSG; this is what we set out to do in this paper. Using these compact representation schemes, a characteristic function is represented by a set of rules, rather than a single blackbox function. It is likely that we can solve the CSG problem more efficiently by applying constraint optimization techniques to the compact representation directly.

We examine three representative compact representation schemes: (i) marginal contribution nets (MC-nets) (Ieong and Shoham, 2005), (ii) synergy coalition groups (SCGs) (Conitzer and Sandholm, 2006), and (iii) SCGs in multi-issue domains (Conitzer and Sandholm, 2004). The optimal choice of a representation scheme depends on the application.

There exist several other compact representation schemes, e.g., logic-based approaches (Wooldridge and Dunne, 2004, 2006) and skill-based approaches (Yokoo et al., 2005; Bachrach and Rosenschein, 2008). In this paper, we restrict our attention to the schemes mentioned earlier, since they are more closely related to the traditional CSG problem.

Quite interestingly, we find that there exists some common structure among these cases: in essence, the problem is to find a subset of rules that maximizes the sum of rule values under certain constraints. For each case, we show that solving the CSG problem is NP-hard, and the size of a problem instance is naturally measured by the number of rules rather than the number of agents.

Furthermore, as an initial step towards developing efficient constraint optimization algorithms for solving the CSG problem, we give a mixed integer programming (MIP) formulation that captures the above mentioned structure. We show that an off-the-shelf optimization package (CPLEX) can solve the resulting MIP problem instances reasonably well, i.e., it can solve instances with a few hundred agents, while the state-of-the-art algorithm (which does not make use of compact representations) can solve instances up to 27 agents.

The rest of this paper is organized as follows. First, we review the model of coalition structure generation (Section 2). Next, we introduce solution algorithms when the characteristic function is represented by MC-nets (Section 3), SCGs (Section 4) and SCGs in multi-issue domains (Section 5). Finally, we show the evaluation results and discussions (Section 6).

2 Model

Let $A = \{1, 2, \ldots, n\}$ be the set of agents. We assume a characteristic function game, i.e., the value of a coalition S is given by a characteristic function v. A characteristic function $v : 2^A \to \Re$ assigns a value to each set of agents (coalition) $S \subseteq A$. We assume that each coalition's value is nonnegative. This is not an unreasonable assumption (Sandholm et al., 1999); even if some coalition's values are negative, as long as each coalition's value is bounded (i.e., not infinitely negative), we can normalize the coalition values so that all values are non-negative. This rescaled game is strategically equivalent to the original game.

A coalition structure CS is a partition of A, into disjoint, exhaustive coalitions. To be more precise, $CS = \{S_1, S_2, \ldots\}$ satisfies the following conditions:

$$\forall i, j\ (i \neq j),\ S_i \cap S_j = \emptyset,\ \bigcup_{S_i \in CS} S_i = A.$$

In other words, in CS, each agent belongs to exactly one coalition, and some agents may be alone in their coalitions.

For example, in a game with three agents a, b, and c, there are seven possible coalitions: $\{a\}$, $\{b\}$, $\{c\}$, $\{a, b\}$,$\{b, c\}$, $\{a, c\}$, $\{a, b, c\}$, and five possible coalition structures: $\{\{a\}, \{b\}, \{c\}\}$, $\{\{a, b\}, \{c\}\}$, $\{\{a\}, \{b, c\}\}$, $\{\{b\}, \{a, c\}\}$, $\{\{a, b, c\}\}$.

The value of a coalition structure CS, denoted as $V(CS)$, is given by:

$$V(CS) = \sum_{S_i \in CS} v(S_i).$$

An optimal coalition structure CS^* is a coalition structure that satisfies the following condition:

$$\forall CS, V(CS^*) \geq V(CS).$$

We say a characteristic function is super-additive, if for any disjoint sets S_i, S_j, $v(S_i \cup S_j) \geq v(S_i) + v(S_j)$ holds. If the characteristic function is super-additive, solving CSG becomes trivial, i.e., the grand coalition (the coalition of all agents) is optimal.

Super-additivity means that any pair of coalitions is better off by merging into one. One might think that super-additivity holds in most of the cases since the agents in the composite coalition can work separately and perform at least as well as the case that they were in different coalitions. However, organizing a large coalition can be costly, e.g., there might be coordination overhead like communication costs, or possible anti-trust penalties. Also, if time is limited, the agents may not have time to carry out the communications and computations required to coordinate effectively within the composite coalition, so component coalitions may be more advantageous. Thus, we assume a characteristic function can be non-super-additive.

3 CSG Using MC-Nets

Ieong and Shoham (2005) develop a concise representation of a characteristic function called *marginal contribution networks (MC-nets).*

Definition 1 (MC-nets). *An* MC-net *consists of a set of* rules R. *Each rule* $r \in R$ *is of the form:* $(P_r, N_r) \rightarrow v_r$, *where* $P_r \subseteq A$, $N_r \subseteq A$, $P_r \cap N_r = \emptyset$, $v_r \in \Re$. *We say that rule* r *is* applicable *to coalition* S *if* $P_r \subseteq S$ *and* $N_r \cap S = \emptyset$, *i.e.,* S *contains all agents in* P_r *(positive literals), and it contains no agent in* N_r *(negative literals). For a coalition* S, $v(S)$ *is given as* $\sum_{r \in R_S} v_r$, *where* R_S *is the set of rules applicable to* S. *Thus, for a coalition structure* CS, $V(CS)$ *is given as* $\sum_{S \in CS} \sum_{r \in R_S} v_r$.

Example 1. Let there be five agents a, b, c, d, e and four rules: $r_1 : (\{b, e\}, \{\}) \rightarrow 3$, $r_2 : (\{a, b, c\}, \{d\}) \rightarrow 2$, $r_3 : (\{a, d\}, \{\}) \rightarrow 1$, and $r_4 : (\{c\}, \{e\}) \rightarrow 1$. In this case, r_1 and r_2 are applicable to coalition $\{a, b, c, e\}$, but r_3 and r_4 are not. Thus, $v(\{a, b, c, e\})$ is equal to $3 + 2 = 5$.

In the original definition from (Ieong and Shoham, 2005), a rule may have a negative value. In this paper, we assume all rules have positive values. Furthermore, we assume each rule has at least one positive literal. Under these restrictions, we can guarantee that having more applicable rules never hurts, and each rule is applicable to only one coalition. Even under these restrictions, MC-nets can represent any characteristic function. This is because, in the worst case, for each coalition $S \subseteq A$, we can create a rule $(S, A \setminus S) \to v(S)$, i.e., each rule is applicable only to S.

Definition 2 (Feasible rule set). *We say a set of rules $R' \subseteq R$ is* feasible *if there exists CS where each rule $r \in R'$ is applicable to some $S \in CS$.*

In Example 1, $\{r_2, r_4\}$ is feasible because each rule is applicable to $CS = \{\{a, b, c\}, \{d, e\}\}$. On the other hand, $\{r_1, r_2, r_4\}$ and $\{r_2, r_3\}$ are infeasible. The problem of finding CS^* is equivalent to finding a feasible rule set R', so that $\sum_{r \in R'} v_r$ is maximized.

Definition 3 (Relations between rules). *The possible relations between two rules r and r' can be classified into the following four nonoverlapping and exhaustive cases:*

Compatible on the same coalition: $P_r \cap P_{r'} \neq \emptyset$ *and* $P_r \cap N_{r'} = P_{r'} \cap N_r = \emptyset$. *For example, in Example 1, r_1 and r_2 are compatible on the same coalition: if r_1 and r_2 are applicable at the same time, there must be a coalition S with $S \supseteq \{a, b, c, e\}$ and $d \notin S$.*

Incompatible: $P_r \cap P_{r'} \neq \emptyset$, *and* $(P_r \cap N_{r'} \neq \emptyset$ *or* $P_{r'} \cap N_r \neq \emptyset)$. *For example, r_2 and r_3 are incompatible: these two rules are not applicable at the same time.*

Compatible on different coalitions: $P_r \cap P_{r'} = \emptyset$, *and* $(P_r \cap N_{r'} \neq \emptyset$ *or* $P_{r'} \cap N_r \neq \emptyset)$. *For example, r_1 and r_4 are compatible on different coalitions: if r_1 and r_4 are applicable at the same time, there must be two different coalitions S_1 and S_2, where $S_1 \supseteq \{b, e\}$ and $S_2 \supseteq \{c\}$.*

Independent: $P_r \cap P_{r'} = \emptyset$, *and* $P_r \cap N_{r'} = P_{r'} \cap N_r = \emptyset$. *For example, r_1 and r_3 are independent. These two rules can be applied to the same coalition or to different coalitions.*

Let us consider a graphical representation of an MC-net in which each vertex is a rule, and between any two vertices, there exists an edge whose type is one of the four cases described above. Figure 1 shows the graphical representation of Example 1 ("independent" edges are not shown).

The following conditions characterize whether a rule set is feasible.

Theorem 1. *A set of rules R' is feasible if and only if it satisfies the following conditions.*

(a) R' includes no pair of rules/vertices connected by an "incompatible" edge, and

(b) if two rules/vertices in R' are connected by a "compatible on different coalitions" edge, then they are not reachable via "compatible on the same coalition" edges within R'.

Proof. First, we prove the "if" part. From (a), there exists no incompatible edge within R'. From (b), R' can be divided into groups $G_1, G_2, \ldots, G_k$ where the rules within G_i are reachable from each other by "compatible on the same coalition"

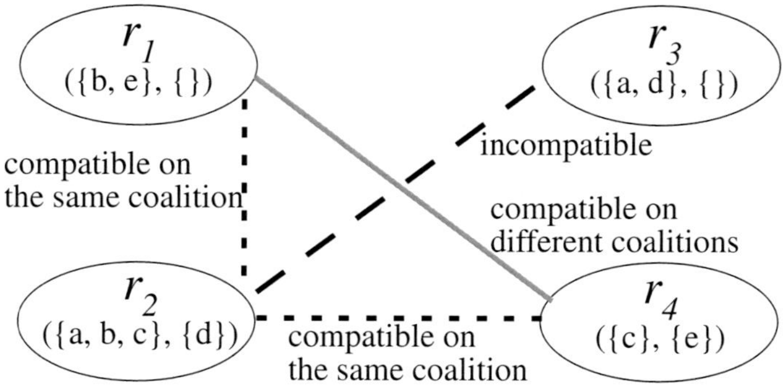

Fig. 1. Graphical representation of Example 1

edges, there exists no "compatible on different coalitions" edge between rules in G_i, and there exists no "compatible on the same coalition" edge between rules that belong to different groups.

Let us choose $CS = \{S_1, S_2, \ldots, S_k\}$ so that S_i is the union of all positive literals of $r \in G_i$. Then, for $i \neq j$, $S_i \cap S_j = \emptyset$ holds. This is because $S_i \cap S_j \neq \emptyset$ would imply that there exists at least one pair $r \in G_i, r' \in G_j$ for which r and r' are connected by a "compatible on the same coalition" edge (since there cannot be an "incompatible" edge between them)—but this is in contradiction with the way in which $G_1, \ldots, G_k$ are chosen. Thus, $\{S_1, \ldots, S_k\}$ is a valid coalition structure.[1]

Now, we show that for any $r \in G_i$, r is applicable to coalition S_i. Clearly, S_i contains all the positive literals of r. It remains to show that S_i does not contain any negative literal of r. For the sake of contradiction, assume S_i contains agent a, where a is a negative literal of r. Then, there exists another rule $r' \in G_i$ for which a is a positive literal. There must be a "compatible on different coalitions" or an "incompatible" edge between r and r'. Either case leads to a contradiction. Hence, R' is feasible.

Next, we prove the "only if" part. We show that if R' does not satisfy the above conditions, then there exists no coalition structure where R' is applicable. Clearly, if (a) is not satisfied, i.e., some $r, r' \in R'$ are connected by an "incompatible" edge, then there exists no coalition structure where r and r' are applicable at the same time.

Now, let us assume (b) is not satisfied, i.e., there exist $r_i, r_j \in R'$ such that r_i and r_j are connected by a "compatible on different coalitions" edge, and they are reachable by "compatible on the same coalition" edges within R'. Assume r_i is applicable to coalition S_i and r_j is applicable to coalition S_j. Since r_i and r_j are connected by a "compatible on different coalitions" edge, S_i and S_j must be different. However, S_i must contain all positive literals of rules reachable from r_i via "compatible on the same coalition" edges: otherwise, some rule in R' is not

[1] If some agent is not included in any S_i, we can assume the agent forms its own coalition.

applicable. Similarly, S_j must contain all positive literals of rules reachable from r_j via "compatible on the same coalition" edges. Since r_i and r_j are reachable from each other via "compatible on the same coalition" edges, S_i and S_j must be the same—but this contradicts the fact that they must be different. □

Theorem 2. *When the characteristic function is represented as an MC-net, finding an optimal coalition structure is NP-hard. Moreover, unless $\mathcal{P} = \mathcal{NP}$, there exists no polynomial-time $O(|R|^{1-\epsilon})$ approximation algorithm for any $\epsilon > 0$, where $|R|$ is the number of rules.*

Proof. The maximum independent set problem is to choose $V' \subseteq V$ for a graph $G = (V, E)$ such that there exists no edge between vertices in V', and $|V'|$ is maximized under this constraint. It is NP-hard and, unless $\mathcal{P} = \mathcal{NP}$, there exists no polynomial-time $O(|V|^{1-\epsilon})$ approximation algorithm for any $\epsilon > 0$ (Håstad, 1999; Zuckerman, 2007). We reduce an arbitrary maximal independent set instance to a CSG problem instance, as follows. For each $v \in V$, let there be an agent a_v; also, for each $e \in E$, let there be an agent a_e. For each $v \in V$, we create a rule r_v where $P_{r_v} = \{a_v\} \cup \{a_e : v \in e\}$, $N_{r_v} = \{a_w : (v, w) \in E\}$, and $v_{r_v} = 1$. Thus, rules are "incompatible" if they correspond to neighboring vertices, and "independent" otherwise. It follows that feasible rule sets correspond exactly to independent sets of vertices. □

The reduction in Theorem 2 relies heavily on "incompatibilities" between rules. If there are no "incompatibilities" then the problem is equivalent to the multi-cut problem (Vazirani, 2001), which is a generalization of the min-cut problem.

Definition 4 (MIP formulation of CSG for MC-nets). *The problem of finding a feasible rule set R' that maximizes $\sum_{r \in R'} v_r$ can be modeled as follows.*

$$
\begin{aligned}
&\mathit{max}\ \textstyle\sum_{r \in R} v_r \cdot x(r) \\
&\quad \mathit{s.t.} \forall e = (r, r'),\ \textit{where } e \textit{ is an "incompatible" edge,} \\
&\qquad x(r) + x(r') \leq 1, \ \text{— } (i) \\
&\quad \forall e = (r_i, r_j),\ \textit{where } e \textit{ is} \\
&\qquad \textit{a "compatible on different coalitions" edge and } i < j, \\
&\qquad dis(e, r_i) = 0,\ dis(e, r_j) \geq 1, \ \text{— } (ii) \\
&\qquad \forall e' = (r_1, r_2),\ \textit{where } e' \textit{ is} \\
&\qquad\quad \textit{a "compatible on the same coalition" edge,} \\
&\qquad\quad dis(e, r_1) \leq dis(e, r_2) + (1 - x(r_1)) + (1 - x(r_2)), \ \text{— } (iii) \\
&\qquad\quad dis(e, r_2) \leq dis(e, r_1) + (1 - x(r_1)) + (1 - x(r_2)), \ \text{— } (iv) \\
&\quad \forall r \in R,\ x(r) \in \{0, 1\}.
\end{aligned}
$$

$x(r) = 1$ means that rule r is selected. The constraint (i) ensures that two rules connected by an "incompatible" edge will not be selected at the same time. Also, for each "compatible on different coalitions" edge $e = (r_i, r_j)$, we define a distance/potential for e, so that $dis(e, r_i) = 0$ and $dis(e, r_j) \geq 1$ (ii). The constraints (iii) and (iv) ensure that if both of r_1 and r_2 are selected, where r_1 and r_2 are connected by a "compatible on the same coalition" edge, then the distance/potential of these two rules for the aforementioned e must be equal.

Then, the facts that $dis(e, r_i) = 0$ and $dis(e, r_j) \geq 1$ ensure that r_i and r_j are not reachable from each other via "compatible on the same coalition" edges. Using such a distance/potential is a standard method for representing connectivity constraints in MIP formalization without enumerating possible paths.

In this formulation, the number of binary variables is equal to the number of rules. The number of constraints is $d_{in} + d_{cd}(2d_{cs} + 1)$, where d_{in}, d_{cd}, d_{cs} are the number of edges with types "incompatible", "compatible on different coalitions", and "compatible on the same coalition", respectively.

4 CSG Using Synergy Coalition Groups

Conitzer and Sandholm (2006) introduce a concise representation of a characteristic function called a *synergy coalition group (SCG)*. The main idea is to explicitly represent the value of a coalition only when there exists some *positive* synergy.

Definition 5 (SCG). *An SCG consists of a set of pairs of the form:* $(S, v(S))$. *For any coalition* S*, the value of the characteristic function is:*

$$v(S) = \max\{ \sum_{S_i \in p_S} v(S_i)\},$$

where p_S *is a partition of* S*, i.e., all the* S_i *are disjoint and* $\bigcup_{S_i \in p_S} S_i = S$*, and for all the* S_i*,* $(S_i, v(S_i)) \in SCG$*. To avoid senseless cases that have no feasible partitions, we require that* $(\{a\}, 0) \in SCG$ *whenever* $\{a\}$ *does not receive a value elsewhere in SCG.*

Thus, if the value of a coalition S is not given explicitly in *SCG*, it is calculated from the possible partitions of S. Using this original definition, we can represent only super-additive characteristic functions, i.e., for any disjoint sets S_i, S_j, $v(S_i \cup S_j) \geq v(S_i) + v(S_j)$ holds. But, as mentioned in Section 2, if the characteristic function is super-additive, solving CSG becomes trivial: the grand coalition is optimal. To allow for characteristic functions that are not super-additive, we add the following requirement on the partition p_S.

- $\forall p'_S \subseteq p_S$, where $|p'_S| \geq 2$, $(\bigcup_{S_i \in p'_S} S_i, v(\bigcup_{S_i \in p'_S} S_i))$ is not an element of *SCG*.

This additional condition requires that if the value of a coalition is explicitly given in *SCG*, then we cannot further divide it into smaller subcoalitions to calculate values. In this way, we can represent *negative* synergies.

Example 2. Let there be five agents a, b, c, d, e and let $SCG = \{(\{a\}, 0), (\{b\}, 0), (\{c\}, 1), (\{d\}, 2), (\{e\}, 3), (\{a, b\}, 3), (\{a, b, c\}, 3)\}$. In this case, $v(\{d, e\}) = v(\{d\}) + v(\{e\}) = 5$, and $v(\{a, b, c, d, e\}) = v(\{a, b, c\}) + v(\{d\}) + v(\{e\}) = 8$. For $v(\{a, b, c, d, e\})$, we cannot use $v(\{a, b\}) + v(\{c\}) + v(\{d\}) + v(\{e\}) = 9$, because $\{a, b\} \cup \{c\} = \{a, b, c\}$ appears in *SCG*.

The (modified) SCG can represent any characteristic function, including characteristic functions that are non-super-additive, or even non-monotone. This is because in the worst case, we can explicitly give the value of every coalition. Due to the additional condition, only these explicit values can then be used to calculate the characteristic function.

We show that when searching for CS^*, we need to consider only the coalitions that are explicitly described in SCG.

Theorem 3. *There exists a coalition structure* CS *for which* $V(CS) = V(CS^*)$ *and* $\forall S \in CS, (S, v(S)) \in SCG$.

Proof. For the sake of contradiction, let us assume there exists some CS^* so that $V(CS^*)$ is strictly larger than any CS that consists of only elements of SCG. Let us examine some coalition $S \in CS^*$ that is not an element of SCG. From the definition of SCG, there exists a partition of S (denoted as p_S) such that $v(S) = \sum_{S_i \in p_S} v(S_i)$, and each S_i is an element of SCG. Then, by replacing each such S by p_S, we obtain a new coalition structure CS that consists of only elements of SCG, and $V(CS) = V(CS^*)$ holds—so we have the desired contradiction. □

Due to Theorem 3, finding CS^* is equivalent to a weighted set packing problem—equivalently, to the winner determination problem in combinatorial auctions (Sandholm, 2002), where each agent is an item and each coalition described in SCG is a bid.

Theorem 4. *When the characteristic function is represented as an SCG, finding an optimal coalition structure is NP-hard. Moreover, unless* $\mathcal{P} = \mathcal{NP}$, *there exists no polynomial-time* $O(|SCG|^{1-\epsilon})$ *approximation algorithm for any* $\epsilon > 0$.

Proof. This follows directly from the corresponding inapproximability for the winner determination problem (Sandholm, 2002) and the maximum independent set problem (Zuckerman, 2007).

Definition 6 (MIP formulation of CSG for SCG). *The problem of finding* CS^* *can be modeled as follows.*

$$\begin{aligned} max \quad & \sum_{(S,v(S)) \in SCG} v(S) \cdot x(S) \\ s.t. \quad & \forall a \in A, \sum_{S \ni a} x(S) = 1, \\ & x(S) \in \{0,1\}. \end{aligned}$$

$x(S)$ *is 1 if* S *is included in* CS^*, *0 otherwise.*

In this formulation (which corresponds to a standard winner determination formulation), the number of binary variables is equal to $|SCG|$, and the number of constraints is equal to the number of agents.

5 CSG in Multi-issue Domain

Conitzer and Sandholm (2004) introduce the concept of a *multi-issue domain.* In a multi-issue domain, there are k independent issues. The overall value of a coalition is the sum of the values of the coalition for the individual issues. More specifically, we assume there are k characteristic functions $v_1, v_2, \ldots, v_k$ such that for any $S \subseteq A$, $v(S) = \sum_{i=1}^{k} v_i(S)$. If each v_i can be represented concisely, then this leads to a concise representation for v. In this paper, we assume that v_i is represented by SCG_i.

Definition 7 (SCGs in multi-issue domains). *We represent the characteristic function by a vector of SCGs* $(SCG_1, \ldots, SCG_k)$. *For any* $S \subseteq A$, $v(S) = \sum_{i=1}^{k} v_i(S)$, *where* v_i *is calculated using* SCG_i. *Also, for a coalition structure* CS, *we denote* $V_i(CS) = \sum_{S \in CS} v_i(S)$. *Thus,* $V(CS) = \sum_{i=1}^{k} V_i(CS)$.

Example 3. Let there be four agents a, b, c, d and two $SCGs$: $SCG_1 = \{(\{a\},0), (\{b\},0), (\{c\},1), (\{d\},0), (\{a,b\},2), (\{a,b,c\},2)\}$, $SCG_2 = \{(\{a\},0), (\{b\},0), (\{c\},0), (\{d\},1), (\{a,b,c\},2)\}$.

In this case, $v(\{a,b,c\})$ is $v_1(\{a,b,c\}) + v_2(\{a,b,c\}) = 2 + 2 = 4$.

When there are multiple issues, an optimal coalition structure CS^* may need to contain a coalition S that is not explicitly described in any SCG_i. For example, assume that in issue i, a and b have a strong positive synergy. Also, in issue j, b and c have a strong positive synergy. Then, coalition $\{a,b,c\}$ may need to be included in CS^*, even though $\{a,b,c\}$ appears in neither SCG_i nor SCG_j.

Definition 8 (Value-producing subset). *Given a coalition structure* CS, *we say that* SCG'_i *(where* $SCG'_i \subseteq SCG_i$*) is a* value-producing subset *of* SCG_i *for* CS, *if* SCG'_i *consists exactly of elements of* SCG_i *that are used to calculate* $V_i(CS)$. *Thus,* $V_i(CS) = \sum_{(S, v_i(S)) \in SCG'_i} v_i(S)$.

In Example 3, $SCG'_1 = \{(\{a,b,c\},2), (\{d\},0)\}$ and $SCG'_2 = \{(\{a,b,c\},2), (\{d\},1)\}$ are value-producing subsets for $CS = \{\{a,b,c\},\{d\}\}$. From this definition, a value-producing subset SCG'_i must contain all agents, and elements of SCG'_i must be disjoint. We call a subset that satisfies these conditions a *valid subset.*

Definition 9 (Valid subset). $SCG'_i \subseteq SCG_i$ *is a valid subset if* $\bigcup_{(S, v_i(S)) \in SCG'_i} S = A$, *and* $\forall (S, v_i(S))$, $(S', v_i(S')) \in SCG'_i$ *where* $S \neq S'$, $S \cap S' = \emptyset$ *holds.*

Theorem 5. *A valid subset* $SCG'_i \subseteq SCG_i$ *is a value-producing subset of* SCG_i *for* CS *if and only if for each* $S \in CS$, *either one of the following conditions holds:*

1. $(S, v_i(S)) \in SCG'_i$,
2. $\exists p_S$, *where* p_S *is a partition of* S, *such that* $|p_S| \geq 2$, $\forall S' \in p_S, (S', v_i(S')) \in SCG'_i$, *and* $\forall p'_S \subseteq p_S$, *where* $|p'_S| \geq 2$, $(\bigcup_{S'' \in p'_S} S'', v_i(\bigcup_{S'' \in p'_S} S'')) \notin SCG_i$.

We omit the proof since it is straightforward from the (modified) definition of the SCG representation. Quite interestingly, we can define the possible relations between elements in SCGs in the same way as we did for MC-nets.

Definition 10 (Relations between coalitions). *The possible relations between two coalitions $(S, v_i(S)) \in SCG_i$ and $(S', v_j(S')) \in SCG_j$ can be classified into the following four cases, which are nonoverlapping and exhaustive:*

Compatible on the same coalition: *$i \neq j$ and $S \cap S' \neq \emptyset$. For example, in Example 3, $(\{a,b\},2) \in SCG_1$ and $(\{a,b,c\},2) \in SCG_2$ are compatible on the same coalition. If these two elements are a part of value-producing subsets at the same time, there must be a coalition S with $S \supseteq \{a,b,c\}$.*

Incompatible: *$i = j$ and $S \cap S' \neq \emptyset$. For example, $(\{a,b\},2) \in SCG_1$ and $(\{a,b,c\},2) \in SCG_1$ are incompatible. They cannot be used simultaneously.*

Compatible on different coalitions: *$i = j$, and there exists $(S \cup S', v_i(S \cup S')) \in SCG_i$. For example, $(\{a,b\},2) \in SCG_1$ and $(\{c\},1) \in SCG_1$ are compatible on different coalitions. If these two elements are included in value-producing subsets at the same time, there must be two coalitions S_1, S_2, where $S_1 \supseteq \{a,b\}$ and $S_2 \supseteq \{c\}$, since if there exists $S \supseteq \{a,b,c\}$, then, we need to use $(\{a,b,c\},2)$ for calculating v_1. To be more precise, this relation must be extended to a hyper-edge. If there exists $(S, v_i(S)) \in SCG_i$, such that $\forall S' \in p_S, (S', v_i(S')) \in SCG_i$ holds, where p_S is a partition of S, then, we create a hyper-edge connecting the elements in p_S. Note that we need to add hyper-edges only for sub-additive cases.*

Independent: *otherwise. For example, $(\{a,b\},2) \in SCG_1$ and $(\{d\},0) \in SCG_1$ are independent. They can be used in both cases.*

The following conditions characterize whether coalitions are value-producing.

Theorem 6. *$(SCG'_1, \ldots, SCG'_k)$, where each SCG'_i is a valid subset of SCG_i, is a vector of value-producing subsets for some CS if and only if the following conditions hold:*

(a) $(SCG'_1, \ldots, SCG'_k)$ include no pair of coalitions connected by an "incompatible" edge, and

(b) if a set of coalitions in $(SCG'_1, \ldots, SCG'_k)$ is connected by a "compatible on different coalitions" hyper-edge, then there exists at least one element that is not reachable from other elements via "compatible on the same coalition" edges.

We omit the proof since it is basically the same as that of Theorem 1.

Definition 11 (MIP formulation in multi-issue domains). *The problem of finding value-producing subsets that maximize the summation of values can be modeled as follows.*

$$
\begin{aligned}
&max \ \textstyle\sum_{p=(S,v_*(S))\in \bigcup_{i=1}^{k} SCG_i} v_*(S) \cdot x(p) \\
&\quad s.t. \forall e = (p, p'), \text{ where } e \text{ is an "incompatible" edge}, \\
&\qquad x(p) + x(p') \leq 1,
\end{aligned}
$$

$\forall e = (p_1, p_2, \ldots, p_l)$, *where* e *is*
a "compatible on different coalitions" hyper-edge,
$dis(e, p_1) = 0,\ dis(e, p_2) + \ldots + dis(e, p_l) \geq 1,$ — (i)
$\forall e' = (p_i, p_j)$, *where* e' *is*
a "compatible on the same coalition" edge,
$dis(e, p_i) \leq dis(e, p_j) + (1 - x(p_i)) + (1 - x(p_j)),$
$dis(e, p_j) \leq dis(e, p_i) + (1 - x(p_i)) + (1 - x(p_j)),$
$\forall p \in \bigcup_{i=1}^{k} SCG_i,\ x(p) \in \{0, 1\}.$

$x(p) = 1$ means the element p in $\bigcup_{i=1}^{k} SCG_i$ is selected. This formulation is basically the same as Definition 4, except for the constraint (i). This constraint means that for a hyper-edge e that connects nodes $p_1, p_2, \ldots, p_l$, at least one element must be unreachable. The number of variables and constraints are basically the same as MC-nets.

Theorem 7. *When the characteristic function is represented as SCGs in a multi-issue domain, finding an optimal coalition structure is NP-hard. Moreover, unless* $\mathcal{P} = \mathcal{NP}$*, there exists no polynomial-time* $O(m^{1-\epsilon})$ *approximation algorithm for any* $\epsilon > 0$*, where* m *is the number of elements in SCGs.*

Proof. We can use the same proof as Theorem 2. □

6 Evaluation and Discussion

We experimentally evaluated the performance of our proposed methods. All the tests were run on a Core 2 Duo E6850 3GHz processor with 8GB RAM. The test machine runs WindowsXP Professional x64 Edition SP2. We used CPLEX version 11.2, a general-purpose mixed integer programming package.

We show results for the following cases: (i) MC-nets, (ii) SCGs, and (iii) SCGs in multi-issue domains. The problem instances are generated slightly differently in each case. For case (ii), we use a decay distribution (Sandholm, 2002) described as follows. Create a coalition with one random agent. Then repeatedly add a new random agent with probability α until an agent is not added or the coalition includes all agents. Choose the value of the coalition between 0 and the number of agents in the coalition uniformly at random. We use $\alpha = 0.55$. For case (i), we first create a rule $(S, \{\}) \to v(S)$ for each SCG in case (ii). Then, we modify each rule by randomly moving an agent from the positive to the negative literals with probability p. We use $p = 0.2$. For case (iii), the way of generating problem instances is basically identical to case (ii), but we create five issues and each issue has the same number of rules. For each issue, we assume 30% of agents are involved.

In Figure 2 (a), we set #rules=#agents[2], and vary #agents from 10 to 120. In Figure 2 (b), (c), (d), we set #rules to 60, 90, and 120, respectively, and vary #agents. Each data point is the median of 50 problem instances.

[2] By #rules, we mean the number of elements in $SCG/SCGs$ in cases (ii) and (iii).

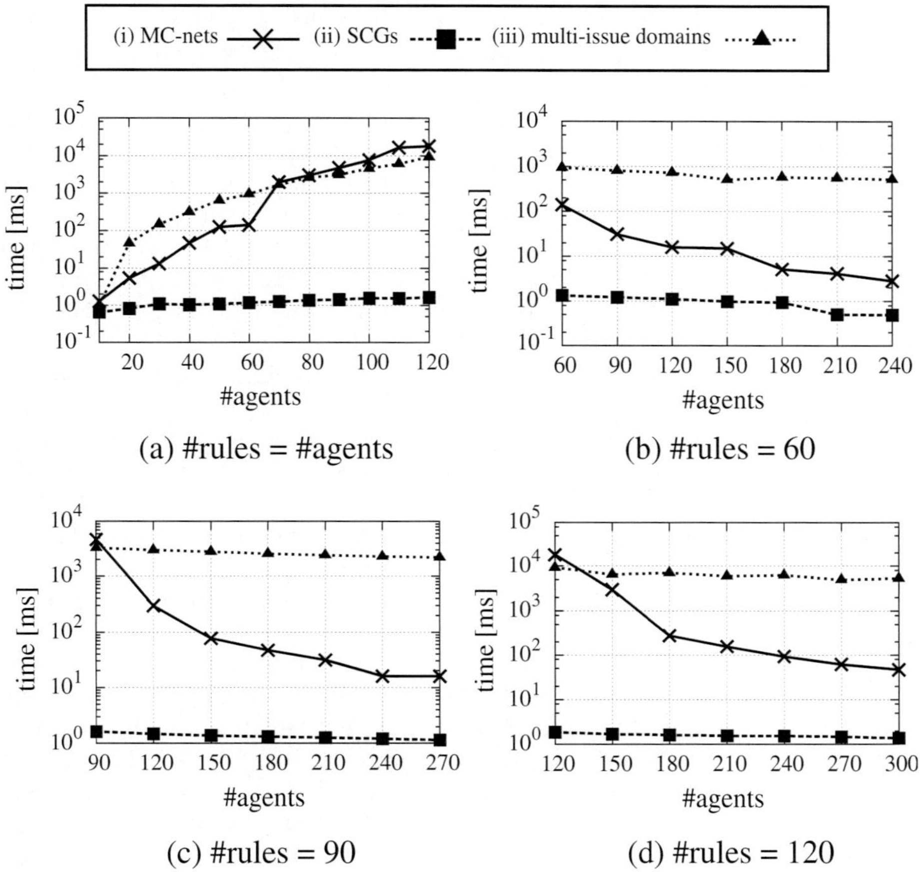

Fig. 2. Computation time for MC-nets, SCGs, SCGs in multi-issue domains

In Figure 2 (b), (c), (d), the CSG problem actually becomes easier when #agents increases. Since #rules, i.e., the number of vertices in the graph, is constant, the graph becomes more sparse by increasing #agents. As long as #rules is the same, case (ii) solves much faster than (i) and (iii). Also, as long as #rules is the same, case (i) and (iii) are about the same, except for the instances where #agents is large (Figure 2 (b), (c), (d)). This is because case (iii) has more constraints since it tends to have more "compatible on different coalitions" hyper-edges. We have tried several different settings and confirmed that the trends are basically similar.

The SCG representation has an advantage since the MIP formulation is simple and the resulting problem instances can be solved quite efficiently. Furthermore, we can leverage existing mature techniques for winner determination problems, including constraint-based approaches such as (Hoos and Boutilier, 2000). When using the MC-net or multi-issue representation, the limiting factor would be the number of edges (excluding "independent" edges) between rules, since we need to use auxiliary variables for representing connectivity constraints. However, in many cases, we can represent a characteristic function much more concisely by using MC-nets or multi-issue domains than by using SCGs.

As discussed in (Sandholm, 2002), specialized algorithms are usually more efficient than CPLEX for solving the winner determination problem. Thus, we can expect that specialized algorithms would be more efficient that CPLEX for solving the CSG problem. Here, we discuss several directions how specialized algorithms can be constructed. Certainly, we can use a depth-first branch and bound procedure. If we relax integer variables to continuous ones in the MIP formalization, we can obtain an admissible estimation. Of course, CPLEX performs a similar procedure, but we can use specialized graph-based heuristics for selecting nodes/rules. Furthermore, it would be possible that we have an efficient algorithm when a graph has some special structure (e.g., the graph is tree, or the graph can be divided into independent subgraphs by removing a small number of nodes). Also, if there exists no "incompatible" edge, then the problem is equivalent to the multi-cut problem (Vazirani, 2001). There exists an efficient approximation algorithm for the multi-cut problem (Vazirani, 2001). We can construct an algorithm that interleaves the selection among incompatible rules and the application of the approximate algorithm for the multi-cut problem.

Rahwan et al. (2007) reports that their IP algorithm can solve problem instances with 27 agents in less than 90 minutes. Also, they report that an extension of the IP algorithm that utilizes DP for preprocessing (Rahwan and Jennings, 2008a) can obtain four-fold speed-up compared to IP. We cannot directly compare our results with these results, since the formalizations of the CSG are different. Here, we are not comparing the efficiency of particular algorithms, but checking the scalability of different formalizations. Their algorithms inevitably evaluate all possible (2^n) coalitions. Thus, it is very unlikely that their approaches can scale up to $n = 100$. On the other hand, the advantage of these approaches is that they do not rely on particular representations.

7 Conclusion

We showed that coalition structure generation can scale up significantly when the characteristic function is represented using recently developed compact representation schemes: MC-nets, SCGs, and SCGs in multi-issue domains, even though we use an off-the-shelf optimization package. For each case, we proved that the problem is NP-hard and inapproximable and developed MIP formulations. Experimental results illustrated that while the state-of-the-art algorithm, which does not make use of compact representations, requires around 90 minutes to solve a problem with 27 agents, our methods can solve a problem with 120 agents and 120 rules in less than 20 seconds. Future work includes developing new algorithms (i) that can find an optimal solution more efficiently, (ii) that can return a suboptimal solution in any time, and (iii) that can find an approximate solution quickly, utilizing constraint optimization techniques.

Acknowledgments

The authors would like to thank Takayoshi Shoudai and Hirotaka Ono for their helpful comments on the earlier version of this paper. This work is partially supported by Japan Society for the Promotion of Science with Grant-in-Aid for Scientific Research (A) 20240015 and 20240003. Conitzer is supported by NSF award number IIS-0812113, a Research Fellowship from the Alfred P. Sloan Foundation, and a Yahoo! Faculty Research Grant.

Bibliography

Bachrach, Y., Rosenschein, J.S.: Coalitional skill games. In: Proceedings of the 7th international joint conference on Autonomous agents and multiagent systems (AAMAS), pp. 1023–1030 (2008)

Conitzer, V., Sandholm, T.: Computing Shapley values, manipulating value division schemes, and checking core membership in multi-issue domains. In: Proceedings of the 19th National Conference on Artificial Intelligence (AAAI), pp. 219–225 (2004)

Conitzer, V., Sandholm, T.: Complexity of constructing solutions in the core based on synergies among coalitions. Artificial Intelligence 170(6), 607–619 (2006)

Dang, V.D., Dash, R.K., Rogers, A., Jennings, N.R.: Overlapping coalition formation for efficient data fusion in multi-sensor networks. In: Proceedings of the 21st National Conference on Artificial Intelligence (AAAI), pp. 635–640 (2006)

Hoos, H.H., Boutilier, C.: Solving combinatorial auctions using stochastic local search. In: Proceedings of the 17th National Conference on Artificial Intelligence (AAAI), pp. 22–29 (2000)

Håstad, J.: Clique is hard to approximate within $n^{1-\epsilon}$. Acta Mathematica 182, 105–142 (1999)

Ieong, S., Shoham, Y.: Marginal contribution nets: a compact representation scheme for coalitional games. In: Proceedings of the 6th ACM Conference on Electronic Commerce (ACM EC), pp. 193–202 (2005)

Rahwan, T., Jennings, N.R.: Coalition structure generation: dynamic programming meets anytime optimisation. In: Proceedings of the 23rd Conference on Artificial Intelligence (AAAI), pp. 156–161 (2008)

Rahwan, T., Jennings, N.R.: An improved dynamic programming algorithm for coalition structure generation. In: Proceedings of the 7th International joint Conference on Autonomous Agents and Multi-agent Systems (AAMAS), pp. 1417–1420 (2008)

Rahwan, T., Ramchurn, S.D., Dang, V.D., Giovannucci, A., Jennings, N.R.: Anytime optimal coalition structure generation. In: Proceedings of the 22nd Conference on Artificial Intelligence (AAAI), pp. 1184–1190 (2007)

Rothkopf, M.H., Pekeč, A., Harstad, R.M.: Computationally manageable combinatorial auctions. Management Science 44(8), 1131–1147 (1998)

Sandholm, T.: Algorithm for optimal winner determination in combinatorial auctions. Artificial Intelligence 135(1-2), 1–54 (2002)

Sandholm, T., Lesser, V.R.: Coalitions among computationally bounded agents. Artificial Intelligence 94(1-2), 99–137 (1997)

Sandholm, T., Larson, K., Andersson, M., Shehory, O., Tohmé, F.: Coalition structure generation with worst case guarantees. Artificial Intelligence 111(1-2), 209–238 (1999)

Shehory, O., Kraus, S.: Methods for task allocation via agent coalition formation. Artificial Intelligence 101(1-2), 165–200 (1998)

Vazirani, V.V.: Approximation Algorithms. Springer, Heidelberg (2001)

Wooldridge, M., Dunne, P.E.: On the computational complexity of qualitative coalitional games. Artificial Intelligence 158(1), 27–73 (2004)

Wooldridge, M., Dunne, P.E.: On the computational complexity of coalitional resource games. Artificial Intelligence 170(10), 835–871 (2006)

Yeh, D.Y.: A dynamic programming approach to the complete set partitioning problem. BIT Numerical Mathematics 26(4), 467–474 (1986)

Yokoo, M., Conitzer, V., Sandholm, T., Ohta, N., Iwasaki, A.: Coalitional games in open anonymous environments. In: Proceedings of the 20th National Conference on Artificial Intelligence (AAAI), pp. 509–515 (2005)

Zuckerman, D.: Linear degree extractors and the inapproximability of max clique and chromatic number. Theory of Computing 3, 103–128 (2007)

Compiling All Possible Conflicts of a CSP*

Alexandre Papadopoulos and Barry O'Sullivan

Cork Constraint Computation Centre
Department of Computer Science, University College Cork, Ireland
{a.papadopoulos,b.osullivan}@4c.ucc.ie

Abstract. In interactive decision-making settings, such as product configuration, users are stating preferences, or foreground constraints, over a set of possible solutions, as defined by background constraints. When the foreground constraints introduce inconsistencies with the background constraints, we wish to find explanations that help the user converge to a solution. In order to provide satisfactory explanations, it can be useful to know one or several subsets of conflicting constraints; such a subset is called a conflict. When computing such conflicts is intractable in an interactive context, we can choose to compile the problem so as to allow faster response times. In this paper we propose a new representation, which implicitly encompasses all conflicts possibly introduced by a user's choices. We claim that it can help in situations where extra information about conflicts is needed, such as when explanations of inconsistency are required.

1 Introduction

We consider a configuration tool with which a user can specify preferences for options. These preferences are expressed as constraints. When preferences conflict, we want to help the user find which preferences to relax. In an iterative process, the user might relax constraints until at least one solution is found. Alternatively, the user might wish to be told which particular subsets of his constraints can be satisfied. Most current approaches to explanation generation in constraint-based settings are based on the notion of a minimal (with respect to inclusion) set of unsatisfiable constraints, known as a minimal conflict set of constraints. To demonstrate the concepts, we provide an example.

Example 1 (Car Configuration). Consider a simple car configuration problem, based on an example in [10], with the following set of options; the Boolean variable $x_i \in \{0,1\}$ indicates whether constraint c_i is in the current set of active constraints or not:

Constraint	Option	Selector	Cost
c_1	Budget	$x_1 = 1$	$\sum_{i\in\{2,\ldots,5\}}(k_i \cdot x_i) \leq 3000$
c_2	Roof Rack	$x_2 = 1$	$k_2 = 500$
c_3	Convertible	$x_3 = 1$	$k_3 = 500$
c_4	CD Player	$x_4 = 1$	$k_4 = 500$
c_5	Leather Seats	$x_5 = 1$	$k_5 = 2600$

Assume that the technical constraints of the configuration problem forbid convertible cars having roof racks, therefore, constraints c_2 and c_3 form a conflict. Note that,

* This work was supported by Science Foundation Ireland (Grant Number 05/IN/I886).

I.P. Gent (Ed.): CP 2009, LNCS 5732, pp. 639–653, 2009.

given the budget constraint, if the user selects option c_5, it is not possible to have any of the options c_2, c_3, c_4. The set of all minimal conflicts for this example are: $\{c_2, c_3\}, \{c_1, c_2, c_5\}, \{c_1, c_3, c_5\}$, and $\{c_1, c_4, c_5\}$. ▲

As explanations, these conflicts are sufficient to explain why all constraints cannot be satisfied simultaneously. Based on the set of minimal conflicts we can compute the set of set-wise maximal relaxations showing which of the user's constraints can be satisfied. Maximal relaxations show how the user can satisfy at least some of his constraints. For example, we can simultaneously satisfy the constraints in $\{c_3, c_4, c_5\}$, but we must exclude c_1 and c_2. The question that is posed then is which relaxation is best to choose. Several answers can be given to it, many of which require some information about all or some of the conflicts. However, computing all conflicts can be intractable in an interactive context, as their number can be very large. In fact, relaxations and conflicts have no direct relationship and the number of either can be exponential in the number of the other. In this paper, we propose to compute in advance all possible conflicts, before a user makes any choices, as a compilation step, thus providing precious information for finding explanations online.

2 Preliminaries

We focus on constraint satisfaction problems in this paper, but the results hold for many other settings in which consistency is monotonic. This property holds whenever the set of solutions to a set of constraints $\mathcal{C}$ is a subset of the solutions to any set of constraints that is a supersubset of $\mathcal{C}$. In addition, we focus on constraint satisfaction problems that are solved in an interactive manner, e.g. product configuration problems. It is useful to distinguish between a background set of constraints, $\mathcal{B}$, that cannot be relaxed, and a set of constraints, $\mathcal{U}$, that are added by the user as he finds a preferred solution to $\mathcal{B}$ by finding a solution to $\mathcal{B} \cup \mathcal{U}$, the constraint problem we denote as $\mathcal{P} \stackrel{\text{def}}{=} \langle \mathcal{B}, \mathcal{U} \rangle$. That way, the set $\mathcal{U}$ can be seen as a *query*. In our context, we will make the assumption that $\mathcal{U}$ contains only unary contraints.

A set of constraints is consistent if it admits a solution. We will assume that the set of background constraints, $\mathcal{B}$, admits at least one solution. If a set of constraints does not admit a solution, we can provide an *explanation* of the inconsistency, by showing a set of conflicting constraints or by excluding constraints in order to recover consistency.

Definition 1 (Minimal Conflict). *Given a constraint problem $\mathcal{P} \stackrel{\text{def}}{=} \langle \mathcal{B}, \mathcal{U} \rangle$ that is inconsistent, a subset C of $\mathcal{U}$ is a* conflict *of $\mathcal{P}$ if $\mathcal{B} \cup C$ is inconsistent. The conflict C is a* minimal conflict *if $\forall C' \subset C$, $\mathcal{B} \cup C'$ admits a solution. The minimal conflict C is a* shortest conflict *if for each other minimal conflict C', $|C| \leq |C'|$.*

Definition 2 (Maximal Relaxation). *Given a constraint problem $\mathcal{P} \stackrel{\text{def}}{=} \langle \mathcal{B}, \mathcal{U} \rangle$ that is inconsistent, a subset R of $\mathcal{U}$ is a* relaxation *of $\mathcal{P}$ if $\mathcal{B} \cup R$ admits a solution. The relaxation R is a* maximal relaxation *if $\forall R' \supset R$, $\mathcal{B} \cup R'$ is inconsistent. The maximal relaxation R is a* longest relaxation *if for each other maximal relaxation R', $|R'| \leq |R|$.*

In some contexts, it might be more convenient to consider the following dual concept.

Definition 3 (Minimal Exclusion Set). *Given a constraint problem $\mathcal{P} \stackrel{\text{def}}{=} \langle \mathcal{B}, \mathcal{U} \rangle$ that is inconsistent, a subset E of $\mathcal{U}$ is an exclusion set (resp. minimal exclusion, shortest exclusion set) of $\mathcal{P}$ if $\mathcal{U} \backslash E$ is a relaxation (resp. maximal relaxation, longest relaxation).*

There is a duality between minimal conflicts and maximal relaxations. Specifically, a minimal conflict is a minimal hitting set of all minimal exclusion sets (i.e. the complement of a maximal relaxation), and a minimal exclusion set is a minimal hitting set of all the minimal conflicts.

In the context we consider, interactivity, implying quick response times, is a key requirements. On the other hand, as a well-known fact, explanation generation can be a highly intractable task in theory and in practice. Therefore, a standard strategy is to employ *compilation* techniques. The key principle is to shift ahead of time, during an *offline phase* (that is before any user interaction) most expensive computations that need to be performed only once, to allow for operations needed during an *online* session to be much faster in practice.

3 Prime Implicates and Explanations

In this section, we recall some standard notions and notations, along with some main results from the literature. Let Φ be a finite set of clauses in Conjunctive Normal Form (CNF) (or any NNF for the purposes of the following definitions), and let C be a disjunction of literals (a *clause*).

Definition 4 (Implicate). *Let C and C' be two clauses.*

- *C is an* implicate *of Φ if $\Phi \models C$.*
- *C* subsumes *C' iff $C \subseteq C'$ iff $C \models C'$.*

Usually we only consider non-trivial implicates, i.e. implicates that are not a tautology. A prime implicate is then defined as follows.

Definition 5 (Prime Implicate). *C is a* prime implicate *of Φ iff:*

- *C is an implicate of Φ and*
- *$\forall C'$ which is an implicate of Φ, $C' \models C \Rightarrow C \models C'$.*
- *Equivalently, $\forall C' \subset C$, $\Phi \not\models C'$.*

$PI(\Phi)$ is the set of all the prime implicates of Φ.

Prime implicates, variants of them, and their generation, are used, amongst other things, for consequence finding (see [14] for a survey). Particularly, they are of great use in the context of explanations. Let us adapt the previously introduced notations for the boolean case as follows. The background constraint are given by a formula Φ, and a user query is a term γ, i.e. a conjunction of literals. The literals involved in γ are denoted by $lit(\gamma)$. Suppose the query is inconsistent, i.e. $\Phi \wedge \gamma \models \bot$. Now, let $C \in PI(\Phi)$ be a prime implicate such that $lit(\neg C) \subseteq \gamma$. Of course, $\neg C$ is inconsistent with Φ. Because C is a prime implicate, $\neg C$ is also a set-wise minimal subset of the literals of γ that is inconsistent with Φ, and so corresponds to a minimal conflict of the query γ.

Example 2. Let Φ be a CNF of the form $(\neg a \vee b) \wedge (a \vee d) \wedge (\ldots)$. $b \vee d$ is a prime implicate. That tells us that if we want $\neg b \wedge c \wedge \neg d \wedge (\ldots)$, it is inconsistent, and a minimal conflict for that query is $\neg b \wedge \neg d$. ▲

Concerning the computation of prime implicates, most existing algorithms rely on repeatedly resolving pairs of clauses. Consider two clauses of the form $x \vee C$ and $\neg x \vee C'$. Then the resolvent $C \vee C'$ is another implicate, which has been obtained by resolving the two clauses on x. If there exists another variable y such that $y \in lit(C)$ and $\neg y \in lit(C')$, then $C \vee C'$ is a tautology, and should not be kept. This is the basis of the Tison method [19]. Note that the method described in [18] is the first one to claim a scalable method for representing and computing the prime implicates of a problem.

Prime implicates present an interest also from a compilation viewpoint. The prime implicates of a given CNF, provide a classic technique to represent this formula in a canonical or minimal way. As such, it can be seen also as a compilation technique. The Compilation Map [3] presents an extensive survey of different compiled representations of boolean formulas. A particular representation type is referred to as a *language*. The **NNF** language is the most general, containing all formulas in negation normal form. Any other language is a subset of **NNF**. Amongst all these languages, the one holding our attention in this context is the **PI** language, characterised as follows:

Definition 6 (PI language). *An NNF Φ belongs to **PI** iff $\Phi = PI(\Phi)$.*

While different languages can be more or less compact, they may or may not support different *queries*. A language is said to *support* a query if, for any formula in this language, there exists an algorithm that can answer this query in polynomial time. For example, counting the number of solutions is not possible in the general case, but is polynomial in the size of a binary decision diagrams. In [17], we defined queries relevant for explanations. **SES** is the query that asks for the size of the shortest exclusion set of a given user query, **MR** is the query that asks for the set of maximal relaxations of a given user query. We can add now the query **SC** that asks for the size of the shortest conflict of a given query. We saw that **PI** supports **SC**. We also have the two following results (answering two questions that were left open in [17]).

Proposition 1. ***PI** supports **MR**.*

Proof. Given a query γ inconsistent with Φ, we first compute the minimal conflicts min-conflicts$(\Phi, \gamma) = \{C \in \Phi / lit(\neg C) \subseteq lit(\gamma)\}$ then compute all the minimal hitting sets of the minimal conflicts $mr(\Phi, \gamma) = \{\gamma \setminus e / e$ is a min hitting set of min-conflicts$(\Phi, \gamma)\}$. Obviously the size of each of these two sets (as well as the time to compute them) is linear in the size of Φ times the number of maximal relaxations. ■

On the other hand, **PI** does not support **SES**. Consider the following result.

Proposition 2. *Let $S = \{S_1, \ldots, S_n\}, S_i \subseteq T$ be a collection of subsets of some set T, let $\Phi(S)$ be the CNF associated with it, where each literal corresponds to an element of T and each clause corresponds to a set of S. Then $\Phi(S) \in$ **PI**.*

Proof. A solution of $\Phi(S)$ corresponds to a hitting set of S. By definition, a (resp. minimal) clause satisfied by every solution of $\Phi(S)$ is a (resp. prime) implicate. A minimal clause satisfied by every solution of $\Phi(S)$ corresponds to a minimal hitting set of all the hitting sets of S. Thus a minimal hitting set of all the hitting sets of S corresponds to a prime implicate of $\Phi(S)$. But the minimal hittings of the hittings sets of S are precisely the sets in S themselves. Thus the clauses of $\Phi(S)$ are all and the only prime implicates of $\Phi(S)$. ■

Proposition 3. ***PI*** *does not support* ***SES*** *unless $P = NP$.*

Proof. Let $S = \{S_1, \ldots, S_n\}, S_i \subseteq T$ be an collection of subsets of T. Consider the sentence $\Phi(S) \in \mathbf{PI}$, and the query $\gamma = \bigwedge_{v \in T} \neg x_v$, where x_v is the literal corresponding to the element $v \in T$. Clearly $\Phi(S) \wedge \gamma$ is inconsistent. A shortest exclusion set gives a smallest subset of literals we must set to true in order to satisfy $\Phi(S)$, *i.e.* a smallest subset of elements T which is a hitting set of S. In other words, the size of the shortest exclusion set of $\Phi \wedge \gamma$ is the size of the smallest hitting set of S, which is NP-Hard to compute. ■

Having given some insights into the application of prime implicates to explanations, the remainder of this paper will therefore deal with generalising these concepts, that is, defining the **PI** language for constraint problems.

4 Domain Consequences

Suppose we have a problem defined on a set of variables $X_1, \ldots, X_n$, taking their values from the domains $D(X_1), \ldots, D(X_n)$. We are interested in simple preferences consisting of domain restrictions, expressed as unary constraints of the form $X_i \in D_i$, with $D_i \subseteq D(X_i)$, holding on a subset of the variables of the problem. Given an inconsistent query, giving one or all or some, according to some criteria, minimal conflicts can be a highly intractable problem, and compilation with any of the classic techniques (automata, BDDs, dDNNF) does not help either. It is, therefore, interesting to have a way to compute in advance all the potential minimal conflicts of a problem, or, to reuse the terminology of the literature, to infer all the consequences of the given problem.

In the lack of any particular query, as we are working in an offline phase, we must define a new, mathematically richer, notion of minimal conflict, which we will call *domain conflicts*.

Definition 7 (Domain Conflict). *The notion of domain conflict is defined as follows:*

- A domain conflict *for given problem is given by the sequence of domains $C = \langle D_1, \ldots, D_n \rangle$, such that imposing $X_i \in D_i$ for each X_i is inconsistent. Such a conflict can be seen as a conjunction of unary constraints, which is inconsistent.*
- *Given two domain conflicts $C_1 = \langle D_1, \ldots, D_n \rangle$ and $C_2 = \langle D'_1, \ldots, D'_n \rangle$, we note that $C_1 \subseteq C_2$ if $\forall X_i$, $D_i \subseteq D'_i$.*
- A maximal *domain conflict is a domain conflict C such that no domain conflict $C' \neq C$ exists with $C \subseteq C'$. In other words, every component D_i of C is maximal.*

Another way of seeing a domain conflict is by considering it defines the Cartesian product $\times\, i \leq nD_i$. This product contains all the tuples that this domain conflict recognises as being inconsistent. However, defining this list of inconsistent tuples as the elements of a given Cartesian product is far more informative than giving an explicit list. Indeed, a sequence of sets tells that *all* the tuples that can be obtained from their product are conflicts of the problem. One implication of that observation is that a domain conflict allows us to take into account conflicts resulting from unary constraints, rather than simply variable assignments.

In this regard, we can observe that the notion of maximal domain conflict recovers the classic one of minimal conflict, in the sense that for a given i, having $D_i = D(X_i)$ is equivalent to having no constraint at all on X_i. Thus, the more values that are in the D_i sets, the fewer constraints there are on the corresponding X_i variables.

As to why this definition can be regarded as being "richer" than the classic one, consider the following example. Suppose that $\langle\{a, b, c\}, \{a, b\}, \{b, c\}\rangle$ is a minimal domain conflict (with the domain of each variable being $\{a, b, c\}$). If the user asks $X_1 \in \{a, b\}$, $X_2 \in \{a\}$, $X_3 \in \{b\}$, it is inconsistent, and a minimal conflict is c_2c_3 (c_i being the constraint holding on X_i). However, seen as a domain conflict, this conflict is not minimal. For example, it does not tell the user that $X_2 \in \{a, b\}$, $X_3 \in \{b, c\}$ is also inconsistent. Also, we can see that a single maximal domain conflict can cover many potential minimal conflicts resulting from a later user query. Conversely, a minimal conflict can be recognised by more than a single maximal domain conflict, thus the potential minimal conflicts covered by different maximal domain conflicts may overlap. Finally, let us point out that this notion is in its definition very close to the one of generalized nogood [12] or global cut seed [9].

With that definition in mind, we can define the consequence of a given problem, thus generalising the concept of prime implicate.

Definition 8 (Domain Consequence). *A domain consequence is defined as follows:*

- *A* domain consequence *of a problem is given by $P = \langle D_1, \ldots, D_n\rangle$ such that $\langle\overline{D_1}, \ldots, \overline{D_n}\rangle$, with $\overline{D_i} = D(X_i) \setminus D_i$, is a domain conflict.*
- *Given two domain consequences P and P', whose corresponding domain conflicts are C and C', we have $P \subseteq P'$ if $C' \subseteq C$. To reuse classic terminology, we may say that P* subsumes *P'.*
- *A domain consequence P is* minimal *if its corresponding domain conflict is maximal. In other words, no domain consequence $P' \neq P$ exists such that $P' \subseteq P$.*

A given consequence of the problem is read as a disjunction. In other words, for any solution of the problem, it must be true that $X_1 \in D_1$ or $X_2 \in D_2$ or... and so on.

Definition 9 (Set of domain consequences). *Let Π be a problem. $Cons(\Pi)$ is the set of all the minimal domain consequences of Π, that is if P is a domain consequence of Π, then $\exists P' \in Cons(\Pi)$ such that $P' \subseteq P$, and $\forall P, P' \in Cons(\Pi)$, $P \subseteq P' \Rightarrow P = P'$.*

A set of domain consequences must be seen as a conjunction, and so $Cons(\Pi) \equiv \Pi$. The strategy will thus be, given a problem, to compute all of its consequences, as efficiently as possible, and represent them as compactly as possible. In other words, we intend to compile a problem to a new representation, in a manner orthogonal to the existing approaches like automata, that supports queries related to conflicts (shortest conflict, minimal conflict, specific subset of all the conflicts with completeness guarantee). By representing a problem by minimal consequences that *cover* all potential consequences (or, equivalently, all potential conflicts), we indeed soundly and completely characterise this problem. However, since this paper opens a new direction in compilation, our main focus is to introduce the concepts and the approaches we propose. How the consequences of a problem can be indeed compactly represented, and how we can operate on this representation to compute the consequences or to perform the subsequent online queries is not addressed in this paper.

Nevertheless, we can still give an idea of how domain consequences can help answer conflict-related queries. Suppose we are given the user query $\mathcal{U}$, where each constraint $c \in \mathcal{U}$ is of the form $X \in D$, with $D \subseteq D(X)$. We additionally assume that every unary constraint holds on a distinct variable. From that query, we can build the domain sequence $S = \langle D_1, \ldots, D_n \rangle$, where $D_i = D$ if there is a constraint $X_i \in D$ in $\mathcal{U}$, or $D_i = D(X_i)$ otherwise. To see whether $\mathcal{U}$ is inconsistent, we simply have to check if S is a domain conflict, by checking if its associated domain consequence is subsumed by a minimal domain consequence of the problem. Let $C = \langle D'_1, \ldots, D'_n \rangle$ be a maximal domain conflict such that S is included in it. The set $\{c \in \mathcal{U}/c$ holds on $X_i \wedge D'_i \neq D(X_i)\}$ is a minimal conflict of the query. If we want a minimum conflict of the query, we only have to iterate over all maximal domain conflicts and pick the one the corresponding minimal conflict of which is of minimal cardinality. Additionally, when looking for specific subsets, i.e. with a given property, of all minimal conflicts of a query (like we did in [16]), reasoning on the set of minimal domain consequences allows us to make efficient consistency checks as well as guarantee the desired property, without generating online all the minimal conflicts of a specific user query.

5 Computation

5.1 Generation

Similar to the way implicates can be deduced from other prime implicates, new domain consequences can be deduced from existing consequences.

Proposition 4. *Let* $P_1 = \langle D_1, \ldots, D_n \rangle$ *and* $P_2 = \langle D'_1, \ldots, D'_n \rangle$ *be two domain consequences of a given problem. Then, for any* X_i, $P = \langle D_1 \cup D'_1, \ldots, D_i \cap D'_i, \ldots, D_n \cup D'_n \rangle$ *is also a domain consequence of the problem. We denote* $P = res_{X_i}(P_1, P_2)$.

Proof. Let us fix, without loss of generality, $i = 1$. P_1 tells us that $X_1 \notin D_1 \Rightarrow X_2 \in D_2 \vee \ldots \vee X_n \in D_n$. Similarly, $X_1 \notin D'_1 \Rightarrow X_2 \in D'_2 \vee \ldots \vee X_n \in D'_n$. In either case, we have $(X_1 \notin D_1 \vee X_1 \notin D'_1) \Rightarrow (X_2 \in D_2 \cup D'_2 \vee \ldots \vee X_n \in D'_n \cup D_n)$, i.e. $(X_1 \in D_1 \cap D_2) \vee (X_2 \in D_2 \cup D'_2) \vee \ldots \vee (X_n \in D'_n \cup D_n)$. ■

We can say, reusing the existing terminology, that P has been inferred on X_i from P_1 and P_2. Note that if $D_i \cap D'_i$ is empty, the consequence becomes independent from the variable X_i, thus retrieving the classical notion of resolvent in the boolean setting.

Some consequences are trivial and therefore useless:

- D_i and D'_i have to be incomparable; if one is included in the other, say $D_i \subseteq D'_i$, the obtained consequence will necessarily be non-minimal, as the one containing D_i will be included in it. For example, suppose we have $\langle ab, ab, c\rangle$ and $\langle a, a, cd\rangle$, we can infer on X_1 the consequence $\langle a, ab, cd\rangle$, which will not be minimal as $\langle a, a, cd\rangle$ is included in it. This condition is equivalent to the fact that the resolvent of two clauses exists only when one contains some literal X and the other $\neg X$.
- if some $D_j \cup D'_j$, for any $j \leq n$, contains all values allowed for X_j, the consequence is always true. For example, suppose we have $\langle ab, ab\rangle$ and $\langle bc, bc\rangle$, with $D_1 = D_2 = abc$, we can infer on X_1 the consequence $\langle b, abc\rangle$, which is trivially true. This condition is equivalent to the fact that two clauses that are resolved on some literal and that also contain another literal in both the negated and non-negated form will result in a trivial clause.

The following example will help illustrate the concepts of conflicts, consequences, minimal consequences, resolution, as well as the representation abilities of these concepts.

Example 3. Suppose we have a constraint on three variables with domain $\{a, b, c\}$, given by the following nogoods. We can notice that a nogood corresponds to a trivial domain conflict. For example, knowing that aaa is forbidden is equivalent to saying that $X \neq a \vee Y \neq a \vee Y \neq a$. The domain consequence thus corresponding to each nogood is given in the second column.

nogood	domain consequence
aaa	$\langle bc, bc, bc\rangle$
aac	$\langle bc, bc, ab\rangle$
abb	$\langle bc, ac, ac\rangle$
abc	$\langle bc, ac, ab\rangle$
bab	$\langle ac, bc, ac\rangle$
bbb	$\langle ac, ac, ac\rangle$

From this set, different new consequences can be inferred, thus resulting in the following set of minimal domain consequences for that constraint, along with their corresponding domain conflict.

minimal consequence	maximal conflict
$\langle c, ac, ac\rangle$	$\langle ab, b, b\rangle$
$\langle bc, c, ab\rangle$	$\langle a, ab, c\rangle$
$\langle ac, c, ac\rangle$	$\langle b, ab, b\rangle$
$\langle bc, bc, b\rangle$	$\langle a, a, ac\rangle$
$\langle bc, ac, a\rangle$	$\langle a, b, bc\rangle$

The consequences that we end up with somehow "factorise" common parts among existing consequences. For example, the consequence $\langle bc, ac, a\rangle$ has been obtained from the two initial consequences $\langle bc, ac, ac\rangle$ and $\langle bc, ac, ab\rangle$ (on the variable X_3). In terms of conflicts, we obtain the conflict $\langle a, b, bc\rangle$, which factorises the common prefix of the two conflicts abb and abc. That is, of course, very similar to what automata perform, with some differences. First, automata merge only common prefixes, while a conflict resulting from two other conflicts might "factorise" the common part of an arbitrary subset of the variables (and thus is independent of any order on the variables). But most importantly, automata are merely just a compact representation of a list of tuples. As previously stressed, this does not allow us to easily answer queries consisting of domain restrictions. Consider a problem where the forbidden tuples are aaa, aab, baa. By inferring domain conflicts, we eventually obtain the two maximal domain conflicts $\langle a, a, ab\rangle$ and $\langle ab, a, a\rangle$. Suppose we required the three variables to be assigned the value a. While an automaton representing these nogoods would easily determine that the query is inconsistent no matter what value we assign to X_3, it would not easily show that the query is also inconsistent regardless of the value we assign to X_1. In either case, it is also not obvious that these two subconflicts are minimal. ▲

Having considered the production rules, we can now present how consequences can be initialised. The basic idea is to start by representing each constraint as the set of its minimal domain consequences. We already saw that trivial domain consequences can be initialised from the nogoods of a constraint. However, most problems that are to be compiled, especially in the context we will eventually address, define constraints as their lists of valid tuples, and it would be intractable to first generate all the corresponding nogoods. Our working assumption will therefore be that we are given the valid tuples of each constraint.

Definition 10. *Let $\mathcal{P}_1, \mathcal{P}_2$ be two sets of minimal domain consequences, $\mathcal{P}_1 \cup \mathcal{P}_2 = \mu\{P_1 \cup P_2 / P_1 \in \mathcal{P}_1, P_2 \in \mathcal{P}_2\}$, where $\mu\mathcal{P}$ is the subset of $\mathcal{P}$ where subsumed elements have been removed.*

It is quite easy to see that $\mathcal{P}_1 \cup \mathcal{P}_2$ contains the minimal domain consequences equivalent to the logical disjunction between $\mathcal{P}_1$ and $\mathcal{P}_2$. The following proposition shows how minimal consequences are built by iteratively taking into account the tuples of a constraint. Slightly abusing notation, for a list of tuples T, we use $Cons(T)$ to denote $Cons(C)$ (which in turn is an abuse of notation), where C is the constraint defined by the valid tuples T.

Definition 11. *Let $t = a_1 \ldots a_n$ be a tuple defined on the variables $X_1, \ldots, X_n$. For each a_i, we define the consequence $P(a_i) = \langle D_1, \ldots, D_n\rangle$ as follows:*

$$D_j = \begin{cases} \{a_i\}, & \text{if } i = j \\ \emptyset, & \text{otherwise.} \end{cases}$$

We thus associate with the tuple t the set of domain consequences $\mathcal{P}(t) = \{P(a_i), i \leq n\}$.

Proposition 5. *For a list T of tuples, we have:*

$$Cons(T) = \begin{cases} \emptyset, & \text{if } T = \emptyset \\ \mu(Cons(T \setminus \{t\}) \cup \mathcal{P}(t)), & \text{otherwise.} \end{cases}$$

Example 4. Consider a constraint with two valid tuples 012 and 021. We have $Cons(012) = \mathcal{P}(012) = \langle\{0\},\emptyset,\emptyset\rangle,\langle\emptyset,\{1\},\emptyset\rangle,\langle\emptyset,\emptyset,\{2\}\rangle$, $Cons(012) \cup \mathcal{P}(021) = \langle\{0\},\emptyset,\emptyset\rangle$, $\langle\{0\},\{2\},\emptyset\rangle$, $\langle\{0\},\emptyset,\{1\}\rangle$, $\langle\{0\},\{1\},\emptyset\rangle$, $\langle\emptyset,\{1,2\},\emptyset\rangle$, $\langle\emptyset,\{1\},\{1\}\rangle$, $\langle\{0\},\emptyset,\{2\}\rangle$, $\langle\emptyset,\{2\},\{2\}\rangle$, $\langle\emptyset,\emptyset,\{1,2\}\rangle$, and finally $Cons(012,021) = \langle\{0\},\emptyset,\emptyset\rangle$, $\langle\emptyset,\{1,2\},\emptyset\rangle$, $\langle\emptyset,\{1\},\{1\}\rangle$, $\langle\emptyset,\{2\},\{2\}\rangle$, $\langle\emptyset,\emptyset,\{1,2\}\rangle$. Although the size of the final set is greater than the number of tuples, it is smaller than the number of nogoods (i.e. 25), and so is the size of the intermediate set, that is before removal of non-minimal domain consequences. ▲

5.2 Algorithm

We present an algorithm for compiling a CSP as its set of all minimal domain consequences, as Algorithm 1. In order to establish the correctness of the algorithm, let us first introduce some notation and results.

Definition 12. *Let $\mathcal{P}$ be a set of domain consequences, and X_i be a variable. P is a domain consequence of $\mathcal{P}$ on X_i if $\exists P_1, P_2$ s.t. $P = res_{X_i}(P_1, P_2)$, where P_1 and P_2 either belong to $\mathcal{P}$ or are themselves consequences of $\mathcal{P}$ on X_i.*

Definition 13. *Let $\mathcal{P}$ be a set of domain consequences, and X_i be a variable. $Res_{X_i}(\mathcal{P})$ is the set containing all the minimal domain consequences that can be inferred on X_i from $\mathcal{P}$. Formally, $\forall P$ that is a domain consequence of $\mathcal{P}$ on X_i, $\exists P' \in Res_{X_i}(\mathcal{P})$ such that $P' \subseteq P$, and $\forall P, P' \in Res_{X_i}(\mathcal{P})$, $P \subseteq P' \Rightarrow P = P'$.*

Example 5. Let $\mathcal{P} = \{\langle ab, bc\rangle, \langle ac, bc\rangle, \langle bc, bc\rangle\}$. $\langle ab, bc\rangle$, $\langle ac, bc\rangle$, $\langle bc, bc\rangle$, $\langle a, bc\rangle$, $\langle b, bc\rangle$, $\langle c, bc\rangle$ and $\langle\emptyset, bc\rangle$ are all domain consequences of $\mathcal{P}$ on X_1, and $Res_{X_1}(\mathcal{P}) = \{\langle\emptyset, bc\rangle\}$. This tells us that given $\mathcal{P}$, we can infer that a is forbidden for X_2. ▲

Definition 14. *Let $\mathcal{P}$ be a set of domain consequences, and $[X_1, \ldots, X_k]$ be a sequence of variables. $Res_{[X_1,\ldots,X_k]}(\mathcal{P}) = Res_{X_k} \circ \cdots \circ Res_{X_2} \circ Res_{X_1}(\mathcal{P})$.*

A very well-know result used in most algorithms for PI generation (as first established in [13]) can be formulated in our context as follows:

Proposition 6. *Let $\mathcal{P}$ be a set of domain consequences, X_i and X_j two variables. Then $Res_{[X_i,X_j,X_i]}(\mathcal{P}) = Res_{[X_i,X_j]}(\mathcal{P})$.*

In other words, it is useless for new consequences inferred on some variable to be tested again with another consequence on an already considered variable. This is a direct consequence of the associativity and the commutativity of the *res* operation. Indeed, $res_{X_j}(P_3, res_{X_i}(P_1, P_2)) = res_{X_i}(P_2, res_{X_j}(P_1, P_3))$. This result is of huge computational importance, as, by fixing in advance an order on the variables, a given consequence will be discovered in a unique way.

Example 6. Let $\mathcal{C} = \{\langle a,\emptyset,a\rangle, \langle b,a,\emptyset\rangle, \langle\emptyset,b,b\rangle\}$, $Res_{X_1}(\mathcal{C}) = \mathcal{C} \cup \{\langle\emptyset,a,a\rangle\}$, and $Res_{X_2}(\mathcal{C}) = Res_{X_1}(\mathcal{C}) \cup \{\langle b,\emptyset,b\rangle, \langle\emptyset,\emptyset,ab\rangle\}$. Take for example $\langle b,\emptyset,b\rangle$, it does not need to be resolved again on X_1 with any other consequence, as it will inevitably give an element already in $Res_{X_2}(\mathcal{C})$. ▲

Algorithm 1. CONSEQUENCECOMPILATION

```
    Data: A problem Π
    Result: N = Cons(Π)
1  N ← ∅
2  foreach C_i ∈ Π do
3      N ← N ∪ Cons(C_i)
4  foreach X_i ∈ Π do
5      S ← N
6      N ← ∅
7      while S ≠ ∅ do
8          D ← first(S)
9          foreach D′ ∈ N do
10             if non-trivial(X_i, D, D′) then
11                 D* ← resolve(X_i, D, D′)
12                 if No element of N ∪ S subsumes D* then
13                     Remove from N ∪ S any element subsumed by D*
14                     Add D* to the end of S
15         if D is still in S then Remove D from S and add to N
16 return N
17 function non-trivial(X_i, ⟨D_1, ..., D_n⟩, ⟨D′_1, ..., D′_n⟩)
18     if D_i ⊆ D′_i ∨ D′_i ⊆ D_i then
19         return false
20     if ∃X_j ≠ X_i s.t. D_j ∪ D′_j = D(X_j) then
21         return false
22     return true
```

Corollary 1. *We have* $Cons(\Pi) = Res_{[X_1,\ldots,X_n]}(\mathcal{P}_0)$ *where* $\mathcal{P}_0$ *is the initial set of domain consequences corresponding to each constraint.*

Proposition 7. *Algorithm 1 computes* $Cons(\Pi)$.

Proof. Lines 6-16 basically perform the Res_{X_i} operation, where S contains the initial domain consequences. During each iteration, the invariant that is maintained is that $N \cup S$ contains only minimal consequences. Every pair of elements is tested exactly once, and for each new inferred consequence, it is added to $N \cup S$ only if it is currently minimal. It is added specifically to S so that it can be in turn tested against other consequences. When a domain consequence D in S is processed, it is resolved against all known minimal consequences contained in N. If D is not subsumed by any new consequence thus inferred, it is added to N. Therefore, when, in the end, S is empty, N contains exactly $Res_{X_i}(S)$. ■

Concerning implementation details, each domain consequence has been implemented using a single bitset, which allows very efficient operations, like the `non-trivial` test, the subsumption test, and the *resolve* operation, which are by far the most frequent operations in this algorithm. Also, some simple optimisations can be performed when computing the disjunction of two sets of domain consequences, while generating the domain consequences of each constraint (line 3). For example, when the union P of two domain consequences is equal to one of them, then this original domain consequence can be discarded, as any subsequently generated consequence will be subsumed by P, as it can be observed in Example 4.

5.3 Complexity

The complexity of Algorithm 1 is determined by the size of $Cons(\Pi)$. We saw that introducing a new valid tuple to the domain consequences of a constraint of arity k can add up to k times new domain consequences. Therefore, we have an upper bound on $|Cons(\Pi)|$ of $n^{|S|}$, where $|S|$ is the number of solutions of the problem. $|S|$ itself is in $O(d^n)$, which gives an idea of the potential number of domain consequences. Additionally, the number of intermediate domain consequences that have been generated during the procedure can be far larger than the final number (as most become eventually subsumed by fewer domain consequences), as we shall see in the next section. These simple facts show that there is very little hope that an efficient way to generate explicitly all the consequences exists, no matter what algorithm we design.

6 Experimental Study

We show how the concepts we introduced are put into practice on some very simple instances. The purpose is not to test the validity of the algorithm, whose implementation has been kept intentionally almost as naive as possible, for the reasons stated in the last section. Indeed, we do not expect that, in the present stage, this algorithm can scale up to any other than trivial problems.

We tested the algorithm on random uniform constraint networks of binary constraints[1], using the seed 100. These instances contained 10 variables, with 10 values, 10 constraints. The tightness of each constraint varied from 1 to 19 allowed tuples, then from 95 to 99. We expected our approach to work better on problems with many conflicts. In Figure 1, we show the average number of consequences produced after the resolution of each variable, as well as the final total. This average number was in direct correspondence with the overall number of iterations. This shows that the tighter the constraints are, the fewer iterations it takes for the algorithm to finish. Not surprisingly, no instance with a tightness outside the extreme values could be resolved in any reasonable amount of time. Note that for the harder instances (1 to 19 allowed tuples per constraint), the final number of consequences is only 1, as they were unsatisfiable. On satisfiable instances, this number ranged from 250000 to 120. It is particularly striking, especially on the unsatisfiable instances, that most of the consequences discovered are not minimal in the end, as they end up being subsumed by the empty consequence.

To further illustrate this point, we show some more detailed figures on two instances, an unsatisfiable instance with 17 allowed tuples per constraint, and a satisfiable one

[1] We used the generator at: `http://www.lirmm.fr/~bessiere/generator.html`

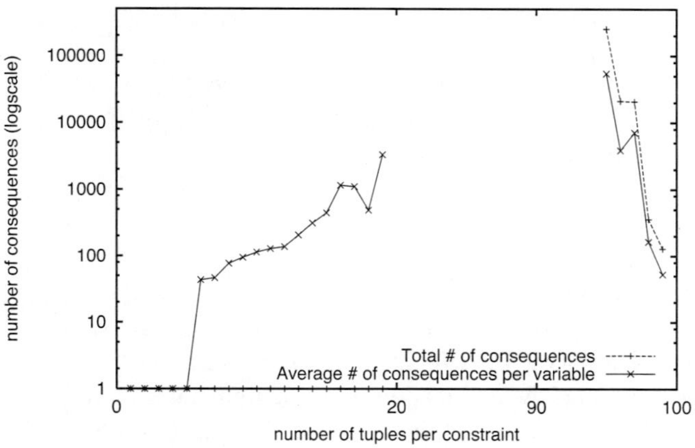

Fig. 1. The total number of consequences and the average for each variable per instance

Table 1. The intermediate number of consequences generated

Variable	17 tuples			95 tuples		
	generated	non-subsumed	variation	generated	non-subsumed	variation
0	239179	6227	730	423	85	85
1	26843	1027	105	0	0	0
2	88841	4894	85	266531	2894	2894
3	16102	609	145	377234	2882	2056
4	397594	21603	-1545	6027194	13708	13708
5	259702	7844	481	13674500	29420	8121
6	58563	2987	-980	51688417	40006	40006
7	1371	225	-263	70889441	50362	18689
8	18	5	-22	63317577	0	0
9	0	0	0	1009075634	263964	165954

with 95. Table 1 shows for each instance and at the end of each basic iteration (i.e. the resolution on each variable) the number of generated consequences (i.e. that passed the `non-trivial` test), the number of non-subsumed consequences (i.e. the part of those that have been successfully added to S), and eventually the final variation of the current number of minimal consequences (i.e. the difference in size between N at the end and at the beginning of the iteration).

These results are encouraging. They show that the size of the final set we aim to produce is tractable on those instances. With an adequately compact representation of the set of minimal consequences, we can manage very satisfactory sizes. As a comparison, it is a notable fact that automata can represent very efficiently a number of solutions several orders of magnitude higher than the figures shown here. The practical ability of such a representation to represent effectively a large number of potential conflicts will depend a lot on different parameters of a problem. These could be parameters like the number of actual conflicts for a user query, the size of such conflicts, value interchangeability, etc. That is of course a fundamental point that will need to be studied in further work. We conjecture that the properties yielding compact consequence compilation will be different from those needed for an automaton representation, where structural considerations (the topological properties of the constraint graph) and variable orderings are crucial. Finally, let us stress again that this work is a "proof of concept", and that these aspects have not yet been dealt with.

7 Related Work

Work on consequence-finding has been an important task in Artificial Intelligence for the past thirty years, with a particular interest for diagnosis, which closely relates to our setting. A very complete survey of the field and its applications has been presented in [14]. Additionally, [5] presents the application of prime implicates to diagnoses. One of the foundational works about prime implicates generation dates from 1967 [19]. Kernel resolution [7, 8] generalises this idea. In 1992, De Kleer proposed a practical implementation, where clause bases are stored using TRIE structures [4]. Several papers present the approach of using prime implicate generation as a compilation technique. In 1995, Marquis introduced the concept of theory prime implicates [15], that allows to reduce the number of generated implicates by defining a stronger notion of entailment. In 1994, del Val introduced the concept of tractable databases [6], which also allows to reduce the number of generated implicates, by only selecting a subset that is refutation complete (a set of clauses is said refutation complete if any clause can be tested to be an implicate only by unit resolution deduction). Both papers show encouraging experimental results. Another approach is to work on the representation of all the prime implicates. The TRIE representation [4] and clause resolution can be extended to a ZBDD representation and multi-resolution [1, 18]. Alternatively, [2] proposes an implicit representation based on meta-products. In particular, the main contribution of [18], is an extensive experimental study; it seems to be the first method for full prime implicates generation that scales to instances of real practical interest. Finally, [11] presents a similar approach to ours (though the concepts are different) in using a pre-analysis of the problem as a compilation technique, in order to facilitate conflict and diagnosis computation.

8 Conclusions

We presented a new approach for compiling a problem into a form that is particularly well-suited for the context of explanations. This approach basically generalises the concept of prime implicate to constraint problems. We showed how domain consequences can be used for computing different types of explanations, and explained the basics of their computation. As a proof of concept, we set up a framework under which further work will be carried out. The main directions will be to examine efficient ways to represent and operate on sets of domain consequences, aiming at scalability to problems of practical interest. We will have to carefully measure the efficiency of explanation computation from this representation, as well as, of course, compare it with other existing representations and algorithms. Another, maybe more realistic, direction will be to define interesting restrictions on the domain conflicts we compute. In the longer term, we intend to investigate other properties that such a canonical representation of a problem might have.

References

1. Chatalic, P., Simon, L.: Multi-resolution on compressed sets of clauses. In: ICTA, pp. 2–10. IEEE Computer Society, Los Alamitos (2000)
2. Coudert, O., Madre, J.C.: A New Method to Compute Prime and Essential Prime Implicants of Boolean Functions. In: Advanced Research in VLSI and Parallel Systems: Proceedings of the 1992 Brown/MIT Conference. MIT Press, Cambridge (1992)
3. Darwiche, A., Marquis, P.: A knowledge compilation map. J. Artif. Intell. Res (JAIR) 17, 229–264 (2002)
4. de Kleer, J.: An improved incremental algorithm for generating prime implicates. In: AAAI, pp. 780–785 (1992)
5. de Kleer, J., Mackworth, A.K., Reiter, R.: Characterizing diagnoses and systems. Artif. Intell. 56(2-3), 197–222 (1992)
6. del Val, A.: Tractable databases: How to make propositional unit resolution complete through compilation. In: KR, pp. 551–561 (1994)
7. del Val, A.: A new method for consequence finding and compilation in restricted languages. In: AAAI/IAAI, pp. 259–264 (1999)
8. del Val, A.: The complexity of restricted consequence finding and abduction. In: AAAI/IAAI, pp. 337–342. AAAI Press/The MIT Press (2000)
9. Focacci, F., Milano, M.: Global cut framework for removing symmetries. In: Walsh, T. (ed.) CP 2001. LNCS, vol. 2239, pp. 77–92. Springer, Heidelberg (2001)
10. Junker, U.: QUICKXPLAIN: Preferred explanations and relaxations for over-constrained problems. In: McGuinness, D.L., Ferguson, G. (eds.) AAAI, pp. 167–172. AAAI Press/The MIT Press (2004)
11. Junquera, B.P., González, C.A.: Possible conflicts: a compilation technique for consistency-based diagnosis. IEEE Transactions on Systems, Man, and Cybernetics, Part B 34(5), 2192–2206 (2004)
12. Katsirelos, G., Bacchus, F.: Generalized nogoods in csps. In: Veloso, M.M., Kambhampati, S. (eds.) AAAI, pp. 390–396. AAAI Press/The MIT Press (2005)
13. Kean, A., Tsiknis, G.K.: An incremental method for generating prime implicants/impicates. J. Symb. Comput. 9(2), 185–206 (1990)
14. Marquis, P.: Consequence finding algorithms. In: Handbook of Defeasible Reasoning and Uncertainty Management Systems, vol. 5, pp. 41–145 (2000)
15. Marquis, P.: Knowledge compilation using theory prime implicates. In: IJCAI (1), pp. 837–845 (1995)
16. O'Sullivan, B., Papadopoulos, A., Faltings, B., Pu, P.: Representative explanations for over-constrained problems. In: AAAI, pp. 323–328. AAAI Press, Menlo Park (2007)
17. Papadopoulos, A., O'Sullivan, B.: Relaxations for compiled over-constrained problems. In: Stuckey, P.J. (ed.) CP 2008. LNCS, vol. 5202, pp. 433–447. Springer, Heidelberg (2008)
18. Simon, L., del Val, A.: Efficient consequence finding. In: Nebel, B. (ed.) IJCAI, pp. 359–370. Morgan Kaufmann, San Francisco (2001)
19. Tison, P.: Generalization of consensus theory and application to the minimization of boolean functions. IEEE Transactions on Electronic Computers 16(4), 446–456 (1967)

On the Power of Clause-Learning SAT Solvers with Restarts

Knot Pipatsrisawat and Adnan Darwiche

University of California, Los Angeles, USA
{thammakn,darwiche}@cs.ucla.edu

Abstract. In this work, we improve on existing work that studied the relationship between the proof system of modern SAT solvers and general resolution. Previous contributions such as those by Beame *et al* (2004), Hertel *et al* (2008), and Buss *et al* (2008) demonstrated that variations on modern clause-learning SAT solvers were as powerful as general resolution. However, the models used in these studies required either extra degrees of non-determinism or a preprocessing step that are not utilized by any state-of-the-art SAT solvers in practice. In this paper, we prove that modern SAT solvers that learn asserting clauses indeed p-simulate general resolution without the need for any additional techniques.

1 Introduction

It is well-known that modern clause-learning SAT solvers (and their original ancestor, the DPLL algorithm [7]) can be interpreted as resolution-based proof systems [4]. For each unsatisfiable formula, these solvers can be viewed as engines that produce refutation proofs. One central question in this research direction is whether the proof system implemented by modern SAT solvers still has enough freedom to generate a short resolution proof (relative to general resolution) for every unsatisfiable formula. The answer to this theoretical question could have important practical implications on the efficiency of modern SAT solvers.

There is much previous work on this subject that demonstrates the strength of variations on modern clause-learning SAT solvers. However, equivalence with respect to general resolution has yet to be proven for modern clause-learning SAT solvers as they are practiced today. In [4], Beame *et al* showed that a proof system based on a more general variation of modern SAT solvers was as powerful as general resolution. The proof presented, however, requires the solver to make decisions on variables that are already implied by unit resolution and to use a learning scheme not utilized by any top-performing SAT solver (e.g., MiniSat [8], Rsat [15]). These modifications "utilize extra degrees of non-determinism that would be very hard to exploit in practice" [10]. Hertel *et al* [10] proved a slightly weaker result that clause-learning solvers *effectively p-simulate* general resolution. This approach allows them to introduce a preprocessing step to transform the CNF into one (with some new variables) that can be efficiently solved by a more practical model of solvers (the model used in [10] is based

I.P. Gent (Ed.): CP 2009, LNCS 5732, pp. 654–668, 2009.

on the one developed by Van Gelder in [19]). Buss *et al* [5] also took a similar approach by modifying the input CNF (and introduced some new variables) to show that a generalized variation of clause-learning algorithm, which allows decision making past conflicts, can effectively p-simulate general resolution.

In this work, we show that modern clause-learning SAT solvers, without any extra modifications, are indeed as powerful as general resolution. In particular, we prove that the proof system implemented by modern clause-learning solvers (which uses unit resolution, asserting clause learning, and restarting) p-simulates general resolution. We show that this result holds for any asserting-clause learning scheme. Our proof does not require any preprocessing or making decisions on implied literal or past a conflict. This result implies that modern SAT solvers in their current form are capable of producing proofs that are as "short" as any resolution proof, given appropriate branching heuristic and restart policy.

The proof of our main result is made possible by the help of two important concepts, namely *1–empowerment* [16] and *1–provability*. Together, they allow us to more accurately capture the power of modern clause-learning SAT solvers and to avoid the need to introduce any technique not present in practice.

The rest of the paper is organized as follows. In the next section, we discuss basic notations and definitions. In Section 3, we present our model of modern clause-learning SAT solvers and the proof system associated with it. Then, in Section 4, we present some interesting key results, which provide some insights on the power of modern SAT solvers and allow us to prove the main result. Next, we present our main result in Section 5. Finally, we conclude in Section 7.

2 Preliminaries

In this section, we review some basic notations related to propositional logic and proof systems. If Δ and α are two boolean formulas and ℓ is a literal, we write $\Delta \models \alpha$ to mean that Δ entails α, and write $\Delta \vdash \ell$ to mean that literal ℓ can be derived from Δ *using unit resolution.* Furthermore, we may treat a clause as the set of literals in the clause and a CNF formula as the set of clauses it contains.

2.1 Proof Systems

A *proof system* is a language for expressing proofs that can be verified in time polynomial in the size of the proof [6]. In this work, we are concerned only with proof systems based on propositional resolution [17]. The *resolution* between clause $\alpha \vee x$ and $\beta \vee \neg x$ is the derivation of clause $\alpha \vee \beta$ (i.e., the *resolvent*). In this case, x is called the *resolved variable.* To make our analysis as related to modern SAT solvers as possible, the *weakening rule*, which allows introduction of arbitrary literals into existing clauses, is not permitted here.

Definition 1. *A resolution proof (or resolution derivation) of the clause* C_k *from the CNF* Δ *is a sequence of clauses* $\Pi = C_1, C_2, ..., C_k$ *where each clause* C_i *is either in* Δ *or is a resolvent of clauses preceding* C_i*.*

We will also treat a resolution proof as the set of clauses in it. The size of a proof is the number of clauses in it. A resolution proof of the empty clause (i.e., false) is called a *refutation proof*. The notion of p-simulation, which was introduced in [6], is used to compare the power of two proof systems. The definition presented here is obtained from [10].

Definition 2 (P-Simulation). *Proof system S p-simulates proof system T, if, for every unsatisfiable formula Δ, the shortest refutation proof of Δ in S is at most polynomially longer than the shortest refutation proof of Δ in T.*

Intuitively, if proof system S p-simulates proof system T, it means that S is unrestricted enough to express proofs that are as short as those expressible in T. As far as resolution proofs are concerned, *general resolution*, which allows any resolution operation to be performed, is the most powerful proof system. Other resolution proof systems that are known to be less powerful (i.e., do not p-simulate general resolution) include *tree-like resolution*, *linear resolution*, and *regular resolution* (see Section 2.3 of [4] for a good review).

3 Modern Clause-Learning SAT Solvers as a Proof System

3.1 Modern Clause-Learning SAT Solvers

In this section, we describe a model of modern clause-learning SAT solvers. Included in our model are the following techniques: unit resolution [7], clause-learning [12,20], restarting [9], and non-chronological backtracking [12,3] (i.e., far-backtracking as termed by [18]). Algorithm 1 shows a pseudo code of a typical clause-learning SAT solver with restarts, which we will refer to as CLR from now on. We will first provide a high-level description of the algorithm before giving formal definitions of its different components.

This algorithm is based on making variable assignments called *decisions*. It starts with an empty decision sequence D and an empty set of learned clauses Γ (Lines 1-2). It then iterates until it either proves the satisfiability or unsatisfiability of the input. In each iteration, the solver has an option of restarting, which amounts to resetting the decision sequence to the empty sequence (Line 5). After that, the conjunction of the input CNF Δ, learned clauses Γ, and decisions D are checked for inconsistency using unit resolution (Line 7). If unit resolution finds an inconsistency, the algorithm does one of two things:

- If the decision sequence is empty, the CNF Δ must be unsatisfiable and the algorithm terminates (Line 9).
- If the decision sequence is not empty, a clause α is generated and a level m is computed based on α. The algorithm then erases all decisions made after level m, adds α to Γ, and moves on to the next iteration (Lines 10-13).[1]

[1] The clause α is known as an asserting clause and m as the assertion level. We will define them formally later.

Algorithm 1. CLR: Clause-learning SAT solver with restarts

```
   input  : CNF formula Δ
   output: A solution of Δ or unsat if Δ is not satisfiable
1  D ← ⟨⟩ // Decision literals
2  Γ ← true // Learned clauses
3  while true do
4      if time to restart then
5          D ← ⟨⟩
6      S ← (Δ, Γ, D)
7      if S is 1–inconsistent then
           // There is a conflict.
8          if D = ⟨⟩ then
9              return unsat
10         α ← an asserting clause of S
11         m ← the assertion level of α
12         D ← D_m // the first m decisions
13         Γ ← Γ ∧ α
14     else
           // There is no conflict.
15         Choose a literal ℓ such that S ⊬ ℓ and S ⊬ ¬ℓ
16         if ℓ = then
17             return D// satisfiable
18         D ← D, ⟨ℓ⟩
```

If unit resolution detects no inconsistency, the solver makes a decision by selecting a literal ℓ whose value is not currently implied or falsified by unit resolution, and adds it to the decision sequence (Line 18). If no such literal is found, the algorithms terminates having proved satisfiability (Line 17). We will now provide the missing definitions.

- A *decision sequence* is an ordered set of literals $D = \langle \ell_1, \ldots, \ell_k \rangle$. Each literal ℓ_i is called the *decision* at *level* i. We write D_m to denote the subsequence $\langle \ell_1, \ldots, \ell_m \rangle$.
- A *SAT state* is a tuple (Δ, Γ, D), where Δ and Γ are CNFs such that $\Delta \models \Gamma$, and D is a decision sequence. We will write S_k to denote the state (Δ, Γ, D_k).
- A CNF Δ is *1–inconsistent* iff $\Delta \vdash \mathsf{false}$. It is *1–consistent* otherwise. A SAT state (Δ, Γ, D) is *1–inconsistent* (*1–consistent*) iff $\Delta \wedge \Gamma \wedge D$ is 1–inconsistent (1–consistent). It is normal for an unsatisfiable CNF to be 1–consistent.
- A literal ℓ is *implied by state* $S = (\Delta, \Gamma, D)$ *at level* k, written $S \vdash_k \ell$, iff k is the smallest integer for which $\Delta \wedge \Gamma \wedge D_k \vdash \ell$. We say that the *implication level* of literals $\ell, \neg\ell$ is k in this case, write $S \vdash \ell$ to mean $S \vdash_i \ell$ for some i, and write $S \not\vdash \ell$ to mean $S \not\vdash_i \ell$ for all i.
- A state $S = (\Delta, \Gamma, \langle \ell_1, \ldots, \ell_k \rangle)$ is *normal* iff for all $1 \leq i \leq k$, S_{i-1} is 1–consistent, $S_{i-1} \not\vdash \ell_i$ and $S_{i-1} \not\vdash \neg\ell_i$.

The notion of normal states prohibits SAT solvers from (1) making a decision in the presence of a conflict and (2) making a decision on a variable that is already assigned a value. By construction, the state S on Line 7 of Algorithm 1 is always normal. Therefore, from now on, we will assume that every SAT state is normal.

We are now ready to define the last two notions used in Algorithm 1: asserting clause and assertion level. An asserting clause is a special type of conflict clause, so we start first by defining the notion of a conflict clause. Our definition of conflict clause closely follows the graphical definition in [20].

Definition 3 (Conflict Clause). *Let $S = (\Delta, \Gamma, D)$ be a 1–inconsistent SAT state. A clause $\alpha = \ell_1 \vee \ldots \vee \ell_m$ is a conflict clause of state S iff:*

1. $\Delta \wedge \Gamma \wedge \neg\alpha \vdash \mathsf{false}$. *That is, we can show that α is implied by $\Delta \wedge \Gamma$ using just unit resolution.*
2. *For each literal ℓ_i, $S \vdash \neg\ell_i$. That is, the literals $\neg\ell_i$ are a subset of the implications (or decisions) discovered by unit resolution in state S.*

In [4], it was shown (in their Proposition 4) that every conflict clause obtained from a cut on an implication graph (or a conflict graph, to be more precise) can be derived from the current knowledge base ($\Delta \wedge \Gamma$) using what is known as *trivial resolution derivation*, which captures the kind of resolution performed by virtually all modern clause-learning SAT solvers [4]. A *trivial resolution derivation* is a resolution derivation in which:

1. Every resolution step (except the very first) is performed between the last resolvent and a clause in the knowledge base.
2. The resolved variables are all distinct.

Our definition of conflict clause is independent of the notion of implication graph and is slightly more general (for example, it encompasses unconventional clauses derived in [2]). Nevertheless, we will later show that all of the conflict clauses that we care to learn can still be "derived" using trivial resolution derivation (to be proven later in Proposition 3).

In any case, modern SAT solvers, in practice, insist on learning conflict clauses that contain *exactly* one literal falsified at the last level.

Definition 4 (Asserting Clause). *A conflict clause α of a SAT state $S = (\Delta, \Gamma, D)$ is an asserting clause iff it has exactly one literal ℓ with implication level $|D|$. The literal ℓ is called the asserted literal of α. Moreover, the assertion level of clause α is defined as the highest implication level $k < |D|$ attained by some literal in α. If α contains only one literal, the assertion level is defined to be zero.*

Given a 1–inconsistent state, there always exists an asserting clause for it [16]. This result ensures that the execution of Line 10 of Algorithm 1 will always succeed. This completes our description of CLR.

3.2 Clause-Learning Schemes

CLR can employ various *learning schemes* to derive conflict clauses (Line 10). Even though we insist on deriving asserting clauses in Algorithm 1, in general, non-asserting clauses may be used.[2] In our context, it is sufficient to view a learning scheme as a function that produces a conflict clause for every 1–inconsistent SAT state. In this work, we will focus on a certain class of learning schemes called *asserting learning schemes*, which always produces asserting clauses. We will later show (in Proposition 3) that every asserting clause can essentially be derived from a trivial resolution proof.

Given a learning scheme LS, we use CLR_{LS} to denote the SAT algorithm obtained by applying LS on Line 10 of Algorithm 1. We use CLR to denote the algorithm with any learning scheme.

3.3 Non-determinism in CLR

Given a learning scheme LS, the only sources of non-determinism remaining in CLR_{LS} are (1) the branching heuristic, (2) the restart policy and (3) the implementation of unit resolution.[3]

In this work, we utilize the notion of *extended branching sequence* (defined by [4]) to capture (1) and (2). An extended branching sequence is simply a sequence of literals and special symbol **R** that is used to control decision making and restarting in CLR. For example, $\sigma = \langle x, \neg y, \mathbf{R}, \neg x \rangle$ indicates that the first decision should be $x = \mathsf{true}$, the second decision should be $y = \mathsf{false}$, then the solver should restart, and set $x = \mathsf{false}$ next (unit resolution and conflict analysis are applied normally between these steps). Given such a sequence σ, $\mathsf{CLR}(\Delta, \sigma)$ refers to the SAT state attained after executing CLR according to the decisions and restarting points specified in σ (i.e., the choices on Lines 5 and 18 should be made based on the next element in the sequence).[4] For simplicity, we will insist that $\mathsf{CLR}(\Delta, \sigma)$ be a 1–consistent state (unless it contains the empty clause).[5] Moreover, we shall use the notation $\mathsf{CLR}(\Delta, \sigma)$ to also refer to the knowledge base (original and learned) of the SAT state $\mathsf{CLR}(\Delta, \sigma)$.

3.4 CLR as a Proof System

Modern clause-learning SAT solvers can be viewed as a proof system that contains all proofs obtainable by executing the solver according to some decision heuristic, restart policy, and implementation of unit resolution. If we view each conflict clause as being derived from a resolution proof, we can combine these

---

[2] Possibly at the expense of completeness.

[3] The implementation of unit resolution may affect, for example, the order and derivations of unit implications, which, in practice, influence which conflict clauses eventually get derived.

[4] A decision in σ should be skipped if its variable is already implied.

[5] This only amounts to letting Algorithm 1 deal with conflicts until it reaches a 1–consistent state.

sub-derivations into a resolution proof that is produced by the SAT solver. In particular, if a given execution of CLR on an unsatisfiable problem produces conflict clauses $C_1, ..., C_k$, we know that $\Sigma = \Delta \wedge C_1 \wedge ... \wedge C_k$ is 1–inconsistent (this is how the algorithm terminates). Let each clause C_i be derived using resolution proof π_i (from original and previously learned clauses), and τ be the unit resolution derivation of false from Σ, then $\Pi = \pi_1, ..., \pi_k, \tau$ is the refutation proof generated by this execution. For the purpose of this work, each π_i can be viewed as a trivial resolution derivation, whose size is at most linear in the number of variables. We will later justify this claim in Proposition 3. In the following definition, which defines the proof system implemented by Algorithm 1, we overload the notation CLR to refer to both the SAT algorithm and the proof system.

Definition 5. *Given a learning scheme LS, proof system CLR_{LS} consists of all refutation proofs that can be generated by Algorithm 1 using learning scheme LS.*

The main result of this work will show that, for all asserting learning schemes LS, CLR_{LS} p-simulates general resolution. Note that the size of τ is always at most linear in the number of variables. Hence, we leave it out from our future discussion and proofs.

4 Ingredients for the Main Result

In this section, we present three key results that allow us to prove the main result. These results, some of which are interesting in their own rights, provide insights on the power of CLR. They are made possible by two important concepts, called *1–empowerment* and *1–provability*, which allow us to formalize the ability of CLR and use it to simulate general resolution. We first give definitions of these notions before presenting the results. The first notion is called 1–empowerment, which is the ability of a clause to allow unit resolution to see a new implication. We present here a slightly modified definition of the one presented in [16].

Definition 6 (1–Empowerment [16]). *Let $\alpha \Rightarrow \ell$ be a clause where ℓ is some literal in the clause and α is a conjunction of literals. The clause is 1–empowering with respect to CNF Δ iff*

1. *$\Delta \models (\alpha \Rightarrow \ell)$: the clause is implied by Δ.*
2. *$\Delta \wedge \alpha$ is 1-consistent: Asserting α does not result in a conflict that is detectable by unit resolution.*
3. *$\Delta \wedge \alpha \not\vdash \ell$: the literal ℓ cannot be derived from $\Delta \wedge \alpha$ using unit resolution.*

In this case, ℓ is called an empowering literal of the clause. On the other hand, a clause implied by Δ that is not 1–empowering is said to be absorbed by Δ.[6]

A clause implied by Δ is 1–empowering if it allows unit resolution to derive a new implication that would be impossible to derive without the clause. For example,

[6] This terminology, "absorbed", was introduced in [1].

consider $\Delta = (a \vee b \vee c) \wedge (a \vee b \vee \neg c) \wedge (a \vee \neg b \vee c) \wedge (a \vee \neg b \vee \neg c) \wedge (\neg c \vee d) \wedge (c \vee e)$. The clause $(a \vee b)$ is 1–empowering with respect to Δ because unit resolution cannot derive a from $\Delta \wedge \neg b$. On the other hand, $(d \vee e)$, which is implied by Δ, is not 1–empowering (i.e., is absorbed), because unit resolution can already derive e from $\Delta \wedge \neg d$ and derive d from $\Delta \wedge \neg e$. Note that if clause C subsumes clause C' (i.e., $C \subseteq C'$), then C absorbes C'. Moreover, adding more clauses to the knowledge base may make a 1–empowering clause become absorbed but can never make an absorbed clause become 1–empowering. Every asserting clause is 1–empowering with respect to the knowledge base at the time of its derivation with its asserted literal as an empowering literal [16].

The second key notion, called 1–provability, is related to the difficulty of deriving a clause from a CNF.

Definition 7 (1–Provability). *Given a CNF Δ, clause C is 1–provable with respect to Δ iff $\Delta \wedge \neg C \vdash \mathsf{false}$.*

If a clause is 1–provable with respect to a given CNF, then we can show that it is implied by the CNF using only unit resolution. For example, consider Δ defined above. The clauses $(a \vee b)$ and (a) are both implied by Δ. In this case, the clause $(a \vee b)$ is 1–provable with respect to Δ, because unit resolution is sufficient to derive a contradiction after we assert the negation of the clause on top of Δ. However, this is not the case for (a) (thus, it is not 1–provable). Notice that, according to Definition 3, every conflict clause is 1–provable with respect to the knowledge base at the time of its derivation. Moreover, a 1–provable clause still remains 1–provable after a clause is added to the knowledge base.

We are now ready to present the results, whose proofs are in the Appendix. The first key result states that, in every refutation proof of a 1–consistent CNF, there is always a clause that is both 1–empowering and 1–provable with respect to the CNF.

Proposition 1. *Let Δ be an unsatisfiable CNF that is 1–consistent and Π be a refutation proof of Δ. There exists a clause $C \in \Pi$ such that C is 1–empowering and 1–provable with respect to Δ.*

The set of clauses which are both 1–empowering and 1–provable plays an important role in our main proof. In the next result, we show that CLR with any asserting learning scheme can absorb such clauses in a quadratic number of decisions. This result essentially states that, given any 1–empowering and 1–provable clause C, we can always come up with a (short) sequence of appropriate decisions (and restarts) to force CLR with an asserting learning scheme to derive clauses that, together with the original knowledge base, allow unit resolution to see any implication that C may allow us to derive (i.e., render C useless, as far as unit resolution is concerned).

Proposition 2. *Let Δ be a CNF with n variables and C be a clause that is 1–empowering and 1–provable with respect to Δ. For any asserting learning scheme AS, there exists an extended branching sequence σ such that*

1. $|\sigma| \in O(n^2)$.
2. *C is absorbed by* $\mathsf{CLR}_{AS}(\Delta, \sigma)$.

Because of this result, we will call any clause that is both 1–empowering and 1–provable with respect to the given CNF Δ, CLR-*derivable* with respect to Δ.

The next result states that every 1–empowering conflict clause can always be derived using a trivial resolution derivation from the original and learned clauses at the time of the conflict.[7]

Proposition 3. *Let* $S = (\Delta, \Gamma, D)$ *be a 1–inconsistent SAT state and* C *be a conflict clause of* S *that is 1–empowering with respect to* $\Delta \wedge \Gamma$. *There exists a trivial resolution proof of some* $C' \subseteq C$ *from* $\Delta \wedge \Gamma$.

Since all resolved variables are distinct in a trivial resolution proof, the size of the proof has to be in $O(n)$, where n is the number of variables of Δ. This result is important as it shows that every empowering learning scheme (including any schemes yet to be conceived) can essentially be "implemented" with the kind of resolution derivation already employed by modern clause-learning SAT solvers. Since every asserting clause is 1–empowering, this result applies to all asserting learning schemes as well.

5 Main Result

In this section, we present our main result, which shows that the proof system implemented by modern clause-learning SAT solvers is as powerful as general resolution. We first present our main result in its most general form, then derive a corollary which is more closely related to modern SAT solvers.

Theorem 1. CLR *with any asserting learning scheme p-simulates general resolution.*

This result is applicable to a class of clause-learning algorithms that is even more general than what is used in practice (for example, it applies to any asserting learning scheme not yet proposed). By restricting the learning scheme, we obtain a more concrete result. Let 1UIP denote the first UIP learning scheme [13,20], which is, by far, the most popular scheme in practice.

Corollary 1. CLR_{1UIP} *p-simulates general resolution.*

We will give an intuitive proof sketch for the main theorem before presenting the actual proof. In contrast to the proofs presented in [4], [10], and [5], we do not try to simulate the derivation of *every* clause in the given resolution proof. Instead, we force CLR to go after CLR-derivable clauses only.

Suppose Δ has n variables. Let Π be a refutation proof of Δ and AS be an asserting learning scheme. If Δ is already 1–inconsistent the proof is trivial.

[7] This result can be viewed as a variation on Proposition 4 of [4] for our definition of conflict clauses and 1–empowering clauses.

Otherwise, we know that we can always find a CLR-derivable clause C in Π. We know that we can force CLR to absorb C while producing proofs whose combined size is only polynomial in n. We can keep repeating this process until the knowledge base becomes 1–inconsistent. Since this can go on for at most $|\Pi|$ times, the combined proof is polynomial in $|\Pi|$ and n.

Proof of Theorem 1. Suppose AS is an asserting learning scheme. Given a CNF Δ with n variables and any refutation proof Π of Δ, we will construct an extended branching sequence σ that will induce CLR_{AS} to derive the empty clause and generate a proof of size $O(n^2|\Pi|)$.

In each iteration, consider $\Sigma = \Delta \wedge \Gamma$, the current knowledge base of CLR_{AS}. We may assume that Σ is 1–consistent. From Proposition 1, we can always find a clause C in Π that is CLR-derivable from Σ. Since we are using an asserting learning scheme, Proposition 2 tells us that such a clause can be absorbed in $O(n^2)$ decisions and restarts. After absorbing C, we force the solver to restart. Let Σ now denote the updated knowledge base. Since Π is still a refutation proof of Σ, we can still find another CLR-derivable clause as long as Σ is still 1–consistent. We repeat this process until Σ is 1–inconsistent, at which point CLR terminates on Line 9 of Algorithm 1. In each iteration, we absorb at least one clause in Π (no absorbed clause can become 1–empowering after we add more clauses to the knowledge base) and perform only $O(n^2)$ decisions and restarts. The proof of Proposition 2 actually shows that at most $O(n)$ conflicts are needed to absorb each CLR-derivable clause. Hence, by Proposition 3, we know that each iteration produces a proof whose length is at most in $O(n^2)$. Therefore, we can use CLR_{AS} to produce a refutation proof of Δ with size $O(n^2|\Pi|)$. □

The next result shows that not only can we construct a CLR refutation proof with length polynomial in the size of any resolution proof, but the construction process can be carried out in polytime as well.

Theorem 2. *The extended branching sequence required for the simulation in Theorem 1 can be constructed in time polynomial in the sizes of the given refutation proof and of the given CNF.*

It suffices to show that finding a CLR-derivable clause in any given refutation proof (with respect to any 1–consistent CNF) can be done in polytime. For each clause C_i in the proof, checking if C_i is 1–provable only requires conditioning and closing the knowledge base under unit resolution. This can be achieved in time linear in the size of the CNF. Checking whether a literal is an empowering literal of a clause can be performed by asserting the negations of the other literals in the clause and see whether unit resolution can detect a conflict or derive the remaining literal from the knowledge base or not. This process, whose time complexity is linear in the size of the CNF, needs to be repeated for each literal in the clause. Therefore, the overall time complexity for finding a CLR-derivable clause is still polynomial in the sizes of the proof and the CNF.

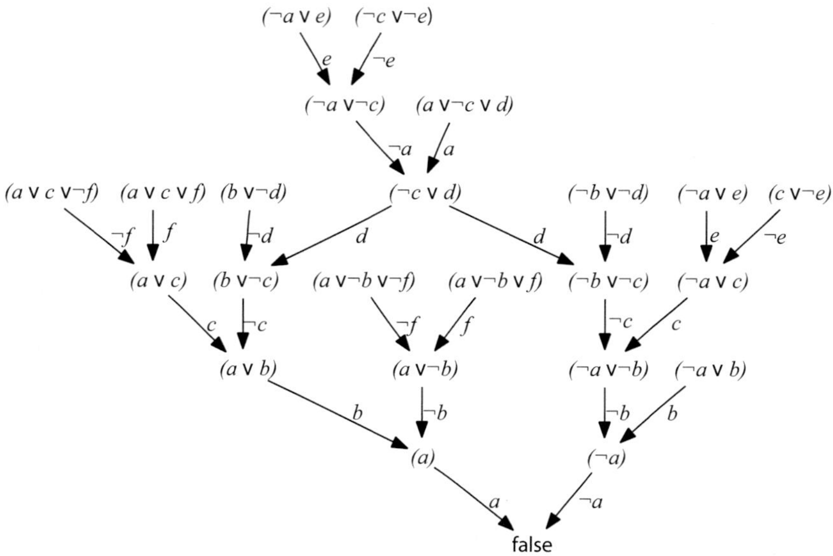

Fig. 1. A refutation proof of Δ. Each resolvent has two incoming edges from its resolved clauses (original clauses have no incoming edges). Each edge is annotated with the resolved literal of the corresponding resolved clause.

5.1 Example

We now show an example of the simulation described in the proof of Theorem 1. Consider the following unsatisfiable CNF:

$$\begin{aligned}\Delta = {} & (\neg a \vee e), (\neg c \vee \neg e), (a \vee \neg c \vee d), (\neg b \vee \neg d), (c \vee \neg e), (\neg a \vee b), \\ & (a \vee \neg b), (a \vee c \vee f), (a \vee c \vee \neg f), (b \vee \neg d), (a \vee \neg b \vee f), (a \vee \neg b \vee \neg f).\end{aligned}$$

Figure 1 shows a refutation proof of Δ. Alternatively, we can write this proof as

$$\begin{aligned}\Pi = {} & (\neg a \vee e), (\neg c \vee \neg e), (\neg a \vee \neg c), (a \vee \neg c \vee d), (\neg c \vee d), \\ & (\neg b \vee \neg d), (\neg b \vee \neg c), (c \vee \neg e), (\neg a \vee c), (\neg a \vee \neg b), (\neg a \vee b), \\ & (\neg a), (b \vee \neg d), (b \vee \neg c), (a \vee c \vee f), (a \vee c \vee \neg f), (a \vee c), (b \vee d), \\ & (a \vee \neg b \vee f), (a \vee \neg b \vee \neg f), (a \vee \neg b), (a), \mathsf{false}.\end{aligned}$$

Initially, one of the CLR-derivable clauses in Π is $(\neg b \vee \neg c)$ and $\neg c$ is the empowering literal. If the solver assigns $b = \mathsf{true}$ and $c = \mathsf{true}$, unit resolution will detect a conflict. In this case, $(\neg b \vee \neg c)$ and $(\neg c \vee d)$ are both asserting clauses. Suppose $(\neg b \vee \neg c)$ is learned. After adding $(\neg b \vee \neg c)$ to the knowledge base, $(\neg c \vee d)$ and $(\neg b \vee \neg c)$, for example, become absorbed. Next, we force the solver to restart. Suppose we choose $(\neg a)$ as the next CLR-derivable clause to absorb. We must now force the solver to set $a = \mathsf{true}$, which will immediately cause a conflict. In this case, $(\neg a)$ is derived. Once again, we force the solver to restart. Suppose we select $(a \vee \neg b)$ as the next CLR-derivable clause. We must now force the solver to set $a = \mathsf{false}, b = \mathsf{true}$. Since $\neg a$ is already implied by the last learned clause, the solver can skip the decision on a and only needs to assert $b = \mathsf{true}$ to cause a conflict. From this conflict, assume that the asserting clause $(a \vee \neg b)$ is learned. Adding this clause into the knowledge base will actually cause it to become 1–inconsistent. Hence, the solver can now terminate, as unit resolution can derive false from the set of original and learned clauses. The whole extended branching sequence used in this process is $\langle b, c, \mathbf{R}, a, \mathbf{R}, \neg a, b \rangle$.

6 Related Work

Early work in this direction was published by Beame *et al* in [4]. In that work, the authors showed that a slight variation of modern SAT solvers can simulate general resolution. However, one key modification required by the proof is to allow the solvers to make decisions on variables that are already assigned. This requirement essentially introduces another degree of freedom, which makes it harder to come up with a good decision heuristic and to actually implement in practice. It is interesting to note that the proof in [4] also requires the solver to restart at every conflict.

Van Gelder proposed a different proof system called POOL for studying modern SAT solvers as resolution engines [19]. In that work, the author focused on understanding the strength of POOL and using it to model modern SAT solvers. The author did not directly compare modern SAT solvers against general resolution.

Nevertheless, POOL later became a basis of the work by Hertel *et al* [10], which proved that modern SAT solvers can *effectively* p-simulate general resolution. In other words, the authors showed that, with an additional preprocessing step, modern solvers can become as strong as general resolution. While the preprocessing is deterministic and independent of the proof being simulated, it can be regarded as an extra component not utilized by any solver in practice.

Buss *et al* [5] also presented a similar argument. The authors showed that with a preprocessing step (different from the one in [10]), a generalized version of clause-learning algorithm can p-simulate general resolution. Apart from requiring an extra preprocessing step, the proof also needed the solver to make decisions on assigned variables.

7 Conclusions and Discussion

In this paper, we proved that clause-learning SAT solvers that utilize restarts correspond to a proof system that is as powerful as general resolution. Our work improves on previous results by avoiding the needs for additional degrees of non-determinism and preprocessing. Our proof is made possible by the notions of 1–empowerment and 1–provability, which allow us to capture the power of modern SAT solvers in a more direct and natural way, and to avoid the need for any special technique. The result presented in this paper essentially shows that modern SAT solvers, as used in practice, are capable of simulating any resolution proof (given the right branching and restarting heuristics, of course).

Note that the our proof requires the solver to restart at every conflict. While no actual solver utilizes this particular restart policy, the proof suggests that a frequent restart policy might be a key to the efficiency of modern solvers. In our proof, restarting gives the solver the freedom to go after any clause necessary for a short refutation. Interestingly, in recent years, there has been a clear trend towards more and more frequent restarts for modern SAT solvers (e.g., [11], [14]).

In spite of our result, more theoretical work still remains to be done in this research direction. The construction of our proof requires the solver to backtrack to the top level upon each conflict (i.e., restart). While it is easy to implement such a strategy, in practice, state-of-the-art solvers only backtrack to the assertion level at each conflict (this type of backtracking is termed far-backtracking in [18]). It still remains an open question whether far-backtracking (or even chronological backtracking) is sufficient to achieve the presented result.

References

1. Atserias, A., Fichte, J.K., Thurley, M.: Clause-learning algorithms with many restarts and bounded-width resolution. In: SAT 2009 (to appear, 2009)
2. Audemard, G., Bordeaux, L., Hamadi, Y., Jabbour, S., Sais, L.: A generalized framework for conflict analysis. In: Kleine Büning, H., Zhao, X. (eds.) SAT 2008. LNCS, vol. 4996, pp. 21–27. Springer, Heidelberg (2008)
3. Bayardo, R.J.J., Schrag, R.C.: Using CSP look-back techniques to solve real-world SAT instances. In: Proceedings of AAAI 1997, pp. 203–208 (1997)
4. Beame, P., Kautz, H., Sabharwal, A.: Towards understanding and harnessing the potential of clause learning. JAIR 22, 319–351 (2004)
5. Buss, S.R., Hoffmann, J., Johannsen, J.: Resolution trees with lemmas: Resolution refinements that characterize DLL algorithms with clause learning. Logical Methods in Computer Science 4, 4 (2008)
6. Cook, S.A., Reckhow, R.A.: The relative efficiency of propositional proof systems. J. Symb. Log. 44(1), 36–50 (1979)
7. Davis, M., Logemann, G., Loveland, D.: A machine program for theorem-proving. Commun. ACM 5(7), 394–397 (1962)
8. Eén, N., Sörensson, N.: An extensible sat-solver. In: Giunchiglia, E., Tacchella, A. (eds.) SAT 2003. LNCS, vol. 2919, pp. 502–518. Springer, Heidelberg (2004)
9. Gomes, C.P., Selman, B., Crato, N.: Heavy-tailed distributions in combinatorial search. In: Smolka, G. (ed.) CP 1997. LNCS, vol. 1330, pp. 121–135. Springer, Heidelberg (1997)
10. Hertel, A.P., Bacchus, F., Pitassi, T., Van Gelder, A.: Clause learning can effectively p-simulate general propositional resolution. In: Proc. of AAAI 2008, pp. 283–290 (2008)
11. Huang, J.: The effect of restarts on the efficiency of clause learning. In: Proc. of IJCAI 2007, pp. 2318–2323 (2007)
12. Marques-Silva, J.P., Sakallah, K.A.: GRASP - A New Search Algorithm for Satisfiability. In: Proceedings of ICCAD 1996, pp. 220–227 (1996)
13. Moskewicz, M., Madigan, C., Zhao, Y., Zhang, L., Malik, S.: Chaff: Engineering an efficient sat solver. In: Proceedings of DAC 2001 (June 2001)

14. Pipatsrisawat, K., Darwiche, A.: Width-based restart policies for clause-learning satisfiability solvers. In: Proceedings of SAT 2009 (to appear, 2009)
15. Pipatsrisawat, K., Darwiche, A.: Rsat 2.0: Sat solver description. Tech. Rep. D–153, Automated Reasoning Group, Comp. Sci. Department, UCLA (2007)
16. Pipatsrisawat, K., Darwiche, A.: A new clause learning scheme for efficient unsatisfiability proofs. In: Proceedings of AAAI 2008, pp. 1481–1484 (2008)
17. Robinson, J.A.: A machine-oriented logic based on the resolution principle. J. ACM 12(1), 23–41 (1965)
18. Sang, T., Beame, P., Kautz, H.: Heuristics for fast exact model counting. In: Bacchus, F., Walsh, T. (eds.) SAT 2005. LNCS, vol. 3569, pp. 226–240. Springer, Heidelberg (2005)
19. Van Gelder, A.: Pool resolution and its relation to regular resolution and dpll with clause learning. In: Sutcliffe, G., Voronkov, A. (eds.) LPAR 2005. LNCS (LNAI), vol. 3835, pp. 580–594. Springer, Heidelberg (2005)
20. Zhang, L., Madigan, C.F., Moskewicz, M.W., Malik, S.: Efficient conflict driven learning in boolean satisfiability solver. In: ICCAD 2001, pp. 279–285 (2001)

A Proofs

In this appendix, we present proofs of Propositions 1, 2, 3. The proof of Proposition 1 is accomplished with the help of two lemmas.

Lemma 1. *Let $C_1 = \alpha \vee \ell$, $C_2 = \beta \vee \neg\ell$. Suppose C_1 and C_2 are not 1–empowering with respect to Δ. Then, $\alpha \vee \beta$ is 1–provable with respect to Δ.*

Proof. Let $C = \alpha \vee \beta$. Since both C_1, C_2 are not 1–empowering, we know that $\Delta \wedge \neg\alpha \vdash \ell$ and $\Delta \wedge \neg\beta \vdash \neg\ell$.[8] Therefore, unit resolution must be able to derive both ℓ and $\neg\ell$ from $\Delta \wedge \neg C = \Delta \wedge \neg\alpha \wedge \neg\beta$. Therefore, C is 1–provable with respect to Δ. □

Lemma 2. *Let C be a clause that is not 1–provable with respect to Δ and Π be a resolution proof of C from Δ. Then, there exists $C' \in \Pi, C' \neq C$, such that C' is* CLR*-derivable from Δ.*

Proof. Let $\Pi = C_1, ..., C_n$ and C_i be the first clause in Π that is not 1–provable with respect to Δ (i may be equal to n). Clearly, C_i must be the resolvent of two 1–provable clauses C_j, C_k for some $j, k < i$. Assume for the sake of contradiction that C_j, C_k are both not 1–empowering. Lemma 1 implies that C_i must be 1–provable, which is a contradiction. Hence, either C_j or C_k must be both 1–provable and 1–empowering. □

Proof of Proposition 1. Given a 1–consistent CNF Δ, it is easy to see that the empty clause (false) is not 1–provable with respect to Δ (otherwise, Δ would be 1–inconsistent by definition). Since every refutation proof contains the empty clause, Lemma 2 implies that the proof Π must contain a clause that is both 1–empowering and 1–provable. □

[8] It is also possible that asserting $\neg\alpha$ or $\neg\beta$ may result in a 1–inconsistent CNF. We omit this case as C is trivially 1–provable.

Next, we present the proof of Proposition 2.

Proof of Proposition 2. Let $C = (\alpha \vee \ell)$ be the clause under consideration and ℓ be an empowering literal. Moreover, let δ be an extended branching sequence consisting of the literals in $\neg\alpha$ in any order. Since C is 1–empowering, the SAT state right after asserting δ must be 1–consistent ($\star$). Moreover, because C is 1–empowering and 1–provable, at this point, neither ℓ nor $\neg\ell$ can be implied by unit resolution.[9] Hence, CLR can select $\neg\ell$ as the next decision. The 1–provability of C ensures that asserting δ together with $\neg\ell$ will result in a 1–inconsistent state S ($\star\star$).

Let D be the asserting clause derived by AS from S. If $\Delta \wedge D$ absorbs C, we are done. Otherwise, we add D to the knowledge base, restart, and repeat this whole process. We will now argue that this can only be repeated $O(n)$ times.

Every asserting clause learned in the process must generate at least one new implication (i.e., its asserted literal) under a subset of δ. In every iteration, at least one more literal that was implied at the conflict level will now be implied by the time that δ is asserted. Because of ($\star$) and ($\star\star$), which hold in every iteration, a conflict only happens after $\neg\ell$ is asserted (thus, after δ is asserted). This implies that, in our proof, once a literal becomes an asserted literal of some asserting clause, it will never be assigned at any future conflict level again. Thus, each literal can only become an asserted literal (of any clause) only once in this whole process.

Since there are only n variables, this can be repeated at most $O(n)$ times before the empowering literal (ℓ) itself is implied by the assertion of some asserting clause. Whenever that happens, it means that the clause C has already been absorbed. The resulting extended branching sequence σ alternates between $\delta, \neg\ell$ and $\mathbf{R}$. Since each iteration takes at most $|C| + 1$ decisions, the size of σ is in $O(n^2)$. □

Note that the above proof only requires $O(n)$ conflicts to absorb a CLR-derivable clause. Next we prove Proposition 3.

Proof of Proposition 3. Let $C = (\alpha \vee \ell)$ and ℓ be its empowering literal. Let σ be a branching sequence consisting of the literals of $\neg\alpha$ (in any order) followed by $\neg\ell$. Let DEC be the decision learning scheme (as defined based on implication graph in Section 2 of [20]). In this scheme, conflict clauses contain only literals of decision variables.

Asserting σ will result in a conflict and DEC will derive an asserting clause C' which consists entirely of the negations of decision literals. Since the decisions in δ are all negations of literals in C, we have $C' \subseteq C$. Since DEC is a learning scheme based on implication graph, Proposition 4 of [4] implies that every conflict clause produced by it, C' in particular, can be derived using a trivial resolution proof. □

[9] That ℓ is not implied is straightforward. If $\neg\ell$ was implied, the current state would be 1–inconsistent, because C is 1–provable. This contradicts ($\star$).

Slice Encoding for Constraint-Based Planning

Cédric Pralet and Gérard Verfaillie

ONERA, 2 avenue Édouard Belin, BP 74025, 31055 Toulouse Cedex 4, France
{Cedric.Pralet,Gerard.Verfaillie}@onera.fr

Abstract. In most of the constraint-based approaches to planning, the problem is unfolded over a given number of steps. Because this unfolded CSP encoding is very time and memory consuming, we propose on top of the CNT framework (*Constraint Networks on Timelines*) a more efficient slice CSP encoding which allows only a limited number of steps to be considered at each step of the search.

1 Introduction

In planning problems [1], we are looking for a plan that satisfies some properties and optimizes some criterion, but whose length *i.e.*, the number of steps it involves, is unknown and possibly unbounded. To overcome such a difficulty, following the seminal work of [2], constraint-based approaches to planning [3,4,5,6] model as a CSP the problem of finding a plan of fixed length H, with H incremented each time no plan of length H has been found. On the other hand, the CNT framework (*Constraint Network on Timelines* [7]) allows the length of the plan to be considered as a variable h, with a domain and with constraints linking it to any other variables. Following the lazy approach of [8], the problem can be unfolded only over the number of steps induced by the minimum value in the domain of h and extended only when this minimum value increases due to constraint propagation or branching choice.

However, what is common to all these approaches is an unfolded CSP encoding of the planning problem over a fixed or variable number of steps. Such an encoding may be very costly in terms of memory required to store all the data structures necessary to define and to manage the CSP and in terms of time required to create the CSP and to perform constraint propagation and search. For example, we observed some minutes or tens of minutes to create the CSP on some instances we wanted to solve. On some of them, there was even not enough memory to create the CSP without memory swaps. This led us to search for more compact CSP encodings of planning problems.

This paper presents the so-called *slice encoding* we developed on top of the CNT framework to allow only a limited number of steps, depending on the constraint graph structure, to be considered at each step of the search. The paper is organized as follows: first, the CNT framework is defined again *via* an illustrative example; second, the slice encoding is defined as well as a generic algorithm able to reason and to search on it; finally, experiments on planning benchmarks show its dramatic impact in terms of memory and time.

I.P. Gent (Ed.): CP 2009, LNCS 5732, pp. 669–683, 2009.

2 The CNT Framework

2.1 An Illustrative Example

Let us consider the following planning problem which is a very simplified version of a real mission management problem for an Earth observation satellite.

We assume a satellite able to perform observations of specified areas of the Earth's surface. We consider a planning horizon $[Ts..Te]$ ($[a..b]$ denotes the set of integers between a and b). We assume N candidate observations on this horizon, numbered from 1 to N. Each observation $k \in [1..N]$ has a starting time $Ts[k]$ and an ending time $Te[k]$, with $Ts < Ts[k] < Te[k] < Te$. Two observations cannot overlap. To perform an observation, the instrument must be switched on at least Δon before starting observing. The instrument is initially off. Solar panels deliver a power Pe. When the instrument is on, a power $Ce > Pe$ is consumed. The initial energy level is Ei. Its minimum and maximum levels are $Emin$ and $Emax$. All data are assumed to be positive integers. The objective is to maximize the number of observations performed.

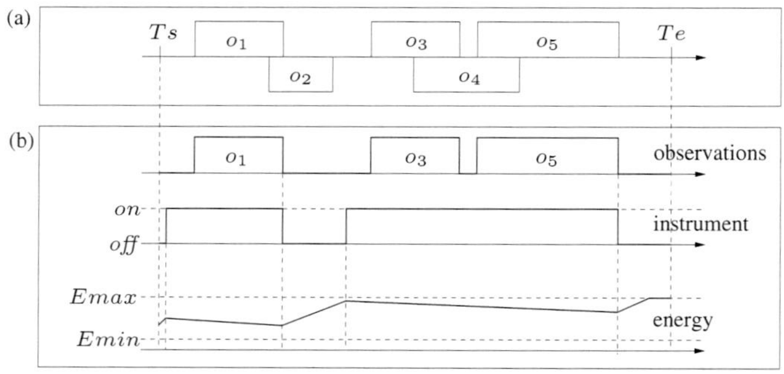

Fig. 1. Graphical representation of an instance (a) and of an optimal solution (b) of the mission management problem for an Earth observation satellite

Figure 1a represents an instance of this problem. Observations o_1 and o_2 are incompatible because of overlapping. Similarly, observation o_4 is incompatible with o_3 and o_5 because of overlapping. Figure 1b is a graphical representation of an optimal solution of this instance. Only observations o_1, o_3, and o_5 are performed. The instrument remains on between o_3 and o_5 because of insufficient time to switch it off. We can observe that the evolution of the energy level depends only on the status of the instrument. Let us use this planning problem to illustrate the basic definitions of the CNT framework.

2.2 Horizon Variables

Horizon variables are used to represent the number of steps to be considered.

Definition 1. *A horizon variable h is a variable whose domain of values is any subset of* $\mathbb{N}$. *We will use* $\mathbf{D}(h)$ *to denote the domain of a horizon variable h.*

In the planning problem described above, it may be useful to consider two horizon variables: a horizon variable *ho* to represent the number of steps in terms of observation and another one *hi* to represent the number of steps in terms of instrument status. If we associate a step with the start and the end of the horizon and with the start and the end of each observation performed, $\mathbf{D}(ho) = [2..2N+2]$: at least 2 steps when no observation is performed and at most $2N+2$ steps when all the candidate observations are performed. In the same way, if we associate a step with the start and the end of the horizon and with each instant at which the status of the instrument changes, $\mathbf{D}(hi) = \mathbf{D}(ho)$.

2.3 Time References and Timelines

Time references represent the temporal positions of the successive steps.

Definition 2. *A time reference t is a pair* $\langle D, h\rangle$ *where D is any subset of* $\mathbb{R}$ *and h a horizon variable. D is the domain of values of t and h is its horizon* i.e., *the number of steps in t. We will use* $\mathbf{D}(t)$ *and* $\mathbf{h}(t)$ *to denote respectively the domain and the horizon of a time reference t.*

In our planning problem, we can consider two time references: one time reference *to* associated with observation, with $\mathbf{h}(to) = ho$ and $\mathbf{D}(to) = \{Ts, Te\} \cup (\cup_{k=1}^{N}\{Ts[k], Te[k]\})$, and one time reference *ti* associated with instrument status, with $\mathbf{h}(ti) = hi$ and $\mathbf{D}(ti) = \{Ts, Te\} \cup (\cup_{k=1}^{N} \{Ts[k] - \Delta on, Te[k]\})$ (it would be counterproductive to switch the instrument on strictly more than Δon before starting observing and to switch it off strictly after ending observing).

Timelines are used to represent the values of the relevant attributes of the system at the successive steps.

Definition 3. *A timeline x is a pair* $\langle D, t\rangle$ *where D is the domain of values of x and t its time reference. We will use* $\mathbf{D}(x)$ *and* $\mathbf{t}(x)$ *to denote respectively the domain and the time reference of a timeline x. For brevity, we will often use* $\mathbf{h}(x)$ *to denote the horizon of the time reference of a timeline x:* $\mathbf{h}(x) = \mathbf{h}(\mathbf{t}(x))$.

In our planning problem, we can consider three timelines: a timeline *no* to represent the number of the starting observation if any, a timeline *in* to represent the current instrument status, and a timeline *en* to represent the current level of energy, with $\mathbf{t}(no) = to$ and $\mathbf{t}(in) = \mathbf{t}(en) = ti$ (timelines *in* and *en* are fully synchronized). Moreover, we have $\mathbf{D}(no) = [0..N]$ (0 when no observation is starting), $\mathbf{D}(in) = \{0, 1\}$ (0 when the instrument is off), and $\mathbf{D}(en) = [Emin..Emax]$.

2.4 Static and Dynamic Variables

Definition 4. *A static variable is either a horizon variable, or any other variable independent from time references and timelines.*

In our planning problem, it may be convenient to associate with each candidate observation $k \in [1..N]$ a variable $st[k]$ of domain $[0..2N+2]$ to represent the observation step at which observation k is performed (0 when k is not performed). It may be also convenient to use a static variable obj of domain $[0..N]$ to represent the number of observations performed.

Dynamic variables are associated with steps of time references and timelines.

Definition 5. *A time reference* t *(resp. timeline* x*) and an assignment* H *of its horizon variable together induce a finite set of variables* $\{t_i \mid i \in [1..H]\}$ *(resp.* $\{x_i \mid i \in [1..H]\}$*). This set is empty when* $H = 0$*. All these variables share the same domain of values* $\mathbf{D}(t)$ *(resp.* $\mathbf{D}(x)$*).*

See Fig. 2 for a graphical representation of time reference ti and timeline en when $\mathbf{h}(ti) = 4$. For example, variable ti_3 represents the temporal position of step 3 and variable en_3 the energy level at step 3.

Fig. 2. Graphical representation of time reference ti and timeline en

2.5 Static and Dynamic Constraints

Static constraints are used to limit the possible assignments of static variables.

Definition 6. *A static constraint* c *is simply a CSP constraint [9] whose scope is limited to static variables.*

In our planning problem, the following static constraints can be defined:

$$obj = card\{k \in [1..N] \mid st[k] \neq 0\} \quad (1)$$
$$(Even(hi)) \wedge (Even(ho)) \wedge (hi \leq ho) \quad (2)$$

Constraint 1 defines the objective to be maximized as the number of observations performed. Constraint 2 limits the values of horizon variables hi and ho.

Dynamic constraints are used to limit the possible combinations of assignments of static and dynamic variables. The difficulty is that the set of dynamic variables associated with time references and timelines is not fixed. It depends on the assignments of the horizon variables. This leads us to a definition of a dynamic constraint which is not as obvious as the previous definitions are. We will use several examples to illustrate it.

Definition 7. *A dynamic constraint c is a tuple* $\langle S, D, f \rangle$ *where S is a finite set of static variables, D is a finite set of time references and timelines, and f is a function which associates a finite set of CSP constraints with each assignment H of the horizon variables of the timelines and time references in D. Variables in the scope of these induced CSP constraints must be either static variables in S or dynamic variables in the set of dynamic variables induced by H for time references and timelines in D.*

In our planning problem, the following dynamic constraints can be defined:

$$(to_1 = Ts) \wedge (no_1 = 0) \wedge (to_{ho} = Te) \wedge (no_{ho} = 0) \tag{3}$$

$$\forall i \in [2..ho-1],\ (no_i = 0) \leftrightarrow (no_{i-1} \neq 0) \tag{4}$$

$$\forall k \in [1..N], \forall i \in [2..ho],\ (no_{i-1} = k) \rightarrow ((to_{i-1} = Ts[k]) \wedge (to_i = Te[k])) \tag{5}$$

$$\forall k \in [1..N], \forall i \in [1..ho],\ (no_i = k) \leftrightarrow (st[k] = i) \tag{6}$$

$$(ti_1 = Ts) \wedge (in_1 = 0) \wedge (en_1 = Ei) \wedge (ti_{hi} = Te) \wedge (in_{hi} = 0) \tag{7}$$

$$\forall i \in [2..hi-1],\ in_i \neq in_{i-1} \tag{8}$$

$$\forall i \in [2..hi],\ en_i = \min(Emax, en_{i-1} + (ti_i - ti_{i-1}) \cdot (Pe - Ce \cdot in_{i-1})) \tag{9}$$

$$\forall i \in [2..ho],\ (no_{i-1} \neq 0) \rightarrow (DuringVal(in, 1, to_{i-1} - \Delta on, to_i)) \tag{10}$$

More precisely, Constraints 3 and 7 define the initial and final states. With regard to Definition 7, a constraint like $to_{ho} = Te$ is a dynamic constraint defined by the tuple $c = \langle \emptyset, \{to\}, f \rangle$ with f the function which associates with each assignment H of ho the unique unary CSP constraint $to_H = Te$ on dynamic variable to_H. Constraint 4 defines the alternation of observation starting and ending for timeline no. It corresponds to a dynamic constraint defined by the tuple $c = \langle \emptyset, \{no\}, f \rangle$ with f the function which associates with each assignment H of ho the set $\{(no_i = 0) \leftrightarrow (no_{i-1} \neq 0) \,|\, i \in [2..H-1]\}$ of $(H-2)$ binary CSP constraints, each one connecting dynamic variables no_i and no_{i-1}.

Constraint 5 links timeline no and time reference to. Together with the basic assumption of the CNT framework according to which temporal positions in any time reference are totally and strictly ordered (see Sect. 2.6 below), it guarantees that there is no overlapping between observations. Constraint 6 enforces consistency between timeline no and static variables $st[k]$. Constraint 8 defines the alternation of values 0 and 1 for timeline in. Constraint 9 describes the way the energy level evolves. Note that the minimum level of energy $Emin$ is guaranteed via the domain of timeline en. Constraint 10 uses a special constraint called *DuringVal* (not detailed here) and guarantees that the instrument is switched on at least Δon before starting observing and that it remains on until ending observing. Such a constraint is called a *synchronization constraint* because it links timeline no and timeline in, which do not share the same time reference.

2.6 Constraint Networks on Timelines (CNTs)

All these definitions can be put together in order to define *constraint networks on timelines.*

Definition 8. *A constraint network on timelines is a tuple* $N = \langle V, CS, T, X, CD\rangle$ *where* V *is a finite set of static variables,* CS *is a finite set of static constraints whose scopes are included in* V*,* T *is a finite set of time references whose horizons belong to* V*,* X *is a finite set of timelines whose time references belong to* T*, and* CD *is a finite set of dynamic constraints whose scopes in terms of static variables, time references, and timelines are respectively included in* V*,* T*, and* X*.*

It is assumed that a default dynamic constraint c_t *is associated with each time reference* $t \in T$*. This constraint enforces that the temporal variables associated with* t *are totally and strictly ordered:* $\forall t \in T, \forall i \in [1..\mathbf{h}(t) - 1]$, $t_i < t_{i+1}$.

The CNT which results from the modeling of our planning problem is defined by the tuple $\langle V, CS, T, X, CD\rangle$, with $V = \{ho, hi, obj\} \cup \{st[k] \mid k \in [1..N]\}$, $CS = \{c_i \mid i \in [1..2]\}$, $T = \{to, ti\}$, $X = \{no, in, en\}$, and $CD = \{c_i \mid i \in [3..10]\}$.

2.7 Assignments, Solutions, Consistency, and Optimality

Definition 9. *An assignment of a CNT* N *is an assignment of all its static variables (including all the horizon variables) and of all the induced dynamic variables. A solution is an assignment of* N *such that all its static constraints and all the CSP constraints induced by all its dynamic constraints are satisfied. A CNT is consistent if and only if it admits a solution. An objective variable* obj *is a static variable which is a function of other static or dynamic variables and whose domain is equipped with a total order* $\succ$*. A solution* Sol *is optimal iff there is no other solution* Sol' *such that* $Sol'[obj] \succ Sol[obj]$*.*[1]

In our planning problem, $obj = npo$ and $\succ =>$. Figure 3 represents an optimal solution of the CNT which results from the modeling of the instance of our planning problem described in Fig. 1a, with $Ts = 0$, $Te = 58$, $\Delta on = 3$, $Pe = 20$, $Ce = 22$, $Ei = 150$, $Emin = 50$, and $Emax = 300$. This is a tabular representation of the solution described in Fig. 1b. Assignments of static variables are omitted.

step	1	2	3	4	5	6	7	8
to	0	4	14	24	34	36	52	58
no	0	1	0	3	0	5	0	0
ti	0	1	14	21	52	58		
in	0	1	0	1	0	0		
en	150	170	144	284	222	300		

Fig. 3. Optimal solution of the CNT resulting from the modeling of the instance of the mission management problem described in Fig. 1a

[1] $A[v]$ denotes the value of variable v in assignment A.

2.8 Comparison with Other Modeling Frameworks

From the constraint programming point of view, a CNT is a kind of dynamic (conditional) CSP [10] whose dynamic aspect comes from the assignment of the horizon variables and from the definition of the dynamic constraints. The CNT framework is in some way more specialized because it aims at modeling specific systems: discrete event dynamic systems. This justifies the presence of the basic notions of horizon and time, which do not appear in standard dynamic CSPs.

From the planning point of view, contrarily to classical frameworks such as PDDL [11], built around the notions of *action*, *precondition*, *effect*, *duration*, and resource *consumption*, the CNT framework is a more basic modeling framework, built around the notions of *time reference*, *timeline*, and *constraint*. This allows complex dynamic phenomena to be modeled. According to several features, the CNT framework is close to the CAIP framework used in the EUROPA planning system [12]. But the CNT framework is a pure CSP framework, which inherits the clear semantics and the flexibility of CSPs.

3 Slice CSP Encoding of Stationary Constraints

We can now present the slice CSP encoding of CNTs we propose. For the moment, this encoding is limited to CNTs where static constraints are of any kind, but dynamic ones are limited to so-called *stationary* dynamic constraints. Although such constraints already allow a wide range of planning problems to be modeled, extension to other dynamic constraints is discussed in Sect. 3.3.

3.1 Stationary Dynamic Constraints

To define stationary dynamic constraints, we first define *constraint generators.*

Definition 10. *A constraint generator c is a tuple $c = \langle S, G, R\rangle$ such that:*

- *S is a finite set of static variables;*
- *G is a finite sequence of pairs $\langle x, -k\rangle$ where x is either a time reference or a timeline and k a positive integer, and such that all time references and timelines x involved in G share the same horizon variable, denoted $\mathbf{h}(c)$;*
- *R is a subset of $\Pi_{v\in S}\mathbf{D}(v) \times \Pi_{(x,-k)\in G}\mathbf{D}(x)$, implicitly or explicitly defined.*

For each x involved in G, we will use $\mathbf{m}(c, x)$ to denote the memory of x in c, defined as $\mathbf{m}(c, x) = \max\{k \mid \langle x, -k\rangle \in G\}$. Finally, we will use $\mathbf{m}(c)$ to denote the memory of c, defined as $\mathbf{m}(c) = \max\{\mathbf{m}(c, x) \mid \langle x, .\rangle \in G\}$.

Definition 11. *Let $c = \langle S, G, R\rangle$ be a constraint generator. Given $i \in [\mathbf{m}(c) + 1..\max(\mathbf{D}(\mathbf{h}(c)))]$, the CSP constraint c_i generated by c at step i is a constraint whose scope is $S \cup \{x_{i-k} \mid \langle x, -k\rangle \in G\}$ and relation is R. The dynamic constraint generated by c is defined by the tuple $\langle S, D, f\rangle$ where $D = \{x \mid \langle x, .\rangle \in G\}$, and f is the function which associates with each assignment H of $\mathbf{h}(c)$ the set $\{c_i \mid i \in [\mathbf{m}(c) + 1..H]\}$ of CSP constraints* i.e., *the dynamic constraint $\forall i \in [\mathbf{m}(c) + 1..\mathbf{h}(c)]$, c_i.*

Such a dynamic constraint is referred to as a *stationary* dynamic constraint because relation R does not depend on i. For example, let us consider the constraint generator $c = \langle \emptyset, \{\langle in, 0\rangle, \langle in, -1\rangle\}, R\rangle$, where in is the timeline introduced in Sect. 2 and R is the relation which includes all pairs (a, b) such that $a \neq b$. Intuitively, this constraint generator states that the value of timeline in at the current step must differ from the value of timeline in one step before. The memory of in in c is $\mathbf{m}(c, in) = 1$, which informally means that one step before the current step needs to be considered for in. In this case, the memory of c is $\mathbf{m}(c) = 1$ too. For each $i \in [\mathbf{m}(c) + 1 .. \max(\mathbf{D}(hi))]$, the CSP constraint c_i generated by c at step i is $in_i \neq in_{i-1}$ and the stationary dynamic constraint generated by c is $\forall i \in [2..hi],\ in_i \neq in_{i-1}$.

When relation R depends on i, it is possible to remain stationary by introducing a timeline $\mathbf{st}(\mathbf{h}(c))$ whose value at each step i is equal to i. This implies introducing the stationary dynamic constraint $\forall i \in [2..\mathbf{h}(c)],\ \mathbf{st}(\mathbf{h}(c))_i = \mathbf{st}(\mathbf{h}(c))_{i-1} + 1$ together with the special initialization constraint $\mathbf{st}(\mathbf{h}(c))_1 = 1$.[2] For example, constraint $ti_1 = Ts$ can be defined as the stationary dynamic constraint $\forall i \in [1..hi],\ (\mathbf{st}(hi)_i = 1) \rightarrow (ti_i = Ts)$ and constraint $ti_{hi} = Te$ as the stationary dynamic constraint $\forall i \in [1..hi],\ (\mathbf{st}(hi)_i = hi) \rightarrow (ti_i = Te)$.

In the following, given the bijection existing between a constraint generator and its associated stationary dynamic constraint, constraint generators are directly considered as stationary dynamic constraints.

3.2 Slice Encoding of Stationary CNTs

It is now possible to define stationary CNTs and their slice encoding:

Definition 12. *A stationary CNT $N = \langle V, CS, T, X, CD\rangle$ is a CNT such that all the dynamic constraints in CD are stationary.*

For each time reference or timeline $x \in T \cup X$, $\mathbf{m}(x)$ denotes the memory of x, defined as $\mathbf{m}(x) = \max\{\mathbf{m}(c, x) \mid c \in CD\}$.

Definition 13. *Let $N = \langle V, CS, T, X, CD\rangle$ be a stationary CNT. The slice encoding of N is the CSP $\langle \mathbf{slV}(N), \mathbf{slC}(N)\rangle$ such that:*

- $\mathbf{slV}(N)$ *is the set of* slice variables *generated by N, defined as $V \cup \{x^{(-k)} \mid (x \in T \cup X) \wedge (k \in [0..\mathbf{m}(x)])\}$ with $\mathbf{D}(x^{(-k)}) = \mathbf{D}(x)$ for each k ($x^{(-k)}$ is just a variable name);*
- $\mathbf{slC}(N)$ *is the set of* slice constraints *generated by N, defined as $CS \cup \{c^{(0)} \mid c \in CD\}$ where, for each constraint generator $c = \langle S, G, R\rangle \in CD$, $c^{(0)}$, called the slice constraint associated with c, is the CSP constraint whose scope is $S \cup \{x^{(-k)} \mid \langle x, -k\rangle \in G\}$ and relation is R.*

Given a stationary CNT, its slice encoding can be built automatically. For example, let us consider the part of our illustrative planning problem limited to

[2] This special initialization constraint, which is not stationary, will be handled specifically by the algorithms defined in Sect. 4 (see function *sliceSolve*).

horizon variable hi, time reference ti, timelines in and en, and Constraints 7 to 9. To get a slice encoding, a timeline $\mathbf{st}(hi)$ is introduced to represent the current step and Constraints 7 to 9 are transformed into the following stationary dynamic constraints (associated constraint generators are omitted):

$$\forall i \in [2..hi],\ \mathbf{st}(hi)_i = \mathbf{st}(hi)_{i-1} + 1 \tag{11}$$
$$\forall i \in [1..hi],\ (\mathbf{st}(hi)_i = 1) \rightarrow ((ti_i = Ts) \wedge (in_i = 0) \wedge (en_i = Ei)) \tag{12}$$
$$\forall i \in [1..hi],\ (\mathbf{st}(hi)_i = hi) \rightarrow ((ti_i = Te) \wedge (in_i = 0)) \tag{13}$$
$$\forall i \in [1..hi],\ (2 \leq \mathbf{st}(hi)_i \leq hi - 1) \rightarrow (in_i \neq in_{i-1}) \tag{14}$$
$$\forall i \in [1..hi],\ (\mathbf{st}(hi)_i \geq 2) \rightarrow (en_i = \min(Emax, en_{i-1} + (ti_i - ti_{i-1}) \cdot (Pe - Ce \cdot in_{i-1}))) \tag{15}$$

In this particular case, all the time references and timelines have a memory of 1. But, in general, time references and timelines can have any memory greater than or equal to 0. According to Def. 13, the resulting slice encoding is as follows:

$$\mathbf{st}(hi)^{(0)} = \mathbf{st}(hi)^{(-1)} + 1 \tag{16}$$
$$(\mathbf{st}(hi)^{(0)} = 1) \rightarrow ((ti^{(0)} = Ts) \wedge (in^{(0)} = 0) \wedge (en^{(0)} = Ei)) \tag{17}$$
$$(\mathbf{st}(hi)^{(0)} = hi) \rightarrow ((ti^{(0)} = Te) \wedge (in^{(0)} = 0)) \tag{18}$$
$$(2 \leq \mathbf{st}(hi)^{(0)} \leq hi - 1) \rightarrow (in^{(0)} \neq in^{(-1)}) \tag{19}$$
$$(\mathbf{st}(hi)^{(0)} \geq 2) \rightarrow (en^{(0)} = \min(Emax, en^{(-1)} + (ti^{(0)} - ti^{(-1)}) \cdot (Pe - Ce \cdot in^{(-1)}))) \tag{20}$$

Informally speaking, the slice encoding of a CNT contains all the constraints that must be checked at each step with regard to previous steps. Contrarily to an unfolded encoding whose size is in the best case a linear function of the number of steps to be considered, the size of a slice encoding is constant, only a function of the time reference and timeline memories. In this case, it involves only 8 variables and 5 constraints.

Step	1	2	3	4	5	6	7	8	9	10
in	in_1	in_2	in_3	in_4	in_5	in_6	in_7	in_8	in_9	in_{10}
en	en_1	en_2	en_3	en_4	en_5	en_6	en_7	en_8	en_9	en_{10}
ti	ti_1	ti_2	ti_3	ti_4	ti_5	ti_6	ti_7	ti_8	ti_9	ti_{10}

Unfolded CSP encoding

$\mathbf{st}(hi)$	$\mathbf{st}(hi)^{(-1)}$	$\mathbf{st}(hi)^{(0)}$
in	$in^{(-1)}$	$in^{(0)}$
en	$en^{(-1)}$	$en^{(0)}$
ti	$ti^{(-1)}$	$ti^{(0)}$

Slice CSP encoding

Fig. 4. Comparison between a unfolded and a slice CSP encoding when $hi = 10$

3.3 Extension to Other Constraints

The slice encoding presented in Sect. 3.2 works only on stationary dynamic constraints, that is on constraints that link dynamic variables that share the same horizon variable. For example, Constraint 10 in our illustrative planning problem, which links time reference to, timeline no, and timeline in in order to synchronize observations and instrument status, cannot be managed.

Fortunately, if each horizon is used by at most one time reference, it is possible to produce an automatic slice encoding (not detailed here) of so-called *synchronization* dynamic constraints, which use functions linking timelines having distinct time references, like $DuringVal$. When an horizon h is shared by two references t, t', it is always possible to create an additional variable h', to take $\mathbf{h}(t') = h'$ instead of $\mathbf{h}(t') = h$, and to add constraint $h = h'$, so that h is used by only one time reference. As a result, it is possible to build a complete slice encoding of our illustrative planning problem and of many similar problems.

However, some constraints, such as $\mathit{allDifferent}(x_i \mid i \in [1..\mathbf{h}(x)])$, remain uncovered. To deal with them, several options could be considered: to exhibit an automatic slice encoding as done with synchronization dynamic constraints, to reformulate the constraint by introducing additional variables in order to get a stationary dynamic constraint, or to come back to an unfolded encoding for x.

4 Reasoning and Searching on a Slice CSP Encoding

Functions 1, 2, 3, and 4 show the pseudo-code of an algorithm able to reason and to search on a slice encoding of any stationary CNT.

Algorithm 1. Main function

$N = \langle V, CS, T, X, CD\rangle$ a stationary CNT, with H the set of horizon variables
sliceSolve(N)
begin
    $slV \leftarrow V \cup \{x^{(-k)} \mid (x \in T \cup X) \wedge (k \in [0..\mathbf{m}(x)])\}$
    $slC \leftarrow CS \cup \{c^{(0)} \mid c \in CD\}$
    $A \leftarrow \{\}$
    foreach $h \in H$ **do** $\mathbf{D}(st(h)^{(0)}) \leftarrow \{1\}$
    $(slV, slC, A) \leftarrow$ **propagateAndShift**(N, slV, slC, A)
    return recSolve(N, slV, slC, A)
end

The main function *sliceSolve* automatically creates the slice encoding from the definition of the static and dynamic constraints, launches the system by setting to 1 the value at slice 0 of the step variable associated with each horizon variable, then calls functions *propagateAndShift* and *recSolve*.

Function *recSolve* is a standard depth-first tree search algorithm. The only restriction is that variables on which to branch must be either static variables

Algorithm 2. Recursive search function

```
recSolve(N, slV, slC, A)
begin
    nasV ← {v ∈ V | (|D(v)| > 1)}
    nadV ← {x^(0) | (x ∈ T∪X) ∧ (|D(x^(0))| > 1) ∧ (st(h(x))^(0) ≤ min(D(h(x))))}
    naV ← nasV ∪ nadV
    if (naV = ∅) then
        return A.{(v, a) | v ∈ V, a ∈ D(v)}
    else
        (A″, opt) ← (null, +∞)
        choose v ∈ naV  and a partition {d_1, d_2} of D(v)
        for i = 1 to 2 do
            D(v) ← d_i
            (slV′, slC′, A′) ← propagateAndShift(N, slV, slC∪{obj < opt}, A)
            if (∀v′ ∈ slV′, |D(v′)| > 0) then  A′ ← recSolve(N, slV′, slC′, A′)
            if (A′ ≠ null) then  (A″, opt) ← (A′, A′[obj])
        return A″
end
```

or dynamic variables at slice 0 when the minimum value in the domain of the associated horizon variable makes this slice mandatory. This allows the search to be performed systematically forward.

Algorithm 3. Propagate and shift function

```
propagateAndShift(N, slV, slC, A)
begin
    (slV, slC) ← propagate(slV, slC)
    shift ← true
    while ((∀v ∈ slV, |D(v)| > 0) ∧ shift)  do
        shift ← false
        shH ← {h ∈ H | (∀x ∈ X ∪ T | h(x) = h, |D(x^(0))| = 1) ∧ (st(h)^(0) ≤
        min(D(h)))}
        if (shH ≠ ∅) then
            foreach h ∈ shH do (slV, slC, A) ← shift(N, h, slV, slC, A)
            (slV, slC) ← propagate(slV, slC)
            shift ← true
    return (slV, slC, A)
end
```

Function *propagateAndShift* alternates classical propagations and shifts. Shifts occur on a horizon variable h each time all the associated dynamic variables at slice 0 are assigned and the minimum value in the domain of h makes a shift mandatory (the horizon is not reached yet).

Algorithm 4. Shift function

shift(N, h, slV, slC, A)
begin
  foreach $c \in CD \mid h(c) = h$ **do**
    $slC \leftarrow slC \cup \{\mathbf{project}(c^{(0)}, V)\}$
    reinit(c)
  foreach $x \in X \cup T \mid \mathbf{h}(x) = h$ **do**
    $A \leftarrow A.\{(x_{\mathbf{st}(h)^{(0)}}, x^{(0)})\}$
    for $k = \mathbf{m}(x)$ *to* 1 **do** $\mathbf{D}(x^{(-k)}) \leftarrow \mathbf{D}(x^{(-k+1)})$
    $\mathbf{D}(x^{(0)}) \leftarrow \mathbf{D}(x)$
  return (slV, slC, A)
end

	(-1)	(0)
st	[0..4]	[0..4]
in	0, 1	0, 1
en	[2..10]	[2..10]
ti	0, 2, 6, 8	0, 2, 6, 8
h : [2..4]		

Slice encoding
(a)

	(-1)	(0)
st	[0..4]	1
in	0, 1	0, 1
en	[2..10]	[2..10]
ti	0, 2, 6, 8	0, 2, 6, 8
h : [2..4]		

Initialization
(b)

	(-1)	(0)
st	0	1
in	0, 1	0
en	[2..10]	4
ti	0, 2, 6, 8	0
h : [2..4]		

Propagation
(c)

	(-1)	(0)
st	1	[0..4]
in	0	0, 1
en	4	[2..10]
ti	0	0, 2, 6, 8
h : [2..4]		

Shift
(d)

	(-1)	(0)
st	1	2
in	0	0, 1
en	4	[5..8]
ti	0	2, 6, 8
h : [2..4]		

Propagation
(e)

	(-1)	(0)
st	1	2
in	0	0, 1
en	4	[5..8]
ti	0	2
h : [2..4]		

Choice $ti^{(0)}$=2
(f)

	(-1)	(0)
st	1	2
in	0	1
en	4	5
ti	0	2
h : 4		

Propagation
(g)

	(-1)	(0)
st	2	[0..4]
in	1	0, 1
en	5	[2..10]
ti	2	0, 2, 6, 8
h : 4		

Shift
(h)

	(-1)	(0)
st	2	3
in	1	0, 1
en	5	∅
ti	2	6, 8
h : 4		

Propagation
Inconsistency
(i)

	(-1)	(0)
st	1	2
in	0	0, 1
en	4	[5..8]
ti	0	6, 8
h : [2..4]		

Backtrack to e
Post $ti^{(0)} \neq 2$
(j)

	(-1)	(0)
st	1	2
in	0	0, 1
en	4	[7..8]
ti	0	6, 8
h : [2..4]		

Propagation
(k)

	(-1)	(0)
st	1	2
in	0	0, 1
en	4	[7..8]
ti	0	8
h : [2..4]		

Choice
$ti^{(0)} = 8$
(l)

	(-1)	(0)
st	1	2
in	0	0
en	4	8
ti	0	8
h : 2		

Propagation
→Solution
(m)

Fig. 5. An example of execution of the algorithm (*st* stands for $\mathbf{st}(hi)$)

As for function *shift*, it handles changes to be performed on the slice encoding when shifting. First, each dynamic constraint is projected to take into account the assignment of dynamic variables before forgetting it. For example, let us assume a constraint $x^{(0)} \leq v$ where x is a timeline and v a static variable. Let us assume that x must be shifted and that $x^{(0)} = 3$. Constraint $3 \leq v$ must be posted before shifting the slice encoding. Second, dynamic variable assignments at slice 0 are recorded and shifting operations are performed. Note that, due to the shifting operations, the domain of a dynamic variable $x^{(-k)}$ may decrease or increase. This contrasts with classical CSP solving where variable domains can only decrease along a search branch.

To make things more concrete, we show in Fig. 5 an example of execution of the algorithm previously described on a very small instance of the slice encoding defined by Constraints 16 to 20 in Sect. 3.2. We assume the following data: $Ts = 0$, $Te = 8$, $N = 1$, $Ts[1] = 5$, $Te[1] = 6$, $\Delta on = 3$, $Pe = 0.5$, $Ce = 1.5$, $Ei = 4$, $Emin = 2$, and $Emax = 10$. In fact, there is only one candidate observation which requires the instrument to be on at least over the interval [2..6]. Figures 5a and 5b show the slice encoding and the initialization. Figures 5c to 5e show an alternation of propagations and shifts. Figure 5f shows a branching choice followed by another alternation of propagations and shifts (Figs. 5g to 5i) which ends with inconsistency detection (there is not enough energy to switch on the instrument). After a backtrack (Fig. 5j) followed by an alternation of propagations and choices (Figs. 5k to 5m), a solution is found (the instrument is not switched on and the candidate observation is not performed).

It can be proved that, if all domains are finite (including the domains of the horizon variables), then this algorithm terminates and returns an optimal solution when the CNT is consistent and no solution when it is inconsistent. The key to the proof is the fact that, thanks to the forward approach, for each stationary dynamic constraint of the form $\forall i \in [\mathbf{m}(c)+1..\mathbf{h}(c)],\ c_i$, the algorithm guarantees at each step the satisfaction of the constraint $\forall i \in [\mathbf{m}(c)+1..\mathbf{st}(\mathbf{h}(c))^{(0)}],\ c_i$.

5 Experiments

The algorithm defined in Sect. 4 has been implemented in a CNT solver called SCOT (*Solver for Constraints On Timelines*) developed on top of the *Choco2* constraint solver (http://choco-solver.net). The implementation includes the management of synchronization dynamic constraints. As SCOT can handle unfolded and slice encodings of a same CNT, we are able to compare both approaches. Experiments were run on a SunUltra45, 1.6GHz, 1GBRAM.

Figure 6 shows the memory and time consumed by CSP creation on three planning problems of the International Planning Competition. The time limit was of 30 minutes. Unsurprisingly, results show the dramatic impact of a slice encoding. On the third problem, which involves synchronization constraints, an unfolded encoding cannot be produced within the time limit except for two instances, whereas a sliced encoding can be built for all instances.

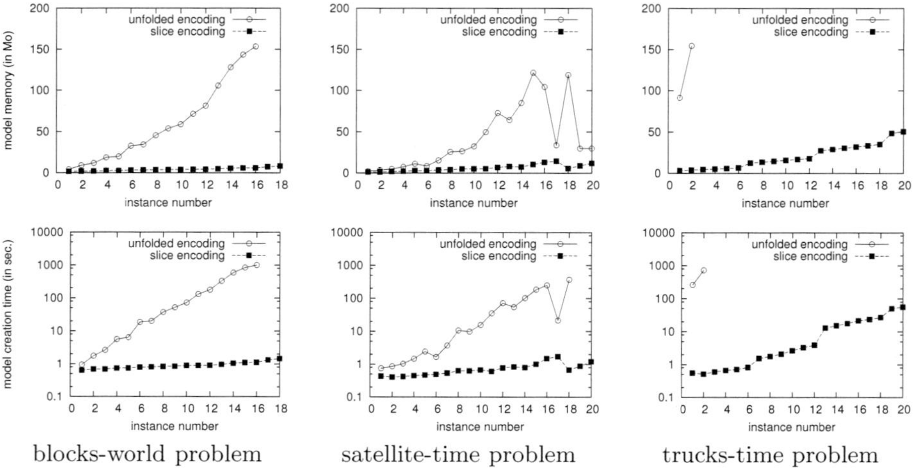

Fig. 6. CSP creation (first row: memory; second row: time)

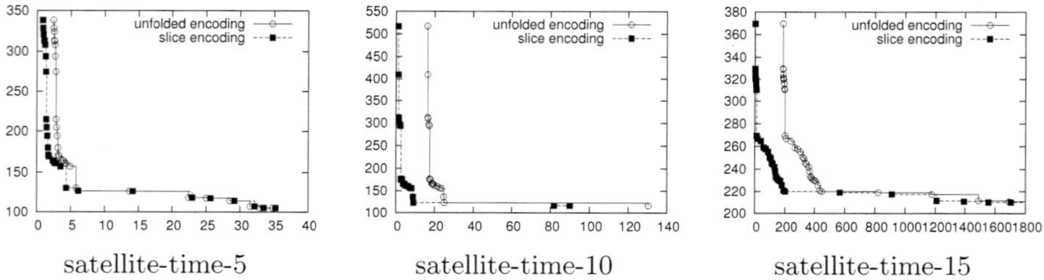

Fig. 7. Evolution of the makespan (y-axis) with cpu-time in seconds (x-axis)

Figure 7 shows the anytime quality profiles, i.e. the way plan quality evolves with cpu-time, on three instances of the second problem (*satellite-time*). On these instances, plan quality corresponds to the makespan, to be minimized. The profiles include the time necessary to create the problem. Search algorithm parameters used are the same for both encodings. Again, results show the positive impact of the slice encoding. On the three instances, the algorithm working on a slice encoding is already producing good quality solutions when the same algorithm working on an unfolded encoding is still creating the CSP. However, the search with a slice encoding can start faster than the search with an unfolded encoding but be overtaken by the latter one. This is due to contradictory effects of the slice encoding: it generates smaller problems, but it limits look-ahead propagation and can result finally in a slower search. To make up for this negative effect, slice encodings with limited look-ahead could be considered.

The comparison of SCOT with other planners is out of the scope of this paper and can be found in [13].

6 Conclusion

This paper shows (1) that the slice encoding is an alternative to the usual unfolded CSP encoding of planning problems, (2) that it is equivalent to the unfolded one thanks to a forward search approach and to changes in standard CSP algorithms, and (3) that it is far more efficient in terms of memory and time required by CSP creation and in terms of anytime quality profile. Finally, we would like to acknowledge the Choco2 developers for fruitful interactions, especially Hadrien Cambazard and Charles Prud'homme.

References

1. Ghallab, M., Nau, D., Traverso, P.: Automated Planning: Theory and Practice. Morgan Kaufmann, San Francisco (2004)
2. Kautz, H., Selman, B.: Planning as Satisfiability. In: Proc. of the 10th European Conference on Artificial Intelligence (ECAI 1992), Vienna, Austria, pp. 359–363 (1992)
3. van Beek, P., Chen, X.: CPlan: A Constraint Programming Approach to Planning. In: Proc. of the 16th National Conference on Artificial Intelligence (AAAI 1999), Orlando, FL, USA, pp. 585–590 (1999)
4. Do, M., Kambhampati, S.: Planning as Constraint Satisfaction: Solving the Planning-Graph by Compiling it into CSP. Artificial Intelligence 132(2), 151–182 (2001)
5. Miguel, I., Shen, Q., Jarvis, P.: Efficient Flexible Planning via Dynamic Flexible Constraint Satisfaction. Engineering Applications of Artificial Intelligence 14(3), 301–327 (2001)
6. Lopez, A., Bacchus, F.: Generalizing GraphPlan by Formulating Planning as a CSP. In: Proc. of the 18th International Joint Conference on Artificial Intelligence (IJCAI 2003), Acapulco, Mexico, pp. 954–960 (2003)
7. Verfaillie, G., Pralet, C., Lemaître, M.: Constraint-based Modeling of Discrete Event Dynamic Systems. Journal of Intelligent Manufacturing (2008) (published online)
8. Pralet, C., Verfaillie, G.: Using Constraint Networks on Timelines to Model and Solve Planning and Scheduling Problems. In: Proc. of the 18th International Conference on Automated Planning and Scheduling (ICAPS 2008), Sydney, Australia, pp. 272–279 (2008)
9. Rossi, R., van Beek, P., Walsh, T. (eds.): Handbook of Constraint Programming. Elsevier, Amsterdam (2006)
10. Mittal, S., Falkenhainer, B.: Dynamic Constraint Satisfaction Problems. In: Proc. of the 8th National Conference on Artificial Intelligence (AAAI 1990), Boston, MA, USA, pp. 25–32 (1990)
11. Fox, M., Long, D.: PDDL2.1: An Extension to PDDL for Expressing Temporal Planning Domains. Journal of Artificial Intelligence Research 20, 61–124 (2003)
12. Frank, J., Jónsson, A.: Constraint-Based Attribute and Interval Planning. Constraints 8(4), 339–364 (2003)
13. Pralet, C., Verfaillie, G.: Forward Constraint-based Algorithms for Anytime Planning. In: Proc. of the 19th International Conference on Automated Planning and Scheduling (ICAPS 2009), Thessaloniki, Greece (2009)

Evolving Parameterised Policies for Stochastic Constraint Programming*

Steven Prestwich[1], S. Armagan Tarim[2], Roberto Rossi[3], and Brahim Hnich[4]

[1]Cork Constraint Computation Centre, University College Cork, Ireland
[2]Operations Management Division, Nottingham University Business School, Nottingham, UK
[3]Logistics, Decision and Information Sciences Group, Wageningen UR, The Netherlands
[4]Faculty of Computer Science, Izmir University of Economics, Turkey
s.prestwich@cs.ucc.ie, armtar@yahoo.com,
roberto.rossi@wur.nl, brahim.hnich@ieu.edu.tr

Abstract. Stochastic Constraint Programming is an extension of Constraint Programming for modelling and solving combinatorial problems involving uncertainty. A solution to such a problem is a policy tree that specifies decision variable assignments in each scenario. Several solution methods have been proposed but none seems practical for large multi-stage problems. We propose an incomplete approach: specifying a policy tree indirectly by a parameterised function, whose parameter values are found by evolutionary search. On some problems this method is orders of magnitude faster than a state-of-the-art scenario-based approach, and it also provides a very compact representation of policy trees.

1 Introduction

Stochastic Constraint Programming (SCP) is a recently proposed extension of Constraint Programming (CP) designed to model and solve complex problems involving uncertainty and probability, a direction of research first proposed in [2]. Stochastic Constraint Satisfaction Problems (SCSPs) are in a higher complexity class than Constraint Satisfaction Problems (CSPs) and usually harder to solve.

An m-stage SCSP is defined as a tuple $(V, S, D, P, C, \Theta, L)$ where V is a set of decision variables, S a set of stochastic variables, D a function mapping each element of $V \cup S$ to a domain of values, P a function mapping each variable in S to a probability distribution, C a set of constraints on $V \cup S$, Θ a function mapping each constraint in C to a threshold value $\theta \in (0, 1]$, and $L = (\langle V_1, S_1 \rangle, \ldots, \langle V_m, S_m \rangle)$ a list of *decision stages* such that the V_i partition V and the S_i partition S. Each constraint must contain at least one V variable, a constraint $h \in C$ containing only V variables is a *hard constraint* with threshold $\Theta(h) = 1$, and one containing at least one S variable is a *chance constraint*. To solve an m-stage SCSP an assignment to the variables in V_1 must be found such that, given random values for S_1, assignments can be found for V_2 such that, given random values for $S_2, \ldots$ assignments can be found for V_m so that, given

* B. Hnich is supported by the Scientific and Technological Research Council of Turkey (TUBITAK) under Grant No. SOBAG-108K027. This material is based in part upon works supported by the Science Foundation Ireland under Grant No. 05/IN/I886.

I.P. Gent (Ed.): CP 2009, LNCS 5732, pp. 684–691, 2009.

random values for S_m, the hard constraints are each satisfied and the chance constraints (containing both decision and stochastic variables) are satisfied in the specified fraction of all possible *scenarios* (set of values for the stochastic variables). A useful concept is that of a *policy tree* of decisions, in which each node represents a value chosen for a decision variable, and each arc from a node represents the value assigned to a stochastic variable. Each path in the tree represents a different possible scenario and the values assigned to decision variables in that scenario. A *satisfying policy (tree)* is a policy tree in which each chance constraint is satisfied with respect to the tree. A chance constraint $h \in C$ is satisfied with respect to a policy tree if it is satisfied under some fraction $\phi \geq \Theta(h)$ of all possible paths in the tree.

As an example, consider a 2-stage SCSP with $V_1 = \{x_1\}$, $S_1 = \{s_1\}$, $V_2 = \{x_2\}$ and $S_2 = \{s_2\}$. Let $\text{dom}(x_1) = [1, 4]$, $\text{dom}(x_2) = [3, 6]$, $\text{dom}(s_1) = [4, 5]$ and $\text{dom}(s_2) = [3, 4]$ where $[a, b]$ represents the discrete interval $\{i \in \mathbf{Z} \,|\, a \leq i \leq b\}$, and the stochastic variable values each have probability 0.5. There are two chance constraints c_1: $(s_1 x_1 + s_2 x_2 \geq 30)$ and c_2: $(s_2 x_1 = 12)$ with $\theta_{c_1} = 0.75$ and $\theta_{c_2} = 0.5$. Decision variable x_1 must be set to a unique value while the value of x_2 depends on that of s_1. A policy for this problem is shown in Figure 1: notice that it is in the form of a tree. The 4 scenarios A, B, C and D each have probability 0.25. Constraint c_1 is satisfied in A, C and D therefore with probability 0.75. Constraint c_2 is satisfied in A and C therefore with probability 0.5. These probabilities satisfy the thresholds $\theta_{c_1}, \theta_{c_2}$ so this is a satisfying policy.

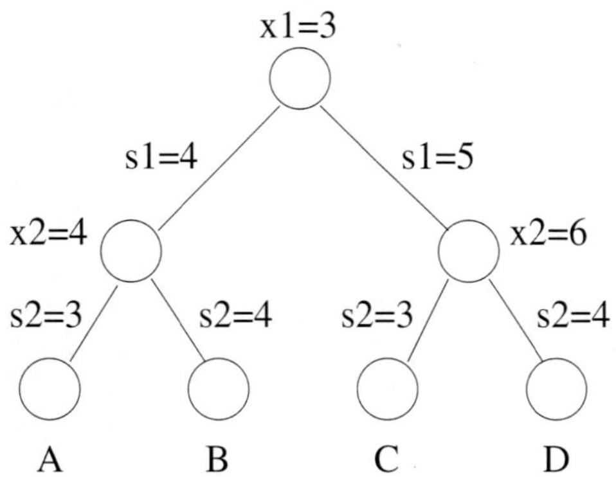

Fig. 1. Example of a satisfying policy tree

No practical way of solving large multi-stage SCSPs has yet been proposed. The design of local search algorithms for SCP has been suggested [20] in order to improve scalability but this idea does not seem to have been pursued, and it does not address the problem of representing large policy trees. We propose a novel approach: using an evolutionary algorithm to choose parameter values for a parameterised function that indirectly specifies a policy tree. The result is an incomplete SCP algorithm that is intended to scale well in two ways: a simple parameterised function can be used to represent a large policy tree, and evolutionary search can handle problems with many decision variables.

The paper is organised as follows. Section 2 describes how to evolve parameterised functions that specify policies. Section 3 shows empirically that an evolutionary algorithm can find a function representing a satisfying policy. Section 4 discusses related work. Section 5 concludes the paper.

2 Evolving Parameterised Policies

Instead of explicitly representing a policy tree we use a parameterised function $\tau_{\underline{w}}$, whose input is the current stochastic variable assignments and a decision variable, and whose output is a domain value for that variable. Its parameters $\underline{w} = (w_1, w_2 \ldots)$ are real-valued numbers which we shall call *weights*. $\tau_{\underline{w}}$ completely defines the policy tree, and if it does not require an exponential number of weights then it avoids the memory problem associated with large trees. For any given function there exist policy trees that cannot be represented, and there is a risk that these are the only satisfying policy trees, but the hope is that relatively simple functions will suffice for most problems of interest.

To simplify the discussion we consider only SCSPs whose decision and stochastic variable domains are intervals $[L, U]$ of integer values, but the method easily generalises to variables with other domains. We assume a fixed ordering of the problem variables (any ordering that conforms to the stage structure will do). First we compute an affine function

$$\alpha_{\underline{w}}(S, x_j) = w_j + \sum_{i \in \sigma_j} w_i s_i$$

where σ_j denotes the set of indices of the stochastic variables S that precede decision variable x_j. This is the simplest possible function that involves all relevant stochastic variables; we do not claim that it will suffice for all SCSPs, but it requires only a linear number of weights and works well in experiments so far (see Section 3). The constant w_j is necessary because in the case of a decision variable x_j that is not preceded by any stochastic variable (so that $\sigma_j = \emptyset$) we require a default value: in the special cases of a deterministic CSP or a 1-stage SCSP no decision variable is preceded by a stochastic variable, so the policy is simply a weight w_j for each decision variable x_j. Note that the stochastic variables S_m (those in the final stage) do not precede any decision variables, and therefore do not appear in $\underline{w}$: thus they do not appear in the policy, though they are used to evaluate it.

However, the value of $\alpha_{\underline{w}}(S, x_j)$ is a real number and not a domain value, so to obtain an integer in $[L, U]$ it is discretised by truncation, then modular arithmetic is used to obtain an integer in the required range:

$$\tau_{\underline{w}}(S, x_j) = L + (\lfloor \alpha_{\underline{w}}(S, x_j) \rfloor \bmod [U - L + 1])$$

Now $\underline{w}$ defines a policy: for each decision variable x_j we choose its value to be $\tau_{\underline{w}}(S, x_j)$.

For example, consider a 3-stage problem with $V_1 = \{x_1, x_2\}$, $S_1 = \{s_1, s_2\}$, $V_2 = \{x_3, x_4\}$, $S_2 = \{s_3, s_4\}$, $V_3 = \{x_5, x_6\}$ and $S_3 = \{s_5, s_6\}$. Suppose we wish to find the value of x_3 given that $s_1 = 5$, $s_2 = 7$, dom$(x_3) = [5, 10]$ and our policy is specified by a weight vector

$$\underline{w} = (0.1, 5.3, 7.1, 9.9, 8.7, 4.1, -0.6, 5.5, -5.2, 2.9)$$

Notice that $\underline{w}$ has 10 components though there are 12 variables in the SCSP: this is because the S_3 variables do not precede any decision variables, as mentioned above. Then

$$\alpha_{\underline{w}}(S, x_3) = 8.7 + 7.1s_1 + 9.9s_2 = 113.5$$

and

$$\tau_{\underline{w}}(S, x_3) = 5 + (\lfloor 113.5 \rfloor \bmod [10 - 5 + 1]) = 5 + (113 \bmod 6) = 5 + 5 = 10$$

So under the policy defined by $\underline{w}$, variable x_3 is set to 10 under any scenario in which $s_1 = 5$ and $s_2 = 7$.

Now that we have defined the form of our policies we can describe how to search for them. The state space to be explored is the Cartesian product $\mathbf{R}^k$ representing the space of real-valued weight vectors $\underline{w}$, where k is the total number of SCP variables not counting those in S_m. To handle the SCP constraints we use *penalty functions* to obtain an unconstrained optimisation problem: this is a standard technique that penalises constraint violations, commonly used when applying genetic algorithms or local search to CSPs. Specifically, the objective function to be minimised is

$$\Phi(\underline{w}) = \sum_{h \in C} \phi(h, \underline{w})$$

where the penalty functions are

$$\phi(h, \underline{w}) = \begin{cases} 0 & \text{if } \pi_h(\underline{w}) \geq \Theta(h) \\ \Theta(h) - \pi_h(\underline{w}) & \text{if } \pi_h(\underline{w}) < \Theta(h) \end{cases}$$

and $\pi_h(\underline{w})$ is the probability that that h is satisfied under the policy defined by $\underline{w}$. Any policy defined by $\underline{w}$ such that $\Phi(\underline{w}) = 0$ is clearly a satisfying policy.

Given the search space and objective function we can apply an evolutionary (or local) search algorithm to solve the problem. In this paper we do not describe the algorithm we used in detail because our emphasis is on showing the feasibility of the approach. Briefly, it is a cellular evolution strategy with Cauchy mutation, plus some additional mutation heuristics designed for this application. Each chromosome is a weight vector $\underline{w}$, and for each chromosome we compute its fitness $\Phi(\underline{w})$ (fitness is conventionally maximised but we minimise Φ). To compute the $\pi_h(\underline{w})$ we check every leaf node in the implied policy tree. The probability associated with a leaf ℓ is the product of the probabilities associated with the stochastic variable assignments in the arcs of the path leading to ℓ. At each leaf a chance constraint $h \in C$ is either satisfied or violated, and by summing the probabilities of the leaves at which h is satisfied we obtain the probability that $\pi_h(\underline{w})$ that h is satisfied under the policy defined by $\underline{w}$. The $\pi_h(\underline{w})$ can also be estimated by sampling the leaves using any of the scenario reduction techniques used in [18], and this is important for problems with many stages. But we can sample many more leaves than [18] because we do not use them to derive a deterministic CSP (in section 3 we use over 1000 scenarios). Chance and hard constraints are treated uniformly: the only difference between them is that a hard constraint h has $\Theta(h) = 1$ while a chance constraint has $\Theta(h) < 1$. We could compute $\pi_h(\underline{w})$ for every chromosome by using all leaves, but to be more efficient we use a number of leaves that depends on how promising the current fitness estimate is: only the fittest chromosomes (including the one representing the satisfying policy tree) sample all leaves. To do this we use the resampling scheme of [15].

We call our method EPP (Evolved Parameterised Policies). EPP transforms a multi-stage SCSP into a noisy numerical optimisation problem. The word "noisy" here refers to the fact that the objective function must be averaged over many scenarios. There are many evolutionary algorithms designed to handle noisy fitness functions: see [3,9] for surveys.

3 Experiments

In this section we show empirically that it is possible to find a satisfying policy using EPP. We use a benchmark set of random SCSPs with 5 chance constraints over 4 decision variables $x_1 \dots x_4$ and 8 stochastic variables $s_1 \dots s_8$. The decision variable domains are the discrete intervals $\text{dom}(x_1) = [5, 10]$, $\text{dom}(x_2) = [4, 10]$, $\text{dom}(x_1) = [3, 10]$ and $\text{dom}(x_2) = [6, 10]$. The domains of stochastic variable s_1, s_3, s_5, s_7 contain 2 values while those of s_2, s_4, s_6, s_8 contain 3 values; in both bases the values are chosen randomly from the discrete interval $[1, 5]$ and have equal probabilities. The chance constraints are:

$$\begin{aligned} x_1 s_1 + x_2 s_2 + x_3 s_3 + x_4 s_4 &= 80 \quad (\theta = \alpha) \\ x_1 s_5 + x_2 s_6 + x_3 s_7 + x_4 s_8 &\leq 100 \quad (\theta = \beta) \\ x_1 s_5 + x_2 s_6 + x_3 s_7 + x_4 s_8 &\geq 60 \quad (\theta = \beta) \\ x_1 s_2 + x_3 s_6 &\geq 30 \quad (\theta = 0.7) \\ x_2 s_4 + x_4 s_8 &= 20 \quad (\theta = 0.05) \end{aligned}$$

where $\alpha \in \{0.005, 0.01, 0.03, 0.05, 0.07, 0.1\}$ and $\beta \in \{0.6, 0.7, 0.8\}$. The problems are 4-stage: $V_1 = \{x_1\}$, $S_1 = \{s_1, s_5\}$, $V_2 = \{x_2\}$, $S_2 = \{s_2, s_6\}$, $V_3 = \{x_3\}$, $S_3 = \{s_3, s_7\}$, $V_4 = \{x_4\}$ and $S_4 = \{s_4, s_8\}$. In total we have 6 α-values and 3 β-values, and we randomly generate 5 different sets of stochastic variable domains, giving 90 instances in total.

The table in Figure 2 compares the scenario-based approach (SBA) of [18] (see Section 4) with EPP. All figures are in seconds and "—" denotes that the time is greater than 200 seconds. All times were obtained on a 2.8 GHz Pentium (R) 4 with 512 MB RAM, or on another machine then normalised to this one. EPP figures are medians over 30 runs. Both methods used all $2^4 \times 3^4 = 1296$ scenarios. Though these are quite small SCSPs they turn out to be non-trivial for SBA, which transforms them into deterministic CSPs with 6739 variables and 6485 constraints. In contrast, EPP transforms them into unconstrained noisy optimisation problems with 10 real-valued variables.

A clear pattern emerges from the results: where SBA solved a problem it was up to 48 times faster than EPP, but EPP solved every problem that SBA solved plus many more, and in some cases EPP was at least 2000 times faster; EPP is on average much faster than SBA. Where SBA and EPP both failed to solve an instance, the instance might be infeasible. However, we do know that in a few cases both SBA and EPP failed to solve a feasible instance (verified by further experiments) so there is room for improvement. It might be that our parameterised policy space does not contain satisfying policies for these problems, and that a more complex parameterised function is required.

problem set 1				problem set 2				problem set 3				problem set 4				problem set 5			
α	β	SBA	EPP	α	β	SBA	EPP	α	β	SBA	EPP	α	β	SBA	EPP	α	β	SBA	EPP
0.6	0.05	—	0.5	0.6	0.05	—	1.6	0.6	0.05	0.7	0.4	0.6	0.05	—	4.2	0.6	0.05	—	0.1
0.6	0.10	—	1.0	0.6	0.10	—	4.8	0.6	0.10	0.5	3.1	0.6	0.10	—	—	0.6	0.10	—	0.5
0.6	0.12	—	0.9	0.6	0.12	—	14	0.6	0.12	0.5	3.1	0.6	0.12	—	—	0.6	0.12	—	0.7
0.6	0.15	—	1.4	0.6	0.15	—	15	0.6	0.15	—	15	0.6	0.15	—	—	0.6	0.15	—	0.8
0.6	0.17	—	1.7	0.6	0.17	—	118	0.6	0.17	—	14	0.6	0.17	—	—	0.6	0.17	—	2.2
0.6	0.20	—	1.6	0.6	0.20	—	—	0.6	0.20	—	49	0.6	0.20	—	—	0.6	0.20	—	1.9
0.7	0.05	—	1.3	0.7	0.05	—	1.7	0.7	0.05	0.6	2.5	0.7	0.05	—	4.9	0.7	0.05	0.2	0.1
0.7	0.10	—	1.2	0.7	0.10	—	4.8	0.7	0.10	0.7	9.1	0.7	0.10	—	—	0.7	0.10	—	0.4
0.7	0.12	—	1.3	0.7	0.12	—	16	0.7	0.12	0.6	12	0.7	0.12	—	—	0.7	0.12	—	0.7
0.7	0.15	—	1.9	0.7	0.15	—	16	0.7	0.15	—	27	0.7	0.15	—	—	0.7	0.15	—	0.8
0.7	0.17	—	2.7	0.7	0.17	—	144	0.7	0.17	—	46	0.7	0.17	—	—	0.7	0.17	—	1.8
0.7	0.20	—	2.8	0.7	0.20	—	—	0.7	0.20	—	159	0.7	0.20	—	—	0.7	0.20	—	3.3
0.8	0.05	—	12	0.8	0.05	—	2.7	0.8	0.05	0.8	9.7	0.8	0.05	—	7.5	0.8	0.05	—	0.2
0.8	0.10	—	9.4	0.8	0.10	—	7.1	0.8	0.10	0.6	17	0.8	0.10	—	—	0.8	0.10	—	0.9
0.8	0.12	—	11	0.8	0.12	—	20	0.8	0.12	0.6	29	0.8	0.12	—	—	0.8	0.12	—	0.8
0.8	0.15	—	12	0.8	0.15	—	13	0.8	0.15	—	58	0.8	0.15	—	—	0.8	0.15	—	1.2
0.8	0.17	—	15	0.8	0.17	—	—	0.8	0.17	—	109	0.8	0.17	—	—	0.8	0.17	—	1.6
0.8	0.20	—	13	0.8	0.20	—	—	0.8	0.20	—	—	0.8	0.20	—	—	0.8	0.20	—	3.3

Fig. 2. Experimental results

4 Related Work

Several SCSP solution methods have been proposed. [20] presented two complete algorithms based on backtracking and forward checking and suggested some approximation procedures, while [1] described an arc-consistency algorithm. In the method of [18] an SCSP is transformed into a *deterministic equivalent* Constraint Satisfaction Problem (CSP) and solved by standard CP methods. It is also extended to handle multiple chance constraints and multiple objective functions. This method gives much better performance on the book production planning problem of [20] compared to the tree search methods. To reduce the size of the CSP *scenario reduction* methods are proposed, as used in Stochastic Programming. These choose a small but representative set of scenarios. However, it might not always be possible to find a small representative set of scenarios, and in some cases choosing an inappropriate set of scenarios can yield an unsolvable CSP. Moreover, using even a modest number of scenarios leads to a CSP that is several times larger than the original SCSP. [4] modify a standard backtracking algorithm to one that can handle multiple chance constraints and uses polynomial space, but is inefficient in time. [16] proposed a cost-based filtering technique for SCP. For the special case of SCP with linear recourse, [19] propose a Bender's decomposition algorithm.

Stochastic Boolean Satisfiability (SSAT) is related to SCP. A recent survey of the SSAT field is given in [13], on which we base this discussion. An SSAT problem can be regarded as an SCSP in which all variable domains are Boolean, all constraints are extensional and may be non-binary, and all constraints are treated as a single chance constraint (there are also restricted and extended versions). Our method therefore applies immediately to SSAT problems. SSAT algorithms fall into three classes: systematic, approximation, and non-systematic. Systematic algorithms are based on the standard SAT backtracking algorithm and correspond roughly to some current SCP algorithms. Approximation algorithms work well on restricted forms of SSAT but less well on general SSAT problems. For example the APPSSAT algorithm [12] considers scenarios in decreasing order of probability to construct a partial tree, but does not work well when all scenarios have similar probability. A non-systematic algorithm for SSAT is randevalssat [10], which applies local search to the decision (existential) variables in a random set of scenarios. This algorithm also suffers from memory problems because it must build a partial tree.

5 Conclusion

We have proposed a method for SCP called EPP, based on the evolution of a parameterised function that indirectly specifies a policy tree. EPP does not suffer from the memory problems of most methods and does not introduce a large number of new variables. It is also the first incomplete algorithm for SCP, and experiments show that on some problems EPP is several orders of magnitude faster than the current best (complete) method. It does not exploit constraint filtering techniques but these could perhaps be used to handle hard constraints. EPP will also require slight modification for handling variable domains that contain arbitrary integers or real numbers, and for handling problems with objective functions. We will explore these issues in future work and test EPP on more interesting SCP problems, and also on SSAT, QBF and QCSP problems which can all be modelled as SCSPs.

EPP is closely related to a machine learning method that has been used for many optimisation problems involving uncertainty: *neuroevolution*. In neuroevolution the parameterised function is an artificial neural network whose parameters are the network weights, which are found by evolutionary search. Unlike our simple function, neural networks are universal function approximators which can in principle approximate any policy. They might turn out to be necessary for harder SCP problems, but on our benchmark set they had no effect other than to make the problem harder to solve, because they must learn more weights. Neuroevolution has been applied to very challenging control problems with good results: see for example [7,8,17]. It has also been used for learning to play games such as Backgammon [14], Go [11], Checkers [5] and Chess [6]. These successes indicate that EPP might work well on real-world SCP problems that are too large to solve by complete methods.

References

1. Balafoutis, T., Stergiou, K.: Algorithms for Stochastic CSPs. In: Benhamou, F. (ed.) CP 2006. LNCS, vol. 4204, pp. 44–58. Springer, Heidelberg (2006)
2. Benoist, T., Bourreau, E., Caseau, Y., Rottembourg, B.: Towards Stochastic Constraint Programming: A Study of On-Line Multi-Choice Knapsack with Deadlines. In: Walsh, T. (ed.) CP 2001. LNCS, vol. 2239, pp. 61–76. Springer, Heidelberg (2001)

3. Beyer, H.-G.: Evolutionary Algorithms in Noisy Environments: Theoretical Issues and Guidelines for Practice. Computer Methods in Applied Mechanics and Engineering 186(2-4), 239–267 (2000)
4. Bordeaux, L., Samulowitz, H.: On the Stochastic Constraint Satisfaction Framework. In: ACM Symposium on Applied Computing, pp. 316–320 (2007)
5. Fogel, D.B., Chellapilla, K.: Verifying Anaconda's Expert Rating by Competing Against Chinook: Experiments in Co-Evolving a Neural Checkers Player. Neurocomputing 42(1-4), 69–86 (2002)
6. Fogel, D.B., Hays, T.J., Hahn, S.L., Quon, J.: A Self-Learning Evolutionary Chess Program. Proceedings of the IEEE 92(12), 1947–1954 (2004)
7. Gomez, F., Schmidhuber, J., Miikkulainen, R.: Efficient Non-Linear Control Through Neuroevolution. Journal of Machine Learning Research 9, 937–965 (2008)
8. Hewahi, N.M.: Engineering Industry Controllers Using Neuroevolution. Artificial Intelligence for Engineering Design, Analysis and Manufacturing 19(1), 49–57 (2005)
9. Jin, Y., Branke, J.: Evolutionary Optimization in Uncertain Environments — a Survey. IEEE Transactions on Evolutionary Computation 9(3), 303–317 (2005)
10. Littman, M.L., Majercik, S.M., Pitassi, T.: Stochastic Boolean Satisfiability. Journal of Automated Reasoning 27(3), 251–296 (2001)
11. Lubberts, A., Miikkulainen, R.: Co-Evolving a Go-Playing Neural Network. In: Genetic and Evolutionary Computation Conference, pp. 14–19. Kaufmann, San Francisco (2001)
12. Majercik, S.M.: APPSSAT: Approximate Probabilistic Planning Using Stochastic Satisfiability. International Journal of Approximate Reasoning 45(2), 402–419 (2007)
13. Majercik, S.M.: Stochastic Boolean Satisfiability. In: Handbook of Satisfiability, ch. 27, pp. 887–925. IOS Press, Amsterdam (2009)
14. Pollack, J.B., Blair, A.D.: Co-Evolution in the Successful Learning of Backgammon Strategy. Machine Learning 32(3), 225–240 (1998)
15. Prestwich, S.D., Tarim, S.A., Rossi, R., Hnich, B.: A Steady-State Genetic Algorithm With Resampling for Noisy Inventory Control. In: Rudolph, G., Jansen, T., Lucas, S., Poloni, C., Beume, N. (eds.) PPSN 2008. LNCS, vol. 5199, pp. 559–568. Springer, Heidelberg (2008)
16. Rossi, R., Tarim, S.A., Hnich, B., Prestwich, S.D.: Cost-Based Domain Filtering for Stochastic Constraint Programming. In: Stuckey, P.J. (ed.) CP 2008. LNCS, vol. 5202, pp. 235–250. Springer, Heidelberg (2008)
17. Stanley, K.O., Miikkulainen, R.: Evolving Neural Networks Through Augmenting Topologies. Evolutionary Computation 10(2), 99–127 (2002)
18. Tarim, S.A., Manandhar, S., Walsh, T.: Stochastic Constraint Programming: A Scenario-Based Approach. Constraints 11(1), 1383–7133 (2006)
19. Tarim, S.A., Miguel, I.: A Hybrid Bender's Decomposition Method for Solving Stochastic Constraint Programs with Linear Recourse. In: Hnich, B., Carlsson, M., Fages, F., Rossi, F. (eds.) CSCLP 2005. LNCS (LNAI), vol. 3978, pp. 133–148. Springer, Heidelberg (2006)
20. Walsh, T.: Stochastic Constraint Programming. In: 15th European Conference on Artificial Intelligence (2002)

Maintaining State in Propagation Solvers

Raphael M. Reischuk[1], Christian Schulte[2], Peter J. Stuckey[3], and Guido Tack[4]

[1] IS&C, Saarland University, Saarbrücken, Germany
reischuk@cs.uni-sb.de
[2] KTH - Royal Institute of Technology, Sweden
cschulte@kth.se
[3] National ICT Australia, Victoria Laboratory, Department of Computer Science and Software Engineering, University of Melbourne, Australia
pjs@cs.mu.oz.au
[4] PS Lab, Saarland University, Saarbrücken, Germany
tack@ps.uni-sb.de

Abstract. Constraint propagation solvers interleave propagation, removing impossible values from variable domains, with search. The solver state is modified during propagation. But search requires the solver to return to a previous state. Hence a propagation solver must determine how to maintain state during propagation and forward and backward search. This paper sets out the possible ways in which a propagation solver can choose to maintain state, and the restrictions that such choices place on the resulting system. Experiments illustrate the result of various choices for the three principle state components of a solver: variables, propagators, and dependencies between them. This paper also provides the first realistic comparison of trailing versus copying for state restoration.

1 Introduction

Constraint propagation solvers interleave propagation (removing impossible values from variable domains) with complete tree search. During propagation and search, these solvers modify their state. Backtracking during search requires the solvers to recover a state that is equivalent to a previous node in the search tree.

The state of a propagation solver, often called the *constraint store*, principally represents three different kinds of information. (1) The information about what *variables* are involved in the problem and what possible values they can take (their domains); (2) the *propagators* implementing the constraints of the problem; and (3) *dependencies* between variables and propagators: which propagators should be (re-)executed on changes to each variable domain.

Each of these kinds of information can change during search. (1) Variables' domain information changes (the key state change made by a propagation solver). New variables may be added by the search (this is uncommon) or by the decomposition of existing propagators. Existing variables may become irrelevant if they are involved in redundant constraints only and are not of interest in the final solution. (2) Propagators can become propagation redundant (they cannot have further effect) and can hence be deleted. New propagators can be added by

I.P. Gent (Ed.): CP 2009, LNCS 5732, pp. 692–706, 2009.

the search. Propagators can replace themselves with more efficient versions (or even split into multiple propagators) as execution proceeds. (3) Dependencies between variables and propagators can change because a variable becomes irrelevant to a propagator (when the variable is fixed, for example). Dependencies may change as execution proceeds, as in watched literals for Boolean clauses.

There are essentially two ways a propagation solver can store and access state information:

globally: all search nodes access the same global state; and
locally: each search node has a local, independent copy of the state.

Given the way in which state information is stored, there may be restrictions on its use. For global state information we have essentially three options:

static: this state may never be modified during the computation.
backtrack-safe: all changes made to the state are such that the state remains equivalent to all previous states on the path from the root node to the current node. The canonical example for this is watched literals for clausal propagators [12]. Although the literals that are watched are modified as search proceeds forward, the resulting state is correct for all previous states.
trailed: when backtracking to a previous node on the path from the root node, any changes to the state between the current and the previous node are undone. The mechanism for storing the undo information is called a *trail*.

For local state, there is only one option. The solver can make arbitrary changes, as they only affect a single search node. Local state is implemented by combining *copying* and *recomputation*: for some nodes in the search tree, a copy of the state of that node is stored; when backtracking, the closest copy on the path from the root is searched, copied, and the remaining search steps to the target node are redone. Finally a fixpoint is computed for the target node.

Global and local storage approaches differ in how optimistic they are. Global trailed state is *optimistic* in that it assumes that only a small part of the state will need restoration upon backtracking; local (copied) state is *pessimistic* in this sense. Both are pessimistic when it comes to search, as they assume that eventually, the solver will have to backtrack. In this sense, recomputation for local state is optimistic, as failure requires recomputation, which may be more expensive than just untrailing the changes or restarting from a copy.

Due to the different nature of the two approaches, each has advantages over the other. Global state using trailing (and other mechanisms where applicable) offers several advantages over local state. First, less copying is required, since the part that has changed can be substantially smaller than the entire state. Second, backtracking is potentially cheaper than when using recomputation. Finally, it is easier to share information (for example objective function values) between search nodes, because information can be global. On the other hand, local state which is copied and recomputed has the advantages of:

- Simplicity of propagator implementation: each propagator has its own local state information that it can arbitrarily adjust;

- Forward simplification, removing useless parts of the state such as propagation-redundant propagators (trailing systems must trail such removals);
- Tuning of required memory by adjusting recomputation parameters;
- Easy parallel search: each thread can work on a local copy independent of the rest of the computation;
- Straightforward support for best-first search: moving to a different part of the search tree just means moving to a different copy. A solver based on global trailed state can only jump between nodes using recomputation [13].

Contributions. In this paper we explore the advantages and disadvantages of different state restoration strategies, we show the interdependence of choices for state restoration within a CP system, and we present experiments that measure the performance for various choices in state restoration.

We give an architecture for a fully hybrid system that uses trailing for propagators and recomputation for domains. The paper is the first to identify and analyze the interdependence of important design decisions with respect to different state restoration strategies.

We provide the first comparison of trailing and recomputation based on a production-quality system, Gecode [18]. Both trailing and copying use the same propagation loop, search control, and (as far as possible) propagators and variable data structures, resulting in meaningful results and a fair comparison.

2 Propagation Solvers

This section reviews the basic design of propagation solvers, and which stateful entities they consist of.

The principle components of a constraint programming problem are: a set of variables $\mathcal{V}$ each with a set of possible values, its domain $D(v), v \in \mathcal{V}$ and a set of constraints C on variables $\mathcal{V}$. A constraint c is implemented by a propagator f which in the abstract is a function from domains to domains, e.g. $f(D) = D'$, removing values from the domains of the variables in c that cannot take part in any solution of c that is possible in the domain D.

In practice, attached to each variable $v \in \mathcal{V}$ is information about its current domain $D(v)$. In addition, for each event changing the variable's domain, a *dependency* list of propagators f (to be woken on this event) is also attached. The usual set of events (for integer variables) consists of: *fix*, the variable becomes fixed; *bnd*, the lower or upper bound of the variable changes; and *dom*, some other change to the domain is made. Each propagator f stores at least the variables $vars(c)$ for the constraint c that it implements, as well as internal state relevant to its implementation.

Search. The propagation solver interleaves propagation with search. A *search tree* is constructed and explored incrementally. Each *node* n in the search tree has an attached state S_n.

Search begins from a *root node* ϵ whose state reflects the original problem after the propagation loop has executed (the state of a node always represents a fixpoint of the propagation loop). Given a *current node* n we determine a set of choice constraints $c_1, \ldots, c_m$ whose disjunction is a consequence of the current state (for example $v = d \lor v \neq d$). The choice defines unexplored *child nodes* $n_i = n.c_i$. To visit node n_i we add propagators representing c_i to the current state S_n and execute the propagation loop to arrive at state S_{n_i}.

If the propagation loop detects failure on the current node n then we choose to visit an unexplored *target node* n' in the search tree. We construct a correct starting state (before propagation) for the target node by copying, recomputing and/or untrailing and then running the propagation loop.

Propagation loop. The propagation loop is the "inner loop" of the propagation solver where it spends the bulk of its computation time. The propagation loop, given a domain D and set of propagators F, computes D' as the greatest fixpoint of all $f \in F$ less than D. The loop starts with an initial queue of propagators $F' \subseteq F$ not guaranteed to be at fixpoint with D. The first propagator from the queue is removed and executed, possibly changing the domains of its variables. These changes create events like *fix*, *bnd* and *dom* which access the dependencies to place more propagators in the queue. Once the propagator has completed, the next propagator from the queue is selected, until the queue is empty. For a complete discussion of the propagation loop, see Schulte and Stuckey [17].

Propagation solver state. There are essentially five kinds of state in a propagation solver: *variable* names; *domain* information for variables (and other variable state); *propagator* names; *internal propagator state*; and *dependencies* between variables and propagators. We separate the names of variables and propagators from their attached state since we may want to use different storage arrangements for each. Note that the propagator queue is always empty at choicepoints, and hence is not part of the state.

Figure 1 shows the access patterns of information in the propagation loop. An arrow $s \rightarrow e$ represents that from data s you can access data e. The importance of this diagram is that it captures interdependencies in choices of storage mechanisms. For an arrow $s \rightarrow e$ if s is local and e is global then this is easily handled since the local object can refer to global names. But if s is global and e is local then we need a way to find which local copy we are referring to, implying

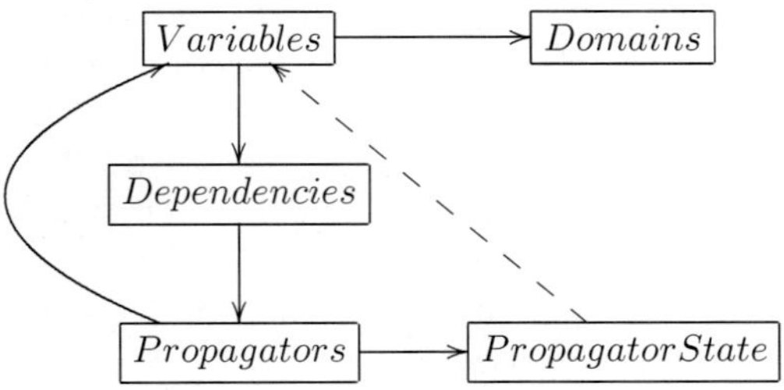

Fig. 1. Diagram of access patterns in a propagation solver

a global name *en* for the local *e* objects and a map $map(en, node) \rightarrow e$. Typically the map is implemented by having the global name be an index in an array or a memory offset, and having each node use the same array/memory block. This of course prevents shrinking the array/memory block in the local state as execution proceeds. Also note that many times propagator state does not refer directly to variables but uses the local names in the propagator (hence the dashed arrow).

Systems. Current constraint programming systems can be roughly categorized as either global state or local state systems, depending on whether their state restoration is predominantly based on trailing or copying with recomputation.

- Constraint solvers based on Prolog such as ECL^iPS^e [2] or SICStus [3] traditionally use global state, and state is recovered by trailing. The same is true for several constraint solvers based on object-oriented programming languages, such as ILOG Solver [9], CHOCO [4], or JaCoP [10].
- Gecode [6, 18] is based on local state, using copying and recomputation.
- Minion [7] is a mainly local state system. Variables and propagators are global static. Dependencies can change in a backtrack safe manner. Most domains are stored locally; for Booleans the domain representation is split: half local and half global in a backtrack safe manner; for one kind of integer variables, the domain is kept in a trailed state. Propagator state may be backtrack-safe or local. In addition to these features described in [7, 8], a look at the source code reveals that recent versions of Minion use trailed state for the internal state of some propagators.
- Figaro [5] allows various choices for state representation. Variables and propagators are global, but domains and propagator state can be: local and copied; lazily copied, i.e., they point to a previous copy until modified; or coarsely trailed, i.e., they trail only the first change to a variable or propagator by trailing the complete old value. Dependencies appear to be static.

3 A Hybrid System

This section briefly reviews three techniques for restoring solver state: trailing, copying, and recomputation. It then presents an architecture for a hybrid system that combines the different approaches.

Trailing. Global state that is neither static nor backtrack-safe can be restored upon backtracking using a *trail* of undo information. In its simplest form, before updating information at address a, a pair of a and the current value v at a is pushed on a stack. When backtracking, the solver restores the memory location at address a to v. This is shown in Fig. 2 (a) and (b): when the solver backtracks from the failed (square) node, it undoes the three trail entries, adds the choice constraint c_5, and computes a fixpoint for the dashed target node.

In principle, this technique can be used to store the necessary undo information for arbitrary state, but it can become inefficient when large parts of the state need to be trailed. To overcome these inefficiencies, *multi-value trailing*

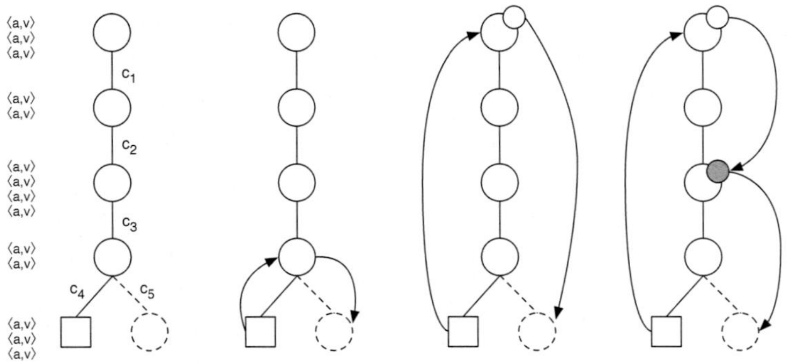

Fig. 2. (a) choices and trail, (b) untrailing, (c) fixed, and (d) adaptive recomputation

allows one to trail an address a and a *vector* of values at consecutive memory cells starting at a, and *time stamping* remembers whether an address a has already been trailed for the current search node and hence does not need trailing again [1]. *Function trailing*, or *semantic trailing*, puts the address of a function and an argument for the function on the trail. Untrailing then simply executes the function, which will cause the undo.

Copying and recomputation. The basis for restoring local state upon backtracking is a *copy* of the state. In the simplest form, a copy is made for each node, this is called *full copying*. To reduce the amount of required memory, solvers based on *recomputation* only store copies at some nodes in the tree. When moving to visit a new target node, the state for the node is computed from the nearest copy above the target node. In Fig. 2 (c), the nearest copy above the target node is at the root (marked as a small circle). When moving from the failed node to the unexplored dashed node, the solver takes the copy, copies it again (it is still needed for later backtracking), adds propagators for the choice constraints c_1, c_2, c_3 and c_5 on the path from the copied node to the target node and computes a fixpoint for the target node. Note this single fixpoint computation is much faster than computing the fixpoint for each intervening node (a technique called *batch recomputation* [5] or *decomposition-based search* [11]).

Schulte [16] describes two strategies for placing the copies in the tree: *fixed* recomputation places a copy every k nodes (k is called the *recomputation distance*); *adaptive* recomputation places a copy in the middle of the path between the current node and the nearest copy (as sketched in Fig. 2 (d)). Here we reconstruct the state of the middle node from a copy of the root node with propagators for c_1 and c_2 added and a fixpoint computed, then store a copy of this state (the grey small circle), before adding c_3 and c_5 and computing a fixpoint to construct the state for the target node.

Combining trailing and recomputation. Combining trailing and copying with recomputation in a single solver is a nontrivial task. This section presents such a hybrid architecture. The hybrid system has both local state and global state. This results in the following issues.

In order to use the same propagation loop, local and global propagators must be able to reside in the same queues. The queues are realized as local state (to support multi-threaded computation). In Gecode, propagators are elements of a doubly-linked list: either they are in the queue, or they are in a list of idle propagators. Global propagators must therefore be removed from the queues when the memory for the local state is deallocated (for example upon failure).

Variables need dependencies to both local and global propagators. The latter should be global themselves, in order to save the copying cost (which is linear in the number of propagators). Similarly the dependencies for local propagators should be local, too; otherwise we require a secondary *map* lookup from the dependency to the propagator. The currently implemented hybrid system therefore does not support global variables with dependencies to local propagators. In summary, dependencies should always be of the same kind of state (global/local) as the propagators they refer to.

Propagators can refer to both global and local variables. Because of the restriction mentioned above, the implemented hybrid system does not support local propagators with references to global variables.

The crucial interaction between the different kinds of state is on backtracking. When moving to a new target node the solver must untrail to the closest copy *above the common ancestor* of the current and target node (instead of just the closest common ancestor as would be usual in pure trailing systems). This is the nearest place where the solver has a consistent view of the overall state.

When performing recomputation (as in Fig. 2 (c)), a single fixpoint is computed for the target node. It is important to note that this implies that the trail entries for all modifications between the copy and the target are now regenerated, but in a different order than during the original exploration. In particular, the sets of trail entries that corresponded to the individual fixpoints during the original exploration may now be arbitrarily mixed. The trail thus cannot recover any state between the copy and the target. But this is not necessary, as recomputation always starts from a copy anyway. Adaptive recomputation (as in Fig. 2 (d)) works exactly like fixed recomputation, except that two fixpoints are computed, one for the copy in the middle of the path, and one for the target.

We have extended Gecode to support both global and local state for variable domains, propagators and dependencies, all coexisting. For the experiments we restrict ourself to one form of hybrid where propagators can be global or local but domains are always local, and to the fully global system. There are other important hybrids to consider. The extended system also supports a hybrid with fully global state but where we can copy the global state at any node in the search tree. This is an important feature for supporting advanced search strategies such as best-first and parallel search in a global state system [13].

4 Local versus Global State

This section presents and evaluates the possible choices of state representation for variables, propagators, and dependencies.

Experimental platform. The experiments in this paper use three different propagation solvers: one based on fully local state, one based on fully global state, and a hybrid using both global and local state. The local state system is Gecode version 3.0.2. For the global state system, we added a trail to the Gecode kernel and reimplemented Boolean and integer variables and several propagators to use backtrack-safe and trailed global state. The hybrid system is based on local variables, but supports both local and global (backtrack-safe and trailed) propagator state and dependencies.

The models are taken from the standard Gecode distribution and were compiled using the `gcc` compiler, version 4.2.1, on a Pentium 4 machine at 2.8 GHz running Linux. All runtime results are given as the arithmetic mean of 20 runs, with a coefficient of deviation less than 2%. Memory measurements represent the peak amount of the system's *overall allocated* memory.

Benchmarking rationale. All three systems are based on the same core solver. They share the same propagation loop, the same propagator scheduling mechanisms, the same data representation for variable domains and dependencies, the same search engines, and (up to the state representation) the same propagation algorithms. That way, the experiments indeed capture solely the difference in state representation. The starting point for the implementation was a copying solver, and some design decisions still reflect that, so there may be potential for optimization. However, the results show that both the global state and the hybrid solver are competitive with Gecode, and previous benchmarks show that Gecode is one of the fastest solvers available. Furthermore, we will see that we can explain the behavior of all three solvers in terms of properties of the *models*. The results are therefore largely independent of concrete implementation details.

4.1 Variable Domains

The first suite of experiments investigates the performance impact of the choice of state representation for the variable domains. The experiments compare the fully global state system with the hybrid system in which all state is global except local variable domains.

Runtime efficiency. Table 1 shows the experimental results. The Knights example has been included for statistics only, the required propagators have not been implemented for the hybrid system. Below the horizontal line are SAT benchmarks. We make the following observations:

- When the average percentage of domains modified in a fixpoint is high ($> 40\%$), copying domains is slightly preferable.
- When the percentage of updated domains is low (and there are many variables) then the optimistic approach of using global trailed domains benefits, and it can be much better than copying.
- The overall results show that trailing versus copying and recomputation of variable domains does not make that much difference except in the extreme cases, illustrated by the Radiotherapy benchmark.

Table 1. Local versus global variable domains. *Var.* is the number of variables in the problem, *Prop.* the number of propagators, *mod %* the average percentage of variables modified per fixpoint.

Benchmark	*Var.*	*Prop.*	*Failures*	*mod %*	*Hybrid* *time (ms)*	*Global* *time %*
Queens (10)	10	3	4992	42.35	31.15	118.46
Queens (100)	100	3	22	32.61	4.40	41.20
Queens (200)	200	3	146 838	3.87	3 033.00	52.51
Golomb Rulers (10)	46	46	173 568	57.27	5 398.50	101.76
Golomb Rulers (bounds, 10)	46	46	30 345	64.10	1 642.00	105.83
Golomb Rulers (bounds, 11)	56	56	689 749	63.05	48 818.00	108.33
Alpha (Z–A)	26	21	10 530 183	42.98	243 556.50	100.00
Alpha (A–Z)	26	21	7 435	30.09	93.15	93.96
Radiotherapy	4 363	2 477	903 982	0.52	360 089.00	4.20
Magic Square (7)	49	19	481 301	19.94	6 075.00	85.37
Sudoku (1)	256	48	9721	8.87	1 007.00	90.07
Sudoku (2)	256	48	14277	8.02	1 281.00	91.76
Failure Stress (500)	2	1 000	1	100.00	3 173.30	171.81
Knights (30)	22188	26880	40	4.25	–	–
Knights (32)	25392	30780	34	4.27	–	–
Knights (34)	28812	34944	20	3.54	–	–
Ramsey (4-4-11)	55	660	211 605	6.65	6 330.00	90.66
Ramsey (4-4-12)	66	990	1 834 459	5.53	56 519.50	85.13
Hanoi (4)	718	4 934	888 424	2.66	95 021.50	58.28
Pigeon Holes (7)	56	204	32 781	10.30	441.00	80.95
Pigeon Holes (8)	72	297	378 344	8.46	6 107.00	69.17
Pigeon Holes (9)	90	415	4 912 515	7.16	88 390.00	77.21
Dubois (20)	60	160	3 145 728	8.05	30 880.00	67.26
Flat (200-1)	600	2 237	167 618	8.45	20 769.00	69.83

Time-stamped trailing. The example Failure Stress exhibits the worst case behavior for trailing. It consists of 1000 propagators for $x < y$ and $y < x$, shrinking the variable domains step by step until detecting failure. Not shown in this table, the memory consumption due to trailing is several orders of magnitude higher than in the local state solver. An implementation therefore must use time-stamped trailing to protect against this pathological behavior, but our experimental system does not use time stamping (to simplify the implementation). However, for all examples we consider for benchmarking here except Failure Stress, each variable is changed at most 2.7 times on average per fixpoint; so the comparisons are fair despite the lack of time-stamped trailing.

Memory consumption. One advantage of local state is the finer control of memory consumption using recomputation. We can confirm the results presented in [15] for the case where only variable domains are copied or trailed, and the remaining state is global. Table 2 compares the peak memory consumption of the global state solver with that of the hybrid solver using different recomputation

Table 2. Percentage of peak memory allocated at different recomputation distances, compared to trailing = 100%. Diagram cut off at 140%. Examples (left to right): Queens (200), Queens (100), Flat (200-1), Pigeon Holes (9), Ramsey (4-4-12), Pigeon Holes (7), Pigeon Holes (8), Ramsey (4-4-11), Hanoi (4), Golomb Rulers (bounds, 11), Golomb Rulers (10), Golomb Rulers (bounds, 10), Radiotherapy, Alpha (Z–A), Alpha (A–Z).

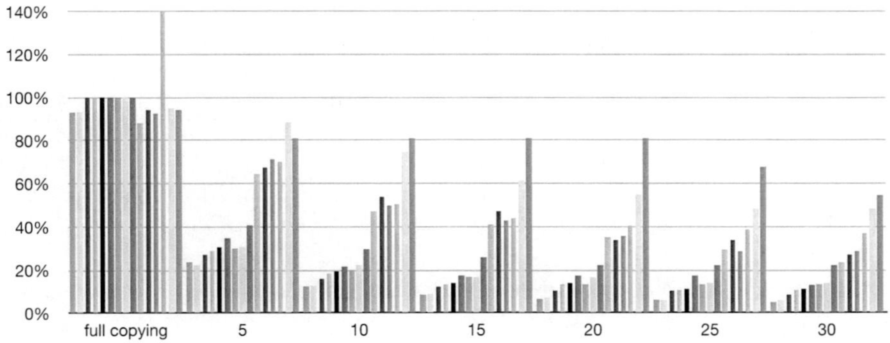

distances. (The cut off example is Radiotherapy at 158%.) Clearly the hybrid solver is advantageous over the trailing solver in memory usage, with median usage of 20% at recomputation distance 10, and 13% at distance 25.

4.2 Dependencies

Dependencies between variables and propagators are a critical part of the propagation solver, as they are accessed frequently during propagation. The key observation about dependencies was made earlier: dependencies should be the same kind of state (global/local) as the propagators they refer to.

Global dependencies. Static dependencies are sufficient for many small constraints, in particular those that are only propagation redundant when all variables are fixed. Backtrack-safe dependencies are sufficient for watched literals (and beneficial for Boolean clauses [12]). It is difficult to compare backtrack-safe with static or dynamic dependencies, as this leads to different propagator scheduling. Table 3 compares watched literals versus dynamic dependencies (two per clause) that are backtracked, and static dependencies (n for a clause of length n). For some SAT instances, watched literals can reduce the runtime in the global state system drastically (left), while for others, there is hardly any difference (right). For some propagators, however, dynamic dependencies are crucial:

- Canceling dependencies is essential for $\neq$ propagators. Using only static dependencies for the Queens example with binary propagators, the number of propagation steps and the runtime are an order of magnitude higher than with dynamic dependencies, independent of which system is used.
- Modifying dependencies (in a non-backtrack-safe way) is important for reified = propagators. Once the Boolean control variable is fixed to false, the propagator should be woken only for *fix* events. For the Knights examples, the global state system shows a 20% runtime overhead otherwise.

Table 3. Relative performance of backtracked and static dependencies for Boolean clauses compared to using watched literals, in the global state solver

Benchmark	*Backtracked* *time %*	*Stateless* *time %*	*Benchmark*	*Backtracked* *time %*	*Stateless* *time %*
Ramsey (4-4-11)	160.57	220.86	Pigeon Holes (7)	91.88	87.54
Ramsey (4-4-12)	193.66	272.75	Pigeon Holes (8)	106.85	103.52
Hanoi (4)	104.23	119.68	Pigeon Holes (9)	100.30	94.05
			Dubois (20)	108.08	103.35
			Flat (200-1)	103.31	105.73

As dependencies are modified much less frequently than domains, the hybrid and global systems use function trailing for dynamic dependencies.

Local dependencies. Local state dependencies are automatically restored upon backtracking. This has the advantage that dynamic dependencies are free, but the disadvantage that watched literals, relying on backtrack-safe state, cannot be implemented directly.

4.3 Propagators and Propagator State

Propagators with internal state are an essential prerequisite for incremental propagation. In the following, the performance impact of the choice of state restoration for propagator state is analyzed.

Runtime efficiency. Table 4 shows the results of comparing the fully local solver with the hybrid solver (differing only in what kind of state the propagators use).

Many simple propagators (e.g. $x \leqslant y$, Boolean clauses) are stateless. The results illustrate that stateless propagators should not be copied. For the SAT examples which are dominated by stateless propagators, the hybrid system improves over the local system. The contrary results for Pigeon Holes are caused by substantially reduced propagation that has occurred (seemingly randomly) due to the use of dynamic dependencies rather than watched literals.

The importance of state in propagators is illustrated by naive `alldifferent`. This propagator has state which eliminates fixed variables from further consideration. If we do not record this state and simply try to remove the values of all fixed variables on each invocation the complexity of the propagator changes. Table 5 shows how a stateless `alldifferent` slows down the hybrid system.

Memory consumption. The memory consumption at different recomputation distances of the local solver compared to the global solver is shown in Table 6. Some examples require significantly more memory in the local solver. (The cut off examples are Radiotherapy at 183% and Knights (30, 32, 34) at more than 300%.) This occurs because all propagators including their complete state must be copied, penalizing examples with many propagators. This is especially true for the added examples Knights (30, 32, 34), which use large numbers of reified

Table 4. Relative runtime of the hybrid and global compared to the local system

Benchmark	*Local*	*Hybrid*	*Global*
	time (ms)	*time %*	*time %*
Queens (10)	36.90	83.33	100.00
Queens (100)	1.50	293.67	120.67
Queens (200)	3 244.00	93.45	49.10
Golomb Rulers (10)	5 925.10	90.98	92.72
Golomb Rulers (bounds, 10)	1 657.75	110.59	104.83
Golomb Rulers (bounds, 11)	51 325.00	105.37	103.04
Alpha (Z–A)	275 562.50	87.68	88.39
Alpha (A–Z)	109.40	86.47	80.00
Radiotherapy	409 870.00	88.25	3.69
Magic Square (7)	6 593.00	92.45	78.66
Sudoku (1)	1 390.25	63.15	65.24
Sudoku (2)	1 743.50	93.36	67.42
Failure Stress (500)	3 173.30	100.00	171.81
Ramsey (4-4-11)	11 486.50	55.11	48.58
Ramsey (4-4-12)	138 305.50	40.87	36.21
Hanoi (4)	112 305.00	84.61	54.19
Pigeon Holes (7)	272.00	162.13	119.85
Pigeon Holes (8)	3 239.50	188.52	145.30
Pigeon Holes (9)	45 846.50	192.80	152.89
Dubois (20)	36 695.00	84.15	60.73
Flat (200-1)	25 205.00	82.40	61.42

Table 5. Relative runtime using a stateless `alldifferent` propagator

Benchmark	*time %*	*Benchmark*	*time %*
Queens (10)	126.83	Golomb Rulers (10)	103.16
Queens (100)	509.93	Alpha (Z–A)	116.80
Queens (200)	1 129.65	Alpha (A–Z)	134.16

propagators. On the other hand, some examples require significantly less memory in the local solver. This is due to the automatic garbage collection made possible by copying: propagation redundant propagators and fixed auxiliary variables can be removed. A trailing solver could partly achieve the same effect by deleting the trail after computing the fixpoint of the root node. In summary, recomputation saves memory of the same order of magnitude as in the hybrid system, while sometimes starting from significantly higher amounts at full copying.

Recomputing propagator state. Some propagator state can be reconstructed from other state information. Its only purpose is to make the propagation more incremental. A well-known example of this is the variable-value graph used in domain consistent `alldifferent` [14]. It can be reconstructed from the domains of the variables. Both local and global propagators can use recomputed state.

Table 6. Percentage of peak memory allocated at different recomputation distances, compared to trailing. The examples are the same as in Table 2, plus three instances of Knights (30,32,34) added on the right.

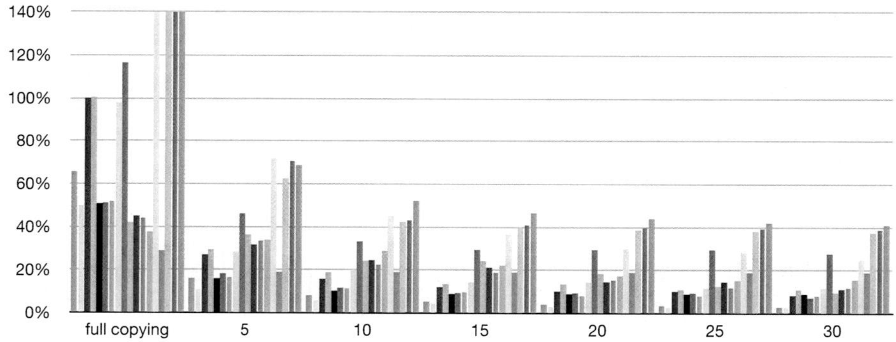

The advantage is that the state does not need to be trailed or copied. Each time the search moves to a non-child target node, this propagator state must be recomputed. This may seem expensive but the number of failures is much smaller than the number of executions of a propagator. Gecode use this approach for the variable-value graph for `alldifferent`, while JaCoP [10] (a trailing solver) also uses the approach for several propagators.

4.4 Summary

Trailing and copying with recomputation have been the dominant restoration techniques. Although there are several papers attempting comparative studies of the two techniques, this is the first time that a realistic comparison based on a production-quality solver is presented. Let us therefore summarize the results.

- Each of the solvers (local, hybrid, and global) is the best on some examples.
- While the trailing solver is generally faster than a copying solver, in almost all cases the difference between them is less than a factor of two.
- The trailing solver is more robust in terms of runtime than either the copying or hybrid solvers for problems with very weak propagation. Robustness for problems with strong propagation can be achieved by time-stamped trailing.
- Copying with recomputation is more robust than either trailing or full copying in terms of memory for any problem.

5 Related Work

There are a couple of papers that describe or compare different techniques for state management in constraint solvers. Schulte [15] introduced the copying and recomputation state maintenance model, and defined fixed and adaptive recomputation. The paper compared the Mozart local state solver versus a number of global state systems and showed that it was comparable or better than each of them in runtime. It also illustrated (but only by simulation of trailing) how the memory requirements of a local state system can be less than global (trailing) state systems when it is using fixed or adaptive recomputation. Here we reinforce the results on memory, but also show that trailing is more robust in runtime than copying and recomputation.

Choi et al. [5] compared different state maintenance approaches in the Figaro system. They compare copying versus lazy copying and coarse-grained trailing. In the system variables and propagators are global, while dependencies appear to be static. Lazy copying is a kind of copy-on-write technique, where there is a level of indirection added to each variable and propagator. Coarse grained trailing works similarly. There is one global *map* which is timestamped. Whenever an object is to be changed, the old value pointed to by the map is copied to the trail and the timestamp updated, and the object can be modified. Their results showed that coarse-grained trailing was faster than full copying which was itself faster than the recomputation approaches, and each of trailing and the recomputations usually gave significant savings in memory; while lazy copying improved on copying and coarse-grained trailing in terms of execution and memory. The results, while interesting, are for a research prototype solver that is several orders of magnitude slower than state of the art solvers.

6 Conclusion

In this paper we set out the possible ways in which a propagation solver can choose to maintain state, and the restrictions that such choices place on the resulting system. We describe how to combine global and local state maintenance in a single solver, and have extended Gecode to support both kinds of state. This allows us to give the first realistic comparison of trailing versus copying solvers, using a state of the art solver. Our results show that while the global state solver is in general faster than the copying and recomputation solver, and avoids some worst case behavior, it uses substantially more memory. As parallelism becomes more important, driven by multi-core CPUs, we foresee the importance of hybrid trailing and copying solvers to support this.

Acknowledgements. We thank Thibaut Feydy and Sebastian Brand for many interesting discussions related to this work. Raphael Reischuk and Guido Tack were supported by the Saarbrücken Graduate School of Computer Science which receives funding from the DFG as part of the Excellence Initiative of the German Federal and State Governments. NICTA is funded by the Australian Government as represented by the Department of Broadband, Communications and the Digital Economy and the Australian Research Council.

References

[1] Aggoun, A., Beldiceanu, N.: Time Stamps Techniques for the Trailed Data in Constraint Logic Programming Systems. In: Actes du Séminaire 1990 de programmation en Logique, pp. 487–509. CNET, Lannion (1990)

[2] Apt, K.R., Wallace, M.: Constraint Logic Programming Using ECLiPSe. Cambridge University Press, Cambridge (2006)
[3] Carlsson, M., Ottosson, G., Carlson, B.: An open-ended finite domain constraint solver. In: Glaser, H., Hartel, P.H., Kuchen, H. (eds.) PLILP 1997. LNCS, vol. 1292, pp. 191–206. Springer, Heidelberg (1997)
[4] CHOCO (2009), http://choco-solver.net
[5] Choi, C.W., Henz, M., Ng, K.B.: Components for state restoration in tree search. In: Walsh, T. (ed.) CP 2001. LNCS, vol. 2239, pp. 240–255. Springer, Heidelberg (2001)
[6] Gecode: generic constraint development environment (2009), http://www.gecode.org
[7] Gent, I.P., Jefferson, C., Miguel, I.: Minion: A fast scalable constraint solver. In: Brewka, G., Coradeschi, S., Perini, A., Traverso, P. (eds.) European Conference on Artificial Intelligence, pp. 98–102. IOS Press, Amsterdam (2006)
[8] Gent, I.P., Jefferson, C., Miguel, I.: Watched literals for constraint propagation in Minion. In: Benhamou, F. (ed.) CP 2006. LNCS, vol. 4204, pp. 182–197. Springer, Heidelberg (2006)
[9] ILOG Solver, part of ILOG CP (2009), http://www.ilog.com/products/cp
[10] JaCoP (2009), http://jacop.osolpro.com/
[11] Michel, L., Van Hentenryck, P.: A decomposition-based implementation of search strategies. ACM Trans. Comput. Logic 5(2), 351–383 (2004)
[12] Moskewicz, M.W., Madigan, C.F., Zhao, Y., Zhang, L., Malik, S.: Chaff: engineering an efficient SAT solver. In: DAC 2001: Proceedings of the 38th conference on Design automation, pp. 530–535. ACM Press, New York (2001)
[13] Perron, L.: Search procedures and parallelism in constraint programming. In: Jaffar, J. (ed.) CP 1999. LNCS, vol. 1713, pp. 346–361. Springer, Heidelberg (1999)
[14] Régin, J.-C.: A filtering algorithm for constraints of difference in CSPs. In: AAAI 1994: Proceedings of the Twelfth National Conference on Artificial intelligence, Menlo Park, CA, USA, vol. 1, pp. 362–367 (1994)
[15] Schulte, C.: Comparing trailing and copying for constraint programming. In: Schreye, D.D. (ed.) Proceedings of the Sixteenth International Conference on Logic Programming, Las Cruces, NM, USA, pp. 275–289. MIT Press, Cambridge (1999)
[16] Schulte, C.: Programming Constraint Services. LNCS (LNAI), vol. 2302. Springer, Heidelberg (2002)
[17] Schulte, C., Stuckey, P.J.: Efficient constraint propagation engines. Transactions on Programming Languages and Systems 31(1), 2:1–2:43 (2008)
[18] Schulte, C., Tack, G., Lagerkvist, M.Z.: Modeling with Gecode (2009), http://www.gecode.org/doc-latest/modeling.pdf

Cost-Driven Interactive CSP with Constraint Relaxation

Yevgeny Schreiber

Intel Corporation, Haifa, Israel
yevgeny.schreiber@intel.com

Abstract. We revisit the Interactive CSP framework (ICSP) and propose a new, somewhat more general model, which we call Cost-Driven Interactive CSP (CICSP). First, we extend the *value acquisition* by a more general concept of *constraint relaxation.* Second, we loosen the basic assumption of ICSP that "value acquisition is expensive" by introducing external *cost functions* of the constraint relaxation and the constraint propagation effort. We also propose a general INTERACTIVE RELAXATION algorithm template that is designated for CICSP. The effectiveness of this approach is illustrated on a real-life scenario from the Functional Test Generation problem domain.

Keywords: Functional Test Generation, Constraint Relaxation, Cost-Driven Interactive CSP, ICSP.

1 Introduction

A Constraint Satisfaction Problem (CSP) is defined as a set $\mathcal{X}$ of variables, a set $\mathcal{D}$ of corresponding value domains, and a set $\mathcal{C}$ of constraints that define the allowed value combinations of the variables. A CSP solution is an assignment of values to variables so that all the constraints in $\mathcal{C}$ are satisfied.

The *Interactive CSP* framework (ICSP) was first introduced by Cucchiara et al. [3] (following the general idea of Sergot [19]). While in the conventional CSP all the variable domains are fully known before the beginning of the solution search, the formulation and the algorithms in [3] allow the process to be interactive, starting with only partially known domains, and acquiring further values when needed during the search. This framework is motivated by real-life problems where the acquisition of all domain values before the search is either inefficient or impossible. In ICSP, the search engine and the external value acquisition engine work *cooperatively*; in particular, the latter can be "guided" by the former in order to produce "better" values. As such, this can be considered as a special case of the *distributed constraint programming framework* [6], where the computation is performed by a network of several engines. This approach was demonstrated in [2,3] on several applications.

A few years later, Faltings and Macho-González [7,13] proposed the *Open CSP* framework (OCSP), which is quite similar to ICSP (see also [14]). This line of work focuses primarily on the minimization of the number of value acquisitions

I.P. Gent (Ed.): CP 2009, LNCS 5732, pp. 707–722, 2009.

during the search, with a somewhat lesser focus on its cooperative nature (in particular, the value acquisition engine in the OCSP framework is not guided).

In the rest of the paper, we refer mainly to the original ICSP model described in [3], to emphasize the interactive nature of the search in their model, as well as in the extension that we propose. When discussing issues that are *common* to both the ICSP and the OCSP models, we will use the name "ICSP" (even in the context of [7,13]), to prevent unnecessary distinction.

While the ICSP framework is mainly oriented to the extension of *variable domains*, it is noted in [7,13] that *constraints* can be handled similarly by using the *hidden variable encoding method* [4]. ICSP can also be considered as a special case of the more general *Dynamic CSP* framework (DCSP) [5,20]. A DCSP is a sequence of constraint satisfaction problems, each of which results from some changes in the definition of its predecessor. These changes may affect any component in the problem definition: additions or removals of variables, domain values, constraints, etc. In this sense, the ICSP can be considered as a special case of the DCSP where all the possible changes result in *extension* of the variable domains (or relaxation of constraints that are encoded as hidden variables).

We choose to analyze the ICSP model from a somewhat different point of view. The condition that a variable x belongs to some "currently known" set of values (its current domain) can be considered as a unary constraint c on x; extending the domain of x is equivalent to a *relaxation* of c.[1] Similarly, we can relax any other (not necessarily unary) constraint *interactively*, during the search process. We can therefore consider the ICSP as a special case of a more general framework with constraints that can be iteratively relaxed. Note that we do not need to encode constraints as hidden variables, since we are no longer bounded by the concept of *variable domain extension*; instead, we can freely operate on any explicit or implicit form of a constraint.

Note that the concept of constraint relaxation was thoroughly examined in several works [8,12,17], where complex *explanation* systems are used in order to determine a constraint that has to be relaxed, and to perform the relaxation efficiently. In this paper we focus on a different angle, using a simple backtracking mechanism instead (as described in Section 3); we leave its enhancement (possibly, using the methods developed in [8,12,17]) for further research.

As in [3], we make the relaxation process efficient by allowing the search engine to *guide* it. The main motivation for this approach is the *cooperative solving*. Often, a large-scale CSP must be decomposed into a set of smaller sub-problems, each of which can be efficiently solved. However, since the sub-problems are not completely disconnected from each other, a solution s_{p_i} of a sub-problem p_i implies a constraint c_i for each sub-problem p_j that is connected to p_i. It is possible that p_j is infeasible with c_i, while it may be feasible with a different (less restrictive) constraint c'_i, implied by a different solution of p_i (if such exists). Note that if there are any other sub-problems connected to p_i that have been successfully solved with c_i before the attempt to solve p_j, then, if c'_i

[1] By "relaxing a constraint" (which is a somewhat overused term) we mean replacing the constraint by another one that is less restrictive.

is *strictly less restrictive* than c_i, replacing c_i by c_i' would not invalidate their solutions. This use case is further discussed in Section 4.

Another aspect that we choose to revisit is the *cost of value acquisition*, or in our (more general) case, of constraint relaxation. The basic assumption in the original ICSP framework in [3], as well as in [7,13], is that each value acquisition is *costly*, and therefore it is never beneficial to try and gather any new information until it is determined that the values gathered so far fail to satisfy all the constraints. However, a "cost" is a real-life term that applies to many facets of the solution process, and should therefore be defined carefully and relatively to other aspects. In particular, many practical constraint satisfaction problems require exponentially long runtime, and so the constraint relaxation cost can be defined, for example, relatively to the cost of the search effort: In some scenarios, it is possible that generating a new tuple that relaxes a constraint, by using the relaxation generator (with correct guidance), is cheaper than searching among the tuples that are already known.

We call the resulting framework *Cost-Driven Interactive CSP* (CICSP). It is mainly intended for problems with one or more "difficult" constraints that involve many variables and imply a large number of allowed and forbidden value tuples. We assume that in these cases it is usually possible to construct an "external" engine that is dedicated for these constraints, and to query it interactively, as in the cooperative solving example described above. A detailed application example is discussed in Section 4; other examples can include a database server that generates table constraints whose size depends on query parameters, or a mathematical equation solver for complex multivariate functions.

Before we continue to describe the details of CICSP, note its differences from the *Valued CSP* framework (VCSP), which deals with *soft constraints* and *penalties* for their violation (see [18]). The constraints in CICSP *are not soft* — the constraint relaxation function is part of the given problem, and there is no way to withdraw a constraint by paying a penalty. Moreover, the constraint relaxation cost is paid each time when a constraint is being relaxed, which is different from the violation penalty that is paid *at most once* for each violated constraint. That being said, in Section 5 we propose a possible extension of CICSP that can support soft constraints and violation penalties.

The rest of the paper is organized as follows. In the next section we describe the CICSP framework more formally; the designated Interactive Relaxation algorithm template is described in Section 3. The address-translation use-case in the field of Functional Test Generation as an application of CICSP is illustrated in Section 4, including some experimental results. In Section 5 we summarize the discussion and suggest several further extensions.

2 Cost-Driven Interactive CSP

Definition 1. *A CICSP is a tuple* $(\mathcal{X}, \mathcal{C}, \mathcal{G}, \mathcal{R}, \mathcal{P})$*, so that:*

(i) *$\mathcal{X}$ is a set of variables $x_1, \ldots, x_n$. The domains of all variables in $\mathcal{X}$ are initially considered* full *— that is, each variable in $\mathcal{X}$ can be assigned* any *value.*[2]

(ii) *Let $\mathcal{S}$ denote the infinite set of all possible constraints on the variables in $\mathcal{X}$. $\mathcal{C} \subset \mathcal{S}$ is the initial set of constraints $c_1, \ldots, c_m$ (of any arity). Each $c_i \in \mathcal{C}$ explicitly or implicitly defines the set S_i of value tuples that satisfy c_i.*

(iii) *$\mathcal{G} : \mathcal{S} \to \mathcal{S}$ is a constraint relaxation function. Given a constraint $c \in \mathcal{C}$, $\mathcal{G}(c) = c'$ is its* extension constraint*: relaxing c replaces it by $c \vee c'$.*[3]

(iv) *$\mathcal{R} : \mathcal{S} \to \mathbb{R}^+$ is a cost function assigning each relaxation step $c \to c \vee c'$ the positive cost $\mathcal{R}(c)$.*

(v) *$\mathcal{P} : \mathcal{S} \to \mathbb{R}^+$ is a cost function assigning each propagation of a constraint c the positive cost $\mathcal{P}(c)$.*

CICSP vs. ICSP. Before discussing the methods of solution of CICSP, we compare the model more formally to the ICSP framework, as formulated in [3]; for the sake of completeness, we summarize the definition of ICSP here. An *interactive domain* D_i of a variable x_i is composed of the *known* set of values $v_{i_1}, \ldots, v_{i_k}$ and the *unknown* component x'_i. The tuples in an *interactive constraint* $c(x_i, x_j) \subseteq D_i \times D_j$ can contain both the known and the unknown parts of the corresponding interactive domains D_i, D_j; if the unknown component x'_i appears in a tuple $(x'_i, v_j) \in c(x_i, x_j)$, then for each possible "future" value v'_i of x'_i, the tuple (v'_i, v_j) satisfies $c(x_i, x_j)$. The ICSP is defined as a finite set of variables $\mathcal{X}$, a set of corresponding interactive domains $\mathcal{D}$, and a set of interactive constraints $\mathcal{C}$.

Theorem 1. *The CICSP extends the ICSP model.*

Proof. Let the tuple $(\mathcal{X}, \mathcal{D}, \mathcal{C})$ be a given ICSP. Then we formulate it as the CICSP tuple $(\mathcal{X}, \mathcal{C}', \mathcal{G}, \mathcal{R}, \mathcal{P})$, where $\mathcal{C}', \mathcal{G}, \mathcal{R}$, and $\mathcal{P}$ are defined as follows.

We start with $\mathcal{C}' = \mathcal{C}$ and then extend it: For each domain $D_i \in \mathcal{D}$ of a variable $x_i \in \mathcal{X}$, let $v_{i_1}, \ldots, v_{i_k}$ be its *known* set of values; we define a unary constraint $c_i(x_i) = \{(v_{i_1}), \ldots, (v_{i_k})\}$, and add it to $\mathcal{C}'$. For each $c \in \mathcal{C}$, we define $\mathcal{G}(c) = \emptyset$. Let the unknown component x'_i of D_i be the set $\{v_{i_{k+1}}, \ldots, v_{i_{k+t}}\}$; for each $j \in [1, \ldots, t]$ denote by $c'_{i,j}$ the constraint that is satisfied by the singleton $(v_{i_{k+j}})$. For each $c_i \in \mathcal{C}'$, we define $\mathcal{G}(c_i) = c'_{i,1}$. For each $j \in [1, \ldots, t-1]$ we define $\mathcal{G}(c_i \vee c'_{i,1} \vee \cdots \vee c'_{i,j}) = c'_{i,j+1}$, and $\mathcal{G}(c_i \vee c'_{i,1} \vee \cdots \vee c'_{i,t}) = \emptyset$.

We define $\mathcal{R}(c) = +\infty$ and $\mathcal{P}(c) = 1$, for any constraint c, implying that c will always be propagated before an attempt to relax it (which will only happen if all other search branches fail). □

Remark 1. It is important to emphasize that the CICSP model does not require all the initial variable domains to be *full*, as well as it is not required that

[2] For practical reasons, each variable can be of a specific *type*, which limits its domain, such as: k-bit integer, k-character string, etc.

[3] Note that this replacement does not change the original CICSP, since $\mathcal{G}$ is given as part of the problem definition.

every constraint in $\mathcal{C}$ should be eligible for relaxation (by $\mathcal{G}$). The case where a variable is defined over some limited domain that is known a priori and does not need to be extended, which is the most common standard CSP scenario, is fully supported by the CICSP framework.

3 Interactive Relaxation Algorithm Template

We propose the following INTERACTIVE RELAXATION (IR) algorithm *template*, which emphasizes the division of functions between the *search engine*, the *relaxation generator*, and the *cost-driven oracle*, and suggests an algorithm for the search engine, as described next — see Fig. 1. The relaxation generator and the cost-driven oracle are treated as black boxes, so that a more specific algorithm can implement each of these entities according to a specific CICSP instance. Moreover, the interface that we suggest for each of these entities is also described in quite general terms, and can easily be extended.

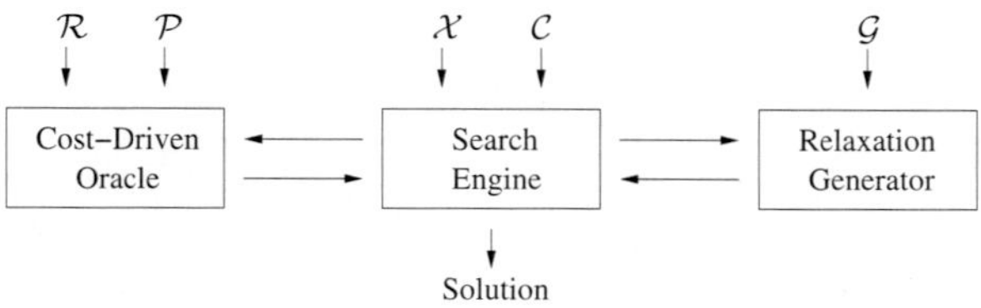

Fig. 1. The entities of the CICSP framework

Relaxation generator. The relaxation generator module (or simply the *generator*) uses the given constraint relaxation function $\mathcal{G}$ to process *relaxation queries* that it receives from the search engine during the solving process. Each relaxation query consists of a constraint c and the *guidance*: the *current state of domains of the variables in* $V(c)$, where $V(c)$ denotes the set of all variables involved in c. Let S' be the set of tuples that satisfy the extension constraint $\mathcal{G}(c) = c'$. When the query is processed, the generator returns a (possibly empty) *subset* $S'' \subseteq S'$ *that does not conflict with the current state of domains of the variables in* $V(c)$, or, equivalently, the corresponding constraint c''.

We assume that all the tuples in S'' have not been known to the search engine before the relaxation query; that is, $S \cap S'' = \emptyset$, where S is the set of tuples that satisfy c. This assumption extends the general idea of the ICSP framework, where the generator returns *new* values at each query. (Note that without this assumption, the search engine can compute $S'' \setminus S$.)

The guidance part of the relaxation query is very important. Consider, for example, the relaxation of a constraint $c(x_i, x_j)$. It is possible that during the solution search, the domains of x_i, x_j (which were considered *full* at the beginning of the search) have been significantly reduced. Without any guidance, the generator, when requested to relax $c(x_i, x_j)$, might repeatedly generate numerous value tuples that conflict with the current state of the domains of x_i, x_j

before it either succeeds to generate a tuple that does not create a conflict or determines that no such tuple can be generated.

Note that the domains do not have to be reported to the generator explicitly (this might be quite costly, since there is no limitation on their size); instead, the state of the domain can be reported implicitly, by reporting only the relevant domain reduction decisions that have been made by the search engine so far. Similarly, the returned constraint can be reported in a compact form.

Cost-driven oracle. In the standard CSP, given a constraint c that defines a set S of tuples that satisfy it, any feasible solution must be consistent with one of the tuples in S. In CICSP, we have a choice: We can either propagate c until either it is satisfied or a conflict is reached, or we can try to relax c and obtain a new tuple (which satisfies at least all the constraints that have been propagated so far). Eventually, to ensure that the search is *complete*, we may need to do both; however, the order in which we perform these operations may have a tremendous effect on the effectiveness of the algorithm in practice.[4] This problem-specific order can be realized for each CICSP instance by implementing the corresponding cost-driven oracle.

The oracle uses the cost functions $\mathcal{R}$ and $\mathcal{P}$ to process *oracle queries* that it receives from the search engine during the solving process. Each oracle query consists of a constraint c and the *search state information* that can be relevant, such as the number of times c has been propagated and / or relaxed so far, etc. Using this information, the oracle selects the next action that should be taken to satisfy c: *propagate* c or *relax* it. The search state information can help the oracle make better choices: For example, if c has been propagated k times (each time reaching a conflict, backtracking, and trying another search branch), the oracle can reach a conclusion that c has a higher chance of being successfully satisfied by relaxation. In this case, the parameter k has to be passed to the oracle as part of the query. Since this kind of knowledge is very application-specific, we leave it without further formalization. In any case, in order to make any kind of nontrivial decision-making possible, the oracle has to compare $\mathcal{R}(c)$ to $\mathcal{P}(c)$.

As shown in Theorem 1, a trivial oracle can implement the policy that is used in [3,7,13], where it is assumed that it is always cheaper to search among the existing values than to generate new ones, without considering $\mathcal{R}$ and $\mathcal{P}$. In Section 4 we describe a real-life problem where this assumption does not hold.

Constraint Propagation. The constraint propagation method used in IR can be considered as a more general version of *forward checking*: Instead of enforcing a constraint c (on variables that have not been instantiated yet) only when a variable in $V(c)$ is instantiated, c can also be enforced each time that the domain of a variable in $V(c)$ is modified (by *registering* c to be triggered when needed — see below). The propagation of c is initiated before any variable in $V(c)$ is instantiated, since the fact that we can start with *full* domains requires us to reduce the domains as much as possible before the instantiation can take place.

[4] This is similar to the effect that may be caused by the order in which variables are instantiated, or the order in which values are selected for each variable instantiation.

INTERACTIVE RELAXATION ALGORITHM

1. **for** each $c \in \mathcal{C}$ **do**
   initialize *unsatisfied*(c) = *extendable*(c) = *enforceable*(c) = *true*;
2. **for** each *unsatisfied* $c \in \mathcal{C}$ **do**
   if not SATISFY(c) **then return** *no solution*;
3. **for** each remaining uninstantiated $x \in \mathcal{X}$ **do** INSTANTIATE(x);
4. **return** *solution found*;

We use the following notations in the rest of the paper. For $x \in \mathcal{X}$, denote by $D(x)$ the *current* domain of x: the full domain minus the reductions that have been made during the search so far. We say that a constraint c is *fully relaxed* by the generator if *relaxation-query*$(c, \{D(x) \mid x \in V(c)\})$ returns $\emptyset$; note that this definition depends on the current state of the relevant domains.

SATISFY(CONSTRAINT c)

1. **while** *unsatisfied*(c) **do**
   (a) **if** *extendable*(c) **and** *enforceable*(c) **then**
       action = *oracle-query*$(c$, {relevant search state data}$)$;
   (b) **else if** *extendable*(c) **then** *action* = *relax*;
   (c) **else if** *enforceable*(c) **then** *action* = *propagate*;
   (d) **else return** *false*;
   (e) **if** (*action* == *relax* **and** RELAX(c)) **or**
       (*action* == *propagate* **and** PROPAGATE(c)) **then**
       i. *unsatisfied*(c) = *false*;
       ii. **return** *true*;

RELAX(CONSTRAINT c)

1. c' = *relaxation-query*$(c, \{D(x) \mid x \in V(c)\})$;
2. **if** $c' == \emptyset$ **then**
   (a) *extendable*(c) = *false*; (∗ c is fully relaxed. ∗)
   (b) **return** *false*;
3. **else return** PROPAGATE(c');

Remark 2. In practice, many changes can be made to improve the efficiency of PROPAGATE(c). For example, it can be split to several sub-procedures, each of which can be triggered separately when the domain of a particular variable in $V(c)$ changes. Moreover, the propagation of the registered constraints does not have to be triggered *each time* when the corresponding domain is reduced. The domain reduction at step 1a can be made implicitly; and so on.

PROPAGATE(CONSTRAINT c)

1. **for** each $x \in V(c)$ **do**
   (a) Remove all values from $D(x)$ that cannot satisfy c;
   (b) **if** $D(x) == \emptyset$ **then goto** step 5;
   (c) **if** $|D(x)| \neq 1$ **then** *register* c for future reductions of $D(x)$;
   (d) **if** $D(x)$ has been reduced by step 1a **then**
       for each *unsatisfied* $c' \neq c$ *registered* for reduction of $D(x)$ **do**
           if not PROPAGATE(c') **then goto** step 5;
2. **for** each *unsatisfied* $c' \in \mathcal{C}$ so that $V(c) \cap V(c') \neq \emptyset$ **do**
   if not SATISFY(c') **then goto** step 5;
3. **for** each $x \in V(c)$ **do**
   if not INSTANTIATE(x) **then goto** step 5;
4. **return** *true*;
5. undo all changes starting from step 1;
6. *enforceable*(c) = *false*;
7. **return** *false*;

INSTANTIATE(VARIABLE x)

1. **if** $|D(x)| == 1$ **then return** *true*;
2. **while** v = next available value in $D(x)$ **do**
   (a) set $D(x) = \{v\}$;
   (b) **for** each *unsatisfied* c *registered* for reduction of $D(x)$ **do**
       if not PROPAGATE(c) **then**
           i. undo changes starting from step 2a;
           ii. remove v from $D(x)$;
           iii. goto step 2;
   (c) **return** *true*;
3. **return** *false*; (* Tried all possible values in $D(x)$. *)

Lemma 1. *Each* INSTANTIATE *procedure called at step 3 of* IR *returns* true.

Proof. Indeed, INSTANTIATE(x) can fail only if it has at least one constraint c *registered* to be triggered when $D(x)$ is modified. However, in this case INSTANTIATE(x) would have been called from step 3 of PROPAGATE(c), and therefore it would not be called at step 3 of IR. □

Lemma 2. *The* IR *algorithm returns* no solution *only if there is a nonempty subset* $\mathcal{C}'$ *of constraints in* $\mathcal{C}$ *that conflict with each other and cannot be satisfied, even if all the possible tuples of each constraint in* $\mathcal{C}'$ *(that can ever be returned by the generator) are known.*

Proof. IR returns *no solution* only if some invocation I of SATISFY(c), called directly from the main loop of IR, returns *false*. This implies that *extendable*(c) = *false* and *enforceable*(c) = *false*. It follows that (a) RELAX(c) was called from I as many times as possible (until the generator returns $\emptyset$), each time returning *false*, and (b) PROPAGATE(c) was called from I at least once, and returned *false*.

We claim that SATISFY(c) was never called before I. Indeed, if SATISFY(c) would have been called from previous calls of PROPAGATE(c') for some constraint $c' \neq c$, it would either result in satisfaction of c, or in failure of PROPAGATE(c'). The former is impossible since c is unsatisfied in I, and the latter would cause SATISFY(c') to fail, which would eventually result in IR returning *no solution* before calling I, a contradiction.

Therefore, at the beginning of I, the domains of the variables in $V(c)$ have never been reduced, and so the generator is not limited when it is requested to relax c, and eventually it returns all the potentially possible tuples of $\mathcal{G}(c)$. Hence, the propagation of both c and of each possible extension of c has failed.

PROPAGATE(c) (where c is either the original constraint or one of its extensions) fails only if it either discovers a conflict (at step 1b or at step 3), or the satisfaction of some constraint c' that has common variables with c. In the latter case we can apply the argument recursively, with only one difference: When *relaxation-query*(c') is called, the generator is limited by the former propagation of c. However, since every possible extension of c is propagated (as shown above), *relaxation-query*(c') eventually returns all the potentially possible tuples.

We can therefore conclude that c, and every constraint c' that was attempted to be satisfied in the function call subtree of I, belong to a subset $\mathcal{C}'$ of constraints that unavoidably conflict with each other. □

Lemma 3. *If* IR *returns a solution, then it satisfies each constraint in* $\mathcal{C}$.

Proof. IR returns a solution only if every call of SATISFY from the main loop succeeds, which implies that each $x \in V(c)$, for all $c \in \mathcal{C}$, is instantiated. INSTANTIATE(x) succeeds only after all constraints that involve x are satisfied. □

Theorem 2. *If* IR *terminates, then it either finds an assignment for each variable in* $\mathcal{X}$ *so that each constraint in* $\mathcal{C}$ *is satisfied, or determines that the problem is infeasible. If the* IR *algorithm does not terminate, then either (a) There is a constraint* $c \in \mathcal{C}$ *that is relaxed an infinite number of times (that is,* relaxation-query(c) *never returns an empty set, given some specific state of the domains of the variables in* $V(c)$*), or (b) There is a constraint* $c \in \mathcal{C}$ *that is enforced on variables with unbounded domains, so that its propagation takes infinite time.*

Proof. The first part of the theorem follows from Lemmas 1–3. To prove the second part, assume that conditions (a) and (b) do not hold. The search tree is composed of two types of nodes: *constraint nodes*, where a constraint can either be propagated or relaxed (possibly more than once), and *variable nodes*, where a variable is instantiated by a value from its current domain. Since conditions (a) and (b) do not hold, both types of nodes take finite number of processing time, so it is only left to show that the total number of nodes is finite.

The depth of the search tree is at most $|\mathcal{C}|+|\mathcal{X}| = m+n$, since in each branch each constraint is satisfied at most once, and each variable is instantiated at most once. Since condition (a) (resp., (b)) does not hold, the maximum rank of a constraint (resp., variable) node is finite, and the theorem follows. □

Remark 3. As it is evident from Theorem 2, IR is only practical when the conditions (a) and (b) do not materialize during the search. One of the most effective ways to prevent this is by selecting a "good" application-specific order of the search tree nodes. Another possible way is, if possible, to configure a "smart" policy for the oracle, preferring to relax constraints whose propagation is a priori suspected to take a long time, and propagate constraints whose relaxation is a priori suspected to be inefficient. It is also possible to extend the problem decomposition and provide the generator not only the current domains of variables in $V(c)$, but also the constraints that interact with c and their variable domains; however, this would require the generator to solve the resulting CSP.

4 CICSP in Functional Test Generation

4.1 Address Generation Problem

In current industrial practice, simulation-based verification techniques play the major role in the functional verification of hardware designs. However, due to the complexity of modern hardware architectures, it is infeasible to cover all possible test scenarios (stimuli) deterministically. Therefore, the common approach is to generate the so called *directed random* tests, using the automated pseudo-random test generation tools [9,15,16]. The input is a set of constraints that reflect the hardware design, and a set of user-given constraints that direct the tool; the output is a sequence of hardware-language instructions that give rise to "interesting" events during the simulation. The randomness is employed to extend the reach (*coverage*) of the generated test. This problem can be formulated as a CSP, and is often (eventually) decomposed into a sequence of sub-problems, each of which generates a *single instruction* in the final instruction sequence.

To demonstrate the CICSP approach, we choose a common sub-problem of the instruction generation problem. An instruction can involve one or more *operands*, each of which can possibly be located at some *memory address*. This address belongs to some *virtual memory address space* that is mapped to the *physical* hardware resources. For example, the IA-32 processor architecture [11] contains two related mechanisms: First, the *segmentation* mechanism translates a *logical* memory address that belongs to some memory *segment* with a set of logical attributes into the so-called *linear address space*. Second, the *paging* mechanism maps chunks (or *pages*) of the linear address space into *physical address* ranges. These mechanisms operate in several different modes that imply page sizes and other parameters; for example, in Fig. 2 the mode implies a page size of 4K bytes. Similar mechanisms also exist in other processor architectures [1].

A common implementation of these mechanisms is a hierarchy of *translation tables*, where each next table is pointed by an entry of the preceding table in the hierarchy. For example, in the paging mechanism in Fig. 2, the location of the required *page table* can be calculated using the relevant entry of the *page directory* (which can be pointed from a higher table in the hierarchy, and so on). The relevant entry within each table is computed using the corresponding field value within the address that is being translated — for example, k specific bits within the linear address are used as an index of the page table entry. Apart from

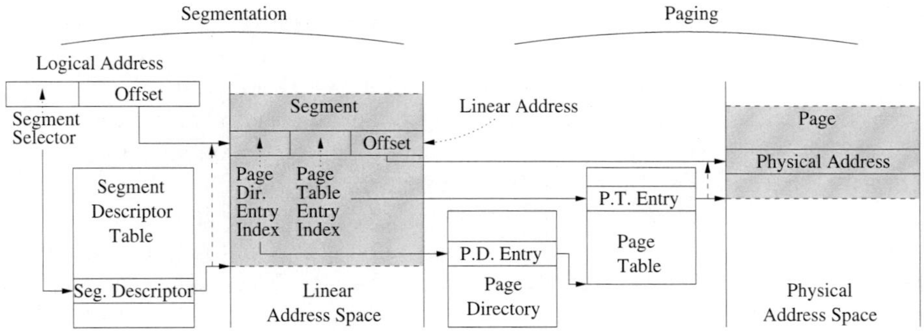

Fig. 2. A general outline of the segmentation and the paging mechanisms in IA-32

pointing to the next node in the hierarchy, each table entry also stores several related *attributes* — for example, the access type of the node (*read* or *write*).

The *Address Generation Problem* (AGP) requires finding an address that (a) has a legal mapping in all the translation mechanisms (that is, all the relevant table entries contain valid data at the time of simulation), and (b) satisfies a set of additional constraints that are implied by a specific test scenario. For instance: Compute a linear address whose least significant bit is 1, most significant bit is 0, it is mapped to a physical address that is greater than 0xFFFF0000, and it belongs to a *read-only* page that is pointed by an even page table entry.

4.2 Solution Approaches

If all the address generation constraints for each required instruction were known at the beginning of the test generation, it would be possible to construct a priori an initial architectural state (that is, all the required translation tables) that satisfies all the constraints. Unfortunately, often this is not the case, since many of the constraints for one instruction are by-products of other instructions, and the problem of generating all the instructions at once is usually too complex to be solved as a single CSP. In the rest of the paper we describe and compare two problem decomposition schemes and the corresponding cooperative solving approaches: a straightforward classic CSP and the CICSP.

Standard CSP approach. The test generation problem can be decomposed into a set of (interconnected) sub-problems $p_0, \ldots, p_k$, where p_0 is the problem of computing the initial architectural state, and p_i is the problem of generating the i-th instruction, for $i \in [1, \ldots, k]$. Here we consider only the AGP aspects of the test generation problem — the solution of p_0 comprises all the address translation tables, and in each further p_i we are interested only in the address of each memory operand of the i-th instruction.

Since the output of each $p_j \neq p_k$ implies an architectural state that restricts p_{j+1}, we can solve all the sub-problems consecutively, until either the whole test is generated, or an infeasible sub-problem is encountered. Obviously, in this approach, the more restrictive is the initial state prepared by p_0, the higher is the probability that some later sub-problem p_i will become infeasible. We can therefore try to initialize the translation tables (as part of the solution of p_0) by as many different entries as possible, in hope that the resulting mechanism will cover all the requirements that may arise during the solving of $p_1, \ldots, p_k$. However, the number of all possible requirement combinations that may arise during the solving of $p_1, \ldots, p_k$ is immensely greater than the number of the translation paths (that is, valid translation table entry combinations) that can be a priori prepared, and therefore, using this approach, we can never be prepared for every possible scenario. To overcome this limitation, we can introduce some randomness into the solving process of p_0, and if the test generation fails on some further p_i, we can retry with a different solution of p_0.

An important practical parameter in the above model is the *initial translation table density* μ. While the maximal number of valid entries in each translation table is constant (defined by the hardware architecture), not every entry has to be part of the CSP model. An entry that is not part of the CSP model is ignored by the solving process, and can never be selected as part of the AGP solution. Using all the possible table entries as part of the CSP model increases the probability of a successful test generation as discussed above, but it also inflates the CSP model and therefore slows down the solution search, as shown below. The initial translation table density $0 \leq \mu \leq 1$ is the probability of each table entry to be added to the CSP model, and so it allows us to control the effective number of the valid translation table entries.

We leave the modeling of p_0 out of the scope of this paper. For each $p_i \neq p_0$, the set $\mathcal{X}_i$ of variables includes the logical, linear, and physical form of a memory address and all their fields, as well as all the fields of the table entries that are involved in the address translation process, for each memory operand of the i-th instruction. The domains of all these variables are completely determined by the initial architectural state: For example, the domain of a variable that represents a specific field f of the involved page table entry contains all the values that are stored at the field f of each page table entry in the initial architectural state. The constraints $\mathcal{C}_i$ include all the constraints that are implied by the hardware architecture, the requirements for the i-th instruction in the test scenario, and the requirement that the table entries that are involved in the generated address translation process are consistent with the architectural state.

CICSP approach. We decompose the problem similarly to the standard CSP approach described above, with the following difference: The initial architectural state is not completely computed before the solving of each $p_1, \ldots, p_k$; rather, it is constructed by an interactive process, where each p_i is responsible to add into the initial architectural state representation the table entries that are required for the i-th instruction. The purpose of p_0 is to construct only a *partial* representation of the initial architectural state, in a way that many "simple" address generation requirements would have a high probability to be satisfied without adding new table entries. See Fig. 3 for an illustration.

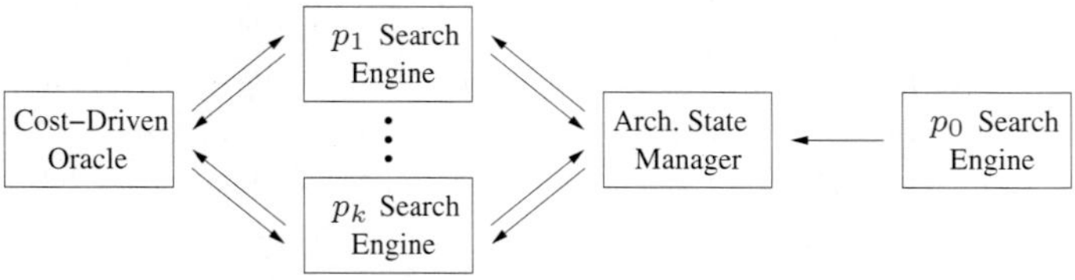

Fig. 3. Cooperative solving of the AGP aspects in the test generation problem

For each $p_i \neq p_0$, $\mathcal{X}_i$ is defined as in the standard CSP approach. The initial domains of most of these variables are *full*: For example, the domain of the 32-bit linear address is $[0, \ldots, 2^{32} - 1]$. The constraints in $\mathcal{C}_i$ can be subdivided into $\mathcal{C}_i^1, \mathcal{C}_i^2$: (a) The constraints $\mathcal{C}_i^1 \subset \mathcal{C}_i$ require that the table entries involved in the generated address translation process are consistent with the current architectural state; these constraints are *extendable*, as described below. (b) The rest of the constraints $\mathcal{C}_i^2 = \mathcal{C}_i \setminus \mathcal{C}_i^1$ are *non-extendable* — this set includes all the constraints that are implied by the IA-32 architecture and the test scenario.

For each constraint in $\mathcal{C}_i^1$, the initially allowed value configurations are based only on the part of the architectural state that has been defined before the beginning of the solution search of p_i. These constraints can be relaxed during the solution search by asking the architectural state manager (which acts as the relaxation generator) to extend the existing architectural state — that is, to add new tables, table entries, pages, etc. The relaxation generator can be guided (by the current domains of all the relevant variables) so that the new generated architectural state elements would satisfy all the constraints propagated so far.

The cost-driven oracle can implement many possible decision policies. The trade-off in these policies is between the amount of the newly generated architectural state elements, and the time of search among all the possibilities bestowed by the existing architectural state. The modification of the existing state is costly since it requires a nontrivial interaction with the architectural state simulator and other practical reasons. On the other hand, the translation table hierarchy can include several layers of tables, where each table can point to thousands of elements in the next layer; searching in all this hierarchy can be very expensive. In our experiments (described below), we have chosen to implement a policy that allocates a *propagation time budget* for each extendable constraint $c \in \mathcal{C}_i^1$. As long as the budget has not run out, we propagate c; when the budget is over, we backtrack to the point before the propagation of c, and ask the generator to relax c. (Note that this specific policy does not guarantee a complete search.) By setting the propagation effort budget of c proportional to $\mathcal{R}(c)$, we can bound the *total* (propagation plus relaxation) effort that is made to satisfy c.

Experiments. We have modeled the AGP both as a standard CSP and as a CICSP using the ILOG Solver [10]. We constrain the table entries involved in the address translation process to be consistent with the architectural state using the *custom lookup constraints* [9], which provide an efficient interaction between the search engine and the external architectural state.

In our experiments we have modeled a specific mode of the IA-32 architecture: 4K-byte and 4M-byte paging mechanism with a two-level translation table hierarchy, and without considering the segmentation mechanism. The test scenario contains a set $\tilde{C}$ of constraints, each of which has an equal probability ρ to be included in C_i^2; each of these constraints is defined on a different subset of paging attributes. We say that ρ is the *test scenario constraint density* (note that except the constraints in $\tilde{C}$ there is also a set of mandatory constraints that describe the IA-32 architecture); we experiment with $\rho \in \{\frac{1}{7}, \frac{1}{5}, \frac{1}{3}\}$.

For each value of ρ we vary the initial translation table density μ, and compare the standard CSP with the CICSP approach. By increasing μ we can increase the probability of success of the standard CSP approach (see Fig. 4(a)) and decrease the average number of constraint relaxations in the CICSP approach (Fig. 4(b)). However, as can be seen in Fig. 4(c), increasing μ inflates the search space, and consequently, the solution search time (here the effect of ρ is minor, and therefore not depicted). According to the results in Fig. 4, increasing μ can be a successful policy only when ρ is very low: Even with ρ as low as $\frac{1}{3}$ the standard CSP approach does not reach the 100% success ratio, in contrast with the CICSP approach, whose success ratio was always 100%. With $\rho = \frac{1}{2}$ most of the standard CSP instances were infeasible (not depicted in Fig. 4).

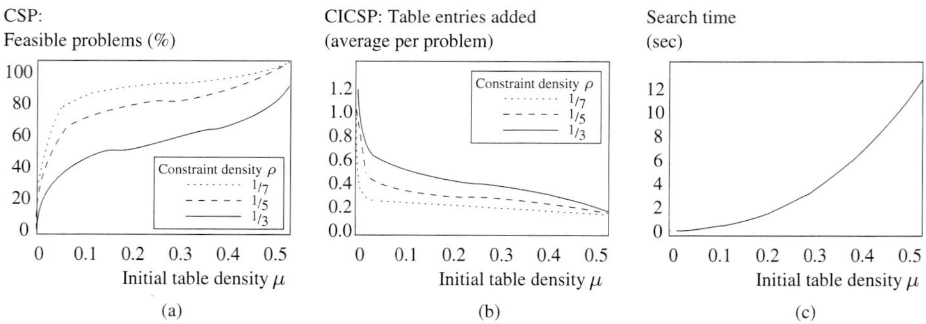

Fig. 4. Experimental results for the AGP in the IA-32 architecture

Moreover, even with low ρ, increasing μ results in a sharp increase of the search algorithm running time: For example, increasing μ from 0.01 to 0.5 increases the average search time from 0.025 to 12.5 seconds. On the other hand, in the CICSP framework, by increasing μ we decrease the average number of times that the relaxation generator is asked to relax a constraint. By analyzing the graphs in Fig. 4(b,c) we can compute the optimal value of μ for the CICSP approach: By setting $\mu = 0.1$ we get a reasonably low running time (about 1 second), while the average number of relaxations is still sufficiently small. Therefore, in order to bring the probability of success close to 100% for $\rho \leq \frac{1}{3}$, we can get $\times 10$ *search time speed-up by using the CICSP*, compared to the standard CSP approach. For higher values of constraint density the gain is even higher.

5 Concluding Remarks

We have presented a new framework that extends the ICSP model of [3] in two directions: First, it allows the relaxation of arbitrary constraints. Second, it does not assume that relaxing a constraint is necessarily much more expensive than its propagation effort. We have also described the IR algorithm template, whose effectiveness was demonstrated on the Address Generation Problem by achieving a $\times 10$ search time speed-up in comparison to the standard CSP approach. Similarly to the original ICSP framework, CICSP is most useful in the presence of a relaxation generator that can be efficiently guided.

The suggested CICSP framework is very flexible — it is not difficult to use it for several additional extensions. For example, a possible extension of CICSP can support soft constraints and violation penalties, simply by extending the interface between the search engine and the cost-driven oracle: When the oracle is asked how to satisfy a constraint c, one of the possible replies for a soft constraint can be "leave it unsatisfied." This decision can be taken by the oracle by considering, among other parameters, the penalty for the violation of c.

It seems that one of the most promising future research directions is the improvement of the search algorithm in the IR template. While its basic version, proposed in Section 3, works very well in the presence of a relaxation generator that can be efficiently guided, its effectiveness can decrease dramatically when the probability of a constraint to be successfully relaxed by the generator is low. This behavior can probably be improved by using more advanced search algorithms. In particular, it would be interesting to combine IR with the *explanation* techniques of [8,12,17].

Acknowledgments. The author is very grateful to Anna Moss and Boris Gutkovich for fruitful discussions and valuable comments, as well as for the usage of their code that demonstrates the lookup constraints concept of [9].

References

1. Adir, A., Emek, R., Katz, Y., Koyfman, A.: DeepTrans — a model-based approach to functional verification of address translation mechanisms. In: Proc. 4th Internat. Worksh. on Microprocessor Test and Verification: Common Challenges and Solutions, pp. 3–6. IEEE CS Press, Los Alamitos (2003)
2. Barruffi, R., Lamma, E., Mello, P., Milano, M.: Least commitment on variable binding in presence of incomplete knowledge. In: Biundo, S., Fox, M. (eds.) ECP 1999. LNCS, vol. 1809, pp. 159–171. Springer, Heidelberg (2000)
3. Cucchiara, R., Gavanelli, M., Lamma, E., Mello, P., Milano, M., Piccardi, M.: Constraint propagation and value acquisition: Why we should do it interactively. In: Proc. 16th Internat. Joint Conf. on Artif. Intell., Stockholm, pp. 468–477 (1999)
4. Dechter, R.: On the expressiveness of networks with hidden variables. In: Proc. 8th National Conf. on Artif. Intell., Boston, pp. 556–562 (1990)
5. Dechter, R., Dechter, A.: Belief maintenance in dynamic constraint networks. In: Proc. 7th National Conf. on Artif. Intell., St. Paul, pp. 37–42 (1988)

6. Faltings, B.: Distributed constraint programming. In: Rossi, F., van Beek, P., Walsh, T. (eds.) Handbook of Constraint Programming, ch. 20, pp. 699–729. Elsevier, Amsterdam (2006)
7. Faltings, B., Macho-González, S.: Open constraint programming. Artif. Intell. 161(1-2), 181–208 (2005)
8. Ferguson, A., O'Sullivan, B.: Quantified constraint satisfaction problems: from relaxations to explanations. In: Proc. Internat. Joint Conf. on Artif. Intell (IJCAI), pp. 74–79. Morgan Kaufmann, San Francisco (2007)
9. Gutkovich, B., Moss, A.: CP with architectural state lookup for functional test generation. In: 11th Annu. IEEE Internat. Worksh. on High Level Design Validation and Test, pp. 111–118 (2006)
10. ILOG Solver 6.5 reference manual (October 2007)
11. Intel® 64 and IA-32 architectures software developer's manual (February 2008)
12. Jussien, N., Boizumault, P.: Implementing constraint relaxation over finite domains using ATMS. In: Jampel, M., Maher, M.J., Freuder, E.C. (eds.) CP-WS 1995. LNCS, vol. 1106, pp. 265–280. Springer, Heidelberg (1996)
13. Macho-González, S., Ansótegui, C., Meseguer, P.: Boosting open CSPs. In: Benhamou, F. (ed.) CP 2006. LNCS, vol. 4204, pp. 314–328. Springer, Heidelberg (2006)
14. Macho-González, S., Meseguer, P.: Open, Interactive and Dynamic CSP. In: Proc. Internat. Worksh. on Constraint Solving under Change and Uncertainty (CP 2005), pp. 13–17 (2005)
15. Moss, A.: Constraint patterns and search procedures for CP-based random test generation. In: Yorav, K. (ed.) HVC 2007. LNCS, vol. 4899, pp. 86–103. Springer, Heidelberg (2008)
16. Naveh, Y., Rimon, M., Jaeger, I., Katz, Y., Vinov, M., Marcus, E., Shurek, G.: Constraint-based random stimuli generation for hardware verification. AI Magazine 28(3), 13–30 (2007)
17. Papadopoulos, A., O'Sullivan, B.: Relaxations for compiled over-constrained problems. In: Stuckey, P.J. (ed.) CP 2008. LNCS, vol. 5202, pp. 433–447. Springer, Heidelberg (2008)
18. Schiex, T., Fargier, H., Verfaillie, G.: Valued constraint satisfaction problems: Hard and easy problems. In: Proc. Internat. Joint Conf. on Artif. Intell (IJCAI), Montreal, pp. 631–637 (1995)
19. Sergot, M.: A query-the-user facility for logic programming. In: Degano, P., Sandewall, E. (eds.) Integrated Interactive Computing Systems, pp. 27–41. North-Holland, Amsterdam (1983)
20. Verfaillie, G., Jussien, N.: Constraint solving in uncertain and dynamic environments: A survey. Constraints 10(33), 253–281 (2005)

Weakly Monotonic Propagators

Christian Schulte[1] and Guido Tack[2]

[1] KTH - Royal Institute of Technology, Sweden
cschulte@kth.se
[2] Programming Systems Lab, Saarland University, Germany
tack@ps.uni-sb.de

Abstract. Today's models for propagation-based constraint solvers require propagators as implementations of constraints to be at least contracting and monotonic. These models do not comply with reality: today's constraint programming systems actually use *non-monotonic* propagators. This paper introduces the first realistic model of constraint propagation by assuming a propagator to be *weakly monotonic* (complying with the constraint it implements). Weak monotonicity is shown to be the minimal property that guarantees constraint propagation to be sound and complete. The important insight is that weak monotonicity makes propagation *in combination* with search well behaved. A case study suggests that non-monotonicity can be seen as an opportunity for more efficient propagation.

1 Introduction

When implementing a propagator for a constraint, the propagator must comply with policies mandated by the underlying constraint programming system such that constraint propagation becomes well behaved. The most obvious property is *contraction*: values are removed but never added. A second property is *monotonicity*: a propagator can perform stronger pruning only when being applied to stronger input (fewer values for variables). Contraction captures pruning as the very essence of constraint propagation, while monotonicity guarantees that the same result (the weakest possible) is computed regardless of propagation order.

However, some propagators are non-monotonic. They may compromise between propagation strength and efficiency, like task intervals in scheduling [1] and propagating the `circuit` (Sect. 5) or `multicost-regular` constraint [2]. Other approaches yield non-monotonic propagators due to delaying or adapting propagation [3,4], using randomization [5,6], or approximation [7]. For some propagators, it may just not be obvious whether they are monotonic.

Systems implement some of these non-monotonic propagators, for example Choco, Gecode, Oz, and SICStus Prolog. It is realistic to assume that many more non-monotonic propagators are used, and that many more systems rely on them. Essentially, many systems use non-monotonic constraint propagation while not spelling out the most basic guarantees: Is the result of propagation unique? Is propagation correct? Do two runs of the same problem return the same solution? Do techniques such as recomputation for search still work?

I.P. Gent (Ed.): CP 2009, LNCS 5732, pp. 723–730, 2009.

This paper attempts to answer these questions. We show that even non-monotonic propagators have to be monotonic to a certain extent, in order to ensure soundness. This leads to the definition of *weakly monotonic propagators*, a class of propagators that covers approximative, heuristic, and randomized algorithms, while yielding strong enough guarantees to keep propagation sound.

After presenting preliminaries in Sect. 2, the paper contributes the first theory of non-monotonic propagators, based on *weak monotonicity*, that fills the gap between models for propagation and reality (Sect. 3). It analyzes the interaction between propagation of weakly monotonic propagators and search, including an analysis of recomputation (Sect. 4). Finally, it provides a case study that suggests that non-monotonicity should be seen as an opportunity rather than a problem (Sect. 5) and concludes with Sect. 6.

2 Preliminaries

We assume a finite set of *variables* $Var = \{x_1, \ldots, x_n\}$ and a finite set of *values* Val. Constraints are characterized by *assignments* $a \in Asn$ that map variables to values: $Asn = Var \to Val$. A *constraint* $c \in Con$ is a relation over the variables: a set of all assignments that satisfy the constraint, $Con = 2^{Asn}$. Constraints are defined for all variables in Var. Typically, only a subset $\mathrm{vars}(c)$ of the variables is *significant*; the constraint is the full relation for all $x \notin \mathrm{vars}(c)$.

A *domain* $d \in Dom$ maps each variable $x \in Var$ to a set of possible values, the *variable domain* $d(x) \subseteq Val$. A domain d can be identified with the set of assignments $\{a \mid \forall x : \ a(x) \in d(x)\}$. We can therefore identify domains with constraints. In particular, $\{a\}$ is a domain and a constraint for any assignment a.

A domain d_1 is *stronger* than a domain d_2 ($d_1 \subseteq d_2$), iff $\forall x \in Var : d_1(x) \subseteq d_2(x)$. By $\mathrm{dom}(c)$ we refer to the *strongest domain* including all valid assignments of a constraint: $\min\{d \in Dom \mid c \subseteq d\} = \{a \mid \forall x \ \exists b \in c.\ a(x) = b(x)\}$. Note that the minimum exists (domains are closed under intersection) and that not every constraint can be captured by a domain. For a constraint c and a domain d, $\mathrm{dom}(c \cap d)$ refers to removing all values from d not supported by c.

A *constraint satisfaction problem* (CSP) is a pair $\langle d, C\rangle$ of a domain d and a set of constraints C. The *solutions* of a CSP $\langle d, C\rangle$ are all assignments that satisfy all constraints: $\mathrm{sol}(d, C) = \{a \in Asn \mid \{a\} \subseteq d,\ \forall c \in C : a \in c\}$.

3 Weakly Monotonic Propagators

Propagators, sometimes also referred to as constraint narrowing operators or filter functions, serve as implementations of constraints. They are usually defined as contracting functions over domains: $p \in Dom \to Dom$, $p(d) \subseteq d$. Requiring propagators to be contracting is uncontroversial, after all it captures the very essence of constraint propagation and guarantees termination.

Many models of propagation additionally require propagators to be idempotent ($p(p(d)) = p(d)$) and monotonic ($d_1 \subseteq d_2 \Rightarrow p(d_1) \subseteq p(d_2)$). Then, propagators are *closure* or *consequence operators* over the lattice of domains [8,9,10]. Examples for definitions of propagators as closure operators are [11,12,13,14].

Theorem 1. *Given a* propagation problem, *a pair* $\langle d, P\rangle$ *of a domain* d *and a set of monotonic propagators* P, *there is a unique weakest simultaneous fixpoint of all* $p \in P$ *that is stronger than* d. *It can be computed by iteration:*

$$\begin{aligned}\mathrm{prop}(d,P) \equiv\ & \textbf{while } \exists p \in P:\ p(d) \neq d \textbf{ do}\\ & \qquad d \leftarrow p(d)\\ & \textbf{return } d\end{aligned}$$

The theorem still holds without idempotency. In practice, it is better to determine the fixpoint status of a propagator dynamically [15]. Similarly, *strength* of propagators is irrelevant (bounds or domain consistency, or forward checking). Consequently, propagators are sometimes defined to be contracting and monotonic. Examples for this definition of propagators are [16,17,15].

In order to relax the definition of propagators, consider how a propagator p can implement a constraint c. The first condition is *correctness*, p must not remove solutions of c: $a \in c \wedge a \in d \Rightarrow a \in p(d)$ for any assignment a and domain d. The second condition is that for an assignment a, p *checks* whether a is a solution of c: $p(\{a\}) = \{a\} \Leftrightarrow a \in c$. The interesting connection between these properties and monotonicity is that *every monotonic propagator implements exactly one constraint.*

Definition 2. *A monotonic propagator* p implements *the constraint defined as the set of assignments* accepted *by the propagator,* $\{a \mid p(\{a\}) = \{a\}\}$. *This is called the* induced constraint c_p of p.

By definition, a propagator p checks whether assignments are solutions of c_p. And by monotonicity, if $a \in c_p$ and $a \in d$, then $p(\{a\}) \subseteq p(d)$, and hence $a \in p(d)$. Thus, p is correct for c_p. Having correct propagators for constraints, it makes sense to define the set of solutions of a propagation problem $\langle d, P\rangle$ for a domain d and a set of propagators P as the set of solutions of the induced constraints: $\mathrm{sol}(d,P) = \mathrm{sol}(d, \{c_p \mid p \in P\})$.

In Def. 2 monotonicity is used to enforce correctness of the propagator. However, monotonicity was *only used for assignments.* This leads to the central definition used in this paper.

Definition 3. *A function* p *over domains is called* weakly monotonic *iff* $a \in d \Rightarrow p(\{a\}) \subseteq p(d)$ *for all assignments* a *and domains* d. *A* propagator *is a contracting and* weakly monotonic *function over domains.*

Every weakly monotonic propagator also induces a single constraint. Weak monotonicity yields a minimal definition of propagators, as every propagator can be made weakly monotonic. Given a non-monotonic propagator that is correct for a constraint c, we can turn it into a weakly monotonic propagator that implements c by composing it with a function that checks c on assignments.

Lemma 4. *A monotonic propagator is weakly monotonic.*

The lemma follows directly from the definitions. Conversely, a weakly monotonic propagator is not necessarily monotonic. Assume a propagator p that only prunes the domain if $|d(x)| \in \{1, 3\}$. A domain with $|d(x)| = 2$ can be stronger than a domain with $|d(x)| = 3$ but yield weaker propagation, so p is not monotonic. But it is weakly monotonic, because p does propagate when $|d(x)| = 1$ and thus still checks assignments.

Lemma 5. *Propagation preserves solutions:* $\mathrm{sol}(d, P) = \mathrm{sol}(p(d), P)$ *for* $p \in P$.

Theorem 1 does not hold for non-monotonic propagators. But, $\mathrm{prop}(d, P)$ still terminates, as propagators are contracting and domains are finite. Thus, $\mathrm{prop}(d, P)$ still produces simultaneous fixpoints for all $p \in P$. However, these fixpoints can now be different for different orders of propagator application. Thus, prop turns into a relation. For convenience, we will continue to write $d' = \mathrm{prop}(d, P)$ instead of $d' \in \mathrm{prop}(d, P)$.

Lemma 6. *Assume that* $\mathrm{prop}(d, P) = d_1$ *and* $\mathrm{prop}(d, P) = d_2$*. Then* d_1 *and* d_2 *may not be comparable:* $d_1 \not\subseteq d_2$ *and* $d_2 \not\subseteq d_1$*.*

Consider propagators p and q with $c_p \equiv (x > 0)$ and $c_q \equiv (x < 2)$. To make them non-monotonic, assume that both p and q only propagate if $|d(x)| \in \{1, 3\}$. Given the domain $d = (x \mapsto \{0, 1, 2\})$, there are two incomparable fixpoints $d_1 = p(q(d)) = (x \mapsto \{0, 1\})$ and $d_2 = q(p(d)) = (x \mapsto \{1, 2\})$.

Although there is no unique weakest fixpoint, the different fixpoints are still well behaved in that they contain all solutions of the original problem. The following lemma will be central for the discussion of search in the next section.

Lemma 7. *If* $\mathrm{prop}(d, P) = d'$*, then* $d' \subseteq d$ *and* $\mathrm{sol}(d, P) = \mathrm{sol}(d', P)$*.*

This follows from the fact that if some $p \in P$ prunes an assignment a from d, then weak monotonicity guarantees that $p(\{a\}) = \emptyset$. Therefore, a is no solution of c_p, and hence not of $\langle d, P \rangle$, either.

But how are these fixpoints related to the unique weakest fixpoint computed by monotonic propagation? We define the *strongest possible propagator* for a constraint c. This so-called *domain propagator* $\hat{p}_c$ establishes *domain consistency* (also known as generalized arc consistency), it removes all values from all variable domains that cannot be extended to a solution of c. That is, $\hat{p}_c$ returns the strongest domain that contains all solutions of c and d: $\hat{p}_c(d) = \mathrm{dom}(c \cap d)$.

Lemma 8. *Any propagator* p *implementing* c *returns a weaker domain than* $\hat{p}_c$*:* $\hat{p}_c(d) \subseteq p(d)$*. For any constraint* c*,* $\hat{p}_c$ *is monotonic.*

Any non-monotonic propagator for a constraint c is weaker than $\hat{p}_c$. Fixpoints are therefore always weaker than those obtained by domain propagation.

The astute reader may have noticed that none of the results depends on propagators being functions, except when being applied to assignments. This is an important insight, as it allows for example *randomized propagation*: on the same domain d, a propagator may return *different* results. In other words, propagators can be relaxed to be contracting and weakly monotonic *relations* over $\mathit{Dom} \times \mathit{Dom}$, as long as they are functional on assignments.

4 Search

A constraint solver interleaves propagation with search. It starts with a propagation problem $\langle d, P\rangle$ and computes a fixpoint. If this fixpoint is neither failed (an empty domain) nor solved (all variables assigned), the solver splits the problem and solves the resulting subproblems recursively. Splitting creates two *branches* (we assume binary search for simplicity), adding propagators to each branch that make the problem simpler (for instance $x = i$ for one branch and $x \neq i$ for the other). Splitting must partition the solution space of the original problem.

Thus, the solver explores a *search tree*. A solver is *sound* if all solutions that it finds by exploring the search tree are solutions of the original problem. It is *complete* if the search tree contains all solutions of the original problem.

If all propagators are monotonic, there is a unique fixpoint for each propagation problem. As long as the addition of propagators to the branches is deterministic, the search tree is completely determined by the initial propagation problem. With non-monotonic propagators, the order of propagation matters. As a result, the shape of the tree also depends on the order of propagation chosen by the solver. As discussed in Lemma 6, the resulting fixpoints may be incomparable, resulting in a different search tree. The good news is that non-monotonic propagation is still correct, it does not remove solutions.

Lemma 9. *A combination of non-monotonic propagation and search is a sound and complete solver for propagation problems. The* set *of solutions found by search is thus determined solely by the original problem to be solved. The* order *of the solutions in the tree depends on the order of propagation.*

For a solver, propagation order may depend on the environment, for instance on memory allocation or other things out of the control of the solver, and may be different for different runs of the same problem. Hence, the first solution may not even be the same between different runs of the same solver. While this may seem inconvenient, non-monotonic propagation is not alone in this respect: parallel search and random restarts share the same properties.

Mozart solves the problem of non-unique fixpoints by fixing the order in which non-monotonic propagators are executed [17]. This technique however clashes with priorities for propagator scheduling, which has proven extremely useful [15].

Recomputation. Recomputation is an important technique for making solvers based on copying efficient [18], and for enabling trailing solvers to perform more advanced search strategies such as best-first search [19].

The main idea is to recompute a node in the search tree from a state further up in the tree, using a *path* that describes the choices that lead to the node. Such a path can consist of a sequence of moves to a child of a node (for example 1.2.2.1.1). Recomputation amounts to redoing the exploration along this path, computing fixpoints for every intermediate step. This is the method described in [18], called *fixpoint recomputation.* Alternatively, a path is a sequence of propagators added by splitting (for example $P_1.P_2.P_3$). Then recomputation adds all propagators to the original propagation problem and computes a single fixpoint.

We will refer to this as *path recomputation* (called *batch recomputation* in [20] and *decomposition based search* in [21]).

Fixpoint recomputation fails in the presence of non-monotonic propagators. Assume that splitting adds $x = 1$ on the left and $x \neq 1$ on the right branch. When recomputing the same node, splitting may make an entirely different decision, choosing $y = 1$ and $y \neq 1$ instead. But recomputation just explores the right branch, resulting in an *incomplete search*: possible solutions with $y = 1$ are lost. The Mozart approach of fixing propagator order solves this problem.

Path recomputation is well-behaved even for non-monotonic propagation. In the above example, the splitting decision would be made only once (during the original exploration), and recomputation just adds $x \neq 1$ on the right branch. As propagation preserves solutions, the set of solutions of the right branch is exactly the same as during original exploration. Interestingly, Choi et al. [20] state that path recomputation requires monotonic propagators. This over-cautious assumption underlines the importance of a theory of non-monotonic propagation.

With non-monotonic propagation, the fixpoint after recomputation can differ from the one during exploration. Instead of failure, the solver may recompute a non-failed node, and instead of a solution, a node where some variables are not assigned yet. However, as the set of solutions is the same, further search will only produce failure or the very same solution, respectively. The solver must thus be able to perform less or additional search when recomputing.

Theorem 10. *Constraint propagation with weakly monotonic propagators, combined with search (possibly based on path recomputation), yields a sound and complete solver for propagation problems.*

5 Case Study: Propagating `circuit`

The `circuit`$(x_1, \ldots, x_n)$ constraint over n finite domain integer variables is true iff the graph with edges $i \rightarrow j$ where $x_i = j$ has a single cycle covering all nodes. Domain-consistent propagation of `circuit` is of course NP-hard.

A simple monotonic propagator for the `circuit` constraint based on the graph G with nodes i and edges $i \rightarrow j$ for all $j \in d(x_i)$ for $1 \leq i \leq n$ is as follows: use a standard `alldifferent` propagation algorithm, as all x_i have to be pairwise distinct; check the mandatory condition that G has only a single strongly connected component. Checking the mandatory condition can be done using DFS on G. However, the DFS spanning tree starting from a node i with $|d(x_i)| > 1$ also offers potential for propagation [22].

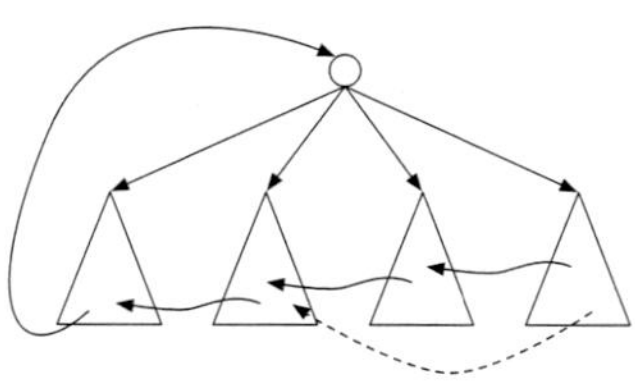

The disjoint subtrees explored by DFS are sketched as triangles. For `circuit`, the following must hold: There must be an edge from each subtree to its predecessor subtree, and an edge from the leftmost subtree to the root (otherwise, there is no covering cycle). There must not be any edges between non-neighbor subtrees (such as the dotted edge: the root node must be visited twice if following this edge). This insight can be propagated: prune edges between non-neighbor subtrees, and if there is a single edge between neighbors or from the leftmost subtree to the root, assign the corresponding variable.

This algorithm is obviously non-monotonic: pruning is likely to increase with the number of subtrees in the DFS spanning tree. The DFS spanning tree has more subtrees, if the variable from which DFS starts has more values. Hence, the more values the variable has from which DFS starts, the more pruning.

The propagator is weakly monotonic, as it checks the constraint on assignments. Propagation depends on where DFS starts, so different heuristics for selecting the start variable can be: the *first* variable; a *random* variable; a variable with the *largest* domain. The experiments below confirm that the propagator is indeed non-monotonic: pruning depends on where DFS starts, as witnessed by the different number of fails during search.

The following table shows the effect of non-monotonicity for `circuit`. We have used Gecode 3.0.2 on a MacPro with 2×2.8 GHz Intel Xeons and 8 GB memory running Windows Vista. The runtimes (average of 20 runs, with a coefficient of deviation less than 2% except for random) and number of failures for pruning are relative to just checking by DFS. The `alldifferent` part of `circuit` uses domain-consistency. `knights-`n finds a Knights tour on an $n \times n$ board and `tsp-*` finds an optimal traveling salesman tour.

Example	check		first		largest		random	
	time (s)	fail	time	fail	time	fail	time	fail
`knights-18`	0.36	6 070	86.1%	83.4%	66.9%	57.4%	11.4%	2.2%
`knights-20`	0.05	39	98.1%	92.3%	101.9%	100.0%	100.0%	87.2%
`knights-22`	45.03	543 384	94.4%	93.2%	81.3%	75.6%	0.9%	0.6%
`knights-24`	4.36	42 260	28.4%	26.1%	67.8%	54.0%	4.9%	1.9%
`tsp-br17`	0.83	48 804	98.2%	97.9%	100.4%	98.1%	102.7%	99.1%
`tsp-ftv33`	1423.71	31 013 229	99.8%	99.5%	99.8%	97.9%	96.9%	92.5%

Random variable selection vastly outperforms the other strategies for difficult `knights` instances and shows some speedup for the medium-sized `tsp-ftv33` instance. The coefficient of deviation for runtime and number of failures for random is less than 5% for `tsp` and around 45% for `knights-`n. That is, for `knights-22` and `knights-24` the speedup is almost always at least one order of magnitude thanks to randomization now legalized by weak monotonicity.

6 Conclusion

This paper has introduced a minimal model of propagation based on weakly monotonic propagators and has clarified the properties of propagation and search based on the model. By this, the paper for the first time gives a model to capture the essential properties of many constraint programming systems that use non-monotonic propagators. The hope is that non-monotonic propagation is seen as a general opportunity for more efficient propagation rather than a problem that is best ignored.

Acknowledgements. We thank Nicolas Beldiceanu, Mats Carlsson, Mikael Lagerkvist, Gert Smolka, and Peter Stuckey for insightful discussions. Part of this work was done while Guido Tack was at NICTA Victoria Laboratory, Melbourne, Australia.

References

1. Baptiste, P., Le Pape, C.: A theoretical and experimental comparison of constraint propagation techniques for disjunctive scheduling. In: IJCAI, pp. 600–606 (1995)
2. Menana, J., Demassey, S.: Sequencing and counting with the multicost-regular constraint. In: van Hoeve, W.-J., Hooker, J.N. (eds.) CPAIOR 2009. LNCS, vol. 5547, pp. 178–192. Springer, Heidelberg (2009)
3. Katriel, I.: Expected-case analysis for delayed filtering. In: Beck, J.C., Smith, B.M. (eds.) CPAIOR 2006. LNCS, vol. 3990, pp. 119–125. Springer, Heidelberg (2006)
4. Stergiou, K.: Heuristics for dynamically adapting propagation. In: ECAI, pp. 485–489 (2008)
5. Katriel, I., Van Hentenryck, P.: Randomized filtering algorithms. Technical Report CS-06-09, Brown University, Providence, RI, USA (2006)
6. Mehta, D., van Dongen, M.R.C.: Probabilistic consistency boosts MAC and SAC. In: IJCAI, pp. 143–148 (2007)
7. Sellmann, M.: Approximated consistency for Knapsack constraints. In: Rossi, F. (ed.) CP 2003. LNCS, vol. 2833, pp. 679–693. Springer, Heidelberg (2003)
8. Ward, M.: The closure operators of a lattice. Annals of Mathematics 43(2), 191–196 (1942)
9. Tarski, A.: Fundamentale Begriffe der Methodologie der deduktiven Wissenschaften. I. Monatshefte für Mathematik 37(1), 361–404 (1930)
10. Tarski, A.: V. In: Logic, semantics, metamathematics, 2nd edn., pp. 60–109. Hackett Publishing Company (1983)
11. Saraswat, V.A., Rinard, M.C., Panangaden, P.: Semantic foundations of concurrent constraint programming. In: POPL, pp. 333–352 (1991)
12. Benhamou, F., McAllester, D.A., Van Hentenryck, P.: CLP(Intervals) revisited. In: ILPS, pp. 124–138. The MIT Press, Cambridge (1994)
13. Van Hentenryck, P., Saraswat, V.A., Deville, Y.: Constraint processing in cc(FD). Technical report, Brown University (1991)
14. Apt, K.R.: Principles of Constraint Programming. Cambridge University Press, Cambridge (2003)
15. Schulte, C., Stuckey, P.J.: Efficient constraint propagation engines. ACM Trans. Program. Lang. Syst. 31(1), 2:1–2:43 (2008)
16. Benhamou, F.: Heterogeneous Constraint Solving. In: Hanus, M., Rodríguez-Artalejo, M. (eds.) ALP 1996. LNCS, vol. 1139, pp. 62–76. Springer, Heidelberg (1996)
17. Müller, T.: Constraint Propagation in Mozart. Doctoral dissertation, Universität des Saarlandes, Saarbrücken, Germany (2001)
18. Schulte, C.: Programming Constraint Services. LNCS (LNAI), vol. 2302. Springer, Heidelberg (2002)
19. Perron, L.: Search procedures and parallelism in constraint programming. In: Jaffar, J. (ed.) CP 1999. LNCS, vol. 1713, pp. 346–361. Springer, Heidelberg (1999)
20. Choi, C.W., Henz, M., Ng, K.B.: Components for state restoration in tree search. In: Walsh, T. (ed.) CP 2001. LNCS, vol. 2239, pp. 240–255. Springer, Heidelberg (2001)
21. Michel, L., Van Hentenryck, P.: A decomposition-based implementation of search strategies. ACM Trans. Comput. Logic 5(2), 351–383 (2004)
22. Carlsson, M.: Personal communication (February 2007)

Constraint-Based Optimal Testing Using DNNF Graphs*

Anika Schumann[1], Martin Sachenbacher[2], and Jinbo Huang[3]

[1] Cork Constraint Computation Centre, University College Cork, Ireland
[2] Institut für Informatik, Technische Universität München, Germany
[3] NICTA and Australian National University, Australia

Abstract. The goal of testing is to distinguish between a number of hypotheses about a system—for example, different diagnoses of faults—by applying input patterns and verifying or falsifying the hypotheses from the observed outputs. Optimal distinguishing tests (ODTs) are those input patterns that are most likely to distinguish between hypotheses about non-deterministic systems. Finding ODTs is practically important, but it amounts in general to determining a ratio of model counts and is therefore computationally very expensive.

In this paper, we present a novel approach to constraint-based ODT generation, which uses structural properties of the system to limit the complexity of computation. We first construct a compact graphical representation of the testing problem via compilation into *decomposable negation normal form*. Based on this compiled representation, we show how one can evaluate distinguishing tests in linear time, which allows us to efficiently determine an ODT. Experimental results from a real-world application show that our method can compute ODTs for instances that were intractable for previous approaches.

Keywords: Algorithms, applications, testing, DNNF graphs.

1 Introduction

Testing asks whether a system can be stimulated with input patterns, such that different hypotheses about the system can be verified or falsified from the observed output patterns. Applications include model-based fault analysis of technical systems, autonomous control (acquiring sensor inputs to discriminate among competing state estimates), and bioinformatics (designing experiments that help distinguish between different possible explanations of biological phenomena).

In many real-world applications of testing, the underlying models are non-deterministic; applying an input can lead to several possible outputs. Different

* This work was supported by Deutsche Forschungsgemeinschaft under grant SA 1000/2-1, the Science Foundation Ireland under the ITOBO Grant No. 05/IN/I886, and NICTA. The latter is funded by the Australian Government as represented by the Department of Broadband, Communications and the Digital Economy and the Australian Research Council through the ICT Centre of Excellence program.

I.P. Gent (Ed.): CP 2009, LNCS 5732, pp. 731–745, 2009.

notions of testing with such non-deterministic models have been introduced. In the area of diagnosis, [13] introduced *definitely* and *possibly discriminating tests* (DDTs and PDTs) for systems modeled as constraints. For a DDT, the sets of possible outputs are disjoint and thus it will necessarily distinguish among hypotheses, whereas for a PDT, the sets partially overlap and thus it may or may not distinguish among hypotheses. In automata theory, [1] studied the analogous problem of generating *strong* and *weak distinguishing sequences* for non-deterministic finite state machines; for sequences of length at most $k \in \mathbb{N}$, this can be reduced to the problem of finding DDTs and PDTs [7].

For example, consider the network shown in Figure 1, which consists of one NOT component and two adders A_{2H} and A_{UL}. The former is high dominant upon receiving input $i_2 = H$ and the latter is low dominant upon receiving input $u = L$. Here we consider the two hypotheses M_1, M_2 that either both adders function normally, i.e. $M_1 = \{\text{NOT}, A_{2H}, A_{UL}\}$ or that both adders are faulty, i.e. $M_2 = \{\text{NOT}, A'_{2H}, A'_{UL}\}$. Then there exists no DDT, but two PDTs: $[-i_1, i_2]$ and $[-i_1, -i_2]$.

[8] introduced *optimal distinguishing tests* (ODTs), which generalize DDTs and PDTs by maximizing the ratio of distinguishing over non-distinguishing possible outcomes. In the example from Figure 1, the PDT $[-i_1, i_2]$ has a better distinguishing ratio than the PDT $[-i_1, -i_2]$, and is therefore an ODT for this example. Finding ODTs is important as it reduces the number of tests to be executed and the overall costs of the testing process. [8] proposed and analyzed a simple greedy-type algorithm to approximate ODTs, which in some real-world applications produces test inputs whose distinguishing ratios are close to those of ODTs.

In this paper, we present a novel search algorithm to compute ODTs (and thus DDTs and PDTs), which exploits structural properties of the model to limit the complexity of optimal test generation. Its main feature is a carefully constructed graph—through manipulation of logical theories and compilation into *decomposable negation normal form* (DNNF) [6]—that allows efficient computation of good upper bounds on ratios of model counts. These upper bounds are used

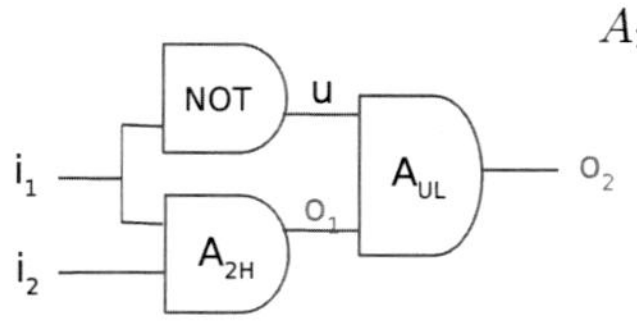

A_{2H}:

i_1	i_2	o_1
L	L	L
L	H	L
L	H	H
H	L	H
H	H	H

A'_{2H}:

i_1	i_2	o_1
L	L	L
L	H	L
H	L	H
H	H	H

NOT:

i_1	u
L	H
H	L

A_{UL}:

u	o_1	o_2
L	L	L
L	H	L
H	L	L
H	L	H
H	H	H

A'_{UL}:

u	o_1	o_2
L	L	L
L	H	L
H	L	L
H	H	L

Fig. 1. Circuit with two possibly faulty adders: A'_{2H} and A'_{UL}

to prune the search in a way motivated by a recent planning algorithm [9]. We show that our method can compute ODTs for instances that were intractable for previous approaches.

2 Background

Following the framework in [8,12,13], we assume that the system can be modeled as a *constraint satisfaction problem* (CSP), which is a triple $M = (\mathcal{V}, \mathcal{D}, \mathcal{C})$, where $\mathcal{D} = D(v_1) \times \ldots \times D(v_n)$ are the finite domains of finitely many variables $v_j \in \mathcal{V}$, $j = 1, \ldots, n$, and $\mathcal{C} = \{C_1, \ldots, C_m\}$ is a finite set of constraints with $C_i \subseteq \mathcal{D}$, $i = 1, \ldots, m$. We denote by X the set of all solutions, that is, assignments $\boldsymbol{x} \in \mathcal{D}$ to the variables such that all constraints are satisfied. That is, $X = \{\boldsymbol{x} \mid \boldsymbol{x} \in \mathcal{D},\ \mathcal{C}(\boldsymbol{x})\}$, where $\mathcal{C}(\boldsymbol{x})$ denotes $\boldsymbol{x} \in C_i$ for all $i = 1, \ldots, m$.

In addition, the system under investigation defines a set of *controllable* (input) variables $\mathcal{I}$ and a set of *observable* (output) variables $\mathcal{O}$. Formally, a hypothesis M for a system is then a CSP whose variables are partitioned into $\mathcal{V} = \mathcal{I} \cup \mathcal{O} \cup \mathcal{L}$, such that $\mathcal{I}$ and $\mathcal{O}$ are the input and output variables of the system, and for all assignments to $\mathcal{I}$, the CSP is satisfiable. The remaining variables $\mathcal{L} = \mathcal{V} \setminus (\mathcal{I} \cup \mathcal{O})$ are called internal state variables. We denote by $\mathcal{D}(\mathcal{I})$ and $\mathcal{D}(\mathcal{O})$ the cross product of the domains of the input and output variables, respectively: $\mathcal{D}(\mathcal{I}) = \times_{v \in \mathcal{I}} D(v)$ and $\mathcal{D}(\mathcal{O}) = \times_{v \in \mathcal{O}} D(v)$.

The goal of testing is then to find assignments of the input variables $\mathcal{I}$ that will cause different assignments of output variables $\mathcal{O}$ for different hypotheses. In the general case of non-deterministic systems, an input assignment can yield more than one possible output assignments. This case is frequent in practice; one reason is that in order to reduce the size of a model, it is common to aggregate the domains of system variables into small sets of values such as 'low', 'med', and 'high'; a side-effect of this abstraction is that the resulting models can no longer be assumed to be deterministic functions, even if the underlying system behavior was deterministic. Another reason is the test situation itself: even in a rigid environment such as an automotive test-bed, there are inevitably variables or parameters that cannot be completely controlled while testing the device.

In order to capture sets of outputs, for a given hypothesis M and an assignment $\boldsymbol{t} \in \mathcal{D}(\mathcal{I})$ to the input variables, we define the *output function* $\mathcal{X} : \mathcal{D}(\mathcal{I}) \to 2^{\mathcal{D}(\mathcal{O})}$ with $\boldsymbol{t} \mapsto \{\boldsymbol{y} \mid \boldsymbol{y} \in \mathcal{D}(\mathcal{O}), \exists \boldsymbol{x} \in X : \boldsymbol{x}[\mathcal{I}] = \boldsymbol{t} \wedge \boldsymbol{x}[\mathcal{O}] = \boldsymbol{y}\}$, where $2^{\mathcal{D}(\mathcal{O})}$ denotes the power set of $\mathcal{D}(\mathcal{O})$, and $\boldsymbol{x}[\mathcal{I}]$, $\boldsymbol{x}[\mathcal{O}]$ denote the restriction of the vector $\boldsymbol{x}$ to the input variables $\mathcal{I}$ and the output variables $\mathcal{O}$, respectively. Note that since M will always yield an output, $\mathcal{X}(\boldsymbol{t})$ is non-empty.

2.1 Distinguishing Tests

Non-deterministic models have given rise to the introduction of so-called *possibly* and *definitely distinguishing tests*, for short PDT and DDT, respectively [13]. The first type of test (PDT) might distinguish between hypotheses, as the sets of possible outputs partially overlap, whereas the second type (DDT) will necessarily do so, as the sets of possible outputs are disjunct:

Definition 1 (Distinguishing Tests). *Consider $k \in \mathbb{N}$ hypotheses $M_1, \ldots, M_k$ with input variables $\mathcal{I}$ and output variables $\mathcal{O}$. Let $\mathcal{X}_i$ be the output function of hypothesis M_i with $i \in \{1, \ldots, k\}$. An assignment $\boldsymbol{t} \in \mathcal{D}(\mathcal{I})$ to $\mathcal{I}$ is a* possibly distinguishing test *(PDT), if there exists an $i \in \{1, \ldots, k\}$ such that $\mathcal{X}_i(\boldsymbol{t}) \setminus \bigcup_{j \neq i} \mathcal{X}_j(\boldsymbol{t}) \neq \varnothing$. An assignment $\boldsymbol{t} \in \mathcal{D}(\mathcal{I})$ is a* definitely distinguishing test *(DDT), if for all $i \in \{1, \ldots, k\}$ it holds that $\mathcal{X}_i(\boldsymbol{t}) \setminus \bigcup_{j \neq i} \mathcal{X}_j(\boldsymbol{t}) = \mathcal{X}_i(\boldsymbol{t})$.*

For testing with non-deterministic automata models instead of logical theories or CSPs, there exists the analogous notion of so-called *weak* and *strong distinguishing sequences* [1,3]. Finding such sequences with a length bounded by some $k \in \mathbb{N}$ can be reduced to the problem of finding PDTs and DDTs, by unrolling automata into a constraint network using k copies of the automata's transition and observation relation [7]. Therefore, in this paper we restrict ourselves to models given as networks of finite-domain constraints.

2.2 Optimal Distinguishing Tests

Due to limited observability or a high degree of non-determinism, it is not uncommon that a DDT for the hypotheses does not exist, and one can instead only find PDTs. This motivates a *quantitative measure* for tests that refines and generalizes the previous notions of PDTs and DDTs. The intuition is that if we assume the possible outcomes (feasible assignments to the output variables) to be (roughly) equally likely, a PDT will be more likely to distinguish among two given hypotheses compared to another PDT, if the ratio of possible outcomes that are unique to a hypothesis versus all possible outcomes is higher. The notion of optimal distinguishing tests introduced in [8] formalizes this goal of finding tests that discriminate among two hypotheses as good as possible:

Definition 2 (Distinguishing Ratio). *Given a test input $\boldsymbol{t} \in \mathcal{D}(\mathcal{I})$ for two hypotheses M_1, M_2 with input variables $\mathcal{I}$ and output variables $\mathcal{O}$, we define $\Gamma(\boldsymbol{t})$ to be the ratio of feasible outputs that distinguish among the hypotheses versus all feasible outputs:*

$$\Gamma(\boldsymbol{t}) := \frac{|\mathcal{X}_1(\boldsymbol{t}) \cup \mathcal{X}_2(\boldsymbol{t})| - |\mathcal{X}_1(\boldsymbol{t}) \cap \mathcal{X}_2(\boldsymbol{t})|}{|\mathcal{X}_1(\boldsymbol{t}) \cup \mathcal{X}_2(\boldsymbol{t})|}$$

Γ is a measure for test quality that refines the notion of PDTs and DDTs: if Γ is 0, then the test does not distinguish at all, as both hypotheses lead to the same observations (output patterns). If the value is 1, then the test is a DDT, since the hypotheses always lead to different observations. If the value is between 0 and 1, then the test is a PDT (there is some non-overlap in the possible observations). Note that Γ is well-defined since for any chosen $\boldsymbol{t} \in \mathcal{D}(\mathcal{I})$, the sets $\mathcal{X}_1(\boldsymbol{t})$ and $\mathcal{X}_2(\boldsymbol{t})$ are non-empty.

An *optimal distinguishing test* (ODT) is one that has the maximal distinguishing ratio. For the example in Figure 1, the two PDTs $[-i_1, i_2]$ and $[-i_1, -i_2]$ for the hypotheses $M_1 = \{\text{NOT}, A_{2H}, A_{UL}\}$ and $M_2 = \{\text{NOT}, A'_{2H}, A'_{UL}\}$ have the following distinguishing ratios:

$\Gamma([-i_1, i_2]) = \frac{|\{o_1,o_2\},\{-o_1,o_2\}|}{|\{o_1,o_2\},\{-o_1,o_2\},\{-o_1,-o_2\}|} = \frac{2}{3}$ and

$\Gamma([-i_1, -i_2]) = \frac{|\{-o_1,o_2\}|}{|\{-o_1,o_2\},\{-o_1,-o_2\}|} = \frac{1}{2}$.

Therefore, the input $[-i_1, i_2]$ is an ODT for this example.

2.3 Deterministic DNNF

We briefly review graph-based representations of propositional theories. A propositional theory f is in negation normal form (NNF) [6] if it only uses conjunction (and, $\wedge$), disjunction (or, $\vee$), and negation (not, $\neg$), and negation only appears next to variables. An NNF is *decomposable* (DNNF) if conjuncts of every conjunction share no variables. A DNNF is *deterministic* (d-DNNF) if disjuncts of every disjunction are pairwise logically inconsistent. A d-DNNF is *smooth* if disjuncts of every OR node mention the same set of variables. In the rest of the paper we also assume that every variable of the logical theory appears in a smooth d-DNNF graph (this can always be ensured in polynomial time [6]). Such graphs represent each of its models by a subgraph G_s that satisfies the properties:

- every OR node in G_s has exactly one child,
- every AND node in G_s has the same children as it has in G, and
- G_s has the same root as G.

Henceforth we will denote subgraphs with these properties as *m-subgraphs*. Those that satisfy only the first two properties we will denote as *s-subgraphs*. Further we say that a subgraph is *labeled* by a literal l if it has a leaf node l, and that it is *consistent* with a partial assignment X to the d-DNNF variables if its labels are consistent with X.

d-DNNF graphs can be generated for propositional theories in conjunctive normal form (CNF) using the publicly available C2D compiler [5]. The complexity of this operation is polynomial in the number of variables and exponential only in the treewidth of the system in the worst case. Furthermore, given a DNNF graph G one can compute its projection $\Pi_\Sigma(G)$ on variable set Σ in linear time. Without impact on the computation time we therefore assume that M_1 and M_2 are defined over input and output variables only.

The left graph of Figure 2 illustrates a smooth d-DNNF graph G_N representing the models for the numerator of Γ for all test vectors for the example illustrated in Figure 1, i.e. $G_N = (M_1 \vee M_2) \wedge \neg(M_1 \wedge M_2)$. Thus it consists of the three models below that are each represented by a m-subgraph:

Model	Nodes of corresponding m-subgraph
$\{-i_1, i_2, o_1, o_2\}$	A1, O2, A3, O5, A6, $-i_1$, i_2, o_1, o_2
$\{-i_1, i_2, -o_1, o_2\}$	A1, O2, A3, O5, A7, $-i_1$, i_2, $-o_1$, o_2
$\{-i_1, -i_2, -o_1, o_2\}$	A1, O2, A4, A7, $-i_1$, $-i_2$, $-o_1$, o_2

Based on a smooth d-DNNF graph G the number of models $|G(X)|$ consistent with a partial assignment X to the d-DNNF variables can be determined by

Algorithm 1. Model counting with respect to instantiation X

$$\Lambda(N) = \begin{cases} 1 & \text{if } N \text{ is a leaf node consistent with X} \\ 0 & \text{if } N \text{ is a leaf node inconsistent with X} \\ \sum_i \Lambda(N_i) & \text{if } N = \bigvee_i N_i \\ \prod_i \Lambda(N_i) & \text{if } N = \bigwedge_i N_i \end{cases}$$

counting the number of consistent m-subgraphs in G. This is done by a bottom-up traversal of the graph that computes for each node N the number of consistent s-subgraphs $\Lambda(N)$ rooted in N. Hence, the $\Lambda(N)$ value of the root of the graph denotes the total number of consistent models represented by G. Algorithm 1 describes this linear time procedure [4]. An example of it is shown on the right of Figure 2. The numbers next to the nodes of that graph denote the $\Lambda(N)$ values computed by Algorithm 1 when running it with $X = [-i_1, i_2]$. Hence for the numerator of $\Gamma([-i_1, i_2])$ we get $|\mathcal{X}_1([-i_1, i_2]) \cup \mathcal{X}_2([-i_1, i_2])| - |\mathcal{X}_1([-i_1, i_2]) \cap \mathcal{X}_2([-i_1, i_2])| = 2$.

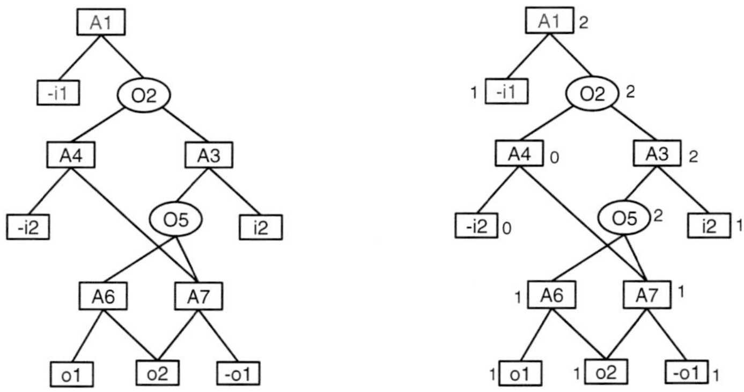

Fig. 2. Smooth d-DNNF graph G_N representing $(M_1 \vee M_2) \wedge \neg(M_1 \cap M_2)$ for the example shown in Figure 1. "A" and "O" indicate an And and an Or node, respectively. Numbers in non-leaf nodes are their identifiers. On the right, the numbers next to the nodes denote the $\Lambda(N)$ values computed by Algorithm 1 when running it with $X = [-i_1, i_2]$.

3 ODT Computation Using a d-DNNF Graph

The last section suggests that we can straightforwardly exploit the linear time d-DNNF based model counting algorithm for our ODT search. Similarly to the generation of a graph G_N representing the models of the numerator of the distinguishing ratio Γ we could also generate a graph representing the models of the denominator of Γ. Based on these two graphs we could then determine the Γ value for any test vector in linear time. However, in order to obtain the test vector with maximal distinguishing ratio, i.e. the ODT, such an approach requires the computation of the Γ values for every complete instantiation of a test vector (CITV). Since the number of test cases is exponential in the number of input variables this procedure would be infeasible for large applications.

This section presents a d-DNNF based branch-and-bound approach that does not require the Γ computation for every test vector. Its main component is a linear time algorithm that computes the upper Γ bound for any partial instantiation of a test vector (PITV). Such a procedure requires the simultaneous count of the models for the numerator and denominator of Γ based on d-DNNF graphs with identical structure. Since it is computationally very expensive to ensure that two independently generated d-DNNF graphs have the same structure we represent the whole ODT problem by a single d-DNNF graph that allows both: the computation of the numerator and that of the denominator of Γ for every test vector. The developed branch-and-bound approach then consists of the following building blocks that are each detailed in the following subsections:

- representation of the ODT problem as single d-DNNF graph $\mathcal{G}$,
- linear time algorithm that computes the Γ value for any CITV based on $\mathcal{G}$,
- linear time algorithm that computes an upper bound for the Γ value for any PITV based on $\mathcal{G}$, and
- an exhaustive search algorithm that iteratively sets input variables until either all variables are set and the Γ value for that CITV is computed or until the upper Γ bound for the PITV is not higher than the Γ value for a previously computed CITV.

3.1 Encoding the ODT Problem as Single d-DNNF Graph

We now describe how we can represent the ODT problem as a single d-DNNF graph $\mathcal{G}$ that allows the distinction of its models into those that belong to the numerator of Γ and those that do not. In order to achieve such a partitioning of nodes we introduce an auxiliary variable d and label every m-subgraph representing a model consistent with the numerator with the literal $-d$ and add the literal d to the remaining m-subgraphs. The latter comprises of the models $G_{\bar{N}} = M_1 \wedge M_2$ as stated in Definition 2. Thus, the propositional formula represented by $\mathcal{G}$ is defined as follows:

$$\begin{aligned}\mathcal{G} &= ((M_1 \vee M_2) \wedge \neg(M_1 \wedge M_2) \wedge \neg d) \vee (M_1 \wedge M_2 \wedge d)\\ &= (G_N \wedge \neg d) \vee (G_{\bar{N}} \wedge d)\end{aligned}$$

For our example we generate a graph $\mathcal{G}$ that represents the following models:

$$\begin{array}{ll}\{-d, -i_1, i_2, o_1, o_2\}, & \{d, -i_1, i_2, -o_1, -o_2\}, \text{ and}\\ \{-d, -i_1, i_2, -o_1, o_2\}, & \{d, -i_1, -i_2, -o_1, -o_2\}.\\ \{-d, -i_1, -i_2, -o_1, o_2\}, & \end{array}$$

The models on the left correspond to the numerator models of Γ and the ones on the right to the denominator models of Γ. The graph is illustrated in Figure 3. Its definition ensures that the distinguishing ratio can be obtained as follows:

$$\begin{aligned}
\Gamma(\boldsymbol{t}) &= \frac{|\mathcal{X}_1(\boldsymbol{t}) \cup \mathcal{X}_2(\boldsymbol{t})| - |\mathcal{X}_1(\boldsymbol{t}) \cap \mathcal{X}_2(\boldsymbol{t})|}{|\mathcal{X}_1(\boldsymbol{t}) \cup \mathcal{X}_2(\boldsymbol{t})|} \\
&= \frac{|\mathcal{X}_1(\boldsymbol{t}) \cup \mathcal{X}_2(\boldsymbol{t})| - |\mathcal{X}_1(\boldsymbol{t}) \cap \mathcal{X}_2(\boldsymbol{t})|}{(|\mathcal{X}_1(\boldsymbol{t}) \cup \mathcal{X}_2(\boldsymbol{t})| - |\mathcal{X}_1(\boldsymbol{t}) \cap \mathcal{X}_2(\boldsymbol{t})|) + |\mathcal{X}_1(\boldsymbol{t}) \cap \mathcal{X}_2(\boldsymbol{t})|} \\
&= \frac{|G_N(\boldsymbol{t})|}{|G_N(\boldsymbol{t})| + |G_{\bar{N}}(\boldsymbol{t})|} && \text{see definition of } G_N \text{ and } G_{\bar{N}} \\
&= \frac{|\mathcal{G}(\boldsymbol{t} \wedge \neg d)|}{|\mathcal{G}(\boldsymbol{t} \wedge \neg d)| + |\mathcal{G}(\boldsymbol{t} \wedge d)|} && \text{see definition of } G \\
&= \frac{|\mathcal{G}(\boldsymbol{t} \wedge \neg d)|}{|\mathcal{G}(\boldsymbol{t})|} && \text{since } d \vee \neg d \text{ evaluates to } \mathit{true}
\end{aligned}$$

The computation of the distinguishing ratio based on the later fraction is the subject of the next subsection.

3.2 Computation of $\Gamma(t)$ Based on DNNF Graph $\mathcal{G}$

We now show that we can compute the distinguishing ratio by resorting to the single graph $\mathcal{G} = (G_N \wedge \neg d) \vee (G_{\bar{N}} \wedge d)$. Although this algorithm has the same complexity than the one based on graphs G_N and $G_{\bar{N}}$, which we described earlier, we chose to present it since it will be the basis for the upper bound algorithm detailed in the next subsection. The latter requires the simultaneous computation of numerator Λ_α and denominator Λ_β values of Γ. This can be done by a single bottom-up traversal of the graph as described in Algorithm 2. The model counting procedure itself is almost identical to the one shown in Algorithm 1. The only difference is that we set the numerator value Λ_α for the leaf node labeled d to 0. This results from the fact that Algorithm 2 is executed with respect to the instantiation $\boldsymbol{t}$ only, but that the numerator of Γ, i.e. $|\mathcal{G}(\boldsymbol{t} \wedge \neg d)|$, is defined with respect to instantiation $\boldsymbol{t} \wedge \neg d$. Hence we have to explicitly add the constraint $\neg d$, i.e. set $\Lambda_\alpha(d)$ to 0.

The bottom node label of graph $\mathcal{G}$ shown on the left of Figure 3 denotes the Λ_α and Λ_β values for $\boldsymbol{t} = [-i_1, -i_2]$. From its root label the distinguishing ratio can be retrieved as formally stated in Theorem 1. Note that the distinguishing ratio of a node is only guaranteed to be correct if it is obtained from the Λ_α and Λ_β values of its children. Resorting to their Γ values would not be sufficient, since the numerator and denominator values for *OR* nodes need to be computed separately (see also Algorithm 1). For instance, consider node $O5$ shown on the left of Figure 3. Its Γ value of $\frac{2}{3}$ cannot be obtained from the Γ values of its children which are 0 (for node $A9$) and 1 (for nodes $A7$ and $A8$).

Theorem 1 (Test Assessment). *Let G be the root node of a smooth d-DNNF graph $\mathcal{G}$ representing the propositional formula $((M_1 \vee M_2) \wedge \neg(M_1 \wedge M_2) \wedge \neg d) \vee (M_1 \wedge M_2 \wedge d)$. Then $\Gamma(\boldsymbol{t}) = \Gamma(G, \boldsymbol{t})$ for any complete instantiation $\boldsymbol{t}$ of input variables.*

The correctness of this Theorem follows from the fact that $\Gamma(\boldsymbol{t}) = \frac{|\mathcal{G}(\boldsymbol{t} \wedge \neg d)|}{|\mathcal{G}(\boldsymbol{t})|}$ as derived in the previous subsection and from the basic d-DNNF model counting procedure shown in Algorithm 1.

Algorithm 2. Test assessment with respect to instantiation $\boldsymbol{t}$

For a leaf N we set:

$$\Lambda_\alpha(N) = \begin{cases} 0 & \text{if } N = d \text{ or} \\ & N \text{ is inconsistent with } \boldsymbol{t} \\ 1 & \text{otherwise} \end{cases} \qquad \Lambda_\beta(N) = \begin{cases} 1 & \text{if } N = d \\ \Lambda_\alpha(N) & \text{otherwise} \end{cases}$$

For remaining nodes we compute:

$$(\Lambda_\alpha(N), \Lambda_\beta(N)) = \begin{cases} (\sum_i \Lambda_\alpha(N_i), \\ \quad \sum_i \Lambda_\beta(N_i)) \text{ if } N = \bigvee_i N_i \\ (\prod_i \Lambda_\alpha(N_i), \\ \quad \prod_i \Lambda_\beta(N_i)) \text{ if } N = \bigwedge_i N_i \end{cases}$$

Then we compute the distinguishing ratio for each node:

$$\Gamma(N, \boldsymbol{t}) = \begin{cases} 0 & \text{if } \Lambda_\beta(N) = 0 \\ \frac{\Lambda_\alpha(N)}{\Lambda_\beta(N)} & \text{otherwise} \end{cases}$$

3.3 Upper Bounds for Partial Test Vectors

While the computation of the distinguishing ratio could have also been done based on two separate d-DNNF graphs, we now show how this single graph is essential to our new method for computing upper bounds on the distinguishing ratio based on a PITV $\boldsymbol{t}_p$. These bounds $\Gamma'(N, \boldsymbol{t}_p)$ are also obtained for each node N by a bottom-up traversal of graph $\mathcal{G}$. We will show that for any CITV $\boldsymbol{t}$ of $\boldsymbol{t}_p$ we have $\Gamma(N, \boldsymbol{t}) \leq \Gamma'(N, \boldsymbol{t}_p)$ and that we can therefore retrieve an upper bound of the distinguishing ratio from the Γ' value of the root of $\mathcal{G}$.

Naturally, the search for an ODT will be the more efficient the tighter the upper bounds. The tightest possible bound for a node N is $\Gamma'(N, \boldsymbol{t}_p) = \Gamma(N, \boldsymbol{t})$ and indeed there is a well defined set of nodes for which we can obtain precisely that bound. This node set $\mathcal{N}_s$ is comprised of all those nodes whose Γ value does not depend on a free input variable, i.e. on a variable in $\mathcal{I}_f$ which is not set by $\boldsymbol{t}_p$. This results from the fact that the Γ value for a node is obtained from a bottom-up traversal of the graph. Hence it depends only on the truth value of its descendant variables, but not on the remaining variables of the graph. Let $\mathcal{N}_f$ denote the set of the remaining nodes, i.e. that are an ancestor of a free input variable.[1] Then resulting from $\mathcal{G}$ being smooth all children $\mathcal{N}_{childN}$ of an *OR* node N belong to the same node set, i.e. either $\mathcal{N}_{childN} \subseteq \mathcal{N}_s$ or $\mathcal{N}_{childN} \subseteq \mathcal{N}_f$. For instance, let us consider the computation of the $\Gamma(O2, [-i_2])$ value for our

[1] Formally, $N \in \mathcal{N}_f$, iff N is a leaf node of a variable $v \in \mathcal{I}_f$, or N is an *AND* or *OR* node that has at least one child $N_i \in \mathcal{N}_f$.

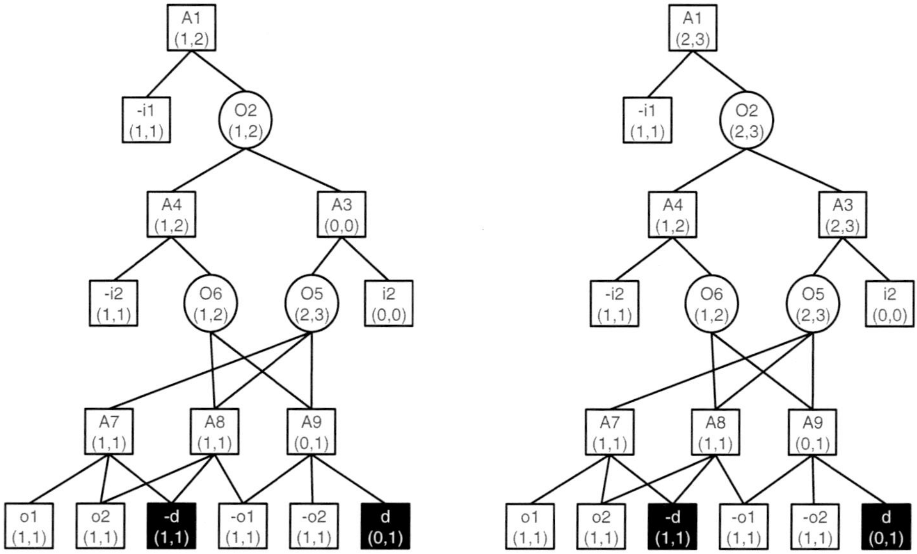

Fig. 3. Graph $\mathcal{G}$ for the example shown in Figure 1. On the left, the bottom labels $(\Lambda_\alpha(N), \Lambda_\beta(N))$ of the nodes refer to the test assessment values computed by Algorithm 2 when running it with $\boldsymbol{t} = [-i_1, -i_2]$ and on the right they to the upper bound values computed by Algorithm 3 when running it with $\boldsymbol{t}_p = [-i_1]$.

example (see left graph of Figure 3). Here we have $\mathcal{I}_f = \{i_1\}$ and thus neither one of the children $A3$ or $A4$ nor $O2$ itself is labeled by a free input variable. This means that $\Gamma(O2, [-i_2])$ is necessarily 1/2 regardless of how $\boldsymbol{t}_p = [-i_2]$ is completed, i.e. regardless of whether the CITV is $[-i_1, -i_2]$ or $[i_1, -i_2]$.

Hence we can compute the $\Gamma'(N, \boldsymbol{t}_p)$ value for any node $N \in \mathcal{N}_s$ using Algorithm 2. In addition, that algorithm can also be used to obtain the $\Gamma'(N, \boldsymbol{t}_p)$ value for any leaf node $N \in \mathcal{N}_f$. This results from the fact that the free variables are not inconsistent with $\boldsymbol{t}_p$ which implies that $\Lambda_\alpha(N)$ and $\Lambda_\beta(N)$ and therefore the $\Gamma(N, \boldsymbol{t}_p)$ value are set to the maximal value 1.

Moreover we show (see proof below) that also the Γ' value of an AND node $N \in \mathcal{N}_f$ can be obtained in analogy to Algorithm 2. Only for OR nodes with more than one child we need to apply a different procedure if the denominator value is larger than 0. This results from the fact that the Γ value of an OR node is retrieved from the separate summation of its children's numerator and denominator values, namely:

$$\Gamma(N, \boldsymbol{t}) = \frac{\Lambda_\alpha(N_1) + \Lambda_\alpha(N_2) \cdots + \Lambda_\alpha(N_j)}{\Lambda_\beta(N_1) + \Lambda_\beta(N_2) \cdots + \Lambda_\beta(N_j)}.$$

This means that the way in which the $\Lambda_\alpha(N_i)$ and $\Lambda_\beta(N_i)$ values of a free child N_i influence the $\Gamma(N, \boldsymbol{t})$ value depends not only on whether a CITV $\boldsymbol{t}$ will turn node N_i into a root of a consistent s-subgraph but also on the specific values of $\Lambda_\alpha(N_i)$ and $\Lambda_\beta(N_i)$. For instance, suppose N has two children N_1 and N_2 with

$\Lambda_\alpha(N_1) = 1$ and $\Lambda_\beta(N_1) = 2$. Depending on the values of N_2 the $\Gamma(N, \boldsymbol{t})$ value could be lower, equal, or higher to the one of $\Gamma(N_1, \boldsymbol{t})$=1/2:

$$\Gamma(N_1, \boldsymbol{t}) > \Gamma(N, \boldsymbol{t}) = \tfrac{1+1}{2+4} = \tfrac{1}{3} \text{ if } \Lambda_\alpha(N_2) = 1 \text{ and } \Lambda_\beta(N_2) = 4$$
$$\Gamma(N_1, \boldsymbol{t}) = \Gamma(N, \boldsymbol{t}) = \tfrac{1+1}{2+2} = \tfrac{1}{2} \text{ if } \Lambda_\alpha(N_2) = 1 \text{ and } \Lambda_\beta(N_2) = 2 \text{ and}$$
$$\Gamma(N_1, \boldsymbol{t}) < \Gamma(N, \boldsymbol{t}) = \tfrac{1+3}{2+4} = \tfrac{2}{3} \text{ if } \Lambda_\alpha(N_2) = 3 \text{ and } \Lambda_\beta(N_2) = 4.$$

Thus it is not possible to determine whether the $\Lambda_\alpha(N_i)$ and $\Lambda_\beta(N_i)$ values of a particular child should be considered for the upper bound computation of $\Gamma' = (N, \boldsymbol{t}_p)$ without looking at the specific values of its other children. Only the values of the child with maximal distinguishing ratio can be safely taken into account for the upper bound computation (see proof below). The procedure is described in Algorithm 3.

Algorithm 3. Upper bound with respect to instantiation $\boldsymbol{t}_p$

For a leaf N we have:

$$\Lambda'_\alpha(N) = \Lambda_\alpha(N) \qquad \text{and} \qquad \Lambda'_\beta(N) = \Lambda_\beta(N).$$

For remaining nodes we compute:

$$(\Lambda'_\alpha(N), \Lambda'_\beta(N)) = \begin{cases} (\Lambda_\alpha(N), \Lambda_\beta(N)) & \text{if } N = \bigvee_i N_i \text{ and } N \in \mathcal{N}_s \\ (\Lambda'_\alpha(N_m), & \text{if } N = \bigvee_i N_i \text{ and } N \in \mathcal{N}_f \text{ and} \\ \quad \Lambda'_\beta(N_m)) & \quad \Gamma(N_m, \boldsymbol{t}_p) = \max_i \Gamma(N_i, \boldsymbol{t}_p) \\ (\prod_i \Lambda'_\alpha(N_i), & \\ \quad \prod_i \Lambda'_\beta(N_i)) & \text{if } N = \bigwedge_i N_i \end{cases}$$

Then we compute the distinguishing ratio for each node:

$$\Gamma'(N, \boldsymbol{t}) = \begin{cases} 0 & \text{if } \Lambda'_\beta(N) = 0 \\ \frac{\Lambda'_\alpha(N)}{\Lambda'_\beta(N)} & \text{otherwise} \end{cases}$$

Note that Algorithm 2 can be regarded as a special case of Algorithm 3 where $\boldsymbol{t}$ is a complete instantiation of a test vector. It is precisely in this case that the computed value is guaranteed to be exact. Otherwise we certainly obtain an upper bound as formally stated in the following theorem:

Theorem 2 (Upper Bound). *Let $\boldsymbol{t}_p$ be a PITV and G the root node of the smooth d-DNNF graph $\mathcal{G}$ representing the propositional formula $((M_1 \vee M_2) \wedge \neg(M_1 \wedge M_2) \wedge \neg d) \vee (M_1 \wedge M_2 \wedge d)$. Then $\Gamma(\boldsymbol{t}) \leq \Gamma(G, \boldsymbol{t}_p)$ for any complete instantiation $\boldsymbol{t}$ of the free variables in $\boldsymbol{t}_p$.*

Proof

We prove this Theorem by showing that $\Gamma(N, \boldsymbol{t}) \leq \Gamma'(N, \boldsymbol{t}_p)$ for every CITV $\boldsymbol{t}$ of $\boldsymbol{t}_p$ and every node N. This is done by induction on the depth of graph $\mathcal{G}$. The base case is straightforward. The distinguishing ratio $\Gamma(N, \boldsymbol{t}_p)$ of a leaf node N can either be 0 or 1. In order to ensure that $\Gamma'(N, \boldsymbol{t}_p)$ is an upper bound for $\Gamma(N, \boldsymbol{t}_p)$ it is therefore sufficient to set the $\Gamma'(N, \boldsymbol{t}_p)$ value for all leaves labeled by a free input variable to 1. This is precisely what Algorithm 3 does by setting the Λ_α and Λ_β values for these nodes to 1. Thus given a node N with children $N_1, \ldots, N_j$ we have the following induction hypothesis for the Λ_α and Λ_β values with respect to any CITV $\boldsymbol{t}$ of $\boldsymbol{t}_p$:

$$\frac{\Lambda_\alpha(N_1)}{\Lambda_\beta(N_1)} \leq \frac{\Lambda'_\alpha(N_1)}{\Lambda'_\beta(N_1)} \quad \frac{\Lambda_\alpha(N_2)}{\Lambda_\beta(N_2)} \leq \frac{\Lambda'_\alpha(N_2)}{\Lambda'_\beta(N_2)} \quad \cdots \quad \frac{\Lambda_\alpha(N_j)}{\Lambda_\beta(N_j)} \leq \frac{\Lambda'_\alpha(N_j)}{\Lambda'_\beta(N_j)} \tag{1}$$

In the induction step we show that for any node whose children satisfy above hypothesis we also have $\Gamma(N, \boldsymbol{t}) \leq \Gamma'(N, \boldsymbol{t}_p)$. Here we distinguish two cases: (i) N is an AND node and (ii) N is an OR node.
(i) For an AND node N Algorithm 2 gives us:[2]

$$\begin{aligned}
\Gamma(N, \boldsymbol{t}) &= \frac{\Lambda_\alpha(N_1) \cdot \Lambda_\alpha(N_2) \cdots \cdot \Lambda_\alpha(N_j)}{\Lambda_\beta(N_1) \cdot \Lambda_\beta(N_2) \cdots \cdot \Lambda_\beta(N_j)} \\
\Rightarrow \Gamma(N, \boldsymbol{t}) &= \frac{\Lambda_\alpha(N_1)}{\Lambda_\beta(N_1)} \cdot \frac{\Lambda_\alpha(N_2)}{\Lambda_\beta(N_2)} \cdots \cdot \frac{\Lambda_\alpha(N_j)}{\Lambda_\beta(N_j)} \\
\Rightarrow \Gamma(N, \boldsymbol{t}) &\leq \frac{\Lambda'_\alpha(N_1)}{\Lambda'_\beta(N_1)} \cdot \frac{\Lambda'_\alpha(N_2)}{\Lambda'_\beta(N_2)} \cdots \cdot \frac{\Lambda'_\alpha(N_j)}{\Lambda'_\beta(N_j)} && \text{see induction hypothesis} \\
\Rightarrow \Gamma(N, \boldsymbol{t}) &\leq \Gamma'(N, \boldsymbol{t}_p) && \text{see Algorithm 3}
\end{aligned}$$

(ii) In case an OR node N is not labeled by a free variable we have $\Gamma(N, \boldsymbol{t}) = \Gamma'(N, \boldsymbol{t}_p)$ and therefore $\Gamma(N, \boldsymbol{t}) \leq \Gamma'(N, \boldsymbol{t}_p)$. To prove the other case we denote with $\Lambda'_{\alpha Max} = \Lambda_\alpha(N_{max})$ and $\Lambda'_{\beta Max} = \Lambda_\beta(N_{max})$ the corresponding values for the node with the maximal distinguishing ratio, i.e. $\Gamma'(N_{max}, \boldsymbol{t}_p) = \frac{\Lambda'_{\alpha Max}}{\Lambda'_{\beta Max}} = \max_i \frac{\Lambda'_\alpha(N_i)}{\Lambda'_\beta(N_i)}$. This gives us:

$$\frac{\Lambda_\alpha(N_1)}{\Lambda_\beta(N_1)} \leq \frac{\Lambda'_{\alpha Max}}{\Lambda'_{\beta Max}} \quad \frac{\Lambda_\alpha(N_2)}{\Lambda_\beta(N_2)} \leq \frac{\Lambda'_{\alpha Max}}{\Lambda'_{\beta Max}} \quad \cdots \quad \frac{\Lambda_\alpha(N_j)}{\Lambda_\beta(N_j)} \leq \frac{\Lambda'_{\alpha Max}}{\Lambda'_{\beta Max}}$$

see induction hypothesis

$$\Rightarrow \quad \begin{aligned} \Lambda_\alpha(N_1) \cdot \Lambda'_{\beta Max} &\leq \Lambda_\beta(N_1) \cdot \Lambda'_{\alpha Max} \\ &\cdots \\ \Lambda_\alpha(N_j) \cdot \Lambda'_{\beta Max} &\leq \Lambda_\beta(N_j) \cdot \Lambda'_{\alpha Max} \end{aligned}$$

$$\Rightarrow \Lambda_\alpha(N_1) \cdot \Lambda'_{\beta Max} \cdots + \Lambda_\alpha(N_j) \cdot \Lambda'_{\beta Max} \leq \Lambda_\beta(N_1) \cdot \Lambda'_{\alpha Max} \cdots + \Lambda_\beta(N_j) \cdot \Lambda'_{\alpha Max}$$

$$\Rightarrow \quad (\Lambda_\alpha(N_1) \cdots + \Lambda_\alpha(N_j)) \cdot \Lambda'_{\beta Max} \leq \Lambda'_{\alpha Max} \cdot (\Lambda_\beta(N_1) \cdots + \Lambda_\beta(N_j))$$

[2] In case the Λ_β value of a child is 0 the $\Gamma(N, \boldsymbol{t})$ value is 0 and hence necessarily less or equal to $\Gamma'(N, \boldsymbol{t}_p)$.

$$\Rightarrow \quad \frac{\Lambda_\alpha(N_1)\cdots+\Lambda_\alpha(N_j)}{\Lambda_\beta(N_1)\cdots+\Lambda_\beta(N_j)} \le \frac{\Lambda'_{\alpha Max}}{\Lambda'_{\beta Max}}$$

$$\Rightarrow \quad \frac{\Lambda_\alpha(N_1)\cdots+\Lambda_\alpha(N_j)}{\Lambda_\beta(N_1)\cdots+\Lambda_\beta(N_j)} \le \Gamma'(N,\boldsymbol{t}_p)$$

$$\Rightarrow \quad \Gamma(N,\boldsymbol{t}) \le \Gamma'(N,\boldsymbol{t}_p)$$

see Algorithm 2 ■

3.4 ODT Computation

With the d-DNNF graph $\mathcal{G}$ and the linear time algorithms to compute the precise distinguishing ratio for a CITV and an upper bound for a PITV we have obtained the basis for our ODT search method. This consists of a branch-and-bound search over the input variables. Iteratively we set the input variables until either all variables are set and the precise Γ value is obtained or until the upper bound of the PITV is lower than the Γ value of a previously computed CITV.

Interestingly, if $\mathcal{G}$ has a certain structure (see below) we can obtain an ODT without a search by making use of the facts (i) that we can compute an upper bound Γ_{UB} for the distinguishing ratio of the ODT by running Algorithm 3 with respect to instantiation *true*, and (ii) that there is exactly one test vector $\boldsymbol{t}$ that is consistent with the resulting subgraph $\mathcal{G}_T$. The latter consists of all nodes from whose Λ'_α and Λ'_β values the upper bound Γ_{UB} was derived.[3]

Note, since $\mathcal{G}_T$ was obtained from running Algorithm 3 with respect to instantiation *true* it means that all input variables belong to the set of free ones, i.e. $\mathcal{I}_F = \mathcal{I}$. Hence every *OR* node $N \in \mathcal{N}_T$ with an input variable as descendant belongs to the set $\mathcal{N}_f$ and has therefore only one child in $\mathcal{G}_T$. Therefore, the labels of the input variables of $\mathcal{G}_T$ form a unique CITV $\boldsymbol{t}$ which is exploited in the following Theorem:

Theorem 3. *Let G be the root node of a smooth d-DNNF graph $\mathcal{G}$ representing the propositional theory $((M_1 \vee M_2) \wedge \neg(M_1 \wedge M_2) \wedge \neg d) \vee (M_1 \wedge M_2 \wedge d)$ and satisfying the following condition: No two children of an OR node are labeled by the same value of an input variable. Then the test vector $\boldsymbol{t}$ consistent with the subgraph $\mathcal{G}_T$ that is obtained from computing $\Gamma'(G, true)$ is an ODT.*

Proof
We prove this Theorem by contradiction. Suppose $\Gamma(G,\boldsymbol{t}) \neq \Gamma'(G, true)$. Since Algorithms 2 and 3 differ only in the Γ computation for an *OR* node in $\mathcal{N}_f$ which has more than one child the assumption implies that there is an *OR* node $N \in \mathcal{N}_f$ in graph $\mathcal{G}$ with at least two children that are both consistent with $\boldsymbol{t}$. This implies that N has at least two children labeled by the same value of an input variable which contradicts the condition of the theorem. ■

[3] Formally, a node $N \in \mathcal{N}_T$ is in $\mathcal{G}_T$, iff one of the following conditions is satisfied: (i) N is the root, (ii) N is a child of an *AND* node in $\mathcal{N}_T$, (iii) $N \in \mathcal{N}_s$ is a child of an *OR* node in $\mathcal{N}_T$, or (iv) $N \in \mathcal{N}_f$ is the child with the maximal Γ' value among the children of an *OR* node in $\mathcal{N}_T$.

For our example the condition of above theorem holds. The right graph of Figure 3 also shows the upper bound values computed by Algorithm 3 when running it with $\boldsymbol{t}_p = true$. Thus we obtained the ODT $\boldsymbol{t} = [-i_1, i_2]$ for our example in linear time. Note that the condition is certainly satisfied if $\mathcal{G}$ is a constrained d-DNNF graph [11]. Unfortunately, the compilation into constrained d-DNNF graphs is more complex and will therefore only be possible for small graphs.

4 Experimental Evaluation

We evaluated our DNNF-based testing method on a model of an automotive engine test-bed [10], which consists of three major components: engine, pipe, and throttle. The goal is to find leaks in the pipe by assigning three to four controllable variables, and observing three to four measurable variables. The problem has been turned into sets of discrete instances of varying sizes by applying different abstractions to the original mixed discrete-continuous model, and using a direct encoding from CSP to SAT [14].

Table 1. Results of ODT (left) and DDT (right) computation

inst.	#nodes	time
1a	58	0.04
2a	103	0.06
3a	161	0.09
4a	205	0.10
5a	329	0.20
6a	245	0.21
7a	362	0.40
8a	4766	5.41
9a	2654	36.6
10a	65063	414

inst.	#nodes	time
1b	66	0.06
2b	124	0.07
3b	191	0.10
4b	4865	14.7
5b	48238	396
6b	102817	1566
7b	–	–
8b	–	–
9b	–	–
10b	–	–

Table 1 shows the experimental results for computing ODTs and DDTs. For each instance, we report the size (number of nodes) of the DNNF graph computed, and the computation time in seconds on a Linux Dual-Core PC with 1GB of RAM. For instances that have DDTs, we compared our method with the most recent specialized DDT approach [12] based on quantified Boolean formulas and the QBF solver sKizzo [2]. That approach was able to solve instances 1b–3b and ran out of memory for the rest. In contrast, our ODT approach could also solve instances 4b–6b.

Our method is the first that computes exact ODTs; hence we cannot compare it directly with previous approaches. However, we used the greedy algorithm of [8] to compute approximate solutions for the same problem instances. This algorithm was only able to solve instances 1a–7a; that is, for instances 8a–10a we were able to compute an optimal solution where the previous approach could not even compute a suboptimal one.

5 Conclusion and Future Work

Optimal distinguishing tests generalize and refine the notion of possibly and definitely distinguishing and strong and weak tests for non-deterministic systems. Since computing ODTs can be computationally very expensive, previous work has focused on approximate solutions. We presented a new method to compute exact ODTs based on innovative ways of compiling the ODT problem into DNNF and computing upper bounds to prune a systematic search. Experimental results show that the method is able to compute both ODTs and DDTs for instances that were too large for previous methods. Thus our approach provides a significant improvement for many applications where output patterns can be assumed to be equally likely or where probabilities are not given. Where probabilistic models are available, our technique provides a baseline from which new techniques can be developed. Note that both model counting and weighted model counting can be done with the same linear complexity on d-DNNF. In fact, this is the basis for the recent compilation based approach to probabilistic reasoning, and will provide the basis also for extending our work to the probabilistic case.

References

1. Alur, R., Courcoubetis, C., Yannakakis, M.: Distinguishing tests for nondeterministic and probabilistic machines. In: Proc. ACM Symposium on Theory of Computing, pp. 363–372 (1995)
2. Benedetti, M.: skizzo: A suite to evaluate and certify QBFs. In: Proc. CADE 2005 (2005)
3. Boroday, S., Petrenko, A., Groz, R.: Can a model checker generate tests for non-deterministic systems? Elec. Notes Theor. Comp. Sci. 190, 3–19 (2007)
4. Darwiche, A.: On the tractable counting of theory models and its application to belief revision and truth maintenance. In: CoRR (2000)
5. Darwiche, A.: The c2d compiler user manual. Technical report, UCLA (2005)
6. Darwiche, A., Marquis, P.: A knowledge compilation map. Journal of Artificial Intelligence Research 17, 229–264 (2002)
7. Esser, M., Struss, P.: Fault-model-based test generation for embedded software. In: Proc. IJCAI 2007, pp. 342–347 (2007)
8. Heinz, S., Sachenbacher, M.: Using model counting to find optimal distinguishing tests. In: Proc. of CPAIOR (2009)
9. Huang, J.: Combining knowledge compilation and search for conformant probabilistic planning. In: Proc. ICAPS 2006, pp. 253–262 (2006)
10. Luo, J., Pattipati, K., Qiao, L., Chigusa, S.: An integrated diagnostic development process for automotive engine control systems. IEEE Trans. on Systems, Man, and Cybernetics 37(6), 1163–1173 (2007)
11. Pipatsrisawat, K., Darwiche, A.: A new d-dnnf-based bound computation algorithm for functional EMAJSAT. In: Proc. of IJCAI 2009 (2009)
12. Sachenbacher, M., Schwoon, S.: Model-based testing using quantified CSPs. In: ECAI 2008 Workshop on Model-based Systems (2008)
13. Struss, P.: Testing physical systems. In: Proc. AAAI 1994, pp. 251–256 (1994)
14. Walsh, T.: SAT vs. CSP. In: Dechter, R. (ed.) CP 2000. LNCS, vol. 1894, pp. 441–456. Springer, Heidelberg (2000)

Why Cumulative Decomposition Is Not as Bad as It Sounds

Andreas Schutt[1], Thibaut Feydy[1], Peter J. Stuckey[1], and Mark G. Wallace[2]

[1] National ICT Australia, Department of Computer Science & Software Engineering, The University of Melbourne, Australia
{aschutt,tfeydy,pjs}@csse.unimelb.edu.au

[2] School of Computer Science & Software Engineering, Monash University, Australia
mark.wallace@infotech.monash.edu.au

Abstract. The global `cumulative` constraint was proposed for modelling cumulative resources in scheduling problems for finite domain (FD) propagation. Since that time a great deal of research has investigated new stronger and faster filtering techniques for `cumulative`, but still most of these techniques only pay off in limited cases or are not scalable. Recently, the "lazy clause generation" hybrid solving approach has been devised which allows a finite domain propagation engine possible to take advantage of advanced SAT technology, by "lazily" creating a SAT model of an FD problem as computation progresses. This allows the solver to make use of SAT nogood learning and autonomous search capabilities. In this paper we show that using lazy clause generation where we model `cumulative` constraint by decomposition gives a very competitive implementation of cumulative resource problems. We are able to close a number of open problems from the well-established PSPlib benchmark library of resource-constrained project scheduling problems.

1 Introduction

Cumulative resources are part of many real-world scheduling problems. A resource can represent not only a machine which is able to run multiple tasks in parallel but also entities such as: electricity, water, consumables or even human skills. Those resources arises for example in the resource-constrained project scheduling problem RCPSP, their variants, their extensions and their specialisations. A RCPSP consists of *tasks* (also called *activities*) consuming one or more resources, *precedences* between some tasks, and *resources*. In this paper we restrict ourselves to case of non-preemptive tasks and renewable resources with a constant resource capacity over the planning horizon. A solution is a schedule of all tasks so that all precedences and resource constraints are satisfied. RCPSP is an NP-hard problem.

Example 1. Consider a simple resource scheduling problem. There are 5 tasks `a`, `b`, `c`, `d` and `e` to be scheduled to end before time 10. The tasks have respective durations 1, 2, 3, 3 and 4, each respective task requiring 1, 1, 2, 2 and 2 units of

I.P. Gent (Ed.): CP 2009, LNCS 5732, pp. 746–761, 2009.

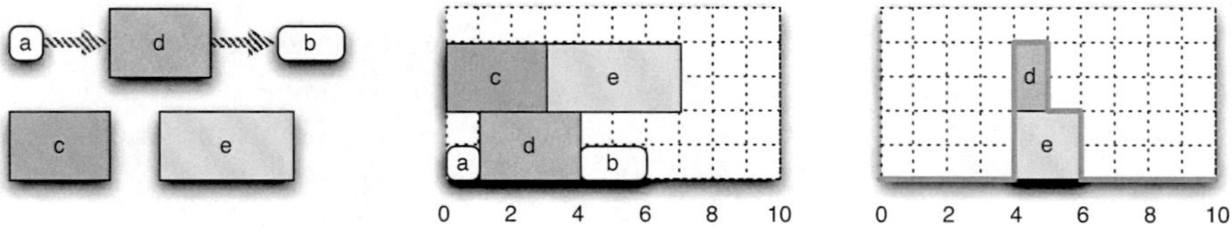

(a) (b) (c)

Fig. 1. (a) A small cumulative resource problem, with 5 tasks to place in the 5x10 box, with task `a` before `d` before `b`, (b) a possible schedule, and (c) a profile under some further conditions

resource, with a resource capacity of 5. Assume further that there are precedence constraints: task `a` must complete before task `d` begins, written `a` $\rightsquigarrow$ `d`, and similarly `d` $\rightsquigarrow$ `b`. Figure 1(a) shows the 5 tasks and precedences, while (b) shows a possible schedule, where the respective start times are: 0, 4, 0, 1, 3.

In 1993 Aggoun and Beldiceanu [2] introduced the global `cumulative` constraint in order to efficiently solve complex scheduling problems in a constraint programming framework. The `cumulative` constraint cannot compete with specific OR methods for restricted forms of scheduling, but since it is applicable whatever the side constraints are it is very valuable. Many improvements have been proposed to the `cumulative` constraint: see e.g. Caseau and Laburthe [5], Carlier and Pinson [4], Nuijten [16] and Baptiste and Le Pape [3].

The best known exact algorithm for solving RCPSP is from Demeulemeester and Herroelen [6]. Their specific method is a branch-and-bound approach relying heavily on dominance rules and cut sets, a kind of problem specific nogoods. They implicitly show the importance of nogoods to fathom the huge search space of RCPSP problems. Unfortunately, the number of cut sets grows exponentially in the number of tasks, so that this method is considered to be efficient only for small problems.

In comparison to Demeulemeester and Herroelen's specific nogoods SAT solvers records general nogoods. Since the introduction of `cumulative` SAT solving has improved drastically. Nowadays, modern SAT solvers can often handle problems with millions of constraints and hundreds of thousands of variables. But problems like RCPSP are difficult to encode into SAT without breaking these implicit limits. Recently, Ohrimenko *et al.* [17] showed how to build a powerful hybrid of SAT solving and FD solving that maintains the advantages of both: the high level modelling and small models of FD solvers, and the efficient nogood recording and conflict driven search of SAT. The key idea in this *lazy clause generation* approach is that finite domain propagators lazily generate a clausal representation of their behaviour. They show that this combination outperforms the best available constraint solvers on Open-Shop-Job problems which is a special case of RCPSP.

Since 1993 little attention has been paid to decompositions of `cumulative` because decomposition cannot compete with the global propagator. But once we consider explanation we have to revisit this. Decomposition of globals means that explanation of behaviour is more fine grained and hence more reusable. Also it avoids the need for complex explanation algorithms to be developed for the global. Note that there is some preliminary work on explanation generation for `cumulative`, in PaLM [9] where (in 2000) it is described as current work, and [18] which restricts attention to the `disjunctive` constraint (resource capacity 1).

In this paper we show how a decomposition based approach for solving complex scheduling problems can be competitive with state-of-the-art specialised methods from the CP and OR community. The G12 Constraint Programming Platform is used for implementation of decomposed cumulative constraint as a lazy clause generator. We evaluate our approach on RCPSP from the well-established and challenging benchmark library PSPLib [1].

2 Lazy Clause Generation

Lazy clause generation is a powerful hybrid of SAT and finite domain solving that inherits advantages of both: high level modelling, and specialised propagation algorithms from FD; nogood recording, and conflict driven search from SAT.

2.1 Finite Domain Propagation

We consider a set of integer variables $\mathcal{V}$. A *domain* D is a complete mapping from $\mathcal{V}$ to finite sets of integers. Let D_1 and D_2 be domains and $V \subseteq \mathcal{V}$. We say that D_1 is *tighter* than D_2, written $D_1 \sqsubseteq D_2$, if $D_1(v) \subseteq D_2(v)$ for all $v \in \mathcal{V}$. We use *range* notation: $[\, l \,..\, u \,]$ denotes the set of integers $\{d \mid l \leq d \leq u, d \in \mathbb{Z}\}$. We assume an *initial domain* D_{init} such that all domains D that occur will be stronger i.e. $D \sqsubseteq D_{init}$.

A *valuation* θ is a mapping of variables to values, written $\{x_1 \mapsto d_1, \ldots, x_n \mapsto d_n\}$. We extend the valuation θ to map expressions or constraints involving the variables in the natural way. Let *vars* be the function that returns the set of variables appearing in an expression, constraint or valuation. In an abuse of notation, we define a valuation θ to be an element of a domain D, written $\theta \in D$, if $\theta(v) \in D(v)$ for all $v \in vars(\theta)$.

A constraint c is a set of valuations over $vars(c)$ which give the allowable values for a set of variables. In finite domain propagation constraints are implemented by propagators. A propagator f for c is a monotonically decreasing function on domains such that for all domains $D \sqsubseteq D_{init}$: $f(D) \sqsubseteq D$ and $\{\theta \in D \mid \theta \in c\} = \{\theta \in f(D) \mid \theta \in c\}$. A *propagation solver* for a set of propagators F and current domain D, $solv(F, D)$, repeatedly applies all the propagators in F starting from domain D until there is no further change in resulting domain. $solv(F, D)$ is the weakest domain $D' \sqsubseteq D$ which is a fixpoint for all $f \in F$.

2.2 SAT Solving

DPLL SAT solvers can be understood as a form of propagation solver where variables are Boolean, and the only constraints are clauses C: $\vee_{l \in C} l$. The difference with an FD solver is that propagation engines are highly specialised and more importantly the reason for propagation is recorded, and on failure used to generate a nogood which explains the failure. This clause is added to the propagators to shortcircuit later search. It also helps direct backtracking to go above the cause of the failure.

2.3 Lazy Clause Generation

The lazy clause generation [17] works as follows. Propagators are considered as clause generators for the SAT solver. Instead of applying propagator f to domain D to obtain $f(D)$, whenever $f(D) \neq D$ we build a clause that encodes the change in domains. In order to do so we must link the integer variables of the finite domain problem to a Boolean representation.

We represent an integer variable x with domain $D_{init}(x) = [\,l\,..\,u\,]$ using the Boolean variables $[\![x = l]\!], \ldots, [\![x = u]\!]$ and $[\![x \leq l]\!], \ldots, [\![x \leq u-1]\!]$ where the former is generated on demand. The variable $[\![x = d]\!]$ is true if x takes the value d, and false for a value different from d. Similarly the variable $[\![x \leq d]\!]$ is true if x takes a value less than or equal to d and false for a value greater than d.

Not every assignment of Boolean variables is consistent with the integer variable x, for example $\{[\![x = 3]\!], [\![x \leq 2]\!]\}$ requires that x is both 3 and ≤ 2. In order to ensure that assignments represent a consistent set of possibilities for the integer variable x we add to the SAT solver clauses $DOM(x)$ that encode $[\![x \leq d]\!] \rightarrow [\![x \leq d+1]\!]$ and $[\![x = d]\!] \leftrightarrow ([\![x \leq d]\!] \wedge \neg[\![x \leq d-1]\!])$. We let $DOM = \cup\{DOM(v) \mid v \in \mathcal{V}\}$.

Any assignment A on these Boolean variables can be converted to a domain: $domain(A)(x) = \{d \in D_{init}(x) \mid \forall [\![c]\!] \in A, vars([\![c]\!]) = \{x\} : x = d \models c\}$ that is the domain includes all values for x that are consistent with all the Boolean variables related to x. Note that the domain may assign no values to some variable.

Example 2. The assignment $A = \{[\![x_1 \leq 10]\!], \neg[\![x_1 \leq 5]\!], \neg[\![x_1 = 7]\!], \neg[\![x_1 = 8]\!], [\![x_2 \leq 11]\!], \neg[\![x_2 \leq 5]\!], [\![x_3 \leq 10]\!], \neg[\![x_3 \leq -2]\!]\}$ is consistent with $x_1 = 6, x_1 = 9$ and $x_1 = 10$. Hence $domain(A)(x_1) = \{6, 9, 10\}$. For the remaining variables $domain(A)(x_2) = [\,6\,..\,11\,]$ and $domain(A)(x_3) = [\,-1\,..\,10\,]$. □

In lazy clause generation a propagator changes from a mapping from domains to domains to a generator of clauses describing propagation. When $f(D) \neq D$ we assume the propagator f can determine a set of clauses C which explain the domain changes.

Example 3. Consider the propagator f for $x_1 \leq x_2 + 1$. When applied to domain $D(x_1) = [\,0\,..\,9\,]$, $D(x_2) = [\,-3\,..\,5\,]$ it obtains $f(D)(x_1) = [\,0\,..\,6\,]$, $f(D)(x_2) = [\,-1\,..\,5\,]$. The clausal explanation of the change in domain of x_1 is $[\![x_2 \leq 5]\!] \rightarrow [\![x_1 \leq 6]\!]$, similarly the change in domain of x_2 is $\neg[\![x_1 \leq -1]\!] \rightarrow \neg[\![x_2 \leq -2]\!]$ $(x_1 \geq 0 \rightarrow x_2 \geq -1)$. These become the clauses $\neg[\![x_2 \leq 5]\!] \vee [\![x_1 \leq 6]\!]$ and $[\![x_1 \leq -1]\!] \vee \neg[\![x_2 \leq -2]\!]$.

Assuming $domain(A) \sqsubseteq D$, then when clauses C that explain the propagation of f are added to the SAT database and unit propagation is performed, then the resulting assignment A' will be such that $domain(A') \sqsubseteq f(D)$.

Using the lazy clause generation we can show that the SAT solver maintains an assignment which is at least as tight as that determined by finite domain propagation [17]. The advantages over a normal FD solver are that we automatically have the nogood recording and backjumping ability of the SAT solver applied to our FD problem. We can also use activity counts from the SAT solver to direct the FD search.

3 Modelling the Cumulative Resource Constraint

In this section we define the `cumulative` constraint and discuss two possible decompositions of it.

The `cumulative` constraint introduced by Aggoun and Beldiceanu [2] in 1993 is a constraint with Zinc [14] type

```
predicate cumulative(list of var int: s, list of var int: d,
                     list of var int: r,        var int: c);
```

Each of the first three arguments are lists of the same length n and indicate information about a set of *tasks*. $s[i]$ is the *start time* of the ith task, $d[i]$ is the *duration* of the ith task, and $r[i]$ is the *resource usage* (per time unit) of the ith task. The last argument c is the *resource capacity*.

The cumulative constraints represent cumulative resources with a constant capacity over the considered planning horizon applied to non-preemptive tasks, *i.e.* if they are started they cannot be interrupted. W.l.o.g. we assume that all values are integral and non-negative and there is a *planning horizon* t_{max} which is the latest time any task can finish.

We also assume for simplicity that each of d, r and c are fixed integers, although this is not important for much of the discussion. This is certainly the most common case of `cumulative`.

The `cumulative` constraint enforces that at all times the sum of resources used by active tasks is no more than the resource capacity.

$$\forall t \in [\,0\,..\,t_{max}-1\,]: \sum_{i\in[\,1\,..\,n\,]:s[i]\leq t<s[i]+d[i]} r[i] \leq c \tag{1}$$

Example 4. Consider the cumulative resource problem defined in Example 1. This can be modelled by the cumulative constraint

$$\texttt{cumulative}\ ([s_{\mathsf{a}}, s_{\mathsf{b}}, s_{\mathsf{c}}, s_{\mathsf{d}}, s_{\mathsf{e}}], [1,2,3,3,4], [1,1,2,2,2], 5)$$

with precedence constraints $\mathsf{a} \rightsquigarrow \mathsf{d}$, $\mathsf{d} \rightsquigarrow \mathsf{b}$, modelled by $s_{\mathsf{a}} + 1 \leq s_{\mathsf{d}}$ and $s_{\mathsf{d}} + 3 \leq s_{\mathsf{b}}$. The propagator for the precedence constraints determines a domain D where $D(s_{\mathsf{a}}) = [\,0\,..\,3\,]$, $D(s_{\mathsf{b}}) = [\,4\,..\,8\,]$, $D(s_{\mathsf{c}}) = [\,0\,..\,7\,]$, $D(s_{\mathsf{d}}) = [\,1\,..\,5\,]$, $D(s_{\mathsf{e}}) = [\,0\,..\,6\,]$. The cumulative constraint does not determine any new information. If we add the constraints $s_{\mathsf{e}} \geq 2$, $s_{\mathsf{e}} \leq 4$, $s_{\mathsf{b}} \leq 7$, $s_{\mathsf{a}} \geq 1$, then precedence determines the domains $D(s_{\mathsf{a}}) = [1..3]$, $D(s_{\mathsf{b}}) = [4..6]$, $D(s_{\mathsf{c}}) = [0..7]$, $D(s_{\mathsf{d}}) = [2..4]$, $D(s_{\mathsf{e}}) = [2..4]$. We can determine that task d must use two resources between times 4 and 5, and task e must use two resources between times 4 and 6 (see Figure 1(c)). Hence task c cannot overlap these between times 4 and 5, and we can determine that $s_{\mathsf{c}} \neq 2$, $s_{\mathsf{c}} \neq 3$, $s_{\mathsf{c}} \neq 4$. If we restrict ourselves to bounds propagation then the cumulative constraint learns nothing. If we then add the constraint that $s_{\mathsf{c}} \geq 2$, then the bounds propagation on the cumulative constraint determines that $s_{\mathsf{c}} \geq 5$ and $D(s_{\mathsf{c}})$ becomes $[5..7]$.

Usually the `cumulative` constraint is implemented as a global propagator, since it can then take more information into account during propagation. In the remainder of this section we give two decompositions.

3.1 Time-Resource Decomposition

The time-resource decomposition (Time-RD) [2] arises from the Formula (1). For every time t the sum of all resource requirements must be less than or equal to the resource capacity. The Zinc encoding of the decomposition is shown below where: $index\_set(a)$ returns the index set of an array a (here $[1..n]$), $lb(x)$ ($ub(x)$) returns the declared lower (resp. upper) bound of a integer variable x, and $bool2int(b)$ is 0 if the Boolean b is false, and 1 if it is true.

```
predicate cumulative(list of var int: s, list of var int: d,
                     list of var int: r,          var int: c) =
  let {set of int: tasks = index_set(s),
       set of int: times = min([lb(s[i]) | i in tasks]) ..
                           max([ub(s[i]) + ub(d[i]) - 1 | i in tasks])
  } in forall( t in times ) (
           c >= sum( i in tasks ) (
             bool2int( s[i] <= t /\ t < s[i] + d[i] ) * r[i]));
```

This decomposition implicitly introduces new Boolean variables B_{it} represents that task i is active at time t:

$$\forall t \in [0..t_{max}-1], \forall i \in [1..n]: \quad B_{it} \leftrightarrow [\![s[i] \leq t]\!] \wedge \neg[\![s[i] \leq t - d[i]]\!]$$

$$\forall t \in [0..t_{max}-1]: \quad \sum_{i \in [1..n]} r[i] \cdot B_{it} \leq c$$

Note that since we are using lazy clause generation, the Booleans for the expressions $[\![s[i] \leq t]\!]$ and $[\![s[i] \leq t - d[i]]\!]$ already exist and that for a task i we only need to construct variables B_{it} where $lb(s[i]) \leq t < ub(s[i]) + ub(d[i])$.

At most nt_{max} new Boolean variables are created, nt_{max} conjunction constraints, and t_{max} sum constraints (of size n). This decomposition implicitly profiles the resource histograms for all times for the resource.

Note that if we have another cumulative constraint for a different resource on the same tasks then we can reuse the Boolean variables, and we just need to create t_{max} new sum constraints.

Example 5. Consider the problem of Example 4 after the addition of $s_{\mathsf{e}} \geq 2$, $s_{\mathsf{e}} \leq 4$, $s_{\mathsf{b}} \leq 7$, $s_{\mathsf{a}} \geq 1$, the decomposition determines that $B_{\mathsf{d}4}$ is true since $s_{\mathsf{d}} \leq 4$ and $\neg(s_{\mathsf{d}} \leq 4 - 4 = 0)$, similarly for $B_{\mathsf{e}4}$ and $B_{\mathsf{e}5}$. Using the sum constraint it determines that $B_{\mathsf{c}4}$ is false, and hence $\neg(s_{\mathsf{c}} \leq 4) \vee s_{\mathsf{c}} \leq 1$. This does not change any bounds. When we add that $s_{\mathsf{c}} \geq 2$ we determine that $\neg(s_{\mathsf{c}} \leq 4)$ or $s_{\mathsf{c}} \geq 5$.

We can expand the model to represent holes in the domains of start times. The literal $[\![s[i] = t]\!]$ is a Boolean representing the start time of the ith task is t. We add the constraint

$$[\![s[i] = t]\!] \rightarrow \bigwedge_{t \leq t' < t + d[i]} B_{it'}$$

which ensures that if $B_{it'}$ becomes false then the values $\{t' - d[i] + 1, t' - d[i] + 2, \ldots, t'\}$ are removed from the domain of $s[i]$. We do not use this constraint for our experiments since it was inferior in solving time to the model without it.

Example 6. Consider again the problem of Example 4, with the extended decomposition we also determine that $s_{\mathsf{c}} \neq 2$, $s_{\mathsf{c}} \neq 3$, $s_{\mathsf{c}} \neq 4$, so the resulting domain is $D(s_{\mathsf{c}}) = \{0, 1, 5, 6, 7\}$.

3.2 Task-Resource Decomposition

The Task-resource decomposition (Task-RD) is a relaxation of the Time-RD. It ensures a non-overload of resources only at the start (or end) times which is sufficient to ensure non-overload at every time for the non-preemptive case. Therefore, the number of variables and linear inequality constraints is independent of the size of the planning horizon t_{max}. It was used by El-Kholy [7] for temporal and resource reasoning in planning. The Zinc code for the decomposition at the start times is below.

```
predicate cumulative(list of var int: s, list of var int: d,
                     list of var int: r,        var int: c) =
   let { set of int: tasks = index_set(s) }
   in forall( j in tasks ) (
         c >= r[j] + sum( i in tasks where i != j ) (
            bool2int( s[i] <= s[j] /\ s[j] < s[i] + d[i] ) * r[i]));
```

The decomposition implicitly introduces new Boolean variables: $B^1_{ij} \equiv$ task j starts at or after task i starts, $B^2_{ij} \equiv$ task j starts before task i ends, and $B_{ij} \equiv$ task j starts when task i is running.

$$\begin{aligned}
\forall j \in [1..n], \forall i \in [1..n] \setminus \{j\}: \quad & B_{ij} \leftrightarrow B^1_{ij} \wedge B^2_{ij} \\
& B^1_{ij} \leftrightarrow s[i] \leq s[j] \\
& B^2_{ij} \leftrightarrow s[j] < s[i] + d[i] \\
\forall j \in [1..n]: \quad & \sum_{i \in [1..n] \setminus \{j\}} r[i] \cdot B_{ij} \leq c - r[j]
\end{aligned}$$

Note not all tasks i must be considered for a task j, only those i which can overlap at the start times $s[j]$ wrt. precedence constraints, resource constraints and the initial domain D_{init}.

Since the SAT solver does not know about the relationship among the B^1_{**} and B^2_{**} the following redundant constraints can be posted for all $i, j \in [1..n]$ where $i < j$ in order to improve the propagation and the learning.

$$B^1_{ij} \vee B^2_{ij} \qquad B^1_{ji} \vee B^2_{ji} \qquad B^1_{ij} \vee B^1_{ji} \qquad B^1_{ij} \rightarrow B^2_{ji} \qquad B^1_{ji} \rightarrow B^2_{ij}$$

The size of this decomposition only depends on n whereas Time-RD depends on n and the number of points in the planning horizon t_{max}. At most $3n(n-1)$ Boolean variables, $3n(n-1)$ equivalence relations, n or relations, $3n$ redundant constraints and n sum constraints are generated. Again adding another cumulative resource constraints can reuse the Boolean variables and requires only adding n new sum constraints.

Example 7. Consider the problem of Example 4 after the addition of $s_{\mathsf{e}} \geq 2$, $s_{\mathsf{e}} \leq 4$, $s_{\mathsf{b}} \leq 7$, $s_{\mathsf{a}} \geq 1$, after the precedence constraints are applied the decomposition learns $\neg B^2_{\mathsf{ad}}$, $\neg B^2_{\mathsf{db}}$ direct from precedence constraints and hence $\neg B_{\mathsf{ad}}$, $\neg B_{\mathsf{db}}$. From the start times it determines that B^1_{ab}, B^1_{db}, B^1_{eb}, $\neg B^2_{\mathsf{ba}}$, $\neg B^2_{\mathsf{bd}}$, $\neg B^2_{\mathsf{be}}$, $\neg B_{\mathsf{ba}}$, $\neg B_{\mathsf{bd}}$, $\neg B_{\mathsf{be}}$. But nothing is determined from the sum constraints and no bounds changes are made by the cumulative. Adding $s_{\mathsf{c}} \geq 2$ does not change this. This illustrates the weaker propagation of the Task-RD.

If we use end time variables $e[i] = s[i] + d[i]$, we can generate a symmetric model to that defined above.

In comparison, to the Time-RD decomposition the Task-RD decomposition is stronger in its ability to relate to task information, but generates a weaker profile of resource usage, since no implicit profile is recorded.

3.3 Explanations

To see the advantage of decomposition in terms of explanations let us revisit Example 4. After the addition of $s_{\mathsf{e}} \geq 2$, $s_{\mathsf{e}} \leq 4$, $s_{\mathsf{b}} \leq 7$, $s_{\mathsf{a}} \geq 1$, together with $s_{\mathsf{c}} \geq 2$ a global (bounds consistent) cumulative constraint propagates that $s_{\mathsf{c}} \geq 5$. A minimal explanation of this is that $s_{\mathsf{e}} \geq 2 \wedge s_{\mathsf{e}} \leq 4 \wedge s_{\mathsf{d}} \geq 2 \wedge s_{\mathsf{d}} \leq 4 \wedge s_{\mathsf{c}} \geq 2 \rightarrow s_{\mathsf{c}} \geq 5$. This is recorded as the clause $[\![s_{\mathsf{e}} \leq 1]\!] \vee \neg[\![s_{\mathsf{e}} \leq 4]\!] \vee [\![s_{\mathsf{d}} \leq 1]\!] \vee [\![s_{\mathsf{d}} \leq 4]\!] \vee [\![s_{\mathsf{c}} \leq 1]\!] \vee \neg[\![s_{\mathsf{c}} \leq 4]\!]$.

Consider what happens in the Time-RD decomposition. After the addition of $s_{\mathsf{e}} \geq 2$, $s_{\mathsf{e}} \leq 4$, $s_{\mathsf{b}} \leq 7$, $s_{\mathsf{a}} \geq 1$, we learn $s_{\mathsf{d}} \leq 4 \wedge s_{\mathsf{d}} \geq 2 \rightarrow B_{\mathsf{d}4}$, $s_{\mathsf{e}} \leq 4 \wedge s_{\mathsf{e}} \geq 1 \rightarrow B_{\mathsf{e}4}$, $s_{\mathsf{e}} \leq 5 \wedge s_{\mathsf{e}} \geq 2 \rightarrow B_{\mathsf{e}5}$, and $B_{\mathsf{d}4} \wedge B_{\mathsf{d}5} \rightarrow \neg B_{\mathsf{c}4}$ With the addition of $s_{\mathsf{c}} \geq 2$ we learn that $s_{\mathsf{c}} \geq 2 \wedge \neg B_{\mathsf{c}4} \rightarrow s_{\mathsf{c}} \geq 5$. Each of the clauses is smaller and more reuseable. For example if we replace the constraint $s_{\mathsf{e}} \geq 2$ by $s_{\mathsf{e}} \geq 1$ then the same reasoning will apply. The crucial benefit of this is that nogoods are more reusable.

4 Resource-Constrained Project Scheduling Problems

Resource-constrained project scheduling problems (RCPSP) appear as variants, extensions and restrictions in many real-world scheduling problems. Therefore we test our decomposition on the well-known RCPSP benchmark library PSPLib [1].

An RCPSP is denoted by a triple (T, A, R) where T is a set of tasks, A a set of precedences between tasks and R is a set of resources. Each task i has a duration $d[i]$ and a resource usage $r[k, i]$ for each resource $k \in R$. Each resource has a resource capacity $c[r]$.

The goal is to find either a schedule or an optimal schedule with respect to an objective function where a schedule s is an assignment which meets following conditions

$$\begin{aligned} \forall i \rightsquigarrow j \in A : \quad & s[i] + d[i] \leq s[j] \\ \forall t \in [\,0\,..\,t_{max} - 1\,], \forall k \in R : \quad & \sum_{i \in T : s[i] \leq t < s[i]+d[i]} r[k, i] \leq c[r] \ , \end{aligned}$$

where t_{max} is the planning horizon. For our experiments we search for a schedule which minimises the makespan (i.e. latest end time). A basic Zinc model is given at `http://www.cs.mu.oz.au/~pjs/rcpsp`

In practice we share the Boolean variables generated inside the cumulative constraints as described in Section 3.1 (by common sub-expression elimination) and add redundant constraints as described in Section 3.2 when using the Task-RD decomposition. We also add redundant non-overlap constraints for each pair of tasks whose resource usages make them unable to overlap. Moreover, the planning horizon t_{max} was determined as the makespan of first solution found by labelling the smallest value of the start time variables in order. The initial domain of each variable $s[i]$ was determined as $D_{init}(s[i]) = [\,p[i]\,..\,t_{max} - q[i]\,]$ where $p[i]$ is the duration of the longest chain of predecessor tasks, and $q[i]$ is the duration of the longest chain of successor tasks.

In the remainder of this section we discuss alternate search strategies.

4.1 Search Using Serial Scheduling Generation

The serial scheduling generation scheme (serial SGS) is one of basic deterministic algorithms to assign stepwise a start time to an unscheduled task. It incrementally extends a partial schedule by choosing an *eligible* task—i.e. all of whose predecessors are fixed in the partial schedule—and assigns it to its earliest start time with respect to the precedence and resource constraints. For more details about SGS, different methods based on it, and computational results in Operations Research see [10, 8, 11].

Baptiste and Le Pape [3] adapt serial SGS for a constraint programming framework. For our experiments we use a form where we do not apply their dominance rules, and where we impose a lower bound on the start time instead of posting the delaying constraint "task i executes after at least one task in S".

1. *Select* an eligible unscheduled task i with the earliest start time $t = lb(s[i])$. If there is a tie between some tasks then select that one with the minimal latest start time $ub(s[i])$. Create a choice point.
2. *Left branch:* Extend the partial schedule by setting $s[i] = t$. If this branch fails then go to the right branch; Otherwise go to step 1.
3. *Right branch:* Delay task i by setting $s[i] \geq t'$ where $t' = \min\{lb(s[j]) + d[j] \mid j \in T : lb(s[j]) + d[j] > lb(s[i])\}$, that is, the earliest end time of the concurrent tasks. If this branch fails then backtrack to the previous choice point; Otherwise go to step 1.

The right branch uses the dominance rule that amongst all optimal schedules there exists one where every task starts either at the first possible time or immediately after the end of another task. Therefore, the imposing of the new lower bound is sound. If we add side constraints then this assumption could be invalid.

Note that we use this search strategy with branch and bound, where whenever a new solution is found, a constraint requiring a better solution is dynamically (globally) added during the search.

4.2 Search Using Variable State Independent Decaying Sum

The SAT decision heuristic Variable State Independent Decaying Sum (VSIDS) [15] is a generic search approach that is currently almost universally used in DPLL SAT solvers. Each variable is associated with a dynamic *activity* counter that is increased when the variable is involved in a failure. Periodically, all counters are reduced, thus *decaying*. The unfixed variable with the highest activity is selected to branch on at each stage. Benchmark results by Moskewicz [15] shows that VSIDS performs better on average on hard problems than other heuristics.

To use VSIDS in a lazy clause generation solver, we ask the SAT solver what its preferred literal for branching on is. This corresponds to an atomic constraint $x \leq d$ or $x = d$ and we branch on $x \leq d \vee x > d$ or $x = d \vee x \neq d$. Note that the search is still controlled by the FD search engine, so that we use its standard approach to implementing branch-and-bound to implement the optimisation search.

Normally SAT solvers use dichotomic restart search for optimisation as the SAT solver itself does not have optimisation search built in, although in some cases it is possible to maintain the nogoods from the previous search. The combination of VSIDS and branch and bound is much stronger since in the continuation of the search with a better bound, the activity counts at the time of finding a new better solution are used in the same part of the search tree.

Restarting is shown to be beneficial in SAT solving (and CSP solving) in speeding up solution finding, and being more robust on hard problems. We also use VSIDS search with restarting, which we denote RESTART.[1]

[1] Note that restarting SGS search while possible is not attractive since the nogoods do not modify the search in most cases.

4.3 Hybrid Search Strategies

One drawback of VSIDS is that at the beginning of the search the activity counters are only related to the clauses occurring in the original model, and not to any conflict. This is exacerbated in lazy clause generation where many of the constraints of the problem may not appear at all in the clause database initially. This can lead to poor decisions in the early stages of the search. Our experiments support this, there are a number of "easy" instances which SGS can solve within a small number of choice points, where VSIDS requires substantially more.

In order to avoid these poor decisions we consider a hybrid search strategy. We use SGS for the first 500 choice points and then restart the search with VSIDS. The SGS search may solve the whole problem if it is easy enough, but otherwise it sets the activity counters to meaningful values so that VSIDS starts concentrating on meaningful decisions. We denote this search as HOT START, and the version where the secondary VSIDS search also restarts as HOT RESTART.

5 Experiments

We carried out extensive experiments on RCPSP instances comparing our approach to decomposition without explanation, global `cumulative` propagators from SICStus and ECLiPSe, as well as a state-of-the-art exact solving algorithm [12]. Detailed results are available at `http://www.cs.mu.oz.au/~pjs/rcpsp`

We use two suites of benchmarks. The library PSPLib [1] contains the four classes J30, J60, J90, and J120 consisting of 480 instances of 30, 60, 90 task and 600 instances of 120 tasks respectively. We also use a suite (BL) of 40 highly cumulative instances with either 20 or 25 tasks constructed by Baptiste and Le Pape [3].

The experiments were run on a X86-64 architecture running GNU/Linux and a 3.4 GHz processor. The code was written in G12 Constraint Programming Platform and compiled with the Mercury Compiler and grade hlc.gc.trseg. Each run was given a 10 minute limit.

5.1 Results on J30 and BL Instances

The first experiment compares different decompositions and search on the smallest instances J30 and BL. We compare SGS, VSIDS, RESTART and the hybrid search approaches using three decompositions Time-RD (t), Task-RD (s), and an equivalent version to Task-RD on end times (e). The results are shown in Table 1. For J30 we show the number of problems solved (#svd), (cmpr(477)) the average solving time in seconds and number of choice points (#cp) on the 477 problems that all approaches solved, and (all(480)) average solving time in seconds and number of choice points on all 480 problems to find the best solution found.[2] Note that we shall use similar comparisons and notation in future

[2] This means that for problems that time out this may be significantly smaller than the number of choice points explored before timeout.

Table 1. Results on J30 and BL instances

search	dec	J30 #svd	J30 cmpr(477) time	J30 cmpr(477) #cp	J30 all(480) time	J30 all(480) #cp	BL #svd	BL all(40) time	BL all(40) #cp	BL #svd	BL cp(4000) time	BL cp(4000) #cp
SGS	s	477	3.25	3069	6.97	4114	**40**	4.18	9628	24	0.22	1261
	e	477	3.31	3054	7.04	4101	**40**	4.41	9443	24	0.19	1144
	t	**480**	1.36	2339	4.09	4230	**40**	1.40	5892	29	**0.05**	781
VSIDS	s	**480**	1.82	2128	2.62	2984	**40**	1.24	4436	31	0.20	1115
	e	**480**	0.85	1504	1.45	2220	**40**	1.27	4104	30	0.20	1025
	t	**480**	0.43	1002	**0.54**	1271	**40**	0.30	2540	34	**0.05**	661
RESTART	s	**480**	0.93	1504	1.73	2339	**40**	1.46	4597	31	0.23	1207
	e	**480**	0.82	1392	1.52	2153	**40**	2.61	5848	32	0.23	1177
	t	**480**	0.39	892	**0.54**	**1212**	**40**	0.17	1670	35	0.06	639
HOT START	t	**480**	**0.34**	**782**	0.56	1223	**40**	**0.13**	**1456**	**36**	**0.05**	688
HOT RESTART	t	**480**	0.42	892	0.59	1241	**40**	0.20	1850	35	0.07	733

tables. For BL we show the number of solved problems, (all(40)) average solving time and number of choice points with 10 minute limit (on all 40 instances), as well as cp(4000) with a 4000 choice point limit.

Clearly the Time-RD decomposition is superior regardless of search, and the best search strategies are RESTART and the hybrid ones.

The results on the BL instances show that approaches using Time-RD and VSIDS could solve between 6 and 8 instances more than the base approach (FE) of Baptiste and Le Pape [3] within 4000 backtracking steps.[3] Their "left-shift/right-shift" approach could solve 40 instances in 30 minutes, with an average of 3634 steps and 39.4 seconds on a 200 MHz machine. All our approaches with Time-RD and VSIDS find the optimal solution faster and in fewer backtracking steps (between a factor of 1.39 and 2.4).

Next we compare the Time-RD decomposition (SGS+t) against implementations of `cumulative` in SICStus v4.0 (default, and with the flag `global`) and ECLiPSe v6.0 (using its 3 cumulative versions from the libraries `cumulative`, `edge_finder` and `edge_finder3`). We also compare against (FD+t) a decomposition without explanation (a normal FD solver) executed in the G12 system. All approaches use the SGS search strategy.

The results are shown in the Table 2. We can see that none of the other approaches compare to the lazy clause generation approach. The best is the SICStus cumulative with `global` flag. Clearly nogoods are very important to fathom search space.

While the Time-RD decomposition clearly outperforms Task-RD on these small examples, as the planning horizon grows at some point Task-RD should be better, since its model size is independent of the planning horizon. To investigate this we took 20 examples from J30 and multiplied the durations and planning horizon by 10 and 100. We compare the Time-RD decomposition versus the (e)

[3] We count the number of choice points which is not smaller than the number of backtracking steps.

Table 2. Results of the FD solvers on the J30 and BL instances

		J30					BL				
	solver	#svd	cmpr(361)		all(480)		#svd	cmpr(6)		all(40)	
SICStus	default	417	0.24	268	89.00	13986	30	2.86	20865	213.59	489218
	global	411	0.43	263	96.85	6661	39	0.32	1265	19.19	10262
ECLiPSe	cumu	365	11.60	19529	158.30	42698	6	149.90	252462	532.31	123839
	ef	361	15.15	15438	161.32	22907	36	8.79	11265	117.21	89034
	ef3	362	13.37	12391	159.41	19186	37	7.17	7717	90.82	49114
G12	FD+t	403	1.93	5665	104.72	156598	30	2.23	34677	217.12	918287
	SGS+t	**480**	**0.02**	**75**	**4.09**	4230	**40**	**0.02**	**293**	**1.40**	5892

Table 3. Results on 20 modified instance from J30 instances

		SGS					VSIDS				
duration	dec	#svd	cmpr(12)		all(20)		#svd	cmpr(12)		all(20)	
1×	e	17	0.74	2383	152.49	67257	**20**	0.25	735	26.89	34563
	t	**20**	**0.44**	**1817**	**87.19**	72888	**20**	**0.11**	**404**	**6.59**	14405
10×	e	13	**4.25**	7493	212.75	47394	**20**	1.98	3971	117.27	68097
	t	**14**	4.74	**2516**	**201.27**	41081	**20**	**1.66**	**1622**	**94.09**	28250
100×	e	**13**	**22.03**	17620	**225.71**	35349	**14**	**10.52**	**6379**	**192.90**	24959
	t	**13**	55.78	**3017**	259.98	7175	**14**	22.42	9836	233.60	12618

end-time Task-RD decomposition (which is slightly better than start-time s). The results are shown in Table 3. First we should note that simply increasing the durations makes the problems significantly more difficult for a decomposed cumulative. While the Time-RD decomposition is still just better than the Task-RD decomposition for the 10× extended examples, it is inferior for scheduling problems with very long durations.

5.2 Results on J60, J90 and J120

We now examine the larger instances J60, J90 and J120 from PSPLib. For J60 we compare the most competitive approaches from the previous subsection: VSIDS + t, RESTART + t, HOT START + t and HOT RESTART + t. For this suite our solvers cannot solve all 480 instances within 10 minutes. The results are presented in the Table 4. For these examples we show the average distance of our best solution found from the best known solution from PSPLib (most of which are generated by specialised heuristic methods), as well as the usual time and number of choice points comparisons. Many of these are currently open problems. Our best approaches close 21 open instances (considering results from PSPlib [1], Laborie [12] and Liess and Michelon [13]). Clearly the hybrid search strategies are superior, although all of these approaches are quite competitive.

For the largest instances J90 and J120 we ran only HOT RESTART + t, since it is the most robust strategy. For J90 we can solve 396 of 480 instances, with

Table 4. Results on J60 instances for Time-RD

solver	#svd	avg. dist.	cmpr(424)		all(480)	
VSIDS + t	424	4.5	5.77	6351	75.07	19781
RESTART + t	428	4.8	5.07	5010	69.70	24333
HOT START + t	**429**	9.3	**3.83**	**4111**	68.27	12072
HOT RESTART + t	**429**	**4.2**	4.66	4617	**68.25**	25810

Table 5. Comparison between Laborie's method and HOT RESTART + t

	J60			J90			J120		
1.4 GHz	45s	300s	1800s	45s	300s	1800s	45s	300s	1800s
Laborie	-	84.2	85.0	-	78.5	79.4	-	41.3	41.7
HOT RESTART + t	85.2	88.1	89.4	79.8	81.3	82.5	42.5	44.8	45.3
3.4 GHz	18s	120s	600s*	18s	120s	600s*	18s	120s	600s*

an average solution distance of 7.6. The average for solved instances is 6.56s with 5077 #cp. We close 13 open instances in J90. For J120 we can solve 272 of 600 instances, with an average solution distance of 9.7, with average (on solved instances) times of 7.52s and 6136 #cp. We close 20 open instances in J120.

We compare our best method HOT RESTART + t to the method by Laborie [12], the best published method on the J60, J90, and J120 instances.

Table 5 shows the percentage of solved instances within a maximal solve time. We give an equivalent time to our solver taking into account the speeds of the processors: 3.4GHz vs. 1.4GHz. At the top of the table is the time cutoff for a 1.4GHz processor, and at the bottom the approximately equivalent cutoff times for a 3.4GHz machine. Note, that all * marked 3.4GHz times are much lower than the equivalent time for the 1.4GHz processor. Clearly this comparison can only be seen as indicative.

Our method clearly outperforms Laborie's method: for every class our method was able to solve more problems within 18s than they could solve in half an hour respectively on their machine. Interestingly, our solver could not solve six instances which were solved by others.

Finally we used HOT START + t to try to improve lower bounds of the remaining open problems, by searching for a solution to the problem with the makespan varying from the best known lower bound to the best known upper bound from PSPLib. In this way we closed 9 more problems and improved 76 lower bounds.

6 Conclusion

We present a new approach solving RCPSP problems by modelling cumulative constraints by decomposition and using lazy clause generation. Benchmarks from the PSPLib show the strong power of nogoods and VSIDS style search to fathom a large part of the search space. Without building complex specific global propagators or highly specialised search algorithms we are able to compete with highly specialised RCPSP solving approaches and close 63 open problems.

Acknowledgments. We would like to thank Phillipe Baptiste for suggesting this line of enquiry. We would also like to thank the anonymous reviewer who pointed out the recent work of Laborie [12] which we had missed, and provided a detailed comparison. NICTA is funded by the Australian Government as represented by the Department of Broadband, Communications and the Digital Economy and the Australian Research Council.

References

[1] PSPLib — project scheduling problem library, `http://129.187.106.231/psplib/` (23.04.2009)
[2] Aggoun, A., Beldiceanu, N.: Extending CHIP in order to solve complex scheduling and placement problems. Mathematical and Computer Modelling 17(7), 57–73 (1993)
[3] Baptiste, P., Le Pape, C.: Constraint propagation and decomposition techniques for highly disjunctive and highly cumulative project scheduling problems. Constraints 5(1-2), 119–139 (2000)
[4] Carlier, J., Pinson, E.: Jackson's pseudo-preemptive schedule and cumulative scheduling problems. Discrete Applied Mathematics 145(1), 80–94 (2004)
[5] Caseau, Y., Laburthe, F.: Cumulative scheduling with task intervals. In: Procs. of the 1996 Joint International Conference and Symposium on Logic Programming, pp. 363–377. MIT Press, Cambridge (1996)
[6] Demeulemeester, E.L., Herroelen, W.S.: New benchmark results for the resource-constrained project scheduling problem. Management Science 43(11), 1485–1492 (1997)
[7] El-Kholy, A.O.: Resource Feasibility in Planning. PhD thesis, Imperial College, University of London (1996)
[8] Hartmann, S., Kolisch, R.: Experimental evaluation of state-of-the-art heuristics for the resource-constrained project scheduling problem. EJOR 127(2), 394–407 (2000)
[9] Jussien, N., Barichard, V.: The PaLM system: explanation-based constraint programming. In: Proceedings of Techniques foR Implementing Constraint programming Systems, pp. 118–133 (2000)
[10] Kolisch, R.: Serial and parallel resource-constrained project scheduling methods revisited: Theory and computation. EJOR 90(2), 320–333 (1996)
[11] Kolisch, R., Hartmann, S.: Experimental investigation of heuristics for resource-constrained project scheduling: An update. EJOR 174(1), 23–37 (2006)
[12] Laborie, P.: Complete MCS-based search: Application to resource constrained project scheduling. In: Kaelbling, L.P., Saffiotti, A. (eds.) Proceedings IJCAI 2005, pp. 181–186. Professional Book Center (2005)
[13] Liess, O., Michelon, P.: A constraint programming approach for the resource-constrained project scheduling problem. Annals of Operations Research 157(1), 25–36 (2008)

[14] Marriott, K., Nethercote, N., Rafeh, R., Stuckey, P.J., Garcia de la Banda, M., Wallace, M.G.: The design of the Zinc modelling language. Constraints 13(3), 229–267 (2008)

[15] Moskewicz, M.W., Madigan, C.F., Zhao, Y., Zhang, L., Malik, S.: Chaff: Engineering an efficient SAT solver. In: Design Automation Conference, pp. 530–535. ACM, New York (2001)

[16] Nuijten, W.P.M.: Time and Resource Constrained Scheduling. PhD thesis, Eindhoven University of Technology (1994)

[17] Ohrimenko, O., Stuckey, P.J., Codish, M.: Propagation = lazy clause generation. In: Bessière, C. (ed.) CP 2007. LNCS, vol. 4741, pp. 544–558. Springer, Heidelberg (2007)

[18] Vilím, P.: Computing explanations for the unary resource constraint. In: Barták, R., Milano, M. (eds.) CPAIOR 2005. LNCS, vol. 3524, pp. 396–409. Springer, Heidelberg (2005)

On Decomposing Knapsack Constraints for Length-Lex Bounds Consistency*

Meinolf Sellmann

Brown University, Department of Computer Science
115 Waterman Street, P.O. Box 1910, Providence, RI 02912

Abstract. The length-lex representation for set variables orders all subsets of a given universe of values according to cardinality and lexicography. To achieve length-lex bounds consistency for Knapsack constraints it has been proposed to decompose the constraint into two sum constraints. We provide theoretical and practical evidence which shows that decomposition increases the problem of computing a fixpoint which is intrinsic to the length-lex representation: 1. The fixpoint problem for this domain representation is NP-hard in general. 2. For a tractable sub-family of Knapsack decomposition takes more time than exponential brute-force enumeration. 3. Experimental results on decomposed Knapsack constraints show that exponential-time fixpoint computation is the rule and not some pathological exception.

1 Introduction

In CP, finite variable domains are commonly given by explicit enumeration of values. Then, when constraints remove values from variables' domains, a fixpoint is trivially reached after a linear number of calls to each filtering routine. The research focus in CP is therefore often concentrated on efficient constraint filtering algorithms. The situation changes fundamentally when domains become exponentially large in the instance input size, as it is, e.g., the case for set variables.

The traditional subset-superset representation for set variables elegantly circumvents the problem by ensuring that the number of potential domain changes is limited by a linear number of steps before the set variable is bound. This is however not the case for the newly proposed length-lex representation for set variables [3]. Consequently, the time until a fixpoint is reached cannot be ignored anymore.

The frequently prohibitively long time before a fixpoint is reached is a well-known practical problem in interval propagation where domains generally also have exponential size. In [1] it was rigorously proven that slow convergence is intrinsic to propagation methods. That is, the decomposition of a problem into individual constraints who exchange information only via their domains can cause NP-hard fixpoint generation problems when domains are exponentially large.

In light of these results, we study the recent proposition to decompose Knapsack constraints to achieve length-lex bounds consistency. We prove a series of negative results. 1. Computing fixpoints at which constraints are length-lex bounds consistent is NP-hard in general. 2. Constraint decomposition may introduce exponential runtimes

* This work was supported by the National Science Foundation through the Career: Cornflower Project (award number 0644113).

I.P. Gent (Ed.): CP 2009, LNCS 5732, pp. 762–770, 2009.

even when the global problem is in P. 3. Experimental results on number partitioning and market split problems show that slow convergence is the rule and not some rare pathological exception.

2 Length-Lex Bounds Consistency

A *set-variable* is a variable whose domain values are sets. We assume that the elements originate from a finite universe of elements. Because the number of possible values of a set-variable can be enormous (the size of a power set, in the worst case), one usually represents the domain of a set-variable S by a 'lower bound' and an 'upper bound' on the values that S can take.

A natural representation for the domain of a set-variable is based on the *subset ordering* of the universe. That is, the lower bound $M(S)$ represents all mandatory elements, while the upper bound $P(S)$ represents all possible elements, i.e., $D_{SS}(S) := \{s \mid M(S) \subseteq s \subseteq P(S)\}$. We refer to this representation as the *subset* representation.

An alternate representation is based on the *length-lexicographic* ordering of the universe [3] where the lower bound $L(S)$ represents the smallest set that can be assigned to S, while the upper bound $U(S)$ represents the largest set, i.e., $D(S) := D_{LL}(S) := [L(S), U(S)] := \{s \mid L(S) \leq_{LL} s \leq_{LL} U(S)\}$. Here $\leq_{LL}$ denotes the length-lexicographic order which sorts sets first according to their cardinality and, as a sub-criterion when the cardinalities are equal, according to their lexicography, whereby we assume that an ordering on the universe is given and elements in a set are sorted accordingly. We refer to this representation as the *length-lex* representation. To keep the presentation simple, throughout the paper we will assume that the length-lex lower bound contains the first value and the upper bound does not as this value could otherwise be removed from consideration (see the length-lex* representation which treats all mandatory and impossible elements separately [4]).

Example 1. Let S be a set-variable over the universe $\{1, \dots, 4\}$ with domain $D(S) = [\{1,3\}, \{2,3,4\}]$. Then, the values that S can take are $D(S) = \{\{1,3\}, \{1,4\}, \{2,3\}, \{2,4\}, \{3,4\}, \{1,2,3\}, \{1,2,4\}, \{1,3,4\}, \{2,3,4\}\}$.

For constraints involving set-variables, the filtering task is to tighten the bounds of the domains. The respective definition of bounds consistency depends on the domain representation used. For the traditional subset representation, bounds consistency of a unary constraint over a set variable means that all values that must be part of the final set according to the constraint are included in the lower bound and all values that cannot be part of the final set are not elements of the upper bound. For the length-lex representation, bounds consistency simply means that both the lower and the upper bound are consistent according to the constraint.

This latter property appears highly desirable. However, as we will see shortly, it is the reason why computing fixpoints of length-lex bounds consistency is intractable.

3 Decomposing Knapsack Constraints

Consider the *Knapsack constraint* $\mathrm{KP}(S, p, B, w, C)$ where we are given a set of items modeled by a set-variable S over a universe of n items, a profit vector p which gives the non-negative profit of each item, a weight vector w which gives the weight of each item, a lower bound B on the total profit, and a capacity C bounding the total weight of the knapsack. The constraint requires that the set s assigned to S fulfills $\sum_{i \in S} p_i \geq B$, and $\sum_{i \in S} w_i \leq C$. In [4] we studied the question whether Knapsack problems are still approximable when conjoined with cardinality and lexicographic constraints. We proved that this is possible and gave an fully polynomial time approximation scheme for the problem. Based on this scheme, we proved that approximated length-lex bounds consistency can be achieved in polynomial time.

While the runtime of our filtering algorithm is polynomial, it is too expensive to be invoked in a constraint propagation engine. Therefore, as a practical alternative to this algorithm, in [7] it is suggested to decompose KP into one sum constraint which enforces the capacity bound and another sum constraint which enforces the profit bound. A complex algorithm for these constraints is devised which requires a preprocessing step in $O(n^3)$ and then runs in amortized time $O(n \log n)$ for each sum constraint. We will now analyze the proposed decomposition theoretically and practically.

3.1 Knapsack Decomposition in Theory

We begin by stating a trivial intractability result which shows that slow-convergence when generating fixpoints is an intrinsic problem of length-lex bounds consistency.

Lemma 1. *It is NP-hard to decide whether a fixpoint exists to a system of only one set-variable and just two unary constraints with associated monotonic and idempotent filtering operators that achieve length-lex bounds consistency.*

Proof. We reduce from Knapsack, i.e., we want to decide whether there exists a subset $I \subseteq \{1, \ldots, n\}$ such that setting $S \leftarrow I$ satisfies constraint $W(S) := \sum_{i \in S} w_i \leq C$ and constraint $P(S) := \sum_{i \in S} p_i \geq B$ for natural numbers w_i, p_i, C, and B. Note that W and P both affect both bounds of the length-lex domain. [7] devised monotonic and idempotent filtering operators for the constraints $W(S)$ and $P(S)$. If there exists a fixpoint where both constraints are length-lex bounds consistent, then, at the bounds found, both constraints are satisfied. Therefore, each bound gives a solution that satisfies both constraints, and thus solves the Knapsack problem. On the other hand, assume that the Knapsack instance is valid, i.e., there exists an $I \subseteq \{1, \ldots, n\}$ such that $W(S)$ and $P(S)$ are satisfied when setting $S \leftarrow I$. Then, $D(S) = [I, I]$ is a fixpoint where both constraints are length-lex bounds consistent. □

Remark 1. The proof above shows that the problem with fixpoint intractability is not actually caused by the specific (in our case: length-lex) ordering of the set variable's domain. The same argument can be made for *any* form of set bounds consistency which requires that constraints are satisfied at the domain bounds.

While the intractability of computing a fixpoint is certainly disconcerting, it is not really surprising since we are tackling an NP-hard problem after all. However, even more concerning is the following:

Lemma 2. *Decomposing Knapsack constraints into two sum constraints in a propagation engine can introduce exponential runtimes even for tractable problems.*

Proof. Consider the following family of Knapsack instances which belong to the large class of Knapsack problems in P for which $\min\{||p||_\infty, ||w||_\infty\}$ is bounded by a polynomial in n. For any even number of variables n we set $p_1 = w_1 = n, p_n = w_n = n$, and $p_i = w_i = 1$ for all $1 < i < n$. Assume we introduce a set variable $S \subseteq \{1, \ldots, n\}$ and pose two separate constraints, $W(S)$ which enforces $\sum_{i \in S} w_i \leq 3n/2$, and $P(S)$ which enforces $\sum_{i \in S} p_i \geq 3n/2$.

Observe how the lower length-lex bound evolves during the fixpoint computation: After filtering W and P twice, we observe the sequence $\{1, 3, 4\}$, $\{1, 3, n\}$, $\{1, 4, 5\}$, $\{1, 4, n\}, \ldots, \{1, n-2, n\}$. That is, we observe *all* sets of cardinality 3 which contain 1 and n and do not contain $n-1$.

The fixpoint computation continues: $\{1, 2, 3, 4\}, \{1, 2, 3, n\}, \{1, 2, 4, 5\}, \{1, 2, 4, n\}$, $\ldots$, $\{1, 2, n-2, n\}$, $\{1, 3, 4, 5\}$, $\{1, 3, 4, n\}$, $\{1, 3, 5, 6\}$, $\{1, 3, 5, n\}$, $\ldots$, $\{1, 3, n-2, n\}$, $\ldots$ $\{1, n-3, n-2, n\}$. That is, among others we visit *all* sets of cardinality 4 which contain 1 and n and do not contain $n-1$.

Next we visit, among other sets, *all* sets of cardinality 5 which contain 1 and n and do not contain $n-1$. And so on and so forth until we finally find the lower bound of cardinality $n/2 + 1$, $\{1, 2, \ldots, n/2 + 1\}$. We observe: by filtering constraints W and P separately and in turn, we implicitly compute a sequence of increasing lower bounds in the length-lex representation of all sets of cardinality from 3 to $n/2$ which contain items 1 and n and not $n-1$. Consequently, for $n \geq 6$, the fixpoint computation requires more than $\sum_{i<n/2-1} \binom{n-3}{i} \geq 2^{n-5}$ calls to the filtering methods, and thus runs in time $\Omega(2^n n \log n)$ – more than brute force enumeration. On the other hand, the pseudo-polynomial approach from [4] can be followed to compute the same bounds in time $O(n^3)$. □

3.2 Knapsack Decomposition in Practice

Number Partitioning. The previous discussion shows that Knapsack decomposition for length-lex bounds consistency is certainly not appealing from a theoretical worst-case point of view. The question arises whether slow convergence occurs only rarely in practice. One of the simplest problems for which we may consider employing a filtering algorithm for Knapsack constraints is number partitioning. Given numbers $a_1, \ldots, a_n$, is there a set $I \subseteq \{1, \ldots, n\}$ such that $\sum_{i \in I} a_i = \frac{\sum_i a_i}{2}$?

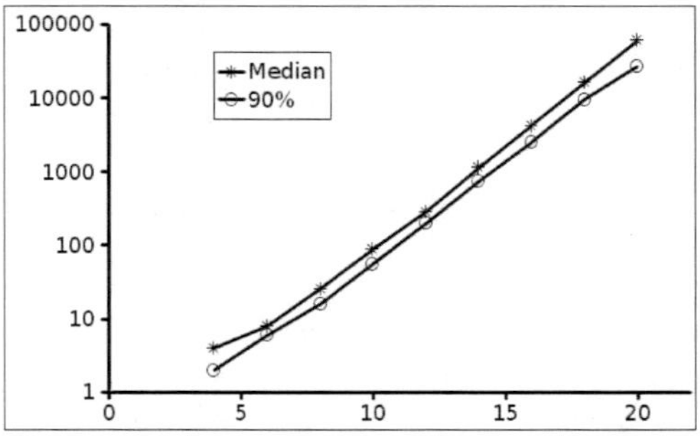

Fig. 1. Filtering calls (log-scale) on number partitioning problems with 4 to 20 numbers (100 instances per data point, numbers drawn uniformly from $[1, 10^6]$). The upper curve shows the minimum number of calls needed by at least 50% of the instances, the lower the minimum number of calls needed by at least 90% of the instances.

Figure 1 shows the number of filtering calls of 100 small, randomly generated number partitioning problems modeled by one set variable with one unary Knapsack constraint that is decomposed into two sum constraints. The two curves (logarithmic scale) depict lower bounds on the number of calls that are required until a fixpoint is reached for 90% and 50% of the instances. We see clearly that exponential-time fixpoint computations occur in the vast majority of test cases. Slow convergence is thus the predominant rule and not some pathological worst-case exception.

Speeding up Convergence. In order to show that Knapsack decomposition is not appealing in practice either, in the following we introduce a heuristic improvement which tries to address the slow convergence problem. We will then show that even this improved algorithm cannot compete with non-decomposition approaches.

To speed up convergence we exploit the method based on dynamic programming (DP) from [4] which simplifies considerably for sum constraints. Let us consider sum constraints of the form $\sum_{i \in I} w_i \leq C$, with $w_i, C \in \mathbb{Q}$. We set up a dynamic program in two dimensions (compare with Figure 2). The first gives the current index of the largest element allowed in the solution set, which is restricted to be a subset of $\emptyset, \{1\}, \{1,2\}, \ldots,$ or $\{1, \ldots, n\}$. The other dimension gives the cardinality $0, 1, \ldots, n$. Then, in W_{ic} we store the weight of the set $I \subseteq \{1, \ldots, i\}$ with cardinality $|I| = c$ that has the smallest weight $w(I) := \sum_{i \in I} w_i$. The ordinary recursion equation $W_{i+1,c} \leftarrow \min\{W_{ic}, W_{i-1,c-1} + w_i\}$ can be used to find the set $I \subseteq \{1, \ldots, n\}$ with cardinality κ that minimizes $w(I) = W_{n\kappa}$.

For the purpose of achieving length-lex bounds consistency for sum constraints, this basic DP can be adapted following the same ideas as given in [4]. Assume that we are given a set variable $S \subseteq \{1, \ldots, n\}$ with domain $D(S) = [L, U]$, $L, U \subseteq \{1, \ldots, n\}$. We are to find new lower and upper bounds $L', U' \subseteq \{1, \ldots, n\}$ such that $w(L'), w(U') \leq C$ and for all L'', U'' with $L \leq_{LL} L'' <_{LL} L'$, $U' <_{LL} U'' \leq_{LL} U$ it holds $w(L''), w(U'') > C$. Following the approach in [4], it is sufficient to devise an algorithm for the case where $\kappa \leftarrow |L| = |U|$.

Note that L and U define paths π_L and π_U in the DP from W_{00} to $W_{n\kappa}$ (see Figure 2). Analogously, every path π from W_{00} to $W_{n\kappa}$ defines a set $S_\pi \subseteq \{1, \ldots, n\}$. We call a path *admissible* iff $L \leq_{LL} S_\pi \leq_{LL} U$. It will be important for us to know the shortest path distance from the root to a given node when the choices implied by that path π already ensure that the resulting set S_π must obey the lexicographic bounds L, U. Conversely, we will also need to argue about paths from nodes in the DP-induced graph to the sink-node $W_{n\kappa}$:

Definition 1. *For a path π from the root to W_{ic}, we write $L <_{lex} S_\pi$ (or $S_\pi <_{lex} U$) if and only if for all $S \subseteq \{1, \ldots, n\}$ such that $S \cap (S_\pi \cup \{i+1, \ldots, n\}) = S$ and $|S| = \kappa$ it holds that $L <_{lex} S$ ($S <_{lex} U$).*

For a path π from W_{ic} to $W_{n\kappa}$, we write $L \leq_{lex} S_\pi$ (or $S_\pi \leq_{lex} U$) if and only if for $T \leftarrow S_\pi \cup (L \cap \{1, \ldots, i\})$ ($T \leftarrow S_\pi \cup (U \cap \{1, \ldots, i\})$) it holds that $|T| = \kappa$ and $L \leq_{lex} T$ ($T \leq_{lex} U$).

Now, for every node W_{ic} in the DP, we compute the following quantities:

1. For $W_{ic} \in \pi_L$, M^1_{ic} gives the distance from the root W_{00} to W_{ic} when following π_L, that is $M^1_{ic} \leftarrow \sum_{i \in L, i \leq k} w_i$. For $W_{ic} \notin \pi_L$, we set $M^1_{ic} \leftarrow \infty$.
2. For $W_{ic} \in \pi_U$, M^2_{ic} gives the distance from the root W_{00} to W_{ic} when following π_U, that is $M^2_{ic} \leftarrow \sum_{i \in U, i \leq k} w_i$. For $W_{ic} \notin \pi_U$, we set $M^2_{ic} \leftarrow \infty$.

3. For arbitrary nodes W_{ic}, M^3_{ic} gives the length of the shortest path π from the root to W_{ic} with $L <_{lex} S_\pi <_{lex} U$.
4. For arbitrary nodes W_{ic}, M^4_{ic} gives the length of the shortest path π from W_{ic} to $W_{n\kappa}$.
5. For $W_{ic} \in \pi_L$, M^5_{ic} gives the length of the shortest path π from W_{ic} to $W_{n\kappa}$ with $L \leq_{lex} S_\pi$.
6. For $W_{ic} \in \pi_U$, M^6_{ic} gives the length of the shortest path π from W_{ic} to $W_{n\kappa}$ with $S_\pi \leq_{lex} U$.

In [4] we devised recursion equations that allow us to compute all these values in time $O(n^2)$. Based on this data, we can compute the desired bounds L', U'. To this end, we propose a new procedure which is more efficient than the one presented in [4].

Let us consider the problem of computing L' (the computation of U' works analogously). We start at node W_{00}. Recall that we may assume that $1 \in L$ and $1 \notin U$. We start at node W_{00} and initialize the already committed cost $g \leftarrow 0$. Clearly $1 \in L'$ if and only if $g + w_1 + M^5_{11} \leq C$. In this case, we move to W_{11} and update $g \leftarrow g + w_1$. While we stay on π_L, we update in this way: at W_{ic}, if $i + 1 \notin L$ or when $g + w_{i+1} + M^5_{i+1,c+1} > C$, we move to $W_{i+1,c}$. Otherwise we add $i + 1$ to L', move to $W_{i+1,c+1}$, and update $g \leftarrow g + w_{i+1}$. At some point it may occur that $i + 1 \in L$ but $g + w_{i+1} + M^5_{i+1,c+1} > C$. Then, we leave π_L and we are already guaranteed that $L <_{LL} L'$. So we continue as follows: If $g + w_{i+1} + M^4_{i+1,c+1} \leq C$, we add $i + 1$ to L', move to $W_{i+1,c+1}$, and update $g \leftarrow g + w_{i+1}$. Otherwise, we move to $W_{i+1,c}$. In this way, we can compute the new lower bound L' in time $O(n)$ once the values M^k_{ic} are known. With this procedure, length-lex bounds consistency for sum constraints can be achieved in time $O(n^2)$.

The previous algorithm is very simple and easy to implement. In contrast to the algorithm presented in [7], it requires no specialized theory, cubic preprocessing, or amortized runtime analysis. On the other hand, it is slower than the $O(n \log n)$ amortized time from [7]. However, as we have seen earlier, the problem with decomposing Knapsack constraints is *not* caused by the inefficiencies when propagating sum constraints. The problem is the very *slow convergence* before a fixpoint is reached. The algorithm presented above has the potential to reduce this time by filtering *all* sum constraints within the *same* dynamic program.

To achieve this practical improvement, we will filter edges from the graph induced by the DP. Since a sum constraint is a Knapsack constraint where the profits are all zero, from Lemma 1 in [4], we know that the following quantities can be used to remove edges from the DP-induced graph:

Lemma 3

- *The length of a shortest admissible path through an edge* $(W_{ic}, W_{i+1,c})$ *is* $\min\{M^3_{ic} + M^4_{i+1,c}, M^1_{ic} + D^1_{ic}, M^2_{ic} + D^2_{ic}\}$, *where* $D^1_{ic} = M^4_{i+1,c}$ *if* $i+1 \in L$ *and* $D^1_{ic} = M^5_{i+1,c}$ *otherwise, and* $D^2_{ic} = M^6_{i+1,c}$ *if* $i+1 \notin U$ *and* $D^2_{ic} = \infty$ *otherwise.*
- *The length of a shortest admissible path through an edge* $(W_{ic}, W_{i+1,c+1})$ *is* $\min\{M^3_{ic} + M^4_{i+1,c+1} + w_{i+1}, M^1_{ic} + E^1_{ic} + w_{i+1}, M^2_{ic} + E^2_{ic} + w_{i+1}\}$, *where* $E^1_{ic} = M^5_{i+1,c+1}$ *if* $i + 1 \in L$ *and* $E^1_{ic} = \infty$ *otherwise, and* $E^2_{ic} = M^4_{i+1,c+1}$ *if* $i + 1 \notin U$ *and* $E^2_{ic} = M^6_{i+1,c+1}$ *otherwise.*

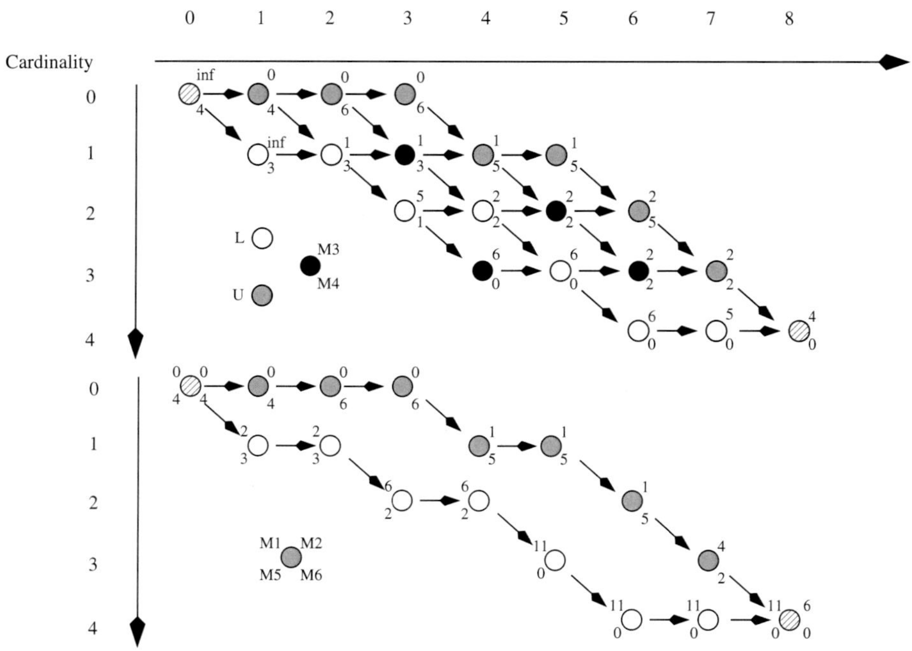

Fig. 2. The DP for the weight constraint in Example 1. The top gives values M^3 and M^4 for all nodes, the bottom M^1, M^2, M^5, M^6 for nodes on L and U. Note that node $W_{4,3}$ is valid as $\{2,3,4,6\}$ obeys the length-lex bounds and defines a path that visits this node. Edge $(W_{6,3}, W_{7,4})$ has been removed previously by the profit constraint.

Example 2. For a set variable $S \subseteq \{1,\ldots,8\}$, consider the Knapsack constraint $KP(S,(0,0,0,0,0,0,-1,1)^T,0,(2,1,4,1,5,0,3,2)^T,7)$. That is, we want a subset of items S such that $x_8 - x_7 \geq 0$ and $2x_1 + x_2 + 4x_3 + x_4 + 5x_5 + 3x_7 + 2x_8 \leq 7$, where $x_i = 1$ if $i \in S$ and $x_i = 0$ otherwise. Assume further that $D(S) = [\{1,3,5,6\},\{4,6,7,8\}]$. Propagating the profit constraint $x_7 - x_8 \leq 0$, we detect that the edge $(W_{6,3}, W_{7,4})$ cannot be used by any admissible path with weight lower or equal 0. Then, we propagate the constraint $2x_1 + x_2 + 4x_3 + x_4 + 5x_5 + 3x_7 + 2x_8 \leq 7$. Figure 2 shows the DP, the lower and upper bounds, as well as all values M_{ic}^k at this point. To compute a new lower bound, we initialize $g \leftarrow 0$ and begin at W_{00}. It holds $g + w_1 + M_{11}^5 = 0 + 2 + 3 \leq 7$, so we include 1 in L' and update $g \leftarrow 2$. Since we are still on L and $2 \notin L$, we cannot include 2 in L' and move to $W_{2,1}$. Then, $2+4+2 > 7$, so we move to $W_{3,1}$, leaving π_L. From now on we will use M^4 instead of M^5. Next, $g + w_4 + M_{4,2}^4 = 2 + 1 + 2 \leq 7$, so we include 4 in L', update $g \leftarrow 3$, and move to $W_{4,2}$. Then, $3 + 5 + 0 > 7$, so we move to $W_{5,2}$. Next, $3 + 0 + 2 \leq 7$, so we include 6 in L', update $g \leftarrow 3$, and move to $W_{6,3}$. Now, the edge $(W_{6,3}, W_{7,4})$ has been removed, so we move to $W_{7,3}$. Finally, we find $3 + 2 + 0 \leq 7$ and include 8 in L'. The new lower bound is $L' = \{1,4,6,8\}$. Note that, without the prior removal of $(W_{6,3}, W_{7,4})$, the new bound would have been $\{1,4,6,7\}$. This bound would disagree with the profit

Table 1. Test results on 100 random market split problems. We give number of variables, constraints, the average number of filtering calls, time in CPU seconds, and, in case where the number of pure bound computations is limited, the average number of choice points.

		LL		LL-red		LL-red-100			LL-red-100-noEdge		
Vars	Cons	Filt. Calls	Time	Filt. Calls	Time	CPs	Filt. Calls	Time	CPs	Filt. Calls	Time
10	2	55.1	0	47.0	0	1	47.0	0	1	109	0
20	3	34.8K	0.58	34.0K	0.57	9.94	31.3K	0.49	31.5	35.0K	0.62
30	4	14.4M	481	14.3M	475	2.9K	12.8M	298	9.3K	13.4M	426
40	5	> 500M	> 5h	> 500M	> 5h	> 100K	> 500M	>5h	> 100K	> 500M	>5h

constraint $x_7 - x_8 \leq 0$, resulting in a need to update the bound again by the profit constraint. The removal of edges and the propagation of sum constraints within the same DP graph can thus contribute to faster convergence in practice.

We test this algorithm on Market Split Problems which consist in the conjunction of several number partitioning constraints. We use randomly generated Cornuejols/ Dawande instances [2] for our tests. Our length-lex approach - which does not need to branch as a fixpoint of length-lex bounds consistency satisfies all constraints, is referred to as "LL". To strengthen the filtering power, we use redundant constraints which result from a weighted sum of the original equalities as they were introduced in [6] and later used successfully in [5]. The respective approach is called "LL-red". In practice it may be beneficial not to wait until a fixpoint of length-lex bounds consistency has been achieved. We test a third approach where we continue propagating sum constraints within the same DP until for 100 iterations only the length-lex bounds have changed, but no edge could be removed. In this case, branching becomes of course necessary. We refer to this approach as "LL-red-100". Finally, we experiment with the latter approach where we turn off the edge-removal heuristic and to which we refer as "LL-red-100-noEdge". Tests were run on on an AMD Athlon 64 3800+ 2.0GHz CPU.

We see that adding redundant constraints helps a little, and limiting the number of pure length-lex bounds improvements helps further reduce the total number of filtering calls. Moreover, our edge-removal heuristic effectively reduces the number of filtering calls, the number of choice points, and the total time needed. However, even with all these improvements the decomposition approach is not competitive at all. The non-decomposition algorithm based on approximated subset-bounds consistency from [5], run on an AMD Athlon 1.2GHz, solves on average instances with 4 constraints in less than a second while visiting less than 6 choice points, and instances with 5 constraints in less than a minute while visiting less than 60 choice points.

4 Conclusion

We rejected the conjecture from [7] that vanilla Knapsack constraint decomposition was appealing for achieving length-lex bounds consistency. In theory, it can lead to exponential filtering times even for tractable filtering problems. In practice, even when using improvements like propagating sum constraints within the same DP, it is vastly outperformed by an approach based on non-decomposed Knapsack constraint filtering.

In general, we argue that filtering algorithms achieving length-lex bounds consistency must not be evaluated based on the filtering-time alone but always have to be viewed as posing a fixpoint problem that is potentially hard in theory and in practice. Decomposing constraints for achieving length-lex bounds consistency appears particularly disadvantageous as this puts even more burden on the fixpoint algorithm.

References

1. Bordeaux, L., Hamadi, Y., Vardi, M.Y.: An Analysis of Slow Convergence in Interval Propagation. In: CP, pp. 790–797 (2007)
2. Cornuejols, G., Dawande, M.: A Class of Hard Small 0-1 Program. In: IPCO, pp. 284–293 (1998)
3. Gervet, C., Van Hentenryck, P.: Length-lex ordering for set CSPs. In: AAAI (2006)
4. Malitsky, Y., Sellmann, M., van Hoeve, W.-J.: Length-Lex Bounds Consistency for Knapsack Constraints. In: Stuckey, P.J. (ed.) CP 2008. LNCS, vol. 5202, pp. 266–281. Springer, Heidelberg (2008)
5. Sellmann, M.: The Practice of Approximated Consistency for Knapsack Constraints. In: AAAI, pp. 179–184 (2004)
6. Trick, M.: A Dynamic Programming Approach for Consistency and Propagation for Knapsack Constraints. In: CPAIOR, pp. 113–124 (2001)
7. Yip, J., Van Hentenryck, P.: Length-Lex Bound Consistency for Knapsack Constraints. In: ACM SAC (2009)

Realtime Online Solving of Quantified CSPs*

David Stynes and Kenneth N. Brown

Cork Constraint Computation Centre,
Dept of Computer Science, University College Cork, Ireland
d.stynes@4c.ucc.ie, k.brown@cs.ucc.ie

Abstract. We define Realtime Online solving of Quantified Constraint Satisfaction Problems (QCSPs) as a model for realtime online CSP solving. We use a combination of propagation, lookahead and heuristics and show how all three improve performance. For adversarial opponents we show that we can achieve promising results through good lookahead and heuristics and that a version of alpha beta pruning performs strongly. For random opponents, we show that we can frequently achieve solutions even on problems which lack a winning strategy and that we can improve our success rate by using Existential Quantified Generalised Arc Consistency, a lower level of consistency than SQGAC, to maximise pruning without removing solutions. We also consider the power of the universal opponent and show that through good heuristic selection we can generate a significantly stronger player than a static heuristic provides.

1 Introduction

Many practical decision problems are not under the control of a single decision maker. For example, in planning under uncertainty, mixed initiative planning, interactive configuration or game playing, either the external environment or another actor refines the detail of the problem as decisions are being made. Such problems can be modeled as *online* constraint satisfaction, where the problem variables must be instantiated in a fixed sequence, but where some of those variables are set externally, and the aim is to achieve a complete satisfying assignment at the end of the process. *Quantified* constraint satisfaction (QCSP) is a generalization of CSP, which also has a fixed sequence of variables, but where some of the variables are universally quantified. The aim is to find a *winning strategy*, which guarantees a complete satisfying assignment for every possible combination of values for the universal variables. QCSP can be regarded as a model for online CSP: the existential variables represent the values under our control, while the universal variables represent the externally assigned variables, and if we can find a winning strategy for the QCSP, then we can guarantee to find a solution to the online CSP. But in many online problems, including for example delivery dispatch, game playing or reservation management, we have

* This work was supported in part by Microsoft Research through the European PhD Scholarship Programme, and by the Embark Initiative of the Irish Research Council for Science, Engineering and Technology (IRCSET).

I.P. Gent (Ed.): CP 2009, LNCS 5732, pp. 771–786, 2009.

limited time in which to make each decision, and so online CSP can be extended to *realtime* online CSP. Given the time limits, it may no longer be feasible to search for a winning strategy for the corresponding QCSP. If we are to continue using QCSP as a model, then we must develop methods for finding solutions to a QCSP interactively, or online, as the universal variables are assigned, and we must do this under time constraints.

Here, we investigate realtime online solving of QCSPs. We continue to use QCSP as an idealised model of online CSP, but at each time-limited step we search partial strategies to find the best decision, in the style of game tree search. We develop the use of constraint propagation, game-tree search and ordering heuristics for finding partial strategies. We consider two types of external actors: (i) adversarial opponents, who try to prevent us finding a solution, and (ii) random solvers, which simply select random values for the variables. We develop existential generalised arc consistency, which does not prune any solutions from a QCSP, and which is particularly effective against random solvers. We develop a version of alpha-beta pruning for adversarial opponents, and a method based on weighted estimates for random solvers. We evaluate our methods empirically. On random binary QCSPs, we show that alpha-beta is most effective on large adversarial problems, while against random solvers we show that weighted estimates frequently finds solutions where no winning strategy exists. We then consider Online Bin Packing problems, requiring non-binary constraints, where the external solver generates the items to be packed into the bins. For these problems we propose some online heuristics tailored to bin packing, and we show that these heuristics outperform static heuristics. Performance is obviously affected by the quality of the opponent, and we show the effect of different strategies and heuristics for the opposing solver.

2 Background

A *Quantified Constraint Satisfaction Problem*[1] is a tuple $\mathcal{P} = (\mathcal{X}, \mathcal{D}, \mathcal{C}, \mathcal{Q})$ where $\mathcal{X} = \{X_1, \ldots, X_n\}$ is a set of variables, $\mathcal{D} = \{D_1, \ldots, D_n\}$ is their domains, $\mathcal{C} = \{C_1, \ldots, C_e\}$ is a finite set of constraints and $\mathcal{Q} = Q_1 X_1 \ldots Q_n X_n$ is a sequence of quantifiers for each variable, where each Q_i is either $\exists$ (existential) or $\forall$ (universal). We define a *Normal Form* QCSP as one with a strictly alternating sequence of quantifiers, starting with $\forall$ and ending with $\exists$. Each constraint C acts on its *scope*, an ordered subset of the variables $S_C = (X_i, \ldots, X_j)$, where $C \subseteq D_i \times \ldots \times D_j$. A *solution* is a tuple containing a value assignment for each variable in $\mathcal{X}$ which satisfies all of the constraints in $\mathcal{C}$. A *strategy* is a tree of value assignments which assigns a value for each existential variable for all possible sequences of assignments to the preceding universal variables. If every path in the strategy tree generates a solution tuple, it is a *winning strategy*. A QCSP is satisfiable if and only if a winning strategy exists, i.e. if we can guarantee to reach a solution no matter what values the universal variables take. Determining whether a winning strategy exists is PSPACE-complete[2]. Note that a QCSP may have solutions which are not part of any winning strategy. When generating binary QCSP problems, it is common[3] to generate only constraints where the second variable is existentially quantified (i.e. $\exists\exists$ and $\forall\exists$ constraints), since other constraints can be removed during preprocessing.

For QCSP, a constraint C is *Strongly Quantified Generalised Arc Consistent* (SQGAC)[4] iff for each variable $X_i \in S_C$ and value $a \in D_i$, a vertex labeled $X_i \leftarrow a$ is contained in M, where M is a multiple winning strategy tree representing the union of all winning strategies for the constraint C. Any values not in the tree are pruned from their domains, and following each domain reduction the tree is updated to be a valid representation of all remaining winning strategies. SQGAC applied to binary QCSPs reduces to arc consistency for QCSPs[5] which we shall refer to as *Quantified Arc Consistency*(QAC), and can be implemented without a complex tree structure.

Gent *et al.*[3] proposed QCSP-Solve as a solver for QCSP, and included a *Pure Value Rule* for binary constraints, later extended to cover non-binary constraints[4]. A value a for variable X_i is *pure* if and only if all possible tuples with $X_i \leftarrow a$ are actual solutions to $\mathcal{P}_i$, the sub-problem containing only all constraints over the variable X_i. As domains are reduced during search, values may dynamically become pure. Existential pure values can be instantly assigned, while universal pure values can be safely pruned from their domains (assuming one other value still remains) to reduce search. Value ordering heuristics have been developed for QCSP, including Dynamic Geelen's Promise[6,7].

A difficulty in modeling using QCSPs is that in many problems some values for the universal variables are only legal depending on preceding decisions, and so no winning strategy is possible. To handle such problem types, Strategic CSPs [8] extend QCSPs by allowing universal variables to adapt their domains to be compatible with previous choices. Alternatively, QCSP$^+$[9] introduces restricted quantification to state when values are legal in the universal domains. [10] proposes backpropagation methods for value ordering in QCSP$^+$. To use QCSP-Solve on such problems, *shadow variables*[4] and associated constraints can be introduced, which make universal values pure when they are no longer legal choices, and thus they are removed from the search.

Dynamic Constraint Satisfaction [11,12] considers problems that change over time, by the addition, retraction or modification of variables, domains and constraints. Formalisms that specifically focus on problems that progress by the assignment of values to variables include Mixed CSP [13] and Stochastic CSP [14], although in the latter case probability distributions are associated with the assignment of each uncontrolled variable. Sampling methods have been used[15,16] to solve Online Stochastic Optimization problems under time constraints, in which problems are gradually revealed, although again with an assumption that there is some model of the likely growth. Finally, Groetschel [17] considers the general problem of Realtime Online Combinatorial Optimization.

In AI game playing [18], the game is represented as a tree of game states. Players take turns making moves, which are a transition from one state in the tree to another. Players perform game-tree search to determine their best option, but it is typically infeasible for the player to search the entire tree. Players are forced to form estimates of which move will lead them to a win. In practice, a subtree of limited depth is generated and the leaf states are evaluated according to a problem specific heuristic function. These values are then propagated back up the tree to the current root state, in order for the player to make a decision. For adversarial zero sum games, most algorithms for performing this propagation are based upon the *minimax* heuristic [18], which states that a player should make the choice which minimises the (estimated) maximum that the opponent can achieve. When propagating up the tree, a state in which it is the player's move will take the value of its child with the highest estimate, since that is the minimum for the opponent. Conversely, a state in which it is the opponent's move will take the value of its child with the lowest estimate. The best known algorithms use variants of *Alpha-Beta Pruning*[19], which uses minimax based reasoning to prune moves which cannot improve on already discovered scores. [20] applied game-tree search in what they call Adversarial CSP, in which solving agents take turns to choose instantiations for variables in a shared CSP.

In a *bin packing* problem [21], we are given a list of items of varying size, and required to place them into a minimal number of fixed capacity bins without causing any of the bins to overflow. In an *online bin packing* problem, we must permanently assign each incoming item to a bin with no knowledge of the future remaining items to come. Two of the simplest heuristics to achieve this are First Fit (FF) and Best Fit (BF). First fit tries to place a packet into the first bin it can fit into. Best Fit places the packet into whatever bin will have the least space remaining after inserting the packet, and is equivalent to ordering the bins in descending fullness and then applying FF. Best Fit is known[22] to have worst case performance of 1.7 times as many bins as an off-line optimal algorithm, where the theoretical best possible by any online algorithm is 1.54, and an average waste of $O(n^{1/2}log^{3/4}n)$ bins.

3 Realtime Online Solving of QCSP

When solving realtime online CSPs using QCSP as a model, we treat the QCSP as a two-player game, in which one player (the existential player) assigns values to the existentially quantified variables, and the other (the universal player) assigns values to the universally quantified variables. The variables are assigned in the order of the quantifier sequence, and a time limit is imposed on each decision. For the existential the objective is to reach a solution, while for an adversarial universal it is to cause a failure (i.e. prevent a solution being reached). Formally, we define Realtime Online solving of QCSP for the existential as:

Definition 1 (Existential RO-QCSP). *Given a normal form QCSP $\mathcal{P}$, an increasing sequence of time points $t_1, t_2, \ldots, t_n$, and a sequence of values $v_1, v_3, v_5, \ldots, v_{n-1}$ such that each value v_j is in D_j and is revealed at time t_j, generate at each time t_k for $k = 2, 4, 6, \ldots, n$ a value $v_k \in D_k$ such that the tuple $(v_1, v_2, \ldots, v_n)$ is a solution for $\mathcal{P}$.*

When choosing each value v_k, the existential player has a known time limit for making the decision. A competent player should reason about the best value to select, taking into account the possible future actions of the other player. If the time limit is sufficiently large, the first player can search for a winning strategy, and if it finds one it can simply execute this for each successive decision. However, we assume that finding a winning strategy will be initially infeasible, and instead we will generate partial strategies. The player looks ahead at possible future moves of both itself and the opponent, performing a partial exploration of the search tree. Different lookahead methods determine which area of the tree is explored. While exploring the tree, constraint propagation prunes unwanted branches. The player heuristically evaluates nodes as they are generated and propagates the evaluations back up to the root node. Once the time limit is reached, it selects the root value with the highest evaluation.

3.1 Constraint Propagation

The strongest level of consistency we consider is SQGAC. However, depending on the type of opponent, maintaining arc consistency can be detrimental, since our aim is simply to find any solution, and not necessarily a winning strategy. Consider a sequence of quantified variables $\exists X_1 \forall X_2 \exists X_3$, with domains $D_1 = D_3 = \{b, c\}$ and $D_2 = \{a, b\}$, and constraints $X_1 = X_3$ and $X_2 \neq X_3$. If we maintain arc consistency, then assigning $X_1 = b$ will remove c from D_3, causing b to be removed from D_2. At this point, we backtrack, since no winning strategy is now possible. For online problem solving with an adversarial opponent this is sensible, since when it reaches variable X_2 after $X_1 = b$, it is simple to detect that $X_2 = b$ removes all options for X_3 and thus the adversary would win. However, when a random opponent reaches X_2 after $X_1 = b$, it may still choose $X_2 = a$, and thus a solution is still possible. In this case, maintaining arc consistency may prevent us finding a solution. Therefore, to avoid losing solutions against random opponents but to keep some of the propagation power of arc consistency, we introduce *Existential Quantified Generalised Arc Consistency* (EQGAC):

Definition 2 (EQGAC). *A QCSP is* Existential Quantified Generalised Arc Consistent *(EQGAC) if for every X_i with $Q_i = \exists$, for all constraints C with $X_i \in S_C, \forall a \in D_i, \exists$ a tuple $t \in C$, s.t. each tuple element $t_j \in D_j$ and $t_i = a$.*

When all constraints are binary, EQGAC reduces to EQAC (*Existential Quantified Arc Consistency)*. Maintaining EQAC or EQGAC does not remove solutions from the problem. When maintaining EQAC, all $\exists\exists$ constraints are propagated like standard constraints in a CSP with MAC, while propagating $\forall\exists$ constraints never prunes values from the universal domain. Maintaining EQGAC uses the SQGAC algorithm[4], except we never place universal values on the remove list, and when removing a universal value from the tree, we do not remove sibling values, maintaining multiple *solution* trees, and not winning strategy trees. Maintaining SQGAC or EQGAC constructs a large tree in a preprocessing step, and as problem sizes increase, it becomes extremely expensive in time and space to construct and maintain these trees. To test the impact of this processing cost, we also implement solvers using forward checking. For non-binary QCSPs we extend nFC0 [23] to QnFC0, which provides much weaker propagation, but requires no preprocessing and no significant extra data structures:

QnFC0: After assigning the current variable, achieve arc consistency on all constraints involving the current variable, past variables and exactly one future variable. If the future variable is existentially quantified, if the domain is not emptied, continue with a new variable, otherwise backtrack. If the future variable is universally quantified, if any value is removed from the domain, backtrack immediately, otherwise continue with a new variable.

3.2 Lookahead and Heuristics

The lookahead strategy and heuristic determine the sub-tree explored for each decision. Each strategy implements the general *lookahead* algorithm (Alg 1). *Nodes* is the main data structure, containing the unexpanded nodes in the search tree: each node represents a partial instantiation of the variables, in order, with the domains of uninstantiated variables reduced by propagation, and is $n_i(X_i, v_i, \sigma_i, p_i)$, recording the last instantiated variable, its value, the domains and propagation information, and its parent node. Initially *Nodes* contains a single node $n_0(\phi, \phi, \sigma_0, \phi)$, where σ_0 is the current state of the QCSP. We also maintain *Store*, a global store of evaluated nodes, initially empty. Different strategies implement *Nodes* differently, and are explained below.

Algorithm 1. lookahead(N,S) - the basic lookahead algorithm

Input: Nodes,Store
Data: τ, ϵ ; // empty data structures
$n_i \leftarrow$ select-and-remove-node(Nodes) ; // select a node to expand
if *$X_j \leftarrow$ next-variable(n_i) is not null* **then**
  $\sigma_i \leftarrow$ pure-value-rule(X_j, σ_i) ; // apply PV rule to X_j
  for *each value w in X_j's domain* **do**
    $\sigma_w \leftarrow$ assign-and-propagate(X_j, w, σ_i) ; // propagate the assignment
    $\tau \leftarrow \tau + n_j(X_j, w, \sigma_w, n_i)$; // record the new state
    $e_w \leftarrow$ evaluate(σ_w) ; // evaluate by inspecting domains
    $\epsilon \leftarrow \epsilon + (X_j, w, e_w)$; // record the evaluation
  Store $\leftarrow$ Store + $n_i(X_i, v_i, \epsilon, p_i)$; // store the old evaluated node
  prop-eval(ϵ,n_i,Store) ; // propagate evaluations upwards
  Nodes $\leftarrow$ Nodes $+\tau$; // add new nodes to data structure
if *Nodes is not empty and time remaining* **then**
  lookahead(Nodes, Store) ; // continue expanding nodes

Depth First(DF): Uses a stack; children are pushed onto the stack in order of generation; the next node to be expanded is popped off the top.

Breadth First(BrF): Uses a queue; children are added to the end in order of generation; the next node to be expanded is taken from the front.

Best First: Uses an ordered list; children are evaluated and inserted into the appropriate position; the next node to be expanded is taken from the front (with highest evaluation).

Partial Best First(PBF): Best First can perform poorly against adversarial opponents, as the best move for a universal variable is the worst for an existential variable and vice versa. PBF is a modification which behaves as best first for existential nodes, but when expanding a universal node the best child for the universal (lowest scored evaluation) will be immediately explored and the remaining children have their estimation negated and then are added to the ordered list.

Alpha Beta Pruning(AB): As depth first but using alpha and beta bounds to prune parts of the search tree which would never be reached, identified when the node is popped off the stack.

Intelligent Depth First(IDF): As DF, but where children are ordered from best to worst before being pushed onto the stack.

Intelligent Alpha Beta(IAB): As AB, but with children ordered as in IDF.

AB and IAB do not search to the bottom of the tree since they are too time consuming and would not finish within the time limit. Instead, we perform an iteratively deepening form of AB which searches to a fixed depth limit and then increases that depth limit before performing AB lookahead again. We continue until we have searched to the final variable (maximum depth) or we have run out of time. In our tests we set the initial depth limit to 2 and at each iteration we increase the depth limit by 1.

For general heuristics to evaluate states, we use *Dynamic Geelen's Promise* (DGP)[7] which is the product of the future existential domain sizes and was shown to be a good value ordering heuristic for solving QCSPs. Since we are looking ahead we need to compare states at different depths of the search tree, where the heuristic evaluations are the product of different numbers of future existential domains, and thus the DGP evaluations are incomparable. We present two different modifications to DGP for achieving that. *Proportional Promise*(PP) is calculated as DGP divided by the product of the original sizes of those future domains. The *Geometric Mean*(GM) of a heuristic is calculated as the nth root of the evaluation given by the heuristic. For random QCSPs, n is the number of future existential domains. For the online bin packing problems, we introduce two heuristics based upon the principals of First Fit (FF) and Best Fit (BF). The *Ordered Fitting* (OF) heuristic is based on First Fit, and prefers states in which the first bin is the most filled, the second bin is the second most filled, etc.. For a problem with a set of k bins, $b = \{b_1, b_2, .., b_k\}$ of maximum capacity c, where f_i is how full the ith bin is, we calculate OF as $\Sigma_{i=1}^{k} f_i * c^{k-i}$. The *Heavily Filled* (HF) heuristic is based on Best Fit, and prefers states in which bins are as highly filled as possible and the rest empty, to those in which many bins are only partially filled. We calculate HF as $\Sigma_{i=1}^{k} \sqrt{f_i/c}$. For applying GM in bin packing, n is the number of packets which have already arrived (since OF/HF evaluations increase with depth, unlike DGP evaluations which decrease).

To propagate heuristic evaluations back up the tree, we use either Minimax, or what we term *Weighted Estimates*(WE) reasoning. At each level, we switch the method of aggregating child scores depending on the quantifier of the parent variable. In a node where an existential variable's valuation is being decided,

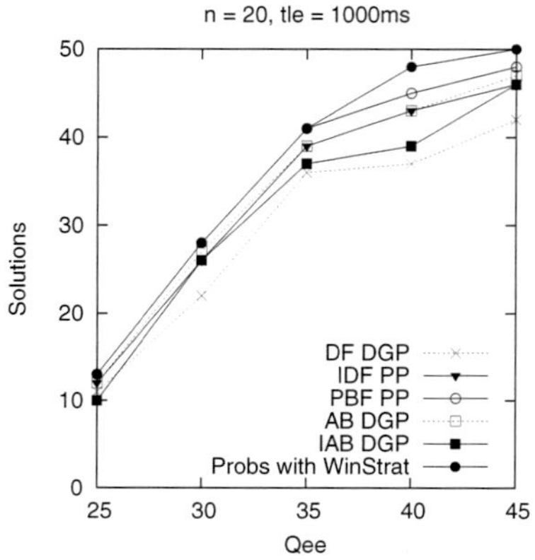

Fig. 1. Random RO-QCSP against universal using AB with no time limit

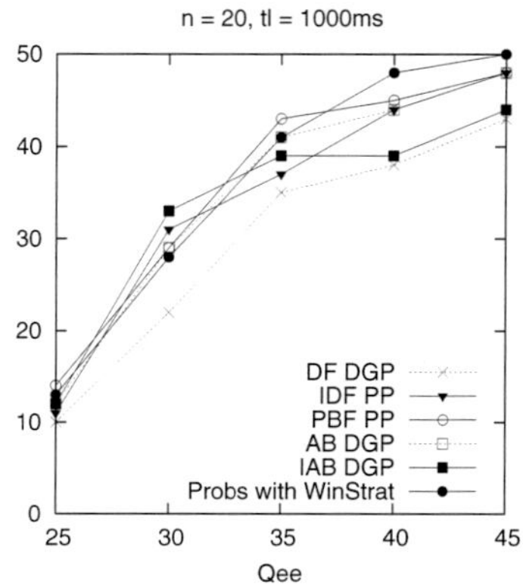

Fig. 2. Random RO-QCSP against universal using DF with time limit

both methods select the highest of the children's valuations. In a node where a universal variable's valuation is being decided, minimax selects the lowest of the children's valuations, while Weighted Estimates computes the average of the children's valuations. While minimax reports that certain values lead to failure, weighted estimates gives an indication of how likely it is an unintelligent opponent can pick a value which causes domain wipe outs.

4 Experiments: Random Binary QCSPs

We tested on randomly generated binary QCSPs with a strictly alternating sequence of $\exists$ and $\forall$ quantifiers, using the flawless generator described in Section 6 of [7]. In all our tests, the block size = 1, domain size = 8, constraint density = 0.20, $q_{\forall\exists} = 1/2$. $q_{\forall\exists}[3]$ is looseness for $\forall\exists$-constraints, $q_{\exists\exists}$ is the looseness of $\exists\exists$-constraints, and n is the number of variables. We denote as tl the time limit for both players, and tle the time limit for the existential. For each value of $Q_{\exists\exists}$ we generated 50 random problems. We measure the number of solutions reached. We omit many of the combinations of lookahead methods, propagation and heuristics from our graphs for clarity. Unless explicitly stated to be using Weighted Estimates(WE), all strategies are using Minimax. Both participants receive the same time limit, tl, per move unless explicitly stated otherwise. For the smaller problems with 20 variables we also show how many problems contained a winning strategy. For larger problems, $n = 51$, finding a winning strategy in reasonable time was not possible. Experiments ran on a 2.0GHz Pentium with 512MB RAM.

Figure 1 shows the number of solutions against an opponent performing a complete lookahead with infinite time on problems with $n = 20$. This is the best a universal player is able to perform - if there is no winning strategy for

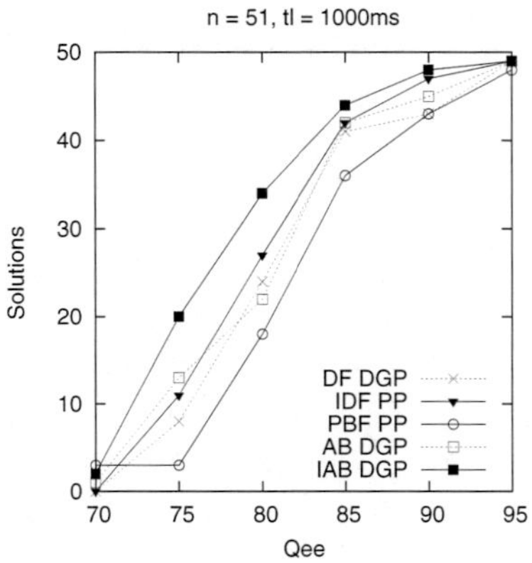

Fig. 3. Random RO-QCSP against universal using DF

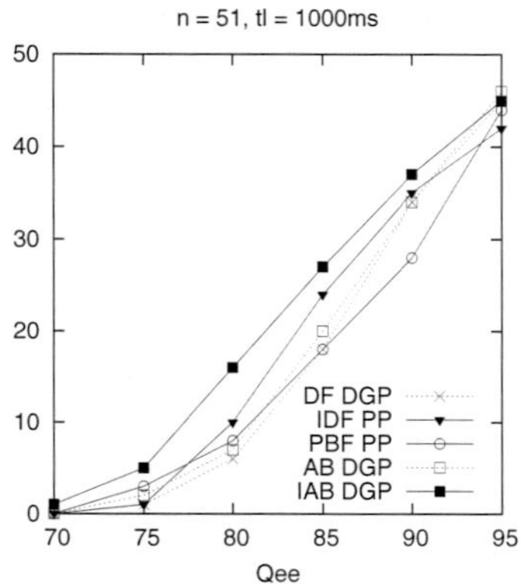

Fig. 4. Random RO-QCSP against universal using DGP AB

the QCSP or if the existential makes any choice which is not part of a winning strategy, the universal will win. We see that our time-limited existential solvers are still able to find most of the winning strategies, with PBF-PP finding them all for lower levels of Q_{EE}. Figure 2 shows performance against a time-limited universal opponent using depth-first lookahead, also with $n = 20$. Against this weaker universal, the existentials improve, and in a number of cases at lower Q_{EE} achieve solutions even when the QCSP has no winning strategy. In both figures, we see that PBF-PP outperforms AB and IAB; we believe that on these relatively small problems PBF-PP is able to search enough of the space to make intelligent decisions, while AB and IAB's higher overhead restricts their search. Figures 3 and 4 show performance on larger problems against weak and strong opponents respectively, and we see that IAB is now outperforming PBF-PP, finding solutions for up to five times as many problems at lower Q_{EE}. The value ordering used by IAB also allows it to prune the search space faster than AB, and thus enables it to search to a deeper level and make more informed decisions.

For problems against a random opponent, again we set $n = 20$, but reduce the time limit to 500ms, as it is significantly easier for the existential player to succeed. In Figure 5 the existential maintains QAC, while in Figure 6 it maintains EQAC. In both cases, IAB and methods using WE find many solutions even when the problem has no winning strategy. The WE approaches, which estimate the likelihood of an opponent picking a particular value, benefit significantly from the use of EQAC, and outperform IAB in Figure 6.

5 Modeling Online Bin Packing

We test on two types of Online Bin Packing problems in which the universal player selects packet sizes, while the existential player attempts to place them in

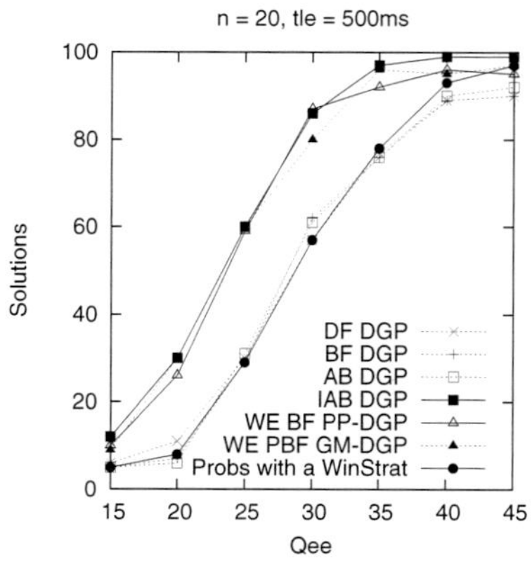

Fig. 5. Random RO-QCSP using QAC against random universal

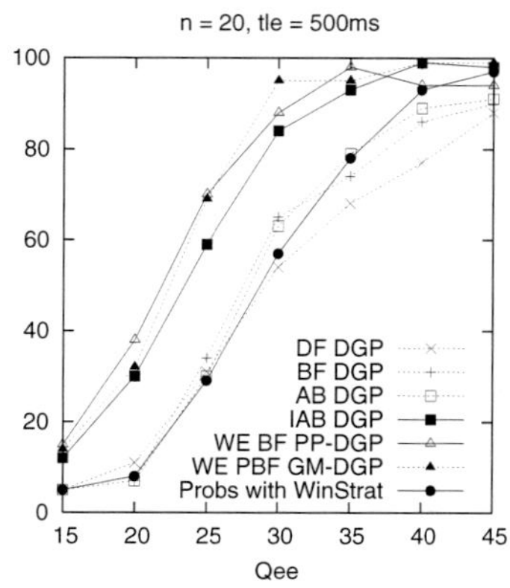

Fig. 6. Random RO-QCSP using EQAC against random universal

the bins. We present the models in an abstract form, in which we assume pruning the universals is unrestricted, to provide a clear and concise description. The footnotes describe how we transform this to a correct quantified form, through use of shadow variables and pure value pruning. The basic model is common to both problems and we describe it first. A known number of packets will be chosen from a limited set, and there is a fixed number of bins, each of the same capacity. We use state variables for each bin to record how much capacity it has left, we have decision variables for the universal which determine the size of each packet, and decision variables for the existential to state into which bin the current packet will be placed. Note that no lookahead is performed before assigning state variables, since they are uniquely determined by previous choices. As an example, the variables for the jth packet choice are

$$\exists a_{(j-1)b_1} \exists a_{(j-1)b_2} \ldots \exists a_{(j-1)b_k} \forall p_j \exists l_j \exists a_{(j)b_1} \exists a_{(j)b_2} \ldots \exists a_{(j)b_k}$$

where $a_{(j-1)b_i}$ is the state of bin b_i before the jth packet arrives, p_j is the size of the jth packet, l_j is the bin the jth packet is placed into, and $a_{(j)b_i}$ is the state of bin b_i after the jth packet has been placed. The following constraints for each j and i ensure the state variables are consistent[1]:

$$(l_j = b_i) \Rightarrow a_{(j)b_i} = a_{(j-1)b_i} - p_j$$
$$(l_j \neq b_i) \Rightarrow a_{(j)b_i} = a_{(j-1)b_i}$$

[1] Note that this describes the abstract model. The limited set of possible packets constrains the universal choices, and so in the implementation we use shadow variables to render illegal universal values pure, and modify the constraints accordingly. Each universal variable p_j has an existential shadow variable sp_j placed immediately after it in the variable sequence (and they will be linked in later constraints). Thus the variables for the jth packet choice become $\ldots \exists a_{(j-1)b_k} \forall p_j \exists sp_j \exists l_j \ldots$, and we replace the first constraint by $(l_j = b_i) \Rightarrow a_{(j)b_i} = a_{(j-1)b_i} - sp_j$.

5.1 Type 1 Problems

In type 1 problems, the universal player has a fixed set of m packets, for which the sum of their sizes is the value B, and must decide on which order to provide them to the existential player. By testing with both a random and an adversarial universal we can evaluate the existential's performance against both average case and worst-case order scenarios for randomly generated sets of packets. The existential player does not know the sizes of the packets before they arrive, but does know the upper bound on the sum of their sizes. Thus the existential is less informed than the universal and the two actually assign values in slightly different synchronised problem models.

We represent the fixed set of packets with a single global cardinality constraint over the p_j variables[2]:

$$gcc(p_1, p_2, \ldots, p_m, c_{s_1}, c_{s_2}, \ldots, c_{s_t}).$$

The domain for each p_j is a set $\{s_1, s_2, ..., s_t\}$ of possible sizes, and the c_{s_i} state exactly how many of the p_j must take each value s_i. Note that the c_{s_i} in this case are constants, rather than constrained variables. The existential player does not see the gcc constraint; instead it sees a less restrictive global sum constraint[3]:

$$\Sigma_{j=1}^{m} p_j \leq B$$

In the Type-1 problems, our aim is to show how the existential player can improve over strategies like First Fit or Best Fit, even when its perception of the problem is more restricted than that of the opponent.

5.2 Type 2 Problems

In the second type of problems, the universal and existential players both share the same problem. This time the list of packets for the universal is *larger* than

---

[2] The universal value must respect the gcc constraint. Instead of posting the gcc constraint, for each p_i we post an extensional constraint with scope $(sp_1, sp_2, ..., sp_{i-1}, p_i, sp_i)$, such that each possible tuple satisfies the following rule: if the values for the sp_j should disallow a value v for p_i ($p_i \leftarrow v$ would cause a violation of the gcc constraint), then $p_i \leftarrow v$ is compatible with all values of sp_i, and otherwise, $p_i = sp_i$. Thus as soon as choices for some p_j disallow a value v for p_i, v becomes pure; immediately before p_i is to be assigned, the pure value rule removes v from its domain.

[3] As before, we replace this with a shadow variable form for each p_i with scope $(sp_1, sp_2, ..., sp_{i-1}, p_i, sp_i)$, as an extensional constraint which implements the rule: if the values of the sp_j plus the number of remaining packets would disallow v for p_i ($p_i \leftarrow v$ would cause the sum to exceed B), then $p_i \leftarrow v$ is compatible with all values of sp_i, and otherwise $p_i = sp_i$. Note that this never makes a value pure if it has not also been made pure in the universal's problem model.

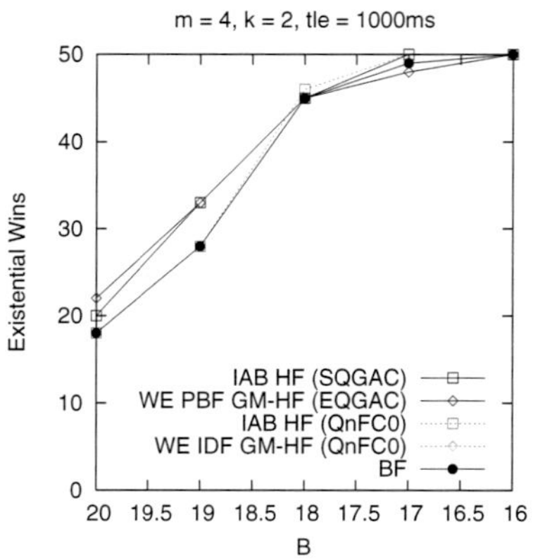

Fig. 7. Bin packing type I against random universal

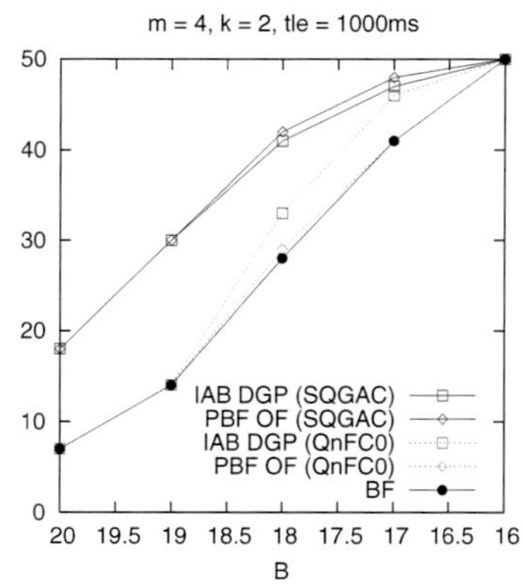

Fig. 8. Bin packing type I against universal using IAB OF

the number of packets it must pick. However, the subset of packets it can pick is restricted by an upper bound, B, on their combined size[4]:

$$gcc(p_1, p_2, \ldots, p_m, v_{s_1}, v_{s_2}, \ldots, v_{s_t})$$
$$\Sigma_{j=1}^{m} p_j \leq \mathrm{B}$$

where the v_{s_i} are now variables, each of which has its own upper bound and a lower bound of 0. In these type-2 problems, the universal has more freedom as to what values to pick and in what order, and our aim is to show that by exploiting constraint propagation, lookahead and heuristics the universal can have a significant effect on the success rate of the existential.

6 Experiments: Online Bin Packing

In our Online Bin Packing problems, packets range in size from 1 to 10 and each bin's capacity is 10. We test 50 problems, each with a different randomly generated list of packets, at each of the different upper bounds on the sum of the sizes of all packets. The size of the list of packets in Type 2 problems is twice the number of incoming packets. The number of incoming packets is m, the number of bins is k, and the time limit per decision for the existential is tle. In all problems the universal has 1000ms per decision. Again, many of the lookaheads and heuristics are omitted from these graphs for clarity. In general, we plot the best AB-based heuristic+lookahead combination, and the best non-AB heuristic+lookahead combination, with other relevant combinations shown

[4] The shadow variable form is an extensional constraint for each p_i with scope $(sp_1, sp_2, ..., sp_{i-1}, p_i, sp_i)$ such that if the sp_j values disallow v for p_i (due to the gcc constraint or the upper bound constraint), then $p_i \leftarrow v$ is compatible with all values of sp_i, and otherwise $p_i = sp_i$.

when appropriate. As a baseline for our tests, we compare our results against an existential using Best Fit, as it is always at least as good as First Fit.

Figures 7 and 8 show the results for small Type 1 problems against Random and IAB OF universals respectively. Against a random opponent we see that the simple BF strategy does well, and our combination of propagation, heuristics and lookahead only achieves a small performance gain over it. Against an adversarial opponent however we see that we can perform significantly better than BF. We also note that the existential using QnFC0 propagation instead of SQGAC or EQGAC performs worse due to the reduced amount of propagation. In all these tests the universal is using SQGAC.

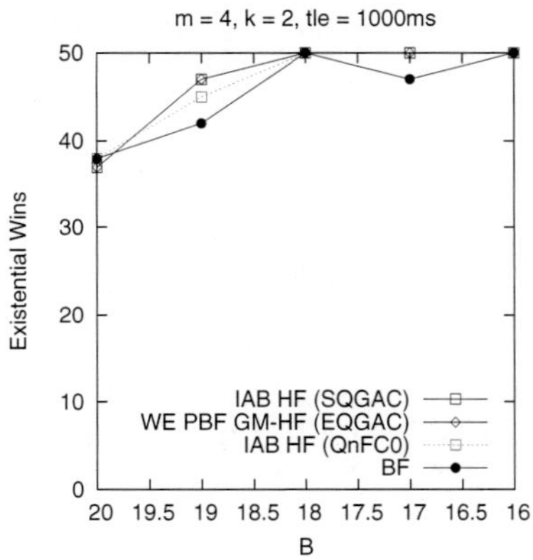

Fig. 9. Bin packing type II against random universal

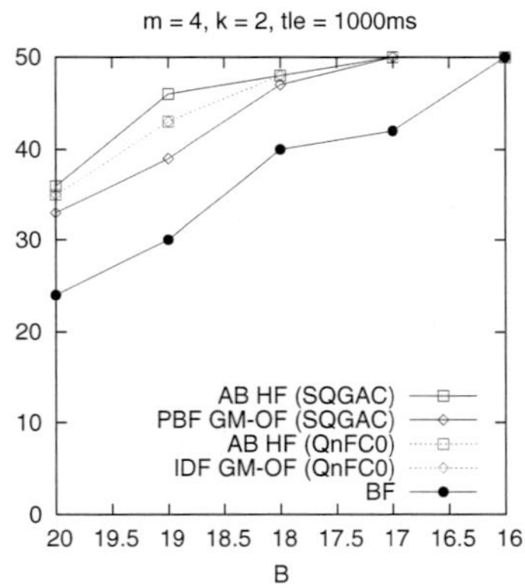

Fig. 10. Bin packing type II against universal using IAB OF

Figures 9 and 10 show the results for small Type 2 problems. The relative performance is similar to Type 1. In Figure 11 we increase the problem size, and use QnFC0 propagation for every player, as SQGAC is too slow on large problems. These larger problems reveal an flaw with IAB OF for the universal, and IAB DGP for the existential player exploits it, obtaining exceptionally good results. A universal using IAB OF essentially assumes the existential wants to maximise the content of the first bin, and so initially picks a small packet to be placed into it. However, IAB DGP's implementation places the first packet into the final bin, so the universal ends up continually feeding small packets expecting them to be placed into the first bin, making it easy for the existential to win.

To overcome this flaw, we develop a new heuristic intended for the universal called *MinSpace*(MS). MS tries to leave a minimal non-zero empty space in each bin. By leaving these small gaps, it makes it hard for the existential to succeed at high upper bounds. We calculate the MS measure using $\Sigma_{i=1}^{k} g(i)$, where $g(i)=0$ when $c - f_i = 0$, and $g(i)=\sqrt{(c - f_i)/c}$ otherwise. Figure 12 shows a universal using MS on the same problems as in Fig. 11. As can be seen the performance of the universal is drastically improved. When both the existential and universal have 1000ms time limits, the existential struggles to do well against MS and only achieves close to the performance of BF, as shown by AB OF (1000ms). The remainder of the plots in Figure 12 show how well we can do when the existential's time limit is raised to 5000ms. With this additional time on these larger problems, we can achieve many more solutions than BF.

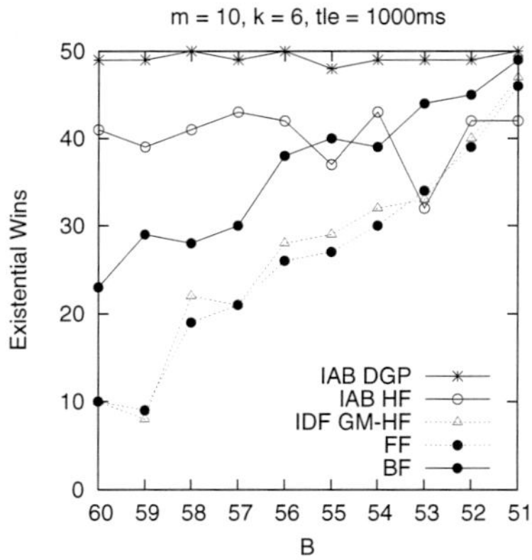

Fig. 11. Bin packing type II against universal using IAB OF

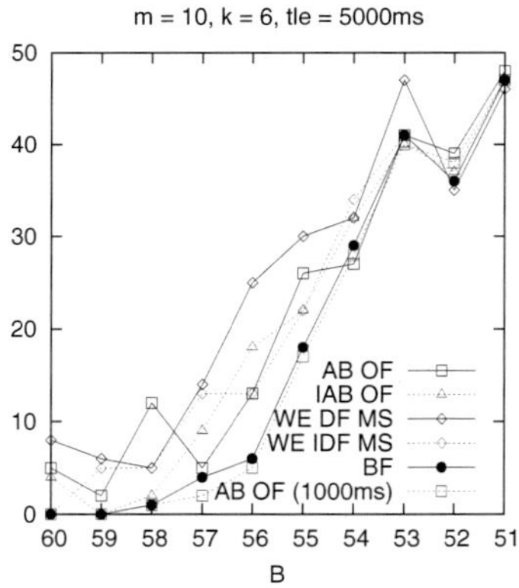

Fig. 12. Bin packing type II against universal using IAB MS

We also compared against a universal using a policy of picking the largest packet it can at each turn. Due to the nature of this universal policy, our choice of existential strategy has no effect until far into the packet stream and BF and most of the heuristics perform almost identically against it. A universal using IAB MS performs consistently significantly better than this Largest First approach at all upper bounds.

7 Conclusions and Future Work

Quantified CSPs can be used as a model for solving online CSPs, generating winning strategies in advance. But when decisions in the online problem have to be made in realtime, complete solving of a QCSP is infeasible. We have developed techniques for realtime online solving of QCSP using a combination of propagation, lookahead and heuristics, for online CSPs involving both adversarial opponents and random external solvers. We have proposed existential quantified generalised arc consistency for handling random solver opponents, which allows us to achieve solutions even when the underlying QCSP has no winning strategy. We have demonstrated that a version of alpha-beta pruning with a constraint-based value-ordering heuristic outperforms other heuristics on large binary QCSPs against adversarial opponents. We have developed a non-binary constraint model of Online Bin Packing, and we have shown that with good heuristic selection, a significantly stronger universal player can be generated using our reasoning, but that against even a strong opponent the existential reasoning can help us reach more solutions.

In future work we will consider weaker consistency levels than SQGAC or EQGAC to avoid the large overhead, we will investigate the use of other forms of quantified constraint problems for realtime online problem solving, and we will consider solution methods based on sampling.

References

1. Bordeaux, L., Cadoli, M., Mancini, T.: CSP Properties for Quantified Constraints: Definitions and Complexity. In: Proceedings of AAAI, pp. 360–365 (2005)
2. Börner, F., Bulatov, A., Jeavons, P., Krokhin, A.: Quantified constraints: Algorithms and complexity. In: Baaz, M., Makowsky, J.A. (eds.) CSL 2003. LNCS, vol. 2803, pp. 58–70. Springer, Heidelberg (2003)
3. Gent, I.P., Nightingale, P., Stergiou, K.: QCSP-Solve: A solver for quantified constraint satisfaction problems. In: Proceedings of IJCAI, pp. 138–143 (2005)
4. Nightingale, P.: Consistency and the Quantified Constraint Satisfaction Problem. PhD thesis, University of St Andrews (2007)
5. Bordeaux, L., Monfroy, E.: Beyond NP: Arc-consistency for quantified constraints. In: Van Hentenryck, P. (ed.) CP 2002. LNCS, vol. 2470, pp. 371–386. Springer, Heidelberg (2002)
6. Stynes, D., Brown, K.N.: Value Ordering for Quantified CSPs. In: Proceedings of CP2007 Doctoral Programme, pp. 157–162 (2007)
7. Stynes, D., Brown, K.N.: Value Ordering for Quantified CSPs. Constraints 14(1), 16–37 (2009)
8. Bessiere, C., Verger, G.: Strategic constraint satisfaction problems. In: Proceedings of CP Workshop on Modelling and Reformulation, pp. 17–29 (2006)
9. Benedetti, M., Lallouet, A., Vautard, J.: QCSP made Practical by Virtue of Restricted Quantification. In: Proceedings of IJCAI, pp. 38–43 (2007)
10. Verger, G., Bessiere, C.: Guiding Search in QCSP^+ with Back-Propagation. In: Stuckey, P.J. (ed.) CP 2008. LNCS, vol. 5202, pp. 175–189. Springer, Heidelberg (2008)
11. Dechter, R., Dechter, A.: Belief maintenance in dynamic constraint networks. In: Proceedings of AAAI, pp. 37–42 (1988)
12. Brown, K.N., Miguel, I.: Uncertainty and Change. In: Handbook of Constraint Programming, ch. 21, pp. 731–760 (2006)
13. Fargier, H., Lang, J., Schiex, T.: Mixed constraint satisfaction: a frameworkfor decision problems under incomplete knowledge. In: Proceedings of AAAI, pp. 175–180 (1996)
14. Walsh, T.: Stochastic constraint programming. In: Proceedings of ECAI, pp. 111–115 (2002)
15. Bent, R., van Hentenryck, P.: Regrets only! online stochastic optimization undertime constraints. In: Proceedings of AAAI, pp. 501–506 (2004)
16. Hentenryck, P.V., Bent, R.: Online Stochastic Combinatorial Optimization. The MIT Press, Cambridge (2006)
17. Grötschel, M., Krumke, S.O., Rambau, J., Winter, T., Zimmermann, U.T.: Combinatorial Online Optimization in Real Time. Online Optimization of Large Scale Systems, 679–704 (2001)

18. Shannon, C.E.: Programming a computer for playing chess. Philosophical Magazine (Series 7), 256–275 (1950)
19. Knuth, D.E., Moore, R.W.: An Analysis of Alpha-Beta Pruning. Artificial Intelligence 6(4), 293–326 (1975)
20. Brown, K.N., Little, J., Creed, P.J., Freuder, E.C.: Adversarial constraint satisfaction by game-tree search. In: Proceedings of ECAI, pp. 151–155 (2004)
21. Johnson, D.S.: Fast Algorithms for Bin Packing. Journal of Computing and System Sciences 8(3), 272–314 (1974)
22. Version, P., Kenyon, C., Rabani, Y., Sinclair, A.: Biased Random Walks, Lyapunov Functions, and Stochastic Analysis of Best Fit Bin Packing. J. Algorithms, 351–358 (1998)
23. Bessière, C., Meseguer, P., Freuder, E.C., Larrosa, J.: On forward checking for non-binary constraint satisfaction. In: Jaffar, J. (ed.) CP 1999. LNCS, vol. 1713, pp. 88–102. Springer, Heidelberg (1999)

Constraint-Based Local Search for the Automatic Generation of Architectural Tests

Pascal Van Hentenryck[1], Carleton Coffrin[1], and Boris Gutkovich[2]

[1] Brown University, Providence RI 02912, USA
[2] Intel Corporation, Haifa, Israel

Abstract. This paper considers the automatic generation of architectural tests (ATGP), a fundamental problem in processor validation. ATGPs are complex conditional constraint satisfaction problems which typically feature both hard and soft constraints and very large domains (e.g., all memory addresses). Moreover, the goal is to generate a large number of diverse solutions under tight runtime constraints. To improve solution diversity, this paper proposes a novel approach to ATGPs by modeling them as MaxDiverse$k$Set problems and solving them with constraint-based local search over conditional variables. The paper presents the semantics and implementation of conditional variables in this context and demonstrates the computational benefits of the approach.

1 Background and Motivation

The automatic generation of architectural tests is a fundamental and complex problem in processor design. It consists of generating random sequences of instructions obeying specified constraints. The complexity of this process prevents the problem from being represented and solved globally. Instead the problem is traditionally solved by an incremental process (see Figure 1) which generates one instruction at a time and transforms the constraints on the sequence into constraints on the test generation of single instructions.

This paper considers the main step in this process: the single instruction generator. This automatic test generation problem (ATGP) can be viewed as a constraint satisfaction problem involving three types of constraints:

1. **Architectural Constraints** that specify the instruction set, i.e., which instructions are valid.
2. **Test Scenario Constraints** that specify the intention of a validation engineer.
3. **State Constraints** that specify the current architectural state maintained by the test generator.

The goal is not to find a single solution or to find all solutions, which is impractical due to the size and complexity of modern architectures. Rather, it is to generate diverse tests which exercise the architecture as thoroughly as possible.

To make the problem concrete, consider the simple example depicted in Figure 2. The left table specifies the instruction set. The first two instruction

I.P. Gent (Ed.): CP 2009, LNCS 5732, pp. 787–801, 2009.

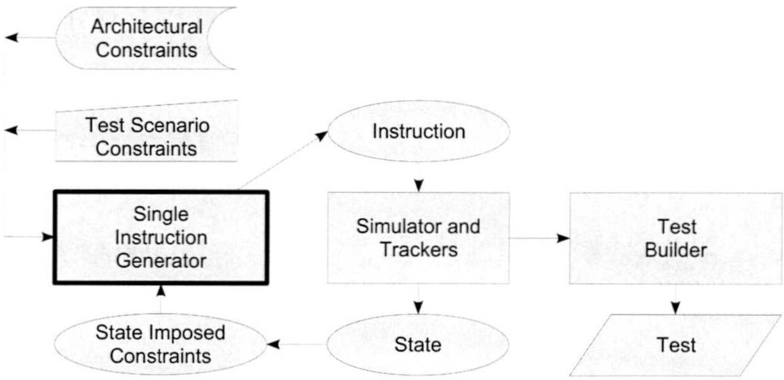

Fig. 1. The Architecture of Automatic Generation of Architectural Tests

Instruction Set				
add	register	r1	r2	r3
sub	register	r1	r2	r3
load	immediate	r1	constant	
load	memory	r1	address	
jump	immediate		constant	
jump	memory		address	

Test Scenario: address = 1234 $\vee$ constant = abcd			
jump	memory		1234
load	immediate	12	abcd
jump	immediate		abcd
load	memory	4b	1234

Fig. 2. A Simple Processor Instruction Example

types are arithmetic operations that operate on registers and have three registers. The third and fourth instruction types are an immediate and a memory load. The last two instruction types are immediate and memory jump. The right table depicts a disjunctive constraint proposed by a validation engineer and the instructions that satisfy it. Observe that the constraint implicitly specifies that the instruction is either a *load* or a *jump*. The CSPs induced by ATGPs are generally quite complex for various reasons:

1. The ATGP is a so-called conditional constraint satisfaction problem [1][1], since the existence of some instruction fields are conditioned on the value of other fields such as the opcode and the addressing mode. In our simple example, the *constant* field is only defined if the instruction is a load or a jump and the mode is immediate.
2. The ATGP typically operates over large instruction sets due to the complexity of modern architectures. Moreover, some of these fields have very large domains, since they represent memory addresses or large constants.
3. The ATGP typically contains both soft and hard constraints, specified by the validation engineer.
4. The validation engineer may specify a desired distribution of instructions to bias toward some specific instructions.

[1] Conditional CSPs were originally called Dynamic CSPs in [1]. Both terms are heavily overloaded; we use CCSPs in this paper.

Earlier approaches (e.g., [2]) address these difficulties by introducing conditional variables to handle the CCSP and use randomization to obtain diverse solutions.

In this paper, we reconsider the issue of producing diverse solutions for ATGPs. We propose to view the ATGP as a MaxDiverse$k$Set problem in the sense of [3] over the CCSP and to use constraint-based local search to obtain high-quality solutions to the model. We show that the resulting approach produces significant improvements in solution diversity compared to pseudo-random solutions, while the running times remain reasonable. The technical contributions can be summarized as follows:

1. The paper proposes a new approach to producing diverse solutions to ATGPs, exploiting jointly earlier results in constraint programming and constraint-based local search.
2. The paper shows that constraint-based local search finds high-quality solutions to MaxDiverse$k$Set problems, even when the number of required solutions increases significantly. The paper also quantifies the quality loss experimentally.
3. The paper presents a new semantics for conditional variables, which is more suited for ATGPs than the original one which was designed for configuration problems.
4. The paper presents the algorithmic foundations for constraint-based local search over conditional variables and shows how constraint-based local search can accommodate conditional variables naturally and compositionally.

The rest of the paper is organized as follows. The next two sections present the building blocks of our approach. Section 2 presents the approach to generate diverse solutions to CSPs and presents experimental results on some simple problems. Section 3 discusses the modeling of the problem as a CCSP and the semantics of conditional variables and how to perform constraint-based local search over conditional variables. Section 4 shows how to model ATGPs in CBLS and search for diverse solutions. Section 5 reports experimental results of our prototype implementation on some benchmark ATGPs to validate the approach, and Section 6 concludes the paper.

2 Generating Diverse Solutions

Modern approaches to ATGPs (e.g., [2]) use a pseudo-random exploration of the search space to generate diverse solutions. This is often sub-optimal however. The goal of this paper is to provide a more principled approach to diversity for ATGPs. For this reason, ATGPs are modeled as MaxDiverse$k$Set problems, which were studied extensively in [3]. More precisely, given a CSP $\mathcal{P}$, its set of solutions $sol(\mathcal{P})$, and a function δ to measure the distance between solutions, the MaxDiverse$k$Set problem for $\mathcal{P}$ consists in finding a set of solutions $S = \{s_1, \ldots, s_k\}$ maximizing

$$\sum_{1 \leq i < j \leq k} \delta(s_i, s_j)$$

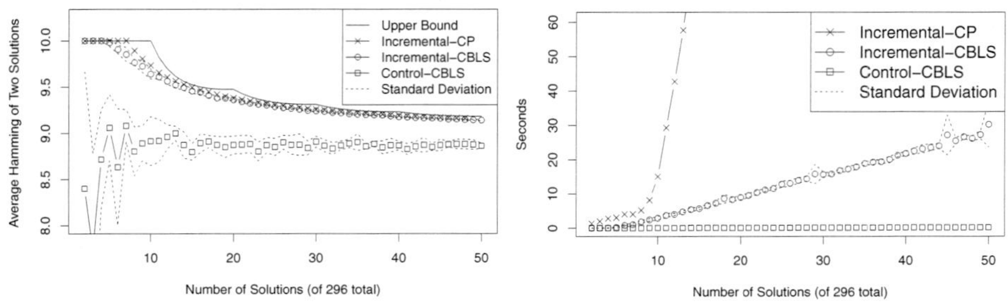

Fig. 3. Diversity Results and Computation Times for the Allinterval Series of Size 10

In ATGPs, it is convenient to use the Hamming distance for the distance function δ between two solutions $s_1 = \langle v_1, \ldots, v_n \rangle$ and $s_2 = \langle w_1, \ldots, w_n \rangle$

$$\delta(s_1, s_2) = \sum_{i=1}^{n} (v_i \neq w_i)$$

This problem is in general quite difficult from a practical standpoint, since it requires to search for k solutions simultaneously and produces very large CSPs (See also [3] for the theoretical complexity which is $FP^{NP[logn]}$-complete). Moreover, in this application domain, it is often desirable to produce the solutions incrementally, one at a time. For this reason, the incremental algorithm in [3] is an excellent candidate for finding approximated solution to MAXDIVERSEkSET problems associated with ATGPs. The algorithm can be formalized as follows:

INCREMENTALGENERATION($\mathcal{P}$)
1 $S \leftarrow \{\}$;
2 **while** $|S| \leq k$
3 **do** FIND $s \in sol(\mathcal{P})$ MAXIMIZING $\sum_{e \in S} \delta(s, e)$;
4 $S \leftarrow S \cup \{s\}$;
5 **return** S;

The algorithm generates one solution at a time until a set of the required cardinality is obtained. The core of the algorithm is in Line 3, which generates the solution maximizing the distance to already generated solutions.

To implement line 3, reference [3] proposed a constraint-programming algorithm using a global constraint for enforcing arc consistency on the distance constraint (derived from the objective). The resulting algorithm was applied to the generation of three solutions to a real-life configuration application, where "three" is considered the "optimal" cardinality to present to users in recommender systems. However, in ATGPs, the cardinality is in general much larger and it is typical to generate 50 to 100 instructions. This led us to approximate the computation in Line 3 with constraint-based local search in order to scale the incremental algorithm effectively.

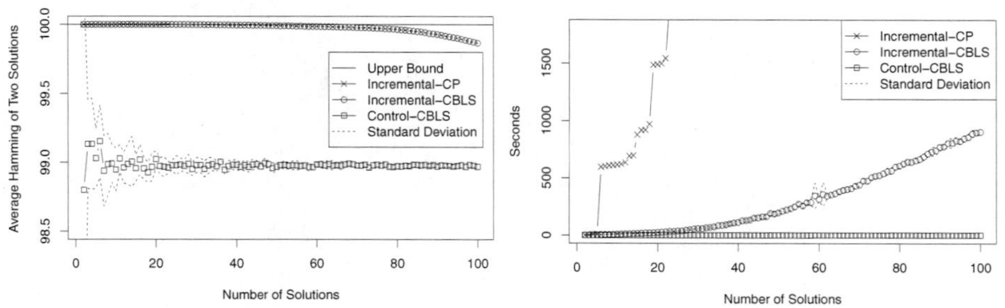

Fig. 4. Diversity Results and Computation Times for the Queens Problem of Size 100

Figures 3 and 4, depict experimental results on the allinterval series of size 10 and the 100-queens problem (with 100 variables). The figures report the average Hamming distance between two generated solutions (y-axis) in sets of increasing cardinality (x-axis) for three methods: the incremental CBLS algorithm (Incremental-CBLS), the incremental CP algorithm (Incremental-CP), and a control CBLS algorithm (Control-CBLS), which generates pseudo-random solutions using CBLS. The CBLS techniques are averaged over ten runs. Incremental-CP uses a global constraint maintaining arc consistency for the Hamming distance and the variable/value heuristic proposed in [3]. Incremental-CBLS uses a generic min-conflict search and a global constraint for the Hamming distance, with value-based violations [4] enabling move evaluation in constant time. The average distance of two solutions is calculated by dividing the MaxDiverse-kSet value by the number of solution pairs in the set S. Specifically, given a distance function δ and a set of solutions $S = \{s_1, \ldots, s_k\}$ the average distance is

$$\frac{\sum\limits_{1 \leq i < j \leq k} \delta(s_i, s_j)}{\binom{k}{2}}$$

The figures also depict a simple upper bound on the maximal diversity and the standard deviation. The upper bound ignores the problem constraints and uses only the domains of the variables, shifting the assignment by one for each successive solution. Assuming that the problem variables are $v_1, \ldots, v_m$ and that $D(v_i)_n$ denotes the n-th value of domain of v_i, the upper bound is computed as follows:

HammingUpperBound(k)
for $i \in 1..m$
  do for $j \in 1..k$
    do $s_{i,j} \leftarrow D(v_i)_{((j-1)\%|D(v_i)|)+1}$
return $\sum_{1 \leq i < j \leq k} \sum_{a=1}^{m} s_{a,i} \neq s_{a,j}$

For consistency with the MaxDiverse$k$Set value, the upper bound is also divided by the number of solution pairs in the set S, i.e. $\binom{k}{2}$, to produce a bound on the average distance between two solutions. The results are particularly

Table 1. The Variables in The Simple ATG Problem

opcode	mode	r1	r2	r3	address	constant
add	register	0..0xff	0..0xff	0..0xff		
sub	register	0..0xff	0..0xff	0..0xff		
load	immediate	0..0xff				0..0xffff
load	memory	0..0xff			0..0xffff	
jump	immediate					0..0xffff
jump	memory				0..0xffff	

interesting. They indicate that Incremental-CBLS is near-optimal in quality on both benchmarks, since the distance to the upper bound is minimal, and outperforms Control-CBLS significantly and asymptotically (the two curves never converge). The quality of Incremental-CP is of course at least as good as the quality of Incremental-CBLS. However, the computational results also show that Incremental-CP does not scale well once the number of solutions is increased. Note also that the Control-CBLS is extremely fast, since it simply finds pseudo-random solutions at each step and never tries to optimize the diversity.

3 Conditional Variables

When modelling an ATGP, it is traditional to introduce a variable for each field which appears in some instruction. As an illustration, Table 1 describes the variables and their domains in our simple ATG example, as well as the instructions they are used in. The first row depicts the variables, while subsequent ones depict the instructions, the variables they use, and the domains of these variables. For instance, the *add* and the *sub* instructions do not use the *address* and *constant* variables. As a result, the ATGP gives rise to a Conditional Constraint Satisfaction Problem (CCSP) in the sense of [1], as observed in [5]. Mittal and Falkenhainer showed how to transform a CCSP into a CSP by introducing additional variables denoting whether a variable is *active*, i.e., whether its condition holds. Subsequent work produces new reformulation techniques (e.g., introducing a dummy value in the domain to express whether the variable is active) and dedicated algorithms to produce significant improvements in efficiency [6,7].

In the context of ATGPs, Moss [2] extended constraint-programming solvers with the concept of conditional variables. A conditional variable y is a pair (x, C), in which x is a regular variable and C is a constraint. Figure 5 depicts the modeling of the simple ATG problem with conditional variables. The variable section declares the variables, their domains, and possibly a condition. The last 5 variables are conditional and depend on the values of the opcode or mode variables. The second section depicts the architectural constraints which, together with the domains, specify the legal instructions. The test scenario is depicted in the third section.

The motivation for introducing conditional variables was twofold. First, Moss argued that the reformulation techniques are not necessarily feasible in ATGPs,

Variables:

$\text{opcode} \in \{\text{add}, \text{sub}, \text{load}, \text{jump}\}$
$\text{mode} \in \{\text{register}, \text{immediate}, \text{memory}\}$
$r_1 \in \{0..0\text{xff}\}$ if $\text{opcode} = \text{add} \vee \text{opcode} = \text{sub} \vee \text{opcode} = \text{load}$
$r_2 \in \{0..0\text{xff}\}$ if $\text{opcode} = \text{add} \vee \text{opcode} = \text{sub}$
$r_3 \in \{0..0\text{xff}\}$ if $\text{opcode} = \text{add} \vee \text{opcode} = \text{sub}$
$\text{address} \in \{0..0\text{xffff}\}$ if $\text{mode} = \text{memory}$
$\text{constant} \in \{0..0\text{xffff}\}$ if $\text{mode} = \text{immediate}$

Architectural Constraints:

$(\text{opcode} \in \{\text{add}, \text{sub}\} \wedge \text{mode} = \text{register}) \vee$
$(\text{opcode} = \text{load} \wedge \text{mode} \in \{\text{immediate}, \text{memory}\}) \vee$
$(\text{opcode} = \text{jump} \wedge \text{mode} \in \{\text{immediate}, \text{memory}\}$

Test Scenario:

$\text{address} = 1234 \vee \text{constant} = abcd$

Fig. 5. Modeling the Simple ATG Example with Conditional Variables

since the domain can already take the entire memory word. She also argued that the more specialized techniques are not general enough for ATGPs. Second, the availability of conditional variables at the modeling level makes it possible to design search algorithms exploiting the semantics of conditional variables in the ATGP context. In particular, the search procedure in [2] nondeterministically decides the active status of each variable and enforces the condition or its negation by adding a new constraint.

This research follows a similar path but for constraint-based local search instead of constraint programming. It uses conditional variables as first-class modeling objects and uses their semantics to guide the search, albeit in a fundamentally different way. The rest of this section will specify the semantics of conditional variables which is only defined informally in [2] and extends the concept of constraint violations in CBLS to conditional variables.

The Semantics of Conditional Variables. There are many possible semantics for conditional variables, each of which may be appropriate for a particular application domain. Mittal and Falkenhainer use what we call a *lenient* semantics in which a constraint holds as soon as one of its conditional variables is inactive. The lenient semantics are appropriate for the configuration problems they consider, but is not suited for ATGPs. Consider the instruction set proposed earlier and the test scenario

$$constant > 10.$$

Using the lenient semantics, the ATGP problem admits as solutions, all the intructions that do not include a constant, i.e., the arithmetic instructions and the memory load and jump instructions, as well as all those for which the constant is greater than 10. Indeed, if the constant variable is inactive, the constraint is ignored in the lenient semantics.

The semantics we propose are strict on the basic constraints: A constraint only holds if all its variables are active. However, the strictness requirement does not carry over logical or threshold connectives. Consider the test scenario

$$address = 1234 \vee constant = abcd.$$

If we require strictness on the disjunction, the resulting ATGP has no solution, since no instruction has both an address and a constant. The intended semantics here is to generate instructions which have either an address with value "1234" or a constant whose value is "abcd". Finally, consider a Hamming distance constraint

$$\sum_{i=1}^{n}(v_i \neq w_i) \geq d$$

which involves reification. The semantics cannot be strict over the entire constraint or it would never be instrumental in comparing two solutions. Rather the reified constraint should only return 1 (true) when it is satisfied and all its variables are active and 0 (false) otherwise. In other words, the strictness is limited to the reified constraint and not the enclosing expression. Note also that a lenient semantics does not make sense for this constraint, since the constraint would hold as soon as a variable is not active. Even a lenient semantics on the reified constraints is not desirable, since similar instructions would have a positive score when many of their variables are undefined.

Figures 7, 8, and 9 describe the semantics of the small language given in Figure 6. The figures use *var(y)* and *cond(y)* to denote the variable and condition part of a conditional variable. The semantics are given for an assignment α of values to the variables. The figures also give the invariants which maintain the truth values of all constraints, showing that the semantics can be implemented compositionally and maintained incrementally, as was the case for differentiable invariants [8]. This indicates that our approach does not require any program transformation and leverages all the functionalities of CBLS. Figure 7 gives the semantics for the evaluation of expressions and should not raise any issue. Observe that the condition of a conditional variable is ignored and is handled at a different level. Figure 8 gives the semantics of constraints. The primitive constraints hold when their traditional semantics hold and when their expressions are well-defined, meaning that their variables are active. The logical connectives simply apply the semantics recursively on their subexpressions. By definition, a conjunction is always strict: all its variables must be active. The rest of the figure specifies when an expression is well-defined. Figure 9 depicts how to handle reification, which is important to give the semantics to the Hamming distance over conditional variables. The evaluation of a reified constraint simply calls the semantic definition for constraints and uses the Kronecker symbol δ to convert Boolean values into 0/1 values:

$$\delta(b) = \begin{cases} 1 & \text{if } b = true; \\ 0 & \text{otherwise.} \end{cases}$$

The rule of well-definedness of a reified constraint simply returns true, meaning that the definedness is local to the reified constraints and does not propagate to the enclosing expression.

$v \in \mathcal{N}$; $x \in \mathit{Variable}$; $y \in \mathit{ConditionalVariable}$; $e \in \mathit{Expression}$; $c \in \mathit{Constraint}$.
$e ::= v \mid x \mid y \mid e + e \mid e - e \mid e \times e \mid c$
$c ::= e = e \mid e \leq e \mid c \vee c \mid c \wedge c$

Fig. 6. The Syntax of Expressions and Constraints (Partial Description)

$$
\begin{array}{ll|ll}
\underset{\alpha}{\mathbb{E}}[v] & = v & i_v & \leftarrow v \\
\underset{\alpha}{\mathbb{E}}[x] & = \alpha(x) & i_x & \leftarrow x \\
\underset{\alpha}{\mathbb{E}}[y] & = \alpha(var(y)) & i_y & \leftarrow i_{var(y)} \\
\underset{\alpha}{\mathbb{E}}[e_1 + e_2] & = \underset{\alpha}{\mathbb{E}}[e_1] + \underset{\alpha}{\mathbb{E}}[e_2] & i_{e_1+e_2} & \leftarrow i_{e_1} + i_{e_2} \\
\underset{\alpha}{\mathbb{E}}[e_1 - e_2] & = \underset{\alpha}{\mathbb{E}}[e_1] - \underset{\alpha}{\mathbb{E}}[e_2] & i_{e_1-e_2} & \leftarrow i_{e_1} - i_{e_2} \\
\underset{\alpha}{\mathbb{E}}[e_1 \times e_2] & = \underset{\alpha}{\mathbb{E}}[e_1] \times \underset{\alpha}{\mathbb{E}}[e_2] & i_{e_1 \times e_2} & \leftarrow i_{e_1} \times i_{e_2}
\end{array}
$$

Fig. 7. The Evaluation of Expressions and their Underlying Invariants

It is worth highlighting that the lenient semantics can be obtained in a very similar way: just replace the conjunction by disjunction and negate the well-definedness condition in the first two lines of Figure 8 and include a recursive call in the definition of well-definedness for reified constraints. So it is possible to accommodate easily both the lenient and the strict semantics in the same system. Observe also that the generated invariants are acyclic by construction since the condition in a conditional variable can only use previously declared variables. Acyclicity is in fact always assumed in CCSPs and is natural in their application domains.

The Definition of Violations. Figure 10 depicts the violations of constraints over conditional variables, as well as the invariants to maintain them. Several points deserve to be highlighted. First, the definition of violations capture the importance of conditions in conditional variables. The violations of a constraint $c(y_1, \ldots, y_n)$ over conditional variables is expressed directly in terms of the violations of the same constraint over traditional variables $c(var(y_1), \ldots, var(y_n))$ but it adds a penalty ϕ for each of its conditional variables whose condition does not hold. The expression $\underset{\alpha}{\mathbb{U}}[e]$ computes the number of inactive conditional variables in e. The penalty is large to focus the search on making the conditional variables active before considering the other violations. Second, observe that conditional variables in reifed constraints are not counted, reflecting the semantics of reification in this context too. Finally, the invariants are once again computed naturally, showing the compositional nature of the implementation.

4 Modeling and Solving the ATGP Problem

We now describe how to model and solve ATGPs.

$\underset{\alpha}{\mathbb{B}}[e_1 = e_2] = \underset{\alpha}{\mathbb{E}}[e_1] = \underset{\alpha}{\mathbb{E}}[e_2] \wedge \underset{\alpha}{\mathbb{D}}[e_1] \wedge \underset{\alpha}{\mathbb{D}}[e_2]$	$b_{e_1=e_2} \leftarrow i_{e_1} = i_{e_2} \wedge d_{e_1} \wedge d_{e_2}$
$\underset{\alpha}{\mathbb{B}}[e_1 \leq e_2] = \underset{\alpha}{\mathbb{E}}[e_1] \leq \underset{\alpha}{\mathbb{E}}[e_2] \wedge \underset{\alpha}{\mathbb{D}}[e_1] \wedge \underset{\alpha}{\mathbb{D}}[e_2]$	$b_{e_1 \leq e_2} \leftarrow i_{e_1} \leq i_{e_2} \wedge d_{e_1} \wedge d_{e_2}$
$\underset{\alpha}{\mathbb{B}}[r_1 \vee r_2] = \underset{\alpha}{\mathbb{B}}[r_1] \vee \underset{\alpha}{\mathbb{B}}[r_2]$	$b_{r_1 \vee r_2} \leftarrow b_{r_1} \vee b_{r_2}$
$\underset{\alpha}{\mathbb{B}}[r_1 \wedge r_2] = \underset{\alpha}{\mathbb{B}}[r_1] \wedge \underset{\alpha}{\mathbb{B}}[r_2]$	$b_{r_1 \wedge r_2} \leftarrow b_{r_1} \wedge b_{r_2}$
$\underset{\alpha}{\mathbb{D}}[v] = true$	$d_v \leftarrow true$
$\underset{\alpha}{\mathbb{D}}[x] = true$	$d_x \leftarrow true$
$\underset{\alpha}{\mathbb{D}}[y] = \underset{\alpha}{\mathbb{B}}[cond(y)]$	$d_y \leftarrow d_{cond(y)}$
$\underset{\alpha}{\mathbb{D}}[e_1 + e_2] = \underset{\alpha}{\mathbb{D}}[e_1] \wedge \underset{\alpha}{\mathbb{D}}[e_2]$	$d_{e_1+e_2} \leftarrow d_{e_1} \wedge d_{e_2}$
$\underset{\alpha}{\mathbb{D}}[e_1 - e_2] = \underset{\alpha}{\mathbb{D}}[e_1] \wedge \underset{\alpha}{\mathbb{D}}[e_2]$	$d_{e_1-e_2} \leftarrow d_{e_1} \wedge d_{e_2}$
$\underset{\alpha}{\mathbb{D}}[e_1 \times e_2] = \underset{\alpha}{\mathbb{D}}[e_1] \wedge \underset{\alpha}{\mathbb{D}}[e_2]$	$d_{e_1 \times e_2} \leftarrow d_{e_1} \wedge d_{e_2}$

Fig. 8. The Evaluation of Constraints and their Corresponding Invariants

$\underset{\alpha}{\mathbb{E}}[c] = \delta(\underset{\alpha}{\mathbb{B}}[c])$	$i_c \leftarrow \delta(b_c)$
$\underset{\alpha}{\mathbb{D}}[c] = true$	$d_c \leftarrow true$

Fig. 9. The Evaluation of Reified Constraints and their Corresponding Invariants

The Model. An ATGP consists of four different components: the objective function to achieve diversity, a hard constraint system, a soft constraint system, and a probabilistic constraint system. The hard constraint system contains the architectural constraints, as well as the hard constraints in the test scenario. The soft constraint system contains the soft constraints of the test scenario. The probabilistic constraint system allows the validation engineer to bias the generated sequence toward some instructions. An entry in a probabilistic constraint system is a tuple $\langle (c_1, p_1), \ldots, (c_k, p_k) \rangle$ where c_i are mutually exclusive constraints and p_i are probabilities satisfying $\sum_{i=1}^{k} p_i = 1$. The intention is to generate a sequence of instructions which satisfy c_i with probability p_i. A typical example would be

$$(opcode = add, 0.7), (opcode = jump, 0.2), (opcode = load, 0.1)$$

which would generate *add*, *jump*, and *load* instructions 70%, 20%, and 10% of the time respectively.

The Search. We experimented with various search procedures sharing a common core. The core has four main features. First, the hard and soft constraint systems H and S are combined into a single constraint system C through weights, i.e., $C = w_h * H + S$. The resulting constraint system is then reified into the objective function, once again using weights

$$O = w_d * HammingDistance - w_c * C$$

$$
\begin{array}{ll|ll}
\underset{\alpha}{\mathbb{V}}[e_1 = e_2] & = \underset{\alpha}{\mathbb{E}}[abs(e_1 - e_2)] + \phi\underset{\alpha}{\mathbb{U}}[e_1] + \phi\underset{\alpha}{\mathbb{U}}[e_2] & v_{e_1=e_2} & \leftarrow i_{abs(e_1-e_2)} + \phi u_{e_1} + \phi u_{e_2} \\
\underset{\alpha}{\mathbb{V}}[e_1 \leq e_2] & = \underset{\alpha}{\mathbb{E}}[\max(e_1 - e_2, 0)] + \phi\underset{\alpha}{\mathbb{U}}[e_1] + \phi\underset{\alpha}{\mathbb{U}}[e_2] & v_{e_1 \leq e_2} & \leftarrow i_{max(e_1-e_2,0)} + \phi u_{e_1} + \phi u_{e_2} \\
\underset{\alpha}{\mathbb{V}}[c_1 \wedge c_2] & = \underset{\alpha}{\mathbb{V}}[c_1] + \mathbb{V}[c_2] & v_{c_1 \wedge c_2} & \leftarrow v_{c_1} + v_{c_2} \\
\underset{\alpha}{\mathbb{V}}[c_1 \vee c_2] & = \min(\underset{\alpha}{\mathbb{V}}[r_1], \underset{\alpha}{\mathbb{V}}[r_2]) & v_{c_1 \wedge c_2} & \leftarrow \min(v_{c_1}, v_{c_2}) \\
 & & & \\
\underset{\alpha}{\mathbb{U}}[v] & = 0 & u_v & \leftarrow 0 \\
\underset{\alpha}{\mathbb{U}}[x] & = 0 & u_x & \leftarrow 0 \\
\underset{\alpha}{\mathbb{U}}[y] & = \delta(\neg\underset{\alpha}{\mathbb{B}}[cond(y)]) & u_y & \leftarrow \delta(\neg d_{cond(y)}) \\
\underset{\alpha}{\mathbb{U}}[e_1 + e_2] & = \underset{\alpha}{\mathbb{U}}[e_1] + \underset{\alpha}{\mathbb{U}}[e_2] & u_{e_1+e_2} & \leftarrow u_{e_1} + u_{e_2} \\
\underset{\alpha}{\mathbb{U}}[e_1 - e_2] & = \underset{\alpha}{\mathbb{U}}[e_1] + \underset{\alpha}{\mathbb{U}}[e_2] & u_{e_1-e_2} & \leftarrow u_{e_1} + u_{e_2} \\
\underset{\alpha}{\mathbb{U}}[e_1 \times e_2] & = \underset{\alpha}{\mathbb{U}}[e_1] + \underset{\alpha}{\mathbb{U}}[e_2] & u_{e_1 \times e_2} & \leftarrow u_{e_1} + u_{e_2} \\
\underset{\alpha}{\mathbb{U}}[c] & = 0 & u_c & \leftarrow 0
\end{array}
$$

Fig. 10. Violations of Constraints over Conditional Variables and their Invariants

Second, the search always selects the variable with the steepest gradient and always assigns to it the value producing the steepest increase in objective O. Third, tabu-search is used as the meta-heuristic. Fourth, the search always includes a restarting strategy. The probabilistic constraint system is handled in an initial step. For each probabilistic constraint, the search flips a coin and imposes the appropriate constraint based on the provided distribution.

This core can be enhanced with a strategic oscillation strategy which adjusts the weights to balance the time spent in the feasible and infeasible region [9]. However, the core procedure achieves the best quality/efficiency tradeoff over the benchmarks presented in the next section. The strategic oscillation offers benefits in solution quality at the expense of an increase in computation times. Since efficiency is a critical factor in ATGPs, this strategy was not retained.

A critical aspect of the search procedure is also its handling of large domains. As mentioned earlier, domains in ATGPs can vary in size considerably. A domain may be small (e.g., the available registers) or very large (e.g., the set of all 32-bit addresses or the set of all 16-bit constants). It is not practical to find the value that decreases the objective the most by differentiation in these cases: It would take too much time to enumerate all the values. Our search handles large domains differently by performing a random sampling of the domain.

It is important to emphasize that ATGPs have many local minima and it is not easy to escape them. This is the main justification for restarts which are critical to achieve a reasonable tradeoff between solution quality and efficiency. This situation is partly due to our modeling which is geared toward feasibility: Violations of the conditions from conditional variables have a significant penalty (the ϕ value in the violation definition). However, without this penalty, the search has difficulty finding feasible solutions and is heavily biased toward the Hamming distance.

```
Test Scenario 2:
   Hard: OpTypeSp1 = imm32 && OpValue1 > 0x12345
   Soft: OpValue1 < 0x22222 || (OpValue1 > 0x33333 && OpValue1 < 0x77777)
Test Scenario 3:
   Prob: InstructionGroups = {arithm%70, logic%20, cmp%10}
Test Scenario 7:
   Hard: OpType1 = immediate || OpType2 = immediate
   Hard: if (OpType1 = immediate) then (OpValue1 > 0xff)
   Hard: if (OpType2 = immediate) then (OpSize2 > 8)
   Soft: OpRole0 = dest
Test Scenario 8:
   Hard: InstructionClass != GP && InstructionGroups = logic
   Soft: OpValue0 = OpValue1
```

Fig. 11. The ATG Test Scenarios

5 Experimental Results

This section presents the experimental results on some benchmarks provided by Intel. The benchmarks consist of an instruction set consisting of 80 instruction types with up to 20 fields and 8 test scenarios. For space reasons, it is not possible to present all the results but Figure 11 depicts some interesting scenarios. Scenario 2 features both hard and soft constraints, including some over large domains. Scenario 3 features a probabilistic constraint. Scenario 7 features hard and soft constraints with implications and disjunctions. Scenario 8 features a soft constraint which restricts the feasible region significantly and creates a significant tension with the Hamming distance. Incremental-CBLS is implemented on top of the COMET system [4] (Significant improvement in speed would result from a native support of conditional variables) and the experiments were run on Intel Xeon CPU 2.80GHz machines running 64-bit Linux Debian.

Figures 12 and 13 depict the experimental results on the above scenarios. Other results are consistent but cannot be included for space reasons. Once again, the graphs on the left give the average Hamming distance between two solutions as a function of the number of solutions and each point corresponds to an average of 10 runs. In general, it is difficult to compute a tight upper bound on ATGPs because of the conditional variables, some of which are not active. The figure reports an upper bound on scenario 8, since the set of feasible instructions is more restricted in this case and the upper bound can exploit that information (a similar result holds for scenario 5 whose results are not shown for space reasons). The results show a significant benefit in solution quality for Incremental-CBLS compared to Control-CBLS. Interestingly, the shape of the results closely follows the queen and allinterval results presented earlier. The curves for the average Hamming distance do not converge, Incremental-CBLS is close to the upper bound on scenarios 5 and 8, and Incremental-CBLS has significantly smaller standard deviations on ATGPs.

The computation times remain reasonable for Incremental-CBLS but obviously the computation times increase compared to Control-CBLS which only

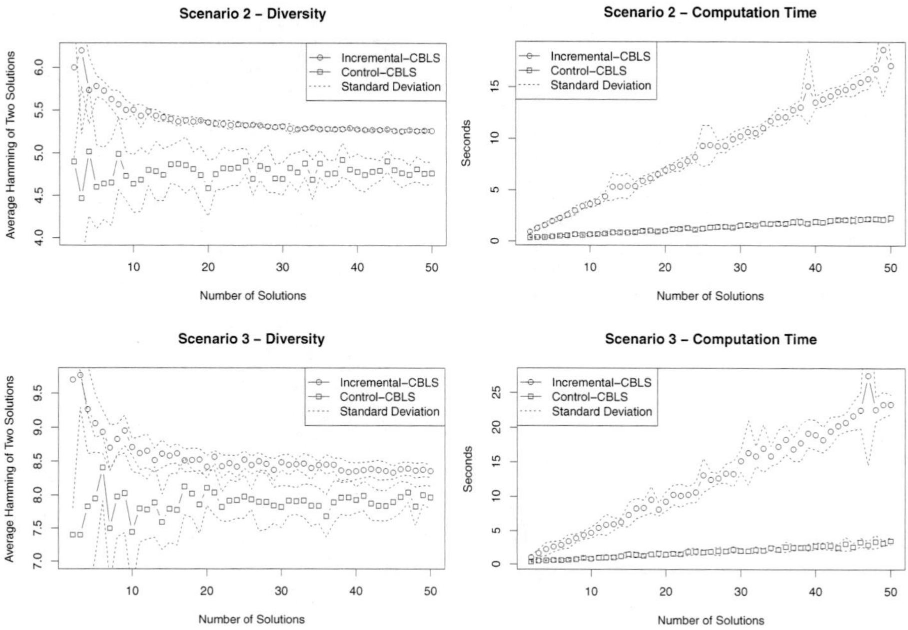

Fig. 12. Diversity Results and Computation Times for the ATGPs

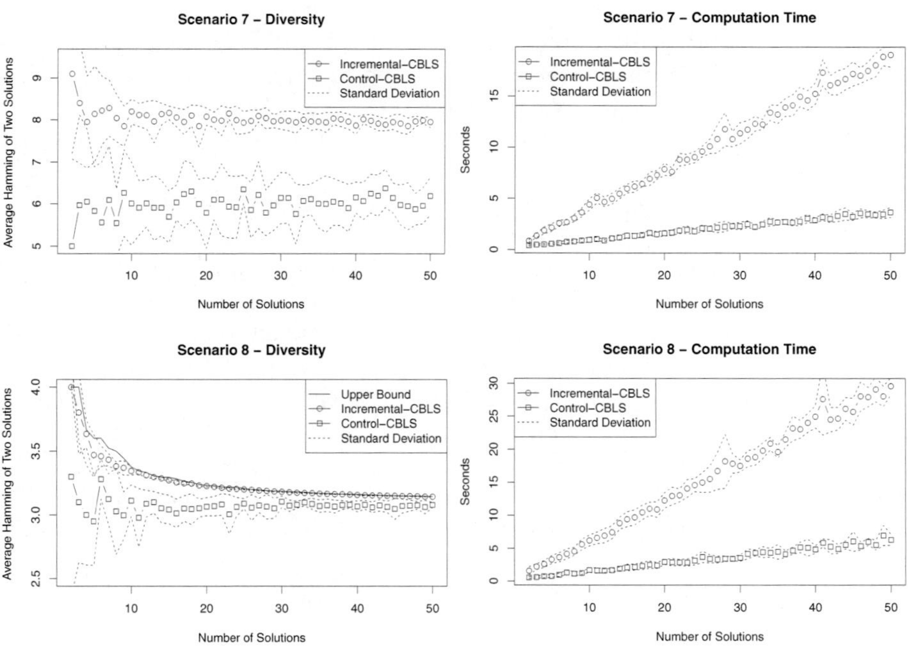

Fig. 13. Diversity Results and Computation Times for the ATGPs

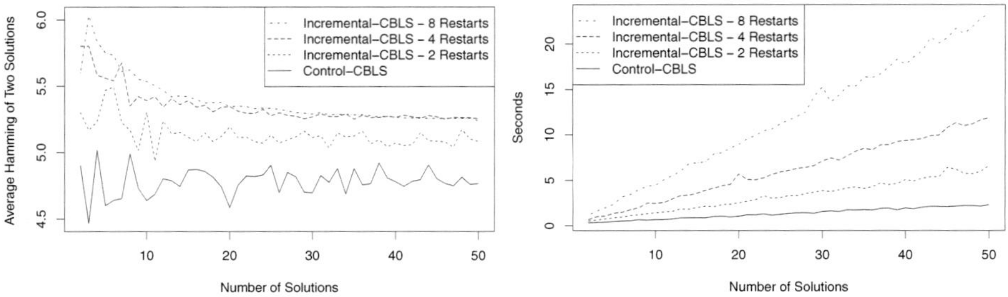

Fig. 14. Tradeoff Between Solution Quality and Efficiency on Scenario 2

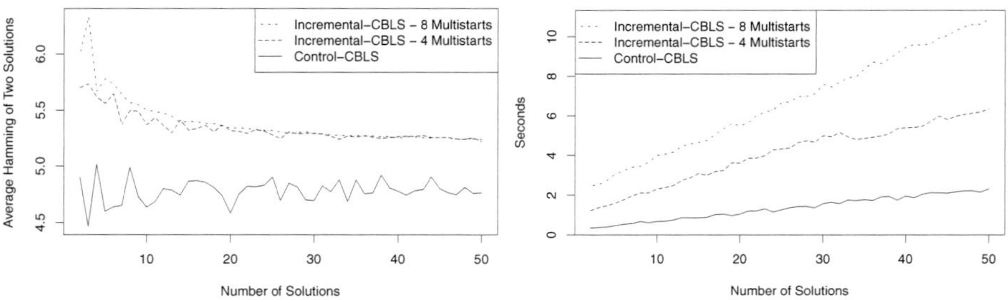

Fig. 15. The Benefits of Parallelism on Scenario 2

searches for random feasible solutions. Scenario 8 is more demanding, since the soft constraint is in direct contradiction with the Hamming distance and restricts the search space significantly when enforced. Only assignments were considered as local moves. An improvement could be gained from considering value swaps as well.

The tradeoffs between solution quality and efficiency was also investigated since both diversity and efficiency are important in ATGPs. Figure 14 shows the results on scenario 2. The number of restarts is reduced from 8 to 4 without significant degradation in quality but with a 50% reduction in computation time. This indicates that Incremental-CBLS can likely be tuned to meet strong timing constraints while still bringing significant benefits.

Finally, Figure 15 shows the benefits of parallelism in ATGPs problems. It transforms the search into a multistart procedure which are executed on 4 processors. The figure shows that the computation times are decreased by 50% again for 8 and 4 multistarts, closing further the gap with Control-CBLS. The generation of 50 diverse solutions now takes less than 6 seconds.

6 Conclusion

This paper reconsidered the automatic generation of architectural tests (ATGP), a fundamental problem in processor validation. It proposed to view ATGPs as

MaxDiverse$k$Set problems to produce more diverse solutions than the random exploration traditionally used. The paper showed that constraint-based local search over conditional variables can provide significant benefits in solution quality, while retaining reasonable efficiency. The paper described a semantics and implementation of constraint-based local search over conditional variables, which is particularly appropriate for ATGPs. It also showed that constraint-based local search brings significant computational benefits over existing techniques as an implementation technique for approximating MaxDiverse$k$Set problems.

There are many directions for future work. The treatment of large domains should be enhanced and the tabu search should be complemented by constraint-based repair techniques to suggest moves that can efficiently reduce the violations of constraints involving large domains. Our prototype implementation should be embedded in the Comet kernel and tested on large scale instances modeling IA-32 and IA-64 processors.

References

1. Mittal, S., Falkenhainer, B.: Dynamic constraint satisfaction problems. In: Proceedings of the Eighth National Conference on Artificial Intelligence (1990)
2. Moss, A.: Constraint patterns and search procedures for cp-based random test generation. In: Yorav, K. (ed.) HVC 2007. LNCS, vol. 4899, pp. 86–103. Springer, Heidelberg (2008)
3. Hebrard, E., Hnich, B., O'Sullivan, B., Walsh, T.: Finding diverse and similar solutions in constraint programming. In: Proceedings of the Twentieth National Conference on Artificial Intelligence (2005)
4. Van Hentenryck, P., Michel, L.: Constraint-Based Local Search. The MIT Press, Cambridge (2005)
5. Bin, E., Emek, R., Shurek, G., Ziv, A.: Using a constraint satisfaction formulation and solution techniques for random test program generation. IBM Systems Journal 41(3) (2002)
6. Geller, F., Veksler, M.: Assumption-based pruning in conditional csp. In: van Beek, P. (ed.) CP 2005. LNCS, vol. 3709, pp. 241–255. Springer, Heidelberg (2005)
7. Sabin, D., Freuder, E.C.: Configuration as composite constraint satisfaction. In: Proceedings of Artificial Intelligence and Manufacturing Research Planning Workshop (1996)
8. Van Hentenryck, P., Michel, L.: Differentiable invariants. In: Benhamou, F. (ed.) CP 2006. LNCS, vol. 4204, pp. 604–619. Springer, Heidelberg (2006)
9. Glover, F., Laguna, M.: Tabu Search. Kluwer, Dordrecht (1997)

Edge Finding Filtering Algorithm for Discrete Cumulative Resources in $\mathcal{O}(kn \log n)$

Petr Vilím

ILOG S.A. an IBM Company, 9, rue de Verdun, BP 85
F-94253 Gentilly Cedex, France
petr_vilim@cz.ibm.com

Abstract. This paper presents new Edge Finding algorithm for discrete cumulative resources, i.e. resources which can process several activities simultaneously up to some maximal capacity C. The algorithm has better time complexity than the current version of this algorithm: $\mathcal{O}(kn \log n)$ versus $\mathcal{O}(kn^2)$ where n is number of activities on the resource and k is number of distinct capacity demands. Moreover the new algorithm is slightly stronger and it is able to handle optional activities. The algorithm is based on the Θ-tree – a binary tree data structure which already proved to be very useful in filtering algorithms for unary resource constraints.

1 Introduction

Nowadays, constraint based scheduling engines like IBM ILOG CP-Optimizer [1] allow to describe and solve very complex scheduling problems involving a variety of different constraints. This paper describes a filtering algorithm called Edge Finding for one of them – for discrete cumulative resource constraint.

Let us demonstrate the problem on a simple example on Figure 1. Note that this example may be just a small part of much more complex problem. There are three equivalent workers (a resource with capacity $C = 3$) who must perform four different activities $T = \{A, B, C, D\}$. Activity A requires all three workers ($c_A = 3$) for one hour ($p_A = 1$), activity B requires only one worker ($c_B = 1$) but

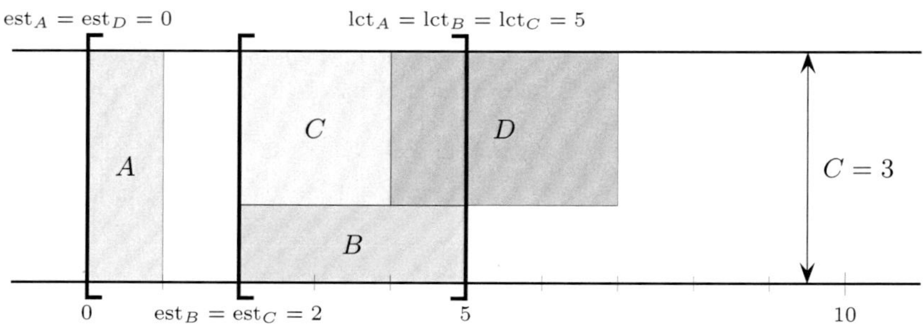

Fig. 1. An example: est_D can be updated from 0 to 4

I.P. Gent (Ed.): CP 2009, LNCS 5732, pp. 802–816, 2009.

for 3 hours ($p_B = 3$) and remaining two activities C and D require two workers each ($c_C = c_D = 2$), activity C for 2 hours ($p_C = 2$) and activity D for 3 hours ($p_D = 3$). Moreover the earliest possible starting time of activities A and D is zero ($\text{est}_A = \text{est}_D = 0$), for activities B and C it is 2 ($\text{est}_B = \text{est}_C = 2$). Latest possible completion time (deadline) for activities A, B and C is 5 ($\text{lct}_A = \text{lct}_B = \text{lct}_C = 5$), activity D does not have a deadline ($\text{lct}_D = \infty$).

Looking closely to the problem we can see that there is no way for D to start before 4. Therefore we can update est_D from 0 to 4 and this way limit the search space of the problem. The rest of this paper describes the algorithm (called Edge Finding) which performs such an update.

Note that there are also filtering algorithms for discrete cumulative resource other than Edge Finding. For example Timetable propagation [2], Not-First/Not-Last [3], Energetic Reasoning [2], Max Energy propagation [4] or Extended Edge Finding [5]. However Timetable and Edge Finding are the ones which are used most of the time.

Let us quickly review existing Edge Finding algorithms for discrete cumulative resources. To the author's knowledge the current state-of-the-art algorithm can be found in [5]. In this paper Luc Mercier and Pascal Van Hentenryck proved that the original cumulative Edge Finding algorithm with time complexity $\mathcal{O}(n^2)$ in [2] is incomplete, and therefore they designed new (complete) algorithm with time complexity $\mathcal{O}(kn^2)$.

This paper further improves the algorithm [5] in several aspects:

- Θ-tree data structure improves time complexity from $\mathcal{O}(kn^2)$ to $\mathcal{O}(kn \log n)$.
- Better usage of relation "ends before end" makes the filtering a little bit stronger. There are cases when the new algorithm propagates while the old one does not (Section 6.2).
- The algorithm can be easily adapted to handle optional activities. Although propagation of optional activities can be further improved [6], it is already pretty strong. We will show how to handle optional activities at the end of the paper (Section 9).
- The algorithm is based on modified Edge Finding rules which are more suitable for propagation. We provide a proof that the new rules are not weaker than the original ones (Section 8).

Like algorithm [5] the new algorithm has two phases:

Detection phase tries to discover necessary relative positions of activities on the resource. The result of this phase is a partial knowledge of a relation "ends before end" (see later), which will be used in the next phase. For the example on Figure 1 the algorithm in this phase detects that activity D must end after the end of $\{A, B, C\}$.

Adjustment phase utilizes results of the previous phase to improve temporal bounds of activities – earliest start times and latest completion times.

In comparison with algorithm [5] the time complexities of both phases are improved. For detection phase from $\mathcal{O}(n^2)$ to $\mathcal{O}(n \log n)$, for adjustment phase from $\mathcal{O}(kn^2)$ to $\mathcal{O}(kn \log n)$. For simplicity we will describe each phase separately.

We present only the algorithm to update earliest start times (not latest completion times) because the algorithm for update of latest completion times is symmetrical.

Note also that Edge Finding algorithm is not idempotent and therefore it is usually repeated until a fixpoint is reached.

2 Basic Notation

Let us formalize the notation already used in the introduction. The input of the algorithm is a discrete cumulative resource with capacity $C \in \mathbb{N}^+$ and a set of activities T ($|T| = n$) which must be processed by the resource. Each activity $i \in T$ is characterized by the following attributes:

- earliest possible start time $\mathrm{est}_i \in \mathbb{N}$,
- latest possible completion time $\mathrm{lct}_i \in \mathbb{N}$,
- processing time (duration) $\mathrm{p}_i \in \mathbb{N}$ – a constant,
- required capacity $\mathrm{c}_i \in \mathbb{N}$ – a constant $\mathrm{c}_i \leq C$.

Activities are not preemptive, that is, if an activity i starts at time t it must continue without interruption until time $t + \mathrm{p}_i$ where it ends. During the whole processing from t to $t + \mathrm{p}_i$ it requires capacity c_i from the resource. At any time, the total capacity required from the resource cannot exceed the maximum capacity C. We define k to be number of distinct capacity demands $k = |\{c_l,\ l \in T\}|$.

Values est_i and lct_i can change – they can be updated by other filtering algorithms or by the Edge Finding algorithm itself. Therefore the input of the Edge Finding algorithm are current bounds est_i and lct_i, the output are new (updated) bounds.

Note that at any time the following inequality must hold for every activity i:

$$\mathrm{est}_i + \mathrm{p}_i \leq \mathrm{lct}_i$$

If it does not hold then the problem does not have any solution and the propagation ends (a *fail*).

From the processing time and required capacity we can compute an energy of an activity i:

$$\mathrm{e}_i = \mathrm{c}_i\, \mathrm{p}_i$$

The energy corresponds to the area occupied by the activity on the resource when depicted like on Figure 1.

The notation for earliest start time, latest completion time and energy can be easily extended for a set of activities $\Omega \subseteq T$:

$$\mathrm{est}_\Omega = \min\{\mathrm{est}_i,\ i \in \Omega\}$$
$$\mathrm{lct}_\Omega = \max\{\mathrm{lct}_i,\ i \in \Omega\}$$
$$\mathrm{e}_\Omega = \sum_{i \in \Omega} \mathrm{e}_i$$

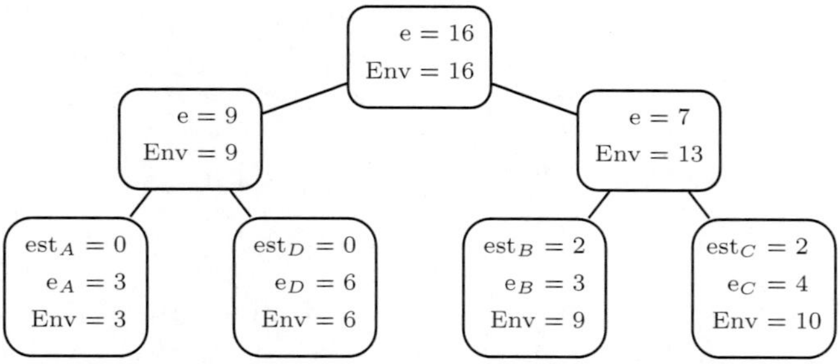

Fig. 2. An example of a Θ-tree for $\Theta = \{A, B, C, D\}$ from Figure 1

3 Earliest Completion Time, Energy Envelope

A critical role in the algorithm plays a computation of possible earliest completion time of a set of intervals $\Theta \subseteq T$. This computation was already described in detail in [4], therefore here we only quickly repeat the main idea. We defined a lower bound $\mathrm{Ect}(\Theta)$ of earliest completion time of a set of activities $\Theta \subseteq T$ as:

$$\mathrm{Ect}(\Theta) = \left\lceil \frac{\mathrm{Env}(\Theta)}{C} \right\rceil$$

where $\mathrm{Env}(\Theta)$ is so-called *energy envelope* of set Θ:

$$\mathrm{Env}(\Theta) = \max_{\Omega \subseteq \Theta} \{C \,\mathrm{est}_\Omega + \mathrm{e}_\Omega\} \tag{1}$$

3.1 Cumulative Θ-Tree

The paper [4] also describes how to compute $\mathrm{Env}(\Theta)$. The idea is to organize set Θ in a balanced binary tree which we call cumulative Θ-tree. Activities are represented by leaf nodes and sorted by est_i from left to right. Each node v of the tree holds the following values:

$$\mathrm{e}_v = \mathrm{e}_{\mathrm{Leaves}(v)} \tag{2}$$

$$\mathrm{Env}_v = \mathrm{Env}\left(\mathrm{Leaves}\left(v\right)\right) \tag{3}$$

Where $\mathrm{Leaves}(v)$ is a set of all activities represented by leaves of the subtree rooted in v.

Figure 2 shows a Θ-tree for $\Theta = \{A, B, C, D\}$ from Figure 1. Notice that the energy envelope of the represented set Θ is equivalent to the value Env of the root node. We can conclude that $\mathrm{Ect}\left(\{A, B, C, D\}\right) = \lceil 16/3 \rceil = 6$.

For a leaf node v representing an activity $i \in T$ the values in the tree are set to:

$$\mathrm{e}_v = \mathrm{e}_i \qquad\qquad \mathrm{Env}_v = \mathrm{Env}\left(\{i\}\right)$$

For internal nodes v these values can be computed recursively from their children nodes $\mathrm{left}(v)$ and $\mathrm{right}(v)$ as shown in the following proposition.

Proposition 1. *For an internal node v, values e_v and Env_v can be computed by the following recursive formulas:*

$$\mathrm{e}_v = \mathrm{e}_{\mathrm{left}(v)} + \mathrm{e}_{\mathrm{right}(v)} \tag{4}$$

$$\mathrm{Env}_v = \max\left\{\mathrm{Env}_{\mathrm{left}(v)} + \mathrm{e}_{\mathrm{right}(v)},\ \mathrm{Env}_{\mathrm{right}(v)}\right\} \tag{5}$$

Proof. See [4]. □

Thanks to formulas (4) and (5), computation of values e_v and Env_v can be integrated within standard operations with balanced binary trees without changing their usual time complexities.

4 Relation "Ends before End"

Before going into details about Edge Finding, let us introduce a notion of *ends before end*. We say that an activity j ends before the end of an activity i (denoted by $j \lessdot i$) if in all solutions $\mathrm{end}(j) \leq \mathrm{end}(i)$. The goal of the Edge Finding algorithm is to discover as much of the relation $\lessdot$ as possible and use it to update temporal bounds. Until a solution is found the relation "ends before end" is known only partially. Therefore in the following we will use the notation $j \lessdot i$ in the sense "it is known that in all solutions j ends before the end of i".

The notation for "ends before end" can be extended also to sets of activities:

$$\forall \Omega \subset T,\ \forall i \in T \setminus \Omega : \quad \Omega \lessdot i \ \Leftrightarrow (\forall j \in \Omega :\ j \lessdot i)$$

5 Edge Finding: Detection Rule

Let us start by definition of a *left cut of T by activity j*:

$$\mathrm{LCut}(T, j) = \{l,\ l \in T \ \&\ \mathrm{lct}_l \leq \mathrm{lct}_j\}$$

To detect the relation $\lessdot$ we will use the following rule:

$$\forall j \in T,\ i \in T \setminus \mathrm{LCut}\,(T, j) :$$
$$(\mathrm{Ect}\,(\mathrm{LCut}\,(T, j) \cup \{i\}) > \mathrm{lct}_j \ \Rightarrow\ \mathrm{LCut}(T, j) \lessdot i)$$

The idea of this rule follows. The set $\mathrm{LCut}(T, j)$ must be processed before lct_j. So if there is not enough time to process i together with $\mathrm{LCut}(T, j)$ then the activity i must end after $\mathrm{LCut}(T, j)$. Note that this rule is different from the original Edge Finding rule. We will show that this new rule is not weaker later in Section 8.

The rule above can be rewritten using energy envelope into a form more suitable for the algorithm:

$$\forall j \in T,\ i \in T \setminus \mathrm{LCut}\,(T, j) :$$
$$(\mathrm{Env}\,(\mathrm{LCut}\,(T, j) \cup \{i\}) > C\,\mathrm{lct}_j \ \Rightarrow\ \mathrm{LCut}(T, j) \lessdot i) \tag{EF1}$$

Our goal is to discover as much of the relation $\lessdot$ as possible. Therefore for each activity $i \in T$ we are looking for an activity $j \in T$ with maximal[1] lct_j such that $\mathrm{LCut}(T, j) \lessdot i$ can be detected by the rule (EF1). This is the task for the first part of the algorithm.

6 Detection Algorithm

Notice that once we prove by (EF1) that $\mathrm{LCut}\,(T, j) \lessdot i$ then it is pointless to evaluate the rule (EF1) for the same activity i and any $j' \in T$ such that $\mathrm{lct}'_j \le \mathrm{lct}_j$ because it cannot bring any new information ($\mathrm{LCut}(T, j') \subseteq \mathrm{LCut}(T, j)$).

The algorithm is similar to the Edge Finding algorithm for unary resource [7]. We iterate over activities j in non-increasing order by lct_j and we maintain a set $\Lambda \subseteq T \setminus \mathrm{LCut}\,(T, j)$ of all activities i for which we still did not find a set which must end before end of i. In each step of the algorithm we check whether there is some activity $i \in \Lambda$ such that the rule (EF1) proves that $\mathrm{LCut}\,(T, j) \lessdot i$. In other words, we test whether the following inequality holds:

$$\max_{i \in \Lambda} \{\mathrm{Env}\,(\mathrm{LCut}\,(T, j) \cup \{i\})\} > C\,\mathrm{lct}_j$$

– If it holds then we find the responsible activity $i \in \Lambda$ and conclude that $\mathrm{LCut}\,(T, j) \lessdot i$. Therefore we can remove i from Λ.
– If it does not hold then we move activity j into Λ and continue by next activity j (because there is no activity i such that $\mathrm{LCut}(j) \lessdot i$ can be proved by (EF1)).

To formalize the algorithm let us define:

$$\mathrm{Env}\,(\Theta, \Lambda) = \max_{i \in \Lambda} \{\mathrm{Env}\,(\Theta \cup \{i\})\}$$

Although we did not show yet how to compute $\mathrm{Env}\,(\Theta, \Lambda)$ we can already present the resulting Algorithm 1. The result of the computation is the array `prec` which has the following meaning:

$$\forall i \in T: \;\; \{l,\ l \in T \;\&\; \mathrm{lct}_l \le \mathtt{prec}\,[i]\} \lessdot i$$

In the algorithm, $\Theta = \mathrm{LCut}(T, j)$ unless there are duplicities in lct_j (the algorithm is correct even with such duplicities). In the following we will concentrate on the computation of $\mathrm{Env}\,(\Theta, \Lambda)$ and the proof that the algorithm 1 has time complexity $\mathcal{O}(n \log n)$.

6.1 Computation of Env (Θ, Λ)

The idea is to extend cumulative Θ-tree into Θ-Λ-tree in a similar way it was done for unary resource in [7]. The cumulative Θ-Λ-tree is a balanced binary tree

[1] Maximality of lct_j assures that for any other $j' \in T$ satisfying $\mathrm{LCut}(T, j') \lessdot i$ by (EF1) we have $\mathrm{LCut}(T, j') \subseteq \mathrm{LCut}(T, j)$.

Algorithm 1. Edge Finding: Detection

```
for i ∈ T do
  prec[i] := −∞;
Θ := T;
Λ := ∅;
for j ∈ T in non-increasing order of lct_j do begin
  while Env(Θ, Λ) > C lct_j do begin
    i := activity from Λ responsible for Env(Θ, Λ);
    prec[i] := lct_j;  // means: LCut(T, j) ⋖ i
    Λ := Λ \ {i};
  end;
  Θ := Θ \ {j};
  Λ := Λ ∪ {j};
end;
```

where each leaf represents one activity from the set Θ or Λ. Leaves are sorted from left to right according to est_i. Each node of the tree holds the following values:

$$\begin{aligned}
\mathrm{e}_v &= \mathrm{e}_{\mathrm{Leaves}(v)\cap\Theta} \\
\mathrm{e}^{\Lambda}_v &= \mathrm{e}_{\mathrm{Leaves}(v)\cap\Theta} + \max_{i\in\mathrm{Leaves}(v)\cap\Lambda}\{\mathrm{e}_i\} \\
\mathrm{Env}_v &= \mathrm{Env}\left(\mathrm{Leaves}\left(v\right)\cap\Theta\right) \\
\mathrm{Env}^{\Lambda}_v &= \mathrm{Env}\left(\mathrm{Leaves}\left(v\right)\cap\Theta, \mathrm{Leaves}\left(v\right)\cap\Lambda\right)
\end{aligned}$$

Notice that $\mathrm{Env}\left(\Theta,\Lambda\right)$ is equivalent to Env^{Λ}_v in the root node. For an example of the cumulative Θ-Λ-tree see Figure 3.

For a leaf node v an activity $i \in \Theta \cup \Lambda$ these values are set to:

$$\mathrm{e}_v = \begin{cases} \mathrm{e}_i & \text{if } i\in\Theta \\ 0 & \text{if } i\in\Lambda \end{cases} \qquad \mathrm{e}^{\Lambda}_v = \begin{cases} -\infty & \text{if } i\in\Theta \\ \mathrm{e}_i & \text{if } i\in\Lambda \end{cases}$$

$$\mathrm{Env}_v = \begin{cases} C\,\mathrm{est}_i + \mathrm{e}_i & \text{if } i\in\Theta \\ -\infty & \text{if } i\in\Lambda \end{cases} \qquad \mathrm{Env}^{\Lambda}_v = \begin{cases} -\infty & \text{if } i\in\Theta \\ C\,\mathrm{est}_i + \mathrm{e}_i & \text{if } i\in\Lambda \end{cases}$$

For internal nodes v these values are computed recursively from their children nodes $\mathrm{left}(v)$ and $\mathrm{right}(v)$:

Proposition 2. *For an internal node v values e_v, e^{Λ}_v, Env_v and Env^{Λ}_v can be computed by the following formulas:*

$$\mathrm{e}_v = \mathrm{e}_{\mathrm{left}(v)} + \mathrm{e}_{\mathrm{right}(v)} \tag{6}$$

$$\mathrm{e}^{\Lambda}_v = \max\left\{\mathrm{e}^{\Lambda}_{\mathrm{left}(v)} + \mathrm{e}_{\mathrm{right}(v)},\ \mathrm{e}_{\mathrm{left}(v)} + \mathrm{e}^{\Lambda}_{\mathrm{right}(v)}\right\} \tag{7}$$

$$\mathrm{Env}_v = \max\left\{\mathrm{Env}_{\mathrm{left}(v)} + \mathrm{e}_{\mathrm{right}(v)},\ \mathrm{Env}_{\mathrm{right}(v)}\right\} \tag{8}$$

$$\mathrm{Env}^{\Lambda}_v = \max\left\{\mathrm{Env}^{\Lambda}_{\mathrm{left}(v)} + \mathrm{e}_{\mathrm{right}(v)},\ \mathrm{Env}_{\mathrm{left}(v)} + \mathrm{e}^{\Lambda}_{\mathrm{right}(v)},\ \mathrm{Env}^{\Lambda}_{\mathrm{right}(v)}\right\} \tag{9}$$

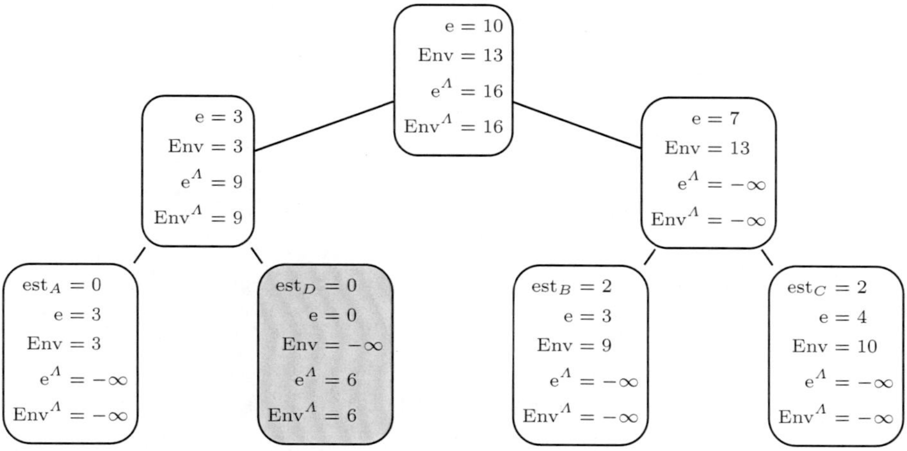

Fig. 3. An example of a Θ-Λ-tree for $\Theta = \text{LCut}(T, A) = \{A, B, C\}$ and $\Lambda = \{D\}$ from Figure 1. We see that $\text{Env}(\Theta, \Lambda) = 16$ which is more than $C\,\text{lct}_A = 15$ and therefore $\{A, B, C\} \lessdot D$.

Proof. First notice that formulas (6) and (8) are the same as formulas (4) and (5) in Proposition 1. Addition of new leaves representing Λ into the tree cannot invalidate these formulas because for these leaves v we have $\text{e}_v = 0$ and $\text{Env}_v = -\infty$. Therefore formulas (6) and (8) hold by Proposition 1.

Formula (7) is simple to prove. It is enough to realize that the difference between computation of e_v by (6) and computation of e^Λ_v is that it is allowed to use one of the activities $i \in \Lambda$. This activity i can be either in the left subtree of v (and in this case we can use $\text{e}^\Lambda_{\text{left}(v)}$ instead of $\text{e}_{\text{left}(v)}$) or in the right subtree of v (and we can use $\text{e}^\Lambda_{\text{right}(v)}$ instead of $\text{e}_{\text{right}(v)}$). Putting this together we transform formula (6) into (7).

It remains to prove formula (9). Again the difference between computation of Env_v and Env^Λ_v is that it is allowed to use one of the activities $i \in \Lambda$. This activity can be either in the left subtree of v (and in this case we can use $\text{Env}^\Lambda_{\text{left}(v)}$ instead of $\text{Env}_{\text{left}(v)}$) or in the right subtree of v (and we can use $\text{Env}^\Lambda_{\text{right}(v)}$ instead of $\text{Env}^\Lambda_{\text{right}(v)}$ or $\text{e}^\Lambda_{\text{right}(v)}$ instead of $\text{e}_{\text{right}(v)}$ but not both). This way we transform formula (8) into (9). □

Thanks to these recursive formulas it is possible to recompute internal values within standard operations with balanced binary trees without changing their time complexity. Therefore lines 9, 11 and 12 of Algorithm 1 has time complexity $\mathcal{O}(\log n)$ and line 6 has time complexity $\mathcal{O}(1)$. To prove that time complexity of the whole Algorithm 1 is $\mathcal{O}(n \log n)$ it remains to show that time complexity of line 7 is also $\mathcal{O}(\log n)$.

The activity $i \in \Lambda$ responsible for $\text{Env}(\Theta, \Lambda)$ can be found by following a path from the root of the tree to the responsible leaf. In each internal node we can recognize in $\mathcal{O}(1)$ whether the responsible activity is in the left or right subtree by analyzing which part of the formulas (9) or (7) was used in the given node:

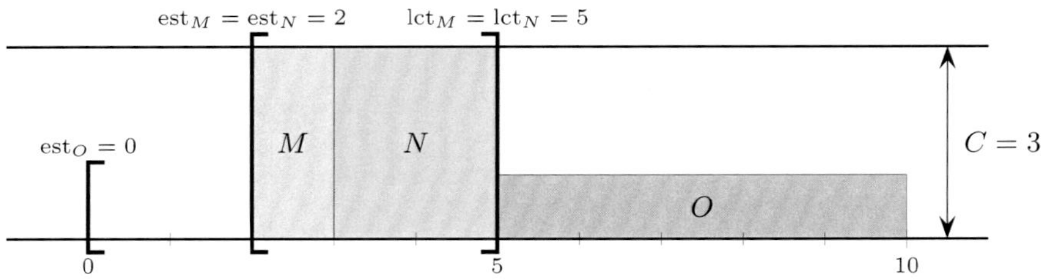

Fig. 4. An example: $\{M, N\} \lessdot O$ but the rule (EF1) is not able to detect it

$$\text{responsible}_{e^\Lambda}(v) = \begin{cases} \text{responsible}_{e^\Lambda}(\text{left}(v)) & \text{if } e^\Lambda(v) = e^\Lambda_{\text{left}(v)} + e_{\text{right}(v)} \\ \text{responsible}_{e^\Lambda}(\text{right}(v)) & \text{if } e^\Lambda(v) = e_{\text{left}(v)} + e^\Lambda_{\text{right}(v)} \end{cases}$$

$$\text{responsible}_{\text{Env}^\Lambda}(v) = \begin{cases} \text{responsible}_{\text{Env}^\Lambda}(\text{right}(v)) & \text{if } \text{Env}^\Lambda(v) = \text{Env}^\Lambda_{\text{right}(v)} \\ \text{responsible}_{e^\Lambda}(\text{right}(v)) & \text{if } \text{Env}^\Lambda(v) = \text{Env}_{\text{left}(v)} + e^\Lambda_{\text{right}(v)} \\ \text{responsible}_{\text{Env}^\Lambda}(\text{left}(v)) & \text{if } \text{Env}^\Lambda(v) = \text{Env}^\Lambda_{\text{left}(v)} + e_{\text{right}(v)} \end{cases}$$

We start the search in the root node r looking for $\text{responsible}_{\text{Env}^\Lambda}(r)$ and continue down the tree using the formulas above (and possibly switching from $\text{responsible}_{\text{Env}^\Lambda}(v)$ to $\text{responsible}_{e^\Lambda}(v)$ on the path) until we reach a leaf.

6.2 Improving Detection

Consider the example on Figure 4. In this example we can see that in every solution $\text{end}(M) \leq \text{end}(O)$ because the maximum possible value for $\text{end}(M)$ is $\text{lct}_M = 5$ and the minimum possible value for $\text{end}(O)$ is $\text{est}_O + \text{p}_O = 5$. Therefore $M \lessdot O$. Similarly $N \lessdot O$. However Edge Finding rule (EF1) is not able to detect that $\{M, N\} \lessdot O$ and we miss the update of est_O from 0 to 5. It is a similar situation to Detectable Precedences for unary resource described in [7].

The idea is to improve the propagation by improving the knowledge of the relation $\lessdot$ stored in the array `prec`:

$$\texttt{prec}[i] := \max\{\texttt{prec}[i], \text{est}_i + \text{p}_i\}$$

It takes time $\mathcal{O}(n)$ to update all `prec[i]` according to the formula above.

7 Time Bound Adjustment

Let us return again to the example on Figure 1. The algorithm presented in the previous chapter detected that $\{A, B, C\} \lessdot D$. We will try to use this knowledge to update est_D. Notice that activity A is actually not important for the update (but it was important in the previous phase to realize that $\{A, B, C\} \lessdot D$), it is a set $\Omega = \{B, C\} \subset \Theta$ which determines new est_D. With this set Ω we can compute new value of est_D denoted as est'_D:

$$\text{est}'_D = \text{est}_\Omega + \left\lceil \frac{\text{e}_\Omega - (C - \text{c}_D)(\text{lct}_\Omega - \text{est}_\Omega)}{\text{c}_D} \right\rceil = 2 + \left\lceil \frac{7 - (3-2)(5-2)}{2} \right\rceil = 4$$

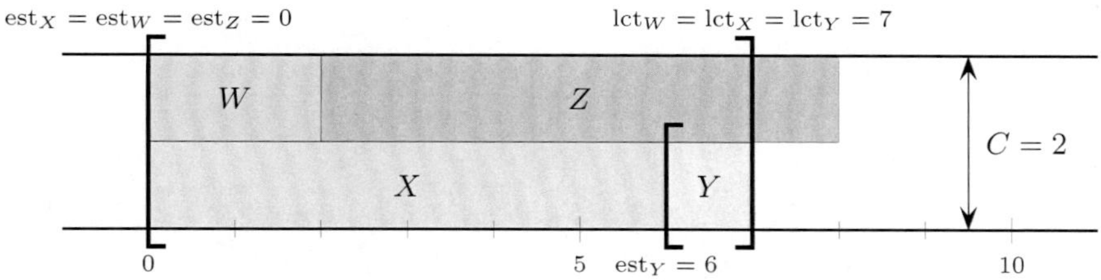

Fig. 5. An example: est_Z can be updated from 0 to 2

However when $\mathrm{LCut}(T, j) \lessdot i$ we cannot use just any subset $\Omega \subseteq \mathrm{LCut}(T, j)$ to compute update of est_i as we did for est_D above. Consider the example on Figure 5. Here $\{W, X, Y\} \lessdot Z$ but we cannot use $\Omega = \{Y\}$ because the result would be invalid:

$$\mathrm{est}_\Omega + \left\lceil \frac{\mathrm{e}_\Omega - (C - \mathrm{c}_Z)(\mathrm{lct}_\Omega - \mathrm{est}_\Omega)}{C_Z} \right\rceil = 6 + \left\lceil \frac{1 - (2-1)(7-6)}{1} \right\rceil = 6$$

The valid update would be to set est_Z to 2, not to 6. The reason that we cannot use $\Omega = \{Y\}$ for update of est_Z is that there is not enough energy in $\Omega = \{Y\}$ to be in potential conflict with Z.

Let us generalize the idea demonstrated on these examples. When $\mathrm{LCut}(T, j) \lessdot i$ then we want to update est_i the following way:

$$\mathrm{LCut}(T, j) \lessdot i \quad \Rightarrow \quad \mathrm{est}'_i := \max \{\mathrm{update}(j, \mathrm{c}_i),\ \mathrm{est}_i\} \qquad \text{(EF2)}$$

where:

$$\mathrm{update}(j, c) = \max_{\substack{\Omega \subseteq \mathrm{LCut}(T,j) \\ \mathrm{e}_\Omega > (C-c)(\mathrm{lct}_\Omega - \mathrm{est}_\Omega)}} \left\{ \mathrm{est}_\Omega + \left\lceil \frac{\mathrm{e}_\Omega - (C-c)(\mathrm{lct}_\Omega - \mathrm{est}_\Omega)}{c} \right\rceil \right\}$$

The condition $\mathrm{e}_\Omega > (C - c)(\mathrm{lct}_\Omega - \mathrm{est}_\Omega)$ makes sure that we do not make any invalid update as described above.

In the following we will describe how to compute values $\mathrm{update}(j, c)$. When all values $\mathrm{update}(j, c)$ are computed then update of est_i using array `prec` and formula (EF2) is trivial.

Let's assume for simplicity that there are no duplicates in the set $\{\mathrm{lct}_j,\ j \in T\}$. Therefore if we sort activities T by increasing lct_j in a sequence $j_1, j_2, \ldots, j_n$ we get:

$$\mathrm{LCut}(T, j_1) \subsetneq \mathrm{LCut}(T, j_2) \subsetneq \cdots \subsetneq \mathrm{LCut}(T, j_n)$$

So when we compute value $\mathrm{update}(j_l, c)$ we do not have to iterate again on all possible subsets $\Omega \subseteq \mathrm{LCut}(T, j_l)$, we can use the fact that we already considered part of them in the computation of $\mathrm{update}(j_{l-1}, c)$. I.e. in the outermost cycle of the algorithm we iterate over all $c \in \{c_m,\ m \in T\}$ and in the inner cycle we iterate over all $j_l \in T$ and compute:

$$\mathrm{update}(j_l, c) = \begin{cases} \mathrm{diff}(j_1, c) & \text{when } l = 1 \\ \max\{\mathrm{update}(j_{l-1}, c),\ \mathrm{diff}(j_l, c)\} & \text{when } l > 1 \end{cases} \qquad (10)$$

where:

$$\text{diff}(j,c) = \max_{\substack{\Omega \subseteq \text{LCut}(T,j) \\ e_\Omega > (C-c)(\text{lct}_j - \text{est}_\Omega)}} \left\{ \text{est}_\Omega + \left\lceil \frac{e_\Omega - (C-c)(\text{lct}_j - \text{est}_\Omega)}{c} \right\rceil \right\} \quad (11)$$

So in the computation of $\text{diff}(j,c)$ we "pretend" that all sets $\Omega \subseteq \text{LCut}(T,j)$ has $\text{lct}_\Omega = \text{lct}_j$. This is not true, there may be sets $\Omega \subsetneq \text{LCut}(T,j)$ such that $\text{lct}_\Omega < \text{lct}_j$. However these sets are correctly considered during computation of $\text{diff}(j',c)$ such that $\text{lct}_{j'} = \text{lct}_\Omega$.

Let's define function $\text{minest}(j,c)$ as:

$$\text{minest}(j,c) = \min \{\text{est}_\Omega,\ \Omega \subseteq \text{LCut}(T,j)\ \&\ e_\Omega > (C-c)(\text{lct}_j - \text{est}_\Omega)\}$$

Notice that for a particular set Ω_m which defines $\text{minest}(j,c)$, i.e. $\text{est}_{\Omega_m} = \text{minest}(j,c)$, we have:

$$\text{est}_{\Omega_m} + \left\lceil \frac{e_{\Omega_m} - (C-c)(\text{lct}_j - \text{est}_\Omega)}{c} \right\rceil > \text{est}_{\Omega_m} = \text{minest}(j,c)$$

and therefore $\text{diff}(j,c) > \text{minest}(j,c)$. Now we will show that:

$$\text{diff}(j,c) = \max_{\substack{\Omega \subseteq \text{LCut}(T,j) \\ \text{est}_\Omega \leq \text{minest}(j,c)}} \left\{ \text{est}_\Omega + \left\lceil \frac{e_\Omega - (C-c)(\text{lct}_j - \text{est}_\Omega)}{c} \right\rceil \right\} \quad (12)$$

The reason follows. The original condition was more restrictive than the new one: in (12) we iterate over more sets Ω than in (11). However for every additional set Ω we have $\text{est}_\Omega \leq \text{minest}(j,c)$ and $e_\Omega \leq (C-c)(\text{lct}_j - \text{est}_\Omega)$ therefore:

$$\text{est}_\Omega + \left\lceil \frac{e_\Omega - (C-c)(\text{lct}_j - \text{est}_\Omega)}{c} \right\rceil \leq \text{est}_\Omega \leq \text{minest}(j,c)$$

And we already know that $\text{diff}(j,c) > \text{minest}(j,c)$. Therefore newly added sets cannot influence the resulting maximum value in formula (12).

Formula (12) is algebraically equivalent to:

$$\text{diff}(j,c) = \left\lceil \frac{\text{Env}(j,c) - (C-c)\,\text{lct}_j}{c} \right\rceil \quad (13)$$

where:

$$\text{Env}(j,c) = \max_{\substack{\Omega \subseteq \text{LCut}(T,j) \\ e_\Omega > (C-c)(\text{lct}_j - \text{est}_\Omega)}} \{C\,\text{est}_\Omega + e_\Omega\} \quad (14)$$

We can split each set Ω by $\text{minest}(j,c)$ into two parts:

$$\Omega_1 = \{l,\ l \in \Omega\ \&\ \text{est}_l \leq \text{minest}\,(j,c)\}$$
$$\Omega_2 = \{l,\ l \in \Omega\ \&\ \text{est}_l > \text{minest}\,(j,c)\}$$

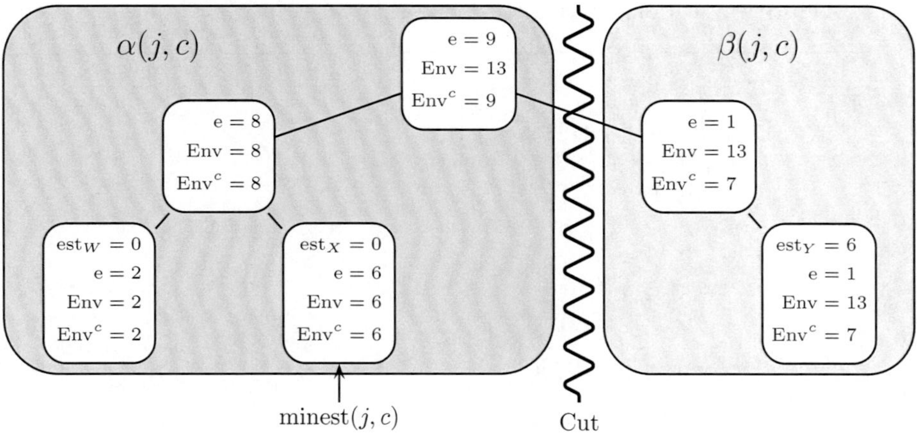

Fig. 6. Example: computation of $\mathrm{Env}(j, c)$ for $c = 1$ and $j = Y$ from example on Figure 5. Therefore $\mathrm{LCut}(T, j) = \{W, X, Y\}$. Situation just before the cut.

And then $C\,\mathrm{est}_\Omega + \mathrm{e}_\Omega = C\,\mathrm{est}_{\Omega_1} + \mathrm{e}_{\Omega_1} + \mathrm{e}_{\Omega_2}$. Let's apply this idea on formula (14). We define:

$$\alpha(j, c) = \{l,\ l \in \mathrm{LCut}\,(T, j)\ \&\ \mathrm{est}_l \le \mathrm{minest}\,(j, c)\}$$
$$\beta(j, c) = \{l,\ l \in \mathrm{LCut}\,(T, j)\ \&\ \mathrm{est}_l > \mathrm{minest}\,(j, c)\}$$

And (14) is equivalent to:

$$\begin{aligned}\mathrm{Env}(j, c) &= \max_{\substack{\Omega_1 \subseteq \alpha(j,c)\\ \Omega_2 \subseteq \beta(j,c)}} \{C\,\mathrm{est}_{\Omega_1} + \mathrm{e}_{\Omega_1} + \mathrm{e}_{\Omega_2}\} = \\ &= \mathrm{e}_{\beta(j,c)} + \mathrm{Env}\,(\alpha\,(j, c))\end{aligned} \tag{15}$$

We can compute $\mathrm{Env}\,(\alpha\,(j, c))$ by building Θ-tree for the set $\alpha(j, c)$ as shown in Proposition 1. However it is more suitable for the algorithm to build Θ-tree for the whole set $\mathrm{LCut}(T, j)$ and cut it into two parts just before the computation of $\mathrm{Env}\,(\alpha\,(j, c))$. The *cut* operation splits the tree into two trees, it is done in such a way that all activities $l \in \mathrm{LCut}(T, j)$ such that $\mathrm{est}_l \le \mathrm{minest}(j, c)$ go to the left part while the others go into the right part. See Figure 6 for an example. The cut operation has time complexity $\mathcal{O}(\log n)$ and it splits the set $\mathrm{LCut}(T, j)$ into sets $\alpha(j, c)$ and $\beta(j, c)$. The value $\mathrm{Env}\,(\alpha\,(j, c))$ can be found in the root node of the Θ-tree for $\alpha(j, c)$, and $\mathrm{e}_{\beta(j,c)}$ can be found in the root node of the Θ-tree for $\beta(j, c)$.

It remains to show how to compute $\mathrm{minest}(j, c)$. The value $\mathrm{minest}(j, c)$ was defined as:

$$\mathrm{minest}(j, c) = \min\{\mathrm{est}_\Omega,\ \Omega \subseteq \mathrm{LCut}\,(T, j)\ \&\ \mathrm{e}_\Omega > (C - c)(\mathrm{lct}_j - \mathrm{est}_\Omega)\}$$

The condition $\mathrm{e}_\Omega > (C - c)(\mathrm{lct}_j - \mathrm{est}_\Omega)$ is algebraically equivalent to:

$$(C - c)\,\mathrm{est}_\Omega + \mathrm{e}_\Omega > (C - c)\,\mathrm{lct}_j \tag{16}$$

Algorithm 2. Computation of $\mathrm{minest}(j, c)$ using Θ-tree for $\mathrm{LCut}(T, j)$

```
v := root;
E := 0;
while v is not a leaf node do begin
   if Env^c(right(v)) + E > (C − c) lct_j then
      v := right(v);
   else begin
      E := E + e_right(v);
      v := left(v);
   end;
end;
l := activity represented by leaf v;
return est_l;
```

Algorithm 3. Computation of all $\mathrm{update}(j, c)$

```
for c ∈ {c_m, m ∈ T} do begin
   Θ := ∅;
   upd := −∞;
   for j ∈ T in non-decreasing order by lct_j do begin
      Θ := Θ ∪ {j};
      minest := minest(j, c);                      // see Algorithm 2
      (α, β) := Cut(Θ, minest);
      Env(j, c) := e(β) + Env(α);                  // see (15)
      diff := ⌈(Env(j,c) + (C − c) lct_j) / c⌉;    // see (13)
      upd := max(upd, diff);                       // see (10)
      update(j, c) := upd;
      Θ := join(α, β);
   end;
end;
```

Notice that the left part of this inequality is very similar to the computation of energy envelope, just C is replaced by $(C - c)$. Let us define a new variant of energy envelope Env^c:

$$\mathrm{Env}^c(\Theta) = \max_{\Omega \subseteq \Theta} \{(C - c)\,\mathrm{est}_\Omega + \mathrm{e}_\Omega\}$$

The computation of Env^c can be done again using Θ-tree by Proposition 1. We can compute Env and Env^c in the same Θ-tree as shown on Figure 6. Now we can see that because of condition (16) it must hold:

$$\mathrm{Env}^c\left(\beta\left(j, c\right)\right) < \left(C - c\right)\mathrm{lct}_j$$

but if we would include activities l with $\mathrm{est}_l = \mathrm{minest}(j, c)$ into the right tree (in other words if we would do the cut more on the left) then this condition would not hold. That allows to find a leaf l with $\mathrm{est}_l = \mathrm{minest}(j, c)$ by following a path from the root the leaf as shown in Algorithm 2. Using this procedure we can compute all values update(j, c) by Algorithm 3 with time complexity $\mathcal{O}(kn \log n)$. Note that once update(j, c) is computed we can trivially update values est_i using (EF2).

8 Relation with Standard Edge Finding

We will show that the algorithm described in the paper does not miss any update done by Edge Finding algorithm described in [5]. It is enough to prove that the original Edge Finding propagation rules are subsumed by the new rules (EF1) and (EF2).

The traditional Edge Finding rule is:

$$\forall i \in T,\ \forall \Theta \subseteq T \backslash \Theta : \ C \left(\mathrm{lct}_\Theta - \mathrm{est}_{\Theta \cup \{i\}}\right) < \mathrm{e}_{\Theta \cup \{i\}} \Rightarrow \mathrm{est}_i := \max\left(\mathrm{est}_i, \mathrm{newest}_i\right)$$

where:

$$\mathrm{newest}_i = \max_{\substack{\Omega \subseteq \Theta \\ \mathrm{e}_\Omega > (C-c)(\mathrm{lct}_\Omega - \mathrm{est}_\Omega)}} \left\{ \mathrm{est}_\Omega + \left\lceil \frac{\mathrm{e}_\Omega - (C-c)\left(\mathrm{lct}_\Omega - \mathrm{est}_\Omega\right)}{c_i} \right\rceil \right\} \tag{17}$$

Let's consider an activity i and sets Θ and Ω which achieves the best update by the rule above. Then we can define j to be an activity from Θ such that $\mathrm{lct}_j = \mathrm{lct}_\Theta$. And because $\Theta \subseteq \mathrm{LCut}(T, j)$ we can see that the rule (EF1) holds for i and j. And because $\Omega \subseteq \Theta \subseteq \mathrm{LCut}(T, j)$ the update by the rule (EF2) must be at least the same as by the original rule (17).

9 Optional Activities

Optional activity is an activity which may or may not be present in the resulting schedule [1]. Optional activities makes modeling of certain types of problems much easier (for example dealing with alternatives) and it also allows the CP engine to propagate better. Therefore it is very important that Edge Finding algorithm can handle optional activities.

To handle optional activities we can use the same idea as suggested in [4]: instead of changing the algorithm we can just change its input data. If an activity j is optional, we set for the algorithm $\mathrm{lct}_j = \infty$ regardless the real value of lct_j. This way the algorithm can never conclude that $j \lessdot i$ for any activity i because from the point of view of the algorithm the activity j can be always scheduled later than i. Therefore optional activities will be influenced by non-optional ones, but non-optional activities will not be influenced by optional ones.

Note that propagation for optional activities could be further improved as suggested for unary resource in [6]. However it would probably result in increase of time complexity of the algorithm.

10 Experimental Results

Speed of the presented algorithm was tested against incomplete algorithm [2] by measuring time needed for initial propagation. These tests was done on cumulative job-shop instances with resources of capacity 2 (note that in this case $k = 1$). For $n = 20$ activities on resource the presented algorithm is on average faster by factor 1.34, for $n = 30$ it is faster by factor 1.60, for $n = 40$ by 1.99, for $n = 60$ by 2.68, for $n = 100$ by 4.15 and for $n = 200$ by factor 7.35.

11 Conclusions

This paper presents a new Edge Finding algorithm for discrete capacity resources. The new algorithm is stronger than the state-of-the-art algorithm [5], it is faster (in term of time complexity) and it can handle optional activities. The algorithm is successfully used by CP-Optimizer [1] starting from version 2.0.

References

1. IBM ILOG CP Optimizer, `http://www.ilog.com/products/cpoptimizer/`
2. Philippe Baptiste, C.L.P., Nuijten, W.: Constraint-Based Scheduling: Applying Constraint Programming to Scheduling Problems. Kluwer Academic Publishers, Dordrecht (2001)
3. Schutt, A., Wolf, A., Schrader, G.: Not-first and not-last detection for cumulative scheduling in $O(n^3 \log n)$. In: Umeda, M., Wolf, A., Bartenstein, O., Geske, U., Seipel, D., Takata, O. (eds.) INAP 2005. LNCS (LNAI), vol. 4369, pp. 66–80. Springer, Heidelberg (2006)
4. Vilím, P.: Max energy filtering algorithm for discrete cumulative resources. In: van Hoeve, W.J., Hooker, J.N. (eds.) CPAIOR 2009. LNCS, vol. 5547, pp. 294–308. Springer, Heidelberg (2009)
5. Mercier, L., Hentenryck, P.V.: Edge finding for cumulative scheduling. Informs Journal of Computing 20, 143–153 (2008)
6. Kuhnert, S.: Efficient edge-finding on unary resources with optional activities. In: Proceedings of INAP 2007 and WLP 2007 (2007)
7. Vilím, P.: Global Constraints in Scheduling. PhD thesis, Charles University in Prague, Faculty of Mathematics and Physics, Department of Theoretical Computer Science and Mathematical Logic (2007)

Evaluation of Length-Lex Set Variables

Justin Yip and Pascal Van Hentenryck

Department of Computer Science, Brown University, USA

Abstract. This paper presents the first experimental evaluation of the length-lex domain for set variables. The implementation is based on bound-consistency algorithms proposed in earlier work and two novel technical contributions: a generic filtering algorithm which automatically pushes ordering constraints into symmetric binary constraints with only a logarithmic overhead and an adaptation of symmetry-breaking constraints from 0/1 matrices to the length-lex ordering. The experimental results indicate that the length-lex representation for set variables is very effective and robust on traditional set-CSPs benchmarks.

1 Introduction

Set variables for constraint programming were proposed in the early 1990s [1] but they have received increasing attention in recent years. The main issue in set variables is that their domain may contain an exponential number of sets and an explicit representation is often too costly. Hence, most of the research has focused on representations that either approximate the set domain or are as compact as possible. The traditional set domain is the subset-bound representation [1,2], which maintains a pair of sets (R, P) to denote the domain $\{s|R \subseteq s \subseteq P\}$. The addition of a cardinality component to the subset-bound representation, and the associated pruning rules, was proposed in [3,4]. This richer domain was further enhanced by a lexicographic component in [5,6] since many set constraint satisfaction problems contain many symmetries which can broken by lexicographic constraints. 0/1 matrix models can also be used to encode the characteristic function of the subset-bound domain, while cardinality and lexicographic restrictions are expressed by posting constraints [7,8]. An exact representation of set domains using binary decision diagrams was proposed in [9]. It enables the solver to enforce arc consistency. The representation may require exponential space in the worst case, but convincing experimental results were given on a variety of benchmarks.

This paper considers the length-lex representation recently proposed in [10]. In theory, the length-lex representation offers three fundamental advantages. First, contrary to the subset-bound representation, it features a total ordering on sets, which makes it possible to enforce bound consistency. Second, the length-lex representation directly captures cardinality and lexicographic information. Finally, the representation often makes it possible to enforce bound consistency in polynomial time. However, despite these potential theoretical benefits, the length-lex domain was never evaluated experimentally.

I.P. Gent (Ed.): CP 2009, LNCS 5732, pp. 817–832, 2009.

The contributions of this paper are threefold. First, the paper presents the first experimental evaluation of the length-lex representation for set variables. The implementation, which consists of about 18,000 lines of code, was integrated inside the COMET system and evaluated on the traditional benchmarks for set variables. The experimental results on this preliminary implementation indicate that the length-lex representation for set variables is extremely effective on traditional benchmarks and the filtering algorithms reduce the search space by orders of magnitude compared to most representations. The models used for the benchmarks also feature two technical contributions. The first technical contribution is algorithmic and shows how to push lexicographic ordering constraints into (symmetric) set constraints. The idea of pushing lexicographic constraints into other constraints was proposed before (e.g., [11,10,20]), as it typically achieves a stronger propagation. The novelty in this paper is a generic routine that pushes the length-lex ordering into any symmetric binary constraint with only an $O(\log n)$ overhead. The second technical contribution pertains to dual modeling and symmetric breaking in primal/dual set-CSPs [7,8,13,14,15,16]. This paper shows that the traditional techniques used in 0/1 matrices to break variables and value symmetries also apply to the length-lex ordering. This is particularly significant since the length-lex representation capture lexicographic information directly, allowing a strong pruning of the search space. The rest of this paper is organized as follows. Section 2 gives an overview of the length-lex representation. Section 3 introduces the filtering algorithms to handle the combination of a symmetric constraint and a length-lex ordering constraint. Section 4 presents how to adapt the dual modeling technique to the length-lex domain. Section 5 presents the empirical results and Section 6 concludes the paper.

2 The Length-Lex Representation of Set Variables

Notations: For simplicity, we assume that sets take their values in a universe $U(n)$ of integers $\{1, \ldots, n\}$ equipped with traditional set operations. n, m are integers denoting the size of a universe, the number of variables, or both. S, T, P and Q, possibly subscripted, are set variables. X and Y are the length-lex domains of S and T respectively. Elements of $U(n)$ are denoted by the letter f, possibly subscripted and modified as $\check{f}$, $\dot{f}$ and $\hat{f}$ to denote the minimum, mean and maximum value respectively. Sets are denoted by lb, ub, s, t, w, x, y, z. A subset s of $U(n)$ of cardinality c is called c-set and is denoted as $\{s_1, s_2, \ldots, s_c\}$ where $(s_1 < s_2 < \ldots < s_c)$. The notation $s_{i..j}$ is a shorthand for $\{s_i, s_{i+1}, \ldots, s_j\}$.

Length-Lex Representation: The length-lex ordering $\preceq$, proposed in [10], totally orders sets first by cardinality and then lexicographically.

Definition 1 (Length-Lex Ordering). *The length-lex ordering is defined by*

$$s \preceq t \textbf{ iff } s = \emptyset \vee |s| < |t| \ \vee |s| = |t| \ \wedge (s_1 < t_1 \vee s_1 = t_1 \wedge s \setminus \{s_1\} \preceq t \setminus \{t_1\})$$

Its strict version is defined by $s \prec t$ ***iff*** $s \preceq t \wedge s \neq t$.

Example 1 (Length-Lex Ordering). Given $U(4) = \{1, \ldots, 4\}$, we have $\emptyset \prec \{1\} \prec \{2\} \prec \{3\} \prec \{4\} \prec \{1,2\} \prec \{1,3\} \prec \{1,4\} \prec \{2,3\} \prec \{2,4\} \prec \{3,4\} \prec \{1,2,3\} \prec \{1,2,4\} \prec \{1,3,4\} \prec \{2,3,4\} \prec \{1,2,3,4\}$.

Definition 2 (Length-Lex Interval). *Given a universe $U(n)$, a length-lex interval is a pair of sets $\langle lb, ub \rangle$. It contains all sets (inclusively) between lb and ub in the length-lex ordering, i.e., $\{s \subseteq U(n) | lb \preceq s \preceq ub\}$.*

Example 2 (Length-Lex Interval). Given $U(6)$, the interval $\langle \{1,3,4\}, \{1,5,6\} \rangle$ denotes the set $\{\{1,3,4\}, \{1,3,5\}, \{1,3,6\}, \{1,4,5\}, \{1,4,6\}, \{1,5,6\}\}$.

Because the length-lex ordering defines a total order on sets, it is possible to enforce bound consistency on set constraints.

Definition 3 (Bound Consistency for Binary Constraint). *A constraint $\mathcal{C}$ over two set variables S and T with respective domains $X = \langle lb_X, ub_X \rangle$ and $Y = \langle lb_Y, ub_Y \rangle$ is bound consistent if*

$$\exists y \in Y : \mathcal{C}(lb_X, y) \wedge \exists y \in Y : \mathcal{C}(ub_X, y) \wedge \exists x \in X : \mathcal{C}(x, lb_Y) \wedge \exists x \in X : \mathcal{C}(x, ub_Y).$$

Consistency algorithms for the length-lex representation have been studied intensively. Reference [10] presented algorithms to enforce bound consistency on many unary constraints in time $\tilde{O}(c)$, where c is the maximum cardinality of the set variable, and provided the first propagator for set domains whose running time is independent of the universe size n (usually $n \gg c$). A generic algorithm that enforces bound consistency on binary constraints and only relies on a simple feasibility routine was proposed in [17]. The generic algorithm can easily be generalized to a propagation routine for a k-arity constraint. That paper also demonstrated a specialized algorithm for binary disjoint constraint running in time $O(c^3)$. An $O(n^4/\epsilon)$ propagator for approximated bound consistency of a combination of two knapsack constraints was proposed in [19], while an algorithm for enforcing bound consistency on a knapsack constraint running in time $\tilde{O}(c)$ with an one-time preprocessing cost $O(c^2 n)$ was proposed in [18]. The same paper also gave an amortization scheme that reduces the running time for the binary disjoint constraint from $O(c^3)$ to $O(c^2)$.

PF-interval: [17] proposed the concept of PF-closed interval (hereinafter called *PF-interval* for short) in order to simplify the design of propagators. The key idea is that any length-lex interval can be partitioned into a linear number of PF-intervals, a special class of length-lex interval that enjoys some elegant properties of the subset-bound representation with cardinality. Informally, a PF-interval denotes all c-sets that begin with the same prefix, immediately followed by one element f of a set F (we call it the F-set), the rest being filled by elements greater than f. The F-set is critical for the efficient inference of lexicographic order, since it determines the most significant element after the required prefix. Moreover, the generic algorithm in [17] guarantees that the F-set is always a range. This paper exploits this property and formally redefines PF-interval.

Definition 4 (PF-Interval). *Let P be a set and $\check{f}$, $\hat{f}$, n and c be integers. A PF-interval $pf\langle P, \check{f}, \hat{f}, n, c\rangle$ satisfies*

$$(\max(P) < \check{f}) \wedge (\check{f} \leq \hat{f}) \wedge (n - \hat{f} + 1 \geq c - |P|)$$

and denotes the set of sets

$$\Big\{P \uplus \{f\} \uplus s \Big| \check{f} \leq f \leq \hat{f} \wedge s \subseteq \{f+1, \ldots, n\} \wedge |P \uplus \{f\} \uplus s| = c\Big\}.$$

Example 3 (PF-interval). Consider the length-lex interval in example 2. It contains all sets that begin with $\{1\}$, and immediately followed by $3, 4$ or 5, and filled by elements in $\{3, ..., 6\}$. It can be expressed as a PF-interval $pf\langle\{1\}, 3, 5, 6, 3\rangle$.

Example 4 (PF-interval). The PF-interval $pf\langle\{1, 2\}, 5, 6, 8, 4\rangle$ denotes of set of sets $\{\{1, 2, 5, 6\}, \{1, 2, 5, 7\}, \{1, 2, 5, 8\}, \{1, 2, 6, 7\}, \{1, 2, 6, 8\}\}$.

The structure of PF-interval makes its inferences almost equivalent to the subset-bound+cardinality representation. The inferences are based on a feasibility routine hs that takes two intervals and return whether there is a solution.

Specification 1 (Feasibility Routine $hs\langle\mathcal{C}\rangle$). *Given a constraint $\mathcal{C}$ and length-lex intervals X and Y, $hs\langle\mathcal{C}\rangle(X, Y) \equiv \exists x \in X, y \in Y : \mathcal{C}(x, y)$.*

Consider the binary disjoint constraint $\mathcal{D}$. Given two PF-intervals, the feasibility routine $hs\langle\mathcal{D}\rangle$ checks whether two prefixes (which corresponds to required sets in the subset-bound domain) are disjoint and whether there are enough *free* elements (possible sets) to satisfy the cardinality requirements. In addition, the routine must check if both PF-intervals can pick different elements from their F-set. If these three conditions hold, there is a solution. The successor(predecessor) construction algorithms are based on this feasibility routine and they greedily pick the smallest(largest) element one at a time that satisfies the feasibility routine. The whole process takes $O(c)$ time.

However, not all length-lex intervals enjoy these elegant properties. This problem can be remedied by observing any length-lex interval can be partitioned into $O(c)$ PF-intervals. With this decomposition technique available, we can enforce bound consistency on binary length-lex constraint by first partitioning both length-lex intervals into $O(c)$ PF-intervals, and then performing pair-wise comparisons. The total runtime is $O(c^3)$. See [17] for more details.

3 Pushing Length-Lex Ordering into Binary Constraints

This section shows how to push a length-lex constraint into a binary symmetric constraint. The idea of pushing lexicographic constraints into other constraints was proposed before (e.g., [11,10,20]). In particular, [11] presented an algorithm to enforce the lexicographic ordering and sum constraint for two vectors of variables simultaneously. [20] proposed an generic algorithm to push the lexicographic order into a global constraint by invoking $O(m)$ calls to the domain-consistent global constraint propagator, m being the number of variables.

Algorithm 1. $bc\langle\mathcal{C}_{\preceq}\rangle(X = \langle lb_X, ub_X\rangle, Y = \langle lb_Y, ub_Y\rangle)$

$(lb_X, ub_X) \leftarrow (succ_X\langle\mathcal{C}_{\preceq}\rangle(X,Y), pred_X\langle\mathcal{C}_{\preceq}\rangle(X,Y))$
$(lb_Y, ub_Y) \leftarrow (succ_Y\langle\mathcal{C}_{\preceq}\rangle(X,Y), pred_Y\langle\mathcal{C}_{\preceq}\rangle(X,Y))$
return $lb_X \neq \bot \wedge ub_X \neq \bot \wedge lb_Y \neq \bot \wedge ub_Y \neq \bot$

Table 1. The Slicing (Left) and Decomposition (Right) of Length-Lex Domains

X:$\langle\{1,4,5\},\{6,7,9\}\rangle$	Y:$\langle\{2,3,4\},\{6,8,9\}\rangle$	$\check{X}^1_{pf}$: $pf\langle\{1\},4,8,9,3\rangle$	:
$\check{X}$:$\langle\{1,4,5\},\{1,8,9\}\rangle$	$\check{Y}$: $\emptyset$	$\dot{X}^1_{pf}$: $pf\langle\{\},2,5,9,3\rangle$	$\dot{Y}^1_{pf}$: $pf\langle\{\},2,5,9,3\rangle$
$\dot{X}$:$\langle\{2,3,4\},\{6,7,9\}\rangle$	$\dot{Y}$:$\langle\{2,3,4\},\{6,7,9\}\rangle$	$\dot{X}^2_{pf}$:$pf\langle\{6,7\},8,9,9,3\rangle$	$\dot{Y}^2_{pf}$:$pf\langle\{6,7\},8,9,9,3\rangle$
$\hat{X}$: $\emptyset$	$\hat{Y}$:$\langle\{6,8,9\},\{6,8,9\}\rangle$	:	$\hat{Y}^1_{pf}$:$pf\langle\{6,8\},9,9,9,3\rangle$

In this section, we give a generic bound consistency algorithm $bc\langle\mathcal{C}_{\preceq}\rangle$ that pushes the length-lex ordering constraint into a symmetric binary constraint[1], only assuming a feasibility routine $hs\langle\mathcal{C}\rangle$. The algorithm $bc\langle\mathcal{C}_{\preceq}\rangle$ enforces bound consistency with time $O(\alpha c^2 \log^2 n)$, which is a $O(\log n)$ overhead incurred over the generic bound consistency algorithm $bc\langle\mathcal{C}\rangle$. Indeed, $bc\langle\mathcal{C}\rangle$ is an algorithm presented in [17] which enforces bound consistency in $O(\alpha c^2 \log n)$ time and only relies on a feasibility routine $hs\langle\mathcal{C}\rangle$ over two PF-intervals(that takes $O(\alpha)$ time). $hs\langle\mathcal{C}\rangle$ is usually computationally inexpensive. For instance, for binary disjoint constraint $\mathcal{D}$, $hs\langle\mathcal{D}\rangle$ takes $O(c)$ and therefore $bc\langle\mathcal{D}\rangle$ takes $O(c^3 \log n)$, which is only a $O(\log n)$ overhead compared to the specialized $O(c^3)$ algorithm.

Definition 5 (Symmetric Constraint). *A binary constraint C over two set variables S and T is* symmetric *if and only if $C(S,T) \Leftrightarrow C(T,S)$.*

The generic algorithm $bc\langle\mathcal{C}_{\preceq}\rangle$ is depicted in Algorithm 1 and simply computes the predecessor and successor for the two sets. What is interesting is their implementation which is based on two key observations. First, if the greatest set in X is smaller than the smallest set in Y, the length-lex ordering constraint is entailed and we can simply apply $bc\langle\mathcal{C}\rangle$. Second, when two PF-intervals are identical ($X = Y$) and C is symmetric, if $hs\langle C\rangle(X,Y)$ holds, then $hs\langle C_{\preceq}\rangle(X,Y)$ also holds. As a consequence of the first property, the algorithms start by slicing the representation in several pieces, which we first illustrate on an example before defining it formally.

Example 5. Suppose we have $U(9)$, $X = \langle\{1,4,5\},\{6,7,9\}\rangle$, $Y = \langle\{2,3,4\},\{6,8,9\}\rangle$. X can be sliced into two length-lex intervals, the first interval $\check{X}$ contains all sets smaller than $\min Y = \{2,3,4\}$ and the second interval $\dot{X}$ contains all sets in Y. Similarly, Y can also be sliced. Table 1 shows the partitions. The table also shows the decomposition of the resulting intervals in PF-intervals. Note that all sets in $\check{X}$ and $\hat{Y}$ satisfy the ordering constraint.

[1] The restriction to symmetric constraint is natural, since otherwise the symmetry would already be broken by the constraint itself.

Specification 2 (3Slices). *Given two length-lex intervals X and $Y = \langle lb_Y, ub_Y \rangle$. $3Slices(X,Y)$ returns three intervals $\check{X}, \dot{X}$, and $\hat{X}$ such that*

$$\check{X} \equiv \{x \in X | x \prec lb_Y\} \text{ and } \dot{X} \equiv \{x \in X | x \in Y\} \text{ and } \hat{X} \equiv \{x \in X | ub_Y \prec x\}$$

Lemma 1. *Given two length-lex intervals X and Y. $3Slices(X,Y) = \check{X}, \dot{X}, \hat{X}$ and $3Slices(Y,X) = \check{Y}, \dot{Y}, \hat{Y}$, we have $\dot{X} = \dot{Y}$.*

The algorithm $bc\langle \mathcal{C}_{\preceq} \rangle$ first decomposes a length-lex interval into two parts. One part can ignore the length-lex ordering because the ordering constraint is already satisfied and calls the existing $bc\langle \mathcal{C} \rangle$ algorithm. The other part deals with two identical intervals and exploits the following symmetric property.

Lemma 2. *If a binary constraint C is symmetric, $hs\langle C \rangle(X,X) = hs\langle C_{\preceq} \rangle(X,X)$.*

Proof. Suppose $C(s,t)$ holds for $s,t \in X$. By symmetry, $C(t,s)$ holds and $s \preceq t \vee t \preceq s$ also holds. □

3.1 $pred_X$ for Length-Lex Intervals

We now demonstrate the generic algorithm (Algorithm 2) for finding the new upper bound for X.

Specification 3 ($pred_X$). *For a constraint $\mathcal{C}$ and two length-lex intervals X and Y, function $pred_X\langle \mathcal{C} \rangle(X,Y)$ returns the greatest set $x \in X$ in the length-lex ordering which belongs to a solution of the constraint or $\perp$ if there is no solution:*

$$pred_X\langle \mathcal{C} \rangle(X,Y) \equiv \begin{cases} \max\{x \in X \mid \exists y \in Y : \mathcal{C}(x,y)\} & \text{if } hs\langle \mathcal{C} \rangle(X,Y) \\ \perp & \text{otherwise} \end{cases}.$$

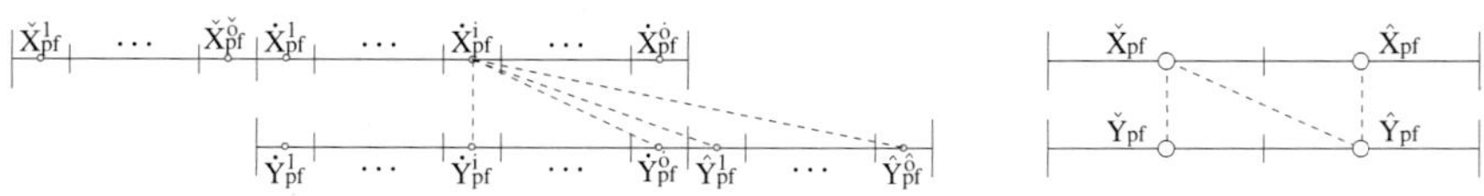

Fig. 1. Key Observations. $pred_X\langle \mathcal{C}_{\preceq} \rangle(X,Y)$ (Left). $pred_X\langle \mathcal{C}_{\preceq} \rangle(X_{pf}, Y_{pf})$ (Right).

The algorithm first slices the length-lex intervals (lines 1–2) and then decompose the results into PF-intervals (lines 3–4). $\hat{X}$ and $\check{Y}$ are not considered as they violate the ordering constraints. The algorithm then locates the largest PF-interval that contains the new bound and constructs the new bound (lines 5–10). The routine for locating the PF-interval is similar to $bc\langle \mathcal{C} \rangle$, except that we must pay some attention to the ordering constraint. The intuition is captured in Figure 1. When considering PF-intervals $\dot{X}^i_{pf}$, we only consider the PF-intervals in Y no smaller than $\dot{X}^i_{pf}$. Two cases must be distinguished. First, there is exactly one PF-interval in Y identical to $\dot{X}^i_{pf}$ and the algorithm must call $pred_X\langle \mathcal{C}_{\preceq} \rangle$ to take into account the ordering constraint (line 8 and the vertical line in Figure 1 (Left)). Second, the remaining PF-intervals are greater than $\dot{X}^i_{pf}$ and the

Algorithm 2. $pred_X\langle\mathcal{C}_{\preceq}\rangle(X = \langle lb_X, ub_X\rangle, Y = \langle lb_Y, ub_Y\rangle)$

1: $\check{X}, \dot{X}, \hat{X} \leftarrow 3Slices(X, Y)$
2: $\check{Y}, \dot{Y}, \hat{Y} \leftarrow 3Slices(Y, X)$
3: $[\dot{X}^1_{pf}, ..., \dot{X}^{\dot{o}}_{pf}] \leftarrow decomp(\dot{X})$
4: $[\hat{Y}^1_{pf}, ..., \hat{Y}^{\hat{o}}_{pf}] \leftarrow decomp(\hat{Y})$
5: $s \leftarrow \top$
6: **for** $i = \dot{o}$ down to 1 **do**
7: $\quad s \leftarrow \max(\max_{j\in\{1..\hat{o}\}}(pred_X\langle\mathcal{C}\rangle(\dot{X}^i_{pf}, \hat{Y}^j_{pf})), \max_{j\in\{i+1..\dot{o}\}}(pred_X\langle\mathcal{C}\rangle(\dot{X}^i_{pf}, \dot{X}^j_{pf})))$
8: $\quad s \leftarrow \max(s, pred_X\langle\mathcal{C}_{\preceq}\rangle(\dot{X}^i_{pf}, \dot{X}^i_{pf}))$
9: $\quad$ **if** $s \neq \top$ **then**
10: $\quad\quad$ **return** s
11: **return** $pred_X\langle\mathcal{C}\rangle(\check{X}, \dot{Y} \uplus \hat{Y})$

algorithm simply calls $pred_X\langle\mathcal{C}\rangle$ (line 7 and the diagonal lines in Figure 1 (Left)). If no PF-interval in $\dot{X}$ contain a new upper bound, we will try $\check{X}$. As all sets in $\check{X}$ are smaller than Y, we use the simple predecessor algorithm $pred_X\langle\mathcal{C}\rangle$ (line 11). Note also that the loop starts with the largest PF-interval (line 6), since we are interested in the largest predecessor.

3.2 $pred_X$ for PF-Interval

It remains to show how to find the predecessor for two identical PF-intervals (line 8). Algorithm 3 constructs the upper bound by performing a binary search in the F-set and the main idea is depicted in Figure 1 (Right). It partitions the F-set into two halves (line 9) and first checks if there is a solution in the upper half. (line 11 and the right vertical line in the figure). Notice that by Lemma 2, we can simply call $hs\langle\mathcal{C}\rangle$. This upper half needs to be compared only with the upper half $\hat{Y}_{pf}$ of Y_{pf}, since the lower half $\check{Y}_{pf}$ violates the ordering constraint. If there is a solution (line 10), the algorithm is called recursively within $\hat{X}_{pf}$ (line 11). Otherwise, the algorithm tries the lower half $\check{X}_{pf}$. Now there are two possible choices ($\check{Y}_{pf}$ and $\hat{Y}_{pf}$) and we do not know which one gives a greater bound. The algorithm tries both (lines 13–14) and returns the largest (line 15). Once the F-set becomes a singleton, it is inserted in the prefix and the algorithm considers the next position (lines 3–8).

Lemma 3. *Suppose* $hs\langle\mathcal{C}\rangle(X_{pf}, Y_{pf})$ *takes* $O(\alpha)$. *Algorithm 3 takes* $O(\alpha c^2 \log^2 n)$.

Proof. The running time of Algorithm 3 is affected by the number of unfixed positions (maximum is c), the number of possible choices r in the F-set (i.e., $\hat{f} - \check{f} + 1$) and the universe size n. Denote the computation time of the algorithm by $T_n(r, c)$. $T_n(r, c)$ can be expressed by the following recurrence relation:

$$T_n(r,c) = \begin{cases} O(\alpha) + \max(T_n(r/2, c), T_n(r/2, c) + O(\alpha c \log n)) & \text{if } r > 1 \\ T_n(n, c-1) & \text{if } r = 1 \wedge c > 1 \\ O(1) & \text{otherwise} \end{cases}$$

Solving the recurrence relation gives $T_n(r, c) = O(\alpha c^2 \log^2 n)$. □

Algorithm 3. $pred_X\langle \mathcal{C}_{\preceq} \rangle(X_{pf} = pf\langle P, \check{f}, \hat{f}, n, c\rangle, Y_{pf} = pf\langle P, \check{f}, \hat{f}, n, c\rangle)$

Require: $X_{pf} == Y_{pf}$
if not $hs\langle \mathcal{C}_{\preceq} \rangle(X_{pf}, Y_{pf})$ **then**
  return $\perp$
if $\check{f} = \hat{f}$ **then**
  if $|P| = c - 1$ **then**
    return $P \uplus \{\check{f}\}$
  else
    $P', \check{f}', \hat{f}' \leftarrow P \uplus \{\check{f}\}, \check{f} + 1, n - c - |P|$
    return $pred_X\langle \mathcal{C}_{\preceq} \rangle(pf\langle P', \check{f}', \hat{f}', n, c\rangle, pf\langle P', \check{f}', \hat{f}', n, c\rangle)$
$\dot{f} \leftarrow (\check{f} + \hat{f})/2$
if $hs\langle \mathcal{C} \rangle(pf\langle P, \dot{f} + 1, \hat{f}, n, c\rangle, pf\langle P, \dot{f} + 1, \hat{f}, n, c\rangle)$ **then**
  return $pred_X\langle \mathcal{C}_{\preceq} \rangle(pf\langle P, \dot{f} + 1, \hat{f}, n, c\rangle, pf\langle P, \dot{f} + 1, \hat{f}, n, c\rangle)$
else
  $s_0 = pred_X\langle \mathcal{C} \rangle(pf\langle P, \check{f}, \dot{f}, n, c\rangle, pf\langle P, \dot{f} + 1, \hat{f}, n, c\rangle)$
  $s_1 = pred_X\langle \mathcal{C}_{\preceq} \rangle(pf\langle P, \check{f}, \dot{f}, n, c\rangle, pf\langle P, \check{f}, \dot{f}, n, c\rangle))$
  return $\max(s_0, s_1)$

Theorem 1. *Assume* $hs\langle \mathcal{C} \rangle(X_{pf}, Y_{pf})$ *takes time* $O(\alpha)$. *Then Algorithm 2 takes* $O(\alpha c^2 \log^2 n)$.

Proof. Locating the first supported X_{pf} takes $O(\alpha c^2)$ time. Once the algorithm locates the X_{pf}, it constructs a predecessor against every possible Y_{pf}. There are at most $O(c)$ possible choices and $O(c)$ of them require only the simple predecessor algorithm $pred_X\langle \mathcal{C} \rangle$ that takes $O(\alpha c \log n)$ [17] . At most one of them must be taken care specially by Algorithm 3 and takes $O(\alpha c^2 \log^2 n)$. Hence, the total run time is $O(\alpha c^2 \log^2 n)$. □

The above generic algorithm pushes the length-lex ordering constraint into arbitrary binary symmetric constraints. Specialized algorithms can of course be designed for specific constraints. For instance, there exists a specialized algorithm for binary disjoint&length-lex constraint with $O(1)$ overhead. Space constraints do not allow us to describe it in detail. The key observation is by disjointness, we include the ordering constraint by considering the first element only.

4 Dual Modeling for Length-Lex Set Variables

This section considers *fully interchangeable* set-CSPs in which both the variables and the values are fully interchangeable. In the 0/1 matrix formulation, these symmetries are broken by imposing a lexicographic ordering on both the rows and columns. It is guaranteed that some solutions of each symmetry class remains after this process. Since the length-lex representation provides a total ordering on its sets, it also provides an ideal vehicle to break symmetries and we would like to use a similar technique with length-lex variables. Variable symmetries can be broken by imposing an ordering on the set variables. If the values

Table 2. Preserving the length-lex ordering by padding dummy elements

	Original	0/1 of Original	Padded	0/1 of Padded
w	{1,3}	{1,0,1,0}	{-4,-3,1,3}	{1,1,0,0,1,0,1,0}
x	{3,4}	{0,0,1,1}	{-4,-3,3,4}	{1,1,0,0,0,0,1,1}
y	{1,2,4}	{1,1,0,1}	{-4,1,2,4}	{1,0,0,0,1,1,0,1}
z	{1,3,4}	{1,0,1,1}	{-4,1,3,4}	{1,0,0,0,1,0,1,1}

are also interchangeable, we can consider the dual problem and impose an ordering on the dual variables. Unfortunately, it is unclear whether enforcing the length-lex ordering on both variables and values will still leave some solutions in each symmetry class, since the length-lex ordering is different from the lex ordering (See Example 6 below). The contribution of this section is to show that imposing a *double length-lex ordering* on a fully interchangeable set-CSP does not eliminate all solutions in each symmetry class.

Definition 6 (Set-CSP). *A set-CSP is a pair $\langle V \times D, C\rangle$, where V denotes the set of variables, D denotes the universe for these variables. An (primal) assignment $\gamma : V \to \mathbb{P}(D)$ maps variables to sets. $C : (V \to \mathbb{P}(D)) \to bool$ is a constraint that specifies which assignments are solutions (i.e. $C(\gamma) = true$).*

Definition 7 (Fully Interchangeable Set-CSP). *A set CSP is fully interchangeable if and only if when γ is a solution, for any bijective $\sigma : V \to V$ and $\tau : D \to D$, and a mapping function $\phi_f(s) = \{f(e)|e \in s\}$, assignment $\gamma' = \phi_\tau \circ \gamma \circ \sigma$ is also a solution.*

The key observation is that, when sets are of the same cardinality, their length-lex order is equivalent to the lexicographic order. By padding some dummy elements to make all sets the same size, we can reduce the length-lex ordering technique to the lexicographic order of 0/1 matrices.

Example 6. Table 2 illustrates difference between length-lex and lex ordering for a universe $U(4)$ and four sets w, x, y, z. The first column shows the sets in length-lex order. These sets are not in lex-order as $x >_{lex} y$. The 0/1 characteristic function (second column) is not in anti-lex-order either since $x <_{lex} y$. By padding dummy elements (third column), the sets are in both length-lex-order and lex-order and their 0/1 characteristic functions are in anti-lex order.

The formal definition of padding is as follows.

Definition 8 (Padding). *We abuse the notation of a universe to allow negative value element, such that $U'(n) = \{-n, ..., -1, 1, ..., n\}$. $pad_n : \mathbb{P}(U(n)) \to \mathbb{P}(U'(n))$ maps a set to a n-set padded by dummy elements. Formally, given $s \subseteq U(n)$ and $c = |s|$, $pad_n(s) \equiv \{-n, ..., -(c+1), s_1, ..., s_c\}$.*

Lemma 4. *Suppose $s, t \subseteq U(n)$. $s \preceq t \Leftrightarrow pad_n(s) \leq_{lex} pad_n(t)$.*

Proof. Trivial when $|s| = |t|$. When $|s| < |t|$, denote $s' = pad_n(s), t' = pad_n(t)$, observe that $s'_{1..n-|t|} = t'_{1..n-|t|}$ as they are dummy elements. $s'_{n-|t|+1}$ is a dummy negative element, whilst $t'_{n-|t|+1} = t_1$, Hence $s' \leq_{lex} t'$. □

Definition 9 (Double Length-Lex Primal/Dual Set-CSP). *Let* $\langle P_M \times N, C\rangle$ *be a CSP where* $P_M = \{P_1, ..., P_m\}$ *are (primal) set variables and* $N = \{1, ..., n\}$*. Its double length-lex primal/dual version is defined as* $\langle P_M \times N \uplus Q_N \times M, C'\rangle$ *where* $Q_N = \{Q_1, ..., Q_n\}$ *are the dual set variables,* $M = \{1, ..., m\}$ *and*

$$C' \equiv C \wedge (P_1 \preceq \ldots \preceq P_m) \wedge (Q_1 \preceq \ldots \preceq Q_n) \wedge \bigwedge_{i \in N, j \in M} (j \in P_i \Leftrightarrow i \in Q_j)$$

Theorem 2. *Given a fully interchangeable CSP, its double length-lex primal/dual version does not eliminate all solutions in each symmetry class.*

Proof. (sketch) Consider a solution γ, transform it to $\gamma' = pad_{\max(n,m)} \circ \gamma$. Every sets in the range of γ' are of the same cardinality. The case reduces to the double anti-lex ordering in the 0/1 matrix model. □

5 Experimental Evaluation

This section compares the length-lex set domain with other set-variable representations in four standard benchmarks: The Social Golfer (SG), Error Correcting Code (EC), Steiner System(SS) and Balanced Incomplete Block Design (BI) problems. The goal is to compare the performance (time and failures) between different domain representations (see Table 3). We try our best to compare different static modeling techniques under the same search strategy (with few explicitly stated exceptions). For some techniques (like the ROBDD domain), a variety of approaches and search strategies were proposed and it is impossible to list all results. Hence we choose the best overall approaches as stated by the authors. Correspondingly, in Table 3, symbol ∘ indicates the compared approaches, while symbol × denotes the fact that the corresponding papers also evaluate this benchmark but the results is not included in our comparison for space reasons. The column *tl* indicates the time limit in seconds (timeout instances are denoted as −1) and the column *cpu* denotes the processor type and speed used. No existing approach, except ours, was evaluated on all benchmarks. On all benchmarks, the search heuristic of our algorithm uses a simple static ordering. Note that dynamic symmetry-breaking techniques are available for some of these benchmarks but such comparison is out of the scope of this paper.

Social Golfer Problem. Figure 2 gives the social golfer model in Comet which takes 3 parameters `nbgroup`, `nbsize`, and `nbweek`. `X[w,i]` is the primal set variable representing the `i`-th group in `w`-th week and `Y[g]` is the dual set variable denoting the groups assigned to golfer `g`. Like most approaches, the search initially fixes some variables due to the symmetries, i.e., the first week and first group of the second week in the primal model and the first `nbsize` players in the dual model. The search first instantiates the first group of every week, then performs a week-wise labeling (except for 6-5-5, where a simple week-wise labeling is used).

The length-lex representation gives very robust results. It quickly solves all instances that other approaches can solve. Compared to ROBDD, length-lex is

Table 3. Evaluation Overview

Name	Abbrev.	SG	EC	SS	BI	tl	cpu
Length-Lex/seq	Length-Lex	○	○	○	○	900	C2D–M 2.53GHz
ROBDD-split/seq [9]	Split/seq	○	×	×		600	P4 2.8GHz
ROBDD-split/minDom [9]	Split/dyn	○				600	P4 2.8GHz
ROBDD-domain/seq [9]	Domain/seq	×	○	○		600	P4 2.8GHz
ROBDD-bound/seq [9]	Bound/seq	×	×	○		600	P4 2.8GHz
Pair-atmost-1/minDom [21]	Pair/dyn	○					Xeon 3.8GHz
Dual-subset-Bound/seq [5][6]	Subset/seq		×	○		240	P4 2GHz
Dual-hybrid/seq [5][6]	Hybrid/seq		×	○		240	P4 2GHz
Cardinal/seq [4]	Card/seq			○		900	P4 2.4GHz
Cardinal/minDom [4]	Card/dyn	○				900	P4 2.4GHz
Dual-set-int/minDom [14]	Set-int/dyn	○		×		7200	Sun Blade 1000
Valprec+dual/minDom [14]	Valprec/dyn	×		○		7200	Sun Blade 1000
0/1-Matrix-lex-sum/seq [8][22]	Lex-sum/seq	×		○	○	3600	PIII 1GHz
0/1-Matrix-lex/seq [8][22]	Lex/seq	×		×	○	3600	PIII 1GHz
Max-Variety/maxDeg [23]	VM/dyn				○		Ultra60 360MHz

```
var<CP>{set{int}} X[Weeks,Groups](cp,Golfers,nbsize);
var<CP>{set{int}} Y[Golfers](cp,WeekGroups,nbweek);
solve<cp> {
  forall(w in Weeks, i in Groups, j in Groups: i < j)
       cp.post(lldisjointLeq(X[w,i],X[w,j]));
  forall(w in Weeks, wo in Weeks: w < wo, i in Groups, j in Groups)
    cp.post(llatmostIntersection(X[w,i],X[wo,j],1));
  forall(w in Weeks: w < Weeks.getUp())
    cp.post(llatmostIntersectionLeq(X[w,1],X[w+1,1],1));
  cp.post(llchanneling(all(w in Weeks, g in Groups) X[w,g],Y));
  forall(g in Golfers: g < Golfers.getUp())
    cp.post(llleq(Y[g],Y[g+1]));
}
```

Fig. 2. Model of Social Golfer Problem in COMET

significantly faster. It dramatically outperforms (in speed and in the explored nodes) Cardinal and Set-int/dyn, a dual-modeling approach using integer dual variables. Pair-at-most-1 uses the subset-bound representation and some decomposition ideas which were inspired by length-lex. In general, length-lex also dominates this approach significantly in time and in explored nodes, except on 10-3-* problems for which it is roughly similar when machines are scaled (We run our tests on an energy efficient mobile processor, while Pair-at-most-1 is on non-consumer-level Xeon processor with a significantly faster processor speed).

Error Correcting Code Problem. The error correcting code problem is a challenging optimization problem which requires to explore the entire search tree to prove the optimality. This benchmark was first proposed in [5] in which only a graph is reported, making it impossible to compare on an instance-by-instance basis. The ROBDD approach was able to prove the optimality of 51 instances

Table 4. Social Golfer Problem

	Length-Lex		Split/seq		Split/dyn		Set-int/dyn		Card/dyn	Pair/dyn	
g,s,w	Time	Fails	Time	Fails	Time	Fails	Time	Fails	Time	Time	Fails
4,2,4	**0.01**	**0**					0.19	266			
4,2,5	**0.01**	**0**					0.52	884			
4,2,6	**0.01**	**0**					0.51	721			
4,2,7	**0.01**	**0**					0.15	97			
4,3,4	**0.01**	**10**					0.81	3222			
4,3,5	**0.40**	**732**	32.1	5165	26	3812			165.63		
4,3,6	**0.02**	**29**	23	2132	15.2	1504			94.67		
4,4,4	**0.06**	**111**					8.11	15759			
4,4,5	**0.05**	**57**					5.63	7510			
5,2,3	**0.01**	**0**					1.07	521			
5,2,4	**0.01**	**0**					78.46	95063			
5,2,5	**0.01**	**0**					1956.71	2554588			
5,2,6	**0.01**	**0**									
5,2,7	**0.02**	**1**									
5,2,8	**0.02**	**1**									
5,2,9	**0.02**	**1**					6841.59	3895064			
5,3,3	**0.01**	**1**					140.06	491452			
5,3,4	**0.01**	**1**									
5,3,5	**0.02**	**5**									
5,3,6	**0.41**	316	1.5	82	1	**34**			-1		
5,3,7	74.59	46117			**13.1**	**528**			-1		
5,4,2	**0.01**	11	0.1	**0**	0.1	**0**			0.83		
5,4,3	**0.02**	24	0.3	**0**	0.3	**0**	98.23	658755	1.89		
5,4,4	**0.14**	194	0.6	**0**	0.6	**0**	4770.94	30802587	3.13		
5,4,5	1.87	1947	2.3	41	**1.7**	**18**			28.65		
5,5,3	**0.06**	**93**					0.46	1418			
5,5,4	4.72	**6876**					**2.1**	13009			
5,5,5	**54.27**	**50623**									
5,5,6	**29.21**	**15769**									
5,5,7	**0.01**	1	0.4	**0**	0.4	**0**			-1		
6,3,2	**0.00**	**0**					1.16	30			
6,3,3	**0.01**	**1**									
6,3,4	**0.01**	**1**									
6,3,5	**0.02**	**6**									
6,3,6	**0.04**	10	1.4	**0**	1.3	7			1.2		
6,4,2	**0.01**	14	0.1	**0**	0.1	**0**	0.99	45	1.75		
6,4,3	**0.03**	42	1.4	**0**	0.9	**0**			4.62		
6,5,2	**0.05**	118					0.1	**60**		17.2	171664
6,5,3	**2.54**	**3351**								29.6	197607
6,5,4	**32.60**	31270	80.7	**0**	40.1	**0**			-1	39.7	197837
6,5,5	**28.76**	**6758**								75.5	239966
6,6,3	**0.82**	**661**					414.16	1521747			
7,2,2	**0.01**	**0**					0.69	21			
7,3,2	**0.01**	**1**					58.94	42			
7,4,2	**0.01**	**0**	0.4	**0**	0.4	**0**	236.73	63	2.82		
7,4,3	**0.03**	21	8.4	**0**	1.7	**0**			6.37		
7,4,4	**0.05**	26	481.6		5.1	**0**			12.46	4.4	27877
7,4,5	**0.36**	152	-1	-1	12.8	**0**			17.18		
7,5,2	**0.31**	574					42.98	**84**			
7,6,2	0.78	1271					**0.53**	**105**			
7,7,2	**0.28**	**0**					0.7	308			
8,3,5	34.52	45477	7.8	**0**	3.9	**0**			**1.01**		
8,4,4	**0.06**	**18**								157.7	738393
8,5,2	**0.25**	307	1.6	**0**	1.6	**0**			-1		
9,4,4	**0.21**	94	-1	-1	107.4	**0**			42.45		
10,3,6	**5.86**	**2941**								17.3	57364
10,3,9	233.80	**45437**								**52.4**	78613
10,3,10	210.80	**25246**								**67.2**	78976
10,4,4	**0.27**	**104**								4	22043
10,4,5	**0.58**	**149**								4.5	22044

Table 5. Error Correcting Code: 11 Difficult Cases(Left), 51 Easy Cases(Right)

	Length-lex		Domain	
l,d,w	Time	Fails	Time	Fails
8,4,4	**0.07**	**110**	1.6	224
9,4,3	**2.05**	**4617**	11.3	5615
9,4,6	**0.40**	**908**	25.4	16554
10,6,5	**0.034**	**158**	26.7	16635
9,4,4	-1	-1	-1	-1
9,4,5	-1	-1	-1	-1
10,4,3	**359.3**	**629822**	-1	-1
10,4,4	-1	-1	-1	-1
10,4,5	-1	-1	-1	-1
10,4,6	-1	-1	-1	-1
10,4,7	**1.99**	**4415**	-1	-1

	Length-Lex		Domain	
	time	fails	time	fails
Mean	**0.0060**	**5.92**	0.2	210.7
Total	**0.31**		11.4	
min	**0.0038**	1	0.03	**0**
25 percentile	**0.0042**	1	0.03	**0**
median	**0.0046**	**1**	0.04	2
75 percentile	**0.0053**	**1**	0.06	25
max	**0.47**	**134**	4.16	3740

Table 6. Experimental Results on the Steiner Triple System

	Length-Lex		Bound/seq		Domain/seq		Lex-sum/seq		Card/seq	Valprec/dyn	
n	Time	Fails	Time	Fails	Time	Fails	Time	Fails	Time	Time	Fail
7	0.002	**0**	**$<$ 0.1**	8	**$<$ 0.1**	8	**0**	1	0.01	**0**	12
9	**0.009**	**1**	0.2	325	0.1	9	0.1	250	0.05	0.03	153
13	**0.05**	**10**	-1	-1	109.2	24723			0.61	1738.24	3935567
15	**0.089**	**0**	0.4	56	1.3	**0**			0.91		
19	**0.459**	**164**							7.94		
21	**1.04**	**448**							39.07		
25	**14.074**	**5100**									
27	**23.548**	**7066**									
31	**5.289**	**0**	23.3	280	-1	-1			48.52		

(compared to 48 in [5]) with a shorter total time. Hence we compare length-lex with the ROBDD approach. The authors claimed there are 11 difficult instances (see Table 5 (Left)) that not all ROBDD-based solvers are able to solve. One of them was able to solve 4. Length-lex solves 6 of the difficult instances and is significantly faster. For the easy instances, length-lex is more than 30 times faster than the ROBDD approach to solve all instances (Table 5 (Right)).

Steiner System. The Steiner triple system is a special class of Steiner system (with $t = 2, k = 3$) which has drawn more attention. Hence, the results are given in two tables. Table 6 depicts the results for the Steiner triple system, while Table 7 shows the results for the remaining instances. For the triple system, the search adopted the labeling technique of [4](and [22]), which assigns the smallest available value to the first possible domain (column-wise labeling). Under our set dual modeling approach, it is equivalent to label the dual variables sequentially. Length-lex solves all instances and outperforms other representations significantly. For the remaining instances, the search uses the standard sequential labeling. Once again, length-lex is the most robust representation and is able to solve all instances efficiently.

Table 7. Experimental Results on the Steiner System

	Length-Lex		Bound/seq		Domain/seq		Hybrid/seq		Subset/seq	
t,k,n	Time	Fails	Time	Fails	Time	Fails	Time	Fails	Time	Fails
2,4,13	**0.008**	**0**	0.1	157	0.1	**0**	0.14	**0**	0.02	1
2,4,16	**0.14**	98	421.4	522706	0.6	**15**				
2,5,21	**0.04**	**0**	0.5	413	1.4	**0**	2.83	**0**	0.1	**0**
2,6,16	**0.005**	**1**	-1	-1	80.7	15205				
3,4,8	**0.006**	**0**	0.1	18	0.1	**0**	0.08	2	0.03	8
3,4,16	**3.43**	**0**	9.7	274	548.7	**0**	54.69	132	7.11	240
3,6,22	2.84	**0**	8.3	1608	-1	-1	54.98	42	**2.47**	92

Table 8. Experimental Results on Balanced Incomplete Block Design

	Length-Lex		VM/dyn		Lex/seq		Lex-sum/seq	
v,b,r,k,l	Time	Fails	Time	Nodes	Time	Fails	Time	Fails
6,20,10,3,4	0.009	**1**	0.033	61	**0**	43	**0**	43
6,30,15,3,6	**0.018**	**1**	0.14	95	0.1	68	0.1	68
6,40,20,3,8	**0.041**	**3**	0.39	128	0.1	108	0.1	108
6,80,40,3,16	**0.82**	**32**	3.6	245	**0.1-1**	100-1000		
7,21,9,3,3	0.008	**0**	0.045	75	**0**	42	**0**	42
7,28,12,3,4	**0.010**	**0**	0.12	86	0.1	64	0.1	64
7,35,15,3,5	**0.012**	**0**	0.27	109	0.1	88	0.1	88
7,42,18,3,6	**0.015**	**0**	0.48	139	0.2	115	0.2	115
7,84,36,3,12	**0.040**	**0**	4.2	254	0.1-1	100-1000		
7,91,39,3,13	**0.047**	**0**	5.4	280	0.1-1	100-1000		
9,24,8,3,2	**0.012**	**1**			0.1	48	0.1	48
9,72,24,3,6	**0.17**	**27**	2.7	252	**0.1-1**	100-1000		
9,84,28,3,7	**0.30**	**43**	4.2	257	1 - 10	1000-10000		
9,96,32,3,8	**0.50**	**66**	6.3	296	1 - 10	1000-10000		
9,108,36,3,9	**0.80**	**97**	14	365	1 - 10	1000-10000		
9,120,40,3,10	**1.27**	**138**	14	268	**1 - 10**	1000-10000		
10,90,27,3,6	**1.05**	**150**	5.3	289	**1 - 10**	**100-1000**		
10,120,36,3,8	**4.77**	**576**	13	377	**1 - 10**	1000-10000		
11,110,30,3,6	**7.66**	1192	16	366	**1 - 10**	**100-1000**		
12,88,22,3,4	**5.65**	1173	5.1	296	**1 - 10**	**100-1000**		
13,52,12,3,2	**0.11**	**15**	2.9	218	**0.1-1**	100-1000		
13,78,18,3,3	**0.47**	**78**	3.5	282	**0.1-1**	100-1000		
13,104,24,3,4	**1.52**	**207**	8.7	344	**1 - 10**	**100-1000**		
15,21,7,5,2	**24.78**	**16891**			**10 - 100**	$10^5 - 10^6$		
15,70,14,3,2	**0.11**	**0**	5.5	383	**0.1-1**	100-1000		
16,32,12,6,4	**3.84**	**980**			10 - 100	$10^6 - 10^7$		
16,80,15,3,2	**0.68**	**148**	4.7	485	1 - 10	**100-1000**		
19,57,9,3,1	**0.13**	**6**	8.2	802	1 - 10	100-1000		
22,22,7,7,2	**74.26**	**21552**			**10 - 100**	$10^5 - 10^6$		

Balanced Incomplete Block Design Problem. For this problem, we give a comprehensive list containing instances from [23] and [8] (some small and trivial instances were removed due to space constraints). VM/dyn is a randomized approach with a maximum node limit 10000. Every instance was run 50 times and not all instances finishes within the limit and only the average time and nodes of the successful instances was reported in [23]. Lex/seq gives only a logarithmic scale for most instances and we can only report the time and number of fails as a range, [22] gives the number of fails for some smaller instances as well. Note that this is the first result of a set representation on the BIBD problem and the length-lex representation is robust and effective.

6 Conclusion

This paper presented the first experimental evaluation of the length-lex domain for set variables. The implementation was based on two novel technical contributions: a generic propagation algorithm which pushes the length-lex ordering constraint into any symmetric binary constraints and an adaptation of the symmetry-breaking technique from 0/1 matrices to the length-lex ordering. The resulting implementation, which consists of 18,000 lines of C++, demonstrates that the length-lex representation for set variables is robust and efficient across the standard benchmarks.

References

1. Puget, J.F.: Pecos a high level constraint programming language. In: Proc. of Spicis (1992)
2. Gervet, C.: Interval propagation to reason about sets: Definition and implementation of a practical language. Constraints 1(3), 191–244 (1997)
3. Azevedo, F., Barahona, P.: Modelling digital circuits problems with set constraints. In: CP 2000, pp. 414–428 (2000)
4. Azevedo, F.: Cardinal: A finite sets constraint solver. Constraints 12(1), 93–129 (2007)
5. Sadler, A., Gervet, C.: Hybrid set domains to strengthen constraint propagation and reduce symmetries. In: Wallace, M. (ed.) CP 2004. LNCS, vol. 3258, pp. 604–618. Springer, Heidelberg (2004)
6. Sadler, A., Gervet, C.: Enhancing set constraint solvers with lexicographic bounds. J. Heuristics 14(1), 23–67 (2008)
7. Flener, P., Frisch, A., Hnich, B., Kiziltan, Z., Miguel, I., Pearson, J., Walsh, T.: Breaking row and column symmetries in matrix models. In: Van Hentenryck, P. (ed.) CP 2002. LNCS, vol. 2470, pp. 462–476. Springer, Heidelberg (2002)
8. Frisch, A.M., Hnich, B., Kiziltan, Z., Miguel, I., Walsh, T.: Propagation algorithms for lexicographic ordering constraints. Artificial Intelligence (170) (2006)
9. Peter Hawkins, V.L., Stuckey, P.J.: Solving set constraint satisfaction problems using robdds. JAIR 24, 109–156 (2005)
10. Gervet, C., Van Hentenryck, P.: Length-lex ordering for set csps. In: AAAI 2006 (2006)
11. Hnich, B., Kiziltan, Z., Walsh, T.: Combining symmetry breaking with other constraints: Lexicographic ordering with sums. In: AMAI 2004 (2004)
12. Cheng, B.M.W., Choi, K.M.F., Lee, J.H.M., Wu, J.C.K.: Increasing constraint propagation by redundant modeling: an experience report. Constraints 4(2), 167–192 (1999)
13. Geelen, P.A.: Dual viewpoint heuristics for binary constraint satisfaction problems. In: ECAI 1992, pp. 31–35 (1992)

14. Law, Y.C., Lee, J.H.M.: Symmetry breaking constraints for value symmetries in constraint satisfaction. Constraints 11 (2006)
15. Puget, J.F.: An efficient way of breaking value symmetries. In: AAAI 2006 (2006)
16. Walsh, T.: General symmetry breaking constraints. In: Benhamou, F. (ed.) CP 2006. LNCS, vol. 4204, pp. 650–664. Springer, Heidelberg (2006)
17. Van Hentenryck, P., Yip, J., Gervet, C., Dooms, G.: Bound consistency for binary length-lex set constraints. In: AAAI 2008, pp. 375–380 (2008)
18. Yip, J., Van Hentenryck, P.: Length-lex bound consistency for knapsack constraints. In: SAC 2009 (2009)
19. Malitsky, Y., Sellmann, M., van Hoeve, W.J.: Length-lex bounds consistency for knapsack constraints. In: Stuckey, P.J. (ed.) CP 2008. LNCS, vol. 5202, pp. 266–281. Springer, Heidelberg (2008)
20. Katsirelos, G., Narodytska, N., Walsh, T.: Combining symmetry breaking and global constraints. In: CSCLP 2008 (2009)
21. van Hoeve, W.J., Sabharwal, A.: Filtering atmost1 on pairs of set variables. In: Perron, L., Trick, M.A. (eds.) CPAIOR 2008. LNCS, vol. 5015, pp. 382–386. Springer, Heidelberg (2008)
22. Kiziltan, Z.: Symmetry breaking ordering constraints. Phd Thesis. Uppsala University
23. Meseguer, P., Torras, C.: Exploiting symmetries within constraint satisfaction search. Artificial Intelligence 129(1-2), 133–163 (2001)

The Complexity of Valued Constraint Models*

Stanislav Živný and Peter G. Jeavons

Computing Laboratory, University of Oxford, Oxford, UK
{stanislav.zivny,peter.jeavons}@comlab.ox.ac.uk

Abstract. The VALUED CONSTRAINT SATISFACTION PROBLEM (VCSP) is a general framework encompassing many optimisation problems. We discuss precisely what it means for a problem to be *modelled* in the VCSP framework. Using our analysis, we show that some optimisation problems, such as (s,t)-MIN-CUT and SUBMODULAR FUNCTION MINIMISATION, can be modelled using a restricted set of valued constraints which are tractable to solve regardless of how they are combined. Hence, these problems can be viewed as special cases of more general problems which include all possible instances using the same forms of valued constraint. However, other, apparently similar, problems such as MIN-CUT and SYMMETRIC SUBMODULAR FUNCTION MINIMISATION, which also have polynomial-time algorithms, can only be naturally modelled in the VCSP framework by using valued constraints which can represent NP-complete problems. This suggests that the reason for tractability in these problems is more subtle; it relies not only on the form of the valued constraints, but also on the precise structure of the problem. Furthermore, our results suggest that allowing constant constraints can significantly alter the complexity of problems in the VCSP framework, in contrast to the CSP framework.

1 Introduction

The study of combinatorial optimisation traditionally considers a range of specific problem types, including integer programming problems, problems on graphs and networks, and Boolean problems [2], such as submodular function minimisation [14]. An important issue for any combinatorial optimisation problem is how to choose an effective representation, which can be crucial to the efficiency of solving the problem.

The VALUED CONSTRAINT SATISFACTION PROBLEM (VCSP) is a single generic framework, for modelling a wide range of optimisation problems [1,11,12].

Our aim in this paper is to investigate which standard combinatorial optimisation problems can be modelled in the VCSP framework, and whether finding such models sheds new light on the complexity of these problems. We will focus on Boolean problems (i.e., each variable can take one of two possible values), which are equivalent to minimising functions defined on sets.

We need to be a little careful in defining what it means for a problem to be modelled in the VCSP framework: we clearly need to exclude modelling

* Stanislav Živný is supported by EPSRC grant EP/F01161X/1.

I.P. Gent (Ed.): CP 2009, LNCS 5732, pp. 833–841, 2009.

procedures that simply obliterate the structure of the problem we are attempting to model. For example, simply finding a solution to each given instance (using some algorithm) and then creating a VCSP instance which allows precisely that solution is not a useful approach to modelling. The standard way of excluding such pathological approaches is to limit the computational resources allowed to transform the problem from one representation to another. However, when dealing with tractable problems, we also need a suitably tight definition of what it means for a problem to be *modelled* in the VCSP framework. (More on modelling via constraints can be found in [15].)

In this paper we shall say that we have a VCSP *model* for a given combinatorial optimisation problem if the entire function to be minimised in that problem is *expressible* using some collection of valued constraints (in a precise sense defined below; note that this notion of expressibility was a major tool in the complexity analysis of a wide variety of Boolean constraint problems carried out by Creignou et al. [5], where it was referred to as *implementation*). We show that many standard problems can be modelled in this way. Moreover, some problems, including for example the (s,t)-MIN-CUT problem and the problem of SUBMODULAR FUNCTION MINIMISATION, can be modelled using very restricted forms of constraints. In fact the forms of constraints needed to model these problems are sufficiently restricted that they can be solved in polynomial time regardless of how they are combined. Hence, these problems can be viewed as special cases of more general problems which include all possible instances using the same forms of valued constraint.

On the other hand, we show that other apparently similar problems, which also have polynomial-time algorithms, can only be modelled using forms of constraint which are powerful enough to represent NP-complete problems. Our examples include the standard MIN-CUT problem. This result indicates that the reason for the tractability of such problems relies on the precise structure of the problem and not just the form of the individual constraints. Such problems provide a fresh incentive to develop the theoretical analysis of the complexity of valued constraint problems, which currently has very little to say about such "hybrid" reasons for tractability.

2 Background

Given some fixed set D, a function from D^k to $\overline{\mathbb{Q}}_+$, where $\overline{\mathbb{Q}}_+$ is the set of all positive rational numbers together with infinity will be called a *cost function.*

Definition 1. *An instance $\mathcal{P}$ of* VCSP *is a triple $\langle V, D, \mathcal{C}\rangle$, where V is a finite set of* variables, *which are to be assigned values from the set D, and $\mathcal{C}$ is a set of* valued constraints. *Each constraint $c \in \mathcal{C}$ is a pair $c = \langle \sigma, \phi\rangle$, where σ is a tuple of variables of length $|\sigma|$, called the* scope *of c, and $\phi : D^{|\sigma|} \to \overline{\mathbb{Q}}_+$ is a cost function. An* assignment *for the instance $\mathcal{P}$ is a mapping s from V to D. The* cost *of an assignment s is defined as follows:*

$$Cost_{\mathcal{P}}(s) = \sum_{\langle\langle v_1,v_2,\ldots,v_m\rangle,\phi\rangle\in\mathcal{C}} \phi(\langle s(v_1), s(v_2), \ldots, s(v_m)\rangle).$$

A solution *to* $\mathcal{P}$ *is an assignment with minimum cost.*

The VCSP is a very general framework which allows us to describe many optimisation problems, including many NP-hard problems [11]. Any set, Γ, of cost functions is called a *valued constraint language.* The class VCSP(Γ) is defined to be the class of all VCSP instances where the cost functions of all valued constraints lie in Γ. The complexity of a valued constraint language Γ is defined as the complexity of VCSP(Γ). A valued constraint language Γ is called *tractable* if VCSP(Γ') is solvable in polynomial time for every finite $\Gamma' \subseteq \Gamma$, and Γ is called *intractable* if VCSP(Γ') is NP-hard for some finite subset $\Gamma' \subseteq \Gamma$. Many examples of tractable valued constraint languages have now been identified [4].

Definition 2. *For any* VCSP *instance* $\mathcal{P} = \langle V, D, \mathcal{C}\rangle$, *and any list* $l = \langle v_1, \ldots, v_m\rangle$ *of variables of* $\mathcal{P}$, *the* projection *of* $\mathcal{P}$ *onto* l, *denoted* $\pi_l(\mathcal{P})$, *is the* m*-ary cost function defined as follows:*

$$\pi_l(\mathcal{P})(x_1, \ldots, x_m) = \min_{\{s:V\rightarrow D \mid \langle s(v_1),\ldots,s(v_m)\rangle=\langle x_1,\ldots,x_m\rangle\}} Cost_{\mathcal{P}}(s).$$

We say that a cost function ϕ *is* expressible *over a valued constraint language* Γ *if there exists an instance* $\mathcal{P} \in$ VCSP(Γ) *and a list* l *of variables of* $\mathcal{P}$ *such that* $\pi_l(\mathcal{P}) = \phi$.

We denote by $\langle\Gamma\rangle$ the *expressive power* of Γ, which is the set of all cost functions expressible over Γ up to additive and multiplicative constants.

Theorem 3 ([4]). *For any valued constraint language* Γ *and any cost function* ϕ *expressible over* Γ, VCSP(Γ) *and* VCSP($\Gamma \cup \{\phi\}$) *are linear-time equivalent.*

3 Boolean Optimisation Problems

In this section we recall some standard Boolean optimisation problems.

(s,t)-Min-Cut: For a directed graph $G = \langle V, E\rangle$ with weights $w : E \rightarrow \mathbb{Q}_+$, $s, t \in V$, C is an (s,t)-*cut* if $C \subseteq V$, $s \in C$ and $t \notin C$. The weight of C is $\sum_{(u,v)\in E, u\in C, v\notin C} w(u,v)$. The (s,t)-Min-Cut problem consists in finding the minimum-weight (s,t)-cut. A cubic-time algorithm (in the number of vertices) based on network flows is known for this problem [6].

Min-Cut: For an undirected graph $G = \langle V, E\rangle$ with weights $w : E \rightarrow \mathbb{Q}_+$, C is a *cut* if $C \subseteq V$. The weight of C is defined as above. The Min-Cut problem consists in finding the minimum-weight cut C such that $C \neq \emptyset$ and $C \neq V$. Using the cubic-time algorithm for the (s,t)-Min-Cut problem [6], one can easily construct an algorithm for the Min-Cut problem of order $O(n^4)$. A purely combinatorial cubic-time algorithm which is not based on network flows is also known for this problem [16].

Submodular Function Minimisation (SFM): A function ψ defined on subsets of a set V is called a *submodular function* [14] if, for all subsets S and T of V, $\psi(S \cap T) + \psi(S \cup T) \leq \psi(S) + \psi(T)$. The problem of Submodular Function Minimisation (SFM) consists in finding a subset S of V for which the value of $\psi(S)$ is minimal. It is a central problem in discrete optimisation, with links to many different areas [14]. The time complexity of the fastest published algorithm for SFM is $O(n^5\mathrm{EO} + n^6)$ where $n = |V|$ and EO is the time to evaluate $\psi(S)$ for some $S \subseteq V$ [9,7].

(s,t)-SFM: Given a submodular function ψ defined on subsets of a set V, and two elements $s, t \in V$, the problem of (s,t)-Submodular Function Minimisation consists in finding a nonempty subset S of V such that $s \in S$, $t \notin S$, and the value of $\psi(S)$ is minimal.

Proposition 4. (s,t)-SFM *is linear-time equivalent to* SFM.

Proof. First we show that (s,t)-SFM is reducible to SFM. Clearly, $S \subset V$ is a solution to an instance $\langle V, \psi \rangle$ of (s,t)-SFM if and only if $S \setminus \{s\}$ is a solution to the instance $\langle V \setminus \{s,t\}, \overline{\psi} \rangle$ of SFM where $\overline{\psi}(U) = \psi(U \cup \{s\})$. On the other hand, $S \subseteq V$ is a solution to an instance $\langle V, \psi \rangle$ of SFM if and only if $S \cup \{s\}$ is a solution to the instance $\langle V \cup \{s,t\}, \overline{\psi} \rangle$ of (s,t)-SFM, where $\overline{\psi}(U) = \psi(U \setminus \{s\})$. □

Symmetric SFM (SSFM): Given a submodular function ψ defined on subsets of a set V, we say that ψ is *symmetric* if for every $U \subseteq V$, $\psi(U) = \psi(V \setminus U)$. Note that $\psi(\emptyset) = \psi(V) \leq \psi(U)$ for every $U \subseteq V$. Symmetric Submodular Function Minimisation consists in finding a nonempty proper subset S of V for which the value of $\psi(S)$ is minimal. Queyranne extended the cubic-time algorithm for the Min-Cut problem mentioned above to obtain a purely combinatorial cubic-time algorithm for SSFM [10].

(s,t)-SSFM: Given a symmetric submodular set function ψ on subsets of V and two elements $s, t \in V$, the problem of (s,t)-Symmetric Submodular Function Minimisation consists in finding a proper nonempty subset S of V, where $s \in S$ and $t \notin S$, for which the value of $\psi(S)$ is minimal.

Similarly to the proof of Proposition 4, it can be easily shown that SSFM is reducible to (s,t)-SSFM.

Proposition 5 ([10]). SFM *is linear-time reducible to* (s,t)-SSFM.

However, using the same proof idea as Proposition 4 does *not* show that (s,t)-SSFM is reducible to SSFM (as it seems difficult to preserve two properties, namely being submodular and symmetric, at the same time).[1] Moreover, Proposition 5 shows that (s,t)-SSFM is as hard as SFM, so a time-complexity-preserving reduction from (s,t)-SSFM to SSFM would make SFM equivalent to SSFM, which would yield a cubic-time algorithm for SFM. This would be a major advance in discrete optimisation.

[1] More on the relationships between SFM, SSFM and (s,t)-SSFM can be found in [8].

4 Modelling in the VCSP Framework

It is easy (and standard) to see that any set function ψ defined on subsets of $V = \{v_1, \dots, v_n\}$ can be associated with a function $\phi : \{0,1\}^n \to \overline{\mathbb{Q}}_+$ defined as follows: for each tuple $t \in \{0,1\}^n$, set $\phi(t) = \psi(T)$, where $T = \{v_i \mid t[i] = 1\}$ (moreover, if $U \subseteq V$ is forbidden, then set $\psi(T) = \infty$).

Note that the submodularity condition on a set function ψ is equivalent to the following condition on the associated Boolean function ϕ: for every two tuples $s, t \in \{0,1\}^n$, $\phi(\text{MIN}(s,t)) + \phi(\text{MAX}(s,t)) \leq \phi(s) + \phi(t)$, where both MIN and MAX are applied coordinate-wise. We therefore call a cost function ϕ satisfying this condition submodular. We now define a precise notion of what it means to model a problem in the VCSP framework. This notion is designed to rule out pathological cases and ensure that the models we allow do provide some insight into the nature of the problem being modelled.

Definition 6. *Let $\mathcal{P}$ be a problem which consists in minimising a given function ψ defined on the subsets of a given set V, and let ϕ be the associated Boolean cost function, as defined above. We say that $\mathcal{P}$ can be* e-modelled *by* VCSP(Γ) *if ϕ can be expressed over Γ.*

In other words, a problem $\mathcal{P}$ can be *e-modelled* by VCSP(Γ) if, for any instance $\langle\{v_1, v_2, \dots, v_n\}, \psi\rangle$ of $\mathcal{P}$, there is an instance $\mathcal{I} = \langle W, \{0,1\}, \mathcal{C}\rangle$ of VCSP(Γ), and a list of variables $\langle w_{v_1}, w_{v_2}, \dots, w_{v_n}\rangle \subseteq W$, such that for any $S \subseteq V$, the minimal cost over all assignments for $\mathcal{I}$ which assign each variable w_{v_i} the value 0 or 1 according to whether or not $v_i \in S$, is equal to $\psi(S)$.

Theorem 7

- (s,t)-MIN-CUT, SFM, (s,t)-SFM, *and* (s,t)-SSFM *can be e-modelled by* VCSP(Γ), *by a suitable choice of tractable valued constraint language* Γ.
- MIN-CUT *and* SSFM *can be e-modelled by* VCSP(Γ), *but only by using an intractable language* Γ.

Proof. Consider first the (s,t)-MIN-CUT problem. We have to prove that there exists a tractable valued constraint language Γ_{cut} such that (s,t)-MIN-CUT can be e-modelled by VCSP(Γ_{cut}). For any $w \in \mathbb{Q}_+$, we define the binary cost function λ^w as $\lambda^w(x,y) = w$ if $\langle x,y\rangle = \langle 0,1\rangle$, and 0 otherwise. For each $d \in \{0,1\}$ and each $c \in \overline{\mathbb{Q}}_+$, we define the unary cost function μ^c_d as $\mu^c_d(x) = c$ if $x \neq d$, and 0 otherwise. Now let Γ_{cut} consist of all λ^w and μ^c_u for $w \in \mathbb{Q}_+$, $c \in \overline{\mathbb{Q}}_+$ and $d \in \{0,1\}$.

Now consider any instance of (s,t)-MIN-CUT with graph $G = \langle V, E\rangle$ and weight function $w : E \to \mathbb{Q}_+$. Define a corresponding instance $\mathcal{I}$ of VCSP(Γ_{cut}) as $\mathcal{I} = \langle V, \{0,1\}, \{\langle\langle i,j\rangle, \lambda^{w(i,j)}\rangle \mid \langle i,j\rangle \in E\} \cup \{\langle s, \mu^\infty_0\rangle, \langle t, \mu^\infty_1\rangle\}\rangle$. Note that in any solution to $\mathcal{I}$ the source and target nodes, s and t, must take the values 0 and 1, respectively. Moreover, the weight of any cut containing s and not containing t is equal to the cost of the corresponding assignment to $\mathcal{I}$. Hence we have shown that (s,t)-MIN-CUT can be e-modelled by VCSP(Γ_{cut}).

On the other hand, we claim that VCSP(Γ_{cut}) can be reduced to (s,t)-MIN-CUT in linear time as follows: any unary constraint on variable v with cost function μ_0^c (respectively μ_1^c) is represented by an edge of weight c from the source node s to node v (respectively, from node v to the target node t). Any binary constraint on variables v_1, v_2 with cost function λ^w is represented by an edge of weight w from nodes v_1 to v_2. Hence VCSP(Γ_{cut}) has the same time complexity as (s,t)-MIN-CUT.

Next we consider the SFM problem. Let Γ_{sub} be the valued constraint language which consists of all submodular cost functions. Because submodular cost functions may take infinite values, instances of VCSP(Γ_{sub}) cannot be simply solved by standard submodular function minimisation algorithms for finite-valued submodular functions. However, Cohen et al. showed [4] that VCSP(Γ_{sub}) is polynomial-time reducible to the problem of SFM over a ring family[2] which is known to be equivalent to SFM [13], so Γ_{sub} is tractable.

Now we consider the (s,t)-SFM problem. A *constant constraint* is a unary constraint μ_d^∞ for an arbitrary $d \in D$. We denote by $\Gamma_{\mathsf{const}} = \{\mu_d^\infty \mid d \in D\}$ the valued constraint language consisting of all constant constraints.

Clearly, (s,t)-SFM can be e-modelled by VCSP($\Gamma_{\mathsf{sub}} \cup \Gamma_{\mathsf{const}}$). Also, as constant constraints are submodular, we have $\Gamma_{\mathsf{const}} \subseteq \Gamma_{\mathsf{sub}}$, so (s,t)-SFM can be e-modelled by VCSP(Γ_{sub}) and we have already shown that Γ_{sub} is tractable.

Now we consider the MIN-CUT problem. To e-model MIN-CUT, we can use the valued constraint language Γ_{cut} defined above, and then forbid the empty and complete cuts by putting a crisp NOT-ALL-EQUAL constraint over the variables $w_1, \ldots, w_n$ which represent V. In other words, MIN-CUT can be e-modelled by VCSP($\Gamma_{\mathsf{cut}} \cup \Gamma_{\mathsf{nae}}$) where Γ_{nae} consists of NOT-ALL-EQUAL constraints of all possible arities.

However, since NOT-ALL-EQUAL SATISFIABILITY is NP-complete, it follows that Γ_{nae} is intractable. We now show that for any valued constraint language Γ which e-models MIN-CUT, VCSP(Γ) must be intractable.

Let Γ be a valued constraint language such that MIN-CUT can be e-modelled by VCSP(Γ). Consider an instance of MIN-CUT which is a triangle with all weights set to zero. The empty and complete cuts are forbidden, any other cut has cost 0, so Γ expresses a NOT-ALL-EQUAL constraint, and hence is intractable.

Now, let Γ_{ssub} be the valued constraint language which consists of all symmetric submodular functions. The SSFM problem can be e-modelled by VCSP($\Gamma_{\mathsf{ssub}} \cup \Gamma_{\mathsf{nae}}$) in a similar way to MIN-CUT. However, since MIN-CUT is just a special case of SSFM, it follows from the argument just given that any suitable choice of Γ will again be intractable.

Finally, we consider the (s,t)-SSFM problem. Similarly to the arguments above, (s,t)-SSFM can be e-modelled by the language $\Gamma_{\mathsf{ssub}} \cup \Gamma_{\mathsf{const}}$ which is a subset of Γ_{sub}. □

We now examine more closely the language $\Gamma_{\mathsf{ssub}} \cup \Gamma_{\mathsf{const}}$ consisting of symmetric submodular constraints and constant constraints, which was introduced to model

[2] A collection of sets $\mathcal{C}$ is called a *ring family* if $\mathcal{C}$ is closed under union and intersection.

the (s,t)-SSFM problem. The following proposition shows that all submodular constraints can be expressed by this language.

Proposition 8. $\Gamma_{\mathsf{sub}} = \langle \Gamma_{\mathsf{ssub}} \cup \Gamma_{\mathsf{const}} \rangle$.

Proof. Since $\Gamma_{\mathsf{ssub}} \cup \Gamma_{\mathsf{const}} \subseteq \Gamma_{\mathsf{sub}}$, it only remains to prove that any submodular function can be expressed by symmetric submodular functions and unary constant constraints. Our proof is adapted from the construction given in [10].

Let ψ be a submodular function defined on subsets of a set V, with $|V| = n$. Let $M = \textsc{Max}(\psi(U) | U \subseteq V, \psi(U) < \infty)$. Define $\overline{V} = V \cup \{s,t\}$ for $s, t \notin V$ and $\overline{\psi}$ as follows:

$$\overline{\psi}(U) = \begin{cases} \psi(U \setminus \{s\}) + M(n+2) & \text{if } s \in U \text{ and } t \notin U, \\ M \mid U \mid & \text{if } s, t \in U, \\ \overline{\psi}(V \setminus U) & \text{if } s \notin U. \end{cases}$$

It can be shown that $\overline{\psi}$ is symmetric and submodular, and that $S \subseteq \overline{V}$ where $s \in S$ and $t \notin S$ minimises $\overline{\psi}$ if and only if $S \setminus \{s\}$ minimises ψ [10]. Hence ψ can be expressed (up to an additive constant) by $\overline{\psi}$ and two elements of Γ_{const}. □

Note that, in the CSP, it has been shown that adding constant constraints to a tractable language which is a core does not change the complexity (i.e., $\text{CSP}(\Gamma)$ is linear-time equivalent to $\text{CSP}(\Gamma \cup \Gamma_{\mathsf{const}})$, provided Γ is a core) [3]. However, in the valued constraint case Proposition 8 suggests that the situation may be rather different: the complexity of the valued language Γ_{ssub}, consisting of all symmetric submodular functions, is cubic, whereas adding constant constraints allows us to express all submodular functions. The best known algorithm for minimising arbitrary submodular functions is $\Omega(n^6)$.

The concept of e-modelling we have defined here is designed to avoid trivial models by requiring the constraints in the model to be capable of expressing the function being minimised. However, more relaxed notions of modelling can also yield non-trivial representations, and may provide more flexibility, as the following result indicates:

Theorem 9. *There is a tractable valued constraint language Γ over an infinite domain such that* Min-Cut *can be reduced to* $\text{VCSP}(\Gamma)$ *in linear-time.*

Proof. Let $D = \{\langle i, d\rangle \mid i \in \mathbb{N}, d \in \{0,1\}\}$.

For any $w \in \mathbb{Q}_+$ and $k \in \mathbb{N}$, we define the binary cost function λ_k^w as follows:

$$\lambda_k^w(\langle i, d_1\rangle, \langle j, d_2\rangle) = \begin{cases} \infty & \text{if } \textsc{Max}(i,j) > k \text{ or } i \neq j, \\ 0 & \text{if } i = j \text{ and } d_1 = d_2, \\ w & \text{otherwise.} \end{cases}$$

Next, for any $d \in \{0,1\}$ and $k \in \mathbb{N}$, we define the unary cost function $\mu_{k \to d}$ as follows:

$$\mu_{k \to d}(\langle i, x\rangle) = \begin{cases} \infty & \text{if } i = k \text{ and } x \neq d, \\ 0 & \text{otherwise.} \end{cases}$$

Let Γ be the valued constraint language over D consisting of the cost functions λ_k^w and $\mu_{k\to d}$ for all $w \in \mathbb{Q}_+$, $k \in \mathbb{N}$ and $d \in \{0,1\}$. We first show how to reduce MIN-CUT to VCSP(Γ).

Consider an instance of MIN-CUT with graph $G = \langle V, E\rangle$ (without loss of generality, without isolated vertices), where $V = \{v_1, \ldots, v_n\}$, and weight $w(i,j)$ for every $\{v_i, v_j\} \in E$. We define a corresponding instance $\mathcal{I}$ of VCSP(Γ) as follows. Let $\mathcal{I} = \langle V, D, \mathcal{C}\rangle$, where $\mathcal{C} = \{\langle\langle v_i, v_j\rangle, \lambda_{n-1}^{w(i,j)}\rangle \mid i < j, \{v_i, v_j\} \in E\} \cup \{\langle v_1, \mu_{(i-1)\to 0}\rangle, \langle v_i, \mu_{(i-1)\to 1}\rangle \mid v_i \in V \setminus \{v_1\}\}$.

Clearly, if $\emptyset \neq U \subsetneq V$ is a minimum cut of G, then so is $V \setminus U$. Therefore, we can assume that $v_1 \in U$. Note that a necessary condition for an assignment of variables of $\mathcal{I}$ to have a finite cost is that there exists some $i \in \{1, \ldots, n-1\}$ such that every variable of $\mathcal{I}$ is assigned a value of the form $\langle i, .\rangle$. Such an i forces the variable v_1 to be assigned $\langle i, 0\rangle$ (i.e., v_1 belongs to the cut), and the variable v_{i+1} to be assigned $\langle i, 1\rangle$ (i.e., v_{i+1} does not belong to the cut). For the minimum cut U, there has to be an $i \in \{2, \ldots, n\}$ such that $v_i \notin U$ and the assignment of $\langle i-1, 0\rangle$ to variables in U, and $\langle i-1, 1\rangle$ to variables not in U, gives the weight of U. (This construction is similar to the one used in the proof of Theorem 7 which shows that (s,t)-MIN-CUT can be e-modelled.)

It remains to show that Γ is tractable. Let Γ' be any finite subset of Γ, and let k be the biggest number which occurs in the subscript of any λ cost function in Γ'. The cost of an assignment to any instance of VCSP(Γ') which is of the form $\langle i, .\rangle$, for $i \in \{1, \ldots, k\}$, corresponds to the cost of an (s,t)-cut in an associated graph. Therefore, VCSP(Γ') can be solved by solving k instances of the (s,t)-MIN-CUT problem, and hence its time complexity is $O(kn^3)$. For a fixed Γ' (and hence a fixed value of k) this is polynomial in the size of the instance. □

The language Γ used in Theorem 9 has a much higher complexity than MIN-CUT. It is an interesting open question whether MIN-CUT can be reduced to VCSP(Γ) for some Γ with the same complexity as MIN-CUT (i.e., such that VCSP(Γ) is linear-time reducible to MIN-CUT).

References

1. Bistarelli, S., Fargier, H., Montanari, U., Rossi, F., Schiex, T., Verfaillie, G.: Semiring-based CSPs and valued CSPs: Frameworks, properties, and comparison. Constraints 4, 199–240 (1999)
2. Boros, E., Hammer, P.L.: Pseudo-boolean optimization. Discrete Applied Mathematics 123(1-3), 155–225 (2002)
3. Bulatov, A., Krokhin, A., Jeavons, P.: Classifying the complexity of constraints using finite algebras. SIAM Journal on Computing 34(3), 720–742 (2005)
4. Cohen, D., Cooper, M., Jeavons, P., Krokhin, A.: The complexity of soft constraint satisfaction. Artificial Intelligence 170, 983–1016 (2006)
5. Creignou, N., Khanna, S., Sudan, M.: Complexity Classification of Boolean Constraint Satisfaction Problems. SIAM Monographs on Discrete Mathematics and Applications, vol. 7. SIAM, Philadelphia (2001)
6. Goldberg, A., Tarjan, R.: A new approach to the maximum flow problem. Journal of the ACM 35, 921–940 (1988)

7. Iwata, S., Orlin, J.B.: A Simple Combinatorial Algorithm for Submodular Function Minimization. In: Proceedings of the 20th SODA, pp. 1230–1237 (2009)
8. Narayanan, H.: A note on the minimization of symmetric and general submodular functions. Discrete Applied Mathematics 131(2), 513–522 (2003)
9. Orlin, J.B.: A faster strongly polynomial time algorithm for submodular function minimization. Mathematical Programming 118, 237–251 (2009)
10. Queyranne, M.: Minimising symmetric submodular functions. Mathematical Programming 82, 3–12 (1998)
11. Rossi, F., van Beek, P., Walsh, T. (eds.): The Handbook of Constraint Programming. Elsevier, Amsterdam (2006)
12. Schiex, T., Fargier, H., Verfaillie, G.: Valued constraint satisfaction problems: hard and easy problems. In: Proceedings of the 14th IJCAI, Montreal, Canada (1995)
13. Schrijver, A.: A combinatorial algorithm minimizing submodular functions in strongly polynomial time. J. of Combinatorial Theory, Series B 80, 346–355 (2000)
14. Schrijver, A.: Combinatorial Optimization: Polyhedra and Efficiency. Algorithms and Combinatorics, vol. 24. Springer, Heidelberg (2003)
15. Smith, B.: Modelling. In: The Handbook of Constraint Programming, ch. 11. Elsevier, Amsterdam (2006)
16. Stoer, M., Wagner, F.: A simple min-cut algorithm. Journal of the ACM 44(4), 585–591 (1997)

Author Index

Kolibri 360 GmbH
Ferdinand-Jühlke-Straße 7
99095 Erfurt, Deutschland
produktsicherheit@kolibri360.de